Student Solutions Manual

Elementary Algebra

FIFTH EDITION

Alan S. Tussy
Citrus College

R. David Gustafson
Rock Valley College

Prepared by

Alexander Lee
Hinds Community College

Australia • Brazil • Japan • Korea • Mexico • Singapore • Spain • United Kingdom • United States

ISBN-13: 978-1-111-98902-6

ISBN-10: 1-111-98902-8

Brooks/Cole
20 Davis Drive
Belmont, CA 94002
USA

Cengage Learning is a leading provider of customized learning solutions with office locations around the globe, including Singapore, the United Kingdom, Australia, Mexico, Brazil, and Japan. Locate your local office at **www.cengage.com/global**

Cengage Learning products are represented in Canada by Nelson Education, Ltd.

To learn more about Brooks/Cole, visit **www.cengage.com/brookscole**

Purchase any of our products at your local college store or at our preferred online store **www.cengagebrain.com**

Printed in the United States of America
2 3 4 5 6 20 19 18 17 16

Student Solutions Manual TABLE OF CONTENTS

SECTION 1.1, STUDY SET
VOCABULARY

Fill in the blanks.

1. A **sum** is the result of an addition.
   A **difference** is the result of a subtraction.
   A **product** is the result of a multiplication.
   A **quotient** is the result of a division.
3. A number, such as 8, is called a **constant** because it does not change.
5. An **equation** is a mathematical sentence that contains an = symbol. An algebraic **expression** does not.
7. The **horizontal** axis of a graph extends left and right and the vertical axis extends up and down.

CONCEPTS

Classify each item as an algebraic expression or an equation.

9. a. **Equation**
   b. **Algebraic expression**

11. a. **Algebraic expression**
    b. **Equation**

13. **Addition, multiplication, division**
    t

15. **Construct a line graph using the data in the table.**

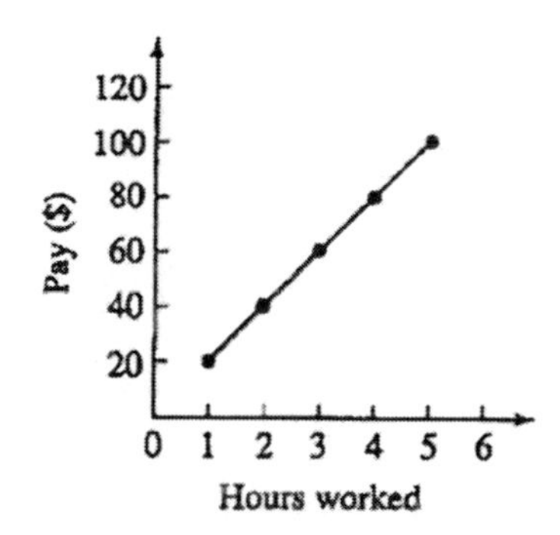

NOTATION

Fill in the blanks.

17. The symbol $\neq$ means **is not equal to**.
19. Write the multiplication 5×6 using a raised dot and then using parentheses. **5•6, 5(6)**

Write each expression without using a multiplication symbol or parentheses.

21. **$4x$** 23. **$2w$**

Write each division using a fraction bar.

25. $\dfrac{32}{x}$ 27. $\dfrac{55}{5}$

GUIDED PRACTICE

29. ACCOUNTING
    15-year-old machinery is worth $35,000.
31. BUSINESS
    30 customers, $250

Express each statement using one of the words sum, product, difference, or quotient.

33. The product of 8 and 2 equals 16.
35. The difference of 11 and 9 equals 2.
37. The sum of x and 2 equals 10.
39. The quotient of 66 and 11 equals 6.

Translate each verbal model into an equation. (Answers may vary, depending on the variables chosen.)

41. $p = 100 - d$
43. $7d = h$
45. $s = 3c$
47. $w = e + 1,200$
49. $p = r - 600$
51. $\dfrac{l}{4} = m$

Use the formula to complete each table.

53. $d = 360 + L$

Lunch Time (minutes) L	School Day (minutes) d
30	$d = 360 + L$ $d = 360 + 30$ $\mathbf{d = 390}$
40	$d = 360 + L$ $d = 360 + 40$ $\mathbf{d = 400}$
45	$d = 360 + L$ $d = 360 + 45$ $\mathbf{d = 405}$

55. $t = 1{,}500 - d$

Deductions d	Take-home pay t
200	$t = 1{,}500 - d$ $t = 1{,}500 - 200$ $\mathbf{t = 1{,}300}$
300	$t = 1{,}500 - d$ $t = 1{,}500 - 300$ $\mathbf{t = 1{,}200}$
400	$t = 1{,}500 - d$ $t = 1{,}500 - 400$ $\mathbf{t = 1{,}100}$

57. $d = \dfrac{e}{\boxed{\mathbf{12}}}$

Eggs e	Dozens d
24	2
36	3
48	4

59. $I = \boxed{\mathbf{2}}\, c$

Couples c	Individuals I
20	40
100	200
200	400

APPLICATIONS

61. EXERCISE

a. The number of calories burned is the product of 3 and the number of minutes cleaning.

b. $c = 3m$

c.

m	10	20	30	40	50	60
c	**30**	**60**	**90**	**120**	**150**	**180**

d.

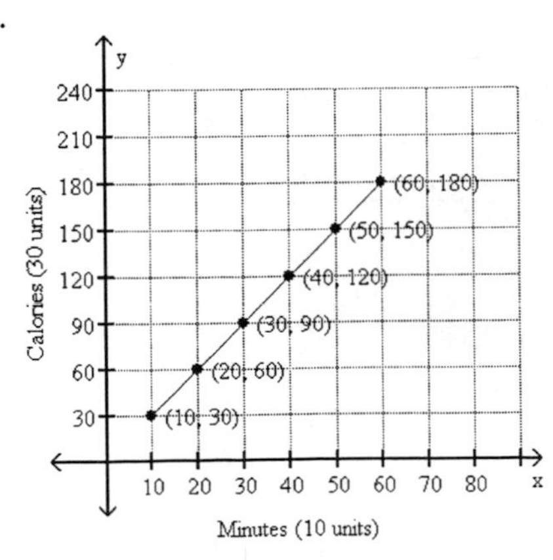

WRITING

63-65. Answers will vary.

CHALLENGE PROBLEMS

67. Complete formula.

$t = \boxed{3}\, s + \boxed{1}$

s	t
18	**55**
33	100
47	142

SECTION 1.2, STUDY SET VOCABULARY

Fill in the blanks.

1. A factor is a number being **multiplied**.
3. When we write 60 as 2·2·3·5, we say that we have written 60 in **prime-factored** form.
5. Two fractions that represent the same number, such as $\frac{1}{2}$ and $\frac{2}{4}$, are called **equivalent** fractions.
7. The **least or lowest** common denominator for a set of fractions is the smallest number each denominator will divide exactly.

CONCEPTS

9. a. **1** b. **a**

c. $\frac{a \cdot c}{b \cdot d}$ d. $\frac{a \cdot d}{b \cdot c}$

e. $\frac{a+b}{d}$ f. $\frac{a-b}{d}$

11. Complete each statement.

a. To simplify a fraction, we remove factors equal to **1** in the form of $\frac{2}{2}$, $\frac{3}{3}$, or $\frac{4}{4}$, and so on.

b. To build a fraction, we multiply it by **1** in the form of $\frac{2}{2}$, $\frac{3}{3}$, or $\frac{4}{4}$, and so on.

NOTATION

13. a. $\frac{5}{6} \cdot \frac{\mathbf{5}}{\mathbf{5}} = \frac{\mathbf{25}}{\mathbf{30}}$ b. $\frac{12}{42} = \frac{\cancel{2}^{1} \cdot \mathbf{2} \cdot \cancel{3}^{1}}{\cancel{2}_{1} \cdot \cancel{3}_{1} \cdot \mathbf{7}} = \frac{\mathbf{2}}{\mathbf{7}}$

GUIDED PRACTICE

Find the prime factorization of each number.

15. $75 = 3 \cdot 5 \cdot 5$

17. $28 = 2 \cdot 2 \cdot 7$

19. $81 = 3 \cdot 3 \cdot 3 \cdot 3$

21. $117 = 3 \cdot 3 \cdot 13$

23. $220 = 2 \cdot 2 \cdot 5 \cdot 11$

25. $1{,}254 = 2 \cdot 3 \cdot 11 \cdot 19$

Perform each operation.

27. $\frac{5}{6} \cdot \frac{1}{8} = \frac{5}{48}$

29. $\frac{7}{11} \cdot \frac{3}{5} = \frac{21}{55}$

31. $\frac{3}{4} \div \frac{2}{5} = \frac{3}{4} \cdot \frac{5}{2} = \frac{15}{8}$

33. $\frac{6}{5} \div \frac{5}{7} = \frac{6}{5} \cdot \frac{7}{5} = \frac{42}{25}$

Build each fraction or whole number to an equivalent fraction with the indicated denominator.

35. $\frac{1}{3} \cdot \frac{\mathbf{3}}{\mathbf{3}} = \frac{3}{9}$

37. $\frac{4}{9} \cdot \frac{\mathbf{6}}{\mathbf{6}} = \frac{24}{54}$

39. $\frac{7}{1} \cdot \frac{\mathbf{5}}{\mathbf{5}} = \frac{35}{5}$

41. $\frac{5}{1} \cdot \frac{\mathbf{7}}{\mathbf{7}} = \frac{35}{7}$

Simplify each fraction, if possible.

43. $\frac{6}{18} = \frac{\cancel{2}^{1} \cdot \cancel{3}^{1}}{\cancel{2}_{1} \cdot \cancel{3}_{1} \cdot 3} = \frac{1}{3}$

45. $\frac{24}{28} = \frac{\cancel{2}^{1} \cdot \cancel{2}^{1} \cdot 6}{\cancel{2}_{1} \cdot \cancel{2}_{1} \cdot 7} = \frac{6}{7}$

47. $\frac{15}{40} = \frac{3 \cdot \cancel{5}^{1}}{2 \cdot 2 \cdot 2 \cancel{5}_{1}} = \frac{3}{8}$

49. $\frac{33}{56} = \frac{3 \cdot 11}{2 \cdot 2 \cdot 2 \cdot 7}$ Simplest form

51. $\frac{26}{39} = \frac{2 \cdot \cancel{13}^{1}}{3 \cdot \cancel{13}_{1}} = \frac{2}{3}$

53. $\frac{36}{225} = \frac{2 \cdot 2 \cdot \cancel{3}^{1} \cdot \cancel{3}^{1}}{5 \cdot 5 \cdot \cancel{3}_{1} \cdot \cancel{3}_{1}} = \frac{4}{25}$

Perform the operations and, if possible, simplify.

55. $\frac{3}{5} + \frac{3}{5} = \frac{3+3}{5} = \frac{6}{5}$

57. $\frac{6}{7} - \frac{2}{7} = \frac{6-2}{7} = \frac{4}{7}$

59. $\frac{1}{6}+\frac{1}{24}=\frac{1}{6}\cdot\frac{\mathbf{4}}{\mathbf{4}}+\frac{1}{24}$

$=\frac{4+1}{24}$

$=\frac{5}{24}$

61. $\frac{7}{10}-\frac{1}{14}=\frac{7}{10}\cdot\frac{\mathbf{7}}{\mathbf{7}}-\frac{1}{14}\cdot\frac{\mathbf{5}}{\mathbf{5}}$

$=\frac{49-5}{70}$

$=\frac{44}{70}$

$=\frac{\overset{1}{\cancel{2}}\cdot 22}{\underset{1}{\cancel{2}}\cdot 35}$

$=\frac{22}{35}$

63. $\frac{2}{15}+\frac{7}{9}=\frac{2}{15}\cdot\frac{\mathbf{3}}{\mathbf{3}}+\frac{7}{9}\cdot\frac{\mathbf{5}}{\mathbf{5}}$

$=\frac{6+35}{45}$

$=\frac{41}{45}$

65. $\frac{13}{28}-\frac{1}{21}=\frac{13}{28}\cdot\frac{\mathbf{3}}{\mathbf{3}}-\frac{1}{21}\cdot\frac{\mathbf{4}}{\mathbf{4}}$

$=\frac{39-4}{84}$

$=\frac{35}{84}$

$=\frac{5\cdot\overset{1}{\cancel{7}}}{\underset{1}{\cancel{7}}\cdot 12}$

$=\frac{5}{12}$

Perform the operations and, if possible, simplify.

67. $16\left(\frac{3}{2}\right)=\frac{16}{1}\cdot\frac{3}{2}$

$=\frac{\overset{1}{\cancel{2}}\cdot 8\cdot 3}{1\cdot\underset{1}{\cancel{2}}}$

$=24$

69. $18\left(\frac{2}{9}\right)=\frac{18}{1}\cdot\frac{2}{9}$

$=\frac{\overset{1}{\cancel{9}}\cdot 2\cdot 2}{1\cdot\underset{1}{\cancel{9}}}$

$=4$

71. $\frac{2}{3}+\frac{5}{18}-\frac{1}{6}=\frac{2}{3}\cdot\frac{\mathbf{6}}{\mathbf{6}}+\frac{5}{18}-\frac{1}{6}\cdot\frac{\mathbf{3}}{\mathbf{3}}$

$=\frac{12+5-3}{18}$

$=\frac{14}{18}$

$=\frac{\overset{1}{\cancel{2}}\cdot 7}{\underset{1}{\cancel{2}}\cdot 9}$

$=\frac{7}{9}$

73. $\frac{5}{12}+\frac{1}{3}-\frac{2}{5}=\frac{5}{12}\cdot\frac{\mathbf{5}}{\mathbf{5}}+\frac{1}{3}\cdot\frac{\mathbf{20}}{\mathbf{20}}-\frac{2}{5}\cdot\frac{\mathbf{12}}{\mathbf{12}}$

$=\frac{25+20-24}{60}$

$=\frac{21}{60}$

$=\frac{\overset{1}{\cancel{3}}\cdot 7}{\underset{1}{\cancel{3}}\cdot 20}$

$=\frac{7}{20}$

Perform the operations and, if possible, simplify.

75. $4\frac{2}{3}\cdot 7=\frac{14}{3}\cdot\frac{7}{1}$

$=\frac{14\cdot 7}{3\cdot 1}$

$=\frac{98}{3}$

$=32\frac{2}{3}$

77. $8\div 3\frac{1}{5}=\frac{8}{1}\div\frac{16}{5}$

$=\frac{8}{1}\cdot\frac{5}{16}$

$=\frac{5\cdot\overset{1}{\cancel{8}}}{1\cdot 2\cdot\underset{1}{\cancel{8}}}$

$=\frac{5}{2}$

$=2\frac{1}{2}$

79. $8\frac{2}{9}-7\frac{2}{3}=\frac{74}{9}-\frac{23}{3}$
$=\frac{74}{9}-\frac{23}{3}\cdot\frac{\mathbf{3}}{\mathbf{3}}$
$=\frac{74-69}{9}$
$=\frac{5}{9}$

81. $3\frac{3}{16}+2\frac{5}{24}=\frac{51}{16}+\frac{53}{24}$
$=\frac{51}{16}\cdot\frac{\mathbf{3}}{\mathbf{3}}+\frac{53}{24}\cdot\frac{\mathbf{2}}{\mathbf{2}}$
$=\frac{153+106}{48}$
$=\frac{259}{48}$
$=5\frac{19}{48}$

TRY IT YOURSELF

Perform the operations and, if possible, simplify.

83. $\frac{3}{5}+\frac{2}{3}=\frac{3}{5}\cdot\frac{\mathbf{3}}{\mathbf{3}}+\frac{2}{3}\cdot\frac{\mathbf{5}}{\mathbf{5}}$
$=\frac{9+10}{15}$
$=\frac{19}{15}$
$=1\frac{4}{15}$

85. $21\left(\frac{10}{3}\right)=\frac{21}{1}\cdot\frac{10}{3}$
$=\frac{\cancel{3}^{1}\cdot 7\cdot 10}{1\cdot\cancel{3}_{1}}$
$=70$

87. $6\cdot 2\frac{7}{24}=\frac{6}{1}\cdot\frac{55}{24}$
$=\frac{\cancel{6}^{1}\cdot 55}{1\cdot\cancel{6}_{1}\cdot 4}$
$=\frac{55}{4}$
$=13\frac{3}{4}$

89. $\frac{2}{3}-\frac{1}{4}+\frac{1}{12}=\frac{2}{3}\cdot\frac{\mathbf{4}}{\mathbf{4}}-\frac{1}{4}\cdot\frac{\mathbf{3}}{\mathbf{3}}+\frac{1}{12}$
$=\frac{8-3+1}{12}$
$=\frac{\cancel{6}^{1}}{2\cdot\cancel{6}_{1}}$
$=\frac{1}{2}$

91. $\frac{21}{35}\div\frac{3}{14}=\frac{21}{35}\cdot\frac{14}{3}$
$=\frac{\cancel{7}^{1}\cdot\cancel{3}^{1}\cdot 14}{5\cdot\cancel{7}_{1}\cdot\cancel{3}_{1}}$
$=\frac{14}{5}$
$=2\frac{4}{5}$

93. $\frac{4}{3}\left(\frac{6}{5}\right)=\frac{4\cdot 6}{3\cdot 5}$
$=\frac{4\cdot 2\cdot\cancel{3}^{1}}{\cancel{3}_{1}\cdot 5}$
$=\frac{8}{5}$
$=1\frac{3}{5}$

95. $\frac{4}{63}+\frac{1}{45}=\frac{4}{63}\cdot\frac{\mathbf{5}}{\mathbf{5}}+\frac{1}{45}\cdot\frac{\mathbf{7}}{\mathbf{7}}$
$=\frac{20+7}{315}$
$=\frac{27}{315}$
$=\frac{3\cdot\cancel{9}^{1}}{\cancel{9}_{1}\cdot 35}$
$=\frac{3}{35}$

97. $3-\frac{3}{4}=\frac{3}{1}\cdot\frac{\mathbf{4}}{\mathbf{4}}-\frac{3}{4}$

$=\frac{12-3}{4}$

$=\frac{9}{4}$

$=2\frac{1}{4}$

99. $\frac{1}{5}\cdot\frac{3}{5}=\frac{1\cdot 3}{5\cdot 5}$

$=\frac{3}{25}$

101. $3\frac{1}{3}\div 1\frac{5}{6}=\frac{10}{3}\div\frac{11}{6}$

$=\frac{10}{3}\cdot\frac{6}{11}$

$=\frac{10\cdot 2\cdot \cancel{3}^{1}}{\cancel{3}_{1}\cdot 11}$

$=\frac{20}{11}$

$=1\frac{9}{11}$

103. $\frac{11}{21}-\frac{8}{21}=\frac{11-8}{21}$

$=\frac{\cancel{3}^{1}}{7\cdot\cancel{3}_{1}}$

$=\frac{1}{7}$

105. $\frac{7}{30}+\frac{1}{50}-\frac{19}{75}=\frac{7}{30}\cdot\frac{\mathbf{5}}{\mathbf{5}}+\frac{1}{50}\cdot\frac{\mathbf{3}}{\mathbf{3}}-\frac{19}{75}\cdot\frac{\mathbf{2}}{\mathbf{2}}$

$=\frac{35+3-38}{150}$

$=\frac{38-38}{150}$

$=\frac{0}{150}$

$=0$

107. $1\frac{31}{32}\cdot 7\frac{1}{9}=\frac{63}{32}\cdot\frac{64}{9}$

$=\frac{7\cdot\cancel{9}^{1}}{\cancel{32}_{1}}\cdot\frac{2\cdot\cancel{32}^{1}}{\cancel{9}_{1}}$

$=\frac{14}{1}$

$=14$

LOOK ALIKES

109. a. $\frac{4}{9}+\frac{3}{7}=\frac{4}{9}\cdot\frac{\mathbf{7}}{\mathbf{7}}+\frac{3}{7}\cdot\frac{\mathbf{9}}{\mathbf{9}}$

$=\frac{28+27}{63}$

$=\frac{55}{63}$

b. $\frac{4}{9}-\frac{3}{7}=\frac{4}{9}\cdot\frac{\mathbf{7}}{\mathbf{7}}-\frac{3}{7}\cdot\frac{\mathbf{9}}{\mathbf{9}}$

$=\frac{28-27}{63}$

$=\frac{1}{63}$

c. $\frac{4}{9}\cdot\frac{3}{7}=\frac{4\cdot\cancel{3}^{1}}{\cancel{3}_{1}\cdot 3\cdot 7}$

$=\frac{4}{27}$

d. $\frac{4}{9}\div\frac{3}{7}=\frac{4}{9}\cdot\frac{7}{3}$

$=\frac{28}{27}$

APPLICATION

111. FORESTRY

a. $\frac{5}{32}+\frac{1}{16}=\frac{5}{32}+\frac{1}{16}\cdot\frac{\mathbf{2}}{\mathbf{2}}$

$=\frac{5+2}{32}$

$=\frac{7}{32}$

The growth was $\frac{7}{32}$ in.

b. $\frac{5}{32}-\frac{1}{16}=\frac{5}{32}-\frac{1}{16}\cdot\frac{\mathbf{2}}{\mathbf{2}}$

$=\frac{5-2}{32}$

$=\frac{3}{32}$

The difference is $\frac{3}{32}$ in.

113. CAMPUS TO CAREERS

$$22\frac{3}{4}+11\frac{1}{2}+28\frac{7}{8}=\frac{91}{4}+\frac{23}{2}+\frac{231}{8}$$
$$=\frac{91}{4}\cdot\frac{\mathbf{2}}{\mathbf{2}}+\frac{23}{2}\cdot\frac{\mathbf{4}}{\mathbf{4}}+\frac{231}{8}$$
$$=\frac{182+92+231}{8}$$
$$=\frac{505}{8}$$
$$=63\frac{1}{8}$$

The total dimensions is $63\frac{1}{8}$ inches.

115. FRAMES

$$4\cdot10\frac{1}{8}=\frac{4}{1}\cdot\frac{81}{8}$$
$$=\frac{\overset{1}{\cancel{4}}\cdot81}{1\cdot2\cdot\underset{1}{\cancel{4}}}$$
$$=\frac{81}{2}$$
$$=40\frac{1}{2}$$

$40\frac{1}{2}$ inches of molding will be needed.

WRITING

117-119. Answers will vary.

REVIEW

121. **Variables** are letters (or symbols) that stand for numbers.

CHALLENGE PROBLEMS

123. $\frac{11}{12}\cdot\frac{9}{9}=\frac{99}{108}$ $\quad\frac{8}{9}\cdot\frac{12}{12}=\frac{96}{108}$

$\frac{11}{12}$ is larger than $\frac{8}{9}$.

SECTION 1.3, STUDY SET
VOCABULARY

Fill in the blanks.

1. The set of **whole** numbers is $\{0, 1, 2, 3, 4, 5, \ldots\}$.
3. The figure [number line from −2 to 2] is called a **number line**.
5. Positive and negative numbers are called **signed** numbers.
7. The symbols < and > are **inequality** symbols.
9. 0.25 is called a **terminating** decimal and 0.333 … is called a **repeating** decimal.
11. An irrational number is a nonterminating, nonrepeating **decimal**.

CONCEPTS

13. Represent each situation using a signed number.
 a. **−$15 million** b. $\frac{5}{16}$ **in. or** $+\frac{5}{16}$ **in.**

15. a. **−20** b. $\frac{2}{3}$

17. **−14 and −4**

NOTATION

Fill in the blanks.

19. $\sqrt{2}$ is read "the **square root** of 2."
21. The symbol ≈ means **is approximately equal to**.
23. The symbol π is a letter from the **Greek** alphabet.
25. $-\frac{4}{5} = \frac{\boxed{-4}}{5} = \frac{4}{\boxed{-5}}$.

GUIDED PRACTICE

27.

	5	0	-3	$\frac{7}{8}$	0.17	$-9\frac{1}{4}$	$\sqrt{2}$	π
Real	✔	✔	✔	✔	✔	✔	✔	✔
Irrational							✔	✔
Rational	✔	✔	✔	✔	✔	✔		
Integer	✔	✔	✔					
Whole	✔	✔						
Natural	✔							

Determine whether each statement is true or false.

29. **True**
31. **False**
33. **True**
35. **True**

Use one of the symbols < or > to make each statement true.

37. $0 > -4$
39. $917 < 971$
41. $-2 > -3$
43. $-\frac{5}{8} < -\frac{3}{8}$
45. $\frac{2}{3} > \frac{3}{5}$
47. $-6.19 < -5.8$

Write each fraction as a decimal. If the result is a repeating decimal, use an over bar.

49. $\frac{5}{8} = 8\overline{)5.000}$ with quotient 0.625
 $= 0.625$

51. $\frac{1}{30} = 30\overline{)1.0000}$ with quotient 0.0333
 $= 0.0\overline{3}$

53. $\frac{1}{60} = 60\overline{)1.0000}$ with quotient 0.0166
 $= 0.01\overline{6}$

55. $\frac{21}{50} = 50\overline{)21.00}$ with quotient 0.42
 $= 0.42$

Graph each set of numbers on a number line.

57. 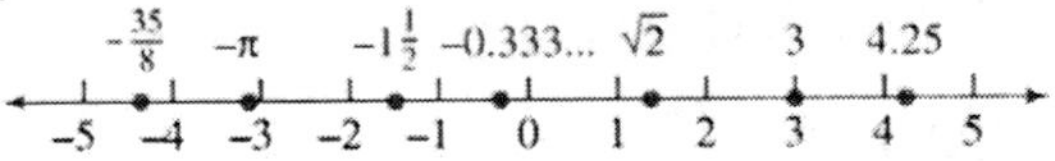 Number line from −5 to 5 with points at $-\frac{35}{8}$, $-\pi$, $-1\frac{1}{2}$, −0.333..., $\sqrt{2}$, 3, 4.25

59. 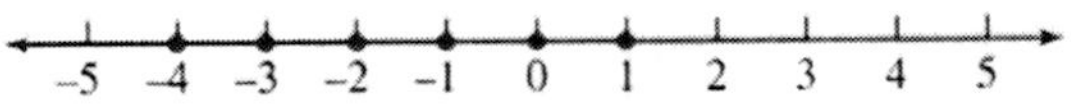 Number line from −5 to 5 with points at −4, −3, −2, −1, 0, 1

Find each absolute value.

61. $|83| = 83$
63. $\left|\frac{4}{3}\right| = \frac{4}{3}$
65. $|-11| = 11$
67. $|-6.1| = 6.1$

Insert one of the symbols <, >, or = in the blank to make each statement true.

69. $|3.4| \boxed{?} -3$

$3.4 \boxed{?} -3$

$3.4 > -3$

Thus, $|3.4| > -3$

71. $|-1.1| \boxed{?} 1.2$

$1.1 \boxed{?} 1.2$

$1.1 < 1.2$

Thus, $|-1.1| < 1.2$

73. $\left|-\frac{15}{2}\right| \boxed{?} 7.5$

$\frac{15}{2} \boxed{?} 7.5$

$7.5 = 7.5$

Thus, $\left|-\frac{15}{2}\right| = 7.5$

75. $\frac{99}{100} = 0.99$

77. $0.3 < 0.333...$

79. $1 \boxed{?} \left|-\frac{15}{16}\right|$

$1 \boxed{?} \frac{15}{16}$

$1 > \frac{15}{16}$

Thus, $1 > \left|-\frac{15}{16}\right|$

APPLICATIONS

81. DRAFTING

natural: 9 whole: 9

integers: 9 rational: $9, \frac{15}{16}, 3\frac{1}{8}, 1.765$

irrational: $2\pi, 3\pi, \sqrt{89}$ real: all

83. iPHONES

iii is the iPhone with the strongest strength of -49.

85. TRADE

a. 2006 had a deficit of $-\$90$ billion.

b. The smallest deficit was in 2009 and it was $-\$45$ billion.

WRITING

87-91. Answers will vary.

REVIEW

93. $\frac{24}{54} = \frac{2\cdot 2\cdot 2\cdot 3}{2\cdot 3\cdot 3\cdot 3}$

$= \frac{\overset{1}{\cancel{2}}\cdot 2\cdot 2\cdot \overset{1}{\cancel{3}}}{\underset{1}{\cancel{2}}\cdot 3\cdot 3\cdot \underset{1}{\cancel{3}}}$

$= \frac{4}{9}$

95. $5\frac{2}{3} \div 2\frac{5}{9} = \frac{17}{3} \div \frac{23}{9}$

$= \frac{17}{3}\cdot\frac{9}{23}$

$= \frac{17\cdot \overset{1}{\cancel{3}}\cdot 3}{\underset{1}{\cancel{3}}\cdot 23}$

$= \frac{51}{23}$

$= 2\frac{5}{23}$

CHALLENGE PROBLEMS

97. 1,999 integers

99. $\frac{1}{8}\cdot\mathbf{\frac{9}{9}}$ $\frac{1}{9}\cdot\mathbf{\frac{8}{8}}$

$\frac{9}{72}$ $\frac{8}{72}$

$\frac{9}{72}\cdot\mathbf{\frac{2}{2}}$ $\frac{8}{72}\cdot\mathbf{\frac{2}{2}}$

$\frac{18}{144}$ $\frac{16}{144}$

$\frac{17}{144}$ is between the 2 fractions.

LOOK ALIKES

103. a. $2.7+(-0.9)=1.8$

b. $2.7-(-0.9)=3.6$

c. $2.7(-0.9)=-2.43$

d. $\dfrac{2.7}{-0.9}=-3$

Use the associative property of multiplication to find each product.

105. $-\frac{1}{2}(2\cdot 67)=\left(-\frac{1}{2}\cdot 2\right)\cdot 67$
$=-1\cdot 67$
$=-67$

107. $-0.2(-10\cdot 3)=(-0.2\cdot -10)\cdot 3$
$=2\cdot 3$
$=6$

APPLICATIONS

109. REAL ESTATE
$\$200,000-\$160,000=\$40,000$
$\$40,000\div 5=\$8,000$
Depreciation per year was $8,000.

111. FLUID FLOW
$-6\cdot 12=-72$
The solution -72° indicates that the temperature dropped 72°.

113. WEIGHT LOSS

a. *ii*

b. $-4\frac{1}{2}\cdot(-8)=\frac{-9}{2}\cdot\frac{-8}{1}$
$=\dfrac{9\cdot 4\cdot \overset{1}{\cancel{2}}}{\underset{1}{\cancel{2}}\cdot 1}$
$=36$

Tom was 36 pounds heavier.

115. CAR RADIATORS
$1\frac{1}{2}(-34)=\frac{3}{2}\cdot\frac{-34}{1}$
$=-\dfrac{3\cdot 17\cdot \overset{1}{\cancel{2}}}{\underset{1}{\cancel{2}}\cdot 1}$
$=-51$

60/40 mixture protects to -51°F.

117. AIRLINES
$-25\cdot 6.4=-160$
2nd qrt losses was $160 million or (160).

$-800\cdot\frac{5}{16}=-250$
3rd qrt losses was $250 million or (250).

$-160\cdot 5=-800$
4th qrt losses was $800 million or (800).

119. PHYSICS

a. 5, -10

b. $5\cdot 0.5=2.5$, $-10\cdot 0.5=-5$

c. $5\cdot 1.5=7.5$, $-10\cdot 1.5=-15$

d. $5\cdot 2=10$, $-10\cdot 2=-20$

WRITING

121-123. Answers will vary.

REVIEW

125. $-3+(-4)+(-5)+4+3=-5$

127. $\frac{1}{2}+\frac{1}{4}+\frac{1}{3}=\frac{1}{2}\cdot\frac{\mathbf{6}}{\mathbf{6}}+\frac{1}{4}\cdot\frac{\mathbf{3}}{\mathbf{3}}+\frac{1}{3}\cdot\frac{\mathbf{4}}{\mathbf{4}}$
$=\dfrac{6+3+4}{12}$
$=\dfrac{13}{12}$

$12\overline{)13.0000}$ = 1.0833

$\dfrac{13}{12}=1.08\overline{3}$

CHALLENGE PROBLEMS

129. An odd number of factors must be negative.

SECTION 1.7, STUDY SET

VOCABULARY

Fill in the blanks.

1. In the exponential expression 7^5, 7 is the **base**, and 5 is the **exponent**. 7^5 is the fifth **power** of seven.

3. An **exponent** is used to represent repeated multiplication.

5. The rules for the **order** of operations guarantee that an evaluation of a numerical expression will result in a single answer.

CONCEPTS

7. a. **Subtraction** b. **Division**
   c. **Addition** d. **Power**

NOTATION

9. a. **Parentheses, brackets, braces, absolute value symbols, fraction bar**
   b. **Innermost : parentheses; outermost : brackets**

11. a. **−5** b. **5**

13. $$\begin{aligned}-19-2[(1+2)^2\cdot 3] &= -19-2[\mathbf{3}^2\cdot 3]\\ &= -19-2[\mathbf{9}\cdot 3]\\ &= -19-2[\mathbf{27}]\\ &= -19-\mathbf{54}\\ &= \mathbf{-73}\end{aligned}$$

GUIDED PRACTICE

Write each product using exponents. See Example 1.

15. $8\cdot 8\cdot 8=8^3$

17. $7\cdot 7\cdot 7\cdot 12\cdot 12=7^3 12^2$

19. $x\cdot x\cdot x=x^3$

21. $r\cdot r\cdot r\cdot r\cdot s\cdot s=r^4 s^2$

Evaluate each expression. See Example 2.

23. $$\begin{aligned}7^2 &= 7\cdot 7\\ &= 49\end{aligned}$$

25. $$\begin{aligned}6^3 &= 6\cdot 6\cdot 6\\ &= 36\cdot 6\\ &= 216\end{aligned}$$

27. $$\begin{aligned}(-5)^4 &= (-5)\cdot(-5)\cdot(-5)\cdot(-5)\\ &= 25\cdot(-5)\cdot(-5)\\ &= -125\cdot(-5)\\ &= 625\end{aligned}$$

29. $$\begin{aligned}(-0.1)^2 &= (-0.1)\cdot(-0.1)\\ &= 0.01\end{aligned}$$

31. $$\begin{aligned}\left(-\frac{1}{4}\right)^3 &= \left(-\frac{1}{4}\right)\cdot\left(-\frac{1}{4}\right)\cdot\left(-\frac{1}{4}\right)\\ &= -\frac{1\cdot 1\cdot 1}{4\cdot 4\cdot 4}\\ &= -\frac{1}{64}\end{aligned}$$

33. $$\begin{aligned}\left(\frac{2}{3}\right)^3 &= \left(\frac{2}{3}\right)\cdot\left(\frac{2}{3}\right)\cdot\left(\frac{2}{3}\right)\\ &= \frac{2\cdot 2\cdot 2}{3\cdot 3\cdot 3}\\ &= \frac{8}{27}\end{aligned}$$

Evaluate each expression. See Example 3.

35. $$\begin{aligned}(-6)^2 &= (-6)\cdot(-6)\\ &= 36\end{aligned}$$

$$\begin{aligned}-6^2 &= -(6\cdot 6)\\ &= -36\end{aligned}$$

37. $$\begin{aligned}(-8)^2 &= (-8)\cdot(-8)\\ &= 64\end{aligned}$$

$$\begin{aligned}-8^2 &= -(8\cdot 8)\\ &= -64\end{aligned}$$

Evaluate each expression. See Example 4.

39. $$\begin{aligned}3-5\cdot 4 &= 3-20\\ &= 3+(-20)\\ &= -17\end{aligned}$$

41. $$\begin{aligned}32-16\div 4+2 &= 32-4+2\\ &= 28+2\\ &= 30\end{aligned}$$

43. $$\begin{aligned}9\cdot 5-6\div 3 &= 45-2\\ &= 45+(-2)\\ &= 43\end{aligned}$$

45. $$\begin{aligned}12\div 3\cdot 2 &= 4\cdot 2\\ &= 8\end{aligned}$$

47. $-22-15+3=-22+(-15)+3$
$=-37+3$
$=-34$

49. $-2(9)-2(5)(10)=-18-100$
$=-18+(-100)$
$=-118$

Evaluate each expression. See Example 5.

51. $-4(6+5)=-4(11)$
$=-44$

53. $(9-3)(9-9)^2=[9+(-3)][9+(-9)]^2$
$=(6)(0)^2$
$=(6)(0)$
$=0$

55. $(-1-3^2\cdot 4)2^2=[-1-(3\cdot 3)\cdot 4](2\cdot 2)$
$=[-1-9\cdot 4]4$
$=[-1-36]4$
$=[-1+(-36)]4$
$=(-37)4$
$=-148$

57. $1+5(10+2\cdot 5)-1=1+5(10+10)-1$
$=1+5(20)-1$
$=1+100-1$
$=0+100$
$=100$

Evaluate each expression. See Example 6.

59. $(-1)^9[-7^2-(-2)^2]=(-1)[-49-4]$
$=(-1)[-49+(-4)]$
$=(-1)(-53)$
$=53$

61. $64-6[15+2(-3+8)]=64-6[15+2(5)]$
$=64-6[15+10]$
$=64-6[25]$
$=64-150$
$=64+(-150)$
$=-86$

63. $-2[2+4^2(8-9)]^2=-2\{2+16[8+(-9)]\}^2$
$=-2[2+16(-1)]^2$
$=-2[2+(-16)]^2$
$=-2[-14]^2$
$=-2(196)$
$=-392$

65. $3+2[-1-(4-5)]=3+2[-1-(-1)]$
$=3+2[-1+1]$
$=3+2[0]$
$=3+0$
$=3$

Evaluate each expression. See Example 7.

67. $\dfrac{-2-5}{-7+(-7)}=\dfrac{-2+(-5)}{-7+(-7)}$
$=\dfrac{-7}{-14}$
$=\dfrac{1}{2}$

69. $\dfrac{2\cdot 2^5-60+(-4)}{5^4-(-4)(-5)}=\dfrac{2\cdot 32-60+(-4)}{625-20}$
$=\dfrac{64-60+(-4)}{625+(-20)}$
$=\dfrac{4+(-4)}{605}$
$=\dfrac{0}{605}$
$=0$

71. $\dfrac{2(-4-2\cdot 2)}{3(-3)(-2)}=\dfrac{2(-4-4)}{3(6)}$
$=\dfrac{2[-4+(-4)]}{18}$
$=\dfrac{2(-8)}{18}$
$=\dfrac{-16}{18}$
$=-\dfrac{\cancel{2}^{1}\cdot 8}{\cancel{2}_{1}\cdot 9}$
$=-\dfrac{8}{9}$

73. $\dfrac{72-(2-2\cdot 4)}{10^2-(9\cdot 10+2^2)}=\dfrac{72-(2-8)}{100-(90+4)}$

$=\dfrac{72-[2+(-8)]}{100-94}$

$=\dfrac{72-(-6)}{100+(-94)}$

$=\dfrac{72+6}{6}$

$=\dfrac{78}{6}$

$=\dfrac{\cancel{6}\cdot 13}{\cancel{6}}$

$=13$

Evaluate each expression.

75. $10-2|4-8|=10-2|4+(-8)|$

$=10-2|-4|$

$=10-2(4)$

$=10-8$

$=2$

77. $-|7-2^3(4-7)|=-|7-8[4+(-7)]|$

$=-|7-8(-3)|$

$=-|7-(-24)|$

$=-|7+24|$

$=-|31|$

$=-31$

79. $\dfrac{(3+5)^2+|-2|}{-2(5-8)}=\dfrac{(8)^2+2}{-2[5+(-8)]}$

$=\dfrac{64+2}{-2(-3)}$

$=\dfrac{66}{6}$

$=\dfrac{\cancel{6}\cdot 11}{\cancel{6}}$

$=11$

81. $\dfrac{|6-4|+2|-4|}{226-6^3}=\dfrac{|6+(-4)|+2(4)}{226-216}$

$=\dfrac{|2|+8}{10}$

$=\dfrac{2+8}{10}$

$=\dfrac{10}{10}$

$=\dfrac{\cancel{10}}{\cancel{10}}$

$=1$

83. $-(2\cdot 3-2^2)^5=-(6-4)^5$

$=-(2)^5$

$=-(2\cdot 2\cdot 2\cdot 2\cdot 2)$

$=-32$

85. $2\cdot 5^2+4\cdot 3^2=2\cdot(5\cdot 5)+4\cdot(3\cdot 3)$

$=2\cdot 25+4\cdot 9$

$=50+36$

$=86$

87. $-2(-1)^2+3(-1)-3=-2(1)+(-3)-3$

$=-2+(-3)+(-3)$

$=-8$

89. $8-3[5^2-(7-3)^2]=8-3[5^2-(4)^2]$

$=8-3(25-16)$

$=8-3[25+(-16)]$

$=8-3(9)$

$=8-27$

$=8+(-27)$

$=-19$

TRY IT YOURSELF

Evaluate each expression.

91. $[6(5)-5(5)]^3(-4)=[30-25]^3(-4)$

$=[30+(-25)]^3(-4)$

$=[5]^3(-4)$

$=125(-4)$

$=-500$

93. $8-6[(130-4^3)-2]$
$$\begin{aligned}&=8-6[(130-64)-2]\\&=8-6[130+(-64)+(-2)]\\&=8-6[64]\\&=8-384\\&=8+(-384)\\&=-376\end{aligned}$$

95. $-2\left(\frac{15}{-5}\right)-\frac{6}{2}+9=-2(-3)-3+9$
$$\begin{aligned}&=6+(-3)+9\\&=3+9\\&=12\end{aligned}$$

97. $-5(-2)^3-|-2+1|=-5(-8)-|-1|$
$$\begin{aligned}&=40-1\\&=39\end{aligned}$$

99. $\frac{18-[2+(1-6)]}{16-(-4)^2}=\frac{18-[2+(1+(-6))]}{16-(16)}$
$$\begin{aligned}&=\frac{18-[2+(-5)]}{16+(-16)}\\&=\frac{18-[-3]}{0}\end{aligned}$$
Undefined

101. $-|-5\cdot 7^2|-30=-|-5\cdot 49|+(-30)$
$$\begin{aligned}&=-|-245|+(-30)\\&=-245+(-30)\\&=-275\end{aligned}$$

103. $(-3)^3\left(\frac{-4}{2}\right)(-1)=-27(-2)(-1)$
$$\begin{aligned}&=54(-1)\\&=-54\end{aligned}$$

105. $\frac{1}{2}\left(\frac{1}{8}\right)+\left(-\frac{1}{4}\right)^2=\frac{1\cdot 1}{2\cdot 8}+\frac{1\cdot 1}{4\cdot 4}$
$$\begin{aligned}&=\frac{1}{16}+\frac{1}{16}\\&=\frac{1+1}{16}\\&=\frac{2}{16}\\&=\frac{\overset{1}{\cancel{2}}}{\underset{1}{\cancel{2}}\cdot 8}\\&=\frac{1}{8}\end{aligned}$$

107. $\frac{-5^2\cdot 10+5\cdot 2^5}{-5-3-1}=\frac{-25\cdot 10+5\cdot 32}{-5+(-3)+(-1)}$
$$\begin{aligned}&=\frac{-250+160}{-9}\\&=\frac{-90}{-9}\\&=\frac{\overset{1}{\cancel{9}}\cdot 10}{\underset{1}{\cancel{9}}}\\&=10\end{aligned}$$

109. $-\left(\frac{40-1^3-2^4}{3(2+5)+2}\right)=-\left(\frac{40-1-16}{3(7)+2}\right)$
$$\begin{aligned}&=-\left(\frac{40+(-1)+(-16)}{21+2}\right)\\&=-\left(\frac{23}{23}\right)\\&=-\frac{\overset{1}{\cancel{23}}}{\underset{1}{\cancel{23}}}\\&=-1\end{aligned}$$

LOOK ALIKES

111. a. $(-7-4)(-2)=(-11)(-2)$
$$=22$$

b. $(-7-4)-2=(-11)+(-2)$
$$=-13$$

113. a. $-100 \div 5 \cdot 2 = -20 \cdot 2$
$= -40$

b. $-100 \div (5 \cdot 2) = -100 \div 10$
$= -10$

APPLICATIONS

115. LIGHT

2 yards = 2^2 square units

3 yards = 3^2 square units

4 yards = 4^2 square units

117. CAMPUS TO CAREERS

Step 1: Find the total minutes.

Gate A: $3:21 - 3:05 = 16$

Gate B: $3:13 - 3:03 = 10$

Gate C: $3:09 - 3:01 = 8$

Gate D: $3:16 - 3:02 = 14$

Step 2: Find the average.

$$\frac{16+10+8+14}{4} = \frac{48}{4} = 12$$

The average wait time is about 12 minutes.

119. CASH AWARDS

a. $2{,}500 + 4(500) + 35(150) + 85(25)$
$= 2{,}500 + 2{,}000 + 5{,}250 + 2{,}125$
$= 4{,}500 + 5{,}250 + 2{,}125$
$= 9{,}750 + 2{,}125$
$= 11{,}875$

The total amount of money is \$11,875.

b. $1 + 4 + 35 + 85 = 125$

125 is the total number of cash prizes.

$$\frac{11{,}875}{125} = 95$$

The average cash prize is \$95.

121. WRAPPING GIFTS

$15 + 4(4) + 2(16) + 2(9)$
$= 15 + 16 + 32 + 18$
$= 31 + 32 + 18$
$= 63 + 18$
$= 81$

The total amount of ribbon is 81 in.

WRITING

123-125. Answers will vary.

REVIEW

127. $-11 + (-6) = -17$

$-11 + 6 = -5$

CHALLENGE PROBLEMS

129. $(3^2)^4 = 9^4$
$= 6{,}561$

Translate the set of instructions to an expression and then evaluate it.

131. $(-3)^3(-4) - (-9+8) = -27(-4) - (-1)$
$= 108 + 1$
$= 109$

SECTION 1.8, STUDY SET
VOCABULARY

Fill in the blanks.

1. Variables and/or numbers can be combined with the operations of arithmetic to create algebraic **expressions**.
3. Addition symbols separate algebraic expressions into parts called **terms**.
5. The **coefficient** of the term $10x$ is 10.

CONCEPTS

7. Number of days in w weeks

   7, 14, 21, $7w$
9. $(\mathbf{12-h})$ inches
11. $(\mathbf{x+20})$ ounces
13. a. $\mathbf{b-15}$ b. $\mathbf{p+15}$

NOTATION

15. $$9a-a^2=9(\mathbf{5})-(5)^2$$
    $$=9(5)-\mathbf{25}$$
    $$=\mathbf{45}-25$$
    $$=20$$
17. a. $\mathbf{8y}$ b. $\mathbf{2cd}$ c. **Commutative**

GUIDED PRACTICE . See Example 1.

19. a. $3x^3+11x^2-x+9$ has 4 terms.

    b. Coefficient of $3x^3$ is 3.

    Coefficient of $11x^2$ is 11.

    Coefficient of $-x$ is -1.

    Coefficient of constant 9 is 9.

Determine whether the variable c is used as a factor or as a term. See Example 2.

21. $c+32$, Term
23. $5c$, Factor

Translate each phrase to an algebraic expression. If no variable is given, use x as the variable. See Example 3.

25. $l+15$
27. $50x$
29. $\frac{w}{l}$
31. $P+\frac{2}{3}p$
33. $k^2-2,005$
35. $2a-1$
37. $\frac{1,000}{n}$
39. $2p+90$
41. $3(35+h+300)$
43. $p-680$
45. $4d-15$
47. $2(200+t)$
49. $|a-2|$
51. $0.1d$ or $\frac{1}{10}d$

Translate each algebraic expression into an English phrase. (Answers may vary.) See Example 3.

53. Three-fourths of r
55. Fifty less than t
57. The product of x, y, and z.
59. Twice m, increased by 5

Answer with an algebraic expression. See Example 4.

61. $(x+2)$ inches
63. $(36-x)$ inches

Answer with an algebraic expression. See Example 8.

65. $60h$ minutes
67. $\frac{i}{12}$ feet

Answer with an algebraic expression. See Example 9.

69. $\$8x$
71. $49x$¢
73. $\$2t$
75. $\$25(x+2)$

Evaluate each expression, for $x = 3$, $y = -2$, and $z = -4$. See Example 10.

77. $$-y=-(-2)$$
    $$=2$$
79. $$-z+3x=-(-4)+3(3)$$
    $$=4+9$$
    $$=13$$
81. $$3y^2-6y-4=3(-2)^2-6(-2)-4$$
    $$=3(4)+12-4$$
    $$=12+12+(-4)$$
    $$=24+(-4)$$
    $$=20$$

83. $(3+x)y = (3+3)(-2)$
$= 6(-2)$
$= -12$

85. $(x+y)^2 - |z+y| = [3+(-2)]^2 - |-4+(-2)|$
$= (1)^2 - |-6|$
$= 1-6$
$= 1+(-6)$
$= -5$

87. $-\dfrac{2x+y^3}{y+2z} = -\dfrac{2(3)+(-2)^3}{(-2)+2(-4)}$
$= -\dfrac{6+(-8)}{(-2)+(-8)}$
$= -\dfrac{-2}{-10}$
$= -\dfrac{\cancel{2}}{\cancel{2}\cdot 5}$
$= -\dfrac{1}{5}$

Evaluate each expression. See Example 10.

89. $b^2 - 4ac = (5)^2 - 4(-1)(-2)$
$= 25-8$
$= 25+(-8)$
$= 17$

91. $a^2 + 2ab + b^2 = (-5)^2 + 2(-5)(-1) + (-1)^2$
$= 25+10+1$
$= 35+1$
$= 36$

93. $\dfrac{n}{2}[2a+(n-1)d]$
$= \dfrac{10}{2}[2(-4.2)+(10-1)6.6]$
$= 5[-8.4+9(6.6)]$
$= 5[-8.4+59.4]$
$= 5(51)$
$= 255$

95. $(27c^2 - 4d^2)^3 = \left[27\left(\dfrac{1}{3}\right)^2 - 4\left(\dfrac{1}{2}\right)^2\right]^3$
$= \left[27\left(\dfrac{1}{9}\right) - 4\left(\dfrac{1}{4}\right)\right]^3$
$= \left(\dfrac{27\cdot 1}{9} - \dfrac{4\cdot 1}{4}\right)^3$
$= \left(\dfrac{\cancel{9}\cdot 3}{\cancel{9}} - \dfrac{\cancel{4}}{\cancel{4}}\right)^3$
$= (3+(-1))^3$
$= (2)^3$
$= 8$

Complete each table. See Example 11.

97.

x	$x^3 - 1$
0	$(\mathbf{0})^3 - 1 = 0 - 1$ $= -1$
-1	$(\mathbf{-1})^3 - 1 = -1 - 1$ $= -2$
-3	$(\mathbf{-3})^3 - 1 = -27 - 1$ $= -28$

99.

s	$\dfrac{5s+36}{s}$
1	$\dfrac{5(\mathbf{1})+36}{\mathbf{1}} = \dfrac{5+36}{1}$ $= \dfrac{41}{1}$ $= 41$
6	$\dfrac{5(\mathbf{6})+36}{\mathbf{6}} = \dfrac{30+36}{6}$ $= \dfrac{66}{6}$ $= 11$
-12	$\dfrac{5(\mathbf{-12})+36}{\mathbf{-12}} = \dfrac{-60+36}{-12}$ $= \dfrac{-24}{-12}$ $= 2$

101.

Input x	Output $2x-\frac{x}{2}$
100	$2(\mathbf{100})-\frac{\mathbf{100}}{2}=200-50$ $=150$
−300	$2(\mathbf{-300})-\frac{\mathbf{-300}}{2}=-600-(-150)$ $=-600+150$ $=-450$

103.

x	$(x+1)(x+5)$
−1	$(\mathbf{-1}+1)(\mathbf{-1}+5)=(0)(4)$ $=0$
−5	$(\mathbf{-5}+1)(\mathbf{-5}+5)=(-4)(0)$ $=0$
−6	$(\mathbf{-6}+1)(\mathbf{-6}+5)=(-5)(-1)$ $=5$

TRY IT YOURSELF

105. a. $x+7$ b. $(x+7)^2$

107. a. $4(x+2)$ b. $4x+2$

APPLICATIONS

109. VEHICLE WEIGHTS

a. Let x = weight of the Element

$2x-340$ = weight of the Hummer

b. Elements weights 3,370 pounds

$$\text{Hummer}=2(3{,}370)+(-340)$$
$$=6{,}740+(-340)$$
$$=6{,}400$$

The Hummer weights 6,400 pounds.

111. COMPUTER COMPANIES

a. Let x = age of Apple

$x+80$ = age of IBM

$x-9$ = age of Dell

b. Let $x=32$ (Apple's age in 2008)

$$\text{IBM's age}=x+80$$
$$=32+80$$
$$=112$$

IBM is 112 years old.

$$\text{Dell's age}=x-9$$
$$=32-9$$
$$=32+(-9)$$
$$=23$$

Dell is 23 years old.

WRITING

113-115. Answers will vary.

REVIEW

117. $12=2\cdot 2\cdot 3$

$15=3\cdot 5$

$\text{LCD}=2\cdot 2\cdot 3\cdot 5=60$

119. $\left(\frac{2}{3}\right)^3=\frac{2\cdot 2\cdot 2}{3\cdot 3\cdot 3}$

$$=\frac{8}{27}$$

CHALLENGE PROBLEMS

121. The answer would be 0, because $(8-8)=0$ is a factor of the expression.

SECTION 1.9, STUDY SET
VOCABULARY

Fill in the blanks.

1. To **simplify** the expression $5(6x)$ means to write it in simpler form: $5(6x) = 30x$.
3. To perform the multiplication $2(x + 8)$, we use the **distributive** property.
5. Terms such as $7x^2$ and $5x^2$, which have the same variables raised to exactly the same power, are called **like** terms.

CONCEPTS

7. a. $4(9t) = (\mathbf{4 \cdot 9})t = \mathbf{36}y$

   b. Associative Property of Multiplication

9. a. $2(x+4) = 2x+8$

   b. $2(x-4) = 2x-8$

   c. $-2(x+4) = -2x-8$

   d. $-2(-x-4) = 2x+8$

11. a. $5(2x) = 10x$

   b. $5+2x$; Can't be simplified

   c. $6(-7x) = -42x$

   d. $6-7x$; Can't be simplified

   e. $2(3x)(3) = 18x$

   f. $2+3x+3 = 3x+5$

NOTATION

13. a. $\mathbf{6(h-4)}$ b. $\mathbf{-(z+16)}$

GUIDED PRACTICE

Use the given property to complete each statement. See Example 1.

15. $8+(7+a) = \underline{(8+7)+a}$

17. $y \cdot 11 = \underline{11y}$

19. $(8d \cdot 2)6 = \underline{8d(2 \cdot 6)}$

21. $9t+(4+t) = 9t+\underline{(t+4)}$

Simplify each expression. See Example 2.

23. $3 \cdot 4t = (3 \cdot 4)t$
$$= 12t$$

25. $5(-7q) = (5 \cdot -7)q$
$$= -35q$$

27. $(-5.6x)(-2) = (-5.6 \cdot -2)x$
$$= 11.2x$$

29. $5(4c)(3) = [5(4)(3)]c$
$$= 60c$$

31. $\frac{5}{3} \cdot \frac{3}{5} g = \frac{5 \cdot 3}{3 \cdot 5} g$
$$= \frac{\overset{1}{\cancel{5}} \cdot \overset{1}{\cancel{3}}}{\underset{1}{\cancel{3}} \cdot \underset{1}{\cancel{5}}} g$$
$$= g$$

33. $12\left(\frac{5}{12}x\right) = \frac{12}{1} \cdot \frac{5}{12} x$
$$= \frac{\overset{1}{\cancel{12}} \cdot 5}{1 \cdot \underset{1}{\cancel{12}}} x$$
$$= 5x$$

Multiply. See Example 3.

35. $5(x+3) = 5 \cdot x + 5 \cdot 3$
$$= 5x+15$$

37. $-3(4x+9) = -3(4x) - 3(9)$
$$= -12x - 27$$

39. $45\left(\frac{x}{5} + \frac{2}{9}\right) = 45\left(\frac{x}{5}\right) + 45\left(\frac{2}{9}\right)$
$$= \frac{45 \cdot 1}{5} x + \left(\frac{45 \cdot 2}{9}\right)$$
$$= \frac{\overset{1}{\cancel{5}} \cdot 9}{\underset{1}{\cancel{5}}} x + \left(\frac{\overset{1}{\cancel{9}} \cdot 5 \cdot 2}{\underset{1}{\cancel{9}}}\right)$$
$$= 9x + 10$$

41. $0.4(x+4) = 0.4(x) + (0.4)(4)$
$$= 0.4x + 1.6$$

Multiply. See Example 4.

43. $6(6c-7) = 6(6c) - 6(7)$
$$= 36c - 42$$

45. $-6(13c-3) = -6(13c) - (-6)(3)$
$$= -78c - (-18)$$
$$= -78c + 18$$

47. $-15(-2t-6)=-15(-2t)-(-15)(6)$
$=30t-(-90)$
$=30t+90$

49. $-1(-4a+1)=-1(-4a)+(-1)(1)$
$=4a+(-1)$
$=4a-1$

Multiply. See Example 5.

51. $(3t+2)8=(3t)8+(2)8$
$=24t+16$

53. $(3w-6)\frac{2}{3}=3w\left(\frac{2}{3}\right)-6\left(\frac{2}{3}\right)$
$=\frac{2\cdot 3}{3}w-\left(\frac{2\cdot 6}{3}\right)$
$=\frac{2\cdot \cancel{3}^1}{\cancel{3}_1}w-\left(\frac{2\cdot 2\cdot \cancel{3}^1}{\cancel{3}_1}\right)$
$=2w-4$

55. $4(7y+4)2=4\cdot 2(7y+4)$
$=8(7y+4)$
$=8(7y)+8(4)$
$=56y+32$

57. $2.5(2a-3b+1)$
$=2.5(2a)-(2.5)(3b)+(2.5)(1)$
$=5a-7.5b+2.5$

Multiply. See Example 6.

59. $-(x-7)=-1(x)-(-1)(7)$
$=-x-(-7)$
$=-x+7$

61. $-(-5.6y+7)=-1(-5.6y)+(-1)(7)$
$=5.6y+(-7)$
$=5.6y-7$

List the like terms in each expression, if any. See Example 7.

63. $3x+2-2x$, $3x$ and $-2x$

65. $-12m^4-3m^3+2m^2-m^3$
$-3m^3$ and $-m^3$

Simplify by combining like terms. See Example 8.

67. $3x+7x=(3+7)x$
$=10x$

69. $-7b^2+27b^2=(-7+27)b^2$
$=20b^2$

Simplify by combining like terms. See Example 9.

71. $36y+y-9y=[36+1+(-9)]y$
$=28y$

73. $\frac{3}{5}t+\frac{1}{5}t=\left(\frac{3}{5}+\frac{1}{5}\right)t$
$=\left(\frac{3+1}{5}\right)t$
$=\frac{4}{5}t$

75. $13r-12r=(13-12)r$
$=1r$
$=r$

77. $43s^3-44s^3=[43+(-44)]s^3$
$=-1s^3$
$=-s^3$

Simplify by combining like terms. See Example 10.

79. $15y-10-y-20y$
$=[15+(-1)+(-20)]y-10$
$=-6y-10$

81. $3x+4-5x+1=(3-5)x+(4+1)$
$=-2x+5$

83. $9m^2-6m+12m-4=9m^2+(-6+12)m-4$
$=9m^2+6m-4$

85. $4x^2+5x-8x+9=4x^2+(5-8)x+9$
$=4x^2-3x+9$

Simplify. See Example 11.

87. $2z+5(z-3)-10=2z+5(z)-(5)(3)-10$
$=2z+5z+(-15-10)$
$=(2+5)z+(-25)$
$=7z-25$

89. $2(s^2-7)-(s^2-2)$
$$=2(s^2)-(2)(7)-1(s^2)-(-1)(2)$$
$$=2s^2-14-s^2+2$$
$$=(2-1)s^2+(-14+2)$$
$$=1s^2+(-12)$$
$$=s^2-12$$

TRY IT YOURSELF

Simplify each expression, if possible.

91. $-\frac{7}{16}x-\frac{3}{16}x=\left(-\frac{7}{16}-\frac{3}{16}\right)x$
$$=\left(\frac{-7+(-3)}{16}\right)x$$
$$=-\frac{10}{16}x$$
$$=-\frac{\cancel{2}^{1}\cdot 5}{\cancel{2}_{1}\cdot 8}x$$
$$=-\frac{5}{8}x$$

93. $-9.8c+6.2c=(-9.8+6.2)c$
$$=-3.6c$$

95. $-4(-6)(-4m)=\left[-4(-6)(-4)\right]m$
$$=-96m$$

97. $-4x+4x=(-4+4)x$
$$=0x$$
$$=0$$

99. $-0.2r-(-0.6r)=[-0.2-(-0.6)]r$
$$=(-0.2+0.6)r$$
$$=0.4r$$

101. $8\left(\frac{3}{4}y\right)=\frac{8}{1}\cdot\frac{3}{4}y$
$$=\frac{\cancel{4}^{1}\cdot 2\cdot 3}{1\cdot\cancel{4}_{1}}y$$
$$=6y$$

103. $-9(3r-9)-7(2r-7)$
$$=-9(3r)-9(-9)-7(2r)-(-7)(7)$$
$$=-27r+81-14r-(-49)$$
$$=-27r+81-14r+49$$
$$=(-27-14)r+(81+49)$$
$$=-41r+130$$

105. $9(7m)=(9\cdot 7)m$
$$=63m$$

107. $6-4(-3c-7)=6-4(-3c)-(-4)(7)$
$$=6+12c+28$$
$$=12c+(6+28)$$
$$=12c+34$$

109. $5t\cdot 60=(5\cdot 60)t$
$$=300t$$

111. $36\left(\frac{2}{9}x-\frac{3}{4}\right)+36\left(\frac{1}{2}\right)$
$$=\frac{36}{1}\left(\frac{2}{9}\right)x-\left(\frac{36}{1}\right)\left(\frac{3}{4}\right)+\left(\frac{36}{1}\right)\left(\frac{1}{2}\right)$$
$$=\frac{36\cdot 2}{1\cdot 9}x-\left(\frac{36\cdot 3}{1\cdot 4}\right)+\left(\frac{36\cdot 1}{1\cdot 2}\right)$$
$$=\frac{\cancel{9}^{1}\cdot 4\cdot 2}{\cancel{9}_{1}}x-\left(\frac{\cancel{4}^{1}\cdot 9\cdot 3}{\cancel{4}_{1}}\right)+\left(\frac{\cancel{2}^{1}\cdot 18}{\cancel{2}_{1}}\right)$$
$$=8x-(27)+18$$
$$=8x+(-27+18)$$
$$=8x-9$$

113. $-4r-7r+2r-r=\left[-4+(-7)+2+(-1)\right]r$
$$=(-10)r$$
$$=-10r$$

115. $24\left(-\frac{5}{6}r\right)=\frac{24}{1}\left(-\frac{5}{6}\right)r$

$$=\left(-\frac{24\cdot 5}{1\cdot 6}\right)r$$

$$=\left(-\frac{\overset{1}{\cancel{6}}\cdot 4\cdot 5}{\underset{1}{\cancel{6}}}\right)r$$

$$=-\frac{20}{1}r$$

$$=-20r$$

117. $a+a+a=(1+1+1)a$

$$=3a$$

119. $60\left(\frac{3}{20}r-\frac{4}{15}\right)=\left(\frac{60}{1}\cdot\frac{3}{20}\right)r-\left(\frac{60}{1}\cdot\frac{4}{15}\right)$

$$=\left(\frac{60\cdot 3}{1\cdot 20}\right)r-\left(\frac{60\cdot 4}{1\cdot 15}\right)$$

$$=\left(\frac{\overset{1}{\cancel{20}}\cdot 3\cdot 3}{\underset{1}{\cancel{20}}}\right)r-\left(\frac{\overset{1}{\cancel{15}}\cdot 4\cdot 4}{\underset{1}{\cancel{15}}}\right)$$

$$=9r-16$$

121. $4a+4b+4c$, Doesn't simplify.

123. $-(c+7)+2(c-3)$

$$=-1(c)+(-1)(7)+2(c)+2(-3)$$

$$=-c+(-7)+2c+(-6)$$

$$=(-1+2)c+(-7-6)$$

$$=c-13$$

125. $a^3+2a^2+4a-2a^2-4a-8$

$$=a^3+(2-2)a^2+(4-4)a-8$$

$$=a^3+(0)a^2+(0)a-8$$

$$=a^3-8$$

LOOK ALIKES

127. a. $2(7x)5=(2\cdot 7\cdot 5)x$

$$=70x$$

b. $2(7x+5)=2\cdot 7x+2\cdot 5$

$$=14x+10$$

APPLICATIONS

129. RED CROSS

$$x+x+x+x+x+x+x+x+x+x+x+x=12x$$

Perimeter of the cross is $12x$ units.

WRITING

131. Answers will vary.

REVIEW

Evaluate each expression for $x=-3$, $y=-5$, *and* $z=0$.

133. $\frac{x-y^2}{2y-1+x}=\frac{-3-(-5)^2}{2(-5)-1+(-3)}$

$$=\frac{-3-(25)}{-10-1+(-3)}$$

$$=\frac{-3+(-25)}{-10+(-1)+(-3)}$$

$$=\frac{-28}{-14}$$

$$=\frac{\overset{1}{\cancel{14}}\cdot 2}{\underset{1}{\cancel{14}}}$$

$$=2$$

CHALLENGE PROBLEMS

135. $\boxed{\mathbf{-17}}(\boxed{\mathbf{11x}}+\boxed{7})=-187x-119$

CHAPTER 1
SUMMARY & REVIEW
SECTION 1.1
Introducing the Language of Algebra

1. 1 hour; 100 cars
2. 100
3. 7 P.M.
4. 12 A.M. (midnight)
5. The difference of 15 and 3 equals 12.
6. The sum of 15 and 3 equals 18.
7. The quotient of 15 and 3 equals 5.
8. The product of 15 and 3 equals 45.
9. a. $4\cdot 9$ and $4(9)$ b. $\frac{9}{3}$
10. a. $8b$ b. Prt
11. a. Equation b. Expression
12. $n = b + 5$

Brackets (b)	Nails (n)
5	**10**
10	**15**
20	**25**

SECTION 1.2
Fractions

13. a. $2\cdot 12, 3\cdot 8$ (Answers may vary)
    b. $2\cdot 2\cdot 6$ (Answers may vary)
    c. 1, 2, 3, 4, 6, 8, 12, 24
14. Equivalent
15. $54 = 2\cdot 3^3$
16. $147 = 3\cdot 7^2$
17. $385 = 5\cdot 7\cdot 11$
18. 41 is prime.
19. $$\frac{20}{35} = \frac{2\cdot 2\cdot \cancel{5}^{1}}{\cancel{5}_{1}\cdot 7} = \frac{4}{7}$$
20. $$\frac{24}{18} = \frac{2\cdot 2\cdot \cancel{2}^{1}\cdot \cancel{3}^{1}}{\cancel{2}_{1}\cdot \cancel{3}_{1}\cdot 3} = \frac{4}{3}$$
21. $$\frac{5}{8}\cdot\frac{\mathbf{8}}{\mathbf{8}} = \frac{40}{64}$$
22. $$\frac{12}{1}\cdot\frac{\mathbf{3}}{\mathbf{3}} = \frac{36}{3}$$
23. $$\begin{aligned} 10 &= 2\cdot 5 \\ 18 &= 2\cdot 3\cdot 3 \\ \text{LCD} &= 2\cdot 3\cdot 3\cdot 5 \\ &= 90 \end{aligned}$$
24. $$\begin{aligned} 21 &= 3\cdot 7 \\ 70 &= 2\cdot 5\cdot 7 \\ \text{LCD} &= 2\cdot 3\cdot 5\cdot 7 \\ &= 210 \end{aligned}$$
25. $$\frac{1}{8}\cdot\frac{7}{8} = \frac{1\cdot 7}{8\cdot 8} = \frac{7}{64}$$
26. $$\frac{16}{35}\cdot\frac{25}{48} = \frac{\cancel{16}^{1}\cdot \cancel{5}^{1}\cdot 5}{\cancel{5}_{1}\cdot 7\cdot \cancel{16}_{1}\cdot 3} = \frac{5}{21}$$
27. $$\frac{1}{3}\div\frac{15}{16} = \frac{1}{3}\cdot\frac{16}{15} = \frac{1\cdot 16}{3\cdot 15} = \frac{16}{45}$$
28. $$16\frac{1}{4}\div 5 = \frac{65}{4}\div\frac{5}{1} = \frac{65}{4}\cdot\frac{1}{5} = \frac{\cancel{5}^{1}\cdot 13}{4\cdot \cancel{5}_{1}} = \frac{13}{4} = 3\frac{1}{4}$$

29. $\frac{17}{25}-\frac{7}{25}=\frac{17-7}{25}$
$=\frac{10}{25}$
$=\frac{2\cdot \cancel{5}}{5\cdot \cancel{5}}$
$=\frac{2}{5}$

30. $\frac{8}{11}-\frac{1}{2}=\frac{8}{11}\cdot\frac{\mathbf{2}}{\mathbf{2}}-\frac{1}{2}\cdot\frac{\mathbf{11}}{\mathbf{11}}$
$=\frac{16-11}{22}$
$=\frac{5}{22}$

31. $\frac{17}{24}+\frac{11}{40}=\frac{17}{24}\cdot\frac{\mathbf{5}}{\mathbf{5}}+\frac{11}{40}\cdot\frac{\mathbf{3}}{\mathbf{3}}$
$=\frac{85+33}{120}$
$=\frac{118}{120}$
$=\frac{59\cdot \cancel{2}}{60\cdot \cancel{2}}$
$=\frac{59}{60}$

32. $4\frac{1}{9}=4\frac{2}{18}=3\frac{18}{18}+\frac{2}{18}=3\frac{20}{18}$
$-3\frac{5}{6}=-3\frac{15}{18}=-3\frac{15}{18}\quad=-3\frac{15}{18}$
$\frac{5}{18}$

33. THE INTERNET

$30\cdot 1\frac{3}{4}=\frac{30}{1}\cdot\frac{7}{4}$
$=\frac{\cancel{2}\cdot 15\cdot 7}{\cancel{2}\cdot 2}$
$=\frac{105}{2}$
$=52\frac{1}{2}$

It received $52\frac{1}{2}$ million hits.

34. MACHINE SHOPS

$\frac{17}{24}-\frac{17}{32}=\frac{17}{24}\cdot\frac{\mathbf{4}}{\mathbf{4}}-\frac{17}{32}\cdot\frac{\mathbf{3}}{\mathbf{3}}$
$=\frac{68-51}{96}$
$=\frac{17}{96}$

$\frac{17}{96}$ of an inch needs to be milled.

SECTION 1.3
The Real Numbers

35. a. 0 b. $\{\ldots,-2,-1,0,1,2,\ldots\}$

36. -206 feet

37. a. $<$ b. $>$

38. a. $\frac{7}{10}$ b. $\frac{14}{3}$

39. $\frac{1}{250}=250\overline{)1.000}$ (quotient 0.004)

$\frac{1}{250}=0.004$

40. $\frac{17}{22}=22\overline{)17.00000}$ (quotient 0.77272)

$\frac{17}{22}=0.7\overline{72}$

41. Number line from -5 to 5 with points plotted at: $-\frac{17}{4}$, -2, $0.333\ldots$, $\frac{7}{8}$, $\sqrt{2}$, π, 3.75

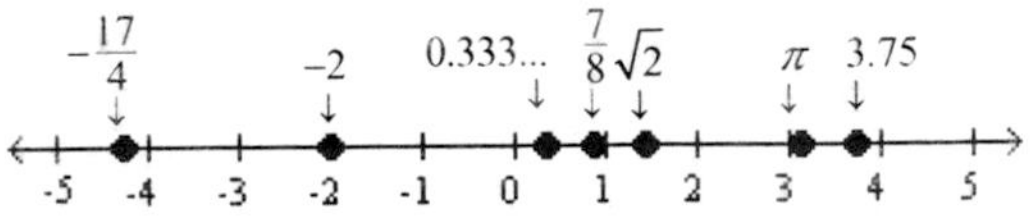

42. Natural: 8

Whole: 0, 8

Integer: 0, −12, 8

Rational: $-\frac{4}{5}, 99.99, 0, -12, 4\frac{1}{2}, 0.666..., 8$

Irrational: $\sqrt{2}$

Real: All

43. False

44. False

45. True

46. True

47. $|-6| \boxed{?} |-5|$

$6 > 5$

$|-6| > |-5|$

48. $-9 \boxed{?} |-10|$

$-9 < 10$

$-9 \boxed{<} |-10|$

SECTION 1.4
Adding Real Numbers; Properties of Addition

49. $-45 + (-37) = -82$

50. $25 + (-13) = 12$

51. $0 + (-7) = -7$

52. $-7 + 7 = 0$

53. $12 + (-8) + (-15) = 4 + (-15)$
$= -11$

54. $-9.9 + (-2.4) = -12.3$

55.
$$\begin{aligned}\frac{5}{16} + \left(-\frac{1}{2}\right) &= \frac{5}{16} + \left(-\frac{1}{2} \cdot \frac{\mathbf{8}}{\mathbf{8}}\right) \\ &= \frac{5 + (-8)}{16} \\ &= -\frac{3}{16}\end{aligned}$$

56. $35 + (-13) + (-17) + 6 = 41 + (-30)$
$= 11$

57. a. Commutative Property of Addition

b. Associative Property of Addition

c. Addition Property of Opposites (Inverse Property of Addition)

d. Addition Property of 0 (Identity Property of Addition)

58. TEMPERATURES

$-48 + 166 = 118$

The record high temperature is 118°F.

SECTION 1.5
Subtracting Real Numbers

59. a. -10 b. 3

60. a. $\frac{9}{16}$ b. -4

61. $45 - 64 = 45 + (-64)$
$= -19$

62.
$$\begin{aligned}-\frac{3}{5} - \frac{1}{3} &= -\frac{3}{5} + \left(-\frac{1}{3}\right) \\ &= -\frac{3}{5} \cdot \frac{\mathbf{3}}{\mathbf{3}} + \left(-\frac{1}{3} \cdot \frac{\mathbf{5}}{\mathbf{5}}\right) \\ &= \frac{-9 + (-5)}{15} \\ &= -\frac{14}{15}\end{aligned}$$

63. $-7 - (-12) = -7 + 12$
$= 5$

64. $3.6 - (-2.1) = 3.6 + 2.1$
$= 5.7$

65. $0 - 10 = 0 + (-10)$
$= -10$

66.
$$\begin{aligned}-33 + 7 - 5 - (-2) &= -33 + 7 + (-5) + 2 \\ &= -26 + (-5) + 2 \\ &= -31 + 2 \\ &= -29\end{aligned}$$

67. GEOGRAPHY

$$29{,}028-(-36{,}205)=29{,}028+36{,}205$$
$$=65{,}233$$

The difference in elevations is 65,233 feet.

Check: $65{,}233+(-36{,}205)=29{,}028$

68. HISTORY

$$-212+(-75)=-287$$

He was born in 287 B.C.

Check: $-287+75=-212$

SECTION 1.6
Multiplying and Dividing Real Numbers
Properties of Multiplication and Division

69. $-8\cdot 7=-56$

70. $$-9\left(-\frac{1}{9}\right)=-\frac{9}{1}\left(-\frac{1}{9}\right)$$
$$=\frac{\not{9}^{1}}{\not{9}_{1}}$$
$$=1$$

71. $$2(-3)(-2)=-6(-2)$$
$$=12$$

72. $$(-4)(-1)(-3)=4(-3)$$
$$=-12$$

73. $-1.2(-5.3)=6.36$

74. $0.002(-1{,}000)=-2$

75. $$-\frac{2}{3}\left(\frac{1}{5}\right)=\frac{-2\cdot 1}{3\cdot 5}$$
$$=-\frac{2}{15}$$

76. $-6(-3)(0)(-1)=0$

77. ELECTRONICS

$2(1.5)=3$

The new high is 3.

$-3(1.5)=-4.5$

The new low is -4.5.

78. a. Associative Property of Multiplication
b. Commutative Property of Multiplication
c. Multiplication Property of 1 (Identity Property of Multiplication)
d. Inverse Property of Multiplication

79. $\frac{44}{-44}=-1$

80. $\frac{-272}{16}=-17$

81. $\frac{-81}{-27}=3$

82. $$-\frac{3}{5}\div\frac{1}{2}=-\frac{3}{5}\cdot\frac{2}{1}$$
$$=-\frac{3\cdot 2}{5\cdot 1}$$
$$=-\frac{6}{5}$$

83. $\frac{-60}{0}$ is undefined

84. $\frac{-4.5}{1}=-4.5$

85. $\frac{0}{18}=\mathbf{0}$ because $\mathbf{0\cdot 18=0}$.

86. GEMSTONES

$$3{,}000-1{,}200=1{,}800$$
$$1{,}800\div 5=360$$

The average annual depreciation is $-\$360$.

SECTION 1.7
Exponents and Order of Operations

87. a. $8\cdot 8\cdot 8\cdot 8\cdot 8=8^5$ b. $9\cdot\pi\cdot r\cdot r=9\pi r^2$

88. a. $9^2 = 9 \cdot 9$
$= 81$

b. $\left(-\frac{2}{3}\right)^3 = \left(-\frac{2}{3}\right)\left(-\frac{2}{3}\right)\left(-\frac{2}{3}\right)$
$= -\frac{2 \cdot 2 \cdot 2}{3 \cdot 3 \cdot 3}$
$= -\frac{8}{27}$

c. $2^5 = 2 \cdot 2 \cdot 2 \cdot 2 \cdot 2$

d. $50^1 = 50$
$= 32$

89. $2 + 5 \cdot 3 = 2 + 15$
$= 17$

90. $-24 \div 2 \cdot 3 = -12 \cdot 3$
$= -36$

91. $-(16-3)^2 = -(13)^2$
$= -(13)(13)$
$= -169$

92. $43 + 2(-6 - 2 \cdot 2) = 43 + 2(-6 - 4)$
$= 43 + 2[-6 + (-4)]$
$= 43 + 2(-10)$
$= 43 + (-20)$
$= 23$

93. $10 - 5[-3 - 2(5 - 7^2)] - 5$
$= 10 - 5[-3 - 2(5 - 49)] - 5$
$= 10 - 5[-3 - 2(-44)] - 5$
$= 10 - 5[-3 + 88] - 5$
$= 10 - 5[85] - 5$
$= 10 - 425 - 5$
$= -415 - 5$
$= -420$

94. $\frac{-4(4+2)-4}{2|-18-4(5)|} = \frac{-4(6)-4}{2|-18-20|}$
$= \frac{-24-4}{2|-38|}$
$= \frac{-28}{2(38)}$
$= -\frac{\overset{1}{\cancel{2}} \cdot \overset{1}{\cancel{2}} \cdot 7}{\underset{1}{\cancel{2}} \cdot \underset{1}{\cancel{2}} \cdot 19}$
$= -\frac{7}{19}$

95. $(-3)^3\left(\frac{-8}{2}\right) + 5 = -27(-4) + 5$
$= 108 + 5$
$= 113$

96. $\frac{2^4 - (4-6)(3-6)}{12 + 4\left[(-1)^8 - 2^2\right]} = \frac{16 - (-2)(-3)}{12 + 4[1-4]}$
$= \frac{16-6}{12 + 4[-3]}$
$= \frac{10}{12 + (-12)}$
$= \frac{10}{0}$
Undefined

97. a. $(-9)^2 = (-9)(-9)$
$= 81$

b. $-9^2 = -(9 \cdot 9)$
$= -81$

98. WALK-A-THONS
$\frac{20 \cdot 5 + 65 \cdot 10 + 25 \cdot 20 + 5 \cdot 50 + 10 \cdot 100}{20 + 65 + 25 + 5 + 10}$
$= \frac{100 + 650 + 500 + 250 + 1{,}000}{125}$
$= \frac{2{,}500}{125}$
$= 20$

The average donation was \$20.

SECTION 1.8
Algebraic Expressions

99. a. 3 terms b. 1 term

100. a. $16x^2-5x+25$; $16, -5, 25$

b. $\frac{x}{2}+y$; $\frac{1}{2}, 1$

101. $h+25$

102. $3s-15$

103. $\frac{1}{2}t-6$

104. $|2-a^2|$

105. HARDWARE

$(n+4)$ inch

106. HARDWARE

$(b-4)$ inch

107. $10d$

108. $(x-5)$ years

109.

Type of coin	Number	· Value (in cents)	= Total value (in cents)
Nickel	6	5	**30**
Dime	d	10	**10*d***

110.

x	$20x-x^3$
0	$20(\mathbf{0})-(\mathbf{0})^3=0-0$ $=\mathbf{0}$
1	$20(\mathbf{1})-(\mathbf{1})^3=20-1$ $=\mathbf{19}$
–4	$20(-\mathbf{4})-(-\mathbf{4})^3=-80-(-64)$ $=-\mathbf{16}$

111. $b^2-4ac=(-10)^2-4(3)(5)$
$=100-60$
$=40$

112. $\frac{x+y}{-x-z}=\frac{19+17}{-(19)-(-18)}$
$=\frac{36}{-19+18}$
$=\frac{36}{-1}$
$=-36$

SECTION 1.9
Simplify Algebraic Expressions Using Properties of Real Numbers

113. $a\cdot 150=\underline{150a}$

114. $9+(1+7y)=\underline{(9+1)+7y}$

115. $2.7(10b)=\underline{(2.7\cdot 10)b}$

116. $x+2x^2=\underline{2x^2+x}$

117. $-4(7w)=-4(7)w$
$=-28w$

118. $3(-2x)(-4)=3(-2)(-4)x$
$=24x$

119. $0.4(5.2f)=0.4(5.2)f$
$=2.08f$

120. $\frac{7}{2}\cdot\frac{2}{7}r=\frac{\overset{1}{\cancel{7}}\cdot\overset{1}{\cancel{2}}}{\underset{1}{\cancel{2}}\cdot\underset{1}{\cancel{7}}}r$
$=r$

121. $5(x+3)=5x+5(3)$
$=5x+15$

122. $-(2x+3-y)=-(2x)+(-1)(3)-(-1)y$
$=-2x-3+y$

123. $\frac{3}{4}(4c-8)=\frac{3}{4}\left(\frac{4}{1}\right)c-\frac{3}{4}\left(\frac{8}{1}\right)$
$=\frac{3\cdot\overset{1}{\cancel{4}}}{\underset{1}{\cancel{4}}\cdot 1}c-\frac{3\cdot\overset{1}{\cancel{4}}\cdot 2}{\underset{1}{\cancel{4}}\cdot 1}$
$=3c-6$

124. $-2(-3c-7)(2.1)=-2(2.1)(-3c-7)$
$=-4.2(-3c-7)$
$=-4.2(-3c)-(-4.2)(7)$
$=12.6c+29.4$

125. $8p+5p-4p=(8+5-4)p$
$=9p$

126. $-5m+2-2m-2=-5m-2m+2-2$
$=(-5-2)m+(2-2)$
$=-7m+0$
$=-7m$

127. $n+n+n+n=(1+1+1+1)n$
$=4n$

128. $5(p-2)-2(3p+4)$
$=5(p)-5(2)-2(3p)+(-2)(4)$
$=5p-10-6p-8$
$=5p-6p-10-8$
$=(5-6)p+(-10-8)$
$=-p-18$

129. $55.7k^2-55.6k^2=(55.7-55.6)k^2$
$=0.1k^2$

130. $8a^3+4a^3+2a-4a^3-2a-1$
$=(8+4-4)a^3+(2-2)a-1$
$=8a^3+0a-1$
$=8a^3-1$

131. $\frac{3}{5}w-\left(-\frac{2}{5}w\right)=\frac{3}{5}w+\frac{2}{5}w$
$=\left(\frac{3+2}{5}\right)w$
$=\frac{\cancel{5}^1}{\cancel{5}_1}w$
$=w$

132. $36\left(\frac{1}{9}h-\frac{3}{4}\right)+36\left(\frac{1}{3}\right)$
$=\frac{36}{1}\left(\frac{1}{9}\right)h-\frac{36}{1}\left(\frac{3}{4}\right)+\frac{36}{1}\left(\frac{1}{3}\right)$
$=\frac{\cancel{9}\cdot 4}{\cancel{9}}h-\frac{\cancel{4}\cdot 9\cdot 3}{\cancel{4}}+\frac{\cancel{3}\cdot 12}{\cancel{3}}$
$=4h-27+12$
$=4h-15$

133. $-(7.6t-1.9)+(1.4t-1.2)8+t$
$=-1(7.6t)-(-1)(1.9)+(1.4t)8-1.2(8)+t$
$=-7.6t+1.9+11.2t-9.6+t$
$=(-7.6t+11.2t+t)+(1.9-9.6)$
$=4.6t-7.7$

134. a. $1x=x$
b. $-1x=-x$
c. $4x-(-1)=4x+1$
d. $4x+(-1)=4x-1$

CHAPTER 1 TEST

1. a. Two fractions, such as $\frac{1}{2}$ and $\frac{5}{10}$, that represent the same number are called **equivalent** fractions.

b. The results of a multiplication is called a **product**.

c. $\frac{8}{7}$ is the **reciprocal** of $\frac{7}{8}$ because $\frac{8}{7}\cdot\frac{7}{8}=1$.

d. $9x^2$ and $7x^2$ are **like terms** because they have the same variable raised to exactly the same power.

e. For any nonzero raeal number a, $\frac{a}{0}$ is **undefined**.

2. SECURITY GUARDS

a. $24 b. 5 hours

3. $f=\frac{a}{5}$

a	$f=\frac{a}{5}$
15	$f=\frac{a}{5}$ $=\frac{15}{5}$ $=\mathbf{3}$
100	$f=\frac{a}{5}$ $=\frac{100}{5}$ $=\mathbf{20}$
350	$f=\frac{a}{5}$ $=\frac{350}{5}$ $=\mathbf{70}$

4. $180 = 2\cdot 2\cdot 3\cdot 3\cdot 5$
$= 2^2\cdot 3^2\cdot 5$

5. $\dfrac{42}{105} = \dfrac{2\cdot 3\cdot 7}{3\cdot 5\cdot 7} = \dfrac{2\cdot \overset{1}{\cancel{3}}\cdot \overset{1}{\cancel{7}}}{\underset{1}{\cancel{3}}\cdot 5\cdot \underset{1}{\cancel{7}}} = \dfrac{2}{5}$

6. $\dfrac{15}{16}\div\dfrac{5}{8} = \dfrac{15}{16}\cdot\dfrac{8}{5} = \dfrac{\overset{1}{\cancel{5}}\cdot 3\cdot \overset{1}{\cancel{8}}}{\underset{1}{\cancel{8}}\cdot 2\cdot \underset{1}{\cancel{5}}} = \dfrac{3}{2} = 1\dfrac{1}{2}$

7. $\dfrac{7}{10}+\dfrac{1}{14} = \dfrac{7}{10}\cdot\dfrac{\mathbf{7}}{\mathbf{7}}+\dfrac{1}{14}\cdot\dfrac{\mathbf{5}}{\mathbf{5}} = \dfrac{49+5}{70} = \dfrac{54}{70} = \dfrac{\overset{1}{\cancel{2}}\cdot 27}{\underset{1}{\cancel{2}}\cdot 35} = \dfrac{27}{35}$

8. $8\dfrac{2}{5} = 8\dfrac{2}{5}\cdot\dfrac{\mathbf{3}}{\mathbf{3}} = 8\dfrac{6}{15} = 7\dfrac{15}{15}+\dfrac{6}{15} = 7\dfrac{21}{15}$

$-1\dfrac{2}{3} = -1\dfrac{2}{3}\cdot\dfrac{\mathbf{5}}{\mathbf{5}} = -1\dfrac{10}{15} = -1\dfrac{10}{15} \qquad = -1\dfrac{10}{15}$

$6\dfrac{11}{15}$

9. SHOPPING

a. $4\dfrac{1}{4}$ pounds

b. $4\dfrac{1}{4}\cdot 84 = \dfrac{17}{4}\cdot\dfrac{84}{1} = \dfrac{17\cdot 84}{4} = \dfrac{17\cdot \overset{1}{\cancel{4}}\cdot 21}{\underset{1}{\cancel{4}}} = 357$

The fruit cost \$3.57.

10. $6\overline{)5.000}$ with quotient 0.833

$\dfrac{5}{6} = 0.8\overline{3}$

11a. Number line from -5 to 5 with points plotted at -3.75, -3, $-1\frac{1}{4}$, 0, 0.5, $\sqrt{2}$, 2, $\frac{7}{2}$.

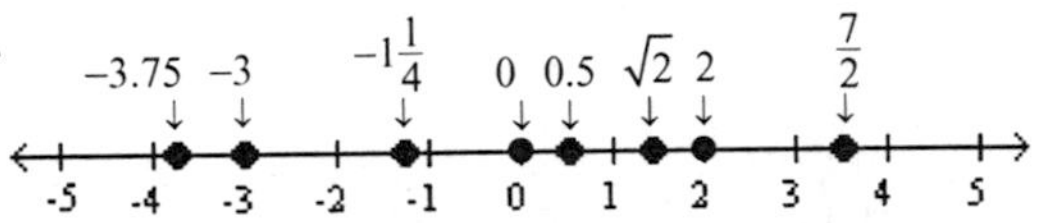

11b. Natural: 2
Whole: 0, 2
Integer: 0, 2, −3
Rational: $-1\dfrac{1}{4}, 0, -3.75, 2, \dfrac{7}{2}, 0.5, -3$
Irrational: $\sqrt{2}$
Real: All

12. a. True b. False c. True
d. True e. True

13. a. $-2 \boxed{>} -3$

b. $-|-9| \ ? \ 8$
$-9 \ ? \ 8$
$-9 < 8$
$-|-9| \boxed{<} 8$

c. $|-4| \ ? \ -(-5)$
$4 \ ? \ 5$
$4 < 5$
$|-4| \boxed{<} -(-5)$

d. $\left|-\dfrac{7}{8}\right| \ ? \ 0.5$
$\dfrac{7}{8} \ ? \ 0.5$
$0.875 > 0.5$
$\left|-\dfrac{7}{8}\right| \boxed{>} 0.5$

14. TELEVISION

$$\frac{0.6+(-0.3)+1.7+1.5+(-0.2)+1.1+(-0.2)}{7}$$
$$=\frac{4.9+(-0.7)}{7}$$
$$=\frac{4.2}{7}$$
$$=0.6$$

A gain of 0.6 of a rating point.

15. $-5.6+(-2)=-7.6$

16. $(-6)+8+(-4)=2+(-4)$
$=-2$

17. $-\frac{1}{2}+\frac{7}{8}=-\frac{1}{2}\cdot\frac{\mathbf{4}}{\mathbf{4}}+\frac{7}{8}$
$=\frac{-4+7}{8}$
$=\frac{3}{8}$

18. a. $-10-(-4)=-10+4$
$=-6$

b. $-6+(-4)=-10$

19. a. $\frac{-12.6}{-0.9}=14$

b. $14(-0.9)=-12.6$

20. $(-2)(-3)(-5)=6(-5)$
$=-30$

21. $-6.1(0.4)=-2.44$

22. $\frac{0}{-3}=0$

23. $\left(-\frac{3}{5}\right)^3=\left(-\frac{3}{5}\right)\left(-\frac{3}{5}\right)\left(-\frac{3}{5}\right)$
$=\frac{-3\cdot-3\cdot-3}{5\cdot5\cdot5}$
$=-\frac{27}{125}$

24. $3+(-3)=0$

25. $0-3=0+(-3)$
$=-3$

26. $-30+50-10-(-40)$
$=-30+50+(-10)+40$
$=-40+90$
$=50$

27. ASTRONOMY

$-12.5-(-26.5)=-12.5+26.5$
$=14$

They differ by a magnitude of 14.

28. GLACIERS

$87-165=87+(-165)$
$=-78$

There was a net lost of 78 inches.

29. INVENTORY

$85\cdot(-15)=-1,275$

There was a financial lost of \$1,275.

30. a. $[-12+(97+3)]$

b. $2x+14$

c. $-2(5)m$

d. 1

e. $15x$

31. a. $9(9)(9)(9)(9)=9^5$

b. $3\cdot x\cdot x\cdot z\cdot z\cdot z=3x^2z^3$

32. $2x-\frac{30}{x}$

x	$2x-\frac{30}{x}$
5	$2x-\frac{30}{x}=2(\mathbf{5})-\frac{30}{\mathbf{5}}$ $=10-6$ $=\mathbf{4}$
10	$2x-\frac{30}{x}=2(\mathbf{10})-\frac{30}{\mathbf{10}}$ $=20-3$ $=\mathbf{17}$
-30	$2x-\frac{30}{x}=2(\mathbf{-30})-\frac{30}{\mathbf{-30}}$ $=-60+1$ $=\mathbf{-59}$

33. $8+2\cdot3^4=8+2\cdot81$
$=8+162$
$=170$

34. $\dfrac{3(40-2^3)}{-2(6-4)^2}=\dfrac{3(40-8)}{-2(2)^2}$
$=\dfrac{3[(40+(-8)]}{-2(4)}$
$=\dfrac{3[32]}{-8}$
$=\dfrac{96}{-8}$
$=-12$

35. $-10^2-5+6=-(10)(10)-5+6$
$=-100-5+6$
$=-105+6$
$=-99$

36. $9-3\left[45-5^2(1^5-4)\right]$
$=9-3[45-25(1-4)]$
$=9-3[45-25[1+(-4)]]$
$=9-3[45-25(-3)]$
$=9-3[45+75]$
$=9-3[120]$
$=9-360$
$=9+(-360)$
$=-351$

37. $|-50\div5\cdot2|=|-10\cdot2|$
$=|-20|$
$=20$

38. $3(10x-y)-5(x+y^2)$
$=3[10(2)-(-5)]-5[2+(-5)^2]$
$=3[20+5]-5[2+25]$
$=3[25]-5[27]$
$=75-135$
$=-60$

39. $2w-7$

40. a. Music
$x-2$
b. Money
$25q$¢

41. 3 terms

42. 1, −6, −1, 10

43. $5(-4x)=5(-4)x$
$=-20x$

44. $-8(-7t)(4)=-8(-7)(4)t$
$=224t$

45. $\dfrac{4}{5}(15a+5)-16a=\dfrac{4}{5}\cdot\dfrac{15}{1}a+\dfrac{4}{5}\cdot\dfrac{5}{1}-16a$
$=\dfrac{4\cdot15}{5\cdot1}a+\dfrac{4\cdot5}{5\cdot1}-16a$
$=\dfrac{4\cdot\overset{1}{\cancel{5}}\cdot3}{\underset{1}{\cancel{5}}}a+\dfrac{4\cdot\overset{1}{\cancel{5}}}{\underset{1}{\cancel{5}}}-16a$
$=12a-16a+4$
$=12a+(-16a)+4$
$=-4a+4$

46. $-1.1d^3-3.8d^3-d^3$
$=[-1.1+(-3.8)+(-1)]d^3$
$=-5.9d^3$

47. $9x+2(7x-3)-9(x-1)$
$=9x+2(7)x+2(-3)-9x-9(-1)$
$=9x+14x-6-9x+9$
$=(9x-9x+14x)+(-6+9)$
$=14x+3$

48. a. True b. False c. False
d. False e. True f. False

49. $m^2+4m^2+5m-2m^2-3m-4$
$=(1+4-2)m^2+(5-3)m-4$
$=3m^2+2m-4$

SECTION 2.1 STUDY SET
VOCABULARY

Fill in the blanks.

1. A statement indicating that two expressions are equal, such as $x + 1 = 7$, is called an **equation**.
3. To **solve** an equation means to find all values of the variable that make the equation true.
5. Equations with the same solutions are called **equivalent** equations.

CONCEPTS

7. a. $x+6$ b. Neither c. No d. Yes

9. a. If $a=b$, then $a+c=b+\boxed{c}$ and $a-c=b-\boxed{c}$.

   b. If $a=b$, then $ca+\boxed{c}b$ and $\frac{a}{c}=\frac{b}{\boxed{c}}$.

11. a. x b. y c. t d. h

NOTATION

Complete each solution to solve the equation.

13. $x-5=45$ Check: $x-5=45$

$$x-5+\boxed{5}=45+\boxed{5} \qquad \boxed{50}-5\overset{?}{=}45$$

$$x=\boxed{50} \qquad \boxed{45}=45 \text{ True}$$

$\boxed{50}$ is the solution.

15. a. Is possibly equal to b. Yes

GUIDED PRACTICE

Check to determine whether the number in red is a solution of the equation. See Example 1.

17. $\mathbf{6}+12\overset{?}{=}28$

$18=28$ False

6 is not a solution.

19. $2(\mathbf{-8})+3\overset{?}{=}-15$

$-16+3\overset{?}{=}-15$

$-13=-15$ False

-8 is not a solution.

21. $0.5(\mathbf{5})\overset{?}{=}2.9$

$2.5=2.9$ False

5 is not a solution.

23. $33-\frac{\mathbf{-6}}{2}\overset{?}{=}30$

$33+3\overset{?}{=}30$

$36=30$ False

-6 is not a solution.

25. $|\mathbf{-2}-8|\overset{?}{=}10$

$|-10|\overset{?}{=}10$

$10=10$ True

-2 is a solution.

27. $3(\mathbf{12})-2\overset{?}{=}4(\mathbf{12})-5$

$36-2\overset{?}{=}48-5$

$34=43$ False

12 is not a solution.

29. $(\mathbf{-3})^2-(\mathbf{-3})-6\overset{?}{=}0$

$9+3-6\overset{?}{=}0$

$12-6\overset{?}{=}0$

$6=0$ False

-3 is not a solution.

31. $\frac{2}{\mathbf{1}+1}+5\overset{?}{=}\frac{12}{\mathbf{1}+1}$

$\frac{2}{2}+5\overset{?}{=}\frac{12}{2}$

$1+5\overset{?}{=}6$

$6=6$ True

1 is a solution.

33. $\frac{\mathbf{3}}{\mathbf{4}}-\frac{1}{8}\overset{?}{=}\frac{5}{8}$

$\frac{6}{8}-\frac{1}{8}\overset{?}{=}\frac{5}{8}$

$\frac{5}{8}=\frac{5}{8}$ True

3/4 is a solution.

35. $(\mathbf{-3}-4)(\mathbf{-3}+3)\overset{?}{=}0$

$(-7)(0)\overset{?}{=}0$

$0=0$ True

-3 is a solution.

Use a property of equality to solve each equation. Then check the result. See Example 2.

37.
$$a-5=66$$
$$a-5+\mathbf{5}=66+\mathbf{5}$$
$$a=71$$
Check: $a-5=66$
$$71-5\overset{?}{=}66$$
$$66=66 \quad \text{True}$$
71 is the solution.

39.
$$9=p-9$$
$$9+\mathbf{9}=p-9+\mathbf{9}$$
$$18=p$$
Check: $9=p-9$
$$9\overset{?}{=}18-9$$
$$9=9 \quad \text{True}$$
18 is the solution.

Use a property of equality to solve each equation. Then check the result. See Example 3.

41.
$$-16=y-4$$
$$-16+\mathbf{4}=y-4+\mathbf{4}$$
$$-12=y$$
Check: $-16=y-4$
$$-16\overset{?}{=}-12-4$$
$$-16=-16 \quad \text{True}$$
−12 is the solution.

43.
$$-3+a=0$$
$$-3+a+\mathbf{3}=0+\mathbf{3}$$
$$a=3$$
Check: $-3+a=0$
$$-3+3\overset{?}{=}0$$
$$0=0 \quad \text{True}$$
3 is the solution.

Use a property of equality to solve each equation. Then check the result. See Example 4.

45.
$$x+\frac{1}{10}=\frac{6}{5}$$
$$x+\frac{1}{10}-\frac{\mathbf{1}}{\mathbf{10}}=\frac{6}{5}-\frac{\mathbf{1}}{\mathbf{10}}$$
$$x=\frac{6}{5}\cdot\frac{\mathbf{2}}{\mathbf{2}}-\frac{1}{10}$$
$$x=\frac{12-1}{10}$$
$$x=\frac{11}{10}$$
Check:
$$x+\frac{1}{10}=\frac{6}{5}$$
$$\frac{11}{10}+\frac{1}{10}\overset{?}{=}\frac{6}{5}$$
$$\frac{12}{10}\overset{?}{=}\frac{6}{5}$$
$$\frac{\cancel{2}^{1}\cdot 6}{\cancel{2}_{1}\cdot 5}\overset{?}{=}\frac{6}{5}$$
$$\frac{6}{5}=\frac{6}{5} \quad \text{True}$$
$\frac{11}{10}$ is the solution.

47.
$$3.5+f=1.2$$
$$3.5+f-\mathbf{3.5}=1.2-\mathbf{3.5}$$
$$f=-2.3$$
Check: $3.5+f=1.2$
$$3.5+(-2.3)\overset{?}{=}1.2$$
$$1.2=1.2 \quad \text{True}$$
−2.3 is the solution.

Use a property of equality to solve each equation. Then check the result. See Example 5.

49.
$$\frac{x}{15}=3$$
$$\mathbf{15}\cdot\frac{x}{15}=\mathbf{15}\cdot 3$$
$$x=45$$
Check: $\frac{x}{15}=3$
$$\frac{45}{15}\overset{?}{=}3$$
$$3=3 \quad \text{True}$$
45 is the solution.

51.
$$\frac{d}{8}=-6$$
$$\mathbf{8}\cdot\frac{d}{8}=\mathbf{8}\cdot-6$$
$$d=-48$$
Check: $\frac{d}{8}=-6$
$$\frac{-48}{8}\overset{?}{=}-6$$
$$-6=-6 \quad \text{True}$$
−48 is the solution.

Use a property of equality to solve each equation. Then check the result. See Example 6.

53. $\frac{4}{5}t = 16$

$\left(\frac{\mathbf{5}}{\mathbf{4}}\right)\cdot\frac{4}{5}t = \left(\frac{\mathbf{5}}{\mathbf{4}}\right)\cdot 16$

$\left(\frac{5}{4}\cdot\frac{4}{5}\right)t = \frac{5\cdot 4\cdot \cancel{4}^{1}}{\cancel{4}_{1}}$

$t = 20$

Check: $\frac{4}{5}t = 16$

$\frac{4}{5}\cdot 20 \stackrel{?}{=} 16$

$\frac{4\cdot 4\cdot \cancel{5}^{1}}{\cancel{5}_{1}} \stackrel{?}{=} 16$

$16 = 16$ True

20 is the solution.

55. $-\frac{7r}{2} = \frac{5}{12}$

$\left(-\frac{\mathbf{2}}{\mathbf{7}}\right)\cdot -\frac{7r}{2} = \left(-\frac{\mathbf{2}}{\mathbf{7}}\right)\cdot\frac{5}{12}$

$\left(-\frac{2}{7}\cdot -\frac{7}{2}\right)r = -\frac{\cancel{2}^{1}\cdot 5}{7\cdot \cancel{2}_{1}\cdot 6}$

$r = -\frac{5}{42}$

Check: $-\frac{7r}{2} = \frac{5}{12}$

$-\frac{7}{2}\cdot\left(-\frac{5}{42}\right) \stackrel{?}{=} \frac{5}{12}$

$\frac{\cancel{7}^{1}\cdot 5}{2\cdot \cancel{7}_{1}\cdot 6} \stackrel{?}{=} \frac{5}{12}$

$\frac{5}{12} = \frac{5}{12}$ True

$-\frac{5}{42}$ is the solution.

Use a property of equality to solve each equation. Then check the result. See Example 7.

57. $4x = 16$

$\frac{4x}{\mathbf{4}} = \frac{16}{\mathbf{4}}$

$x = 4$

Check: $4x = 16$

$4\cdot 4 \stackrel{?}{=} 16$

$16 = 16$ True

4 is the solution.

59. $-1.7 = -3.4y$

$\frac{-1.7}{\mathbf{-3.4}} = \frac{-3.4y}{\mathbf{-3.4}}$

$0.5 = y$

Check: $-1.7 = -3.4y$

$-1.7 \stackrel{?}{=} -3.4\cdot 0.5$

$-1.7 = -1.7$ True

0.5 is the solution.

Use a property of equality to solve each equation. Then check the result. See Example 8.

61. $-x = 18$

$\mathbf{(-1)} - x = \mathbf{(-1)}18$

$x = -18$

Check: $-x = 18$

$-(-18) \stackrel{?}{=} 18$

$18 = 18$ True

-18 is the solution.

63. $-n = \frac{4}{21}$

$\mathbf{(-1)} - n = \mathbf{(-1)}\frac{4}{21}$

$n = -\frac{4}{21}$

Check: $-n = \frac{4}{21}$

$-(-\frac{4}{21}) \stackrel{?}{=} \frac{4}{21}$

$\frac{4}{21} = \frac{4}{21}$ True

$-\frac{4}{21}$ is the solution.

TRY IT YOURSELF

Solve each equation. Then check the results.

65. $63 = 9c$

$$\frac{63}{\mathbf{9}} = \frac{9c}{\mathbf{9}}$$

$$7 = c$$

Check: $63 = 9c$

$$63 \stackrel{?}{=} 9 \cdot 7$$

$$63 = 63 \quad \text{True}$$

7 is the solution.

67. $d - \frac{1}{9} = \frac{7}{9}$

$$d - \frac{1}{9} + \mathbf{\frac{1}{9}} = \frac{7}{9} + \mathbf{\frac{1}{9}}$$

$$d = \frac{8}{9}$$

Check: $d - \frac{1}{9} = \frac{7}{9}$

$$\frac{8}{9} - \frac{1}{9} \stackrel{?}{=} \frac{7}{9}$$

$$\frac{7}{9} = \frac{7}{9} \quad \text{True}$$

$\frac{8}{9}$ is the solution.

69. $0 = \frac{v}{11}$

$$\mathbf{11} \cdot 0 = \mathbf{11} \cdot \frac{v}{11}$$

$$0 = v$$

Check: $0 = \frac{v}{11}$

$$0 \stackrel{?}{=} \frac{0}{11}$$

$$0 = 0 \quad \text{True}$$

0 is the solution.

71. $x - 1.6 = -2.5$

$$x - 1.6 + \mathbf{1.6} = -2.5 + \mathbf{1.6}$$

$$x = -0.9$$

Check: $x - 1.6 = -2.5$

$$-0.9 - 1.6 \stackrel{?}{=} -2.5$$

$$-2.5 = -2.5$$

-0.9 is the solution.

73. $\frac{2}{3}c = 10$

$$\left(\mathbf{\frac{3}{2}}\right) \cdot \frac{2}{3}c = \left(\mathbf{\frac{3}{2}}\right) \cdot 10$$

$$\left(\frac{3}{2} \cdot \frac{2}{3}c\right) = \frac{3 \cdot 5 \cdot \cancel{2}^{1}}{\cancel{2}_{1}}$$

$$c = 15$$

Check: $\frac{2}{3}c = 10$

$$\frac{2}{3} \cdot 15 \stackrel{?}{=} 10$$

$$\frac{2 \cdot 5 \cdot \cancel{3}^{1}}{\cancel{3}_{1}} \stackrel{?}{=} 10$$

$$10 = 10 \quad \text{True}$$

15 is the solution.

75. $-100 = -5g$

$$\frac{-100}{\mathbf{-5}} = \frac{-5g}{\mathbf{-5}}$$

$$20 = g$$

Check: $-100 = -5g$

$$-100 \stackrel{?}{=} -5 \cdot 20$$

$$-100 = -100 \quad \text{True}$$

20 is the solution.

77. $s + \frac{1}{5} = \frac{4}{25}$

$$s + \frac{1}{5} - \mathbf{\frac{1}{5}} = \frac{4}{25} - \mathbf{\frac{1}{5}}$$

$$s = \frac{4}{25} - \frac{1}{5} \cdot \mathbf{\frac{5}{5}}$$

$$s = \frac{4}{25} - \frac{5}{25}$$

$$s = -\frac{1}{25}$$

Check: $s+\frac{1}{5}=\frac{4}{25}$

$$-\frac{1}{25}+\frac{1}{5}\cdot\frac{\mathbf{5}}{\mathbf{5}}\stackrel{?}{=}\frac{4}{25}$$

$$-\frac{1}{25}+\frac{5}{25}\stackrel{?}{=}\frac{4}{25}$$

$$\frac{4}{25}=\frac{4}{25} \quad \text{True}$$

$-\frac{1}{25}$ is the solution.

79. $$\frac{d}{-7}=-3$$

$$(\mathbf{-7})\frac{d}{-7}=(\mathbf{-7})\cdot -3$$

$$d=21$$

Check: $\frac{d}{-7}=-3$

$$\frac{21}{-7}\stackrel{?}{=}-3$$

$$-3=-3 \quad \text{True}$$

21 is the solution.

81. $$8h=0$$

$$\frac{8h}{\mathbf{8}}=\frac{0}{\mathbf{8}}$$

$$h=0$$

Check: $8h=0$

$$8\cdot 0\stackrel{?}{=}0$$

$$0=0 \quad \text{True}$$

0 is the solution.

83. $$\frac{y}{0.6}=-4.4$$

$$(\mathbf{0.6})\cdot\frac{y}{0.6}=(\mathbf{0.6})\cdot -4.4$$

$$y=-2.64$$

Check: $\frac{y}{0.6}=-4.4$

$$\frac{-2.64}{0.6}\stackrel{?}{=}-4.4$$

$$-4.4=-4.4 \quad \text{True}$$

-2.64 is the solution.

85. $$23b=23$$

$$\frac{23b}{\mathbf{23}}=\frac{23}{\mathbf{23}}$$

$$b=1$$

Check: $23b=23$

$$23\cdot 1\stackrel{?}{=}23$$

$$23=23 \quad \text{True}$$

1 is the solution.

87. $$-\frac{5}{4}h=-5$$

$$\left(-\frac{\mathbf{4}}{\mathbf{5}}\right)\cdot -\frac{5}{4}h=\left(-\frac{\mathbf{4}}{\mathbf{5}}\right)\cdot -5$$

$$\left(-\frac{4}{5}\cdot -\frac{5}{4}\right)h=\frac{-4\cdot -1\cdot \cancel{5}^{1}}{\cancel{5}_{1}}$$

$$h=4$$

Check: $-\frac{5}{4}h=-5$

$$-\frac{5}{4}\cdot 4\stackrel{?}{=}-5$$

$$\frac{-5\cdot \cancel{4}^{1}}{\cancel{4}_{1}}\stackrel{?}{=}-5$$

$$-5=-5 \quad \text{True}$$

4 is the solution.

89. $$8.9=-4.1+t$$

$$8.9+\mathbf{4.1}=-4.1+t+\mathbf{4.1}$$

$$13=t$$

Check: $8.9=-4.1+t$

$$8.9\stackrel{?}{=}-4.1+13$$

$$8.9=8.9 \quad \text{True}$$

13 is the solution.

91. $$-2.5=-m$$

$$(\mathbf{-1})-2.5=(\mathbf{-1})(-m)$$

$$2.5=m$$

Check: $-2.5=-m$

$$-2.5\stackrel{?}{=}(-1)2.5$$

$$-2.5=-2.5 \quad \text{True}$$

2.5 is the solution.

93.
$$-\frac{9}{8}x = 3$$
$$\left(-\mathbf{\frac{8}{9}}\right)\cdot -\frac{9}{8}x = \left(-\mathbf{\frac{8}{9}}\right)\cdot 3$$
$$\left(-\frac{8}{9}\cdot -\frac{9}{8}\right)x = \frac{-8\cdot \cancel{3}^{1}}{3\cdot \cancel{3}_{1}}$$
$$x = -\frac{8}{3}$$

Check:
$$-\frac{9}{8}x = 3$$
$$-\frac{9}{8}\cdot -\frac{8}{3} \stackrel{?}{=} 3$$
$$\frac{3\cdot \cancel{3}\cdot \cancel{8}}{\cancel{8}\cdot \cancel{3}} \stackrel{?}{=} 3$$
$$3 = 3 \quad \text{True}$$

$-\frac{8}{3}$ is the solution.

95.
$$\frac{2}{3}n = -\frac{7}{8}$$
$$\left(\mathbf{\frac{3}{2}}\right)\cdot \frac{2}{3}n = \left(\mathbf{\frac{3}{2}}\right)\cdot -\frac{7}{8}$$
$$\left(\frac{3}{2}\cdot \frac{2}{3}\right)n = \frac{3\cdot -7}{2\cdot 8}$$
$$n = -\frac{21}{16}$$

Check:
$$\frac{2}{3}n = -\frac{7}{8}$$
$$\frac{2}{3}\cdot\left(-\frac{21}{16}\right) \stackrel{?}{=} -\frac{7}{8}$$
$$\frac{\cancel{2}\cdot \cancel{3}\cdot -7}{\cancel{3}\cdot \cancel{2}\cdot 8} \stackrel{?}{=} -\frac{7}{8}$$
$$-\frac{7}{8} = -\frac{7}{8} \quad \text{True}$$

$-\frac{21}{16}$ is the solution.

97.
$$-10 = n - 5$$
$$-10 + \mathbf{5} = n - 5 + \mathbf{5}$$
$$-5 = n$$

Check:
$$-10 = n - 5$$
$$-10 \stackrel{?}{=} -5 - 5$$
$$-10 = -10 \quad \text{True}$$

-5 is the solution.

99.
$$\frac{h}{-40} = 5$$
$$(\mathbf{-40})\cdot \frac{h}{-40} = (\mathbf{-40})\cdot 5$$
$$h = -200$$

Check:
$$\frac{h}{-40} = 5$$
$$\frac{-200}{-40} \stackrel{?}{=} 5$$
$$5 = 5 \quad \text{True}$$

-200 is the solution.

101.
$$-\frac{15}{16}a = -\frac{5}{4}$$
$$\left(-\mathbf{\frac{16}{15}}\right)\cdot -\frac{15}{16}a = \left(-\mathbf{\frac{16}{15}}\right)\cdot -\frac{5}{4}$$
$$\left(-\frac{16}{15}\cdot -\frac{15}{16}\right)a = \frac{-4\cdot \cancel{4}\cdot -1\cdot \cancel{5}}{3\cdot \cancel{5}\cdot \cancel{4}}$$
$$a = \frac{4}{3}$$

Check:
$$-\frac{15}{16}a = -\frac{5}{4}$$
$$-\frac{15}{16}\cdot\left(\frac{4}{3}\right) \stackrel{?}{=} -\frac{5}{4}$$
$$\frac{\cancel{3}\cdot -5\cdot \cancel{4}}{4\cdot \cancel{4}\cdot \cancel{3}} \stackrel{?}{=} -\frac{5}{4}$$
$$-\frac{5}{4} = -\frac{5}{4} \quad \text{True}$$

$\frac{4}{3}$ is the solution.

103. $-15x=-60$

$$\frac{-15x}{\mathbf{-15}}=\frac{-60}{\mathbf{-15}}$$

$$x=4$$

Check: $-15x=-60$

$$-15\cdot 4 \stackrel{?}{=} -60$$

$$-60=-60 \quad \text{True}$$

4 is the solution.

LOOK ALIKES

105. a. $d+\frac{1}{10}=\frac{3}{4}$

$$d+\frac{1}{10}-\frac{\mathbf{1}}{\mathbf{10}}=\frac{3}{4}-\frac{\mathbf{1}}{\mathbf{10}}$$

$$d=\frac{3}{4}\left(\frac{\mathbf{5}}{\mathbf{5}}\right)-\frac{1}{10}\left(\frac{\mathbf{2}}{\mathbf{2}}\right)$$

$$d=\frac{15-2}{20}$$

$$d=\frac{13}{20}$$

Check: $d+\frac{1}{10}=\frac{3}{4}$

$$\frac{13}{20}+\frac{1}{10}\cdot\frac{\mathbf{2}}{\mathbf{2}}\stackrel{?}{=}\frac{3}{4}$$

$$\frac{13}{20}+\frac{2}{20}\stackrel{?}{=}\frac{3}{4}$$

$$\frac{15}{20}\stackrel{?}{=}\frac{3}{4}$$

$$\frac{\overset{1}{\cancel{5}}\cdot 3}{\underset{1}{\cancel{5}}\cdot 4}\stackrel{?}{=}\frac{3}{4}$$

$$\frac{3}{4}=\frac{3}{4} \quad \text{True}$$

$\frac{13}{20}$ is the solution.

105. b. $d-\frac{1}{10}=\frac{3}{4}$

$$d-\frac{1}{10}+\frac{\mathbf{1}}{\mathbf{10}}=\frac{3}{4}+\frac{\mathbf{1}}{\mathbf{10}}$$

$$d=\frac{3}{4}\left(\frac{\mathbf{5}}{\mathbf{5}}\right)+\frac{1}{10}\left(\frac{\mathbf{2}}{\mathbf{2}}\right)$$

$$d=\frac{15+2}{20}$$

$$d=\frac{17}{20}$$

Check: $d-\frac{1}{10}=\frac{3}{4}$

$$\frac{17}{20}-\frac{1}{10}\cdot\frac{\mathbf{2}}{\mathbf{2}}\stackrel{?}{=}\frac{3}{4}$$

$$\frac{17}{20}-\frac{2}{20}\stackrel{?}{=}\frac{3}{4}$$

$$\frac{15}{20}\stackrel{?}{=}\frac{3}{4}$$

$$\frac{\overset{1}{\cancel{5}}\cdot 3}{\underset{1}{\cancel{5}}\cdot 4}\stackrel{?}{=}\frac{3}{4}$$

$$\frac{3}{4}=\frac{3}{4} \quad \text{True}$$

$\frac{17}{20}$ is the solution.

105. c. $\frac{1}{10}d=\frac{3}{4}$

$$\frac{\mathbf{10}}{\mathbf{1}}\left(\frac{1}{10}d\right)=\frac{\mathbf{10}}{\mathbf{1}}\cdot\frac{3}{4}$$

$$d=\frac{\overset{1}{\cancel{2}}\cdot 5\cdot 3}{1\cdot\underset{1}{\cancel{2}}\cdot 2}$$

$$d=\frac{15}{2}$$

Check: $\frac{1}{10}d = \frac{3}{4}$

$$\frac{1}{10} \cdot \frac{15}{2} \stackrel{?}{=} \frac{3}{4}$$

$$\frac{15}{20} \stackrel{?}{=} \frac{3}{4}$$

$$\frac{\overset{1}{\cancel{5}} \cdot 3}{\underset{1}{\cancel{5}} \cdot 4} \stackrel{?}{=} \frac{3}{4}$$

$$\frac{3}{4} = \frac{3}{4} \quad \text{True}$$

$\frac{15}{2}$ is the solution.

105. d. $10d = \frac{3}{4}$

$$\mathbf{\frac{1}{10}}\left(\frac{10}{1}d\right) = \mathbf{\frac{1}{10}} \cdot \frac{3}{4}$$

$$d = \frac{3}{40}$$

Check: $10d = \frac{3}{4}$

$$\frac{10}{1} \cdot \frac{3}{40} \stackrel{?}{=} \frac{3}{4}$$

$$\frac{30}{40} \stackrel{?}{=} \frac{3}{4}$$

$$\frac{\overset{1}{\cancel{10}} \cdot 3}{\underset{1}{\cancel{10}} \cdot 4} \stackrel{?}{=} \frac{3}{4}$$

$$\frac{3}{4} = \frac{3}{4} \quad \text{True}$$

$\frac{3}{40}$ is the solution.

APPLICATIONS

107. SYNTHESIZERS

$$x + 115 = 180$$

$$x + 115 - \mathbf{115} = 180 - \mathbf{115}$$

$$x = 65$$

Check: $x + 115 = 180$

$$65 + 115 \stackrel{?}{=} 180$$

$$180 = 180 \quad \text{True}$$

65° is the measure of the angle.

109. SHARING THE WINNING TICKET

$$\frac{x}{16} = 375{,}000$$

$$\mathbf{16} \cdot \frac{x}{16} = \mathbf{16} \cdot 375{,}000$$

$$x = 6{,}000{,}000$$

Check: $\frac{x}{16} = 375{,}000$

$$\frac{6{,}000{,}000}{16} \stackrel{?}{=} 375{,}000$$

$$375{,}000 = 375{,}000 \quad \text{True}$$

$6,000,000 is the amount of the jackpot.

WRITING

111-113. Answers will vary.

REVIEW

115. Evaluate:

$$-9 - 3x = -9 - 3(-3)$$

$$= -9 + 9$$

$$= 0$$

117. Translate to symbols:

Subtract x from 45.

$$45 - x$$

CHALLENGE PROBLEMS

119. If $a + 81 = 49$, what is $a - 81$?

$$a + 81 = 49$$

$$a + 81 - \mathbf{81} = 49 - \mathbf{81}$$

$$a = -32$$

$$a - 81 = -32 - 81$$

$$= -113$$

SECTION 2.2 STUDY SET

VOCABULARY

Fill in the blanks.

1. $3x + 8 = 10$ is an example of a linear **equation** in one variable.

3. A linear equation that is true for any permissible replacement value for the variable is called an **identity**.

CONCEPTS

Fill in the blanks.

5. a. To solve $3x - 5 = 1$, we first undo the **subtraction** of 5 by adding 5 to both sides. Then we undo the **multiplication** by 3 by dividing both sides by 3.

   b. To solve $\frac{x}{2} + 3 = 5$, we can undo the **addition** of 3 by subtracting 3 from both sides. Then we undo the **division** by 2 by multiplying both sides by 2.

7. a.
$$6x + 5 = 7$$
$$6(\mathbf{-2}) + 5 \stackrel{?}{=} 7$$
$$-12 + 5 \stackrel{?}{=} 7$$
$$-7 = 7$$
No

9. a. 6 b. 10

LOOK ALIKES…

10. a. $2x + 5$ b. $\frac{4}{3}$ c. 23 d. No

NOTATION

Complete the solution.

11. Solve: $2x - 7 = 21$
$$2x - 7\boxed{+7} = 21\boxed{+7}$$
$$2x = 28$$
$$\frac{2x}{\boxed{2}} = \frac{28}{\boxed{2}}$$
$$x = 14$$

Check: $2x - 7 = 21$
$$2(\boxed{14}) - 7 \boxed{\stackrel{?}{=}} 21$$
$$\boxed{28} - 7 \stackrel{?}{=} 21$$
$$\boxed{21} = 21$$
$\boxed{14}$ is the solution.

GUIDED PRACTICE

Solve each equation and check the result.

13.
$$-8x + 1 = 73$$
$$-8x + 1 \mathbf{- 1} = 73 \mathbf{- 1}$$
$$-8x = 72$$
$$\frac{-8x}{\mathbf{-8}} = \frac{72}{\mathbf{-8}}$$
$$x = -9$$

Check: $-8x + 1 = 73$
$$-8(-9) + 1 \stackrel{?}{=} 73$$
$$72 + 1 \stackrel{?}{=} 73$$
$$73 = 73$$
-9 is the solution.

-8 is the solution.

15.
$$-5q - 2 = 23$$
$$-5q - 2 \mathbf{+ 2} = 23 \mathbf{+ 2}$$
$$-5q = 25$$
$$\frac{-5q}{\mathbf{-5}} = \frac{25}{\mathbf{-5}}$$
$$q = -5$$
-10 is the solution.

Check: $-5q - 2 = 23$
$$-5(-5) - 2 \stackrel{?}{=} 23$$
$$25 - 2 \stackrel{?}{=} 23$$
$$23 = 23$$
-5 is the solution.

17.
$$\frac{5}{6}k - 5 = 10$$
$$\frac{5}{6}k - 5 \mathbf{+ 5} = 10 \mathbf{+ 5}$$
$$\frac{\mathbf{6}}{\mathbf{5}}\left(\frac{5}{6}k\right) = \frac{\mathbf{6}}{\mathbf{5}}(15)$$
$$k = \frac{6 \cdot 3 \cdot \cancel{5}}{\cancel{5}}$$
$$k = 18$$

Check: $\frac{5}{6}k - 5 = 10$
$$\frac{5}{6}(18) - 5 \stackrel{?}{=} 10$$
$$\frac{5 \cdot 3 \cdot \cancel{6}}{\cancel{6}} - 5 \stackrel{?}{=} 10$$
$$15 - 5 \stackrel{?}{=} 10$$
$$10 = 10$$
18 is the solution.

19.
$$-\frac{7}{16}h + 28 = 21$$
$$-\frac{7}{16}h + 28 \mathbf{- 28} = 21 \mathbf{- 28}$$
$$\mathbf{-\frac{16}{7}}\left(-\frac{7}{16}h\right) = \mathbf{-\frac{16}{7}}(-7)$$
$$h = \frac{-16 \cdot \cancel{-7}}{\cancel{7}}$$
$$h = 16$$

Check: $-\frac{7}{16}h + 28 = 21$
$$-\frac{7}{16}(16) + 28 \stackrel{?}{=} 21$$
$$\frac{-7 \cdot \cancel{16}}{\cancel{16}} + 28 \stackrel{?}{=} 21$$
$$-7 + 28 \stackrel{?}{=} 21$$
$$21 = 21$$
16 is the solution.

21.
$$-6 - y = -2$$
$$-6 - y \mathbf{+ 6} = -2 \mathbf{+ 6}$$
$$-y = 4$$
$$(\mathbf{-1})(-y) = (\mathbf{-1})4$$
$$y = -4$$

Check: $-6 - y = -2$
$$-6 - (-4) \stackrel{?}{=} -2$$
$$-6 + 4 \stackrel{?}{=} -2$$
$$-2 = -2$$
-4 is the solution.

23.
$$-1.7 = 1.2 - x$$
$$-1.7 \mathbf{- 1.2} = 1.2 - x \mathbf{- 1.2}$$
$$-2.9 = -x$$
$$(\mathbf{-1})(-2.9) = (\mathbf{-1})(-x)$$
$$2.9 = x$$

Check: $-1.7 = 1.2 - x$
$$-1.7 \stackrel{?}{=} 1.2 - (2.9)$$
$$-1.7 \stackrel{?}{=} 1.2 + (-2.9)$$
$$-1.7 = -1.7$$
2.9 is the solution.

25.
$$3(2y - 2) - y = 5$$
$$6y - 6 - y = 5$$
$$5y - 6 = 5$$
$$5y - 6 \mathbf{+ 6} = 5 \mathbf{+ 6}$$
$$5y = 11$$
$$\frac{5y}{\mathbf{5}} = \frac{11}{\mathbf{5}}$$
$$y = \frac{11}{5}$$

Check: $3(2y - 2) - y = 5$
$$6y - 6 - y = 5$$
$$5y - 6 = 5$$
$$5 \cdot \frac{11}{5} - 6 \stackrel{?}{=} 5$$
$$11 - 6 \stackrel{?}{=} 5$$
$$5 = 5 \quad \text{True}$$
$\frac{11}{5}$ is the solution.

27.
$$\begin{aligned} 6a-3(3a-4)&=30\\ 6a-9a+12&=30\\ -3a+12&=30\\ -3a+12-\mathbf{12}&=30-\mathbf{12}\\ -3a&=18\\ \frac{-3a}{\mathbf{-3}}&=\frac{18}{\mathbf{-3}}\\ a&=-6 \end{aligned}$$

Check: $6a-3(3a-4)=30$

$$\begin{aligned} 6(-6)-3(3\cdot-6-4)&\overset{?}{=}30\\ -36-3(-22)&\overset{?}{=}30\\ -36+66&\overset{?}{=}30\\ 30&=30 \quad \text{True} \end{aligned}$$

-6 is the solution.

29.
$$\begin{aligned} 7a-12&=8a+9\\ 7a-12-\mathbf{7a}&=8a+9-\mathbf{7a}\\ -12&=a+9\\ -12-\mathbf{9}&=a+9-\mathbf{9}\\ -21&=a \end{aligned}$$

Check: $7a-12=8a+9$

$$\begin{aligned} 7(-21)-12&\overset{?}{=}8(-21)+9\\ -147-12&\overset{?}{=}-168+9\\ -159&=-159 \end{aligned}$$

-21 is the solution.

31.
$$\begin{aligned} 60r-50&=15r-5\\ 60r-50-\mathbf{15r}&=15r-5-\mathbf{15r}\\ 45r-50&=-5\\ 45r-50+\mathbf{50}&=-5+\mathbf{50}\\ 45r&=45\\ \frac{45r}{\mathbf{45}}&=\frac{45}{\mathbf{45}}\\ r&=1 \end{aligned}$$

Check: $60r-50=15r-5$

$$\begin{aligned} 60\cdot1-50&\overset{?}{=}15\cdot1-5\\ 60-50&\overset{?}{=}15-5\\ 10&=10 \quad \text{True} \end{aligned}$$

1 is the solution.

33. LCD $=18$
$$\begin{aligned} \frac{5}{6}x+\frac{2}{9}&=\frac{1}{3}\\ \mathbf{18}\left(\frac{5}{6}x\right)+\mathbf{18}\left(\frac{2}{9}\right)&=\mathbf{18}\left(\frac{1}{3}\right)\\ 15x+4&=6\\ 15x+4-\mathbf{4}&=6-\mathbf{4}\\ 15x&=2\\ \frac{15x}{\mathbf{15}}&=\frac{2}{\mathbf{15}}\\ x&=\frac{2}{15} \end{aligned}$$

Check: $\frac{1}{3}=\frac{5}{6}x+\frac{2}{9}$

$$\begin{aligned} 18\left(\frac{1}{3}\right)&=18\left(\frac{5}{6}x\right)+18\left(\frac{2}{9}\right)\\ 6&=15x+4\\ 6&\overset{?}{=}15\cdot\left(\frac{2}{15}\right)+4\\ 6&\overset{?}{=}2+4\\ 6&=6 \quad \text{True} \end{aligned}$$

$\frac{2}{15}$ is the solution.

35. LCD $=8$
$$\begin{aligned} \frac{1}{8}y-\frac{1}{2}&=\frac{1}{4}\\ \mathbf{8}\left(\frac{1}{8}y\right)-\mathbf{8}\left(\frac{1}{2}\right)&=\mathbf{8}\left(\frac{1}{4}\right)\\ y-4&=2\\ y-4+\mathbf{4}&=2+\mathbf{4}\\ y&=6 \end{aligned}$$

Check: $\frac{1}{8}y-\frac{1}{2}=\frac{1}{4}$

$$\begin{aligned} \frac{1}{8}(6)-\frac{1}{2}&\overset{?}{=}\frac{1}{4}\\ \frac{6}{8}-\frac{1}{2}&\overset{?}{=}\frac{1}{4}\\ \frac{3}{4}-\frac{2}{4}&\overset{?}{=}\frac{1}{4}\\ \frac{1}{4}&=\frac{1}{4} \end{aligned}$$

6 is the solution.

Solve each equation and check the result.

37.
$$\begin{aligned} 0.02(62)-0.08s&=0.06(s+9)\\ \mathbf{100}\left[0.02(62)-0.08s\right]&=\mathbf{100}\left[0.06(s+9)\right]\\ 100\left[0.02(62)\right]-100\left[0.08s\right]&=100\left[0.06(s+9)\right]\\ 2(62)-8s&=6(s+9)\\ 124-8s&=6s+54\\ 124-8s+\mathbf{8s}&=6s+\mathbf{8s}+54\\ 124&=14s+54\\ 124-\mathbf{54}&=14s+54-\mathbf{54}\\ 70&=14s\\ \frac{70}{\mathbf{14}}&=\frac{14s}{\mathbf{14}}\\ 5&=s \end{aligned}$$

Check: $0.02(62)-0.08s=0.06(s+9)$

$$\begin{aligned} 0.02(62)-0.08(5)&\overset{?}{=}0.06(5+9)\\ 1.24-0.4&\overset{?}{=}0.06(14)\\ 0.84&=0.84 \end{aligned}$$

5 is the solution.

39.
$$\begin{aligned} 0.09(t+50)+0.15t&=52.5\\ \mathbf{100}\left[0.09(t+50)\right]+\mathbf{100}\left[0.15t\right]&=\mathbf{100}(52.5)\\ 19(t+50)+15t&=5,250\\ 9t+450+15t&=5,250\\ (9+15)t+450&=5,250\\ 24t+450&=5,250\\ 24t+450-\mathbf{450}&=5,250-\mathbf{450}\\ 24t&=4,800\\ \frac{24t}{\mathbf{24}}&=\frac{4,800}{\mathbf{24}}\\ t&=200 \end{aligned}$$

Check: $0.09(t+50)+0.15t=52.5$

$$\begin{aligned} 0.09(200+50)+0.15(200)&\overset{?}{=}52.5\\ 0.09(250)+30&\overset{?}{=}52.5\\ 22.5+30&\overset{?}{=}52.5\\ 52.5&=52.5 \quad \text{True} \end{aligned}$$

200 is the solution.

Solve each equation and check the result.

41. LCD $=3$
$$\begin{aligned} \frac{10-5s}{3}&=-s+6\\ \mathbf{3}\left(\frac{10-5s}{3}\right)&=\mathbf{3}(-s+6)\\ 10-5s&=-3s+18\\ 10-5s+\mathbf{5s}&=-3s+18+\mathbf{5s}\\ 10&=2s+18\\ 10-\mathbf{18}&=2s+18-\mathbf{18}\\ -8&=2s\\ \frac{-8}{\mathbf{2}}&=\frac{2s}{\mathbf{2}}\\ -4&=s \end{aligned}$$

Check:

$$\begin{aligned} \frac{10-5s}{3}&=-s+6\\ \frac{10-5(-4)}{3}&\overset{?}{=}-(-4)+6\\ \frac{10+20}{3}&\overset{?}{=}4+6\\ \frac{30}{3}&\overset{?}{=}10\\ 10&=10 \end{aligned}$$

-4 is the solution.

43. LCD = 16

$$\frac{7t-9}{16}=t$$

$$\mathbf{16}\left(\frac{7t-9}{16}\right)=\mathbf{16}t$$

$$7t-9=16$$

$$7t-9-\mathbf{7t}=16t-\mathbf{7t}$$

$$-9=9t$$

$$\frac{-9}{\mathbf{9}}=\frac{9t}{\mathbf{9}}$$

$$-1=t$$

Check: $\frac{7t-9}{16}=t$

$$\frac{7(-1)-9}{16}\overset{?}{=}-1$$

$$\frac{-7-9}{16}\overset{?}{=}-1$$

$$\frac{-16}{16}\overset{?}{=}-1$$

$$-1=-1 \text{ True}$$

-1 is the solution.

45. $8x+3(2-x)=5x+6$

$$8x+6-3x=5x+6$$

$$5x+6=5x+6$$

$$5x+6-\mathbf{5x}=5x+6-\mathbf{5x}$$

$$6=6$$

The terms involving x drop out and the result is true. This means that *all real numbers* are solutions and this equation is an identity.

47. $-3(s+2)=-2(s+4)-s$

$$-3s-6=-2s-8-s$$

$$-3s-6=-3s-8$$

$$-3s-6+\mathbf{3s}=-3s-8+\mathbf{3s}$$

$$-6=-8$$

The terms involving s drop out and the result is false. This means the equation has *no solution* and it is a contradiction.

TRY IT YOURSELF

Solve each equation, if possible.

49. $3x-8-4x-7x=-2-8$

$$-8x-8=-10$$

$$-8x-8+\mathbf{8}=-10+\mathbf{8}$$

$$-8x=-2$$

$$\frac{-8x}{\mathbf{-8}}=\frac{-2}{\mathbf{-8}}$$

$$x=\frac{\cancel{2}^{1}}{4\cdot\cancel{2}_{1}}$$

$$x=\frac{1}{4}$$

Check: $3x-8-4x-7x=-2-8$

let $\frac{1}{4}=0.25$

$$3(0.25)-8-4(0.25)-7(0.25)\overset{?}{=}-2-8$$

$$0.75-8-1-1.75\overset{?}{=}-10$$

$$-8-1-1\overset{?}{=}-10$$

$$-10=-10 \text{ True}$$

$\frac{1}{4}$ or 0.25 is the solution.

51. $\frac{t}{3}+2=6$

$$\frac{t}{3}+2-\mathbf{2}=6-\mathbf{2}$$

$$\frac{\mathbf{3}}{\mathbf{1}}\left(\frac{t}{3}\right)=\frac{\mathbf{3}}{\mathbf{1}}(4)$$

$$t=12$$

Check: $\frac{t}{3}+2=6$

$$\frac{12}{3}+2\overset{?}{=}6$$

$$4+2\overset{?}{=}6$$

$$6=6$$

12 is the solution.

53. $4(5b)+2(6b-1)=-34$

$$20b+12b-2=-34$$

$$32b-2=-34$$

$$32b-2-\mathbf{2}=-34-\mathbf{2}$$

$$32b=-32$$

$$\frac{32b}{\mathbf{32}}=\frac{-32}{\mathbf{32}}$$

$$b=-1$$

Check: $4(5b)+2(6b-1)=-34$

$$4(5\cdot-1)+2(6\cdot-1-1)\overset{?}{=}-34$$

$$4(-5)+2(-7)\overset{?}{=}-34$$

$$-20+(-14)\overset{?}{=}-34$$

$$-34=-34 \text{ True}$$

-1 is the solution.

55. $2x+5=17$

$$2x+5-\mathbf{5}=17-\mathbf{5}$$

$$2x=12$$

$$\frac{2x}{\mathbf{2}}=\frac{12}{\mathbf{2}}$$

$$x=6$$

Check: $2x+5=17$

$$2(6)+5\overset{?}{=}17$$

$$12+5\overset{?}{=}17$$

$$17=17$$

6 is the solution.

57. LCD = 6

$$\frac{5}{6}(1-x)=-x+1$$

$$\mathbf{6}\left(\frac{5}{6}\right)(1-x)=\mathbf{6}(-x+1)$$

$$5(1-x)=6(-x+1)$$

$$5-5x=-6x+6$$

$$5-5x+\mathbf{6x}=-6x+6+\mathbf{6x}$$

$$x+5=6$$

$$x+5-\mathbf{5}=6-\mathbf{5}$$

$$x=1$$

Check: $\frac{5}{6}(1-x)=-x+1$

$$\frac{5}{6}(1-1)\overset{?}{=}-1+1$$

$$\frac{5}{6}(0)\overset{?}{=}0$$

$$0=0$$

1 is the solution.

59. $0.05a+0.01(90)=0.02(a+90)$

$$\mathbf{100}[0.05a+0.01(90)]=\mathbf{100}[0.02(a+90)]$$

$$100(0.05a)+100[0.01(90)]=100[0.02(a+90)]$$

$$5a+1(90)=2(a+90)$$

$$5a+90=2a+180$$

$$5a+90-\mathbf{2a}=2a+180-\mathbf{2a}$$

$$3a+90=180$$

$$3a+90-\mathbf{90}=180-\mathbf{90}$$

$$3a=90$$

$$\frac{3a}{\mathbf{3}}=\frac{90}{\mathbf{3}}$$

$$a=30$$

Check: $0.05a+0.01(90)=0.02(a+90)$

$$0.05(30)+0.01(90)\stackrel{?}{=}0.02(30+90)$$
$$1.5+0.9\stackrel{?}{=}0.02(120)$$
$$2.4=2.4 \text{ True}$$

30 is the solution.

61.
$$\frac{7}{2}+\frac{3}{2}d=-9+1.5d$$
$$3.5+1.5d=-9+1.5d$$
$$3.5+1.5d-\mathbf{1.5d}=-9+1.5d-\mathbf{1.5d}$$
$$3.5=-9$$

The terms involving d drop out and the result is false. This means the equation has *no solution* and is a contration.

63.
$$-(19-3s)-(8s+1)=35$$
$$-19+3s-8s-1=35$$
$$-5s-20=35$$
$$-5s-20+\mathbf{20}=35+\mathbf{20}$$
$$-5s=55$$
$$\frac{-5s}{\mathbf{-5}}=\frac{55}{\mathbf{-5}}$$
$$s=-11$$

Check:
$$-(19-3s)-(8s+1)\stackrel{?}{=}35$$
$$-(19-3\cdot\mathbf{-11})-(8\cdot\mathbf{-11}+1)\stackrel{?}{=}35$$
$$-(19+33)-(-88+1)\stackrel{?}{=}35$$
$$-(52)-(-87)\stackrel{?}{=}35$$
$$-52+87\stackrel{?}{=}35$$
$$35=35 \text{ True}$$

-11 is the solution.

65.
$$5x=4x+7$$
$$5x-\mathbf{4x}=4x+7-\mathbf{4x}$$
$$x=7$$

Check: $5x=4x+7$
$$5(7)\stackrel{?}{=}4(7)+7$$
$$35\stackrel{?}{=}28+7$$
$$35=35$$

7 is the solution.

67. LCD = 4
$$\frac{3(b+2)}{2}=\frac{4b-10}{4}$$
$$\mathbf{4}\left(\frac{3(b+2)}{2}\right)=\mathbf{4}\left(\frac{4b-10}{4}\right)$$
$$2[3(b+2)]=4b-10$$
$$2(3b+6)=4b-10$$
$$6b+12=4b-10$$
$$6b+12-\mathbf{4b}=4b-10-\mathbf{4b}$$
$$2b+12=-10$$
$$2b+12-\mathbf{12}=-10-\mathbf{12}$$
$$2b=-22$$
$$\frac{2b}{\mathbf{2}}=\frac{-22}{\mathbf{2}}$$
$$b=-11$$

Check:
$$\frac{3(b+2)}{2}=\frac{4b-10}{4}$$
$$\frac{3(-11+2)}{2}\stackrel{?}{=}\frac{4(-11)-10}{4}$$
$$\frac{3(-9)}{2}\stackrel{?}{=}\frac{-44-10}{4}$$
$$\frac{-27}{2}\stackrel{?}{=}\frac{-54}{4}$$
$$\frac{-27}{2}\stackrel{?}{=}\frac{\cancel{2}^{1}\cdot-27}{\cancel{2}_{1}\cdot 2}$$
$$-\frac{27}{2}=-\frac{27}{2} \quad \text{True}$$

-11 is the solution.

69.
$$8y-2=4y+16$$
$$8y-2-\mathbf{4y}=4y+16-\mathbf{4y}$$
$$4y-2=16$$
$$4y-2+\mathbf{2}=16+\mathbf{2}$$
$$4y=18$$
$$\frac{4y}{\mathbf{4}}=\frac{18}{\mathbf{4}}$$
$$y=\frac{9\cdot\cancel{2}^{1}}{2\cdot\cancel{2}_{1}}$$
$$y=\frac{9}{2}$$

Check: $8y-2=4y+16$
$$8\cdot\frac{9}{2}-2\stackrel{?}{=}4\cdot\frac{9}{2}+16$$
$$36-2\stackrel{?}{=}18+16$$
$$34=34 \quad \text{True}$$

$\frac{9}{2}$ is the solution.

71. LCD = 12
$$\frac{1}{6}y+\frac{1}{4}y=-1$$
$$\mathbf{12}\left(\frac{1}{6}y\right)+\mathbf{12}\left(\frac{1}{4}y\right)=\mathbf{12}(-1)$$
$$2y+3y=-12$$
$$5y=-12$$
$$\frac{5y}{\mathbf{5}}=\frac{-12}{\mathbf{5}}$$
$$y=-\frac{12}{5}$$

Check: $\frac{1}{6}y+\frac{1}{4}y=-1$

$$\left(\frac{1}{6}\cdot-\frac{12}{5}\right)+\left(\frac{1}{4}\cdot-\frac{12}{5}\right)\stackrel{?}{=}-1$$

$$-\frac{2}{5}+\left(-\frac{3}{5}\right)\stackrel{?}{=}-1$$

$$-\frac{5}{5}\stackrel{?}{=}-1$$

$$-1=-1$$

$-\frac{12}{5}$ is the solution.

73. $0.7-4y=1.7$

$$0.7-4y-\mathbf{0.7}=1.7-\mathbf{0.7}$$

$$-4y=1$$

$$\frac{-4y}{\mathbf{-4}}=\frac{1}{\mathbf{-4}}$$

$$y=-0.25$$

Check: $0.7-4y=1.7$

$$0.7-4(-0.25)\stackrel{?}{=}1.7$$

$$0.7+1\stackrel{?}{=}1.7$$

$$1.7=1.7$$

-0.25 is the solution.

75. $-33=5t+2$

$$-33-\mathbf{2}=5t+2-\mathbf{2}$$

$$-35=5t$$

$$\frac{-35}{\mathbf{5}}=\frac{5t}{\mathbf{5}}$$

$$-7=t$$

Check: $-33=5t+2$

$$-33\stackrel{?}{=}5(-7)+2$$

$$-33\stackrel{?}{=}-35+2$$

$$-33=-33$$

-7 is the solution.

77. $-3p+7=-3$

$$-3p+7-\mathbf{7}=-3-\mathbf{7}$$

$$-3p=-10$$

$$\frac{-3p}{\mathbf{-3}}=\frac{-10}{\mathbf{-3}}$$

$$p=\frac{10}{3}$$

Check: $-3p+7=-3$

$$-3\left(\frac{10}{3}\right)+7\stackrel{?}{=}-3$$

$$-10+7\stackrel{?}{=}-3$$

$$-3=-3$$

$\frac{10}{3}$ is the solution.

79. $2(-3)+4y=14$

$$-6+4y=14$$

$$-6+4y+\mathbf{6}=14+\mathbf{6}$$

$$4y=20$$

$$\frac{4y}{\mathbf{4}}=\frac{20}{\mathbf{4}}$$

$$y=5$$

Check: $2(-3)+4y=14$

$$-6+4(5)\stackrel{?}{=}14$$

$$-6+20\stackrel{?}{=}14$$

$$14=14 \quad \text{True}$$

5 is the solution.

81. $0.06(a+200)+0.1a=172$

$$\mathbf{100}[0.06(a+200)+0.1a]=\mathbf{100}(172)$$

$$\mathbf{100}[0.06(a+200)]+\mathbf{100}[0.1a]=\mathbf{100}(172)$$

$$6(a+200)+100(0.1a)=100(172)$$

$$6a+1,200+10a=17,200$$

$$16a+1,200=17,200$$

$$16a+1,200-\mathbf{1,200}=17,200-\mathbf{1,200}$$

$$16a=16,000$$

$$\frac{16a}{\mathbf{16}}=\frac{16,000}{\mathbf{16}}$$

$$a=1,000$$

Check: $0.06(a+200)+0.1a=172$

$$0.06(1,000+200)+(0.1)(1,000)\stackrel{?}{=}172$$

$$0.06(1,200)+100\stackrel{?}{=}172$$

$$72+100\stackrel{?}{=}172$$

$$172=172 \quad \text{True}$$

1,000 is the solution.

83. $8.6y+3.4=4.2y-9.8$

$$8.6y+3.4-\mathbf{4.2y}=4.2y-9.8-\mathbf{4.2y}$$

$$4.4y+3.4=-9.8$$

$$4.4y+3.4-\mathbf{3.4}=-9.8-\mathbf{3.4}$$

$$4.4y=-13.2$$

$$\frac{4.4y}{\mathbf{4.4}}=\frac{-13.2}{\mathbf{4.4}}$$

$$y=-3$$

Check: $8.6y+3.4=4.2y-9.8$

$$8.6(\mathbf{-3})+3.4\stackrel{?}{=}4.2(\mathbf{-3})-9.8$$

$$-25.8+3.4\stackrel{?}{=}-12.6-9.8$$

$$-22.4=-22.4$$

-3 is the solution.

85. LCD $=15$

$$\frac{2}{3}y+2=\frac{1}{5}+y$$

$$\mathbf{15}\left(\frac{2}{3}y\right)+\mathbf{15}(2)=\mathbf{15}\left(\frac{1}{5}\right)+\mathbf{15}(y)$$

$$10y+30=3+15y$$

$$10y+30-\mathbf{10y}=3+15y-\mathbf{10y}$$

$$30=3+5y$$

$$30-\mathbf{3}=3+5y-\mathbf{3}$$

$$27=5y$$

$$\frac{27}{\mathbf{5}}=\frac{5y}{\mathbf{5}}$$

$$\frac{27}{5}=y$$

Check: $\frac{2}{3}y+2=\frac{1}{5}+y$

$$15\left(\frac{2}{3}y\right)+15(2)=15\left(\frac{1}{5}\right)+15(y)$$

$$10y+30=3+15y$$

$$10\cdot\left(\frac{27}{5}\right)+30\stackrel{?}{=}3+15\cdot\left(\frac{27}{5}\right)$$

$$54+30\stackrel{?}{=}3+81$$

$$84=84 \quad \text{True}$$

$\frac{27}{5}$ is the solution.

87. $$0.4b-0.1(b-100)=70$$
$$\mathbf{10}(0.4b)-\mathbf{10}[0.1(b-100)]=\mathbf{10}(70)$$
$$4b-1(b-100)=700$$
$$4b-b+100=700$$
$$3b+100=700$$
$$3b+100-\mathbf{100}=700-\mathbf{100}$$
$$3b=600$$
$$\frac{3b}{\mathbf{3}}=\frac{600}{\mathbf{3}}$$
$$b=200$$

Check: $$0.4b-0.1(b-100)=70$$
$$0.4(200)-0.1(200-100)\stackrel{?}{=}70$$
$$80-0.1(100)\stackrel{?}{=}70$$
$$80-10\stackrel{?}{=}70$$
$$70=70 \quad \text{True}$$

200 is the solution.

89. LCD $=4$
$$\frac{1}{4}(10-2y)=8$$
$$\mathbf{4}\left[\frac{1}{4}(10-2y)\right]=\mathbf{4}(8)$$
$$10-2y=32$$
$$10-2y-\mathbf{10}=32-\mathbf{10}$$
$$-2y=22$$
$$\frac{-2y}{\mathbf{-2}}=\frac{22}{\mathbf{-2}}$$
$$y=-11$$

Check: $$\frac{1}{4}(10-2y)=8$$
$$\frac{1}{4}[10-2(-11)]\stackrel{?}{=}8$$
$$\frac{1}{4}[10+22]\stackrel{?}{=}8$$
$$\frac{1}{4}[32]\stackrel{?}{=}8$$
$$8=8$$

-11 is the solution.

91. $$2-3(x-5)=4(x-1)$$
$$2-3x+15=4x-4$$
$$-3x+17=4x-4$$
$$-3x+17+\mathbf{3x}=4x-4+\mathbf{3x}$$
$$17=7x-4$$
$$17+4=7x-4+\mathbf{4}$$
$$21=7x$$
$$\frac{21}{\mathbf{7}}=\frac{7x}{\mathbf{7}}$$
$$3=x$$

Check: $$2-3(x-5)=4(x-1)$$
$$2-3(3-5)\stackrel{?}{=}4(3-1)$$
$$2-3(-2)\stackrel{?}{=}4(2)$$
$$2+6\stackrel{?}{=}8$$
$$8=8 \quad \text{True}$$

3 is the solution.

93. LCD $=12$
$$2n-\frac{3}{4}n=\frac{1}{2}n+\frac{13}{3}$$
$$\mathbf{12}\left(2n-\frac{3}{4}n\right)=\mathbf{12}\left(\frac{1}{2}n+\frac{13}{3}\right)$$
$$12(2n)-12\left(\frac{3}{4}n\right)=12\left(\frac{1}{2}n\right)+12\left(\frac{13}{3}\right)$$
$$24n-3(3n)=6n+4(13)$$
$$24n-9n=6n+52$$
$$15n-\mathbf{6n}=6n+52-\mathbf{6n}$$
$$9n=52$$
$$\frac{9n}{\mathbf{9}}=\frac{52}{\mathbf{9}}$$
$$n=\frac{52}{9}$$

Check: $$2n-\frac{3}{4}n=\frac{1}{2}n+\frac{13}{3}$$
$$2\left(\frac{52}{9}\right)-\frac{3}{4}\left(\frac{52}{9}\right)\stackrel{?}{=}\frac{1}{2}\left(\frac{52}{9}\right)+\frac{13}{3}$$
$$\frac{104}{9}-\frac{39}{9}\stackrel{?}{=}\frac{26}{9}+\frac{39}{9}$$
$$\frac{65}{9}=\frac{65}{9} \quad \text{True}$$

$\frac{52}{9}$ is the solution.

95. $$10.08=4(0.5x+2.5)$$
$$10.08=2x+10$$
$$10.08-\mathbf{10}=2x+10-\mathbf{10}$$
$$0.08=2x$$
$$\frac{0.08}{\mathbf{2}}=\frac{2x}{\mathbf{2}}$$
$$0.04=x$$

Check:
$$10.08=4(0.5x+2.5)$$
$$10.08\stackrel{?}{=}4(0.5\cdot 0.04+2.5)$$
$$10.08\stackrel{?}{=}4(0.02+2.5)$$
$$10.08\stackrel{?}{=}4(2.52)$$
$$10.08=10.08 \quad \text{True}$$

0.04 is the solution.

97. LCD $=12$

$$\frac{3}{4}(d-8)=\frac{2}{3}(d+1)$$
$$\mathbf{12}\left(\frac{3}{4}\right)(d-8)=\mathbf{12}\left(\frac{2}{3}\right)(d+1)$$
$$\mathbf{3}[3(d-8)]=\mathbf{4}[2(d+1)]$$
$$9(d-8)=8(d+1)$$
$$9d-72=8d+8$$
$$9d-72-\mathbf{8d}=8d+8-\mathbf{8d}$$
$$d-72=8$$
$$d-72+\mathbf{72}=8+\mathbf{72}$$
$$d=80$$

Check: $\frac{3(d-8)}{4}=\frac{2(d+1)}{3}$

$$\frac{3(80-8)}{4}\stackrel{?}{=}\frac{2(80+1)}{3}$$
$$\frac{3(72)}{4}\stackrel{?}{=}\frac{2(81)}{3}$$
$$54=54 \quad \text{True}$$

80 is the solution.

99. $2d+5=0$

$$2d+5-\mathbf{5}=0-\mathbf{5}$$
$$2d=-5$$
$$\frac{2d}{\mathbf{2}}=\frac{-5}{\mathbf{2}}$$
$$d=-\frac{5}{2}$$

Check: $2d+5=0$

$$2\left(-\frac{5}{2}\right)+5\stackrel{?}{=}0$$
$$-5+5\stackrel{?}{=}0$$
$$0=0$$

$-\frac{5}{2}$ is the solution.

101. $3(A+2)=2(A-7)$

$$3A+6=2A-14$$
$$3A+6-\mathbf{2A}=2A-14-\mathbf{2A}$$
$$A+6=-14$$
$$A+6-\mathbf{6}=-14-\mathbf{6}$$
$$A=-20$$

Check: $3(A+2)=2(A-7)$

$$3(-20+2)\stackrel{?}{=}2(-20-7)$$
$$3(-18)\stackrel{?}{=}2(-27)$$
$$-54=-54 \quad \text{True}$$

-20 is the solution.

103. $4(a-3)=-2(a-6)+6a$

$$4a-12=-2a+12+6a$$
$$4a-12=4a+12$$
$$4a-12-\mathbf{4a}=4a+12-\mathbf{4a}$$
$$-12=+12$$

The terms involving a drop out and the result is false. This means the equation has *no solution* and it is a contradiction.

105. $4(y-3)-y=3(y-4)$

$$4y-12-y=3y-12$$
$$3y-12=3y-12$$
$$3y-12-\mathbf{3y}=3y-12-\mathbf{3y}$$
$$-12=-12$$

The terms involving y drop out and the result is true. This means that *all real numbers* are solutions and this equation is an identity.

107. $-(4-m)=-10$

$$-4+m=-10$$
$$-4+m+\mathbf{4}=-10+\mathbf{4}$$
$$m=-6$$

Check: $-(4-m)=-10$

$$-[4-(-6)]\stackrel{?}{=}-10$$
$$-(4+6)\stackrel{?}{=}-10$$
$$-10=-10$$

-6 is the solution.

109. LCD $=3$

$$-\frac{2}{3}z+4=8$$
$$\mathbf{3}\left(-\frac{2}{3}z+4\right)=\mathbf{3}(8)$$
$$3\left(-\frac{2}{3}z\right)+3(4)=3(8)$$
$$-2z+12=24$$
$$-2z+12-\mathbf{12}=24-\mathbf{12}$$
$$-2z=12$$
$$\frac{-2z}{\mathbf{-2}}=\frac{12}{\mathbf{-2}}$$
$$z=-6$$

Check: $-\frac{2}{3}z+4=8$

$$-\frac{2}{3}(-6)+4\stackrel{?}{=}8$$
$$4+4\stackrel{?}{=}8$$
$$8=8 \quad \text{True}$$

-6 is the solution.

Look Alikes . . .

Simplify each expression and solve each equation.

111. a. $-2(9-3x)-(5x+2)$

$$=\mathbf{-2}(9)-\mathbf{2}(-3x)-\mathbf{1}(5x)-\mathbf{1}(2)$$
$$=-18+6x-5x-2$$
$$=(6x-5x)+(-18-2)$$
$$=x-20$$

111. b.
$$-2(9-3x)-(5x+2)=-25$$
$$-\mathbf{2}(9)-\mathbf{2}(-3x)-\mathbf{1}(5x)-\mathbf{1}(2)=-25$$
$$-18+6x-5x-2=-25$$
$$(6x-5x)+(-18-2)=-25$$
$$x-20=-25$$
$$x-20+\mathbf{20}=-25+\mathbf{20}$$
$$x=-5$$

Check:
$$-2(9-3x)-(5x+2)=-25$$
$$-2[9-3(-5)]-[5(-5)+2]\overset{?}{=}-25$$
$$-2(9+15)-(-25+2)\overset{?}{=}-25$$
$$-2(24)-(-23)\overset{?}{=}-25$$
$$-48+23\overset{?}{=}-25$$
$$-25=-25$$

-5 is the solution.

113. a.
$$0.6-0.2(x+1)=0.6-\mathbf{0.2}(x)-\mathbf{0.2}(1)$$
$$=0.6-0.2x-0.2$$
$$=(0.6-0.2)-0.2x$$
$$=0.4-0.2x$$

113. b.
$$0.6-0.2(x+1)=0.4$$
$$0.6-\mathbf{0.2}(x)-\mathbf{0.2}(1)=0.4$$
$$0.6-0.2x-0.2=0.4$$
$$(0.6-0.2)-0.2x=0.4$$
$$0.4-0.2x=0.4$$
$$0.4-0.2x-\mathbf{0.4}=0.4-\mathbf{0.4}$$
$$-0.2x=0$$
$$\frac{-0.2x}{\mathbf{-0.2}}=\frac{0}{\mathbf{-0.2}}$$
$$x=0$$

Check:
$$0.6-0.2(x+1)=0.4$$
$$0.6-0.2(0+1)\overset{?}{=}0.4$$
$$0.6-0.2(1)\overset{?}{=}0.4$$
$$0.6-0.2\overset{?}{=}0.4$$
$$0.4=0.4$$

0 is the solution.

WRITING

115.–117. Answers will vary.

REVIEW

Name the property that is used.

119. $x\cdot 9=9x$; **Com. Prop. Mult.**

121. $(x+1)+2=x+(1+2)$; **Assoc. Prop. Add.**

CHALLENGE PROBLEMS

123. Solve: LCD = 6
$$\frac{5}{6}\left(-\frac{3}{4}m+1\right)=-\frac{2}{3}\left(\frac{1}{2}m-1\right)$$
$$\mathbf{6}\left[\frac{5}{6}\left(-\frac{3}{4}m+1\right)\right]=\mathbf{6}\left[-\frac{2}{3}\left(\frac{1}{2}m-1\right)\right]$$
$$5\left(-\frac{3}{4}m+1\right)=-4\left(\frac{1}{2}m-1\right)$$
$$\mathbf{5}\left(-\frac{3}{4}m\right)+\mathbf{5}(1)=\mathbf{-4}\left(\frac{1}{2}m\right)\mathbf{-4}(-1)$$
$$-\frac{15}{4}m+5=-2m+4$$
LCD = 4
$$\mathbf{4}\left(-\frac{15}{4}m+5\right)=\mathbf{4}(-2m+4)$$
$$\mathbf{4}\left(-\frac{15}{4}m\right)+\mathbf{4}(5)=\mathbf{4}(-2m)+\mathbf{4}(4)$$
$$-15m+20=-8m+16$$
$$-15m+20+\mathbf{15m}=-8m+16+\mathbf{15m}$$
$$20=7m+16$$
$$20-\mathbf{16}=7m+16-\mathbf{16}$$
$$4=7m$$
$$\frac{4}{\mathbf{7}}=\frac{7m}{\mathbf{7}}$$
$$\frac{4}{7}=m$$

Check:
$$\frac{5}{6}\left(-\frac{3}{4}m+1\right)=-\frac{2}{3}\left(\frac{1}{2}m-1\right)$$
$$\frac{5}{6}\left(-\frac{3}{4}\cdot\frac{\mathbf{4}}{\mathbf{7}}+1\right)\overset{?}{=}-\frac{2}{3}\left(\frac{1}{2}\cdot\frac{\mathbf{4}}{\mathbf{7}}-1\right)$$
$$\frac{5}{6}\left(-\frac{3}{7}+1\right)\overset{?}{=}-\frac{2}{3}\left(\frac{2}{7}-1\right)$$
$$\frac{5}{6}\left(-\frac{3}{7}+\frac{7}{7}\right)\overset{?}{=}-\frac{2}{3}\left(\frac{2}{7}-\frac{7}{7}\right)$$
$$\frac{5}{6}\left(\frac{4}{7}\right)\overset{?}{=}-\frac{2}{3}\left(-\frac{5}{7}\right)$$
$$\frac{10}{21}=\frac{10}{21}$$

$\frac{4}{7}$ is the solution.

SECTION 2.3 STUDY SET

VOCABULARY

Fill in the blanks.

1. **Percent** means parts per one hundred.

3. In percent questions, the word *of* means **multiplication**, and the word **is** means equals.

CONCEPTS

5. $\frac{51}{100}$, 0.51 , 51%

7. Fill in the blanks using the words percent, amount, and base.

   Amount = **percent** · **base**

9. a. Find the amount of increase in the number of earthquakes.

   $4,257 - 3,618 = 639$

   b. Fill in blanks to find the percent of increase in earthquakes:

   639 is **what** % of **3,618** ?

NOTATION

11. Change each percent to a decimal.

   a. 35% **0.35** b. 8.5% **0.085**

   c. 150% **1.5** d. $2\frac{3}{4}\%$ **0.0275**

   e. 9.25% **0.0925** f. $1\frac{1}{2}\%$ **0.015**

GUIDED PRACTICE

13. What number is 48% of 650?

$x = 0.48 \cdot 650$

$x = 312$

312 is the solution.

15. What number is 92.4% of 50?

$x = 0.924 \cdot 50$

$x = 46.2$

46.2 is the solution.

17. 75 is 25% of what number?

$75 = 0.25 \cdot x$

$\frac{75}{\mathbf{0.25}} = \frac{0.25x}{\mathbf{0.25}}$

$300 = x$

300 is the solution.

19. 128.1 is 8.75% of what number?

$128.1 = 0.0875 \cdot x$

$\frac{128.1}{\mathbf{0.0875}} = \frac{0.0875x}{\mathbf{0.0875}}$

$1,464 = x$

1,464 is the solution.

21. 78 is what percent of 300?

$78 = x \cdot 300$

$\frac{78}{\mathbf{300}} = \frac{300x}{\mathbf{300}}$

$0.26 = x$

$26\% = x$

26% is the solution.

23. 0.42 is what percent of 16.8?

$0.42 = x \cdot 16.8$

$\frac{0.42}{\mathbf{16.8}} = \frac{16.8x}{\mathbf{16.8}}$

$0.025 = x$

$2.5\% = x$

2.5% is the solution.

APPLICATIONS

25. ANTISEPTICS

Let x = amt of pure hydrogen peroxide

What number is 3% of 16?

$x = 0.03 \cdot 16$

$x = 0.48$

0.48 oz is pure hydrogen peroxide.

27. U.S. FEDERAL BUDGET

a. Let x = amt spent on SS, Medicare

What number is 37% of 2,980 billion?

$x = 0.37 \cdot 2,980$ billion

$x = 1,102.6$ billion

$1,102.6 billion was spent on SS, Medicare.

---

b. Let y = amt spent on Nat Def, vet

What number is 24% of 2,980 billion?

$y = 0.24 \cdot 2,980$ billion

$y = 715.2$ billion

$715.2 billion was spent on Nat Def, Vet.

29. PAYPAL

Step 1: Let x = fee charged by PayPal

What number is 2.9% of 350?

$x = 0.029 \cdot 350$

$x = 10.15$

$10.15 is the fee charged by PayPal.

Step 2: $10.15 + $0.30 = $10.45

$10.45 is the final fee paid to PayPal.

31. PRICE GUARANTEES

Step 1: Let x = amt of difference saved

$\$120 - \$98 = \$22$

What amount is 10% of 22?

$x = 0.1 \cdot 22$

$x = 2.2$

\$2.20 is the amount reimbursed on \$22.

Step 2: $\$120 - \$98 = \$22$

\$22 is the difference in prices.

Step 3: $\$22.00 + \$2.20 = \$24.20$

\$24.20 is the total reimbursement.

33. COMPUTER MEMORY

Step 1: Let x = percent of used disk space

44.7 is what percent of 74.5 GB?

$44.7 = x \cdot 74.5$ GB

$$\frac{44.7}{\mathbf{74.5}} = \frac{74.5x}{\mathbf{74.5}}$$

$0.6 \approx x$

$60\% \approx x$

60% of the disk is being used.

Step 2: Let y = amt of free disk space

29.8 is what percent of 74.5 GB?

$29.8 = y \cdot 74.5$ GB

$$\frac{29.8}{\mathbf{74.5}} = \frac{74.5y}{\mathbf{74.5}}$$

$0.4 \approx y$

$40\% \approx y$

40% of the disk space is free.

35. DENTISTRY

Let x = percent of teeth have filling

6 is what percent of 32?

$6 = x \cdot 32$

$$\frac{6}{\mathbf{32}} = \frac{32x}{\mathbf{32}}$$

$0.1875 = x$

$19\% = x$

19% of the teeth have filling.

37. DMV WRITTEN TEST

Let x = test score of teenager

33 is what percent of 50?

$33 = x \cdot 50$

$$\frac{33}{\mathbf{50}} = \frac{50x}{\mathbf{50}}$$

$0.66 = x$

$66\% = x$

66% is the test score.

No, the teenager did not pass.

39. CHILD CARE

Let x = maximum number of children

84 is 70% of what number?

$84 = 0.70 \cdot x$

$$\frac{84}{\mathbf{0.70}} = \frac{0.70x}{\mathbf{0.70}}$$

$120 = x$

120 is the maximum number of children.

41. NUTRITION

a. 5 grams and 25%

b. Let x = total amount of fat in grams

5 is 25% of what number?

$5 = 0.25 \cdot x$

$$\frac{5}{\mathbf{0.25}} = \frac{0.25x}{\mathbf{0.25}}$$

$20 = x$

20 grams is the total amount of fat.

43. RAINFOREST

Part a:

Step 1: Let x = amt of increase

$4{,}984 - 4{,}490 = 494$

494 is the amount of increase.

Step 2: 494 is what percent of 4,490?

$494 = x \cdot 4{,}490$

$$\frac{494}{\mathbf{4{,}490}} = \frac{4{,}490x}{\mathbf{4{,}490}}$$

$0.11 \approx x$

There was a 11% increase from year 2007 to 2008.

Part b:

Step 1: Let x = amt of decrease

$4{,}984 - 2{,}705 = 2{,}279$

2,279 is the amount of decrease.

Step 2: 2,279 is what percent of 4,984?

$2{,}279 = x \cdot 4{,}984$

$$\frac{2{,}279}{\mathbf{4{,}984}} = \frac{4{,}984x}{\mathbf{4{,}984}}$$

$0.457 \approx x$

There was a 46% decrease from year 2008 to 2009.

45. INSURANCE COSTS

Step 1: $\$1,050-\$925=\$125$

This is the amount of decrease.

Step 2: 125 is what percent of 1,050?

$125 = x \cdot 1,050$

$\frac{125}{\mathbf{1,050}}=\frac{1,050x}{\mathbf{1,050}}$

$0.119 \approx x$

12% is the percent of decrease.

from Campus To Careers

47. AUTO SERVICE TECHNICIAN

a. 4 tires times 1.5% equals what number?

$4 \cdot 0.015 = x$

$0.06 = x$

6% is the percent that the mileage decreases.

b. 6% of 25 miles equals what number?

$0.06 \cdot 25 = x$

$1.5 = x$

$25-1.5=$ 23.5 the mileage per gallon.

23.5 is the mileage per gallon of gas.

49. TV SHOPPING

Let $x=$ the original price of the toy

15 is 20% of what number?

$15 = 0.20 \cdot x$

$\frac{15}{\mathbf{0.20}}=\frac{0.20x}{\mathbf{0.20}}$

$75 = x$

\$75 is the original price of the toy.

51. SALES

Step 1: $\$210 \div 2=\$105$

\$105 is the savings for one set.

Step 2: Let $x=$ the original price of one set

105 is 35% of what number?

$105 = 0.35 \cdot x$

$\frac{105}{\mathbf{0.35}}=\frac{0.35x}{\mathbf{0.35}}$

$300 = x$

\$300 is the original price of one set.

53. REAL ESTATE

Let $x=$ the selling price of the condo

3,325 is $3\frac{1}{2}$% of what number?

$3,325 = 0.035 \cdot x$

$\frac{3,325}{\mathbf{0.035}}=\frac{0.035x}{\mathbf{0.035}}$

$95,000 = x$

\$95,000 is the selling price of the condo.

54. CONSIGNMENT

Let $x=$ the selling price of the sculpture

13,500 is 45% of what number?

$13,500 = 0.45 \cdot x$

$\frac{13,500}{\mathbf{0.45}}=\frac{0.45x}{\mathbf{0.45}}$

$30,000 = x$

\$30,000 is the selling price of the sculpture.

56. AGENTS

Let $x=$ the amount of the contract

1,000,000 is 12.5% of what number?

$1.000,000 = 0.125 \cdot x$

$\frac{1,000,000}{\mathbf{0.125}}=\frac{0.125x}{\mathbf{0.125}}$

$8,000,000 = x$

\$8 million is the amount of the contract.

WRITING

57 - 59. Answers will vary.

REVIEW

61. $-\frac{16}{25}\div\left(-\frac{4}{15}\right)=-\frac{16}{25}\cdot\left(-\frac{15}{4}\right)$

$=\frac{16\cdot 15}{25\cdot 4}$

$=\frac{\cancel{2}\cdot\cancel{2}\cdot 3\cdot 4\cdot\cancel{5}}{\cancel{2}\cdot\cancel{2}\cdot\cancel{5}\cdot 5}$

$=\frac{12}{5}$

$=2\frac{2}{5}$

63. $x+15=-49$

$-34+15\overset{?}{=}-49$

$-19\overset{?}{=}-49$ False

-34 is not a solution.

CHALLENGE PROBLEMS

65. SOAPS

Step 1: $100\%-99\frac{44}{100}\%=\frac{56}{100}\%$

$=\frac{14}{25}\%$

This is the amount of impurities.

Step 2: $\frac{56}{100}\%=0.56\%$

$=0.0056$

SECTION 2.4 STUDY SET
VOCABULARY

Fill in the blanks.

1. A **formula** is an equation that states a mathematical relationship between two or more variables.

3. The **volume** of a three-dimensional geometric solid is the amount of space it encloses.

CONCEPTS

5. a. Time, distance, rate $\mathbf{d = rt}$

   b. Markup, retail price, cost $\mathbf{r = c + m}$

   c. Costs, revenue, profit $\mathbf{p = r - c}$

   d. Interest rate, time, interest, principal $\mathbf{I = Prt}$

7. Complete the table.

	rate	• time =	distance
Light	186,282 mi/sec	60 sec	**11,176,920 mi**
Sound	1,088 ft/sec	60 sec	**65,280 ft**

NOTATION

9. Solve $Ax + By = C$ for y.

$$Ax + By = C$$
$$Ax + By - \boxed{\mathbf{Ax}} = C - \boxed{\mathbf{Ax}}$$
$$By = C - Ax$$
$$\frac{By}{\boxed{\mathbf{B}}} = \frac{C - Ax}{\boxed{\mathbf{B}}}$$
$$y = \frac{C - Ax}{\boxed{\mathbf{B}}}$$

11. a. Write $\pi \cdot r^2 \cdot h$ in simpler form. $\boldsymbol{\pi r^2 h}$

    b. In the formula $V = \pi r^2 h$, what does r represent?

    The *r* represents the radius of the cylinder.

    What does h represent?

    The *h* represents the height of the cylinder.

GUIDED PRACTICE

Use a formula to solve each problem.

13. HOLLYWOOD

Let $c =$ the cost of the movie in millions

profit = revenue − cost

$$p = r - c$$
$$1,602 = 1,842 - c$$
$$1,602 - \mathbf{1,842} = 1,842 - c - \mathbf{1,842}$$
$$-240 = -c$$
$$\frac{-240}{\mathbf{-1}} = \frac{-c}{\mathbf{-1}}$$
$$240 = c$$

\$240 million is the cost of the movie.

15. SERVICE CLUBS

Let $r =$ the gross profit

profit = revenue − cost

$$p = r - c$$
$$875.85 = r - 55.15$$
$$875.85 - \mathbf{55.15} = r - 55.15 - \mathbf{55.15}$$
$$931 = r$$

\$931 is the gross profit made.

17. ENTREPRENEURS

Let $r =$ the simple interest rate

Principal •	rate •	time =	interest
2,500	***r***	2	175

$$2,500 \cdot r \cdot 2 = 175$$
$$5,000r = 175$$
$$\frac{5,000r}{\mathbf{5,000}} = \frac{175}{\mathbf{5,000}}$$
$$r = 0.035$$
$$r = 3.5\%$$

3.5% is the simple interest rate.

19. LOANS

Let $P =$ the amount borrowed

Principal •	rate •	time =	interest
P	0.02	3	360

$$p \cdot 0.02 \cdot 3 = 360$$
$$0.06p = 360$$
$$\frac{0.06p}{\mathbf{0.06}} = \frac{360}{\mathbf{0.06}}$$
$$p = 6,000$$

He borrowed \$6,000.

21. SWIMMING

Let $r =$ the average rate in mph

rate •	time =	distance
r	742	1,826

$$r \cdot 742 = 1{,}826$$
$$\frac{742r}{\mathbf{742}} = \frac{1{,}826}{\mathbf{742}}$$
$$r \approx 2.46$$
$$r \approx 2.5$$

2.5 mph was his average rate.

23. HOT AIR BALLOONS

Let $t =$ the number of hours

rate •	time =	distance
37	t	166.5

$$37t = 166.5$$
$$\frac{37t}{\mathbf{37}} = \frac{166.5}{\mathbf{37}}$$
$$t = 4.5$$

It will take 4.5 hours.

25. FRYING FOODS

$$C = \frac{5}{9}(F - 32)$$
$$C = \frac{5}{9}(365 - 32)$$
$$C = \frac{5}{9}(333)$$
$$C = \frac{\cancel{3} \cdot \cancel{3} \cdot 5 \cdot 37}{\cancel{3} \cdot \cancel{3}}$$
$$C = 185$$

The temperature is 185° Celsius.

27. BIOLOGY

$$C = \frac{5}{9}(F - 32)$$
$$-270 = \frac{5}{9}(F - 32)$$
$$\frac{\mathbf{9}}{\mathbf{5}} \cdot (-270) = \frac{\mathbf{9}}{\mathbf{5}} \cdot \frac{5}{9}(F - 32)$$
$$-486 = F - 32$$
$$-486 + \mathbf{32} = F - 32 + \mathbf{32}$$
$$-454 = F$$

The temperature is $-454°$ Fahrenheit.

29. ENERGY SAVINGS

$$P = 2l + 2w$$
$$100 = 2 \cdot 30 + 2w$$
$$100 = 60 + 2w$$
$$100 - \mathbf{60} = 60 + 2w - \mathbf{60}$$
$$40 = 2w$$
$$\frac{40}{\mathbf{2}} = \frac{2w}{\mathbf{2}}$$
$$20 = w$$

The width is 20 inches.

31. STRAWS

$$V = \pi r^2 h \quad , \pi = 3.141592654$$
$$= 3.141592654 \cdot 2^2 \cdot 150$$
$$= 3.141592654 \cdot 4 \cdot 150$$
$$\approx 1{,}884.9$$
$$\approx 1{,}885$$

The volume is about 1,885 mm^3.

GUIDED PRACTICE

Solve for the specified variable.

33. $r = c + m$ for c

$$r = c + m$$
$$r - \boldsymbol{m} = c + m - \boldsymbol{m}$$
$$r - m = c$$
$$c = r - m$$

35. $P = a + b + c$ for b

$$P = a + b + c$$
$$P - \boldsymbol{a} = a + b + c - \boldsymbol{a}$$
$$P - a = b + c$$
$$P - a - \boldsymbol{c} = b + c - \boldsymbol{c}$$
$$P - a - c = b$$
$$b = P - a - c$$

37. $V = \frac{1}{3}Bh$ for h

$$V = \frac{1}{3}Bh$$
$$\mathbf{3} \cdot V = \mathbf{3} \cdot \frac{1}{3}Bh$$
$$3V = Bh$$
$$\frac{3V}{\boldsymbol{B}} = \frac{Bh}{\boldsymbol{B}}$$
$$\frac{3V}{B} = h$$
$$h = \frac{3V}{B}$$

39. $E = IR$ for R

$$E = IR$$
$$\frac{E}{\boldsymbol{I}} = \frac{IR}{\boldsymbol{I}}$$
$$\frac{E}{I} = R$$
$$R = \frac{E}{I}$$

41. $T = 2r + 2t$ for r

$$T = 2r + 2t$$
$$T - \mathbf{2t} = 2r + 2t - \mathbf{2t}$$
$$T - 2t = 2r$$
$$\frac{T - 2t}{\mathbf{2}} = \frac{2r}{\mathbf{2}}$$
$$\frac{T - 2t}{2} = r$$
$$r = \frac{T - 2t}{2}$$

43. $Ax + By = C$ for x

$$Ax + By = C$$
$$Ax + By - \mathbf{By} = C - \mathbf{By}$$
$$Ax = C - By$$
$$\frac{Ax}{\mathbf{A}} = \frac{C - By}{\mathbf{A}}$$
$$x = \frac{C - By}{A}$$

45. $2x + 7y = 21$ for y

$$2x + 7y = 21$$
$$2x + 7y - \mathbf{2x} = 21 - \mathbf{2x}$$
$$7y = -2x + 21$$
$$\frac{7y}{\mathbf{7}} = \frac{-2x}{\mathbf{7}} + \frac{21}{\mathbf{7}}$$
$$y = -\frac{2}{7}x + 3$$

47. $9x - 2y = -8$ for y

$$9x - 2y = -8$$
$$9x - 2y - \mathbf{9x} = -8 - \mathbf{9x}$$
$$-2y = -9x - 8$$
$$\frac{-2y}{\mathbf{-2}} = \frac{-9x}{\mathbf{-2}} - \frac{8}{\mathbf{-2}}$$
$$y = \frac{9}{2}x + 4$$

49. $T = 4b(a + am)$ for m

$$T = 4b(a + am)$$
$$T = \mathbf{4b}(a) + \mathbf{4b}(am)$$
$$T = 4ab + 4abm$$
$$T - \mathbf{4ab} = 4ab + 4abm - \mathbf{4ab}$$
$$T - 4ab = 4abm$$
$$\frac{T - 4ab}{\mathbf{4ab}} = \frac{4abm}{\mathbf{4ab}}$$
$$\frac{T - 4ab}{4ab} = m$$
$$m = \frac{T - 4ab}{4ab}$$

51. $G = g(4r - 1)$ for r

$$G = g(4r - 1)$$
$$G = \mathbf{g}(4r) - \mathbf{g}(1)$$
$$G = 4gr - g$$
$$G + \mathbf{g} = 4gr - gz - \mathbf{g}$$
$$G + g = 4gr$$
$$\frac{G + g}{\mathbf{4g}} = \frac{4gr}{\mathbf{4g}}$$
$$\frac{G + g}{4g} = r$$
$$r = \frac{G + g}{4g}$$

TRY IT YOURSELF

Solve for the specified variable or expression.

53. $A = \dfrac{a + b + c}{3}$ for c

$$A = \frac{a + b + c}{3}$$
$$\mathbf{3} \cdot A = \mathbf{3} \cdot \left(\frac{a + b + c}{3}\right)$$
$$3A = a + b + c$$
$$3A - \mathbf{a} = a + b + c - \mathbf{a}$$
$$3A - a = b + c$$
$$3A - a - \mathbf{b} = b + c - \mathbf{b}$$
$$3A - a - b = c$$
$$c = 3A - a - b$$

55. $3x + y = 9$ for y

$$3x + y = 9$$
$$3x + y - \mathbf{3x} = 9 - \mathbf{3x}$$
$$y = -3x + 9$$

57. $K = \dfrac{1}{2}mv^2$ for m

$$K = \frac{1}{2}mv^2$$
$$\mathbf{2} \cdot K = \mathbf{2} \cdot \left(\frac{1}{2}mv^2\right)$$
$$2K = mv^2$$
$$\frac{2K}{\mathbf{v^2}} = \frac{mv^2}{\mathbf{v^2}}$$
$$\frac{2K}{v^2} = m$$
$$m = \frac{2K}{v^2}$$

59. $C = 2\pi r$ for r

$$C = 2\pi r$$
$$\frac{C}{\mathbf{2\pi}} = \frac{2\pi r}{\mathbf{2\pi}}$$
$$\frac{C}{2\pi} = r$$
$$r = \frac{C}{2\pi}$$

61. $\dfrac{M}{2}-9.9=2.1B$ for M

$$\frac{M}{2}-9.9=2.1B$$
$$\mathbf{2}\cdot\left(\frac{M}{2}-9.9\right)=\mathbf{2}\cdot(2.1B)$$
$$2\cdot\frac{M}{2}-2\cdot 9.9=2(2.1B)$$
$$M-19.8=4.2B$$
$$M-19.8+\mathbf{19.8}=4.2B+\mathbf{19.8}$$
$$M=4.2B+19.8$$

63. $w=\dfrac{s}{f}$ for f

$$w=\frac{s}{f}$$
$$\boldsymbol{f}\cdot w=\boldsymbol{f}\cdot\frac{s}{f}$$
$$fw=s$$
$$\frac{fw}{\boldsymbol{w}}=\frac{s}{\boldsymbol{w}}$$
$$f=\frac{s}{w}$$

65. $-x+3y=9$ for y

$$-x+3y=9$$
$$-x+3y+\boldsymbol{x}=9+\boldsymbol{x}$$
$$3y=x+9$$
$$\frac{3y}{\mathbf{3}}=\frac{x+9}{\mathbf{3}}$$
$$\frac{3y}{3}=\frac{x}{3}+\frac{9}{3}$$
$$y=\frac{1}{3}x+3$$

67. $A=\dfrac{1}{2}h(b+d)$ for b

$$A=\frac{1}{2}h(b+d)$$
$$\mathbf{2}\cdot A=\mathbf{2}\cdot\left(\frac{1}{2}h(b+d)\right)$$
$$2A=h(b+d)$$
$$2A=bh+dh$$
$$2A-\boldsymbol{dh}=bh+dh-\boldsymbol{dh}$$
$$2A-dh=bh$$
$$\frac{2A-dh}{\boldsymbol{h}}=\frac{bh}{\boldsymbol{h}}$$
$$\frac{2A-dh}{h}=b$$
$$b=\frac{2A-dh}{h}\text{ or }\frac{2A}{h}-d$$

69. $c^2=a^2+b^2$ for a^2

$$c^2=a^2+b^2$$
$$c^2-\boldsymbol{b^2}=a^2+b^2-\boldsymbol{b^2}$$
$$c^2-b^2=a^2$$
$$a^2=c^2-b^2$$

71. $\dfrac{7}{8}c+w=9$ for c

$$\frac{7}{8}c+w=9$$
$$\mathbf{8}\cdot\left(\frac{7}{8}c+w\right)=\mathbf{8}\cdot 9$$
$$8\cdot\left(\frac{7}{8}c\right)+8\cdot w=8\cdot 9$$
$$7c+8w=72$$
$$7c+8w-\mathbf{8}\boldsymbol{w}=72-\mathbf{8}\boldsymbol{w}$$
$$7c=72-8w$$
$$\frac{7c}{\mathbf{7}}=\frac{72-8w}{\mathbf{7}}$$
$$c=\frac{72-8w}{7}$$

73. $m=70+t(a+b)$ for b

$$m=70+t(a+b)$$
$$m=70+\boldsymbol{t}(a)+\boldsymbol{t}(b)$$
$$m=70+at+bt$$
$$m-\mathbf{70}-\boldsymbol{at}=70+at+bt-\mathbf{70}-\boldsymbol{at}$$
$$m-70-at=bt$$
$$\frac{m-70-at}{\boldsymbol{t}}=\frac{bt}{\boldsymbol{t}}$$
$$\frac{m-70-at}{t}=t$$

75. $V = lwh$ for l

$$V = lwh$$
$$\frac{V}{\mathbf{wh}} = \frac{lwh}{\mathbf{wh}}$$
$$\frac{V}{wh} = l$$
$$l = \frac{V}{wh}$$

77. $2E = \frac{T-t}{9}$ for t

$$2E = \frac{T-t}{9}$$
$$9 \cdot 2E = \mathbf{9} \cdot \left(\frac{T-t}{9}\right)$$
$$18E = T - t$$
$$18E + \mathbf{t} = T - t + \mathbf{t}$$
$$18E + t = T$$
$$18E + t - \mathbf{18E} = T - \mathbf{18E}$$
$$t = T - 18E$$

79. $s = 4\pi r^2$ for r^2

$$s = 4\pi r^2$$
$$\frac{s}{\mathbf{4\pi}} = \frac{4\pi r^2}{\mathbf{4\pi}}$$
$$\frac{s}{4\pi} = r^2$$
$$r^2 = \frac{s}{4\pi}$$

Look Alikes . . .

81. Solve $A = R + ab$

a. for R

$$A = R + ab$$
$$A - \mathbf{ab} = R + ab - \mathbf{ab}$$
$$A - ab = R$$
$$R = A - ab$$

b. for a

$$A = R + ab$$
$$A - \mathbf{R} = R + ab - \mathbf{R}$$
$$A - R = ab$$
$$\frac{A-R}{\mathbf{b}} = \frac{ab}{\mathbf{b}}$$
$$\frac{A-R}{b} = a$$
$$a = \frac{A-R}{b}$$

83. Solve $S = 2(2lw + wh)$

a. for h

$$S = 2(2lw + wh)$$
$$S = \mathbf{2}(2lw) + \mathbf{2}(wh)$$
$$S = 4lw + 2wh$$
$$S - \mathbf{4lw} = 4lw + 2wh - \mathbf{4lw}$$
$$S - 4lw = 2wh$$
$$\frac{S-4lw}{\mathbf{2w}} = \frac{2wh}{\mathbf{2w}}$$
$$\frac{S-4lw}{2w} = h$$
$$h = \frac{S-4lw}{2w}$$

b. for l

$$S = 2(2lw + wh)$$
$$S = \mathbf{2}(2lw) + \mathbf{2}(wh)$$
$$S = 4lw + 2wh$$
$$S - \mathbf{2wh} = 4lw + 2wh - \mathbf{2wh}$$
$$S - 2wh = 4lw$$
$$\frac{S-2wh}{\mathbf{4w}} = \frac{4lw}{\mathbf{4w}}$$
$$\frac{S-2wh}{4w} = l$$
$$l = \frac{S-2wh}{4w}$$

APPLICATIONS
from Campus To Careers

85. Solve $\text{Torque} = \frac{5,252 \cdot \text{Horsepower}}{\text{RPM}}$ for Horsepower

$$\text{Torque} = \frac{5,252 \cdot \text{Horsepower}}{\text{RPM}}$$
$$\mathbf{RPM} \cdot \text{Torque} = \mathbf{RPM}\left(\frac{5,252 \cdot \text{Horsepower}}{\text{RPM}}\right)$$
$$\text{RPM} \cdot \text{Torque} = 5,252 \cdot \text{Horsepower}$$
$$\frac{\text{RPM} \cdot \text{Torque}}{\mathbf{5,252}} = \frac{5,252 \cdot \text{Horsepower}}{\mathbf{5,252}}$$
$$\frac{\text{RPM} \cdot \text{Torque}}{5,252} = \text{Horsepower}$$
$$\text{Horsepower} = \frac{\text{RPM} \cdot \text{Torque}}{5,252}$$

87. AVON PRODUCTS

Let p = the operating profit ending Mar. '09

operating profit = revenue – cost

$$p = r - c$$
$$= 2{,}186.9 - 2{,}018.5$$
$$= 168.4$$

\$168.4 million is the operating profit for the Quarter ending Mar. '09.

Let p = the operating profit ending Mar. '10

operating profit = revenue – cost

$$p = r - c$$
$$= 2{,}490.4 - 2{,}297.6$$
$$= 192.8$$

\$192.8 million is the operating profit f or the Quarter ending Mar. '10.

89. CAMPERS

Let w = the length of the shorter side in inches

$$P = 2l + 2w$$
$$140 = 2 \cdot 56 + 2w$$
$$140 = 112 + 2w$$
$$140 - \mathbf{112} = 112 + 2w - \mathbf{112}$$
$$28 = 2w$$
$$\frac{28}{\mathbf{2}} = \frac{2w}{\mathbf{2}}$$
$$14 = w$$

The length of the shorter side is 14 inches.

91. KITES

Let b = the wing span of the kite in inches

$$A = \frac{1}{2}bh$$

$A = 650 \text{ in}^2$, $h = 26$ inches

$$A = \frac{1}{2}bh$$
$$650 = \frac{1}{2}(26b)$$
$$650 = 13b$$
$$\frac{650}{\mathbf{13}} = \frac{13b}{\mathbf{13}}$$
$$50 = b$$

The wing span of the kite is 50 in.

93. WHEELCHAIRS

Find the diameter of the rear wheel.

$$12.5 \cdot 2 = 25$$

The diameter of the wheel is 25 inches.

Find the radius of the front wheel.

$$5 \div 2 = 2.5$$

The radius of the wheel is 2.5 inches.

95. BULLS-EYE

Step 1: Find the radius of the bulls-eye.

$$4.8 \div 2 = 2.4$$

Step 2: Let A = the area of bulls-eye

$$A = \pi r^2 \quad , \quad \pi = 3.141592654$$
$$A = 3.141592654 \cdot (2.4)^2$$
$$A = 3.141592654 \cdot 5.76$$
$$A \approx 18.09$$
$$A \approx 18.1$$

The area of the bulls-eye is about 18.1 in^2.

97. HORSES

Let A = the area of the circle

$$A = \pi r^2 \quad , \quad \pi = 3.141592654$$
$$A = 3.141592654 \cdot (28)^2$$
$$A = 3.141592654 \cdot 784$$
$$A \approx 2{,}463.0$$
$$A \approx 2{,}463$$

The area of the circle is about $2{,}463 \text{ ft}^2$.

99. WORLD HISTORY

Let A = the area of the window

$$A = \frac{1}{2}h(B + b)$$
$$A = \frac{1}{2} \cdot 70 \cdot (50 + 40)$$
$$A = \frac{1}{2} \cdot 70 \cdot 90$$
$$A = 35 \cdot 90$$
$$A = 3{,}150$$

The area of the window is $3{,}150 \text{ cm}^2$.

101. TIRES

Let w = the width of the tire

$$A = lw$$
$$45 = 7\frac{1}{2} \cdot w$$
$$45 = 7.5 \cdot w$$
$$\frac{45}{\mathbf{7.5}} = \frac{7.5w}{\mathbf{7.5}}$$
$$6 = w$$

The width of the tire is 6 inches.

103. FIREWOOD

Let l = the length of the cord of wood

$$V = lwh$$
$$128 = l \cdot 4 \cdot 4$$
$$128 = 16l$$
$$\frac{128}{\mathbf{16}} = \frac{16l}{\mathbf{16}}$$
$$8 = l$$

The length of the cord of wood is 8 feet.

105. IGLOOS

Let V = the volume of a sphere in ft^3

Step 1: $V = \frac{4}{3}\pi r^3$, $\pi = 3.141592654$

$V = \frac{4}{3} \cdot 3.141592654 \cdot (5.5)^3$

$V = \frac{4}{3} \cdot 3.141592654 \cdot (166.375)$

$V \approx 696.9099703$

Step 2: Divide the answer from Step 1 by 2.

$\frac{696.9099703}{2} = 348.4$

The volume of the igloo is 348 ft^3.

107. COOKING

Let A = the area of the grill

Find the radius: $r = \frac{18}{2}$

$= 9$

$A = \pi r^2$, $\pi = 3.141592654$

$A = 3.141592654 \cdot (9)^2$

$A = 3.141592654 \cdot 81$

$A \approx 254.4$

$A = 254$

The area of the grill is about 254 in^2.

109. PULLEYS

$L = 2D + 3.25(r + R)$ Solve for R.

$$L = 2D + 3.25(r + R)$$
$$L = 2D + 3.25r + 3.25R$$
$$L - \mathbf{3.25r} = 2D + 3.25r + 3.25R - \mathbf{3.25r}$$
$$L - 3.25r = 2D + 3.25R$$
$$L - 3.25r - \mathbf{2D} = 2D + 3.25R - \mathbf{2D}$$
$$L - 2D - 3.25r = 3.25R$$
$$\frac{L - 2D - 3.25r}{\mathbf{3.25}} = \frac{3.25R}{\mathbf{3.25}}$$
$$\frac{L - 2D - 3.25r}{3.25} = R$$
$$R = \frac{L - 2D - 3.25r}{3.25}$$

or

$$R = \frac{L - 2D}{3.25} - \frac{3.25r}{3.25}$$
$$R = \frac{L - 2D}{3.25} - r$$

WRITING

111 - 113. Answers will vary.

REVIEW

115. What number is 82% of 168?

$$x = 0.82 \cdot 168$$
$$x = 137.76$$

137.76 is the solution.

117. What percent of 200 is 30?

$$x \cdot 200 = 30$$
$$\frac{200x}{\mathbf{200}} = \frac{30}{\mathbf{200}}$$
$$x = 0.15$$
$$x = 15\%$$

15% is the solution.

CHALLENGE PROBLEMS

119. Solve $-7(\alpha - \beta) - (4\alpha - \theta) = \frac{\alpha}{2}$ for α

$$-7(\alpha - \beta) - (4\alpha - \theta) = \frac{\alpha}{2}$$
$$\mathbf{2}[-7(\alpha - \beta) - (4\alpha - \theta)] = \mathbf{2}\left(\frac{\alpha}{2}\right)$$
$$2[-7(\alpha - \beta)] - 2[(4\alpha - \theta)] = 2\left(\frac{\alpha}{2}\right)$$
$$-14(\alpha - \beta) - 2(4\alpha - \theta) = \alpha$$
$$-14\alpha + 14\beta - 8\alpha + 2\theta = \alpha$$
$$-22\alpha + 14\beta + 2\theta = \alpha$$
$$-22\alpha + 14\beta + 2\theta + \mathbf{22\alpha} = \alpha + \mathbf{22\alpha}$$
$$14\beta + 2\theta = 23\alpha$$
$$\frac{14\beta + 2\theta}{\mathbf{23}} = \frac{23\alpha}{\mathbf{23}}$$
$$\frac{14\beta + 2\theta}{23} = \alpha$$
$$\alpha = \frac{14\beta + 2\theta}{23}$$

SECTION 2.5 STUDY SET
VOCABULARY

Fill in the blanks.

1. Integers that follow one another, such as 7 and 8, are called **consecutive** integers.
3. The equal sides of an isosceles triangle meet to form the **vertex** angle. The angles opposite the equal sides are called **base** angles, and they have equal measures.

CONCEPTS

5. $\boxed{\textbf{17 feet}}$ = total length
   $x =$ length of shortest section
   $\boxed{x+2}$ = length of middle-sized section
   $\boxed{3x}$ = length of longest section

7. **0.03x$**

9. **180°**

NOTATION

11. a. $x+1$ b. $x+2$ c. $x+2$, $x+4$

GUIDED PRACTICE

13. **Analyze**

- 12 ft board is cut into 2 pieces.
- Longer piece is twice as long as shorter piece.
- Find the length of each piece.

Assign

Let x = length of shorter piece in feet
$2x$ = length of longer piece in feet

Form

The length of the shorter board	plus	the length of the longer board	equals	the total length.
x	$+$	$2x$	$=$	12

Solve

$$x + 2x = 12$$
$$3x = 12$$
$$\frac{3x}{\mathbf{3}} = \frac{12}{\mathbf{3}}$$
$$x = 4$$

Longer length board

$$2x = 2(4)$$
$$= 8$$

State

Length of shorter piece is 4 ft.
Length of longer piece is 8 ft.

Check

$4 \text{ ft} + 8 \text{ ft} = 12 \text{ ft}$
The results check.

APPLICATIONS

15. NATIONAL PARKS

Analyze

- Total route is 444 miles.
- It took 4 days to drive.
- Each day they drove 6 miles further than the previous day.
- How many miles did they drive each day?

Assign

Let x = miles driven the first day
$x+6$ = miles driven the second day
$(x+6)+6$ = miles driven the third day
$(x+12)+6$ = miles driven the fourth day

Form

The number of miles driven on the 1st day	plus	the number of miles driven on the 2nd day	plus	the number of miles driven on the 3rd day	plus	the number of miles driven on the 4th day	equals	the total distance.
x	$+$	$x+6$	$+$	$x+12$	$+$	$x+18$	$=$	444

Solve

$$x + x + 6 + x + 12 + x + 18 = 444$$
$$4x + 36 = 444$$
$$4x + 36 - \mathbf{36} = 444 - \mathbf{36}$$
$$4x = 408$$
$$\frac{4x}{\mathbf{4}} = \frac{408}{\mathbf{4}}$$
$$x = 102$$

2nd day
$$x + 6 = 102 + 6$$
$$= 108$$

3rd day
$$x + 12 = 102 + 12$$
$$= 114$$

4th day
$$x + 18 = 102 + 18$$
$$= 120$$

State

They drove 102 miles the 1st day.
They drove 108 miles the 2nd day.
They drove 114 miles the 3rd day.
They drove 120 miles the 4th day.

Check

$102 + 108 + 114 + 120 = 444$

17. SOLAR HEATING

Analyze

- The width of both is 18 feet.
- One panel is 3.4 feet wider than the other.
- What is the width of each panel?

Assign

Let x = width of smaller panel in feet

$x+3.4$ = width of larger panel in feet

Form

The width of the smaller panel	plus	the width of the larger panel	equals	the total lengths.
x	$+$	$x+3.4$	$=$	18

Solve

$$x+x+3.4=18$$
$$2x+3.4=18$$
$$2x+3.4-\mathbf{3.4}=18-\mathbf{3.4}$$
$$2x=14.6$$
$$\frac{2x}{\mathbf{2}}=\frac{14.6}{\mathbf{2}}$$
$$x=7.3$$

Width of larger panel

$$x+3.4=7.3+3.4$$
$$=10.7$$

State

The width of the smaller panel is 7.3 feet.
The width of the larger panel is 10.7 feet.

Check

$7.3+10.7=18$

19. iPHONE APPS

Analyze

- Total of $52.97 spent on 3 applications.
- Call of Duty & Zombies II cost $7 more than Guitar Hero.
- Tom Tom cost $4.11 more than twelve times Guitar Hero.
- Find the cost of each application.

Assign

Let x = cost of Guitar Hero

$x+7$ = cost of Call of Duty: Zombies

$12x+4.11$ = cost of Tom Tom

Form

Guitar Hero cost	plus	Call of Duty and Zombies cost	plus	Tom Tom cost	equals	the total cost.
x	$+$	$(x+7)$	$+$	$(12x+4.11)$	$=$	52.97

Solve

$$x+(x+7)+(12x+4.11)=52.97$$
$$14x+11.11=52.97$$
$$14x+11.1-\mathbf{11.11}=52.97-\mathbf{11.11}$$
$$14x=41.86$$
$$\frac{14x}{\mathbf{14}}=\frac{41.86}{\mathbf{14}}$$
$$x=2.99$$

WW & Zombie	Tom Tom
$x+7=2.99+7$	$12x+4.11=12(2.99)+4.11$
$=9.99$	$=35.88+4.11$
	$=39.99$

State

The cost of Guitar Hero is $2.99
The cost of Call of Duty & Zombies is $9.99
The cost of Tom Tom is $39.99.

Check

$\$2.99+\$9.99+\$39.99=\$52.97$

21. COUNTING CALORIES

Analyze

- Both have a total of 850 calories.
- The pie has 100 more than twice the calories in ice cream.
- How many calories are in each food?

Assign

Let x = number of calories in the ice cream

$2x+100$ = number of calories in the pie

Form

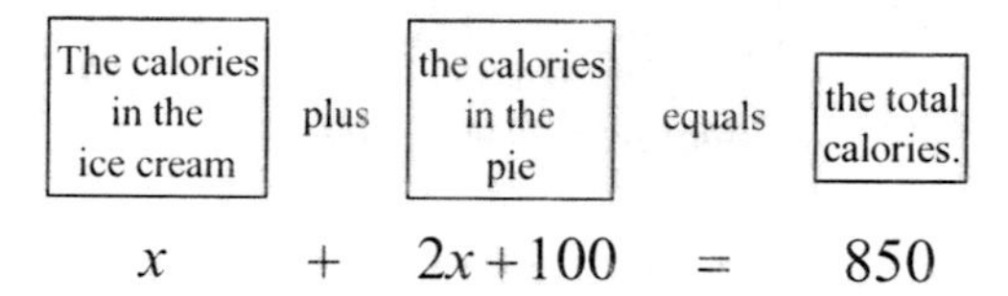

$$x \quad + \quad 2x+100 \quad = \quad 850$$

Solve

$$x+2x+100=850$$
$$3x+100=850$$
$$3x+100-\mathbf{100}=850-\mathbf{100}$$
$$3x=750$$
$$\frac{3x}{\mathbf{3}}=\frac{750}{\mathbf{3}}$$
$$x=250$$

Calories in the pie

$$2x+100=2(250)+100$$
$$=500+100$$
$$=600$$

State

The ice cream has 250 calories.
The pie has 600 calories.

Check

$250+600=850$

23. CONCERTS

Analyze

- Fee to rent the concert hall is $2,250.
- $150 an hour to pay support staff.
- Budget is $3,300.
- How many hours can the hall be rented?

Assign

Let x = number of hours

Form

The amount to pay the support staff	plus	the amount to rent the hall	equals	the total budget.
$150x$	$+$	$2{,}250$	$=$	$3{,}300$

Solve

$$150x+2,250=3,300$$
$$150x+2,250-\mathbf{2,250}=3,300-\mathbf{2,250}$$
$$150x=1,050$$
$$\frac{150x}{\mathbf{150}}=\frac{1,050}{\mathbf{150}}$$
$$x=7$$

State

The concert hall can be rented for 7 hours.

Check

$\$150(7)+\$2,250=\$3,300$

$\$1050+\$2,250=\$3,300$

25. FIELD TRIPS

Analyze

- Budget is $275.
- Charged $0.25 per mile to rent van.
- Cost $65 a day to rent van (2 days).
- How many miles can be driven?

Assign

Let x = number of miles

Form

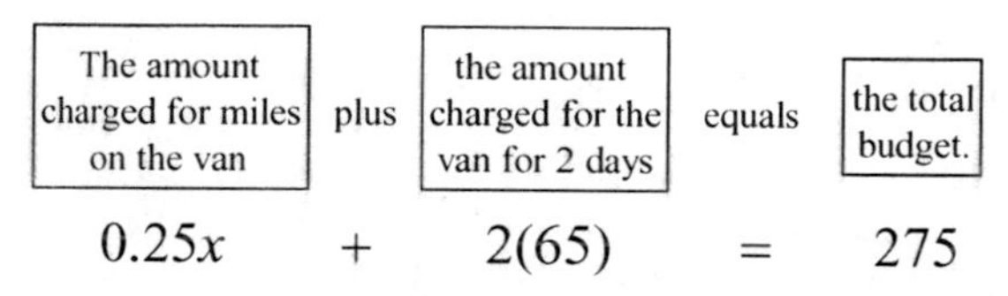

$$0.25x \quad + \quad 2(65) \quad = \quad 275$$

Solve

$$0.25x+2(65)=275$$
$$0.25x+130=275$$
$$0.25x+130-\mathbf{130}=275-\mathbf{130}$$
$$0.25x=145$$
$$\frac{0.25x}{\mathbf{0.25}}=\frac{145}{\mathbf{0.25}}$$
$$x=580$$

State

The van can be driven for 580 miles.

Check

$\$0.25(580)+2(\$65)=\$275$

$\$145+\$130=\$275$

27. TUTORING

Analyze

- Money set aside for tutoring is \$400.
- Placement test cost \$25.
- Tutoring cost \$18.75 per hour.
- How many hours of tutoring can be had?

Assign

Let x = number of hours

Form

The cost of the test	plus	the amount charged for tutoring	equals	the total money.
25	+	$18.75x$	=	400

Solve

$$25+18.75x=400$$
$$25+18.75x-\mathbf{25}=400-\mathbf{25}$$
$$18.75x=375$$
$$\frac{18.75x}{\mathbf{18.75}}=\frac{375}{\mathbf{18.75}}$$
$$x=20$$

State

They can afford 20 hours of tutoring.

Check

$$\$25+\$18.75(20)=\$400$$
$$\$25+\$375=\$400$$

29. CATTLE AUCTIONS

Analyze

- Wants to make \$45,500 off the auction.
- Charged 9% commision of the sale.
- What is the selling price?

Assign

Let x = the selling price of the bull

Form

The selling price of the bull	minus	the amount of the commission	equals	the amount made on the auction.
x	−	$0.09x$	=	45,500

Solve

$$x-0.09x=45{,}500$$
$$0.91x=45{,}500$$
$$\frac{0.91x}{\mathbf{0.91}}=\frac{45{,}500}{\mathbf{0.91}}$$
$$x=50{,}000$$

State

The selling price was \$50,000.

Check

$$\$50{,}000-\$50{,}000(0.09)=\$45{,}500$$
$$\$50{,}000-\$4{,}500=\$45{,}500$$

31. SELLING USED CLOTHING

Analyze

- Wants to make \$210 on the sale of the coat.
- Consignment charge is $12\frac{1}{2}\%$ of the cost.
- What is the price of the coat?

Assign

Let x = the price of the coat

Form

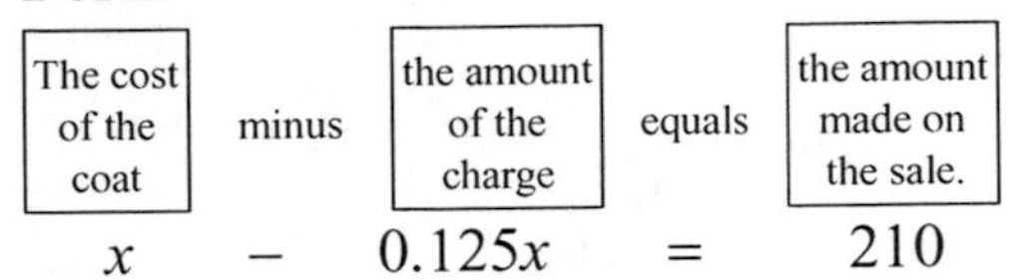

$x \quad - \quad 0.125x \quad = \quad 210$

Solve

$$x-0.125x=210$$
$$0.875x=210$$
$$\frac{0.875x}{\mathbf{0.875}}=\frac{210}{\mathbf{0.875}}$$
$$x=240$$

State

The price of the coat should be \$240.

Check

$$\$240-\$240(0.125)=\$210$$
$$\$240-\$30=\$210$$

33. SAVINGS ACCOUNTS

Analyze

- Balance after one year was \$5,512.50.
- Interest rate was 5%.
- What was the beginning balance?

Assign

Let x = the beginning balance

Form

The beginning balance	plus	the amount of the interest	equals	the ending balance.
x	+	$0.05x$	=	5,512.50

Solve

$$x+0.05x=5{,}512.50$$
$$1.05x=5{,}512.50$$
$$\frac{1.05x}{\mathbf{1.05}}=\frac{5{,}512.50}{\mathbf{1.05}}$$
$$x=5{,}250$$

State

The beginning balance was \$5,250.

Check

$$\$5{,}250+\$5{,}250(0.05)=\$5{,}512.50$$
$$\$5{,}250+\$262.50=\$5{,}512.50$$

CONSECUTIVE INTEGER PROBLEMS

35. SOCCER

Analyze

- Total number of goals is 29.
- Ronaldo scored one more goal than Mueller.
- The respective number of goals made by each are consecutive integers.
- How many goals did each make?

Assign

Let $x =$ number of goals made by Mueller
$x+1 =$ number of goals made by Ronaldo

Form

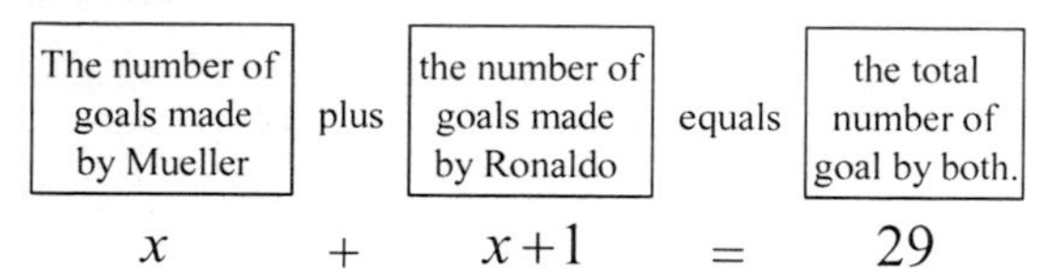

$$x \quad + \quad x+1 \quad = \quad 29$$

Solve

$$\begin{aligned} x+x+1 &= 29 \\ 2x+1 &= 29 \\ 2x+1-\mathbf{1} &= 29-\mathbf{1} \\ 2x &= 28 \\ \frac{2x}{\mathbf{2}} &= \frac{28}{2} \\ x &= 14 \end{aligned}$$

Ronaldo

$$\begin{aligned} x+1 &= 14+1 \\ &= 15 \end{aligned}$$

State

Mueller scored 14 goals.
Ronaldo scored 15 goals.

Check

$14+15=29$

37. TV HISTORY

Analyze

- Total number of episodes is 470.
- *Friends* has two more episodes than *Leave it to Beaver.*
- The respective number of eposides each are consecutive even integers.
- How many episodes of each are there?

Assign

Let $x =$ number of *Leave It to Beaver* episodes
$x+2 =$ number of *Friends* episodes

Form

The number of *Leave it to Beaver* episodes made	plus	the number of *Friends* episodes made	equals	the total number of episodes.
x	$+$	$x+2$	$=$	470

Solve

$$\begin{aligned} x+x+2 &= 470 \\ 2x+2 &= 470 \\ 2x+2-\mathbf{2} &= 470-\mathbf{2} \\ 2x &= 468 \\ \frac{2x}{\mathbf{2}} &= \frac{468}{\mathbf{2}} \\ x &= 234 \end{aligned}$$

Friends

$$\begin{aligned} x+2 &= 234+2 \\ &= 236 \end{aligned}$$

State

Number of *Leave It to Beaver* eposides is 234.
Number *Friends* eposides is 236.

Check

$234+236=470$

39. CELEBRITY BIRTHDAYS

Analyze

- Total of the three calendar dates is 72.
- The respective value for each birthday is a consecutive even integers.
- What is the birth date of each person?

Assign

Let $x =$ birth date of Selena
$x+2 =$ birth date of Jennifer
$x+4 =$ birth date of Sandra

Form

The date of Selena's birthday	plus	the date of Jennifer's birthday	plus	the date of Sandra's birthday	equals	the total of all three dates.
x	$+$	$x+2$	$+$	$x+4$	$=$	72

Solve

$$\begin{aligned} x+x+2+x+4 &= 72 \\ 3x+6 &= 72 \\ 3x+6-\mathbf{6} &= 72-\mathbf{6} \\ 3x &= 66 \\ \frac{3x}{\mathbf{3}} &= \frac{66}{\mathbf{3}} \\ x &= 22 \end{aligned}$$

Jennifer

$$\begin{aligned} x+2 &= 22+2 \\ &= 24 \end{aligned}$$

Sandra

$$\begin{aligned} x+4 &= 22+4 \\ &= 26 \end{aligned}$$

State

The birthdate of Selena is July 22th.
The birthdate of Jennifer is July 24th.
The birthdate of Sandra is July 26th.

Check

$22+24+26=72$

GEOMETRY PROBLEMS

41. TENNIS

Analyze

- Perimeter of the rectangular court is 210 feet.
- The length is 3 feet less than three times the width.
- What are the dimensions of the court?

Assign

Let x = width of the court in feet

$3x-3$ = length of the court in feet

Form

Two widths of the court	plus	two lengths of the court	equals	the perimeter of the court.
$2x$	$+$	$2(3x-3)$	$=$	210

Solve

$$
\begin{aligned}
2x+2(3x-3) &= 210\\
2x+6x-6 &= 210\\
8x-6 &= 210\\
8x-6+\mathbf{6} &= 210+\mathbf{6}\\
8x &= 216\\
\frac{8x}{\mathbf{8}} &= \frac{216}{\mathbf{8}}\\
x &= 27
\end{aligned}
$$

length

$$
\begin{aligned}
3x-3 &= 3(27)-3\\
&= 81-3\\
&= 78
\end{aligned}
$$

State

The width of the court is 27 feet.
The length of the court is 78 feet.

Check

$27+78+27+78=210$

43. ART

Analyze

- Perimeter of the rectangular picture is 102.5 inches.
- The length is 11.75 inches shorter than twice the width.
- What are the dimensions of the picture?

Assign

Let x = width of the picture in inches

$2x-11.75$ = length of the picture in inches

Form

Two widths of the picture	plus	two lengths of the picture	equals	the perimeter of the picture.
$2x$	$+$	$2(2x-11.75)$	$=$	102.5

Solve

$$
\begin{aligned}
2x+2(2x-11.75) &= 102.5\\
2x+4x-23.5 &= 102.5\\
6x-23.5 &= 102.5\\
6x-23.5+\mathbf{23.5} &= 102.5+\mathbf{23.5}\\
6x &= 126\\
\frac{6x}{\mathbf{6}} &= \frac{126}{\mathbf{6}}\\
x &= 21
\end{aligned}
$$

length

$$
\begin{aligned}
2x-11.75 &= 2(21)-11.75\\
&= 42-11.75\\
&= 30.25
\end{aligned}
$$

State

The width of the picture is 21 inches.
The length of the picture is 30.25 inches.

Check

$21+30.25+21+30.25=102.5$

45. ENGINEERING

Analyze

- Perimeter of the isosceles triangular truss is 25 feet.
- Each of the two equal sides is 4 feet shorter than the third side
- What are the lengths of the sides?

Assign

Let x = length of third side in feet

$x-4$ = length of 1 of the 2 equal sides in feet

Form

The length of the third side	plus	the 2 equal lengths	equals	the perimeter of the truss.
x	$+$	$2(x-4)$	$=$	25

Solve

$$x+2(x-4)=25$$
$$x+2x-8=25$$
$$3x-8+\mathbf{8}=25+\mathbf{8}$$
$$3x=33$$
$$\frac{3x}{\mathbf{3}}=\frac{33}{\mathbf{3}}$$
$$x=11$$

1 equal side

$$x-4=11-4$$
$$-7$$

State

Length of the third side is 11 feet.
Length of each of the 2 equal sides is 7 feet.

Check

$11+7+7=25$

47. TV TOWERS

Analyze

- Internal measure of the 3 angles of a triangle is 180°.
- Each base angle is 4 times the vertex angle.
- What is the measure of the vertex angle?

Assign

Let x = measure of the vertex angle

$4x$ = measure of each base angle

Form

The measure of the vertex angle	plus	the measure of the 2 base angles	equals	the measure of the angles of a triangle.
x	$+$	$2(4x)$	$=$	180

Solve

$$x+2(4x)=180$$
$$x+8x=180$$
$$9x=180$$
$$\frac{9x}{\mathbf{9}}=\frac{180}{\mathbf{9}}$$
$$x=20$$

State

Measure of the vertex angle is 20°.

Check

$$20^\circ+4(20^\circ)+4(20^\circ)=180^\circ$$
$$20^\circ+80^\circ+80^\circ=180^\circ$$

49. MOUNTAIN BICYCLE

Analyze

- Internal measure of the 3 angles of a triangle is 180°.
- The angle the crossbar makes with the seat is 15° less than twice the angle at the steering.
- The angle at the pedal is 25° more than the angle at the steering column.
- What is the measures of the three angle?

Assign

Let $x =$ measure of the steering column angle

$2x - 15 =$ measure of crossbar angle

$x + 25 =$ measure of pedal gear angle

Form

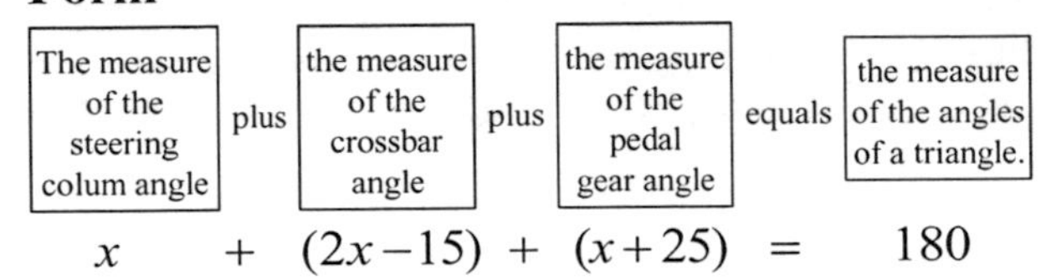

$$x \quad + \quad (2x-15) \quad + \quad (x+25) \quad = \quad 180$$

Solve

$$x + (2x - 15) + (x + 25) = 180$$
$$4x + 10 = 180$$
$$4x + 10 - \mathbf{10} = 180 - \mathbf{10}$$
$$4x = 170$$
$$\frac{4x}{4} = \frac{170}{4}$$
$$x = 42.5$$

crossbar angle

$$2x - 15 = 2(42.5) - 15 = 85 - 15 = 70$$

pedal gear angle

$$x + 25 = 42.5 + 25 = 67.5$$

State

Measure of the steering colum angle is 42.5°.

Measure of the crossbar angle is 70°.

Measure of the pedal gear angle is 67.5°.

Check

$42.5^\circ + 70^\circ + 67.5^\circ = 180^\circ$

51. ANGLES

Analyze

- The measure of complementary angles is 90°.
- The measure of the first angle is $2x$.
- The measure of the second angle is $(6x + 2)^\circ$.
- What are the measure of each angles?

Assign

Let $2x =$ measure of $\angle 1$

$6x + 2 =$ measure of $\angle 2$

Form

The measure of $\angle 1$	plus	the measure of $\angle 2$	equals	the measure of the complementary angles.
$2x$	$+$	$6x+2$	$=$	90

Solve

$$2x + 6x + 2 = 90$$
$$8x + 2 = 90$$
$$8x + 2 - \mathbf{2} = 90 - \mathbf{2}$$
$$8x = 88$$
$$\frac{8x}{\mathbf{8}} = \frac{88}{\mathbf{8}}$$
$$x = 11$$

$\angle 1$

$$2x = 2(11) = 22$$

$\angle 2$

$$6x + 2 = 6(11) + 2 = 66 + 2 = 68$$

State

$x = 11$

Measure of $\angle 1$ is 22°.

Measure of $\angle 2$ is 68°.

Check

$22^\circ + 68^\circ = 90^\circ$

53. "LIGHTNING BOLT"

Analyze

- The measure of supplementary angles is 180°.
- The measure of the maximum stride angle is 18° less than twice its supplement.
- What is the angle of the maximum stride?

Assign

Let $x =$ measure of the supplement angle

$2x - 18 =$ measure of the maximum stide angle

Form

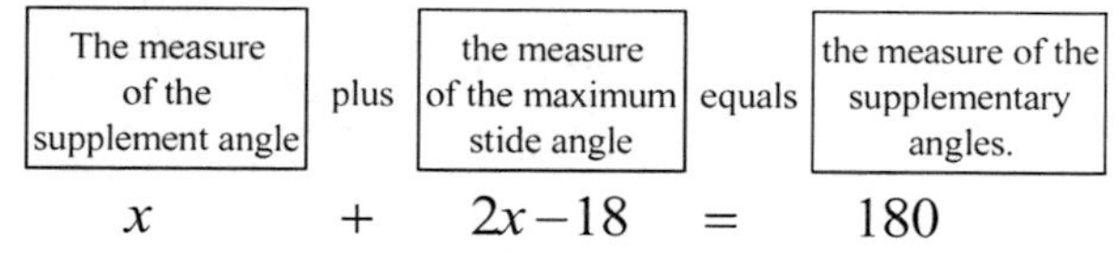

$$x \quad + \quad 2x - 18 \quad = \quad 180$$

Solve

$$x + 2x - 18 = 180$$
$$3x - 18 = 180$$
$$3x - 18 + \mathbf{18} = 180 + \mathbf{18}$$
$$3x = 198$$
$$\frac{3x}{\mathbf{3}} = \frac{198}{\mathbf{3}}$$
$$x = 66$$

stide angle

$$2x - 18 = 2(66) - 18$$
$$= 132 - 18$$
$$= 114$$

State

Measure of supplemental angle is 66°.

Measure of stride angle is 114°.

Check

$66^\circ + 114^\circ = 180^\circ$

WRITING

55 - 57. Answers will vary.

REVIEW

Solve.

59. $$\frac{5}{8}x = -15$$
$$\frac{\mathbf{8}}{\mathbf{5}}\left(\frac{5}{8}x\right) = \frac{\mathbf{8}}{\mathbf{5}} \cdot -15$$
$$x = -\frac{8 \cdot 3 \cdot \overset{1}{\cancel{5}}}{\underset{1}{\cancel{5}}}$$
$$x = -24$$

-24 is the solution.

The solution checks.

61. $$\frac{3}{4}y = \frac{2}{5}y - \frac{3}{2}y - 2$$
$$\mathbf{20} \cdot \left(\frac{3}{4}y\right) = \mathbf{20} \cdot \left(\frac{2}{5}y - \frac{3}{2}y - 2\right)$$
$$20 \cdot \left(\frac{3}{4}y\right) = 20 \cdot \left(\frac{2}{5}y\right) - 20 \cdot \left(\frac{3}{2}y\right) - 20(2)$$
$$15y = 8y - 30y - 40$$
$$15y = -22y - 40$$
$$15y + \mathbf{22y} = -22y - 40 + \mathbf{22y}$$
$$37y = -40$$
$$\frac{37y}{\mathbf{37}} = \frac{-40}{\mathbf{37}}$$
$$y = -\frac{40}{37}$$

$-\frac{40}{37}$ is the solution.

The solution checks.

CHALLENGE PROBLEMS

63. Consecutive integers

01:02:03 04/05/06

SECTION 2.6 STUDY SET
VOCABULARY

Fill in the blanks.

1. Problems that involve depositing money are called **investment** problems, and problems that involve moving vehicles are called uniform **motion** problems.

CONCEPTS

3. Complete the table.

	Principal	• rate	• time	= interest
Stocks	x			
Art	**30,000 – x**			

5. Complete the table.

	rate	• time	= distance
West	r		
East	**r – 150**		

7. Complete the table.

	rate	• time	= distance
Husband	35	t	**$35t$**
Wife	45	t	**$45t$**
		Total:	**80**

Form

$$\mathbf{35t + 45t = 80}$$

9a. Complete the table.

	Amount	• Strength	= Amount of pure antifreeze
Strong	6	0.50	**0.50(6)**
Weak	x	0.25	**$0.25x$**
Mixture	**6 + x**	0.30	**0.30(6 + x)**

Form

$$\mathbf{0.50(6) + 0.25x = 0.30(6 + x)}$$

9b. Complete the table.

	Amount	• Strength	= Amount of pure vinegar
Strong	x	0.06	**$0.06x$**
Weak	**10 – x**	0.03	**0.03(10 – x)**
Mixture	10	0.05	**0.05(10)**

Form

$$\mathbf{0.06x + 0.03(10 - x) = 0.05(10)}$$

NOTATION

11. $6\% = \mathbf{0.06}$, $15.2\% = \mathbf{0.152}$

GUIDED PRACTICE

Solve each equation.

13.
$$0.18x + 0.45(12 - x) = 0.36(12)$$
$$\mathbf{100}[0.18x + 0.45(12 - x)] = \mathbf{100}[0.36(12)]$$
$$100(0.18x) + 100[0.45(12 - x)] = 100[0.36(12)]$$
$$18x + 45(12 - x) = 36(12)$$
$$18x + 540 - 45x = 432$$
$$-27x + 540 = 432$$
$$-27x + 540 - \mathbf{540} = 432 - \mathbf{540}$$
$$-27x = -108$$
$$\frac{-27x}{\mathbf{-27}} = \frac{-108}{\mathbf{-27}}$$
$$x = 4$$

$$0.18x + 0.45(12 - x) = 0.36(12)$$
$$0.18(4) + 0.45(12 - 4) \stackrel{?}{=} 0.36(12)$$
$$0.72 + 0.45(8) \stackrel{?}{=} 4.32$$
$$0.72 + 3.6 \stackrel{?}{=} 4.32$$
$$4.32 = 4.32$$

4 is the solution.

15.
$$0.08x + 0.07(15{,}000 - x) = 1{,}110$$
$$\mathbf{100}[0.08x + 0.07(15{,}000 - x)] = \mathbf{100}(1{,}110)$$
$$100(0.08x) + 100[0.07(15{,}000 - x)] = 100(1{,}110)$$
$$8x + 7(15{,}000 - x) = 111{,}000$$
$$8x + 105{,}000 - 7x = 111{,}000$$
$$x + 105{,}000 = 111{,}000$$
$$x + 105{,}000 - \mathbf{105{,}000} = 111{,}000 - \mathbf{105{,}000}$$
$$x = 6{,}000$$

$$0.08x + 0.07(15{,}000 - x) = 1{,}110$$
$$0.08(6{,}000) + 0.07(15{,}000 - 6{,}000) \stackrel{?}{=} 1{,}110$$
$$480 + 0.07(9{,}000) \stackrel{?}{=} 1{,}110$$
$$480 + 630 \stackrel{?}{=} 1{,}110$$
$$1{,}110 = 1{,}110$$

6,000 is the solution.

APPLICATIONS

Investment problems

17. CORPORATE INVESTMENTS

Analyze

- \$25,000 is the total investment.
- 1st part at 4% annual interest.
- 2nd part at 7% annual interest.
- \$1,300 is the first-year combined income.
- How much is invested at each rate?

Assign

Let x = amount invested at 4%

$25,000 - x$ = amount invested at 7%

Form

	Principal •	rate •	time =	interest
1st part	x	0.04	1	**0.04x**
2nd part	$25,000 - x$	0.07	1	**0.07(25,000 – x)**
			Total:	**1,300**

The interest earned at 4%	plus	the interest earned at 7%	equals	the total interest.
$0.04x$	$+$	$0.07(25,000-x)$	$=$	$1,300$

Solve

$$0.04x + 0.07(25,000 - x) = 1,300$$
$$\mathbf{100}[0.04x + 0.07(25,000 - x)] = \mathbf{100}(1,300)$$
$$100(0.04x) + 100[0.07(25,000 - x)] = 100(1,300)$$
$$4x + 7(25,000 - x) = 130,000$$
$$4x + 175,000 - 7x = 130,000$$
$$-3x + 175,000 = 130,000$$
$$-3x + 175,000 - \mathbf{175,000} = 130,000 - \mathbf{175,000}$$
$$-3x = -45,000$$
$$\frac{-3x}{\mathbf{-3}} = \frac{-45,000}{\mathbf{-3}}$$
$$x = 15,000$$

7%

$$25,000 - x = 25,000 - 15,000$$
$$= 10,000$$

State

\$15,000 was invested at 4%.

\$10,000 was invested at 7%.

Check

The first investment earned 0.04(\$15,000), or \$600. The second earned 0.07(\$10,000), or \$700. Since the total return was \$600 + \$700 = \$1,300, the results check.

19. OLD COINS

Analyze

- \$3,500 was used to buy coins.
- Gold coins earning 15% annual interest.
- Silver coins earning 12% annual interest.
- \$480 was the return after 1 year.
- How much was invested in each?

Assign

Let x = amount invested in gold at 15%

$3,500 - x$ = amount invested in silver at12%

Form

	Principal •	rate •	time =	interest
Gold	x	0.15	1	**0.15x**
Silver	$3,500 - x$	0.12	1	**0.12(3,500 – x)**
			Total:	**480**

The return on gold coin at 15%	plus	the return on silver coins at 12%	equals	the total return.
$0.15x$	$+$	$0.12(3,500-x)$	$=$	480

Solve

$$0.15x + 0.12(3,500 - x) = 480$$
$$\mathbf{100}\left[0.15x + 0.12(3,500 - x)\right] = \mathbf{100}(480)$$
$$100(0.15x) + 100[0.12(3,500 - x)] = 100(480)$$
$$15x + 12(3,500 - x) = 48,000$$
$$15x + 42,000 - 12x = 48,000$$
$$3x + 42,000 = 48,000$$
$$3x + 42,000 - \mathbf{42,000} = 48,000 - \mathbf{42,000}$$
$$3x = 6,000$$
$$\frac{3x}{\mathbf{3}} = \frac{6,000}{\mathbf{3}}$$
$$x = 2,000$$

12%

$$3,500 - x = 3,500 - 2,000$$
$$= 1,500$$

State

\$2,000 was invested in gold coins at 15%.

\$1,500 was invested in silver coins at 12%.

Check

The gold coins earned 0.15(\$2,000), or \$300. The silver coins earned 0.12(\$1,500), or \$180. Since the total interest was \$300 + \$180 = \$480, the results check.

21. RETIREMENT

Analyze

- $28,000 invested at 6% annual interest.
- $3,500 is the return goal.
- What amount is invested at 7% annual interest.

Assign

Let x = amount invested at 7%

Form

	Principal	• rate	• time =	interest
1st part	28,000	0.06	1	**1,680**
2nd part	x	0.07	1	**0.07x**
		Total:		**3,500**

The interest on 6% investment	plus	the interest on 7% investment	equals	the total return goal.
1,680	+	$0.07x$	=	3,500

Solve

$$1,680+0.07x=3,500$$
$$\mathbf{100}(1,680+0.07x)=\mathbf{100}(3,500)$$
$$100(1,680)+100(0.07x)=100(3,500)$$
$$168,000+7x=350,000$$
$$168,000+7x-\mathbf{168,000}=350,000-\mathbf{168,000}$$
$$7x=182,000$$
$$\frac{7x}{\mathbf{7}}=\frac{182,000}{\mathbf{7}}$$
$$x=26,000$$

State

$26,000 is needed to be invested at 7%.

Check

The 1st investment earned 0.06($28,000), or $1,680.
The 2nd investment earned 0.07($26,000), or $1,820.
Since the total interest was $1,680 + $1,820 = $3,500, the results check.

23. 1099 FORMS

Analyze

- $15,000 is the total deposit.
- 822 part at 5% annual interest.
- 721 part at 4.5% annual interest.
- $720 is the total interest income.
- How much is invested at each rate?

Assign

Let x = amount invested at 5%
$15,000-x$ = amount invested at 4.5%

Form

	Principal	• rate	• time =	interest
822	x	0.05	1	**0.05x**
721	$15,000-x$	0.045	1	**0.045(15,000 − x)**
		Total:		**720**

The interest earned at 5%	plus	the interest earned at 4.5%	equals	the total interest.
$0.05x$	+	$0.045(15,000-x)$	=	720

Solve

$$0.05x+0.045(15,000-x)=720$$
$$\mathbf{1,000}[0.05x+0.045(15,000-x)]=\mathbf{1,000}(720)$$
$$1,000(0.05x)+1,000[0.045(15,000-x)]=1,000(720)$$
$$50x+45(15,000-x)=720,000$$
$$50x+675,000-45x=720,000$$
$$5x+675,000=720,000$$
$$5x+675,000-\mathbf{675,000}=720,000-\mathbf{675,000}$$
$$5x=45,000$$
$$\frac{5x}{\mathbf{5}}=\frac{45,000}{\mathbf{5}}$$
$$x=9,000$$

4.5%

$$15,000-x=15,000-9,000$$
$$=6,000$$

State

$9,000 was invested in the 822 at 5%.
$6,000 was invested in the 721 at 4.5%.

Check

The 1st investment earned 0.05($9,000), or $450.
The 2nd investment earned 0.045($6,000), or $270.
Since the total interest was $450 + $270 = $720, the results check.

25. INVESTMENTS

Analyze

- 1st equal amount invested at 7%.
- 2nd equal amount invested at 8%.
- 3rd equal amount invested at 10.5%.
- $1,249.50 is the total interest income.
- How much is invested at each rate?

Assign

Let x = equal amount invested at all rates

Form

	Principal •	rate •	time =	interest
1st part	x	0.07	1	**$0.07x$**
2nd part	x	0.08	1	**$0.08x$**
3rd part	x	0.105	1	**$0.105x$**
			Total:	**1,249.50**

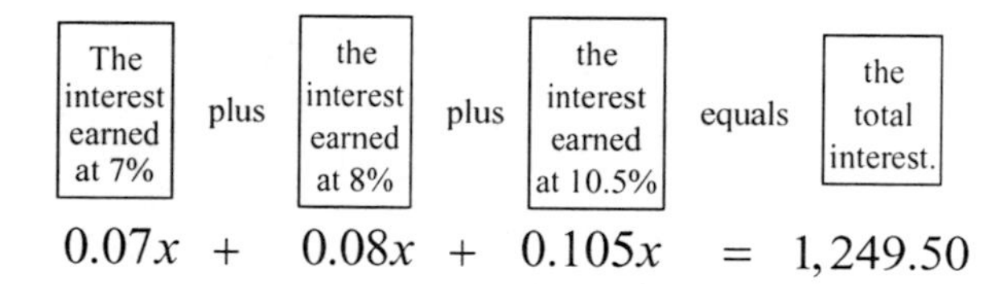

$$0.07x + 0.08x + 0.105x = 1,249.50$$

Solve

$$0.07x + 0.08x + 0.105x = 1,249.50$$
$$0.255x = 1,249.50$$
$$\mathbf{1,000}(0.255x) = \mathbf{1,000}(1,249.50)$$
$$255x = 1,249,500$$
$$\frac{255x}{\mathbf{255}} = \frac{1,249,500}{\mathbf{255}}$$
$$x = 4,900$$

State

$4,900 was invested at each of the rates.

Check

The 1st investment earned 0.07($4,900), or $343.

The 2nd investment earned 0.08($4,900), or $392.

The 3rd investment earned 0.105($4,900), or $514.50.

Since the total interest was $\$343 + \$392 + \$514.50 = \$1,249.50$, the results check.

27. BAD INVESTMENTS

Analyze

- $18,000 is the total investment.
- Credit union earns 3% annual interest.
- Utility stocks earns 7% annual interest, but suffered a loss.
- $90 was the net income after first year..
- How much was invested in each investment?

Assign

Let x = amount invested at 3%

$18,000 - x$ = amount invested at 7%

Form

	Principal •	rate •	time =	interest
CU	x	0.03	1	**$0.03x$**
US	$18,000 - x$	0.07	1	**$0.07(18,000 - x)$**
			Total:	**90**

The interest earned at 3% | minus | the interest loss at 7% | equals | the total interest.

$$0.03x - 0.07(18,000 - x) = 90$$

Solve

$$0.03x - 0.07(18,000 - x) = 90$$
$$\mathbf{100}[0.03x - 0.07(18,000 - x)] = \mathbf{100}(90)$$
$$100(0.03x) - 100[0.07(18,000 - x)] = 100(90)$$
$$3x - 7(18,000 - x) = 9,000$$
$$3x - 126,000 + 7x = 9,000$$
$$10x - 126,000 = 9,000$$
$$10x - 126,000 + \mathbf{126,000} = 9,000 + \mathbf{126,000}$$
$$10x = 135,000$$
$$\frac{10x}{\mathbf{10}} = \frac{135,000}{\mathbf{10}}$$
$$x = 13,500$$

7%

$$18,000 - x = 18,000 - 13,500$$
$$= 4,500$$

State

$13,500 was invested in the credit union.

$4,500 was invested in utility stock.

Check

The CU investment earned 0.03($13,500), or $405.

The US investment lost value 0.07($4,500), or $315.

Since the net interest was $\$405 - \$315 = \$90$, the results check.

Uniform Motion Problems

29. TORNADOES

Analyze

- One team travels east at 20 mph.
- One team travels west at 25 mph.
- Both leave from same place at the same time.
- Radios have a range of 90 miles.
- How long before they lose radio contact?

Assign

Let t = time in hours before they lose contact

Form

	rate	• time	= distance
East	20	t	$\mathbf{20t}$
West	25	t	$\mathbf{25t}$
		Total:	**90**

The distance traveled east	plus	the distance traveled west	equals	the total distance.
$20t$	+	$25t$	=	90

Solve

$$\begin{aligned} 20t + 25t &= 90 \\ 45t &= 90 \\ \frac{45t}{\mathbf{45}} &= \frac{90}{\mathbf{45}} \\ t &= 2 \end{aligned}$$

State

After 2 hours they will be out of contact with each other.

Check

The distance traveled east is 20(2), or 40 mi.
The distance traveled west is 25(2), or 50 mi.
Since the total was $40\text{ mi} + 50\text{ mi} = 90\text{ mi}$, the result checks.

31. HELLO/GOOD-BYE

Analyze

- Husband travels towards home at 45 mph.
- Wife travels towards work at 35 mph.
- Both leave at the same time.
- Combined distance is 20 miles.
- How long before they wave at each other?

Assign

Let t = time in hours they wave to each other

Form

	rate	• time	= distance
Husband	45	t	$\mathbf{45t}$
Wife	35	t	$\mathbf{35t}$
		Total:	**20**

The distance traveled by the husband	plus	the distance traveled by the wife	equals	the total distance.
$45t$	+	$35t$	=	20

Solve

$$\begin{aligned} 45t + 35t &= 20 \\ 80t &= 20 \\ \frac{80t}{\mathbf{80}} &= \frac{20}{\mathbf{80}} \\ t &= 0.25 \\ t &= \frac{1}{4} \end{aligned}$$

State

After $\frac{1}{4}$ hour or 15 minutes they will wave.

Check

The distance traveled by husband is 45(0.25), or 11.25 mi.
The distance traveled by wife is 35(0.25), or 8.75 mi.
Since the total was $11.25\text{ mi} + 8.75\text{ mi} = 20\text{ mi}$, the result checks.

33. CYCLING

Analyze

- Cyclist leaves base at 18 mph.
- Staff leaves 1.5 hours later in the same direction at 45 mph.
- Both leave from the same place.
- Both travel the same distances.
- How long before the staff catches the cylist?

Assign

Let t = time in hrs before car catches cyclist

Form

	rate •	time =	distance
Cyclist	18	$t + 1.5$	$\mathbf{18(t + 1.5)}$
Staff	45	t	$\mathbf{45t}$
		Distance:	**same**

The distance traveled by the cyclist	equals	the distance traveled by the staff.
$18(t+1.5)$	$=$	$45t$

Solve

$$18(t+1.5) = 45t$$
$$18t + 27 = 45t$$
$$18t + 27 - \mathbf{18t} = 45t - \mathbf{18t}$$
$$27 = 27t$$
$$\frac{27}{\mathbf{27}} = \frac{27t}{\mathbf{27}}$$
$$1 = t$$

State

After 1 hour, the staff car will catch the cyclist.

Check

The distance traveled by cyclist is 18(2.5), or 45 mi.
The distance traveled by car is 45(1), or 45 mi.
Since the each distance is the same 45 mi = 45 mi, the result checks.

35. ROAD TRIPS

Analyze

- 1st part of the trip averaged 40 mph.
- 2nd part of the trip averaged 50 mph.
- Total time of the trip was 5 hours.
- Total distance covered is 210 miles.
- How long did the car average 40 mph?

Assign

Let t = time in hours at 40 mph

Form

	rate •	time =	distance
1st part	40	t	$\mathbf{40t}$
2nd part	50	$5 - t$	$\mathbf{50(5 - t)}$
		Total:	**210**

The distance traveled at 40 mph	plus	the distance traveled at 50 mph	equals	the total distance.
$40t$	$+$	$50(5-t)$	$=$	210

Solve

$$40t + 50(5-t) = 210$$
$$40t + 250 - 50t = 210$$
$$-10t + 250 = 210$$
$$-10t + 250 - \mathbf{250} = 210 - \mathbf{250}$$
$$-10t = -40$$
$$\frac{-10t}{\mathbf{-10}} = \frac{-40}{\mathbf{-10}}$$
$$t = 4$$

State

The 40 mph part of the trip averaged 4 hours.

Check

The distance traveled at 40 mph is 40(4), or 160 mi.
The distance traveled at 50 mph is 50(1), or 50 mi.
Since the total of the distances is 160 mi + 50 mi, or 210 miles, the result checks.

37. WINTER DRIVING

Analyze

- 1st part of the trip was for 4 hours.
- 2nd part of the trip was for 3 hours at 20 mph less.
- Total distance of the trip was 325 miles.
- What was the average speed for the 1st part of the trip?

Assign

Let r = rate of 1st part of the trip in mph

Form

	rate	• time	= distance
1st part	r	4	$\mathbf{4r}$
2nd part	$r-20$	3	$\mathbf{3(t-20)}$
		Total:	**325**

The distance traveled by the 1st part	plus	the distance traveled by the 2nd part	equals	the total distance.
$4r$	$+$	$3(r-20)$	$=$	325

Solve

$$4r+3(r-20)=325$$
$$4r+3r-60=325$$
$$7r-60=325$$
$$7r-60+\mathbf{60}=325+\mathbf{60}$$
$$7r=385$$
$$\frac{7r}{\mathbf{7}}=\frac{385}{\mathbf{7}}$$
$$r=55$$

2nd part

$$r-20=55-20$$
$$=35$$

State

1st part was averaged at 55 mph.

2nd part was averaged at 35 mph.

Check

The distance traveled at 55 mph is 55(4), or 220 mi.
The distance traveled at 33 mph is 35(3), or 105 mi.
Since the total of the distances is $220\text{ mi}+105\text{ mi}$, or 325 miles, the results check.

Liquid Mixture Problems

39. SALT SOLUTION

Analyze

- Weak salt solution is 3%.
- Strong salt solution is 7%.
- A mixture of the two is to be 5%.
- Start with 50 gallons of 7% solution.
- How many gallons of 3% solution are needed?

Assign

Let x = the amount of 3% solution in gallons

Form

	Amount •	Strength =	Amount of pure salt
Weak	x	0.03	$\mathbf{0.03x}$
Strong	50	0.07	**3.5**
Mixture	$\mathbf{x+50}$	0.05	$\mathbf{0.05(x+50)}$

The salt in the 3% solution	plus	the salt in the 7% solution	equals	the salt in the 5% solution.
$0.03x$	$+$	3.5	$=$	$0.05(x+50)$

Solve

$$0.03x+3.5=0.05(x+50)$$
$$\mathbf{100}[0.03x+3.5]=\mathbf{100}[0.05(x+50)]$$
$$100(0.03x)+100(3.5)=100[0.05(x+50)]$$
$$3x+350=5(x+50)$$
$$3x+350=5x+250$$
$$3x+350-\mathbf{3x}=5x+250-\mathbf{3x}$$
$$350=2x+250$$
$$350-\mathbf{250}=2x+250-\mathbf{250}$$
$$100=2x$$
$$\frac{100}{\mathbf{2}}=\frac{2x}{\mathbf{2}}$$
$$50=x$$

State

50 gallons of the 3% salt solution will be needed.

Check

The salt in the 3% solution is 50(0.03), or 1.5
The salt in the 7% solution is 50(0.07), or 3.5
The salt in the 5% solution is 100(0.05), or 5.0
Since the total amount of salt in the ingredients is $1.5+3.5$, or 5.0, and that equals the amount of salt in the mixture, 5.0, the result checks.

41. MAKING CHEESE

Analyze

- Milk contains 4% butterfat.
- Milk contains 1% butterfat.
- A mixture of 15 gallons is to be 2%.
- How many gallons of each are needed?

Assign

Let $x =$ the amount of 4% milk in gal

$15 - x =$ the amount of 1% milk in gal

Form

	Amount	• Strength =	Amount of pure butterfat
4% Milk	x	0.04	**0.04*x***
1% Milk	**15 – *x***	0.01	**0.01(15 – *x*)**
Mixture	15	0.02	**0.3**

The butterfat in the 4% solution	plus	the butterfat in the 1% solution	equals	the butterfat in the 2% solution.
$0.04x$	$+$	$0.01(15-x)$	$=$	0.3

Solve

$$0.04x + 0.01(15 - x) = 0.3$$
$$\mathbf{100}[0.04x + 0.01(15 - x)] = \mathbf{100}(0.3)$$
$$100(0.04x) + 100[0.01(15 - x)] = 100(0.3)$$
$$4x + 1(15 - x) = 30$$
$$4x + 15 - x = 30$$
$$3x + 15 = 30$$
$$3x + 15 - \mathbf{15} = 30 - \mathbf{15}$$
$$3x = 15$$
$$\frac{3x}{\mathbf{3}} = \frac{15}{\mathbf{3}}$$
$$x = 5$$

State

5 gallons of the 4% milk is needed.

$15 - 5 = 10$ gallons of 1% milk is needed.

Check

The butterfat in the 4% solution is 5(0.04), or 0.2
The butterfat in the 1% solution is 10(0.01), or 0.1
The butterfat in the 2% solution is 15(0.02), or 0.3
Since the total amount of butterfat in the ingredients is $0.2 + 0.1$, or 0.3 and that equals the amount of butterfat in the mixture, 0.3, the result checks.

43. PRINTING

Analyze

- 8% cobalt blue color ink.
- 22% cobalt blue color ink.
- A mixture of 64 ounces is to be 15%.
- How many ounces of each are needed?

Assign

Let $x =$ the amount of 8% ink in ounces

$64 - x =$ the amount of 22% ink in ounces

Form

	Amount	• Strength =	Amount of pure cobalt blue
Weaker	x	0.08	**0.08*x***
Stronger	**64 – *x***	0.22	**0.22(64 – *x*)**
Mixture	64	0.15	**9.6**

The cobalt blue in the 8% solution	plus	the cobalt blue in the 22% solution	equals	the cobalt blue in the 15% solution.
$0.08x$	$+$	$0.22(64-x)$	$=$	9.6

Solve

$$0.08x + 0.22(64 - x) = 9.6$$
$$\mathbf{100}[0.08x + 0.22(64 - x)] = \mathbf{100}(9.6)$$
$$100(0.08x) + 100[0.22(64 - x)] = 100(9.6)$$
$$8x + 22(64 - x) = 960$$
$$8x + 1{,}408 - 22x = 960$$
$$-14x + 1{,}408 = 960$$
$$-14x + 1{,}408 - \mathbf{1{,}408} = 960 - \mathbf{1{,}408}$$
$$-14x = -448$$
$$\frac{-14x}{\mathbf{-14}} = \frac{-448}{\mathbf{-14}}$$
$$x = 32$$

22%

$$64 - x = 64 - 32$$
$$= 32$$

State

32 ounces of the 8% cobalt blue is needed.

32 ounces of the 22% cobalt blue is needed.

Check

The coblt blue in the 8% solution is 32(0.08), or 2.56 ounces.

The cobalt blue in the 22% solution is 32(0.22), or 7.04 ounces.

The cobalt blue in the 15% solution is 64(0.15), or 9.6 ounces.

Since the total amount of C B in the ingredients is 2.56 oz $+ 7.04$ oz, or 9.6 oz and that equals the amount of C B in the mixture, 9.6 oz, the result checks.

45. INTERIOR DECORATING

Analyze

- 7% Desert Sunrise tint.
- 1 gal of 18.2% Bright Pumpkin tint.
- Mixture is Cool Cantaloupe of 8.6% tint.
- How many gal of Desert tint is needed?

Assign

Let $x =$ the gal of 3% Desert tint

Form

	Amount	• Strength =	Amount of pure tint
Desert	x	0.07	**0.07x**
Bright	1	0.182	**0.182(1)**
Cool	$x+1$	0.086	**0.086(x + 1)**

The tint in the 7% Desert plus the tint in the 18.2% Bright equals the tint in the 8.6% Cool.

$$0.07x + 0.182(1) = 0.086(x+1)$$

Solve

$$0.07x + 0.182 = 0.086(x+1)$$
$$\mathbf{1,000}[0.07x + 0.182] = \mathbf{1,000}[0.086(x+1)]$$
$$1,000(0.07x) + 1,000(0.182) = 1,000[0.086(x+1)]$$
$$70x + 182 = 86(x+1)$$
$$70x + 182 = 86x + 86$$
$$70x + 182 - \mathbf{70x} = 86x + 86 - \mathbf{70x}$$
$$182 = 16x + 86$$
$$182 - \mathbf{86} = 16x + 86 - \mathbf{86}$$
$$96 = 16x$$
$$\frac{96}{\mathbf{16}} = \frac{16x}{\mathbf{16}}$$
$$6 = x$$

State

6 gal of 7% Desert Sunrise is needed.

Check

The tint in the 7% paint is 6(0.07), or 0.42 gal.
The tint in the 18.2% paint is 1(0.182), or 0.182 gal.
The tint in the 8.6% paint is 7(0.086), or 0.602 gal.
Since the total amount of tint in the ingredients is 0.42 gal + 0.182 gal, or 0.602 gal and that equals the amount of tint in the mixture, 0.602 gal, the result checks.

Dry Mixture Problems

47. LAWN SEED

Analyze

- Bluegrass seed sells for \$6 per lb.
- Ryegrass seed sells for \$3 per lb.
- A blend of the two is to sell for \$5 per lb.
- Start with 100 lb of bluegrass.
- How many pounds of ryegrass are needed?

Assign

Let $x =$ the lb of ryegrass needed

Form

	Number	• Value	= Total value
Bluegrass	100	6	**600**
Ryegrass	x	3	**3x**
Blend	**x + 100**	5	**5(x + 100)**

The value of the bluegrass plus the value of the ryegrass equals the total value of the blend.

$$600 + 3x = 5(x+100)$$

Solve

$$600 + 3x = 5(x+100)$$
$$600 + 3x = 5x + 500$$
$$600 + 3x - \mathbf{3x} = 5x + 500 - \mathbf{3x}$$
$$600 = 2x + 500$$
$$600 - \mathbf{500} = 2x + 500 - \mathbf{500}$$
$$100 = 2x$$
$$\frac{100}{\mathbf{2}} = \frac{2x}{\mathbf{2}}$$
$$50 = x$$

State

50 lb of the ryegrass seed will be needed.

Check

The value of the bluegrass is 100(\$6), or \$600.
The value of the ryegrass is 50(\$3), or \$150.
The value of the blend is 150(\$5), or \$750.
Since the total was \$600 + \$150 = \$750, the result checks.

49. RAISINS

Analyze

- **Natural** raisins sells for \$3.45 per scoop.
- **Golden** raisins sells for \$2.55 per scoop.
- **A blend** of the two is to sell for \$3 per scoop.
- **Start with** 20 scoops of golden raisins.
- How many scoops of natural raisins are needed?

Assign

Let x = the scoops of natural raisins needed

Form

	Number	• Value	= Total value
Golden	20	2.55	**51**
Natural	x	3.45	**3.45x**
Blend	**x + 20**	3	**3(x + 20)**

The value of the golden	plus	the value of the natural	equals	the total value of the blend.
51	+	$3.45x$	=	$3(x+20)$

Solve

$$51+3.45x=3(x+20)$$
$$51+3.45x=3x+60$$
$$51+3.45x-\mathbf{3x}=3x+60-\mathbf{3x}$$
$$51+0.45x=60$$
$$51+0.45x-\mathbf{51}=60-\mathbf{51}$$
$$0.45x=9$$
$$\frac{0.45x}{\mathbf{0.45}}=\frac{9}{\mathbf{0.45}}$$
$$x=20$$

State

20 scoops of the golden raisins will be needed.

Check

The value of the golden is 20(\$2.55), or \$51.
The value of the natural is 20(\$3.45), or \$69.
The value of the blend is 40(\$3), or \$120.
Since the total was $\$51+\$69=\$120$, the result checks.

51. PACKAGED SALAD

Analyze

- **Iceberg** sells for \$2.20 per 10 oz bag.
- **Romaine** sells for \$3.50 per 10 oz bag.
- **Start with** 50 bags of Iceberg.
- How many bags of Romaine are needed a blend sells for \$2.50 per 10 oz bag.?

Assign

Let x = the bags of Roamine needed

Form

	Number	• Value	= Total value
Romaine	x	3.50	**3.50x**
Iceberg	50	2.20	**50(2.20)**
Blend	$x + 50$	2.50	**2.50(x + 50)**

The value of the Romaine	plus	the value of the Iceberg	equals	the total value of the blend.
$3.50x$	+	$50(2.20)$	=	$2.50(x+50)$

Solve

$$3.50x+50(2.20)=2.50(x+50)$$
$$3.50x+110=2.50x+125$$
$$3.50x+110-\mathbf{2.50x}=2.50x+125-\mathbf{2.50x}$$
$$x+110=125$$
$$x+110-\mathbf{110}=125-\mathbf{110}$$
$$x=15$$

State

15 bags of Romaine will be needed.
Optional: Needed for checking
50 bags of Iceberg will be needed.

Check

The value of Romaine is 15(\$3.50), or \$52.50.
The value of Iceberg is 50(\$2.20), or \$110.
The value of blend is 65(\$2.50), or \$162.50.
Since the total was $\$52.50+\$110=\$162.50$, the result checks.

53. BRONZE

Analyze

- 4 pounds of tin.
- 6 pounds of copper.
- Tin cost $1 more than copper.
- The bronze is 10 pounds.
- How much is a pound of tin?

Assign

Let x = the cost of tin

$x-1$ = the cost of copper

Form

	Number	• Value	= Total value
Tin	4	x	$4x$
Copper	6	$x-1$	$6(x-1)$
Bronze	10	3.65	36.5

The value of the tin	plus	the value of the copper	equals	the value of the bronze.
$4x$	$+$	$6(x-1)$	$=$	36.5

Solve

$$4x+6(x-1)=36.5$$
$$4x+6x-6=36.5$$
$$10x-6=36.5$$
$$10x-6+\mathbf{6}=36.5+\mathbf{6}$$
$$10x=42.5$$
$$\frac{10x}{\mathbf{10}}=\frac{42.5}{\mathbf{10}}$$
$$x=4.25$$

copper

$$x-1=4.25-1$$
$$=3.25$$

State

Tin cost $4.25 per pound.

Optional: Needed for checking

Copper cost $3.25 per pound.

Check

The value of the tin is 4($4.25), or $17.
The value of the copper is 6($3.25), or $19.50.
The value of the blend is 10($3.65), or $36.50.
Since the total was $\$17+\$19.50=\$36.50$, the results check.

Number - Value Problems

55. RENTALS

Analyze

- 1-room rents for $550.
- 2-room rents for $700.
- 3-room rents for $900.
- An equal number is rented.
- Total monthly rent is $36,550.
- How many of each apartment are rented?

Assign

Let x = the number of each apartment rented

Form

	Number	• Value	= Total value
1-room	x	550	$550x$
2-room	x	700	$700x$
3-room	x	900	$900x$
		Total:	**36,550**

The value of the 1-room	plus	the value of the 2-room	plus	the value of the 3-room	equals	the total rent.
$550x$	$+$	$700x$	$+$	$900x$	$=$	$36,550$

Solve

$$550x+700x+900x=36,550$$
$$2,150x=36,550$$
$$\frac{2,150x}{\mathbf{2,150}}=\frac{36,550}{\mathbf{2,150}}$$
$$x=17$$

State

17 of each type of bedroom was rented.

Check

The value of the 1-room is 17($550), or $9,350.
The value of the 2-room is 17($700), or $11,900.
The value of the 3-room is 17($900), or $15,300.
Since the total was $\$9,350+\$11,000+\$15,300$ $=\$36,550,$ the result checks.

57. SOFTWARE

Analyze

- Spreadsheet cost \$150. Database cost \$195.
- Word processing cost \$210.
- An equal number of SS and data were sold.
- 15 more Word than the other two combined were sold.
- \$72,000 in sales were generated by all 3.
- How many spreadsheets were sold?

Assign

Let x = the number of spreadsheets and database sold

Form

	Number	• Value	= Total value
Spreadsheet	x	150	$\mathbf{150x}$
Database	x	195	$\mathbf{195x}$
Word	$\mathbf{2x+15}$	210	$\mathbf{210(2x+15)}$
		Total:	**72,000**

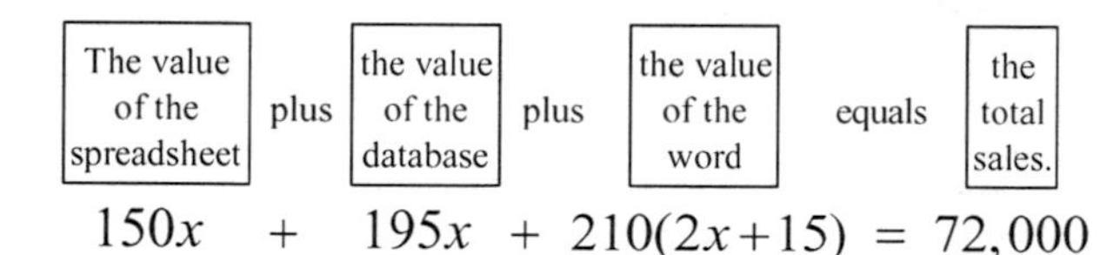

$$150x \quad + \quad 195x \quad + \quad 210(2x+15) \quad = \quad 72,000$$

Solve

$$150x+195x+210(2x+15)=72,000$$
$$150x+195x+420x+3,150=72,000$$
$$765x+3,150=72,000$$
$$765x+3,150-\mathbf{3,150}=72,000-\mathbf{3,150}$$
$$765x=68,850$$
$$\frac{765x}{\mathbf{765}}=\frac{68,850}{\mathbf{765}}$$
$$x=90$$

word processing

$$2x+15=2(90)+15$$
$$=180+15$$
$$=195$$

State

90 spreadsheets were sold.
Optional: Needed for checking
90 database were sold.
195 word processing were sold.

Check

The value of the SS is 90(\$150), or \$13,500.
The value of the data is 90(\$195), or \$17,550.
The value of the word is 195(\$210), or \$40,950.
Since the total was \$13,500 + \$17,550 + \$40,950 = \$72,000, the result checks.

59. PIGGY BANKS

Analyze

- Pennies are worth \$0.01.
- Nickels are worth \$0.05.
- Dimes are worth \$0.10.
- 20 more pennies than dimes.
- Nickels triple the number of dimes.
- \$5.40 is the values of all the coins.
- How many of each type coins are there?

Assign

Let x = the number of dimes
$x+20$ = the number of pennies
$3x$ = the number of nickels

Form

	Number	• Value	= Total value
Dimes	x	0.10	$\mathbf{0.10x}$
Pennies	$\mathbf{x+20}$	0.01	$\mathbf{0.01(x+20)}$
Nickels	$\mathbf{3x}$	0.05	$\mathbf{0.15x}$
		Total:	**5.40**

The value of the dimes	plus	the value of the pennies	plus	the value of the nickels	equals	the value of all coins.

$$0.10x \quad + \quad 0.01(x+20) \quad + \quad 0.15x \quad = \quad 5.40$$

Solve

$$0.10x+0.01(x+20)+0.15x=5.40$$
$$0.10x+0.01x+0.20+0.15x=5.40$$
$$0.26x+0.20=5.40$$
$$0.26x+0.20-\mathbf{0.20}=5.40-\mathbf{0.20}$$
$$0.26x=5.20$$
$$\frac{0.26x}{\mathbf{0.26}}=\frac{5.20}{\mathbf{0.26}}$$
$$x=20$$

pennies

$$x+20=20+20$$
$$=40$$

nickels

$$3x=3(20)$$
$$=60$$

State

There are 20 dimes.
There are 40 pennies.
There are 60 nickels.

Check

The value of the dimes is 20(\$0.10), or \$2.
The value of the pennies is 40(\$0.01), or \$0.40.
The value of the nickels is 60(\$0.05), or \$3.
Since the total was \$2 + \$0.40 + \$3 = \$5.40, the results check.

61. BASKETBALL

Analyze

- Made 46 more 2-point baskets than 3-point.
- Made only 1 free throw.
- 113 total points were scored.
- How many 2-point and 3-points baskets were made?

Assign

Let $x =$ the number of 3-point baskets
$x + 46 =$ the number of 2-point baskets

Form

	Number	• Value	= Total value
3-point	**x**	3	**$3x$**
2-point	**$x + 46$**	2	**$2(x + 46)$**
Free throw	1	1	**1**
		Total:	**113**

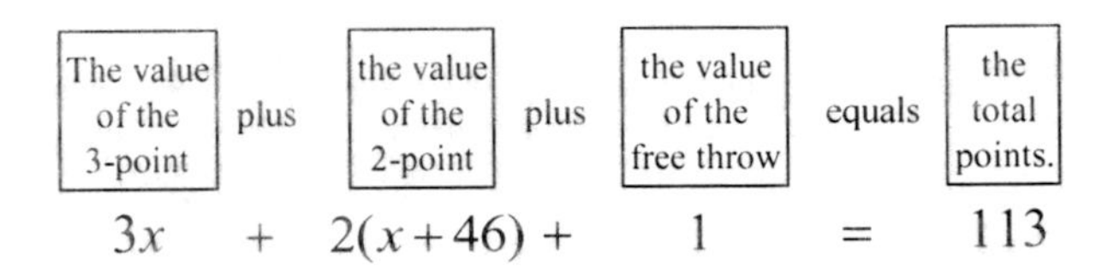

Solve

$$3x + 2(x+46) + 1 = 113$$
$$3x + 2x + 92 + 1 = 113$$
$$5x + 93 = 113$$
$$5x + 93 - \mathbf{93} = 113 - \mathbf{93}$$
$$5x = 20$$
$$\frac{5x}{\mathbf{5}} = \frac{20}{\mathbf{5}}$$
$$x = 4$$

2-point

$$x + 46 = 4 + 46$$
$$= 50$$

State

4 3-point baskets were made.
50 2-point baskets were made.
1 free throw.

Check

The value of the 3-point is 4(3), or 12.
The value of the 2-point is 50(2), or 100.
The value of the free throw is 1(1), or 1.
Since the total was $12 + 100 + 1 = 113$, the results check.

WRITING

63-65. Answers will vary.

REVIEW

67. $-12(3a + 4b - 32)$
$$= -12(3a) - 12(4b) - 12(-32)$$
$$= -36a - 48b + 384$$

69. $3(5t+1)2 = [3(5t) + 3(1)]2$
$$= (15t + 3)2$$
$$= 2(15t) + 2(3)$$
$$= 30t + 6$$

CHALLENGE

71. EVAPORATION

Analyze

- 300 ml of a 2% salt solution.
- Reduce to a 3% salt solution.
- How much water needs to be boiled away?

Assign

Let $x =$ the amount of water in ml

Form

	Amount	• Strength =	Amount of pure salt
Weak	300	0.02	**6**
Strong	**300 − x**	0.03	**30.0(300 − x)**

The salt in the 2% solution equals the salt in the 3% solution.

$$6 = 0.03(300 - x)$$

Solve

$$6 = 0.03(100 - x)$$
$$\mathbf{100}(6) = \mathbf{100}[0.03(100 - x)]$$
$$100(6) = 100[0.03(100 - x)]$$
$$600 = 3(100 - x)$$
$$600 = 300 + 3x$$
$$600 - \mathbf{300} = 300 + 3x - \mathbf{300}$$
$$300 = 3x$$
$$\frac{300}{\mathbf{3}} = \frac{3x}{\mathbf{3}}$$
$$100 = x$$

State

100 milliliters of water needs to be boiled off.

Check

The salt in the 2% solution is 300(0.02), or 6 ml.
The salt in the 3% solution is 200(0.03), or 6 ml.
Since the salt in the weaker is 6 ml and the salt in the stronger is 6 ml, the result checks.

73. FINANCIAL PLANNING

Analyze

- Insured fund pays 11% interest.
- Risky investment pays 13% interest.
- Same amount invested in both.
- Higher rate generates $150 extra.
- How much is invested at each rate?

Assign

Let $x =$ amount invested at 11% and 13%

Form

	Principal •	rate	• time =	interest
Insured	$\boldsymbol{x}$	0.11	1	$\mathbf{0.11}\boldsymbol{x}$
Risky	$\boldsymbol{x}$	0.13	1	$\mathbf{0.13}\boldsymbol{x}$

In order for the two amounts of interest to be equal, one must add $150 to the Insured interest before solving.

The interest earned at 11%	+	$150	equals	the interest earned at 13%
$0.11x$	$+$	150	$=$	$0.13x$

Solve

$$0.11x + 150 = 0.13x$$
$$\mathbf{100}[(0.11x)+150] - \mathbf{100}(0.13x)$$
$$100(0.11x)+100(150) = 100(0.13x)$$
$$11x + 15,000 = 13x$$
$$11x + 15,000 - \mathbf{11x} = 13x - \mathbf{11x}$$
$$15,000 = 2x$$
$$\frac{15,000}{\mathbf{2}} = \frac{2x}{\mathbf{2}}$$
$$7,500 = x$$

State

$7,500 was invested in each.

Check

The 1st investment earned $0.11(\$7,500) + 150$, or $975.

The 2nd investment earned $0.13(\$7,500)$, or $975.

Since the 2 interests are equal, $\$975 = \$975$, the result checks.

SECTION 2.7 STUDY SET
VOCABULARY

Fill in the blanks.

1. An **inequality** is a statement that contains one of the symbols: $>$, $\geq$, $<$, or $\leq$. An equation is a statement that contains an **=** symbol.

3. The solution set of $x > 2$ can be expressed in **interval** notation as $(2, \infty)$.

CONCEPTS

5. a. Adding the same number to **both** sides of an inequality does not change the solutions.
   b. Multiplying or dividing both sides of an inequality by the same **positive** number does not change the solutions.
   c. If we multiply or divide both sides of an inequality by a **negative** number, the direction of the inequality symbol must be reversed for the inequalities to have the same solutions.

7. Rewrite the inequality $32 < x$ in an equivalent form with the variable on the left side. $\mathbf{x > 32}$

9. a. $x < \mathbf{-1}$, $(\mathbf{-\infty}, -1)$ b. $x \geq \mathbf{2}$, $[2, \mathbf{\infty})$

NOTATION

11. a. $\leq$ b. ∞ c. [or] d. $>$

13.
$$4x - 5 \geq 7$$
$$4x - 5 + \boxed{5} \geq 7 + \boxed{5}$$
$$4x \geq \boxed{12}$$
$$\frac{4x}{\boxed{4}} \geq \frac{12}{\boxed{4}}$$
$$x \geq 3$$
Solution set: $[\boxed{3}, \infty)$

GUIDED PRACTICE

15. Determine whether each number is a solution of $3x - 2 > 5$.

a.
$$3x - 2 > 5$$
$$3(\mathbf{5}) - 2 \overset{?}{>} 5$$
$$15 - 2 \overset{?}{>} 5$$
$$13 > 5 \text{ True}$$
5 is a solution.

b.
$$3x - 2 > 5$$
$$3(\mathbf{-4}) - 2 \overset{?}{>} 5$$
$$-12 - 2 \overset{?}{>} 5$$
$$-14 > 5 \text{ False}$$
-4 is not a solution.

17. Determine whether each number is a solution of $-5(x-1) \geq 2x + 12$.

a.
$$-5(x-1) \geq 2x + 12$$
$$-5(\mathbf{1} - 1) \overset{?}{\geq} 2(\mathbf{1}) + 12$$
$$-5(0) \overset{?}{\geq} 2 + 12$$
$$0 \geq 14 \text{ False}$$
1 is not a solution.

b.
$$-5(x-1) \geq 2x + 12$$
$$-5(\mathbf{-1} - 1) \overset{?}{\geq} 2(\mathbf{-1}) + 12$$
$$-5(-2) \overset{?}{\geq} -2 + 12$$
$$10 \geq 10 \text{ True}$$
-1 is a solution.

Graph each inequality and describe the graph using interval notation.

19. $x < 5$

5

$(-\infty, 5)$

21. $-3 < x \leq 1$

-3 1

$(-3, 1]$

Solve each inequality. Write the solution set in interval notation and graph it.

23.
$$x + 2 > 5$$
$$x + 2 - \mathbf{2} > 5 - \mathbf{2}$$
$$x > 3$$
$$(3, \infty)$$

3

25.
$$g - 30 \geq -20$$
$$g - 30 + \mathbf{30} \geq -20 + \mathbf{30}$$
$$g \geq 10$$
$$[10, \infty)$$

10

27.
$$-\frac{3}{16}x \geq -9$$
$$-\frac{\mathbf{16}}{\mathbf{3}}\left(-\frac{3}{16}x\right) \leq -\frac{\mathbf{16}}{\mathbf{3}}(-9)$$
$$x \leq 48$$
$$(-\infty, 48]$$

48

29.
$$\frac{2}{3}x \ge 2$$
$$\frac{\mathbf{3}}{\mathbf{2}}\left(\frac{2}{3}x\right) \ge \frac{\mathbf{3}}{\mathbf{2}}(2)$$
$$x \ge 3$$
$$[3, \infty)$$

3

31.
$$-3y \le -6$$
$$\frac{-3y}{\mathbf{-3}} \ge \frac{-6}{\mathbf{-3}}$$
$$y \ge 2$$
$$[2, \infty)$$

2

33.
$$8h < 48$$
$$\frac{8h}{\mathbf{8}} < \frac{48}{\mathbf{8}}$$
$$h < 6$$
$$(-\infty, 6)$$

6

35.
$$64 < 9x + 1$$
$$64 - \mathbf{1} < 9x + 1 - \mathbf{1}$$
$$63 < 9x$$
$$\frac{63}{\mathbf{9}} < \frac{9x}{\mathbf{9}}$$
$$7 < x$$
$$x > 7$$
$$(7, \infty)$$

7

37.
$$-20 \ge 3m - 5$$
$$-20 + \mathbf{5} \ge 3m - 5 + \mathbf{5}$$
$$-15 \ge 3m$$
$$\frac{-15}{\mathbf{3}} \ge \frac{3m}{\mathbf{3}}$$
$$-5 \ge m$$
$$m \le -5$$
$$(-\infty, -5]$$

-5

39.
$$1.3 - 2x \ge 0.5$$
$$1.3 - 2x - \mathbf{1.3} \ge 0.5 - \mathbf{1.3}$$
$$-2x \ge -0.8$$
$$\frac{-2x}{\mathbf{-2}} \le \frac{-0.8}{\mathbf{-2}}$$
$$x \le 0.4$$
$$(-\infty, 0.4]$$

0.4

41.
$$24.9 - 12a < -3.9$$
$$24.9 - 12a - \mathbf{24.9} < -3.9 - \mathbf{24.9}$$
$$-12a < -28.8$$
$$\frac{-12a}{\mathbf{-12}} > \frac{-28.8}{\mathbf{-12}}$$
$$a > 2.4$$
$$(2.4, \infty)$$

2.4

43.
$$9a + 4 > 5a - 16$$
$$9a + 4 - \mathbf{5a} > 5a - 16 - \mathbf{5a}$$
$$4a + 4 > -16$$
$$4a + 4 - \mathbf{4} > -16 - \mathbf{4}$$
$$4a > -20$$
$$\frac{4a}{\mathbf{4}} > \frac{-20}{\mathbf{4}}$$
$$x > -5$$
$$(-5, \infty)$$

−5

45.
$$8(2n+1) \le 4(6n+7) + 4n$$
$$\mathbf{8}(2n) + \mathbf{8}(1) \le \mathbf{4}(6n) + \mathbf{4}(7) + 4n$$
$$16n + 8 \le 24n + 28 + 4n$$
$$16n + 8 \le 28n + 28$$
$$16n + 8 - \mathbf{28n} \le 28n + 28 - \mathbf{28n}$$
$$-12n + 8 \le 28$$
$$-12n + 8 - \mathbf{8} \le 28 - \mathbf{8}$$
$$-12n \le 20$$
$$\frac{-12n}{\mathbf{-12}} \ge \frac{20}{\mathbf{-12}}$$
$$n \ge -\frac{5}{3}$$
$$\left[-\frac{5}{3}, \infty\right)$$

$-\frac{5}{3}$

47. $\frac{1}{2}+\frac{n}{5}>\frac{3}{4}$, LCD = 20

$$\mathbf{20}\left(\frac{1}{2}+\frac{n}{5}\right)>\mathbf{20}\left(\frac{3}{4}\right)$$
$$20\left(\frac{1}{2}\right)+20\left(\frac{n}{5}\right)>20\left(\frac{3}{4}\right)$$
$$10+4n>15$$
$$10+4n-\mathbf{10}>15-\mathbf{10}$$
$$4n>5$$
$$\frac{4n}{\mathbf{4}}>\frac{5}{\mathbf{4}}$$
$$n>\frac{5}{4}$$
$$\left(\frac{5}{4},\infty\right)$$

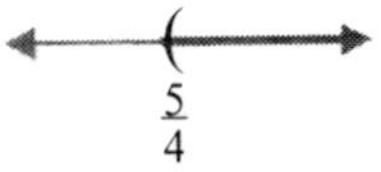

$\frac{5}{4}$

49. $\frac{1}{2}-\frac{x}{24}\geq-\frac{1}{8}$, LCD = 24

$$\mathbf{24}\left(\frac{1}{2}-\frac{x}{24}\right)\geq\mathbf{24}\left(-\frac{1}{8}\right)$$
$$24\left(\frac{1}{2}\right)-24\left(\frac{x}{24}\right)\geq24\left(-\frac{1}{8}\right)$$
$$12-x\geq-3$$
$$12-x-\mathbf{12}\geq-3-\mathbf{12}$$
$$-x\geq-15$$
$$\frac{-x}{\mathbf{-1}}\leq\frac{-15}{\mathbf{-1}}$$
$$x\leq15$$
$$(-\infty,15]$$

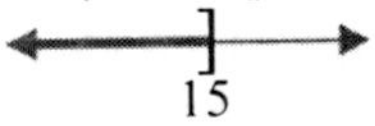

15

Graph each inequality and describe the graph using interval notation.

51. $-2\leq x<3$

$[-2,3)$

−2 3

53. $-\frac{7}{4}<x<2$

$$\left(-\frac{7}{4},2\right)$$

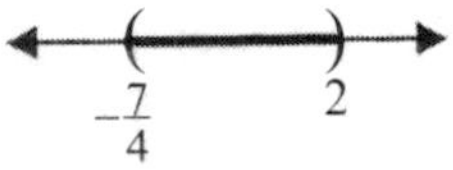

$-\frac{7}{4}$ 2

Solve each compound inequality. Write the solution set in interval notation and graph it.

55. $2<x-5<5$

$$2+\mathbf{5}<x-5+\mathbf{5}<5+\mathbf{5}$$
$$7<x<10$$
$$(7,10)$$

7 10

57. $0\leq x+10\leq10$

$$0-\mathbf{10}\leq x+10-\mathbf{10}\leq10-\mathbf{10}$$
$$-10\leq x\leq0$$
$$[-10,0]$$

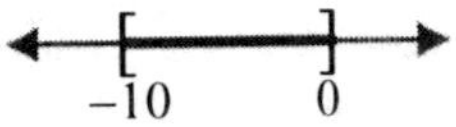

−10 0

59. $3\leq2x-1<5$

$$3+\mathbf{1}\leq2x-1+\mathbf{1}<5+\mathbf{1}$$
$$4\leq2x<6$$
$$\frac{4}{\mathbf{2}}\leq\frac{2x}{\mathbf{2}}<\frac{6}{\mathbf{2}}$$
$$2\leq x<3$$
$$[2,3)$$

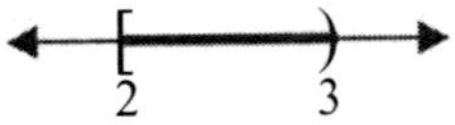

2 3

61. $-9<6x+9\leq45$

$$-9-\mathbf{9}<6x+9-\mathbf{9}\leq45-\mathbf{9}$$
$$-18<6x\leq36$$
$$\frac{-18}{\mathbf{6}}<\frac{6x}{\mathbf{6}}\leq\frac{36}{\mathbf{6}}$$
$$-3<x\leq6$$
$$(-3,6]$$

−3 6

TRY IT YOURSELF

Solve each inequality or compound inequality. Write the solution set in interval notation and graph it.

63. $\frac{6x+1}{4} \le x+1$, LCD = 4

$$\mathbf{4}\left(\frac{6x+1}{4}\right) \le \mathbf{4}(x+1)$$
$$6x+1 \le 4x+4$$
$$6x+1-\mathbf{4x} \le 4x+4-\mathbf{4x}$$
$$2x+1 \le 4$$
$$2x+1-\mathbf{1} \le 4-\mathbf{1}$$
$$2x \le 3$$
$$\frac{2x}{\mathbf{2}} \le \frac{3}{\mathbf{2}}$$
$$x \le \frac{3}{2}$$
$$\left(-\infty, \frac{3}{2}\right]$$

$\frac{3}{2}$

65. $17(3-x) \ge 3-13x$

$$51-17x \ge 3-13x$$
$$51-17x+\mathbf{13x} \ge 3-13x+\mathbf{13x}$$
$$51-4x \ge 3$$
$$51-4x-\mathbf{51} \ge 3-\mathbf{51}$$
$$-4x \ge -48$$
$$\frac{-4x}{\mathbf{-4}} \le \frac{-48}{\mathbf{-4}}$$
$$x \le 12$$
$$(-\infty, 12]$$

12

67. $0 < 5(x+2) \le 15$

$$0 < 5x+10 \le 15$$
$$0-\mathbf{10} < 5x+10-\mathbf{10} \le 15-\mathbf{10}$$
$$-10 < 5x \le 5$$
$$\frac{-10}{\mathbf{5}} < \frac{5x}{\mathbf{5}} \le \frac{5}{\mathbf{5}}$$
$$-2 < x \le 1$$
$$(-2, 1]$$

−2 1

69. $0.4x \le 0.1x+0.45$

$$0.4x-\mathbf{0.1x} \le 0.1x+0.45-\mathbf{0.1x}$$
$$0.3x \le 0.45$$
$$\frac{0.3x}{\mathbf{0.3}} \le \frac{0.45}{\mathbf{0.3}}$$
$$x \le 1.5$$
$$(-\infty, 1.5]$$

1.5

71. $-\frac{2}{3} \ge \frac{2y}{3} - \frac{3}{4}$, LCD = 12

$$\mathbf{12}\left(-\frac{2}{3}\right) \ge \mathbf{12}\left(\frac{2y}{3} - \frac{3}{4}\right)$$
$$12\left(-\frac{2}{3}\right) \ge 12\left(\frac{2y}{3}\right) - 12\left(\frac{3}{4}\right)$$
$$-8 \ge 8y-9$$
$$-8+\mathbf{9} \ge 8y-9+\mathbf{9}$$
$$1 \ge 8y$$
$$\frac{1}{\mathbf{8}} \ge \frac{8y}{\mathbf{8}}$$
$$\frac{1}{8} \ge y$$
$$y \le \frac{1}{8}$$
$$\left(-\infty, \frac{1}{8}\right]$$

$\frac{1}{8}$

73. $\frac{m}{-42} - 1 > -1$

$$\frac{m}{-42} - 1 + \mathbf{1} > -1 + \mathbf{1}$$
$$\frac{m}{-42} > 0$$
$$\mathbf{-42}\left(\frac{m}{-42}\right) < \mathbf{-42}(0)$$
$$m < 0$$
$$(-\infty, 0)$$

0

75. $$\begin{aligned} 6-x &\le 3(x-1) \\ 6-x &\le 3(x)-3(1) \\ 6-x &\le 3x-3 \\ 6-x-\mathbf{3x} &\le 3x-3-\mathbf{3x} \\ 6-4x &\le -3 \\ 6-4x-\mathbf{6} &\le -3-\mathbf{6} \\ -4x &\le -9 \\ \frac{-4x}{\mathbf{-4}} &\mathbf{\ge} \frac{-9}{\mathbf{-4}} \\ x &\ge \frac{9}{4} \\ &\left[\frac{9}{4},\infty\right) \end{aligned}$$

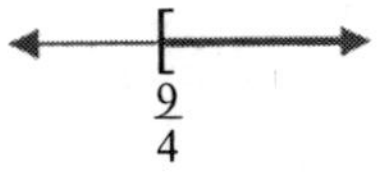

77. $$\begin{aligned} 6 < -2(x-1) &< 12 \\ 6 < -2x+2 &< 12 \\ 6-\mathbf{2} < -2x+2-\mathbf{2} &< 12-\mathbf{2} \\ 4 < -2x &< 10 \\ \frac{4}{\mathbf{-2}} \mathbf{>} \frac{-2x}{\mathbf{-2}} &\mathbf{>} \frac{10}{\mathbf{-2}} \\ -2 > x &> -5 \\ -5 < x &< -2 \\ &(-5,-2) \end{aligned}$$

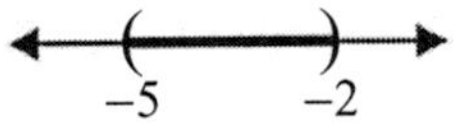

79. $$\begin{aligned} -1 &\le -\frac{1}{2}n \quad ,\ \text{LCD} = -2 \\ \mathbf{-2}(-1) &\mathbf{\ge} \mathbf{-2}\left(-\frac{1}{2}n\right) \\ 2 &\ge n \\ n &\le 2 \\ &(-\infty,2] \end{aligned}$$

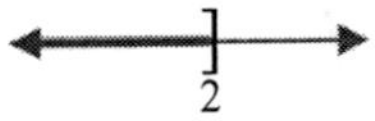

81. $$\begin{aligned} -m-12 &> 15 \\ -m-12+\mathbf{12} &> 15+\mathbf{12} \\ -m &> 27 \\ \frac{-m}{\mathbf{-1}} &\mathbf{<} \frac{27}{\mathbf{-1}} \\ m &< -27 \\ &(-\infty,-27) \end{aligned}$$

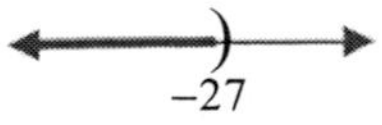

83. $$\begin{aligned} y-\frac{1}{7} &\le \frac{2}{3} \quad ,\ \text{LCD} = 21 \\ \mathbf{21}\left(y-\frac{1}{7}\right) &\le \mathbf{21}\left(\frac{2}{3}\right) \\ 21(y)-21\left(\frac{1}{7}\right) &\le 21\left(\frac{2}{3}\right) \\ 21y-3 &\le 14 \\ 21y-3+\mathbf{3} &\le 14+\mathbf{3} \\ 21y &\le 17 \\ \frac{21y}{\mathbf{21}} &\le \frac{17}{\mathbf{21}} \\ y &\le \frac{17}{21} \\ &\left(-\infty,\frac{17}{21}\right] \end{aligned}$$

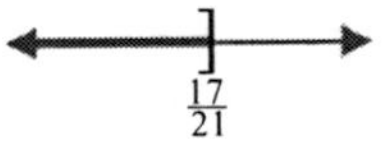

85. $$\begin{aligned} 9x+13 &\ge 2x+6x \\ 9x+13 &\ge 8x \\ 9x+13-\mathbf{8x} &\ge 8x-\mathbf{8x} \\ x+13 &\ge 0 \\ x+13-\mathbf{13} &\ge 0-\mathbf{13} \\ x &\ge -13 \\ &[-13,\infty) \end{aligned}$$

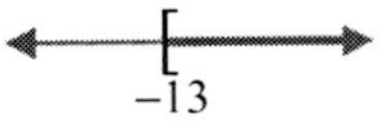

87.

$$7 < \frac{5}{3}a + (-3)$$
$$7 + \mathbf{3} < \frac{5}{3}a - 3 + \mathbf{3}$$
$$10 < \frac{5}{3}a$$
$$\frac{\mathbf{3}}{\mathbf{5}} \cdot 10 < \frac{\mathbf{3}}{\mathbf{5}} \cdot \left(\frac{5}{3}a\right)$$
$$6 < a$$
$$a > 6$$
$$(6, \infty)$$

6

89.

$$-8 \le \frac{y}{8} - 4 \le 2$$
$$-8 + \mathbf{4} \le \frac{y}{8} - 4 + \mathbf{4} \le 2 + \mathbf{4}$$
$$-4 \le \frac{y}{8} \le 6$$
$$\mathbf{8}(-4) \le \mathbf{8}\left(\frac{y}{8}\right) \le \mathbf{8}(6)$$
$$-32 \le y \le 48$$
$$[-32, 48]$$

−32 48

91.

$$0.04x + 1.04 \le 0.01x + 1.085$$
$$0.04x + 1.04 - \mathbf{0.01x} \le 0.01x + 1.085 - \mathbf{0.01x}$$
$$0.03x + 1.04 \le 1.085$$
$$0.03x + 1.04 - \mathbf{1.04} \le 1.085 - \mathbf{1.04}$$
$$0.03x \le 0.045$$
$$\frac{0.03x}{\mathbf{0.03}} \le \frac{0.045}{\mathbf{0.03}}$$
$$x \le 1.5$$
$$(-\infty, 1.5]$$

1.5

93.

$$\frac{5}{3}(x+1) \ge -x + \frac{2}{3} \quad , \text{ LCD} = 3$$
$$\mathbf{3}\left(\frac{5}{3}(x+1)\right) \ge \mathbf{3}\left(-x + \frac{2}{3}\right)$$
$$3\left(\frac{5}{3}(x+1)\right) \ge 3(-x) + 3\left(\frac{2}{3}\right)$$
$$5(x+1) \ge -3x + 2$$
$$5x + 5 \ge -3x + 2$$
$$5x + 5 + \mathbf{3x} \ge -3x + 2 + \mathbf{3x}$$
$$8x + 5 \ge 2$$
$$8x + 5 - \mathbf{5} \ge 2 - \mathbf{5}$$
$$8x \ge -3$$
$$\frac{8x}{\mathbf{8}} \ge \frac{-3}{\mathbf{8}}$$
$$x \ge -\frac{3}{8}$$
$$\left[-\frac{3}{8}, \infty\right)$$

$-\frac{3}{8}$

95.

$$\frac{4}{5}x < \frac{2}{5} \quad , \text{ LCD} = 5$$
$$\mathbf{5}\left(\frac{4}{5}x\right) < \mathbf{5}\left(\frac{2}{5}\right)$$
$$4x < 2$$
$$\frac{4x}{\mathbf{4}} < \frac{2}{\mathbf{4}}$$
$$x < \frac{1}{2}$$
$$\left(-\infty, \frac{1}{2}\right)$$

$\frac{1}{2}$

97.
$$2x+3(2x+3)\le 7(x+1)+1$$
$$2x+\mathbf{3}(2x)+\mathbf{3}(3)\le \mathbf{7}(x)+\mathbf{7}(1)+1$$
$$2x+6x+9\le 7x+7+1$$
$$8x+9\le 7x+8$$
$$8x+9-\mathbf{7x}\le 7x+8-\mathbf{7x}$$
$$x+9\le 8$$
$$x+9-\mathbf{9}\le 8-\mathbf{9}$$
$$x\le -1$$
$$(-\infty,-1]$$

Look Alikes . . .

Solve each equation and inequality. Write the solution set of each inequality in interval notation and graph it.

99. a.
$$\frac{3}{8}+\frac{b}{3}>\frac{5}{12}\quad \text{, LCD} = 24$$
$$\mathbf{24}\left(\frac{3}{8}\right)+\mathbf{24}\left(\frac{b}{3}\right)>\mathbf{24}\left(\frac{5}{12}\right)$$
$$3(3)+8b>2(5)$$
$$9+8b>10$$
$$9+8b-\mathbf{9}>10-\mathbf{9}$$
$$8b>1$$
$$\frac{8b}{\mathbf{8}}>\frac{1}{\mathbf{8}}$$
$$b>\frac{1}{8}$$
$$\left(\frac{1}{8},\infty\right)$$

99. b.
$$\frac{3}{8}+\frac{b}{3}=\frac{5}{12}\quad \text{, LCD} = 24$$
$$\mathbf{24}\left(\frac{3}{8}\right)+\mathbf{24}\left(\frac{b}{3}\right)=\mathbf{24}\left(\frac{5}{12}\right)$$
$$3(3)+8b=2(5)$$
$$9+8b=10$$
$$9+8b-\mathbf{9}=10-\mathbf{9}$$
$$8b=1$$
$$\frac{8b}{\mathbf{8}}=\frac{1}{\mathbf{8}}$$
$$b=\frac{1}{8}$$

101. a.
$$4\le 2x-6$$
$$2x-6+\mathbf{6}\ge 4+\mathbf{6}$$
$$2x\ge 10$$
$$\frac{2x}{\mathbf{2}}\ge\frac{10}{\mathbf{2}}$$
$$x\ge 5$$
$$[5,\infty)$$

101. b.
$$4\le 2x-6<18$$
$$4+\mathbf{6}\le 2x-6+\mathbf{6}<18+\mathbf{6}$$
$$10\le 2x<24$$
$$\frac{10}{\mathbf{2}}\le\frac{2x}{\mathbf{2}}<\frac{24}{\mathbf{2}}$$
$$5\le x<12$$
$$[5,12)$$

APPLICATIONS

103. GRADES

Analyze

- A student has test scores of 68%, 75% and 79%.
- What should be the score of the next test to have a B (80%) or better?

Assign

Let $x =$ the 4th test score

Form

The average of the four test grades	must be no less than	80.
$\frac{68+75+79+x}{4}$	$\ge$	80

Solve
$$\frac{68+75+79+x}{4}\ge 80$$
$$\mathbf{4}\left(\frac{68+75+79+x}{4}\right)\ge \mathbf{4}(80)$$
$$x+222\ge 320$$
$$x+228-\mathbf{222}\ge 320-\mathbf{222}$$
$$x\ge 98$$

State

The 4th test has to be a 98% or better.

Check
$$\frac{68+75+79+98}{4}\ge 80$$
$$\frac{320}{4}\ge 80$$
$$80\ge 80$$

105. GAS MILEAGE

Analyze

- Two cars average 17 and 19 mpg.
- What should be the mpg of the 3rd car to have an average of 21 mpg or better?

Assign

Let x = the mpg of the 3rd car

Form

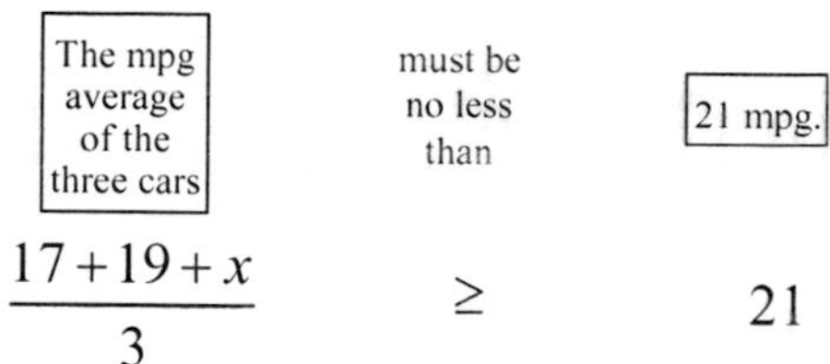

$$\frac{17+19+x}{3} \geq 21$$

Solve

$$\frac{17+19+x}{3} \geq 21$$
$$\mathbf{3}\left(\frac{17+19+x}{3}\right) \geq \mathbf{3}(21)$$
$$x+36 \geq 63$$
$$x+36-\mathbf{36} \geq 63-\mathbf{36}$$
$$x \geq 27$$

State

The 3rd car will need to average 27 mpg or better.

Check

$$\frac{17+19+27}{3} \geq 21$$
$$\frac{63}{3} \geq 21$$
$$21 \geq 21$$

107. GEOMETRY

Analyze

- The perimeter of an equilateral triangle is at most 57 feet.
- What could be the length of a side?

Assign

Let x = the length of one side in feet

Form

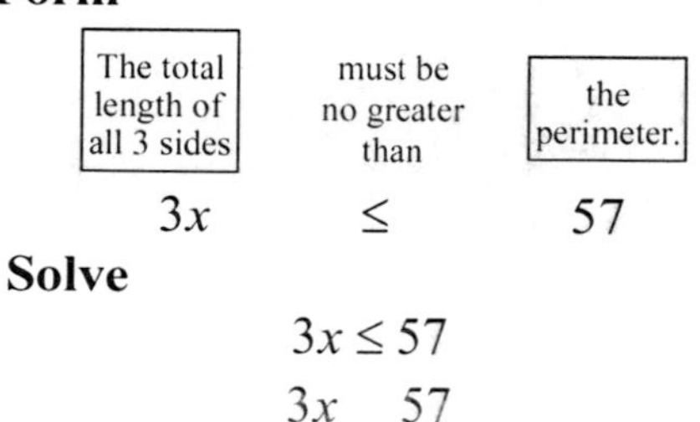

$$3x \leq 57$$

Solve

$$3x \leq 57$$
$$\frac{3x}{\mathbf{3}} \leq \frac{57}{\mathbf{3}}$$
$$x \leq 19$$

State

The length of one side is 19 feet or less.

Check

$$3x \leq 57$$
$$3(19) \leq 57$$
$$57 \leq 57$$

109. COUNTER SPACE

Analyze

- The width is x feet.
- The length is $x+5$ feet.
- The outside perimeter needs to exceed 30 feet.
- What is the value of x?

Assign

Let x = the value

Form

Two widths of the counter	plus	two lengths of the counter	must be no less than	the perimeter of the counter.
$2x$	$+$	$2(x+5)$	$\geq$	30

Solve

$$2x+2(x+5) \geq 30$$
$$2x+2x+10 \geq 30$$
$$4x+10 \geq 30$$
$$4x+10-\mathbf{10} \geq 30-\mathbf{10}$$
$$4x \geq 20$$
$$\frac{4x}{\mathbf{4}} \geq \frac{20}{\mathbf{4}}$$
$$x \geq 5$$

State

The value of x is 5 feet or more.

Check

$$2x+2(x+5) \geq 30$$
$$2(5)+2(5+5) \geq 30$$
$$10+20 \geq 30$$
$$30 \geq 30$$

111. GRADUATIONS

Analyze

- Cost \$18 to rent a cap and gown.
- \$0.80 per announcement.
- Does not want to spend over \$50.
- How many announcements can she order?

Assign

Let $x =$ the number of announcements

Form

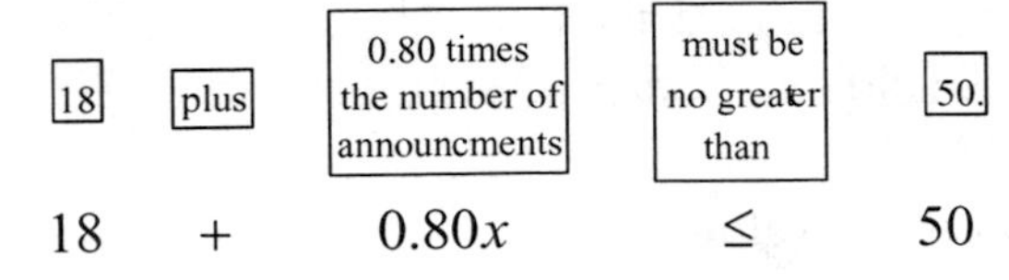

$$18 \quad + \quad 0.80x \quad \leq \quad 50$$

Solve

$$18+0.80x \leq 50$$
$$18+0.80x-\mathbf{18} \leq 50-\mathbf{18}$$
$$0.80x \leq 32$$
$$\frac{0.80x}{\mathbf{0.80}} \leq \frac{32}{\mathbf{0.80}}$$
$$x \leq 40$$

State

She can buy 40 or less announcements.

Check

$$18+0.80x \leq 50$$
$$18+0.80(40) \leq 50$$
$$18+32 \leq 50$$
$$50 \leq 50$$

113. WINDOWS

Analyze

- Triangular window area is no greater than 100 in^2.
- Base must be 16 inches.
- Formula: $A = \frac{1}{2}bh$ or $A = 0.5bh$.
- What is the height of the window?

Assign

Let $h =$ the height of the window in inches

Form

$$A \geq 0.5bh$$

Solve

$$A \geq 0.5bh$$
$$100 \geq 0.5(16)h$$
$$100 \geq 8h$$
$$\frac{100}{\mathbf{8}} \geq \frac{8h}{\mathbf{8}}$$
$$12.5 \geq h$$
$$h \leq 12.5$$

State

The height can be 12.5 inches or less.

Check

$$100 \geq 0.5(16)h$$
$$100 \geq 0.5(16)(12.5)$$
$$100 \geq 100$$

115. NUMBER PUZZLES

Analyze

- What *whole* numbers satisfy the condition: Twice the number decreased by 1 is between 50 and 60?

Assign

Let $x =$ whole # between the 2 numbers

Form

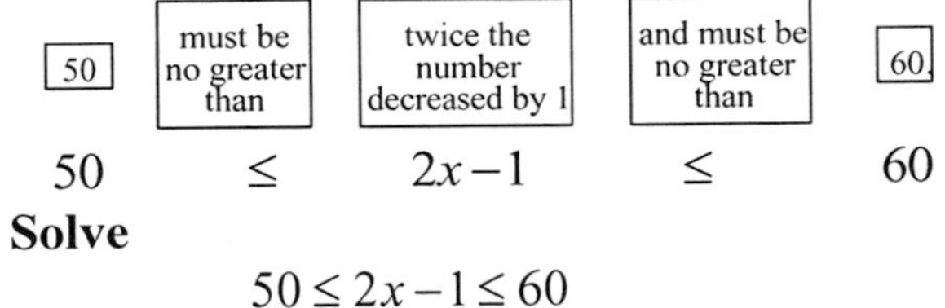

$$50 \quad \leq \quad 2x-1 \quad \leq \quad 60$$

Solve

$$50 \leq 2x-1 \leq 60$$
$$50+\mathbf{1} \leq 2x-1+\mathbf{1} \leq 60+\mathbf{1}$$
$$51 \leq 2x \leq 61$$
$$\frac{51}{\mathbf{2}} \leq \frac{2x}{\mathbf{2}} \leq \frac{61}{\mathbf{2}}$$
$$25.5 \leq x \leq 30.5$$

State

The whole numbers are 26, 27, 28, 29, 30.

Check

$$50 \leq 2x-1 \leq 60$$
$$50 \leq 2(\mathbf{27})-1 \leq 60$$
$$50 \leq 53 \leq 60$$

WRITING

117. Answers will vary.

REVIEW

119.

x	x^2-3
-2	$(-2)^2-3$ $=4-3$ $=\mathbf{1}$
0	0^2-3 $=0-3$ $=\mathbf{-3}$
3	3^2-3 $=9-3$ $=\mathbf{6}$

CHALLENGE PROBLEMS

121. Solve the inequality. Write the solution set in interval notation and graph it.

$$3-x<5<7-x$$
$$3-x+\mathbf{x}<5+\mathbf{x}<7-x+\mathbf{x}$$
$$3<5+x<7$$
$$3-\mathbf{5}<5+x-\mathbf{5}<7-\mathbf{5}$$
$$-2<x<2$$
$$(-2,2)$$

−2 2

CHAPTER 2 REVIEW

SECTION 2.1

Solving Equations Using Properties of Equality

Determine whether the given number is a solution of the equation.

1. 84, $x-34=50$

$$84-34\overset{?}{=}50$$
$$50=50 \text{ True}$$

Since the resulting statement is true, 84 is a solution.

2. 3, $5y+2=12$

$$5(3)+2\overset{?}{=}12$$
$$15+2\overset{?}{=}12$$
$$17=12 \text{ False}$$

Since the resulting statement is false, 3 is not a solution.

3. -30, $\frac{x}{5}=6$

$$\frac{-30}{5}\overset{?}{=}6$$
$$-6=6 \text{ False}$$

Since the resulting statement is false, –30 is not a solution.

4. 2, $|a^2-a-1|=0$

$$|2^2-2-1|\overset{?}{=}0$$
$$|4-2-1|\overset{?}{=}0$$
$$|2-1|\overset{?}{=}0$$
$$|1|\overset{?}{=}0$$
$$1=0 \text{ False}$$

Since the resulting statement is false, 2 is not a solution.

5. -3, $5b-2=3b-8$

$$5(-3)-2\overset{?}{=}3(-3)-8$$
$$-15-2\overset{?}{=}-9-8$$
$$-17=-17 \text{ True}$$

Since the resulting statement is true, –3 is a solution.

6. 1, $\frac{2}{y+1}=\frac{12}{y+1}-5$

$$\frac{2}{1+1}\overset{?}{=}\frac{12}{1+1}-5$$
$$\frac{2}{2}\overset{?}{=}\frac{12}{2}-5$$
$$1\overset{?}{=}6-5$$
$$1=1 \text{ True}$$

Since the resulting statement is true, 1 is a solution.

Fill in the blanks.

7. An **equation** is a statement indicating that two expressions are equal.

8. To solve $x-8=10$ means to find all the values of the variable that make the equation a **true** statement.

Solve each equation and check the result.

9. $x-9=12$

$$x-9\mathbf{+9}=12\mathbf{+9}$$
$$x=21$$

$$x-9=12$$
$$21-9\overset{?}{=}12$$
$$12=12$$

The solution is 21. The result checks.

10. $-y=-32$

$$\frac{-y}{-1}=\frac{-32}{-1}$$
$$y=32$$

$$-y=-32$$
$$-\mathbf{(32)}\overset{?}{=}-32$$
$$-32=-32$$

The solution is 32. The result checks.

11. $a+3.7=-16.9$

$$a+3.7\mathbf{-3.7}=-16.9\mathbf{-3.7}$$
$$a=-20.6$$

$$a+3.7=-16.9$$
$$\mathbf{-20.6}+3.7\overset{?}{=}-16.9$$
$$-16.9=-16.9$$

The solution is -2.6. The result checks.

12. $100=-7+r$

$$100\mathbf{+7}=-7+r\mathbf{+7}$$
$$107=r$$

$$100=-7+r$$
$$100\overset{?}{=}-7+\mathbf{107}$$
$$100=100$$

The solution is 107. The result checks.

13. $120 = 5c$

$$\frac{120}{\mathbf{5}} = \frac{5c}{\mathbf{5}}$$

$$24 = c$$

$$120 = 5c$$

$$120 \stackrel{?}{=} 5(\mathbf{24})$$

$$120 = 120$$

The solution is 24. The result checks.

14.
$$t - \frac{1}{3} = \frac{3}{7}$$

$$t - \frac{1}{3} + \frac{\mathbf{1}}{\mathbf{3}} = \frac{3}{7} + \frac{\mathbf{1}}{\mathbf{3}}$$

$$t = \frac{3}{7} \cdot \frac{\mathbf{3}}{\mathbf{3}} + \frac{1}{3} \cdot \frac{\mathbf{7}}{\mathbf{7}}$$

$$t = \frac{9+7}{21}$$

$$t = \frac{16}{21}$$

The solution is $\frac{16}{21}$. The result checks.

15.
$$\frac{4}{3}t = -12$$

$$\frac{\mathbf{3}}{\mathbf{4}} \cdot \frac{4}{3}t = \frac{\mathbf{3}}{\mathbf{4}} \cdot (-12)$$

$$t = -\frac{3 \cdot 3 \cdot \cancel{4}^{1}}{\cancel{4}_{1}}$$

$$t = -9$$

$$\frac{4}{3}t = -12$$

$$\frac{4}{3}(\mathbf{-9}) \stackrel{?}{=} -12$$

$$\frac{4}{\cancel{3}_{1}} \cdot \frac{-3 \cdot \cancel{3}^{1}}{1} \stackrel{?}{=} -12$$

$$-12 = -12$$

The solution is -9. The result checks.

16.
$$3 = \frac{q}{-2.6}$$

$$\frac{\mathbf{-2.6}}{\mathbf{1}} \cdot 3 = \frac{\mathbf{-2.6}}{\mathbf{1}} \cdot \frac{q}{-2.6}$$

$$-7.8 = q$$

The solution is -7.8. The result checks.

17. $6b = 0$

$$\frac{6b}{\mathbf{6}} = \frac{0}{\mathbf{6}}$$

$$b = 0$$

$$6b = 0$$

$$6 \cdot \mathbf{0} \stackrel{?}{=} 0$$

$$0 = 0$$

The solution is 0. The result checks.

18.
$$\frac{15}{16}s = -3$$

$$\frac{\mathbf{16}}{\mathbf{15}} \cdot \left(\frac{15}{16}s\right) = \frac{\mathbf{16}}{\mathbf{15}} \cdot (-3)$$

$$s = -\frac{\cancel{3}^{1} \cdot 16}{\cancel{3}_{1} \cdot 5}$$

$$s = -\frac{16}{5}$$

The solution is $\frac{16}{5}$. The result checks.

SECTION 2.2
More about Solving Equations

Solve each equation and check the answer.

19.
$$5x + 4 = 14$$

$$5x + 4 - \mathbf{4} = 14 - \mathbf{4}$$

$$5x = 10$$

$$\frac{5x}{\mathbf{5}} = \frac{10}{\mathbf{5}}$$

$$x = 2$$

$$5x + 4 = 14$$

$$5(2) + 4 \stackrel{?}{=} 14$$

$$10 + 4 \stackrel{?}{=} 14$$

$$14 = 14$$

The solution is 2. The result checks.

20.
$$98.6 - t = 129.2$$

$$98.6 - t - \mathbf{98.6} = 129.2 - \mathbf{98.6}$$

$$-t = 30.6$$

$$\frac{-t}{\mathbf{-1}} = \frac{30.6}{\mathbf{-1}}$$

$$t = -30.6$$

The solution is -30.6. The result checks.

21.
$$\frac{n}{5} + (-2) = 4$$

$$\frac{n}{5} + (-2) + \mathbf{2} = 4 + \mathbf{2}$$

$$\frac{n}{5} = 6$$

$$\frac{\mathbf{5}}{\mathbf{1}} \cdot \frac{n}{5} = \frac{\mathbf{5}}{\mathbf{1}} \cdot (6)$$

$$n = 30$$

$$\frac{n}{5} + (-2) = 4$$

$$\frac{\mathbf{30}}{5} + (-2) \stackrel{?}{=} 4$$

$$6 + (-2) \stackrel{?}{=} 4$$

$$4 = 4$$

The solution is 30. The result checks.

22. $$\frac{b-5}{4}=-6$$
$$\left(\frac{\mathbf{4}}{\mathbf{1}}\right)\cdot\frac{b-5}{4}=\left(\frac{\mathbf{4}}{\mathbf{1}}\right)\cdot(-6)$$
$$b-5=-24$$
$$b-5+\mathbf{5}=-24+\mathbf{5}$$
$$b=-19$$
The solution is -19. The solution checks.

23. $$5(2x-4)-5x=0$$
$$10x-20-5x=0$$
$$5x-20=0$$
$$5x-20+\mathbf{20}=0+\mathbf{20}$$
$$5x=20$$
$$\frac{5x}{\mathbf{5}}=\frac{20}{\mathbf{5}}$$
$$x=4$$

$$5(2x-4)-5x=0$$
$$5(2\cdot4-4)-5\cdot4\overset{?}{=}0$$
$$5(4)-20\overset{?}{=}0$$
$$20-20\overset{?}{=}0$$
$$0=0$$
The solution is 4. The solution checks.

24. $$-2(x-5)=5(-3x+4)+3$$
$$-2x+10=-15x+20+3$$
$$-2x+10=-15x+23$$
$$-2x+10-\mathbf{10}=-15x+23-\mathbf{10}$$
$$-2x=-15x+13$$
$$-2x+\mathbf{15x}=-15x+13+\mathbf{15x}$$
$$13x=13$$
$$\frac{13x}{\mathbf{13}}=\frac{13}{\mathbf{13}}$$
$$x=1$$

The solution is 1. The solution checks.

25. $$\frac{3}{4}=\frac{1}{2}+\frac{d}{5} \qquad \text{LCD}=20$$
$$\mathbf{20}\cdot\frac{3}{4}=\mathbf{20}\cdot\left(\frac{1}{2}+\frac{d}{5}\right)$$
$$20\cdot\frac{3}{4}=20\cdot\left(\frac{1}{2}\right)+20\cdot\left(\frac{d}{5}\right)$$
$$15=10+4d$$
$$15-\mathbf{10}=10+4d-\mathbf{10}$$
$$5=4d$$
$$\frac{5}{\mathbf{4}}=\frac{4d}{\mathbf{4}}$$
$$\frac{5}{4}=d$$

$$\frac{3}{4}=\frac{1}{2}+\frac{d}{5}$$
$$\frac{3}{4}\overset{?}{=}\frac{1}{2}+\frac{\frac{5}{4}}{5}$$
$$\frac{3}{4}\overset{?}{=}\frac{1}{2}+\frac{5}{4}\cdot\frac{1}{5}$$
$$\frac{3}{4}\overset{?}{=}\frac{1}{2}+\frac{1}{4}$$
$$\frac{3}{4}=\frac{3}{4}$$

The solution is $\frac{5}{4}$. The solution checks.

26. $$\frac{5(7-x)}{4} = 2x-3 \quad \text{LCD} = 4$$

$$\mathbf{4}\cdot\left(\frac{5(7-x)}{4}\right) = \mathbf{4}\cdot(2x-3)$$

$$4\cdot\left(\frac{5(7-x)}{4}\right) = 4\cdot(2x) - 4\cdot(3)$$

$$5(7-x) = 8x-12$$

$$35-5x = 8x-12$$

$$35-5x+\mathbf{12} = 8x-12+\mathbf{12}$$

$$47-5x = 8x$$

$$47-5x+5\boldsymbol{x} = 8x+5\boldsymbol{x}$$

$$47 = 13x$$

$$\frac{47}{\mathbf{13}} = \frac{47x}{\mathbf{13}}$$

$$\frac{47}{13} = x$$

The solution is $\frac{47}{13}$. The solution checks.

27. $$\frac{3(2-c)}{2} = \frac{-2(2c+3)}{5}, \quad \text{LCD} = 10$$

$$\mathbf{10}\cdot\left(\frac{3(2-c)}{2}\right) = \mathbf{10}\cdot\left(\frac{-2(2c+3)}{5}\right)$$

$$5[3(2-c)] = 2[-2(2c+3)]$$

$$15(2-c) = -4(2c+3)$$

$$30-15c = -8c-12$$

$$30-15c+\mathbf{12} = -8c-12+\mathbf{12}$$

$$42-15c = -8c$$

$$42-15c+\mathbf{15c} = -8c+\mathbf{15c}$$

$$42 = 7c$$

$$\frac{42}{\mathbf{7}} = \frac{7c}{\mathbf{7}}$$

$$6 = c$$

$$\frac{3(2-c)}{2} = \frac{-2(2c+3)}{5}$$

$$\frac{3(2-6)}{2} \stackrel{?}{=} \frac{-2(2\cdot 6+3)}{5}$$

$$\frac{3(-4)}{2} \stackrel{?}{=} \frac{-2(15)}{5}$$

$$\frac{-12}{2} \stackrel{?}{=} \frac{-30}{5}$$

$$-6 = -6$$

The solution is 6. The solution checks.

28. $$\frac{b}{3}+\frac{11}{9}+3b = -\frac{5}{6}b, \quad \text{LCD} = 18$$

$$\mathbf{18}\cdot\left(\frac{b}{3}+\frac{11}{9}+3b\right) = \mathbf{18}\cdot\left(-\frac{5}{6}b\right)$$

$$18\cdot\left(\frac{b}{3}\right)+18\cdot\left(\frac{11}{9}\right)+18\cdot(3b) = 18\cdot\left(-\frac{5}{6}b\right)$$

$$6b+22+54b = -15b$$

$$60b+22 = -15b$$

$$60b+22-\mathbf{60b} = -15b-\mathbf{60b}$$

$$22 = -75b$$

$$\frac{22}{\mathbf{-75}} = \frac{-75b}{\mathbf{-75}}$$

$$-\frac{22}{75} = b$$

The solution is $-\frac{22}{75}$. The solution checks.

29. $$0.15(x+2)+0.3 = 0.35x-0.4$$

$$\mathbf{100}[0.15(x+2)+0.3] = \mathbf{100}[0.35x-0.4]$$

$$100[0.15(x+2)]+100(0.3) = 100(0.35x)-100(0.4)$$

$$15(x+2)+30 = 35x-40$$

$$15x+30+30 = 35x-40$$

$$15x+60 = 35x-40$$

$$15x+60-\mathbf{15x} = 35x-40-\mathbf{15x}$$

$$60 = 20x-40$$

$$60+\mathbf{40} = 20x-40+\mathbf{40}$$

$$100 = 20x$$

$$\frac{100}{\mathbf{20}} = \frac{20x}{\mathbf{20}}$$

$$5 = x$$

$$0.15(x+2)+0.3 = 0.35x-0.4$$

$$0.15(5+2)+0.3 \stackrel{?}{=} 0.35(5)-0.4$$

$$0.15(7)+0.3 \stackrel{?}{=} 1.75-0.4$$

$$1.05+0.3 \stackrel{?}{=} 1.35$$

$$1.35 = 1.35$$

The solution is 5. The solution checks.

30. $$0.5-0.02(y-2)=0.16+0.36y$$
$$\mathbf{100}[0.5-0.02(y-2)]=\mathbf{100}[0.16+0.36y]$$
$$100(0.5)-100[0.02(y-2)]=100(0.16)+100(0.36y)$$
$$50-2(y-2)=16+36y$$
$$50-2y+4=16+36y$$
$$54-2y=16+36y$$
$$54-2y+\mathbf{2y}=16+36y+\mathbf{2y}$$
$$54=16+38y$$
$$54-\mathbf{16}=16+38y-\mathbf{16}$$
$$38=38y$$
$$\frac{38}{\mathbf{38}}=\frac{38y}{\mathbf{38}}$$
$$1=y$$

The solution is 1. The solution checks.

31. $$3(a+8)=6(a+4)-3a$$
$$3a+24=6a+24-3a$$
$$3a+24=3a+24$$
$$3a+24-\mathbf{3a}=3a+24-\mathbf{3a}$$
$$24=24$$

The terms involving a drop out and the result is true. This means that *all real numbers* are solutions and this equation is called an identity.

32. $$2(y+10)+y=3(y+8)$$
$$2y+20+y=3y+24$$
$$3y+20=3y+24$$
$$3y+20-\mathbf{3y}=3y+24-\mathbf{3y}$$
$$20=24$$

The terms involving x drop out and the result is false. This means the equation has *no solution* and it is a contradiction.

SECTION 2.3
Application of Percent

33. Fill in the blanks
   a. **Percent** means parts per one hundred.
   b. When the price of an item is reduced, we call the amount of the reduction a **discount** .
   c. An employee who is paid a **commission** is paid a percent of the goods or services that he or she sells.

34. 4.81 is 2.5% of what number?
$$4.81=0.025\cdot x$$
$$\frac{4.81}{\mathbf{0.025}}=\frac{0.025x}{\mathbf{0.025}}$$
$$192.4=x$$
192.4 is the solution.

35. What number is 15% of 950?
$$x=0.15\cdot 950$$
$$x=142.5$$
142.5 is the solution.

36. What percent of 410 is 49.2?
$$x\cdot 410=49.2$$
$$\frac{410x}{\mathbf{410}}=\frac{49.2}{\mathbf{410}}$$
$$x=0.12$$
$$x=12\%$$
12% is the solution.

37. INTERNET USERS
   a. What percent did not use the Internet?
$$100\%-71.2\%=28.8\%$$
   The percent of non-users is 28.8%.
   b. How many people used the Internet?
   Let x = number of people using the Internet
   What number is 71.2% of 310 million?
$$x=0.712\cdot 310$$
$$x=220.72$$
$$x\approx 221$$
   About 221 million people used the Internet.

38. COST OF LIVING
   Let x = amt of cost of living
   What number is 3.5% of $764?
$$x=0.035\cdot 764$$
$$x=26.74$$
   The increase is $26.74.

39. FAMILY BUDGETS

Let $x =$ percent of housing cost

625 is what percent of 1,890?

$625 = x \cdot 1{,}890$

$$\frac{625}{\mathbf{1{,}890}} = \frac{1{,}890x}{\mathbf{1{,}890}}$$

$0.33068 \approx x$

$33\% \approx x$

About 33% of the income is spent on housing. No, they are not within the range.

40. DISCOUNTS

Let $x =$ the original cost of food processor

148.50 is 33% of what number?

$148.50 = 0.33 \cdot x$

$$\frac{148.50}{\mathbf{0.33}} = \frac{0.33x}{\mathbf{0.33}}$$

$450 = x$

$450 was the original cost of the food processor.

41. TUPPERWARE

Let $x =$ amt of the commission

What number is 25% of $600?

$x = 0.25 \cdot 600$

$x = 150$

The commission is $150.

42. COLLECTIBLES

Step 1: $\$100 - \$6 = \$94$

This is the amount of increase.

Step 2: 94 is what percent of 6?

$94 = x \cdot 6$

$$\frac{94}{\mathbf{6}} = \frac{6x}{\mathbf{6}}$$

$15.\overline{6} \approx x$

$15.67 \approx x$

The percent of increase is 1,567%.

SECTION 2.4
Formulas

43. SHOPPING

Let $m =$ the amount of markup

$r = c + m$

$395 = 219 + m$

$395 - \mathbf{219} = 219 + m - \mathbf{219}$

$176 = m$

$176 is the amount of the markup.

44. RESTAURANTS

Let $r =$ the revenue

$r = p + e$

$13{,}500 = p + e$

$13{,}500 - \mathbf{1{,}700} = 1{,}700 + e - \mathbf{1{,}700}$

$11{,}800 = e$

$11,800 is the amount of expenses.

45. SNAILS

Let $t =$ the number of minutes

Rate •	time =	distance
2.5	t	20

$2.5 \cdot t = 20$

$$\frac{2.5t}{\mathbf{2.5}} = \frac{20}{\mathbf{2.5}}$$

$t = 8$

It will take 8 minutes.

46. CERTIFICATES OF DEPOSIT

Let $r =$ the annual interest rate

Principal •	rate	• time =	interest
26,000	r	1	1,170

$26{,}000 \cdot r \cdot 1 = 1{,}170$

$26{,}000r = 1{,}170$

$$\frac{26{,}000r}{\mathbf{26{,}000}} = \frac{1{,}170}{\mathbf{26{,}000}}$$

$r = 0.045$

$r\% = 4.5\%$

The rate is 4.5% annually.

47. JEWELRY

$$C = \frac{5}{9}(F - 32)$$
$$1,065 = \frac{5}{9}(F - 32)$$
$$\frac{\mathbf{9}}{\mathbf{5}} \cdot (1,065) = \frac{\mathbf{9}}{\mathbf{5}} \cdot \frac{5}{9}(F - 32)$$
$$1,917 = F - 32$$
$$1,917 + \mathbf{32} = F - 32 + \mathbf{32}$$
$$1,949 = F$$

The temperature is $1,949^{\circ}$ Fahrenheit.

48. CAMPING

a. $P = 2l + 2w$
$$= 2 \cdot 60 + 2 \cdot 24$$
$$= 120 + 48$$
$$= 168$$
The perimeter is 168 inches.

b. $A = lw$
$$= 60 \cdot 24$$
$$= 1,440$$
The area is 1,440 square inches.

c. $V = lwh$
$$= 60 \cdot 24 \cdot 3$$
$$= 4,320$$
The volume is 4,320 cubic inches.

49. $A = \frac{1}{2}bh$
$$= \frac{1}{2} \cdot 17 \cdot 9$$
$$= \frac{1}{2} \cdot 153$$
$$= 76.5$$
The area is 76.5 square meters.

50. $A = \frac{1}{2}h(b + d)$
$$= \frac{1}{2} \cdot 12 \cdot (11 + 13)$$
$$= \frac{1}{2} \cdot 12 \cdot 24$$
$$= 6 \cdot 24$$
$$= 144$$
The area is 144 square inches.

51. a. $C = 2\pi r \quad , \pi = 3.141592654$
$$= 2 \cdot 3.141592654 \cdot 8$$
$$\approx 50.265$$
$$\approx 50.27$$
The circumference is about 50.27 cm.

b. $A = \pi r^2 \quad , \pi = 3.141592654$
$$= 3.141592654 \cdot 8 \cdot 8$$
$$\approx 201.061$$
$$\approx 201$$
The area is about 201 sq cm.

52. $V = \pi r^2 h \quad , \pi = 3.141592654$
$$= 3.141592654 \cdot (0.5)^2 \cdot 12$$
$$= 3.141592654 \cdot 0.25 \cdot 12$$
$$\approx 9.42$$
$$\approx 9.4$$
The volume is about 9.4 cubic feet.

53. HALLOWEEN

$$V = \frac{4}{3}\pi r^3 \quad , \pi = 3.1415926542$$
$$= \frac{4}{3} \cdot 3.1415926542 \cdot (4.5)^3$$
$$= \frac{4}{3} \cdot 3.1415926542 \cdot (91.125)$$
$$\approx 381.70$$
$$\approx 381.7$$
The volume is about 381.7 cubic inches.

54. $V = \frac{1}{3}Bh$
$$= \frac{1}{3} \cdot 6 \cdot 6 \cdot 10$$
$$= 120$$
The volume is 120 cubic feet.

Solve each formula for the specified variable.

55. $A = 2\pi rh$ for h
$$A = 2\pi rh$$
$$\frac{A}{\mathbf{2\pi r}} = \frac{2\pi rh}{\mathbf{2\pi r}}$$
$$\frac{A}{2\pi r} = h$$
$$h = \frac{A}{2\pi r}$$

56. $$A - BC = \frac{G-K}{3} \quad \text{for } G$$

$$3(A - BC) = \mathbf{3}\left(\frac{G-K}{3}\right)$$

$$3(A - BC) = G - K$$

$$3(A - BC) + \boldsymbol{K} = G - K + \boldsymbol{K}$$

$$3(A - BC) + K = G$$

$$G = 3(A - BC) + K$$

or

$$G = 3A - 3BC + K$$

57. $$C = \frac{1}{4}s(t-d) \quad \text{for } t$$

$$\mathbf{4} \cdot C = \mathbf{4} \cdot \left[\frac{1}{4}s(t-d)\right]$$

$$4C = s(t-d)$$

$$\frac{4C}{\boldsymbol{s}} = \frac{s(t-d)}{\boldsymbol{s}}$$

$$\frac{4C}{s} = t - d$$

$$\frac{4C}{s} + \boldsymbol{d} = t - d + \boldsymbol{d}$$

$$\frac{4C}{s} + d = t$$

$$t = \frac{4C}{s} + d \text{ or } \frac{4C + ds}{s}$$

58. $$4y - 3x = 16 \quad \text{for } y$$

$$4y - 3x + \mathbf{3x} = 16 + \mathbf{3x}$$

$$4y = 3x + 16$$

$$\frac{4y}{\mathbf{4}} = \frac{3x}{\mathbf{4}} + \frac{16}{\mathbf{4}}$$

$$y = \frac{3}{4}x + 4$$

SECTION 2.5

Problem Solving

59. SOUND SYSTEMS

Analyze

- A 45-foot wire is cut into 3 pieces.
- One piece is 15 feet long.
- 2nd piece is 2 feet less than 3 times the 3rd
- Find the shorter piece.

Assign

Let x = length of 3rd piece in feet

$3x - 2$ = length of 2nd piece in feet

Form

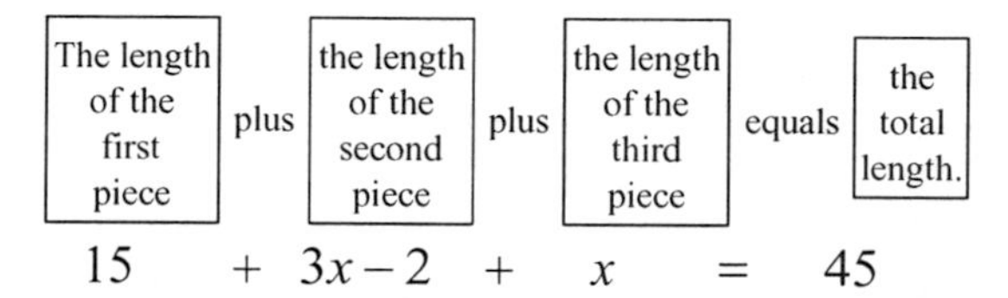

$$15 \quad + \quad 3x - 2 \quad + \quad x \quad = \quad 45$$

Solve

$$15 + 3x - 2 + x = 45$$

$$4x + 13 = 45$$

$$4x + 13 - \mathbf{13} = 45 - \mathbf{13}$$

$$4x = 32$$

$$\frac{4x}{\mathbf{4}} = \frac{32}{\mathbf{4}}$$

$$x = 8$$

Second length

$$3x - 2 = 3(8) - 2$$

$$= 24 - 2$$

$$= 22$$

State

The length of the shorter piece is 8 ft.

Optional answers for checking.
Length of second piece is 22 ft.
Length of first piece is 15 ft.

Check

$15 \text{ ft} + 22 \text{ ft} + 8 \text{ ft} = 45 \text{ ft}$

The results check.

60. SIGNING PETITIONS

Analyze

- Collector makes \$50 a day.
- Collector makes \$2.25 per verified signature.
- How many signature are needed to earn \$500 a day?

Assign

Let x = number of signatures

Form

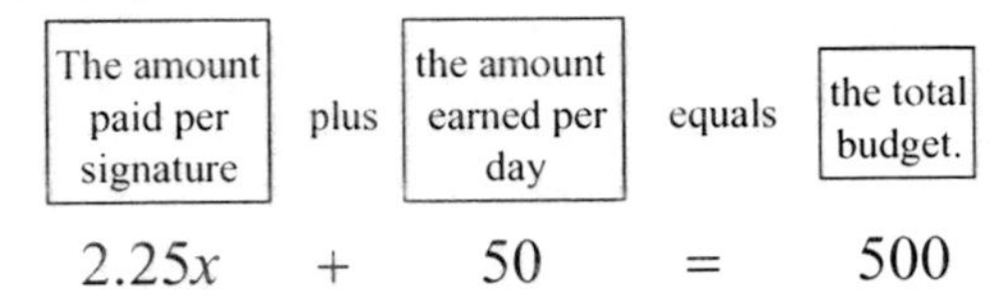

$$2.25x \quad + \quad 50 \quad = \quad 500$$

Solve

$$2.25x + 50 = 500$$
$$2.25x + 50 - \mathbf{50} = 500 - \mathbf{50}$$
$$2.25x = 450$$
$$\frac{2.25x}{\mathbf{2.25}} = \frac{450}{\mathbf{2.25}}$$
$$x = 200$$

State

200 signatures are needed.

Check

$$2.25x + 50 = 500$$
$$2.25(200) + 50 = 500$$
$$450 + 50 = 500$$
$$500 = 500$$

The result checks.

61. LOTTERY WINNINGS

Analyze

- \$1,800,000 was the lump sum after taxes.
- 28% of the original prize was federal income tax.
- What was the original cash prize?

Assign

Let x = the original amount

Form

The original amount	minus	the amount of federal taxes	equals	the lum sum.
x	$-$	$0.28x$	$=$	1,800,000

Solve

$$x - 0.28x = 1,800,000$$
$$0.72x = 1,800,000$$
$$\frac{0.72x}{\mathbf{0.72}} = \frac{1,800,000}{\mathbf{0.72}}$$
$$x = 2,500,000$$

State

The original price was \$2,500,000.

Check

$$\$2,500,000 - 0.28(\$2,500,000)$$
$$= \$2,500,000 - \$700,000$$
$$= \$1,800,000$$

The result checks.

62. NASCAR

Analyze

- The sum of the two consecutive odd numbers is 88.
- The smaller number belongs to Labonte.
- Find the two numbers.

Assign

Let $x = 1^{st}$ consecutive odd integer

$x + 2 = 2^{nd}$ consecutive odd integer

Form

The first consecutive odd integer	plus	the second consecutive odd integer	equals	the total of the two odd integers.
x	$+$	$x+2$	$=$	88

Solve

$$x + x + 2 = 88$$
$$2x + 2 = 88$$
$$2x + 2 - \mathbf{2} = 88 - \mathbf{2}$$
$$2x = 86$$
$$\frac{2x}{\mathbf{2}} = \frac{86}{\mathbf{2}}$$
$$x = 43$$

State

43 is Labonte's number.

$43 + 2 = 45$ is Petty's number.

Check

$$x + x + 2 = 88$$
$$43 + 43 + 2 = 88$$
$$88 = 88$$

The results check.

63. ART HISTORY

Analyze

- Perimeter of the rectangular painting is 109.5 ($109\frac{1}{2}$) inches.
- The length is 5 inches more than its width.
- What are the dimensions of the painting?

Assign

Let x = width of the painting in inches

$x+5$ = length of the painting in inches

Form

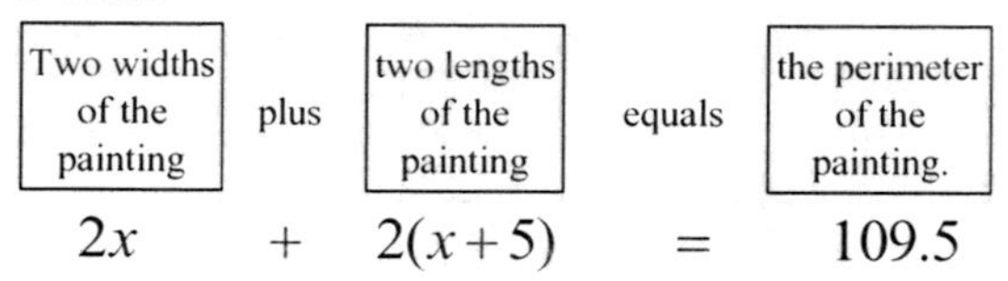

$$2x \quad + \quad 2(x+5) \quad = \quad 109.5$$

Solve

$$2x+2(x+5)=109.5$$
$$2x+2x+10=109.5$$
$$4x+10=109.5$$
$$4x+10-\mathbf{10}=109.5-\mathbf{10}$$
$$4x=99.5$$
$$\frac{4x}{\mathbf{4}}=\frac{99.5}{\mathbf{4}}$$
$$x=24.875$$
or
$$x=24\tfrac{7}{8}$$

length
$$x+5=24.875+5$$
$$=29.875$$
or
$$=29\tfrac{7}{8}$$

State

The width of the painting is 24.875 inches.
The length of the painting is 29.875 inches.

Check

$$2x+2(x+5)=109.5$$
$$2(24.875)+2(29.875)=109.5$$
$$49.75+59.75=109.5$$
$$109.5=109.5$$

The results check.

64. GEOMETRY

Analyze

- Internal measure of the 3 angles of a triangle is 180°.
- The measure of the vertex angle is 27°?
- What is the measure of each base angle which are equal to each other.

Assign

Let x = measure of each base angle

Form

The measure of the vertex angle	plus	the measure of the 2 base angles	equals	the measure of the angles of a triangle.
27	+	$2x$	=	180

Solve

$$27+2x=180$$
$$27+2x-\mathbf{27}=180-\mathbf{27}$$
$$2x=153$$
$$\frac{2x}{\mathbf{2}}=\frac{153}{\mathbf{2}}$$
$$x=76.5$$

State

Measure of each base angle is 76.5°.

Check

$$27+2x=180$$
$$27+2(76.5)=180$$
$$27+153=180$$
$$180=180$$

The result checks.

SECTION 2.6
More on Problem Solving

65. INVESTMENT INCOME

Analyze

- \$27,000 is the total investment.
- CD investment at 7% annual interest.
- CMF investment at 9% annual interest.
- \$2,110 is the first-year combined interest.
- How much is invested at each rate?

Assign

Let x = amount invested at 7%

$27{,}000 - x$ = amount invested at 9%

Form

	Principal	• rate	• time =	interest
CD	x	0.07	1	**0.07x**
CMF	$27{,}000 - x$	0.09	1	**0.09(27,000 − x)**
		Total:		**2,110**

The interest earned at 7%	plus	the interest earned at 9%	equals	the total interest.
$0.07x$	$+$	$0.09(27{,}000-x)$	$=$	$2{,}110$

Solve

$$0.07x + 0.09(27{,}000 - x) = 2{,}110$$
$$\mathbf{100}[0.07x + 0.09(27{,}000 - x)] = \mathbf{100}(2{,}110)$$
$$100(0.07x) + 100[0.09(27{,}000 - x)] = 100(2{,}110)$$
$$7x + 9(27{,}000 - x) = 211{,}000$$
$$7x + 243{,}000 - 9x = 211{,}000$$
$$-2x + 243{,}000 = 211{,}000$$
$$-2x + 243{,}000 - \mathbf{243{,}000} = 211{,}000 - \mathbf{243{,}000}$$
$$-2x = -32{,}000$$
$$\frac{-2x}{\mathbf{-2}} = \frac{-32{,}000}{\mathbf{-2}}$$
$$x = 16{,}000$$

State

\$16,000 was invested at 7%.
$\$27{,}000 - \$16{,}000 = \$11{,}000$ was the amount invested at 9%.

Check

The CD investment earned 0.07(\$16,000), or \$1,120.
The CMF earned 0.09(\$11,000), or \$990. Since the total return was $\$1{,}120 + \$990 = \$2{,}110$, the results check.

66. WALKING AND BICYCLING

Analyze

- Man walks toward biker at 3 mph.
- Friend bikes toward walker at 12 mph.
- Both leave at the same time.
- Combined distance is 5 miles.
- How long before they meet at each other?

Assign

Let t = time in hours they meet to each other

Form

	rate	• time	= distance
Walker	3	t	**3t**
Biker	12	t	**12t**
		Total:	**5**

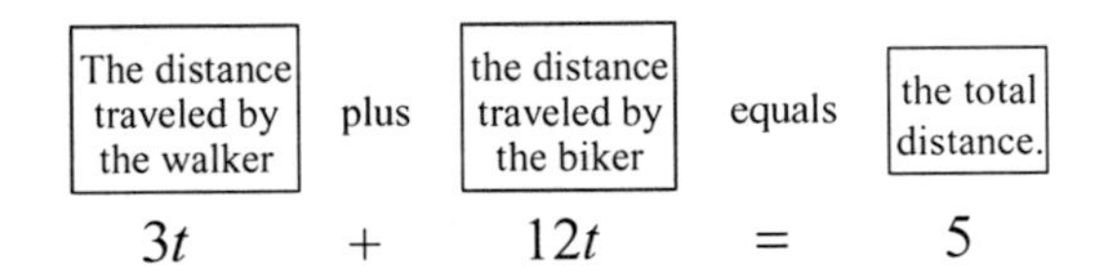

$3t \quad + \quad 12t \quad = \quad 5$

Solve

$$3t + 12t = 5$$
$$15t = 5$$
$$\frac{15t}{\mathbf{15}} = \frac{5}{\mathbf{15}}$$
$$t = \frac{1}{3}$$

State

After $\frac{1}{3}$ hour or 20 minutes they will meet.

Check

The distance traveled by walker is $3\left(\frac{1}{3}\right)$, or 1 mi.

The distance traveled by biker is $12\left(\frac{1}{3}\right)$, or 4 mi.

Since the total was $1 \text{ mi} + 4 \text{ mi} = 5 \text{ mi}$, the result checks.

67. AIRPLANES

Analyze

- Propeller plane's rate is 180 mph.
- Propeller plane leaves base 2.5 hours earlier.
- Jet's rate is 450 mph.
- Both leave from the same place.
- Both travel in the same direction.
- How long before the jet catches the prop plane?

Assign

Let t = time in hours before jet catches propeller plane

Form

	rate	• time	= distance
Prop	180	$t+2.5$	$\mathbf{180(t+2.5)}$
Jet	450	t	$\mathbf{450t}$
		Distance:	**same**

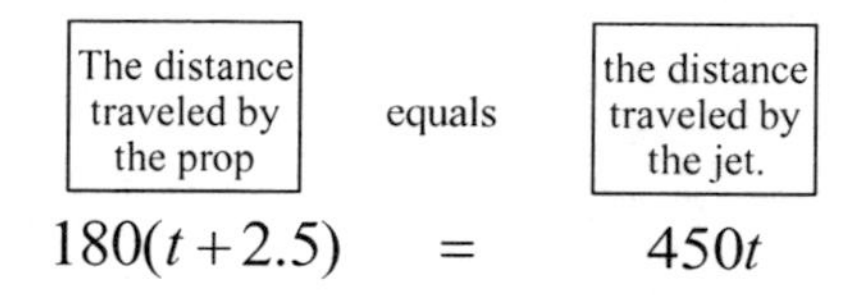

$$180(t+2.5) \quad = \quad 450t$$

Solve

$$180(t+2.5)=450t$$
$$180t+450=450t$$
$$180t+450-\mathbf{180t}=450t-\mathbf{180t}$$
$$450=270t$$
$$\frac{450}{\mathbf{270}}=\frac{270t}{\mathbf{270}}$$
$$1\tfrac{2}{3}=t$$

or

$$1\text{ hr }40\text{ minutes}=t$$

State

After $1\frac{2}{3}$ hours, the jet will catch the prop plane.

Check

Distance traveled by prop is $180(2\frac{1}{2}+1\frac{2}{3}=4\frac{1}{6})$, or 750 mi.

Distance traveled by jet is $450(1\frac{2}{3})$, or 750 mi.

Since the each distance is the same 750 mi = 750 mi, the result checks.

68. AUTOGRAPHS

Analyze

- Kesha has 8 more TV autographs than movie star autographs.
- Each TV autograph is worth \$75.
- Each movie star autograph is worth \$250.
- Total collection is worth \$1,900.
- How many of each does she have?

Assign

Let x = the # of movie star autographs

$x+8$ = the # of TV autographs

Form

	Number	• Value	= Total value
Celebrity	x	250	$\mathbf{250x}$
TV	$x+8$	75	$\mathbf{75(x+8)}$
		Total:	**1,900**

The value of the celebrity autographs	plus	the value of the TV autographs	equals	the total value of the autographs.
$250x$	$+$	$75(x+8)$	$=$	$1,900$

Solve

$$250x+75(x+8)=1,900$$
$$250x+75x+600=1,900$$
$$325x+600=1,900$$
$$325x+600-\mathbf{600}=1,900-\mathbf{600}$$
$$325x=1,300$$
$$\frac{325x}{\mathbf{325}}=\frac{1,300}{\mathbf{325}}$$
$$x=4$$

TV

$$x+8=4+8$$
$$=12$$

State

She has 4 movie star autographs.

She has 12 TV celebrity autographs.

Check

The value of the movie is 4(\$250), or \$1,000.
The value of the TV is 12(\$75), or \$900.
Since the total was \$1,000 + \$900 = \$1,900, the results check.

69. MIXTURES

Analyze

- Candy sells for $0.90 per lb.
- Gumdrops sells for $1.50 per lb.
- A mixture of the two sells for $1.20 per lb.
- Blend of 20 lb is needed.
- How many pounds of each is needed?

Assign

Let $x =$ the pounds of candy

$20 - x =$ the pounds of gumdrops

Form

	Number	• Value	= Total value
Candy	x	0.90	$\mathbf{0.90x}$
Gumdrops	$\mathbf{20 - x}$	1.50	$\mathbf{1.50(20 - x)}$
Mixture	20	1.20	**24**

The value of the candy plus the value of the gumdrops equals the total value of the mixture.

$$0.90x + 1.50(20 - x) = 24$$

Solve

$$0.90x + 1.50(20 - x) = 24$$
$$0.90x + 30 - 1.50x = 24$$
$$-0.60x + 30 = 24$$
$$-0.60x + 30 - \mathbf{30} = 24 - \mathbf{30}$$
$$-0.60x = -6$$
$$\frac{-0.60x}{\mathbf{-0.60}} = \frac{-6}{\mathbf{-0.60}}$$
$$x = 10$$

gumdrops

$$20 - x = 20 - 10$$
$$= 10$$

State

10 lb of the candy will be needed.

10 lb of gumdrops will be needed.

Check

The value of the candy is 10($0.90), or $9.
The value of the gumdrops is 10($1.50), or $15.
The value of the blend is 20($1.20), or $24.
Since the total was $\$9 + \$15 = \$24$,
the results check.

70. ELIMINATING MILDEW

Analyze

- Weak solution is 2% fungicide.
- Start with 4 gallons of 5% fungicide, the strong solution.
- How many gallons 2% fungicide is needed to make a solution of 4% fungicide?

Assign

Let $x =$ Number of gallons of 2% fungicide

Form

	Amount	• Strength	= Amount of pure fungicide
Weak	x	0.02	$\mathbf{0.02x}$
Strong	4	0.05	**0.20**
4% solution	$\mathbf{x + 4}$	0.04	$\mathbf{0.04(x + 4)}$

The pure fungicide in the 2% solution plus the pure fungicide in the 5% solution equals the pure fungicide in the 4% solution.

$$0.02x + 0.20 = 0.04(x + 4)$$

Solve

$$0.02x + 0.20 = 0.04(x + 4)$$
$$\mathbf{100}(0.02x + 0.20) = \mathbf{100}[0.04(x + 4)]$$
$$100(0.02x) + 100(0.20) = 100[(0.04(x + 4)]$$
$$2x + 20 = 4(x + 4)$$
$$2x + 20 = 4x + 16$$
$$2x + 20 - \mathbf{20} = 4x + 16 - \mathbf{20}$$
$$2x = 4x - 4$$
$$2x - \mathbf{4x} = 4x - 4 - \mathbf{4x}$$
$$-2x = -4$$
$$\frac{-2x}{\mathbf{-2}} = \frac{-4}{\mathbf{-2}}$$
$$x = 2$$

State

2 gallons of the 2% fungicide is needed.

Check

The fungicide in the 2% solution is 2(0.02), or 0.04.
The fungicide in the 5% solution is 4(0.05), or 0.20.
The fungicide in the 4% solution is 6(0.04), or 0.24.
Since the total was $0.04 + 0.20 = 0.24$,
the results check.

SECTION 2.7
Solving Inequalities

Solve each inequality. Write the solution set in interval notation and graph it.

71. $3x+2<5$

$3x+2-\mathbf{2}<5-\mathbf{2}$

$3x<3$

$\frac{3x}{\mathbf{3}}<\frac{3}{\mathbf{3}}$

$x<1$

$(-\infty,1)$

1

72. $-\frac{3}{4}x\geq -9$

$-\frac{\mathbf{4}}{\mathbf{3}}\left(-\frac{3}{4}x\right)\leq -\frac{\mathbf{4}}{\mathbf{3}}(-9)$

$x\leq 12$

$(-\infty,12]$

12

73. $\frac{3}{4}<\frac{d}{5}+\frac{1}{2}$, LCD = 20

$\mathbf{20}\cdot\left(\frac{3}{4}\right)<\mathbf{20}\cdot\left(\frac{d}{5}+\frac{1}{2}\right)$

$20\cdot\left(\frac{3}{4}\right)<20\cdot\left(\frac{d}{5}\right)+20\cdot\left(\frac{1}{2}\right)$

$15<4d+10$

$15-\mathbf{10}<4d+10-\mathbf{10}$

$5<4d$

$\frac{5}{\mathbf{4}}<\frac{4d}{\mathbf{4}}$

$\frac{5}{4}<d$

$d>\frac{5}{4}$

$\left(\frac{5}{4},\infty\right)$

$\frac{5}{4}$

74. $5(3-x)\leq 3(x-3)$

$15-5x\leq 3x-9$

$15-5x-\mathbf{3x}\leq 3x-9-\mathbf{3x}$

$15-8x\leq -9$

$15-8x-\mathbf{15}\leq -9-\mathbf{15}$

$-8x\leq -24$

$\frac{-8x}{\mathbf{-8}}\geq\frac{-24}{\mathbf{-8}}$

$x\geq 3$

$[3,\infty)$

3

75. $\frac{t}{-5}-(-1.8)\geq -6.2$

$\frac{t}{-5}+1.8\geq -6.2$

$\frac{t}{-5}+1.8-\mathbf{1.8}\geq -6.2-\mathbf{1.8}$

$\frac{t}{-5}\geq -8$

$\mathbf{-5}\cdot\frac{t}{-5}\leq \mathbf{-5}\cdot(-8)$

$t\leq 40$

$(-\infty,40]$

40

76. $a+5-2(10-a)>6$

$a+5-\mathbf{2}(10)-\mathbf{2}(-a)>6$

$a+5-20+2a>6$

$3a-15>6$

$3a-15+\mathbf{15}>6+\mathbf{15}$

$3a>21$

$\frac{3a}{\mathbf{3}}>\frac{21}{\mathbf{3}}$

$a>7$

$(7,\infty)$

7

77.
$$24 < 3(x+2) < 39$$
$$24 < \mathbf{3}(x) + \mathbf{3}(2) < 39$$
$$24 < 3x + 6 < 39$$
$$24 - \mathbf{6} < 3x + 6 - \mathbf{6} < 39 - \mathbf{6}$$
$$18 < 3x < 33$$
$$\frac{18}{\mathbf{3}} < \frac{3x}{\mathbf{3}} < \frac{33}{\mathbf{3}}$$
$$6 < x < 11$$
$$(6, 11)$$

6 11

78.
$$0 \le 3 - 2x < 10$$
$$0 - \mathbf{3} \le 3 - 2x - \mathbf{3} < 10 - \mathbf{3}$$
$$-3 \le -2x < 7$$
$$\frac{-3}{\mathbf{-2}} \ge \frac{-2x}{\mathbf{-2}} > \frac{7}{\mathbf{-2}}$$
$$\frac{3}{2} \ge x > -\frac{7}{2}$$
$$-\frac{7}{2} < x \le \frac{3}{2}$$
$$\left(-\frac{7}{2}, \frac{3}{2}\right]$$

$-\frac{7}{2}$ $\frac{3}{2}$

79. SPORTS EQUIPMENT

$2.40 \text{ g} \le w \le 2.53 \text{ g}$

80. SIGNS

Analyze

- The width must be 18 inches.
- The perimeter must not exceed 132 inches.
- What are the possible lengths?

Assign

Let x = the length in inches

Form

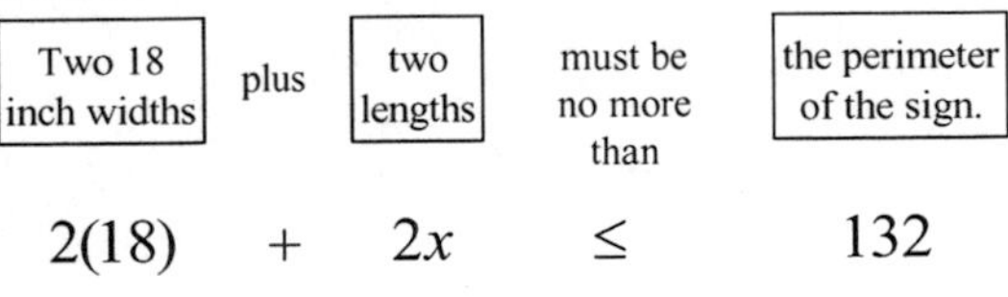

$$2(18) \quad + \quad 2x \quad \le \quad 132$$

Solve
$$2(18) + 2x \le 132$$
$$36 + 2x \le 132$$
$$36 + 2x - \mathbf{36} \le 132 - \mathbf{36}$$
$$2x \le 96$$
$$\frac{2x}{\mathbf{2}} \le \frac{96}{\mathbf{2}}$$
$$x \le 48$$

State

The lengths must be 48 inches or less or $0 \text{ inches} < l \le 48 \text{ inches}$.

Check
$$2(18) + 2x \le 132$$
$$2(18) + 2(48) \le 132$$
$$36 + 96 \le 132$$
$$132 \le 132$$
The results check.

CHAPTER 2 TEST

1. Fill in the blanks.

a. To **solve** an equation means to find all of the values of the variable that make the equation true.

b. **Percent** means parts per one hundred.

c. The distance around a circle is called its **circumference** .

d. An **inequality** is a statement that contains one of the symbols $>, \ge, <, \text{or} \le$.

e. The **multiplication** property of **equality** says that multiplying both sides of an equation by the same nonzero number does not change its solution.

2.
$$5y + 2 = 12$$
$$5(3) + 2 \stackrel{?}{=} 12$$
$$15 + 2 \stackrel{?}{=} 12$$
$$17 = 12 \quad \text{False}$$
3 is not a solution.

Solve each equation.

3.
$$3h + 2 = 8$$
$$3h + 2 \mathbf{- 2} = 8 \mathbf{- 2}$$
$$3h = 6$$
$$\frac{3h}{\mathbf{3}} = \frac{6}{\mathbf{3}}$$
$$h = 2$$

Checking
$$3h + 2 = 8$$
$$3(2) + 2 \stackrel{?}{=} 8$$
$$6 + 2 \stackrel{?}{=} 8$$
$$8 = 8$$

The solution is 2. The result checks.

4. $-22 = -x$

$$\frac{-22}{-1} = \frac{-x}{-1}$$

$$22 = x$$

Checking

$$-22 = -x$$

$$-22 \stackrel{?}{=} -(22)$$

$$-22 = -22$$

The solution is 22. The result checks.

5. $\frac{4}{5}t = -4$

$$\frac{\mathbf{5}}{\mathbf{4}} \cdot \frac{4}{5}t = \frac{\mathbf{5}}{\mathbf{4}} \cdot (-4)$$

$$t = -\frac{5 \cdot \overset{1}{\cancel{4}}}{\underset{1}{\cancel{4}}}$$

$$t = -5$$

Checking

$$\frac{4}{5}t = -4$$

$$\frac{4}{5}(-5) \stackrel{?}{=} -4$$

$$-4 = -4$$

The solution is -5. The result checks.

6. $\frac{11b-11}{5} = \frac{3b-2}{2}$

$$\left(\frac{\mathbf{10}}{\mathbf{1}}\right)\left(\frac{11b-11}{5}\right) = \left(\frac{\mathbf{10}}{\mathbf{1}}\right)\left(\frac{3b-2}{2}\right)$$

$$2(11b-11) = 5(3b-2)$$

$$22b - 22 = 15b - 10$$

$$22b - 22 - \mathbf{15b} = 15b - 10 - \mathbf{15b}$$

$$7b - 22 = -10$$

$$7b - 22 + \mathbf{22} = -10 + \mathbf{22}$$

$$7b = 12$$

$$\frac{7b}{\mathbf{7}} = \frac{12}{\mathbf{7}}$$

$$b = \frac{12}{7}$$

This problem was not checked due to its complexity.

The solution is $\frac{12}{7}$. The result checks.

7. $0.8(x - 1{,}000) + 1.3 = 2.9 + 0.2x$

$$\mathbf{10}[0.8(x - 1{,}000) + 1.3] = \mathbf{10}[2.9 + 0.2x]$$

$$10[0.8(x - 1{,}000)] + 10(1.3) = 10(2.9) + 10(0.2x)$$

$$8(x - 1{,}000) + 13 = 29 + 2x$$

$$8x - 8{,}000 + 13 = 29 + 2x$$

$$8x - 7{,}987 = 29 + 2x$$

$$8x - 7{,}987 - \mathbf{2x} = 29 + 2x - \mathbf{2x}$$

$$6x - 7{,}987 = 29$$

$$6x - 7{,}987 + \mathbf{7{,}987} = 29 + \mathbf{7{,}987}$$

$$6x = 8{,}016$$

$$\frac{6x}{\mathbf{6}} = \frac{8{,}016}{\mathbf{6}}$$

$$x = 1{,}336$$

$$0.8(x - 1{,}000) + 1.3 = 2.9 + 0.2x$$

$$0.8(\mathbf{1{,}336} - 1{,}000) + 1.3 \stackrel{?}{=} 2.9 + 0.2(\mathbf{1{,}336})$$

$$0.8(336) + 1.3 \stackrel{?}{=} 2.9 + 267.2$$

$$268.8 + 1.3 \stackrel{?}{=} 270.1$$

$$270.1 = 270.1$$

The solution is 1,336. The result checks.

8. $2(y-7) - 3y = -(y-3) - 17$

$$2y - 14 - 3y = -y + 3 - 17$$

$$-y - 14 = -y - 14$$

$$-y - 14 + \mathbf{y} = -y - 14 + \mathbf{y}$$

$$-14 = -14 \text{ True}$$

The terms involving y drop out and the result is true. This means that *all real numbers* are solutions and this equation is an identity.

9. $$\frac{m}{2}-\frac{1}{3}=\frac{1}{4}+\frac{m}{6}\quad,\ \text{LCD}=12$$

$$\mathbf{12}\cdot\left(\frac{m}{2}-\frac{1}{3}\right)=\mathbf{12}\cdot\left(\frac{1}{4}+\frac{m}{6}\right)$$

$$12\cdot\frac{m}{2}-12\cdot\frac{1}{3}=12\cdot\left(\frac{1}{4}\right)+12\cdot\left(\frac{m}{6}\right)$$

$$6m-4=3+2m$$

$$6m-4-\mathbf{2m}=3+2m-\mathbf{2m}$$

$$4m-4=3$$

$$4m-4+\mathbf{4}=3+\mathbf{4}$$

$$4m=7$$

$$\frac{4m}{\mathbf{4}}=\frac{7}{\mathbf{4}}$$

$$m=\frac{7}{4}$$

This problem was not checked due to its complexity.

The solution is $\frac{7}{4}$. The solution checks.

10. $$\frac{3}{4}(6n-2)=246$$

$$\mathbf{4}\left[\frac{3}{4}(6n-2)\right]=\mathbf{4}(246)$$

$$3(6n-2)=984$$

$$\mathbf{3}(6n)-\mathbf{3}(2)=984$$

$$18n-6=984$$

$$18n-6+\mathbf{6}=984+\mathbf{6}$$

$$18n=990$$

$$\frac{18n}{\mathbf{18}}=\frac{990}{\mathbf{18}}$$

$$n=55$$

$$\frac{3}{4}(6n-2)=246$$

$$\frac{3}{4}(6\cdot\mathbf{55}-2)\overset{?}{=}246$$

$$\frac{3}{4}(330-2)\overset{?}{=}246$$

$$\frac{3}{4}(328)\overset{?}{=}246$$

$$\frac{3\cdot\overset{82}{\cancel{328}}}{\underset{1}{\cancel{4}}}\overset{?}{=}246$$

$$246=246$$

The solution is 55. The solution checks .

11. $$5x=0$$

$$\frac{5x}{\mathbf{5}}=\frac{0}{\mathbf{5}}$$

$$x=0$$

Checking

$$5x=0$$

$$5(0)\overset{?}{=}0$$

$$0=0$$

The solution is 0. The solution checks.

12. $$6a+(-7)=3a-7+2a$$

$$6a-7=5a-7$$

$$6a-7+\mathbf{7}=5a-7+\mathbf{7}$$

$$6a=5a$$

$$6a-\mathbf{5a}=5a-\mathbf{5a}$$

$$a=0$$

$$6a+(-7)=3a-7+2a$$

$$6(0)+(-7)\overset{?}{=}3(0)-7+2(0)$$

$$-7=-7$$

The solution is 0. The solution checks.

13. $$9-5(2x+10)=-1$$

$$9-10x-50=-1$$

$$-10x-41=-1$$

$$-10x-41+\mathbf{41}=-1+\mathbf{41}$$

$$-10x=40$$

$$\frac{-10x}{\mathbf{-10}}=\frac{40}{\mathbf{-10}}$$

$$x=-4$$

$$9-5(2x+10)=-1$$

$$9-5[2(-4)+10]\overset{?}{=}-1$$

$$9-5(-8+10)\overset{?}{=}-1$$

$$9-5(2)\overset{?}{=}-1$$

$$9-10\overset{?}{=}-1$$

$$-1=-1$$

The solution is -4. The solution checks.

14. $$24t=-6(8-4t)$$

$$24t=-48+24t$$

$$24t-\mathbf{24t}=-48+24t-\mathbf{24t}$$

$$0=-48\ \text{ False}$$

The terms involving t drop out and the result is false. This means the equation has *no solution* and it is a contradiction.

15. What number is 15.2% of 80?

$$x = 0.152\cdot 80$$

$$x=12.16$$

12.16 is the solution.

16. DOWN PAYMENTS

Let $x =$ the selling price of the house

11,400 is 15% of what number?

$11{,}400 = 0.15 \cdot x$

$\frac{11{,}400}{\mathbf{0.15}} = \frac{0.15x}{\mathbf{0.15}}$

$76{,}000 = x$

$76,000 is the selling price.

17. BODY TEMPERATURES

Step 1: $105 - 98.6 = 6.4$

This is the amount of increase.

Step 2: 6.4 is what percent of 98.6?

$6.4 = x \cdot 98.6$

$\frac{6.4}{\mathbf{98.6}} = \frac{98.6x}{\mathbf{98.6}}$

$0.064 \approx x$

$6\% \approx x$

The percent of increase is about 6%.

18. COMMISSIONS

Let $x =$ amt of the commission

What number is 5% of $599.99?

$x = 0.05 \cdot 599.99$

$x = 30$

$30 is amount of the commission.

19. GRAND OPENINGS

Let $c =$ the cost

$r = p + c$

$445 = 150 + c$

$445 - \mathbf{150} = 150 + c - \mathbf{150}$

$295 = c$

$295 is the amount of the cost.

20. Find the Celsius temperature.

$C = \frac{5}{9}(F - 32)$

$= \frac{5}{9}(14 - 32)$

$= \frac{5}{9}(-18)$

$= -10$

The temperature is -10° Celsius.

21. SOUND

$d = rt$

$= 1{,}108\text{ft/second (60 seconds)}$

$= 66{,}480$

In one minute it will travel 66,480 ft.

22. PETS

$V = \frac{4}{3}\pi r^3$, Use this formula to find the volume of water in the bowl that is $\frac{3}{4}$ full.

$V = \frac{4}{3}\pi r^3$, $\pi = 3.141592654$

$= \frac{3}{4} \cdot \frac{4}{3} \cdot 3.1415926542 \cdot (5)^3$

$= 1 \cdot 3.1415926542 \cdot (125)$

≈ 392.6

≈ 393

The volume of water is about 393 in^3.

Solve for the specified variable.

23. $V = \pi r^2 h$ for h.

$V = \pi r^2 h$

$\frac{V}{\boldsymbol{\pi r^2}} = \frac{\pi r^2 h}{\boldsymbol{\pi r^2}}$

$\frac{V}{\pi r^2} = h$

$h = \frac{V}{\pi r^2}$

24. $A = P + Prt$ for r.

$$A = P + Prt$$
$$A - \mathbf{P} = P + Prt - \mathbf{P}$$
$$A - P = Prt$$
$$\frac{A-P}{\mathbf{Pt}} = \frac{Prt}{\mathbf{Pt}}$$
$$\frac{A-P}{Pt} = r$$
$$r = \frac{A-P}{Pt}$$

N25. $A = \dfrac{a+b+c+d}{4}$ for c.

$$A = \frac{a+b+c+d}{4}$$
$$\mathbf{4} \cdot A = \mathbf{4} \cdot \frac{a+b+c+d}{4}$$
$$4A = a+b+c+d$$
$$4A - \boldsymbol{a} - \boldsymbol{b} - \boldsymbol{d} = a+b+c+d - \boldsymbol{a} - \boldsymbol{b} - \boldsymbol{d}$$
$$4A - a - b - d = c$$
$$c = 4A - a - b - d$$

N26. $2x - 3y = 9$ for y.

$$2x - 3y = 9$$
$$2x - 3y - \mathbf{2x} = 9 - \mathbf{2x}$$
$$-3y = 9 - 2x$$
$$\frac{-3y}{\mathbf{-3}} = -\frac{2x}{\mathbf{-3}} + \frac{9}{\mathbf{-3}}$$
$$y = \frac{2x}{3} - 3$$

27. IRONS

$$A = \frac{1}{2}bh$$
$$= \frac{1}{2} \cdot 8 \cdot 5$$
$$= 4 \cdot 5$$
$$= 20$$

The area is 20 square inches.

28. TELEVISION

Analyze

- 30-minute block of TV time
- Programming minutes are 2 less than three times the number of commercials minutes.
- Find the number of minutes for both.

Assign

Let x = number of commercial minutes
$3x - 2$ = number of programming minutes

Form

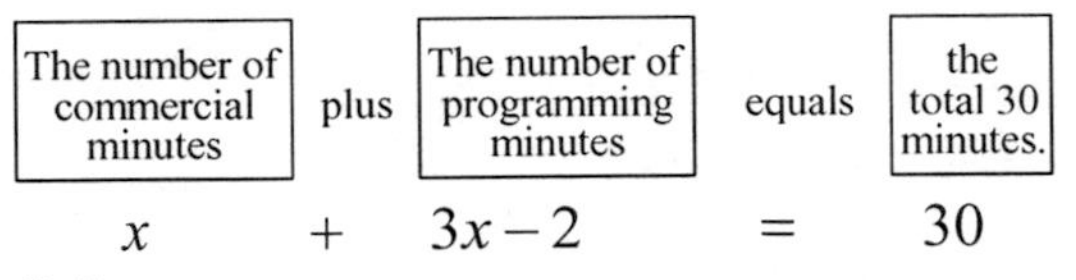

$$x \quad + \quad 3x - 2 \quad = \quad 30$$

Solve

$$x + 3x - 2 = 30$$
$$4x - 2 = 30$$
$$4x - 2 + \mathbf{2} = 30 + \mathbf{2}$$
$$4x = 32$$
$$\frac{4x}{\mathbf{4}} = \frac{32}{\mathbf{4}}$$
$$x = 8$$

Programming

$$3x - 2 = 3(8) - 2$$
$$= 24 - 2$$
$$= 22$$

State

8 minutes of commercials.
22 minutes of programming.

Check

8 minutes + 22 minutes = 30 minutes
The results check.

29. PLUMBING BILLS

Analyze

- Standard service charge is \$25.75
- Parts cost \$38.75.
- 4 hours of labor.
- Total charges are \$226.70.
- Find the cost per hour for the labor.

Assign

Let x = cost per hour for labor

Form

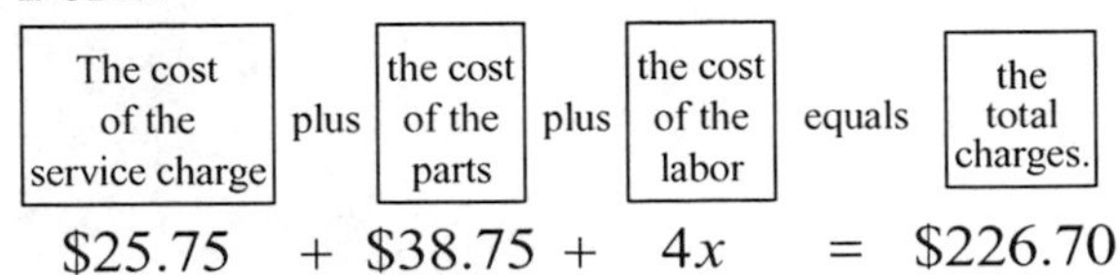

$$\$25.75 \quad + \quad \$38.75 \quad + \quad 4x \quad = \quad \$226.70$$

Solve

$$25.75 + 38.75 + 4x = 226.70$$
$$4x + 64.50 = 226.70$$
$$4x + 64.50 - \mathbf{64.50} = 226.70 - \mathbf{64.50}$$
$$4x = 162.20$$
$$\frac{4x}{\mathbf{4}} = \frac{162.20}{\mathbf{4}}$$
$$x = 40.55$$

State

The cost per hour for the labor is $40.55.

Check

$$\$25.75 + \$38.75 + 4x = \$226.70$$
$$\$25.75 + \$38.75 + 4(\$40.55) = \$226.70$$
$$\$25.75 + \$38.75 + \$162.55 = \$226.70$$
$$\$226.70 = \$226.70$$

The result checks.

30. CONCERT SEATING

Analyze

- Floor seats cost $12.50 each.
- Balcony seats cost $20.50 each.
- Ten times as many floor seats were sold as balcony seats (smaller quanity of the two).
- Total receipts is $11,640.
- How many of each type of tickets were sold?

Assign

Let x = the number of balcony seats sold
$10x$ = the number of floor seats sold

Form

	Number •	Value =	Total value
Balcony	x	20.50	$\mathbf{20.50x}$
Floor	$10x$	12.50	$\mathbf{12.50(10x)}$
		Total:	**11,640**

The value of the balcony seats	plus	the value of the floor seats	equals	the total value of the receipts.
$20.50x$	$+$	$12.50(10x)$	$=$	$11,640$

Solve

$$20.50x + 12.50(10x) = 11,640$$
$$20.50x + 125x = 11,640$$
$$145.50x = 11,640$$
$$\frac{145.50x}{\mathbf{145.50}} = \frac{11,640}{\mathbf{145.50}}$$
$$x = 80$$

Floor seats

$$10x = 10(80)$$
$$= 800$$

State

80 balcony seats tickets were sold.

800 floor seats tickets were sold.

Check

The value of the balcony tickets is 80($20.50), or $1,640.
The value of the floor seat tickets is 800($12.50), or $10,000.
Since the total was $\$1,640 + 10,000 = \$11,640$, the results check.

31. HOME SALES

Analyze

- Cleared $114,600 on the sale of a house.
- Paid agent 4.5% of the selling price.
- What was the selling price of the house?

Assign

Let x = the selling price of the house

Form

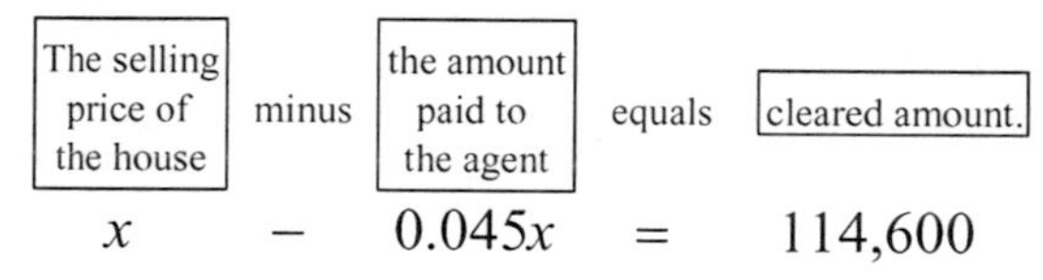

$$x \quad - \quad 0.045x \quad = \quad 114,600$$

Solve

$$x - 0.045x = 114,600$$
$$0.955x = 114,600$$
$$\frac{0.955x}{\mathbf{0.955}} = \frac{114,600}{\mathbf{0.955}}$$
$$x = 120,000$$

State

The selling price was $120,000.

Check

$$x - 0.045x = 120,000 - 0.045(120,000)$$
$$= 120,000 - 5,400$$
$$= 114,600$$

The result checks.

32. COLORADO

Analyze

- Perimeter of the state is 1,320 miles.
- Length is 100 miles longer than the width.
- What are the dimensions of the state?

Assign

Let $x =$ width of the state in miles

$x+100 =$ length of the state in miles

Form

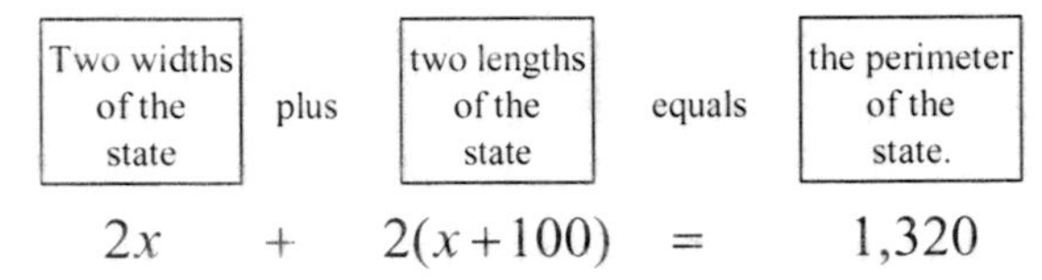

$$2x \quad + \quad 2(x+100) \quad = \quad 1{,}320$$

Solve

$$2x+2(x+100)=1{,}320$$
$$2x+2x+200=1{,}320$$
$$4x+200=1{,}320$$
$$4x+200-\mathbf{200}=1{,}320-\mathbf{200}$$
$$4x=1{,}120$$
$$\frac{4x}{\mathbf{4}}=\frac{1{,}120}{\mathbf{4}}$$
$$x=280$$

length

$$x+100=280+100$$
$$=380$$

State

Width is 280 miles.
Length is 380 miles.

Check

$$2(280)+2(380)=560+760$$
$$=1{,}320$$

The results check.

33. TEA

Analyze

- Green tea is worth \$40 a lb.
- Herbal tea is worth \$50 a lb.
- A 20 pound blend is worth \$42 a lb.
- How many pounds of each is needed?

Assign

Let $x =$ the pounds of green tea

$20-x =$ the pounds of herbal tea

Form

	Number	• Value	= Total value
Green	$\mathbf{x}$	40	$\mathbf{40x}$
Herbal	$\mathbf{20-x}$	50	$\mathbf{50(20-x)}$
Blend	20	42	**840**

The value of the green tea	plus	the value of the herbal tea	equals	the total value of the blend.
$40x$	$+$	$50(20-x)$	$=$	840

Solve

$$40x+50(20-x)=840$$
$$40x+1{,}000-50x=840$$
$$-10x+1{,}000=840$$
$$-10x+1{,}000-\mathbf{1{,}000}=840-\mathbf{1{,}000}$$
$$-10x=-160$$
$$\frac{-10x}{\mathbf{-10}}=\frac{-160}{\mathbf{-10}}$$
$$x=16$$

herbal

$$20-x=20-16$$
$$=4$$

State

16 lb of green tea will be needed.
4 lb of herbal tea will be needed

Check

$$40x+50(20-x)=840$$
$$40(16)+50(20-16)=840$$
$$640+200=840$$
$$840=840$$

The results check.

34. READING

Analyze

- Total of the two page numbers is 825.
- The respective number for each page is a consecutive integers.
- What are the two page numbers?

Assign

Let $x = 1^{st}$ page number

$x+1 = 2^{nd}$ page number

Form

The 1st page number	plus	the 2nd page number	equals	the total of both page numbers.
x	$+$	$x+1$	$=$	825

Solve

$$x + x + 1 = 825$$
$$2x + 1 = 825$$
$$2x + 1 - \mathbf{1} = 825 - \mathbf{1}$$
$$2x = 824$$
$$\frac{2x}{\mathbf{2}} = \frac{824}{\mathbf{2}}$$
$$x = 412$$

2nd page

$$x + 1 = 412 + 1$$
$$= 413$$

State

The 1st page number is 412.
The 2nd page number is 413.

Check

$$x + x + 1 = 825$$
$$412 + 412 + 1 = 825$$
$$825 = 825$$

The results check.

35. TRAVEL TIMES

Analyze

- A car drives towards Madison at 65 mph.
- A truck drives towards Rockford at 55 mph.
- Both leave at the same time using the same road.
- The distance between the 2 cities is 72 miles.
- How long before they meet at each other?

Assign

Let t = time in hours they meet to each other

	rate	• time	= distance
Car	65	t	$\mathbf{65t}$
Truck	55	t	$\mathbf{55t}$
		Total:	**72**

Form

The distance traveled by the car	plus	the distance traveled by the truck	equals	the total distance.
$65t$	$+$	$55t$	$=$	72

Solve

$$65t + 55t = 72$$
$$120t = 72$$
$$\frac{120t}{\mathbf{120}} = \frac{72}{\mathbf{120}}$$
$$t = \frac{3}{5}$$

or

$$t = 0.6$$

State

After $\frac{3}{5}$ hour or 36 minutes they will meet.

Check

$$65t + 55t = 72$$
$$65\left(\frac{3}{5}\right) + 55\left(\frac{3}{5}\right) = 72$$
$$13(3) + 11(3) = 72$$
$$39 + 33 = 72$$
$$72 = 72$$

The result checks.

36. PICKLES

Analyze

- Start with 30 liters of 10% brine
- Want to dilute 10% brine to 8%
- How many liters of 2% is needed?

Assign

Let x = the number of liters of 2% brine

	Amount •	Strength =	Amount of pure salt
2%	x	0.02	$\mathbf{0.02x}$
10%	30	0.10	**3**
8%	$\mathbf{x+30}$	0.08	$\mathbf{0.08(x+30)}$

Form

The salt in the 2% brine plus the salt in the 10% brine equals the salt in the 8% brine.

$$0.02x \quad + \quad 3 \quad = \quad 0.08(x+30)$$

Solve

$$0.02x+3=0.08(x+30)$$
$$\mathbf{100}(0.02x+3)=\mathbf{100}[0.08(x+30)]$$
$$100(0.02x)+100(3)=100[(0.08(x+30)]$$
$$2x+300=8(x+30)$$
$$2x+300=8x+240$$
$$2x+300-\mathbf{300}=8x+240-\mathbf{300}$$
$$2x=8x-60$$
$$2x-\mathbf{8x}=8x-60-\mathbf{8x}$$
$$-6x=-60$$
$$\frac{-6x}{\mathbf{-6}}=\frac{-60}{\mathbf{-6}}$$
$$x=10$$

State

10 liters of 2% brine is needed.

Check

$$0.02x+3=0.08(x+30)$$
$$0.02(10)+3=0.08(10+30)$$
$$0.2+3=0.08(40)$$
$$3.2=3.2$$

The result checks.

37. EXERCISE

Analyze

- Jogger jogs at 8 mph with a half-hour head start.
- Bicyclist travels at 20 mph.
- How long before the biker catches the jogger?

Assign

Let t = time in hours before the biker catches the jogger

Form

	rate •	time =	distance
Jogger	8	$t+0.5$	$\mathbf{8(t+0.5)}$
Biker	20	t	$\mathbf{20t}$

Distance same

The distance traveled by the biker equals the distance traveled by the jogger.

$$20t \quad = \quad 8(t+0.5)$$

Solve

$$20t=8(t+0.5)$$
$$20t=\mathbf{8}(t)+\mathbf{8}(0.5)$$
$$20t=8t+4$$
$$20t-\mathbf{8t}=8t+4-\mathbf{8t}$$
$$12t=4$$
$$\frac{12t}{\mathbf{12}}=\frac{4}{\mathbf{12}}$$
$$t=\frac{1}{3}$$

State

The biker catches the jogger after $\frac{1}{3}$ hour or 20 minutes.

Check

$$20t=8(t+0.5)$$
$$20\left(\frac{1}{3}\right)=8\left(\frac{1}{3}+\frac{1}{2}\right),$$
$$\frac{20}{3}=8\left(\frac{5}{6}\right)$$
$$\frac{20}{3}=\frac{20}{3}$$

The distances are equal, the result checks.

38. GEOMETRY

Analyze

- Internal measure of the 3 angles of a triangle is 180°.
- The measure of the vertex angle is 44°?
- What is the measure of each base angle which are equal to each other.

Assign

Let x = measure of each base angle

Form

The measure of the vertex angle	plus	the measure of the 2 base angles	equals	the measure of the angles of a triangle.
44	+	$2x$	=	180

Solve

$$44+2x=180$$
$$44+2x-\mathbf{44}=180-\mathbf{44}$$
$$2x=136$$
$$\frac{2x}{\mathbf{2}}=\frac{136}{\mathbf{2}}$$
$$x=68$$

State

The measure of each base angle is 68°.

Check

$44+2x=180$
$44+2(68)=180$
$44+136=180$
$180=180$

The result checks.

39. INVESTMENTS

Analyze

- \$13,750 is the total investment.
- 1st part invested at 9% annual interest.
- 2nd part invested at 8% annual interest.
- \$1,185 is the first-year combined interest.
- How much is invested at the lower rate?

Assign

Let x = amount invested at 8%

$13{,}750-x$ = amount invested at 9%

Form

	Principal	• rate	• time =	interest
1st	x	0.08	1	**$0.08x$**
2nd	$13{,}750-x$	0.09	1	**$0.09(13{,}750-x)$**
		Total:		**1,185**

The interest earned at 8%	plus	the interest earned at 9%	equals	the total interest.
$0.08x$	+	$0.09(13,750-x)$	=	1,185

Solve

$$0.08x+0.09(13,750-x)=1,185$$
$$\mathbf{100}[0.08x+0.09(13,750-x)]=\mathbf{100}(1,185)$$
$$100(0.08x)+100[0.09(13,750-x)]=100(1,185)$$
$$8x+9(13,750-x)=118,500$$
$$8x+123,750-9x=118,500$$
$$-x+123,750=118,500$$
$$-x+123,750-\mathbf{123,750}=118,500-\mathbf{123,750}$$
$$-x=-5,250$$
$$\frac{-x}{\mathbf{-1}}=\frac{-5,250}{\mathbf{-1}}$$
$$x=5,250$$

State

\$5,250 was invested at 8%.

For checking purposes the other amount is needed. $\$13,750-\$5,250=\$8,500$.

Check

$0.08(5,250)+0.09(8,509)=1,185$

$420+765.81=1,185$

The results check.

Determine if -3 is a solution.

40. $4-9w<-4w+19$, Is -3 a solution?

$$4-9(-3)\overset{?}{<}-4(-3)+19$$
$$4+27\overset{?}{<}12+19$$
$$31\overset{?}{<}31$$

False

Solve each inequality. Write the solution set in interval notation and graph it.

41. $-8x-20\le 4$

$$-8x-20+\mathbf{20}\le 4+\mathbf{20}$$
$$-8x\le 24$$
$$\frac{-8x}{\mathbf{-8}}\ge\frac{24}{\mathbf{-8}}$$
$$x\ge -3$$
$$[-3,\infty)$$

−3

42.
$$-8.1 > \frac{t}{2} + (-11.3)$$
$$-8.1 > \frac{t}{2} - 11.3$$
$$-8.1 + \mathbf{11.3} > \frac{t}{2} - 11.3 + \mathbf{11.3}$$
$$3.2 > \frac{t}{2}$$
$$\mathbf{2} \cdot 3.2 > \mathbf{2} \cdot \left(\frac{t}{2}\right)$$
$$6.4 > t$$
$$t < 6.4$$
$$(-\infty, 6.4)$$

6.4

43.
$$-12 \le 2(x+1) < 10$$
$$-12 \le 2x + 2 < 10$$
$$-12 - \mathbf{2} \le 2x + 2 - \mathbf{2} < 10 - \mathbf{2}$$
$$-14 \le 2x < 8$$
$$\frac{-14}{\mathbf{2}} \le \frac{2x}{\mathbf{2}} < \frac{8}{\mathbf{2}}$$
$$-7 \le x < 4$$
$$[-7, 4)$$

−7 4

44.
$$\frac{1}{3}(a-5) > \frac{1}{2}(a+1)$$
$$\mathbf{6}\left[\frac{1}{3}(a-5)\right] > \mathbf{6}\left[\frac{1}{2}(a+1)\right]$$
$$2(a-5) > 3(a+1)$$
$$\mathbf{2}(a) - \mathbf{2}(5) > \mathbf{3}(a) + \mathbf{3}(1)$$
$$2a - 10 > 3a + 3$$
$$2a - 10 \mathbf{- 3a} > 3a + 3 \mathbf{- 3a}$$
$$-a - 10 > 3$$
$$-a - 10 \mathbf{+ 10} > 3 \mathbf{+ 10}$$
$$-a > 13$$
$$\frac{-a}{\mathbf{-1}} \mathbf{<} \frac{13}{\mathbf{-1}}$$
$$a < -13$$
$$(-\infty, -13)$$

−13

45.
$$-9(h-3) + 2h \le 8(4-h)$$
$$\mathbf{-9}(h) - \mathbf{9}(-3) + 2h \le \mathbf{8}(4) - \mathbf{8}(h)$$
$$-9h + 27 + 2h \le 32 - 8h$$
$$-7h + 27 \le 32 - 8h$$
$$-7h + 27 \mathbf{+ 8h} \le 32 - 8h \mathbf{+ 8h}$$
$$h + 27 \le 32$$
$$h + 27 \mathbf{- 27} \le 32 \mathbf{- 27}$$
$$h \le 5$$
$$(-\infty, 5]$$

5

46. AWARDS

Analyze

- The frame cost \$15.
- Each word cost \$0.75.
- Total cost cannot exceed \$150.
- What is the maximum number of words?

Assign

Let x = the number of words

Form

The cost of the frame	plus	the cost of the words	must be no more than	the maximum amount.
15	+	$0.75x$	$\le$	150

Solve

$$15 + 0.75x \le 150$$
$$15 + 0.75x \le 150$$
$$15 + 0.75x \mathbf{- 15} \le 150 \mathbf{- 15}$$
$$0.75x \le 135$$
$$\frac{0.75x}{\mathbf{0.75}} \le \frac{135}{\mathbf{0.75}}$$
$$x \le 180$$

State

The maximum numer of words is 180.

Check

$$15 + 0.75(180) \le 150$$
$$15 + 135 \le 150$$
$$150 \le 150$$

The result checks.

SECTION 3.1 STUDY SET
VOCABULARY

Fill in the blanks.

1. (7, 1) is called an **ordered** pair.
3. A rectangular coordinate system is formed by two perpendicular number lines called the x-**axis** and the y-**axis**. The point where the axes cross is called the **origin**.
5. The point with coordinates (4, 2) can be graphed on a **rectangular** coordinate system.

CONCEPTS

7. **Fill in the blanks.**
   a. To plot (–5, 4), we start at the **origin** and move 5 units to the **left** and then move 4 units **up**.
   b. To plot $\left(6, -\frac{3}{2}\right)$, we start at the **origin** and move 6 units to the **right** and then move $\frac{3}{2}$ units **down**.
9. a. I and II b. II and III c. IV d. The y-axis

NOTATION

11. Explain the difference between (3, 5) and 3(5).
    (3, 5) is an ordered pair, 3(5) = 3 • 5.
13. Do these ordered pairs name the same point?
    $\left(2.5, -\frac{7}{2}\right), \left(2\frac{1}{2}, -3.5\right), \left(2.5, -3\frac{1}{2}\right)$
    Yes
15. In the ordered pair (4, 5), is the number 4 associated with the horizontal or the vertical axis?
    Horizontal

GUIDED PRACTICE

17. Graph each point: (–3, 4), (4, 3.5), $\left(-2, -\frac{5}{2}\right)$ (0, –4), $\left(\frac{3}{2}, 0\right)$, (2.7, –4.1).

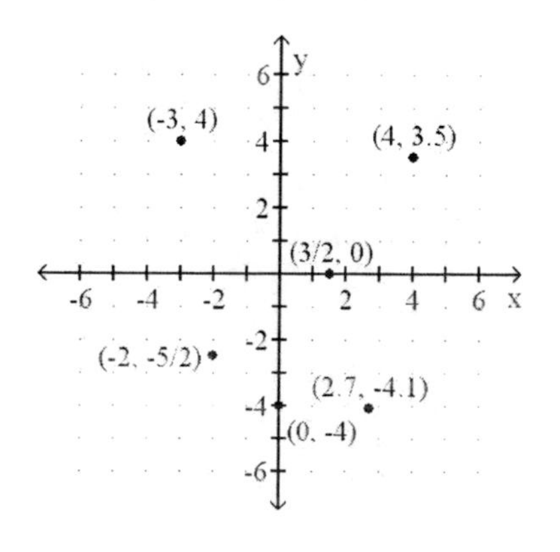

19. Complete the coordinates for each point.
    (4, 3), (0, 4), (–5, 0), (–4, –5), (3, –3)
21. a. **60 beats/min** b. **10 min**
23. a. **5 min and 50 min after starting**
    b. **20 min**
25. a **2 hrs** b. **–1,000 ft**
27. a **It ascends (rises) 500 ft** b. **–500 ft**

APPLICATIONS

29. BRIDGE CONSTRUCTION
    Rivets: $(-60, 0), (-20, 0), (20, 0), (60, 0)$
    Welds: $(-40, 30), (0, 30), (40, 30)$
    Anchors: $(-60, -30), (60, -30)$
31. GAMES
    (G, 2), (G, 3), (G, 4)

from Campus to Careers

33. DENTAL ASSISTANT
    a. 8
    b. It represents the patient's left side.
35. WATER PRESSURE
    a. 60°; 4 ft b. 30°; 4 ft
37. AREA

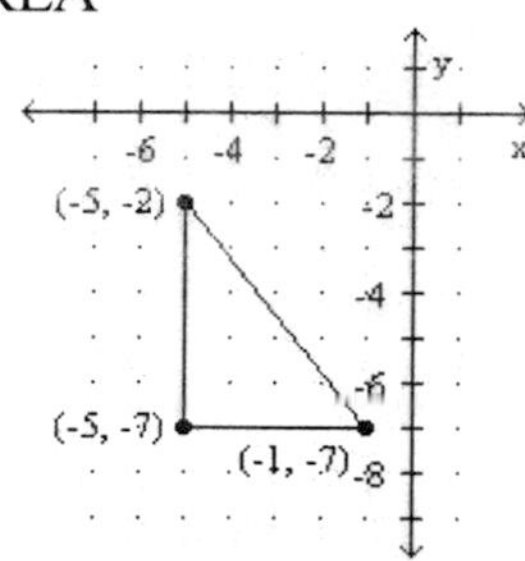

Area: $A = 0.5bh$
$= 0.5(4)(5)$
$= 10$
The solution is 10 sq. units.

39. TRUCKS

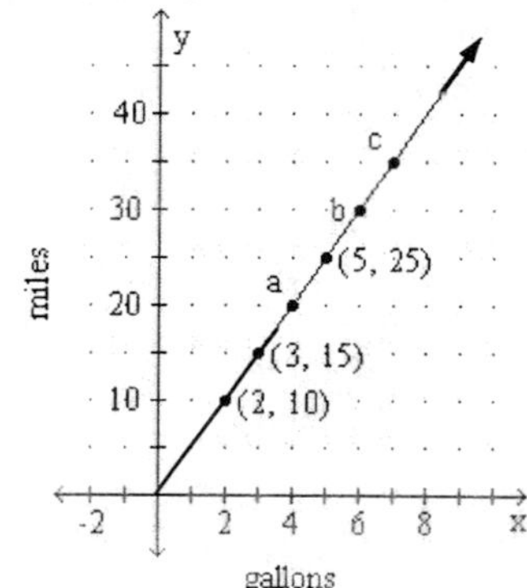

a. 20 miles b. 6 gallons c. 35 miles

41. DEPRECIATION

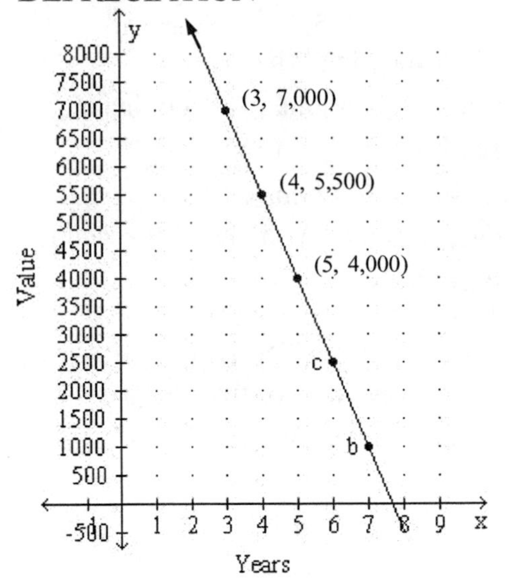

a. A 3-year old copier is worth $7,000.

b. $1,000

c. 6 years

WRITING

43- 45. Answers will vary.

REVIEW

47. Solve $AC = \frac{2}{3}h - T$ for h.

$$AC = \frac{2}{3}h - T$$

$$AC + \boldsymbol{T} = \frac{2}{3}h - T + \boldsymbol{T}$$

$$AC + T = \frac{2}{3}h$$

$$AC + T = \frac{2}{3}h$$

$$\frac{\mathbf{3}}{\mathbf{2}}(AC + T) = \frac{\mathbf{3}}{\mathbf{2}}\left(\frac{2}{3}h\right)$$

$$\frac{3(AC + T)}{2} = h$$

$$h = \frac{3(AC + T)}{2}$$

49. $\dfrac{-4(4+2)-2^3}{|-12-4(5)|} = \dfrac{-4(6)-8}{|-12-20|}$

$$= \frac{-24-8}{|-32|}$$

$$= \frac{-32}{32}$$

$$= -1$$

CHALLENGE PROBLEMS

51. Quadrant III is the quadrant.

Example point. $(-3, -5)$

Sum: $-3 + (-5) = -8$

Product: $-3(-5) = 15$

SECTION 3.2 STUDY SET
VOCABULARY

Fill in the blanks.

1. $y = 9x + 5$ is an equation in **two** variables, x and y.
3. Solutions of equations in two variables are often listed in a **table** of solutions.
5. $y = 3x + 8$ is called a **linear** equation because its graph is a line.

CONCEPTS

7. Consider $y = -3x + 6$.
   a. How many variables does the equation contain? **2**
   b. Does $(4, -6)$ satisfy the equation? **Yes**
   c. Is $(-2, 0)$ a solution? **No**
   d. How many solutions does this equation have? **Infinitely many**
9. Every point on the graph represents an ordered-pair **solution** of $y = -2x - 3$ and every ordered-pair solution is a **point** on the graph.
11. a. One could choose $\mathbf{-5, 0, 5}$. These numbers are multiples of the denominator. There are many others to choose from.
    b. One could choose $\mathbf{-10, 0, 10}$. These numbers would simplify the decimal. There are many others to choose from.

NOTATION

13. Verify that $(-2, 6)$ is a solution of $y = -x + 4$.

$$y = -x + 4$$
$$\boxed{6} \stackrel{?}{=} -(\boxed{-2}) + 4$$
$$6 \stackrel{?}{=} \boxed{2} + 4$$
$$6 = \boxed{6}$$

15. a. In the linear equation $y = \frac{1}{2}x + 7$ what are the understood exponents on the variables.
    They are 1's.
    b. Explain why $y = x^2 + 2$ and $y = x^3 - 4$ are not linear equations.
    The exponent on x is not 1.

GUIDED PRACTICE

Determine whether each equation has the given ordered pair as a solution.

17. $y = 5x - 4$; $(1, 1)$

$$1 \stackrel{?}{=} 5(1) - 4$$
$$1 \stackrel{?}{=} 5 - 4$$
$$1 = 1 \text{ True}$$

$(1, 1)$ is a solution.

19. $7x - 2y = 3$; $(2, 6)$

$$7(2) - 2(6) \stackrel{?}{=} 3$$
$$14 - 12 \stackrel{?}{=} 3$$
$$2 = 3 \text{ False}$$

$(2, 6)$ is not a solution.

21. $x + 12y = -12$; $(0, -1)$

$$0 + 12(-1) \stackrel{?}{=} -12$$
$$0 - 12 \stackrel{?}{=} -12$$
$$-12 = -12 \text{ True}$$

$(0, -1)$ is a solution.

23. $3x - 6y = 12$; $(-3.6, -3.8)$

$$3(-3.6) - 6(-3.8) \stackrel{?}{=} 12$$
$$-10.8 + 22.8 \stackrel{?}{=} 12$$
$$12 = 12 \text{ True}$$

$(-3.6, -3.8)$ is a solution.

25. $y - 6x = 12$; $\left(\frac{5}{6}, 7\right)$

$$7 - 6\left(\frac{5}{6}\right) \stackrel{?}{=} 12$$
$$7 - 5 \stackrel{?}{=} 12$$
$$2 = 12 \text{ False}$$

$\left(\frac{5}{6}, 7\right)$ is not a solution.

27. $y = -\frac{3}{4}x + 8$; $(-8, 12)$

$$12 \stackrel{?}{=} -\frac{3}{4}(-8) + 8$$
$$12 \stackrel{?}{=} 6 + 8$$
$$12 = 14 \text{ False}$$

$(-8, 12)$ is not a solution.

For each equation, complete the solution.

29. $y=-5x-4$; $(-3,?)$

$y=-5(-3)-4$

$y=15-4$

$y=11$

$(-3, \boxed{11})$ is the completed ordered pair.

31. $4x-5y=-4$; $(?,4)$

$4x-5(4)=-4$

$4x-20=-4$

$4x-20+\mathbf{20}=-4+\mathbf{20}$

$4x=16$

$\frac{4x}{\mathbf{4}}=\frac{16}{\mathbf{4}}$

$x=4$

$(\boxed{4}, 4)$ is the completed ordered pair.

33. $y=\frac{x}{4}+9$; $(16,?)$

$y=\frac{16}{4}+9$

$y=4+9$

$y=13$

$(16, \boxed{13})$ is the completed ordered pair.

35. $7x=4y$; $(?,-2)$

$7x=4(-2)$

$7x=-8$

$\frac{7x}{\mathbf{7}}=\frac{-8}{\mathbf{7}}$

$x=-\frac{8}{7}$

$\left(\boxed{-\frac{8}{7}}, -2\right)$ is the completed ordered pair.

Complete each table of solutions.

37. $y=2x-4$

$(8,?)$

$y=2(8)-4$

$y=16-4$

$y=12$

$y=2x-4$

$(?,8)$

$8=2x-4$

$8+\mathbf{4}=2x-4+\mathbf{4}$

$12=2x$

$\frac{12}{\mathbf{2}}=\frac{2x}{\mathbf{2}}$

$6=x$

x	y	(x, y)
8	**12**	**(8,12)**
6	8	**(6,8)**

39. $3x-y=-2$

$(-5,?)$

$3(-5)-y=-2$

$-15-y=-2$

$-15-y+\mathbf{15}=-2+\mathbf{15}$

$-y=13$

$\frac{-y}{\mathbf{-1}}=\frac{13}{\mathbf{-1}}$

$y=-13$

$3x-y=-2$

$(?,-1)$

$3x-(-1)=-2$

$3x+1-\mathbf{1}=-2-\mathbf{1}$

$3x=-3$

$\frac{3x}{\mathbf{3}}=\frac{-3}{\mathbf{3}}$

$x=-1$

x	y	(x, y)
−5	**−13**	**(−5,−13)**
−1	−1	**(−1,−1)**

Complete a table of solutions and then graph each equation.

41. $y=2x-3$

x	y	(x, y)
−2	$y=2x-3$ $=2(-2)-3$ $=-4-3$ $=-7$	(−2, −7)
0	$y=2x-3$ $=2(0)-3$ $=0-3$ $=-3$	(0, −3)
2	$y=2x-3$ $=2(2)-3$ $=4-3$ $=1$	(2, 1)

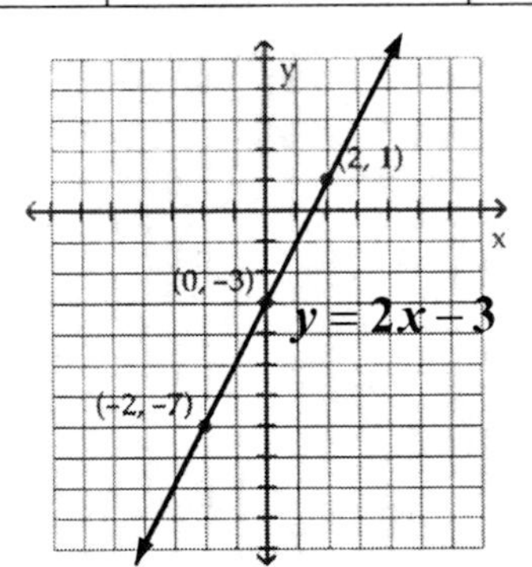

43. $y = 5x - 4$

x	y	(x, y)
-1	$y = 5x - 4$ $= 5(-1) - 4$ $= -5 - 4$ $= -9$	$(-1, -9)$
0	$y = 5x - 4$ $= 5(0) - 4$ $= 0 - 4$ $= -4$	$(0, -4)$
1	$y = 5x - 4$ $= 5(1) - 4$ $= 5 - 4$ $= 1$	$(1, 1)$

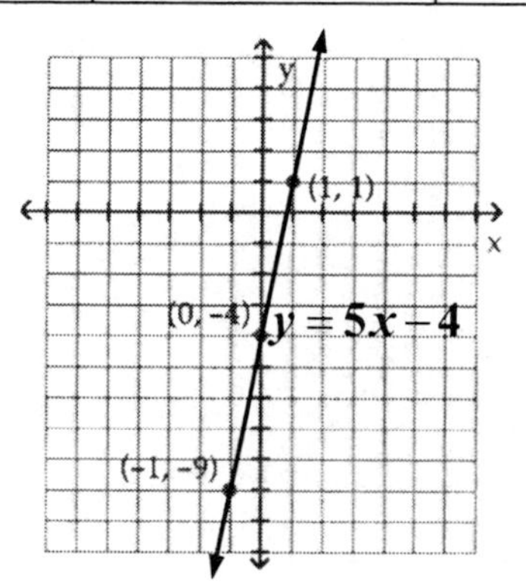

45. $y = -6x$

x	y	(x, y)
-1	$y = -6x$ $= -6(-1)$ $= 6$	$(-1, 6)$
0	$y = -6x$ $= -6(0)$ $= 0$	$(0, 0)$
1	$y = -6x$ $= -6(1)$ $= -6$	$(1, -6)$

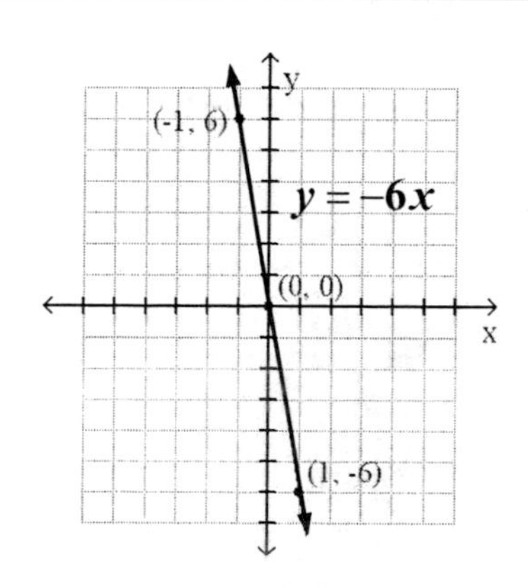

47. $y = -7x$

x	y	(x, y)
-1	$y = -7x$ $= -7(-1)$ $= 7$	$(-1, 7)$
0	$y = -7x$ $= -7(0)$ $= 0$	$(0, 0)$
1	$y = -7x$ $= -7(1)$ $= -7$	$(1, -7)$

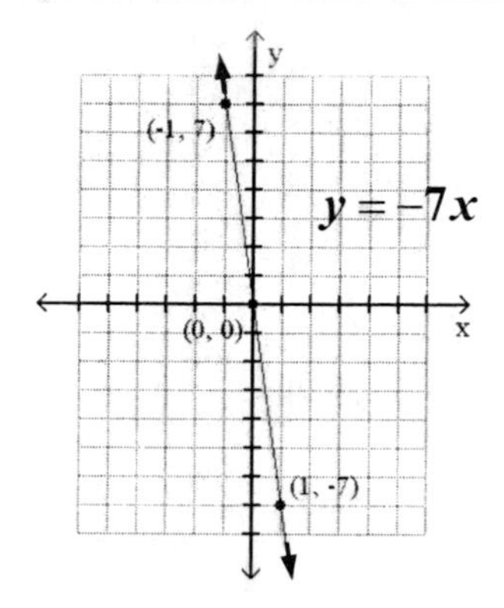

49.

$$\begin{aligned} 2x + 3y &= -3 \\ 2x + 3y - \mathbf{2x} &= -3 - \mathbf{2x} \\ 3y &= -2x - 3 \\ \frac{3y}{\mathbf{3}} &= \frac{-2x}{\mathbf{3}} - \frac{3}{\mathbf{3}} \\ y &= -\tfrac{2}{3}x - 1 \end{aligned}$$

x	y	(x, y)
-3	$y = -\frac{2}{3}x - 1$ $= -\frac{2}{3}(-3) - 1$ $= 2 - 1$ $= 1$	$(-3, 1)$
0	$y = -\frac{2}{3}x - 1$ $= -\frac{2}{3}(0) - 1$ $= -1$	$(0, -1)$
3	$y = -\frac{2}{3}x - 1$ $= -\frac{2}{3}(3) - 1$ $= -2 - 1$ $= -3$	$(3, -3)$

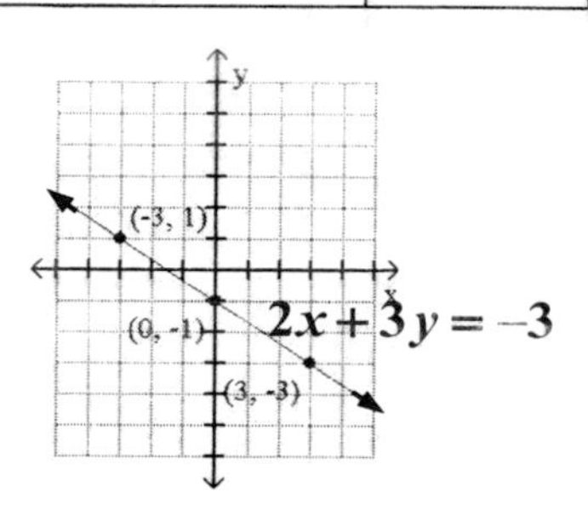

51.
$$5y - x = 20$$
$$5y - x + \mathbf{x} = 20 + \mathbf{x}$$
$$5y = x + 20$$
$$\frac{5y}{\mathbf{5}} = \frac{x}{\mathbf{5}} + \frac{20}{\mathbf{5}}$$
$$y = \tfrac{1}{5}x + 4$$

x	y	(x, y)
–5	$y = \frac{1}{5}x + 4$ $= \frac{1}{5}(-5) + 4$ $= -1 + 4$ $= 3$	(–5, 3)
0	$y = \frac{1}{5}x + 4$ $= \frac{1}{5}(0) + 4$ $= 4$	(0, 4)
5	$y = \frac{1}{5}x + 4$ $= \frac{1}{5}(5) + 4$ $= 1 + 4$ $= 5$	(5, 5)

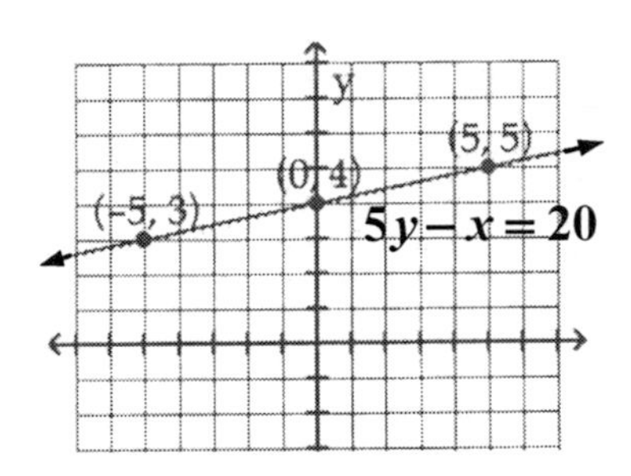

53. $y = x$

x	y	(x, y)
–3	–3	(–3, –3)
0	0	(0, 0)
4	4	(4, 4)

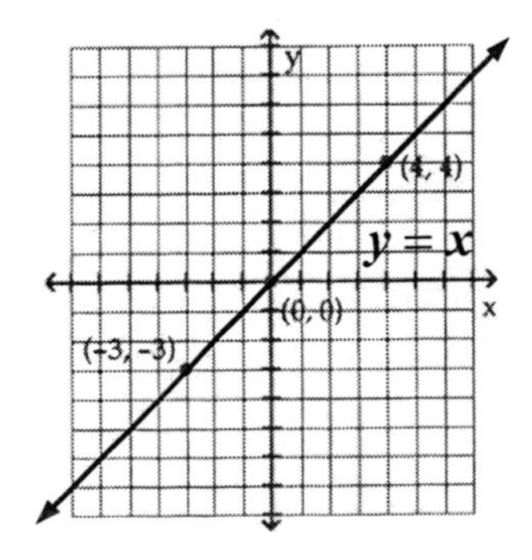

55. $y = -x - 1$

x	y	(x, y)
–2	$y = -x - 1$ $= -(-2) - 1$ $= 2 - 1$ $= 1$	(–2, 1)
0	$y = -x - 1$ $= -(0) - 1$ $= 0 - 1$ $= -1$	(0, –1)
2	$y = -x - 1$ $= -(2) - 1$ $= -2 - 1$ $= -3$	(2, –3)

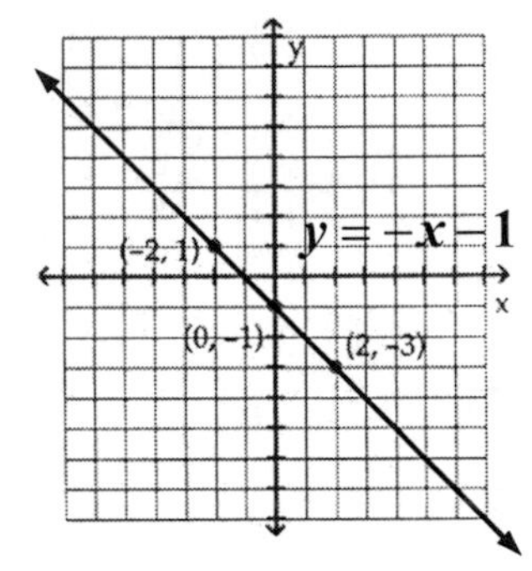

57.
$$3y = 12x + 15$$
$$\frac{3y}{\mathbf{3}} = \frac{12x}{\mathbf{3}} + \frac{15}{\mathbf{3}}$$
$$y = 4x + 5$$

x	y	(x, y)
–1	$y = 4x + 5$ $= 4(-1) + 5$ $= -4 + 5$ $= 1$	(–1, 1)
0	$y = 4x + 5$ $= 4(0) + 5$ $= 0 + 5$ $= 5$	(0, 5)
1	$y = 4x + 5$ $= 4(1) + 5$ $= 4 + 5$ $= 9$	(1, 9)

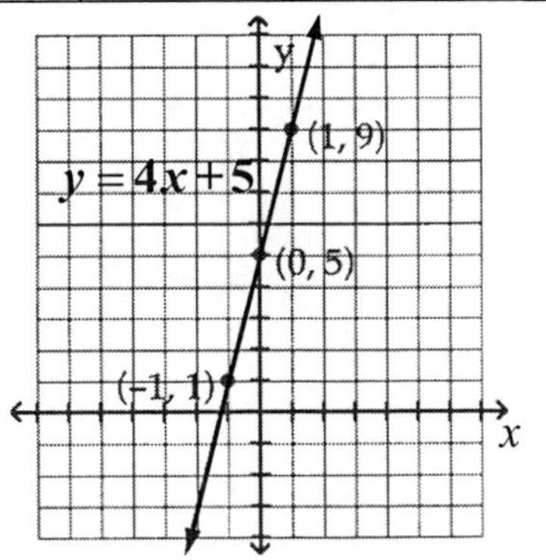

59. $y = \frac{3}{8}x - 6$

x	y	(x, y)
–8	$y = \frac{3}{8}x - 6$ $= \frac{3}{8}(-8) - 6$ $= -3 - 6$ $= -9$	(–8, –9)
0	$y = \frac{3}{8}x - 6$ $= \frac{3}{8}(0) - 6$ $= 0 - 6$ $= -6$	(0, –6)
8	$y = \frac{3}{8}x - 6$ $= \frac{3}{8}(8) - 6$ $= 3 - 6$ $= -3$	(8, –3)

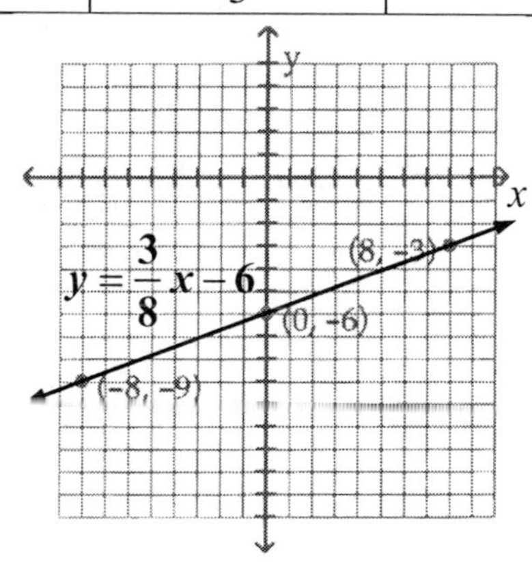

61. $y = 1.5x - 4$

x	y	(x, y)
–2	$y = 1.5x - 4$ $= 1.5(-2) - 4$ $= -3 - 4$ $= -7$	(–2, –7)
0	$y = 1.5x - 4$ $= 1.5(0) - 4$ $= 0 - 4$ $= -4$	(0, –4)
2	$y = 1.5x - 4$ $= 1.5(2) - 4$ $= 3 - 4$ $= -1$	(2, –1)

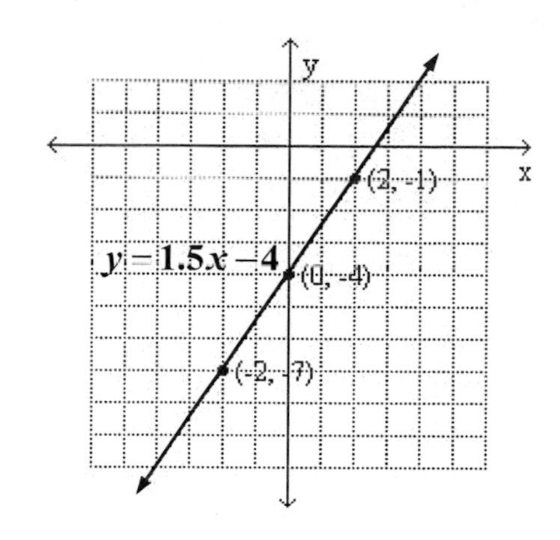

63.

$$8x + 4y = 16$$
$$8x + 4y - \mathbf{8x} = -\mathbf{8x} + 16$$
$$4y = -8x + 16$$
$$\frac{4y}{\mathbf{4}} = \frac{-8x}{\mathbf{4}} + \frac{16}{\mathbf{4}}$$
$$y = -2x + 4$$

x	y	(x, y)
–1	$y = -2x + 4$ $= -2(-1) + 4$ $= 2 + 4$ $= 6$	(–1, 6)
0	$y = -2x + 4$ $= -2(0) + 4$ $= 0 + 4$ $= 4$	(0, 4)
1	$y = -2x + 4$ $= -2(1) + 4$ $= -2 + 4$ $= 2$	(1, 2)

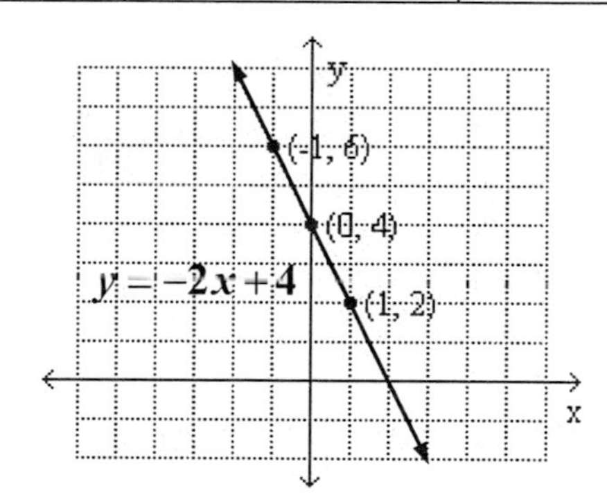

65. $y = -\frac{1}{2}x$

x	y	(x, y)
-2	$y = -\frac{1}{2}x$ $= -\frac{1}{2}(-2)$ $= 1$	$(-2, 1)$
0	$y = -\frac{1}{2}x$ $= -\frac{1}{2}(0)$ $= 0$	$(0, 0)$
2	$y = -\frac{1}{2}x$ $= -\frac{1}{2}(2)$ $= -1$	$(2, -1)$

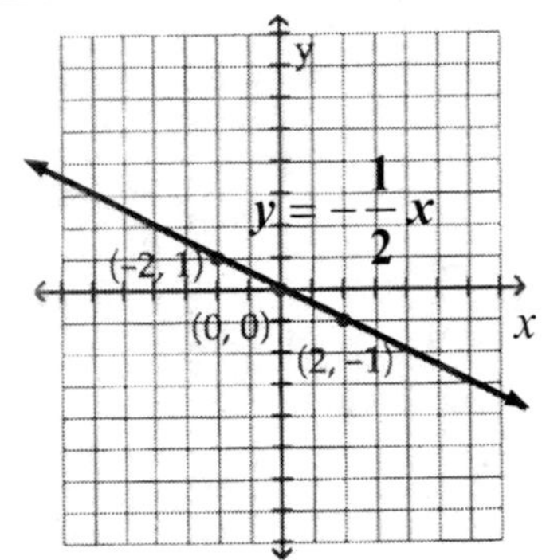

67. $y = \frac{5}{6}x - 5$

x	y	(x, y)
-6	$y = \frac{5}{6}x - 5$ $= \frac{5}{6}(-6) - 5$ $= -5 - 5$ $= -10$	$(-6, -10)$
0	$y = \frac{5}{6}x - 5$ $= \frac{5}{6}(0) - 5$ $= 0 - 5$ $= -5$	$(0, -5)$
6	$y = \frac{5}{6}x - 5$ $= \frac{5}{6}(6) - 5$ $= 5 - 5$ $= 0$	$(6, 0)$

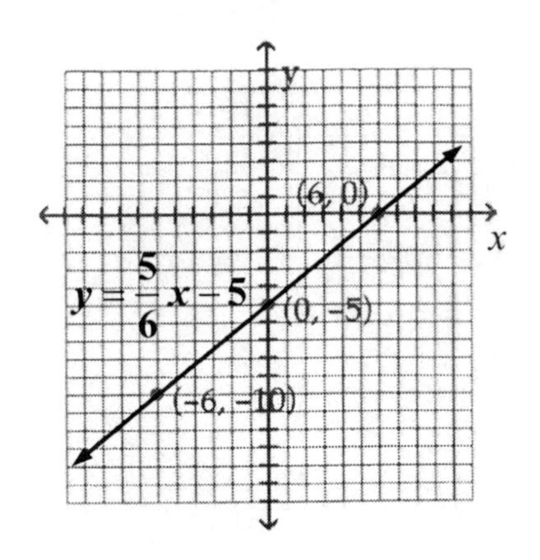

69. $$-6y = 30x + 12$$
$$\frac{-6y}{-6} = \frac{30x}{-6} + \frac{12}{-6}$$
$$y = -5x - 2$$

x	y	(x, y)
-1	$y = -5x - 2$ $= -5(-1) - 2$ $= 5 - 2$ $= 3$	$(-1, 3)$
0	$y = -5x - 2$ $= -5(0) - 2$ $= 0 - 2$ $= -2$	$(0, -2)$
1	$y = -5x - 2$ $= -5(1) - 2$ $= -5 - 2$ $= -7$	$(1, -7)$

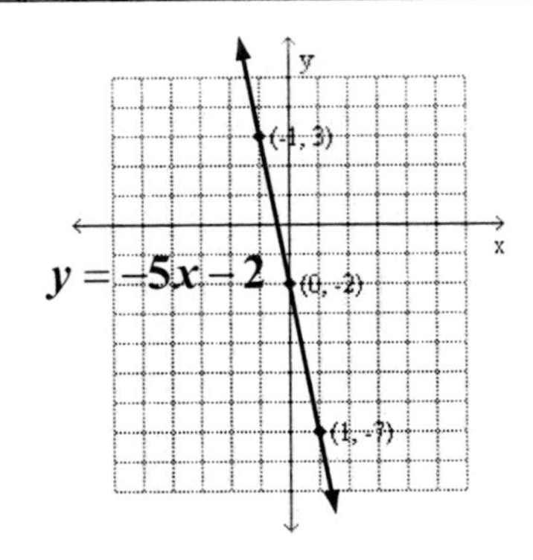

71. $y = \dfrac{x}{3}$

x	y	(x, y)
-3	$y = \frac{x}{3}$ $= \frac{-3}{3}$ $= -1$	$(-3, -1)$
0	$y = \frac{x}{3}$ $= \frac{0}{3}$ $= 0$	$(0, 0)$
3	$y = \frac{x}{3}$ $= \frac{3}{3}$ $= 1$	$(3, 1)$

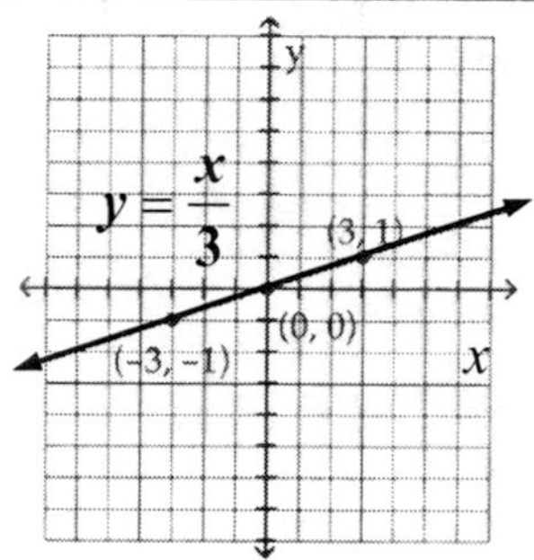

73. $y = -2x + 1$

x	y	(x, y)
-1	$y = -2x + 1$ $= -2(-1) + 1$ $= 2 + 1$ $= 3$	$(-1, 3)$
0	$y = -2x + 1$ $= -2(0) + 1$ $= 0 + 1$ $= 1$	$(0, 1)$
1	$y = -2x + 1$ $= -2(1) + 1$ $= -2 + 1$ $= -1$	$(1, -1)$

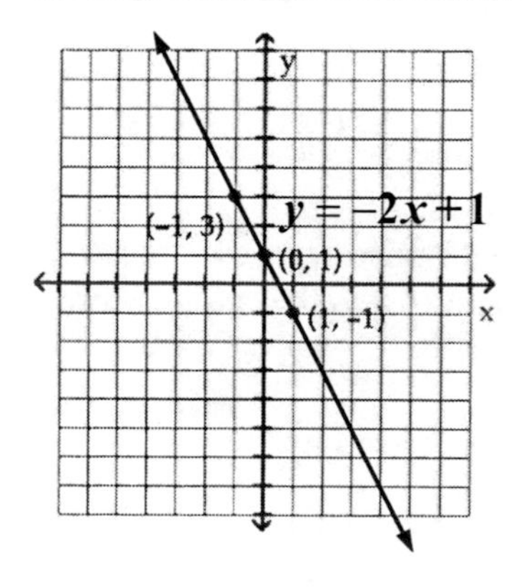

75.
$$7x - y = 1$$
$$7x - y + \mathbf{y} = \mathbf{y} + 1$$
$$7x = y + 1$$
$$7x - \mathbf{1} = y + 1 - \mathbf{1}$$
$$y = 7x - 1$$

x	y	(x, y)
-1	$y = 7x - 1$ $= 7(-1) - 1$ $= -7 - 1$ $= -8$	$(-1, -8)$
0	$y = 7x - 1$ $= 7(0) - 1$ $= 0 - 1$ $= -1$	$(0, -1)$
1	$y = 7x - 1$ $= 7(1) - 1$ $= 7 - 1$ $= 6$	$(1, 6)$

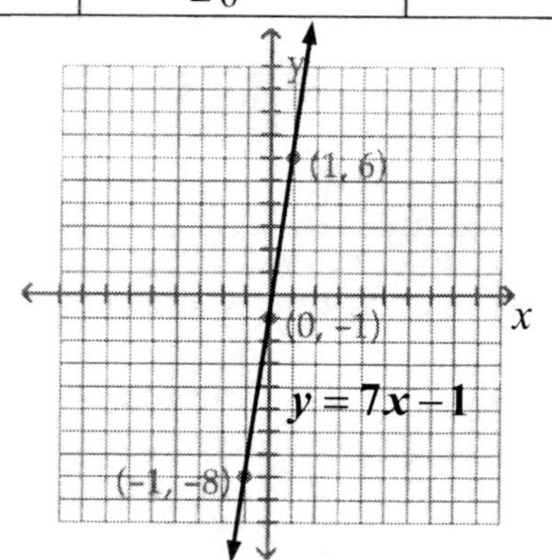

77.
$$7y = -2x$$
$$\frac{7y}{\mathbf{7}} = \frac{-2x}{\mathbf{7}}$$
$$y = -\frac{2}{7}x$$

x	y	(x, y)
-7	$y = -\frac{2}{7}x$ $= -\frac{2}{7}(-7)$ $= 2$	$(-7, 2)$
0	$y = -\frac{2}{7}x$ $= -\frac{2}{7}(0)$ $= 0$	$(0, 0)$
7	$y = -\frac{2}{7}x$ $= -\frac{2}{7}(7)$ $= -2$	$(7, -2)$

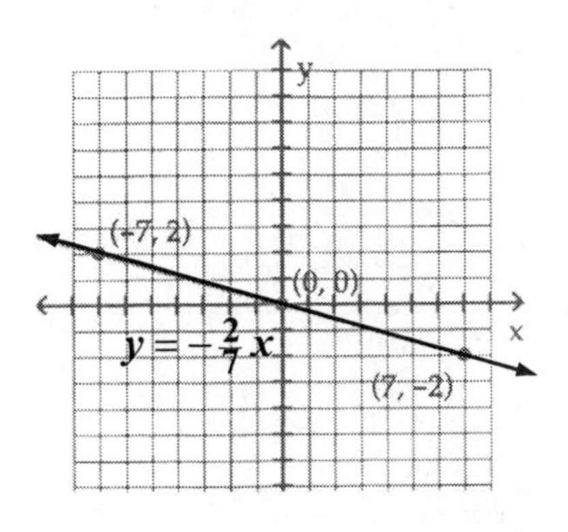

79. $y=-2.5x+5$

x	y	(x, y)
–2	$y=-2.5x+5$ $=-2.5(-2)+5$ $=5+5$ $=10$	(–2, 10)
0	$y=-2.5x+5$ $=-2.5(0)+5$ $=5$	(0, 5)
2	$y=-2.5x+5$ $=-2.5(2)+5$ $=-5+5$ $=0$	(2, 0)

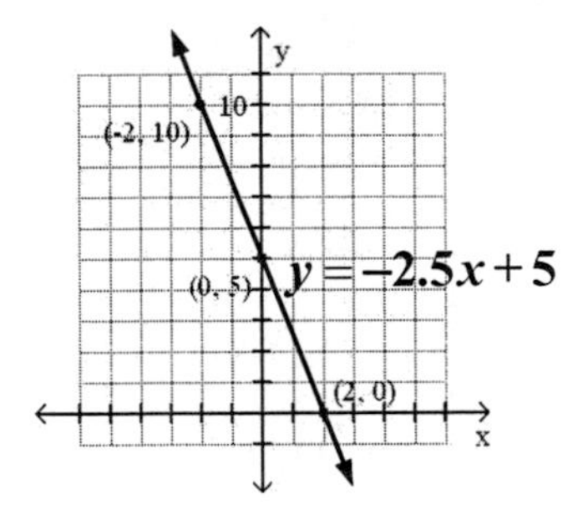

APPLICATIONS

81. BILLARDS

$y=2x-4$

x	y	(x, y)
1	–2	(1, –2)
2	0	(2, 0)
4	4	(4, 4)

$y=-2x+12$

x	y	(x, y)
4	4	(4, 4)
6	0	(6, 0)
8	–4	(8, –4)

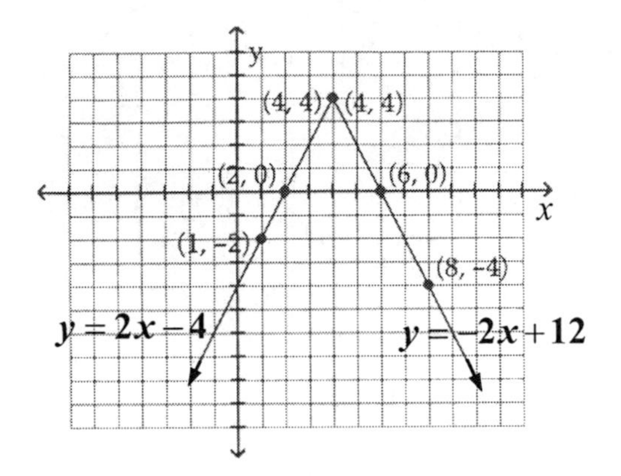

83. DEFROSTING POULTRY

$h=5p$

p	h	(p, h)
5	25	(5, 25)
15	75	(15, 75)
20	100	(20, 100)
25	?	(25, ?)

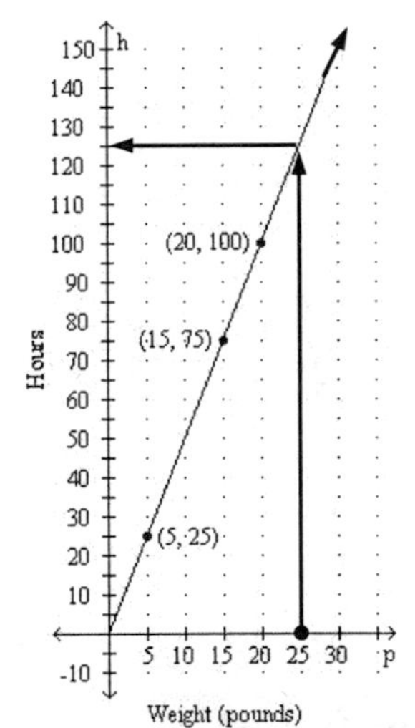

It will take a 25-pound turkey
125 hours to defrost in the refrigerator.

85. HOUSEKEEPING

The problem wants you to estimate (not calculate) the amount of polish that is left in the bottle after 650 sprays. Use a graph to find the answer.

n = number of sprays (x-axis) scale of 50
A = ounces remaining (y-axis) scale of 2

$$A = -0.02n + 16$$

n	A	(n, A)
100	$A = -0.02(100) + 16$ $= -2 + 16$ $= 14$	(100, 14)
200	12	(200, 12)
300	$A = -0.02(300) + 16$ $= -6 + 16$ $= 10$	(300, 10)
500	6	(500, 6)
650	?	(650, ?)

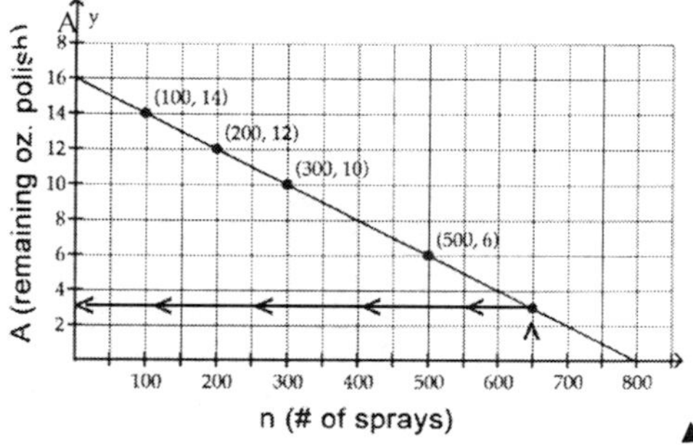

The amount of polish left in the bottle after 650 sprays is 3 ounces.

87. NFL TICKETS

The problem wants you to estimate (not calculate) the price of a ticket in 2020. Use a graph to find the answer.

t = the year (x-axis) scale of 2
p = price of the ticket (y-axis) scale of 5

$$p = 2.7t + 20$$

t	p	(t, p)
0 (1990)	20	(0, 20)
5 (1995)	33.50	(5, 33.50)
10 (2000)	47	(10, 47)
20(2010)	74	(20, 74)
30(2020)	?	(30, ?)

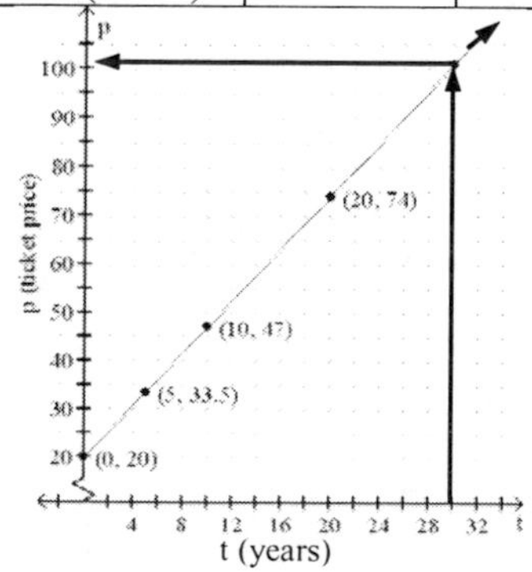

In the year 2020, the price of a NFL football ticket will be about $101.

89. RAFFLES

The problem wants you to predict (not calculate) the number of raffle tickets that will be sold at $6. Use a graph to find the answer.

p = the price (x-axis) scale of 1
n = the number of tickets (y-axis) scale of 10

$$n = -20p + 300$$

p	n	(p, n)
1	280	(1, 280)
3	240	(3, 240)
5	200	(5, 200)
6	?	(6, ?)

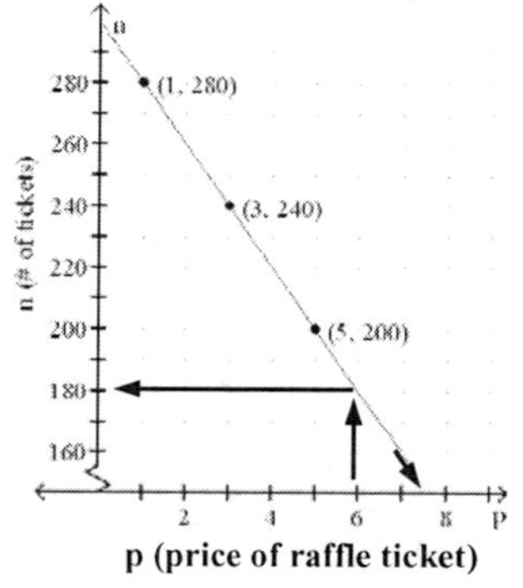

The predicted number of $6 raffle tickets to be sold is around 180 tickets.

91. U.S.SPACE PROGRAM

The problem wants you to predict (not calculate) when 70% will say "yes". Use a graph to find the answer.

n = the year (x-axis) scale of 2
p = the percent of "yes" responses (y-axis) scale of 5

$$p = \frac{3}{5}n + 40$$

n	p	(n, p)
0(1980)	40	(0, 40)
5(1985)	43	(5, 43)
10(1990)	46	(10, 46)
20(2000)	52	(20, 52)
?	70	(? ,70)

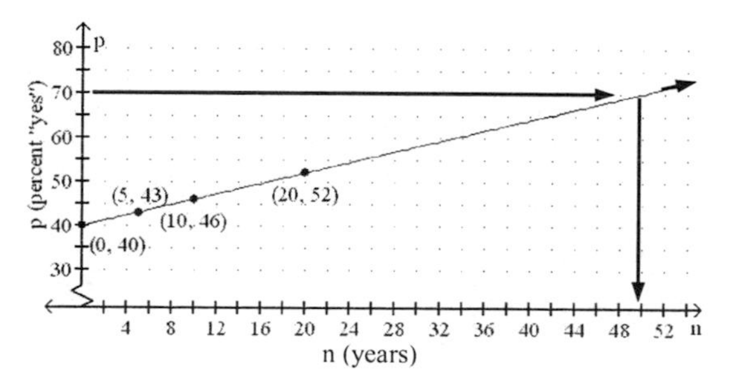

The predicted year that 70% will respond with a "yes" is the year 2030 (1980 + 50).

WRITING

93 – 99. Answers will vary.

REVIEW

101. Simplify: $-(-5-4c) = -(-5)-(-4c)$
$= 5+4c$

103. $V = \frac{4}{3}\pi r^3, \quad \pi = 3.141592654$
$= \frac{4}{3}(3.141592654)(6)(6)(6)$
≈ 904.77
≈ 904.8
The volume of the sphere is approximately 904.8 ft^3.

CHALLENGE PROBLEMS

105. Graph $y = x^2 + 1$.

x	$y = x^2 + 1$	(x, y)
–3	$y = (-3)^2 + 1$ $= 9+1$ $= 10$	(–3, 10)
–2	$y = (-2)^2 + 1$ $= 4+1$ $= 5$	(–2, 5)
–1	$y = (-1)^2 + 1$ $= 1+1$ $= 2$	(–1, 2)
0	$y = (0)^2 + 1$ $= 0+1$ $= 1$	(0, 1)
1	$y = (1)^2 + 1$ $= 1+1$ $= 2$	(1, 2)
2	$y = (2)^2 + 1$ $= 4+1$ $= 5$	(2, 5)
3	$y = (3)^2 + 1$ $= 9+1$ $= 10$	(3, 10)

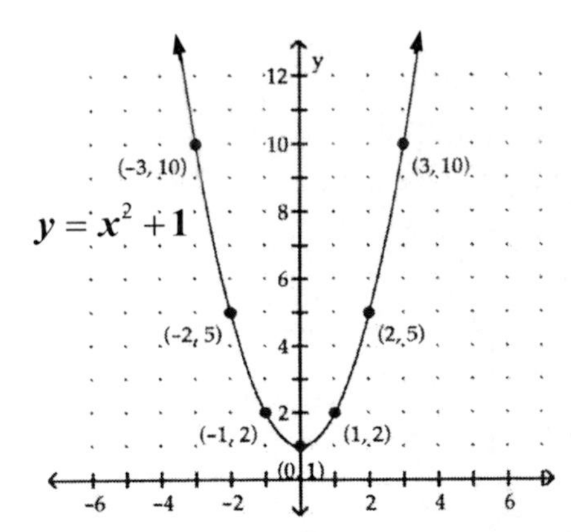

This type of graph is called a **Parabola**.

107. Graph $y = |x| - 2$.

x	$y = \|x\| - 2$	(x, y)
–3	$y = \|-3\| - 2$ $= 3-2$ $= 1$	(–3, 1)
–2	$y = \|-2\| - 2$ $= 2-2$ $= 0$	(–2, 0)
–1	$y = \|-1\| - 2$ $= 1-2$ $= -1$	(–1, –1)
0	$y = \|0\| - 2$ $= 0-2$ $= -2$	(0, –2)
1	$y = \|1\| - 2$ $= 1-2$ $= -1$	(1, –1)
2	$y = \|2\| - 2$ $= 2-2$ $= 0$	(2, 0)
3	$y = \|3\| - 2$ $= 3-2$ $= 1$	(3, 1)

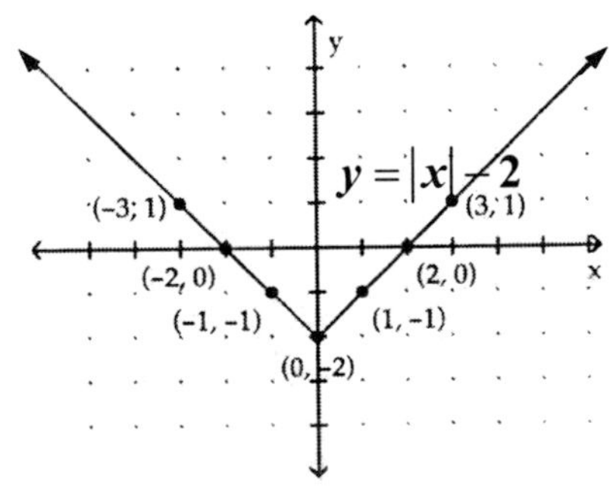

This kind of graph is called the **Absolute Value** graph.

SECTION 3.3 STUDY SET
VOCABULARY

Fill in the blanks.

1. The **x-intercept** of a line is the point where the line intersects/crosses the x-axis.

3. The graph of $y = 4$ is a **horizontal** line and the graph of $x = 6$ is a **vertical** line.

CONCEPTS

5. **Fill in the blanks.**
   a. To find the y-intercept of the graph of a line, substitute **0** for x in the equation and solve for **y**.
   b. To find the x-intercept of the graph of a line, substitute **0** for y in the equation and solve for **x**.

7. a. Which intercept tells the purchase price of the machinery? ***y*-intercept: (0, 80,000)**
   What was that price? **$80,000**
   b. Which intercept indicates when the machinery will have lost all of its value?
   ***x*-intercept: (30, 0)**
   When is that? **30 years after purchase**

NOTATION

9. What is the equation of the x-axis? $\mathbf{y = 0}$
   What is the equation of the y-axis? $\mathbf{x = 0}$

GUIDED PRACTICE

Give the coordinates of the intercepts of each graph.

11. The x-intercept is (4, 0) and the y-intercept is (0, 3).
13. The x-intercept is (–5, 0) and the y-intercept is (0, –4).
15. The y-intercept is (0, 2). There is no x-intercept because the line does not cross the x-axis.

Estimate the coordinates of the intercepts of each graph.

17. The x-intercept is $\left(-2\frac{1}{2}, 0\right)$ and the y-intercept is $\left(0, \frac{2}{3}\right)$.

Find the x- and y-intercepts of the graph of each equation. *Do not graph the line.*

19. x – intercept: $y = 0$

$$8x + 3y = 24$$
$$8x + 3(0) = 24$$
$$8x = 24$$
$$x = 3$$

The x – intercept is $(3, 0)$.

y – intercept: $x = 0$

$$8x + 3y = 24$$
$$8(0) + 3y = 24$$
$$3y = 24$$
$$y = 8$$

The y – intercept is $(0, 8)$.

21. x – intercept: $y = 0$

$$7x - 2y = 28$$
$$7x - 2(0) = 28$$
$$7x = 28$$
$$x = 4$$

The x – intercept is $(4, 0)$.

y – intercept: $x = 0$

$$7x - 2y = 28$$
$$7(0) - 2y = 28$$
$$-2y = 28$$
$$y = -14$$

The y – intercept is $(0, -14)$.

23. x – intercept: $y = 0$

$$-5x - 3y = 10$$
$$-5x - 3(0) = 10$$
$$-5x = 10$$
$$x = -2$$

The x – intercept is $(-2, 0)$.

y – intercept: $x = 0$

$$-5x - 3y = 10$$
$$-5(0) - 3y = 10$$
$$-3y = 10$$
$$y = -\frac{10}{3}$$

The y – intercept is $\left(0, -\frac{10}{3}\right)$.

25. x – intercept: $y = 0$

$$6x + y = 9$$
$$6x + (0) = 9$$
$$6x = 9$$
$$x = \frac{9}{6}$$
$$x = \frac{3}{2}$$

The x – intercept is $\left(\frac{3}{2}, 0\right)$.

y – intercept: $x = 0$

$$6x + y = 9$$
$$6(0) + y = 9$$
$$y = 9$$

The y – intercept is $(0, 9)$.

Use the intercept method to graph each equation.

27. $4x + 5y = 20$

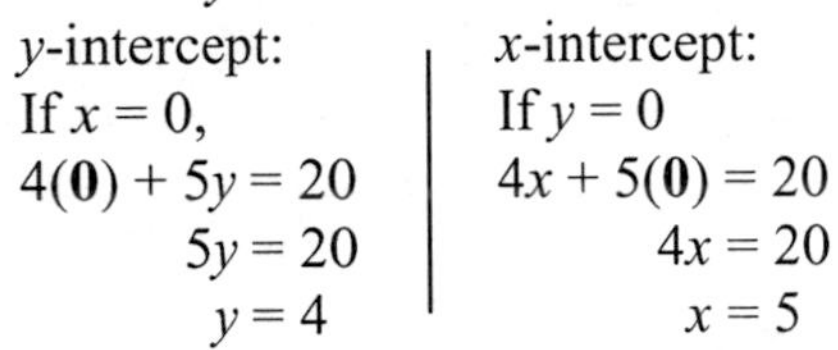

y-intercept:	x-intercept:
If $x = 0$,	If $y = 0$
$4(\mathbf{0}) + 5y = 20$	$4x + 5(\mathbf{0}) = 20$
$5y = 20$	$4x = 20$
$y = 4$	$x = 5$

The y-intercept is (0, 4), and the x-intercept is (5, 0).

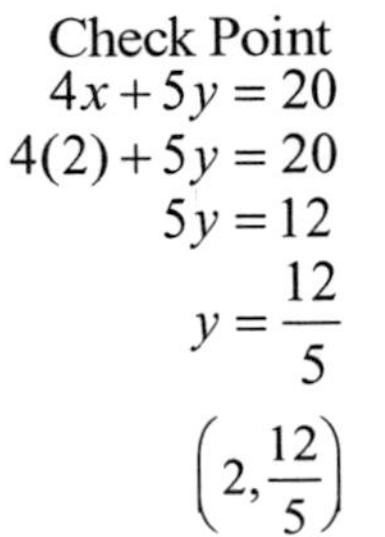

Check Point
$$4x + 5y = 20$$
$$4(2) + 5y = 20$$
$$5y = 12$$
$$y = \frac{12}{5}$$
$$\left(2, \frac{12}{5}\right)$$

29. $5x + 15y = -15$

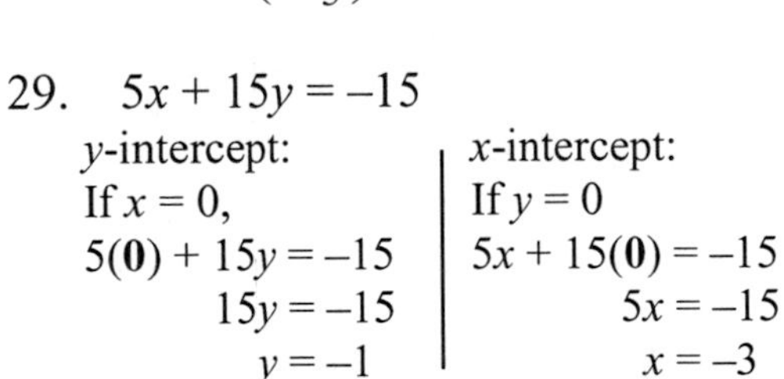

y-intercept:	x-intercept:
If $x = 0$,	If $y = 0$
$5(\mathbf{0}) + 15y = -15$	$5x + 15(\mathbf{0}) = -15$
$15y = -15$	$5x = -15$
$y = -1$	$x = -3$

The y-intercept is (0, –1), and the x-intercept is (–3, 0).

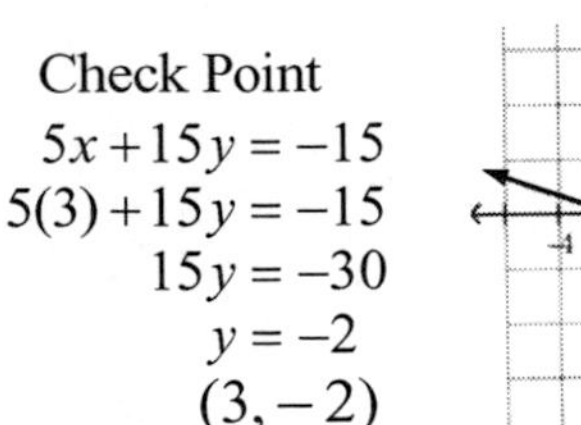

Check Point
$$5x + 15y = -15$$
$$5(3) + 15y = -15$$
$$15y = -30$$
$$y = -2$$
$$(3, -2)$$

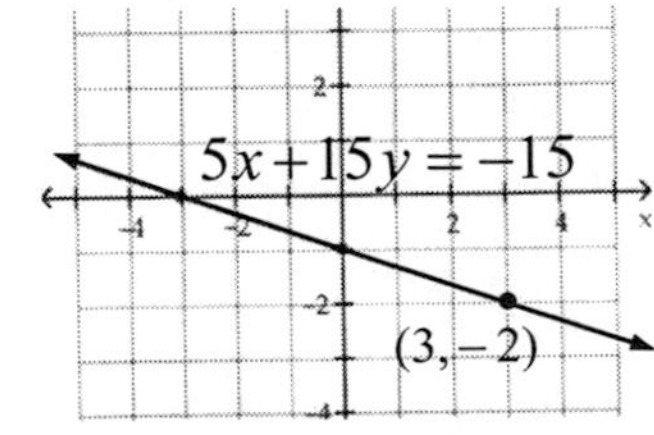

31. $x - y = -3$

y-intercept:	x-intercept:
If $x = 0$,	If $y = 0$
$(\mathbf{0}) - y = -3$	$x - (\mathbf{0}) = -3$
$-y = -3$	$x = -3$
$y = 3$	

The y-intercept is (0, 3), and the x-intercept is (–3, 0).

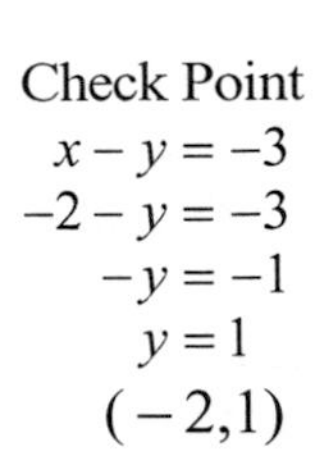

Check Point
$$x - y = -3$$
$$-2 - y = -3$$
$$-y = -1$$
$$y = 1$$
$$(-2, 1)$$

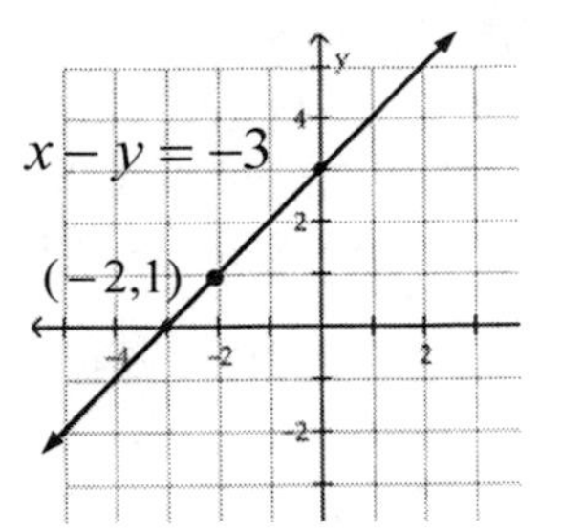

33. $x + 2y = -2$

y-intercept:	x-intercept:
If $x = 0$,	If $y = 0$
$(\mathbf{0}) + 2y = -2$	$x + 2(\mathbf{0}) = -2$
$2y = -2$	$x = -2$
$y = -1$	

The y-intercept is (0, –1), and the x-intercept is (–2, 0).

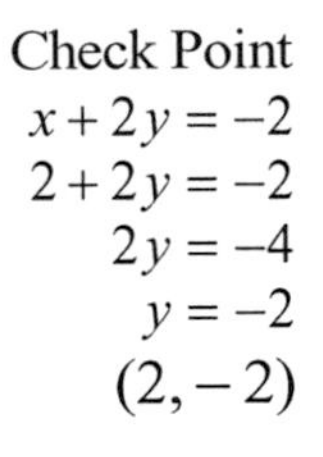

Check Point
$$x + 2y = -2$$
$$2 + 2y = -2$$
$$2y = -4$$
$$y = -2$$
$$(2, -2)$$

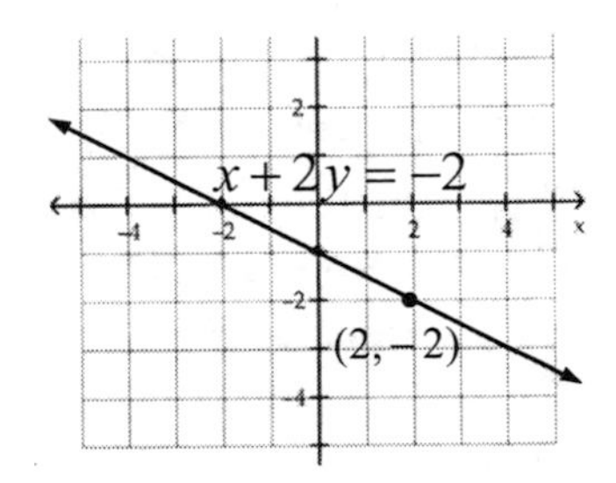

Use the intercept method to graph each equation.

35. $30x + y = -30$

y-intercept:	x-intercept:
If $x = 0$,	If $y = 0$
$30(\mathbf{0}) + y = -30$	$30x + (\mathbf{0}) = -30$
$y = -30$	$30x = -30$
	$x = -1$

The y-intercept is (0, –30), and the x-intercept is (–1, 0).

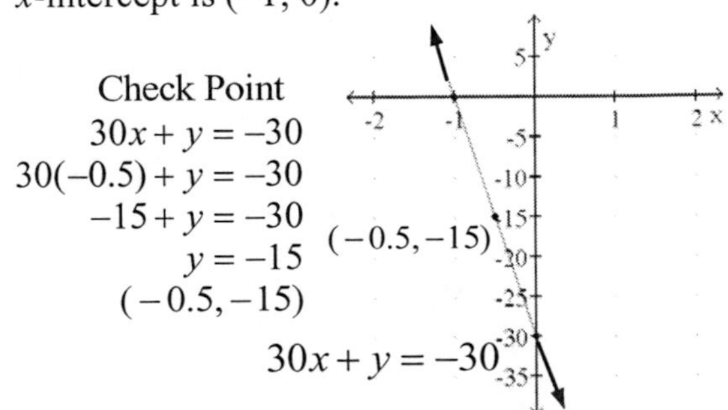

Check Point
$$30x + y = -30$$
$$30(-0.5) + y = -30$$
$$-15 + y = -30$$
$$y = -15$$
$$(-0.5, -15)$$

37. $4x - 20y = 60$

y-intercept:	x-intercept:
If $x = 0$,	If $y = 0$
$4(\mathbf{0}) - 20y = 60$	$4x - 20(\mathbf{0}) = 60$
$-20y = 60$	$4x = 60$
$y = -3$	$x = 15$

The y-intercept is $(0, -3)$, and the x-intercept is $(15, 0)$.

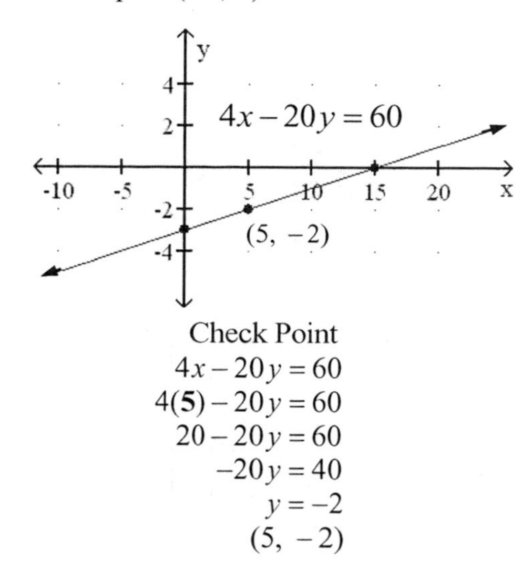

Check Point

$$4x - 20y = 60$$
$$4(\mathbf{5}) - 20y = 60$$
$$20 - 20y = 60$$
$$-20y = 40$$
$$y = -2$$
$$(5, -2)$$

Use the intercept method to graph each equation.

39. $3x + 4y = 8$

y-intercept:	x-intercept:
If $x = 0$,	If $y = 0$
$3(\mathbf{0}) + 4y = 8$	$3x + 4(\mathbf{0}) = 8$
$4y = 8$	$3x = 8$
$y = 2$	$x = \frac{8}{3}$

The y-intercept is $(0, 2)$, and the x-intercept is $(\frac{8}{3}, 0)$.

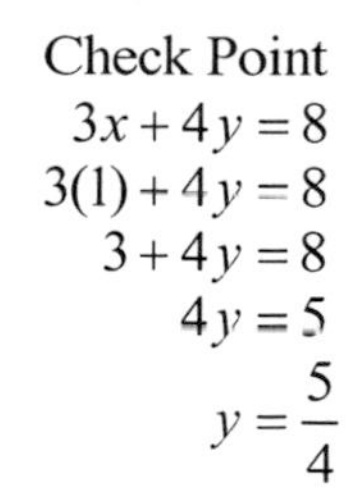

Check Point

$$3x + 4y = 8$$
$$3(1) + 4y = 8$$
$$3 + 4y = 8$$
$$4y = 5$$
$$y = \frac{5}{4}$$

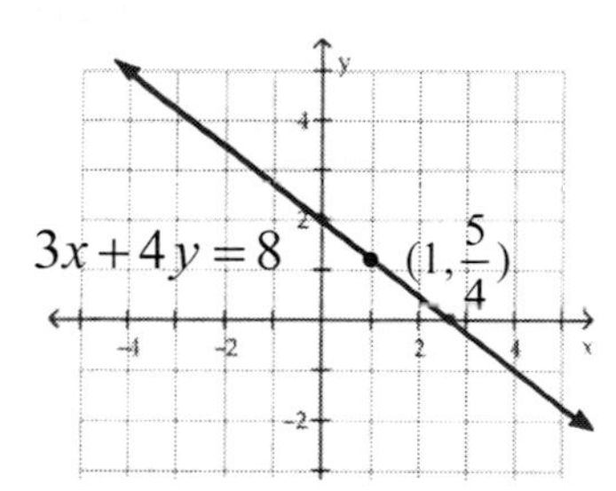

41. $-9x + 4y = 9$

y-intercept:	x-intercept:
If $x = 0$,	If $y = 0$
$-9(\mathbf{0}) + 4y = 9$	$-9x + 4(\mathbf{0}) = 9$
$4y = 9$	$-9x = 9$
$y = \frac{9}{4}$	$x = -1$

The y-intercept is $(0, \frac{9}{4})$, and the x-intercept is $(-1, 0)$.

Check Point

$$-9x + 4y = 9$$
$$-9(-2) + 4y = 9$$
$$18 + 4y = 9$$
$$4y = -9$$
$$y = -\frac{9}{4}$$

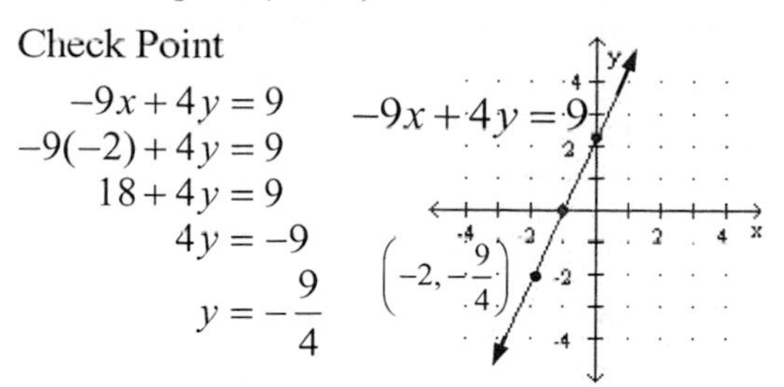

43. $3x - 4y = 11$

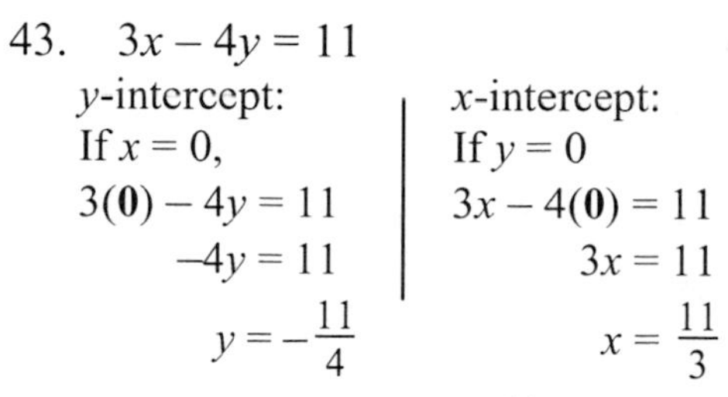

y-intercept:	x-intercept:
If $x = 0$,	If $y = 0$
$3(\mathbf{0}) - 4y = 11$	$3x - 4(\mathbf{0}) = 11$
$-4y = 11$	$3x = 11$
$y = -\frac{11}{4}$	$x = \frac{11}{3}$

The y-intercept is $(0, -\frac{11}{4})$, and the x-intercept is $(\frac{11}{3}, 0)$.

Check Point

$$3x - 4y = 11$$
$$3(1) - 4y = 11$$
$$3 - 4y = 11$$
$$-4y = 8$$
$$y = -2$$
$$(1, -2)$$

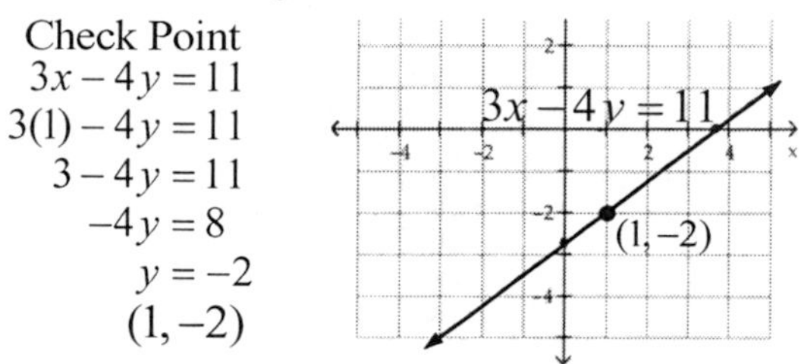

45. $9x + 3y = 10$

y-intercept:	x-intercept:
If $x = 0$,	If $y = 0$
$9(\mathbf{0}) + 3y = 10$	$9x + 3(\mathbf{0}) = 10$
$3y = 10$	$9x = 10$
$y = \frac{10}{3}$	$x = \frac{10}{9}$

The y-intercept is $(0, \frac{10}{3})$, and the x-intercept is $(\frac{10}{9}, 0)$.

Check Point

$$9x + 3y = 10$$
$$9(2) + 3y = 10$$
$$18 + 3y = 10$$
$$3y = -8$$
$$y = -\frac{8}{3}$$
$$\left(2, -\frac{8}{3}\right)$$

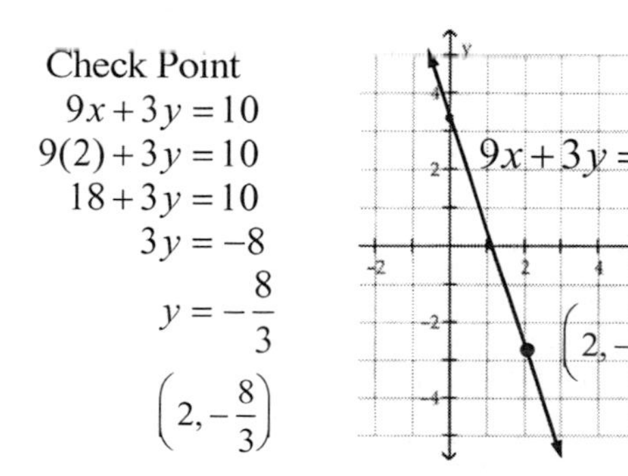

Use the intercept method to graph each equation.

47. $3x + 5y = 0$

y-intercept:	x-intercept:
If $x = 0$,	If $y = 0$
$3(\mathbf{0}) + 5y = 0$	$3x + 5(\mathbf{0}) = 0$
$5y = 0$	$3x = 0$
$y = 0$	$x = 0$

The y-intercept is (0, 0), and the x-intercept is (0, 0).

Find two other points.

If $x = 5$,	If $y = 3$
$3(\mathbf{5}) + 5y = 0$	$3x + 5(\mathbf{3}) = 0$
$15 + 5y = 0$	$3x + 15 = 0$
$5y = -15$	$3x = -15$
$y = -3$	$x = -5$
$(5, -3)$	$(-5, 3)$

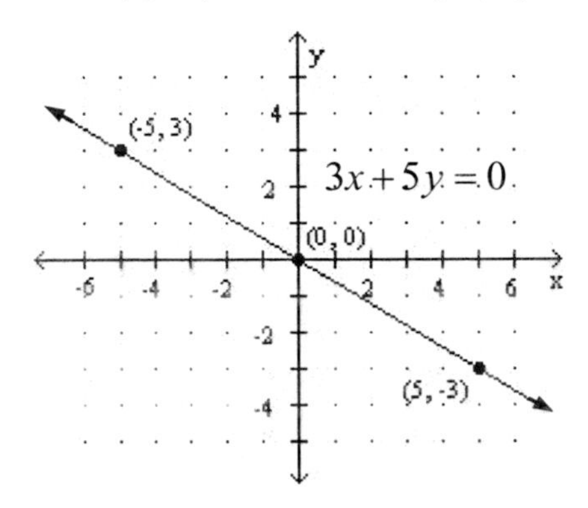

49. $2x - 7y = 0$

y-intercept:	x-intercept:
If $x = 0$,	If $y = 0$
$2(\mathbf{0}) - 7y = 0$	$2x - 7(\mathbf{0}) = 0$
$-7y = 0$	$2x = 0$
$y = 0$	$x = 0$

The y-intercept is (0, 0), and the x-intercept is (0, 0).

Find two other points.

If $x = 7$,	If $y = -2$
$2(\mathbf{7}) - 7y = 0$	$2x - 7(\mathbf{-2}) = 0$
$14 - 7y = 0$	$2x + 14 = 0$
$-7y = -14$	$2x = -14$
$y = 2$	$x = -7$
$(7, 2)$	$(-7, -2)$

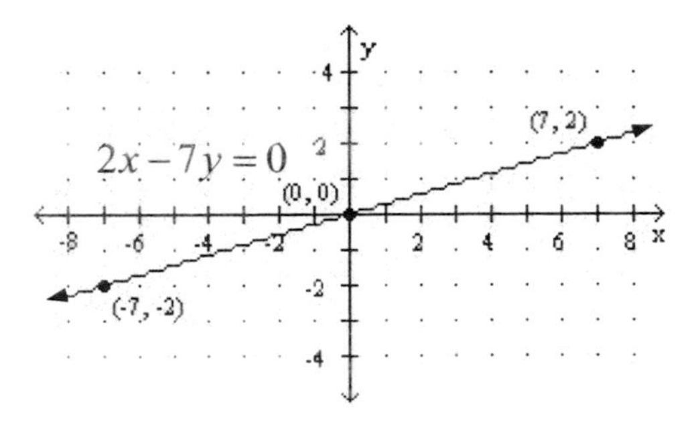

Graph each equation.

51. $y = 5$

x	y	(x, y)
−2	5	(−2,5)
0	5	(0,5)
3	5	(3,5)

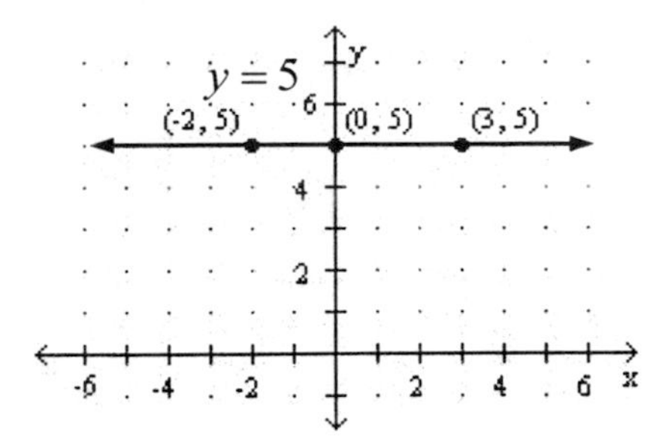

53. $y = 0$

x	y	(x, y)
−2	0	(−2,0)
0	0	(0,0)
3	0	(3,0)

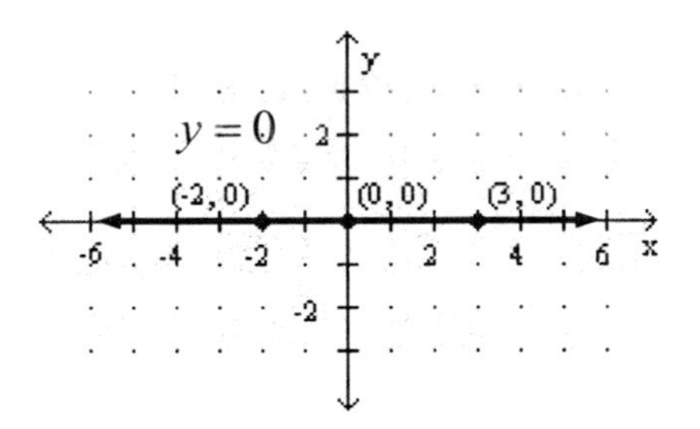

55. $x = -2$

x	y	(x, y)
−2	−3	(−2,−3)
−2	0	(−2,0)
−2	2	(−2,2)

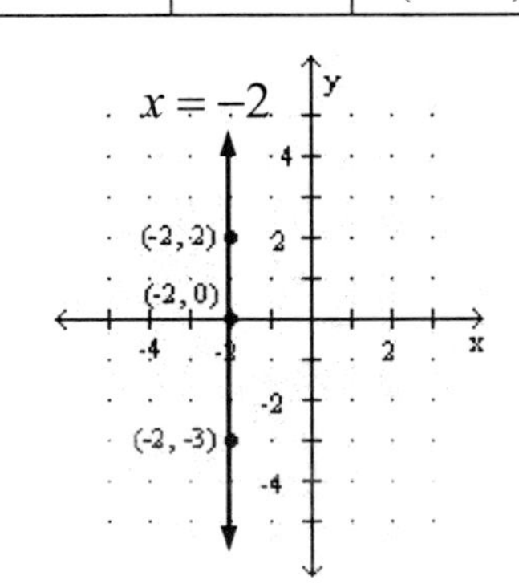

57. $x = \frac{4}{3}$

x	y	(x, y)
$\frac{4}{3}$	-3	$\left(\frac{4}{3}, -3\right)$
$\frac{4}{3}$	0	$\left(\frac{4}{3}, 0\right)$
$\frac{4}{3}$	2	$\left(\frac{4}{3}, 2\right)$

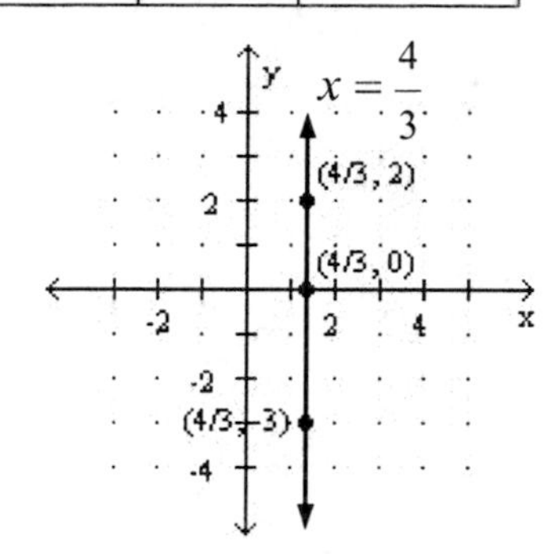

59. $y - 2 = 0$

$y = 2$

x	y	(x, y)
-2	2	$(-2, 2)$
0	2	$(0, 2)$
3	2	$(3, 2)$

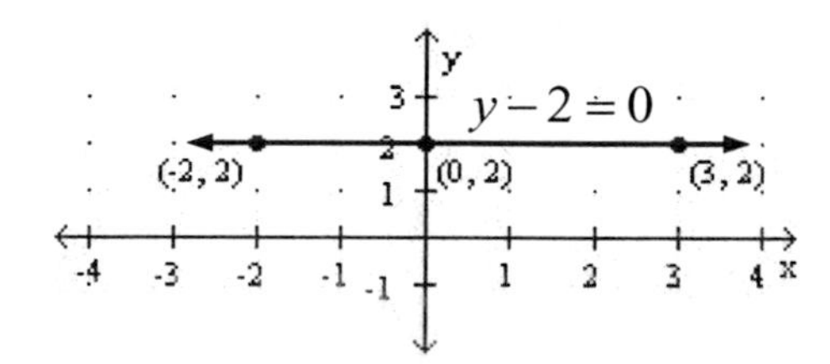

61. $5x = 7.5$

$\frac{5x}{\mathbf{5}} = \frac{7.5}{\mathbf{5}}$

$x = 1.5$

x	y	(x, y)
1.5	-3	$(1.5, -3)$
1.5	0	$(1.5, 0)$
1.5	2	$(1.5, 2)$

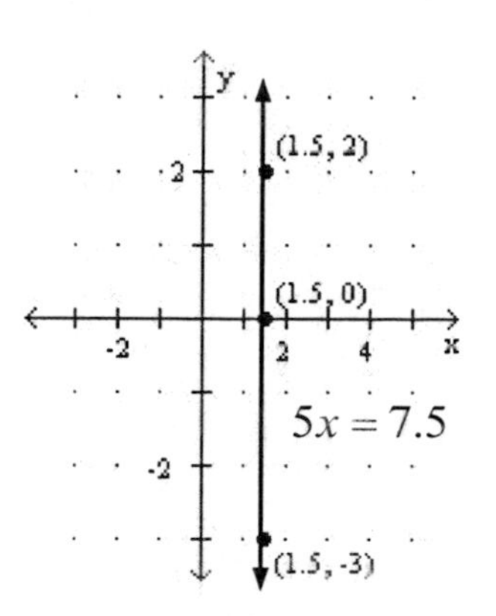

TRY IT YOURSELF

Graph each equation.

63. $7x = 4y - 12$

y-intercept:
If $x = 0$,
$7(\mathbf{0}) = 4y - 12$
$0 = 4y - 12$
$-4y = -12$
$y = 3$

x-intercept:
If $y = 0$
$7x = 4(\mathbf{0}) - 12$
$7x = -12$
$x = -\frac{12}{7}$

The y-intercept is $(0, 3)$, and the x-intercept is $(-\frac{12}{7}, 0)$.

Check Point
$7x = 4y - 12$
$7(-4) = 4y - 12$
$-28 = 4y - 12$
$-16 = 4y$
$-4 = y$

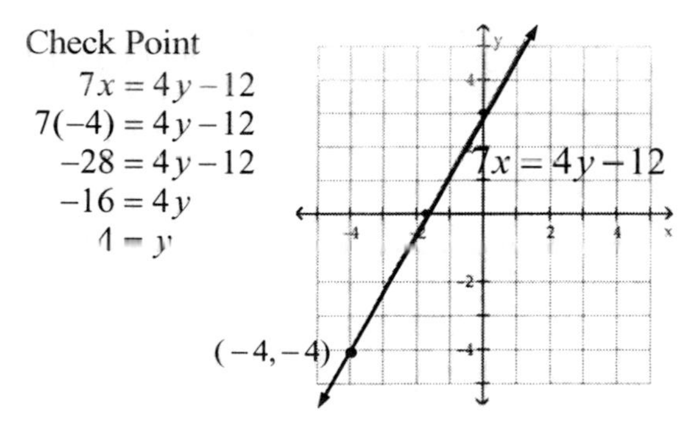

65. $4x - 3y = 12$

y-intercept:
If $x = 0$,
$4(\mathbf{0}) - 3y = 12$
$-3y = 12$
$y = -4$

x-intercept:
If $y = 0$
$4x - 3(\mathbf{0}) = 12$
$4x = 12$
$x = 3$

The y-intercept is $(0, -4)$, and the x-intercept is $(3, 0)$.

Check Point

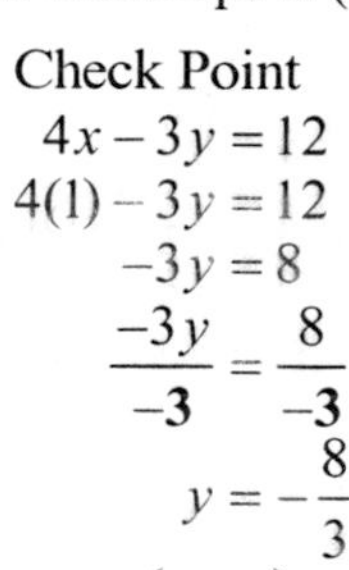

$4x - 3y = 12$
$4(1) - 3y = 12$
$-3y = 8$
$\frac{-3y}{\mathbf{-3}} = \frac{8}{\mathbf{-3}}$
$y = -\frac{8}{3}$
$\left(1, -\frac{8}{3}\right)$

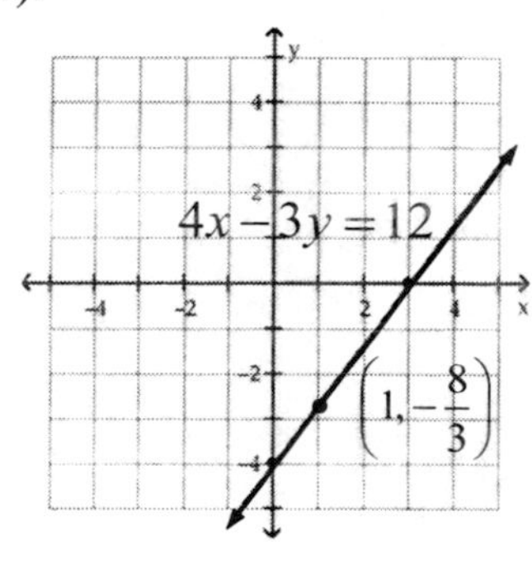

67. $x = -\frac{5}{3}$

x	y	(x, y)
$-\frac{5}{3}$	-3	$\left(-\frac{5}{3}, -3\right)$
$-\frac{5}{3}$	0	$\left(-\frac{5}{3}, 0\right)$
$-\frac{5}{3}$	2	$\left(-\frac{5}{3}, 2\right)$

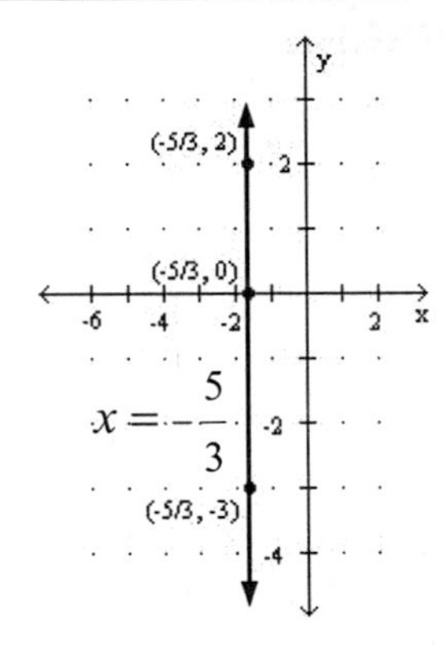

69. $y - 3x = -\frac{4}{3}$

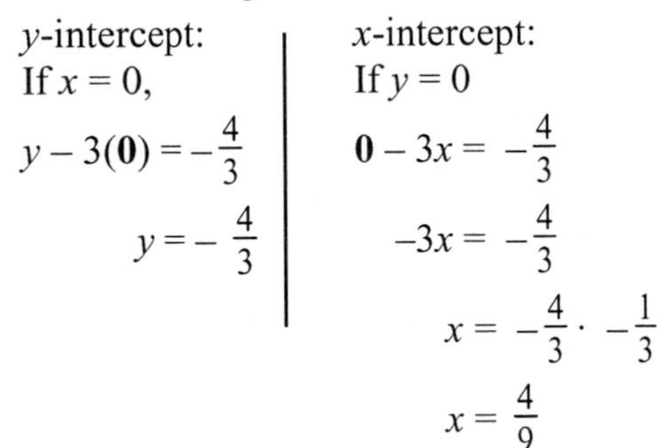

y-intercept:	x-intercept:
If $x = 0$,	If $y = 0$
$y - 3(\mathbf{0}) = -\frac{4}{3}$	$\mathbf{0} - 3x = -\frac{4}{3}$
$y = -\frac{4}{3}$	$-3x = -\frac{4}{3}$
	$x = -\frac{4}{3} \cdot -\frac{1}{3}$
	$x = \frac{4}{9}$

The y-intercept is $(0, -\frac{4}{3})$, and the x-intercept is $(\frac{4}{9}, 0)$.

Check Point

$y - 3x = -\frac{4}{3}$

$y - 3(\frac{3}{3}) = -\frac{4}{3}$

$y - \frac{9}{3} = -\frac{4}{3}$

$y = \frac{5}{3}$

$\left(1, \frac{5}{3}\right)$

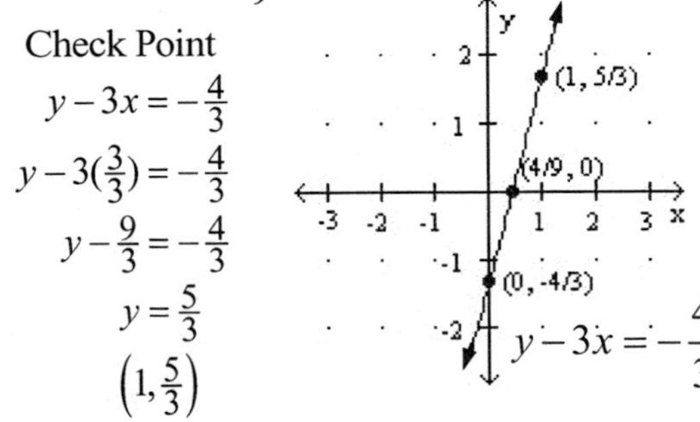

71. $7x + 3y = 0$

y-intercept:	x-intercept:
If $x = 0$,	If $y = 0$
$7(\mathbf{0}) + 3y = 0$	$7x + 3(\mathbf{0}) = 0$
$3y = 0$	$7x = 0$
$y = 0$	$x = 0$

The y-intercept is (0, 0), and the x-intercept is (0, 0).

Find two other points.

If $x = 3$,	If $y = 7$
$7(\mathbf{3}) + 3y = 0$	$7x + 3(\mathbf{7}) = 0$
$21 + 3y = 0$	$7x + 21 = 0$
$3y = -21$	$7x = -21$
$y = -7$	$x = -3$
$(3, -7)$	$(-3, 7)$

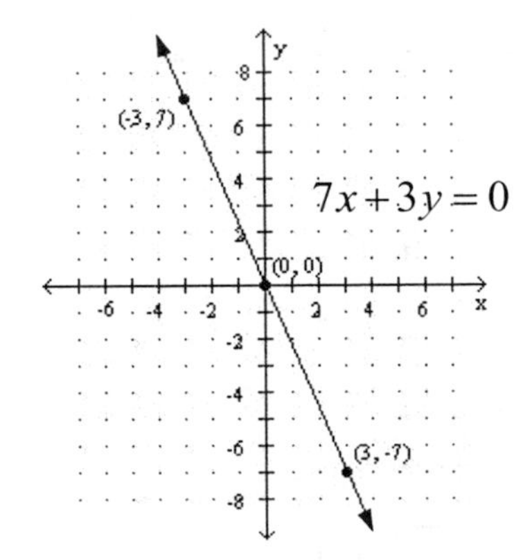

73. $-4x = 8 - 2y$

y-intercept:	x-intercept:
If $x = 0$,	If $y = 0$
$-4(\mathbf{0}) = 8 - 2y$	$-4x = 8 - 2(\mathbf{0})$
$0 = 8 - 2y$	$-4x = 8$
$2y = 8$	$x = -2$
$y = 4$	

The y-intercept is (0, 4), and the x-intercept is (–2, 0).

Check Point

$-4x = 8 - 2y$

$-4(-1) = 8 - 2y$

$4 = 8 - 2y$

$-4 = -2y$

$2 = y$

$(-1, 2)$

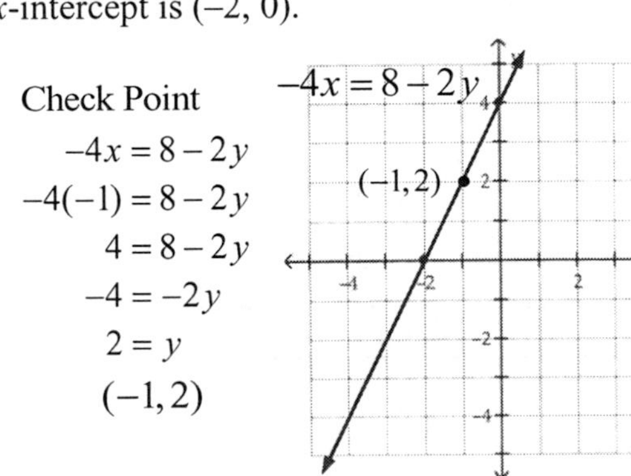

75. $3x = -150 - 5y$

y-intercept:	x-intercept:
If $x = 0$,	If $y = 0$
$3(\mathbf{0}) = -150 - 5y$	$3x = -150 - 5(\mathbf{0})$
$0 = -150 - 5y$	$3x = -150$
$5y = -150$	$x = -50$
$y = -30$	

The y-intercept is $(0, -30)$, and the x-intercept is $(-50, 0)$.

Check Point

$$3x = -150 - 5y$$
$$3(-25) = -150 - 5y$$
$$-75 = -150 - 5y$$
$$75 = -5y$$
$$-15 = y$$

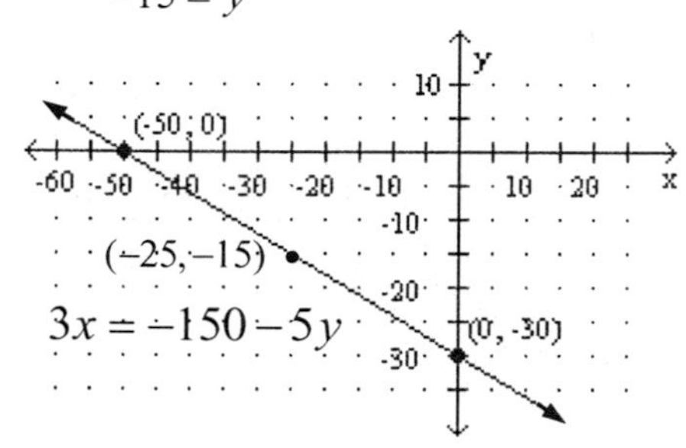

77. $-3y = 3$

y-intercept:	x-intercept:
If $x = 0$,	none
$-3y = 3$	
$y = -1$	

The y-intercept is $(0, -1)$, and the x-intercept does not exist.

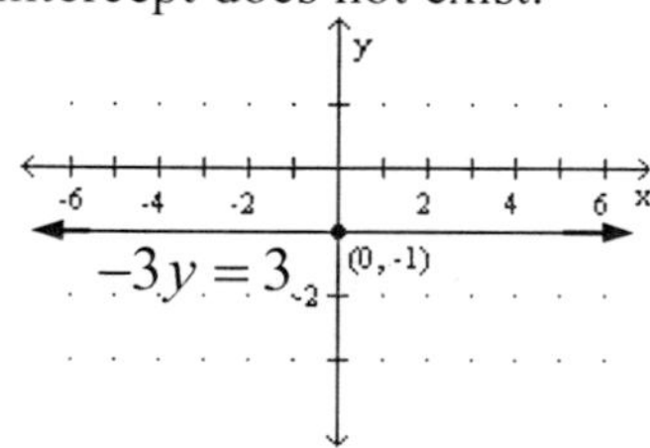

APPLICATIONS

79. CHEMISTRY

a. Estimate absolute zero. **about –270°C**

b. What is the volume of the gas when the temperature is absolute zero? **0 milliliters**

81. BOTTLED WATER DISPENSER

The g-intercept is $(0, 5)$: Before any cups of water have been served from the bottle, it contains 5 gallons of water. The c-intercept is $(106\frac{2}{3}, 0)$: The bottle will be empty after $106\frac{2}{3}$ 6-ounce cups of water have been served from it.

83. LANDSCAPING

x – axis represents the number of trees

y – axis represents the number of shrubs

$$50x + 25y = 5{,}000$$

y-intercept:	x-intercept:
If $x = 0$,	If $y = 0$
$50(\mathbf{0}) + 25y = 5{,}000$	$50x + 25(\mathbf{0}) = 5{,}000$
$25y = 5{,}000$	$50x = 5{,}000$
$y = 200$	$x = 100$

The y-intercept is $(0, 200)$, and the x-intercept is $(100, 0)$.

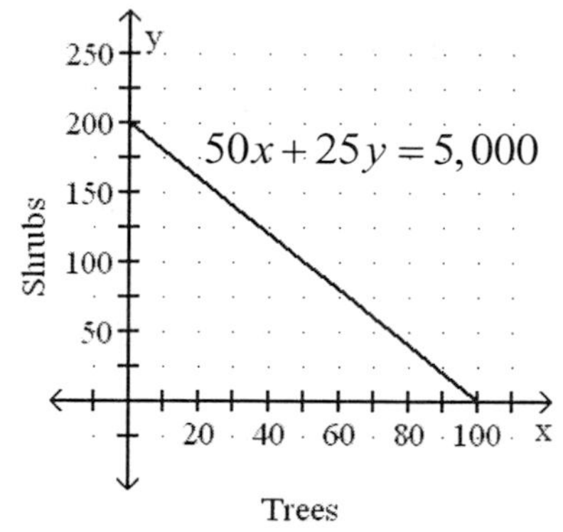

a. What information is given by the y-intercept?

If only shrubs are purchased, he can buy 200.

b. What information is given by the x-intercept?

If only trees are purchased, he can buy 100.

WRITING

85-87. Answers will vary.

REVIEW

89. $$\frac{3\cdot5\cdot5}{3\cdot5\cdot5\cdot5} = \frac{\cancel{3}\cdot\cancel{5}\cdot\cancel{5}}{\cancel{3}\cdot\cancel{5}\cdot\cancel{5}\cdot5} = \frac{1}{5}$$

91. Translate: Six less than twice x.

$$2x - 6$$

CHALLENGE PROBLEMS

93. Where will the line y = b intersect the line $x = a$? **Intersect at (a, b).**

95. What is the least number of intercepts a line can have? **1;** What is the greatest number a line can have? **Infinitely many**

SECTION 3.4 STUDY SET
VOCABULARY

Fill in the blanks.

1. The **slope** of a line is a measure of the line's steepness. It is the **ratio** of the vertical change to the horizontal change.
3. The rate of **change** of a linear relationship can be found by finding the slope of the graph of the line and attaching the proper units.

CONCEPTS

5. Which line graphed has
   a. a positive slope? **Line 2**
   b. a negative slope? **Line 1**
   c. zero slope? **Line 4**
   d. undefined slope? **Line 3**
7. For each graph, determine which line has the greater slope.
   a. **Line 1** b. **Line 1** c. **Line 2** d. **Line 2**
9. a. Rise of 1 and a run of 3, slope is $\frac{1}{3}$
   b. Rise of 4 and a run of 12, slope is $\frac{4}{12} = \frac{1}{3}$
   c. When calculating the slope of a line, the **same** value will be obtained no matter which two points on the line are used.
11. Write each slope in a better way.
   a. $m = \frac{0}{6}$, $m = 0$ b. $m = \frac{8}{0}$, **Undefined**
   c. $m = \frac{3}{12}$, $m = \frac{1}{4}$ d. $m = \frac{-10}{-5}$, $m = 2$
13. The grade of an incline is its slope expressed as a percent. Express the slope $\frac{2}{5}$ as a grade.
   40%
15. Find the negative reciprocal of each number.
   a. 6, $-\frac{1}{6}$ b. $-\frac{7}{8}$, $\frac{8}{7}$ c. −1, **1**

NOTATION

17. a. What is the formula used to find the slope of a line passing through (x_1, y_1) and (x_2, y_2)?

$$m = \frac{y_2 - y_1}{x_2 - x_1}$$

   b. Fill in the blanks to state the slope formula in words: m equals y **sub** two minus y **sub** one **over (divided by)** x sub **two** minus x sub **one**.
19. Consider the points (7, 2) and (−4, 1). If we let $x_1 = 7$, then what is y_2? **1**

GUIDED PRACTICE

21. $m = \frac{2}{3}$
23. $m = \frac{4}{3}$
25. $m = -2$
27. $m = 0$
29. $m = -\frac{1}{5}$
31. $m = \frac{1}{2}$

Find the slope of the line passing through the given points, when possible.

33. Let $(x_1, y_1) = (1,3)$
   Let $(x_2, y_2) = (2,4)$
$$m = \frac{y_2 - y_1}{x_2 - x_1} = \frac{4-3}{2-1} = \frac{1}{1} = 1$$

35. Let $(x_1, y_1) = (3,4)$
   Let $(x_2, y_2) = (2,7)$
$$m = \frac{y_2 - y_1}{x_2 - x_1} = \frac{7-4}{2-3} = \frac{3}{-1} = -3$$

37. Let $(x_1, y_1) = (0,0)$
   Let $(x_2, y_2) = (4,5)$
$$m = \frac{y_2 - y_1}{x_2 - x_1} = \frac{5-0}{4-0} = \frac{5}{4}$$

39. Let $(x_1, y_1) = (-3,5)$
   Let $(x_2, y_2) = (-5,6)$
$$m = \frac{y_2 - y_1}{x_2 - x_1} = \frac{6-5}{-5-(-3)} = \frac{1}{-5+3} = -\frac{1}{2}$$

41. Let $(x_1, y_1) = (-2,-2)$
   Let $(x_2, y_2) = (-12,-8)$
$$m = \frac{y_2 - y_1}{x_2 - x_1} = \frac{-8-(-2)}{-12-(-2)} = \frac{-8+2}{-12+2} = \frac{-6}{-10} = \frac{3}{5}$$

43. Let $(x_1, y_1) = (5,7)$
   Let $(x_2, y_2) = (-4,7)$
$$m = \frac{y_2 - y_1}{x_2 - x_1} = \frac{7-7}{-4-5} = \frac{0}{-9} = 0$$

45. Let $(x_1, y_1) = (8, -4)$
Let $(x_2, y_2) = (8, -3)$

$$m = \frac{y_2 - y_1}{x_2 - x_1} = \frac{-3-(-4)}{8-8} = \frac{-3+4}{8-8} = \frac{1}{0}$$

Undefined

47. Let $(x_1, y_1) = (-6, 0)$
Let $(x_2, y_2) = (0, -4)$

$$m = \frac{y_2 - y_1}{x_2 - x_1} = \frac{-4-0}{0\ \ (\ \ 6)} = \frac{-4}{6} = -\frac{2}{3}$$

49. Let $(x_1, y_1) = (-2.5,\ 1.75)$
Let $(x_2, y_2) = (-0.5,\ -7.75)$

$$m = \frac{y_2 - y_1}{x_2 - x_1} = \frac{-7.75-1.75}{-0.5-(-2.5)} = \frac{-9.5}{-0.5+2.5} = \frac{-9.5}{2} = -4.75$$

51. Let $(x_1, y_1) = (-2.2,\ 18.6)$
Let $(x_2, y_2) = (-1.7,\ 18.6)$

$$m = \frac{y_2 - y_1}{x_2 - x_1} = \frac{18.6-18.6}{-1.7-(-2.2)} = \frac{0}{-1.7+2.2} = 0$$

53. Let $(x_1, y_1) = \left(-\frac{4}{7}, -\frac{1}{5}\right)$
Let $(x_2, y_2) = \left(\frac{3}{7}, \frac{6}{5}\right)$

$$m = \frac{y_2 - y_1}{x_2 - x_1} = \frac{\frac{6}{5} - \left(-\frac{1}{5}\right)}{\frac{3}{7} - \left(-\frac{4}{7}\right)} = \frac{\frac{6}{5} + \frac{1}{5}}{\frac{3}{7} + \frac{4}{7}} = \frac{\frac{7}{5}}{\frac{7}{7}} = \frac{7}{5}$$

55. Let $(x_1, y_1) = \left(-\frac{3}{4}, \frac{2}{3}\right)$
Let $(x_2, y_2) = \left(\frac{4}{3}, -\frac{1}{6}\right)$

$$m = \frac{y_2 - y_1}{x_2 - x_1} = \frac{-\frac{1}{6} - \frac{2}{3}}{\frac{4}{3} - \left(-\frac{3}{4}\right)} = \frac{-\frac{1}{6} - \frac{4}{6}}{\frac{16}{12} + \frac{9}{12}} = \frac{-\frac{5}{6}}{\frac{25}{12}} = -\frac{5}{6} \cdot \frac{12}{25} = -\frac{5 \cdot 12}{6 \cdot 25} = -\frac{\overset{1}{\cancel{5}} \cdot 2 \cdot \overset{1}{\cancel{6}}}{\underset{1}{\cancel{6}} \cdot \underset{1}{\cancel{5}} \cdot 5} = -\frac{2}{5}$$

Determine the slope of the graph of the line that has the given table of solutions.

57. Let $(x_1, y_1) = (-3, -1)$
Let $(x_2, y_2) = (1, 2)$

$$m = \frac{y_2 - y_1}{x_2 - x_1} = \frac{2-(-1)}{1-(-3)} = \frac{2+1}{1+3} = \frac{3}{4}$$

59. Let $(x_1, y_1) = (-3, 6)$
Let $(x_2, y_2) = (0, 6)$

$$m = \frac{y_2 - y_1}{x_2 - x_1} = \frac{6-6}{0-(-3)} = \frac{0}{3} = 0$$

Find the slope of each line, if possible.

61. $y = -11$ Select any two points that lie on the horizontal line.
Let $(x_1, y_1) = (3, -11)$
Let $(x_2, y_2) = (5, -11)$

$$m = \frac{y_2 - y_1}{x_2 - x_1} = \frac{-11-(-11)}{5-3} = \frac{-11+11}{2} = \frac{0}{2} = 0$$

63. $y = 0$ Select any two points that lie on the horizontal line.
Let $(x_1, y_1) = (3, 0)$
Let $(x_2, y_2) = (5, 0)$

$$m = \frac{y_2 - y_1}{x_2 - x_1} = \frac{0-0}{5-3} = \frac{0}{2} = 0$$

65. $x=6$ Select any two points that lie on the vertical line.

Let $(x_1, y_1) = (6,4)$
Let $(x_2, y_2) = (6,5)$

$$m = \frac{y_2 - y_1}{x_2 - x_1} = \frac{5-4}{6-6} = \frac{1}{0}$$

Undefined

67. $x=-10$ Select any two points that lie on the vertical line.

Let $(x_1, y_1) = (-10,4)$
Let $(x_2, y_2) = (-10,5)$

$$m = \frac{y_2 - y_1}{x_2 - x_1} = \frac{5-4}{-10-(-10)} = \frac{1}{-10+10} = \frac{1}{0}$$

Undefined

69. $y-9=0$ Solve for y.
$y=9$ Select any two points that lie on the horizontal line.

Let $(x_1, y_1) = (3,9)$
Let $(x_2, y_2) = (5,9)$

$$m = \frac{y_2 - y_1}{x_2 - x_1} = \frac{9-9}{5-3} = \frac{0}{2} = 0$$

71. $3x=-12$ Solve for x.
$x=-4$ Select any two points that lie on the vertical line.

Let $(x_1, y_1) = (-4,4)$
Let $(x_2, y_2) = (-4,5)$

$$m = \frac{y_2 - y_1}{x_2 - x_1} = \frac{5-4}{-4-(-4)} = \frac{1}{-4+4} = \frac{1}{0}$$

Undefined

Determine whether the lines through each pair of points are parallel, perpendicular, or neither.

73. Step 1: Determine the slopes.

Let $(x_1, y_1) = (5,3)$
Let $(x_2, y_2) = (1,4)$

$$m_1 = \frac{y_2 - y_1}{x_2 - x_1} = \frac{4-3}{1-5} = -\frac{1}{4}$$

Let $(x_1, y_1) = (-3,-4)$
Let $(x_2, y_2) = (1,-5)$

$$m_2 = \frac{y_2 - y_1}{x_2 - x_1} = \frac{-5-(-4)}{1-(-3)} = \frac{-5+4}{1+3} = -\frac{1}{4}$$

Step 2: Compare the slopes.
The slopes are equal.
The lines are parallel.

75. Step 1: Determine the slopes.

Let $(x_1, y_1) = (-4,-2)$
Let $(x_2, y_2) = (2,-3)$

$$m_1 = \frac{y_2 - y_1}{x_2 - x_1} = \frac{-3-(-2)}{2-(-4)} = \frac{-3+2}{2+4} = -\frac{1}{6}$$

Let $(x_1, y_1) = (7,1)$
Let $(x_2, y_2) = (8,7)$

$$m_2 = \frac{y_2 - y_1}{x_2 - x_1} = \frac{7-1}{8-7} = \frac{6}{1} = 6$$

Step 2: Compare the slopes.
The slopes are negative reciprocals.
The lines are perpendicular.

77. Step 1: Determine the slopes.

Let $(x_1, y_1) = (2,2)$	Let $(x_1, y_1) = (-3,4)$
Let $(x_2, y_2) = (4,-3)$	Let $(x_2, y_2) = (-1,9)$
$m_1 = \frac{y_2 - y_1}{x_2 - x_1} = \frac{-3-2}{4-2} = -\frac{5}{2}$	$m_2 = \frac{y_2 - y_1}{x_2 - x_1} = \frac{9-4}{-1-(-3)} = \frac{5}{-1+3} = \frac{5}{2}$

Step 2: Compare the slopes.

The slopes are not equal.
The slopes are not negative reciprocals.
The lines are neither parallel nor perpendicular.

79. Step 1: Determine the slopes.

Let $(x_1, y_1) = (-1,8)$	Let $(x_1, y_1) = (3,3)$
Let $(x_2, y_2) = (-6,8)$	Let $(x_2, y_2) = (3,7)$
$m_1 = \frac{y_2 - y_1}{x_2 - x_1} = \frac{8-8}{-6-(-1)} = \frac{0}{-6+1} = 0$	$m_2 = \frac{y_2 - y_1}{x_2 - x_1} = \frac{7-3}{3-3} = \frac{4}{0}$ Undefined

Step 2: Compare the slopes.

The slopes are of a horizontal and a vertical line. The lines are perpendicular.

81. Step 1: Determine the slopes.

Let $(x_1, y_1) = (6,4)$	Let $(x_1, y_1) = (-2,-3)$
Let $(x_2, y_2) = (2,5)$	Let $(x_2, y_2) = (2,-4)$
$m_1 = \frac{y_2 - y_1}{x_2 - x_1} = \frac{5-4}{2-6} = -\frac{1}{4}$	$m_2 = \frac{y_2 - y_1}{x_2 - x_1} = \frac{-4-(-3)}{2-(-2)} = \frac{-4+3}{2+2} = -\frac{1}{4}$

Step 2: Compare the slopes.

The slopes are equal.
The lines are parallel.

83. Step 1: Determine the slopes.

Let $(x_1, y_1) = (4,2)$	Let $(x_1, y_1) = (-5,3)$
Let $(x_2, y_2) = (5,-3)$	Let $(x_2, y_2) = (-2,9)$
$m_1 = \frac{y_2 - y_1}{x_2 - x_1} = \frac{-3-2}{5-4} = -5$	$m_2 = \frac{y_2 - y_1}{x_2 - x_1} = \frac{9-3}{-2-(-5)} = \frac{6}{-2+5} = \frac{6}{3} = 2$

Step 2: Compare the slopes.

The slopes are not equal.
The slopes are not negative reciprocals.
The lines are neither parallel nor perpendicular.

Find the slope of a line perpendicular to the line passing through the given two points.

85. Step 1: Determine the slope.

Let $(x_1, y_1) = (0,0)$
Let $(x_2, y_2) = (5,-9)$

$$m = \frac{y_2 - y_1}{x_2 - x_1} = \frac{-9-0}{5-0} = -\frac{9}{5}$$

Step 2: Determine the negative reciprocal.

The negative reciprocal is $\frac{5}{9}$.

87. Step 1: Determine the slope.

Let $(x_1, y_1) = (-1, 7)$
Let $(x_2, y_2) = (1, 10)$

$$\begin{aligned} m &= \frac{y_2 - y_1}{x_2 - x_1} \\ &= \frac{10 - 7}{1 - (-1)} \\ &= \frac{3}{1 + 1} \\ &= \frac{3}{2} \end{aligned}$$

Step 2: Determine the negative reciprocal.

The negative reciprocal is $-\frac{2}{3}$.

89. Step 1: Determine the slope.

Let $(x_1, y_1) = \left(-2, \frac{1}{2}\right)$

Let $(x_2, y_2) = \left(-1, \frac{3}{2}\right)$

$$\begin{aligned} m &= \frac{y_2 - y_1}{x_2 - x_1} \\ &= \frac{\frac{3}{2} - \frac{1}{2}}{-1 - (-2)} \\ &= \frac{\frac{2}{2}}{-1 + 2} \\ &= \frac{1}{1} \\ &= 1 \end{aligned}$$

Step 2: Determine the negative reciprocal.
The negative reciprocal is -1.

91. Step 1: Determine the slope.

Let $(x_1, y_1) = (-1, 2)$
Let $(x_2, y_2) = (-3, 6)$

$$\begin{aligned} m &= \frac{y_2 - y_1}{x_2 - x_1} \\ &= \frac{6 - 2}{-3 - (-1)} \\ &= \frac{4}{-3 + 1} \\ &= \frac{4}{-2} \\ &= -2 \end{aligned}$$

Step 2: Determine the negative reciprocal.

The negative reciprocal is $\frac{1}{2}$.

APPLICATIONS

93. POOL DESIGN

Step 1: Determine the rise of the slope by finding the difference of the two given vertical heights. **9 ft – 3 ft = 6 ft**

Step 2: Determine the run of the slope by looking at the distance between the two depths. **15 ft**

Compute the ratio.

$$\begin{aligned} m &= \frac{\text{rise}}{\text{run}} \\ &= \frac{6}{15} \\ &= \frac{2}{5} \end{aligned}$$

Step 3: The slope falls from left to right thus indicating it is negative.

$$m = -\frac{2}{5}$$

95. GRADE OF A ROAD

$$m = \frac{\text{rise}}{\text{run}}$$
$$= \frac{264}{5,280}$$
$$= \frac{1}{20}$$

Change the fraction to a percent by dividing:

$$\frac{1}{20} = 0.05$$
$$= 5\%$$

97. TREADMILLS

The height setting is the change in y and the length of the treadmill is the change in x.

For 6 inches:

$$m = \frac{\text{rise}}{\text{run}}$$
$$= \frac{6}{50}$$
$$= \frac{3}{25}$$
$$= 0.12$$
$$= 12\%$$

99. CARPENTRY

The pitch of the front of the roof is

$$\frac{9}{6} = \frac{3}{2}$$

The pitch of the side of the roof is

$$\frac{9}{15} = \frac{3}{5}$$

101. IRRIGATION

Find an order pair for each of the endpoints: (0, 8,000) and (8, 1,000). Now apply the slope formula:

Let $(x_1, y_1) = (0,\ 8{,}000)$ Let $(x_2, y_2) = (8,\ 1{,}000)$

$$m = \frac{y_2 - y_1}{x_2 - x_1}$$
$$= \frac{1{,}000 - 8{,}000}{8 - 0}$$
$$= \frac{-7{,}000}{8}$$
$$= -875$$

The rate of change would be –875 gal per hour.

103. MILK PRODUCTION

Find an order pair for each of the endpoints: (1996, 16,433) and (2009, 20,580). Now apply the slope formula:

Let $(x_1, y_1) = (1996,\ 16{,}433)$

Let $(x_2, y_2) = (2009,\ 20{,}580)$

$$m = \frac{y_2 - y_1}{x_2 - x_1}$$
$$= \frac{20{,}580 - 16{,}433}{2009 - 1996}$$
$$= \frac{4{,}147}{13}$$
$$= 319$$

The rate of change would be 319 lb per year.

From Campus to Careers

105. DENTAL ASSISTANT

Find an order pair for each of the endpoints: (2000, 6,600) and (2008, 9,200). Now apply the slope formula:

Let $(x_1, y_1) = (2000,\ 6{,}600)$

Let $(x_2, y_2) = (2008,\ 9{,}200)$

$$m = \frac{y_2 - y_1}{x_2 - x_1}$$
$$= \frac{9{,}200 - 6{,}600}{2008 - 2000}$$
$$= \frac{2{,}600}{8}$$
$$= 325$$

There is an increase of 325 students per year.

WRITING

107 – 109. Answers will vary.

REVIEW

111. HALLOWEEN CANDY

Analyze

- Black licorice sells for $1.90 per lb.
- Orange gumdrops sells for $2.20 per lb.
- A mixture of the two is to sell for $2 per lb.
- A mixture of 60 pounds is needed.
- How many pounds of each is needed?

Assign

Let $x =$ the # of lbs of black licorice

$60 - x =$ the # of lbs of orange gumdrops

Form

	Number	• Value	= Total value
Licorice	x	1.90	**1.90*x***
Gumdrops	60 - x	2.20	**2.20(60 - *x*)**
Blend	60	2.00	**60(2)**

The value of the black licorice	plus	the value of the orange gumdrops	equals	the total value of the blend.
$1.90x$	+	$2.20(60-x)$	=	$60(2)$

Solve

$$1.90x + 2.20(60 - x) = 60(2)$$
$$1.90x + 132 - 2.20x = 120$$
$$-0.30x + 132 = 120$$
$$-0.30x + 132 - \mathbf{132} = 120 - \mathbf{132}$$
$$-0.30x = -12$$
$$\frac{-0.30x}{\mathbf{-0.30}} = \frac{-12}{\mathbf{-0.30}}$$
$$x = 40$$

gumdrops

$$60 - x = 60 - 40$$
$$= 20$$

State

40 lb of black licorice will be needed.
20 lb of orange gumdrops will be needed.

Check

The value of the licorice is 40($1.90), or $76.
The value of the gumdrops is 20($2.20), or $44.
The value of the blend is 60($2), or $120.
Since the total was $\$76 + \$44 = \$120$, the results check.

CHALLENGE

113. Find the slope from A to B.

Let $(x_1, y_1) = A(-50, -10)$
Let $(x_2, y_2) = B(20, 0)$

$$m_{\overline{AB}} = \frac{y_2 - y_1}{x_2 - x_1} = \frac{0-(-10)}{20-(-50)} = \frac{10}{70} = \frac{1}{7}$$

Find the slope from A to C.

Let $(x_1, y_1) = A(-50, -10)$
Let $(x_2, y_2) = C(34, 2)$

$$m_{\overline{AC}} = \frac{y_2 - y_1}{x_2 - x_1} = \frac{2-(-10)}{34-(-50)} = \frac{12}{84} = \frac{1}{7}$$

Find the slope from C to B.

Let $(x_1, y_1) = C(34, 2)$
Let $(x_2, y_2) = B(20, 0)$

$$m_{\overline{CB}} = \frac{y_2 - y_1}{x_2 - x_1} = \frac{0-2}{20-34} = \frac{-2}{-14} = \frac{1}{7}$$

The slopes are the same; therefore all three points do lie on the same line.

SECTION 3.5 STUDY SET
VOCABULARY

Fill in the blanks.

1. The equation $y = mx + b$ is called the **slope–intercept** form of the equation of a line.

CONCEPTS

3. a. $7x + 4y = 2$ **No** b. $5y = 2x - 3$ **No**

c. $y = 6x + 1$ **Yes** d. $x = 4y - 8$ **No**

5. **Simplify the right side of each equation.**

a. $y = \frac{4x}{2} + \frac{16}{2}$

$\mathbf{y = 2x + 8}$

b. $y = \frac{15x}{-3} + \frac{9}{-3}$

$\mathbf{y = -5x - 3}$

c. $y = \frac{2x}{6} - \frac{6}{6}$

$\mathbf{y = \frac{1}{3}x - 1}$

d. $y = \frac{-9x}{-5} - \frac{20}{-5}$

$\mathbf{y = \frac{9}{5}x + 4}$

NOTATION

7.
$$2x + 5y = 15$$
$$2x + 5y - 2x = \boxed{\mathbf{-2x}} + 15$$
$$\boxed{5y} = -2x + 15$$
$$\frac{5y}{\boxed{5}} = \frac{-2x}{\boxed{5}} + \frac{15}{\boxed{5}}$$
$$y = \boxed{-\frac{2}{5}}x + \boxed{3}$$

The slope is $\boxed{-\frac{2}{5}}$ and the y-intercept is $\boxed{(\boxed{0}, \boxed{3})}$.

9. $-\frac{3}{2} = \frac{3}{\boxed{-2}} = \frac{\boxed{-3}}{2}$

GUIDED PRACTICE

Find the slope and the y-intercept of the line with the given equation.

11. The slope is 4. The y-intercept is $(0,2)$.

13. The slope is -5. The y-intercept is $(0,-8)$.

15. The slope is 25. The y-intercept is $(0,-9)$.

17. The slope is -1. The y-intercept is $(0,11)$.

19. The slope is $\frac{1}{2}$. The y-intercept is $(0,6)$.

21. The slope is $\frac{1}{4}$. The y-intercept is $\left(0, -\frac{1}{2}\right)$.

23. The slope is -5. The y-intercept is $(0,0)$.

25. The slope is 1. The y-intercept is $(0,0)$.

27. The slope is 0. The y-intercept is $(0,-2)$.

29.
$$-5y - 2 = 0$$
$$-5y - 2 + \mathbf{2} = 0 + \mathbf{2}$$
$$-5y = 2$$
$$\frac{-5y}{\mathbf{-5}} = \frac{2}{\mathbf{-5}}$$
$$y = -\frac{2}{5}$$

The slope is 0. The y-intercept is $\left(0, -\frac{2}{5}\right)$.

31.
$$x + y = 8$$
$$x + y - \mathbf{x} = 8 - \mathbf{x}$$
$$y = -x + 8$$

The slope is -1. The y-intercept is $(0,8)$.

33.
$$6y = x - 6$$
$$\frac{6y}{\mathbf{6}} = \frac{x}{\mathbf{6}} - \frac{6}{\mathbf{6}}$$
$$y = \frac{x}{6} - 1$$

The slope is $\frac{1}{6}$. The y-intercept is $(0,-1)$.

35.
$$-4y = 6x - 4$$
$$\frac{-4y}{\mathbf{-4}} = \frac{6x}{\mathbf{-4}} - \frac{4}{\mathbf{-4}}$$
$$y = -\frac{3}{2}x + 1$$

The slope is $-\frac{3}{2}$. The y-intercept is $(0,1)$.

37.
$$2x + 3y = 6$$
$$2x + 3y - \mathbf{2x} = 6 - \mathbf{2x}$$
$$3y = -2x + 6$$
$$\frac{3y}{\mathbf{3}} = \frac{-2x}{\mathbf{3}} + \frac{6}{\mathbf{3}}$$
$$y = -\frac{2}{3}x + 2$$

The slope is $-\frac{2}{3}$. The y-intercept is $(0,2)$.

39.
$$3x - 5y = 15$$
$$3x - 5y - \mathbf{3x} = 15 - \mathbf{3x}$$
$$-5y = -3x + 15$$
$$\frac{-5y}{\mathbf{-5}} = \frac{-3x}{\mathbf{-5}} + \frac{15}{\mathbf{-5}}$$
$$y = \frac{3}{5}x - 3$$

The slope is $\frac{3}{5}$. The y-intercept is $(0,-3)$.

41.
$$\begin{aligned} -6x+6y &= -11 \\ -6x+6y+\mathbf{6x} &= -11+\mathbf{6x} \\ 6y &= 6x-11 \\ \frac{6y}{\mathbf{6}} &= \frac{6x}{\mathbf{6}}-\frac{11}{\mathbf{6}} \\ y &= x-\frac{11}{6} \end{aligned}$$

The slope is 1. The y-intercept is $\left(0, -\frac{11}{6}\right)$

Write an equation of the line with the given slope and *y*-intercept and graph it.

43. Slope 5, y-intercept $(0, -3)$

The equation of the line is $y = 5x - 3$.

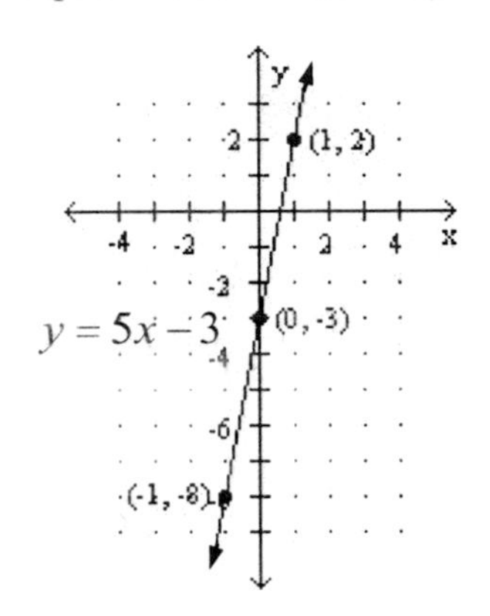

45. Slope -3, y-intercept $(0, 6)$

The equation of the line is $y = -3x + 6$.

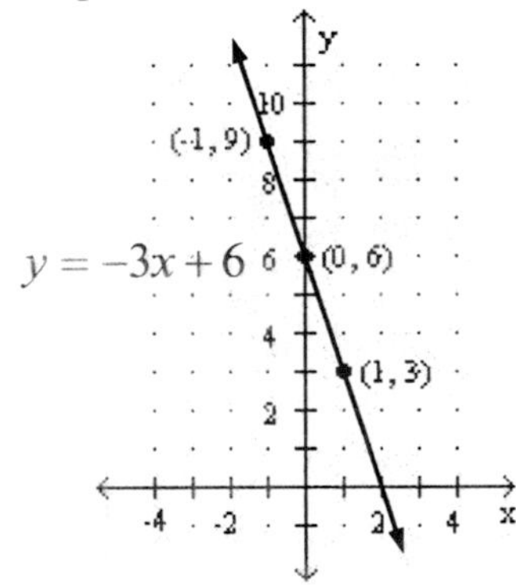

47. Slope $\frac{1}{4}$, y-intercept $(0, -2)$

The equation of the line is $y = \frac{1}{4}x - 2$.

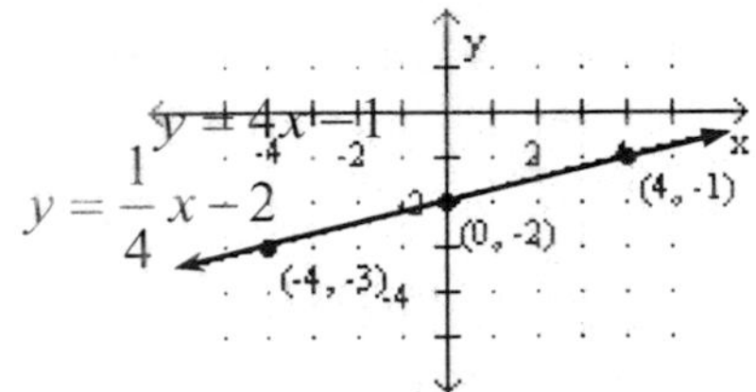

49. Slope $-\frac{8}{3}$, y-intercept $(0, 5)$

The equation of the line is $y = -\frac{8}{3}x + 5$.

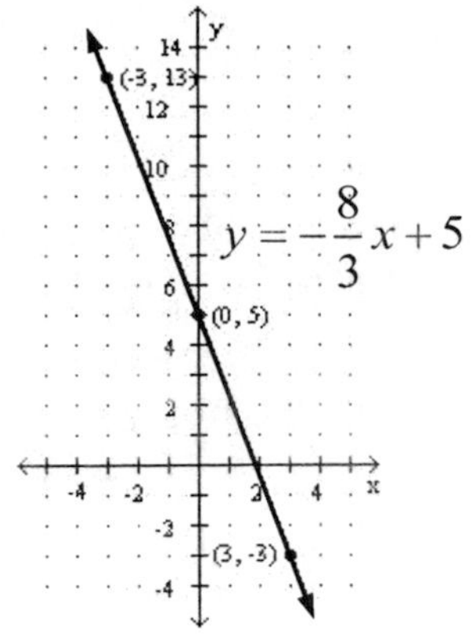

51. Slope $\frac{6}{5}$, y-intercept $(0, 0)$

The equation of the line is $y = \frac{6}{5}x$.

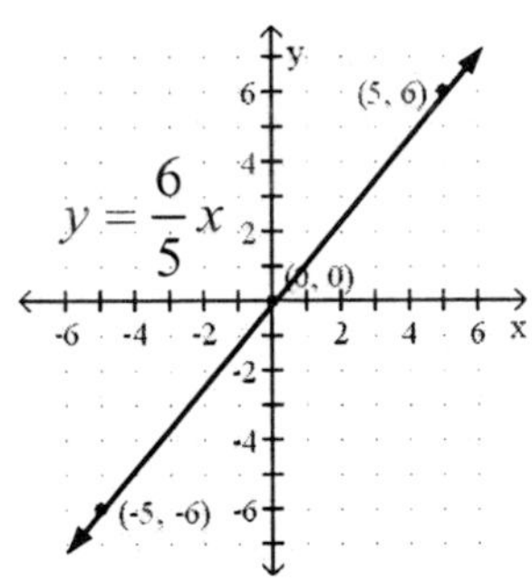

53. Slope -2, y-intercept $\left(0, \frac{1}{2}\right)$

The equation of the line is $y = -2x + \frac{1}{2}$.

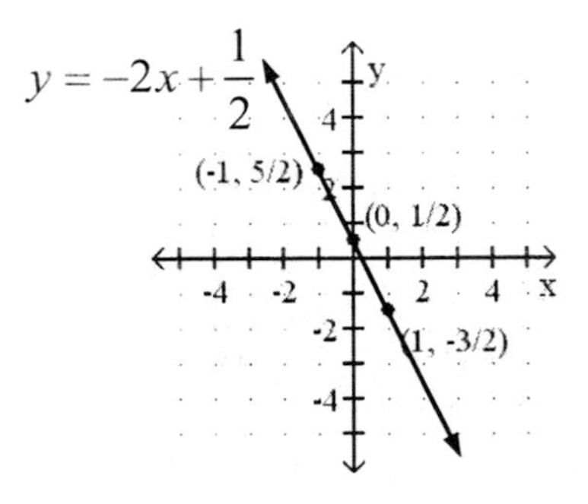

Write an equation for each line shown.

55. Slope 5, y-intercept $(0, -1)$

The equation of the line is $y = 5x - 1$.

57. Slope -2, y-intercept $(0, 3)$

The equation of the line is $y = -2x + 3$.

59. Slope $\frac{4}{5}$, y-intercept $(0, -2)$

The equation of the line is $y = \frac{4}{5}x - 2$.

61. Slope $-\frac{5}{3}$, y-intercept $(0, 2)$

The equation of the line is $y = -\frac{5}{3}x + 2$.

Find the slope and the *y*-intercept of the graph of each equation and graph it.

63. $y = 3x + 3$

$m = 3$, y-intercept $= (0, 3)$

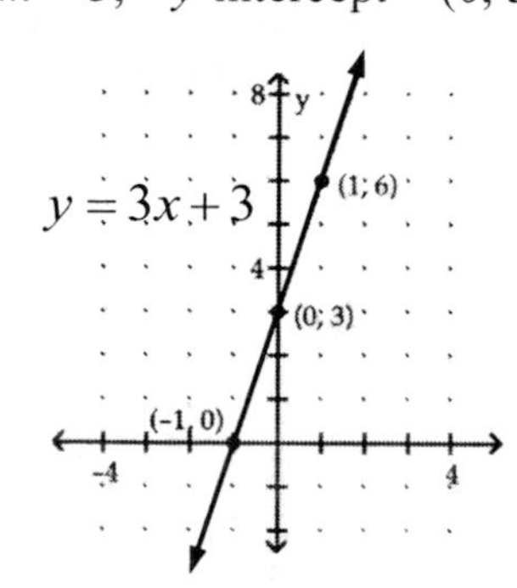

65. $y = \frac{1}{2}x + 2$

$m = \frac{1}{2}$, y-intercept $= (0, 2)$

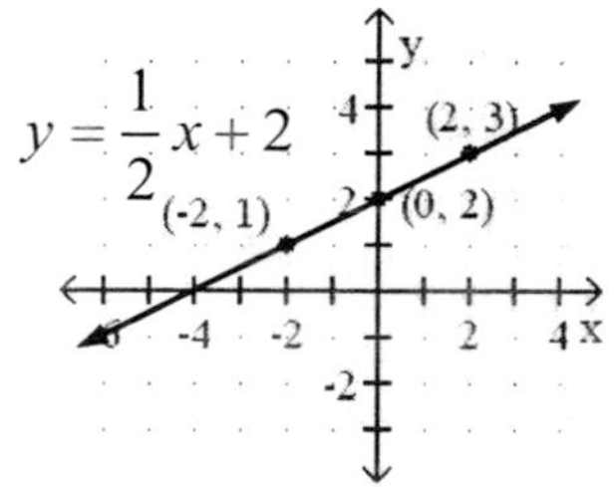

67. $y = -3x$

$m = -3$, y-intercept $= (0, 0)$

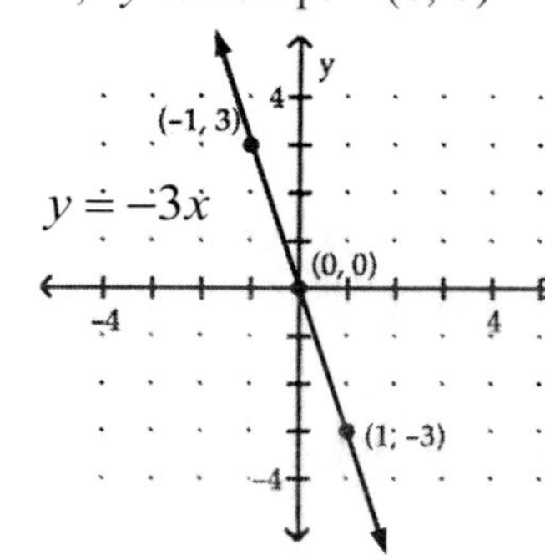

69.

$$4x + y = -4$$
$$4x + y - \mathbf{4x} = -4 - \mathbf{4x}$$
$$y = -4x - 4$$

$m = -4$, y-intercept $= (0, -4)$

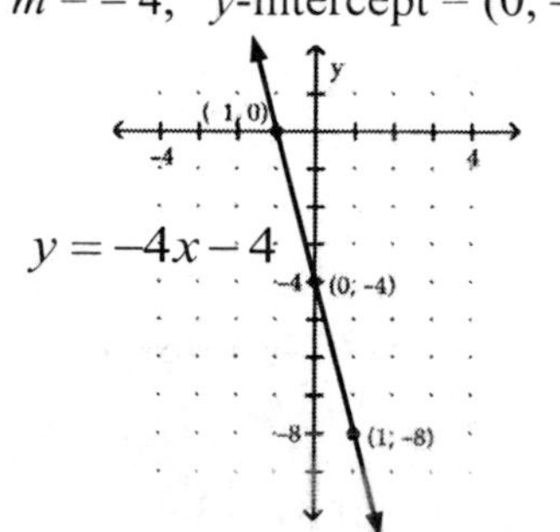

71.

$$3x + 4y = 16$$
$$3x + 4y - \mathbf{3x} = 16 - \mathbf{3x}$$
$$4y = -3x + 16$$
$$\frac{4y}{\mathbf{4}} = \frac{-3x}{\mathbf{4}} + \frac{16}{\mathbf{4}}$$
$$y = -\frac{3}{4}x + 4$$

$m = -\frac{3}{4}$, y-intercept $= (0, 4)$

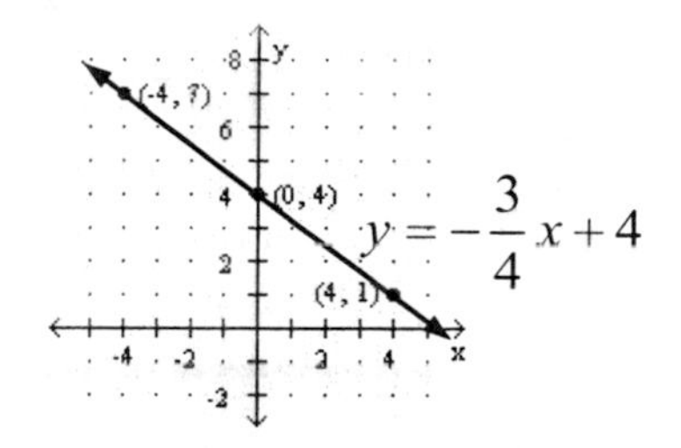

73.

$$10x - 5y = 5$$
$$10x - 5y - \mathbf{10x} = 5 - \mathbf{10x}$$
$$-5y = -10x + 5$$
$$\frac{-5y}{\mathbf{-5}} = \frac{-10x}{\mathbf{-5}} + \frac{5}{\mathbf{-5}}$$
$$y = 2x - 1$$

$m = 2$, y-intercept $= (0, -1)$

y
2
(1, 1)
-4
-2
2
4
x
-2
(0, -1)
$y = 2x - 1$
-4
(-2, -5)
-6

For each pair of equations, determine whether their graphs are parallel, perpendicular, or neither.

75. $y = 6x + 8$ | $y = 6x$

$m = 6$ | $m = 6$

The slopes are equal.
The lines are parallel.

77. $y = x$ | $y = -x$

$m = 1$ | $m = -1$

The slopes are negative reciprocals.
The lines are perpendicular.

79. $y = \frac{1}{2}x - \frac{4}{5}$ | $y = 0.5x + 3$

$m = \frac{1}{2}$ | $m = 0.5$

The slopes are equal.
The lines are parallel.

81. $y = -2x - 9$, $m = -2$

$$\begin{aligned} 2x - y &= 9 \\ 2x - y + \mathbf{y} &= 9 + \mathbf{y} \\ 2x &= y + 9 \\ 2x - \mathbf{9} &= y + 9 - \mathbf{9} \\ y &= 2x - 9 \\ m &= 2 \end{aligned}$$

The slopes are not equal.
The slopes are not negative reciprocals.
The lines are neither parallel nor perpendicular.

83.

$$\begin{aligned} 3x &= 5y - 10 \\ 3x + \mathbf{10} &= 5y - 10 + \mathbf{10} \\ 3x + 10 &= 5y \\ \frac{3x}{\mathbf{5}} + \frac{10}{\mathbf{5}} &= \frac{5y}{\mathbf{5}} \\ y &= \frac{3}{5}x + 2 \\ m &= \frac{3}{5} \end{aligned}$$

$$\begin{aligned} 5x &= 1 - 3y \\ 5x - \mathbf{1} &= 1 - 3y - \mathbf{1} \\ 5x - 1 &= -3y \\ \frac{5x}{\mathbf{-3}} - \frac{1}{\mathbf{-3}} &= \frac{-3y}{\mathbf{-3}} \\ y &= -\frac{5}{3}x + \frac{1}{3} \\ m &= -\frac{5}{3} \end{aligned}$$

The slopes are negative reciprocals.
The lines are perpendicular.

85.

$$\begin{aligned} x - y &= 12 \\ x - y - \mathbf{x} &= 12 - \mathbf{x} \\ -y &= -x + 12 \\ \frac{-y}{\mathbf{-1}} &= \frac{-x}{\mathbf{-1}} + \frac{12}{\mathbf{-1}} \\ y &= x - 12 \\ m &= 1 \end{aligned}$$

$$\begin{aligned} -2x + 2y &= -23 \\ -2x + 2y + \mathbf{2x} &= -23 + \mathbf{2x} \\ 2y &= 2x - 23 \\ \frac{2y}{\mathbf{2}} &= \frac{2x}{\mathbf{2}} - \frac{23}{\mathbf{2}} \\ y &= x - \frac{23}{2} \\ m &= 1 \end{aligned}$$

The slopes are equal.
The lines are parallel.

87. $x = 9$ | $y = 8$

$m =$ undefined | $m = 0$

The slopes are of a horizontal and a vertical line. The lines are perpendicular.

89.

$$\begin{aligned} -4x + 3y &= -12 \\ -4x + 3y + \mathbf{4x} &= -12 + \mathbf{4x} \\ 3y &= 4x - 12 \\ \frac{3y}{\mathbf{3}} &= \frac{4x}{\mathbf{3}} - \frac{12}{\mathbf{3}} \\ y &= \frac{4}{3}x - 4 \\ m &= \frac{4}{3} \end{aligned}$$

$$\begin{aligned} 8x + 6y &= 54 \\ 8x + 6y - \mathbf{8x} &= 54 - \mathbf{8x} \\ 6y &= -8x + 54 \\ \frac{6y}{\mathbf{6}} &= \frac{-8x}{\mathbf{6}} + \frac{54}{\mathbf{6}} \\ y &= -\frac{4}{3}x + 9 \\ m &= -\frac{4}{3} \end{aligned}$$

The slopes are not equal.
The slopes are not negative reciprocals.
The lines are neither parallel nor perpendicular.

APPLICATIONS

91. PRODUCTION COSTS

a. basic fee is $5,000 (constant)
extra cost is $2,000/hr (not constant)
total hours used in not known (h)
total cost is not known (c)

Total product cost	equals	extra cost of $2,000/hr	plus	basic fee of $5,000.
c	$=$	$2{,}000h$	$+$	$5{,}000$

b. The cost for 8 hours of filming is found by replacing h with 8 and then calculating the results.

$$\begin{aligned} c &= 2{,}000h + 5{,}000 \\ c &= 2{,}000(\mathbf{8}) + 5{,}000 \\ c &= 16{,}000 + 5{,}000 \\ c &= 21{,}000 \end{aligned}$$

The total cost for the 8 hours is $21,000.

93. CHEMISTRY
$F = 5t - 10$

95. EMPLOYMENT SERVICE
$c = -20m + 500$

97. SEWING COSTS
$c = 5x + 20$

99. iPADS

a. $c = 14.95m + 629.99$

b. $c = 14.95m + 629.99$

$m = 24$ months (2 years)

$c = 14.95(24) + 629.99$

$c = 358.80 + 629.99$

$c = 988.79$

The total cost for the iPad after 2 years is \$988.79.

101. NAVIGATION

a. When there are no head waves, the ship can travel at 18 knots.

b. Points (0,18) and (4,16) are to be used to calculate the rate of change.

$$m = \frac{y_2 - y_1}{x_2 - x_1}$$

$$m = \frac{16-18}{4-0}$$

$$m = \frac{-2}{4}$$

$$m = -\frac{1}{2}$$

The rate of change is $-\frac{1}{2}$ knot per foot

c. $y = -\frac{1}{2}x + 18$

WRITING

103. Answers will vary.

REVIEW

105. CABLE TV

Analyze

- A 186-foot cable is cut into four pieces.
- Each successive piece is 3 feet longer than the previous one.
- Find the length of each piece.

Assign

Let x = length of 1st piece in feet

$x+3$ = length of 2nd piece in feet

$x+6$ = length of 3rd piece in feet

$x+9$ = length of 4th piece in feet

Form

The length of the 1st piece	plus	the length of the 2nd piece	plus	the length of the 3rd piece	plus	the length of the 4th piece	equals	the total length.
x	+	$x+3$	+	$x+6$	+	$x+9$	=	186

Solve

$$x + (x+3) + (x+6) + (x+9) = 186$$

$$4x + 18 = 186$$

$$4x + 18 - \mathbf{18} = 186 - \mathbf{18}$$

$$4x = 168$$

$$\frac{4x}{\mathbf{4}} = \frac{168}{\mathbf{4}}$$

$$x = 42$$

2nd piece	3rd piece	4th piece
$x+3 = \mathbf{42}+3$	$x+6 = \mathbf{42}+6$	$x+9 = \mathbf{42}+9$
$= 45$	$= 48$	$= 51$

State

Length of 1st piece is 42 ft.
Length of 2nd piece is 45 ft.
Length of 3rd piece is 48 ft.
Length of 4th piece is 51 ft.

Check

$42 \text{ ft} + 45 \text{ ft} + 48 \text{ ft} + 51 \text{ ft} = 186 \text{ ft}$

$186 \text{ ft} = 186 \text{ ft}$

The results check.

CHALLENGE PROBLEMS

107. If one would draw several lines such that they went through quadrants I, II, and IV, one would see that the **slopes** of all the lines would be **negative** and all the ***y*-intercepts** would be **positive**.

$$m < 0;\ b > 0$$

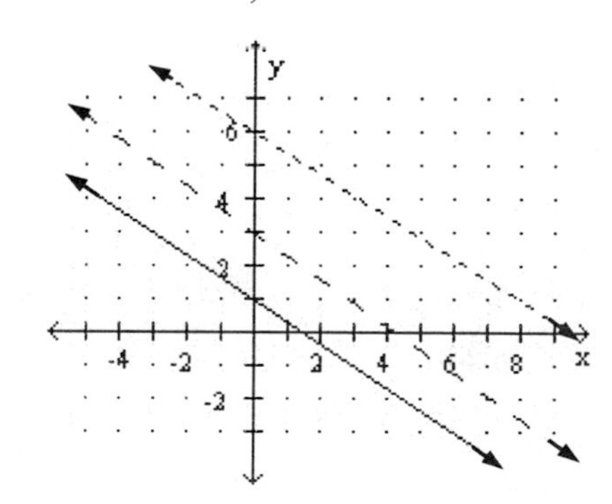

SECTION 3.6 STUDY SET
VOCABULARY

Fill in the blanks.

1. $y - y_1 = m(x - x_1)$ is called the **point–slope** form of the equation of a line. In words, we read this as y minus y **sub** one equals m **times** the quantity of x **minus** x sub **one**.

CONCEPTS

3. Determine in what form each equation is written.
   a. $y - 4 = 2(x - 5)$ **Point–slope**
   b. $y = 2x + 15$ **Slope–intercept**

5. Refer to the following graph of a line.
   a. What highlighted point does the line pass through? **(– 2, – 3)**
   b. What is the slope of the line? $\frac{5}{6}$
   c. Write an equation of the line in point–slope form. $y + 3 = \frac{5}{6}(x + 2)$

7. Suppose you are asked to write an equation of the line in the scatter diagram below. What two points would you use to write the point–slope equation?

   (67, 170), (79, 220)

NOTATION

Complete the solution.

9. Find an equation of the line with slope –2 that passes through the point (–1, 5). Write the answer in slope–intercept form.

$$y - y_1 = m(x - x_1)$$
$$y - \mathbf{5} = -2[x - (\mathbf{-1})]$$
$$y - 5 = -2[x \mathbf{+} 1]$$
$$y - 5 = -2x - \mathbf{2}$$
$$y = -2x + \mathbf{3}$$

11. Consider the steps below and then fill in the blanks:

$$y - 3 = 2(x + 1)$$
$$y - 3 = 2x + 2$$
$$y = 2x + 5$$

The original equation was in **point–slope** form. After solving for y, we obtain an equation in **slope–intercept** form.

GUIDED PRACTICE

Use the point - slope form to write an equation of the line with the given slope and point.

13. Passes through (2, 1) , $m = 3$

$$x_1 = 2 \text{ and } y_1 = 1$$
$$y - y_1 = m(x - x_1)$$
$$y - 1 = 3(x - 2)$$

15. Passes through (–5, –1) , $m = \frac{4}{5}$

$$x_1 = -5 \text{ and } y_1 = -1$$
$$y - y_1 = m(x - x_1)$$
$$y - (-1) = \frac{4}{5}(x - (-5))$$
$$y + 1 = \frac{4}{5}(x + 5)$$

Use the point - slope form to find an equation of the line with the given slope and point. Then write the equation in slope - intercept form. See Example 1.

17. Passes through (3, 5) , $m = 2$

$$x_1 = 3 \text{ and } y_1 = 5$$
$$y - y_1 = m(x - x_1)$$
$$y - 5 = 2(x - 3) \text{ point-slope form}$$

$$y - 5 = 2(x - 3)$$
$$y - 5 = 2x - 6$$
$$y - 5 + \mathbf{5} = 2x - 6 + \mathbf{5}$$
$$y = 2x - 1 \text{ slope-intercept form}$$

19. Passes through (–9, 8) , $m = -5$

$$x_1 = -9 \text{ and } y_1 = 8$$
$$y - y_1 = m(x - x_1)$$
$$y - 8 = -5(x - (-9))$$
$$y - 8 = -5(x + 9) \text{ point-slope form}$$

$$y - 8 = -5(x + 9)$$
$$y - 8 = -5x - 45$$
$$y - 8 + \mathbf{8} = -5x - 45 + \mathbf{8}$$
$$y = -5x - 37 \text{ slope-intercept form}$$

21. Passes through (0, 0) , $m = -3$

$$x_1 = 0 \text{ and } y_1 = 0$$
$$y - y_1 = m(x - x_1)$$
$$y - 0 = -3(x - 0)$$
$$y = -3x \text{ slope-intercept form}$$

23. Passes through (10, 1) , $m=\frac{1}{5}$

$x_1=10$ and $y_1=1$

$$y-y_1=m(x-x_1)$$

$$y-1=\frac{1}{5}(x-10) \text{ point-slope form}$$

$$y-1=\frac{1}{5}(x-10)$$

$$y-1=\frac{1}{5}x-2$$

$$y-1+\mathbf{1}=\frac{1}{5}x-2+\mathbf{1}$$

$$y=\frac{1}{5}x-1 \text{ slope-intercept form}$$

25. Passes through $(6,-4)$, $m=-\frac{4}{3}$

$x_1=6$ and $y_1=-4$

$$y-y_1=m(x-x_1)$$

$$y-(-4)=-\frac{4}{3}(x-6)$$

$$y+4=-\frac{4}{3}(x-6) \text{ point-slope form}$$

$$y+4=-\frac{4}{3}(x-6)$$

$$y+4=-\frac{\mathbf{4}}{\mathbf{3}}x-\left(-\frac{\mathbf{4}}{\mathbf{3}}\right)6$$

$$y+4=-\frac{4}{3}x+8$$

$$y+4-\mathbf{4}=-\frac{4}{3}x+8-\mathbf{4}$$

$$y=-\frac{4}{3}x+4 \text{ slope intercept form}$$

27. Passes through $(2,-6)$, $m=-\frac{11}{6}$

$x_1=2$ and $y_1=-6$

$$y-y_1=m(x-x_1)$$

$$y-(-6)=-\frac{11}{6}(x-2)$$

$$y+6=-\frac{11}{6}(x-2) \text{ point-slope form}$$

$$y+6=-\frac{11}{6}(x-2)$$

$$y+6=-\frac{\mathbf{11}}{\mathbf{6}}x-\left(-\frac{\mathbf{11}}{\mathbf{6}}\right)2$$

$$y+6=-\frac{11}{6}x+\frac{11}{3}$$

$$y+6-\mathbf{6}=-\frac{11}{6}x+\frac{11}{3}-\frac{\mathbf{18}}{\mathbf{3}}$$

$$y=-\frac{11}{6}x-\frac{7}{3} \text{ slope-intercept form}$$

Find an equation of the line that passes through the two given points. Write the equation in slope - intercept form. See Example 2.

29. $(1,7)$ and $(-2,1)$

(x_1,y_1) and (x_2,y_2)

$$m=\frac{y_2-y_1}{x_2-x_1}=\frac{1-7}{-2-1}=\frac{-6}{-3}=2$$

Passes through (1, 7) , $m=2$

$x_1=1$ and $y_1=7$

$$y-y_1=m(x-x_1)$$

$$y-7=2(x-1)$$

$$y-7=2x-2$$

$$y-7+\mathbf{7}=2x-2+\mathbf{7}$$

$$y=2x+5 \text{ slope-intercept form}$$

31. $(-4,3)$ and $(2,0)$

(x_1, y_1) and (x_2, y_2)

$$m = \frac{y_2 - y_1}{x_2 - x_1}$$
$$= \frac{0-3}{2-(-4)}$$
$$= \frac{-3}{6}$$
$$= -\frac{1}{2}$$

Passes through $(2,0)$, $m = -\frac{1}{2}$

$x_1 = 2$ and $y_1 = 0$

$$y - y_1 = m(x - x_1)$$
$$y - 0 = -\frac{1}{2}(x-2)$$
$$y = -\frac{\mathbf{1}}{\mathbf{2}}x - \frac{\mathbf{1}}{\mathbf{2}}(-2)$$
$$y = -\frac{1}{2}x + 1 \text{ slope-intercept form}$$

33. $(5,5)$ and $(7,5)$

(x_1, y_1) and (x_2, y_2)

$$m = \frac{y_2 - y_1}{x_2 - x_1}$$
$$= \frac{5-5}{7-5}$$
$$= \frac{0}{2}$$
$$= 0$$

Passes through $(5,5)$, $m = 0$

$x_1 = 5$ and $y_1 = 5$

$$y - y_1 = m(x - x_1)$$
$$y - 5 = 0(x-5)$$
$$y - 5 = 0$$
$$y - 5 + \mathbf{5} = 0 + \mathbf{5}$$
$$y = 5$$

35. $(5,1)$ and $(-5,0)$

(x_1, y_1) and (x_2, y_2)

$$m = \frac{y_2 - y_1}{x_2 - x_1}$$
$$= \frac{0-1}{-5-5}$$
$$= \frac{-1}{-10}$$
$$= \frac{1}{10}$$

Passes through $(5,1)$, $m = \frac{1}{10}$

$x_1 = 5$ and $y_1 = 1$

$$y - y_1 = m(x - x_1)$$
$$y - 1 = \frac{1}{10}(x-5)$$
$$y - 1 = \frac{\mathbf{1}}{\mathbf{10}}x - \frac{\mathbf{1}}{\mathbf{10}}(5)$$
$$y - 1 = \frac{1}{10}x - \frac{1}{2}$$
$$y - 1 + \mathbf{1} = \frac{1}{10}x - \frac{1}{2} + \frac{\mathbf{2}}{\mathbf{2}}$$
$$y = \frac{1}{10}x + \frac{1}{2} \text{ slope-intercept form}$$

37. $(-8,2)$ and $(-8,17)$

(x_1, y_1) and (x_2, y_2)

$$m = \frac{y_2 - y_1}{x_2 - x_1}$$
$$= \frac{17-2}{-8-(-8)}$$
$$= \frac{15}{0}$$

Undefined

Passes through $(-8,2)$, $m = \text{undefined}$

$x_1 = -8$ and $y_1 = 2$

$$x = -8$$

39. $\left(\frac{2}{3},\frac{1}{3}\right)$ and $(0,0)$

(x_1,y_1) and (x_2,y_2)

$$m=\frac{y_2-y_1}{x_2-x_1}$$
$$=\frac{0-\frac{1}{3}}{0-\frac{2}{3}}$$
$$=\frac{-\frac{1}{3}}{-\frac{2}{3}}$$
$$=-\frac{1}{\not{3}_1}\left(-\frac{\not{3}^1}{2}\right)$$
$$=\frac{1}{2}$$

Passes through $(0,0)$, $m=\frac{1}{2}$

$x_1=0$ and $y_1=0$

$$y-y_1=m(x-x_1)$$
$$y-0=\frac{1}{2}(x-0)$$
$$y=\frac{1}{2}x \text{ slope-intercept form}$$

Write an equation of each line. See Example 3.

41. Vertical, passes through (4,5)
The equation of a vertical line can be written in the form $x=a$.
$x=4$

43. Horizontal, passes through (4,5)
The equation of a horizontal line can be written in the form $y=b$.
$y=5$

Graph the line that passes through the given point and has the given slope. See Example 4.

45. $(1,-2)$, slope $-1=\frac{-1}{1}$

Start at $(1,-2)$, go down 1, right 1

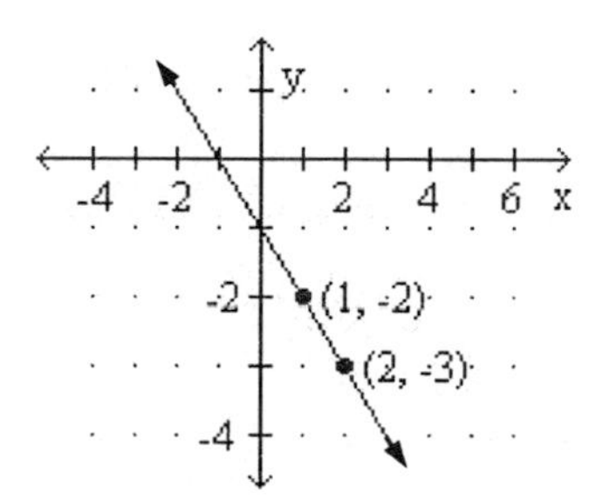

47. $(5,-3)$, $m=\frac{3}{4}$

Start at $(5,-3)$, go up 3, right 4

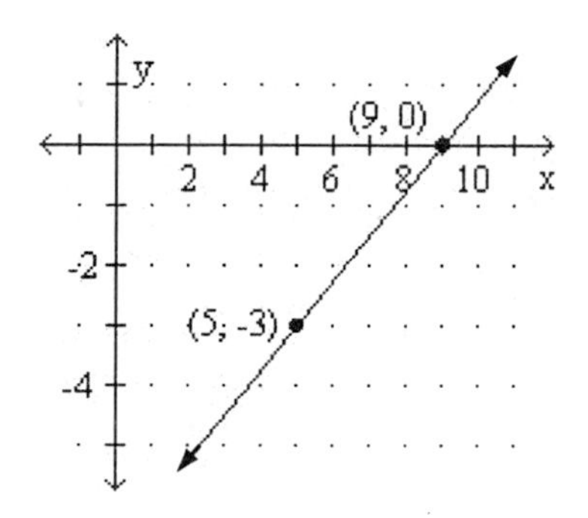

49. $(-2,-3)$, $m=2=\frac{2}{1}$

Start at $(-2,-3)$, go up 2, right 1

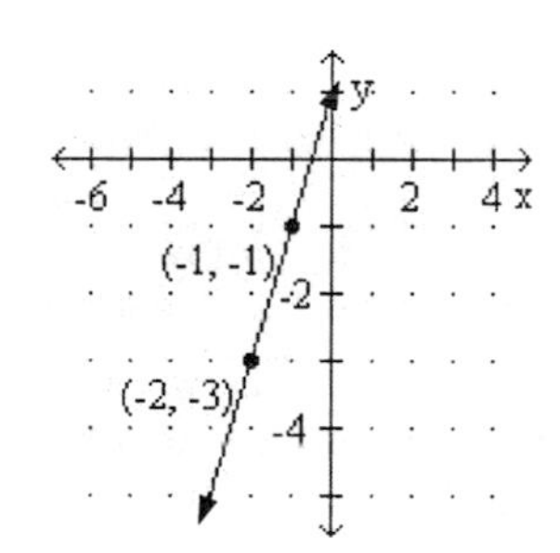

51. $(4,-3)$, $m=-\frac{7}{8}=\frac{-7}{8}$

Start at $(4,-3)$, go down 7, right 8

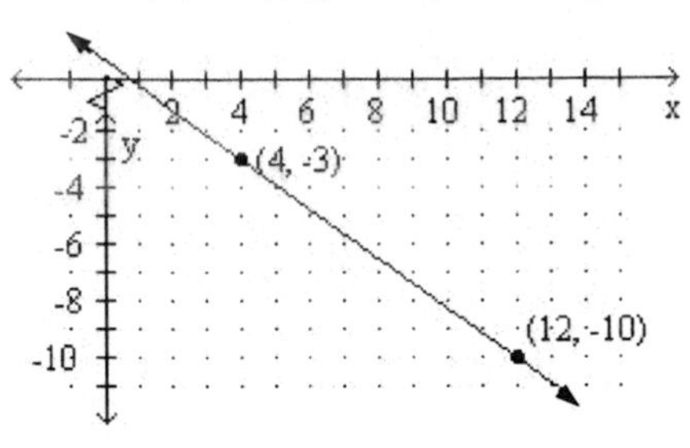

TRY IT YOURSELF

Use either the slope - intercept form (from Section 3.5) or the point - slope form (from Section 3.6) to find an equation of each line. Write each result in slope - intercept form, if possible.

53. $(5,0)$ and $(-11,-4)$

(x_1,y_1) and (x_2,y_2)

$$m=\frac{y_2-y_1}{x_2-x_1}$$
$$=\frac{-4-0}{-11-5}$$
$$=\frac{-4}{-16}$$
$$=\frac{1}{4}$$

Passes through $(5,0)$, $m=\frac{1}{4}$

$x_1=5$ and $y_1=0$

$$y-y_1=m(x-x_1)$$
$$y-0=\frac{1}{4}(x-5)$$
$$y=\mathbf{\frac{1}{4}}x-\mathbf{\frac{1}{4}}(5)$$
$$y=\frac{1}{4}x-\frac{5}{4} \text{ slope-intercept form}$$

55. Horizontal, passes through $(-8,12)$

The equation of a horizontal line can be written in the form $y=b$.

$$y=12$$

57. Passes through y-intercept $\left(0,\frac{7}{8}\right)$, $m=-\frac{1}{4}$

$x_1=0$ and $y_1=\frac{7}{8}$

$$y-y_1=m(x-x_1)$$
$$y-\frac{7}{8}=-\frac{1}{4}(x-0)$$
$$y-\frac{7}{8}=-\frac{1}{4}x$$
$$y-\frac{7}{8}+\mathbf{\frac{7}{8}}=-\frac{1}{4}x+\mathbf{\frac{7}{8}}$$
$$y=-\frac{1}{4}x+\frac{7}{8} \text{ slope-intercept form}$$

59. Passes through $(3,0)$, $m=-\frac{2}{3}$

$x_1=3$ and $y_1=0$

$$y-y_1=m(x-x_1)$$
$$y-0=-\frac{2}{3}(x-3)$$
$$y=-\frac{2}{3}(x-3)$$
$$y=-\mathbf{\frac{2}{3}}x-\left(-\mathbf{\frac{2}{3}}\right)3$$
$$y=-\frac{2}{3}x+2 \text{ slope-intercept form}$$

61. Passes through $(2,20)$, $m=8$

$x_1=2$ and $y_1=20$

$$y-y_1=m(x-x_1)$$
$$y-20=8(x-2)$$
$$y-20=\mathbf{8}x-\mathbf{8}(2)$$
$$y-20=8x-16$$
$$y-20+\mathbf{20}=8x-16+\mathbf{20}$$
$$y=8x+4 \text{ slope-intercept form}$$

63. Vertical, passes through $(-3,7)$

The equation of a vertical line can be written in the form $x=a$.

$$x=-3$$

65. Passes through y-intercept $(0,-11)$, $m=7$

$b=-11$

$$y=mx+b$$
$$y=7x+(-11)$$
$$y=7x-11$$

67. $(-2,-1)$ and $(-1,-5)$

(x_1, y_1) and (x_2, y_2)

$$m = \frac{y_2 - y_1}{x_2 - x_1}$$
$$= \frac{-5-(-1)}{-1-(-2)}$$
$$= \frac{-4}{1}$$
$$= -4$$

Passes through $(-2,-1)$, $m = -4$

$x_1 = -2$ and $y_1 = -1$

$$y - y_1 = m(x - x_1)$$
$$y - (-1) = -4(x - (-2))$$
$$y + 1 = -4(x + 2)$$
$$y + 1 = -4x - 8$$
$$y + 1 - \mathbf{1} = -4x - 8 - \mathbf{1}$$
$$y = -4x - 9 \text{ slope-intercept form}$$

69. x-intercept $(7,0)$ and y-intercept $(0,-2)$

$(7,0)$ and $(0,-2)$

(x_1, y_1) and (x_2, y_2)

$$m = \frac{y_2 - y_1}{x_2 - x_1}$$
$$= \frac{-2-0}{0-7}$$
$$= \frac{-2}{-7}$$
$$= \frac{2}{7}$$

Passes through $(7,0)$, $m = \frac{2}{7}$

$x_1 = 7$ and $y_1 = 0$

$$y - y_1 = m(x - x_1)$$
$$y - 0 = \frac{2}{7}(x - 7)$$
$$y = \frac{\mathbf{2}}{\mathbf{7}}(x) - \frac{\mathbf{2}}{\mathbf{7}}(7)$$
$$y = \frac{2}{7}x - 2 \text{ slope-intercept form}$$

71. Passes through origin $(0,0)$, $m = \frac{1}{10}$

$x_1 = 0$ and $y_1 = 0$

$$y - y_1 = m(x - x_1)$$
$$y - 0 = \frac{1}{10}(x - 0)$$
$$y = \frac{1}{10}x \text{ slope-intercept form}$$

73. Passes through $\left(-\frac{1}{8}, 12\right)$, m is undefined

$x_1 = -\frac{1}{8}$ and $y_1 = 12$

$$x = -\frac{1}{8}$$

75. Passes through y-intercept $(0,-2.8)$, $m = 1.7$

$b = -2.8$

$$y = mx + b$$
$$y = 1.7x + (-2.8)$$
$$y = 1.7x - 2.8$$

APPLICATIONS

77. ANATOMY

In this problem, we are given two different letters r and h to use other than x and y. One has to determine which letter correlates with which letter. $x = r$ and $y = h$.

Find the slope which is given as 3.9 inches for each 1-inch for the radius.

$$m = \frac{\text{rise}}{\text{run}}$$
$$= \frac{3.9}{1}$$
$$= 3.9$$

Given: 64-inch tall woman is h_1.

Given: 9-inch-long radius bone is r_1.

Now use the point-slope formula.

$$h - h_1 = m(r - r_1)$$
$$h - 64 = 3.9(r - 9)$$
$$h - 64 = \mathbf{3.9}(r) - \mathbf{3.9}(9)$$
$$h - 64 = 3.9r - 35.1$$
$$h - 64 + \mathbf{64} = 3.9r - 35.1 + \mathbf{64}$$
$$h = 3.9r + 28.9 \text{ slope-intercept form}$$

79. POLE VAULTING

Part 1: Given (5, 2) and (10, 0)

Find the slope:

$$m = \frac{y_2 - y_1}{x_2 - x_1}$$
$$= \frac{2-0}{5-10}$$
$$= \frac{2}{-5}$$
$$= -\frac{2}{5}$$

Part 2: Now pick a point (10, 0) and use

$$y - y_1 = m(x - x_1)$$
$$y - 0 = -\frac{2}{5}(x-10)$$
$$y = -\frac{2}{5}x + 4 \text{ slope-intercept form}$$

Part 3: Given (9, 7) and (10, 0)

Find the slope:

$$m = \frac{y_2 - y_1}{x_2 - x_1}$$
$$= \frac{7-0}{9-10}$$
$$= \frac{7}{-1}$$
$$= -7$$

Now pick a point (10, 0) and use

$$y - y_1 = m(x - x_1)$$
$$y - 0 = -7(x-10)$$
$$y = -7x + 70 \text{ slope-intercept form}$$

Part 4: Given (10, 7) and (10, 0)

Find the slope:

$$m = \frac{y_2 - y_1}{x_2 - x_1}$$
$$= \frac{7-0}{10-10}$$
$$= \frac{7}{0}$$

Slope is undefined.

Undefined slope means the line is vertical and $x = a$ is the equation. $x = 10$

81. TOXIC CLEANUP

a. In this problem, we are given two different letters m (month) and y (yard) to use for x and y. One has to determine which letter correlates with which letter, $x = m$ and $y = y$.

Find the slope which is given as two ordered pairs (3, 800) and (5, 720). The reason for the numbers 3 and 5 is because after "3 months" and then "2 months later".

$$m = \frac{y_2 - y_1}{m_2 - m_1}$$
$$= \frac{800-720}{3-5}$$
$$= \frac{80}{-2}$$
$$= -40$$

Now pick one point (3, 800) and use

$$y - y_1 = m(m - m_1)$$
$$y - 800 = -40(m-3)$$
$$y - 800 = -40m + 120$$
$$y - 800 + \mathbf{800} = -40m + 120 + \mathbf{800}$$
$$y = -40m + 920 \text{ slope-intercept form}$$

The remainder of the solution is on the next page.

b. To predict the number of cubic yards of waste on the site one year after the cleanup project began,

let $m = 1$ yr, 1 year = 12 months.

$$y = -40(\mathbf{12}) + 920$$
$$= -480 + 920$$
$$= 440$$

440 cubic yards of waste that will still be on the site after 1 year.

83. TRAMPOLINES

In this problem, we are given two different letters r and l to use for x and y. One has to determine which letter correlates with which letter, $x = r$ and $y = l$.

Find the slope which is given as two ordered pairs (3, 19) and (7, 44).

$$m = \frac{l_2 - l_1}{r_2 - r_1} = \frac{44 - 19}{7 - 3} = \frac{25}{4}$$

Now pick one point (3, 19) and use

$$l - l_1 = m(r - r_1)$$

$$l - 19 = \frac{25}{4}(r - 3)$$

$$l - 19 = \frac{25}{4}r - \frac{75}{4}$$

$$l - 19 + \mathbf{19} = \frac{25}{4}r - \frac{75}{4} + \mathbf{19}$$

$$l = \frac{25}{4}r - \frac{75}{4} + \frac{\mathbf{76}}{\mathbf{4}}$$

$$l = \frac{25}{4}r + \frac{1}{4} \quad \text{slope-intercept form}$$

85. GOT MILK

a. Two points must be selected (yr, gal). (6, 26.5) and (21, 22.0) were selected because they both lie on the line. Now you must find the slope of the two selected ordered pairs.

$$m = \frac{y_2 - y_1}{x_2 - x_1} = \frac{26.5 - 22.0}{6 - 21} = \frac{4.5}{-15} = -\frac{4.5 \cdot 10}{15 \cdot 10} = -\frac{45}{150} = -\frac{3}{10}$$

Now pick one point (21, 22.0) and use

$$y - y_1 = m(x - x_1)$$

$$y - 22.0 = -\frac{3}{10}(x - 21)$$

$$y - 22.0 = -\frac{3}{10}x + \frac{63}{10}$$

$$y - 22.0 + 22.0 = -\frac{3}{10}x + 6.3 + 22.0$$

$$y = -\frac{3}{10}x + 28.3 \quad \text{slope-intercept form}$$

or

$$y = -0.3x + 28.3 \quad \text{slope-intercept form}$$

or

$$y = -\frac{3}{10}x + \frac{283}{10} \quad \text{slope-intercept form}$$

b. Use $x = 40$ (the year 2020 correlates with 40) to calculate the number of gallons of milk to be consumed.

$$y = -0.3x + 28.3$$

$$y = -0.3(\mathbf{40}) + 28.3$$

$$y = -12 + 28.3$$

$$y = 16.3$$

The amount of milk that an average American will drink in 2020 is 16.3 gal.

WRITING

87-89. Answers will vary.

REVIEW

91. FRAMES

Analyze

- The length of a rectangular picture is 5 inches greater than twice the width.
- The perimeter is 112 inches.
- Find the dimensions of the frame.

Assign

Let x = width of frame in inches

$2x+5$ = length of frame in inches

Form

2•	the width of the frame	plus	2•	the length of the frame	equals	the perimeter.
	$2x$	$+$		$2(2x+5)$	$=$	112

Solve

$$2x+2(2x+5)=112$$
$$2x+4x+10=112$$
$$6x+10=112$$
$$6x+10-\mathbf{10}=112-\mathbf{10}$$
$$6x=102$$
$$\frac{6x}{\mathbf{6}}=\frac{102}{\mathbf{6}}$$
$$x=17$$

length

$$2x+5=2(\mathbf{17})+5$$
$$=34+5$$
$$=39$$

State

The dimensions of the frame are 17 inches by 39 inches.

Check

$$2x+2(2x+5)=112$$
$$2(\mathbf{17})+2(\mathbf{39})=112$$
$$34+78=112$$
$$112=112$$

The results check.

CHALLENGE PROBLEMS

93. Given (2, 5) Find the equation of a line parallel to $y = 4x - 7$.

First, find the slope from $y = 4x - 7$, $m = 4$.

The two lines are parallel and have the same slope, so use $m = 4$ and (2, 5) to write the equation of the line.

$$y-y_1=m(x-x_1)$$
$$y-5=4(x-2)$$
$$y-5=4x-8$$
$$y-5+\mathbf{5}=4x-8+\mathbf{5}$$
$$y=4x-3 \text{ slope-intercept form}$$

SECTION 3.7 STUDY SET
VOCABULARY

Fill in the blanks.

1. $2x - y \leq 4$ is a linear **inequality** in two variables.
3. (7, 2) is a solution of $x - y > 1$. We say that (7, 2) **satisfies** the inequality.
5. In the graph, the line $2x - y = 4$ divides the coordinate plane into two **half-planes**.

CONCEPTS

7. Determine whether (–3, –5) is a solution of $5x - 3y \geq 0$. **Yes**

$$5x - 3y \overset{?}{\geq} 0$$
$$5(-3) - 3(-5) \overset{?}{\geq} 0$$
$$-15 + 15 \overset{?}{\geq} 0$$
$$0 \geq 0 \text{ True}$$

9. Fill in the blanks: A **dashed** line indicates that points on the boundary are not solutions and a **solid** line indicates that points on the boundary are solutions.
11. If a false statement results when the coordinates of a test point are substituted into a linear inequality, which half-plane should be shaded to represent the solution of the inequality?

    The half-plane opposite that in which the test point lies

13. A linear inequality has been graphed. Determine whether each point satisfies the inequality.
    a. (2, 1) **Yes**
    b. (–2, –4) **No**
    c. (4, –2) **No**
    d. (–3, 4) **Yes**

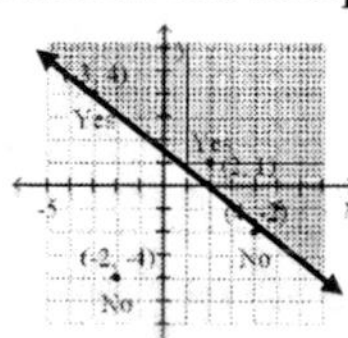

NOTATION

15. Write the meaning of each symbol in words.
    a. < **Is less than**
    b. ≥ **Is greater than or equal to**
    c. ≤ **Is less than or equal to**
    d. $\overset{?}{>}$ **Is possibly greater than**
17. Fill in the blanks: The inequality $4x + 2y \leq 9$ means $4x + 2y$ **=** 9 or $4x + 2y$ **<** 9.

GUIDED PRACTICE

Determine whether each ordered pair is a solution of the given inequality.

19. $2x + y > 6;\ (3,2)$

$$2(\mathbf{3}) + \mathbf{2} \overset{?}{>} 6$$
$$6 + 2 \overset{?}{>} 6$$
$$8 > 6 \text{ True}$$

(3,2) is a solution.

20. $4x - 2y \geq -6;\ (-2,1)$

$$4(\mathbf{-2}) - 2(\mathbf{1}) \overset{?}{\geq} -6$$
$$-8 - 2 \overset{?}{\geq} -6$$
$$-10 \geq -6 \text{ False}$$

(–2,1) is not a solution.

21. $-5x - 8y < 8;\ (-8,4)$

$$-5(\mathbf{-8}) - 8(\mathbf{4}) \overset{?}{<} 8$$
$$40 - 32 \overset{?}{<} 8$$
$$8 < 8 \text{ False}$$

(–8, 4) is not a solution.

22. $x + 3y > 14;\ (-3,8)$

$$\mathbf{-3} + 3(\mathbf{8}) \overset{?}{>} 14$$
$$-3 + 24 \overset{?}{>} 14$$
$$21 > 14 \text{ True}$$

(–3,8) is a solution.

23. $4x - y \leq 0;\ \left(\frac{1}{2}, 1\right)$

$$4\left(\frac{\mathbf{1}}{\mathbf{2}}\right) - \mathbf{1} \overset{?}{\leq} 0$$
$$2 - 1 \overset{?}{\leq} 0$$
$$1 \leq 0 \text{ False}$$

$\left(\frac{1}{2}, 1\right)$ is not a solution.

25. $-5x + 2y > -4;\ (0.8, 0.6)$

$$-5(\mathbf{0.8}) + 2(\mathbf{0.6}) \overset{?}{>} -4$$
$$-4 + 1.2 \overset{?}{>} -4$$
$$-2.8 > -4 \text{ True}$$

(0.8, 0.6) is a solution.

Complete the graph by shading the correct side of the boundary.

27. $x - y \geq -2$, Boundary line is <u>solid</u> because of the $\geq$ symbol.

Select test point (0,0) and substitute into

$x - y \geq -2$

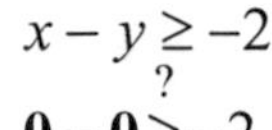

$\mathbf{0} - \mathbf{0} \overset{?}{\geq} -2$

$0 \geq -2$

True

The coordinates of every point on the same side of the line as the origin satisfy the inequality. To indicate this, we shade the half-plane that contains the test point (0,0).

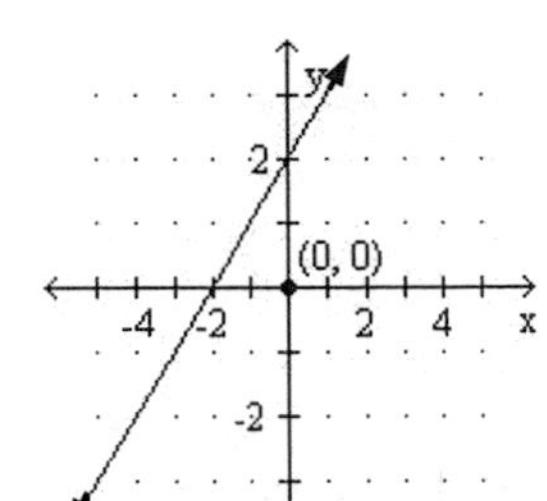

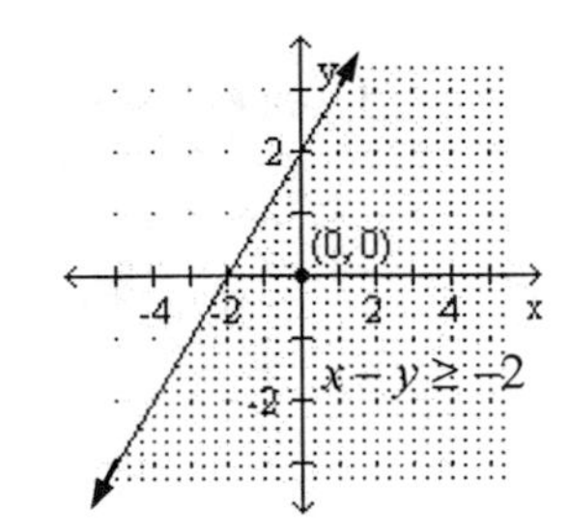

29. $y > 2x - 4$, Boundary line is dashed because of the $>$ symbol.

Select test point (0,0) and substitute into

$y > 2x - 4$

$\mathbf{0} \overset{?}{>} 2(\mathbf{0}) - 4$

$0 > -4$

True

The coordinates of every point on the same side of the line as the origin satisfy the inequality. To indicate this, we shade the half-plane that contains the test point (0,0).

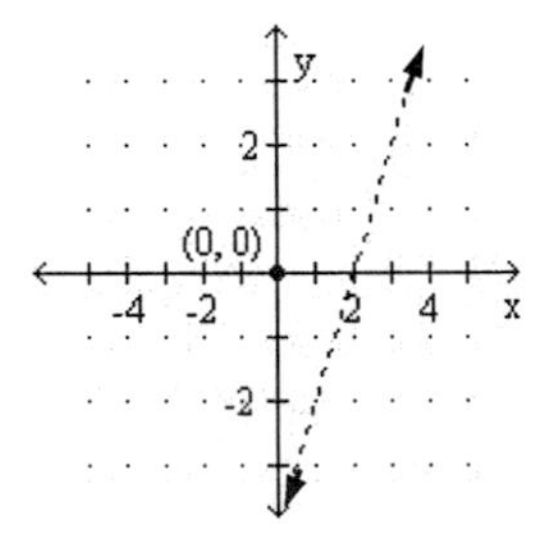

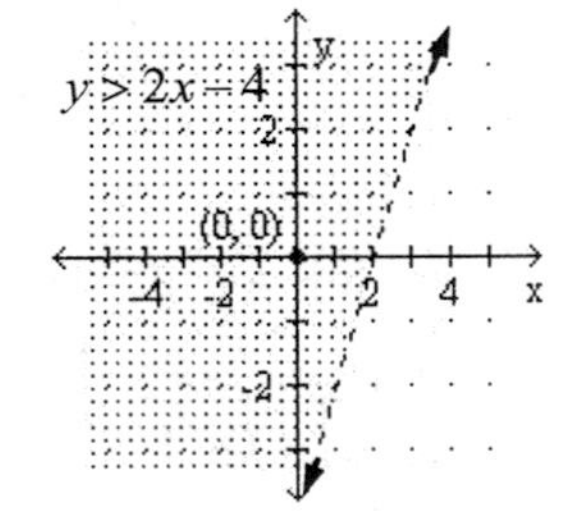

31. $x - 2y \geq 4$, Boundary line is <u>solid</u> because of the $\geq$ symbol.

Select test point (0,0) and substitute into

$x - 2y \geq 4$

$\mathbf{0} - 2(\mathbf{0}) \overset{?}{\geq} 4$

$0 \geq 4$

False

The coordinates of every point on the same side of the line as the origin ***do not*** satisfy the inequality. To indicate this, we shade the half-plane that ***does not*** contain the test point (0, 0).

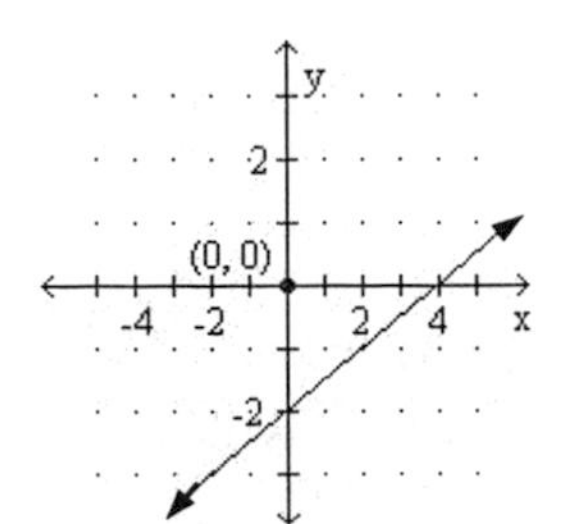

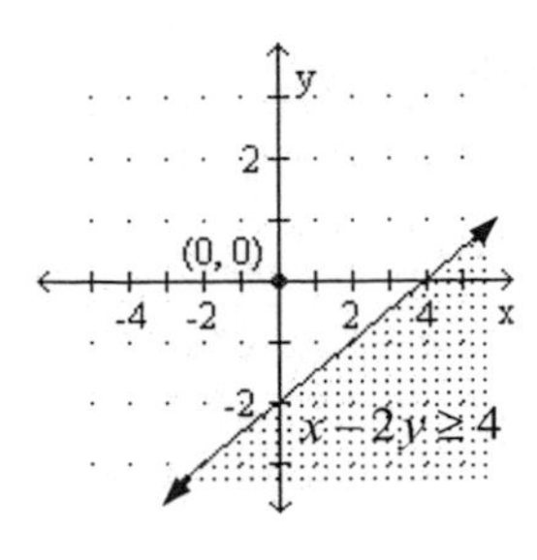

33. $y \leq 4x$, Boundary line is <u>solid</u> because of the $\leq$ symbol.

The line passes through the origin so select test point (1,1) and substitute into

$y \leq 4x$

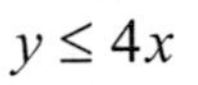

$\mathbf{1} \overset{?}{\leq} 4(\mathbf{1})$

$1 \leq 4$

True

The coordinates of every point on the same side of the line as (1,1) satisfy the inequality. To indicate this, we shade the half-plane that contains the test point (1,1).

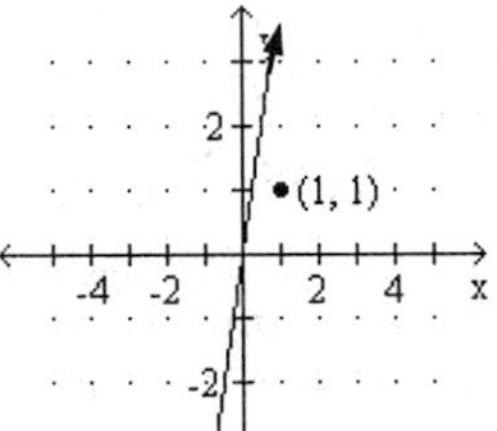

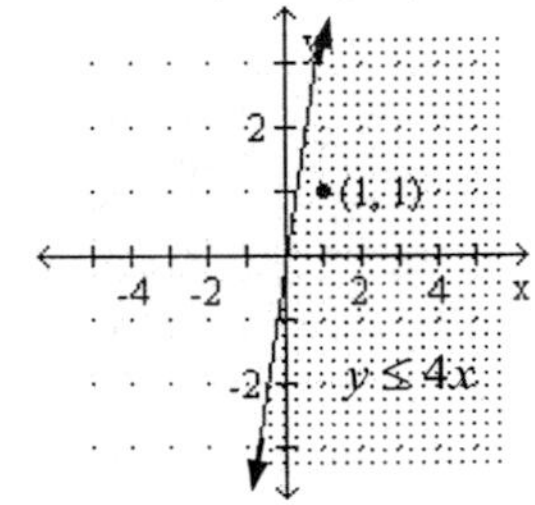

Graph each inequality.

35. $x + y \geq 3$, Boundary line is <u>solid</u> because of the $\geq$ symbol.

Graph as $x + y = 3$.

y-intercept:	x-intercept:
If $x = 0$,	If $y = 0$,
$\mathbf{0} + y = 3$	$x + \mathbf{0} = 3$
$y = 3$	$x = 3$

The y-int is (0, 3), and the x-int is (3, 0).

Select test point (0,0) and substitute into

$x + y \geq 3$

$\mathbf{0} - 2(\mathbf{0}) \overset{?}{\geq} 3$

$0 \geq 3$

False

The coordinates of every point on the same side of the line as the origin ***do not*** satisfy the inequality. To indicate this, we shade the half-plane that ***does not*** contain the test point (0, 0).

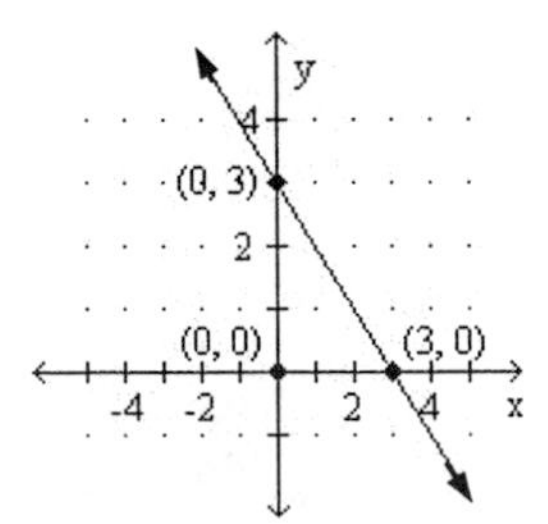

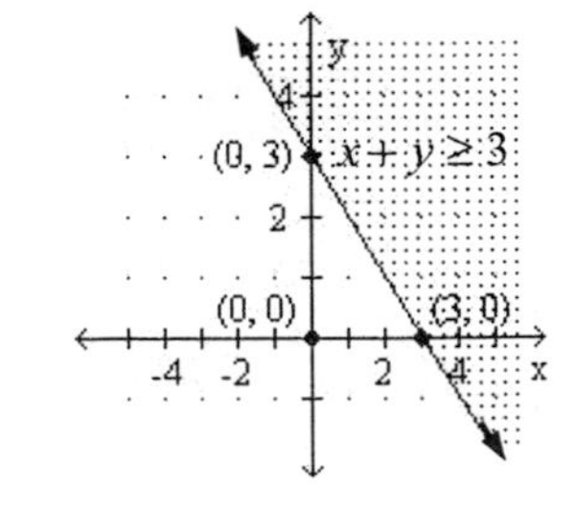

37. $3x-4y>12$, Boundary line is dashed because of the $>$ symbol.
Graph as $3x-4y=12$.

y-intercept:	x-intercept:
If $x=0$,	If $y=0$,
$3(\mathbf{0})-4y=12$	$3x+4(\mathbf{0})=12$
$-4y=12$	$3x=12$
$\frac{-4y}{\mathbf{-4}}=\frac{12}{\mathbf{-4}}$	$\frac{3x}{\mathbf{3}}=\frac{12}{\mathbf{3}}$
$y=-3$	$x=4$

The y-int is $(0,\ -3)$, and the x-int is $(4,\ 0)$.
Select test point (0,0) and substitute into

$$3x-4y>12$$
$$3(\mathbf{0})-4(\mathbf{0})\overset{?}{>}12$$
$$0>12$$
False

The coordinates of every point on the same side of the line as the origin ***do not*** satisfy the inequality. To indicate this, we shade the half-plane that ***does not*** contain the test point (0, 0).

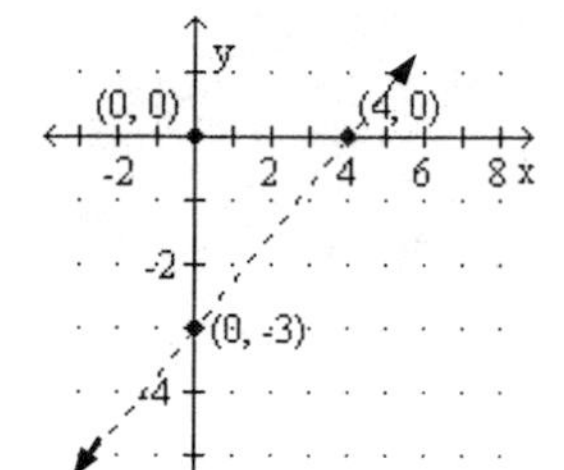

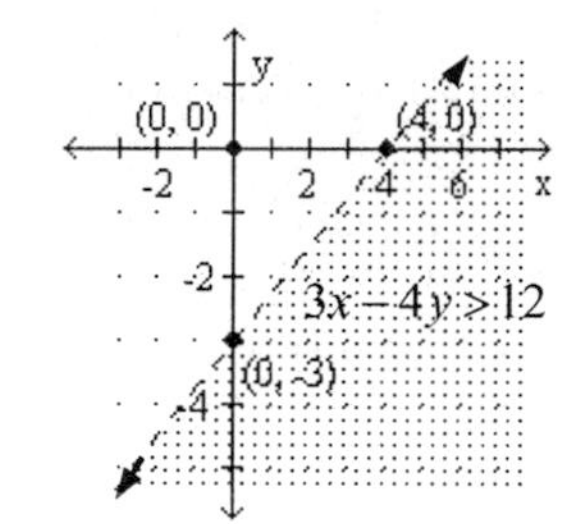

39. $2x+3y\le -12$. Boundary line is solid because of the $\le$ symbol.
Graph as $2x+3y=-12$.

y-intercept:	x-intercept:
If $x=0$,	If $y=0$,
$2(\mathbf{0})+3y=-12$	$2x+3(\mathbf{0})=-12$
$3y=-12$	$2x=-12$
$\frac{3y}{\mathbf{3}}=\frac{-12}{\mathbf{3}}$	$\frac{2x}{\mathbf{2}}=\frac{-12}{\mathbf{2}}$
$y=-4$	$x=-6$

The y-int is $(0,\ -4)$, and the x-int is $(-6,\ 0)$.
Select test point (0,0) and substitute into

$$2x+3y\le -12$$
$$2(\mathbf{0})+3(\mathbf{0})\overset{?}{\le}-12$$
$$0\le -12$$
False

The coordinates of every point on the same side of the line as the origin ***do not*** satisfy the inequality. To indicate this, we shade the half-plane that ***does not*** contain the test point (0, 0).

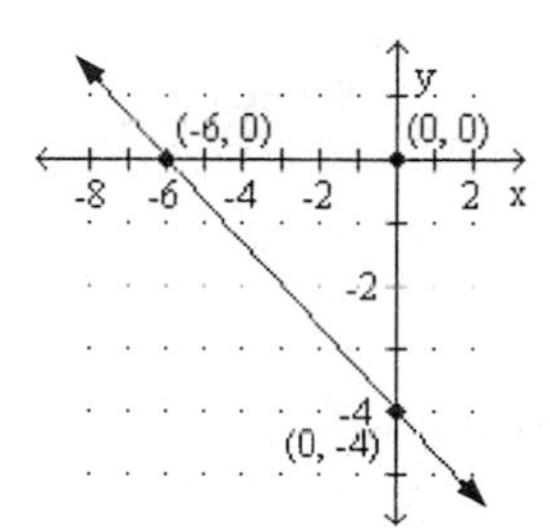

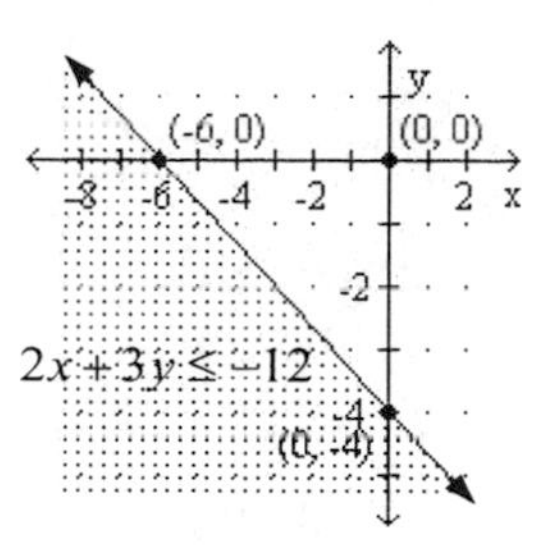

41. $y<2x-1$, Boundary line is dashed because of the $<$ symbol.
Graph as $y=2x-1$.

y-intercept:	x-intercept:
If $x=0$,	If $y=0$,
$y=2(\mathbf{0})-1$	$\mathbf{0}=2x-1$
$y=-1$	$1=2x$
	$\frac{1}{\mathbf{2}}=\frac{2x}{\mathbf{2}}$
	$\frac{1}{2}=x$

The y-int is $(0,\ -1)$, and the x-int is $\left(\frac{1}{2},0\right)$.

Select test point (0,0) and substitute into

$$y<2x-1$$
$$\mathbf{0}\overset{?}{<}2(\mathbf{0})-1$$
$$0<-1$$
False

The coordinates of every point on the same side of the line as the origin ***do not*** satisfy the inequality. To indicate this, we shade the half-plane that ***does not*** contain the test point (0, 0).

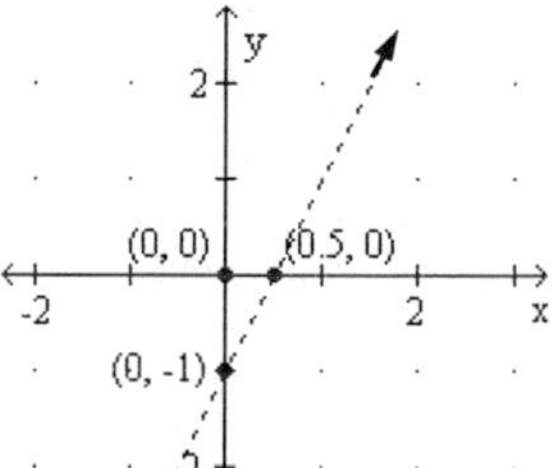

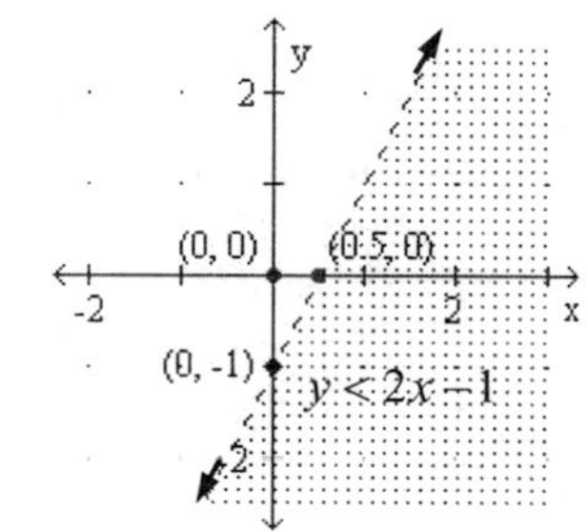

43. $y < -3x + 2$, Boundary line is dashed because of the $<$ symbol.
Graph as $y = -3x + 2$.

y-intercept:	x-intercept:
If $x = 0$,	If $y = 0$,
$y = -3(\mathbf{0}) + 2$	$\mathbf{0} = -3x + 2$
$y = 2$	$0 - \mathbf{2} = -3x + 2 - \mathbf{2}$
	$-2 = -3x$
	$\dfrac{-2}{\mathbf{-3}} = \dfrac{-3x}{\mathbf{-3}}$
	$\dfrac{2}{3} = x$

The y-int is (0, 2), and the x-int is $\left(\dfrac{2}{3}, 0\right)$.

Select test point (0,0) and substitute into

$y < -3x + 2$

$\mathbf{0} \overset{?}{<} -3(\mathbf{0}) + 2$

$0 < 2$

True

The coordinates of every point on the same side of the line as the origin satisfy the inequality. To indicate this, we shade the half-plane that contains the test point (0,0).

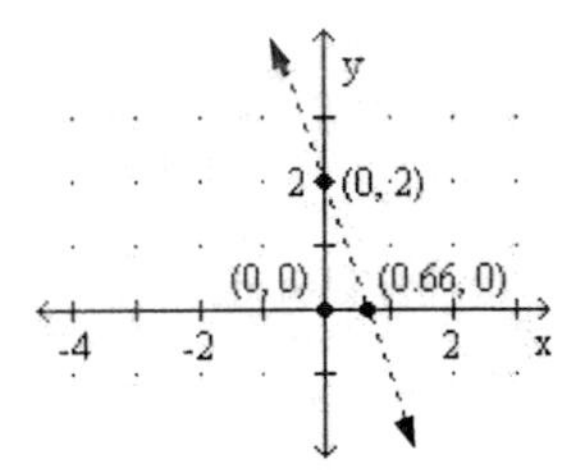

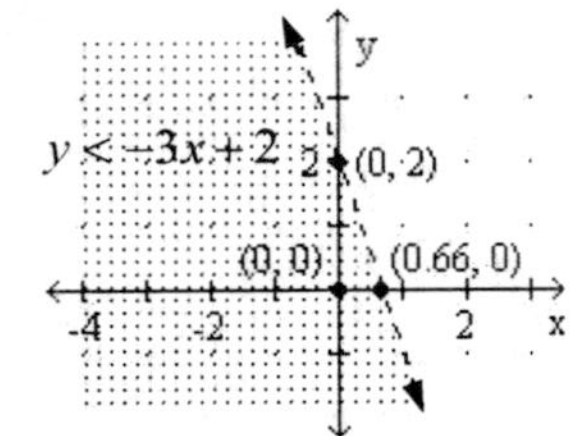

45. $y \geq -\dfrac{3}{2}x + 1$, Boundary line is solid because of the $\geq$ symbol.
Graph as $y = -\dfrac{3}{2}x + 1$.

y-intercept:	x-intercept:
If $x = 0$,	If $y = 0$,
$y = -\dfrac{3}{2}(\mathbf{0}) + 1$	$\mathbf{0} = -\dfrac{3}{2}x + 1$
$y = 1$	$0 - \mathbf{1} = -\dfrac{3}{2}x + 1 - \mathbf{1}$
	$-1 = -\dfrac{3}{2}x$
	$\left(-\dfrac{\mathbf{2}}{\mathbf{3}}\right)\cdot(-1) = \left(-\dfrac{\mathbf{2}}{\mathbf{3}}\right) - \dfrac{3}{2}x$
	$\dfrac{2}{3} = x$

The y-int is (0, 1), and the x-int is $\left(\dfrac{2}{3}, 0\right)$.

Select test point (0,0) and substitute into

$y \geq -\dfrac{3}{2}x + 1$

$\mathbf{0} \overset{?}{\geq} -\dfrac{3}{2}(\mathbf{0}) + 1$

$0 \geq 1$

False

The coordinates of every point on the same side of the line as the origin ***do not*** satisfy the inequality. To indicate this, we shade the half-plane that ***does not*** contain the test point (0, 0).

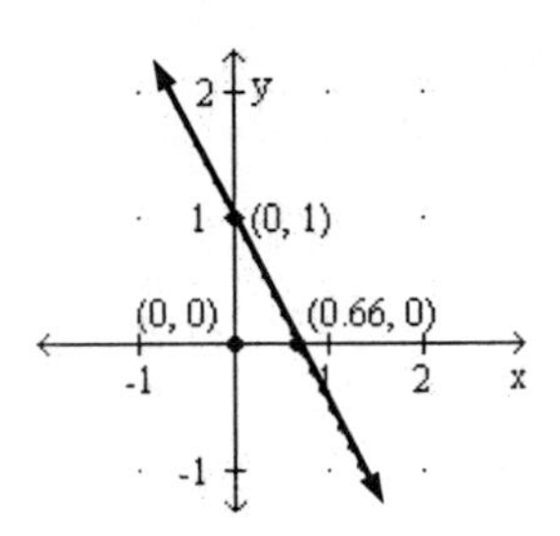

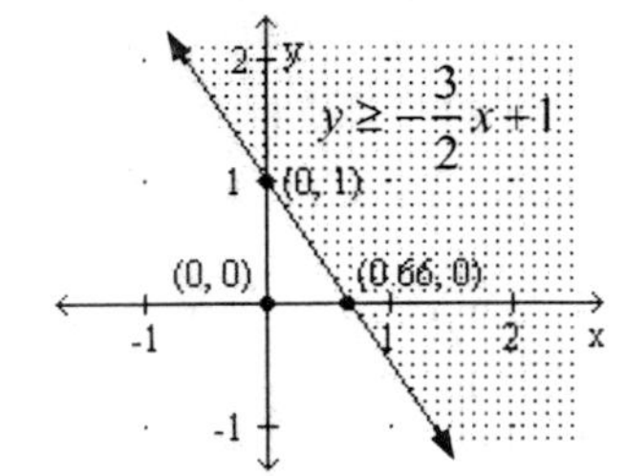

47. $x - 2y \geq 4$, Boundary line is solid because of the $\geq$ symbol.
Graph as $x - 2y = 4$.

y-intercept:	x-intercept:
If $x = 0$,	If $y = 0$,
$x - 2y = 4$	$x - 2y = 4$
$\mathbf{0} - 2y = 4$	$x - 2(\mathbf{0}) = 4$
$-2y = 4$	$x = 4$
$\dfrac{-2y}{\mathbf{-2}} = \dfrac{4}{\mathbf{-2}}$	
$y = -2$	

The y-int is $(0, -2)$, and the x-int is (4,0).

Select test point (0,0) and substitute into

$x - 2y \geq 4$

$\mathbf{0} - 2(\mathbf{0}) \overset{?}{\geq} 4$

$0 \geq 4$

False

The coordinates of every point on the same side of the line as the origin ***do not*** satisfy the inequality. To indicate this, we shade the half-plane that ***does not*** contain the test point (0, 0).

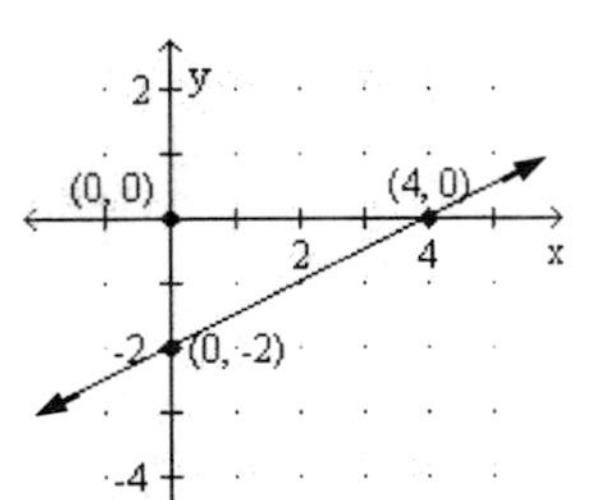

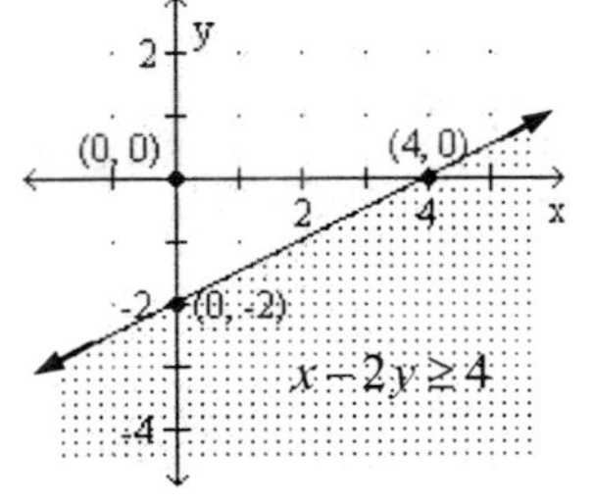

49. $2y - x < 8$, Boundary line is dashed because of the $<$ symbol.
Graph as $2y - x = 8$.

y-intercept:	x-intercept:
If $x = 0$,	If $y = 0$,
$2y - \mathbf{0} = 8$	$2(\mathbf{0}) - x = 8$
$2y = 8$	$-x = 8$
$y = 4$	$x = -8$

The y-int is $(0,4)$, and the x-int is $(-8,0)$.

Select test point $(0,0)$ and substitute into

$$2y - x < 8$$
$$2(\mathbf{0}) - \mathbf{0} \overset{?}{<} 8$$
$$0 < 8$$
True

The coordinates of every point on the same side of the line as the origin satisfy the inequality. To indicate this, we shade the half-plane that contains the test point (0,0).

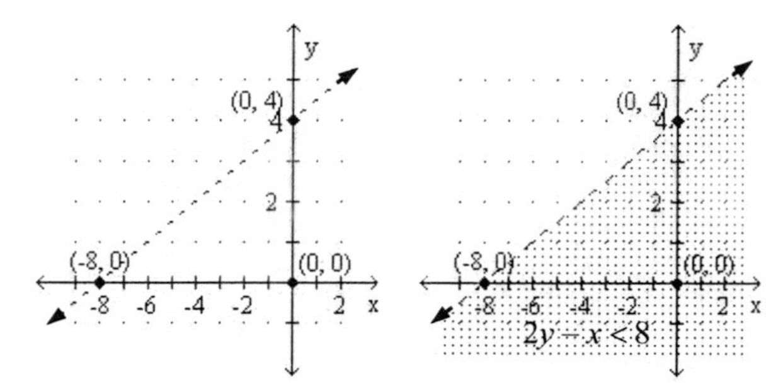

51. $7x - 2y < 21$, Boundary line is dashed because of the $<$ symbol.
Graph as $7x - 2y = 21$.

y-intercept:	x-intercept:
If $x = 0$,	If $y = 0$,
$7(\mathbf{0}) - 2y = 21$	$7x - 2(\mathbf{0}) = 21$
$-2y = 21$	$7x = 21$
$\dfrac{-2y}{\mathbf{-2}} = \dfrac{21}{\mathbf{-2}}$	$\dfrac{7x}{\mathbf{7}} = \dfrac{21}{\mathbf{7}}$
$y = -\dfrac{21}{2}$	$x = 3$

The y-int is $\left(0, -\dfrac{21}{2}\right)$, and the x-int is $(3,0)$.

Select test point $(0,0)$ and substitute into

$$7(\mathbf{0}) - 2(\mathbf{0}) < 21$$
$$0 < 21$$
True

The coordinates of every point on the same side of the line as (0,0) satisfy the inequality. To indicate this, we shade the half-plane that contains the test point (0,0).

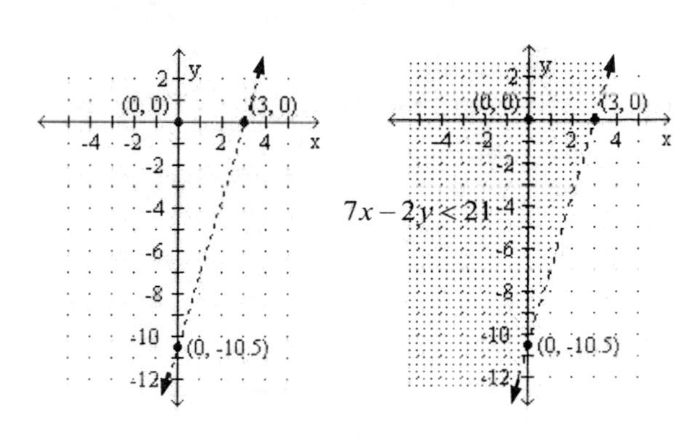

53. $2x - 3y \geq 4$, Boundary line is solid because of the $\geq$ symbol.
Graph as $2x - 3y = 4$.

y-intercept:	x-intercept:
If $x = 0$,	If $y = 0$,
$2(\mathbf{0}) - 3y = 4$	$2x - 3(\mathbf{0}) = 4$
$-3y = 4$	$2x = 4$
$\dfrac{-3y}{\mathbf{-3}} = \dfrac{4}{\mathbf{-3}}$	$\dfrac{2x}{\mathbf{2}} = \dfrac{4}{\mathbf{2}}$
$y = -\dfrac{4}{3}$	$x = 2$

The y-int is $\left(0, -\dfrac{4}{3}\right)$, and the x-int is $(2,0)$.

Select test point $(0,0)$ and substitute into

$$2x - 3y \geq 4$$
$$2(\mathbf{0}) - 3(\mathbf{0}) \overset{?}{\geq} 4$$
$$0 \geq 4$$
False

The coordinates of every point on the same side of the line as the origin ***do not*** satisfy the inequality. To indicate this, we shade the half-plane that ***does not*** contain the test point (0, 0).

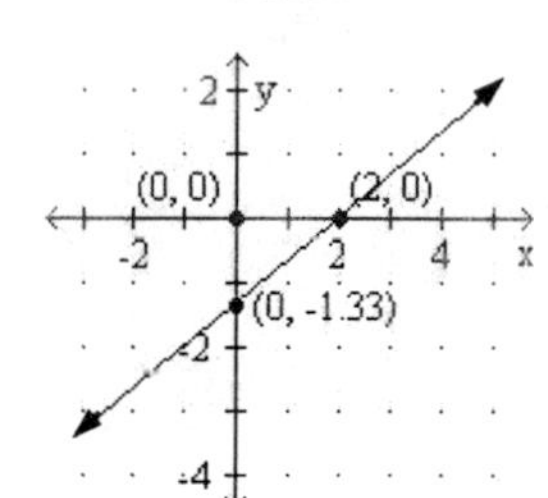

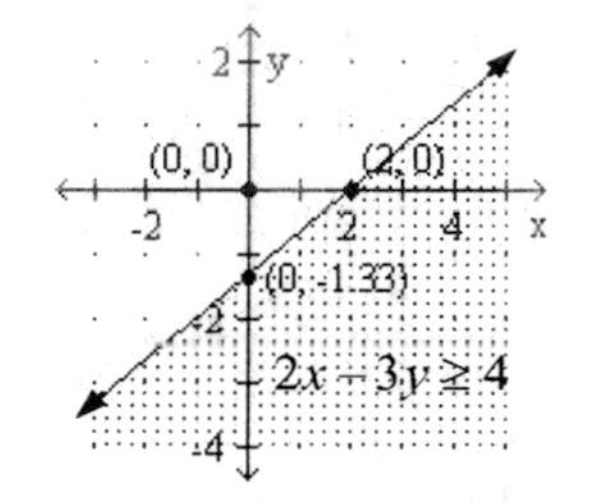

Graph each inequality.

55. $y \geq 2x$, Boundary line is solid because of the $\geq$ symbol. Graph as $y = 2x$.

y-intercept:	x-intercept:
If $x = 0$,	If $y = 0$,
$y = 2(\mathbf{0})$	$\mathbf{0} = 2x$
$y = 0$	$0 = x$

The y-int is (0,0), and the x-int is $(0,0)$.

The line passes through the origin so select test point (0,2) and substitute into

$y > 2x$

$\mathbf{2} \overset{?}{>} 2(\mathbf{0})$

$2 > 0$

True

The coordinates of every point on the same side of the line as (0,2) satisfy the inequality. To indicate this, we shade the half-plane that contains the test point (0,2).

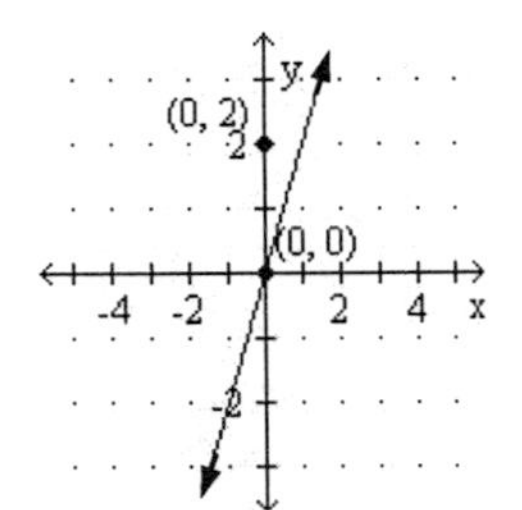

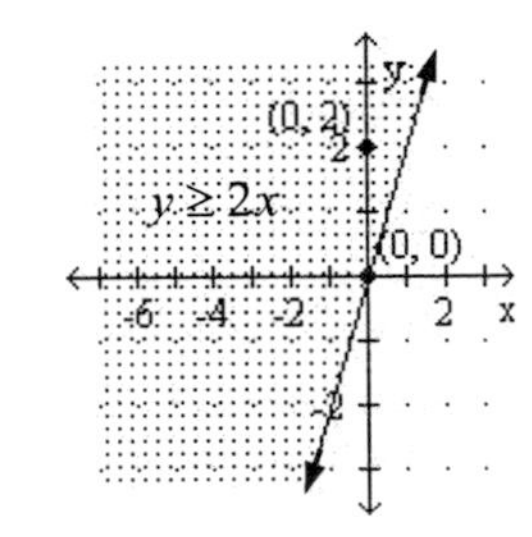

57. $y < -\frac{x}{2}$, Boundary line is dashed because of the $<$ symbol.

Graph as $y = -\frac{x}{2}$.

y-intercept:	x-intercept:
If $x = 0$,	If $y = 0$,
$y = -\frac{\mathbf{0}}{2}$	$\mathbf{0} = -\frac{x}{2}$
$y = 0$	$0 = x$

The y-int is (0,0), and the x-int is $(0,0)$.

The line passes through the origin so select test point (0,2) and substitute into

$y < -\frac{x}{2}$

$\mathbf{2} \overset{?}{<} -\frac{\mathbf{0}}{2}$

$2 < 0$

False

The coordinates of every point on the same side of the line as (0, 2) ***do not*** satisfy the inequality. To indicate this, we shade the half-plane that ***does not*** contain the test point (0, 2).

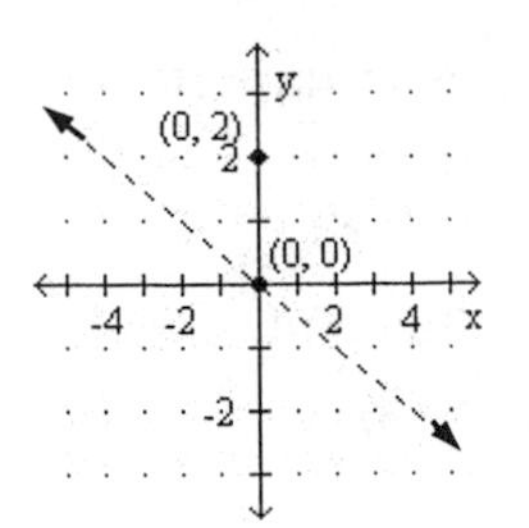

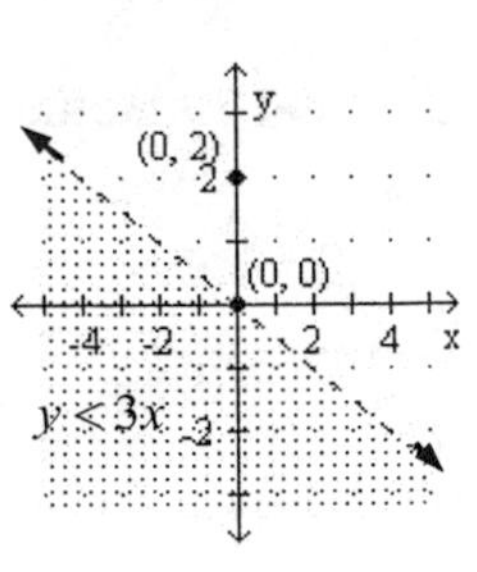

59. $y + x < 0$, Boundary line is dashed because of the $<$ symbol. Graph as $y = -x$.

y-intercept:	x-intercept:
If $x = 0$,	If $y = 0$,
$y = -\mathbf{0}$	$\mathbf{0} = -x$
$y = 0$	$0 = x$

The y-int is (0,0), and the x-int is $(0,0)$.

The line passes through the origin so select test point (0,2) and substitute into

$y + x < 0$

$\mathbf{2} + \mathbf{0} \overset{?}{<} 0$

$2 < 0$

False

The coordinates of every point on the same side of the line as (0, 2) ***do not*** satisfy the inequality. To indicate this, we shade the half-plane that ***does not*** contain the test point (0, 2).

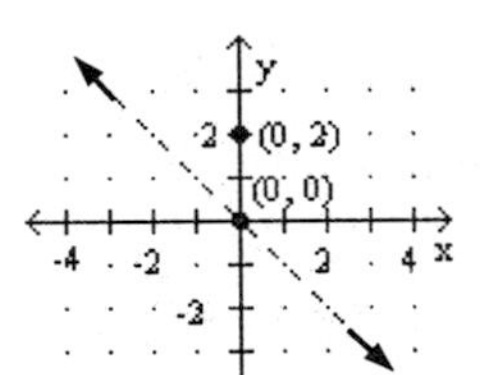

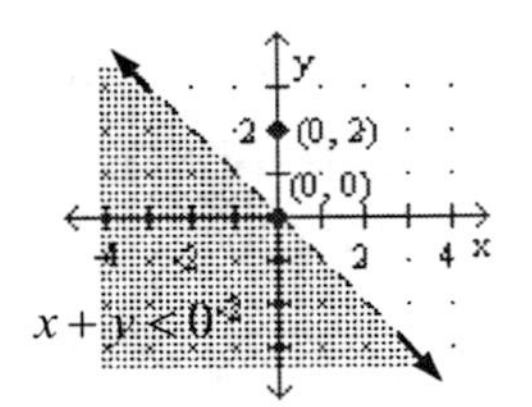

61. $5x + 3y < 0$, Boundary line is dashed because of the $<$ symbol. Graph as $5x + 3y = 0$.

y-intercept:	x-intercept:
If $x = 0$,	If $y = 0$,
$5(\mathbf{0}) + 3y = 0$	$5x + 3(\mathbf{0}) = 0$
$3y = 0$	$5x = 0$
$\frac{3y}{\mathbf{3}} = \frac{0}{\mathbf{3}}$	$\frac{5x}{\mathbf{5}} = \frac{0}{\mathbf{5}}$
$y = 0$	$x = 0$

The y-int is (0,0), and the x-int is $(0,0)$.

The line goes through the origin so select test point (0,2) and substitute into

$$5x + 3y < 0$$

$$5(\mathbf{0}) + 3(\mathbf{2}) \overset{?}{<} 0$$

$$6 < 0$$

False

The coordinates of every point on the same side of the line as (0, 2) ***do not*** satisfy the inequality. To indicate this, we shade the half-plane that ***does not*** contain the test point (0, 2).

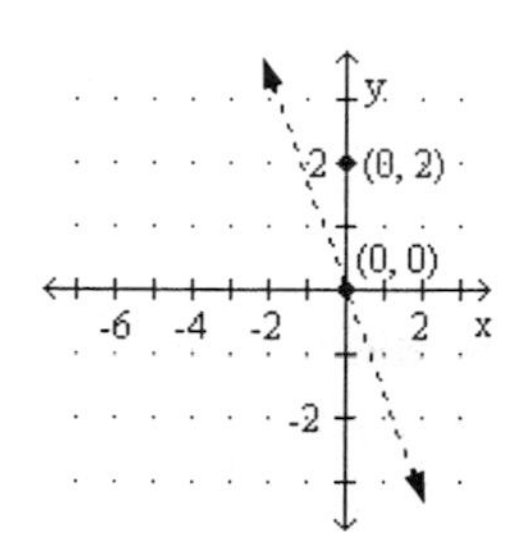

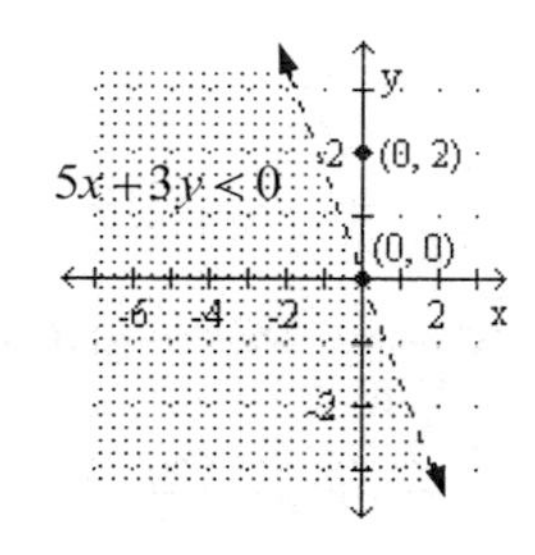

Graph each inequality.

63. $x < 2$, Boundary line is dashed because of the $<$ symbol.
Graph as $x = 2$.
There is no y-int, and the x-int is $(2,0)$.
$x = 2$ is a line parallel to the y-axis.
Select test point (0,0) and substitute into

$x < 2$

$\mathbf{0} < 2$

True

The coordinates of every point on the same side of the line as (0,0) satisfy the inequality. To indicate this, we shade the half-plane that contains the test point (0,0).

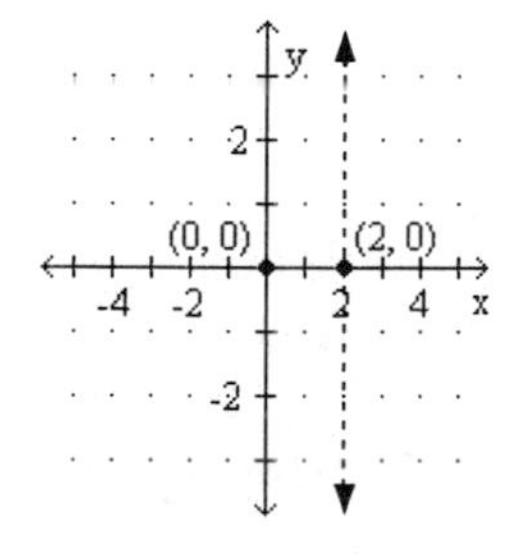

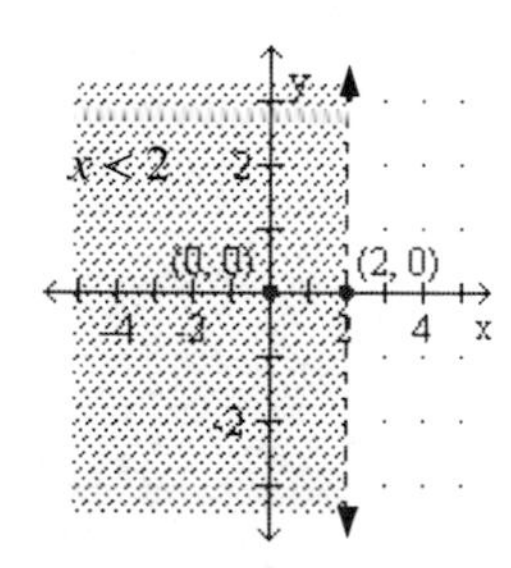

65. $y \le 1$, Boundary line is solid because of the $\le$ symbol.
Graph as $y = 1$.
The y-int is $(0,1)$, and there is no x-int.
$y = 1$ is a line parallel to the x-axis.
Select test point (0,0) and substitute into

$y \le 1$

$\mathbf{0} \le 1$

True

The coordinates of every point on the same side of the line as (0,0) satisfy the inequality. To indicate this, we shade the half-plane that contains the test point (0,0).

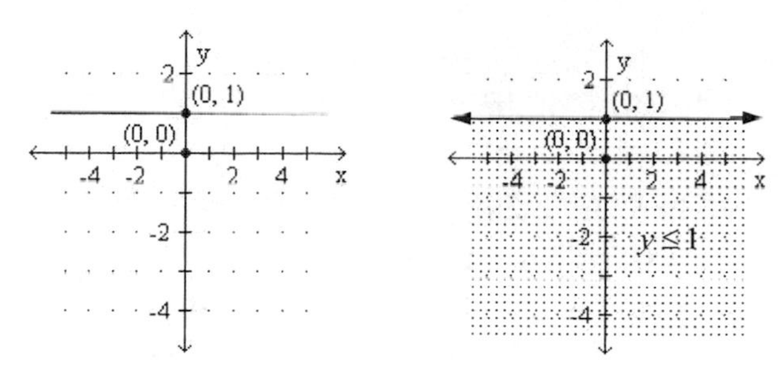

67. $y + 2.5 > 0$, Boundary line is dashed because of the $>$ symbol.
Graph as $y + 2.5 = 0$.
y-intercept:
If $x = 0$,

$$y + 2.5 = 0$$

$$y = -2.5$$

The y-int is $(0, -2.5)$, and no x-int.
$y = -2.5$ is parallel to the x-axis.
Select test point $(1,1)$ and substitute into

$$y + 2.5 > 0$$

$$\mathbf{1} + 2.5 \overset{?}{>} 0$$

$$3.5 > 0$$

True

The coordinates of every point on the same side of the line as (1, 1) satisfy the inequality. To indicate this, we shade the half-plane that contains the test point (1, 1).

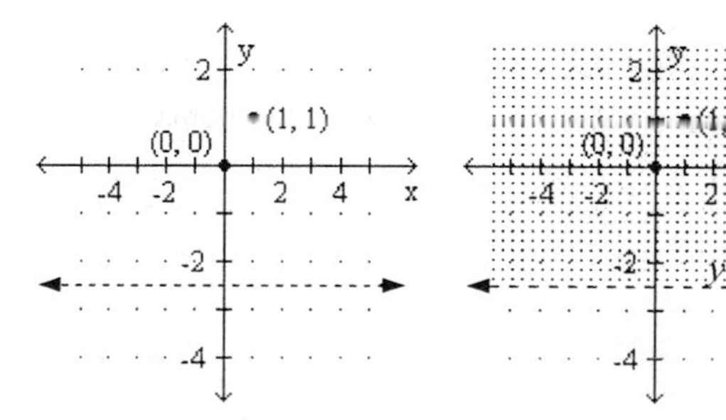

69. $x \le 0$, Boundary line is solid because of the $\le$ symbol.
Graph as $x = 0$.

There are many y-int, and the x-int is $(0,0)$.
$x = 0$ is the same line as the y-axis.
Select test point $(1,1)$ and substitute into

$x \le 0$
$1 \le 0$
False

The coordinates of every point on the same side of the line as (1, 1) ***do not*** satisfy the inequality. To indicate this, we shade the half-plane that ***does not*** contain the test point (1, 1).

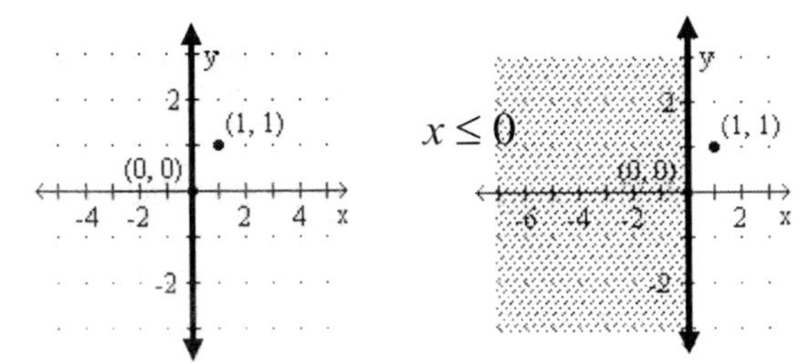

TRY IT YOURSELF

71. a. $5x - 3y \ge -15$, Boundary line is solid because of the $\ge$ symbol.
Graph as $5x - 3y = -15$.

y-intercept:	x-intercept:
If $x = 0$,	If $y = 0$,
$5(\mathbf{0}) - 3y = -15$	$5x - 3(\mathbf{0}) = -15$
$-3y = -15$	$5x = -15$
$\frac{-3y}{\mathbf{-3}} = \frac{-15}{\mathbf{-3}}$	$\frac{5x}{\mathbf{5}} = \frac{-15}{\mathbf{5}}$
$y = 5$	$x = -3$

The y-int is $(0,5)$, and the x-int is $(-3,0)$.
Select test point (0,0) and substitute into

$5x - 3y \ge -15$
$5(\mathbf{0}) - 3(\mathbf{0}) \overset{?}{\ge} -15$
$0 \ge -15$
True

The coordinates of every point on the same side of the line as (0,0) satisfy the inequality. To indicate this, we shade the half-plane that contains the test point (0,0).

$5x - 3y \ge -15$

b. $5x - 3y < -15$, Boundary line is dashed because of the $<$ symbol.
Graph as $5x - 3y = -15$.

If one would notice the inequality symbol is the opposite of the original, then logically one would shade the opposite side of the boundary line and make it dashed instead of solid.

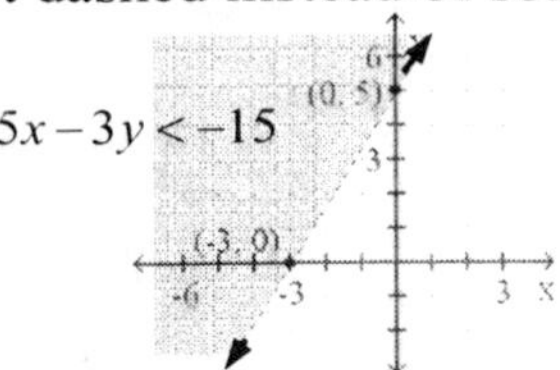

73. a. $y + 2x < 0$, Boundary line is dashed because of the $<$ symbol.
Graph as $y = -2x$.

y-intercept:	x-intercept:
If $x = 0$,	If $y = 0$,
$y = -2(\mathbf{0})$	$\mathbf{0} = -2x$
$y = 0$	$0 = x$

The y-int is (0,0), and the x-int is $(0,0)$.
The line passes through the origin so select test point (0,2) and substitute into

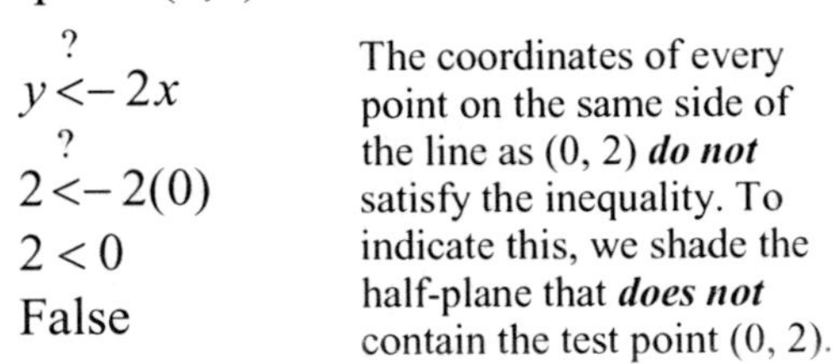

$y \overset{?}{<} -2x$
$2 \overset{?}{<} -2(0)$
$2 < 0$
False

The coordinates of every point on the same side of the line as (0, 2) ***do not*** satisfy the inequality. To indicate this, we shade the half-plane that ***does not*** contain the test point (0, 2).

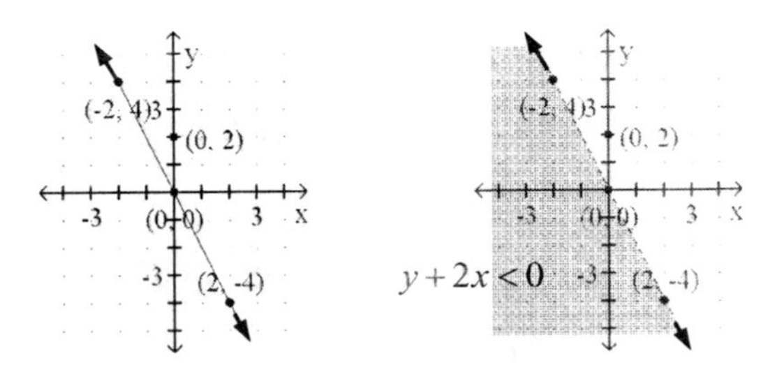

b. $y + 2x \ge 0$, Boundary line is solid because of the $\ge$ symbol.
Graph as $y = -2x$.

If one would notice the inequality symbol is the opposite of the original, then logically one would shade the opposite side of the boundary line and make it solid instead of dashed.

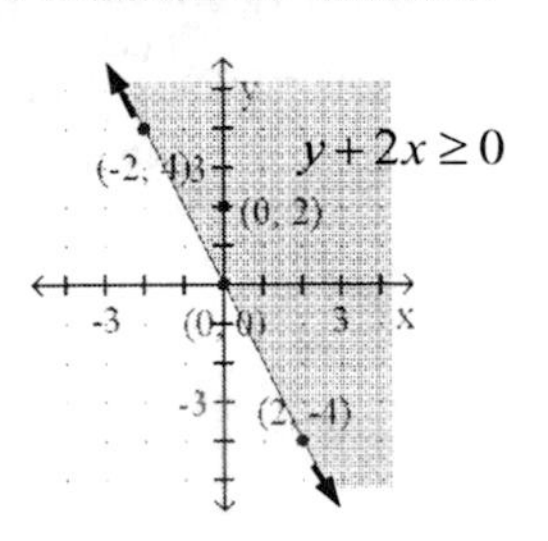

APPLICATIONS

75. DELIVERIES

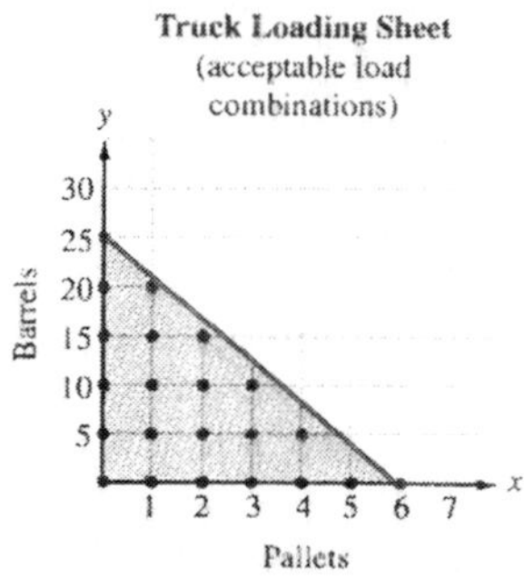

No, the truck can not deliver 4 pallets and 10 barrels in one trip.

FROM CAMPUS TO CAREERS

77. DENTAL ASSISTANT

Let x represents the number children and y represents the number of adults so let $c = x$ and $a = y$

$$\frac{3}{4}c + a \le 9$$

$$a \le -\frac{3}{4}c + 9$$

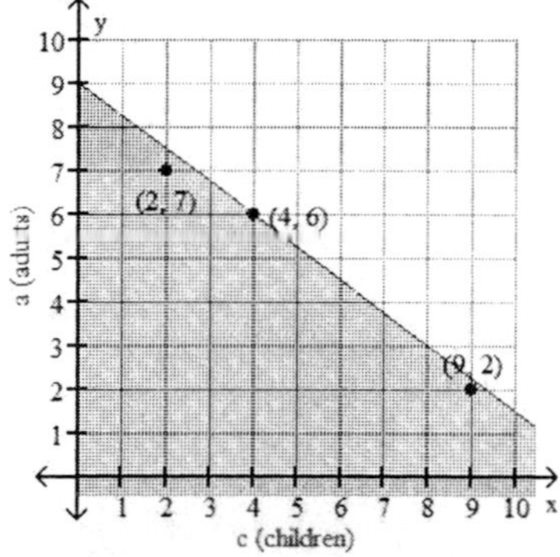

(2,7), (4,6), (9,2); answers may vary.

79. PRODUCTION PLANNING

$3x + 4y \le 120$, Graph as $3x + 4y = 120$.
Boundary line is solid.

y-intercept:	x-intercept:
If $x = 0$,	If $y = 0$,
$3(\mathbf{0}) + 4y = 120$	$3x + 4(\mathbf{0}) = 120$
$4y = 120$	$3x = 120$
$\frac{4y}{\mathbf{4}} = \frac{120}{\mathbf{4}}$	$\frac{3x}{\mathbf{3}} = \frac{120}{\mathbf{3}}$
$y = 30$	$x = 40$

The y-int is $(0,30)$, and the x-int is (40,0).
Select test point (0,0) and substitute into

$3(\mathbf{0}) + 4(\mathbf{0}) \le 120$
$0 \le 120$
True

The coordinates of every point on the same side of the line as (0,0) satisfy the inequality. To indicate this, we shade the half-plane that contains the test point (0,0).

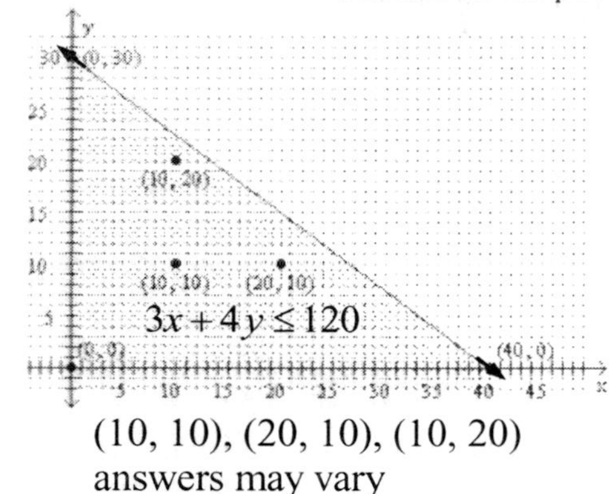

(10, 10), (20, 10), (10, 20)
answers may vary

81. INVENTORIES

$100x + 88y \ge 4,400$,

Graph as $100x + 88y = 4,400$.
Boundary line is solid.

y-intercept:	x-intercept:
If $x = 0$,	If $y = 0$,
$100(\mathbf{0}) + 88y = 4,400$	$100x + 88(\mathbf{0}) = 4,400$
$88y = 4,400$	$100x = 4,400$
$\frac{88y}{\mathbf{88}} = \frac{4,400}{\mathbf{88}}$	$\frac{100x}{\mathbf{100}} = \frac{4,400}{\mathbf{100}}$
$y = 50$	$x = 44$

The y-int is $(0,50)$, and the x-int is (44,0).
Select test point (0,0) and substitute into

$100(\mathbf{0}) + 88(\mathbf{0}) \ge 4,400$
$0 \ge 4,400$
False

The coordinates of every point on the same side of the line as (0,0) ***does not*** satisfy the inequality. To indicate this, we shade the half-plane that ***does not*** contain the test point (0,0).

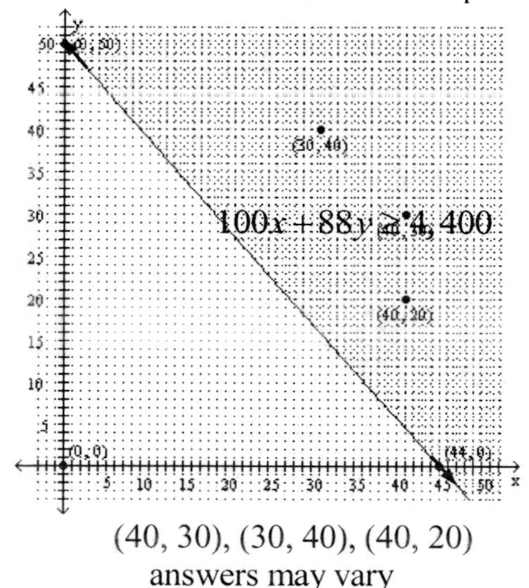

(40, 30), (30, 40), (40, 20)
answers may vary

WRITING

83-85. Answers will vary.

REVIEW

87. Solve $A = P + Prt$ for t.

$$A = P + Prt$$
$$A - \mathbf{P} = P + Prt - \mathbf{P}$$
$$A - P = Prt$$
$$\frac{A-P}{\mathbf{Pr}} = \frac{Prt}{\mathbf{Pr}}$$
$$\frac{A-P}{Pr} = t$$
$$t = \frac{A-P}{Pr}$$

89. Simplify:

$$40\left(\frac{3}{8}x - \frac{1}{4}\right) + 40\left(\frac{4}{5}\right) = \mathbf{40}\left(\frac{3}{8}x\right) - \mathbf{40}\left(\frac{1}{4}\right) + 40\left(\frac{4}{5}\right)$$
$$= 15x - 10 + 32$$
$$= 15x + 22$$

CHALLENGE PROBLEMS

91. Find a linear inequality that has the graph shown.

slope $= \frac{3}{2}$

y-intercept $= (0, -3)$

Line is solid indicating $\le$ or $\ge$.

Start with a line equation using the found information.

$$y = mx + b$$
$$y = \frac{3}{2}x - 3$$
$$\mathbf{2}y = \mathbf{2}\left(\frac{3}{2}x\right) - \mathbf{2}(3)$$
$$2y = 3x - 6$$
$$2y - \mathbf{3x} = 3x - 6 - \mathbf{3x}$$
$$-3x + 2y = -6$$

Now test the inequality: $-3x + 2y \le -6$

Substitute test point (0,0) in it.

$$-3x + 2y \le -6$$
$$-3(\mathbf{0}) + 2(\mathbf{0}) \overset{?}{\le} -6$$
$$0 \le -6$$

False

The above false statement indicates this is the inequality with the given shaded half:

$$-3x + 2y \le -6$$

or

$$3x - 2y \ge 6$$

SECTION 3.8 STUDY SET

VOCABULARY

Fill in the blanks.

1. A set of ordered pairs is called a **relation** .

3. The set of all input values for a function is called the **domain**, and the set of all output values is called the **range**.

5. If $f(2) = -3$, we call -3 a function **value**.

CONCEPTS

7. **N** FEDERAL MINIMUM HOURLY

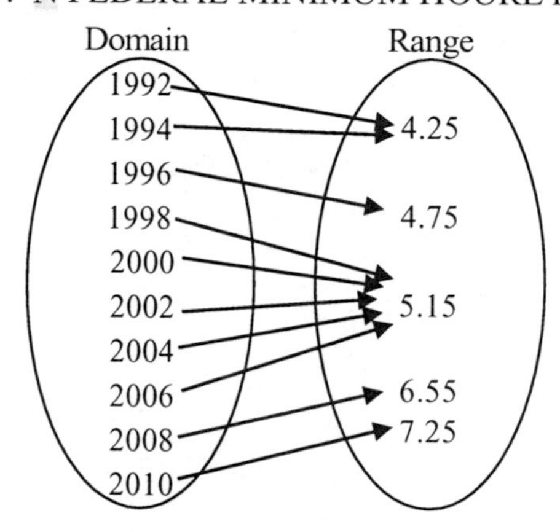

9. For the given input, what value will the function machine output?

$$f(x) = x^2 + 8, \ x = -5$$
$$f(-5) = (-5)^2 + 8$$
$$f(-5) = 25 + 8$$
$$f(-5) = \mathbf{33}$$

NOTATION

Fill in the blanks.

11. We read $f(x) = 5x - 6$ as "f **of** x is $5x$ minus 6."

13. The notation $f(4) = 5$ indicates that when the x-value **4** is input into a function rule, the output is **5**. This fact can be shown graphically by plotting the ordered pair (**4** , **5**).

GUIDED PRACTICE

Find the domain and range of each relation.

15. **Domain:{–6, –1, 6, 8}; Range:{–10, –5, –1, 2}**

17. **Domain:{–8, 0, 6}; Range: {9, 50}**

Determine whether each arrow diagram or table defines *y* as a function of *x*. If a function is defined, give its domain and range. If it does not define a function, find two ordered pairs that show a value of *x* that is assigned more than one value of *y*.

19. **Yes; domain: {10,20,30}; range: {20,40,60}**

21. **No; (4, 2), (4, 4), (4, 6)** (Answers may vary)

23. **Yes; domain: {1, 2, 3, 4, 5}; range: {7, 8, 15, 16, 23}**

25. **No; (–1, 0), (–1, 2)**

27. **No; (3, 4), (3, –4) or (4, 3), (4, –3)**

29. **Yes; domain: {–3, 1, 5, 6}; range: {–8, 0, 4, 9}**

31. **No; (3, 4), (3, –4) or (4, 3), (4, –3)**

33. **Yes; domain: {–2, –1, 0, 1}; range: {7, 10, 13, 16}**

Find each function value.

35. $f(x) = 4x - 1$

a. $f(x) = 4x - 1$

$$f(\mathbf{1}) = 4(\mathbf{1}) - 1$$
$$= 4 - 1$$
$$= 3$$

Thus, $f(1) = 3$

b. $f(x) = 4x - 1$

$$f(\mathbf{-2}) = 4(\mathbf{-2}) - 1$$
$$= -8 - 1$$
$$= -9$$

Thus, $f(-2) = -9$

c. $f(x) = 4x - 1$

$$f\left(\mathbf{\frac{1}{4}}\right) = 4\left(\mathbf{\frac{1}{4}}\right) - 1$$
$$= 1 - 1$$
$$= 0$$

Thus, $f\left(\frac{1}{4}\right) = 0$

d. $f(x) = 4x - 1$

$$f(\mathbf{50}) = 4(\mathbf{50}) - 1$$
$$= 200 - 1$$
$$= 199$$

Thus, $f(50) = 199$

37. $f(x) = 2x^2$

a. $f(x) = 2x^2$

$$f(\mathbf{0.4}) = 2(\mathbf{0.4})^2$$
$$= 2(0.16)$$
$$= 0.32$$

Thus, $f(0.4) = 0.32$

b. $f(x)=2x^2$

$f(-\mathbf{3})=2(-\mathbf{3})^2$
$=2(9)$
$=18$

Thus, $f(-3)=18$

c. $f(x)=2x^2$

$f(\mathbf{1,000})=2(\mathbf{1,000})^2$
$=2(1,000,000)$
$=2,000,000$

Thus, $f(1,000)=2,000,000$

d. $f(x)=2x^2$

$f\left(\frac{\mathbf{1}}{\mathbf{8}}\right)=2\left(\frac{\mathbf{1}}{\mathbf{8}}\right)^2$
$=2\left(\frac{1}{64}\right)$
$=\frac{1}{32}$

Thus, $f\left(\frac{1}{8}\right)=\left(\frac{1}{32}\right)$

39. $h(x)=|x-7|$

a. $h(x)=|x-7|$

$h(\mathbf{0})=|\mathbf{0}-7|$
$=|-7|$
$=7$

Thus, $h(0)=7$

b. $h(x)=|x-7|$

$h(-\mathbf{7})=|-\mathbf{7}-7|$
$=|-14|$
$=14$

Thus, $h(-7)=14$

c. $h(x)=|x-7|$

$h(\mathbf{7})=|\mathbf{7}-7|$
$=|0|$
$=0$

Thus, $h(7)=0$

d. $h(x)=|x-7|$

$h(\mathbf{8})=|\mathbf{8}-7|$
$=|1|$
$=1$

Thus, $h(8)=1$

41. $g(x)=x^3-x$

a. $g(x)=x^3-x$

$g(\mathbf{1})=\mathbf{1}^3-\mathbf{1}$
$=1-1$
$=0$

Thus, $g(1)=0$

b. $g(x)=x^3-x$

$g(\mathbf{10})=\mathbf{10}^3-\mathbf{10}$
$=1,000-10$
$=990$

Thus, $g(10)=990$

c. $g(x)=x^3-x$

$g(-\mathbf{3})=(-\mathbf{3})^3-(-\mathbf{3})$
$=-27+3$
$=-24$

Thus, $g(-3)=-24$

d. $g(x)=x^3-x$

$g(\mathbf{6})=(\mathbf{6})^3-\mathbf{6}$
$=216-6$
$=210$

Thus, $g(6)=210$

43. $s(x)=(x+3)^2$

a. $s(x)=(x+3)^2$

$s(\mathbf{3})=(\mathbf{3}+3)^2$
$=(6)^2$
$=36$

Thus, $s(3)=36$

b. $s(x)=(x+3)^2$

$s(-\mathbf{3})=(-\mathbf{3}+3)^2$
$=(0)^2$
$=0$

Thus, $s(-3)=0$

c. $s(x)=(x+3)^2$

$s(\mathbf{0})=(\mathbf{0}+3)^2$
$=(3)^2$
$=9$

Thus, $s(0)=9$

d. $s(x)=(x+3)^2$

$s(-\mathbf{5})=(-\mathbf{5}+3)^2$
$=(-2)^2$
$=4$

Thus, $s(-5)=4$

45. $f(x)=3.4x^2-1.2x+0.5$, find $f(-0.3)$

$f(-\mathbf{0.3})=3.4(-\mathbf{0.3})^2-1.2(-\mathbf{0.3})+0.5$
$=3.4(0.09)-1.2(-0.3)+0.5$
$=0.306+0.36+0.5$
$=0.666+0.5$
$=1.166$

Thus, $f(-0.3)=1.166$

Complete each table of function values and then graph each function.

47. $f(x) = -3x - 2$

x	$f(x)$
-2	$f(-2) = -3(-2) - 2$ $= 6 - 2$ $= 4$
-1	$f(-1) = -3(-1) - 2$ $= 3 - 2$ $= 1$
0	$f(0) = -3(0) - 2$ $= 0 - 2$ $= -2$
1	$f(1) = -3(1) - 2$ $= -3 - 2$ $= -5$

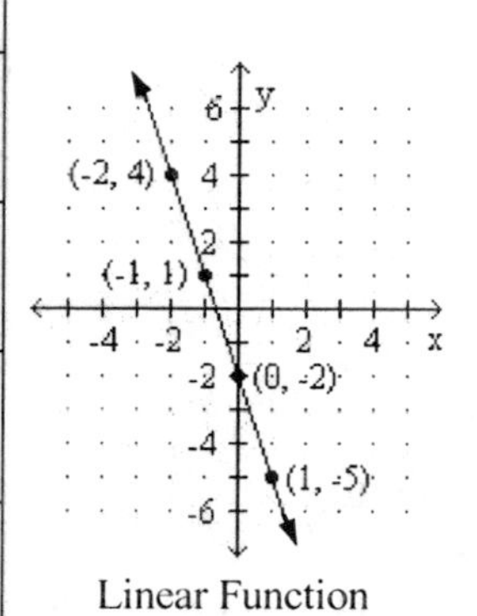

Linear Function

49. $h(x) = |1 - x|$

x	$h(x)$
-2	$h(-2) = \|1-(-2)\|$ $= \|1+2\|$ $= \|3\|$ $= 3$
-1	$h(-1) = \|1-(-1)\|$ $= \|1+1\|$ $= \|2\|$ $= 2$
0	$h(0) = \|1-0\|$ $= \|1\|$ $= 1$
1	$h(1) = \|1-1\|$ $= \|0\|$ $= 0$
2	$h(2) = \|1-2\|$ $= \|-1\|$ $= 1$
3	$h(3) = \|1-3\|$ $= \|-2\|$ $= 2$
4	$h(4) = \|1-4\|$ $= \|-3\|$ $= 3$

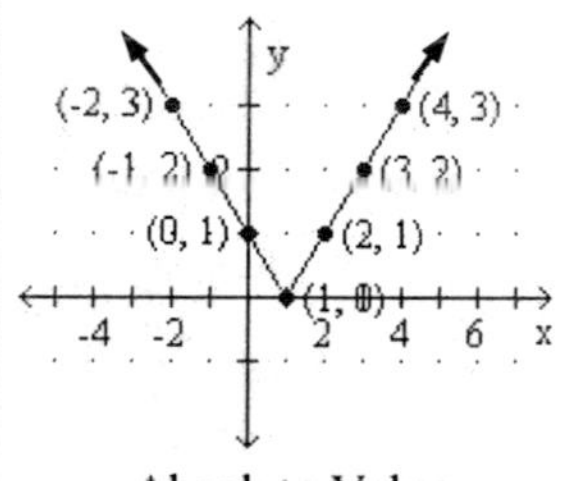

Absolute Value Function

Graph each function.

51. $f(x) = \frac{1}{2}x - 2$

x	$f(x)$
-2	$f(-2) = \frac{1}{2}(-2) - 2$ $= -1 - 2$ $= -3$
0	$f(0) = \frac{1}{2}(0) - 2$ $= 0 - 2$ $= -2$
2	$f(2) = \frac{1}{2}(2) - 2$ $= 1 - 2$ $= -1$

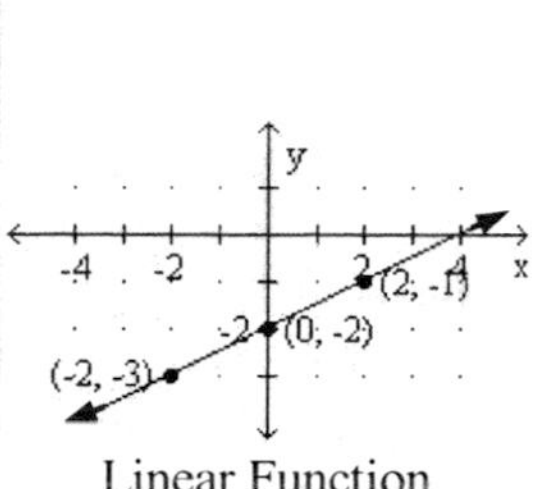

Linear Function

53. $h(x) = -|x|$

x	$h(x)$
-3	$h(-3) = -\|-3\|$ $= -(3)$ $= -3$
-2	$h(-2) = -\|-2\|$ $= -(2)$ $= -2$
-1	$h(-1) = -\|-1\|$ $= -(1)$ $= -1$
0	$h(0) = -\|0\|$ $= 0$
1	$h(1) = -\|1\|$ $= -(1)$ $= -1$
2	$h(2) = -\|2\|$ $= -(2)$ $= -2$
3	$h(3) = -\|3\|$ $= -(3)$ $= -3$

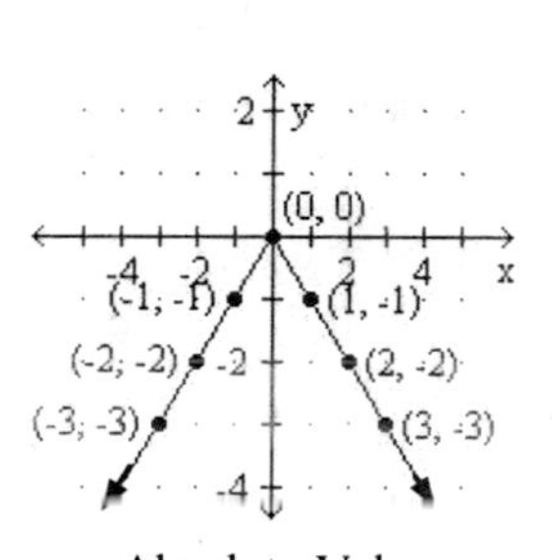

Absolute Value Function

Determine whether each graph is the graph of a function. If it is not, find ordered pairs that show a value of x that is assigned more than one value of y.

55. **Yes**

57. **No**; (3, 4), (3, −1); Answers will vary.

59. **No**; (0, 2), (0, −4); Answers will vary.

61. **No**; (3, 0), (3, 1); Answers will vary.

APPLICATIONS

63. REFLECTIONS

$f(x) = |x|$

65. VACATIONING

$$C(d) = 500 + 100(d - 3)$$
$$C(7) = 500 + 100(\mathbf{7} - 3)$$
$$= 500 + 100(4)$$
$$= 500 + 400$$
$$= 900$$

The cost will be \$900.

67. LAWN SPRINKLERS

$$A(r) = \pi r^2, \quad \pi = 3.141592654$$
$$A(\mathbf{5}) \approx (3.141592654)(\mathbf{5})^2$$
$$\approx (3.141592654)(25)$$
$$\approx 78.53$$
$$\approx 78.5$$

The area covered is about 78.5 ft^2.

$$A(r) = \pi r^2, \quad \pi = 3.141592654$$
$$A(\mathbf{20}) \approx (3.141592654)(\mathbf{20})^2$$
$$\approx (3.141592654)(400)$$
$$\approx 1,256.63$$
$$\approx 1,256.6$$

The area covered is about $1{,}256.6 \text{ ft}^2$.

69. POSTAGE

Yes

WRITING

71-75. Answers will vary.

REVIEW PROBLEMS

77. COFFEE BLENDS

Analyze

- Regular coffee sells for \$4/lb.
- Gourmet coffee sells for \$7/lb.
- 40 lbs of gourmet coffee on hand.
- Blend to sell for \$5/lb.
- How much regular coffee is needed?

Assign

Let x = the amount of regular coffee in lb.

Form

	Amount •	Value =	Total value
Regular	$\mathbf{x}$	4	$\mathbf{4x}$
Gourmet	40	7	**7(40)**
Blend	$x + 40$	5	$\mathbf{5(x + 40)}$

The value of the regular coffee	plus	the value of the gourmet coffee	equals	the value of the blend.
$4x$	$+$	280	$=$	$5(x+40)$

Solve

$$4x + 280 = 5(x + 40)$$
$$4x + 280 = 5x + 200$$
$$4x + 280 - \mathbf{4x} = 5x + 200 - \mathbf{4x}$$
$$280 = x + 200$$
$$280 - \mathbf{200} = x + 200 - \mathbf{200}$$
$$80 = x$$

State

80 pounds of regular coffee will be needed.

Check

The value of regular is 80(\$4), or \$320.
The value of gourmet is 40(\$7), or \$280.
The value of blend is 120(\$5), or \$600.
Since the total was \$320 + \$280 = \$600, the result checks.

CHALLENGE PROBLEMS

79. No. The graph contains many values of x that are assigned more than one value of y, such as (2, 1) and (2, 2).

81. Let $f(x) = -2x + 5$. For what value of x is $f(x) = -7$?

If $f(x) = -2x + 5$ and $f(x) = -7$, then $-2x + 5 = -7$.

$$-2x + 5 = -7$$
$$-2x + 5 - \mathbf{5} = -7 - \mathbf{5}$$
$$-2x = -12$$
$$\frac{-2x}{\mathbf{-2}} = \frac{-12}{\mathbf{-2}}$$
$$x = 6$$

The value of x is 6.

CHAPTER 3 REVIEW

SECTION 3.1

Graphing Using the Rectangular Coordinate System

1. Graph the points with coordinates (–1, 3), (0, 1.5), (–4, –4), $\left(2, \frac{7}{2}\right)$, and (4, 0).

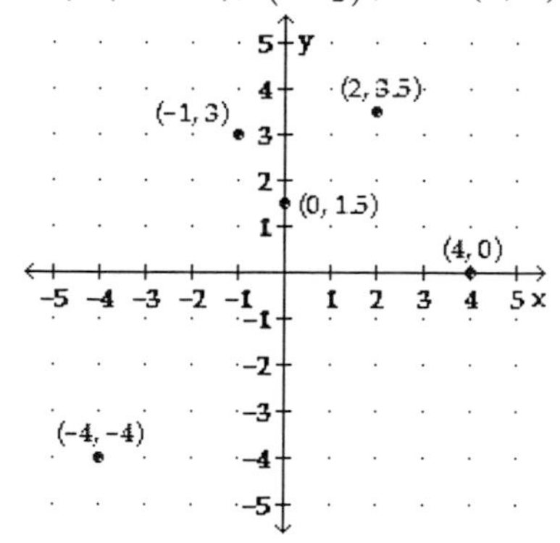

2. HAWAII
   Estimate the coordinates of Oahu using an ordered pair of the form (longitude, latitude).
   (158, 21.5)

3. In what quadrant does the point (–3, –4) lie?
   Quadrant III

4. What are the coordinates of the origin? **(0, 0)**

5. GEOMETRY
   Three vertices (corners) of a square are (–5, 4), (–5, –2), and (1, –2). Find the coordinates of the fourth vertex and then find the area of the square.

   The coordinates of the fourth vertex would be **(1, 4)**. Each side of the square is 6 units long.

$$A = lw$$
$$= (6)(6)$$
$$= 36$$

   The area of the square is 36 unit^2.

6. COLLEGE ENROLLMENTS
   The graph gives the number of students enrolled at a college for the period from 4 weeks before to 5 weeks after the semester began.
   a. What was the maximum enrollment and when did it occur? **2,500; week 2**
   b. How many students had enrolled 2 weeks before the semester began? **1,000**
   c. When was the enrollment 2,250?
      1st week and 5th week

SECTION 3.2

Graphing Linear Equations

7. Is (–3, –2) a solution of y = 2x + 4?

$$y = 2x + 4$$
$$(\mathbf{-2}) \stackrel{?}{=} 2(\mathbf{-3}) + 4$$
$$-2 \stackrel{?}{=} -6 + 4$$
$$-2 = -2$$
$$\text{True}$$

$(-3, -2)$ is a solution of $y = 2x + 4$.

8. Complete the table of solutions.
   $3x + 2y = -18$

x	y	(x, y)
–2	**–6**	**(–2, –6)**
–8	3	**(–8, 3)**

$$3x + 2y = -18$$
$$3(\mathbf{-2}) + 2y = -18$$
$$-6 + 2y = -18$$
$$-6 + 2y + \mathbf{6} = -18 + \mathbf{6}$$
$$2y = -12$$
$$\frac{2y}{\mathbf{2}} = \frac{-12}{\mathbf{2}}$$
$$y = -6$$

$$3x + 2y = -18$$
$$3x + 2(\mathbf{3}) = -18$$
$$3x + 6 = -18$$
$$3x + 6 - \mathbf{6} = -18 - \mathbf{6}$$
$$3x = -24$$
$$\frac{3x}{\mathbf{3}} = \frac{-24}{\mathbf{3}}$$
$$x = -8$$

9. Which of the following equations are not linear equations?

   $8x - 2y = 6$ $\quad$ $y = x^2 + 1$ $\quad$ $y = x$

   $3y = -x + 4$ $\quad$ $y - x^3 = 0$

   $\boldsymbol{y = x^2 + 1}$ **and** $\boldsymbol{y - x^3 = 0}$

10. The graph of a linear equation is shown.
    a. When the coordinates of point *A* are substituted into the equation, will a true or false statement result?
       True, because point *A* lies on the line and therefore is a solution of the equation.
    b. When the coordinates of point *B* are substituted into the equation, will a true or false statement result?
       False, because point *B* does not lie on the line and therefore is not a solution of the equation.

Graph each equation by constructing a table of solutions.

11. $y = 4x - 2$

x	y	(x, y)
-1	$y = 4x - 2$ $y = 4(\mathbf{-1}) - 2$ $y = -4 - 2$ $y = -6$	$(-1, -6)$
0	$y = 4x - 2$ $y = 4(\mathbf{0}) - 2$ $y = 0 - 2$ $y = -2$	$(0, -2)$
1	$y = 4x - 2$ $y = 4(\mathbf{1}) - 2$ $y = 4 - 2$ $y = 2$	$(1, 2)$

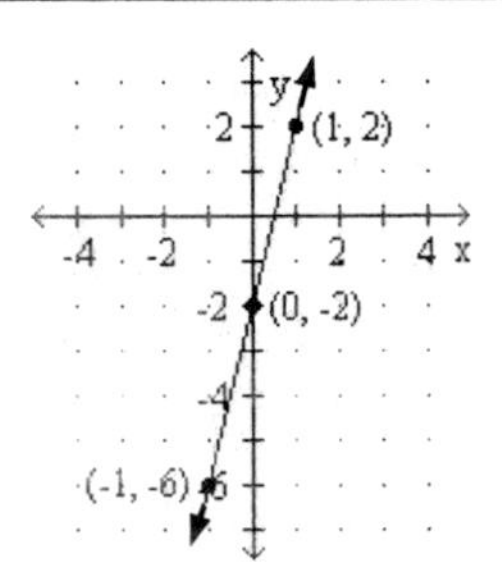

12. $y = \frac{3}{4}x$

x	y	(x, y)
-4	$y = \frac{3}{4}x$ $y = \frac{3}{4}(\mathbf{-4})$ $y = -3$	$(-4, -3)$
0	$y = \frac{3}{4}x$ $y = \frac{3}{4}(\mathbf{0})$ $y = 0$	$(0, 0)$
4	$y = \frac{3}{4}x$ $y = \frac{3}{4}(\mathbf{4})$ $y = 3$	$(4, 3)$

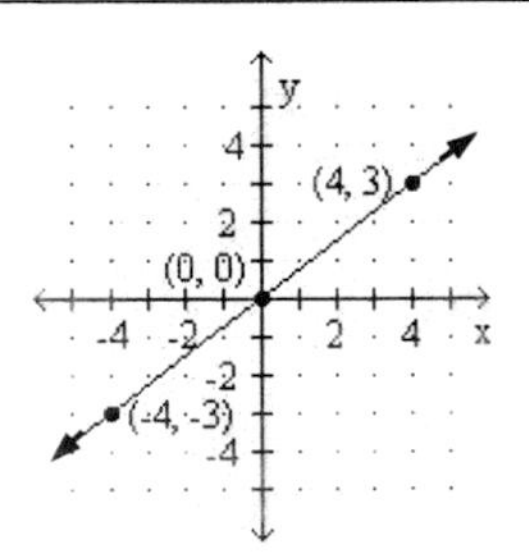

13. $5y = -5x + 15$

$$\frac{5y}{\mathbf{5}} = \frac{-5x}{\mathbf{5}} + \frac{15}{\mathbf{5}}$$

$$y = -x + 3$$

x	y	(x, y)
-1	$y = -x + 3$ $y = -(\mathbf{-1}) + 3$ $y = 1 + 3$ $y = 4$	$(-1, 4)$
0	$y = -x + 3$ $y = -(\mathbf{0}) + 3$ $y = 0 + 3$ $y = 3$	$(0, 3)$
1	$y = -x + 3$ $y = -(\mathbf{1}) + 3$ $y = -1 + 3$ $y = 2$	$(1, 2)$

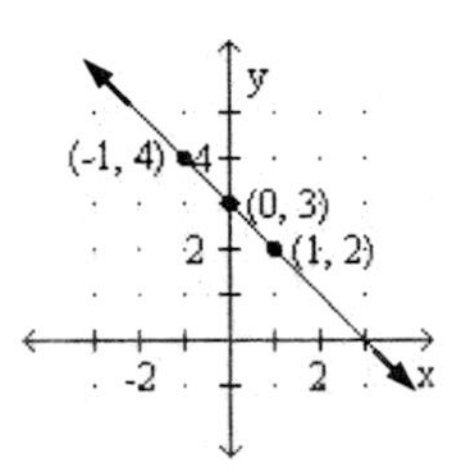

14. $6y = -4x$

$$\frac{6y}{\mathbf{6}} = \frac{-4x}{\mathbf{6}}$$

$$y = -\frac{2}{3}x$$

x	y	(x, y)
-3	$y = -\frac{2}{3}x$ $y = -\frac{2}{3}(\mathbf{-3})$ $y = 2$	$(-3, 2)$
0	$y = -\frac{2}{3}x$ $y = -\frac{2}{3}(\mathbf{0})$ $y = 0$	$(0, 0)$
3	$y = -\frac{2}{3}x$ $y = -\frac{2}{3}(\mathbf{3})$ $y = -2$	$(3, -2)$

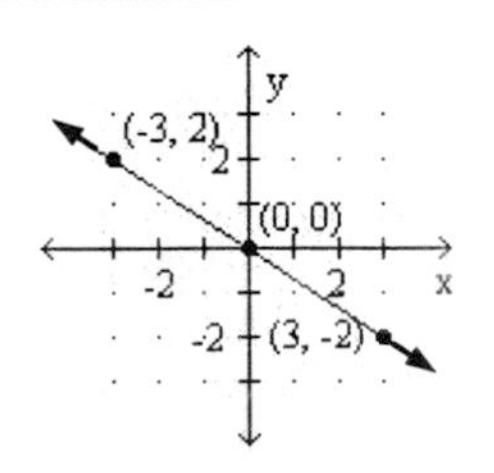

15. BIRTHDAY PARTIES

$c = 8n + 50$

Do not calculate the answer.

n	c	(n, c)
1	$c = 8n + 50$ $c = 8(\mathbf{1}) + 50$ $c = 8 + 50$ $c = 58$	(1, 58)
5	$c = 8n + 50$ $c = 8(\mathbf{5}) + 50$ $c = 40 + 50$ $c = 90$	(5, 90)
10	$c = 8n + 50$ $c = 8(\mathbf{10}) + 50$ $c = 80 + 50$ $c = 130$	(10, 130)

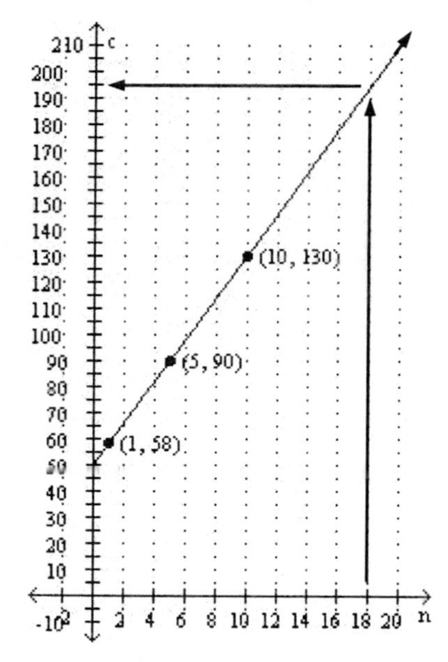

According to the graph, the cost of the party for 18 children is about $195.

16. Determine whether each statement is true or false.

a. It takes three or more points to determine a line. **False**

b. A linear equation in two variables has infinitely many solutions. **True**

SECTION 3.3 REVIEW

Intercepts

17. Identify the x- and y-intercepts of the graph.

The x-intercept is $(-3, 0)$.

The y-intercept is $(0, 2.5)$.

18. DEPRECIATION

The y-intercept (0, 25,000) indicates the equipment was originally valued at $25,000.

The x-intercept (10, 0) indicates in 10 years the sound equipment has no value.

Use the intercept method to graph each equation.

19. $-4x + 2y = 8$

y-intercept:	x-intercept:
If $x = 0$,	If $y = 0$
$-4(\mathbf{0}) + 2y = 8$	$-4x + 2(\mathbf{0}) = 8$
$2y = 8$	$-4x = 8$
$y = 4$	$x = -2$

The y-intercept is $(0, 4)$, and the x-intercept is $(-2, 0)$.

Check Point

$-4x + 2y = 8$

$-4(\mathbf{-1}) + 2y = 8$

$4 + 2y - \mathbf{4} = 8 - \mathbf{4}$

$2y = 4$

$y = 2$

$(-1, 2)$

$-4x + 2y = 8$

(0, 4)

(-1, 2)

(-2, 0)

20. $5x - 4y = 13$

y-intercept:	x-intercept:
If $x = 0$,	If $y = 0$
$5(\mathbf{0}) - 4y = 13$	$5x - 4(\mathbf{0}) = 13$
$-4x = 13$	$5x = 13$
$y = -13/4$	$x = 13/5$

The y-intercept is $\left(0, -\frac{13}{4}\right)$, and the x-intercept is $\left(\frac{13}{5}, 0\right)$.

Check Point

$5x - 4y = 13$

$5(\mathbf{1}) - 4y = 13$

$5 - 4y - \mathbf{5} = 13 - \mathbf{5}$

$-4y = 8$

$y = -2$

$(1, -2)$

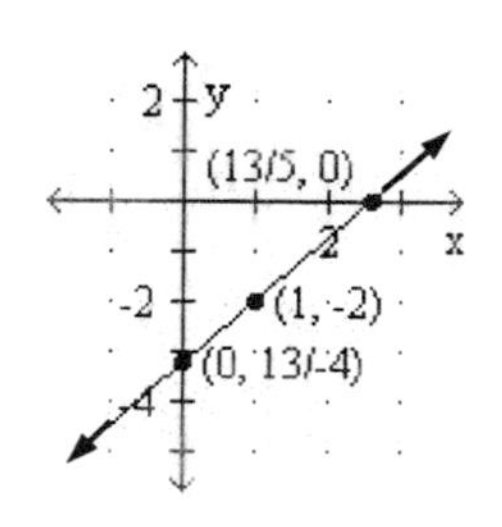

21. $y = 4$

Let $x = -2$	Let $x = 0$	Let $x = 3$
$0x + y = 4$	$0x + y = 4$	$0x + y = 4$
$0(\mathbf{-2}) + y = 4$	$0(\mathbf{0}) + y = 4$	$0(\mathbf{3}) + y = 4$
$y = 4$	$y = 4$	$y = 4$
$(-2, 4)$	$(0, 4)$	$(3, 4)$

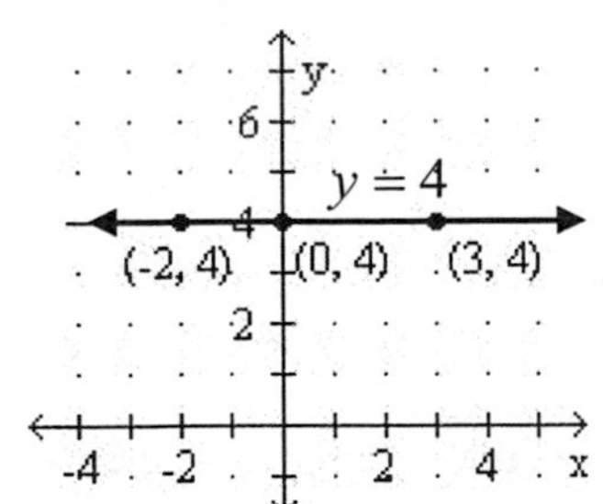

22. $x = -1$

Let $y = -3$	Let $y = 0$	Let $y = 2$
$x + 0y = -1$	$x + 0y = -1$	$x + 0y = -1$
$x + 0(\mathbf{-3}) = -1$	$x + 0(\mathbf{0}) = -1$	$x + 0(\mathbf{2}) = -1$
$x = -1$	$x = -1$	$x = -1$
$(-1, -3)$	$(-1, 0)$	$(-1, 2)$

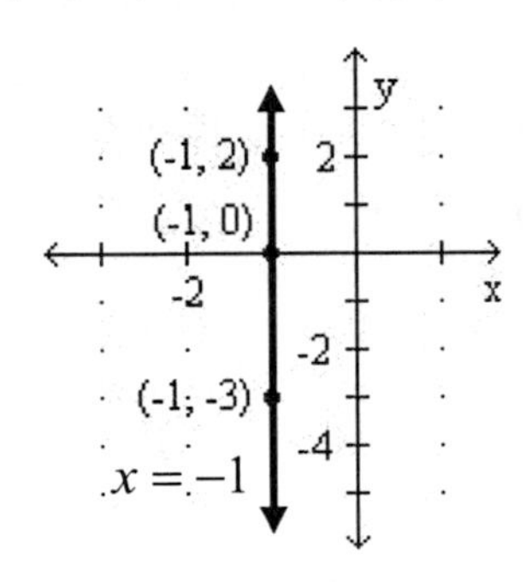

SECTION 3.4 REVIEW
Slope and Rate of Change

In each case, find the slope of the line.

23. $m = \dfrac{1}{4}$

24. $m = -\dfrac{7}{8}$

25. The line with this table of solutions

x	*y*	*(x, y)*
2	–3	(2, –3)
4	–17	(4, –17)

Let $(x_1, y_1) = (2, -3)$

Let $(x_2, y_2) = (4, -17)$

$$m = \frac{y_2 - y_1}{x_2 - x_1} = \frac{-17 - (-3)}{4 - 2} = \frac{-17 + 3}{4 - 2} = \frac{-14}{2} = -7$$

26. The line passing through the points (1, –4) and (3, –7)

Let $(x_1, y_1) = (1, -4)$

Let $(x_2, y_2) = (3, -7)$

$$m = \frac{y_2 - y_1}{x_2 - x_1} = \frac{-7 - (-4)}{3 - 1} = \frac{-7 + 4}{3 - 1} = -\frac{3}{2}$$

27. Draw a line having a slope that is

a. Positive

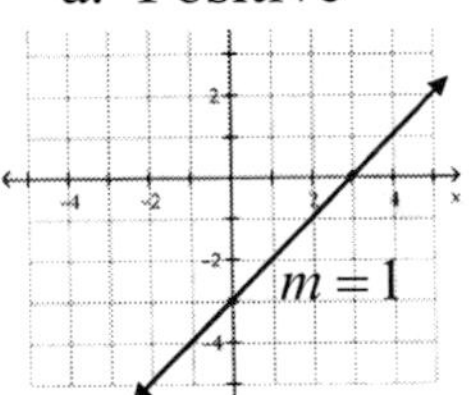

b. Negative

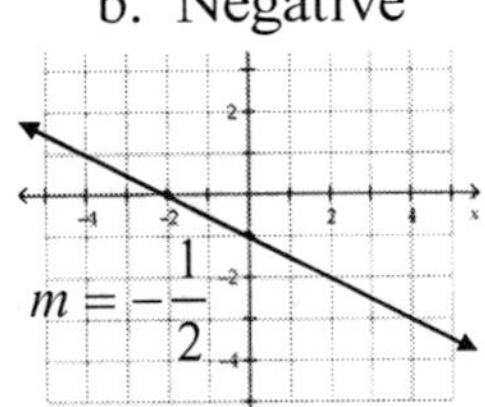

c. 0

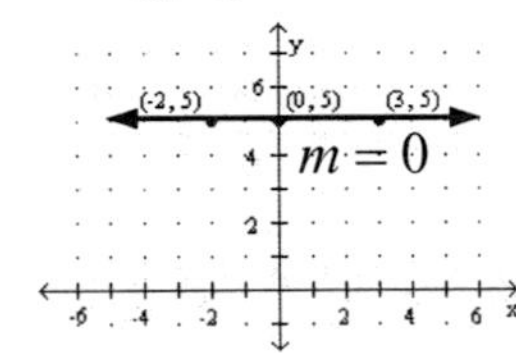

d. Undefined

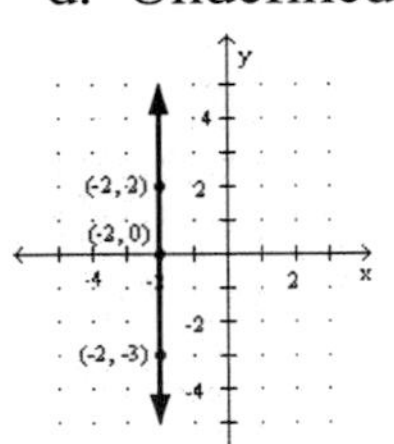

28. CARPENTRY

$$m = \frac{\text{rise}}{\text{run}} = \frac{6}{8} = \frac{\overset{1}{\cancel{2}} \cdot 3}{\underset{1}{\cancel{2}} \cdot 4} = \frac{3}{4}$$

29. RAMPS

$$m = \frac{\text{rise}}{\text{run}} = \frac{2}{24} \approx 0.083 \approx 8.3\%$$

The grade is about 8.3%.

30. BOTTLED WATER

a. The y values would be the number of gallons and the x values would be the years.

$$m = \frac{\text{change in } y}{\text{change in } x} = \frac{29-17}{2008-2000} = \frac{12}{8} = 1.5$$

The rate of change was 1.5 gallons per year.

31. Step 1: Find each slope.

Let $(x_1, y_1) = (6,6)$	Let $(x_1, y_1) = (2,-10)$
Let $(x_2, y_2) = (4,2)$	Let $(x_2, y_2) = (-2,-2)$
$m_1 = \frac{y_2 - y_1}{x_2 - x_1} = \frac{2-6}{4-6} = \frac{-4}{-2} = 2$	$m_2 = \frac{y_2 - y_1}{x_2 - x_1} = \frac{-2-(-10)}{-2-2} = \frac{-2+10}{-2-2} = \frac{8}{-4} = -2$

Step 2: Compare the 2 slopes.

The slopes are not equal.
The slopes are not negative reciprocals.
The lines are neither parallel nor perpendicular.

32. Step 1: Determine the slope.

Let $(x_1, y_1) = (-1,9)$

Let $(x_2, y_2) = (-8,4)$

$$m = \frac{y_2 - y_1}{x_2 - x_1} = \frac{4-9}{-8-(-1)} = \frac{4-9}{-8+1} = \frac{-5}{-7} = \frac{5}{7}$$

Step 2: Determine the negative reciprocal.

The negative reciprocal is $-\frac{7}{5}$.

This is the slope of the line perpendicular and passing through the given points.

SECTION 3.5 REVIEW

Slope - Intercept Form

Find the slope and the y - intercept of each line.

33. $y = \frac{3}{4}x - 2$

Since $m = \frac{3}{4}$ and $b = -2$, the slope is $\frac{3}{4}$ and the y-intercept is $(0,-2)$.

34. $y = -4x$

$y = -4x + 0$

Since $m = -4$ and $b = 0$, the slope is -4 and the y-intercept is $(0,0)$.

35. $y = \frac{x}{8} + 10$

$y = \frac{1}{8}x + 10$

Since $m = \frac{1}{8}$ and $b = 10$, the slope is $\frac{1}{8}$ and the y-intercept is $(0,10)$.

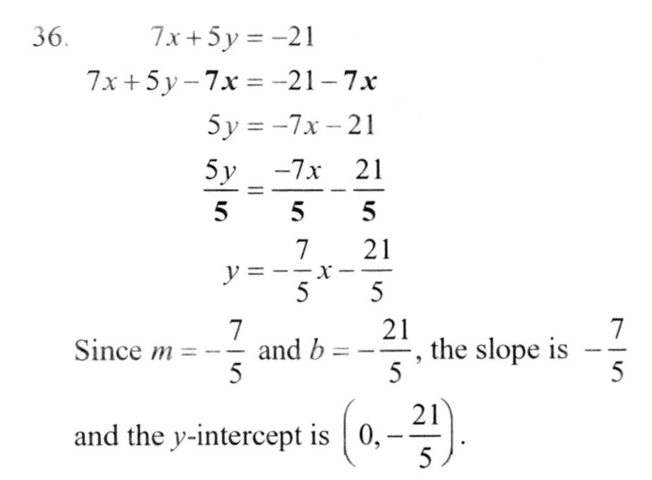

36. $7x + 5y = -21$

$7x + 5y - \mathbf{7x} = -21 - \mathbf{7x}$

$5y = -7x - 21$

$\frac{5y}{\mathbf{5}} = \frac{-7x}{\mathbf{5}} - \frac{21}{\mathbf{5}}$

$y = -\frac{7}{5}x - \frac{21}{5}$

Since $m = -\frac{7}{5}$ and $b = -\frac{21}{5}$, the slope is $-\frac{7}{5}$ and the y-intercept is $\left(0, -\frac{21}{5}\right)$.

37. Slope 4, y-intercept $(0, -1)$

The equation of the line is $y = 4x - 1$.

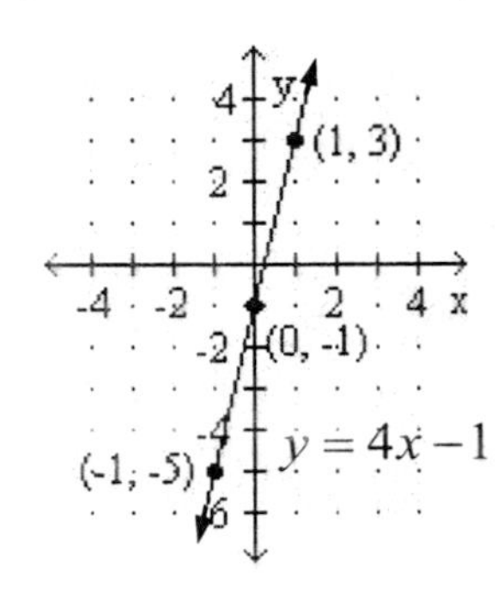

38. $m = \frac{3}{2}$, y-intercept is $(0, -3)$

Equation of the line is $y = \frac{3}{2}x - 3$.

39. $9x - 3y = 15$

$9x - 3y - \mathbf{9x} = 15 - \mathbf{9x}$

$-3y = -9x + 15$

$\frac{-3y}{\mathbf{-3}} = \frac{-9x}{\mathbf{-3}} + \frac{15}{\mathbf{-3}}$

$y = 3x - 5$

$m = 3$; y-intercept is $(0, -5)$.

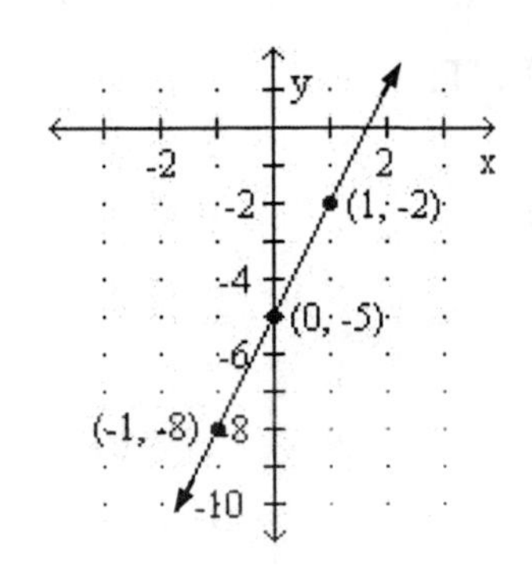

40. COPIERS

a. • Copies already made, 75,000
• New copies, 300 per wk
• Total copies is not known (c)

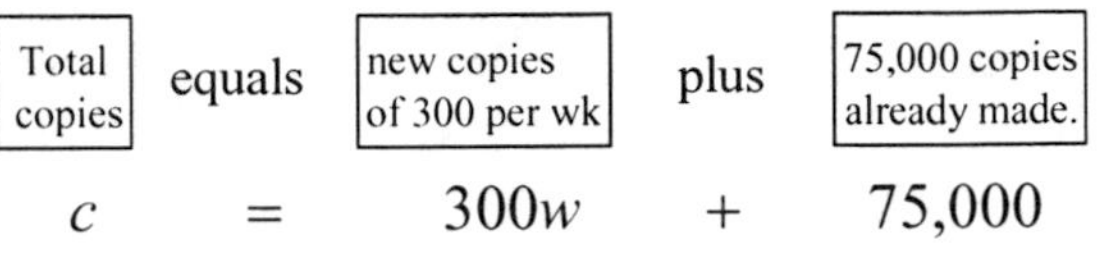

Total copies	equals	new copies of 300 per wk	plus	75,000 copies already made.
c	$=$	$300w$	$+$	$75,000$

b. Find the predicted copies after 52 weeks, replace w with 52 and then calculate the result.

$c = 300w + 75,000$

$c = 300(\mathbf{52}) + 75,000$

$c = 15,600 + 75,000$

$c = 90,600$

The total copies for 52 weeks is 90,600.

Without graphing, determine whether graphs of the given pairs of lines are be parallel, perpendicular, or neither.

41. $y = -\frac{2}{3}x + 6$ | $y = -\frac{2}{3}x - 6$

$m = -\frac{2}{3}$ | $m = -\frac{2}{3}$

The slopes are equal.
The lines are parallel.

42. $x + 5y = -10$

$x + 5y - \mathbf{x} = -10 - \mathbf{x}$

$5y = -x - 10$

$\frac{5y}{\mathbf{5}} = \frac{-x}{\mathbf{5}} - \frac{10}{\mathbf{5}}$

$y = -\frac{1}{5}x - 2$

$m = -\frac{1}{5}$

$y - 5x = 0$

$y - 5x + \mathbf{5x} = 0 + \mathbf{5x}$

$y = 5x$

$m = 5$

The slopes are negative reciprocals.
The lines are perpendicular.

SECTION 3.6 REVIEW
Point - Slope Form

Find an equation of a line with the given slope that passes through the given point. Write the equation in slope - intercept form and graph it.

43. Passes through (1, 5), $m = 3$

$x_1 = 1$ and $y_1 = 5$

$$y - y_1 = m(x - x_1)$$
$$y - 5 = 3(x - 1) \text{ point-slope form}$$

$$y - 5 = 3(x - 1)$$
$$y - 5 = 3x - 3$$
$$y - 5 + \mathbf{5} = 3x - 3 + \mathbf{5}$$
$$y = 3x + 2 \text{ slope-intercept form}$$

y-int (0, 2), $m = 3 = \frac{3}{1}$

Start at (0, 2), go up 3, right 1

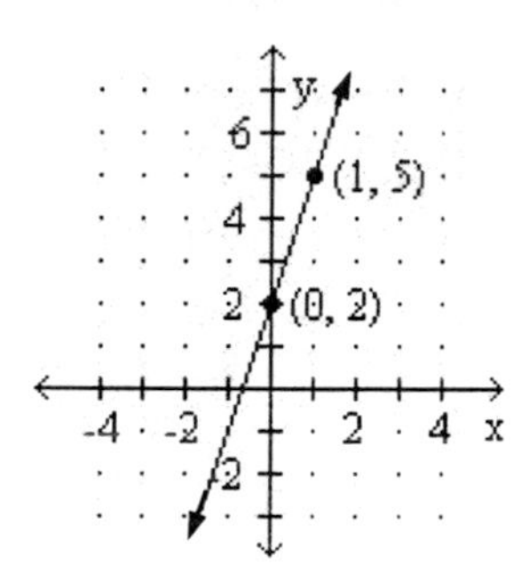

44. Passes through $(-4, -1)$, $m = -\frac{1}{2}$

$x_1 = -4$ and $y_1 = -1$

$$y - y_1 = m(x - x_1)$$
$$y - (-1) = -\frac{1}{2}(x - (-4))$$
$$y + 1 = -\frac{1}{2}(x + 4) \text{ point-slope form}$$

$$y + 1 = -\frac{1}{2}(x + 4)$$
$$y + 1 = -\frac{\mathbf{1}}{\mathbf{2}}(x) - \frac{\mathbf{1}}{\mathbf{2}}(4)$$
$$y + 1 = -\frac{1}{2}x - 2$$
$$y + 1 - \mathbf{1} = -\frac{1}{2}x - 2 - \mathbf{1}$$
$$y = -\frac{1}{2}x - 3 \text{ slope-intercept form}$$

y-int $(0, -3)$, $m = -\frac{1}{2} = \frac{-1}{2}$

Start at $(0, -3)$, go down 1, right 2

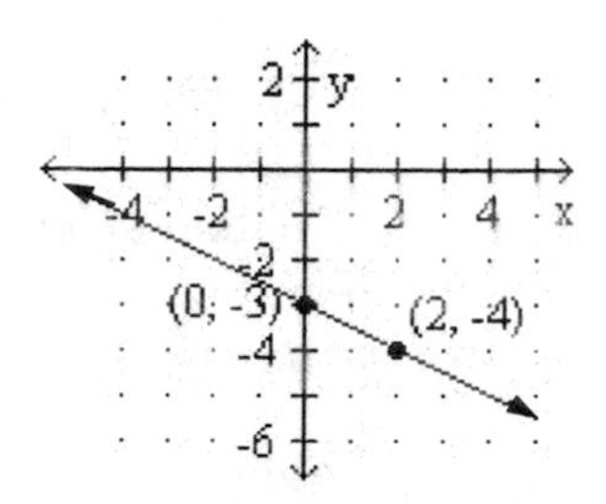

Write an equation of the line with the following characteristics. Write the equation in slope - intercept form.

45. Let $(x_1, y_1) = (3, 7)$

Let $(x_2, y_2) = (-6, 1)$

Step 1: Find the slope.

$$m = \frac{y_2 - y_1}{x_2 - x_1}$$
$$= \frac{1 - 7}{-6 - 3}$$
$$= \frac{-6}{-9}$$
$$= \frac{2}{3}$$

Passes through (3, 7), $m = \frac{2}{3}$

Step 2: $x_1 = 3$ and $y_1 = 7$

$$y - y_1 = m(x - x_1)$$
$$y - 7 = \frac{2}{3}(x - 3)$$
$$y - 7 = \frac{\mathbf{2}}{\mathbf{3}}x - \frac{\mathbf{2}}{\mathbf{3}}(3)$$
$$y - 7 = \frac{2}{3}x - 2$$
$$y - 7 + \mathbf{7} = \frac{2}{3}x - 2 + \mathbf{7}$$
$$y = \frac{2}{3}x + 5 \text{ slope-intercept form}$$

46. Horizontal, passes through $(6,-8)$

The equation of a horizontal line can be written in the form $y=b$.

$$y=-8$$

47. CAR REGISTRATION

a. In this problem, we are told to let x = # of years and y = value. Now you must find the slope from the two ordered pairs: (2, 380) and (4, 310)

Let $(x_1, y_1)=(2,380)$

Let $(x_2, y_2)=(4,310)$

$$m=\frac{y_2-y_1}{x_2-x_1}$$
$$=\frac{310-380}{4-2}$$
$$=\frac{-70}{2}$$
$$=-35$$

Now pick one point (2, 380) and use

$$y-y_1=m(x-x_1)$$
$$y-380=-35(x-2)$$
$$y-380=-35x+70$$
$$y-380+\mathbf{380}=-35x+70+\mathbf{380}$$
$$y=-35x+450 \text{ slope-intercept form}$$

48. THE ATMOSPHERE

a. Let $(x_1, y_1)=(20,\ 340)$

Let $(x_2, y_2)=(40,\ 370)$

Step 1: Find the slope.

$$m=\frac{y_2-y_1}{x_2-x_1}$$
$$=\frac{370-340}{40-20}$$
$$=\frac{30}{20}$$
$$=\frac{3}{2}$$

Passes through $(20,\ 340)$, $m=\frac{3}{2}$

Step 2: $x_1=20$ and $y_1=340$

$$y-y_1=m(x-x_1)$$
$$y-340=\frac{3}{2}(x-20)$$
$$y-340=\frac{3}{2}x-30$$
$$y-340+\mathbf{340}=\frac{3}{2}x-30+\mathbf{340}$$
$$y=\frac{3}{2}x+310 \text{ slope-intercept form}$$

b. Use 60 for x (the year 2020-1960 = 60) to calculate the amount of carbon dioxide.

$$y=\frac{3}{2}x+310$$
$$y=\frac{3}{2}(\mathbf{60})+310$$
$$y=90+310$$
$$y=400$$

The amount of carbon dioxide in 2020 will be 400 parts per million.

SECTION 3.7 REVIEW

Graphing Linear Inequalities

49. Determine whether each ordered pair is a solution of $2x-y\le-4$.

a. $2x-y\le-4$; (0, 5)

$$2(\mathbf{0})-(\mathbf{5})\overset{?}{\le}-4$$
$$0-5\overset{?}{\le}-4$$
$$-5\le-4 \text{ True}$$

$(0,5)$ is a solution.

b. $2x-y\le-4$; (2, 8)

$$2(\mathbf{2})-(\mathbf{8})\overset{?}{\le}-4$$
$$4-8\overset{?}{\le}-4$$
$$-4\le-4 \text{ True}$$

$(2,8)$ is a solution.

c. $2x-y\le-4$; $(-3,-2)$

$$2(\mathbf{-3})-(\mathbf{-2})\overset{?}{\le}-4$$
$$-6+2\overset{?}{\le}-4$$
$$-4\le-4 \text{ True}$$

$(-3,-2)$ is a solution.

d. $2x - y \le -4;\ \left(\frac{1}{2}, -5\right)$

$$2\left(\mathbf{\frac{1}{2}}\right) - (\mathbf{-5}) \overset{?}{\le} -4$$

$$1 + 5 \overset{?}{\le} -4$$

$$6 \le -4 \quad \text{False}$$

$\left(\frac{1}{2}, -5\right)$ is not a solution.

50. Fill in the blanks: $2x - 3y \ge 6$ means $2x - 3y \boxed{=} 6$ or $2x - 3y \boxed{>} 6$.

Graph each inequality.

51. $x - y < 5$, Boundary line is dashed because of the $<$ symbol.
Graph as $x - y = 5$.

y-intercept:	x-intercept:
If $x = 0$,	If $y = 0$,
$\mathbf{0} - y = 5$	$x - \mathbf{0} = 5$
$-y = 5$	$x = 5$
$\frac{-y}{\mathbf{-1}} = \frac{5}{\mathbf{-1}}$	
$y = -5$	

The y-int is $(0, -5)$, and the x-int is $(5, 0)$.

Select test point (0,0) and substitute into

$$x - y < 5$$
$$\mathbf{0} - \mathbf{0} \overset{?}{<} 5$$
$$0 < 5$$
True

The coordinates of every point on the same side of the line as the origin satisfy the inequality. To indicate this, we shade the half-plane that contains the test point (0,0).

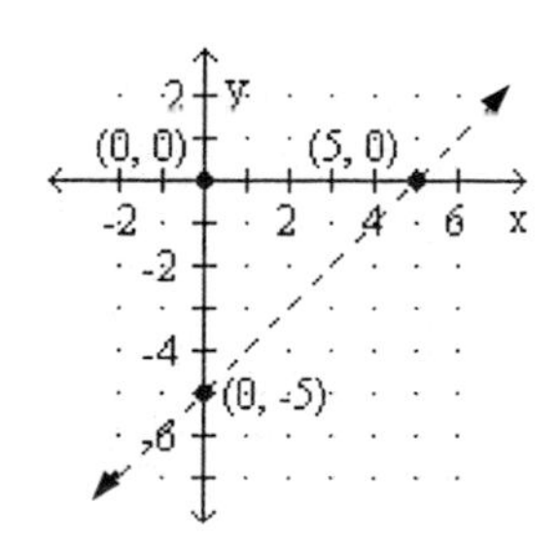

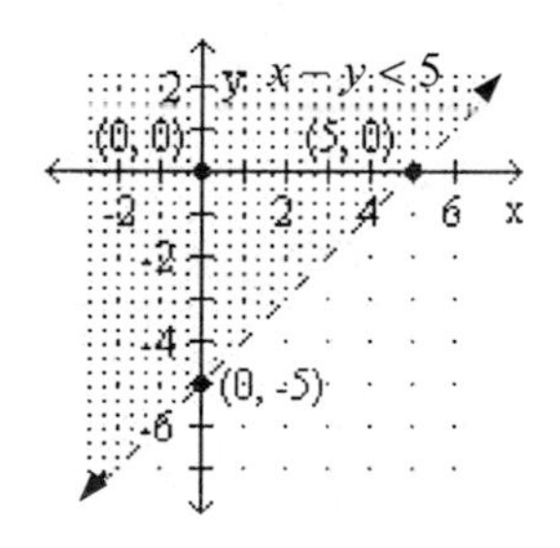

52. $2x - 3y \ge 6$, Boundary line is solid because of the $\ge$ symbol.
Graph as $2x - 3y = 6$.
The y-int is $(0, -2)$, and the x-int is $(3, 0)$.
Select test point (0,0) and substitute into

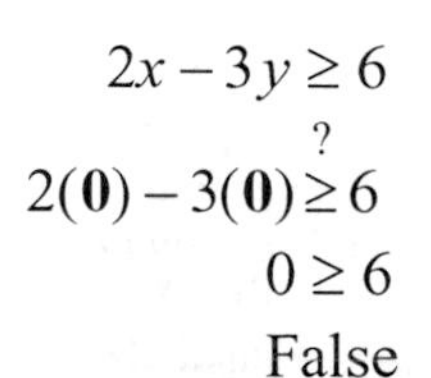

$$2x - 3y \ge 6$$
$$2(\mathbf{0}) - 3(\mathbf{0}) \overset{?}{\ge} 6$$
$$0 \ge 6$$
False

The coordinates of every point on the same side of the line as the origin ***does not*** satisfy the inequality. To indicate this, we shade the half-plane that ***does not*** contains the test point (0,0).

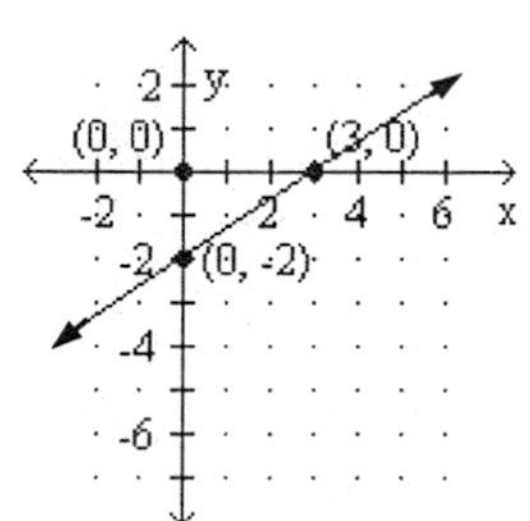

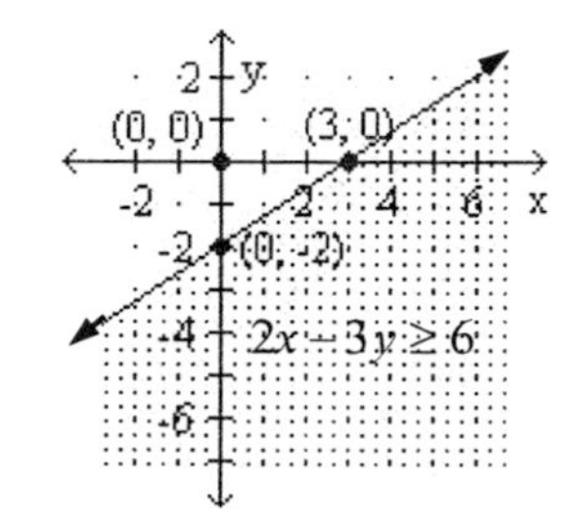

53. $y \le -2x$, Boundary line is solid because of the $\le$ symbol.
Graph as $y = -2x$.

y-intercept:	x-intercept:
If $x = 0$,	If $y = 0$,
$y = -2(\mathbf{0})$	$\mathbf{0} = -2x$
$y = 0$	$0 = x$

The y-int is $(0, 0)$, and the x int is $(0, 0)$.
The line goes through the origin so select test point (0,2) and substitute into

$$y \le -2x$$
$$\mathbf{2} \overset{?}{\le} -2(\mathbf{0})$$
$$2 \le 0$$
False

The coordinates of every point on the same side of the line as (0,2) ***does not*** satisfy the inequality. To indicate this, we shade the half-plane that ***does not*** contain the test point (0,2).

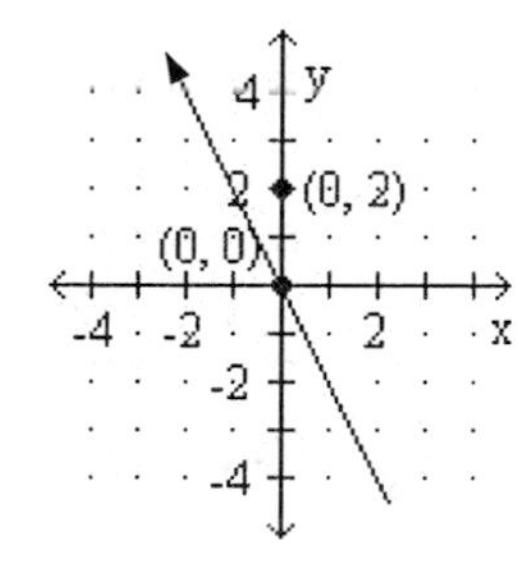

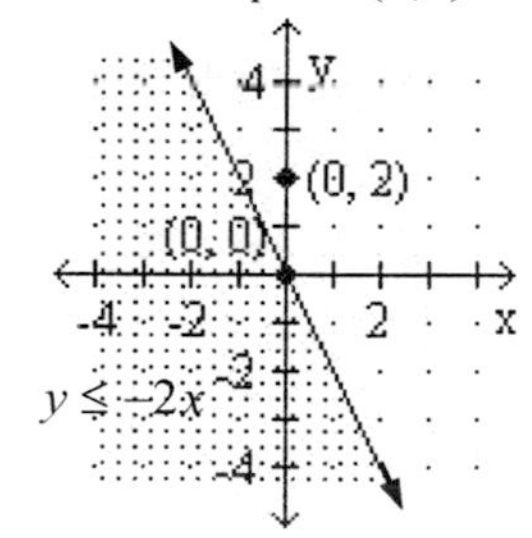

54. $y < -4$, Boundary line is dashed because of the $<$ symbol.

Graph as $y = -4$.

The y-int is $(0, -4)$, and there is no x-int.

$y = -4$ is a line parallel to the x-axis.

Select test point (0,0) and substitute into

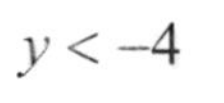

$y < -4$

$0 < -4$

False

The coordinates of every point on the same side of the line as the origin ***do not*** satisfy the inequality. To indicate this, we shade the half-plane that ***does not*** contain the test point (0, 0).

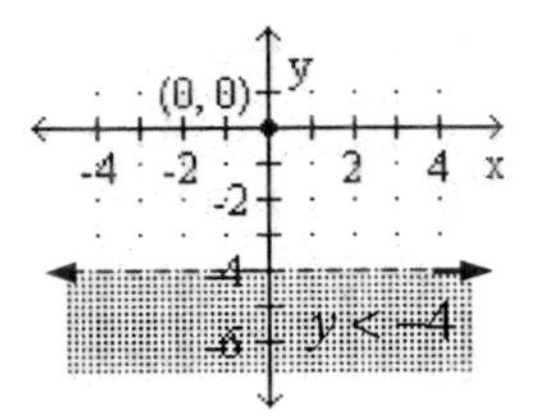

55. a. True b. False c. False

56. WORK SCHEDULES

$3x + 5y \leq 30$

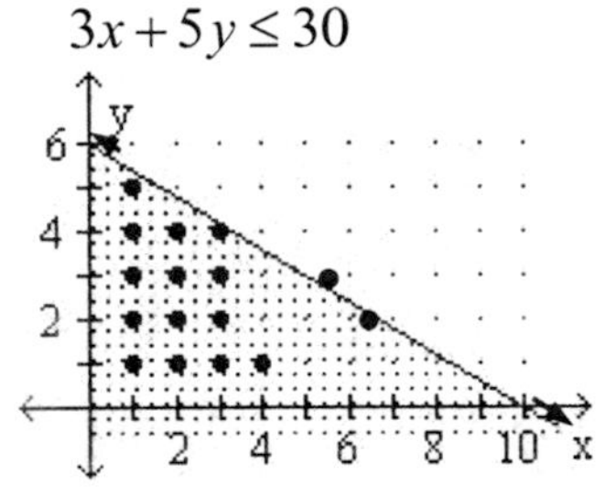

$(2,4)$, $(5,3)$, $(6,2)$

Answers will vary.

SECTION 3.8 REVIEW

An Introduction to Functions

Find the domain and range of each relation.

57. $\{(7, -3), (-5, 9), (4, 4), (0, -11)\}$

Domain: $\{-5, 0, 4, 7\}$

Range: $\{-11, -3, 4, 9\}$

58. $\{(2, -2), (15, -8), (-6, 9), (1, -8)\}$

Domain: $\{-6, 1, 2, 15\}$

Range: $\{-8, -2, 9\}$

Determine whether each arrow diagram or table defines y as a function of x. If a function is defined, give its domain and range. If it does not define a function, find ordered pairs that show a value of x that is assigned more than one value of y.

59. Yes; domain: $\{1, 4, 8\}$; range: $\{0, 6, 9\}$

60. Yes; domain: $\{2,3,5,6\}$; range: $\{1,4\}$

61. Yes; domain: $\{3,5,7,9\}$; range: $\{9,25,49,81\}$

62. No; $(-1, 2)$, $(-1, 4)$

63. Yes; domain: $\{-1, 0, 1, 2\}$; range: $\{5, 6\}$

64. No; $(4, 4)$, $(4, 6)$

Fill in the blanks.

65. The set of all input values for a function is called the **domain**, and the set of all output values is called the **range**.

66. Since $y =$ **$f(x)$** , the equations $y = 2x - 8$ and $f(x) = 2x - 8$ are equivalent.

For $f(x) = x^2 - 4x$, find each of the following function values.

67. $f(x) = x^2 - 4x$

$$f(\mathbf{1}) = \mathbf{1}^2 - 4(\mathbf{1}) = 1 - 4 = -3$$

Thus, $f(1) = -3$

68. $f(x) = x^2 - 4x$

$$f(\mathbf{0}) = \mathbf{0}^2 - 4(\mathbf{0}) = 0 - 0 = 0$$

Thus, $f(0) = 0$

69. $f(x) = x^2 - 4x$

$$f(\mathbf{-3}) = (\mathbf{-3})^2 - 4(\mathbf{-3}) = 9 + 12 = 21$$

Thus, $f(-3) = 21$

70. $f(x) = x^2 - 4x$

$$f\left(\frac{\mathbf{1}}{\mathbf{2}}\right) = \left(\frac{\mathbf{1}}{\mathbf{2}}\right)^2 - 4\left(\frac{\mathbf{1}}{\mathbf{2}}\right) = \frac{1}{4} - \frac{4}{2} = \frac{1}{4} - \frac{8}{4} = -\frac{7}{4}$$

Thus, $f\left(\frac{1}{2}\right) = -\frac{7}{4}$

For $g(x) = 1 - 6x$, find each of the following function values.

71. $g(x) = 1 - 6x$

$$\begin{aligned} g(\mathbf{1}) &= 1 - 6(\mathbf{1}) \\ &= 1 - 6 \\ &= -5 \end{aligned}$$

Thus, $g(1) = -5$

72. $g(x) = 1 - 6x$

$$\begin{aligned} g(\mathbf{-6}) &= 1 - 6(\mathbf{-6}) \\ &= 1 + 36 \\ &= 37 \end{aligned}$$

Thus, $g(-6) = 37$

73. $g(x) = 1 - 6x$

$$\begin{aligned} g(\mathbf{0.5}) &= 1 - 6(\mathbf{0.5}) \\ &= 1 - 3 \\ &= -2 \end{aligned}$$

Thus, $g(0.5) = -2$

74. $g(x) = 1 - 6x$

$$\begin{aligned} g\left(\frac{\mathbf{3}}{\mathbf{2}}\right) &= 1 - 6\left(\frac{\mathbf{3}}{\mathbf{2}}\right) \\ &= 1 - 9 \\ &= -8 \end{aligned}$$

Thus, $g\left(\frac{3}{2}\right) = -8$

Determine whether each graph is the graph of a function. If it is not, find two ordered pairs that show a value of *x* that is assigned more than one value of *y*.

75. No: (1, 0.5), (1, 4) (answers may vary)

76. Yes.

Complete the table of function values. Then graph the function.

77. $f(x) = 1 - |x|$

x	$f(x)$
0	$f(x) = 1 - \lvert x \rvert$ $f(\mathbf{0}) = 1 - \lvert \mathbf{0} \rvert$ $= 1 - 0$ $= 1$
1	$f(x) = 1 - \lvert x \rvert$ $f(\mathbf{1}) = 1 - \lvert \mathbf{1} \rvert$ $= 1 - 1$ $= 0$
3	$f(x) = 1 - \lvert x \rvert$ $f(\mathbf{3}) = 1 - \lvert \mathbf{3} \rvert$ $= 1 - 3$ $= -2$
−1	$f(x) = 1 - \lvert x \rvert$ $f(\mathbf{-1}) = 1 - \lvert \mathbf{-1} \rvert$ $= 1 - 1$ $= 0$
−3	$f(x) = 1 - \lvert x \rvert$ $f(\mathbf{-3}) = 1 - \lvert \mathbf{-3} \rvert$ $= 1 - 3$ $= -2$

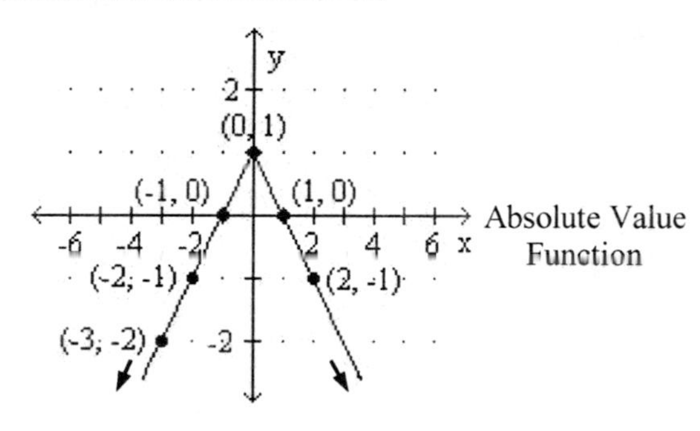

78. ALUMINUM CANS

$$\begin{aligned} V(r) &= 15.7r^2 \\ V(\mathbf{8}) &= 15.7(\mathbf{8})^2 \\ &= 15.7(64) \\ &= 1{,}004.8 \end{aligned}$$

The volume of the can is 1,004.8 in^3.

CHAPTER 3 TEST

1. **Fill in the blanks.**

a. A rectangular coordinate system is formed by two perpendicular number lines called the x- **axis** and the y- **axis** .

b. A **solution** of an equation in two variables is an ordered pair of numbers that makes the equation a true statement.

c. $3x + y = 10$ is called a **linear** equation in two variables because its graph is a line.

d. The **slope** of a line is a measure of steepness.

e. A **function** is a set of ordered pairs in which to each first component there corresponds exactly one second component.

The graph shows the number of dogs being boarded in a kennel over a 3-day holiday weekend.

2. How many dogs were in the kennel 2 days before the holiday? **10**

3. What is the maximum number of dogs that were boarded on the holiday weekend? **60**

4. When were there 30 dogs in the kennel? **1 day before and the 3rd day of the holiday**

5. What information does the -intercept of the graph give? **50 dogs were in the kennel when the holiday began.**

6. Plot each point on a rectangular coordinate system:

$$(1,3),(-2,4),(-3,-2),(3,-2),(-1,0),$$

$$(0,-1), \text{ and } \left(-\frac{1}{2},\frac{7}{2}\right)$$

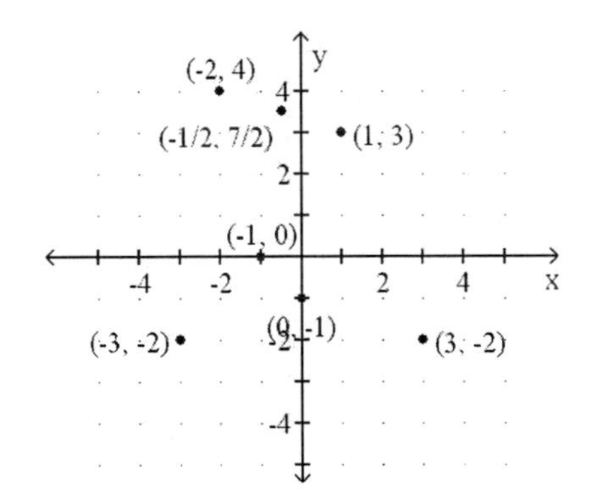

7. Find the coordinates of each point shown in the graph.

***A*(2, 4), *B*(–3, 3), *C*(–2, –3), *D*(4, –3), *E*(–4, 0), *F*(3.5, 1.5)**

8. In which quadrant is each point located?

a. $(-1,-5)$ **III** b. $\left(6,-2\frac{3}{4}\right)$ **IV**

9. Is (–3, –4) a solution of $3x - 4y = 7$?

$$3x - 4y = 7$$

$$3(\mathbf{-3}) - 4(\mathbf{-4}) \stackrel{?}{=} 7$$

$$-9 + 16 \stackrel{?}{=} 7$$

$$7 = 7$$

True

(–3, –4) is a solution of $3x - 4y = 7$.

10. Complete the table of solutions for the linear equation.

x	y	(x, y)
2	$x+4y=6$ $\mathbf{2}+4y=6$ $2+4y-\mathbf{2}=6-\mathbf{2}$ $4y=4$ $\frac{4y}{\mathbf{4}}=\frac{4}{\mathbf{4}}$ $y=1$	(2,1)
$x+4y=6$ $x+4(\mathbf{3})=6$ $x+12=6$ $x+12-\mathbf{12}=6-\mathbf{12}$ $x=-6$	3	(–6,3)

11. a. False, because point C does not lie on the line and therefore is not a solution of the equation.

b. True, because point D lies on the line and therefore is a solution of the equation.

12. Graph: $y = \frac{x}{3}$

x	y	(x, y)
–3	$y=\frac{x}{3}$ $y=\frac{\mathbf{-3}}{3}$ $y=-1$	(–3, –1)
0	$y=\frac{x}{3}$ $y=\frac{\mathbf{0}}{3}$ $y=0$	(0,0)
3	$y=\frac{x}{3}$ $y=\frac{\mathbf{3}}{3}$ $y=1$	(3, 1)

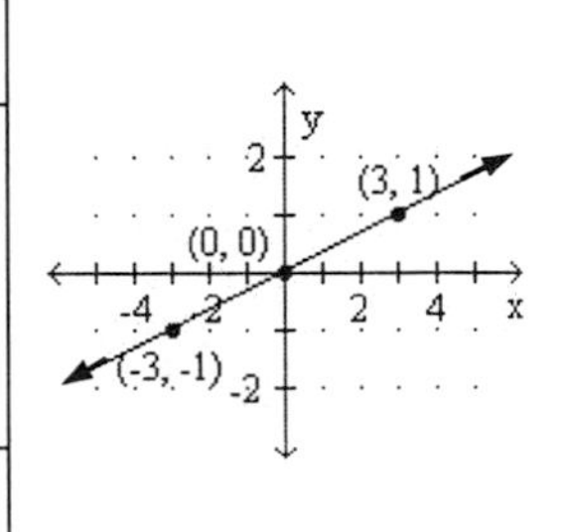

13. **What are the x- and y-intercepts of the graph of $2x - 3y = 6$?**

let $x = 0$	let $y = 0$
$2x - 3y = 6$	$2x - 3y = 6$
$2(\mathbf{0}) - 3y = 6$	$2x - 3(\mathbf{0}) = 6$
$-3y = 6$	$2x = 6$
$\frac{-3y}{\mathbf{-3}} = \frac{6}{\mathbf{-3}}$	$\frac{2x}{\mathbf{2}} = \frac{6}{\mathbf{2}}$
$y = -2$	$x = 3$
y-int: $(0, -2)$	x-int: $(3, 0)$

14. Graph: $8x + 4y = -24$.

let $x = 0$	let $y = 0$
$8x + 4y = -24$	$8x + 4y = -24$
$8(\mathbf{0}) + 4y = -24$	$8x + 4(\mathbf{0}) = -24$
$4y = -24$	$8x = -24$
$\frac{4y}{\mathbf{4}} = \frac{-24}{\mathbf{4}}$	$\frac{8x}{\mathbf{8}} = \frac{-24}{\mathbf{8}}$
$y = -6$	$x = -3$
y-int: $(0, -6)$	x-int: $(-3, 0)$

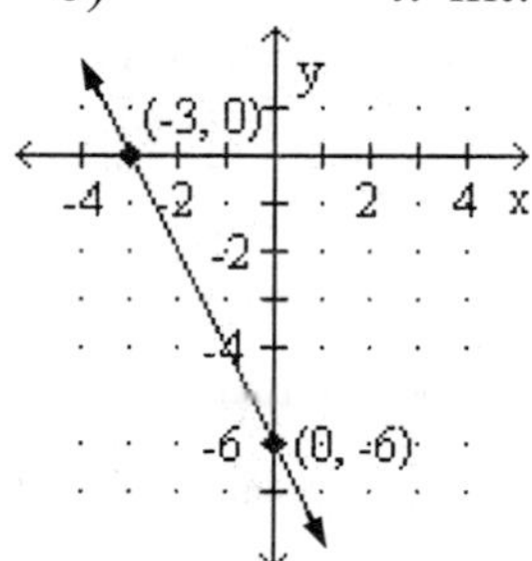

15. Find the slope of the line.

$$m = \frac{8}{7}$$

16. Find the slope of the line passing through

Let $(x_1, y_1) = (-1, 3)$

Let $(x_2, y_2) = (3, -1)$

$$\begin{aligned} m_1 &= \frac{y_2 - y_1}{x_2 - x_1} \\ &= \frac{-1 - 3}{3 - (-1)} \\ &= \frac{-4}{3 + 1} \\ &= \frac{-4}{4} \\ &= -1 \end{aligned}$$

The slope of the line is -1.

17. What is the slope of a horizontal line?

The slope is zero or $m = 0$.

18. RAMPS

$$\begin{aligned} m &= \frac{\text{rise}}{\text{run}} \\ &= \frac{2}{20} \\ &= 0.1 \\ &= 10\% \end{aligned}$$

The grade is 10%.

19. **One line passes through (9, 2) and (6, 4). Another line passes through (0, 7) and (2, 10). Without graphing, determine whether the lines are parallel, perpendicular, or neither.**

Step 1: Find each slope.

Let $(x_1, y_1) = (9, 2)$	Let $(x_1, y_1) = (0, 7)$
Let $(x_2, y_2) = (6, 4)$	Let $(x_2, y_2) = (2, 10)$
$m_1 = \frac{y_2 - y_1}{x_2 - x_1}$	$m_2 = \frac{y_2 - y_1}{x_2 - x_1}$
$= \frac{4 - 2}{6 - 9}$	$= \frac{10 - 7}{2 - 0}$
$= \frac{2}{-3}$	$= \frac{3}{2}$
$= -\frac{2}{3}$	

Step 2: Compare the slopes.

The slopes are negative reciprocals.
The lines are perpendicular.

When graphed, are the lines $y = 2x + 6$ and $2x - y = 0$ parallel, perpendicular, or neither?

20.

$y = 2x + 6$

x	y
-3	$y = 2x + 6$ $y = 2(-3) + 6$ $y = -6 + 6$ $y = 0$
-2	$y = 2x + 6$ $y = 2(-2) + 6$ $y = -4 + 6$ $y = 2$
-1	$y = 2x + 6$ $y = 2(-1) + 6$ $y = -2 + 6$ $y = 4$

$2x - y = 0$

$2x = y$

x	y
-1	$y = 2x$ $y = 2(-1)$ $y = -2$
0	$y = 2x$ $y = 2(0)$ $y = 0$
1	$y = 2x$ $y = 2(1)$ $y = 2$

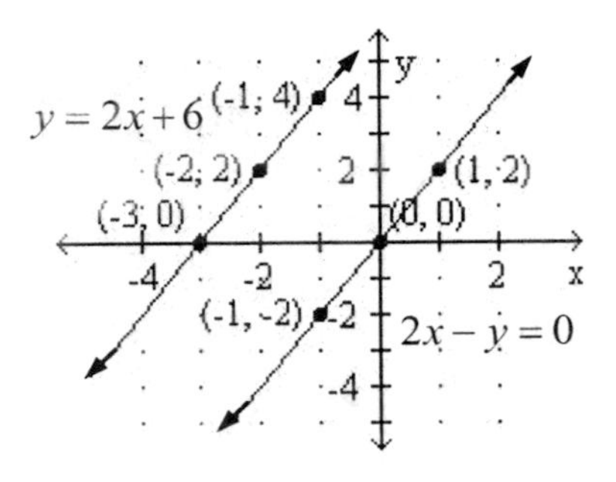

The lines are parallel.

21. Find the rate of change of the decline on which the woman is running.

(2, 250) and (12, 100)

(x_1, y_1) and (x_2, y_2)

$$m_{\text{woman}} = \frac{y_2 - y_1}{x_2 - x_1} = \frac{100 - 250}{12 - 2} = \frac{-150}{10} = -15$$

Her rate of change is -15 feet per mi.

22. Find the rate of change of the incline on which the man is running.

(12, 100) and (20, 300)

(x_1, y_1) and (x_2, y_2)

$$m_{\text{man}} = \frac{y_2 - y_1}{x_2 - x_1} = \frac{300 - 100}{20 - 12} = \frac{200}{8} = 25$$

His rate of change is 25 feet per mi.

23. Graph: $x = -4$.

x	y	(x, y)
-4	-3	$(-4, -3)$
-4	0	$(-4, 0)$
-4	2	$(-4, 2)$

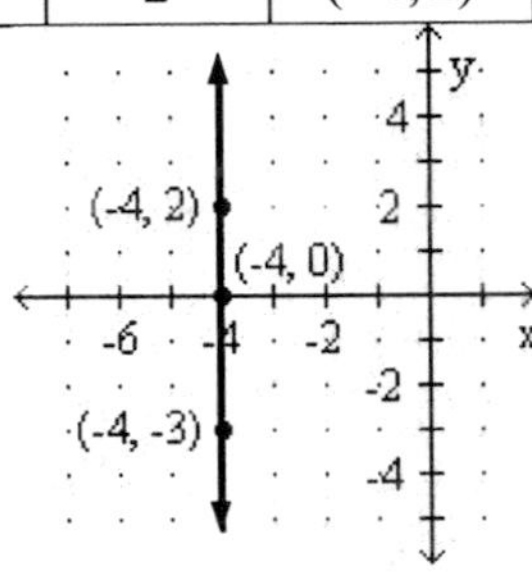

24. Graph the line passing through $(-2, -4)$ having slope $\frac{2}{3}$.

Start at the given point $(-2, -4)$ and $m = \frac{2}{3}$.

Move up 2 spaces, then move right 3 spaces.

Start at the given point $(-2, -4)$ and $m = \frac{-2}{-3}$.

Move down 2 spaces, then move left 3 spaces.

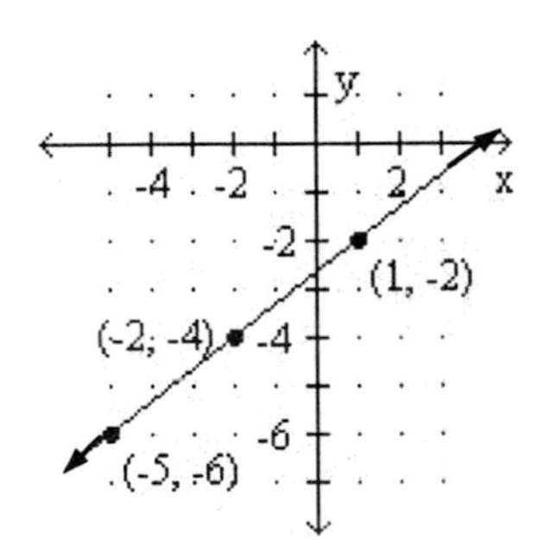

25. Find the slope and the y-intercept of the graph of $x+2y=8$.

$$x+2y=8$$
$$x+2y-\mathbf{x}=8-\mathbf{x}$$
$$2y=-x+8$$
$$\frac{2y}{\mathbf{2}}=\frac{-x}{\mathbf{2}}+\frac{8}{\mathbf{2}}$$
$$y=-\frac{x}{2}+4$$

The slope is $-\frac{1}{2}$. The y-intercept is (0,4).

Find an equation of the line passing through $(-2,5)$ with slope 7. Write the equation in slope - intercept form.

26. Passes through $(-2,5)$, $m=7$

$x_1=-2$ and $y_1=5$

$$y-y_1=m(x-x_1)$$
$$y-5=7(x-(-2))$$
$$y-5=7(x+2) \text{ point-slope form}$$
$$y-5=7(x+2)$$
$$y-5=\mathbf{7}x+\mathbf{7}(2)$$
$$y-5=7x+14$$
$$y-5+\mathbf{5}=7x+14+\mathbf{5}$$
$$y=7x+19 \text{ slope-intercept form}$$

27. Find an equation for the line shown. Write the equation in slope-intercept form.

Slope -2, y-intercept $(0, -5)$

The equation of the line is $y=-2x-5$.

28. DEPRECIATION

a. computer cost new \$15,000
loses value of \$1,500 per yr
Write a linear equation that gives the resale value v of the computer x years after being purchased.

Resale value	equals	depreciation of \$1,500 per yr	plus	the cost of \$15,000.
v	$=$	$-1,500x$	$+$	$15,000$

b. Use your equation from part 'a' to predict the value of the computer 8 years after it is purchased.

$$v=-1,500x+15,000$$
$$v=-1,500(\mathbf{8})+15,000$$
$$v=-12,000+15,000$$
$$v=3,000$$

The value of the computer after 8 years is \$3,000.

29. **Determine whether (6, 1) is a solution of** $2x-4y\geq 8$.

$$2x-4y\geq 8$$
$$2(\mathbf{6})-4(\mathbf{1})\overset{?}{\geq}8$$
$$12-4\overset{?}{\geq}8$$
$$8\geq 8$$
True

(6,1) is a solution.

30. WATER HEATERS

Choose two ordered pairs.

(140, 13) and (180, 5)

(T_1, y_1) and (T_2, y_2)

$$m=\frac{y_2-y_1}{T_2-T_1}=\frac{5-13}{180-140}=\frac{-8}{40}=-\frac{1}{5}$$

Passes through $(180,5)$, $m=-\frac{1}{5}$

$T_1=180$ and $y_1=5$

$$y-y_1=m(T-T_1)$$
$$y-5=-\frac{1}{5}(T-180)$$
$$y-5=-\frac{\mathbf{1}}{\mathbf{5}}(T)-\frac{\mathbf{1}}{\mathbf{5}}(-180)$$
$$y-5=-\frac{1}{5}T+36 \text{ point-slope form}$$
$$y-5+\mathbf{5}=-\frac{1}{5}T+36+\mathbf{5}$$
$$y=-\frac{1}{5}T+41 \text{ slope-intercept form}$$

31. **Determine whether each point satisfies the inequality.**

a. $(-2,3)$, Yes, because the point lies within the shaded region of the graph and therefore is a solution of the inequality.

b. $(3,-4)$, No, because point does not lie within the shaded region of the graph and therefore is not a solution of the inequality.

c. $(0,0)$, No, because the point lies on the the dashed boundary line of the graph and therefore is not a solution of the inequality.

32. Is $(-20,-2)$ a solution of

$$\frac{1}{2}x-3y\geq -4?$$

$$\frac{1}{2}(\mathbf{-20})-3(\mathbf{-2})\overset{?}{\geq}-4$$

$$-10+6\overset{?}{\geq}-4$$

$$-4\geq -4 \text{ Ture}$$

$(-20,-2)$ is a solution.

Graph the linear inequality $2x-5y\leq -10$.

33. $2x-5y\leq -10$, Boundary line is <u>solid</u> because of the $\leq$ symbol.

Graph as $2x-5y=-10$.

y-intercept:	x-intercept:
If $x=0$,	If $y=0$,
$2(\mathbf{0})-5y=-10$	$2x-5(\mathbf{0})=-10$
$-5y=-10$	$2x=-10$
$\frac{-5y}{\mathbf{-5}}=\frac{-10}{\mathbf{-5}}$	$\frac{2x}{\mathbf{2}}=\frac{-10}{\mathbf{2}}$
$y=2$	$x=-5$

The y-int is $(0,2)$, and the x-int is $(-5,0)$.

Select test point $(0,0)$ and substitute into

$$2x-5y\leq -10$$

$$5(\mathbf{0})-3(\mathbf{0})\overset{?}{\leq}-10$$

$$0\leq -10$$

False

The coordinates of every point on the same side of the line as (0, 0) ***do not*** satisfy the inequality. To indicate this, we shade the half-plane that ***does not*** contain the test point (0, 0).

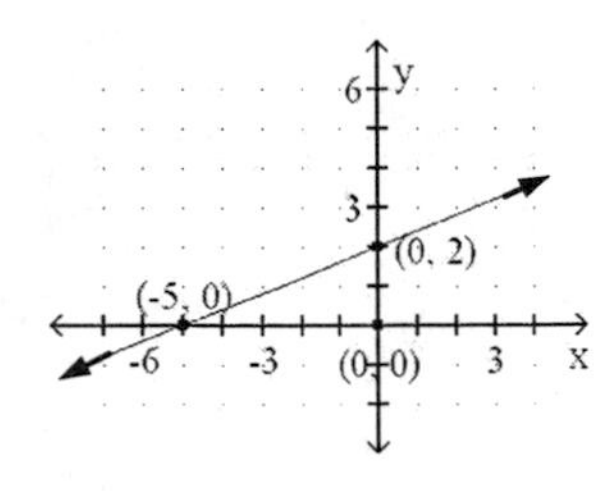

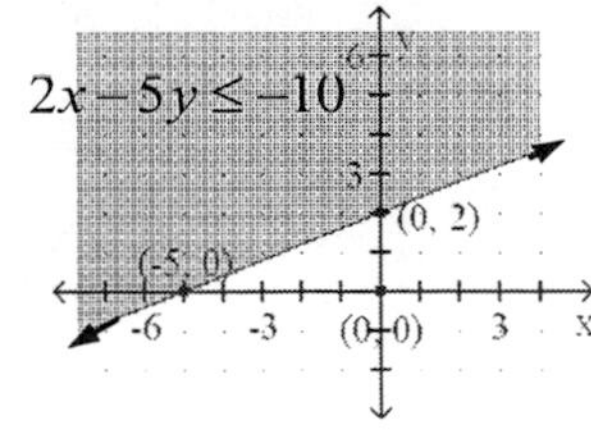

34. Find the domain and range of the relation: $\{(5, 3), (1, 12), (-4, 3), (0, -8)\}$.

Domain: $\{-4, 0, 1, 5\}$; Range: $\{-8, 3, 12\}$

Determine whether the relation defines y as a function of x. If a function is defined, give its domain and range. If it does not define a function, find ordered pairs that show a value of x that is assigned more than one value of y.

35. Yes; domain: $\{1, 2, 3, 4\}$; range: $\{1, 2, 3, 4\}$

36. No; domain: $(-3,9),(-3,-7)$

37. Yes; domain: $\{6, 7, 8, 9, 10\}$; range: $\{5\}$

38. No; domain: $(2,6),(2,2)$

39. No; $(2,3.5),(2,-3.5)$, (answers may vary)

40. No; $(-2,2),(-2,-1)$, (answers may vary)

41. If $f(x)=2x-7$, find: $f(-3)$

$$f(\mathbf{-3})=2(\mathbf{-3})-7$$
$$=-6-7$$
$$=-13$$

Thus, $f(-3)=-13$

42. If $g(x)=3.5x^3$, find $g(6)$.

$$g(x)=3.5x^3$$
$$g(\mathbf{6})=3.5(\mathbf{6})^3$$
$$=3.5(216)$$
$$=756$$

Thus, $g(6)=756$

43. TELEPHONE CALLS

$C(n)=0.30n+15$, find $C(45)$.

$$C(n)=0.30n+15$$
$$C(\mathbf{45})=0.30(\mathbf{45})+15$$
$$=13.5+15$$
$$=28.5$$

It costs $28.50 to make 45 calls.

44. Graph: $f(x)=|x|-1$.

x	$f(x)$
-3	$f(x)=\|x\|-1$ $f(-3)=\|-3\|-1$ $=3-1$ $=2$
-2	$f(x)=\|x\|-1$ $f(-2)=\|-2\|-1$ $=2-1$ $=1$
1	$f(x)=\|x\|-1$ $f(-1)=\|-1\|-1$ $=1-1$ $=0$
0	$f(x)=\|x\|-1$ $f(0)=\|0\|-1$ $=0-1$ $=-1$
2	$f(x)=\|x\|-1$ $f(2)=\|2\|-1$ $=2-1$ $=1$
3	$f(x)=\|x\|-1$ $f(3)=\|3\|-1$ $=3-1$ $=2$

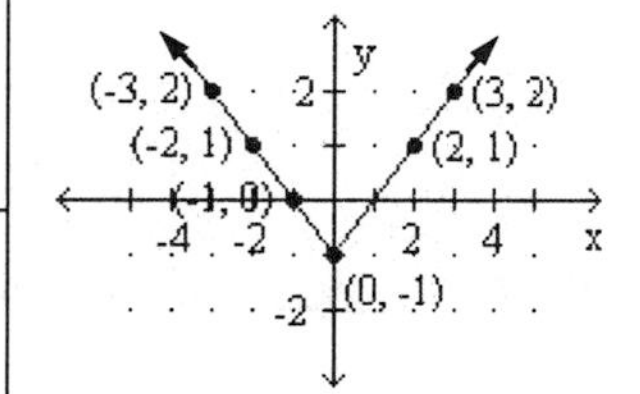

Absolute Value Function

CUMULATIVE REVIEW CHAPTERS 1 - 3

1. Prime Factorization: [Section 1.2]

$$\begin{aligned}108 &= 2\cdot 54\\ &= 2\cdot 2\cdot 27\\ &= 2\cdot 2\cdot 3\cdot 9\\ &= 2\cdot 2\cdot 3\cdot 3\cdot 3\\ &= 2^2\cdot 3^3\end{aligned}$$

2. Write $\frac{1}{250}$ as a decimal. [Section 1.3]

$$\frac{1}{250} = 250\overline{)1.000} = 0.004 \quad (\text{quotient } 0.004)$$

3. Determine whether each statement is true or false. [Section 1.3]

   a. Every whole number is an integer. **True**

   b. Every integer is a real number. **True**

   c. 0 is a whole number, an integer, and a rational number. **True**

Perform the operations.

4. $-27+21+(-9)$ [Section 1.4]

$$\begin{aligned}-27+21+(-9) &= -6+(-9)\\ &= -15\end{aligned}$$

5. $-1.57-(-0.8)$ [Section 1.5]

$$\begin{aligned}-1.57-(-0.8) &= -1.57+0.8\\ &= -0.77\end{aligned}$$

6. $-9(-7)(5)(-3)$ [Section 1.6]

$$\begin{aligned}-9(-7)(5)(-3) &= 63(5)(-3)\\ &= 63(5)(-3)\\ &= 315(-3)\\ &= -945\end{aligned}$$

7. $\frac{-180}{-6}$ [Section 1.6]

$$\begin{aligned}\frac{-180}{-6} &= \frac{\overset{1}{\cancel{-6}}\cdot 30}{\underset{1}{\cancel{-6}}}\\ &= 30\end{aligned}$$

8. Evaluate: $\left|\frac{(6-5)^4-(-21)}{-27+4^2}\right|$. [Section 1.7]

$$\begin{aligned}\left|\frac{(6-5)^4-(-21)}{-27+4^2}\right| &= \left|\frac{(1)^4+21}{-27+16}\right|\\ &= \left|\frac{1+21}{-11}\right|\\ &= \left|\frac{22}{-11}\right|\\ &= |-2|\\ &= 2\end{aligned}$$

9. Evaluate: b^2-4ac for $a=2$, $b=-8$, and $c=4$. [Section 1.8]

$$\begin{aligned}b^2-4ac &= (-8)^2-4(2)(4)\\ &= 64-32\\ &= 32\end{aligned}$$

10. Suppose x sheets from a 500-sheet ream of paper have been used. How many sheets are left? [Section 1.8]

    $500-x$

11. How many terms does the algebraic expression $3x^2-2x+1$ have? What is the coefficient of the second term? [Section 1.8]

    3 terms

    -2 is the coefficient of the second term

12. Use the distributive property to remove parentheses. [Section 1.9]

    a. $2(x+4) = \mathbf{2}x+\mathbf{2}(4)$
    $= 2x+8$

    b. $-2(x-4) = \mathbf{-2}x-\mathbf{2}(-4)$
    $= -2x+8$

Simplify each expression. [Section 1.9]

13. $5a+10-a = (5-1)a+10$
    $= 4a+10$

14. $-7(9t) = -63t$

15. $-2b^2+6b^2 = (-2+6)b^2$
    $= 4b^2$

16. $5(-17)(0)(2)=0$

17. $$\begin{aligned}(a+2)-(a-2)&=a+2-\mathbf{1}(a)-\mathbf{1}(-2)\\&=a+2-a+2\\&=(1-1)a+(2+2)\\&=4\end{aligned}$$

18. $$\begin{aligned}-4(-5)(-8a)&=20(-8a)\\&=20(-8)a\\&=-160a\end{aligned}$$

19. $$\begin{aligned}-y-y-y&=[-1+(-1)+(-1)]y\\&=-3y\end{aligned}$$

20. $$\begin{aligned}\frac{3}{2}(4x-8)+x&=\frac{\mathbf{3}}{\mathbf{2}}(4x)-\frac{\mathbf{3}}{\mathbf{2}}(8)+x\\&=6x-12+x\\&=(6+1)x-12\\&=7x-12\end{aligned}$$

Solve each equation. [Sections 2.1 and 2.2]

21. $$\begin{aligned}3x-5&=13\\3x-5+\mathbf{5}&=13+\mathbf{5}\\3x&=18\\\frac{3x}{\mathbf{3}}&=\frac{18}{\mathbf{3}}\\x&=6\end{aligned}$$

The solution is 6.

22. $$\begin{aligned}1.2-x&=-1.7\\1.2-x-\mathbf{1.2}&=-1.7-\mathbf{1.2}\\-x&=-2.9\\\frac{-x}{\mathbf{-1}}&=\frac{-2.9}{\mathbf{-1}}\\x&=2.9\end{aligned}$$

The solution is 2.9.

23. $$\begin{aligned}\frac{2}{3}x-2&=4\\\frac{2}{3}x-2+\mathbf{2}&=4+\mathbf{2}\\\frac{2}{3}x&=6\\\left(\frac{\mathbf{3}}{\mathbf{2}}\right)\frac{2}{3}x&=\left(\frac{\mathbf{3}}{\mathbf{2}}\right)6\\x&=9\end{aligned}$$

The solution is 9.

24. $$\begin{aligned}\frac{y-2}{7}&=-3\\\mathbf{7}\cdot\left(\frac{y-2}{7}\right)&=\mathbf{7}(-3)\\y-2&=-21\\y-2+\mathbf{2}&=-21+\mathbf{2}\\y&=-19\end{aligned}$$

The solution is -19.

25. $$\begin{aligned}-3(2y-2)-y&=5\\-\mathbf{3}(2y)-\mathbf{3}(-2)-y&=5\\-6y+6-y&=5\\-7y+6&=5\\-7y+6-\mathbf{6}&=5-\mathbf{6}\\-7y&=-1\\\frac{-7y}{\mathbf{-7}}&=\frac{-1}{\mathbf{-7}}\\y&=\frac{1}{7}\end{aligned}$$

The solution is $\frac{1}{7}$.

26. $$\begin{aligned}9y-3&=6y\\9y-3-\mathbf{9y}&=6y-\mathbf{9y}\\-3&=-3y\\\frac{-3}{\mathbf{-3}}&=\frac{-3y}{\mathbf{-3}}\\1&=y\end{aligned}$$

The solution is 1.

27. $$\frac{1}{3}+\frac{c}{5}=-\frac{3}{2}$$
$$\mathbf{30}\left(\frac{1}{3}\right)+\mathbf{30}\left(\frac{c}{5}\right)=\mathbf{30}\left(-\frac{3}{2}\right)$$
$$10+6c=-45$$
$$10+6c-\mathbf{10}=-45-\mathbf{10}$$
$$6c=-55$$
$$\frac{6c}{\mathbf{6}}=\frac{-55}{\mathbf{6}}$$
$$c=-\frac{55}{6}$$
The solution is $-\frac{55}{6}$.

28. $$5(x+2)=5x-2$$
$$\mathbf{5}x+\mathbf{5}(2)=5x-2$$
$$5x+10=5x-2$$
$$5x+10-\mathbf{5x}=5x-2-\mathbf{5x}$$
$$10=-2$$
No solution, contradiction

29. $$-x=99$$
$$\frac{-x}{\mathbf{-1}}=\frac{99}{\mathbf{-1}}$$
$$x=-99$$
The solution is -99.

30. $$3c-2=\frac{11(c-1)}{5}$$
$$\mathbf{5}(3c)-\mathbf{5}(2)=\mathbf{5}\left(\frac{11(c-1)}{5}\right)$$
$$15c-10=11(c-1)$$
$$15c-10=\mathbf{11}c-\mathbf{11}(1)$$
$$15c-10=11c-11$$
$$15c-10+\mathbf{10}=11c-11+\mathbf{10}$$
$$15c=11c-1$$
$$15c-\mathbf{11c}=11c-1-\mathbf{11c}$$
$$4c=-1$$
$$\frac{4c}{\mathbf{4}}=\frac{-1}{\mathbf{4}}$$
$$c=-\frac{1}{4}$$
The solution is $-\frac{1}{4}$.

31. PENNIES [Section 2.3]

79% of what number is 869?

$0.79 \bullet x = 869$
$$0.79x=869$$
$$\frac{0.79x}{\mathbf{0.79}}=\frac{869}{\mathbf{0.79}}$$
$$x=1{,}100$$
1,100 people were surveyed.

32. Solve for h: $S=2\pi rh+2\pi r^2$. [Section 2.4]
$$S=2\pi rh+2\pi r^2$$
$$S-\mathbf{2\pi r^2}=2\pi rh+2\pi r^2-\mathbf{2\pi r^2}$$
$$S-2\pi r^2=2\pi rh$$
$$\frac{S-2\pi r^2}{\mathbf{2\pi r}}=\frac{2\pi rh}{\mathbf{2\pi r}}$$
$$\frac{S-2\pi r^2}{2\pi r}=h$$
$$h=\frac{S-2\pi r^2}{2\pi r}$$

33. BAND AIDS [Section 2.4]

$$P = 2w + 2l$$
$$= 2\left(\frac{13}{16}\right) + 2\left(\frac{3}{4}\right)$$
$$= \frac{13}{8} + \frac{3}{2}$$
$$= \frac{13}{8} + \frac{3}{2} \cdot \mathbf{\frac{4}{4}}$$
$$= \frac{13 + 12}{8}$$
$$= \frac{25}{8}$$
$$= 3\frac{1}{8}$$

The perimeter is $3\frac{1}{8}$ inches.

$$A = lw$$
$$= \left(\frac{13}{16}\right)\left(\frac{3}{4}\right)$$
$$= \frac{39}{64}$$

The area is $\frac{39}{64}$ in^2.

34. HIGH HEELS [Section 2.5]

Analyze

- 3 angles of a triangle measure $180°$.
- Right angle is $90°$.
- Find the value of x.

Assign

Let x = the measure of one angle in degrees

Form

2 times	the measure of one angle	plus	the measure of the right angle	equals	the total degrees of the triangle.
	$2x$	$+$	90	$=$	180

Solve

$$2x + 90 = 180$$
$$2x + 90 - \mathbf{90} = 180 - \mathbf{90}$$
$$2x = 90$$
$$\frac{2x}{\mathbf{2}} = \frac{90}{\mathbf{2}}$$
$$x = 45$$

State

$45°$ is the measure of each angle.

Check

The measure of each angle is $45°$.
The measure of the right angle is $90°$.
Since the total was $45° + 45° + 90° = 180°$, the solution checks.

35. Complete the table. [Section 2.6]

	% acid •	Liters =	Amt of acid
50% solution	0.50	x	**0.50*x***
25% solution	0.25	$13 - x$	**0.25(13–x)**
30% solution	0.30	13	**0.30(13)**

36. ROAD TRIPS [Section 2.6]

Analyze

- Bus and truck leave at the same time and from the same place.
- Bus travels at 60 mph.
- Truck travels at 50 mph.
- How long will it take for them to be 75 miles apart?

Assign

Let t = the number of hours traveled by each

	rate •	time =	distance
Bus	60	***t***	**60*t***
Truck	50	***t***	**50*t***
		Total:	**75**

Form

The distance traveled by the bus	minus	the distance traveled by the truck	equals	the total of 75 miles.
$60t$	$-$	$50t$	$=$	75

Solve

$$60t - 50t = 75$$
$$10t = 75$$
$$\frac{10t}{\mathbf{10}} = \frac{75}{\mathbf{10}}$$
$$t = 7.5$$

State

It will take 7.5 hours before the two are 75 miles apart.

Check

The bus's distance is $60(7.5) = 450$.
The truck's distance is $50(7.5) = 375$.
Since the difference was $450 - 375 = 75$, the solution checks.

37. MIXING CANDY [Section 2.6]

Analyze

- Candy corn sells for $2.85 per lb.
- Black gumdrop sells for $1.80 per lb.
- A blend of the two is to sell for $2.22 per lb.
- Mixture of 200 lb is needed.
- How many pounds of each is needed?

Assign

Let x = the pounds candy corn

$200 - x$ = the pounds black gumdrops

Form

	Number •	Value =	Total value
Candy corn	$\mathbf{x}$	2.85	**2.85*x***
Black gumdrops	$\mathbf{200 - x}$	1.80	$\mathbf{1.80(200 - x)}$
Blend	200	2.22	**200(2.22)**

The value of the candy corn	plus	the value of the gumdrops	equals	the total value of the blend.
$2.85x$	+	$1.80(200 - x)$	=	$200(2.22)$

Solve

$$2.85x + 1.80(200 - x) = 200(2.22)$$
$$2.85x + 360 - 1.80x = 444$$
$$1.05x + 360 = 444$$
$$1.05x + 360 - \mathbf{360} = 444 - \mathbf{360}$$
$$1.05x = 84$$
$$\frac{1.05x}{\mathbf{1.05}} = \frac{84}{\mathbf{1.05}}$$
$$x = 80$$

gumdrops

$$200 - x = 200 - \mathbf{80}$$
$$= 120$$

State

80 lb of candy corn will be needed.

120 lb of black gumdrops will be needed.

Check

The value of corn is 80($2.85), or $228.
The value of dumdrops is 120($1.80), or $216.
The value of the blend is 200($2.22), or $444.
Since the total was $228 + $216 = $444,
the results check.

Solve each inequality. Write the solution set in interval notation and graph it. [Section 2.7]

38. $$-\frac{3}{16}x \geq -9$$

$$\left(-\frac{\mathbf{16}}{\mathbf{3}}\right)\left(-\frac{3}{16}x\right) \leq \left(-\frac{\mathbf{16}}{\mathbf{3}}\right)(-9)$$
$$x \leq 48$$
$$(-\infty, 48]$$

48

39. $$8x + 4 > 3x + 4$$

$$8x + 4 - \mathbf{4} > 3x + 4 - \mathbf{4}$$
$$8x > 3x$$
$$8x - \mathbf{3x} > 3x - \mathbf{3x}$$
$$5x > 0$$
$$\frac{5x}{\mathbf{5}} > \frac{0}{\mathbf{5}}$$
$$x > 0$$
$$(0, \infty)$$

0

40. In which quadrants are the second coordinates of points positive? [Section 3.1] **Quadrants I and II**

41. Is $(-2, 4)$ a solution of $y = 2x - 8$? [Section 3.2]

$$y = 2x - 8$$
$$\mathbf{4} \stackrel{?}{=} -2(\mathbf{-2}) - 8$$
$$4 \stackrel{?}{=} 4 - 8$$
$$4 = -4 \quad \text{False}$$

$(-2, 4)$ is not a solution.

Graph each equation.

42. $y = x$ [Section 3.2]

x	y	(x, y)
–2	–2	(–2, –2)
0	0	(0, 0)
2	2	(2, 2)

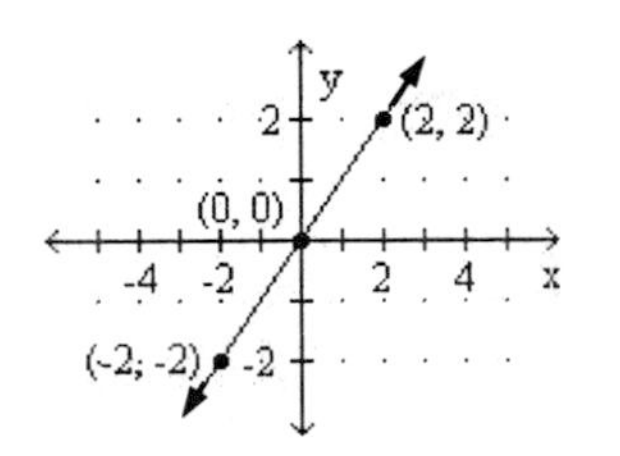

43. $4y+2x=-8$ [Section 3.3]

y-intercept:	x-intercept:
If $x=0$,	If $y=0$
$4y+2(\mathbf{0})=-8$	$4(\mathbf{0})+2x=-8$
$4y=-8$	$2x=-8$
$y=-2$	$x=-4$

The y-intercept is (0, –2), and the x-intercept is (–4, 0).

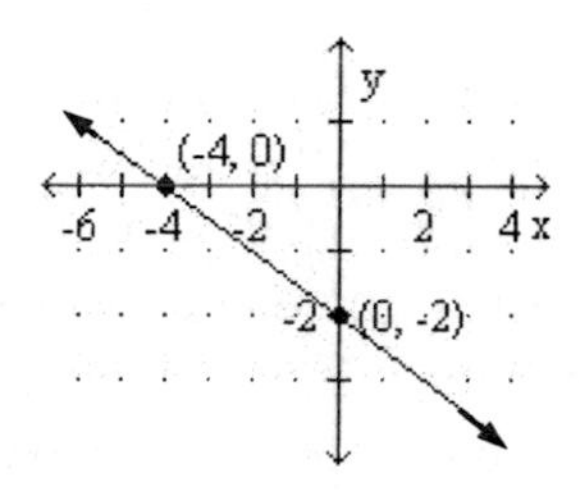

44. What is the slope of the graph of the line $y=5$? [Section 3.4]

$y=5$ is the graph of a horizontial line.

The slope of the line is zero.

45. What is the slope of the line passing through $(-2, 4)$ and $(5, -6)$? [Section 3.4]

Let $(x_1, y_1)=(-2,4)$. Let $(x_2, y_2)=(5,-6)$.

$$m=\frac{y_2-y_1}{x_2-x_1}=\frac{-6-4}{5-(-2)}=\frac{-6-4}{5+2}=\frac{-10}{7}=-\frac{10}{7}$$

46. ROOFING

Find the pitch of the roof. [Section 3.4]

$$m=\frac{\text{rise}}{\text{run}}=\frac{7}{12}$$

47. Find the slope and the y-intercept of the graph of the line described by $4x-6y=-12$. [Section 3.5]

Write the equation in $y=mx+b$ form.

$$4x-6y=-12$$
$$4x-6y-\mathbf{4x}=-12-\mathbf{4x}$$
$$-6y=-4x-12$$
$$\frac{-6y}{\mathbf{-6}}=\frac{-4x}{\mathbf{-6}}-\frac{12}{\mathbf{-6}}$$
$$y=\frac{2}{3}x+2$$

The slope is $\frac{2}{3}$. The y-intercept is $(0,2)$.

48. Write an equation of the line that has slope -2 and y-int (0, 1). [Section 3.5]

$$y=mx+b$$
$$y=-2x+1$$

49. Write an equation of the line that has slope $-\frac{7}{8}$ and passes through $(2, -9)$. Express the answer in point-slope form and slope-intercept form. [Section 3.6]

Passes through $(2,-9)$, $m=-\frac{7}{8}$

$x_1=2$ and $y_1=-9$

$$y-y_1=m(x-x_1)$$
$$y-(-9)=-\frac{7}{8}(x-2)$$
$$y+9=-\frac{7}{8}(x-2) \quad \text{point-slope form}$$
$$y+9=-\frac{7}{8}(x-2)$$
$$y+9=-\frac{\mathbf{7}}{\mathbf{8}}x-\frac{\mathbf{7}}{\mathbf{8}}(-2)$$
$$y+9-\mathbf{9}=-\frac{7}{8}x+\frac{7}{4}-\frac{\mathbf{36}}{\mathbf{4}}$$
$$y=-\frac{7}{8}x-\frac{29}{4} \quad \text{slope-intercept form}$$

50. Is $(-2, -4)$ a solution of $x+y\le -6$? [Section 3.7]

$$x+y\le -6$$
$$\mathbf{-2}+(\mathbf{-4})\overset{?}{\le}-6$$
$$-6\le -6$$

True

(–2, –4) is a solution.

51. $y \geq x+1$, Boundary line is solid because of the $\geq$ symbol.

y-intercept:	x-intercept:
If $x = 0$,	If $y = 0$,
$y = \mathbf{0}+1$	$\mathbf{0} = x+1$
$y = 1$	$0-\mathbf{1} = x+1-\mathbf{1}$
	$-1 = x$

The y-int is (0, 1), and the x-int is $(-1, 0)$.

Select test point (0,0) and substitute into

$y \geq x+1$

$\mathbf{0} \overset{?}{\geq} \mathbf{0}+1$

$0 \geq 1$

False

The coordinates of every point on the same side of the line as the origin ***do not*** satisfy the inequality. To indicate this, we shade the half-plane that ***does not*** contain the test point (0, 0).

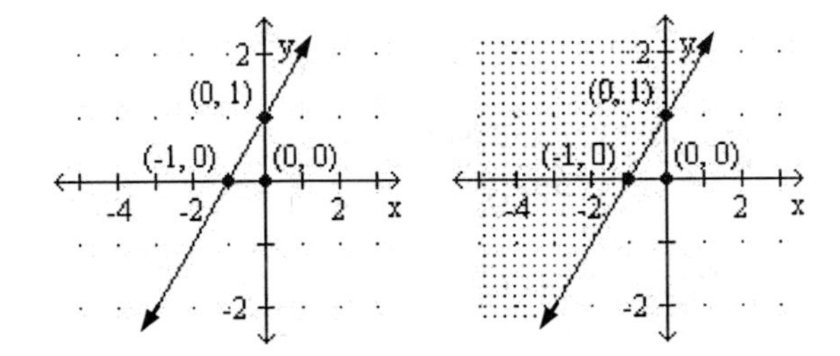

52. Graph $x < 4$ on a rectangular coordinate system. [Section 3.7]
Boundary line is dashed because of the $<$ symbol. Graph as $x = 4$.
There is no y-int, and the x-int is $(4,0)$.
$x = 4$ is a line parallel to the y-axis.
Select test point (0,0) and substitute into

$x < 4$

$\mathbf{0} < 4$

True

The coordinates of every point on the same side of the line as (0,0) satisfy the inequality. To indicate this, we shade the half-plane that contains the test point (0,0).

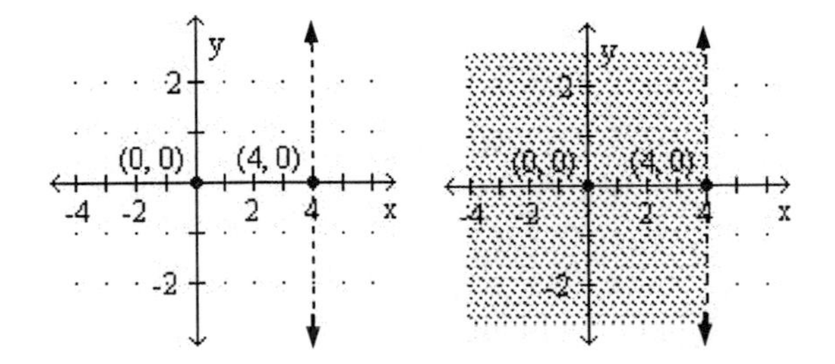

53. If $f(x) = x^4 + x$, find $f(-3)$. [Section 3.8]

$f(x) = x^4 + x$

$f(-\mathbf{3}) = (-\mathbf{3})^4 + (-\mathbf{3})$

$= 81 - 3$

$= 78$

Thus, $f(-3) = 78$

54. Is this the graph of a function? [Section 3.8] **No, because it will not pass the vertical line test since the following points** $(1,2)$ and $(1,-2)$ **have the same x - coordinate and different y - coordinates.**

SECTION 4.1 STUDY SET
VOCABULARY

Fill in the blanks.

1. The pair of equations $\begin{cases} x - y = -1 \\ 2x - y = 1 \end{cases}$ is called a **system** of equations.

3. The point of **intersection** of the lines graphed in part (a) of the following figure is (1, 2).

5. A system of equations that has at least one solution is called a **consistent** system. A system with no solution is called an **inconsistent** system.

CONCEPTS

7. Refer to the illustration.
   a. **True, because the point lies on the line.**
   b. **True, because the point lies on the line.**

9. a. To graph $5x - 2y = 10$, we can use the intercept method.

x	y	(x, y)
0	$5x - 2y = 10$ $5(0) - 2y = 10$ $-2y = 10$ $y = -5$	**(0, –5)**
$5x - 2y = 10$ $5x - 2(0) = 10$ $5x = 10$ $x = 2$	0	**(2,0)**

   b. To graph $y = 3x - 2$, we can use the slope and y-intercept.

   slope: $\mathbf{3 = \frac{3}{1}}$

   y-intercept: **(0, –2)**

11. **No solution; independent, there are different graphs**

GUIDED PRACTICE

Determine whether the ordered pair is a solution of the given system.

13. $(1,1)$, $\begin{cases} x + y = 2 \\ 2x - y = 1 \end{cases}$

$x + y = 2$	$2x - y = 1$
$1 + 1 \stackrel{?}{=} 2$	$2(1) - 1 \stackrel{?}{=} 1$
$2 = 2$ True	$2 - 1 \stackrel{?}{=} 1$
	$1 = 1$ True

Since (1, 1) satisfies both equations, it is a solution of the system.

15. $(3,-2)$, $\begin{cases} 2x + y = 4 \\ y = 1 - x \end{cases}$

$2x + y = 4$	$y = 1 - x$
$2(3) + (-2) \stackrel{?}{=} 4$	$-2 \stackrel{?}{=} 1 - 3$
$6 - 2 \stackrel{?}{=} 4$	$-2 = -2$ True
$4 = 4$ True	

Since (3, – 2) satisfies both equations, it is a solution of the system.

17. $(12,0)$, $\begin{cases} x - 9y = 12 \\ y = 10 - x \end{cases}$

$x - 9y = 12$	$y = 10 - x$
$12 - 9(0) \stackrel{?}{=} 12$	$0 \stackrel{?}{=} 10 - (12)$
$12 - 0 \stackrel{?}{=} 12$	$0 = -2$ False
$12 = 12$ True	

Since (12, 0) does not satisfy both equations, it is not a solution of the system.

19. $(-2,-4)$, $\begin{cases} 4x + 5y = -23 \\ -3x + 2y = 0 \end{cases}$

$4x + 5y = -23$	$-3x + 2y = 0$
$4(-2) + 5(-4) \stackrel{?}{=} -23$	$-3(-2) + 2(-4) \stackrel{?}{=} 0$
$-8 - 20 \stackrel{?}{=} -23$	$6 - 8 \stackrel{?}{=} 0$
$-28 = -23$	$-2 = 0$
False	False

Since (–2, –4) does not satisfy both equations, it is not a solution of the system.

21. $\left(\frac{1}{2}, 3\right)$, $\begin{cases} 2x + y = 4 \\ 4x - 11 = 3y \end{cases}$

$2x + y = 4$	$4x - 11 = 3y$
$2\left(\frac{1}{2}\right) + 3 \stackrel{?}{=} 4$	$4\left(\frac{1}{2}\right) - 11 \stackrel{?}{=} 3(3)$
$1 + 3 \stackrel{?}{=} 4$	$2 - 11 \stackrel{?}{=} 9$
$4 = 4$	$-9 = 9$
True	False

Since $(\frac{1}{2}, 3)$ does not satisfy both equations, it is not a solution of the system.

23. $(2.5, 3.5)$, $\begin{cases} 4x - 3 = 2y \\ 4y + 1 = 6x \end{cases}$

$4x - 3 = 2y$	$4y + 1 = 6x$
$4(2.5) - 3 \stackrel{?}{=} 2(3.5)$	$4(3.5) + 1 \stackrel{?}{=} 6(2.5)$
$10 - 3 \stackrel{?}{=} 7$	$14 + 1 \stackrel{?}{=} 15$
$7 = 7$	$15 = 15$
True	True

Since (2.5, 3.5) satisfies both equations, it is a solution of the system.

Solve each system by graphing. If a system has no solution or infinitely many, so state.

25. $\begin{cases} 2x + 3y = 12; \; x\text{-int } (6,0); \; y\text{-int } (0,4) \\ 2x - y = 4; \quad x\text{-int } (2,0); \; y\text{-int } (0,-4) \end{cases}$

The solution is (3,2).

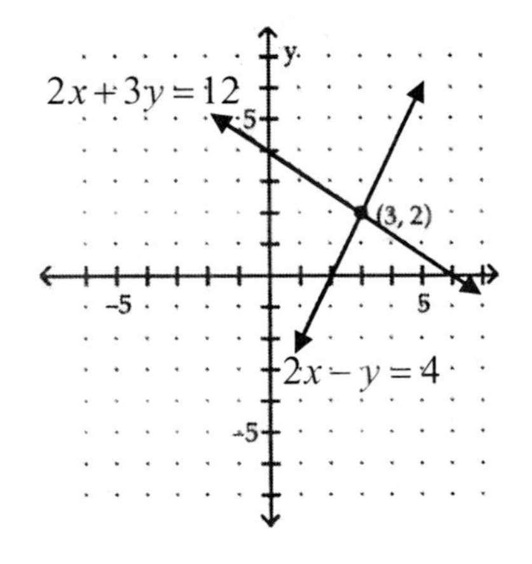

27. $\begin{cases} x + y = 4; \; x\text{-int } (4,0); \; y\text{-int } (0,4) \\ x - y = -6; \; x\text{-int } (-6,0); \; y\text{-int } (0,6) \end{cases}$

The solution is $(-1,5)$.

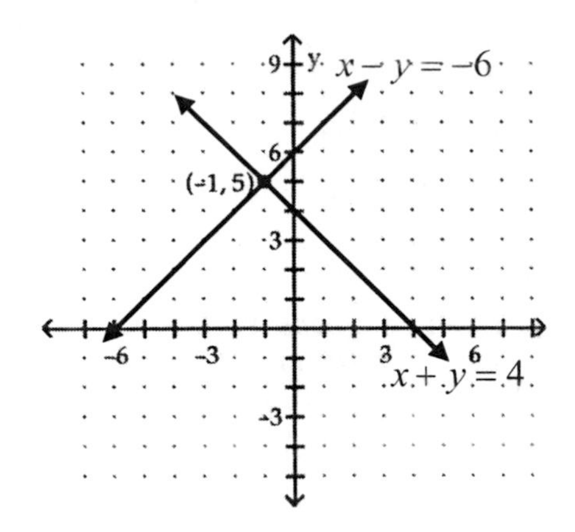

29. $\begin{cases} y = -\frac{1}{3}x - 4; \; m = -\frac{1}{3}; \; y\text{-int } (0,-4) \\ x + 3y = 6; \quad x\text{-int } (6,0); \; y\text{-int } (0,2) \end{cases}$

The lines are parallel.
The system has no solution.

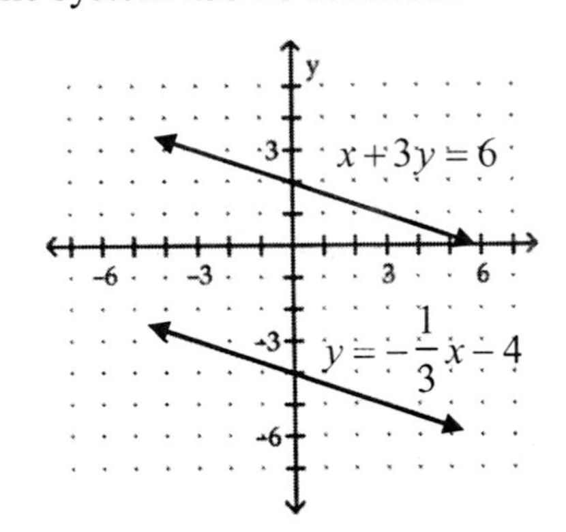

$N31.$ $\begin{cases} y = 3x; \qquad m = 3; \; y\text{-int } (0,0) \\ y - 3x = -3; \; x\text{-int } (1,0); \; y\text{-int } (0,-3) \end{cases}$

The lines are parallel.
The system has no solution.

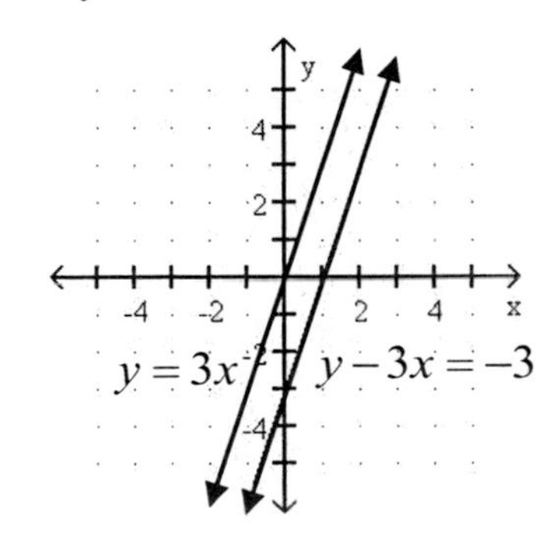

33. $\begin{cases} y = x-1; & m=1;\ y\text{-int } (0,-1) \\ 3x-3y=3; & x\text{-int } (1,0);\ y\text{-int } (0,-1) \end{cases}$

The graphs are the same line.
The system has infinitely many solutions.

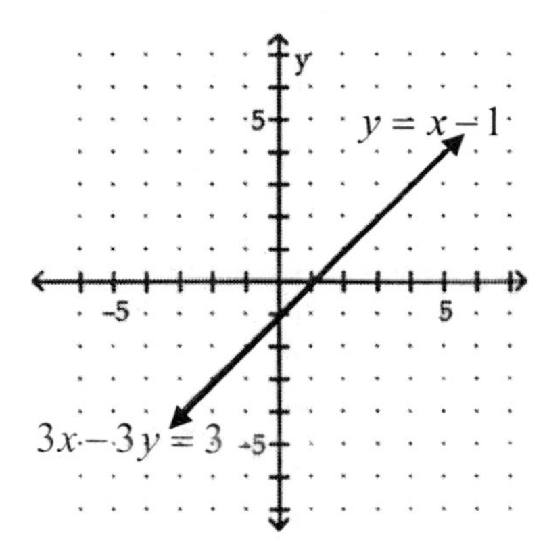

N35. $\begin{cases} 4x+6y=12; & x\text{-int } (3,0);\ y\text{-int } (0,2) \\ 2x+3y=6; & x\text{-int } (3,0);\ y\text{-int } (0,2) \end{cases}$

The graphs are the same line.
The system has infinitely many solutions.

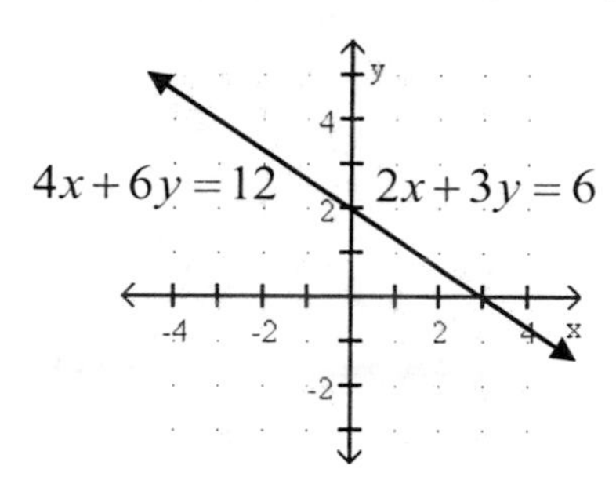

Find the slope and the y - intercept of the graph of each line in the system of equations. Then, use that information to determine the number of solutions of the system.

37. $\begin{cases} y = 6x-7 \\ y = -2x+1 \end{cases}$

$y = 6x - 7$	$y = -2x+1$
$m = 6$	$m = -2$
y-int $= (0, -7)$	y-int $= (0, 1)$

Since the line graphs have different slopes, the lines are neither parallel nor identical. Therefore, they will intersect at one point and the system will have 1 solution.

39. $\begin{cases} 3x - y = -3 \\ y - 3x = 3 \end{cases}$

$3x - y = -3$	$y - 3x = 3$
$3x - y + y = -3 + y$	$y - 3x + 3x = 3 + 3x$
$3x = -3 + y$	$y = 3x + 3$
$3x + 3 = -3 + y + 3$	$m = 3$
$y = 3x + 3$	y-int $= (0, 3)$
$m = 3$	
y-int $= (0, 3)$	

Since the line graphs are exactly the same line, the system has infinitely many solutions.

41. $\begin{cases} x + y = 6 \\ x + y = 8 \end{cases}$

$x + y = 6$	$x + y = 8$
$x + y - x = 6 - x$	$x + y - x = 8 - x$
$y = -x + 6$	$y = -x + 8$
$m = -1$	$m = -1$
y-int $= (0, 6)$	y-int $= (0, 8)$

Since the line graphs have the same slope, they are parallel and there is no solution.

43. $\begin{cases} 6x + y = 0 \\ 2x + 2y = 0 \end{cases}$

$6x + y = 0$	$2x + 2y = 0$
$6x + y - 6x = 0 - 6x$	$2x + 2y - 2x = 0 - 2x$
$y = -6x + 0$	$2y = -2x + 0$
$m = -6$	$\frac{2y}{2} = \frac{-2x}{2} + \frac{0}{2}$
y-int $= (0, 0)$	$y = -x + 0$
	$m = -1$
	y-int $= (0, 0)$

Since the line graphs have different slopes, the lines are neither parallel nor identical. Therefore, they will intersect at one point and the system will have 1 solution.

Use a graphing calculator to solve each system, if possible.

45. $\begin{cases} y = 4 - x \\ y = 2 + x \end{cases}$

The solution is $(1,3)$.

47. $\begin{cases} 6x - 2y = 5 \\ 3x = y + 10 \end{cases}$

The lines are parallel. There is no solution.

TRY IT YOURSELF

49. $\begin{cases} y = 3x + 6; & m = 3;\ y\text{-int } (0,6) \\ y = -2x - 4; & m = -2;\ y\text{-int } (0,-4) \end{cases}$

The solution is $(-2,0)$.

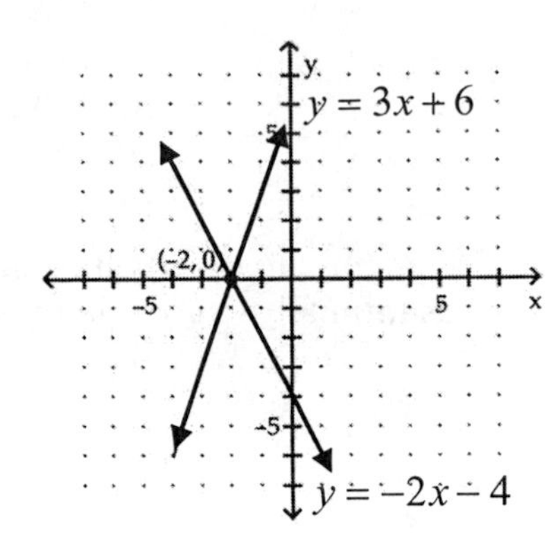

51. $\begin{cases} 2y = 3x + 2; & m = \frac{3}{2};\ y\text{-int } (0,1) \\ 3x - 2y = 6; & m = \frac{3}{2};\ y\text{-int } (0,-3) \end{cases}$

The lines are parallel.
The system has no solution.

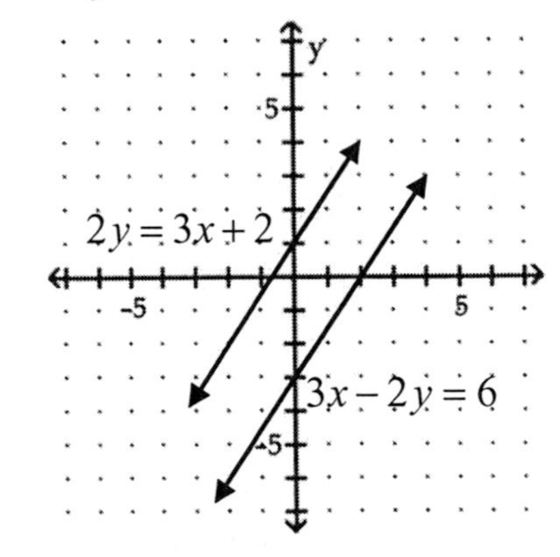

53. $\begin{cases} x + y = 2; & x\text{-int } (2,0);\ y\text{-int } (0,2) \\ y = x - 4; & m = 1;\ y\text{-int } (0,-4) \end{cases}$

The solution is $(3,-1)$.

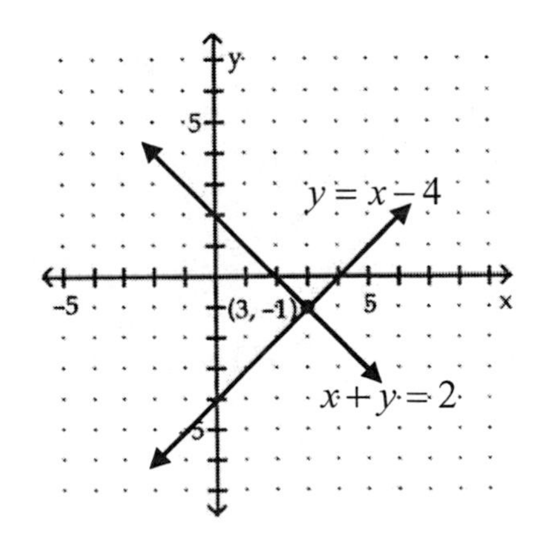

55. $\begin{cases} x = 3; & x\text{-int } (3,0);\ \text{no } y\text{-int} \\ 3y = 6 - 2x; & m = -\frac{2}{3};\ y\text{-int } (0,2) \end{cases}$

The solution is $(3,0)$.

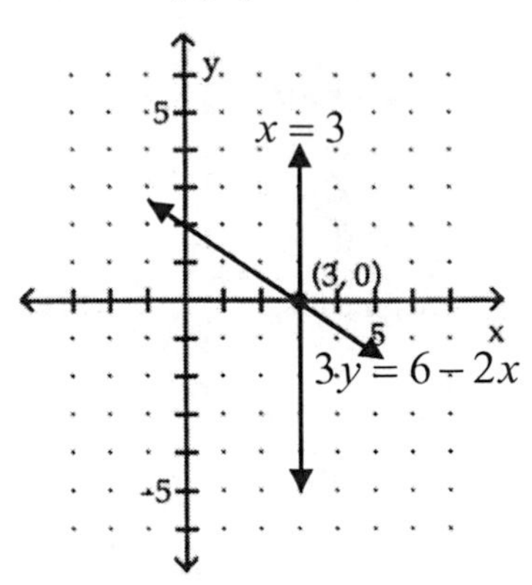

57. $\begin{cases} y = \frac{3}{4}x + 3; & m = \frac{3}{4};\ y\text{-int } (0,3) \\ y = -\frac{x}{4} - 1; & m = -\frac{1}{4};\ y\text{-int } (0,-1) \end{cases}$

The solution is $(-4,0)$.

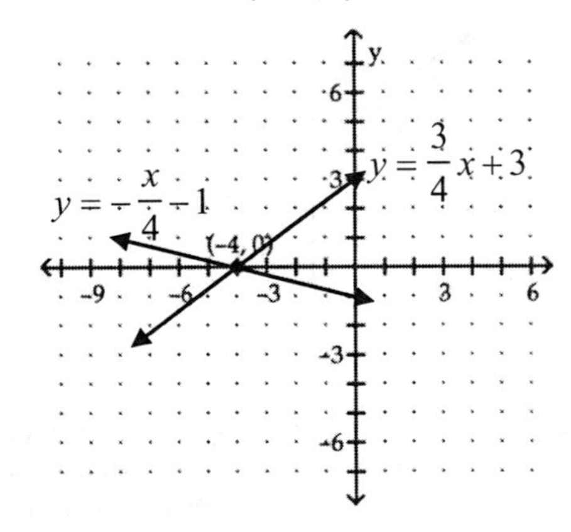

59. $\begin{cases} y = -x - 2; & m = -1;\ y\text{-int } (0,-2) \\ y = -3x + 6; & m = -3;\ y\text{-int } (0,6) \end{cases}$

The solution is $(4,-6)$.

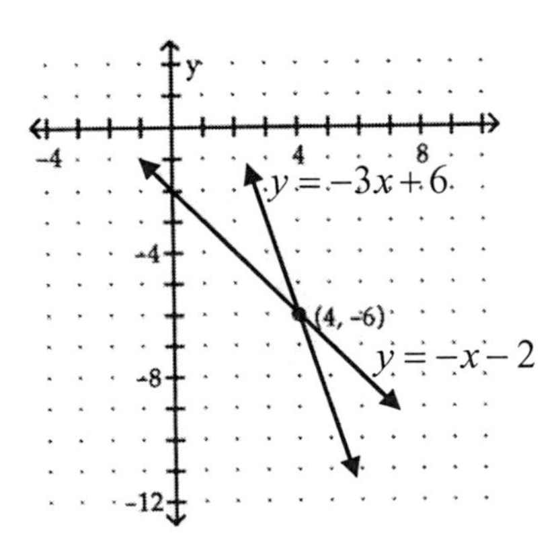

61. $\begin{cases} -x+3y=-11; \; x\text{-int } (11,0); \; y\text{-int } (0,-\frac{11}{3}) \\ 3x-y=17; \; x\text{-int } (\frac{17}{3},0); \; y\text{-int } (0,-17) \end{cases}$

The solution is $(5,-2)$.

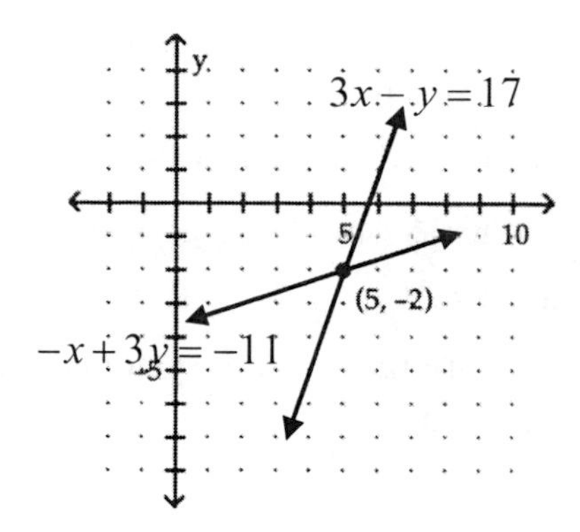

63. $\begin{cases} x+y=2; \; x\text{-int } (2,0); \; y\text{-int } (0,2) \\ y=x; \quad m=1; \; y\text{-int } (0,0) \end{cases}$

The solution is $(1,1)$.

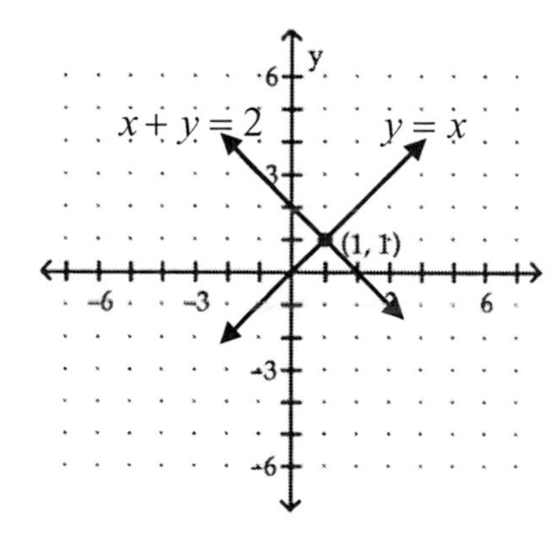

65. $\begin{cases} 4x-2y=8; \; x\text{-int } (2,0); \; y\text{-int } (0,-4) \\ y=2x-4; \quad m=2; \; y\text{-int } (0,-4) \end{cases}$

The graphs are the same line.
The system has infinitely many solutions.

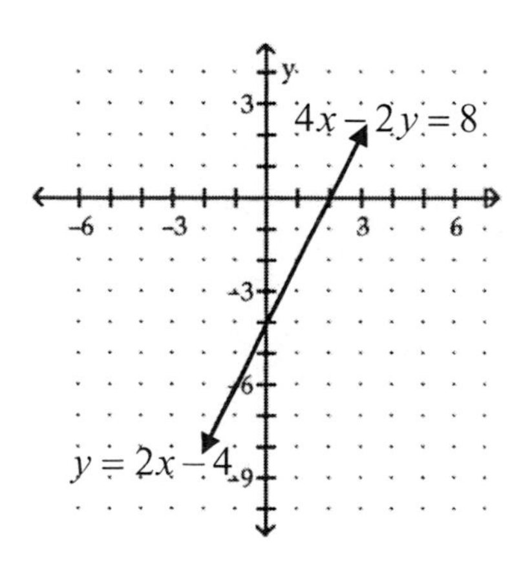

67. $\begin{cases} x+4y=-2; \; x\text{-int } (-2,0); \; y\text{-int } (0,-\frac{1}{2}) \\ y=-x-5; \quad m=-1; \; y\text{-int } (0,-5) \end{cases}$

The solution is $(-6,1)$.

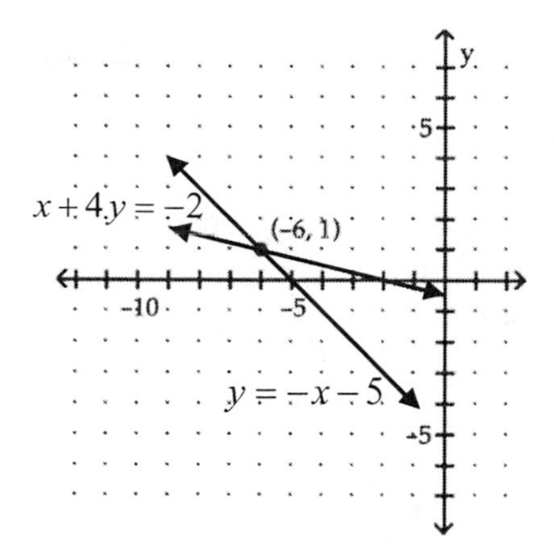

69. $\begin{cases} y=-3; \quad \text{no } x\text{-int }; \; y\text{-int } (0,-3) \\ -x+2y=-4; \; x\text{-int } (4,0); \; y\text{-int } (0,-2) \end{cases}$

The solution is $(-2,-3)$.

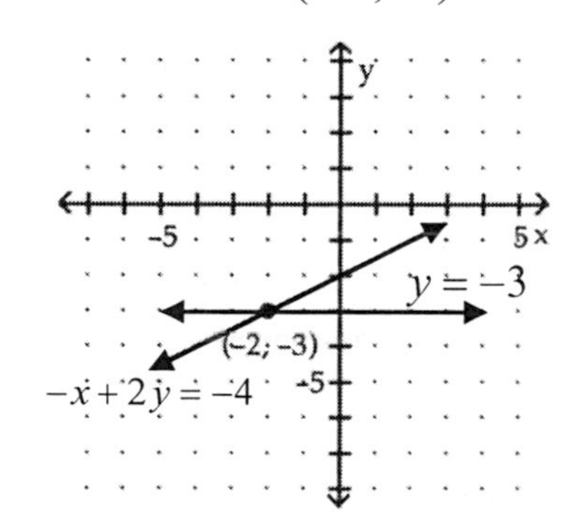

71. $\begin{cases} x+2y=-4; \; x\text{-int } (-4,0); \; y\text{-int } (0,-2) \\ x-\frac{1}{2}y=6; \; x\text{-int } (6,0); \; y\text{-int } (0,-12) \end{cases}$

The solution is $(4,-4)$.

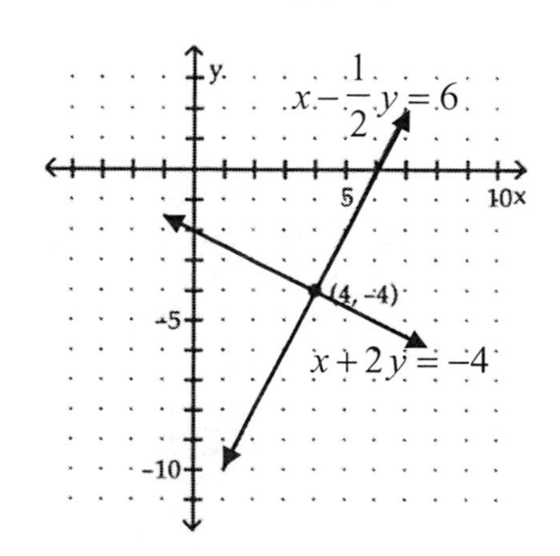

APPLICATIONS

73. SOCIAL NETWORKS
    The same number of visitors was in May '09 and it was about 70 million.

75. LATITUDE AND LONGITUDE
    a. Houston, New Orleans, St. Augustine
    b. St. Louis, Memphis, New Orleans
    c. New Orleans

77. DAILY TRACKING POLL
    a. The incumbent; 7%
    b. November 2
    c. The challenger; 3

FROM CAMPUS TO CAREERS

79. PHOTOGRAPHERS
    $(6,4),(6,8),(12,4),(12,8)$; yes

WRITING

81-86. Answers will vary.

REVIEW

Solve each inequality. Write the solution set in interval notation and graph it.

87.
$$\begin{aligned}-4(2y+2) &\le 4y+28\\ -8y-8-\mathbf{4y} &\le 4y+28-\mathbf{4y}\\ -12y-8 &\le 28\\ -12y-8+\mathbf{8} &\le 28+\mathbf{8}\\ -12y &\le 36\\ \frac{-12y}{\mathbf{-12}} &\ge \frac{36}{\mathbf{-12}}\\ y &\ge -3\end{aligned}$$
$[-3,\infty)$

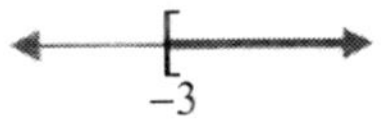

89.
$$\begin{aligned}-1 &\le -\frac{1}{2}n\\ -\frac{1}{2}n &\ge -1\\ -\frac{\mathbf{2}}{\mathbf{1}}\cdot\left(-\frac{1}{2}n\right) &\le -\frac{\mathbf{2}}{\mathbf{1}}\cdot(-1)\\ n &\le 2\end{aligned}$$
$(-\infty,2]$

CHALLENGE PROBLEMS

91. No. Suppose the equations are dependent, then they have the same graph, and would intersect at an infinite number of points. Therefore, the system would have at least one solution, and be consistent.

93. Answers will vary.
    Given (2, 3) is a solution.
$$\begin{cases}x+y=5\\ x-y=10\end{cases}$$
    There are many more systems than these two.

SECTION 4.2 STUDY SET
VOCABULARY

Fill in the blanks.

1. To solve the system $\begin{cases} x = y+1 \\ 3x+2y=8 \end{cases}$ using the method discussed in this section, we begin by **substituting** $y + 1$ for x in the second equation.

CONCEPTS

3. If the substitution method is used to solve $\begin{cases} 5x+y=2 \\ y=-3x \end{cases}$ which equation should be used as the substitution equation? $\mathbf{y = -3x}$

5. Suppose $x - 4$ is substituted for y in the equation $x + 3y = 8$. Insert parentheses in $x + 3x - 4 = 8$ to show the substitution.

   $\mathbf{x + 3(x - 4) = 8}$

7. A student uses the substitution method to solve the system $\begin{cases} 4a+5b=2 \\ b=3a-11 \end{cases}$ and finds that $a = 3$. What is the easiest way for her to determine the value of b?

   Substitute 3 for *a* in the second equation.

9. Suppose $-2 = 1$ is obtained when a system is solved by the substitution method.
   a. Does the system have a solution? **No**
   b. Which of the following is a possible graph of the system? **ii, parallel lines have no solution point**

NOTATION

Complete the solution of the system.

11. Solve: $\begin{cases} y=3x \\ x-y=4 \end{cases}$

$x - y = 4$ This is the second equation. Substitute for y.

$x - (\mathbf{3x}) = 4$
$-2x = \mathbf{4}$
$x = \mathbf{-2}$

$y = 3x$ This is the first equation.
$y = 3(\mathbf{-2})$
$y = \mathbf{-6}$

The solution is $(\mathbf{-2}, \mathbf{-6})$.

GUIDED PRACTICE

Solve each system by substituion. See Example 1.

13. $\begin{cases} y=2x \\ x+y=6 \end{cases}$

$x + y = 6$ The 2nd equation. Substitute for y.

$x + (\mathbf{2x}) = 6$
$3x = 6$
$\frac{3x}{\mathbf{3}} = \frac{6}{\mathbf{3}}$
$x = 2$

$y = 2x$ The 1st equation
$y = 2(\mathbf{2})$
$y = 4$

The solution is (2,4).

Check:	Check:
$y = 2x$	$x + y = 6$
$4 \stackrel{?}{=} 2(2)$	$2 + 4 \stackrel{?}{=} 6$
$4 = 4$	$6 = 6$
True	True

15. $\begin{cases} y=2x-6 \\ 2x+y=6 \end{cases}$

$2x + y = 6$ The 2nd equation. Substitute for y.

$2x + (\mathbf{2x - 6}) = 6$
$4x - 6 = 6$
$4x - 6 + \mathbf{6} = 6 + \mathbf{6}$
$4x = 12$
$\frac{4x}{\mathbf{4}} = \frac{12}{\mathbf{4}}$
$x = 3$

$y = 2x - 6$ The 1st equation
$y = 2(\mathbf{3}) - 6$
$y = 6 - 6$
$y = 0$

The solution is (3, 0).

Check:	Check:
$y = 2x - 6$	$2x + y = 6$
$0 \stackrel{?}{=} 2(3) - 6$	$2(3) + 0 \stackrel{?}{=} 6$
$0 = 0$	$6 = 6$
True	True

Solve each system by substituion.
See Example 2.

17. $\begin{cases} x+3y=-4 \\ x=-5y \end{cases}$

$x+3y=-4$ The 1st equation Substitute for x.

$$\begin{aligned} -\mathbf{5y}+3y&=-4 \\ -2y&=-4 \\ \frac{-2y}{\mathbf{-2}}&=\frac{-4}{\mathbf{-2}} \\ y&=2 \end{aligned}$$

$x=-5y$ The 2nd equation

$x=-5(\mathbf{2})$

$x=-10$

The solution is $(-10,2)$.

Check:	Check:
$x+3y=-4$	$x=-5y$
$-10+3(2)\stackrel{?}{=}-4$	$-10\stackrel{?}{=}-5(2)$
$-4=-4$	$-10=-10$
True	True

19. $\begin{cases} y=-5x \\ x+3y=-28 \end{cases}$

$x+3y=-28$ The 2nd equation Substitute for y.

$$\begin{aligned} x+3(\mathbf{-5x})&=-28 \\ x-15x&=-28 \\ -14x&=-28 \\ \frac{-14x}{\mathbf{-14}}&=\frac{-28}{\mathbf{-14}} \\ x&=2 \end{aligned}$$

$y=-5x$ The 1st equation

$y=-5(\mathbf{2})$

$y=-10$

The solution is $(2,-10)$.

Check:	Check:
$y=-5x$	$x+3y=-28$
$-10\stackrel{?}{=}-5(2)$	$2+3(-10)\stackrel{?}{=}-28$
$-10=-10$	$2-30\stackrel{?}{=}-28$
True	$-28=-28$
	True

Solve each system by substituion.
See Example 3.

21. $\begin{cases} r+3s=9 \\ 3r+2s=13 \end{cases}$

Solve for r in the 1st equation.

$\begin{cases} r=9-3s \\ 3r+2s=13 \end{cases}$

$3r+2s=13$ The 2nd equation Substitute for x.

$$\begin{aligned} 3(\mathbf{9-3s})+2s&=13 \\ 27-9s+2s&=13 \\ 27-7s&=13 \\ 27-7s-\mathbf{27}&=13-\mathbf{27} \\ -7s&=-14 \\ \frac{-7s}{\mathbf{-7}}&=\frac{-14}{\mathbf{-7}} \\ s&=2 \end{aligned}$$

$r=9-3s$ The 1st equation

$r=9-3(\mathbf{2})$

$r=9-6$

$r=3$

The solution is (3, 2).

Check:	Check:
$r+3s=9$	$3r+2s=13$
$3+3(2)\stackrel{?}{=}9$	$3(3)+2(2)\stackrel{?}{=}13$
$3+6\stackrel{?}{=}9$	$9+4\stackrel{?}{=}13$
$9=9$	$13=13$
True	True

23. $\begin{cases} 4x+y=-15 \\ 2x+3y=5 \end{cases}$

Solve for y in the 1st equation.

$\begin{cases} y=-4x-15 \\ 2x+3y=5 \end{cases}$

$2x+3y=5$ The 2nd equation Substitute for y.

$$\begin{aligned} 2x+3(\mathbf{-4x-15})&=5 \\ 2x-12x-45&=5 \\ -10x-45&=5 \\ -10x-45+\mathbf{45}&=5+\mathbf{45} \\ -10x&=50 \\ \frac{-10x}{\mathbf{-10}}&=\frac{50}{\mathbf{-10}} \\ x&=-5 \end{aligned}$$

$y=-4x-15$ The 1st equation

$y=-4(\mathbf{-5})-15$

$y=20-15$

$y=5$

The solution is $(-5, 5)$.

Check:	Check:
$4x+y=-15$	$2x+3y=5$
$4(-5)+5\stackrel{?}{=}-15$	$2(-5)+3(5)\stackrel{?}{=}5$
$-20+5\stackrel{?}{=}-15$	$-10+15\stackrel{?}{=}5$
$-15=-15$	$5=5$
True	True

Solve each system by substituion.
See Example 4.

25. $\begin{cases} 8x - 6y = 4 \\ 2x - y = -2 \end{cases}$

Solve for y in the 2nd equation.

$\begin{cases} 8x - 6y = 4 \\ y = 2x + 2 \end{cases}$

$8x - 6y = 4$ The 1st equation Substitute for y.

$$8x - 6(\mathbf{2x+2}) = 4$$
$$8x - 12x - 12 = 4$$
$$-4x - 12 = 4$$
$$-4x - 12 + \mathbf{12} = 4 + \mathbf{12}$$
$$-4x = 16$$
$$\frac{-4x}{\mathbf{-4}} = \frac{16}{\mathbf{-4}}$$
$$x = -4$$

$y = 2x + 2$ The 2nd equation
$$y = 2(\mathbf{-4}) + 2$$
$$y = -8 + 2$$
$$y = -6$$

The solution is $(-4, -6)$.

Check:
$$8x - 6y = 4$$
$$8(\mathbf{-4}) - 6(\mathbf{-6}) \overset{?}{=} 4$$
$$-32 + 36 \overset{?}{=} 4$$
$$4 = 4$$
True

Check:
$$2x - y = -2$$
$$2(-4) - (-6) \overset{?}{=} -2$$
$$-8 + 6 \overset{?}{=} -2$$
$$-2 = -2$$
True

27. $\begin{cases} 4x + 5y = 2 \\ 3x - y = 11 \end{cases}$

Solve for y in the 2nd equation.

$\begin{cases} 4x + 5y = 2 \\ y = 3x - 11 \end{cases}$

$4x + 5y = 2$ The 1st equation Substitute for y.

$$4x + 5(\mathbf{3x - 11}) = 2$$
$$4x + 15x - 55 = 2$$
$$19x - 55 = 2$$
$$19x - 55 + \mathbf{55} = 2 + \mathbf{55}$$
$$19x = 57$$
$$\frac{19x}{\mathbf{19}} = \frac{57}{\mathbf{19}}$$
$$x = 3$$

$y = 3x - 11$ The 2nd equation
$$y = 3(\mathbf{3}) - 11$$
$$y = 9 - 11$$
$$y = -2$$

The solution is $(3, -2)$.

Check:
$$4x + 5y = 2$$
$$4(3) + 5(-2) \overset{?}{=} 2$$
$$12 - 10 \overset{?}{=} 2$$
$$2 = 2$$
True

Check:
$$3x - y = 11$$
$$3(3) - (-2) \overset{?}{=} 11$$
$$9 + 2 \overset{?}{=} 11$$
$$11 = 11$$
True

Solve each system by substituion.
See Example 5.

29. $\begin{cases} \frac{x}{4} + \frac{y}{4} = -\frac{1}{2} \\ x - 2y = y - 24 - x \end{cases}$

Clear 1st equation of fractions.
Write 2nd equation in standard form.

$\begin{cases} \mathbf{4}\left(\frac{x}{4}\right) + \mathbf{4}\left(\frac{y}{4}\right) = \mathbf{4}\left(-\frac{1}{2}\right) \\ x - 2y + \mathbf{x} - \mathbf{y} = y - 24 - x + \mathbf{x} - \mathbf{y} \end{cases}$

$\begin{cases} x + y = -2 \\ 2x - 3y = -24 \end{cases}$

Solve for x in the 1st equation.

$\begin{cases} x = -y - 2 \\ 2x - 3y = -24 \end{cases}$

$2x - 3y = -24$ The 2nd equation Substitute for x.

$$2(\mathbf{-y - 2}) - 3y = -24$$
$$-2y - 4 - 3y = -24$$
$$-5y - 4 = -24$$
$$-5y - 4 + \mathbf{4} = -24 + \mathbf{4}$$
$$-5y = -20$$
$$\frac{-5y}{\mathbf{-5}} = \frac{-20}{\mathbf{-5}}$$
$$y = 4$$

$x = -y - 2$ The 1st equation
$$x = -(\mathbf{4}) - 2$$
$$x = -6$$

The solution is $(-6, 4)$.

Check:
$$\frac{x}{4} + \frac{y}{4} = -\frac{1}{2}$$
$$\frac{-6}{4} + \frac{4}{4} \overset{?}{=} -\frac{1}{2}$$
$$\frac{-2}{4} \overset{?}{=} -\frac{1}{2}$$
$$-\frac{1}{2} = -\frac{1}{2}$$
True

Check:
$$x - 2y = y - 24 - x$$
$$-6 - 2(4) \overset{?}{=} 4 - 24 - (-6)$$
$$-6 - 8 \overset{?}{=} 4 - 24 + 6$$
$$-14 = -14$$
True

31. $\begin{cases} x-5y=20-4x-y \\ \dfrac{y}{3}=\dfrac{x}{2}-\dfrac{5}{2} \end{cases}$

Write 1st equation in standard form.
Clear 2nd equation of fractions.

$\begin{cases} x-5y+\mathbf{4x}+\mathbf{y}=20-4x-y+\mathbf{4x}+\mathbf{y} \\ \mathbf{6}\left(\dfrac{y}{3}\right)=\mathbf{6}\left(\dfrac{x}{2}\right)-\mathbf{6}\left(\dfrac{5}{2}\right) \end{cases}$

$\begin{cases} 5x-4y=20 \\ 2y=3x-15 \end{cases}$

Solve for y in the 2nd equation.

$\begin{cases} 5x-4y=20 \\ y=\dfrac{3x}{2}-\dfrac{15}{2} \end{cases}$

$5x-4y=20$ The 1st equation Substitute for y.

$5x-4\left(\dfrac{3x}{2}-\dfrac{15}{2}\right)=20$

$5x-6x+30=20$

$-x+30=20$

$-x+30-\mathbf{30}=20-\mathbf{30}$

$-x=-10$

$\dfrac{-x}{\mathbf{-1}}=\dfrac{-10}{\mathbf{-1}}$

$x=10$

$2y=3x-15$ The 2nd equation

$2y=3(\mathbf{10})-15$

$2y=30-15$

$2y=15$

$\dfrac{2y}{\mathbf{2}}=\dfrac{15}{\mathbf{2}}$

$y=\dfrac{15}{2}$

The solution is $\left(10,\dfrac{15}{2}\right)$.

Solve each system by substituion. See Example 6.

33. $\begin{cases} 2a+4b=-24 \\ a=20-2b \end{cases}$

$2a+4b=-24$ The 1st equation Substitute for a.

$2(\mathbf{20}-\mathbf{2b})+4b=-24$

$40-4b+4b=-24$ The variables drop out.

$40=-24$

False

No Solution.

35. $\begin{cases} 6-y=4x \\ 2y=-8x-20 \end{cases}$

Solve for y in the 1st equation.

$\begin{cases} 6-4x=y \\ 2y=-8x-20 \end{cases}$

$2y=-8x-20$ The 2nd equation Substitute for y.

$2(6-4x)=-8x-20$

$12-8x=-8x-20$

$12-8x+\mathbf{8x}=-8x-20+\mathbf{8x}$ The variables drop out.

$12=-20$

False

No Solution.

Solve each system by substituion. See Example 7.

37. $\begin{cases} y-3x=-5 \\ 21x=7y+35 \end{cases}$

Solve for y in the 1st equation.

$\begin{cases} y=3x-5 \\ 21x=7y+35 \end{cases}$

$21x=7y+35$ The 2nd equation Substitute for y.

$21x=7(\mathbf{3x}-\mathbf{5})+35$

$21x=21x-35+35$

$21x=21x$

$21x-\mathbf{21x}=21x-\mathbf{21x}$ The variables drop out.

$0=0$

True

The system has infinitely many solutions.

39. $\begin{cases} x=-3y+6 \\ 2x+4y=6+x+y \end{cases}$

Write the 2nd equation in standard form.

$\begin{cases} x=-3y+6 \\ 2x+4y-\mathbf{x}-\mathbf{y}=6+x+y-\mathbf{x}-\mathbf{y} \end{cases}$

$\begin{cases} x=-3y+6 \\ x+3y=6 \end{cases}$

$x+3y=6$ The 2nd equation Substitute for x.

$(\mathbf{-3y}+\mathbf{6})+3y=6$ The variables drop out.

$6=6$

True

The system has infinitely many solutions.

TRY IT YOURSELF

Solve each system by substitution. If a system has no solution or infinitely many solutions, so state.

41. $\begin{cases} -y = 11 - 3x \\ 2x + 5y = -4 \end{cases}$

Solve for y in the 1st equation.

$\begin{cases} y = 3x - 11 \\ 2x + 5y = -4 \end{cases}$

$2x + 5y = -4$ The 2nd equation Substitute for y.

$$\begin{aligned} 2x + 5(\mathbf{3x - 11}) &= -4 \\ 2x + 15x - 55 &= -4 \\ 17x - 55 &= -4 \\ 17x - 55 + \mathbf{55} &= -4 + \mathbf{55} \\ 17x &= 51 \\ \frac{17x}{\mathbf{17}} &= \frac{51}{\mathbf{17}} \\ x &= 3 \end{aligned}$$

$y = 3x - 11$ The 1st equation

$y = 3(\mathbf{3}) - 11$

$y = 9 - 11$

$y = -2$

The solution is $(3, -2)$.

Check:

$-y = 11 - 3x$

$-(-2) \stackrel{?}{=} 11 - 3(3)$

$2 \stackrel{?}{=} 11 - 9$

$2 = 2$

True

Check:

$2x + 5y = -4$

$2(3) + 5(-2) \stackrel{?}{=} -4$

$6 - 10 \stackrel{?}{=} -4$

$-4 = -4$

True

43. $\begin{cases} \dfrac{x}{2} + \dfrac{y}{6} = \dfrac{2}{3} \\ \dfrac{x}{3} - \dfrac{y}{4} = \dfrac{1}{12} \end{cases}$

Clear both equations of fractions.

$\begin{cases} 6\left(\dfrac{x}{2}\right) + 6\left(\dfrac{y}{6}\right) = 6\left(\dfrac{2}{3}\right) \\ 12\left(\dfrac{x}{3}\right) - 12\left(\dfrac{y}{4}\right) = 12\left(\dfrac{1}{12}\right) \end{cases}$

$\begin{cases} 3x + y = 4 \\ 4x - 3y = 1 \end{cases}$

Solve for y in the 1st equation.

$\begin{cases} y = 4 - 3x \\ 4x - 3y = 1 \end{cases}$

$4x - 3y = 1$ The 2nd equation Substitute for y.

$$\begin{aligned} 4x - 3(\mathbf{4 - 3x}) &= 1 \\ 4x - 12 + 9x &= 1 \\ 13x - 12 &= 1 \\ 13x - 12 + \mathbf{12} &= 1 + \mathbf{12} \\ 13x &= 13 \\ \frac{13x}{\mathbf{13}} &= \frac{13}{\mathbf{13}} \\ x &= 1 \end{aligned}$$

$y = 4 - 3x$ The 1st equation

$y = 4 - 3(\mathbf{1})$

$y = 4 - 3$

$y = 1$

The solution is $(1,1)$.

Check:

$\dfrac{x}{2} + \dfrac{y}{6} = \dfrac{2}{3}$

$\dfrac{1}{2} + \dfrac{1}{6} \stackrel{?}{=} \dfrac{2}{3}$

$\dfrac{3}{6} + \dfrac{1}{6} \stackrel{?}{=} \dfrac{2}{3}$

$\dfrac{4}{6} \stackrel{?}{=} \dfrac{2}{3}$

$\dfrac{2}{3} = \dfrac{2}{3}$

True

Check:

$\dfrac{x}{3} - \dfrac{y}{4} = \dfrac{1}{12}$

$\dfrac{1}{3} - \dfrac{1}{4} \stackrel{?}{=} \dfrac{1}{12}$

$\dfrac{4}{12} - \dfrac{3}{12} \stackrel{?}{=} \dfrac{1}{12}$

$\dfrac{1}{12} = \dfrac{1}{12}$

True

45. $\begin{cases} y + x = 2x + 2 \\ 6x - 4y = 21 - y \end{cases}$

Write both equations in standard form.

$\begin{cases} y + x - \mathbf{2x} = 2x + 2 - \mathbf{2x} \\ 6x - 4y + \mathbf{y} = 21 - y + \mathbf{y} \end{cases}$

$\begin{cases} -x + y = 2 \\ 6x - 3y = 21 \end{cases}$

Solve for y in the 1st equation.

$\begin{cases} y = 2 + x \\ 6x - 3y = 21 \end{cases}$

$6x - 3y = 21$ The 2nd equation Substitute for y.

$$\begin{aligned} 6x - 3(\mathbf{2 + x}) &= 21 \\ 6x - 6 - 3x &= 21 \\ 3x - 6 &= 21 \\ 3x - 6 + \mathbf{6} &= 21 + \mathbf{6} \\ 3x &= 27 \\ \frac{3x}{\mathbf{3}} &= \frac{27}{\mathbf{3}} \\ x &= 9 \end{aligned}$$

$y = 2 + x$ The 1st equation

$y = 2 + \mathbf{9}$

$y = 11$

The solution is (9, 11).

Check:	Check:
$y + x = 2x + 2$	$6x - 4y = 21 - y$
$11 + 9 \stackrel{?}{=} 2(9) + 2$	$6(9) - 4(11) \stackrel{?}{=} 21 - 11$
$20 \stackrel{?}{=} 18 + 2$	$54 - 44 \stackrel{?}{=} 10$
$20 = 20$	$10 = 10$
True	True

47. $\begin{cases} y - 4 = 2x \\ y = 2x + 2 \end{cases}$

$y - 4 = 2x$ The 1st equation Substitute for y.

$(\mathbf{2x + 2}) - 4 = 2x$

$2x - 2 = 2x$

$2x - 2 - \mathbf{2x} = 2x - \mathbf{2x}$ The variables drop out.

$-2 = 0$

False

No Solution.

49. $\begin{cases} a + b = 1 \\ a - 2b = -1 \end{cases}$

Solve for a in the 1st equation.

$\begin{cases} a = 1 - b \\ a - 2b = -1 \end{cases}$

$a - 2b = -1$ The 2nd equation Substitute for a.

$(\mathbf{1 - b}) - 2b = -1$

$1 - 3b = -1$

$1 - 3b - \mathbf{1} = -1 - \mathbf{1}$

$-3b = -2$

$\dfrac{-3b}{\mathbf{-3}} = \dfrac{-2}{\mathbf{-3}}$

$b = \dfrac{2}{3}$

$a = 1 - b$ The 1st equation

$a = 1 - \mathbf{\dfrac{2}{3}}$

$a = \dfrac{3}{3} - \dfrac{2}{3}$

$a = \dfrac{1}{3}$

The solution is $\left(\dfrac{1}{3}, \dfrac{2}{3}\right)$.

Check:	Check:
$a + b = 1$	$a - 2b = -1$
$\dfrac{1}{3} + \dfrac{2}{3} \stackrel{?}{=} 1$	$\dfrac{1}{3} - 2\left(\dfrac{2}{3}\right) \stackrel{?}{=} -1$
$\dfrac{3}{3} \stackrel{?}{=} 1$	$\dfrac{1}{3} - \dfrac{4}{3} \stackrel{?}{=} -1$
$1 = 1$	$-\dfrac{3}{3} \stackrel{?}{=} -1$
True	$-1 = -1$
	True

51. $\begin{cases} 5x = \dfrac{1}{2}y - 1 \\ \dfrac{1}{4}y = 10x - 1 \end{cases}$

Clear both equations of fractions.

$\begin{cases} \mathbf{2}(5x) = \mathbf{2}\left(\dfrac{1}{2}y\right) - \mathbf{2}(1) \\ \mathbf{4}\left(\dfrac{1}{4}y\right) = \mathbf{4}(10x) - \mathbf{4}(1) \end{cases}$

$\begin{cases} 10x = y - 2 \\ y = 40x - 4 \end{cases}$

$10x = y - 2$ The 1st equation Substitute for y.

$10x = (\mathbf{40x - 4}) - 2$

$10x = 40x - 6$

$10x - \mathbf{40x} = 40x - 6 - \mathbf{40x}$

$-30x = -6$

$\dfrac{-30x}{\mathbf{-30}} = \dfrac{-6}{\mathbf{-30}}$

$x = \dfrac{1}{5}$

$y = 40x - 4$ The 2nd equation

$y = 40\left(\mathbf{\dfrac{1}{5}}\right) - 4$

$y = 8 - 4$

$y = 4$

The solution is $\left(\dfrac{1}{5}, 4\right)$

Check:	Check:
$5x = \dfrac{1}{2}y - 1$	$\dfrac{1}{4}y = 10x - 1$
$5\left(\dfrac{1}{5}\right) \stackrel{?}{=} \dfrac{1}{2}(4) - 1$	$\dfrac{1}{4}(4) \stackrel{?}{=} 10\left(\dfrac{1}{5}\right) - 1$
$1 \stackrel{?}{=} 2 - 1$	$1 \stackrel{?}{=} 2 - 1$
$1 = 1$	$1 = 1$
True	True

53. $\begin{cases} x+2y=-6 \\ x=y \end{cases}$

$x+2y=-6$ The 1st equation Substitute for x.

$$y+2y=-6$$
$$3y=-6$$
$$\frac{3y}{\mathbf{3}}=\frac{-6}{\mathbf{3}}$$
$$y=-2$$

$x=y$ The 2nd equation

$x=\mathbf{-2}$

The solution is $(-2,-2)$.

Check:
$$x+2y=-6$$
$$-2+2(-2)\overset{?}{=}-6$$
$$-2+(-4)\overset{?}{=}-6$$
$$-6=-6$$
True

Check:
$$x=y$$
$$-2=-2$$
True

55. $\begin{cases} b=\frac{2}{3}a \\ 8a-3b=3 \end{cases}$

$8a-3b=3$ The 2nd equation Substitute for b.

$$8a-3\left(\frac{\mathbf{2}}{\mathbf{3}}\mathbf{a}\right)=3$$
$$8a-2a=3$$
$$6a=3$$
$$\frac{6a}{\mathbf{6}}=\frac{3}{\mathbf{6}}$$
$$a=\frac{1}{2}$$

$b=\frac{2}{3}a$ The 1st equation

$$b=\frac{2}{3}\left(\frac{\mathbf{1}}{\mathbf{2}}\right)$$
$$b=\frac{1}{3}$$

The solution is $\left(\frac{1}{2},\frac{1}{3}\right)$.

Check:
$$b=\frac{2}{3}a$$
$$\frac{1}{3}\overset{?}{=}\frac{2}{3}\left(\frac{1}{2}\right)$$
$$\frac{1}{3}=\frac{1}{3}$$
True

Check:
$$8a-3b=3$$
$$8\left(\frac{1}{2}\right)-3\left(\frac{1}{3}\right)\overset{?}{=}3$$
$$4-1\overset{?}{=}3$$
$$3=3$$
True

57. $\begin{cases} 6x-3y=5 \\ x+2y=0 \end{cases}$

Solve for x in the 2nd equation.

$\begin{cases} 6x-3y=5 \\ x=-2y \end{cases}$

$6x-3y=5$ The 1st equation Substitute for x.

$$6(\mathbf{-2y})-3y=5$$
$$-12y-3y=5$$
$$-15y=5$$
$$\frac{-15y}{\mathbf{-15}}=\frac{5}{\mathbf{-15}}$$
$$y=-\frac{1}{3}$$

$x=-2y$ The 2nd equation

$$x=-2\left(-\frac{\mathbf{1}}{\mathbf{3}}\right)$$
$$x=\frac{2}{3}$$

The solution is $\left(\frac{2}{3},-\frac{1}{3}\right)$.

Check:
$$6x-3y=5$$
$$6\left(\frac{2}{3}\right)-3\left(-\frac{1}{3}\right)\overset{?}{=}5$$
$$4+1\overset{?}{=}5$$
$$5=5$$
True

Check:
$$x+2y=0$$
$$\frac{2}{3}+2\left(-\frac{1}{3}\right)\overset{?}{=}0$$
$$\frac{2}{3}-\frac{2}{3}\overset{?}{=}0$$
$$0=0$$
True

59. $\begin{cases} 2x+3=-4y \\ x-6=-8y \end{cases}$

Solve for x in 2nd equation.

$\begin{cases} 2x+3=-4y \\ x=6-8y \end{cases}$

$2x+3=-4y$ The 1st equation Substitute for x.

$$2(\mathbf{6-8y})+3=-4y$$
$$12-16y+3=-4y$$
$$-16y+15=-4y$$
$$-16y+15+\mathbf{16y}=-4y+\mathbf{16y}$$
$$15=12y$$
$$\frac{15}{\mathbf{12}}=\frac{12y}{\mathbf{12}}$$
$$\frac{5}{4}=y$$

$x=6-8y$ The 2nd equation

$$x=6-8\left(\frac{\mathbf{5}}{\mathbf{4}}\right)$$
$$x=6-10$$
$$x=-4$$

The solution is $\left(-4,\frac{5}{4}\right)$.

Check:
$$2x+3=-4y$$
$$2(-4)+3\overset{?}{=}-4\left(\frac{5}{4}\right)$$
$$-8+3\overset{?}{=}-5$$
$$-5=-5$$
True

Check:
$$x-6=-8y$$
$$-4-6\overset{?}{=}-8\left(\frac{5}{4}\right)$$
$$-10=-10$$
True

61. $\begin{cases} 2x+5y=-2 \\ y=-\dfrac{x}{2} \end{cases}$

Clear the fraction from the 2nd equation.

$$\begin{cases} 2x+5y=-2 \\ \mathbf{-2}(y)=\mathbf{-2}\left(-\dfrac{x}{2}\right) \end{cases}$$

$$\begin{cases} 2x+5y=-2 \\ -2y=x \end{cases}$$

$$2x+5y=-2 \quad \text{The 1}^{\text{st}}\text{ equation Substitute for } x.$$
$$2(\mathbf{-2y})+5y=-2$$
$$-4y+5y=-2$$
$$y=-2$$

$$y=-\frac{x}{2} \quad \text{The 2}^{\text{nd}}\text{ equation}$$
$$\mathbf{-2}=-\frac{x}{2}$$
$$-2(-2)=-2\left(-\frac{x}{2}\right)$$
$$4=x$$

The solution is $(4,-2)$.

Check:
$$2x+5y=-2$$
$$2(4)+5(-2)\overset{?}{=}-2$$
$$8-10\overset{?}{=}-2$$
$$-2=-2$$
True

Check:
$$y=-\frac{x}{2}$$
$$-2\overset{?}{=}-\left(\frac{4}{2}\right)$$
$$-2=-2$$
True

63. $\begin{cases} 3x+4y=-19 \\ 2y-x=3 \end{cases}$

Solve for x in the 2nd equation.

$$\begin{cases} 3x+4y=-19 \\ x=2y-3 \end{cases}$$

$$3x+4y=-19 \quad \text{The 1}^{\text{st}}\text{ equation Substitute for } x.$$
$$3(\mathbf{2y}\ \ \mathbf{3})+4y=-19$$
$$6y-9+4y=-19$$
$$10y-9=-19$$
$$10y-9+\mathbf{9}=-19+\mathbf{9}$$
$$10y=-10$$
$$\frac{10y}{\mathbf{10}}=\frac{-10}{\mathbf{10}}$$
$$y=-1$$

$$x=2y-3 \quad \text{The 2}^{\text{nd}}\text{ equation}$$
$$x=2(\mathbf{-1})-3$$
$$x=-2-3$$
$$x=-5$$

The solution is $(-5,-1)$.

Check:
$$3x+4y=-19$$
$$3(-5)+4(-1)\overset{?}{=}-19$$
$$-15-4\overset{?}{=}-19$$
$$-19=-19$$
True

Check:
$$2y-x=3$$
$$2(-1)-(-5)\overset{?}{=}3$$
$$-2+5\overset{?}{=}3$$
$$3=3$$
True

65. $\begin{cases} 3(x-1)+3=8+2y \\ 2(x+1)=8+y \end{cases}$

Distribute in both equations.

$$\begin{cases} 3x-3+3=8+2y \\ 2x+2=8+y \end{cases}$$

Write the 1st equation in standard form.
Solve for y in the 2nd equation.

$$\begin{cases} 3x-\mathbf{2y}=8+2y-\mathbf{2y} \\ 2x+2-\mathbf{8}=8+y-\mathbf{8} \end{cases}$$

$$\begin{cases} 3x-2y=8 \\ 2x-6=y \end{cases}$$

$$3x-2y=8 \quad \text{The 1}^{\text{st}}\text{ equation Substitute for } y.$$
$$3x-2(\mathbf{2x-6})=8$$
$$3x-4x+12=8$$
$$-x+12=8$$
$$-x+12-\mathbf{12}=8-\mathbf{12}$$
$$-x=-4$$
$$\frac{-x}{\mathbf{-1}}=\frac{-4}{\mathbf{-1}}$$
$$x=4$$

$$2x-6=y \quad \text{The 2}^{\text{nd}}\text{ equation}$$
$$2(\mathbf{4})-6=y$$
$$8-6=y$$
$$2=y$$

The solution is $(4, 2)$.

Check:
$$3(x-1)+3=8+2y$$
$$3(4-1)+3\overset{?}{=}8+2(2)$$
$$3(3)+3\overset{?}{=}8+4$$
$$9+3\overset{?}{=}12$$
$$12=12$$
True

Check:
$$2(x+1)=8+y$$
$$2(4+1)\overset{?}{=}8+2$$
$$2(5)\overset{?}{=}10$$
$$10=10$$
True

67. $\begin{cases} x = \frac{1}{3}y - 1 \\ x = y + 5 \end{cases}$

Clear the fraction from the 1st equation.

$\begin{cases} \mathbf{3}(x) = \mathbf{3}\left(\frac{1}{3}y\right) - \mathbf{3}(1) \\ x = y + 5 \end{cases}$

$\begin{cases} 3x = y - 3 \\ x = y + 5 \end{cases}$

$3x = y - 3$ The 1st equation Substitute for x.

$3(\boldsymbol{y+5}) = y - 3$

$3y + 15 = y - 3$

$3y + 15 - \mathbf{15} = y - 3 - \mathbf{15}$

$3y = y - 18$

$3y - \boldsymbol{y} = y - 18 - \boldsymbol{y}$

$2y = -18$

$\frac{2y}{\mathbf{2}} = \frac{-18}{\mathbf{2}}$

$y = -9$

$x = y + 5$ The 2nd equation

$x = \mathbf{-9} + 5$

$x = -4$

The solution is $(-4, -9)$.

Check:

$x = \frac{1}{3}y - 1$

$-4 \overset{?}{=} \frac{1}{3}(-9) - 1$

$-4 \overset{?}{=} -3 - 1$

$-4 = -4$

True

Check:

$x = y + 5$

$-4 \overset{?}{=} -9 + 5$

$-4 = -4$

True

69. $\begin{cases} 2a - 3b = -13 \\ -b = -2a - 7 \end{cases}$

Solve for b in 2nd equation.

$\begin{cases} 2a - 3b = -13 \\ b = 2a + 7 \end{cases}$

$2a - 3b = -13$ The 1st equation Substitute for b.

$2a - 3(\boldsymbol{2a+7}) = -13$

$2a - 6a - 21 = -13$

$-4a - 21 = -13$

$-4a - 21 + \mathbf{21} = -13 + \mathbf{21}$

$-4a = 8$

$\frac{-4a}{\mathbf{-4}} = \frac{8}{\mathbf{-4}}$

$a = -2$

$b = 2a + 7$ The 2nd equation

$b = 2(\mathbf{-2}) + 7$

$b = -4 + 7$

$b = 3$

The solution is $(-2, 3)$.

Check:

$2a - 3b = -13$

$2(-2) - 3(3) \overset{?}{=} -13$

$-4 - 9 \overset{?}{=} -13$

$-13 = -13$

True

Check:

$-b = -2a - 7$

$-(3) \overset{?}{=} -2(-2) - 7$

$-3 \overset{?}{=} 4 - 7$

$-3 = -3$

True

71. $\begin{cases} x = 7y - 10 \\ 2x - 14y + 20 = 0 \end{cases}$

Put the 2nd equation in standard form.

$\begin{cases} x = 7y - 10 \\ 2x - 14y + 20 - \mathbf{20} = 0 - \mathbf{20} \end{cases}$

$\begin{cases} x = 7y - 10 \\ 2x - 14y = -20 \end{cases}$

$2x - 14y = -20$ The 2nd equation Substitute for x.

$2(\boldsymbol{7y - 10}) - 14y = -20$

$14y - 20 - 14y = -20$ The variables drop out.

$-20 = -20$

True

The system has infinitely many solutions.

73. $\begin{cases} 4x + 1 = 2x + 5 + y \\ 2x + 2y = 5x + y + 6 \end{cases}$

Write both equation in standard form.

$\begin{cases} 4x + 1 - \mathbf{2x} - \boldsymbol{y} - \mathbf{1} = 2x + 5 + y - \mathbf{2x} - \boldsymbol{y} - \mathbf{1} \\ 2x + 2y - \mathbf{5x} - \boldsymbol{y} = 5x + y + 6 - \mathbf{5x} - \boldsymbol{y} \end{cases}$

$\begin{cases} 2x - y = 4 \\ -3x + y = 6 \end{cases}$

Solve for y in the 1st equation.

$\begin{cases} 2x - y + \boldsymbol{y} - \mathbf{4} = 4 + \boldsymbol{y} - \mathbf{4} \\ -3x + y = 6 \end{cases}$

$\begin{cases} 2x - 4 = y \\ -3x + y = 6 \end{cases}$

$-3x + y = 6$ The 2nd equation Substitute for y.

$-3x + (\mathbf{2x - 4}) = 6$

$-x - 4 = 6$

$-x - 4 + \mathbf{4} = 6 + \mathbf{4}$

$-x = 10$

$\frac{-x}{\mathbf{-1}} = \frac{10}{\mathbf{-1}}$

$x = -10$

$2x - 4 = y$ The 1st equation

$2(\mathbf{-10}) - 4 = y$

$-20 - 4 = y$

$-24 = y$

The solution is $(-10, -24)$.

Check:

$4x + 1 = 2x + 5 + y$

$4(-10) + 1 \overset{?}{=} 2(-10) + 5 + (-24)$

$-40 + 1 \overset{?}{=} -20 + 5 - 24$

$-39 = -39$

True

Check:

$2x + 2y = 5x + y + 6$

$2(-10) + 2(-24) \overset{?}{=} 5(-10) + (-24) + 6$

$-20 - 48 \overset{?}{=} -50 - 24 + 6$

$-68 = -68$

True

75. $\begin{cases} 2a+3b=7 \\ 6a-b=1 \end{cases}$

Solve for b in the 2nd equation.

$\begin{cases} 2a+3b=7 \\ 6a-b+\mathbf{b}-\mathbf{1}=1+\mathbf{b}-\mathbf{1} \end{cases}$

$\begin{cases} 2a+3b=7 \\ 6a-1=b \end{cases}$

$2a+3b=7$ The 1st equation. Substitute for b.

$$2a+3(\mathbf{6a-1})=7$$
$$2a+18a-3=7$$
$$20a-3=7$$
$$20a-3+\mathbf{3}=7+\mathbf{3}$$
$$20a=10$$
$$\frac{20a}{\mathbf{20}}=\frac{10}{\mathbf{20}}$$
$$a=\frac{1}{2}$$

$6a-1=b$ The 2nd equation

$$6\left(\mathbf{\frac{1}{2}}\right)-1=b$$
$$3-1=b$$
$$2=b$$

The solution is $\left(\frac{1}{2},2\right)$.

Check:
$$2a+3b=7$$
$$2\left(\frac{1}{2}\right)+3(2)\overset{?}{=}7$$
$$1+6\overset{?}{=}7$$
$$7=7$$
True

Check:
$$6a-b=1$$
$$6\left(\frac{1}{2}\right)-2\overset{?}{=}1$$
$$3-2\overset{?}{=}1$$
$$1=1$$
True

77. $\begin{cases} 2x-3y=-4 \\ x=-\frac{3}{2}y \end{cases}$

$2x-3y=-4$ The 1st equation. Substitute for x.

$$2\left(-\mathbf{\frac{3}{2}}y\right)-3y=-4$$
$$-3y-3y=-4$$
$$-6y=-4$$
$$\frac{-6y}{\mathbf{-6}}=\frac{-4}{\mathbf{-6}}$$
$$y=\frac{2}{3}$$

$x=-\frac{3}{2}y$ The 2nd equation

$$x=-\frac{3}{2}\left(\mathbf{\frac{2}{3}}\right)$$
$$x=-1$$

The solution is $\left(-1,\frac{2}{3}\right)$.

Check:
$$2x-3y=-4$$
$$2(-1)-3\left(\frac{2}{3}\right)\overset{?}{=}-4$$
$$-2-2\overset{?}{=}-4$$
$$-4=-4$$
True

Check:
$$x=-\frac{3}{2}y$$
$$-1\overset{?}{=}-\frac{3}{2}\left(\frac{2}{3}\right)$$
$$-1=-1$$
True

79. $\begin{cases} \frac{9x}{7}-\frac{3y}{7}=\frac{12}{7} \\ y-3x=-4 \end{cases}$

Clear 1st equations of fractions.
Solve for y in the 2nd equation.

$\begin{cases} \mathbf{7}\left(\frac{9x}{7}\right)-\mathbf{7}\left(\frac{3y}{7}\right)=\mathbf{7}\left(\frac{12}{7}\right) \\ y-3x+\mathbf{3x}=-4+\mathbf{3x} \end{cases}$

$\begin{cases} 9x-3y=12 \\ y=3x-4 \end{cases}$

$9x-3y=12$ The 1st equation. Substitute for y.

$$9x-3(\mathbf{3x-4})=12$$

$9x-9x+12=12$ The variables drop out.

$$12=12$$

True

The system has infinitely many solutions.

81. $\begin{cases} 4x+5y+1=-16+x \\ x-3y+2=-3-x \end{cases}$

Write both equations in standard form.

$\begin{cases} 4x+5y+1-\mathbf{x}-\mathbf{1}=-16+x-\mathbf{x}-\mathbf{1} \\ x-3y+2+\mathbf{x}-\mathbf{2}=-3-x+\mathbf{x}-\mathbf{2} \end{cases}$

$\begin{cases} 3x+5y=-17 \\ 2x-3y=-5 \end{cases}$

Solve for x in the 1st equation.

$\begin{cases} 3x+5y=-17 \\ 2x-3y=-5 \end{cases}$

$$3x+5y=-17$$
$$3x+5y-\mathbf{5y}=-17-\mathbf{5y}$$
$$3x=-17-5y$$
$$\frac{3x}{\mathbf{3}}=\frac{-17}{\mathbf{3}}-\frac{5y}{\mathbf{3}}$$
$$x=\frac{-17}{3}-\frac{5y}{3}$$
$$\begin{cases} x=\frac{-17}{3}-\frac{5y}{3} \\ 2x-3y=-5 \end{cases}$$

$2x-3y=-5$ The 2nd equation. Substitute for x.

$$2\left(\frac{\mathbf{-17}}{\mathbf{3}}-\frac{\mathbf{5y}}{\mathbf{3}}\right)-3y=-5$$
$$\frac{-34}{3}-\frac{10y}{3}-3y=-5$$
$$\mathbf{3}\left(\frac{-34}{3}\right)-\mathbf{3}\left(\frac{10y}{3}\right)-\mathbf{3}(3y)=\mathbf{3}(-5)$$
$$-34-10y-9y=-15$$
$$-34-19y=-15$$
$$-34-19y+\mathbf{34}=-15+\mathbf{34}$$
$$-19y=19$$
$$\frac{-19y}{\mathbf{-19}}=\frac{19}{\mathbf{-19}}$$
$$y=-1$$

$2x=-13-5y$ The 1st equation

$$2x=-13-5(\mathbf{-1})$$
$$2x=-13+5$$
$$2x=-8$$
$$\frac{2x}{2}=\frac{-8}{2}$$
$$x=-4$$

The solution is $(-4,-1)$.

Check:

$$4x+5y+1=-12+2x$$
$$4(-4)+5(-1)+1\overset{?}{=}-12+2(-4)$$
$$-16-5+1\overset{?}{=}-12-8$$
$$-20=-20$$

True

Check:

$$x-3y+2=-3-x$$
$$-4-3(-1)+2\overset{?}{=}-3-(-4)$$
$$-4+3+2\overset{?}{=}-3+4$$
$$1=1$$

True

APPLICATIONS

83. OFFROADING

Let a = angle of approach in degrees
d = angle of departure in degrees

$$\begin{cases} a+d=77 \\ a=d+3 \end{cases}$$

$a+d=77$ The 1st equation. Substitute for a.

$$\mathbf{d+3}+d=77$$
$$2d+3=77$$
$$2d+3-\mathbf{3}=77-\mathbf{3}$$
$$2d=74$$
$$\frac{2d}{\mathbf{2}}=\frac{74}{\mathbf{2}}$$
$$d=37$$

$a=d+3$ The 2nd equation

$$a=\mathbf{37}+3$$
$$a=40$$

The angle of approach is 40°.
The angle of departure is 37°.

Check:	Check:
$a+d=77$	$a=d+3$
$40+37\overset{?}{=}77$	$40\overset{?}{=}37+3$
$77=77$	$40=40$
True	True

85. GEOMETRY

Let x = measure of 1st angle in degrees
y = measure of 2nd angle in degrees

$$\begin{cases} x+y=90 \\ y=3x \end{cases}$$

$x+y=90$ The 1st equation. Substitute for y.

$$x+\mathbf{3x}=90$$
$$4x=90$$
$$\frac{4x}{\mathbf{4}}=\frac{90}{\mathbf{4}}$$
$$x=22.5$$

$y=3x$ The 2nd equation

$$y=3(\mathbf{22.5})$$
$$y=67.5$$

The measure of angle x is 22.5°.
The measure of angle y is 67.5°.

Check:	Check:
$x+y=90$	$y=3x$
$22.5+67.5\overset{?}{=}90$	$67.5\overset{?}{=}3(22.5)$
$90=90$	$67.5=67.5$
True	True

WRITING

87-91. Answers will vary.

REVIEW

93. Find the prime factors of 189.

$$\begin{aligned}189 &= 3\cdot 63\\ &= 3\cdot 7\cdot 9\\ &= 3\cdot 7\cdot 3\cdot 3\\ &= 3^3\cdot 7\end{aligned}$$

95. Add:

$$\begin{aligned}\frac{5}{12}+\frac{1}{4} &= \frac{5}{12}+\frac{1}{4}\left(\frac{\mathbf{3}}{\mathbf{3}}\right)\\ &= \frac{5+3}{12}\\ &= \frac{8}{12}\\ &= \frac{\not{2}\cdot\not{2}\cdot 2}{\not{2}\cdot\not{2}\cdot 3}\\ &= \frac{2}{3}\end{aligned}$$

CHALLENGE PROBLEMS

97. $\begin{cases}\dfrac{6x-1}{3}-\dfrac{5}{3}=\dfrac{3y+1}{2}\\[2mm] \dfrac{1+5y}{4}+\dfrac{x+3}{4}=\dfrac{17}{2}\end{cases}$

Clear both equations of fractions.

$$\begin{cases}\mathbf{6}\left(\dfrac{6x-1}{3}\right)-\mathbf{6}\left(\dfrac{5}{3}\right)=\mathbf{6}\left(\dfrac{3y+1}{2}\right)\\[2mm] \mathbf{4}\left(\dfrac{1+5y}{4}\right)+\mathbf{4}\left(\dfrac{x+3}{4}\right)=\mathbf{4}\left(\dfrac{17}{2}\right)\end{cases}$$

$$\begin{cases}12x-2-10=9y+3\\ 1+5y+x+3=34\end{cases}$$

Write the 1st equation in standard form.
Solve for x in the 2nd equation.

$$\begin{cases}12x-9y=15\\ x=30-5y\end{cases}$$

$12x-9y=15$ The 1st equation Substitute for x.

$$\begin{aligned}12(\mathbf{30-5y})-9y&=15\\ 360-60y-9y&=15\\ 360-69y&=15\\ 360-69y-\mathbf{360}&=15-\mathbf{360}\\ -69y&=-345\\ \frac{-69y}{\mathbf{-69}}&=\frac{-345}{\mathbf{-69}}\\ y&=5\end{aligned}$$

$x=30-5y$ The 2nd equation

$$\begin{aligned}x&=30-5(\mathbf{5})\\ x&=30-25\\ x&=5\end{aligned}$$

The solution is (5, 5).

Check:

$$\begin{aligned}\frac{6x-1}{3}-\frac{5}{3}&=\frac{3y+1}{2}\\ \frac{6(5)-1}{3}-\frac{5}{3}&\stackrel{?}{=}\frac{3(5)+1}{2}\\ \frac{30-1}{3}-\frac{5}{3}&\stackrel{?}{=}\frac{15+1}{2}\\ \frac{29}{3}-\frac{5}{3}&\stackrel{?}{=}\frac{16}{2}\\ \frac{24}{3}&\stackrel{?}{=}\frac{16}{2}\\ 8&=8\end{aligned}$$

True

Check:

$$\begin{aligned}\frac{1+5y}{4}+\frac{x+3}{4}&=\frac{17}{2}\\ \frac{1+5(5)}{4}+\frac{5+3}{4}&\stackrel{?}{=}\frac{17}{2}\\ \frac{1+25}{4}+\frac{8}{4}&\stackrel{?}{=}\frac{17}{2}\\ \frac{26}{4}+\frac{8}{4}&\stackrel{?}{=}\frac{17}{2}\\ \frac{34}{4}&\stackrel{?}{=}\frac{17}{2}\\ \frac{17}{2}&=\frac{17}{2}\end{aligned}$$

True

99. $\begin{cases} 2(2x+3y)=5 \\ 8x=3(1+3y) \end{cases}$

Distribute in both equations.

$\begin{cases} 4x+6y=5 \\ 8x=3+9y \end{cases}$

Solve 2nd for x.

$\begin{cases} 4x+6y=5 \\ x=\frac{9}{8}y+\frac{3}{8} \end{cases}$

$$4x+6y=5 \quad \text{The 1}^{\text{st}} \text{ equation} \quad \text{Substitute for } x.$$

$$4\left(\mathbf{\frac{9}{8}}y+\mathbf{\frac{3}{8}}\right)+6y=5$$

$$\frac{9}{2}y+\frac{3}{2}+6y=5$$

$$\mathbf{2}\left(\frac{9}{2}y\right)+\mathbf{2}\left(\frac{3}{2}\right)+\mathbf{2}(6y)=\mathbf{2}(5)$$

$$9y+3+12y=10$$

$$21y+3=10$$

$$21y+3-\mathbf{3}=10-\mathbf{3}$$

$$21y=7$$

$$\frac{21y}{\mathbf{21}}=\frac{7}{\mathbf{21}}$$

$$y=\frac{\cancel{7}^{1}}{3\cdot\cancel{7}_{1}}$$

$$y=\frac{1}{3}$$

$$x=\frac{9}{8}y+\frac{3}{8} \quad \text{The 2}^{\text{nd}} \text{ equation}$$

$$x=\frac{\cancel{9}^{3}}{8}\left(\frac{1}{\cancel{3}_{1}}\right)+\frac{3}{8}$$

$$x=\frac{3}{8}+\frac{3}{8}$$

$$x=\frac{6}{8}$$

$$x=\frac{\cancel{2}^{1}\cdot 3}{\cancel{2}_{1}\cdot 4}$$

$$x=\frac{3}{4}$$

The solution is $\left(\frac{3}{4},\frac{1}{3}\right)$.

Check:

$$2(2x+3y)=5$$

$$2\left(2\cdot\frac{3}{4}+3\cdot\frac{1}{3}\right)\stackrel{?}{=}5$$

$$2\left(\frac{3}{2}+1\right)\stackrel{?}{=}5$$

$$2\left(\frac{5}{2}\right)\stackrel{?}{=}5$$

$$5=5$$

True

Check:

$$8x=3(1+3y)$$

$$8\left(\frac{3}{4}\right)\stackrel{?}{=}3\left(1+3\cdot\frac{1}{3}\right)$$

$$2(3)\stackrel{?}{=}3(1+1)$$

$$6=6$$

True

SECTION 4.3 STUDY SET

VOCABULARY

Fill in the blanks.

1. The coefficients of $3x$ and $-3x$ are **opposites**.

CONCEPTS

3. In the following system, which terms have coefficients that are opposites? **7y and –7y**

$$\begin{cases} 3x+7y=-25 \\ 4x-7y=12 \end{cases}$$

5. Add each pair of equations.

a. $\begin{array}{r} 2a+2b=-6 \\ \underline{3a-2b=\ \ 2} \\ \mathbf{5a\qquad=-4} \end{array}$

b. $\begin{array}{r} x-3y=15 \\ \underline{-x-y=-14} \\ \mathbf{-4y=1} \end{array}$

7. If the elimination method is used to solve

$$\begin{cases} 3x+12y=4 \\ 6x-4y=7 \end{cases}$$

a. By what would we multiply the first equation to eliminate x? **–2**

b. By what would we multiply the second equation to eliminate y? **3**

9. What algebraic step should be performed to

a. Clear $\frac{2}{3}x+4y=-\frac{4}{5}$ of fractions?

Multiply both sides by 15.

b. Clear $0.2x - 0.9y = 6.4$ of decimals?

Multiply both sides by 10.

NOTATION

Complete the solution to solve the system.

11. Solve: $\begin{cases} x+y=5 \\ x-y=-3 \end{cases}$.

$$\begin{array}{r} x+y=5 \\ \underline{x-y=-3} \\ \boxed{2x}=2 \end{array}$$ Add the equations.

$$x=\boxed{1}$$

$x+y=5$ This is the first equation.

$$\boxed{1}+y=5$$

$$y=\boxed{4}$$

The solution is $(\boxed{1},\boxed{4})$.

GUIDED PRACTICE

Use the elimination method to solve each system. See Example 1.

13. $\begin{cases} x+y=5 \\ x-y=1 \end{cases}$

Eliminate y.

$$\begin{array}{r} x+y=5 \\ \underline{x-y=1} \\ 2x\qquad=6 \end{array}$$

$$\frac{2x}{\mathbf{2}}=\frac{6}{\mathbf{2}}$$

$$x=3$$

$x+y=5$ The 1st equation

$$\mathbf{3}+y=5$$

$$3+y-\mathbf{3}=5-\mathbf{3}$$

$$y=2$$

The solution is (3, 2).

15. $\begin{cases} x+y=-5 \\ -x+y=-1 \end{cases}$

Eliminate x.

$$\begin{array}{r} x+y=-5 \\ \underline{-x+y=-1} \\ 2y=-6 \end{array}$$

$$\frac{2y}{\mathbf{2}}=\frac{-6}{\mathbf{2}}$$

$$y=-3$$

$x+y=-5$ The 1st equation

$$x+(\mathbf{-3})=-5$$

$$x-3+\mathbf{3}=-5+\mathbf{3}$$

$$x=-2$$

The solution is $(-2,-3)$.

17. $\begin{cases} 4x+3y=24 \\ 4x-3y=-24 \end{cases}$

Eliminate y.

$$\begin{array}{r} 4x+3y=\ \ 24 \\ \underline{4x-3y=-24} \\ 8x\qquad=0 \end{array}$$

$$\frac{8x}{\mathbf{8}}=\frac{0}{\mathbf{8}}$$

$$x=0$$

$4x+3y=24$ The 1st equation

$$4(\mathbf{0})+3y=24$$

$$3y=24$$

$$\frac{3y}{\mathbf{3}}=\frac{24}{\mathbf{3}}$$

$$y=8$$

The solution is (0, 8).

19. $\begin{cases} 2s+t=-2 \\ -2s-3t=-6 \end{cases}$

Eliminate s.

$$\begin{array}{r} 2s+\ t=-2 \\ -2s-3t=\ 6 \\ \hline -2t=-8 \end{array}$$

$$\frac{-2t}{\mathbf{-2}}=\frac{-8}{\mathbf{-2}}$$

$$t=4$$

$2s+t=-2$ The 1st equation

$$2s+\mathbf{4}=-2$$

$$2s+4-\mathbf{4}=-2-\mathbf{4}$$

$$2s=-6$$

$$\frac{2s}{\mathbf{2}}=\frac{-6}{\mathbf{2}}$$

$$s=-3$$

The solution is $(-3, 4)$.

23. $\begin{cases} 7x-y=10 \\ 8x-y=13 \end{cases}$

Eliminate y.

Multiply both sides of the 1st equation by -1.

$\begin{cases} \mathbf{-1}(7x)\mathbf{-1}(-y)=\mathbf{-1}(10) \\ 8x-y=13 \end{cases}$

$$\begin{array}{r} -7x+y=-10 \\ 8x-y=\ 13 \\ \hline x\qquad =3 \end{array}$$

$7x-y=10$ The 1st equation

$$7(\mathbf{3})-y=10$$

$$21-y=10$$

$$21-y+\mathbf{y}=10+\mathbf{y}$$

$$21-\mathbf{10}=10+y-\mathbf{10}$$

$$11=y$$

The solution is $(3, 11)$

Use the elimination method to solve each system. See Example 2.

21. $\begin{cases} x+3y=-9 \\ x+8y=-4 \end{cases}$

Eliminate x.

Multiply both sides of the 1st equation by -1.

$\begin{cases} \mathbf{-1}(x)\mathbf{-1}(3y)=\mathbf{-1}(-9) \\ x+8y=-4 \end{cases}$

$$\begin{array}{r} -x-3y=\ 9 \\ x+8y=-4 \\ \hline 5y=5 \end{array}$$

$$\frac{5y}{\mathbf{5}}=\frac{5}{\mathbf{5}}$$

$$y=1$$

$x+3y=-9$ The 1st equation

$$x+3(\mathbf{1})=-9$$

$$x+3-\mathbf{3}=-9-\mathbf{3}$$

$$x=-12$$

The solution is $(-12, 1)$.

Use the elimination method to solve each system. See Example 3.

25. $\begin{cases} 7x+4y-14=0 \\ 3x=2y-20 \end{cases}$

Write both equations in standard form.

$\begin{cases} 7x+4y-14\mathbf{+14}=0\mathbf{+14} \\ 3x\mathbf{-2y}=2y-20\mathbf{-2y} \end{cases}$

$\begin{cases} 7x+4y=14 \\ 3x-2y=-20 \end{cases}$

Eliminate y.

Multiply both sides of the 2nd equation by 2.

$\begin{cases} 7x+4y=14 \\ \mathbf{2}(3x)-\mathbf{2}(2y)=\mathbf{2}(-20) \end{cases}$

$$\begin{array}{r} 7x+4y=\ 14 \\ 6x-4y=-40 \\ \hline 13x\qquad =-26 \end{array}$$

$$\frac{13x}{\mathbf{13}}=\frac{-26}{\mathbf{13}}$$

$$x=-2$$

$7x+4y=14$ The 1st equation

$$7(\mathbf{-2})+4y=14$$

$$-14+4y=14$$

$$-14+4y+\mathbf{14}=14+\mathbf{14}$$

$$4y=28$$

$$\frac{4y}{\mathbf{4}}=\frac{28}{\mathbf{4}}$$

$$y=7$$

The solution is $(-2, 7)$.

27. $\begin{cases} 7x-50y+43=0 \\ x=4-3y \end{cases}$

Write both equations in standard form.

$\begin{cases} 7x-50y+43-\mathbf{43}=0-\mathbf{43} \\ x+\mathbf{3y}=4-3y+\mathbf{3y} \end{cases}$

$\begin{cases} 7x-50y=-43 \\ x+3y=4 \end{cases}$

Eliminate x.

Multiply both sides of the 2nd equation by -7.

$\begin{cases} 7x-50y=-43 \\ \mathbf{-7}(x)-\mathbf{7}(3y)=\mathbf{-7}(4) \end{cases}$

$$\begin{aligned} 7x-50y&=-43 \\ -7x-21y&=-28 \\ \hline -71y&=-71 \\ \frac{-71y}{\mathbf{-71}}&=\frac{-71}{\mathbf{-71}} \\ y&=1 \end{aligned}$$

$x+3y=4$ The 2nd equation

$$\begin{aligned} x+3(\mathbf{1})&=4 \\ x+3-\mathbf{3}&=4-\mathbf{3} \\ x&=1 \end{aligned}$$

The solution is (1, 1).

Use the elimination method to solve each system. See Example 4.

29. $\begin{cases} 4x+3y=7 \\ 3x-2y=-16 \end{cases}$

Eliminate y.

Multiply both sides of the 1st equation by 2.

Multiply both sides of the 2nd equation by 3.

$\begin{cases} \mathbf{2}(4x)+\mathbf{2}(3y)=\mathbf{2}(7) \\ \mathbf{3}(3x)-\mathbf{3}(2y)=\mathbf{3}(-16) \end{cases}$

$$\begin{aligned} 8x+6y&=14 \\ 9x-6y&=-48 \\ \hline 17x\quad&=-34 \\ \frac{17x}{\mathbf{17}}&=\frac{-34}{\mathbf{17}} \\ x&=-2 \end{aligned}$$

$4x+3y=7$ The 1st equation

$$\begin{aligned} 4(-\mathbf{2})+3y&=7 \\ -8+3y+\mathbf{8}&=7+\mathbf{8} \\ 3y&=15 \\ \frac{3y}{\mathbf{3}}&=\frac{15}{\mathbf{3}} \\ y&=5 \end{aligned}$$

The solution is $(-2, 5)$

31. $\begin{cases} 5a+8b=2 \\ 11a-3b=25 \end{cases}$

Eliminate b.

Multiply both sides of the 1st equation by 3.

Multiply both sides of the 2nd equation by 8.

$\begin{cases} \mathbf{3}(5a+\mathbf{3}(8b)=\mathbf{3}(2) \\ \mathbf{8}(11a)-\mathbf{8}(3b)=\mathbf{8}(25) \end{cases}$

$$\begin{aligned} 15a+24b&=6 \\ 88a-24b&=200 \\ \hline 103a\quad&=206 \\ \frac{103a}{\mathbf{103}}&=\frac{206}{\mathbf{103}} \\ a&=2 \end{aligned}$$

$5a+8b=2$ The 1st equation

$$\begin{aligned} 5(\mathbf{2})+8b&=2 \\ 10+8b-\mathbf{10}&=2-\mathbf{10} \\ 8b&=-8 \\ \frac{8b}{\mathbf{8}}&=\frac{-8}{\mathbf{8}} \\ b&=-1 \end{aligned}$$

The solution is $(2, -1)$

Use the elimination method to solve each system. See Example 5.

33. $\begin{cases} \frac{3}{5}s+\frac{4}{5}t=1 \\ -\frac{1}{4}s+\frac{3}{8}t=1 \end{cases}$

Clear both equations of fractions.

$\begin{cases} \mathbf{5}\left(\frac{3}{5}s\right)+\mathbf{5}\left(\frac{4}{5}t\right)=\mathbf{5}(1) \\ \mathbf{8}\left(-\frac{1}{4}s\right)+\mathbf{8}\left(\frac{3}{8}t\right)=\mathbf{8}(1) \end{cases}$

$\begin{cases} 3s+4t=5 \\ -2s+3t=8 \end{cases}$

Eliminate s.

Multiply both sides of the 1st equation by 2.

Multiply both sides of the 2nd equation by 3..

$\begin{cases} \mathbf{2}(3s)+\mathbf{2}(4t)=\mathbf{2}(5) \\ \mathbf{3}(-2s)+\mathbf{3}(3t)=\mathbf{3}(8) \end{cases}$

$\begin{cases} 6s+8t=10 \\ -6s+9t=24 \end{cases}$

$$\begin{aligned} 6s+8t&=10 \\ -6s+9t&=24 \\ \hline 17t&=34 \\ \frac{17t}{\mathbf{17}}&=\frac{34}{\mathbf{17}} \\ t&=2 \end{aligned}$$

$3s + 4t = 5$ The 1st equation

$3s + 4(\mathbf{2}) = 5$

$3s + 8 - \mathbf{8} = 5 - \mathbf{8}$

$3s = -3$

$\frac{3s}{\mathbf{3}} = \frac{-3}{\mathbf{3}}$

$s = -1$

The solution is $(-1, 2)$.

35. $\begin{cases} \frac{1}{2}s - \frac{1}{4}t = 1 \\ \frac{1}{3}s + t = 3 \end{cases}$

Clear both equations of fractions.

Multiply both sides of the 1st equation by 4.

Multiply both sides of the 2nd equation by 3.

$\begin{cases} \mathbf{4}\left(\frac{1}{2}s\right) - \mathbf{4}\left(\frac{1}{4}t\right) = \mathbf{4}(1) \\ \mathbf{3}\left(\frac{1}{3}s\right) + \mathbf{3}(t) = \mathbf{3}(3) \end{cases}$

$\begin{cases} 2s - t = 4 \\ s + 3t = 9 \end{cases}$

Eliminate s.

Multiply both sides of the 2nd equation by -2.

$$\begin{array}{r} 2s - t = 4 \\ -2s - 6t = -18 \\ \hline -7t = -14 \end{array}$$

$\frac{-7t}{\mathbf{-7}} = \frac{-14}{\mathbf{-7}}$

$t = 2$

$\frac{1}{3}s + t = 3$ The 2nd equation

$\frac{1}{3}s + (\mathbf{2}) = 3$

$\frac{1}{3}s + 2 - \mathbf{2} = 3 - \mathbf{2}$

$\frac{1}{3}s = 1$

$\frac{\mathbf{3}}{\mathbf{1}}\left(\frac{1}{3}s\right) = \mathbf{3}(1)$

$s = 3$

The solution is $(3, 2)$.

Use the elimination method to solve each system. If there is no solution, or infinitely many solutions, so indicate. See Example 6.

37. $\begin{cases} 3x - 5y = -29 \\ 3x - 5y = 15 \end{cases}$

Eliminate x.

Multiply both sides of the 1st equation by -1.

$\begin{cases} \mathbf{-1}(3x) - \mathbf{1}(-5y) = \mathbf{-1}(-29) \\ 3x - 5y = 15 \end{cases}$

$$\begin{array}{r} -3x + 5y = 29 \\ 3x - 5y = 15 \\ \hline 0 = 44 \end{array}$$

Both variables are eliminated.

False

No solution.

39. $\begin{cases} 3x - 16 = 5y \\ -3x + 5y - 33 = 0 \end{cases}$

Write both equations in standard form.

$\begin{cases} 3x - 16 - \mathbf{5y} + \mathbf{16} = 5y - \mathbf{5y} + \mathbf{16} \\ -3x + 5y - 33 + \mathbf{33} = 0 + \mathbf{33} \end{cases}$

$\begin{cases} 3x - 5y = 16 \\ -3x + 5y = 33 \end{cases}$

Eliminate x.

$$\begin{array}{r} 3x - 5y = 16 \\ -3x + 5y = 33 \\ \hline 0 = 49 \end{array}$$

Both variables are eliminated.

False

No solution.

Use the elimination method to solve each system. If there is no solution, or infinitely many solutions, so indicate. See Example 7.

41. $\begin{cases} 0.4x - 0.7y = -1.9 \\ -x + \frac{7}{4}y = \frac{19}{4} \end{cases}$

Multiply the 1st equation by 10.

Multiply the 2nd equation by 4.

$\begin{cases} \mathbf{10}(0.4x) - \mathbf{10}(0.7y) = \mathbf{10}(-1.9) \\ \mathbf{4}(-x) + \mathbf{4}\left(\frac{7}{4}y\right) = \mathbf{4}\left(\frac{19}{4}\right) \end{cases}$

$\begin{cases} 4x - 7y = -19 \\ -4x + 7y = 19 \end{cases}$

$$\begin{array}{r} 4x - 7y = -19 \\ -4x + 7y = 19 \\ \hline 0 = 0 \end{array}$$

Both variables are eliminated.

True

The system has infinitely many solutions.

43. $\begin{cases} \dfrac{x-6y}{2}=7 \\ -x+6y+14=0 \end{cases}$

Write the 2nd equations in standard form.

$$\begin{cases} \dfrac{x-6y}{2}=7 \\ -x+6y+14-\mathbf{14}=0-\mathbf{14} \end{cases}$$

$$\begin{cases} \dfrac{x-6y}{2}=7 \\ -x+6y=-14 \end{cases}$$

Multiply both sides of the 1st equation by 2.

$$\begin{cases} \mathbf{2}\left(\dfrac{x-6y}{2}\right)=\mathbf{2}(7) \\ -x+6y=-14 \end{cases}$$

$$\begin{cases} x-6y=14 \\ -x+6y=-14 \end{cases}$$

$$\begin{array}{r} x-6y=14 \\ -x+6y=-14 \\ \hline 0=0 \end{array}$$

Both variables are eliminated.

True

The system has infinitely many solutions.

TRY IT YOURSELF

Solve the system by either the substitution or the elimination method if possible.

45. $\begin{cases} y=-3x+9 \\ y=x+1 \end{cases}$

$y=-3x+9$ The 1st equation
Substitute for y.

$$\begin{aligned} (\mathbf{x+1}) &= -3x+9 \\ x+1+\mathbf{3x} &= -3x+9+\mathbf{3x} \\ 4x+1 &= 9 \\ 4x+1-\mathbf{1} &= 9-\mathbf{1} \\ 4x &= 8 \\ \frac{4x}{\mathbf{4}} &= \frac{8}{\mathbf{4}} \\ x &= 2 \end{aligned}$$

$y=x+1$ The 2nd equation

$y=\mathbf{2}+1$

$y=3$

The solution is (2, 3).

47. $\begin{cases} 4x+6y=5 \\ 8x-9y=3 \end{cases}$

Eliminate y.

Multiply both sides of the 1st equation by 3.

Multiply both sides of the 2nd equation by 2.

$$\begin{array}{r} 12x+18y=15 \\ 16x-18y=6 \\ \hline 28x \qquad =21 \end{array}$$

$$\frac{28x}{\mathbf{28}}=\frac{21}{\mathbf{28}}$$

$$x=\frac{3}{4}$$

$4x+6y=5$ The 1st equation

$$\begin{aligned} 4\left(\frac{\mathbf{3}}{\mathbf{4}}\right)+6y &= 5 \\ 3+6y &= 5 \\ 3+6y-\mathbf{3} &= 5-\mathbf{3} \\ 6y &= 2 \\ \frac{6y}{\mathbf{6}} &= \frac{2}{\mathbf{6}} \\ y &= \frac{1}{3} \end{aligned}$$

The solution is $\left(\frac{3}{4}, \frac{1}{3}\right)$.

49. $\begin{cases} 6x-3y=-7 \\ y+9x=6 \end{cases}$

Write the 2nd equation in standard form.

$$\begin{cases} 6x-3y=-7 \\ 9x+y=6 \end{cases}$$

Eliminate y.

Multiply both sides of the 2nd equation by 3.

$$\begin{array}{r} 6x-3y=-7 \\ 27x+3y=18 \\ \hline 33x \qquad =11 \end{array}$$

$$\frac{33x}{\mathbf{33}}=\frac{11}{\mathbf{33}}$$

$$x=\frac{1}{3}$$

$$6x-3y=-7 \quad 1^{st} \text{ equation}$$
$$6\left(\frac{\mathbf{1}}{\mathbf{3}}\right)-3y=-7$$
$$2-3y=-7$$
$$2-3y-\mathbf{2}=-7-\mathbf{2}$$
$$-3y=-9$$
$$\frac{-3y}{\mathbf{-3}}=\frac{-9}{\mathbf{-3}}$$
$$y=3$$

The solution is $\left(\frac{1}{3}, 3\right)$.

51. $\begin{cases} x+y=1 \\ x-y=5 \end{cases}$

Eliminate y.

$$x+y=1$$
$$\underline{x-y=5}$$
$$2x \quad =6$$
$$\frac{2x}{\mathbf{2}}=\frac{6}{\mathbf{2}}$$
$$x=3$$

$$x+y=1 \quad \text{The } 1^{st} \text{ equation}$$
$$\mathbf{3}+y=1$$
$$3+y-\mathbf{3}=1-\mathbf{3}$$
$$y=-2$$

The solution is $(3,-2)$.

53. $\begin{cases} 4(x-2y)=36 \\ 3x-6y=27 \end{cases}$

Distribute in 1st equation.

$\begin{cases} 4x-8y=36 \\ 3x-6y=27 \end{cases}$

Eliminate x.

Multiply both sides of the 1st equation by 3.

Multiply both sides of the 2nd equation by -4.

$\begin{cases} \mathbf{3}(4x)-\mathbf{3}(8y)=\mathbf{3}(36) \\ \mathbf{-4}(x)-\mathbf{4}(-6y)=\mathbf{-4}(27) \end{cases}$

$$12x-24y= \ 108$$
$$\underline{-12x+24y=-108}$$
$$0=0 \quad \text{Both variables are eliminated.}$$

True

The system has infinitely many solutions.

55. $\begin{cases} x=y \\ 0.1x+0.2y=1.0 \end{cases}$

Multiply both sides of the 2nd equation by 10.

$\begin{cases} x=y \\ x+2y=10 \end{cases}$

$$x+2y=10 \quad \text{The } 2^{nd} \text{ equation}$$
Substitute for y.
$$x+2(\mathbf{x})=10$$
$$3x=10$$
$$\frac{3x}{\mathbf{3}}=\frac{10}{\mathbf{3}}$$
$$x=\frac{10}{3}$$

$$x=y \quad \text{The } 1^{st} \text{ equation}$$
$$\frac{\mathbf{10}}{\mathbf{3}}=y$$

The solution is $\left(\frac{10}{3}, \frac{10}{3}\right)$.

57. $\begin{cases} 2x+11y=-10 \\ 5x+4y=22 \end{cases}$

Eliminate x.

Multiply both sides of the 1st equation by 5.

Multiply both sides of the 2nd equation by -2.

$\begin{cases} \mathbf{5}(2x)+\mathbf{5}(11y)=\mathbf{5}(-10) \\ \mathbf{-2}(5x)+\mathbf{-2}(4y)=\mathbf{-2}(22) \end{cases}$

$$10x+55y=-50$$
$$\underline{-10x-8y \ =-44}$$
$$47y=-94$$
$$\frac{47y}{\mathbf{47}}=\frac{-94}{\mathbf{47}}$$
$$y=-2$$

$$5x+4y=22 \quad \text{The } 2^{nd} \text{ equation}$$
$$5x+4(\mathbf{-2})=22$$
$$5x-8+\mathbf{8}=22+\mathbf{8}$$
$$5x=30$$
$$\frac{5x}{\mathbf{5}}=\frac{30}{\mathbf{5}}$$
$$x=6$$

The solution is $(6, -2)$

59. $\begin{cases} 7x = 21 - 6y \\ 4x + 5y = 12 \end{cases}$

Write the 1st equations in standard form.

$$\begin{cases} 7x + \mathbf{6y} = 21 - 6y + \mathbf{6y} \\ 4x + 5y = 12 \end{cases}$$

$$\begin{cases} 7x + 6y = 21 \\ 4x + 5y = 12 \end{cases}$$

Eliminate x.

Multiply both sides of the 1st equation by 4.

Multiply both sides of the 2nd equation by -7.

$$\begin{cases} \mathbf{4}(7x) + \mathbf{4}(6y) = \mathbf{4}(21) \\ \mathbf{-7}(4x) - \mathbf{7}(5y) = \mathbf{-7}(12) \end{cases}$$

$$\begin{array}{r} 28x + 24y = 84 \\ -28x - 35y = -84 \\ \hline -11y = 0 \end{array}$$

$$\frac{-11y}{\mathbf{-11}} = \frac{0}{\mathbf{-11}}$$

$$y = 0$$

$7x = 21 - 6y$ The 1st equation

$$7x = 21 - 6(\mathbf{0})$$

$$7x = 21$$

$$\frac{7x}{\mathbf{7}} = \frac{21}{\mathbf{7}}$$

$$x = 3$$

The solution is (3, 0).

61. $\begin{cases} 9x - 10y = 0 \\ \dfrac{9x - 3y}{63} = 1 \end{cases}$

Multiply both sides of the 2nd equation by 63.

$$\begin{cases} 9x - 10y = 0 \\ \mathbf{63}\left(\dfrac{9x - 3y}{63}\right) = \mathbf{63}(1) \end{cases}$$

$$\begin{cases} 9x - 10y = 0 \\ 9x - 3y = 63 \end{cases}$$

Eliminate x.

Multiply both sides of the 1st equation by -1.

$$\begin{cases} (\mathbf{-1})9x - (\mathbf{-1})(10y) = (\mathbf{-1})0 \\ 9x - 3y = 63 \end{cases}$$

$$\begin{array}{r} -9x + 10y = 0 \\ 9x - 3y = 63 \\ \hline 7y = 63 \end{array}$$

$$\frac{7y}{\mathbf{7}} = \frac{63}{\mathbf{7}}$$

$$y = 9$$

$9x - 10y = 0$ The 1st equation

$$9x - 10(\mathbf{9}) = 0$$

$$9x - 90 + \mathbf{90} = 0 + \mathbf{90}$$

$$9x = 90$$

$$\frac{9x}{\mathbf{9}} = \frac{90}{\mathbf{9}}$$

$$x = 10$$

The solution is (10, 9).

63. $\begin{cases} \dfrac{m}{4} + \dfrac{n}{3} = -\dfrac{1}{12} \\ \dfrac{m}{2} - \dfrac{5}{4}n = \dfrac{7}{4} \end{cases}$

Clear both equations of fractions.

Multiply both sides of the 1st equation by 12.

Multiply both sides of the 2nd equation by 4.

$$\begin{cases} \mathbf{12}\left(\dfrac{m}{4}\right) + \mathbf{12}\left(\dfrac{n}{3}\right) = \mathbf{12}\left(-\dfrac{1}{12}\right) \\ \mathbf{4}\left(\dfrac{m}{2}\right) - \mathbf{4}\left(\dfrac{5}{4}n\right) = \mathbf{4}\left(\dfrac{7}{4}\right) \end{cases}$$

$$\begin{cases} 3m + 4n = -1 \\ 2m - 5n = 7 \end{cases}$$

Eliminate n.

Multiply both sides of the 1st equation by 5.

Multiply both sides of the 2nd equation by 4.

$$\begin{cases} (\mathbf{5})3m + (\mathbf{5})4n = (\mathbf{5})(-1) \\ (\mathbf{4})2m - (\mathbf{4})5n = (\mathbf{4})7 \end{cases}$$

$$\begin{array}{r} 15m + 20n = -5 \\ 8m - 20n = 28 \\ \hline 23m = 23 \end{array}$$

$$\frac{23m}{\mathbf{23}} = \frac{23}{\mathbf{23}}$$

$$m = 1$$

$2m - 5n = 7$ The 2nd equation

$$2(\mathbf{1}) - 5n = 7$$

$$2 - 5n - \mathbf{2} = 7 - \mathbf{2}$$

$$-5n = 5$$

$$\frac{-5n}{\mathbf{-5}} = \frac{5}{\mathbf{-5}}$$

$$n = -1$$

The solution is $(1, -1)$.

65. $\begin{cases} x - \dfrac{4}{3}y = \dfrac{1}{3} \\ 2x + \dfrac{3}{2}y = \dfrac{1}{2} \end{cases}$

Clear both equations of fractions.

Multiply both sides of the 1st equation by 3.

Multiply both sides of the 2nd equation by 2.

$$\begin{cases} \mathbf{3}(x) - \mathbf{3}\left(\dfrac{4}{3}y\right) = \mathbf{3}\left(\dfrac{1}{3}\right) \\ \mathbf{2}(2x) + \mathbf{2}\left(\dfrac{3}{2}y\right) = \mathbf{2}\left(\dfrac{1}{2}\right) \end{cases}$$

$$\begin{cases} 3x - 4y = 1 \\ 4x + 3y = 1 \end{cases}$$

Eliminate y.

Multiply both sides of the 1st equation by 3.

Multiply both sides of the 2nd equation by 4.

$$\begin{cases} (\mathbf{3})3x - (\mathbf{3})4y = (\mathbf{3})1 \\ (\mathbf{4})4x + (\mathbf{4})3y = (\mathbf{4})1 \end{cases}$$

$$\begin{array}{l} 9x - 12y = 3 \\ \underline{16x + 12y = 4} \\ 25x \qquad\quad = 7 \end{array}$$

$$\frac{25x}{\mathbf{25}} = \frac{7}{\mathbf{25}}$$

$$x = \frac{7}{25}$$

$4x + 3y = 1$ The 2nd equation

$$4\left(\frac{\mathbf{7}}{\mathbf{25}}\right) + 3y = 1$$

$$\frac{28}{25} + 3y = 1$$

$$\frac{28}{25} + 3y - \frac{\mathbf{28}}{\mathbf{25}} = \frac{25}{25} - \frac{\mathbf{28}}{\mathbf{25}}$$

$$3y = -\frac{3}{25}$$

$$\frac{\mathbf{1}}{\mathbf{3}}(3y) = -\frac{3}{25}\left(\frac{\mathbf{1}}{\mathbf{3}}\right)$$

$$y = -\frac{1}{25}$$

The solution is $\left(\dfrac{7}{25}, -\dfrac{1}{25}\right)$.

67. $\begin{cases} 4x - 7y + 32 = 0 \\ 5x = 4y - 2 \end{cases}$

Write both equations in standard form.

$$\begin{cases} 4x - 7y + 32 - \mathbf{32} = 0 - \mathbf{32} \\ 5x - \mathbf{4y} = 4y - 2 - \mathbf{4y} \end{cases}$$

$$\begin{cases} 4x - 7y = -32 \\ 5x - 4y = -2 \end{cases}$$

Eliminate x.

Multiply both sides of the 1st equation by 5.

Multiply both sides of the 2nd equation by -4.

$$\begin{cases} \mathbf{5}(4x) - \mathbf{5}(7y) = \mathbf{5}(-32) \\ (\mathbf{-4})5x(\mathbf{-4})(-4y) = (\mathbf{-4})(-2) \end{cases}$$

$$\begin{array}{r} 20x - 35y = -160 \\ \underline{-20x + 16y = \quad\;\; 8} \\ -19y = -152 \end{array}$$

$$\frac{-19y}{\mathbf{-19}} = \frac{-152}{\mathbf{-19}}$$

$$y = 8$$

$5x = 4y - 2$ The 2nd equation

$$5x = 4(\mathbf{8}) - 2$$

$$5x = 32 - 2$$

$$5x = 30$$

$$\frac{5x}{\mathbf{5}} = \frac{30}{\mathbf{5}}$$

$$x = 6$$

The solution is (6, 8).

69. $\begin{cases} 3(x + 4y) = -12 \\ x = 3y + 10 \end{cases}$

Distribute in 1st equation.

Write the 2nd equation in standard form.

$$\begin{cases} \mathbf{3}(x) + \mathbf{3}(12y) = -12 \\ x - \mathbf{3y} = 3y + 10 - \mathbf{3y} \end{cases}$$

$$\begin{cases} 3x + 12y = -12 \\ x - 3y = 10 \end{cases}$$

Eliminate y.

Multiply both sides of the 2nd equation by 4.

$$\begin{cases} 3x + 12y = -12 \\ \mathbf{4}(x) - \mathbf{4}(3y) = \mathbf{4}(10) \end{cases}$$

$$\begin{array}{l} 3x + 12y = -12 \\ \underline{4x - 12y = \;\; 40} \\ 7x \qquad\quad = \;\; 28 \end{array}$$

$$\frac{7x}{\mathbf{7}} = \frac{28}{\mathbf{7}}$$

$$x = 4$$

$$x-3y=10 \quad \text{The 2}^{\text{nd}} \text{ equation.}$$
$$\mathbf{4}-3y=10$$
$$4-3y-\mathbf{4}=10-\mathbf{4}$$
$$-3y=6$$
$$\frac{-3y}{\mathbf{-3}}=\frac{6}{\mathbf{-3}}$$
$$y=-2$$

The solution is $(4, -2)$.

71. $\begin{cases} 4a+7b=2 \\ 9a-3b=1 \end{cases}$

Eliminate b.

Multiply both sides of the 1st equation by 3.

Multiply both sides of the 2nd equation by 7.

$$\begin{cases} \mathbf{3}(4a)+\mathbf{3}(7b)=\mathbf{3}(2) \\ \mathbf{7}(9a)-\mathbf{7}(3b)=\mathbf{7}(1) \end{cases}$$

$$\begin{array}{r} 12a+21b=6 \\ 63a-21b=7 \\ \hline 75a \qquad =13 \end{array}$$

$$\frac{75a}{\mathbf{75}}=\frac{13}{\mathbf{75}}$$
$$a=\frac{13}{75}$$

$$9a-3b=1 \quad \text{The 2}^{\text{nd}} \text{ equation}$$
$$9\left(\frac{\mathbf{13}}{\mathbf{75}}\right)-3b=1$$
$$\frac{39}{25}-3b=1$$
$$\frac{39}{25}-3b-\frac{\mathbf{39}}{\mathbf{25}}=\frac{25}{25}-\frac{\mathbf{39}}{\mathbf{25}}$$
$$-3b=-\frac{14}{25}$$
$$-\frac{\mathbf{1}}{\mathbf{3}}(-3b)=-\frac{\mathbf{1}}{\mathbf{3}}\left(-\frac{14}{25}\right)$$
$$b=\frac{14}{75}$$

The solution is $\left(\frac{13}{75}, \frac{14}{75}\right)$.

73. $\begin{cases} 3a-b=12.3 \\ 4a-b=14.9 \end{cases}$

Eliminate b.

Multiply both sides of the 1st equation by -1.

$$\begin{cases} \mathbf{-1}(3a)-\mathbf{1}(-b)=\mathbf{-1}(12.3) \\ 4a-b=14.9 \end{cases}$$

$$\begin{array}{r} -3a+b=-12.3 \\ 4a-b=14.9 \\ \hline a \qquad =2.6 \end{array}$$

$$3a-b=12.3 \quad \text{The 1}^{\text{st}} \text{ equation}$$
$$3(\mathbf{2.6})-b=12.3$$
$$7.8-b=12.3$$
$$7.8-b-\mathbf{7.8}=12.3-\mathbf{7.8}$$
$$-b=4.5$$
$$\frac{-b}{\mathbf{-1}}=\frac{4.5}{\mathbf{-1}}$$
$$b=-4.5$$

The solution is $(2.6, -4.5)$.

75. $\begin{cases} 5x-4y=8 \\ -5x-4y=8 \end{cases}$

Eliminate x.

$$\begin{array}{r} 5x-4y=8 \\ -5x-4y=8 \\ \hline -8y=16 \end{array}$$

$$\frac{-8y}{\mathbf{-8}}=\frac{16}{\mathbf{-8}}$$
$$y=-2$$

$$5x-4y=8 \quad \text{1}^{\text{st}} \text{ equation}$$
$$5x-4(\mathbf{-2})=8$$
$$5x+8=8$$
$$5x+8-\mathbf{8}=8-\mathbf{8}$$
$$5x=0$$
$$\frac{5x}{\mathbf{5}}=\frac{0}{\mathbf{5}}$$
$$x=0$$

The solution is $(0, -2)$.

77. $\begin{cases} 9a+16b=-36 \\ 7a+4b=48 \end{cases}$

Eliminate b.

Multiply both sides of the 2nd equation by -4.

$\begin{cases} 9a+16b=-36 \\ \mathbf{-4}(7a)-\mathbf{4}(4b)=\mathbf{-4}(48) \end{cases}$

$$\begin{array}{r} 9a+16b= -36 \\ -28a-16b=-192 \\ \hline -19a \qquad =-228 \end{array}$$

$$\frac{-19a}{\mathbf{-19}}=\frac{-228}{\mathbf{-19}}$$

$$a=12$$

$$7a+4b=48 \quad \text{The 2}^{\text{nd}}\text{ equation}$$

$$7(\mathbf{12})+4b=48$$

$$84+4b=48$$

$$84+4b-\mathbf{84}=48-\mathbf{84}$$

$$4b=-36$$

$$\frac{4b}{\mathbf{4}}=\frac{-36}{\mathbf{4}}$$

$$b=-9$$

The solution is $(12,-9)$.

79. $\begin{cases} 8x+12y=-22 \\ 3x-2y=8 \end{cases}$

Eliminate y.

Multiply both sides of the 2nd equation by 6.

$\begin{cases} 8x+12y=-22 \\ (\mathbf{6})3x-(\mathbf{6})2y=(\mathbf{6})8 \end{cases}$

$$\begin{array}{r} 8x+12y=-22 \\ 18x-12y= \ \ 48 \\ \hline 26x \qquad =26 \end{array}$$

$$\frac{26x}{\mathbf{26}}=\frac{26}{\mathbf{26}}$$

$$x=1$$

$$3x-2y=8 \quad \text{The 2}^{\text{nd}}\text{ equation}$$

$$3(\mathbf{1})-2y=8$$

$$3-2y=8$$

$$3-2y-\mathbf{3}=8-\mathbf{3}$$

$$-2y=5$$

$$\frac{-2y}{\mathbf{-2}}=\frac{5}{\mathbf{-2}}$$

$$y=-\frac{5}{2}$$

The solution is $\left(1,-\frac{5}{2}\right)$.

81. $\begin{cases} 6x+5y+29=0 \\ 0.02x=0.03y-0.05 \end{cases}$

Write both equations in standard form.

$\begin{cases} 6x+5y=-29 \\ 0.02x-0.03y=-0.05 \end{cases}$

Multiply both sides of the 2nd equation by 100.

$\begin{cases} 6x+5y=-29 \\ \mathbf{100}(0.02x)-\mathbf{100}(0.03y)=\mathbf{100}(-0.05) \end{cases}$

$\begin{cases} 6x+5y=-29 \\ 2x-3y=-5 \end{cases}$

$\begin{cases} 6x+5y=-29 \\ (\mathbf{-3})2x-(\mathbf{-3})3y=(\mathbf{-3})(-5) \end{cases}$

$$\begin{array}{r} 6x+5y=-29 \\ -6x+9y= \ \ 15 \\ \hline 14y=-14 \end{array}$$

$$\frac{14y}{\mathbf{14}}=\frac{-14}{\mathbf{14}}$$

$$y=-1$$

$$2x-3y=-5 \quad \text{The 2}^{\text{nd}}\text{ equation}$$

$$2x-3(\mathbf{-1})=-5$$

$$2x+3=-5$$

$$2x+3-\mathbf{3}=-5-\mathbf{3}$$

$$2x=-8$$

$$\frac{2x}{\mathbf{2}}=\frac{-8}{\mathbf{2}}$$

$$x=-4$$

The solution is $(-4,-1)$.

83. $\begin{cases} c=d-9 \\ 5c=3d-35 \end{cases}$

Write both equations in standard form.

$\begin{cases} c-d=-9 \\ 5c-3d=-35 \end{cases}$

Eliminate c.

Multiply both sides of the 1st equation by -5.

$\begin{cases} (\mathbf{-5})c-(\mathbf{-5})d=(\mathbf{-5})(-9) \\ 5c-3d=-35 \end{cases}$

$$\begin{array}{r} -5c+5d= \ \ 45 \\ 5c-3d=-35 \\ \hline 2d=10 \end{array}$$

$$\frac{2d}{\mathbf{2}}=\frac{10}{\mathbf{2}}$$

$$d=5$$

$$c=d-9 \quad \text{The 1}^{\text{st}}\text{ equation}$$

$$c=\mathbf{5}-9$$

$$c=-4$$

The solution is $(-4,5)$.

85. $\begin{cases} 0.9x + 2.1 = 0.3y \\ 0.4x = 0.7y + 1.9 \end{cases}$

Write both equations in standard form.

$\begin{cases} 0.9x - 0.3y = -2.1 \\ 0.4x - 0.7y = 1.9 \end{cases}$

Multiply both sides of both equations by 10.

$\begin{cases} \mathbf{10}(0.9x) - \mathbf{10}(0.3y) = \mathbf{10}(-2.1) \\ \mathbf{10}(0.4x) - \mathbf{10}(0.7y) = \mathbf{10}(1.9) \end{cases}$

$\begin{cases} 9x - 3y = -21 \\ 4x - 7y = 19 \end{cases}$

Eliminate y.

Multiply both sides of the 1st equation by 7.

Multiply both sides of the 2nd equation by -3.

$\begin{cases} \mathbf{7}(9x) - \mathbf{7}(3y) = \mathbf{7}(-21) \\ \mathbf{-3}(4x) - \mathbf{3}(-7y) = \mathbf{-3}(19) \end{cases}$

$$\begin{array}{rl} 63x - 21y &= -147 \\ -12x + 21y &= -57 \\ \hline 51x &= -204 \end{array}$$

$$\frac{51x}{\mathbf{51}} = \frac{-204}{\mathbf{51}}$$

$$x = -4$$

$9x + 21 = 3y$ The 1st equation

$$9(\mathbf{-4}) + 21 = 3y$$
$$-36 + 21 = 3y$$
$$-15 = 3y$$
$$\frac{-15}{\mathbf{3}} = \frac{3y}{\mathbf{3}}$$
$$-5 = y$$

The solution is $(-4, -5)$.

87. $\begin{cases} 5c + 2d = -5 \\ 6c + 2d = -10 \end{cases}$

Eliminate d.

Multiply both sides of the 1st equation by -1.

$\begin{cases} \mathbf{-1}(5c) - \mathbf{1}(2d) = \mathbf{-1}(-5) \\ 6c + 2d = -10 \end{cases}$

$$\begin{array}{rl} -5c - 2d &= 5 \\ 6c + 2d &= -10 \\ \hline c &= -5 \end{array}$$

$5c + 2d = -5$ The 1st equation

$$5(\mathbf{-5}) + 2d = -5$$
$$-25 + 2d = -5$$
$$-25 + 2d + \mathbf{25} = -5 + \mathbf{25}$$
$$2d = 20$$
$$\frac{2d}{\mathbf{2}} = \frac{20}{\mathbf{2}}$$
$$d = 10$$

The solution is $(-5, 10)$.

89. $\begin{cases} \frac{2}{15}x - \frac{1}{5}y = \frac{1}{3} \\ \frac{2}{15}x - \frac{1}{5}y = \frac{1}{10} \end{cases}$

Clear both equations of fractions.

Multiply both sides of the 1st equation by 30.

Multiply both sides of the 2nd equation by -30.

$\begin{cases} \mathbf{30}\left(\frac{2}{15}x\right) - \mathbf{30}\left(\frac{1}{5}y\right) = \mathbf{30}\left(\frac{1}{3}\right) \\ \mathbf{-30}\left(\frac{2}{15}x\right) - \mathbf{30}\left(-\frac{1}{5}y\right) = \mathbf{-30}\left(\frac{1}{10}\right) \end{cases}$

$$\begin{array}{rl} 4x - 6y &= 10 \\ -4x + 6y &= -3 \\ \hline 0 &= 7 \end{array}$$

Both variables are eliminated.

False

No solution.

APPLICATIONS

91. EDUCATION

Let x = the number of years since 1980

y = the % less than high school

$\begin{cases} 9x + 11y = 352 \\ 5x - 11y = -198 \end{cases}$

Eliminate y.

$$\begin{array}{rl} 9x + 11y &= 352 \\ 5x - 11y &= -198 \\ \hline 14x &= 154 \end{array}$$

$$\frac{14x}{\mathbf{14}} = \frac{154}{\mathbf{14}}$$
$$x = 11$$

$x = 0$ implies 1980

$x = 11$ implies $1980 + 11$

The year when the percents are equal is 1991.

93. CFL BULBS

Let c = the operating cost of a CFL bulb

d = the number of days before savings start for the CFL bulb

$$\begin{cases} 60c - d = 96 \\ 15c - d = 6 \end{cases}$$

Eliminate d.

Multiply both sides of the 2nd equation by -1.

$$\begin{cases} 60c - d = 96 \\ \mathbf{-1}(15c) - \mathbf{1}(-d) = \mathbf{-1}(6) \end{cases}$$

$$\begin{aligned} 60c - d &= 96 \\ -15c + d &= -6 \\ \hline 45c \quad &= 90 \end{aligned}$$

$$\frac{45c}{\mathbf{45}} = \frac{90}{\mathbf{45}}$$

$$c = 2$$

$60c - d = 96$ The 1st equation

$$\begin{aligned} 60(\mathbf{2}) - d &= 96 \\ 120 - d - \mathbf{120} &= 96 - \mathbf{120} \\ -d &= -24 \\ -1(-d) &= -1(-24) \\ d &= 24 \end{aligned}$$

After 24 days the CFL bulb will begin to save money.

WRITING

95 - 97 Answers will vary.

REVIEW

99. Find an equation of the line with slope $-\frac{11}{6}$ that passes through $(2, -6)$. Write the equation in slope-intercept form.

$$y - y_1 = m(x - x_1)$$

$$y - (-6) = -\frac{11}{6}(x - 2)$$

$$y + 6 = -\frac{11}{6}x + \frac{22}{6}$$

$$y + 6 - \mathbf{6} = -\frac{11}{6}x + \frac{22}{6} - \frac{\mathbf{36}}{\mathbf{6}}$$

$$y = -\frac{11}{6}x - \frac{14}{6}$$

$$y = -\frac{11}{6}x - \frac{7}{3} \quad \text{slope-intercept form}$$

101. Evaluate: $-10(18 - 4^2)^3$

$$\begin{aligned} -10\left(18 - 4^2\right)^3 &= -10\left(18 - 16\right)^3 \\ &= -10\left(2\right)^3 \\ &= -10(8) \\ &= -80 \end{aligned}$$

CHALLENGE PROBLEMS

Use the elimination method to solve each system.

103. $\begin{cases} \dfrac{x-3}{2} = \dfrac{11}{6} - \dfrac{y+5}{3} \\ \dfrac{x+3}{3} - \dfrac{y+3}{4} = \dfrac{5}{12} \end{cases}$

Clear both equations of fractions

$$\begin{cases} \mathbf{6}\left(\dfrac{x-3}{2}\right) = \mathbf{6}\left(\dfrac{11}{6}\right) - \mathbf{6}\left(\dfrac{y+5}{3}\right) \\ \mathbf{12}\left(\dfrac{x+3}{3}\right) - \mathbf{12}\left(\dfrac{y+3}{4}\right) = \mathbf{12}\left(\dfrac{5}{12}\right) \end{cases}$$

$$\begin{cases} 3(x-3) = 11 - 2(y+5) \\ 4(x+3) - 3(y+3) = 5 \end{cases}$$

Distribute twice in both equations.

Collect like terms

$$\begin{cases} 3x - 9 = 1 - 2y \\ 4x - 3y + 3 = 5 \end{cases}$$

Write both equations in standard form.

$$\begin{cases} 3x - 9 + \mathbf{2y + 9} = 1 - 2y + \mathbf{2y + 9} \\ 4x - 3y + 3 - \mathbf{3} = 5 - \mathbf{3} \end{cases}$$

$$\begin{cases} 3x + 2y = 10 \\ 4x - 3y = 2 \end{cases}$$

Eliminate y.

Multiply both sides of the 1st equation by 3.

Multiply both sides of the 2nd equation by 2.

$$\begin{cases} \mathbf{3}(3x) + \mathbf{3}(2y) = \mathbf{3}(10) \\ \mathbf{2}(4x) - \mathbf{2}(3y) = \mathbf{2}(2) \end{cases}$$

$$\begin{aligned} 9x + 6y &= 30 \\ 8x - 6y &= \ \ 4 \\ \hline 17x \quad &= 34 \end{aligned}$$

$$\frac{17x}{\mathbf{17}} = \frac{34}{\mathbf{17}}$$

$$x = 2$$

$3x + 2y = 10$ The 1st equation

$$\begin{aligned} 3(\mathbf{2}) + 2y &= 10 \\ 6 + 2y - \mathbf{6} &= 10 - \mathbf{6} \\ 2y &= 4 \\ \frac{2y}{\mathbf{2}} &= \frac{4}{\mathbf{2}} \\ y &= 2 \end{aligned}$$

The solution is (2, 2).

SECTION 4.4 STUDY SET

VOCABULARY

Fill in the blanks.

1. Two angles are said to be **complementary** if the sum of their measures is 90°. Two angles are said to be **supplementary** if the sum of their measures is 180°.

CONCEPTS

3. A length of pipe is to be cut into two pieces. The longer piece is to be 1 foot less than twice the shorter piece. Write two equations that model the situation. $\mathbf{x + y = 20,\ y = 2x - 1}$

5. Two angles are supplementary. The measure of the smaller angle is 25° less than the measure of the larger angle. Write two equations that model the situation. $\mathbf{x + y = 180, y = x - 25}$

7. Let x = the cost of a chicken taco and y = the cost of a beef taco. Write an equation that models the offer shown in the advertisement.

   $\mathbf{5x + 2y = 15}$

9. For each case below, write an algebraic expression that represents the speed of the canoe in miles per hour if its speed in still water is x mph. $\mathbf{x + c,\ x - c}$

11. a. If the contents of the two test tubes are poured into a third tube, how much solution will the third tube contain? (mL stands for milliliter. A milliliter is about two drops from an eyedropper.) **(x + y) mL**

    b. Which of the following strengths could the mixture possibly be: 27%, 33%, or 44% acid solution? **33%**

GUIDED PRACTICE

13. COMPLEMENTARY ANGLES

Analyze

- Two angles are complementary (90°).
- One angle is 10° more than three times the measure of the other.
- Find the measure of each angle.

Assign

Let x = the measure of one ∠

y = the measure of the other ∠

Form

The measure of one angle	plus	the measure of the other angle	is	complementary.
x	$+$	y	$=$	90

One angle	is	10°	more than	3 times the other angle.
x	$=$	10	$+$	$3y$

Solve

$$\begin{cases} x + y = 90 \\ x = 10 + 3y \end{cases}$$

Write the 2nd equation in standard form.

$$\begin{cases} x + y = 90 \\ x - 3y = 10 \end{cases}$$

Eliminate x.

Multply both sides of the 2nd equation by -1.

$$\begin{aligned} x + \ \ y &= \ \ 90 \\ -x + 3y &= -10 \\ \hline 4y &= 80 \end{aligned}$$

$$\frac{4y}{\mathbf{4}} = \frac{80}{\mathbf{4}}$$

$$y = 20$$

$x = 10 + 3y$ The 2nd equation

$x = 10 + 3(20)$

$x = 10 + 60$

$x = 70$

State

The measure of one angle is 70°.

The measure of the other angle is 20°.

Check

$70 + 20 = 90$ | $70 = 10 + 3(20)$

$70 = 70$

The results check.

15. SUPPLEMENTARY ANGLES

Analyze

- Two angles are supplementary (180°).
- The difference of the measures of two supplementary angles is 80°.
- Find the measure of each angle.

Assign

Let x = the measure of one $\angle$ (larger$\angle$)

y = the measure of the other $\angle$ (smaller$\angle$)

Form

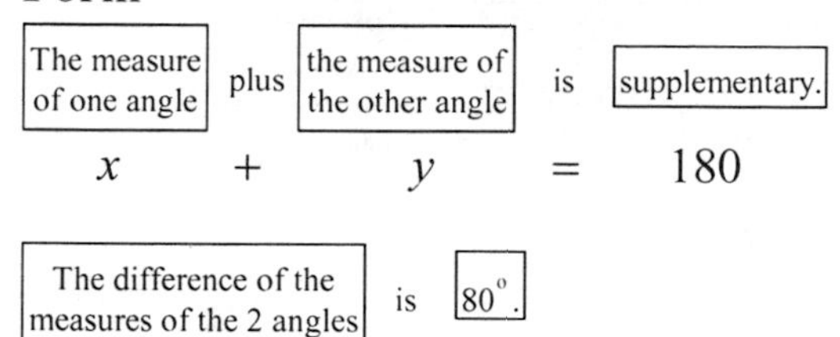

Solve

$$\begin{cases} x + y = 180 \\ x - y = 80 \end{cases}$$

Eliminate y.

$$\begin{aligned} x + y &= 180 \\ x - y &= 80 \\ \hline 2x &= 260 \end{aligned}$$

$$\frac{2x}{\mathbf{2}} = \frac{260}{\mathbf{2}}$$

$$x = 130$$

$x + y = 180$ The 1st equation

$$\mathbf{130} + y = 180$$

$$130 + y - \mathbf{130} = 180 - \mathbf{130}$$

$$y = 50$$

State

The measure of one angle is 130°.
The measure of the other angle is 50°.

Check

$130 + 50 = 180$	$130 - 50 = 80$
$180 = 180$	$80 = 80$

The results check.

APPLICATION

Write a system of two equations in two variables to solve each problem.

17. TREE TRIMMING

Analyze

- The total length of both arms is 51 ft.
- The upper arm is 7 ft. shorter than the lower arm.
- Find the length of each arm.

Assign

Let x = length of upper arm in feet (shorter length)

y = length of lower arm in feet (longer length)

The length of the upper arm	plus	the length of the lower arm	is	51 feet.
x	$+$	y	$=$	51

Form

The upper arm	is	7 feet	shorter than	the lower arm.
x	$=$	y	$-$	7

Solve

$$\begin{cases} x + y = 51 \\ x = y - 7 \end{cases}$$

Write the 2nd equation in standard form.

$$\begin{cases} x + y = 51 \\ x - y = -7 \end{cases}$$

Eliminate y.

$$\begin{aligned} x + y &= 51 \\ x - y &= -7 \\ \hline 2x &= 44 \end{aligned}$$

$$\frac{2x}{\mathbf{2}} = \frac{44}{\mathbf{2}}$$

$$x = 22$$

$x + y = 51$ The 1st equation

$$\mathbf{22} + y = 51$$

$$22 + y - \mathbf{22} = 51 - \mathbf{22}$$

$$y = 29$$

State

The measure of the upper arm is 22 ft.
The measure of the lower arm is 29 ft.

Check

$22 + 29 = 51$	$22 = 29 - 7$
$51 = 51$	$22 = 22$

The results check.

19. GOVERNMENT

Analyze

- Both salaries total \$627,300.
- The president's makes \$172,700 more than VP.
- Find the salary of each.

Assign

Let x = president's salary (larger salary)

y = vice-president salary (smaller salary)

Form

The salary of the President	plus	the salary of the VP	is	\$627,300.
x	$+$	y	$=$	627,300

The President	makes	\$172,700	more than	the VP.
x	$=$	172,700	$+$	y

Solve

$$\begin{cases} x+y=627,300 \\ x=172,700+y \end{cases}$$

Write the 2nd equation in standard form

$$\begin{cases} x+y=627,300 \\ x-y=172,700 \end{cases}$$

Eliminate y.

$$x+y=627,300$$
$$\underline{x-y=172,700}$$
$$2x \quad =800,000$$
$$\frac{2x}{\mathbf{2}}=\frac{800,000}{\mathbf{2}}$$
$$x=400,000$$

$$x+y=627,300$$
$$\mathbf{400,000}+y=627,300$$
$$400,000+y-\mathbf{400,000}=627,300-\mathbf{400,000}$$
$$y=227,300$$

State

The President's salary is \$400,000.
The Vice-President's salary is \$227,300.

Check

$400,000+227,300=627,300$	$400,000=172,700+227,300$
$627,300=627,300$	$400,000=400,000$

The results check.

GEOMETRY PROBLEMS

21. MONUMENTS

Analyze

- Two angles are supplementary (180º).
- The measure of $\angle 2$ is 15º less than twice the measure of $\angle 1$.
- Find the measure of each angle.

Assign

Let x = the measure of $\angle 1$ (larger$\angle$)

y = the measure of $\angle 2$ (smaller$\angle$)

Form

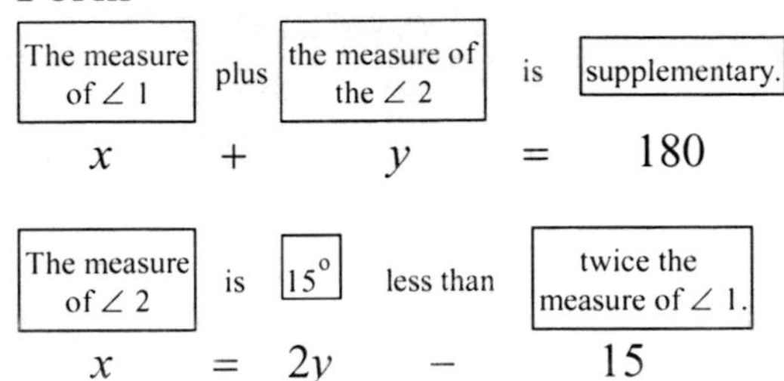

Solve

$$\begin{cases} x+y=180 \\ y=2x-15 \end{cases}$$

$x+y=180$ The 1st equation. Substitute for x.

$$x+(\mathbf{2x-15})=180$$
$$3x-15=180$$
$$3x-15+\mathbf{15}=180+\mathbf{15}$$
$$\frac{3x}{\mathbf{3}}=\frac{195}{\mathbf{3}}$$
$$x=65$$

$y=2x-15$ The 2nd equation

$$y=2(\mathbf{65})-15$$
$$y=130-15$$
$$y=115$$

State

The measure of $\angle 1$ is 65º.
The measure of $\angle 2$ is 115º.

Check

$65+115=180$	$115=2(65)-15$
$180=180$	$115=130-15$
	$115=115$

The results check.

23. THEATER SCREENS

Analyze

- A giant rectangular movie screen has a width 26 feet less than its length.
- Its perimeter is 332 feet.
- Find the measures of the width and length.

Assign

Let w = width of screen in ft (shorter length)
l = length of screen in ft (longer length)

Form

The measure of two widths	plus	the measure of two lengths	is	the perimerter of 332 feet.
$2w$	$+$	$2l$	$=$	332

The width	is	26 ft	less than	the length.
w	$=$	l	$-$	26

Solve

$$\begin{cases} 2l+2w=332 \\ w=l-26 \end{cases}$$

$2l+2w=332$ The 1st equation Substitute for w.

$$2l+2(\mathbf{l-26})=332$$
$$2l+2l-52=332$$
$$4l-52=332$$
$$4l-52+\mathbf{52}=332+\mathbf{52}$$
$$4l=384$$
$$\frac{4l}{\mathbf{4}}=\frac{384}{\mathbf{4}}$$
$$l=96$$

$w=l-26$ The 2nd equation

$$w=\mathbf{96}-26$$
$$w=70$$

State

The width of the screen is 70 feet.
The length of the screen is 96 feet.

Check

$2(70)+2(96)=332$	$70=96-26$
$140+192=332$	$70=70$
$332=332$	

The results check.

25. GEOMETRY

Analyze

- A 50-meter path surrounds a rectangular garden.
- The width of the garden is two-thirds its length.
- Find the measures of the width and length.

Assign

Let w = width of garden in m (shorter length)
l = length of garden in m (longer length)

Form

The measure of two widths	plus	the measure of two lengths	is	the perimeter of 50 meters.
$2w$	$+$	$2l$	$=$	50

The width	is	two-thirds	of	the length.
w	$=$	$\frac{2}{3}$	$\bullet$	l

Solve

$$\begin{cases} 2l+2w=50 \\ w=\frac{2}{3}l \end{cases}$$

Clear the fraction from the 2nd equation by multiplying both sides by 3.

$$\begin{cases} 2l+2w=50 \\ 2l=3w \end{cases}$$

$2l+2w=50$ The 1st equation Substitute for $2l$.

$$\mathbf{3w}+2w=50$$
$$5w=50$$
$$\frac{5w}{\mathbf{5}}=\frac{50}{\mathbf{5}}$$
$$w=10$$

$2l+2w=50$ The 1st equation

$$2l+2(\mathbf{10})=50$$
$$2l+20=50$$
$$2l+20-\mathbf{20}=50-\mathbf{20}$$
$$2l=30$$
$$\frac{2l}{\mathbf{2}}=\frac{30}{\mathbf{2}}$$
$$l=15$$

State

The width of the garden is 10 m.
The length of the garden is 15 m.

Check

$2(15)+2(10)=50$	$3(10)=2(15)$
$30+20=50$	$30=30$
$50=50$	

The results check.

NUMBER/VALUE PROBLEMS

27. EMPTY CARTRIDGES

Analyze

- \$40 for 5 printer and 2 copier cartridges.
- \$57 for 6 printer and 3 copier cartridges.
- How much is each type of cartridge?

Assign

Let x = value for a printer cartridge

y = value for a copier cartridge

	Number •	Value	= Total Value
Printer	5	x	**5x**
Copier	2	y	**2y**
		Total	**40**

Form

The value of 5 printer cartridges	plus	the value of 2 copier cartridges	is	\$40.
$5x$	$+$	$2y$	$=$	40

	Number •	Value	= Total Value
Printer	6	x	**6x**
Copier	3	y	**3y**
		Total	**57**

The value of 6 printer cartridges	plus	the value of 3 copier cartridges	is	\$57.
$6x$	$+$	$3y$	$=$	57

Solve

$$\begin{cases} 5x+2y=40 \\ 6x+3y=57 \end{cases}$$

Eliminate y.

Multiply both sides of the 1st equation by 3.

Multiply both sides of the 2nd equation by -2.

$$\begin{array}{r} 15x+6y=\ \ 120 \\ -12x-6y=-114 \\ \hline 3x\ \ \ \ \ \ \ =6 \end{array}$$

$$\frac{3x}{\mathbf{3}}=\frac{6}{\mathbf{3}}$$

$$x=2$$

$$5x+2y=40 \quad \text{The 1st equation}$$

$$5(\mathbf{2})+2y=40$$

$$10+2y-\mathbf{10}=40-\mathbf{10}$$

$$2y=30$$

$$\frac{2y}{\mathbf{2}}=\frac{30}{\mathbf{2}}$$

$$y=15$$

State

Bank was paid \$2 for each printer cartridge.

Bank was paid \$15 for each copier cartridge.

Check

$5(\$2)+2(\$15)=\$40$	$6(\$2)+3(\$15)=\$57$
$\$10+\$30=\$40$	$\$12+\$45=\$57$
$\$40=\$40$	$\$57=\$57$

The results check.

from CAMPUS TO CAREERS

29. PORTRAIT PHOTOGRAPHER

Analyze

- Pk 1: 1 10×14 and 10 8×10 photos cost \$239.50.
- Pk 2: 1 10×14 and 5 8×10 photos cost \$134.50.
- Find the cost of each size photo.

Assign

Let x = the cost of one 10x14 photo

y = the cost of one 8x10 photo

	Number •	Value	= Total Value
10x14	1	x	***x***
8x10	10	y	**10y**
		Total	**239.50**

Form

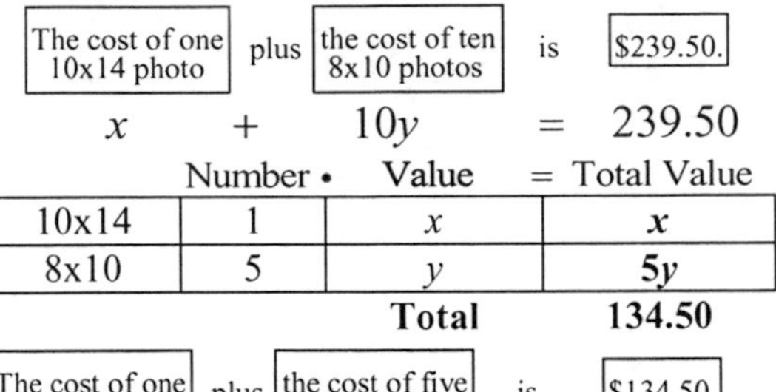

The cost of one 10x14 photo	plus	the cost of ten 8x10 photos	is	\$239.50.
x	$+$	$10y$	$=$	239.50

	Number •	Value	= Total Value
10x14	1	x	***x***
8x10	5	y	**5y**
		Total	**134.50**

The cost of one 10x14 photo	plus	the cost of five 8x10 photos	is	\$134.50.
x	$+$	$5y$	$=$	134.50

Solve

$$\begin{cases} x+10y=239.50 \\ x+5y=134.50 \end{cases}$$

Eliminate y.

Multiply both sides of the 2nd equation by -1.

$$\begin{array}{r} x+10y=\ \ 239.50 \\ -x-5y\ =-134.50 \\ \hline \end{array}$$

$$5y=105$$

$$\frac{5y}{\mathbf{5}}=\frac{105}{\mathbf{5}}$$

$$y=21$$

$$x+5y=134.50 \quad \text{The 2nd equation}$$

$$x+5(\mathbf{21})=134.50$$

$$x+105-\mathbf{105}=134.50-\mathbf{105}$$

$$x=29.50$$

State

The cost of one 10×14 photo is \$29.50.

The cost of one 8×10 photo is \$21.

Check

$\mathbf{\$29.50}+10(\mathbf{\$21})=\$239.50$	$\mathbf{\$29.50}+5(\mathbf{\$21})=\$134.50$
$\$29.50+\$210=\$239.50$	$\$29.50+\$105=\$134.50$
$\$239.50=\$239.50$	$\$134.50=\$134.50$

The results check.

31. COLLECTING STAMPS

Analyze

- One Elvis stamp and one Liberty stamp cost a total of $0.63.
- A sheet of 40 Elvis stamps and a sheet of 20 Liberty stamps cost a total of $18.40.
- How much is each stamp?

Assign

Let x = the cost of one Elvis stamp
y = the cost of one Liberty stamp

	Number •	Value	= Total Value
Elvis	1	x	x
Liberty	1	y	y
		Total	**0.63**

Form

The value of one Elvis stamp plus the value of one Liberty stamp is $0.63.

$$x + y = 0.63$$

	Number •	Value	= Total Value
Elvis	40	x	**40x**
Liberty	20	y	**20y**
		Total	**18.40**

The value of 40 Elvis stamps plus the value of 20 Liberty stamps is $18.40.

$$40x + 20y = 18.40$$

Solve

$$\begin{cases} x + y = 0.63 \\ 40x + 20y = 18.40 \end{cases}$$

Eliminate y.
Multiply both sides of the 1st equation by -20.

$$-20x - 20y = -12.60$$
$$40x + 20y = 18.40$$
$$20x = 5.80$$
$$\frac{20x}{\mathbf{20}} = \frac{5.80}{\mathbf{20}}$$
$$x = 0.29$$

$x + y = 0.63$ The 1st equation
$$(\mathbf{0.29}) + y = 0.63$$
$$0.29 + y - \mathbf{0.29} = 0.63 - \mathbf{0.29}$$
$$y = 0.34$$

State

The cost of one Elvis stamp is $0.29.
The cost of one Liberty stamp is $0.34.

Check

$\$0.29 + \$0.34 = \$.63$
$\$.63 = \$.63$

$40(\$0.29) + 20(\$0.34) = \$18.40$
$\$11.60 + \$6.80 = \$18.40$
$\$18.40 = \$18.40$

The results check.

33. SELLING ICE CREAM

Analyze

- Ice cream cones cost $1.80 and sundaes cost $3.30.
- The receipts for a total of 148 cones and sundaes were $360.90.
- How many of each was sold?

Assign

Let x = number of cones sold
y = number of sundaes sold

Form

The number of cones plus the number of sundaes is 148.

$$x + y = 148$$

	Number •	Value	= Total Value
Cones	x	1.80	**1.80x**
Sundaes	y	3.30	**3.30y**
		Total	**360.90**

The value of x cones plus the value of y sundaes is $360.90.

$$1.80x + 3.30y = 360.90$$

Solve

$$\begin{cases} x + y = 148 \\ 1.80x + 3.30y = 360.90 \end{cases}$$

Solve the 1st equation for x.

$$\begin{cases} x = 148 - y \\ 1.80x + 3.30y = 360.90 \end{cases}$$

$1.80x + 3.30y = 360.90$ The 2nd equation
$1.80(\mathbf{148 - y}) + 3.30y = 360.90$ Substitute for x.
$$266.40 - 1.80y + 3.30y = 360.90$$
$$266.40 + 1.50y = 360.90$$
$$266.40 + 1.50y - \mathbf{266.40} = 360.90 - \mathbf{266.40}$$
$$1.50y = 94.50$$
$$\frac{1.50y}{\mathbf{1.50}} = \frac{94.50}{\mathbf{1.50}}$$
$$y = 63$$

$x + y = 148$ The 1st equation
$$x + \mathbf{63} = 148$$
$$x + 63 - \mathbf{63} = 148 - \mathbf{63}$$
$$x = 85$$

State

The number of cones sold is 85.
The number of sunades sold is 63.

Check

$85 + 63 = 148$
$148 = 148$

$\$1.80(\mathbf{85}) + \$3.30(\mathbf{63}) = \$360.90$
$\$153 + \$207.90 = \$360.90$
$\$360.90 = \$360.90$

The results check.

INTEREST PROBLEMS

35. STUDENT LOANS

Analyze

- A college used \$5,000 to make two student loans.
- The first at 5% annual interest to a nursing student.
- The second at 7% to a business major.
- College collected \$310 in interest the first year.
- How much was loaned to each student?

Assign

Let x = amount loaned to nursing student

y = amount loaned to business student

Form

The amount loaned to nursing student	plus	the amount loaned to business student	is	\$5,000.
x	$+$	y	$=$	5,000

	Principal	· Rate	· Time	= Interest
Nurse	x	5%	1 yr	$\mathbf{0.05x}$
Business	y	7%	1 yr	$\mathbf{0.07y}$

Total Interest = \$310

The amount of interest of 5% loan	plus	the amount of interest of 7% loan	is	\$310.
$0.05x$	$+$	$0.07y$	$=$	310

Solve

$$\begin{cases} x+y=5,000 \\ 0.05x+0.07y=310 \end{cases}$$

Eliminate x.

Multiply both sides of the 1st equation by -5.

Multiply both sides of the 2nd equation by 100.

$$\begin{aligned} -5x-5y &= -25,000 \\ 5x+7y &= 31,000 \\ \hline 2y &= 6,000 \\ \frac{2y}{\mathbf{2}} &= \frac{6,000}{\mathbf{2}} \\ y &= 3,000 \end{aligned}$$

$$\begin{aligned} x+y &= 5,000 \quad \text{The 1st equation} \\ x+\mathbf{3,000} &= 5,000 \\ x+3,000-\mathbf{3,000} &= 5,000-\mathbf{3,000} \\ x &= 2,000 \end{aligned}$$

State

The nursing student loan is \$2,000.

The business student loan is \$3,000

Check

$\$2,000+\$3,000=\$5,000$

$\$5,000=\$5,000$

$0.05(\$2,000)+0.07(\$3,000)=\$310$

$\$100+\$210=\$310$

$\$310=\$310$

The results check.

37. INVESTING A BONUS

Analyze

- Bonus amount is \$40,000.
- Invested part in a fund at 8% annual interest.
- The rest in an offshore bank at 9% annual interest.
- \$3,415 in interest was earned the first year.
- How much was invested at each rate?

Assign

Let x = amount invested int fund at 8%

y = amount invested offshore at 9%

Form

The amount invested in an int. fund	plus	the amount invested in an offshore bank	is	\$40,000.
x	$+$	y	$=$	40,000

	Principal	· Rate	· Time	= Interest
Int fund	x	0.08	1 yr	$\mathbf{0.08x}$
Offshore bank	y	0.09	1 yr	$\mathbf{0.09y}$

Total Interest = \$3,415

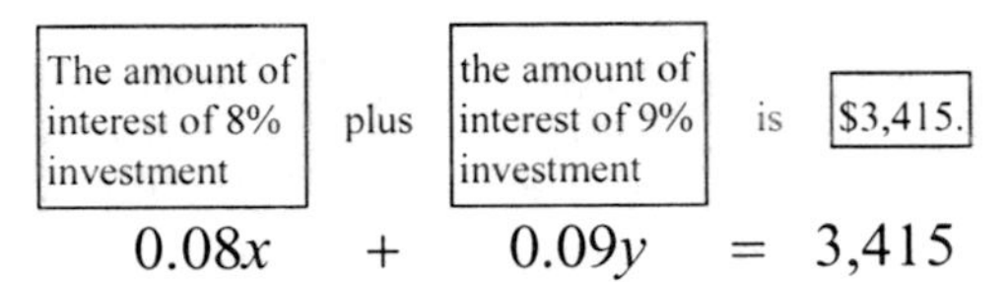

$0.08x \quad + \quad 0.09y \quad = \quad 3,415$

Solve

$$\begin{cases} x+y=40,000 \\ 0.08x+0.09y=3,415 \end{cases}$$

Eliminate x.

Multiply both sides of the 1st equation by -8.

Multiply both sides of the 2nd equation by 100.

$$\begin{aligned} -8x-8y &= -320,000 \\ 8x+9y &= 341,500 \\ \hline y &= 21,500 \end{aligned}$$

$$\begin{aligned} x+y &= 40,000 \quad \text{The 1st equation} \\ x+\mathbf{21,500} &= 40,000 \\ x+21,500-\mathbf{21,500} &= 40,000-\mathbf{21,500} \\ x &= 18,500 \end{aligned}$$

State

\$18,500 was invested at 8% with Int fund.

\$21,500 was invested at 9% with offshore.

Check

$\$18,500+\$21,500=\$40,000$

$\$40,000=\$40,000$

$0.08(\$18,500)+0.09(\$21,500)=\$3,415$

$\$1,480+\$1,935=\$3,415$

$\$3,415=\$3,415$

The results check.

39. LOSSES

Analyze the Problem

- CEO has $22,000 to invest.
- Invested part at 4% annual interest.
- Invested in biotech at 3% annual interest.
- Biotech showed a loss after first year.
- $110 in interest was earned the first year.
- Find the amount of each investment.

Assign

Let $x =$ amount invested at 4%

$y =$ amount invested at 3%

Form

The amount invested in the 4% account	plus	the amount invested in the 3% biotech	is	$22,000.
x	$+$	y	$=$	22,000

	Principal •	Rate •	Time =	Interest
account	x	0.04	1 yr	**0.04x**
biotech	y	0.03	1 yr	**– 0.03y**

Total Interest = $110

The amount of interest of 4% investment	plus	the amount of interest of 3% investment lost	is	$110.
$0.04x$	$-$	$0.03y$	$=$	110

Solve

$$\begin{cases} x + y = 22,000 \\ 0.04x - 0.03y = 110 \end{cases}$$

Eliminate y.

Multiply both sides of the 1st equation by 3.

Multiply both sides of the 2nd equation by 100.

$$\begin{aligned} 3x + 3y &= 66,000 \\ 4x - 3y &= 11,000 \\ \hline 7x \quad &= 77,000 \end{aligned}$$

$$\frac{7x}{\mathbf{7}} = \frac{77,000}{\mathbf{7}}$$

$$x = 11,000$$

$$x + y = 22,000 \quad \text{The 1st equation}$$

$$x + \mathbf{11,000} = 22,000$$

$$x + 11,000 - \mathbf{11,000} = 22,000 - \mathbf{11,000}$$

$$x = 11,000$$

State

$11,000 was invested at 4%.

$11,000 was invested at 3%.

Check

$\$11,000 + \$11,000 = \$22,000$

$\$22,000 = \$22,000$

$0.04(\$11,000) - 0.03(\$11,000) = \$110$

$\$440 - \$330 = \$110$

$\$110 = \$110$

The results check.

UNIFORM MOTION PROBLEMS

41. THE GULF STREAM

Analyze

- With the current, a ship traveled 300 mi in 10 hrs.
- Against the current, it took 15 hours for return trip.
- Find the speed of the ship in still water and the speed of the current.

Assign

Let $x =$ speed of ship in still water in mph

$y =$ speed of current in mph

	Rate •	Time =	Distance
With current	$x + y$	10	$\mathbf{10(x+y)}$

Total Distance = 300

Form

The rate of the ship with the current	times	10 hours traveled	is	300 miles.
$(x + y)$	•	10	$=$	300

	Rate •	Time =	Distance
Against current	$x - y$	15	$\mathbf{15(x-y)}$

Total Distance = 300

The rate of the ship against the current	times	15 hours traveled	is	300 miles.
$(x - y)$	•	15	$=$	300

Solve

$$\begin{cases} 10(x + y) = 300 \\ 15(x - y) = 300 \end{cases}$$

Distribute in both equations.

$$\begin{cases} 10x + 10y = 300 \\ 15x - 15y = 300 \end{cases}$$

Eliminate y.

Multiply both sides of the 1st equation by 3.

Multiply both sides of the 2nd equation by 2.

$$\begin{aligned} 30x + 30y &= 900 \\ 30x - 30y &= 600 \\ \hline 60x \quad &= 1,500 \end{aligned}$$

$$\frac{60x}{\mathbf{60}} = \frac{1,500}{\mathbf{60}}$$

$$x = 25$$

$$10(x + y) = 300 \quad \text{The 1st equation}$$

$$10(\mathbf{25} + y) = 300$$

$$250 + 10y = 300$$

$$250 + 10y - \mathbf{250} = 300 - \mathbf{250}$$

$$10y = 50$$

$$\frac{10y}{\mathbf{10}} = \frac{50}{\mathbf{10}}$$

$$y = 5$$

State

The speed of the ship in still water is 25 mph.

The speed of the current is 5 mph.

Check

$10(25 + 5) = 300$

$10(30) = 300$

$300 = 300$

$15(25 - 5) = 300$

$15(20) = 300$

$300 = 300$

The results check.

43. AVIATION

Analyze

- Airplane flys with the wind 800 mi in 4 hrs.
- The return trip against the wind takes 5 hours.
- Find the speed of the plane in still air and the speed of the wind current.

Assign

Let x = speed of plane in still air in mph

y = speed of wind in mph

	Rate •	Time =	Distance
With wind	$x+y$	4	$\mathbf{4(x+y)}$

Total Distance = 800

Form

The rate of the plane with the wind	times	4 hours traveled	is	800 miles.
$(x+y)$	•	4	=	800

	Rate •	Time =	Distance
Against wind	$x-y$	5	$\mathbf{5(x-y)}$

Total Distance = 800

The rate of the plane against the wind	times	5 hours traveled	is	800 miles.
$(x-y)$	•	5	=	800

Solve

$$\begin{cases} 4(x+y)=800 \\ 5(x-y)=800 \end{cases}$$

Distribute in both equations.

$$\begin{cases} 4x+4y=800 \\ 5x-5y=800 \end{cases}$$

Eliminate y.

Multiply both sides of the 1st equation by 5.

Multiply both sides of the 2nd equation by 4.

$$\begin{array}{r} 20x+20y=4,000 \\ 20x-20y=3,200 \\ \hline 40x \quad\quad =7,200 \end{array}$$

$$\frac{40x}{\mathbf{40}}=\frac{7,200}{\mathbf{40}}$$

$$x=180$$

$$4(x+y)=800 \quad \text{The 1st equation}$$

$$4(\mathbf{180}+y)=800$$

$$720+4y=800$$

$$720+4y-\mathbf{720}=800-\mathbf{720}$$

$$4y=80$$

$$\frac{4y}{\mathbf{4}}=\frac{80}{\mathbf{4}}$$

$$y=20$$

State

The speed of the plane in still air is 180 mph.
The speed of the wind is 20 mph.

Check

$4(180+20)=800$	$5(180-20)=800$
$4(200)=800$	$5(160)=800$
$800=800$	$800=800$

The results check.

45. MARINE BIOLOGY

Analyze

- Set up an aquarium containing 3% salt water.
- 2 tanks contain 6% and 2% salt water.
- How much from each tank must he use to fill a 32-gallon aquarium with a 3% mixture?

Assign

Let x = amount of 2% salt water in gallons

y = amount of 6% salt water in gallons

Form

The # of gallons of 2% salt water	plus	the # of gallons of 6% salt water	is	32 gallons.
x	+	y	=	32

	Amt •	Strength =	Amt of salt
Weak	x	0.02	$\mathbf{0.02x}$
Strong	y	0.06	$\mathbf{0.06y}$
Mix	32	0.03	**32(0.03)**

The amount of salt in the 2% salt water	plus	the amount of salt in the 6% salt water	is equal to	the amount of salt in the 3% salt water.
$0.02x$	+	$0.06y$	=	$32(0.03)$

Solve

$$\begin{cases} x+y=32 \\ 0.02x+0.06y=0.96 \end{cases}$$

Eliminate x.

Multiply both sides of the 1st equation by -2.

Multiply both sides of the 2nd equation by 100.

$$\begin{array}{r} -2x-2y=-64 \\ 2x+6y=\ \ 96 \\ \hline 4y=32 \end{array}$$

$$\frac{4y}{\mathbf{4}}=\frac{32}{\mathbf{4}}$$

$$y=8$$

$$x+y=32 \quad \text{The 1st equation}$$

$$x+\mathbf{8}=32$$

$$x+8-\mathbf{8}=32-\mathbf{8}$$

$$x=24$$

State

24 gallons of 2% salt water will be needed.
8 gallons of 6% salt water will be needed.

Check

$24+8=32$	$0.02(24)+0.06(8)=0.96$
$32=32$	$0.48+0.48=0.96$
	$0.96=0.96$

The results check.

47. CLEANING FLOORS

Analyze

- Custodian mixes a 4% ammonia solution and a 12% ammonia solution to get 1 gallon (128 fluid ounces) of a 9% ammonia solution.
- How many fluid ounces of the 4% solution and the 12% solution should be used?

Assign

Let $x =$ amount of 4% ammonia solution in oz

$y =$ amount of 12% ammonia solution in oz

Form

The # of ounces of 4% ammonia	plus	the # of ounces of 12% ammonia	is	128 ounces.
x	$+$	y	$=$	128

	Amt •	Strength =	Amt of salt
Weak	x	0.04	**$0.04x$**
Strong	y	0.12	**$0.12y$**
Mix	128	0.09	**128(0.09)**

The amount of ammonia in the 4% solution	plus	the amount of ammonia in the 12% solution	is equal to	the amount of ammonia in the 9% solution.
$0.04x$	$+$	$0.12y$	$=$	$128(0.09)$

Solve

$$\begin{cases} x + y = 128 \\ 0.04x + 0.12y = 11.52 \end{cases}$$

Eliminate x.

Multiply both sides of the 1st equation by -4.

Multiply both sides of the 2nd equation by 100.

$$\begin{array}{r} -4x - 4y = -512 \\ 4x + 12y = 1{,}152 \\ \hline 8y = 640 \\ \frac{8y}{\mathbf{8}} = \frac{640}{\mathbf{8}} \\ y = 80 \end{array}$$

$x + y = 128$ The 1st equation

$x + \mathbf{80} = 128$

$x + 80 - \mathbf{80} = 128 - \mathbf{80}$

$x = 48$

State

48 ounces of 4% solution will be needed.

80 ounces of12% solution will be needed.

Check

$48 + 80 = 128$

$128 = 128$

$0.04(48) + 0.12(80) = 11.52$

$1.92 + 9.6 = 11.52$

$11.52 = 11.52$

The results check.

49. COFFEE SALES

Analyze

- A 100 pound blend sells for \$6.35 a pound.
- Brazilian coffee selling for \$3.75 a pound.
- Columbian coffee selling for \$8.75 a pound.
- How much of each type should be used?

Assign

Let $x =$ amount of \$8.75/lb coffee in lbs

$y =$ amount of \$3.75/lb coffee in lbs

Form

The # of lbs of \$8.75/lb coffee	plus	the # of lbs of \$3.75/lb coffee	is	100 pounds.
x	$+$	y	$=$	100

	Amt •	Cost/lb =	Total Value
Columbian	x	8.75	**$8.75x$**
Brazilian	y	3.75	**$3.75y$**
Mixture	100	6.35	**100(6.35)**

The value of \$8.75/lb coffee	plus	the value of \$3.75/lb coffee	is equal to	the value of \$6.35/lb coffee.
$8.75x$	$+$	$3.75y$	$=$	$6.35(100)$

Solve

$$\begin{cases} x + y = 100 \\ 8.75x + 3.75y = 635 \end{cases}$$

Eliminate y.

Multiply both sides of the 1st equation by -375.

Multiply both sides of the 2nd equation by 100.

$$\begin{array}{r} -375x - 375y = -37{,}500 \\ 875x + 375y = 63{,}500 \\ \hline 500x = 26{,}000 \\ \frac{500x}{\mathbf{500}} = \frac{26{,}000}{\mathbf{500}} \\ x = 52 \end{array}$$

$x + y = 100$ The 1st equation

$\mathbf{52} + y = 100$

$52 + y - \mathbf{52} = 100 - \mathbf{52}$

$y = 48$

State

52 lb of \$8.75/lb coffee will be needed.

48 lb of \$3.75/lb coffee will be needed.

Check

$52 + 48 = 100$

$100 = 100$

$\$8.75(52) + \$3.75(48) = \$635$

$\$455 + \$180 = \$635$

$\$635 = \$635$

The results check.

51. GOURMET FOODS

Analyze

- Stuffed Kalamata olives sell for \$9 a pint.
- Marinated mushrooms sell for \$12 a pint.
- How many pints of each are needed to get 20 pints of a mixture that will sell for \$10 a pint?

Assign

Let x = pints of \$9.00/pt olives
y = pints of \$12.00/pt mushrooms

Form

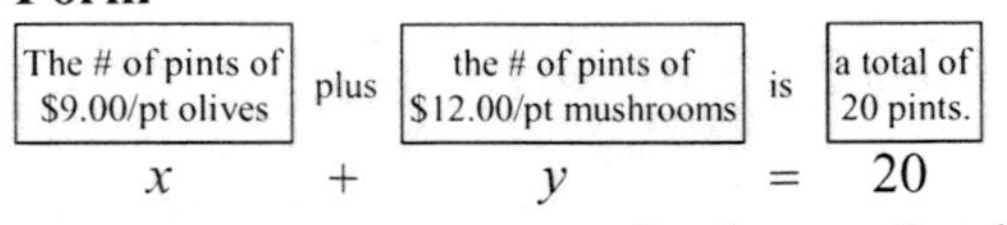

$x \quad + \quad y \quad = \quad 20$

	Amt •	Cost/pt =	Total Value
Olives	x	9.00	**9.00x**
Mushrooms	y	12.00	**12.00y**
Mixture	20	10.00	**10.00(20)**

The value of \$9/pt olives plus the value of \$12/pt mushrooms is equal to the value of \$10/pt blend.

$9x \quad + \quad 12y \quad = \quad 10(20)$

Solve

$$\begin{cases} x + y = 20 \\ 9x + 12y = 200 \end{cases}$$

Eliminate x.

Multiply both sides of the 1st equation by -9.

$$\begin{aligned} -9x - 9y &= -180 \\ 9x + 12y &= 200 \\ \hline 3y &= 20 \\ \frac{3y}{\mathbf{3}} &= \frac{20}{\mathbf{3}} \\ y &= 6\frac{2}{3} \end{aligned}$$

$x + y = 20$ The 1st equation

$$x + \mathbf{6\frac{2}{3}} = 20$$

$$x + 6\frac{2}{3} - \mathbf{6\frac{2}{3}} = 19\frac{3}{3} - \mathbf{6\frac{2}{3}}$$

$$x = 13\frac{1}{3}$$

State

$13\frac{1}{3}$ pt of \$9.00/pt olives will be needed.

$6\frac{2}{3}$ lb of \$12.0/pt mushrooms will be needed.

Check

$$13\frac{1}{3} + 6\frac{2}{3} = 20$$
$$20 = 20$$

$$\$9\left(\frac{40}{3}\right) + \$12\left(\frac{20}{3}\right) = \$200$$
$$\$120 + \$80 = \$200$$
$$\$200 = \$200$$

The results check.

WRITING

53. Answers will vary.

REVIEW

Graph each inequality. Then describe the graph using interval notation.

55. $x < 4$

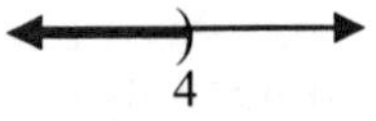

(number line: arrow to the left, open parenthesis at 4)

$(-\infty, 4)$

57. $-1 < x \le 2$

(number line: open parenthesis at -1, bracket at 2)

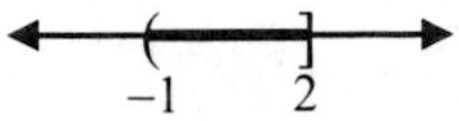

$(-1, 2]$

CHALLENGE PROBLEMS

59. SCALE

Analyze

- 3 nails and 1 bolt is the same as 3 nuts.
- 1 bolt and 1 nut is the same as 5 nails.
- How many nails will it take to balance 1 nut?

Assign

Let bolt = number of bolts
nut = number of nuts
nail = number of nails

Form

Three nails plus one bolt is 3 nuts.

3 nails + 1 bolt = 3 nuts

One bolt plus one nut is 5 nails.

1 bolt + 1 nut = 5 nails

Solve

$$\begin{cases} 3 \text{ nails} + 1 \text{ bolt} = 3 \text{ nuts} \\ 1 \text{ bolt} + 1 \text{ nut} = 5 \text{ nails} \end{cases}$$

Solve the 2nd equation for 1 bolt.

$$\begin{cases} 3 \text{ nails} + 1 \text{ bolt} = 3 \text{ nuts} \\ 1 \text{ bolt} = 5 \text{ nails} - 1 \text{ nut} \end{cases}$$

$3 \text{ nails} + 1 \text{ bolt} = 3 \text{ nuts}$ The 1st equation

$3 \text{ nails} + (\mathbf{5\ nails - 1\ nut}) = 3 \text{ nuts}$ Substitute for 1 bolt.

$$\begin{aligned} 8 \text{ nails} - 1 \text{ nut} &= 3 \text{ nuts} \\ 8 \text{ nails} - 1 \text{ nut} + \mathbf{1\ nut} &= 3 \text{ nut} + \mathbf{1\ nut} \\ 8 \text{ nails} &= 4 \text{ nuts} \\ \frac{8 \text{ nails}}{\mathbf{4}} &= \frac{4 \text{ nuts}}{\mathbf{4}} \\ 2 \text{ nails} &= 1 \text{ nut} \end{aligned}$$

State

It will take 2 nails to balance 1 nut.

Check

The results check.

SECTION 4.5 STUDY SET

VOCABULARY

Fill in the blanks.

1. $\begin{cases} x+y>2 \\ x+y<4 \end{cases}$ is a system of linear **inequalities**.

3. To find the solutions of a system of two linear inequalities graphically, look for the **intersection**, or overlap, of the two shaded regions.

CONCEPTS

5. a. What is the equation of the boundary line of the graph of $3x-y<5$? **$3x-y=5$**

   b. Is the boundary a solid or dashed line? **Dashed**

7. Find the slope and the y-intercept of the line whose equation is $y=4x-3$. **Slope: $4=\frac{4}{1}$, y-intercept: (0, –3)**

9. The boundary of the graph of $2x+y>4$ is shown.

   a. Does the point (0, 0) make the inequality true? **No**

   b. Should the region above or below the boundary be shaded? **Above**

11. The graph of a system of two linear inequalities is shown. Determine whether each point is a solution of the system.

   a. (4, –2) **Yes**

   b. (1, 3) **No**

   c. the origin **No**

13. Match each equation, inequality, or system with the graph of its solution.

   a. $x+y=2$ **ii** b. $x+y\geq 2$ **iii**

   c. $\begin{cases} x+y=2 \\ x-y=2 \end{cases}$ **iv** d. $\begin{cases} x+y\geq 2 \\ x-y\leq 2 \end{cases}$ **i**

GUIDED PRACTICE

Graph the solutions of each system.

See Example 1.

15. $\begin{cases} x+2y\leq 3 \\ 2x-y\geq 1 \end{cases}$

Step 1: Graph $x+2y\leq 3$. We begin by graphing the boundary line $x+2y=3$. Since the inequality contains a $\leq$ symbol, the boundary is a solid line. We check test point (0,0) to determine which side of the line to shade.

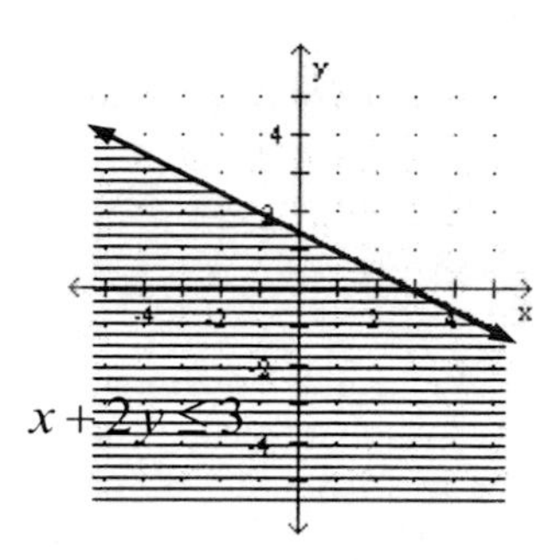

Step 2: Graph $2x-y\geq 1$. We begin by graphing the boundary line $2x-y=1$. Since the inequality contains a $\geq$ symbol, the boundary is a solid line. We check test point (0,0) to determine which side of the line to shade.

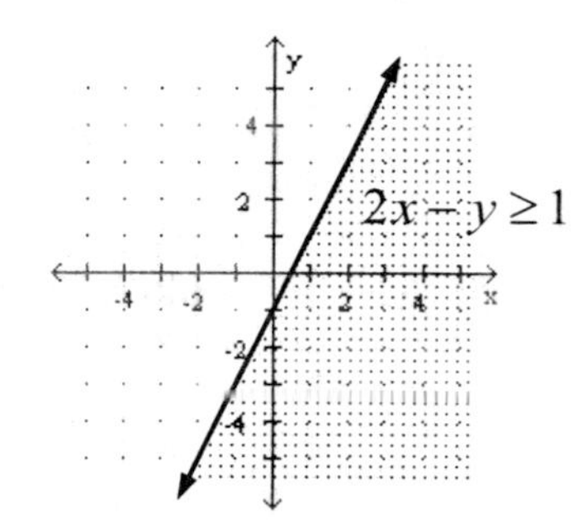

Step 3: We now superimpose the two graphs so that we can determine the region that the graphs have in common.

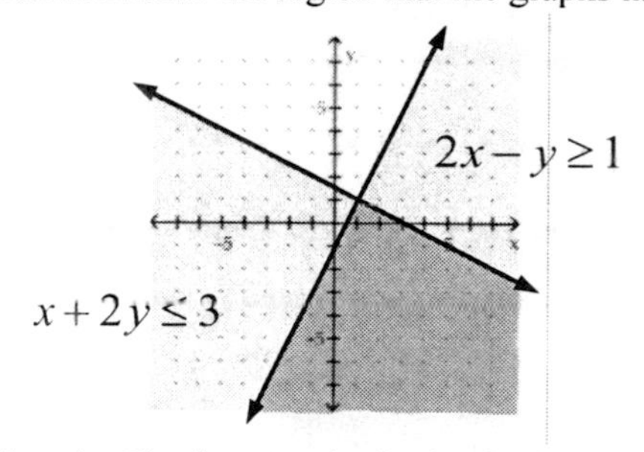

Step 4: Check one point in the double shaded region.

Check: (2,0)	Check: (2,0)
$x+2y\leq 3$	$2x-y\geq 1$
$2+2(0)\leq 3$	$2(2)-0\geq 1$
$2\leq 3$	$4\geq 1$
True	True

17. $\begin{cases} x+y<-1 \\ x-y>-1 \end{cases}$

Step 1: Graph $x+y<-1$. We begin by graphing the boundary line $x+y=-1$. Since the inequality contains a $<$ symbol, the boundary is a dashed line. We check test point (0,0) to determine which side of the line to shade.

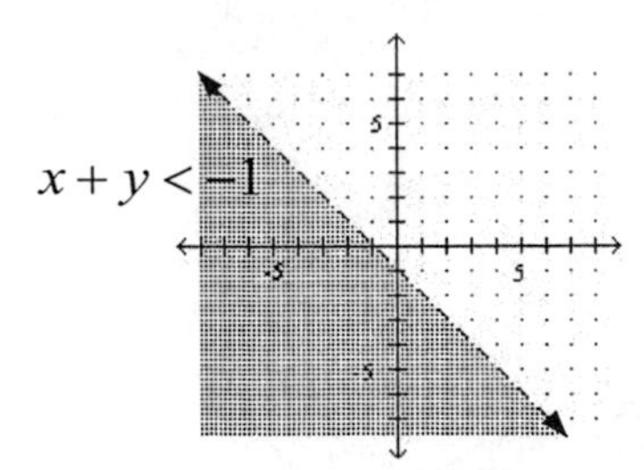

Step 2: Graph $x-y>-1$. We begin by graphing the boundary line $x-y=-1$. Since the inequality contains a $>$ symbol, the boundary is a dashed line. We check test point (0,0) to determine which side of the line to shade.

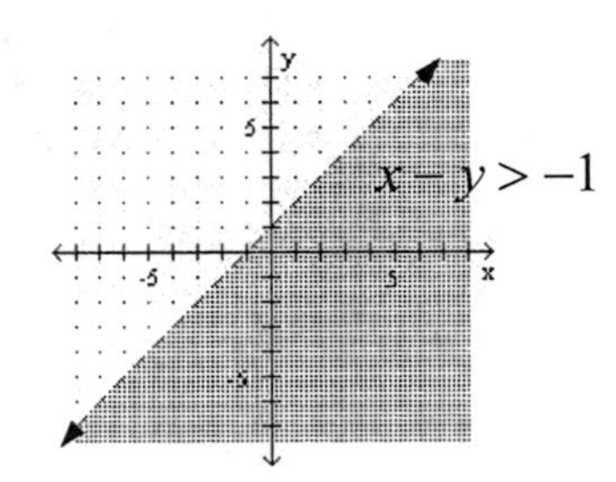

Step 3: We now superimpose the two graphs so that we can determine the region that the graphs have in common.

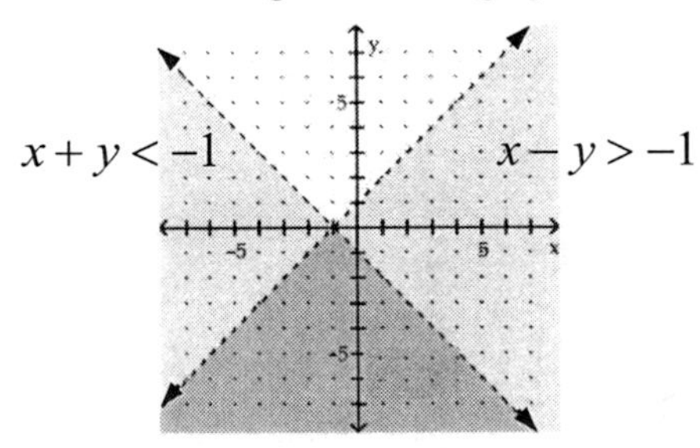

Step 4: Check one point in the double shaded region.

Check: $(0,-2)$	Check: $(0,-2)$
$x+y<-1$	$x-y>-1$
$0+(-2)<-1$	$0-(-2)>-1$
$-2<-1$	$2>-1$
True	True

Graph the solutions of each system. See Example 2.

19. $\begin{cases} y>2x \\ x+2y<6 \end{cases}$

21. $\begin{cases} y \ge x \\ y \le \frac{1}{3}x+1 \end{cases}$

Graph the solutions of each system. See Example 3.

23. $\begin{cases} x \ge 2 \\ y \le 3 \end{cases}$

25. $\begin{cases} x>0 \\ y>0 \end{cases}$

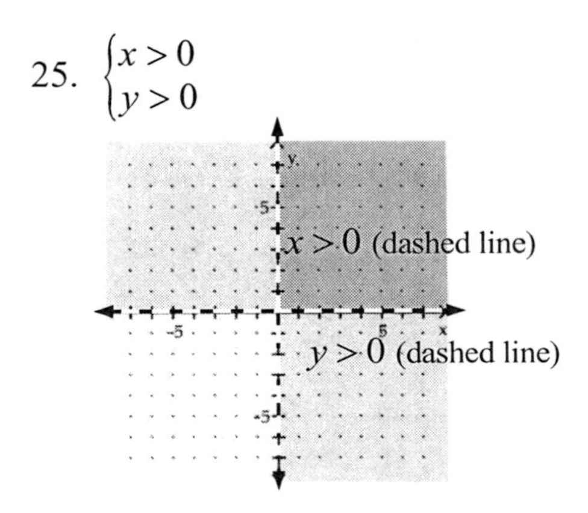

Graph the solutions of each system. See Example 4. With these systems, one extra boundary is drawn before identifying the intersection of the system.

27. $\begin{cases} x \geq 0 \\ y \geq 0 \\ x + y \leq 3 \end{cases}$

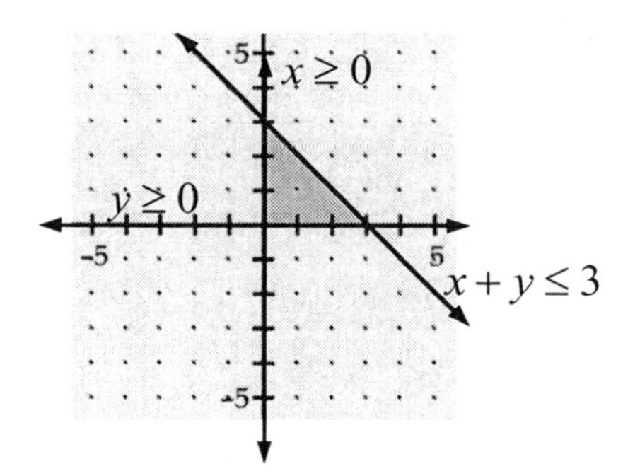

29. $\begin{cases} x - y < 4 \\ y \leq 0 \\ x \geq 0 \end{cases}$

TRY IT YOURSELF

Graph the solutions of each system.

31. $\begin{cases} 2x - 3y \leq 0 \\ y \geq x - 1 \end{cases}$

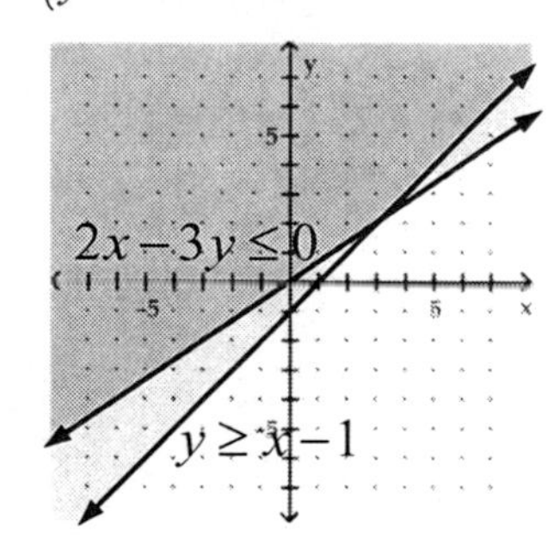

33. $\begin{cases} x + y < 2 \\ x + y \leq 1 \end{cases}$

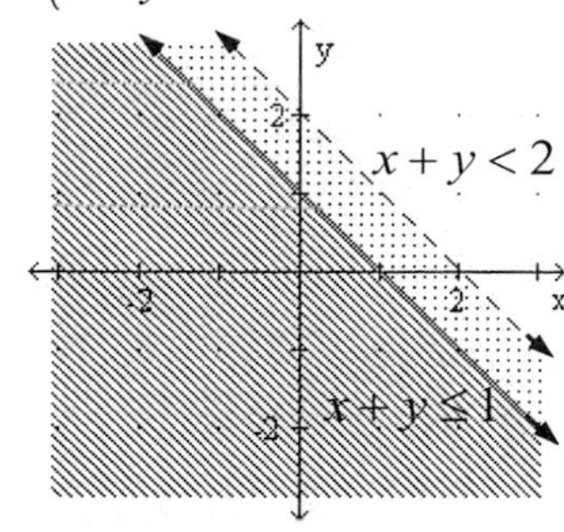

35. $\begin{cases} 3x + 4y \geq -7 \\ 2x - 3y \geq 1 \end{cases}$

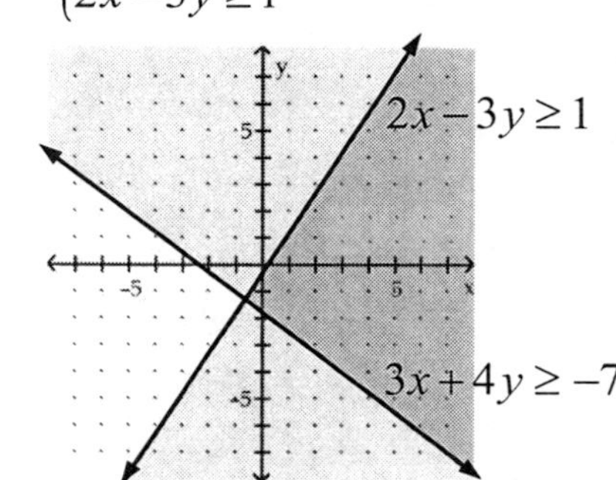

37. $\begin{cases} 2x + y < 7 \\ y > 2 - 2x \end{cases}$

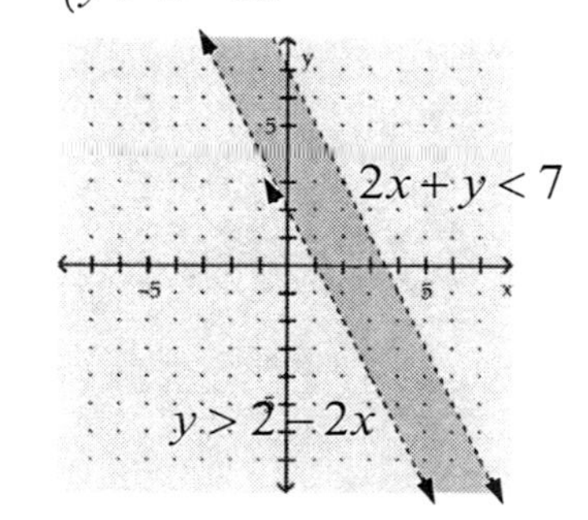

39. $\begin{cases} 2(x - 2y) > -6 \\ 3x + y \geq 5 \end{cases}$

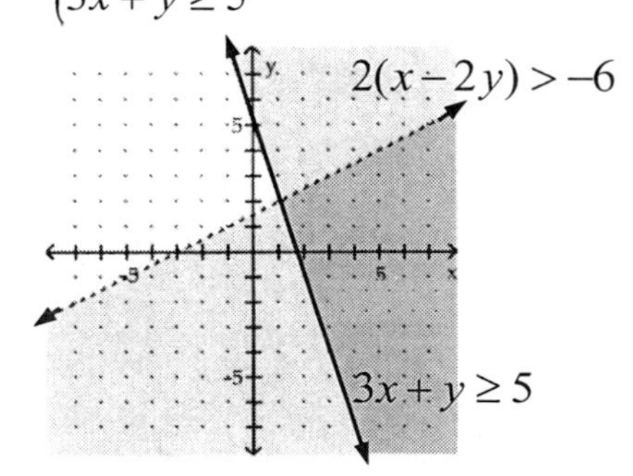

41. $\begin{cases} 3x - y + 4 \le 0 \\ 3y > -2x - 10 \end{cases}$

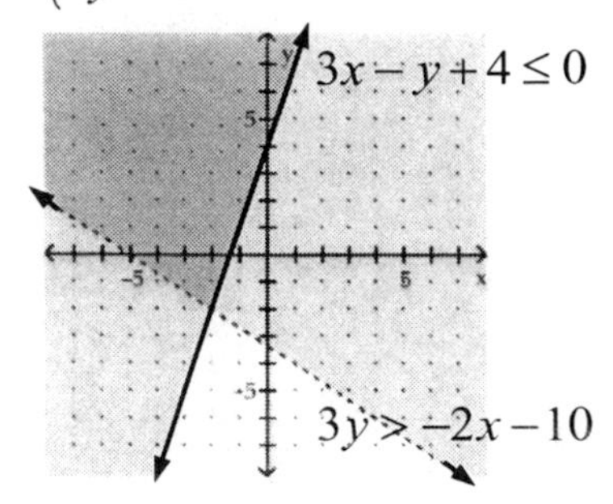

43. $\begin{cases} x \ge -1 \\ y \le -x \\ x - y \le 3 \end{cases}$

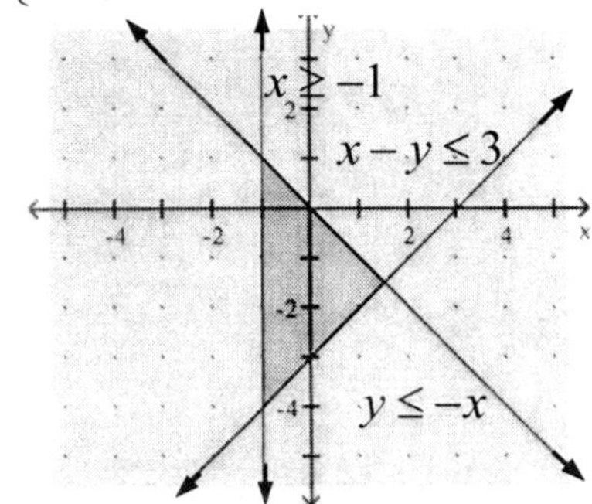

45. $\begin{cases} x + y > 0 \\ y - x < -2 \end{cases}$

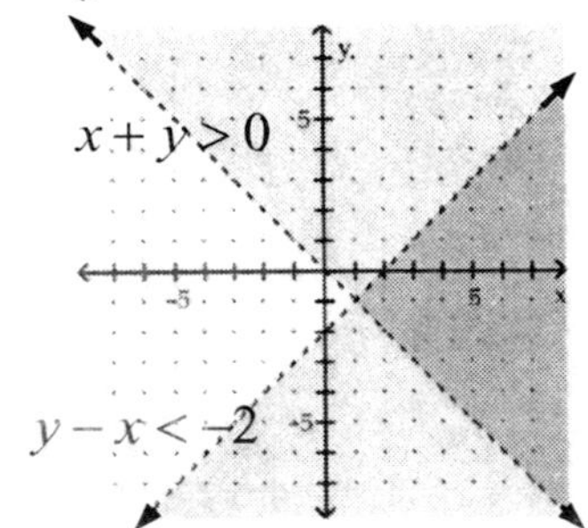

LOOK ALIKES

In part a, graph the solution of each system. Use your answer to part a to determine the solution of the system of equations in part b.

47. a. $\begin{cases} x + y > -1 \\ y \ge x - 3 \end{cases}$ b. $\begin{cases} x + y = -1 \\ y = x - 3 \end{cases}$

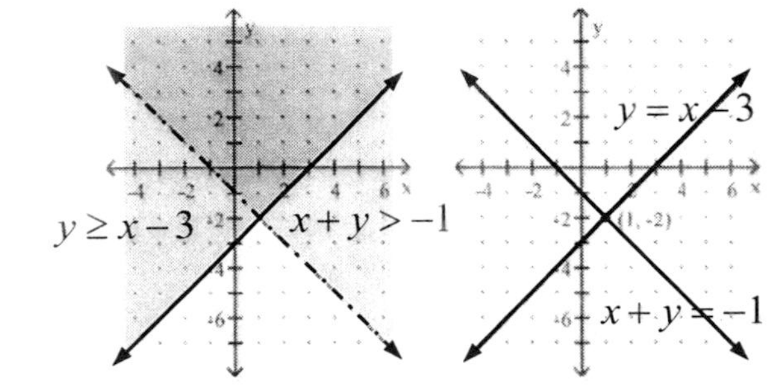

APPLICATIONS

49. BIRDS OF PREY

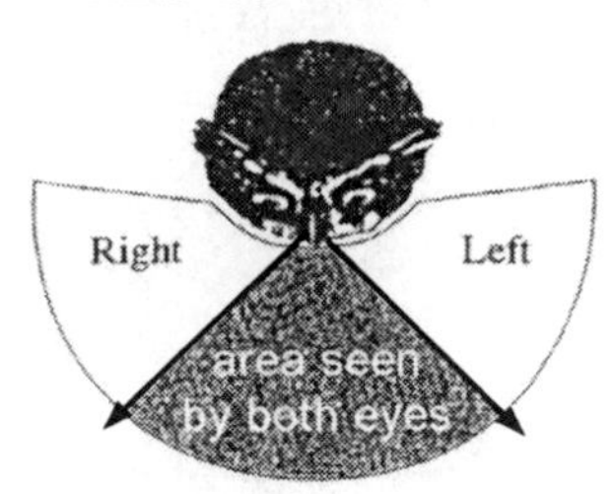

Graph each system of inequalities and give two possible solutions. See Example 5.

51. BUYING COMPACT DISCS

$\begin{cases} 10x + 15y \ge 3(\\ 10x + 15y \le 6(\end{cases}$

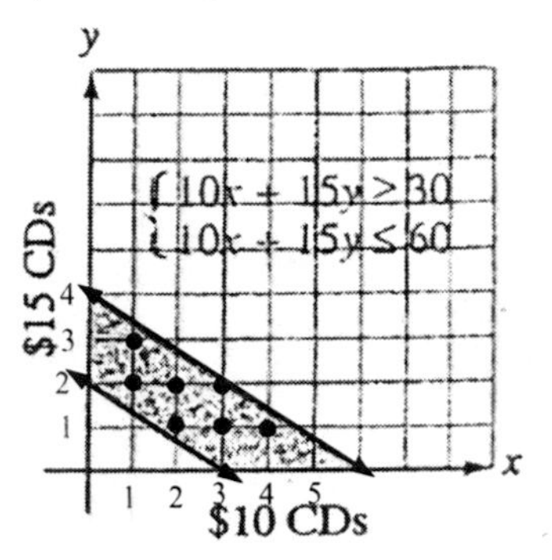

1 $10 CD and 2 $15 CD's
1 $10 CD and 3 $15 CD's
2 $10 CD's and 1 $15 CD
2 $10 CD's and 2 $15 CD's
3 $10 CD's and 1 $15 CD
3 $10 CD's and 2 $15 CD's
4 $10 CD's and 1 $15 CD

53. FURNITURE

$$\begin{cases} 150x + 100y \le 900 \\ y > x \end{cases}$$

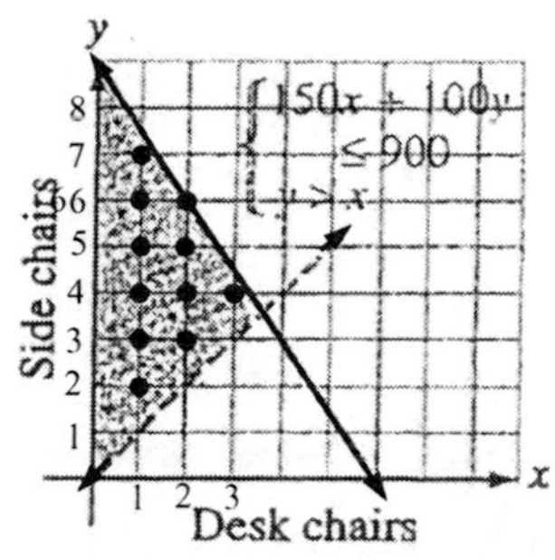

1 desk chair and 2 side chairs
1 desk chair and 3 side chairs
1 desk chair and 4 side chairs
1 desk chair and 5 side chairs
1 desk chair and 6 side chairs
1 desk chair and 7 side chairs
2 desk chair and 3 side chairs
2 desk chair and 4 side chairs
2 desk chair and 5 side chairs
2 desk chair and 6 side chairs
3 desk chair and 4 side chairs

from CAMPUS TO CAREERS

55. PHOTOGRAPHERS

$$\begin{cases} y \le \frac{1}{4}x + 2 \\ y \ge -\frac{1}{4}x + 2 \end{cases}$$

$y \le \frac{1}{4}x + 2$

$y \ge -\frac{1}{4}x + 2$

WRITING

57-59. Answers will vary.

REVIEW

Simplify each expression.

61. $8\left(\frac{3}{4}t\right) = \frac{8 \cdot 3t}{4}$

$= \frac{2 \cdot \cancel{4} \cdot 3t}{\cancel{4}}$

$= 6t$

63. $-\frac{7}{16}x - \frac{3}{16}x = \frac{-7x - 3x}{16}$

$= \frac{-10x}{16}$

$= -\frac{\cancel{2} \cdot 5x}{\cancel{2} \cdot 8}$

$= -\frac{5}{8}x$

CHALLENGE PROBLEMS

Graph the solutions of each system.

65. $\begin{cases} \frac{x}{3} - \frac{y}{2} < -3 \\ \frac{x}{3} + \frac{y}{2} > -1 \end{cases}$

Clear both inequalities of fractions.

$$\begin{cases} 2x - 3y < -18 \\ 2x + 3y > -6 \end{cases}$$

Graph the boundaries of both.

Identify the intersection.

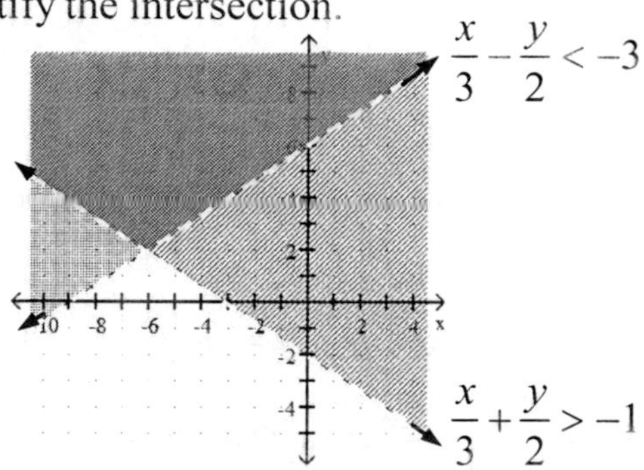

67. $\begin{cases} 2x + 3y \le 6 \\ 3x + y \le 1 \\ x \le 0 \end{cases}$

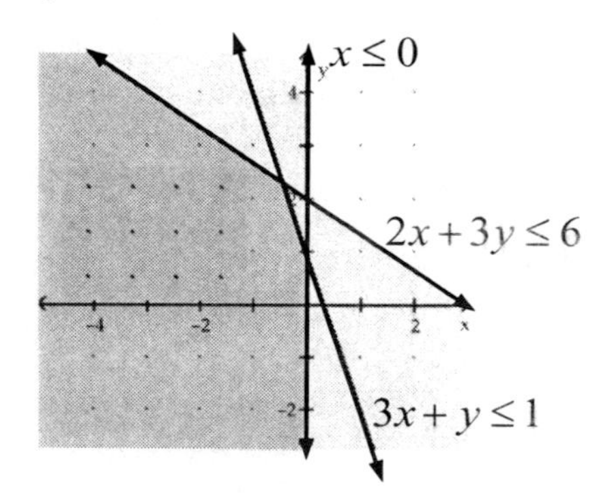

CHAPTER 4

SECTION 4.1 REVIEW EXERCISES

Solving System Equations by Graphing

Determine whether the ordered pair is a solution of the system.

1. $(2,-3)$ $\begin{cases} 3x-2y=12 \\ 2x+3y=-5 \end{cases}$

$3x-2y=12$	$2x+3y=-5$
$3(2)-2(-3)\overset{?}{=}12$	$2(2)+3(-3)\overset{?}{=}-5$
$6+6\overset{?}{=}12$	$4-9\overset{?}{=}-5$
$12=12$	$-5=-5$
True	True

Since $(2,-3)$ satisfies both equations, it is a solution of the system.

2. $\left(\frac{7}{2},-\frac{2}{3}\right)$ $\begin{cases} 3y=2x-9 \\ 2x+3y=6 \end{cases}$

$3y=2x-9$	$2x+3y=6$
$3\left(\frac{-2}{3}\right)\overset{?}{=}2\left(\frac{7}{2}\right)-9$	$2\left(\frac{7}{2}\right)+3\left(\frac{-2}{3}\right)\overset{?}{=}6$
$-2\overset{?}{=}7-9$	$7-2\overset{?}{=}6$
$-2=-2$	$5=6$
True	False

Since $\left(\frac{7}{2},-\frac{2}{3}\right)$ does not satisfy both equations, it is not a solution of the system.

Use the graphing method to solve each system.

3. $\begin{cases} x+y=7 \\ 2x-y=5 \end{cases}$

The solution is (4, 3).

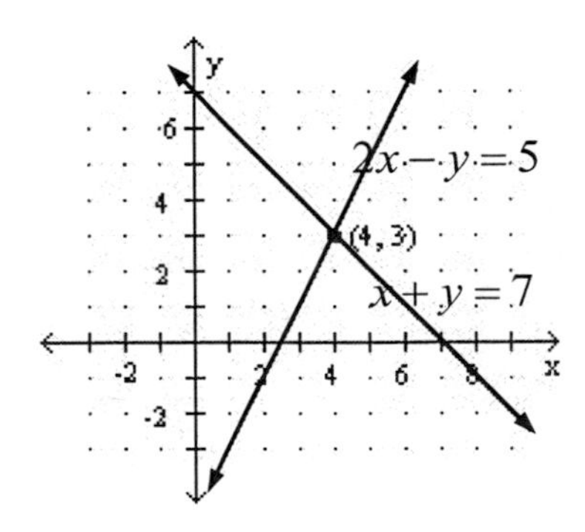

4. $\begin{cases} 2x+y=5 \\ y=-\frac{x}{3} \end{cases}$

The solution is $(3,-1)$.

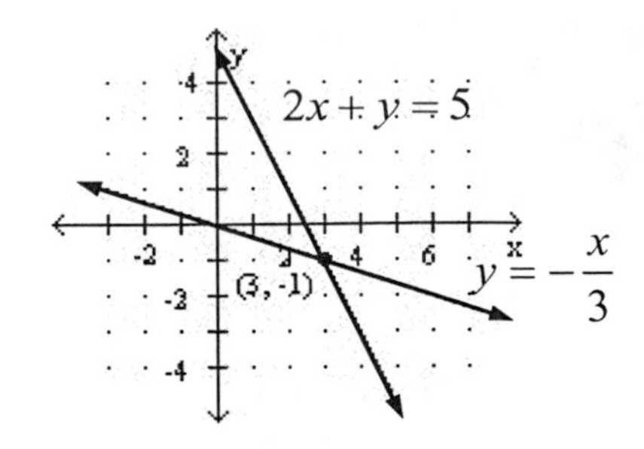

$y=-\frac{x}{3}$

5. $\begin{cases} 3x+6y=6 \\ x+2y-2=0 \end{cases}$

The graphs are the same line.
The system has infinitely many solutions.

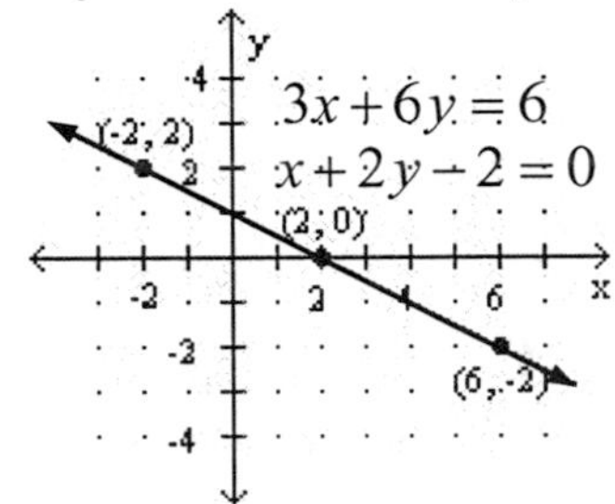

6. $\begin{cases} 6x+3y=12 \\ y=-2x+2 \end{cases}$

The lines are parallel.
The system has no solution.

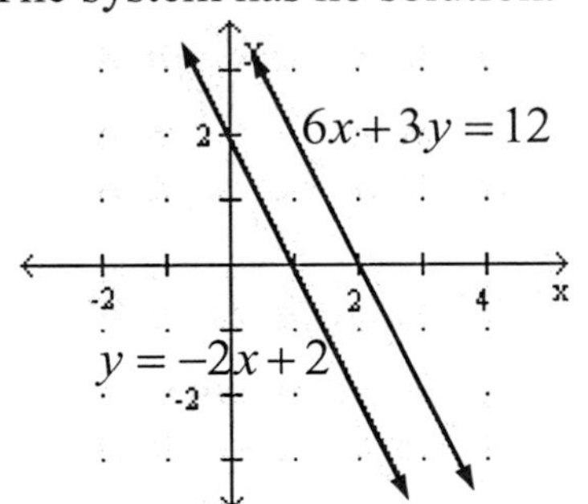

Find the slope and the y - intercept of the graph of each line in the system. Then, use that information to determine the number of solutions of the system.

7. $\begin{cases} y = -2x + 1 \\ 8x + 4y = 3 \end{cases}$

$y = -2x + 1$
$m = -2$
$y\text{-int} = (0, 1)$

$8x + 4y = 3$
$8x + 4y - \mathbf{8x} = 3 - \mathbf{8x}$
$4y = 3 - 8x$
$y = -2x + \frac{3}{4}$
$m = -2$
$y\text{-int} = \left(0, \frac{3}{4}\right)$

Since the line graphs have the same slope, they are parallel and there is no solution.

8. BACHELOR'S DEGREE
Estimate the point of intersection of the graphs. Explain its significance.
(1980, 480,000)
In 1980, the same number of men as women awarded Bachelor' Degrees in the U.S. ws the same 480,000.

SECTION 4.2 REVIEW EXERCISES SOLVING SYSTEMS OF EQUATIONS BY SUBSTITUTION

Use the substitution method to solve each system.

9. $\begin{cases} y = 15 - 3x \\ 7y + 3x = 15 \end{cases}$

$7y + 3x = 15$ The 2nd equation. Substitute for y.
$7(\mathbf{15 - 3x}) + 3x = 15$
$105 - 21x + 3x = 15$
$105 - 18x = 15$
$105 - 18x - \mathbf{105} = 15 - \mathbf{105}$
$-18x = -90$
$\frac{-18x}{\mathbf{-18}} = \frac{-90}{\mathbf{-18}}$
$x = 5$

$y = 15 - 3x$ The 1st equation
$y = 15 - 3(\mathbf{5})$
$y = 15 - 15$
$y = 0$
The solution is (5,0).

10. $\begin{cases} x = y \\ 5x - 4y = 3 \end{cases}$

$5x - 4y = 3$ The 2nd equation. Substitute for x.
$5(\mathbf{y}) - 4y = 3$
$y - 3$
$x = y$
$x = \mathbf{3}$
The solution is (3,3).

11. $\begin{cases} 6x + 2y = 8 - y + x \\ 3x = 2 - y \end{cases}$

Write the 1st equation in standard form.
Solve the 2nd equation for y.

$\begin{cases} 5x + 3y = 8 \\ y = 2 - 3x \end{cases}$

$5x + 3y = 8$ The 1st equation. Substitute for y.
$5x + 3(\mathbf{2 - 3x}) = 8$
$5x + 6 - 9x = 8$
$-4x + 6 = 8$
$-4x + 6 - \mathbf{6} = 8 - \mathbf{6}$
$-4x = 2$
$\frac{-4x}{\mathbf{-4}} = \frac{2}{\mathbf{-4}}$
$x = -\frac{1}{2}$

$y = 2 - 3x$ The 2nd equation
$y = 2 - 3\left(-\frac{\mathbf{1}}{\mathbf{2}}\right)$
$y = 2 + \frac{3}{2}$
$y = \frac{4}{2} + \frac{3}{2}$
$y = \frac{7}{2}$

The solution is $\left(-\frac{1}{2}, \frac{7}{2}\right)$.

12. $\begin{cases} r = 3s+7 \\ r = 2s+5 \end{cases}$

$$r = 2s+5 \quad \text{The 2}^{\text{nd}} \text{ equation. Substitute for } r.$$
$$(\mathbf{3s+7}) = 2s+5$$
$$3s+7-\mathbf{2s} = 2s+5-\mathbf{2s}$$
$$s+7=5$$
$$s+7-\mathbf{7}=5-\mathbf{7}$$
$$s=-2$$

$r = 3s+7$ The 1st equation
$r = 3(\mathbf{-2})+7$
$r = -6+7$
$r = 1$
The solution is $(1,-2)$.

13. $\begin{cases} 9x+3y-5=0 \\ 3x+y=\dfrac{5}{3} \end{cases}$

Solve the 2nd equation for y.

$\begin{cases} 9x+3y-5=0 \\ y=\dfrac{5}{3}-3x \end{cases}$

$$9x+3y-5=0 \quad \text{The 1}^{\text{st}} \text{ equation. Substitute for } y.$$
$$9x+3\left(\mathbf{\frac{5}{3}-3x}\right)-5=0$$
$$9x+5-9x-5=0 \quad \text{The variables drop out.}$$
$$0=0$$

True
The system has infinitely many solutions.

14. $\begin{cases} \dfrac{x}{2}+\dfrac{y}{2}=11 \\ \dfrac{5x}{16}-\dfrac{3y}{16}=\dfrac{15}{8} \end{cases}$

Clear both equations of fractions.

$\begin{cases} \mathbf{2}\left(\dfrac{x}{2}\right)+\mathbf{2}\left(\dfrac{y}{2}\right)=\mathbf{2}(11) \\ \mathbf{16}\left(\dfrac{5x}{16}\right)-\mathbf{16}\left(\dfrac{3y}{16}\right)=\mathbf{16}\left(\dfrac{15}{8}\right) \end{cases}$

$\begin{cases} x+y=22 \\ 5x-3y=30 \end{cases}$

Solve the 1st equation for x.

$\begin{cases} x=22-y \\ 5x-3y=30 \end{cases}$

$$5x-3y=30 \quad \text{The 2}^{\text{nd}} \text{ equation. Substitute for } x.$$
$$5(\mathbf{22-y})-3y=30$$
$$110-5y-3y=30$$
$$-8y+110=30$$
$$-8y+110-\mathbf{110}=30-\mathbf{110}$$
$$-8y=-80$$
$$\frac{-8y}{\mathbf{-8}}=\frac{-80}{\mathbf{-8}}$$
$$y=10$$

$x = 22-y$ The 1st equation
$x = 22-(\mathbf{10})$
$x = 12$
The solution is $(12,10)$.

15. When solving a system using the substitution method, suppose you obtain the result $8 = 9$.
   a. How many solutions does the system have?
   No solution
   b. Describe the graph of the system.
   Two parallel lines
   c. What term is used to describe the system?
   Inconsistent system

16. Fill in the blank. With the substitution method, the objective is to use an appropriate substitution to obtain one equation in **one** variable.

SECTION 4.3 REVIEW EXERCISES SOLVING SYSTEMS OF EQUATIONS BY ELIMINATION (ADDITION)

17. Write each equation of the system in general $Ax + By = C$ form.

$\begin{cases} 4x+2y-7=0 \\ 3y=5x+6 \end{cases} \rightarrow \begin{cases} \mathbf{4x+2y=7} \\ \mathbf{5x-3y=-6} \end{cases}$

18. Fill in the blank. With the elimination method, the basic objective is to obtain two equations whose sum will be one equation in **one** variable.

Solve each system using the elimination (addition) method.

19. $\begin{cases} 2x+y=1 \\ 5x-y=20 \end{cases}$

Eliminate y.

$$\begin{array}{r} 2x+y=1 \\ 5x-y=20 \\ \hline 7x \quad =21 \end{array}$$

$$\frac{7x}{\mathbf{7}}=\frac{21}{\mathbf{7}}$$

$$x=3$$

$2x+y=1$ The 1st equation

$$2(\mathbf{3})+y=1$$

$$6+y-\mathbf{6}=1-\mathbf{6}$$

$$y=-5$$

The solution is $(3, -5)$.

20. $\begin{cases} x+8y=7 \\ x-4y=1 \end{cases}$

Eliminate x.

Multiply both sides of the 2nd equation by -1.

$$\begin{array}{r} x+8y=7 \\ -x+4y=-1 \\ \hline 12y=6 \end{array}$$

$$\frac{12y}{\mathbf{12}}=\frac{6}{\mathbf{12}}$$

$$y=\frac{1}{2}$$

$x+8y=7$ The 1st equation

$$x+8\left(\frac{\mathbf{1}}{\mathbf{2}}\right)=7$$

$$x+4-\mathbf{4}=7-\mathbf{4}$$

$$x=3$$

The solution is $\left(3, \frac{1}{2}\right)$.

21. $\begin{cases} 5a+b=2 \\ 3a+2b=11 \end{cases}$

Eliminate b.

Multiply both sides of the 1st equation by -2.

$$\begin{array}{r} -10a-2b=-4 \\ 3a+2b=11 \\ \hline -7a \quad =7 \end{array}$$

$$\frac{-7a}{\mathbf{-7}}=\frac{7}{\mathbf{-7}}$$

$$a=-1$$

$5a+b=2$ The 1st equation

$$5(\mathbf{-1})+b=2$$

$$-5+b+\mathbf{5}=2+\mathbf{5}$$

$$b=7$$

The solution is $(-1, 7)$.

22. $\begin{cases} 11x+3y=27 \\ 8x+4y=36 \end{cases}$

Eliminate y.

Multiply both sides of the 1st equation by 4.

Multiply both sides of the 2nd equation by -3.

$$\begin{array}{r} 44x+12y=108 \\ -24x-12y=-108 \\ \hline 20x \quad =0 \end{array}$$

$$\frac{20x}{\mathbf{20}}=\frac{0}{\mathbf{20}}$$

$$x=0$$

$8x+4y=36$ The 2nd equation

$$8(\mathbf{0})+4y=36$$

$$4y=36$$

$$\frac{4y}{\mathbf{4}}=\frac{36}{\mathbf{4}}$$

$$y=9$$

The solution is $(0, 9)$.

23. $\begin{cases} 9x+3y=15 \\ 3x=5-y \end{cases}$

Write the 2nd equation in standard form.

$$\begin{cases} 9x+3y=15 \\ 3x+y=5 \end{cases}$$

Eliminate y.

Multiply both sides of the 2nd equation by -3.

$$\begin{array}{r} 9x+3y=15 \\ -9x-3y=-15 \\ \hline 0=0 \end{array}$$

Both variables are eliminated.

True

The system has infinitely many solutions.

24. $\begin{cases} 0.02x+0.05y=0 \\ 0.3x-0.2y=-1.9 \end{cases}$

Clear both equations of decimals.
Multiply both sides of the 1st equation by 100.
Multiply both sides of the 2nd equation by 10.

$$\begin{cases} 100(0.02x)+100(0.05y)=100(0) \\ 10(0.3x)-10(0.2y)=10(-1.9) \end{cases}$$

$$\begin{cases} 2x+5y=0 \\ 3x-2y=-19 \end{cases}$$

Eliminate y.
Multiply both sides of the 1st equation by 2.
Multiply both sides of the 2nd equation by 5.

$$\begin{aligned} 4x+10y &= 0 \\ 15x-10y &= -95 \\ \hline 19x \quad &= -95 \\ \frac{19x}{\mathbf{19}} &= \frac{-95}{\mathbf{19}} \\ x &= -5 \end{aligned}$$

$4x+10y=0$ The 1st equation

$$\begin{aligned} 4(\mathbf{-5})+10y &= 0 \\ -20+10y+\mathbf{20} &= 0+\mathbf{20} \\ 10y &= 20 \\ \frac{10y}{\mathbf{10}} &= \frac{20}{\mathbf{10}} \\ y &= 2 \end{aligned}$$

The solution is $(-5, 2)$.

25. $\begin{cases} -\dfrac{a}{4}-\dfrac{b}{3}=\dfrac{1}{12} \\ \dfrac{a}{2}-\dfrac{5b}{4}=\dfrac{7}{4} \end{cases}$

Clear both equations of fractions.

$$\begin{cases} \mathbf{12}\left(-\dfrac{a}{4}\right)-\mathbf{12}\left(\dfrac{b}{3}\right)=\mathbf{12}\left(\dfrac{1}{12}\right) \\ \mathbf{4}\left(\dfrac{a}{2}\right)-\mathbf{4}\left(\dfrac{5b}{4}\right)=\mathbf{4}\left(\dfrac{7}{4}\right) \end{cases}$$

$$\begin{cases} -3a-4b=1 \\ 2a-5b=7 \end{cases}$$

Eliminate a.
Multiply both sides of the 1st equation by 2.
Multiply both sides of the 2nd equation by 3.

$$\begin{aligned} -6a-\ 8b &= \ 2 \\ 6a-15b &= 21 \\ \hline -23b &= 23 \\ \frac{-23b}{\mathbf{-23}} &= \frac{23}{\mathbf{-23}} \\ b &= -1 \end{aligned}$$

$2a-5b=7$ The 2nd equation

$$\begin{aligned} 2a-5(\mathbf{-1}) &= 7 \\ 2a+5 &= 7 \\ 2a+5-\mathbf{5} &= 7-\mathbf{5} \\ 2a &= 2 \\ \frac{2a}{\mathbf{2}} &= \frac{2}{\mathbf{2}} \\ a &= 1 \end{aligned}$$

The solution is $(1, -1)$.

26. $\begin{cases} -\dfrac{1}{4}x=1-\dfrac{2}{3}y \\ 6x-18y=5-2y \end{cases}$

Clear the 1st equation of fractions.

$$\begin{cases} \mathbf{12}\left(-\dfrac{1}{4}x\right)=\mathbf{12}(1)-\mathbf{12}\left(\dfrac{2}{3}y\right) \\ 6x-18y=5-2y \end{cases}$$

$$\begin{cases} -3x=12-8y \\ 6x-18y=5-2y \end{cases}$$

Write both equations in standard form.

$$\begin{cases} -3x+8y=12 \\ 6x-16y=5 \end{cases}$$

Eliminate x.
Multiply both sides of the 1st equation by 2.

$$\begin{aligned} -6x+16y &= 24 \\ 6x-16y &= \ 5 \\ \hline 0 &= 29 \end{aligned}$$

Both variables are eliminated.
False
No solution.

For each system, determine which method, substitution or elimination (addition), would be easier to use to solve the system and explain why.

27. $\begin{cases} 6x+2y=5 \\ 3x-3y=-4 \end{cases}$ **Elimination; no variables have a coefficient of 1 or -1.**

28. $\begin{cases} x=5-7y \\ 3x-3y=-4 \end{cases}$ **Substitution; equation 1 is solved for x.**

SECTION 4.4 REVIEW EXERCISES PROBLEM SOLVING USING SYSTEMS OF EQUATIONS

29. ELEVATIONS

Analyze

- Elevation of Las Vegas is 20 times greater than that of Baltimore.
- The sum of their elevations is 2,100 feet.
- Find the elevation of each city.

Assign

Let x = the elevation of Las Vegas in feet
y = the elevation of Baltimore in feet

Form

The elevation of Las Vegas	is	twenty	times greater than	the elevation of Baltimore.
x	$=$	20	$\cdot$	y

The elevation of Las Vegas	plus	the elevation of Baltimore	is	2,100 feet.
x	$+$	y	$=$	$2{,}100$

Solve

$$\begin{cases} x = 20y \\ x + y = 2{,}100 \end{cases}$$

$x + y = 2{,}100$ The 2nd equation. Substitute for x.

$$(\mathbf{20y}) + y = 2{,}100$$
$$21y = 2{,}100$$
$$\frac{21y}{\mathbf{21}} = \frac{2{,}100}{\mathbf{21}}$$
$$y = 100$$

$x = 20y$ The 1st equation

$$x = 20(\mathbf{100})$$
$$x = 2{,}000$$

State

The elevation of Las Vegas is 2,000 feet.
The elevation of Baltimore is 100 feet.

Check

$2{,}000 = 20(100)$	$2{,}000 + 100 = 2{,}100$
$2{,}000 = 2{,}000$	$2{,}100 = 2{,}100$

The results check.

30. PAINTING EQUIPMENT

Analyze

- Fully extended, a ladder is 35 feet in length.
- The extension is 7 feet shorter than the base.
- How long is each part of the ladder?

Assign

Let x = the base in feet (longer section)
y = the extension in feet (shorter section)

Form

The length of the bases	plus	the length of the bases extension	is	35 feet.
x	$+$	y	$=$	35

The extension	is	7 feet	shorter than	the base.
y	$=$	x	$-$	7

Solve

$$\begin{cases} x + y = 35 \\ y = x - 7 \end{cases}$$

$x + y = 35$ The 1st equation. Substitute for y.

$$x + (\mathbf{x - 7}) = 35$$
$$2x - 7 = 35$$
$$2x - 7 + \mathbf{7} = 35 + \mathbf{7}$$
$$2x = 42$$
$$\frac{2x}{\mathbf{2}} = \frac{42}{\mathbf{2}}$$
$$x = 21$$

$y = x - 7$ The 2nd equation

$$y = \mathbf{21} - 7$$
$$y = 14$$

State

The length of the base is 21 feet.
The length of the extension is 14 feet.

Check

The results check.

31. GEOMETRY

Analyze

- Two angles are complementary (90°).
- One is 15° more than twice the measure of the other.
- Find the measure of each angle.

Assign

Let x = the measure of one $\angle$

y = the measure of the other $\angle$

Form

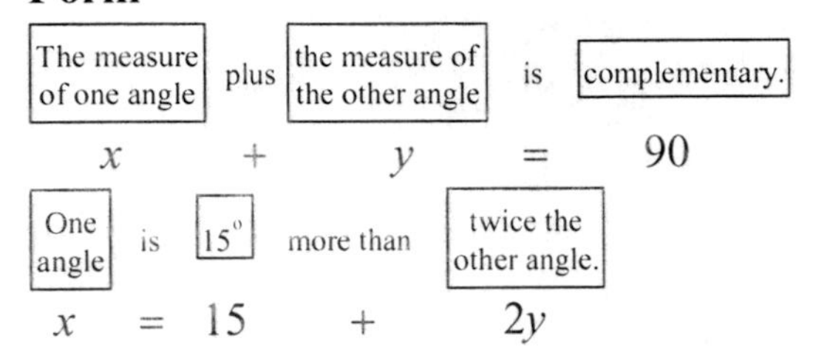

Solve

$$\begin{cases} x + y = 90 \\ x = 15 + 2y \end{cases}$$

$$x + y = 90 \quad \text{The 1}^{\text{st}} \text{ equation. Substitute for } x.$$
$$(\mathbf{15} + \mathbf{2y}) + y = 90$$
$$3y + 15 - \mathbf{15} = 90 - \mathbf{15}$$
$$3y = 75$$
$$\frac{3y}{\mathbf{3}} = \frac{75}{\mathbf{3}}$$
$$y = 25$$

$$x + y = 90 \quad \text{The 1}^{\text{st}} \text{ equation}$$
$$x + \mathbf{25} = 90$$
$$x + 25 - 25 = 90 - 25$$
$$x = 65$$

State

The measure of one angle is 65°.
The measure of the other angle is 25°.

Check

$65 + 25 = 90$	$65 = 15 + 2(25)$
	$65 = 15 + 50$
	$65 = 65$

The results check.

32. CRASH INVESTIGATION

Analyze

- 420 yards of tape is used to seal off a rectangular-shaped area.
- Width is three-fourths of the length.
- What is the area of the rectangle?

Assign

Let w = width of rectangle in yds (shorter length)

l = length of rectangle in yds (longer length)

Form

The measure of two widths	plus	the measure of two lengths	is	the perimerter of 420 feet.
$2w$	$+$	$2l$	$=$	420

The width	is	three-fourth	of	the length.
w	$=$	$\frac{3}{4}$	$\bullet$	l

Solve

$$\begin{cases} 2l + 2w = 420 \\ w = \frac{3}{4}l \end{cases}$$

$$2w + 2l = 420 \quad \text{The 1}^{\text{st}} \text{ equation. Substitute for } w.$$
$$2\left(\frac{3}{4}\boldsymbol{l}\right) + 2l = 420$$
$$\frac{3}{2}l + 2l = 420$$
$$\frac{3}{2}l + \frac{4}{2}l = 420$$
$$\frac{7}{2}l = 420$$
$$\frac{\mathbf{2}}{\mathbf{7}}\left(\frac{7}{2}l\right) = \frac{\mathbf{2}}{\mathbf{7}}(420)$$
$$l = 120$$

$$w = \frac{3}{4}l \quad \text{The 2}^{\text{nd}} \text{ equation}$$
$$w = \frac{3}{4}(\mathbf{120})$$
$$w = 3(30)$$
$$w = 90$$

State

The area of the rectangle is
$120 \text{ yd} \bullet 90 \text{ yd} = 10{,}800 \text{ yd}^2$.

Check

The results check.

33. Complete each table.

a.

	Amt •	Strength =	Amt of antifreeze
Weak	x	0.02	**$0.02x$**
Strong	y	0.09	**$0.09y$**
Mixture	100	0.08	**0.08(100)**

b.

	Rate •	Time =	Distance
With the wind	$s+w$	5	**$5(s+w)$**
Against the wind	$s-w$	7	**$7(s-w)$**

c.

	Principal •	Rate •	Time =	Interest
Mack Financial	x	0.11	1 yr	**$0.11x$**
Union Savings	y	0.06	1 yr	**$0.06y$**

d.

	Amount •	Price =	Total Value
Carmel corn	x	4	**$4x$**
Peanuts	y	8	**$8y$**
Mixture	10	5	**5(10)**

34. CANDY STORE

Analyze

- Gummy bears worth \$3 per pound.
- Gummy worms worth \$6 per pound.
- 30 pound mixture sells for \$4.20 per pound.
- How much of each type should be used?

Assign

Let x = amount of \$3/lb bears in lbs

y = amount of \$6/lb worms in lbs

Form

The # of lbs of \$3/lb bears	plus	the # of lbs of \$6/lb worms	is	30 pounds.
x	$+$	y	$=$	30

	Amt •	Cost/lb =	Total Value
Bears	x	3	**$3x$**
Worms	y	6	**$6y$**
Mixture	30	4.20	**4.20(30)**

The value of \$3/lb bears	plus	the value of \$6/lb worms	is equal to	the value of \$4.20/lb mixture.
$3x$	$+$	$6y$	$=$	4.20(30)

Solve

$$\begin{cases} x+y=30 \\ 3x+6y=126 \end{cases}$$

Eliminate x.

Multiply both sides of the 1st equation by -3.

$$-3x-3y=-90$$
$$3x+6y=126$$
$$3y=36$$
$$\frac{3y}{\mathbf{3}}=\frac{36}{\mathbf{3}}$$
$$y=12$$

$x+y=30$ The 1st equation

$$x+\mathbf{12}=30$$
$$x+12-\mathbf{12}=30-\mathbf{12}$$
$$x=18$$

State

18 lb of \$3/lb gummy bears will be needed.
12 lb of \$6/lb gummy worms will be needed.

Check

The results check.

35. BOATING

Analyze

- A boat can travel 56 miles downriver in 4 hours.
- The return trip takes 3 hours longer (7 hours).
- Find the speed of the current.

Assign

Let x = speed of boat in still water in mph

y = speed of river in mph

Form

	Rate •	Time =	Distance
With current	$x+y$	4	$\mathbf{4(x+y)}$

Total Distance = 56

The rate of the boat with the current times 4 hours traveled is 56 miles.

$$(x+y) \cdot 4 = 56$$

	Rate •	Time =	Distance
Against current	$x-y$	7	$\mathbf{7(x-y)}$

Total Distance = 56

The rate of the boat against the current times 7 hours traveled is 56 miles.

$$(x-y) \cdot 7 = 56$$

Solve

$$\begin{cases} 4(x+y)=56 \\ 7(x-y)=56 \end{cases}$$

Distribute in both equations.

$$\begin{cases} 4x+4y=56 \\ 7x-7y=56 \end{cases}$$

Eliminate x.

Multiply both sides of the 1st equation by 7.

Multiply both sides of the 2nd equation by -4.

$$\begin{array}{r} 28x+28y= 392 \\ -28x+28y=-224 \\ \hline 56y=168 \end{array}$$

$$\frac{56y}{\mathbf{56}}=\frac{168}{\mathbf{56}}$$

$$y=3$$

State

The speed of the river is 3 mph.

Check

The results check.

36. SHOPPING

Analyze

- 2 bottles of cleaner and 3 bottles of soaking solution cost $63.40
- 3 bottles of cleaner and 3 bottles of soaking solution cost $69.60.
- How much is each bottle of solution?

Assign

Let x = cost of a bottle of cleaner

y = cost of a bottle of soaking solution

Form

	Amount •	Price	= Total Value
Cleaner	2	x	$\mathbf{2x}$
Soaking	3	y	$\mathbf{3y}$
		Total	**63.40**

The value of 2 bottles of cleaner plus the value of 3 bottles of soaking is $63.40.

$$2x + 3y = 63.40$$

	Amount •	Price	= Total Value
Cleaner	3	x	$\mathbf{3x}$
Soaking	2	y	$\mathbf{2y}$
		Total	**69.60**

The value of 3 bottles of cleaner plus the value of 2 bottles of soaking is $69.60.

$$3x + 2y = 69.60$$

Solve

$$\begin{cases} 2x+3y=63.40 \\ 3x+2y=69.60 \end{cases}$$

Eliminate x.

Multiply both sides of the 1st equation by 3.

Multiply both sides of the 2nd equation by -2.

$$\begin{array}{r} 6x+9y= 190.20 \\ -6x-4y=-139.20 \\ \hline 5y=51 \end{array}$$

$$\frac{5y}{\mathbf{5}}=\frac{51}{\mathbf{5}}$$

$$y=10.20$$

$$2x+3y=63.40 \quad \text{The 1st equation}$$

$$2x+3(\mathbf{10.20})=63.40$$

$$2x+30.60-\mathbf{30.60}=63.40-\mathbf{30.60}$$

$$2x=32.80$$

$$\frac{2x}{\mathbf{2}}=\frac{32.80}{\mathbf{2}}$$

$$x=16.40$$

State

A bottle of cleaning solution cost $16.40.

A bottle of soaking solution cost $10.20.

Check

The results check.

37. INVESTING

Analyze

- Investing \$3,000.
- Passbook account paying 6% annual interest.
- Certificate account paying 10% annually.
- Total interest for first year was \$270.
- How much was invested at 6%?

Assign

Let x = amt invested in passbook account

y = amt invested in certificate account

Form

The amount invested in the passbook	plus	the amount invested in the certificate	is	\$3,000.
x	$+$	y	$=$	3,000

	Principal •	Rate •	Time =	Interest
Passbook	x	0.06	1 yr	**0.06x**
Certificate	y	0.10	1 yr	**0.10y**
			Total Interest =	**\$270**

The interest earned by the passbook	plus	the interest earned by the certificate	is	\$270.
$0.06x$	$+$	$0.10y$	$=$	270

Solve

$$\begin{cases} x + y = 3{,}000 \\ 0.06x + 0.10y = 270 \end{cases}$$

Eliminate y.

Multiply both sides of the 1st equation by -10.

Multiply both sides of the 2nd equation by 100.

$$\begin{aligned} -10x - 10y &= -30{,}000 \\ 6x + 10y &= 27{,}000 \\ \hline -4x &= -3{,}000 \end{aligned}$$

$$\frac{-4x}{\mathbf{-4}} = \frac{-3{,}000}{\mathbf{-4}}$$

$$x = 750$$

State

\$750 was invested at 6%.

Check

The results check.

38. ANTIFREEZE

Analyze

- A 40% antifreeze solution.
- A 70% antifreeze solution.
- How much of each is needed to make a 20 gal mixture that is 50% antifreeze solution?

Assign

Let x = amount of 40% antifreeze in gallons

y = amount of 70% antifreeze in gallons

Form

The # of gallons of 40% antifreeze	plus	the # of gallons of 70% antifreezes	is	20 gallons.
x	$+$	y	$=$	20

	Amt •	Strength =	Amt of antifreeze
Weak	x	0.40	**0.4x**
Strong	y	0.70	**0.7y**
Mix	20	0.50	**20(0.5)**

The amount of antifreeze in the 40% solution	plus	the amount of antifreeze in the 70% solution	is equal to	the amount of anitfreeze 50% solution.
$0.4x$	$+$	$0.7y$	$=$	$20(0.5)$

Solve

$$\begin{cases} x + y = 20 \\ 0.4x + 0.7y = 10 \end{cases}$$

Eliminate x.

Multiply both sides of the 1st equation by -4.

Multiply both sides of the 2nd equation by 10.

$$\begin{aligned} -4x - 4y &= -80 \\ 4x + 7y &= 100 \\ \hline 3y &= 20 \end{aligned}$$

$$\frac{3y}{\mathbf{3}} = \frac{20}{\mathbf{3}}$$

$$y = 6\frac{2}{3}$$

$x + y = 20$ The 1st equation

$$x + \mathbf{6\frac{2}{3}} = 20$$

$$x + \frac{20}{3} - \frac{\mathbf{20}}{\mathbf{3}} = \frac{60}{3} - \frac{\mathbf{20}}{\mathbf{3}}$$

$$x = \frac{40}{3}$$

$$x = 13\frac{1}{3}$$

State

$13\frac{1}{3}$ gallons of 40% antifreeze will be needed.

$6\frac{2}{3}$ gallons of 70% antifreeze will be needed.

Check

The results check.

SECTION 4.5 REVIEW EXERCISES

Solving Systems of Linear Inequalities

39. $\begin{cases} 5x+3y<15 \\ 3x-y>3 \end{cases}$

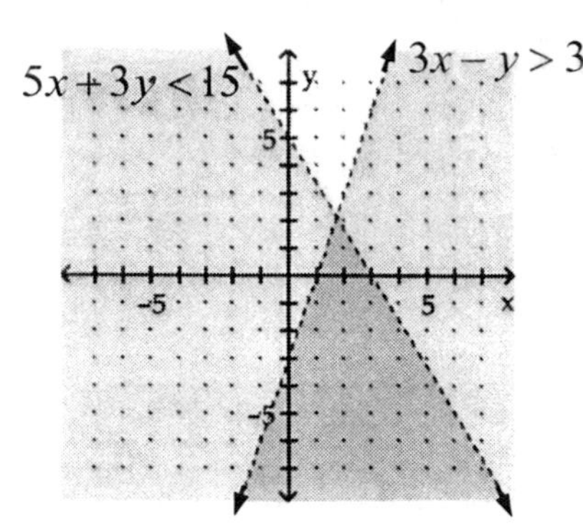

40. $\begin{cases} 3y \le x \\ y>3x \end{cases}$

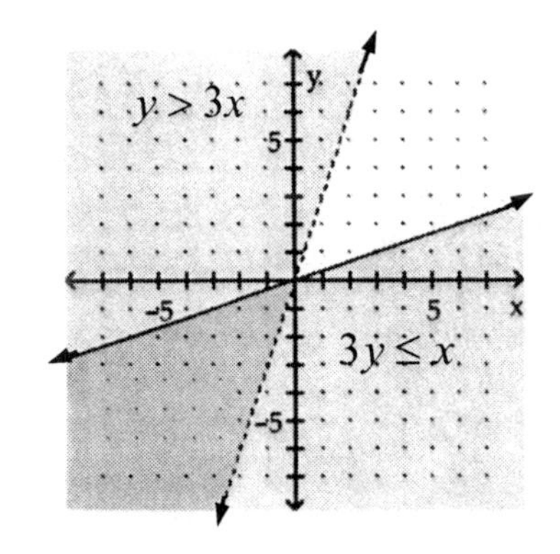

41. $\begin{cases} x \le 0 \\ y<0 \end{cases}$

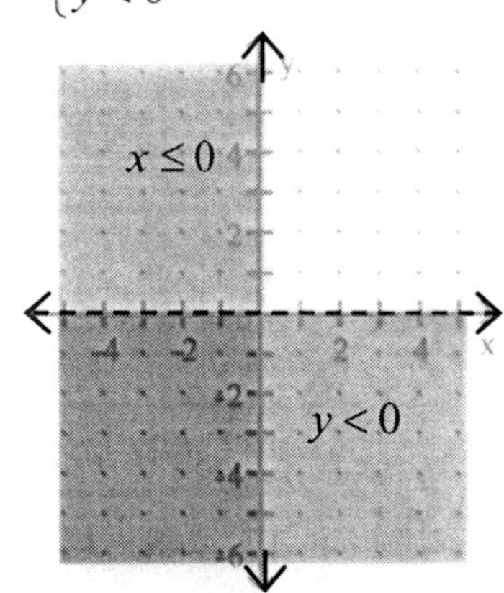

42. $\begin{cases} y \ge x \\ y \le \frac{1}{3}x+1 \\ x>-3 \end{cases}$

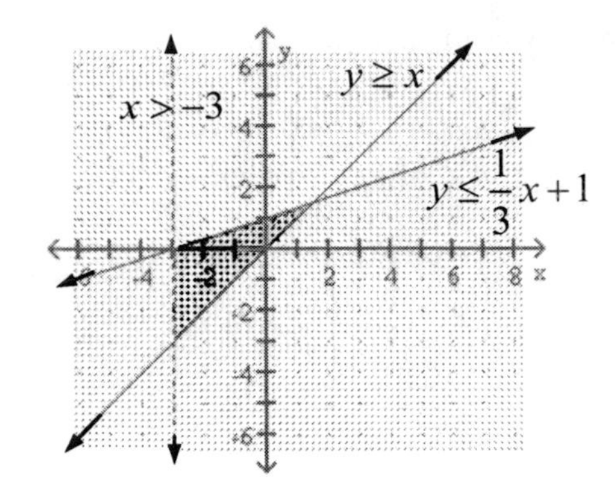

Use a check to determine whether each ordered pair is a solution of the system.

43. a. $(5, -4)$ $\begin{cases} x+2y \le 3 \\ 2x-y>1 \end{cases}$

$$\begin{array}{cc} x+2y \le 3 & 2x-y>1 \\ \mathbf{5}+2(\mathbf{-4}) \overset{?}{\le} 3 & 2(\mathbf{5})-(\mathbf{-4}) \overset{?}{>} 1 \\ 5-8 \overset{?}{\le} 3 & 10+4 \overset{?}{>} 1 \\ -3 \le 3 & 14>3 \\ \text{True} & \text{True} \end{array}$$

$(5,-4)$ is a solution of the system.

b. $(-1, -3)$ $\begin{cases} x+2y \le 3 \\ 2x-y>1 \end{cases}$

$$\begin{array}{cc} x+2y \le 3 & 2x-y>1 \\ \mathbf{-1}+2(\mathbf{-3}) \overset{?}{\le} 3 & 2(\mathbf{-1})-(\mathbf{-3}) \overset{?}{>} 1 \\ -1-6 \overset{?}{\le} 3 & -2+3 \overset{?}{>} 1 \\ -7 \le 3 & 1>3 \\ \text{True} & \text{False} \end{array}$$

$(-1,-3)$ is not a solution of the system.

44. GIFT SHOPPING

$$\begin{cases} 10x+20y \ge 40 \\ 10x+20y \le 60 \end{cases}$$

Let $x = \#$ of T-shirts

$y = \#$ of pants

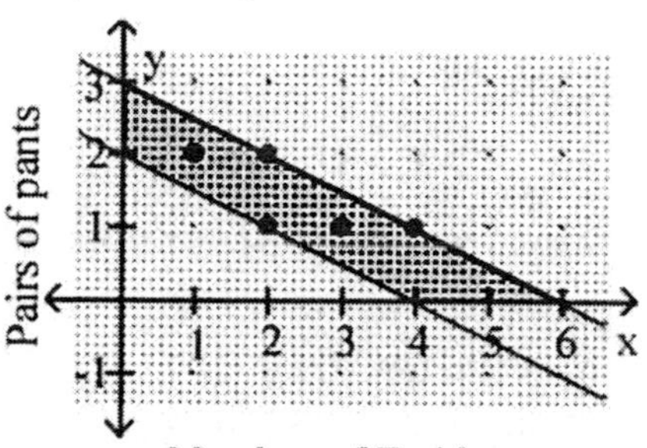

1 T-shirt and 2 pairs of pants

3 T-shirts and 1 pair of pants

There are other answers.

CHAPTER 4 TEST

Determine whether the ordered pair is a solution of the system.

1. $(5,3)$ $\begin{cases} 3x+2y=21 \\ x+y=8 \end{cases}$

$3x+2y=21$	$x+y=8$
$3(5)+2(3)\overset{?}{=}21$	$5+3\overset{?}{=}8$
$15+6\overset{?}{=}21$	$8=8$
$21=21$	True
True	

Since (5,3) satisfies both equations, it is a solution of the system.

2. $(-2,-1)$ $\begin{cases} 4x+y=-9 \\ 2x-3y=-7 \end{cases}$

$4x+y=-9$	$2x-3y=-7$
$4(-2)+(-1)\overset{?}{=}-9$	$2(-2)-3(-1)\overset{?}{=}-7$
$-8-1\overset{?}{=}-9$	$-4+3\overset{?}{=}-7$
$-9=-9$	$-1=-7$
True	False

Since $(-2,-1)$ does not satisfies both equations, it is not a solution of the system.

3. **Fill in the blanks.**

a. A **solution** of a system of linear equations is an ordered pair that satisfies each equation.

b. A system of equations that has at least one solution is called a **consistent** system.

c. A system of equations that has no solution is called an **inconsistent** system.

d. Equations with different graphs are called **independent** equations.

e. A system of **dependent** equations has an infinite number of solutions.

f. Two angles are said to be **supplementary** if the sum of their measures is 180°.

4. **ENERGY**

(2005, 770); In 2005, the amount of electricity generated by natural gas and nuclear sources was the same, about 770 billion kilowatts hours each.

Solve each system by graphing.

5. $\begin{cases} y=2x-1 \\ x-2y=-4 \end{cases}$

The solution is (2,3).

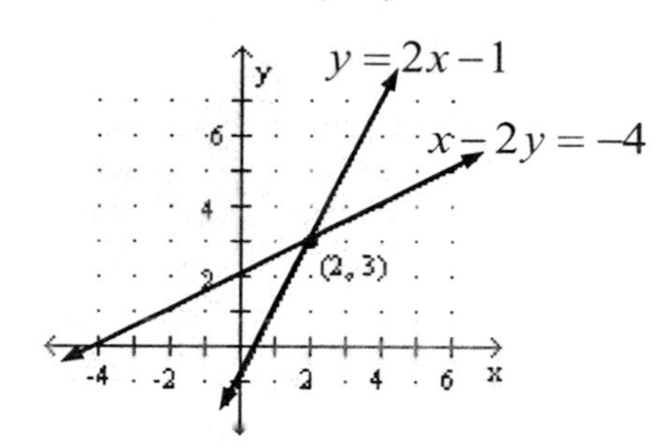

6. $\begin{cases} x+y=5 \\ y=-x \end{cases}$

The lines are parallel.
The system has no solution.

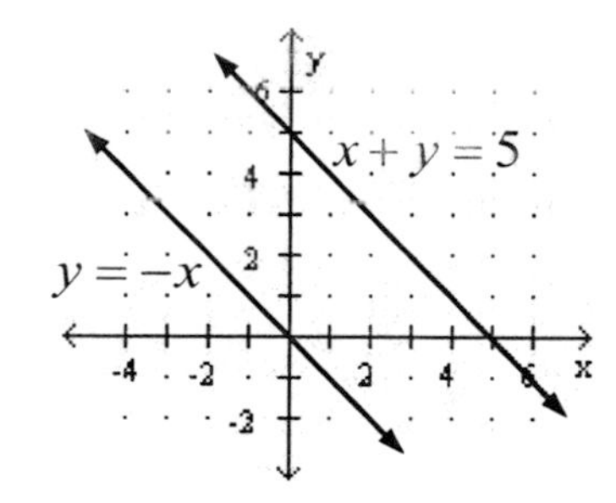

7. Find the slope and the y-intercept of the graph of each line in the system $\begin{cases} y = 4x - 10 \\ x - 2y = -16 \end{cases}$.

Then, use that information to determine the *number of solutions* of the system. **Do not solve the system.**

$$\begin{cases} y = 4x - 10 \\ x - 2y = -16 \end{cases}$$

$y = 4x - 10$	$x - 2y = -16$
$m = 4$	$x - 2y - x = -16 - x$
$y\text{-int} = (0, \ -10)$	$\dfrac{-2y}{-2} = \dfrac{-x}{-2} - \dfrac{16}{-2}$
	$y = \dfrac{1}{2}x + 8$
	$m = \dfrac{1}{2}$
	$y\text{-int} = (0, 8)$

Since the lines have different slopes, they will intersect at one point. The system has 1 solution.

8. Infinitely many solutions. **(0, 2), (3, 1), (6, 0).**

Solve each system by substitution.

9. $\begin{cases} y = x - 1 \\ 2x + y = -7 \end{cases}$

$$2x + y = -7 \quad \text{The 2}^{\text{nd}} \text{ equation. Substitute for } y.$$
$$2x + (\mathbf{x - 1}) = -7$$
$$3x - 1 = -7$$
$$3x - 1 + \mathbf{1} = -7 + \mathbf{1}$$
$$3x = -6$$
$$\frac{3x}{\mathbf{3}} = \frac{-6}{\mathbf{3}}$$
$$x = -2$$

$y = x - 1$ The 1st equation

$y = \mathbf{-2} - 1$

$y = -3$

The solution is $(-2, -3)$.

10. $\begin{cases} 3x + 6y = -15 \\ x + 2y = -5 \end{cases}$

Solve the 2nd equation for x.

$$\begin{cases} 3x + 6y = -15 \\ x = -5 - 2y \end{cases}$$

$$3x + 6y = -15 \quad \text{The 1}^{\text{st}} \text{ equation. Substitute for } x.$$
$$3(\mathbf{-5 - 2y}) + 6y = -15$$
$$-15 - 6y + 6y = -15$$
$$-15 = -15 \quad \text{The variables drop out.}$$

True

The system has infinitely many solutions.

Solve each system using elimination (addition).

11. $\begin{cases} 3x - y = 2 \\ 2x + y = 8 \end{cases}$

Eliminate y.

$$\begin{aligned} 3x - y &= 2 \\ 2x + y &= 8 \\ \hline 5x \quad &= 10 \end{aligned}$$
$$\frac{5x}{\mathbf{5}} = \frac{10}{\mathbf{5}}$$
$$x = 2$$

$2x + y = 8$ The 2nd equation

$2(\mathbf{2}) + y = 8$

$4 + y - \mathbf{4} = 8 - \mathbf{4}$

$y = 4$

The solution is (2, 4).

12. $\begin{cases} 4x + 3y = -3 \\ -3x = -4y + 21 \end{cases}$

Write the 2nd equation in standard form.

$$\begin{cases} 4x + 3y = -3 \\ -3x + 4y = 21 \end{cases}$$

Eliminate x.

Multiply both sides of the 1st equation by 3.

Multiply both sides of the 2nd equation by 4.

$$\begin{aligned} 12x + 9y &= -9 \\ -12x + 16y &= 84 \\ \hline 25y &= 75 \end{aligned}$$
$$\frac{25y}{\mathbf{25}} = \frac{75}{\mathbf{25}}$$
$$y = 3$$

$$4x+3y=-3 \text{ The 1}^{\text{st}} \text{ equation}$$
$$4x+3(\mathbf{3})=-3$$
$$4x+9-\mathbf{9}=-3-\mathbf{9}$$
$$4x=-12$$
$$\frac{4x}{\mathbf{4}}=\frac{-12}{\mathbf{4}}$$
$$x=-3$$
The solution is $(-3,3)$.

$$-a+2b=-1 \text{ The 2}^{\text{nd}} \text{ equation}$$
$$-a+2(\mathbf{-1})=-1$$
$$-a-2+\mathbf{2}=-1+\mathbf{2}$$
$$-a=1$$
$$\frac{-a}{\mathbf{-1}}=\frac{1}{\mathbf{-1}}$$
$$x=-1$$
The solution is $(-1,-1)$.

Solve each system using substitution or elimination (addition).

13. $\begin{cases} 3x-5y-16=0 \\ \frac{x}{2}-\frac{5}{6}y=\frac{1}{3} \end{cases}$

Write the 1st equation in standard form.
Clear the 2nd equation of fractions.

$$\begin{cases} 3x-5y-16\mathbf{+16}=0\mathbf{+16} \\ \mathbf{6}\left(\frac{x}{2}\right)-\mathbf{6}\left(\frac{5}{6}y\right)=\mathbf{6}\left(\frac{1}{3}\right) \end{cases}$$

$$\begin{cases} 3x-5y=16 \\ 3x-5y=2 \end{cases}$$

Eliminate x.
Multiply both sides of the 1st equation by -1.

$$-3x+5y=-16$$
$$3x-5y=\quad 2$$
$$0=-14 \text{ Both variables are eliminated.}$$
False
No solution.

14. $\begin{cases} 3a+4b=-7 \\ 2b-a=-1 \end{cases}$

Rewrite the 2nd equation in standard form.

$$\begin{cases} 3a+4b=-7 \\ -a+2b=-1 \end{cases}$$

Eliminate a.
Multiply both sides of the 2nd equation by 3.

$$3a+4b=-7$$
$$-3a+6b=-3$$
$$10b=-10$$
$$\frac{10b}{\mathbf{10}}=\frac{-10}{\mathbf{10}}$$
$$b=-1$$

15. $\begin{cases} y=3x-1 \\ y=2x+4 \end{cases}$

$$y=2x+4 \text{ The 2}^{\text{nd}} \text{ equation. Substitute for } y.$$
$$(\mathbf{3x-1})=2x+4$$
$$3x-1+\mathbf{1}=2x+4+\mathbf{1}$$
$$3x=2x+5$$
$$3x-\mathbf{2x}=2x+5-\mathbf{2x}$$
$$x=5$$

$$y=3x-1 \text{ The 1}^{\text{st}} \text{ equation}$$
$$y=3(\mathbf{5})-1$$
$$y=15-1$$
$$y=14$$
The solution is $(5,14)$.

16. $\begin{cases} 0.6c+0.5d=0 \\ 0.02c+0.09d=0 \end{cases}$

Eliminate c.
Multiply both sides of the 1st equation by -10.
Multiply both sides of the 2nd equation by 300.

$$-6c-5d=0$$
$$6c+27d=0$$
$$22d=0$$
$$\frac{22d}{\mathbf{22}}=\frac{0}{\mathbf{22}}$$
$$d=0$$

$$0.6c+0.5d=0 \text{ The 1}^{\text{st}} \text{ equation}$$
$$0.6c+0.5(\mathbf{0})=0$$
$$0.6c=0$$
$$\frac{0.6c}{\mathbf{0.6}}=\frac{0}{\mathbf{0.6}}$$
$$c=0$$
The solution is $(0,0)$.

17. $\begin{cases} a-1=2b \\ 3a+1=-10b \end{cases}$

Rewrite both equations in standard form.

$\begin{cases} a-2b=1 \\ 3a+10b=-1 \end{cases}$

Eliminate b.

Multiply both sides of the 1st equation by 5.

$\begin{cases} 5a-10b=5 \\ 3a+10b=-1 \end{cases}$

$$\begin{aligned} 5a-10b&=5 \\ 3a+10b&=-1 \\ \hline 8a&=4 \\ \frac{8a}{\mathbf{8}}&=\frac{4}{\mathbf{8}} \\ a&=\frac{1}{2} \end{aligned}$$

$a-2b=1$ The 1st equation

$$\frac{\mathbf{1}}{\mathbf{2}}-2b=1$$

$$\frac{1}{2}-2b-\frac{\mathbf{1}}{\mathbf{2}}=1-\frac{\mathbf{1}}{\mathbf{2}}$$

$$-2b=\frac{2}{2}-\frac{1}{2}$$

$$-2b=\frac{1}{2}$$

$$-\frac{\mathbf{1}}{\mathbf{2}}(-2b)=-\frac{\mathbf{1}}{\mathbf{2}}\left(\frac{1}{2}\right)$$

$$b=-\frac{1}{4}$$

The solution is $\left(\frac{1}{2},-\frac{1}{4}\right)$.

18. $\begin{cases} \frac{x+2}{6}=\frac{y+3}{4} \\ \frac{x}{5}=\frac{3y-3}{6} \end{cases}$

Clear both equations of fractions.

$\begin{cases} \mathbf{12}\left(\frac{x+2}{6}\right)=\mathbf{12}\left(\frac{y+3}{4}\right) \\ \mathbf{30}\left(\frac{x}{5}\right)=\mathbf{30}\left(\frac{3y-3}{6}\right) \end{cases}$

$\begin{cases} 2(x+2)=3(y+3) \\ 6x=5(3y-3) \end{cases}$

Distribute both equations.

$\begin{cases} 2x+4=3y+9 \\ 6x=15y-15 \end{cases}$

Rewrite both equations in standard form.

$\begin{cases} 2x-3y=5 \\ 6x-15y=-15 \end{cases}$

Eliminate x.

Multiply both sides of the 1st equation by -3.

$\begin{cases} -6x+9y=-15 \\ 6x-15y=-15 \end{cases}$

$$\begin{aligned} -6x+9y&=-15 \\ 6x-15y&=-15 \\ \hline -6y&=-30 \\ \frac{-6y}{\mathbf{-6}}&=\frac{-30}{\mathbf{-6}} \\ y&=5 \end{aligned}$$

$2x-3y=5$ The 1st equation

$$2x-3(\mathbf{5})=5$$

$$2x-15=5$$

$$2x-15\mathbf{+15}=5\mathbf{+15}$$

$$2x=20$$

$$\frac{2x}{\mathbf{2}}=\frac{20}{\mathbf{2}}$$

$$x=10$$

The solution is (10, 5).

19. $\begin{cases} 4(a-2)+5y=19 \\ 3(a-2)-y=0 \end{cases}$

Distribute both equations.

$\begin{cases} 4a-8+5y=19 \\ 3a-6-y=0 \end{cases}$

Rewrite both equations in standard form.

$\begin{cases} 4a-8+5y+\mathbf{8}=19+\mathbf{8} \\ 3a-6-y+\mathbf{6}=0+\mathbf{6} \end{cases}$

$\begin{cases} 4a+5y=27 \\ 3a-y=6 \end{cases}$

Eliminate y.

Multiply both sides of the 2nd equation by 5.

$\begin{cases} 4a+5y=27 \\ 15a-5y=30 \end{cases}$

$$\begin{aligned} 4a+5y&=27 \\ 15a-5y&=30 \\ \hline 19a&=57 \\ \frac{19a}{\mathbf{19}}&=\frac{57}{\mathbf{19}} \\ a&=3 \end{aligned}$$

$$\begin{aligned} 3a-y&=6 \quad \text{The 2nd equation. Substitute for } x. \\ 3(\mathbf{3})-y&=6 \\ 9-y&=6 \\ 9-y-\mathbf{9}&=6-\mathbf{9} \\ -y&=-3 \\ \frac{-y}{\mathbf{-1}}&=\frac{-3}{\mathbf{-1}} \\ y&=3 \end{aligned}$$

The solution is (3, 3).

20. $\begin{cases} 3x+1=2x-4y+2 \\ 7x-y-1=9y+5x+10 \end{cases}$

Rewrite both equations in standard form.

$\begin{cases} 3x+1\mathbf{-1-2x+4y}=2x-4y+2\mathbf{-1-2x+4y} \\ 7x-y-1\mathbf{+1-5x-9y}=9y+5x+10\mathbf{+1-5x-9y} \end{cases}$

$\begin{cases} x+4y=1 \\ 2x-10y=11 \end{cases}$

Solve the 1st equation for x.

Substitute for x in the 2nd equation.

$\begin{cases} x=1-4y \\ 2x-10y=11 \end{cases}$

$$\begin{aligned} 2(\mathbf{1-4y})-10y&=11 \quad \text{The 2nd equation. Substitute for } x. \\ 2-8y-10y&=11 \\ 2-18y&=11 \\ 2-18y\mathbf{-2}&=11\mathbf{-2} \\ -18y&=9 \\ \frac{-18y}{\mathbf{-18}}&=\frac{9}{\mathbf{-18}} \\ y&=-\frac{1}{2} \end{aligned}$$

$$\begin{aligned} x&=1-4y \\ x&=1-4\left(\mathbf{-\frac{1}{2}}\right) \\ x&=1+2 \\ x&=3 \end{aligned}$$

The solution is $\left(3,-\frac{1}{2}\right)$.

Write a system of two equations in two variables to solve each problem.

21. CHILD CARE

Analyze

- 22-mile commute, one way.
- First part of the trip is 6 miles less than the second part.
- How long is each part of her commute?

Assign

Let x = the 1st part in miles

y = the 2nd part in miles

Form

The first part	plus	the second part	is	22 miles.
x	$+$	y	$=$	22

The first part	is	6 miles	less than	the second part.
x	$=$	y	$-$	6

Solve

$$\begin{cases} x+y=22 \\ x=y-6 \end{cases}$$

$x+y=22$ The 1st equation. Substitute for x.

$(\mathbf{y-6})+y=22$

$2y-6=22$

$2y-6+\mathbf{6}=22+\mathbf{6}$

$2y=28$

$\frac{2y}{\mathbf{2}}=\frac{28}{\mathbf{2}}$

$y=14$

$x=y-6$ The 2nd equation

$x=(\mathbf{14})-6$

$x=8$

State

The first part is 8 miles.
The second part is 14 miles.

Check

$8+14=22$ | $8=14-6$

$22=22$ | $8=8$

The results check.

22. VACATIONING

Analyze

- Cost \$219 for a family of 7.
- Adult tickets cost \$37.00.
- Child tickets cost \$27.00.
- How many of each type were bought?

Assign

Let x = number of child tickets

y = number of adult tickets

Form

The number of children tickets	plus	the number of adult tickets	is	7.
x	$+$	y	$=$	7

	Number •	Value	= Total Value
Child	x	\$27	$\mathbf{27x}$
Adult	y	\$37	$\mathbf{37y}$
		Total	**219**

The value of children tickets	plus	the value of adult tickets	is	\$219.
$27x$	$+$	$37y$	$=$	219

Solve

$$\begin{cases} x+y=7 \\ 27x+33y=219 \end{cases}$$

Eliminate x.
Multiply both sides of the 1st equation by -27.

$$\begin{aligned} -27x-27y&=-189 \\ 27x+37y&=\ \ 219 \\ \hline 10y&=30 \end{aligned}$$

$\frac{10y}{\mathbf{10}}=\frac{30}{\mathbf{10}}$

$y=3$

$x+y=7$ The 1st equation

$x+\mathbf{3}=7$

$x+3-\mathbf{3}=7-\mathbf{3}$

$x=4$

State

4 child tickets were bought.
3 adult tickets were bought.

Check

The results check.

23. FINANCIAL PLANNING

Analyze

- Investing $10,000.
- Invest part at 8% annual interest.
- Invest the rest at 9% annual interest.
- Combined interest for first year was $840.
- How much was invested at each rate?

Assign

Let x = amt invested at 8%

y = amt invested at 9%

Form

The amount invested at 8%	plus	the amount invested at 9%	is	$10,000.
x	$+$	y	$=$	10,000

	Principal ·	Rate ·	Time =	Interest
8%	x	8%	1 yr	**$0.08x$**
9%	y	9%	1 yr	**$0.09y$**

Total Interest = $840

The interest earned by 8% investment	plus	the interest earned by 9% investment	is	$840.
$0.08x$	$+$	$0.09y$	$=$	840

Solve

$$\begin{cases} x + y = 10{,}000 \\ 0.08x + 0.09y = 840 \end{cases}$$

Eliminate x.

Multiply both sides of the 1st equation by -8.

Multiply both sides of the 2nd equation by 100.

$$\begin{array}{r} -8x - 8y = -80{,}000 \\ \underline{8x + 9y = 84{,}000} \\ y = 4{,}000 \end{array}$$

$$x + y = 10{,}000 \quad \text{The 1}^{\text{st}} \text{ equation}$$

$$x + \mathbf{4{,}000} = 10{,}000$$

$$x + 4{,}000 - \mathbf{4{,}000} = 10{,}000 - \mathbf{4{,}000}$$

$$x = 6{,}000$$

State

$6,000 was invested at 8%.

$4,000 was invested at 9%.

Check

$\$6{,}000 + \$4{,}000 = \$10{,}000$	$0.08(\$6{,}000) + 0.09(\$4{,}000) = \$840$
$\$10{,}000 = \$10{,}000$	$\$480 + \$360 = \$840$
	$\$840 = \$840$

The results check.

24. TAILWINDS/HEADWINDS

Analyze

- Plane flies 450 miles in 2.5 hours with a tailwind.
- Plane flies 450 miles in 3 hours into a headwind.
- Find the speed of the plane in calm air and the speed of the wind.

Assign

Let x = speed of plane in calm air in mph

y = speed of wind in mph

Form

	Rate ·	Time =	Distance
With tailwind	$x + y$	2.5	**$2.5(x + y)$**

Total Distance = 450

The rate of the plane with tailwind	times	2.5 hours traveled	is	450 miles.
$(x + y)$	$\cdot$	2.5	$=$	450

	Rate ·	Time =	Distance
Into headwind	$x - y$	3	**$3(x - y)$**

Total Distance = 450

The rate of the plnae into headwind	times	3 hours traveled	is	450 miles.
$(x - y)$	$\cdot$	3	$=$	450

Solve

$$\begin{cases} 2.5(x + y) = 450 \\ 3(x - y) = 450 \end{cases}$$

Distribute in both equations.

$$\begin{cases} 2.5x + 2.5y = 450 \\ 3x - 3y = 450 \end{cases}$$

Eliminate y.

Multiply both sides of the 1st equation by 30.

Multiply both sides of the 2nd equation by 25.

$$\begin{array}{l} 75x + 75y = 13{,}500 \\ \underline{75x - 75y = 11{,}250} \\ 150x \qquad = 24{,}750 \end{array}$$

$$\frac{150x}{\mathbf{150}} = \frac{24{,}750}{\mathbf{150}}$$

$$x = 165$$

$$3x - 3y = 450 \quad \text{The 2}^{\text{nd}} \text{ equation}$$

$$3(\mathbf{165}) - 3y = 450$$

$$495 - 3y - \mathbf{495} = 450 + \mathbf{495}$$

$$-3y = -45$$

$$\frac{-3y}{\mathbf{-3}} = \frac{-45}{\mathbf{-3}}$$

$$y = 15$$

State

The speed of the plane is 165 mph.

The speed of the wind is 15 mph.

Check

The results check.

25. TETHER BALL

Analyze

- Two angles are complementary (90°).
- Larger angle is 10° more than 3 times the measure of the smaller angle.
- Find the measure of each angle.

Assign

Let x = the measure of smaller ∠

y = the measure of larger ∠

Form

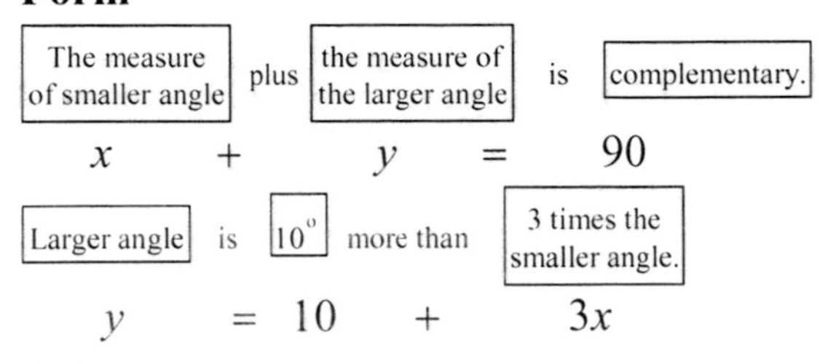

Solve

$$\begin{cases} x+y=90 \\ y=10+3x \end{cases}$$

$x+y=90$ The 1st equation. Substitute for y.

$$x+(\mathbf{10+3x})=90$$

$$4x+10=90$$

$$4x+10-\mathbf{10}=90-\mathbf{10}$$

$$4x=80$$

$$\frac{4x}{\mathbf{4}}=\frac{80}{\mathbf{4}}$$

$$x=20$$

$y=10+3x$ The 2nd equation

$$y=10+3(\mathbf{20})$$

$$y=10+60$$

$$y=70$$

State

The measure of the smaller angle is 20°.
The measure of the larger angle is 70°.

Check

$20+70=90$	$70=10+3(20)$
$90=90$	$70=70$

The results check.

26. ANTIFREEZE

Analyze

- A 5% antifreeze solution.
- A 20% antifreeze solution.
- How much of each is needed to make a 12 pint solution that is 15% antifreeze?

Assign

Let x = amount of 5% antifreeze in pints

y = amount of 20% antifreeze in pints

Form

The # of pints of 5% antifreeze	plus	the # of pints of 20% antifreezes	is	12 pints.
x	+	y	=	12

	Amt •	Strength =	Amt of pure antifreeze
Weak	x	0.05	**0.05*x***
Strong	y	0.20	**0.20*y***
Mixture	12	0.15	**12(0.15)**

The amount of pure antifreeze in the 5% solution	plus	the amount of pure antifreeze in the 20% solution	is equal to	the amount of pure anitfreeze in the 15% solution.
$0.05x$	+	$0.20y$	=	$12(0.15)$

Solve

$$\begin{cases} x+y=12 \\ 0.05x+0.20y=1.8 \end{cases}$$

Eliminate x.
Multiply both sides of the 1st equation by −5.
Multiply both sides of the 2nd equation by 100.

$$\begin{aligned} -5x-\ 5y &= -60 \\ 5x+20y &= \ 180 \\ \hline 15y &= \ 120 \end{aligned}$$

$$\frac{15y}{\mathbf{15}}=\frac{120}{\mathbf{15}}$$

$$y=8$$

$x+y=12$ The 1st equation

$$x+\mathbf{8}=12$$

$$x+8-\mathbf{8}=12-\mathbf{8}$$

$$x=4$$

State

4 pints of 5% antifreeze will be needed.
8 pints of 20% antifreeze will be needed.

Check

The results check.

27. SUNSCREEN

Analyze

- \$1.50/ounce sunscreen.
- \$0.80/ounce sunscreen.
- How much of each is needed to make 10 ounces that sells for \$1.01?

Assign

Let x = amount of \$1.50/oz sunscreen in oz

y = amount of \$0.80/oz sunscreen in oz

Form

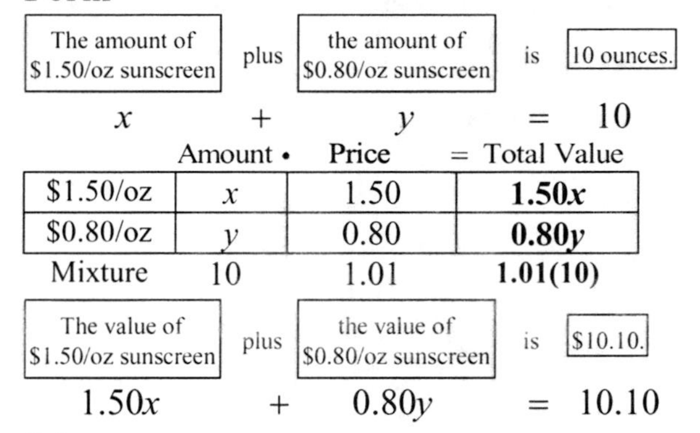

The amount of \$1.50/oz sunscreen	plus	the amount of \$0.80/oz sunscreen	is	10 ounces.
x	$+$	y	$=$	10

	Amount •	Price	= Total Value
\$1.50/oz	x	1.50	**1.50*x***
\$0.80/oz	y	0.80	**0.80*y***
Mixture	10	1.01	**1.01(10)**

The value of \$1.50/oz sunscreen	plus	the value of \$0.80/oz sunscreen	is	\$10.10.
$1.50x$	$+$	$0.80y$	$=$	10.10

Solve

$$\begin{cases} x+y=10 \\ 1.50x+0.80y=10.10 \end{cases}$$

Eliminate y.

Multiply both sides of the 1st equation by -80.

Multiply both sides of the 2nd equation by 100.

$$\begin{array}{l} -80x-80y=-800 \\ \underline{150x+80y=1010} \\ 70x \qquad\quad = 210 \end{array}$$

$$\frac{70x}{\mathbf{70}}=\frac{210}{\mathbf{70}}$$

$$x=3$$

$x+y=10$ The 1st equation

$\mathbf{3}+y=10$

$3+y-\mathbf{3}=10-\mathbf{3}$

$y=7$

State

3 ounces of \$1.50/oz sunscreen is needed.

7 ounces of \$0.80/oz sunscreen is needed.

Check

$3+7=10$	$\$1.50(3)+\$0.80(7)=\$10.10$
$10=10$	$\$4.50+\$5.60=\$10.10$
	$\$10.10=\$10.10$

The results check.

Use a check to determine whether (3, 1) is a solution of the system :

28. $\begin{cases} y \le 2x-1 \\ x+3y>6 \end{cases}$

$y \le 2x-1$	$x+3y>6$
$\mathbf{1} \overset{?}{\le} 2(\mathbf{3})-1$	$(\mathbf{3})+3(\mathbf{1}) \overset{?}{>} 6$
$1 \overset{?}{\le} 6-1$	$3+3 \overset{?}{>} 6$
$1 \le 5$	$6>6$
True	False

(3,1) is not a solution of the system.

Solve the system by graphing.

29. $\begin{cases} 3x+2y \le 6 \\ y \ge x+1 \end{cases}$

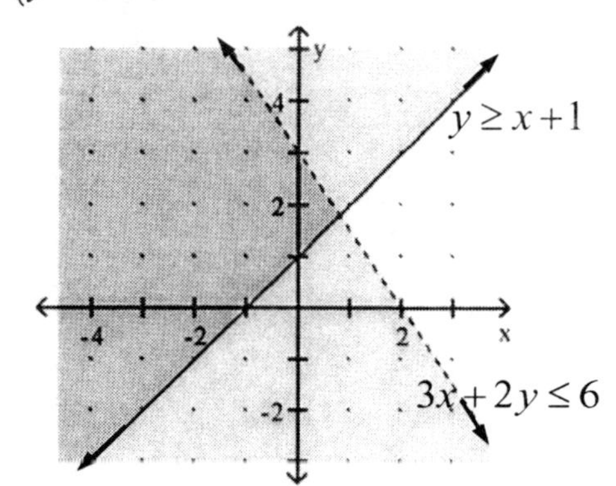

30. $\begin{cases} x-y<3 \\ y \le 0 \\ x \ge 0 \end{cases}$

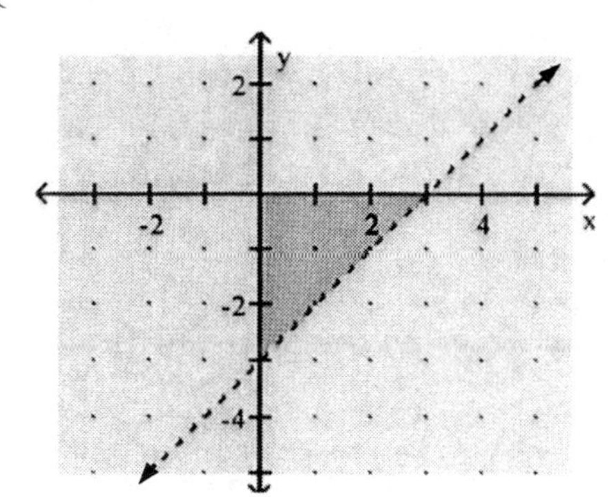

31. CLOTHES SHOPPING

$$\begin{cases} 20x+40y \ge 80 \\ 20x+40y \le 120 \end{cases}$$

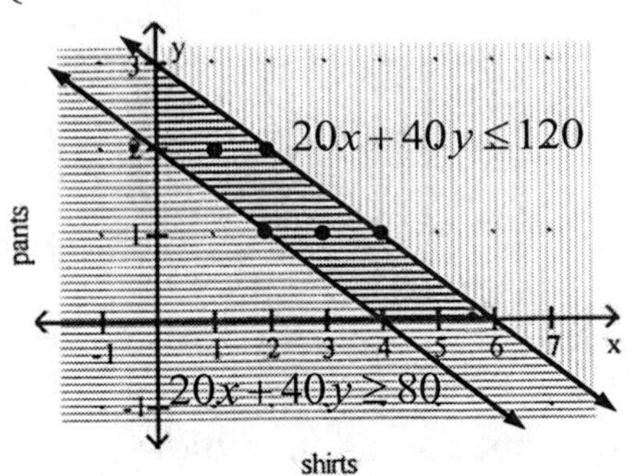

1 shirt, 2 pants

2 shirts, 1 pants

2 shirts, 2 pants

32. Match each equation, inequality, or system with the graph of its solution.

a. $2x+y=2$ **iii** b. $2x+y \ge 2$ **ii**

c. $\begin{cases} 2x+y=2 \\ 2x-y=2 \end{cases}$ **i** d. $\begin{cases} 2x+y \ge 2 \\ 2x-y \le 2 \end{cases}$ **iv**

CUMULATIVE REVIEW CHAPTERS 1 - 4

Fill in the blanks.

1. The answer to an addition problem is called a **sum**. The answer to a subtraction problem is called a **difference**. The answer to a multiplication problem is called a **product**. The answer to a division problem is called a **quotient**. [Section 1.1]

2. Prime Factorization: [Section 1.2]

$$\begin{aligned}100 &= 2\cdot 50\\ &= 2\cdot 2\cdot 25\\ &= 2\cdot 2\cdot 5\cdot 5\\ &= 2^2\cdot 5^2\end{aligned}$$

3. Divide: [Section 1.2]

$$\begin{aligned}\frac{3}{4}\div\frac{6}{5} &= \frac{3}{4}\cdot\frac{5}{6}\\ &= \frac{\overset{1}{\cancel{3}}\cdot 5}{4\cdot 2\cdot \underset{1}{\cancel{3}}}\\ &= \frac{5}{8}\end{aligned}$$

4. Subtract: [Section 1.2]

$$\text{LCD} = 2\cdot 5\cdot 7 = 70$$

$$\begin{aligned}\frac{7}{10}-\frac{1}{14} &= \frac{7}{10}\cdot\left(\frac{7}{7}\right)-\frac{1}{14}\cdot\left(\frac{5}{5}\right)\\ &= \frac{49-5}{70}\\ &= \frac{44}{70}\\ &= \frac{\overset{1}{\cancel{2}}\cdot 2\cdot 11}{\underset{1}{\cancel{2}}\cdot 5\cdot 7}\\ &= \frac{22}{35}\end{aligned}$$

5. Rational or Irrational: [Section 1.3]
π is irrational.

6. Graph: [Section 1.3]

$$\left\{-2\frac{1}{4}, \sqrt{2}, -1.75, \frac{7}{2}, 0.5\right\}$$

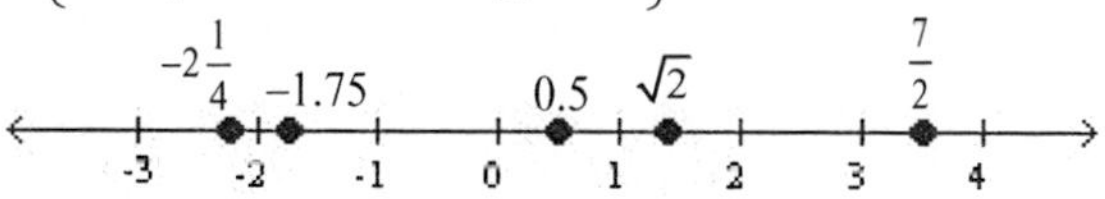

7. Write as a decimal: [Section 1.3]

$$\frac{2}{3} = 3\overline{)2.000}\ \ (0.666) = 0.6\overline{6}$$

8. What property of real numbers is illustrated? [Section 1.6]
$3(2x) = (3\cdot 2)x$
Associative property of multiplication

Evaluate each expression.

9. $-3^2+\left|4^2-5^2\right|$ [Section 1.7]

$$\begin{aligned}-3^2+\left|4^2-5^2\right| &= -(3)(3)+\left|16-25\right|\\ &= -9+\left|-9\right|\\ &= -9+9\\ &= 0\end{aligned}$$

10. $(4-5)^{20}$ [Section 1.7]

$$\begin{aligned}(4-5)^{20} &= (-1)^{20}\\ &= 1\end{aligned}$$

11. $\dfrac{-3-(-7)}{2^2-3}$ [Section 1.7]

$$\begin{aligned}\frac{-3-(-7)}{2^2-3} &= \frac{-3+7}{4-3}\\ &= \frac{4}{1}\\ &= 4\end{aligned}$$

12. $12-2[1-(-8+2)]$ [Section 1.7]

$$\begin{aligned}12-2[1-(-8+2)] &= 12-2[1-(-6)]\\ &= 12-2[1+6]\\ &= 12-2(7)\\ &= 12-14\\ &= -2\end{aligned}$$

13. RACING [Section 1.8]
$250-x$

14. Value of d dimes? [Section 1.8]
$10d$ cents or $\$0.10d$

Simplify each expression.

15. $13r - 12r$ [Section 1.9]

$$13r - 12r = (13 - 12)r$$
$$= r$$

16. $27\left(\frac{2}{3}x\right)$ [Section 1.9]

$$27\left(\frac{2}{3}x\right) = \frac{27 \cdot 2}{3}x$$
$$= \frac{\overset{1}{\cancel{3}} \cdot 9 \cdot 2}{\underset{1}{\cancel{3}}}x$$
$$= 18x$$

17. $4(d-3)-(d-1)$ [Section 1.9]

$$4(d-3)-(d-1) = \mathbf{4}d - \mathbf{4}(3) - d + 1$$
$$= 4d - 12 - d + 1$$
$$= (4-1)d + (-12+1)$$
$$= 3d - 11$$

18. $(13c-3)(-6)$ [Section 1.9]

$$(13c-3)(-6) = -\mathbf{6}(13c) - \mathbf{6}(-3)$$
$$= -78c + 18$$

Solve each equation. Check each result.

19. $3(x-5)+2=2x$ [Section 2.2]

$$3(x-5)+2 = 2x$$
$$\mathbf{3}(x) - \mathbf{3}(5) + 2 = 2x$$
$$3x - 15 + 2 = 2x$$
$$3x - 13 = 2x$$
$$3x - 13 + \mathbf{13} = 2x + \mathbf{13}$$
$$3x = 2x + 13$$
$$3x - \mathbf{2x} = 2x + 13 - \mathbf{2x}$$
$$x = 13$$

Check: $x = 13$

$$3(x-5)+2 = 2x$$
$$3(\mathbf{13}-5)+2 \stackrel{?}{=} 2(\mathbf{13})$$
$$3(8)+2 \stackrel{?}{=} 26$$
$$24+2 \stackrel{?}{=} 26$$
$$26 = 26 \text{ True}$$

The solution is 13.

20. $\frac{x-5}{3} - 5 = 7$ [Section 2.2]

$$\frac{x-5}{3} - 5 = 7$$
$$\mathbf{3}\left(\frac{x-5}{3} - 5\right) = \mathbf{3}(7)$$
$$\mathbf{3}\left(\frac{x-5}{3}\right) - \mathbf{3}(5) = \mathbf{3}(7)$$
$$x - 5 - 15 = 21$$
$$x - 20 = 21$$
$$x - 20 + \mathbf{20} = 21 + \mathbf{20}$$
$$x = 41$$

Check: $x = 41$

$$\frac{x-5}{3} - 5 = 7$$
$$\frac{\mathbf{41}-5}{3} - 5 \stackrel{?}{=} 7$$
$$\frac{36}{3} - 5 \stackrel{?}{=} 7$$
$$12 - 5 \stackrel{?}{=} 7$$
$$7 = 7 \text{ True}$$

The solution is 41.

21. $\frac{2}{5}x + 1 = \frac{1}{3} + x$ [Section 2.2]

$$\frac{2}{5}x + 1 = \frac{1}{3} + x$$
$$\mathbf{15}\left(\frac{2}{5}x + 1\right) = \mathbf{15}\left(\frac{1}{3} + x\right)$$
$$\mathbf{15}\left(\frac{2}{5}x\right) + \mathbf{15}(1) = \mathbf{15}\left(\frac{1}{3}\right) + \mathbf{15}(x)$$
$$6x + 15 = 5 + 15x$$
$$6x + 15 - \mathbf{5} = 5 + 15x - \mathbf{5}$$
$$6x + 10 = 15x$$
$$6x + 10 - \mathbf{6x} = 15x - \mathbf{6x}$$
$$10 = 9x$$
$$\frac{10}{\mathbf{9}} = \frac{9x}{\mathbf{9}}$$
$$\frac{10}{9} = x$$

Check: $x = \frac{10}{9}$

$$\frac{2}{5}\left(\mathbf{\frac{10}{9}}\right)+1\stackrel{?}{=}\frac{1}{3}+\mathbf{\frac{10}{9}}$$

$$\frac{4}{9}+\frac{9}{9}\stackrel{?}{=}\frac{3}{9}+\frac{10}{9}$$

$$\frac{13}{9}=\frac{13}{9} \quad \text{True}$$

The solution is $\frac{10}{9}$.

22. $-\frac{5}{8}h=15$ [Section 2.2]

$$-\frac{5}{8}h=15$$

$$-\mathbf{\frac{8}{5}}\left(-\frac{5}{8}h\right)=-\mathbf{\frac{8}{5}}(15)$$

$$h=-24$$

Check: $h=-24$

$$-\frac{5}{8}h=15$$

$$-\frac{5}{8}(\mathbf{-24})\stackrel{?}{=}15$$

$$15=15 \quad \text{True}$$

The solution is -24.

23. GYMNASTICS [Section 2.3]

85% of what number is 119.

$$0.85 \bullet x = 119$$

$$0.85x=119$$

$$\frac{0.85x}{\mathbf{0.85}}=\frac{119}{\mathbf{0.85}}$$

$$x=140$$

Check: $x=140$

$$0.85x=119$$

$$0.85(\mathbf{140})\stackrel{?}{=}119$$

$$119=119 \quad \text{True}$$

The maximum number of children is 140.

24. GREEN HOUSE [Section 2.3]

Let x = amt of emissions from transportation

What number is 27.9% of 6,957 teragrams?

$$x = 0.279 \cdot 6{,}957 \text{ teragrams}$$

$$x = 1{,}941.003 \text{ teragrams}$$

1,941 teragrams comes from transportation.

25. Solve: $A=\frac{1}{2}h(b+B)$ for h. [Section 2.4]

$$A=\frac{1}{2}h(b+B)$$

$$\mathbf{2}(A)=\mathbf{2}\left[\frac{1}{2}h(b+B)\right]$$

$$2A=h(b+B)$$

$$\frac{2A}{\mathbf{b+B}}=\frac{h(b+B)}{\mathbf{b+B}}$$

$$\frac{2A}{b+B}=h$$

$$h=\frac{2A}{b+B}$$

26. CANCER [Section 2.5]

Analyze

- Total number of people with cancer is 219,000.
- 13,000 more cases dealing with men than women.
- How many cases for each gender are there?

Assign

Let x = number of cases for women

$x+13{,}000$ = number of cases for men

Form

The number of cancer cases for women	plus	the number of cancer cases for men	equals	the total number of cases.
x	$+$	$x+13{,}000$	$=$	$219{,}000$

Solve

$$x+x+13{,}000=219{,}000$$

$$2x+13{,}000=219{,}000$$

$$2x+13{,}000-\mathbf{13{,}000}=219{,}000-\mathbf{13{,}000}$$

$$2x=206{,}000$$

$$\frac{2x}{\mathbf{2}}=\frac{206{,}000}{\mathbf{2}}$$

$$x=103{,}000$$

men

$$x+13{,}000=103{,}000+13{,}000$$

$$=116{,}000$$

State

Number of women with cancer is 103,000.

Number of men with cancer is 116,000.

Check

$$103{,}000+113{,}000=219{,}000$$

27. MIXING CANDY [Section 2.6]

Analyze

- Lemon gumdrops sell for \$4.40 per lb.
- Red licorice bits sell for \$3.80 per lb.
- The blend is to sell for \$4 per lb.
- A blend of 30 lb is needed.
- How many pounds of each is needed?

Assign

Let $x =$ the number of pounds lemon gumdrops
$30 - x =$ the number of pounds red licorice

	Number	• Value	= Total value
Lemon	x	4.40	**4.40*x***
Licorice	**30 – *x***	3.80	**3.80(30 – *x*)**
Blend	30	4.00	**120**

Form

The value of the lemon gumdrops	plus	the value of the licorice	equals	the total value of the blend.
$4.40x$	$+$	$3.80(30-x)$	$=$	120

Solve

$$4.40x + 3.80(30 - x) = 120$$
$$4.40x + \mathbf{3.80}(30) - \mathbf{3.80}(x) = 120$$
$$4.40x + 114 - 3.80x = 120$$
$$0.60x + 114 = 120$$
$$0.60x + 114 - \mathbf{114} = 120 - \mathbf{114}$$
$$0.60x = 6$$
$$\frac{0.60x}{\mathbf{0.60}} = \frac{6}{\mathbf{0.60}}$$
$$x = 10$$

licorice

$$30 - x = 30 - \mathbf{10}$$
$$= 20$$

State

10 lb of the \$4.40/lb lemon drops will be needed.

20 lb of the \$3.80/lb licorice will be needed.

Check

The value of the lemon drops is 10(\$4.40), or \$44.
The value of the licorice is 20(\$3.80), or \$76.
The value of the blend is 30(\$4), or \$120.
Since the total was \$44 + \$76 = \$120, the answers check.

28. Solve: $8(4+x) > 10(6+x)$ [Section 2.7]

$$8(4+x) > 10(6+x)$$
$$\mathbf{8}(4) + \mathbf{8}(x) > \mathbf{10}(6) + \mathbf{10}(x)$$
$$32 + 8x > 60 + 10x$$
$$32 + 8x - \mathbf{32} > 60 + 10x - \mathbf{32}$$
$$8x > 10x + 28$$
$$8x - \mathbf{10x} > 10x + 28 - \mathbf{10x}$$
$$-2x > 28$$
$$\frac{-2x}{\mathbf{-2}} < \frac{28}{\mathbf{-2}}$$
$$x < -14$$
$$(-\infty, -14)$$

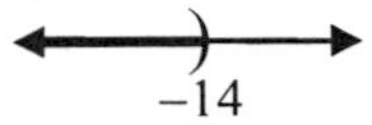

29. NEW YORK [Section 3.1]
(Ninth Avenue, 44th Street)

30. PERIMETER [Section 3.1]

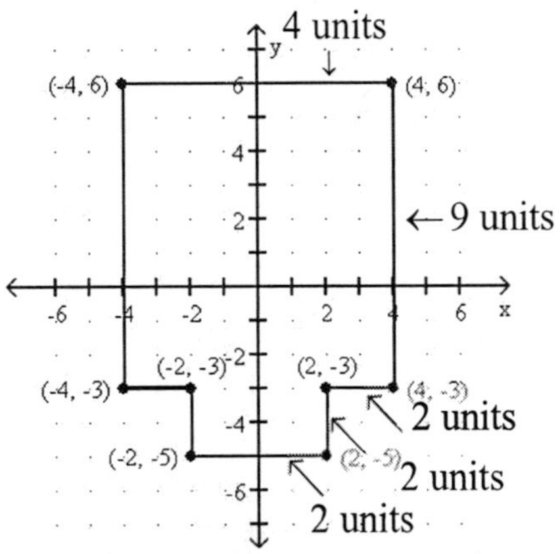

$$\text{Perimeter} = 2(4+9+2+2+2)$$
$$= 2(19)$$
$$= 38 \text{ units}$$

The perimeter is 38 units.

31. In what quadrant does $(-3.5, 6)$ lie? [Section 3.1] **Quadrant II**

32. Is $(-2, 8)$ a solution of $y = -2x + 3$? [Section 3.2]

$$y = -2x + 3$$
$$\mathbf{8} \stackrel{?}{=} -2(\mathbf{-2}) + 3$$
$$8 \stackrel{?}{=} 4 + 3$$
$$8 = 7 \text{ False}$$

$(-2, 8)$ is not a solution.

Graph each equation.

33. $x = 4$ [Section 3.3]

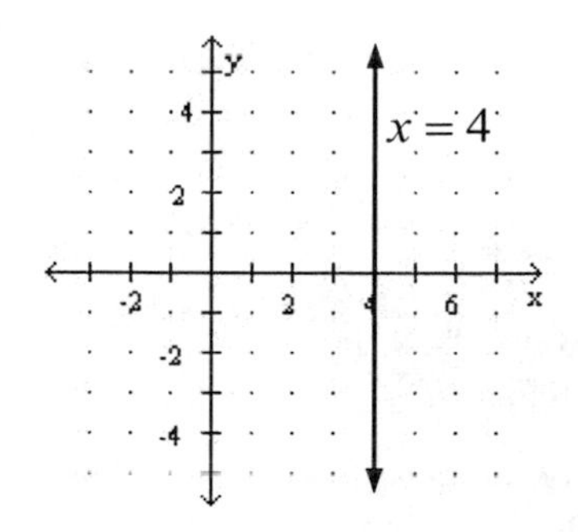

34. $4x - 3y = 12$ [Section 3.3]

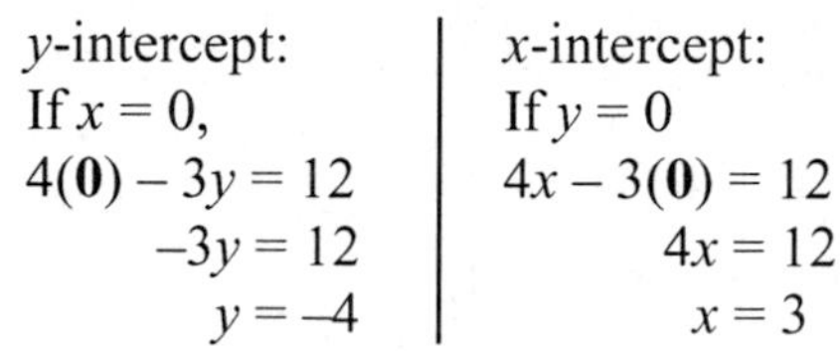

y-intercept:	x-intercept:
If $x = 0$,	If $y = 0$
$4(\mathbf{0}) - 3y = 12$	$4x - 3(\mathbf{0}) = 12$
$-3y = 12$	$4x = 12$
$y = -4$	$x = 3$

The y-intercept is (0, –4), and the x-intercept is (3, 0).

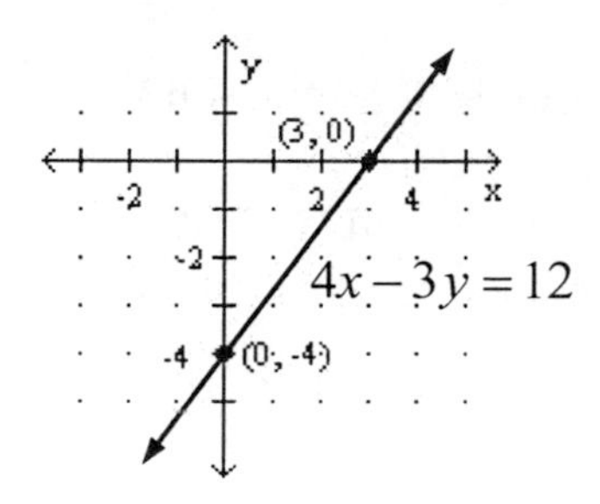

Find the slope of the line with the given properties.

35. Passing through $(-2, 4)$ and $(6, 8)$ [Section 3.4]

$(-2, 4)$ and $(6, 8)$

(x_1, y_1) and (x_2, y_2)

$$m = \frac{y_2 - y_1}{x_2 - x_1} = \frac{8-4}{6-(-2)} = \frac{4}{6+2} = \frac{4}{8} = \frac{1}{2}$$

36. A line that is horizontial [Section 3.4]

$m = 0$

37. An equation of $2x - 3y = 12$ [Section 3.5]

$$2x - 3y = 12$$
$$2x - 3y - \mathbf{2x} = 12 - \mathbf{2x}$$
$$-3y = -2x + 12$$
$$\frac{-3y}{\mathbf{-3}} = \frac{-2x}{\mathbf{-3}} + \frac{12}{\mathbf{-3}}$$
$$y = \frac{2}{3}x - 4$$

The slope is $\frac{2}{3}$.

38. Are the graphs of the lines parallel or perpendicular? [Section 3.5]

$$y = -\frac{3}{4}x + \frac{15}{4}$$
$$m = -\frac{3}{4}$$

$$4x - 3y = 25$$
$$4x - 3y - \mathbf{4x} = 25 - \mathbf{4x}$$
$$-3y = -4x + 25$$
$$\frac{-3y}{\mathbf{-3}} = \frac{-4x}{\mathbf{-3}} + \frac{25}{\mathbf{-3}}$$
$$y = \frac{4}{3}x - \frac{25}{3}$$
$$m = \frac{4}{3}$$

The slopes are negative reciprocals. The lines are perpendicular.

39. Find the slope and the x- and y-intercepts of the line.

The slope is 2.
The x-intercept is $(-1, 0)$.
The y-intercept is (0, 2).

40. NEWSPAPERS [Section 3.5]

$A(1999,\ 44\%)$ and $B(2009,\ 28\%)$
(x_1, y_1) (x_2, y_2)

$$m = \frac{y_2 - y_1}{x_2 - x_1}$$
$$m = \frac{28\% - 44\%}{2009 - 1999}$$
$$m = \frac{-16\%}{10}$$
$$m = -1.6\%$$

There is a decrease of 1.6% per year.

Find an equation of the line with the following properties. Write the equation in slope - intercept form.

41. Slope $= \frac{2}{3}$, y-intercept $= (0,5)$
[Section 3.5]

$$y = mx + b$$
$$y = \frac{2}{3}x + 5$$

42. Passing through $(-2,4)$ and $(6,10)$
[Section 3.6]

Step 1: $(-2,4)$ and $(6,10)$
(x_1, y_1) and (x_2, y_2)

$$m = \frac{y_2 - y_1}{x_2 - x_1} = \frac{10-4}{6-(-2)} = \frac{6}{6+2} = \frac{6}{8} = \frac{3}{4}$$

Step 2: $y - y_1 = m(x - x_1)$

$$y - 10 = \frac{3}{4}(x-6)$$
$$y - 10 = \frac{3}{4}x - \frac{3}{4}(6)$$
$$y - 10 = \frac{3}{4}x - \frac{9}{2}$$
$$y - 10 + \mathbf{10} = \frac{3}{4}x - \frac{9}{2} + \mathbf{\frac{20}{2}}$$
$$y = \frac{3}{4}x + \frac{11}{2}$$

43. A horizontial line passing through (2,4)
[Section 3.6]

$$y = 4$$

44. Graph: $y < \frac{x}{3} - 1$ [Section 3.7]

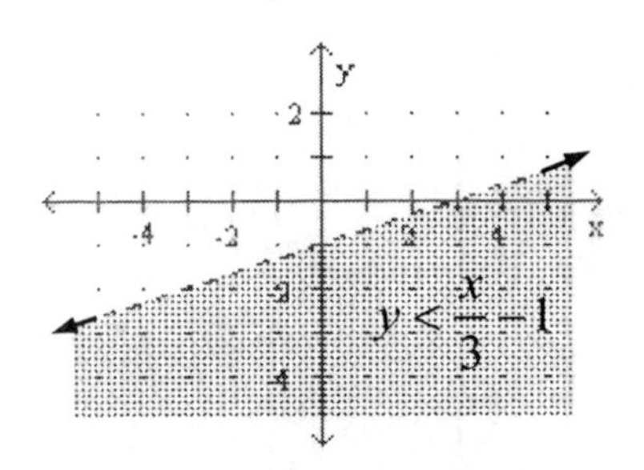

45. If $f(x) = -2x^2 - 3x^3$, find $f(-1)$.
[Section 3.8]

$$f(x) = -2x^2 - 3x^3$$
$$f(-1) = -2(-1)^2 - 3(-1)^3$$
$$f(-1) = -2(1) - 3(-1)$$
$$f(-1) = -2 + 3$$
$$f(-1) = 1$$

46. Is this a graph of a function?
[Section 3.8]

No ; (1, 2) and (1, −2)
There are other points.

Solve each system by graphing.

47. $\begin{cases} x + 4y = -2 \\ y = -x - 5 \end{cases}$ [Section 4.1]

The solution is (−6, 1)

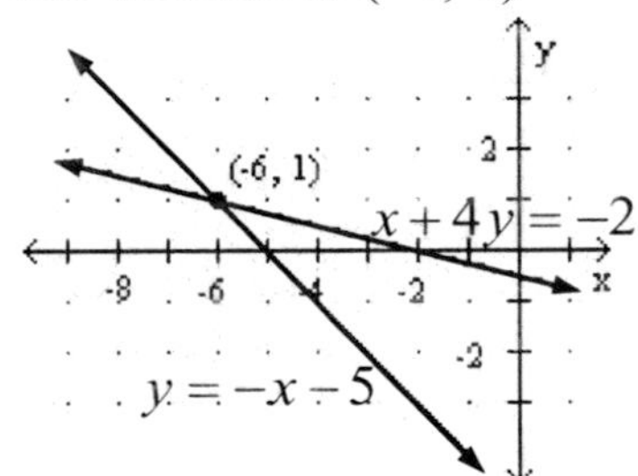

48. $\begin{cases} 2x - 3y < 0 \\ y > x - 1 \end{cases}$ [Section 4.5]

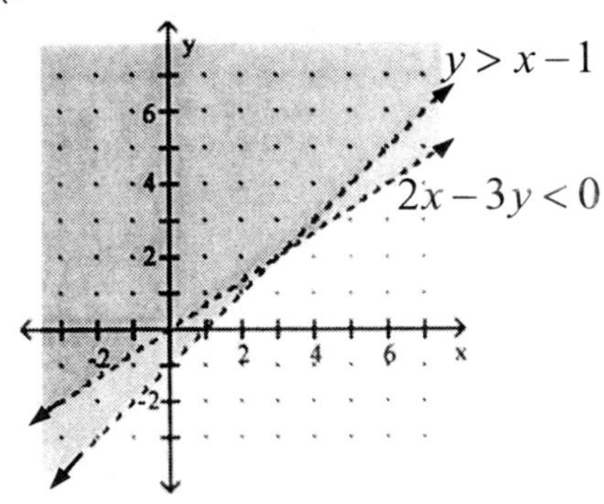

49. Solve $\begin{cases} x-2y=2 \\ 2x+3y=11 \end{cases}$ by substitution

[Section 4.2]

$$\begin{cases} x-2y=2 \\ 2x+3y=11 \end{cases}$$

Solve the 1st equation for x.

$$\begin{cases} x=2y+2 \\ 2x+3y=11 \end{cases}$$

$$2x+3y=11 \quad \text{The 2nd equation. Substitute for } x.$$

$$2(\mathbf{2y+2})+3y=11$$

$$4y+4+3y=11$$

$$7y+4=11$$

$$7y+4-\mathbf{4}=11-\mathbf{4}$$

$$7y=7$$

$$\frac{7y}{\mathbf{7}}=\frac{7}{\mathbf{7}}$$

$$y=1$$

$x=2y+2$ The 1st equation

$x=2(1)+2$

$x=4$

The solution is $(4,1)$.

50. $\begin{cases} \frac{3}{2}x-\frac{2}{3}y=0 \\ \frac{3}{4}x+\frac{4}{3}y=\frac{5}{2} \end{cases}$

Clear both equations of fractions.

$$\begin{cases} \mathbf{6}\left(\frac{3}{2}x\right)-\mathbf{6}\left(\frac{2}{3}y\right)=\mathbf{6}(0) \\ \mathbf{12}\left(\frac{3}{4}x\right)+\mathbf{12}\left(\frac{4}{3}y\right)=\mathbf{12}\left(\frac{5}{2}\right) \end{cases}$$

$$\begin{cases} 9x-4y=0 \\ 9x+16y=30 \end{cases}$$

Eliminate x.

Multiply both sides of the 1st equation by -1.

$$\begin{array}{r} -9x+4y=0 \\ 9x+16y=30 \\ \hline 20y=30 \end{array}$$

$$\frac{20y}{\mathbf{20}}=\frac{30}{\mathbf{20}}$$

$$y=\frac{3}{2}$$

$9x-4y=0$ The 1st equation

$$9x-4\left(\frac{\mathbf{3}}{\mathbf{2}}\right)=0$$

$$9x-6=0$$

$$9x-6+\mathbf{6}=0+\mathbf{6}$$

$$9x=6$$

$$\frac{9x}{\mathbf{9}}=\frac{6}{\mathbf{9}}$$

$$x=\frac{2}{3}$$

The solution is $\left(\frac{2}{3},\frac{3}{2}\right)$.

51. NUTRITION [Section 4.4]

Analyze

- Protein in 1 serving of egg noodles: 5 g
- Protein in 1 serving of rice pilaf: 4 g
- Wants to consume only 22 g of protein.
- Fat in 1 serving of egg noodles: 3 g
- Fat in 1 serving of rice pilaf: 5 g
- Wants to consume only 21 g of fat.

Assign

Let x = # of servings of egg noodles
y = # of servings of rice pilaf

Form

	Grams •	Servings =	Total Protein
Noodles	5	x	$\mathbf{5x}$
Rice	4	y	$\mathbf{4y}$
		Total	**22**

The grams of protein in egg noodles plus the grams of protein in rice pilaf is 22 grams.

$$5x + 4y = 22$$

	Grams •	Servings =	Total Fat
Noodles	3	x	$\mathbf{3x}$
Rice	5	y	$\mathbf{5y}$
		Total	**21**

The grams of fat in egg noodles plus the grams of fat in rice pilaf is 21 grams.

$$3x + 5y = 21$$

Solve

$$\begin{cases} 5x+4y=22 \\ 3x+5y=21 \end{cases}$$

Eliminate x.

Multiply both sides of the 1st equation by -3.

Multiply both sides of the 2nd equation by 5.

$$\begin{aligned} -15x-12y &= -66 \\ 15x+25y &= 105 \\ \hline 13y &= 39 \\ \frac{13y}{\mathbf{13}} &= \frac{39}{\mathbf{13}} \\ y &= 3 \end{aligned}$$

$$\begin{aligned} 5x+4y &= 22 \quad \text{The 1st equation} \\ 5x+4(\mathbf{3}) &= 22 \\ 5x+12-\mathbf{12} &= 22-\mathbf{12} \\ \frac{5x}{\mathbf{5}} &= \frac{10}{\mathbf{5}} \\ x &= 2 \end{aligned}$$

State

2 servings of egg noodles.
3 servings of rice pilaf.

Check

The results check.

52. INVESTMENT CLUBS

Analyze

- $8,000 to invest.
- Invested part at 10% annual interest.
- The rest at 12% annual interest.
- $900 in interest was earned the first year.
- How much was invested at each rate?

Assign

Let x = amount invested at 10%
y = amount invested at 12%

Form

The amount invested at 10% plus the amount invested at 12% is $8,000.

$$x + y = 8{,}000$$

	Principal •	Rate •	Time =	Interest
10% part	x	0.10	1 yr	$\mathbf{0.10x}$
12% part	y	0.12	1 yr	$\mathbf{0.12y}$
			Total Interest =	**$900**

The amount of interest of 10% investment plus the amount of interest of 12% investment is $900.

$$0.10x + 0.12y = 900$$

Solve

$$\begin{cases} x+y=8{,}000 \\ 0.10x+0.12y=900 \end{cases}$$

Eliminate x.

Multiply both sides of the 1st equation by -10.

Multiply both sides of the 2nd equation by 100.

$$\begin{aligned} -10x-10y &= -80{,}000 \\ 10x+12y &= 90{,}000 \\ \hline 2y &= 10{,}000 \\ \frac{2y}{\mathbf{2}} &= \frac{10{,}000}{\mathbf{2}} \\ y &= 5{,}000 \end{aligned}$$

$$\begin{aligned} x+y &= 8{,}000 \quad \text{The 1st equation} \\ x+\mathbf{5{,}000} &= 8{,}000 \\ x+5{,}000-\mathbf{5{,}000} &= 8{,}000-\mathbf{5{,}000} \\ x &= 3{,}000 \end{aligned}$$

State

$3,000 was invested at 10%.
$5,000 was invested at 12%.

Check

The results check.

SECTION 5.1 STUDY SET
VOCABULARY

Fill in the blanks.

1. Expressions such as x^4, 10^3, and $(5t)^2$ are called **exponential** expressions.

CONCEPTS

Fill in the blanks.

3. a. $(3x)^4 = \mathbf{3x \cdot 3x \cdot 3x \cdot 3x}$

   b. $(-5y)(-5y)(-5y) = \mathbf{(-5y)^3}$

5. To simplify each expression, determine whether you add, subtract, multiply, or divide the exponents.

   a. $\dfrac{x^8}{x^2}$ **Subtract** b. $b^6 \cdot b^9$ **Add**

   c. $(n^8)^4$ **Multiply** d. $(a^4b^2)^5$ **Multiply**

Simplify each expression, if possible.

7. *a.* $x^2 + x^2 = (1+1)x^2$
   $= \mathbf{2x^2}$

   b. $x^2 \cdot x^2 = x^{2+2}$
   $= \mathbf{x^4}$

   c. $x^2 + x$, Doesn't simplify.

   d. $x^2 \cdot x = x^{2+1}$
   $= \mathbf{x^3}$

NOTATION

Complete each solution to simplify each expression.

9. $\left(x^4x^2\right)^3 = \left(\mathbf{x^6}\right)^3 = x^{\mathbf{18}}$

11. a. We read 9^4 as "nine to the fourth **power**."

    b. We read $(a^2b^6)(a^4b^5)$ as "the **quantity** of of a^2b^6 times the **quantity** of a^4b^5."

    c. We read $(3^6)^9$ as "3 to the **sixth** power, raised to the **ninth** power."

GUIDED PRACTICE

Identify the base and the exponent in each expression. See Example 1.

Look Alikes . . .

13. a. 4^3; 4, 3 b. -4^3; 4, 3 c. $(-4)^3$; -4, 3

15. a. $(-3x)^2$; $-3x$, 2 b. -3^2; x, 2

    c. $-(-3x)^2$; $-3x$, 2

17. a. $9m^{12}$; m, 12 b. $(9m)^{12}$; $9m$, 12

    c. $-9m^{12}$; , 12

Write each expression in an equivalent form using an exponent. See Example 2.

19. $4t \cdot 4t \cdot 4t \cdot 4t = (4t)^4$

21. $-4 \cdot t \cdot t \cdot t \cdot t \cdot t = -4t^5$

23. $\dfrac{t}{2} \cdot \dfrac{t}{2} \cdot \dfrac{t}{2} = \left(\dfrac{t}{2}\right)^3$

25. $(x-y)(x-y) = (x-y)^2$

Use the product rule for exponents to simplify each expression. Write the results using exponents. See Example 3.

27. $5^3 \cdot 5^4 = 5^{3+4}$
    $= 5^7$

29. $bb^2b^3 = b^{1+2+3}$
    $= b^6$

31. $(y-2)^5(y-2)^2 = (y-2)^{5+2}$
    $= (y-2)^7$

33. $(a^2b^3)(a^3b^3) = (a^2a^3)(b^3b^3)$
    $= a^{2+3}b^{3+3}$
    $= a^5b^6$

Find an expression that represents the area or volume of each figure. Recall that the formula for the volume of a rectangular solid is V = length • width • height. See Example 4.

35. $A = lw$
    $= a^5 \cdot a^5$
    $= a^{5+5}$
    $= a^{10}$

    The area is a^{10} mi^2.

37. $V = lwh$
    $= x^4 \cdot x^3 \cdot x^2$
    $= x^{4+3+2}$
    $= x^9$

    The volume is x^9 ft^3.

Use the quotient rule for exponents to simplify each expression. Write the results using exponents. See Example 5.

39. $\dfrac{8^{12}}{8^4} = 8^{12-4}$
    $= 8^8$

41. $\dfrac{x^{15}}{x^3} = x^{15-3}$
    $= x^{12}$

43. $\dfrac{(3.7p)^7}{(3.7p)^2} = (3.7p)^{7-2}$
    $= (3.7p)^5$

45. $\dfrac{c^3d^7}{cd} = \dfrac{c^3}{c} \cdot \dfrac{d^7}{d}$
    $= c^{3-1}d^{7-1}$
    $= c^2d^6$

Use the product and quotient rules for exponents to simplify each expression. See Example 6.

47. $\dfrac{y^3y^4}{yy^2} = \dfrac{y^{3+4}}{y^{1+2}} = \dfrac{y^7}{y^3} = y^{7-3} = y^4$

49. $\dfrac{a^2a^3a^4}{a^8} = \dfrac{a^{2+3+4}}{a^8} = \dfrac{a^9}{a^8} = a^{9-8} = a^1 = a$

Use the power rule for exponents to simplify each expression. Write the results using exponents. See Example 7.

51. $(3^2)^4 = 3^{2\cdot4} = 3^8$

53. $[(-4.3)^3]^8 = (-4.3)^{3\cdot8} = (-4.3)^{24}$

55. $(m^{50})^{10} = m^{50\cdot10} = m^{500}$

57. $(y^5)^3 = y^{5\cdot3} = y^{15}$

Use the product and power rules for exponents to simplify each expression. See Example 8.

59. $(x^2x^3)^5 = (x^{2+3})^5 = (x^5)^5 = x^{5\cdot5} = x^{25}$

61. $(p^2p^3)^5 = (p^{2+3})^5 = (p^5)^5 = p^{5\cdot5} = p^{25}$

63. $(t^3)^4(t^2)^3 = (t^{3\cdot4})(t^{2\cdot3}) = (t^{12})(t^6) = t^{12+6} = t^{18}$

65. $(u^4)^2(u^3)^2 = (u^{4\cdot2})(u^{3\cdot2}) = (u^8)(u^6) = u^{8+6} = u^{14}$

Use the power of a product rule for exponents to simplify each expression. See Example 8.

67. $(6a)^2 = 6^2a^2 = 36a^2$

69. $(5y)^4 = 5^4y^4 = 625y^4$

71. $(-2r^2s^3)^3 = (-2)^3(r^2)^3(s^3)^3 = -8r^{2\cdot3}s^{3\cdot3} = -8r^6s^9$

73. $\left(-\dfrac{1}{3}y^2z^4\right)^5 = \left(-\dfrac{1}{3}\right)^5(y^2)^5(z^4)^5 = -\dfrac{1}{243}y^{2\cdot5}z^{4\cdot5} = -\dfrac{1}{243}y^{10}z^{20}$

Use rules for exponents to simplify each expression. See Example 10.

75. $\dfrac{(ab^2)^3}{a^2b^2} = \dfrac{(a^1)^3(b^2)^3}{a^2b^2} = \dfrac{(a^{1\cdot3})(b^{2\cdot3})}{a^2b^2} = \dfrac{a^3b^6}{a^2b^2} = a^{3-2}b^{6-2} = a^1b^4 = ab^4$

77. $\dfrac{(r^4s^3)^4}{r^3s^9} = \dfrac{(r^4)^4(s^3)^4}{r^3s^9} = \dfrac{(r^{4\cdot4})(s^{3\cdot4})}{r^3s^9} = \dfrac{r^{16}s^{12}}{r^3s^9} = r^{16-3}s^{12-9} = r^{13}s^3$

Use rules for exponents to simplify each expression. See Example 11.

79. $\dfrac{(6k)^7}{(6k)^4} = (6k)^{7-4} = (6k)^3 = 6^3k^3 = 216k^3$

81. $\dfrac{(3q)^5}{(3q)^3} = (3q)^{5-3} = (3q)^2 = 3^3q^2 = 9q^2$

Use the power of a quotient rule for exponents to simplify each expression. See Example 12.

83. $\left(\dfrac{a}{b}\right)^3 = \dfrac{a^3}{b^3}$

85. $\left(\frac{8a^2}{11b^5}\right)^2 = \frac{(8a^2)^2}{(11b^5)^2}$

$= \frac{(8)^2(a^2)^2}{(11)^2(b^5)^2}$

$= \frac{8^2 a^{2\cdot2}}{11^2 b^{5\cdot2}}$

$= \frac{64a^4}{121b^{10}}$

99. $\left(\frac{y^3y^5}{yy^2}\right)^3 = \left(\frac{y^{3+5}}{y^{1+2}}\right)^3$

$= \left(\frac{y^8}{y^3}\right)^3$

$= (y^{8-3})^3$

$= (y^5)^3$

$= y^{5\cdot3}$

$= y^{15}$

101. $\frac{s^2s^2s^2}{s^3s} = \frac{s^2s^2s^2}{s^3s^1}$

$= \frac{s^{2+2+2}}{s^{3+1}}$

$= \frac{s^6}{s^4}$

$= s^{6-4}$

$= s^2$

TRY IT YOURSELF

Simplify each expression, if possible

87. $\left(\frac{x^2}{y^3}\right)^5 = \frac{(x^2)^5}{(y^3)^5}$

$= \frac{x^{2\cdot5}}{y^{3\cdot5}}$

$= \frac{x^{10}}{y^{15}}$

89. $y^3y^2y^4 = y^{3+2+4}$

$= y^9$

91. $\frac{15^9}{15^6} = 15^{9-6}$

$= 15^3$

93. $\frac{t^5t^6t}{t^2t^3} = \frac{t^5t^6t^1}{t^2t^3}$

$= \frac{t^{5+6+1}}{t^{2+3}}$

$= \frac{t^{12}}{t^5}$

$= t^{12-5}$

$= t^7$

95. $\frac{(k-2)^{15}}{(k-2)} = \frac{(k-2)^{15}}{(k-2)^1}$

$= (k-2)^{15-1}$

$= (k-2)^{14}$

97. $cd^4 \cdot cd = cc \cdot dd^4$

$= c^{1+1} \cdot d^{1+4}$

$= c^2d^5$

103. $(-6a^3b^2)^3 = (-6)^3(a^3)^3(b^2)^3$

$= -216a^{3\cdot3}b^{2\cdot3}$

$= -216a^9b^6$

105. $\left(\frac{3m^4}{2n^5}\right)^5 = \frac{(3m^4)^5}{(2n^5)^5}$

$= \frac{(3)^5(m^4)^5}{(2)^5(n^5)^5}$

$= \frac{3^5 m^{4\cdot5}}{2^5 n^{5\cdot5}}$

$= \frac{243m^{20}}{32n^{25}}$

107. $\frac{(a^2b^2)^{15}}{(ab)^9} = \frac{(a^2)^{15}(b^2)^{15}}{(a^1)^9(b^1)^9}$

$= \frac{(a^{2\cdot15})(b^{2\cdot15})}{(a^{1\cdot9})(b^{1\cdot9})}$

$= \frac{a^{30}b^{30}}{a^9b^9}$

$= a^{30-9}b^{30-9}$

$= a^{21}b^{21}$

109. $\left(n^4n\right)^3\left(n^3\right)^6=\left(n^{4+1}\right)^3\left(n^3\right)^6$
$=\left(n^5\right)^3\left(n^3\right)^6$
$=\left(n^{5\cdot3}\right)\left(n^{3\cdot6}\right)$
$=\left(n^{15}\right)\left(n^{18}\right)$
$=n^{15+18}$
$=n^{33}$

111. $\frac{(6h)^8}{(6h)^6}=(6h)^{8-6}$
$=(6h)^2$
$=6^2h^2$
$=36h^2$

113. $\frac{x^4y^7}{xy^3}=\frac{x^4}{x^1}\cdot\frac{y^7}{y^3}$
$=x^{4-1}y^{7-3}$
$=x^3y^4$

115. $\left(\frac{m}{3}\right)^4=\frac{m^4}{3^4}$
$=\frac{m^4}{81}$

(All new) LOOK ALIKES . . .

117. a. $a^3\cdot a^3=a^{3+3}$
$=a^6$

b. $\left(a^3\right)^3=a^{3\cdot3}$
$=a^9$

c. $a^3+a^3=1a^3+1a^3$
$=2a^3$

119. a. $b^3b^2b^4=b^{3+2+4}$
$=b^9$

b. $\left(b^3b^2\right)^4=\left(b^{3+2}\right)^4$
$=\left(b^5\right)^4$
$=b^{5\cdot4}$
$=b^{20}$

c. $\frac{b^3b^2}{b^4}=\frac{b^{3+2}}{b^4}$
$=\frac{b^5}{b^4}$
$=b^{5-4}$
$=b$

APPLICATIONS

121. ART HISTORY

a. $A=lw$
$=(5x)(5x)$
$=25x^{1+1}$
$=25x^2$

The area of the square is $25x^2$ ft^2.

b. $A=\pi r^2$
$=(3a)(3a)\pi$
$=9a^{1+1}\pi$
$=9a^2\pi$

The area of the circle is $9a^2\pi$ ft^2.

123. CHILDBIRTH

$\left(\frac{1}{2}\right)^{13}=\frac{1^{13}}{2^{13}}$
$=\frac{1}{8{,}192}$

The probability is $\frac{1}{8{,}192}$.

WRITING

125. Answers will vary.

REVIEW

Match each equation with its graph below.

127. $y=2x-1$ **c** 129. $y=3$ **d**

CHALLENGE PROBLEMS

131. **Simplify each expression. The variables represent natural numbers.**

a. $x^{2m}x^{3m}=x^{2m+3m}$
$=x^{5m}$

b. $\left(y^{5c}\right)^4=y^{5c\cdot4}$
$=y^{20c}$

c. $\frac{m^{8x}}{m^{4x}}=m^{8x-4x}$
$=m^{4x}$

d. $\left(2a^{6y}\right)^4=\left(2^1\right)^4\left(a^{6y}\right)^4$
$=\left(2^{1\cdot4}\right)\left(a^{6y\cdot4}\right)$
$=2^4a^{24y}$
$=16a^{24y}$

SECTION 5.2 STUDY SET

VOCABULARY

Fill in the blanks.

1. In the expression 5^{-1}, the exponent is a **negative** integer.

3. We read a^0 as "a to the **zero** power".

CONCEPTS

5. Complete the table.

Expression	Base	Exponent
4^{-2}	**4**	**–2**
$6x^{-5}$	x	**–5**
$\left(\frac{3}{y}\right)^{-8}$	$\frac{3}{y}$	**–8**
-7^{-1}	**7**	**–1**
$(-2)^{-3}$	**–2**	**–3**
$10a^0$	a	**0**

7. Complete the table.

x	3^x
2	$3^2 = 3\cdot 3 = 9$
1	$3^1 = 3$
0	$3^0 = 1$
-1	$3^{-1} = \frac{1}{3}$
-2	$3^{-2} = \frac{1}{3^2} = \frac{1}{3\cdot 3} = \frac{1}{9}$

9. Fill in the blanks.

a. $2^{-3} = \dfrac{1}{2^3}$

b. $\dfrac{1}{t^{-6}} = t^6$

c. A factor can be moved from the denominator to the numerator or from the numerator to the denominator of a fraction if the **sign** of its exponent is changed.

$$\frac{5^{-2}}{6^{-3}} = \frac{6^3}{5^2}$$

d. A fraction raised to a power is equal to the **reciprocal** of the fraction raise to the opposite power.

$$\left(\frac{3}{d}\right)^{-2} = \left(\frac{d}{3}\right)^2$$

NOTATION

Complete each solution to simplify each expression.

11. $(y^5y^3)^{-5} = (y^8)^{-5}$
$= y^{-40}$
$= \dfrac{1}{y^{40}}$

GUIDED PRACTICE

Simplify each expression. See Example 1.

13. $7^0 = 1$

15. $\left(\dfrac{1}{4}\right)^0 = 1$

17. $2x^0 = 2\cdot 1$
$= 2$

19. $\dfrac{5}{2x^0} = \dfrac{5}{2\cdot 1}$
$= \dfrac{5}{2}$

Express using positive exponents and simplify, if possible. See Example 2.

21. $2^{-2} = \dfrac{1}{2^2}$
$= \dfrac{1}{4}$

23. $6^{-1} = \dfrac{1}{6^1}$
$= \dfrac{1}{6}$

25. $b^{-5} - \dfrac{1}{b^5}$

27. $(-5)^{-1} = \dfrac{1}{(-5)^1}$
$= -\dfrac{1}{5}$

29. $2^{-2} + 4^{-1} = \dfrac{1}{2^2} + \dfrac{1}{4^1}$
$= \dfrac{1}{4} + \dfrac{1}{4}$
$= \dfrac{2}{4}$
$= \dfrac{1}{2}$

31. $9^0 - 9^{-1} = 1 - \dfrac{1}{9^1}$
$= \dfrac{9}{9} - \dfrac{1}{9}$
$= \dfrac{8}{9}$

Simplify. Do not use negative exponents in the answer. See Example 3.

33. $15g^{-6} = 15\cdot g^{-6}$
$= 15\cdot\dfrac{1}{g^6}$
$= \dfrac{15}{g^6}$

35. $5x^{-3} = 5\cdot x^{-3}$
$= 5\cdot\dfrac{1}{x^3}$
$= \dfrac{5}{x^3}$

37. $-3^{-3}=-1\cdot 3^{-3}$
$=-1\cdot\frac{1}{3^3}$
$=-\frac{1}{27}$

39. $-8^{-2}=-1\cdot 8^{-2}$
$=-1\cdot\frac{1}{8^2}$
$=-\frac{1}{64}$

Simplify. Do not use negative exponents in the answer. See Example 4.

41. $\frac{1}{5^{-3}}=5^3$
$=125$

43. $\frac{8}{s^{-1}}=8s$

45. $\frac{2^{-4}}{3^{-1}}=\frac{3^1}{2^4}$
$=\frac{3}{16}$

47. $\frac{-4d^{-1}}{p^{-10}}=\frac{-4p^{10}}{d^1}$
$=-\frac{4p^{10}}{d}$

Simplify. See Example 5.

49. $\left(\frac{1}{6}\right)^{-2}=\left(\frac{6^1}{1^1}\right)^2$
$=\frac{6^{1\cdot 2}}{1^{1\cdot 2}}$
$=\frac{6^2}{1}$
$=36$

51. $\left(\frac{1}{2}\right)^{-3}=\left(\frac{2^1}{1^1}\right)^3$
$=\frac{2^{1\cdot 3}}{1^{1\cdot 3}}$
$=\frac{2^3}{1}$
$=8$

53. $\left(\frac{c}{d}\right)^{-8}=\left(\frac{d^1}{c^1}\right)^8$
$=\frac{d^{1\cdot 8}}{c^{1\cdot 8}}$
$=\frac{d^8}{c^8}$

55. $\left(\frac{3}{m}\right)^{-4}=\left(\frac{m^1}{3^1}\right)^4$
$=\frac{m^{1\cdot 4}}{3^{1\cdot 4}}$
$=\frac{m^4}{3^4}$
$=\frac{m^4}{81}$

Simplify. Do not use negative exponents in the answer. See Example 6.

57. $y^8\cdot y^{-2}=y^{8+(-2)}$
$=y^6$

59. $b^{-7}\cdot b^{14}=b^{-7+14}$
$=b^7$

61. $\frac{y^4}{y^5}=y^{4-5}$
$=y^{-1}$
$=\frac{1}{y}$

63. $\frac{h^{-5}}{h^2}=h^{-5-2}$
$=h^{-7}$
$=\frac{1}{h^7}$

65. $(x^4)^{-3}=x^{-12}$
$=\frac{1}{x^{12}}$

67. $(b^2)^{-4}=b^{-8}$
$=\frac{1}{b^8}$

69. $\left(6s^4t^{-7}\right)^2=\left(6^1s^4t^{-7}\right)^2$
$=6^2\left(s^4\right)^2\left(t^{-7}\right)^2$
$=36s^{4\cdot 2}t^{-7\cdot 2}$
$=36s^8t^{-14}$
$=\frac{36s^8}{t^{14}}$

71. $\left(\frac{4}{x^3}\right)^{-3}=\left(\frac{x^3}{4^1}\right)^3$
$=\frac{\left(x^3\right)^3}{4^3}$
$=\frac{x^{3\cdot 3}}{64}$
$=\frac{x^9}{64}$

Simplify. Do not use negative exponents in the answer. See Example 7.

73. $\frac{y^{-3}}{y^{-4}y^{-2}}=\frac{y^{-3}}{y^{-4+(-2)}}$
$=\frac{y^{-3}}{y^{-6}}$
$=y^{-3-(-6)}$
$=y^{-3+6}$
$=y^3$

75. $\frac{a^{-5}a^{-9}}{a^{-8}}=\frac{a^{-5+(-9)}}{a^{-8}}$
$=\frac{a^{-14}}{a^{-8}}$
$=a^{-14-(-8)}$
$=a^{-14+8}$
$=a^{-6}$
$=\frac{1}{a^6}$

77. $\dfrac{2^{-1}a^4b^2}{3^{-2}a^2b^4} = \dfrac{3^2a^4b^2}{2a^2b^4}$

$= \dfrac{9a^{4-2}b^{2-4}}{2}$

$= \dfrac{9a^2b^{-2}}{2}$

$= \dfrac{9a^2}{2b^2}$

79. $\left(\dfrac{y^3z^{-2}}{y^{-4}z^3}\right)^2 = \left(y^{3-(-4)}z^{-2-3}\right)^2$

$= \left(y^{3+4}z^{-5}\right)^2$

$= \left(y^7z^{-5}\right)^2$

$= y^{7\cdot 2}z^{-5\cdot 2}$

$= y^{14}z^{-10}$

$= \dfrac{y^{14}}{z^{10}}$

TRY IT YOURSELF

Simplify. Do not use negative exponents in the answer. Assume that no variables are 0.

81. $\left(\dfrac{a^4}{2b}\right)^{-3} = \left(\dfrac{2^1b^1}{a^4}\right)^3$

$= \dfrac{\left(2^1b^1\right)^3}{\left(a^4\right)^3}$

$= \dfrac{2^3b^3}{a^{4\cdot 3}}$

$= \dfrac{8b^3}{a^{12}}$

83. $\dfrac{r^{-50}}{r^{-70}} = \dfrac{r^{70}}{r^{50}}$

$= r^{70-50}$

$= r^{20}$

85. $\dfrac{a^{-5}}{b^{-2}} = \dfrac{b^2}{a^5}$

87. $(-10)^{-3} = \dfrac{1}{(-10)^3}$

$= \dfrac{1}{(-10)(-10)(-10)}$

$= -\dfrac{1}{1{,}000}$

89. $\left(\dfrac{a^2b^3}{ab^4}\right)^0 = 1$

91. $\dfrac{9^{-2}s^6t}{4^{-3}s^4t^5} = \dfrac{4^3s^6t^1}{9^2s^4t^5}$

$= \dfrac{64s^{6-4}t^{1-5}}{81}$

$= \dfrac{64s^2t^{-4}}{81}$

$= \dfrac{64s^2}{81t^4}$

93. $\left(2u^{-2}v^5\right)^5 = \left(2^1u^{-2}v^5\right)^5$

$= 2^5\left(u^{-2}\right)^5\left(v^5\right)^5$

$= 32u^{-2\cdot 5}v^{5\cdot 5}$

$= 32u^{-10}v^{25}$

$= \dfrac{32v^{25}}{u^{10}}$

95. $\left(\dfrac{y^4}{3}\right)^{-2} = \left(\dfrac{3^1}{y^4}\right)^2$

$= \dfrac{3^2}{\left(y^4\right)^2}$

$= \dfrac{9}{y^{4\cdot 2}}$

$= \dfrac{9}{y^8}$

97. $-15x^0y = -15\cdot 1y$

$= -15y$

99. $\left(\dfrac{4}{h^{10}}\right)^{-2} = \left(\dfrac{h^{10}}{4^1}\right)^2$

$= \dfrac{h^{10\cdot 2}}{4^{1\cdot 2}}$

$= \dfrac{h^{10\cdot 2}}{4^2}$

$= \dfrac{h^{20}}{16}$

101. $x^{-3} \cdot x^{-3} = x^{-3+(-3)}$
$= x^{-6}$
$= \frac{1}{x^6}$

103. $\left(\frac{c^3 d^{-4}}{c^{-1} d^5}\right)^3 = \left(c^{3-(-1)} d^{-4-5}\right)^3$
$= \left(c^{3+1} d^{-9}\right)^3$
$= \left(c^4 d^{-9}\right)^3$
$= c^{4 \cdot 3} d^{-9 \cdot 3}$
$= c^{12} d^{-27}$
$= \frac{c^{12}}{d^{27}}$

105. $15(-6x)^0 = 15 \cdot 1$
$= 15$

107. $\frac{2^{-2} g^{-2} h^{-3}}{9^{-1} h^{-3}} = \frac{9^1 h^{-3-(-3)}}{2^2 g^2}$
$= \frac{9h^{-3+3}}{4g^2}$
$= \frac{9h^0}{4g^2}$
$= \frac{9 \cdot 1}{4g^2}$
$= \frac{9}{4g^2}$

109. $\left(5d^{-2}\right)^3 = \left(5^1 d^{-2}\right)^3$
$= 5^3 \left(d^{-2}\right)^3$
$= 125d^{-2 \cdot 3}$
$= 125d^{-6}$
$= \frac{125}{d^6}$

111. $\left(\frac{x^2 y^{-2}}{x^{-5} y^3}\right)^4 = \left(x^{2-(-5)} y^{-2-3}\right)^4$
$= \left(x^{2+5} y^{-5}\right)^4$
$= \left(x^7 y^{-5}\right)^4$
$= x^{7 \cdot 4} y^{-5 \cdot 4}$
$= x^{28} y^{-20}$
$= \frac{x^{28}}{y^{20}}$

113. $\left(2x^3 y^{-2}\right)^5 = \left(2^1 x^3 y^{-2}\right)^5$
$= 2^5 \left(x^3\right)^5 \left(y^{-2}\right)^5$
$= 32x^{3 \cdot 5} y^{-2 \cdot 5}$
$= 32x^{15} y^{-10}$
$= \frac{32x^{15}}{y^{10}}$

115. $\frac{t(t^{-2})^{-2}}{t^{-5}} = \frac{t^1 (t^{-2 \cdot -2})}{t^{-5}}$
$= \frac{t^1 (t^4)}{t^{-5}}$
$= \frac{t^{1+4}}{t^{-5}}$
$= \frac{t^5}{t^{-5}}$
$= t^{5-(-5)}$
$= t^{5+5}$
$= t^{10}$

117. $\frac{-4s^{-5}}{t^{-2}} = \frac{-4t^2}{s^5}$
$= -\frac{4t^2}{s^5}$

119. $\left(x^{-4} x^3\right)^3 = \left(x^{-4+3}\right)^3$
$= \left(x^{-1}\right)^3$
$= x^{-1 \cdot 3}$
$= x^{-3}$
$= \frac{1}{x^3}$

(All new) LOOK ALIKES . . .

121. a. $8^{-1} = \frac{1}{8}$ b. $-8^{-1} = -\frac{1}{8}$

c. $(-8)^{-1} = \frac{1}{(-8)^1} = -\frac{1}{8}$ d. $-(-8)^{-1} = -\frac{1}{(-8)^1} = \frac{1}{8}$

123. a. $4xy^{-2} = \frac{4x}{y^2}$ b. $(4xy)^{-2} = \frac{1}{(4xy)^2} = \frac{1}{16x^2y^2}$

c. $4x^{-2}y = \frac{4y}{x^2}$ d. $4^{-2}xy = \frac{xy}{16}$

APPLICATIONS
from Campus to Careers

125. SOUND ENGINEERING TECHNICIAN

Type of sound	Intensity
Front row rock concert	10^{-1}
Normal conversation	10^{-6}
Vacuum cleaner	10^{-4}
Military jet takeoff	10^{2}
Whisper	10^{-10}

WRITING

127. Answers will vary.

REVIEW

Find the slope of the line that passes through the given points.

129. Let $(x_1, y_1) = (1, -4)$
Let $(x_2, y_2) = (3, -7)$

$$m = \frac{y_2 - y_1}{x_2 - x_1} = \frac{-7-(-4)}{3-1} = \frac{-7+4}{2} = \frac{-3}{2} = -\frac{3}{2}$$

131. Write an equation of the line having slope $\frac{3}{4}$ and y-intercept -5.

$$y = mx + b$$
$$y = \frac{3}{4}x - 5$$

CHALLENGE PROBLEMS

133. **Simplify each expression. Do not use negative exponents in the answer. The variable *m* represents a positive integer.**

a. $r^{5m}r^{-6m} = r^{5m+(-6m)} = r^{-m} = \frac{1}{r^m}$

b. $\frac{x^{3m}}{x^{6m}} = x^{3m-(6m)} = x^{3m+(-6m)} = x^{-3m} = \frac{1}{x^{3m}}$

SECTION 5.3 STUDY SET

VOCABULARY

Fill in the blanks.

1. 4.84×10^5 is written in **scientific** notation. 484,000 is written in **standard** notation.

CONCEPTS

Fill in the blanks.

3. When we multiply a decimal by 10^5, the decimal point moves 5 places to the **right**. When we multiply a decimal by 10^{-7}, the decimal point moves 7 places to the **left**.

5. a. When a real number greater than 10 is written in scientific notation, the exponent on 10 is a **positive** integer.
   b. When a real number between 0 and 1 is written in scientific notation, the exponent on 10 is a **negative** integer.

Fill in the blanks to write each number in scientific notation.

7. a. $7{,}700 = \mathbf{7.7} \times 10^3$
   b. $500{,}000 = \mathbf{5.0} \times 10^5$
   c. $114{,}000{,}000 = 1.44 \times 10^{\mathbf{8}}$

9. Write each expression so that the decimal numbers are grouped together and the powers of ten are grouped together.
   a. $(5.1\times10^9)(1.5\times10^{22}) = (\mathbf{5.1\times1.5})(\mathbf{10^9\times10^{22}})$
   b. $\dfrac{8.8\times10^{30}}{2.2\times10^{19}} = \dfrac{\mathbf{8.8}}{\mathbf{2.2}}\times\dfrac{\mathbf{10^{30}}}{\mathbf{10^{19}}}$

NOTATION

11. Fill in the blanks. A positive number is written in scientific notation when it is written in the form $N \times 10^n$, where $\mathbf{1 \le N < 10}$ and n is an **integer**.

GUIDED PRACTICE

Convert each number to standard notation. See Example 1.

13. The exponent is 2.
    Move the decimal 2 places to the right.
    $2.3 \times 10^2 = 230$

15. The exponent is 5.
    Move the decimal 5 places to the right.
    $8.12 \times 10^5 = 812{,}000$

17. The exponent is -3.
    Move the decimal 3 places to the left.
    $1.15 \times 10^{-3} = 0.00115$

19. The exponent is -4.
    Move the decimal 4 places to the left.
    $9.76\times10^{-4} = 0.000976$

21. The exponent is 6.
    Move the decimal 6 places to the right.
    $6.001\times10^6 = 6{,}001{,}000$

23. The exponent is 0.
    Do not move the decimal.
    $2.718\times10^0 = 2.718$

25. The exponent is -2.
    Move the decimal 2 places to the left.
    $6.798\times10^{-2} = 0.06798$

27. The exponent is -5.
    Move the decimal 5 places to the left.
    $2.0\times10^{-5} = 0.00002$

Write each number in scientific notation. See Example 2.

29. Move the decimal 4 places to the left.
    $23{,}000 = 2.3\times10^4$

31. Move the decimal 6 places to the left.
    $1{,}700{,}000 = 1.7\times10^6$

33. Move the decimal 2 places to the right.
    $0.062 = 6.2\times10^{-2}$

35. Move the decimal 6 places to the right.
    $0.0000051 = 5.1\times10^{-6}$

37. Move the decimal 9 places to the left.
    $5{,}000{,}000{,}000 = 5.0\times10^9$

39. Move the decimal 7 places to the right.
    $0.0000003 = 3.0\times10^{-7}$

41. Move the decimal 8 places to the left.
    $909{,}000{,}000 = 9.09\times10^8$

43. Move the decimal 2 places to the right.
    $0.0345 = 3.45\times10^{-2}$

45. Do not move the decimal.
    $9 = 9.0\times10^0$

47. Move the decimal 1 place to the left.
    $11 = 1.1\times10^1$

49. Move the decimal 18 places to the left.
$1{,}718{,}000{,}000{,}000{,}000{,}000 = 1.718\times10^{18}$

51. Move the decimal 14 places to the right.
$0.0000000000000123 = 1.23\times10^{-14}$

53. Move the decimal 1 place to the left.
$$73\times10^4 = 7.3\times10^{4+1}$$
$$= 7.3\times10^5$$

55. Move the decimal 2 places to the left.
$$201.8\times10^{15} = 2.018\times10^{15+2}$$
$$= 2.018\times10^{17}$$

57. Move the decimal 2 places to the right.
$$0.073\times10^{-3} = 7.3\times10^{-3+(-2)}$$
$$= 7.3\times10^{-5}$$

59. Move the decimal 1 place to the left.
$$36.02\times10^{-20} = 3.602\times10^{-20+1}$$
$$= 3.602\times10^{-19}$$

Use scientific notation to perform the calculations. Give all answers in scientific notation and standard notation. See Examples 3 and 4.

61. $(3.4\times10^2)(2.1\times10^3)$
$$= (3.4\cdot2.1)\times(10^2\times10^3)$$
$$= (3.4\cdot2.1)\times(10^{2+3})$$
$$= 7.14\times10^5$$
$$= 714{,}000$$

63. $(8.4\times10^{-13})(4.8\times10^9)$
$$= (8.4\cdot4.8)\times(10^{-13}\times10^9)$$
$$= (8.4\cdot4.8)\times(10^{-13+9})$$
$$= 40.32\times10^{-4}$$
$$= 4.032\times10^{-4+1}$$
$$= 4.032\times10^{-3}$$
$$= 0.004032$$

65. $$\frac{2.24\times10^4}{5.6\times10^7} = \frac{2.24}{5.6}\times\frac{10^4}{10^7}$$
$$= \frac{2.24}{5.6}\times10^{4-7}$$
$$= 0.4\times10^{-3}$$
$$= 4.0\times10^{-3+(-1)}$$
$$= 4.0\times10^{-4}$$
$$= 0.0004$$

67. $$\frac{9.3\times10^2}{3.1\times10^{-2}} = \frac{9.3}{3.1}\times\frac{10^2}{10^{-2}}$$
$$= \frac{9.3}{3.1}\times10^{2-(-2)}$$
$$= 3\times10^{2+2}$$
$$= 3.0\times10^4$$
$$= 30{,}000$$

69. $$\frac{0.00000129}{0.0003} = \frac{1.29\times10^{-6}}{3.0\times10^{-4}}$$
$$= \frac{1.29}{3.0}\times\frac{10^{-6}}{10^{-4}}$$
$$= \frac{1.29}{3.0}\times10^{-6-(-4)}$$
$$= 0.43\times10^{-6+4}$$
$$= 0.43\times10^{-2}$$
$$= 4.3\times10^{-2+(-1)}$$
$$= 4.3\times10^{-3}$$
$$= 0.0043$$

71. $(0.0000000056)(5{,}500{,}000)$
$$= (5.6\times10^{-9})(5.5\times10^6)$$
$$= (5.6\cdot5.5)\times(10^{-9}\times10^6)$$
$$= (5.6\cdot5.5)\times(10^{-9+6})$$
$$= 30.8\times10^{-3}$$
$$= 3.08\times10^{-3+1}$$
$$= 3.08\times10^{-2}$$
$$= 0.0308$$

73. $$\frac{96{,}000}{(12{,}000)(0.00004)} = \frac{9.6\times10^4}{(1.2\times10^4)(4.0\times10^{-5})}$$
$$= \frac{9.6\times10^4}{(1.2\cdot4.0)\times(10^4\times10^{-5})}$$
$$= \frac{9.6\times10^4}{(1.2\cdot4.0)\times(10^{4+(-5)})}$$
$$= \frac{9.6\times10^4}{4.8\times10^{-1}}$$
$$= \frac{9.6}{4.8}\times\frac{10^4}{10^{-1}}$$
$$= \frac{9.6}{4.8}\times10^{4-(-1)}$$
$$= \frac{9.6}{4.8}\times10^{4+1}$$
$$= 2.0\times10^5$$
$$= 200{,}000$$

75. $\dfrac{2,475}{(132,000,000,000,000)(0.25)}$

$$= \frac{2.475 \times 10^3}{(1.32 \times 10^{14})(2.5 \times 10^{-1})}$$
$$= \frac{2.475 \times 10^3}{(1.32 \cdot 2.5) \times (10^{14} \times 10^{-1})}$$
$$= \frac{2.475 \times 10^3}{(1.32 \cdot 2.5) \times (10^{14+(-1)})}$$
$$= \frac{2.475 \times 10^3}{3.3 \times 10^{13}}$$
$$= \frac{2.475}{3.3} \times \frac{10^3}{10^{13}}$$
$$= \frac{2.475}{3.3} \times 10^{3-13}$$
$$= \frac{2.475}{3.3} \times 10^{3+(-13)}$$
$$= 0.75 \times 10^{-10}$$
$$= 7.5 \times 10^{-10+(-1)}$$
$$= 7.5 \times 10^{-11}$$
$$= 0.000000000075$$

Find each power. These exact answers were computed using MS Excel.

77. $(456.4)^6 = 9{,}038{,}030{,}747{,}579{,}110$
$= 9.03803074757911 \times 10^{15}$

79. $225^{-5} = 0.00000000000173415299158326$
$= 1.73415299158326 \times 10^{-12}$

APPLICATIONS

81. ASTRONOMY
Move the decimal 13 places to the left.
$25{,}700{,}000{,}000{,}000 = 2.57 \times 10^{13}$
The distance is about 2.57×10^{13} miles.

83. EARTH, SUN, MOON
Surface area of the Earth
The exponent is 8.
Move the decimal 8 places to the right.
$1.97 \times 10^8 = 197{,}000{,}000$
The surface area is $197{,}000{,}000$ mi^2.

Surface area of the sun
The exponent is 17.
Move the decimal 17 places to the right.
$1.09 \times 10^{17} = 109{,}000{,}000{,}000{,}000{,}000$
The surface area is
$109{,}000{,}000{,}000{,}000{,}000$ mi^2.

Surface area of the moon
The exponent is 7.
Move the decimal 7 places to the right.
$1.46 \times 10^7 = 14{,}600{,}000$
The surface area is $14{,}600{,}000$ mi^2.

85. SAND
Move the decimal 10 places to the right.
$0.00000000045 = 4.5 \times 10^{-10}$
The mass of one grain of beach sand is approximately 4.5×10^{-10} oz.

87. WAVELENGTHS

gamma ray	treating cancer	8.9×10^{-14}
x-ray	medical	2.3×10^{-11}
ultraviolet	sun lamp	6.1×10^{-8}
visible light	lighting	9.3×10^{-6}
infrared	photography	3.7×10^{-5}
microwave	cooking	1.1×10^{-2}
radio wave	communication	3.0×10^{2}

89. **from CAMPUS TO CAREERS**
SOUND ENGINEERING TECHNICIAN

$$\frac{3.3\times10^4}{(100)(1000)} = \frac{3.3\times10^4}{(1.0\times10^2)(1.0\times10^3)}$$
$$= \frac{3.3\times10^4}{(1.0\times1.0)(10^2\times10^3)}$$
$$= \frac{3.3}{1.0}\cdot\frac{10^4}{10^{2+3}}$$
$$= 3.3\times10^{4-5}$$
$$= 3.3\times10^{-1}$$

The speed of sound in air is approximately 3.3×10^{-1} km/sec.

91. LIGHT YEARS

$$\begin{aligned}(5.87\times10^{12})(5.28\times10^{3}) &= (5.87\times5.28)(10^{12}\times10^{3})\\ &= 30.9936\times10^{12+3}\\ &= 30.9936\times10^{15}\\ &= 3.09936\times10^{15+1}\\ &= 3.09936\times10^{16}\end{aligned}$$

One light year is about 3.09936×10^{16} feet.

93. INSURED DEPOSITS

$$\begin{aligned}I &= Prt\\ &= (7.6\times10^{12})(0.04)(1)\\ &= (7.6\times10^{12})(4.0\times10^{-2})(1)\\ &= (7.6\times4.0\times1)(10^{12}\times10^{-2})\\ &= 30.4\times10^{12+(-2)}\\ &= 30.4\times10^{10}\\ &= 3.04\times10^{10+1}\\ &= 3.04\times10^{11}\end{aligned}$$

The amount of interest would be about 3.04×10^{11} dollars.

95. POWERS OF 10

One million:
$1{,}000{,}000 = 1.0\times10^{6}$

One billion:
$1{,}000{,}000{,}000 = 1.0\times10^{9}$

One trillion:
$1{,}000{,}000{,}000{,}000 = 1.0\times10^{12}$

One quadrillion:
$1{,}000{,}000{,}000{,}000{,}000 = 1.0\times10^{15}$

One quintillion:
$1{,}000{,}000{,}000{,}000{,}000{,}000 = 1.0\times10^{18}$

WRITING

97-99. Answers will vary.

REVIEW

101. If $y=-1$, find the value of $-5y^{55}$.

$$\begin{aligned}-5y^{55} &= -5(-1)^{55}\\ &= -5(-1)\\ &= 5\end{aligned}$$

103. COUNSELING

1st year, counselor had 75 clients.

2nd year, she had 105 clients.

t years is the x-axis.

c clints is the y-axis.

Write a linear equation.

In this problem, we are given two different letters t(years) and c(clients) to use for x and y. One has to determine which letter correlates with which letter, $x=t$ and $y=c$.

Find the slope which is given as two ordered pairs (1, 75) and (2, 105). The reason for the numbers 1 and 2 is because after "1 year" and then after "2 years".

Let $(t_1, c_1) = (1, 75)$

Let $(t_2, c_2) = (2, 105)$

$$\begin{aligned}m &= \frac{c_2-c_1}{t_2-t_1}\\ &= \frac{105-75}{2-1}\\ &= \frac{30}{1}\\ &= 30\end{aligned}$$

Now pick one point (1, 75) and use

$$\begin{aligned}c-c_1 &= m(t-t_1)\\ c-75 &= 30(t-1)\\ c-75 &= 30t-30\\ c-75-\mathbf{75} &= 30t-30+\mathbf{75}\\ c &= 30t+45 \text{ slope-intercept form}\end{aligned}$$

CHALLENGE PROBLEMS

105. Consider 2.5×10^{-4}. Answer the following questions in scientific notation form.

a. What is its opposite?

Its opposite is -2.5×10^{-4}.

b. What is its reciprocal?

$$\begin{aligned}\frac{1}{2.5\times10^{-4}} &= \frac{1}{2.5}\times\frac{1}{10^{-4}}\\ &= 0.4\times10^{4}\\ &= 4.0\times10^{4+(-1)}\\ &= 4.0\times10^{3}\end{aligned}$$

Its reciprocal is 4.0×10^{3}.

SECTION 5.4 STUDY SET
VOCABULARY

Fill in the blanks.

1. A **polynomial** is a term or a sum of terms in which all variables have whole-number exponents and no variable appears in a denominator.
3. $x^3 - 6x^2 + 9x - 2$ is a polynomial in **one** variable, and is written in **descending** powers of x and $c^3 + 2c^2d - d^2$ is a polynomial in **two** variables and is written in **ascending** powers of d.
5. A **monomial** is a polynomial with exactly one term. A **binomial** is a polynomial with exactly two terms. A **trinomial** is a polynomial with exactly three terms.
7. To **evaluate** the polynomial $x^2 - 2x + 1$ for $x = 6$, we substitute 6 for x and follow the rules for the order of operations.

CONCEPTS

Determine whether each expression is a polynomial.

9. a. $x^3 - 5x^2 - 2$ **Yes** b. $x^{-4} - 5x$ **No**

c. $x^2 - \dfrac{1}{2x} + 3$ **No** d. $x^3 - 1$ **Yes**

e. $x^2 - y^2$ **Yes** f. $a^4 + a^3 + a^2 + a$ **Yes**

Make a term-coefficient-degree table like that shown in Example 1 for each polynomial.

11. $8x^2 + x - 7$

Term	Coefficient	Degree
$8x^2$	8	2
x	1	1
−7	−7	0

Degree of the polynomial: **2**

13. $8a^6b^3 - 27ab$

Term	Coefficient	Degree
$8a^6b^3$	8	9
$-27ab$	−27	2

Degree of the polynomial: **9**

NOTATION

15. a. Write $x - 9 + 3x^2 + 5x^3$ in descending powers of x.

$5x^3 + 3x^2 + x - 9$

b. Write $-2xy + y^2 + x^2$ in ascending powers of y.

$x^2 - 2xy + y^2$

GUIDED PRACTICE

Classify each polynomial as a monomial, a binomial, a trinomial, or none of these. See Example 1.

	Monomial one term	Binomial two terms	Trinomial three terms	none of these
17. $3x + 7$		X		
19. $y^2 + 4y + 3$			X	
21. $\frac{3}{2}z^2$	X			
23. $t - 32$		X		
25. $s^2 - 23s + 31$			X	
27. $6x^5 - x^4 - 3x^3 + 7$				X
29. $3m^3n - 4m^2n^2 + mn - 1$				X
31. $2a^2 - 3ab + b^2$			X	

Find the degree of each polynomial. See Example 1.

33. $3x^4$ **4th**
35. $-2x^2 + 3x + 1$ **2nd**
37. $\frac{1}{3}x - 5$ **1st**
39. $-5r^2s^2 - r^3s + 3$ **4th**
41. $x^{12} + 3x^2y^3$ **12th**
43. 38 **0th**
45. $\frac{3}{2}m^7 - \frac{3}{4}m^{18}$ **18th**
47. $5.5tw - 6.5t^2w - 7.5t^3$ **3rd**

Evaluate each expression. See Examples 2 and 3.

49. $x^2 - x + 1$

a. For $x = 2$

$$\begin{aligned} x^2 - x + 1 &= (\mathbf{2})^2 - (\mathbf{2}) + 1 \\ &= 4 - 2 + 1 \\ &= 2 + 1 \\ &= 3 \end{aligned}$$

b. For $x = -3$

$$\begin{aligned} x^2 - x + 1 &= (\mathbf{-3})^2 - (\mathbf{-3}) + 1 \\ &= 9 + 3 + 1 \\ &= 12 + 1 \\ &= 13 \end{aligned}$$

51. $4t^2 + 2t - 8$

a. For $t = -1$

$$\begin{aligned} 4t^2 + 2t - 8 &= 4(\mathbf{-1})^2 + 2(\mathbf{-1}) - 8 \\ &= 4(1) + 2(-1) - 8 \\ &= 4 - 2 - 8 \\ &= 2 - 8 \\ &= 2 + (-8) \\ &= -6 \end{aligned}$$

b. For $t = 0$

$$\begin{aligned} 4t^2 + 2t - 8 &= 4(\mathbf{0})^2 + 2(\mathbf{0}) - 8 \\ &= 4(0) + 2(0) - 8 \\ &= 0 + 0 - 8 \\ &= -8 \end{aligned}$$

53. $\frac{1}{2}a^2 - \frac{1}{4}a$

a. For $a = 4$

$$\begin{aligned} \frac{1}{2}a^2 - \frac{1}{4}a &= \frac{1}{2}(\mathbf{4})^2 - \frac{1}{4}(\mathbf{4}) \\ &= \frac{1}{2}(16) - \frac{1}{4}(4) \\ &= 8 - 1 \\ &= 7 \end{aligned}$$

b. For $a = -8$

$$\begin{aligned} \frac{1}{2}a^2 - \frac{1}{4}a &= \frac{1}{2}(\mathbf{-8})^2 - \frac{1}{4}(\mathbf{-8}) \\ &= \frac{1}{2}(64) - \frac{1}{4}(-8) \\ &= 32 + 2 \\ &= 34 \end{aligned}$$

55. $-9.2x^2 + x - 1.4$

a. For $x = -1$

$$\begin{aligned} -9.2x^2 + x - 1.4 &= -9.2(\mathbf{-1})^2 + (\mathbf{-1}) - 1.4 \\ &= -9.2(1) - 1 - 1.4 \\ &= -9.2 - 1 - 1.4 \\ &= -10.2 - 1.4 \\ &= -11.6 \end{aligned}$$

b. For $x = -2$

$$\begin{aligned} -9.2x^2 + x - 1.4 &= -9.2(\mathbf{-2})^2 + (\mathbf{-2}) - 1.4 \\ &= -9.2(4) - 2 - 1.4 \\ &= -36.8 - 2 - 1.4 \\ &= -38.8 - 1.4 \\ &= -40.2 \end{aligned}$$

57. $x^3 + 3x^2 + 2x + 4$

a. For $x = 2$

$$\begin{aligned} x^3 + 3x^2 + 2x + 4 &= (\mathbf{2})^3 + 3(\mathbf{2})^2 + 2(\mathbf{2}) + 4 \\ &= 8 + 3(4) + 4 + 4 \\ &= 8 + 12 + 4 + 4 \\ &= 20 + 4 + 4 \\ &= 24 + 4 \\ &= 28 \end{aligned}$$

b. For $x = -2$

$$\begin{aligned} &x^3 + 3x^2 + 2x + 4 \\ &\quad = (\mathbf{-2})^3 + 3(\mathbf{-2})^2 + 2(\mathbf{-2}) + 4 \\ &\quad = -8 + 3(4) - 4 + 4 \\ &\quad = -8 + 12 - 4 + 4 \\ &\quad = 4 - 4 + 4 \\ &\quad = 0 + 4 \\ &\quad = 4 \end{aligned}$$

59. $y^4 - y^3 + y^2 + 2y - 1$

a. For $y = 1$

$$\begin{aligned} &y^4 - y^3 + y^2 + 2y - 1 \\ &\quad = (\mathbf{1})^4 - (\mathbf{1})^3 + (\mathbf{1})^2 + 2(\mathbf{1}) - 1 \\ &\quad = 1 - 1 + 1 + 2 - 1 \\ &\quad = 0 + 1 + 2 - 1 \\ &\quad = 1 + 2 - 1 \\ &\quad = 3 - 1 \\ &\quad = 2 \end{aligned}$$

b. For $y = -1$

$$\begin{aligned} &y^4 - y^3 + y^2 + 2y - 1 \\ &\quad = (\mathbf{-1})^4 - (\mathbf{-1})^3 + (\mathbf{-1})^2 + 2(\mathbf{-1}) - 1 \\ &\quad = 1 - (-1) + 1 + 2(-1) - 1 \\ &\quad = 1 + 1 + 1 - 2 - 1 \\ &\quad = 2 + 1 - 2 - 1 \\ &\quad = 3 - 2 - 1 \\ &\quad = 1 - 1 \\ &\quad = 0 \end{aligned}$$

Evaluate each polynomial for a = –2 and b = 3. See Example 4.

61. $6a^2b$, For $a = -2,\ b = 3$.

$$\begin{aligned} 6a^2b &= 6(\mathbf{-2})^2(\mathbf{3}) \\ &= 6(4)(3) \\ &= 72 \end{aligned}$$

63. $a^3 + b^3$, For $a = -2,\ b = 3$.

$$\begin{aligned} a^3 + b^3 &= (\mathbf{-2})^3 + (\mathbf{3})^3 \\ &= -8 + 27 \\ &= 19 \end{aligned}$$

65. $a^2 + 5ab - b^2$, For $a = -2,\ b = 3$.

$$\begin{aligned} a^2 + 5ab - b^2 &= (\mathbf{-2})^2 + 5(\mathbf{-2})(\mathbf{3}) - (\mathbf{3})^2 \\ &= 4 - 30 - 9 \\ &= -26 - 9 \\ &= -35 \end{aligned}$$

67. $5ab^3 - ab - b + 10$, For $a = -2$, $b = 3$.

$$\begin{aligned} 5ab^3 - ab - b + 10 &= 5(\mathbf{-2})(\mathbf{3})^3 - (\mathbf{-2})(\mathbf{3}) - (\mathbf{3}) + 10 \\ &= 5(-2)27 + 6 - 3 + 10 \\ &= -270 + 6 - 3 + 10 \\ &= -270 + 6 - 3 + 10 \\ &= -264 - 3 + 10 \\ &= -267 + 10 \\ &= -257 \end{aligned}$$

Construct a table of solutions and then graph the equation. See Examples 5-7.

69. $y = x^2 + 1$

x	y	(x, y)
-3	$y = (-3)^2 + 1$ $= 9 + 1$ $= 10$	$(-3, 10)$
-2	$y = (-2)^2 + 1$ $= 4 + 1$ $= 5$	$(-2, 5)$
-1	$y = (-1)^2 + 1$ $= 1 + 1$ $= 2$	$(-1, 2)$
0	$y = (0)^2 + 1$ $= 0 + 1$ $= 1$	$(0, 1)$
1	$y = (1)^2 + 1$ $= 1 + 1$ $= 2$	$(1, 2)$
2	$y = (2)^2 + 1$ $= 4 + 1$ $= 5$	$(2, 5)$
3	$y = (3)^2 + 1$ $= 9 + 1$ $= 10$	$(3, 10)$

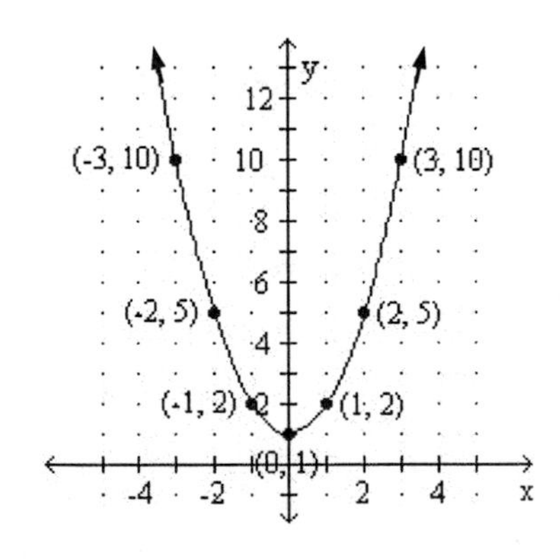

71. $y = -x^2 - 2$

x	y	(x, y)
-3	$y = -(-3)^2 - 2$ $= -(9) - 2$ $= -9 + (-2)$ $= -11$	$(-3, -11)$
-2	$y = -(-2)^2 - 2$ $= -(4) - 2$ $= -4 + (-2)$ $= -6$	$(-2, -6)$
-1	$y = -(-1)^2 - 2$ $= -(1) - 2$ $= -1 + (-2)$ $= -3$	$(-1, -3)$
0	$y = -(0)^2 - 2$ $= -(0) - 2$ $= 0 + (-2)$ $= -2$	$(0, -2)$
1	$y = -(1)^2 - 2$ $= -(1) - 2$ $= -1 + (-2)$ $= -3$	$(1, -3)$
2	$y = -(2)^2 - 2$ $= -(4) - 2$ $= -4 + (-2)$ $= -6$	$(2, -6)$
3	$y = -(3)^2 - 2$ $= -(9) - 2$ $= -9 + (-2)$ $= -11$	$(3, -11)$

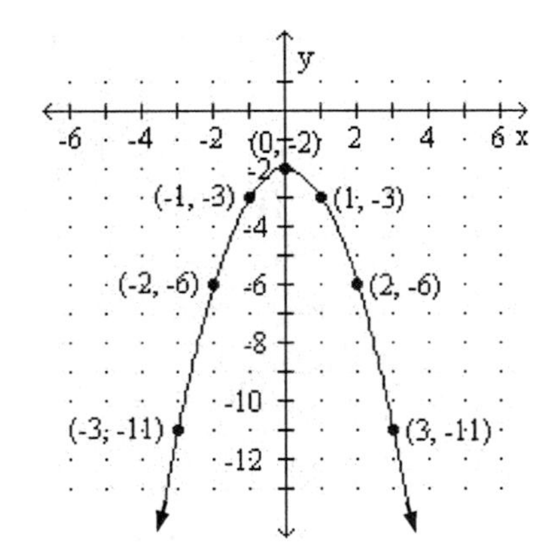

73. $y = 2x^2 - 3$

x	y	(x, y)
−3	$y = 2(-3)^2 - 3$ $= 2(9) - 3$ $= 18 + (-3)$ $= 15$	(−3, 15)
−2	$y = 2(-2)^2 - 3$ $= 2(4) - 3$ $= 8 + (-3)$ $= 5$	(−2, 5)
−1	$y = 2(-1)^2 - 3$ $= 2(1) - 3$ $= 2 + (-3)$ $= -1$	(−1, −1)
0	$y = 2(0)^2 - 3$ $= 2(0) - 3$ $= 0 + (-3)$ $= -3$	(0, −3)
1	$y = 2(1)^2 - 3$ $= 2(1) - 3$ $= 2 + (-3)$ $= -1$	(1, −1)
2	$y = 2(2)^2 - 3$ $= 2(4) - 3$ $= 8 + (-3)$ $= 5$	(2, 5)
3	$y = 2(3)^2 - 3$ $= 2(9) - 3$ $= 18 + (-3)$ $= 15$	(3, 15)

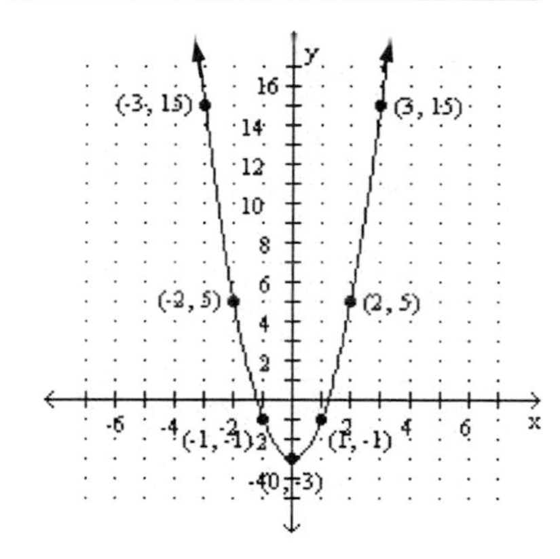

75. $y = x^3 + 2$

x	y	(x, y)
−3	$y = (-3)^3 + 2$ $= -27 + 2$ $= -25$	(−3, −25)
−2	$y = (-2)^3 + 2$ $= -8 + 2$ $= -6$	(−2, −6)
−1	$y = (-1)^3 + 2$ $= -1 + 2$ $= 1$	(−1, 1)
0	$y = (0)^3 + 2$ $= 0 + 2$ $= 2$	(0, 2)
1	$y = (1)^3 + 2$ $= 1 + 2$ $= 3$	(1, 3)
2	$y = (2)^3 + 2$ $= 8 + 2$ $= 10$	(2, 10)
3	$y = (3)^3 + 2$ $= 27 + 2$ $= 29$	(3, 29)

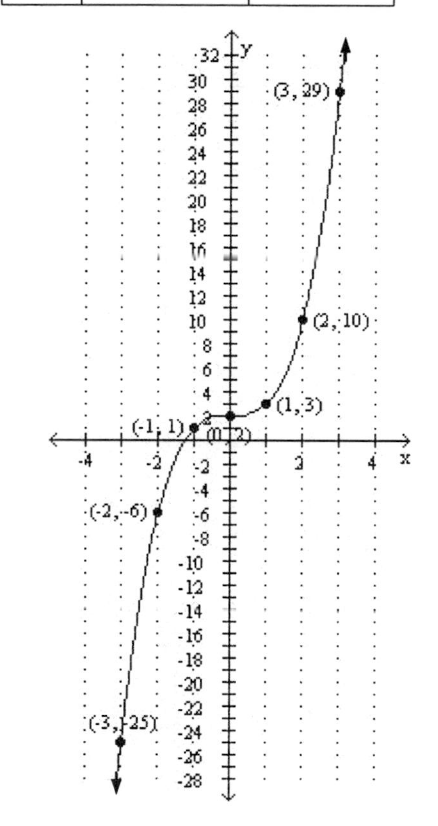

77. $y = x^3 - 3$

x	y	(x, y)
-3	$y = (-3)^3 - 3$ $= -27 + (-3)$ $= -30$	$(-3, -30)$
-2	$y = (-2)^3 - 3$ $= -8 + (-3)$ $= -11$	$(-2, -11)$
-1	$y = (-1)^3 - 3$ $= -1 + (-3)$ $= -4$	$(-1, -4)$
0	$y = (0)^3 - 3$ $= 0 + (-3)$ $= -3$	$(0, -3)$
1	$y = (1)^3 - 3$ $= 1 + (-3)$ $= -2$	$(1, -2)$
2	$y = (2)^3 - 3$ $= 8 + (-3)$ $= 5$	$(2, 5)$
3	$y = (3)^3 - 3$ $= 27 + (-3)$ $= 24$	$(3, 24)$

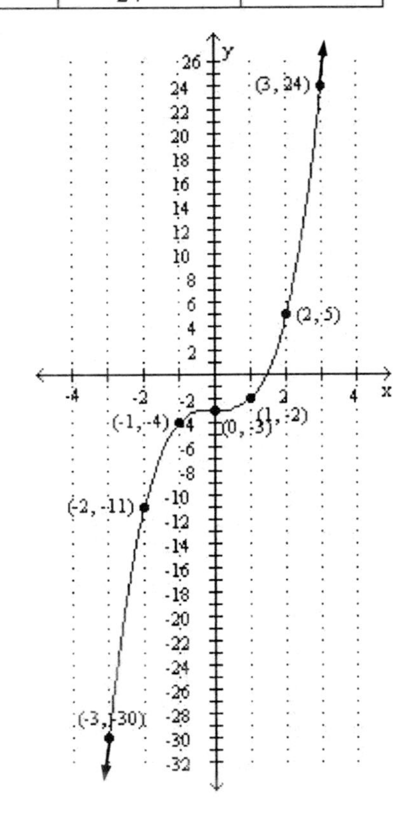

79. $y = -x^3 - 1$

x	y	(x, y)
-3	$y = -(-3)^3 - 1$ $= -(-27) - 1$ $= 27 + (-1)$ $= 26$	$(-3, 26)$
-2	$y = -(-2)^3 - 1$ $= -(-8) - 1$ $= 8 + (-1)$ $= 7$	$(-2, 7)$
-1	$y = -(-1)^3 - 1$ $= -(-1) - 1$ $= 1 + (-1)$ $= 0$	$(-1, 0)$
0	$y = -(0)^3 - 1$ $= -(0) - 1$ $= 0 + (-1)$ $= -1$	$(0, -1)$
1	$y = -(1)^3 - 1$ $= -(1) - 1$ $= -1 + (-1)$ $= -2$	$(1, -2)$
2	$y = -(2)^3 - 1$ $= -(8) - 1$ $= -8 + (-1)$ $= -9$	$(2, -9)$
3	$y = -(3)^3 - 1$ $= -(27) - 1$ $= -27 + (-1)$ $= -28$	$(3, -28)$

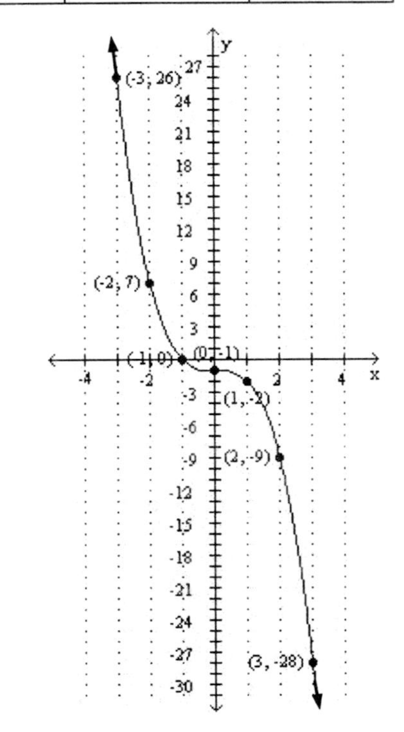

APPLICATION

81. SUPERMARKETS

$$\frac{1}{3}c^3+\frac{1}{2}c^2+\frac{1}{6}c=\frac{1}{3}(\mathbf{6})^3+\frac{1}{2}(\mathbf{6})^2+\frac{1}{6}(\mathbf{6})$$
$$=\frac{1}{3}(216)+\frac{1}{2}(36)+\frac{1}{6}(6)$$
$$=72+18+1$$
$$=91$$

91 cantaloupes will be used.

83. STOPPING DISTANCE

$$0.04v^2+0.9v=0.04(\mathbf{30})^2+0.9(\mathbf{30})$$
$$=0.04(900)+0.9(30)$$
$$=36+27$$
$$=63$$

The stopping distance is 63 feet.

from CAMPUS TO CAREERS

85. SOUND ENGINEERING TECHNICIAN

Let $x=0$, when the year is 2004.

Let $x=1$, when the year is 2005.

Let $x=10$, when the year is 2014.

$$0.32x^2-0.36x+0.21$$
$$=0.32(\mathbf{10})^2-0.36(\mathbf{10})+0.21$$
$$=0.32(100)-0.36(10)+0.21$$
$$=32-3.6+0.21$$
$$=28.61$$

The approximate number of iTunes downloads as of January 2014 will be about 28.6 billion.

87. SCIENCE HISTORY

Time (seconds)	Distance (ft)	(s, f)
0	0	(0, 0)
$\frac{1}{4}$	$\frac{1}{2}$	$\left(\frac{1}{4},\frac{1}{2}\right)$
$\frac{1}{2}$	2	$\left(\frac{1}{2},2\right)$
$\frac{3}{4}$	$4\frac{1}{2}$	$\left(\frac{3}{4},4\frac{1}{2}\right)$
1	8	(1, 8)
$1\frac{1}{4}$	$12\frac{1}{2}$	$\left(1\frac{1}{4},12\frac{1}{2}\right)$
$1\frac{1}{2}$	18	$\left(1\frac{1}{2},18\right)$

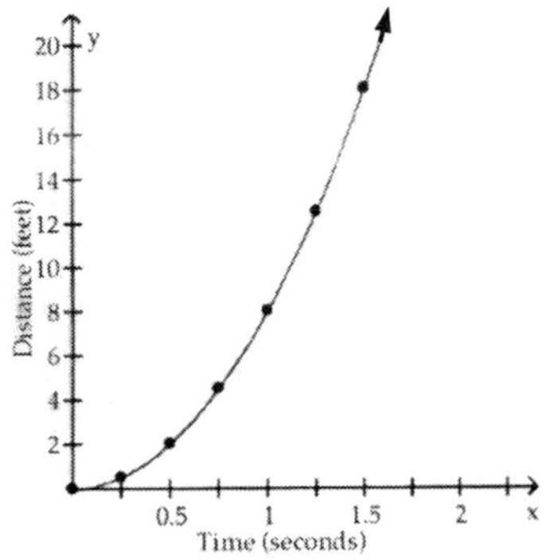

WRITING

89-91. Answers will vary.

REVIEW

Solve each inequality. Write the solution set in interval notation and graph it.

93.
$$-4(3y+2)\le 28$$
$$-4(\mathbf{3y})-4(\mathbf{2})\le 28$$
$$-12y-8\le 28$$
$$-12y-8+\mathbf{8}\le 28+\mathbf{8}$$
$$-12y\le 36$$
$$\frac{-12y}{\mathbf{-12}}\ge\frac{36}{\mathbf{-12}}$$
$$y\ge -3$$
$$[-3,\infty)$$

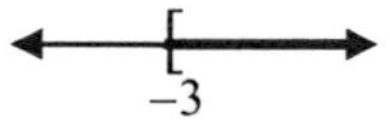

Simplify each expression. Do not use negative exponents in the answer.

95. $\left(x^2x^4\right)^3 = \left(x^{2+4}\right)^3$

$= \left(x^6\right)^3$

$= x^{6\cdot 3}$

$= x^{18}$

97. $\left(\dfrac{y^2y^5}{y^4}\right)^3 = \left(\dfrac{y^{2+5}}{y^4}\right)^3$

$= \left(\dfrac{y^7}{y^4}\right)^3$

$= \left(y^{7-4}\right)^3$

$= \left(y^3\right)^3$

$= y^{3\cdot 3}$

$= y^9$

CHALLENGE PROBLEMS

99. Find a three-term polynomial of degree 2 whose value will be 1 when it is evaluated for $x = 2$. Answers will vary.

Example: $x^2 - x - 1 = (\mathbf{2})^2 - (\mathbf{2}) - 1$

$= 4 - 2 - 1$

$= 2 - 1$

$= 1$

SECTION 5.5 STUDY SET

VOCABULARY

Fill in the blanks.

1. $(b^3 - b^2 - 9b + 1) + (b^3 - b^2 - 9b + 1)$ is the sum of two **polynomials**.

3. **Like** terms have the same variables with the same exponents.

CONCEPTS

Fill in the blanks.

5. To add polynomials, **combine** their like terms.

7. Simplify each polynomial, if possible.

a. $2x^2 + 3x^2 = (2+3)x^2$
$= 5x^2$

b. $15m^3 - m^3 = (15-1)m^3$
$= 14m^3$

c. $8a^3b - a^3b = (8-1)a^3b$
$= 7a^3b$

d. $6cd + 4c^2d$; Will not simplify.

9. Write without parentheses.

a. $-(5x^2 - 8x + 23)$
$\mathbf{-5x^2 + 8x - 23}$

b. $-(-5y^4 + 3y^2 - 7)$
$\mathbf{5y^4 - 3y^2 + 7}$

NOTATION

Fill in the blanks to add (subtract) the polynomials.

11. $(6x^2 + 2x + 3) + (4x^2 - 7x + 1)$
$= (6x^2 + \mathbf{4x^2}) + (\mathbf{2x} - 7x) + (3 + \mathbf{1})$
$= \mathbf{10x^2} - 5x + \mathbf{4}$

GUIDED PRACTICE

Simplify each polynomial and write it in descending powers of one variable. See Example 1.

13. $8t^2 + 4t^2 = (8+4)t^2$
$= 12t^2$

15. $18x^2 - 19x + 2x^2 = (18+2)x^2 - 19x$
$= 20x^2 - 19x$

17. $10x^2 - 8x + 9x - 9x^2 = (10-9)x^2 + (-8+9)x$
$= x^2 + x$

19. $\frac{1}{5}x^2 - \frac{3}{8}x + \frac{2}{3}x^2 + \frac{1}{4}x$
$= \left(\frac{1}{5} + \frac{2}{3}\right)x^2 + \left(-\frac{3}{8} + \frac{1}{4}\right)x$
$= \left(\frac{1}{5}\cdot\frac{\mathbf{3}}{\mathbf{3}} + \frac{2}{3}\cdot\frac{\mathbf{5}}{\mathbf{5}}\right)x^2 + \left(-\frac{3}{8} + \frac{1}{4}\cdot\frac{\mathbf{2}}{\mathbf{2}}\right)x$
$= \left(\frac{3+10}{15}\right)x^2 + \left(\frac{-3+2}{8}\right)x$
$= \left(\frac{13}{15}\right)x^2 + \left(\frac{-1}{8}\right)x$
$= \frac{13}{15}x^2 - \frac{1}{8}x$

21. $0.6x^3 + 0.8x^4 + 0.7x^3 + (-0.8x^4)$
$= [0.8 + (-0.8)]x^4 + (0.6 + 0.7)x^3$
$= 0x^4 + (1.3)x^3$
$= 1.3x^3$

23. $\frac{1}{2}st + \frac{3}{2}st = \left(\frac{1}{2} + \frac{3}{2}\right)st$
$= \frac{4}{2}st$
$= 2st$

25. $-4ab + 4ab - ab = (-4 + 4 - 1)ab$
$= -ab$

27. $4x^2y + 5 - 6x^3y - 3x^2y + 2x^3y$
$= (-6+2)x^3y + (4-3)x^2y + 5$
$= -4x^3y + x^2y + 5$

Add the polynomials. See Example 2.

29. $(3q^2 - 5q + 7) + (2q^2 + q - 12)$
$= (3q^2 + 2q^2) + (-5q + q) + (7 - 12)$
$= 5q^2 - 4q - 5$

31. $\left(\frac{2}{3}y^3+\frac{3}{4}y^2+\frac{1}{2}\right)+\left(\frac{1}{3}y^3+\frac{1}{5}y^2-\frac{1}{6}\right)$

$=\left(\frac{2}{3}+\frac{1}{3}\right)y^3+\left(\frac{3}{4}+\frac{1}{5}\right)y^2+\left(\frac{1}{2}-\frac{1}{6}\right)$

$=\left(\frac{2+1}{3}\right)y^3+\left(\frac{3}{4}\cdot\frac{\mathbf{5}}{\mathbf{5}}+\frac{1}{5}\cdot\frac{\mathbf{4}}{\mathbf{4}}\right)y^2+\left(\frac{1}{2}\cdot\frac{\mathbf{3}}{\mathbf{3}}-\frac{1}{6}\right)$

$=\left(\frac{3}{3}\right)y^3+\left(\frac{15+4}{20}\right)y^2+\left(\frac{3-1}{6}\right)$

$=y^3+\frac{19}{20}y^2+\frac{\overset{1}{\cancel{2}}}{\underset{1}{\cancel{2}}\cdot 3}$

$=y^3+\frac{19}{20}y^2+\frac{1}{3}$

33. $(0.3p+2.1q)+(0.4p-3q)$

$=(0.3+0.4)p+(2.1-3)q$

$=0.7p-0.9q$

35. $(2x^2+xy+3y^2)+(5x^2-y^2)$

$=(2+5)x^2+xy+(3-1)y^2$

$=7x^2+xy+2y^2$

Find a polynomial that represents the perimeter of the figure. See Example 3.

37. $(x^2+3x+1)+(x^2+3x+1)+(x^2-4)$

$=(1+1+1)x^2+(3+3)x+(1+1-4)$

$=3x^2+6x-2$

The perimeter is $(3x^2+6x-2)$ yd.

39. $(2x^2-7)+(x+6)+(5x^2+3x+1)+(x+6)$

$=(2+5)x^2+(1+3+1)x+(-7+6+1+6)$

$=7x^2+5x+6$

The perimeter is $(7x^2+5x+6)$ mi.

Use vertical form to add the polynomials. See Example 4.

41.
$$\begin{array}{r} 3x^2+4x+5 \\ \underline{\mathbf{2x^2-3x+6}} \\ 5x^2+x+11 \end{array}$$

43.
$$\begin{array}{r} 6a^2+7a+9 \\ \underline{-\ 9a^2\qquad -2} \\ -3a^2+7a+7 \end{array}$$

45.
$$\begin{array}{r} z^3+6z^2-7z+16 \\ \underline{9z^3-6z^2+8z-18} \\ 10z^3\qquad +z-2 \end{array}$$

47.
$$\begin{array}{r} -3x^3y^2+4x^2y-4x+9 \\ \underline{2x^3y^2\qquad\qquad +9x-3} \\ -x^3y^2+4x^2y+5x+6 \end{array}$$

Subtract the polynomials. See Example 5.

49. $(3a^2-2a+4)-(a^2-3a+7)$

$=3a^2-2a+4-a^2+3a-7$

$=(3-1)a^2+(-2+3)a+(4-7)$

$=2a^2+a-3$

51. $(-4h^3+5h^2+15)-(h^3-15)$

$=-4h^3+5h^2+15-h^3+15$

$=(-4-1)h^3+5h^2+(15+15)$

$=-5h^3+5h^2+30$

53. $\left(\frac{3}{8}s^8-\frac{3}{4}s^7\right)-\left(\frac{1}{3}s^8+\frac{1}{5}s^7\right)$

$=\frac{3}{8}s^8-\frac{3}{4}s^7-\frac{1}{3}s^8-\frac{1}{5}s^7$

$=\left(\frac{3}{8}-\frac{1}{3}\right)s^8+\left(-\frac{3}{4}-\frac{1}{5}\right)s^7$

$=\left(\frac{3}{8}\cdot\frac{\mathbf{3}}{\mathbf{3}}-\frac{1}{3}\cdot\frac{\mathbf{8}}{\mathbf{8}}\right)s^8+\left(-\frac{3}{4}\cdot\frac{\mathbf{5}}{\mathbf{5}}-\frac{1}{5}\cdot\frac{\mathbf{4}}{\mathbf{4}}\right)s^7$

$=\frac{9-8}{24}s^8+\frac{-15-4}{20}s^7$

$=\frac{1}{24}s^8-\frac{19}{20}s^7$

55. $(5ab+2b^2)-(2+ab+b^2)$

$=5ab+2b^2-2-ab-b^2$

$=(2-1)b^2+(5-1)ab-2$

$=b^2+4ab-2$

Use vertical form to subtract the polynomials. See Example 6.

57.
$$\begin{array}{r} 3x^2+4x+5 \\ \underline{-(2x^2-2x+3)} \end{array} \xrightarrow[\text{and add}]{\text{Change signs}} \begin{array}{r} 3x^2+4x+5 \\ \underline{-2x^2+2x-3} \\ x^2+6x+2 \end{array}$$

59.
$$\begin{array}{r} 5s^2\qquad +9 \\ \underline{-(s^2+4s+2)} \end{array} \xrightarrow[\text{and add}]{\text{Change signs}} \begin{array}{r} 5s^2\qquad +9 \\ \underline{-s^2-4s-2} \\ 4s^2-4s+7 \end{array}$$

61.
$$\begin{array}{r} 17a^3\qquad +25a-10 \\ \underline{-(8a^3+8a^2-3a+1)} \end{array} \xrightarrow[\text{and add}]{\text{Change signs}} \begin{array}{r} 17a^3\qquad +25a-10 \\ \underline{-8a^3-8a^2+3a-1} \\ 9a^3-8a^2+28a-11 \end{array}$$

63. $$\begin{array}{r} 0.8x^3 \qquad\qquad -2.3x+0.6 \\ -\underline{(0.2x^3-1.2x^2-3.6x+0.9)} \end{array}$$

Change the signs of the 2nd line and add

$$\begin{array}{r} 0.8x^3 \qquad\qquad -2.3x+0.6 \\ -\underline{0.2x^3+1.2x^2+3.6x-0.9} \\ 0.6x^3+1.2x^2+1.3x-0.3 \end{array}$$

Perform the operations. See Example 7.

65. $(-2x^2-7x+1)+(-4x^2+8x-1)$

$=(-2-4)x^2+(-7+8)x+(1-1)$

$=-6x^2+x$

$(-6x^2+x)-(3x^2+4x-7)$

$=-6x^2+x-3x^2-4x+7$

$=(-6-3)x^2+(1-4)x+7$

$=-9x^2-3x+7$

67. $(3t^3+t^2)+(-t^3+6t-3)$

$=(3-1)t^3+t^2+6t-3$

$=2t^3+t^2+6t-3$

$(2t^3+t^2+6t-3)-(t^3-2t^2+2)$

$=2t^3+t^2+6t-3-t^3+2t^2-2$

$=(2-1)t^3+(1+2)t^2+6t+(-3-2)$

$=t^3+3t^2+6t-5$

TRY IT YOURSELF

Perform the operations.

69. $(9a^2+3a)-(2a-4a^2)=9a^2+3a-2a+4a^2$

$=(9+4)a^2+(3-2)a$

$=13a^2+a$

71. $(2y^5-y^4)-(-y^5+5y^4-1.2)$

$=2y^5-y^4+y^5-5y^4+1.2$

$=(2+1)y^5+(-1-5)y^4+1.2$

$=3y^5-6y^4+1.2$

73. $3r^4-4r+7r^4=(3+7)r^4-4r$

$=10r^4-4r$

75. $(0.03f^2+0.25f+0.91)-(0.17f^2-1.18)$

$=0.03f^2+0.25f+0.91-0.17f^2+1.18$

$=(0.03-0.17)f^2+0.25f+(0.91+1.18)$

$=-0.14f^2+0.25f+2.09$

77. $\left(\frac{7}{8}r^4+\frac{5}{9}r^2-\frac{9}{4}\right)-\left(-\frac{3}{8}r^4-\frac{2}{3}r^2-\frac{1}{4}\right)$

$=\frac{7}{8}r^4+\frac{5}{9}r^2-\frac{9}{4}+\frac{3}{8}r^4+\frac{2}{3}r^2+\frac{1}{4}$

$=\left(\frac{7}{8}+\frac{3}{8}\right)r^4+\left(\frac{5}{9}+\frac{2}{3}\right)r^2+\left(-\frac{9}{4}+\frac{1}{4}\right)$

$=\left(\frac{7}{8}+\frac{3}{8}\right)r^4+\left(\frac{5}{9}+\frac{2}{3}\cdot\frac{3}{3}\right)r^2+\left(-\frac{9}{4}+\frac{1}{4}\right)$

$=\frac{7+3}{8}r^4+\frac{5+6}{9}r^2+\frac{-9+1}{4}$

$=\frac{10}{8}r^4+\frac{11}{9}r^2+\frac{-8}{4}$

$=\frac{\cancel{2}\cdot 5}{\cancel{2}\cdot 4}r^4+\frac{11}{9}r^2-2$

$=\frac{5}{4}r^4+\frac{11}{9}r^2-2$

79. $$\begin{array}{r} 8c^2-4c-5 \\ -\underline{(-c^2+2c+9)} \end{array} \xrightarrow[\text{and add}]{\text{Change signs}} \begin{array}{r} 8c^2-4c-5 \\ \underline{c^2-2c-9} \\ 9c^2-6c-14 \end{array}$$

81. $(12.1h^3+9.9h^2)+(7.3h^3+1.1h^2)$

$=(12.1+7.3)h^3+(9.9+1.1)h^2$

$=19.4h^3+11h^2$

83. $(20-4rt-5r^2t)+(10-5rt)$

$=-5r^2t+(-4rt-5rt)+(20+10)$

$=-5r^2t+(-4-5)rt+(30)$

$=-5r^2t-9rt+30$

85. $(3x^2-3x-2)+(3x^2+4x-3)$

$=3x^2-3x-2+3x^2+4x-3$

$=(3+3)x^2+(-3+4)x+(-2-3)$

$=6x^2+x-5$

87. $\frac{2}{3}d^2-\frac{1}{4}c^2+\frac{5}{6}c^2-\frac{1}{2}cd+\frac{1}{3}d^2$

$=\left(-\frac{1}{4}+\frac{5}{6}\right)c^2-\frac{1}{2}cd+\left(\frac{2}{3}+\frac{1}{3}\right)d^2$

$=\left(-\frac{1}{4}\cdot\frac{\mathbf{3}}{\mathbf{3}}+\frac{5}{6}\cdot\frac{\mathbf{2}}{\mathbf{2}}\right)c^2-\frac{1}{2}cd+\left(\frac{2}{3}+\frac{1}{3}\right)d^2$

$=\left(\frac{-3+10}{12}\right)c^2-\frac{1}{2}cd+\left(\frac{2+1}{3}\right)d^2$

$=\frac{7}{12}c^2-\frac{1}{2}cd+\frac{3}{3}d^2$

$=\frac{7}{12}c^2-\frac{1}{2}cd+d^2$

89. $(3x+7)+(4x-3)=(3+4)x+(7-3)$

$=7x+4$

91. $(-2.7t^2+2.1t-1.7)+(3.1t^2-2.5t+2.3)$

$=(-2.7+3.1)t^2+(2.1-2.5)t+(-1.7+2.3)$

$=0.4t^2-0.4t+0.6$

$(0.4t^2-0.4t+0.6)-(1.7t^2-1.1t)$

$=0.4t^2-0.4t+0.6-1.7t^2+1.1t$

$=(0.4-1.7)t^2+(-0.4+1.1)t+0.6$

$=-1.3t^2+0.7t+0.6$

93. $-32u^3-16u^3=(-32-16)u^3$

$=-48u^3$

95. $(9d^2+6d)+(8d-4d^2)$

$=(9-4)d^2+(6+8)d$

$=5d^2+14d$

97.
$$\begin{array}{r} 3x^3y^2+4x^2y+7x+12 \\ -\underline{(-4x^3y^2+6x^2y+9x-3)} \end{array}$$

Change the signs of the 2nd line and add

$$\begin{array}{r} 3x^3y^2+4x^2y+7x+12 \\ \underline{4x^3y^2-6x^2y-9x\quad+3} \\ 7x^3y^2-2x^2y-2x+15 \end{array}$$

99. $(2x^2-3x+1)-(4x^2-3x+2)+(2x^2+3x+2)$

$=2x^2-3x+1-4x^2+3x-2+2x^2+3x+2$

$=(2-4+2)x^2+(-3+3+3)x+(1-2+2)$

$=0x^2+3x+1$

$=3x+1$

101.
$$\begin{array}{r} 4x^3+4x^2-3x+10 \\ +\underline{(5x^3-2x^2-4x-4)} \\ 9x^3+2x^2-7x+\ 6 \end{array}$$

LOOK ALIKES . . .

103. a. $(-8x^2-3x)+(-11x^2+6x+10)$

$=(-8-11)x^2+(-3+6)x+10$

$=-19x^2+3x+10$

b. $(-8x^2-3x)-(-11x^2+6x+10)$

$=-8x^2-3x+11x^2-6x-10$

$=(-8+11)x^2+(-3-6)x-10$

$=3x^2-9x-10$

APPLICATIONS

105. GREEK ARCHITECTURE

a. $(x^2-3x+2)-(5x-10)$

$=x^2-3x+2-5x+10$

$=x^2+(-3-5)x+(2+10)$

$=x^2-8x+12$

The difference in the heights is $(x^2-8x+12)$ ft.

b. $(x^2-3x+2)+(5x-10)$

$=x^2-3x+2+5x-10$

$=x^2+(-3+5)x+(2-10)$

$=x^2+2x-8$

The sum of the heights is (x^2+2x-8) ft.

107. PIÑATAS

$(4a^2+6a-1)-(2a^2-6)$

$=4a^2+6a-1-2a^2+6$

$=(4-2)a^2+6a+(-1+6)$

$=2a^2+6a+5$

The length of the rope is $(2a^2+6a+5)$ in.

109. NAVAL OPERATIONS

a. $(-16t^2+150t+40)-(-16t^2+128t+20)$

$=-16t^2+150t+40+16t^2-128t-20$

$=(-16+16)t^2+(150-128)t+(40-20)$

$=0t^2+22t+20$

$=22t+20$

The difference in the heights is $(22t+20)$ ft.

b. $22t+20=22(\mathbf{4})+20$

$=88+20$

$=108$

The second flare is 108 ft higher.

WRITING

111-115. Answers will vary.

REVIEW

117. What is the sum of the measures of the angles of a triangle? **180°**

119. Graph: $y=-\frac{1}{2}x+2$

y-intercept is (0,2), $m=\frac{-1}{2}$

Start at (0,2), go down 1, right 2

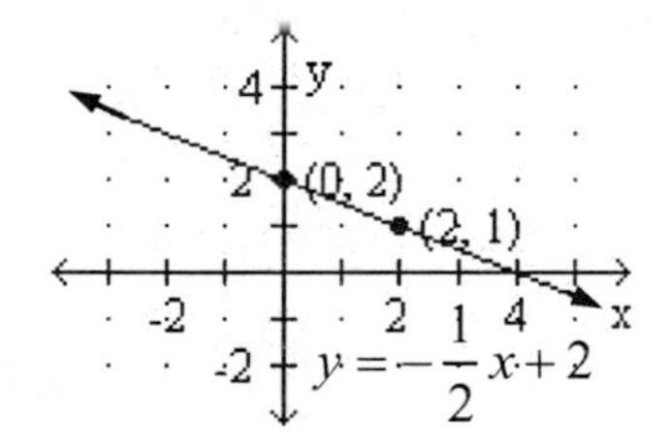

CHALLENGE PROBLEMS

121. $(6x^2-7x-8)-(2x^2-x+3)$

$=6x^2-7x-8-2x^2+x-3$

$=(6-2)x^2+(-7+1)x+(-8-3)$

$=4x^2-6x-11$

The needed polynomial is $(4x^2-6x-11)$.

SECTION 5.6 STUDY SET
VOCABULARY

Fill in the blanks.

1. $(2x^3)(3x^4)$ is the product of two **monomials** and $(2a-4)(3a+5)$ is the product of two **binomials**.

3. In the acronym FOIL, F stands for **first** terms, O for **outer** terms, I for **inner** terms, and L for **last** terms.

CONCEPTS

Fill in the blanks.

5. a. To multiply two polynomials, multiply **each** term of one polynomial by **each** term of the other polynomial, and then combine like terms.

   b. When multiplying three polynomials, we begin by multiplying **any** two of them, and then we multiply that result by the **third** polynomial.

7. Simplify each polynomial by combining like terms.

   a. $6x^2-8x+9x-12$ $\quad \mathbf{6x^2+x-12}$

   b. $5x^4+3ax^2+5ax^2+3a^2$ $\quad \mathbf{5x^4+8ax^2+3a^2}$

NOTATION

Complete each solution.

9. $(9n^3)(8n^2)=(9\cdot\mathbf{8})(\mathbf{n^3}\cdot n^2)=\mathbf{72n^5}$

11. $(2x+5)(3x-2)=2x(3x)-\mathbf{2x}(2)+\mathbf{5}(3x)-\mathbf{5}(2)$
$=6x^2-\mathbf{4x}+\mathbf{15x}-10$
$=6x^2+\mathbf{11x}-10$

GUIDED PRACTICE

Multiply. See Example 1.

13. $5m(m)=5m^{1+1}$
$=5m^2$

15. $(3x^2)(4x^3)=(3\cdot4)(x^2x^3)$
$=12(x^{2+3})$
$=12x^5$

17. $(1.2c^3)(5c^3)=(1.2\cdot5)(c^3c^3)$
$=6(c^{3+3})$
$=6c^6$

19. $(3b^2)(-2b)(4b^3)=(3\cdot-2\cdot4)(b^2bb^3)$
$=-24(b^{2+1+3})$
$=-24b^6$

21. $(2x^2y^3)(4x^3y^2)=(2\cdot4)(x^2x^3)(y^3y^2)$
$=8(x^{2+3})(y^{3+2})$
$=8x^5y^5$

23. $(8a^5)\left(-\frac{1}{4}a^6\right)=\left(8\cdot-\frac{1}{4}\right)(a^5a^6)$
$=-2(a^{5+6})$
$=-2a^{11}$

Multiply. See Example 2.

25. $3x(x+4)=\mathbf{3x}(x)+\mathbf{3x}(4)$
$=3(x^{1+1})+12x$
$=3x^2+12x$

27. $-4t(t^2-7)=\mathbf{-4t}(t^2)-\mathbf{4t}(-7)$
$=-4(t^{1+2})+28t$
$=-4t^3+28t$

29. $-2x^3(3x^2-x+1)$
$=\mathbf{-2x^3}(3x^2)-\mathbf{2x^3}(-x)-\mathbf{2x^3}(1)$
$=-6(x^{3+2})+2(x^{3+1})-2x^3$
$=-6x^5+2x^4-2x^3$

31. $\frac{5}{8}t^2(t^6+8t^2)=\mathbf{\frac{5}{8}t^2}(t^6)+\mathbf{\frac{5}{8}t^2}(8t^2)$
$=\frac{5}{8}(t^{2+6})+5(t^{2+2})$
$=\frac{5}{8}t^8+5t^4$

33. $-4x^2z(3x^2+z^2+xz-1)$
$=\mathbf{-4x^2z}(3x^2)-\mathbf{4x^2z}(z^2)-\mathbf{4x^2z}(xz)-\mathbf{4x^2z}(-1)$
$=-12(x^{2+2}z)-4(x^2z^{1+2})-4(x^{2+1}z^{1+1})+4x^2z$
$=-12x^4z-4x^2z^3-4x^3z^2+4x^2z$

35. $(x^2-12x)(6x^{12})=x^2(\mathbf{6x^{12}})-12x(\mathbf{6x^{12}})$
$=6(x^{2+12})-72(x^{1+12})$
$=6x^{14}-72x^{13}$

Find a polynomial that represents the area of the parallelogram or rectangle. See Example 3.

37. $A=bh$
$=(7h+3)h$
$=(7h)\mathbf{h}+(3)\mathbf{h}$
$=7h^{1+1}+3h$
$=7h^2+3h$

The area is $(7h^2+3h)$ in^2.

39. $A = bh$
$= (4w-2)w$
$= (4w)\mathbf{w} - (2)\mathbf{w}$
$= 4w^{1+1} - 2w$
$= 4w^2 - 2w$

The area is $(4w^2 - 2w)$ ft^2.

Multiply. See Examples 4 and 5.

41. $(y+3)(y+5) = \mathbf{y}(y) + \mathbf{y}(5) + \mathbf{3}(y) + \mathbf{3}(5)$
$= y^2 + 5y + 3y + 15$
$= y^2 + 8y + 15$

43. $(m+6)(m-9) = \mathbf{m}(m) + \mathbf{m}(-9) + \mathbf{6}(m) + \mathbf{6}(-9)$
$= m^2 - 9m + 6m - 54$
$= m^2 - 3m - 54$

45. $(4y-5)(y+7) = \mathbf{4y}(y) + \mathbf{4y}(7) - \mathbf{5}(y) - \mathbf{5}(7)$
$= 4y^2 + 28y - 5y - 35$
$= 4y^2 + 23y - 35$

47. $(2x-3)(6x-5)$
$= \mathbf{2x}(6x) + \mathbf{2x}(-5) - \mathbf{3}(6x) - \mathbf{3}(-5)$
$= 12x^2 - 10x - 18x + 15$
$= 12x^2 - 28x + 15$

49. $(3.8y-1)(2y-1)$
$= \mathbf{3.8y}(2y) + \mathbf{3.8y}(-1) - \mathbf{1}(2y) - \mathbf{1}(-1)$
$= 7.6y^2 - 3.8y - 2y + 1$
$= 7.6y^2 - 5.8y + 1$

51. $\left(6m - \frac{2}{3}\right)\left(3m - \frac{4}{3}\right)$

$= \mathbf{6m}(3m) + \mathbf{6m}\left(-\frac{4}{3}\right) + \left(-\frac{\mathbf{2}}{\mathbf{3}}\right)3m + \left(-\frac{\mathbf{2}}{\mathbf{3}}\right)\left(-\frac{4}{3}\right)$

$= 18m^2 - 8m - 2m + \frac{8}{9}$

$= 18m^2 - 10m + \frac{8}{9}$

53. $(t^2-3)(t^2-4) = \mathbf{t^2}(t^2) + \mathbf{t^2}(-4) - \mathbf{3}(t^2) - \mathbf{3}(-4)$
$= t^4 - 4t^2 - 3t^2 + 12$
$= t^4 - 7t^2 + 12$

55. $(3a-2b)(4a+b)$
$= \mathbf{3a}(4a) + \mathbf{3a}(b) - \mathbf{2b}(4a) - \mathbf{2b}(b)$
$= 12a^2 + 3ab - 8ab - 2b^2$
$= 12a^2 - 5ab - 2b^2$

Multiply. See Example 6.

57. $(x+2)(x^2-2x+3)$
$= \mathbf{x}(x^2-2x+3) + \mathbf{2}(x^2-2x+3)$
$= \mathbf{x}(x^2) + \mathbf{x}(-2x) + \mathbf{x}(3) + \mathbf{2}(x^2) + \mathbf{2}(-2x) + \mathbf{2}(3)$
$= x^3 - 2x^2 + 3x + 2x^2 - 4x + 6$
$= x^3 + (-2+2)x^2 + (3-4)x + 6$
$= x^3 + 0x^2 - x + 6$
$= x^3 - x + 6$

59. $(4t+3)(t^2+2t+3)$
$= \mathbf{4t}(t^2+2t+3) + \mathbf{3}(t^2+2t+3)$
$= \mathbf{4t}(t^2) + \mathbf{4t}(2t) + \mathbf{4t}(3) + \mathbf{3}(t^2) + \mathbf{3}(2t) + \mathbf{3}(3)$
$= 4t^3 + 8t^2 + 12t + 3t^2 + 6t + 9$
$= 4t^3 + (8+3)t^2 + (12+6)t + 9$
$= 4t^3 + 11t^2 + 18t + 9$

61. $(x^2+6x+7)(2x-5)$
$= (\mathbf{x^2} + \mathbf{6x} + \mathbf{7})2x - (\mathbf{x^2} + \mathbf{6x} + \mathbf{7})5$
$= \mathbf{x^2}(2x) + \mathbf{6x}(2x) + \mathbf{7}(2x) - [\mathbf{x^2}(5) + \mathbf{6x}(5) + \mathbf{7}(5)]$
$= 2x^3 + 12x^2 + 14x - [5x^2 + 30x + 35]$
$= 2x^3 + 12x^2 + 14x - 5x^2 - 30x - 35$
$= 2x^3 + (12-5)x^2 + (14-30)x - 35$
$= 2x^3 + 7x^2 - 16x - 35$

63. $(r^2 - r + 3)(r^2 - 4r - 5)$
$= \mathbf{r^2}(r^2 - 4r - 5) - \mathbf{r}(r^2 - 4r - 5) + \mathbf{3}(r^2 - 4r - 5)$
$= \mathbf{r^2}(r^2) + \mathbf{r^2}(-4r) + \mathbf{r^2}(-5) - \mathbf{r}(r^2) - \mathbf{r}(-4r) - \mathbf{r}(-5)$
$+ \mathbf{3}(r^2) + \mathbf{3}(-4r) + \mathbf{3}(-5)$
$= r^4 - 4r^3 - 5r^2 - r^3 + 4r^2 + 5r + 3r^2 - 12r - 15$
$= r^4 + (-4-1)r^3 + (-5+4+3)r^2 + (5-12)r - 15$
$= r^4 - 5r^3 + 2r^2 - 7r - 15$

Multiply using vertical form. See Example 7.

65.
$$\begin{array}{r} x^2 - 2x + 1 \\ \underline{x + 2} \\ 2x^2 - 4x + 2 \\ \underline{x^3 - 2x^2 + x \quad\;\;} \\ x^3 \qquad\quad - 3x + 2 \end{array}$$

67.
$$\begin{array}{r} 4x^2 + 3x - 4 \\ \underline{3x + 2} \\ 8x^2 + 6x - 8 \\ \underline{12x^3 + 9x^2 - 12x \quad\;\;} \\ 12x^3 + 17x^2 - 6x - 8 \end{array}$$

Multiply. See Example 8.

69. $4x(2x+1)(x-2)$
$= 4x[\mathbf{2x}(x) + \mathbf{2x}(-2) + \mathbf{1}(x) + \mathbf{1}(-2)]$
$= 4x[2x^2 - 4x + x - 2]$
$= 4x[2x^2 - 3x - 2]$
$= \mathbf{4x}(2x^2) + \mathbf{4x}(-3x) + \mathbf{4x}(-2)$
$= 8x^3 - 12x^2 - 8x$

71. $-3a(a+b)(a-b)$
$= -3a[\mathbf{a}(a) + \mathbf{a}(-b) + \mathbf{b}(a) + \mathbf{b}(-b)]$
$= -3a[a^2 - ab + ab - b^2]$
$= -3a[a^2 + 0ab - b^2]$
$= \mathbf{-3a}(a^2) \mathbf{-3a}(-b^2)$
$= -3a^3 + 3ab^2$

73. $(-2a^2)(-3a^3)(3a-2)$
$= (\mathbf{-2})(\mathbf{-3})(\boldsymbol{a}^{\mathbf{2+3}})(3a-2)$
$= 6a^5(3a-2)$
$= \mathbf{6a^5}(3a) + \mathbf{6a^5}(-2)$
$= (6)(3)a^{5+1} + (6)(-2)a^5$
$= 18a^6 - 12a^5$

75. $(x-4)(x+1)(x-3)$
$= (x-4)[\mathbf{x}(x) + \mathbf{x}(-3) + \mathbf{1}(x) + \mathbf{1}(-3)]$
$= (x-4)[x^2 - 3x + x - 3]$
$= (x-4)[x^2 - 2x - 3]$
$= \mathbf{x}(x^2) - \mathbf{x}(2x) - \mathbf{x}(3) - \mathbf{4}(x^2) - \mathbf{4}(-2x) - \mathbf{4}(-3)$
$= x^{2+1} - 2x^{1+1} - 3x - 4x^2 + 8x + 12$
$= x^3 - 2x^2 - 3x - 4x^2 + 8x + 12$
$= x^3 + (-2-4)x^2 + (-3+8)x + 12$
$= x^3 - 6x^2 + 5x + 12$

Try It Yourself

Multiply.

77. $(5x-2)(6x-1)$
$= \mathbf{5x}(6x) + \mathbf{5x}(-1) - \mathbf{2}(6x) - \mathbf{2}(-1)$
$= 30x^2 - 5x - 12x + 2$
$= 30x^2 + (-5-12)x + 2$
$= 30x^2 - 17x + 2$

79. $(3x^2 + 4x - 7)(2x^2)$
$= 3x^2(\mathbf{2x^2}) + 4x(\mathbf{2x^2}) - 7(\mathbf{2x^2})$
$= 6x^{2+2} + 8x^{1+2} - 14x^2$
$= 6x^4 + 8x^3 - 14x^2$

81. $2(t+4)(t-3) = 2[\,\mathbf{t}(t) + \mathbf{t}(-3) + \mathbf{4}(t) + \mathbf{4}(-3)]$
$= 2(t^2 - 3t + 4t - 12)$
$= 2(t^2 + t - 12)$
$= 2t^2 + 2t - 24$

83.
$$\begin{array}{r} 2a^2 + 3a + 1 \\ \underline{3a^2 - 2a + 4} \\ 8a^2 + 12a + 4 \\ -4a^3 - 6a^2 - 2a \\ \underline{6a^4 + 9a^3 + 3a^2} \\ 6a^4 + 5a^3 + 5a^2 + 10a + 4 \end{array}$$

85. $(t+2s)(9t-3s)$
$= \mathbf{t}(9t) + \mathbf{t}(-3s) + \mathbf{2s}(9t) + \mathbf{2s}(-3s)$
$= 9t^2 - 3st + 18st - 6s^2$
$= 9t^2 + 15st - 6s^2$

87. $\left(\frac{1}{2}a\right)(4a^4)(a^5) = \left(\frac{1}{2}\cdot 4\right)(a \cdot a^4 \cdot a^5)$
$= 2a^{1+4+5}$
$= 2a^{10}$

89. $\left(4a - \frac{5}{4}r\right)\left(4a + \frac{3}{4}r\right)$
$= \mathbf{4a}(4a) + \mathbf{4a}\left(\frac{3}{4}r\right) + \left(-\frac{\mathbf{5}}{\mathbf{4}}\mathbf{r}\right)4a + \left(-\frac{\mathbf{5}}{\mathbf{4}}\mathbf{r}\right)\left(\frac{3}{4}r\right)$
$= 16a^2 + 3ar - 5ar - \frac{15}{16}r^2$
$= 16a^2 + (3-5)ar - \frac{15}{16}r^2$
$= 16a^2 - 2ar - \frac{15}{16}r^2$

91. $(a+b)(a+b) = \mathbf{a}(a) + \mathbf{a}(b) + \mathbf{b}(a) + \mathbf{b}(b)$
$= a^2 + ab + ab + b^2$
$= a^2 + 2ab + b^2$

93. $(x+6)(x^3 + 5x^2 - 4x - 4)$
$= \mathbf{x}(x^3 + 5x^2 - 4x - 4) + \mathbf{6}(x^3 + 5x^2 - 4x - 4)$
$= \mathbf{x}(x^3) + \mathbf{x}(5x^2) + \mathbf{x}(-4x) + \mathbf{x}(-4)$
$+ \mathbf{6}(x^3) + \mathbf{6}(5x^2) + \mathbf{6}(-4x) + \mathbf{6}(-4)$
$= x^4 + 5x^3 - 4x^2 - 4x + 6x^3 + 30x^2 - 24x - 24$
$= x^4 + (5+6)x^3 + (-4+30)x^2 + (-4-24)x - 24$
$= x^4 + 11x^3 + 26x^2 - 28x - 24$

95. $9x^2(x^2 - 2x + 6)$
$= \mathbf{9x^2}(x^2) + \mathbf{9x^2}(-2x) + \mathbf{9x^2}(6)$
$= 9(x^{2+2}) - 18(x^{2+1}) + 54x^2$
$= 9x^4 - 18x^3 + 54x^2$

97. $4y(y+3)(y+7)$
$= 4y[\mathbf{y}(y) + \mathbf{y}(7) + \mathbf{3}(y) + \mathbf{3}(7)]$
$= 4y[y^2 + 7y + 3y + 21]$
$= 4y[y^2 + 10y + 21]$
$= \mathbf{4y}(y^2) + \mathbf{4y}(10y) + \mathbf{4y}(21)$
$= 4y^3 + 40y^2 + 84y$

99. $0.3p^5(0.4p^4 - 6p^2)$
$= \mathbf{0.3p^5}(0.4p^4) + \mathbf{0.3p^5}(-6p^2)$
$= 0.12(p^{5+4}) - 1.8(p^{5+2})$
$= 0.12p^9 - 1.8p^7$

101. $8.2pq(2pq-3p+5q)$
$= \mathbf{8.2pq}(2pq)-\mathbf{8.2pq}(3p)+\mathbf{8.2pq}(5q)$
$= 16.4p^2q^2-24.6p^2q+41pq^2$

103. $(-3x+y)(x^2-8xy+16y^2)$
$= -\mathbf{3x}(x^2-8xy+16y^2)+\mathbf{y}(x^2-8xy+16y^2)$
$= -\mathbf{3x}(x^2)-\mathbf{3x}(-8xy)-\mathbf{3x}(16y^2)$
$\quad +\mathbf{y}(x^2)+\mathbf{y}(-8xy)+\mathbf{y}(16y^2)$
$= -3x^3+24x^2y-48xy^2+x^2y-8xy^2+16y^3$
$= -3x^3+(24+1)x^2y+(-48-8)xy^2+16y^3$
$= -3x^3+25x^2y-56xy^2+16y^3$

LOOK ALIKES . . .

Perform the indicated operations to simplify each expression, if possible.

105. a. $(x-2)+(x^2+2x+4)$
$= x^2+(1+2)x+(-2+4)$
$= x^2+3x+2$

105. b.
$$\begin{array}{r} x^2+2x+4 \\ \underline{x-2} \\ -2x^2-4x-8 \\ \underline{x^3+2x^2+4x \quad\quad} \\ x^3 \quad\quad\quad\quad -8 \end{array}$$

107. a. $(6x^2z^5)-(-3xz^3)$; Does not simplify.

b. $(6x^2z^5)(-3xz^3)=(6\cdot -3)(x^2x)(z^5z^3)$
$= -18x^{2+1}z^{5+3}$
$= -18x^3z^8$

109. a. $(2x^2-x)-(3x^2-3x)$
$= 2x^2-x-3x^2+3x$
$= (2-3)x^2+(-1+3)x$
$= -x^2+2x$

b. $(2x^2-x)(3x^2-3x)$
$= \mathbf{2x^2}(3x^2-3x)-\mathbf{x}(3x^2-3x)$
$= \mathbf{2x^2}(3x^2)+\mathbf{2x^2}(-3x)-\mathbf{x}(3x^2)-\mathbf{x}(-3x)$
$= 6x^{2+2}-6x^{2+1}-3x^{2+1}+3x^{1+1}$
$= 6x^4-6x^3-3x^3+3x^2$
$= 6x^4-9x^3+3x^2$

111. a. $3a+(4a-1)+(6a+2)$
$= (3+4+6)a+(-1+2)$
$= 13a+1$

b. $3a(4a-1)(6a+2)$
$= 3a\left[\mathbf{4a}(6a+2)-\mathbf{1}(6a+2)\right]$
$= 3a\left[\mathbf{4a}(6a)+\mathbf{4a}(2)-\mathbf{1}(6a)-\mathbf{1}(2)\right]$
$= 3a\left[24a^2+8a-6a-2\right]$
$= 3a\left[24a^2+2a-2\right]$
$= \mathbf{3a}(24a^2)+\mathbf{3a}(2a)-\mathbf{3a}(2)$
$= 72a^3+6a^2-6a$

APPLICATIONS

113. STAMPS
$A = lw$
$= (3x-1)(2x+1)$
$= \mathbf{3x}(2x)+\mathbf{3x}(1)-\mathbf{1}(2x)-\mathbf{1}(1)$
$= 6x^2+3x-2x-1$
$= 6x^2+(3-2)x-1$
$= 6x^2+x-1$
The area is $(6x^2+x-1)$ cm^2.

115. SUNGLASSES
$A = 0.785ab$
$= 0.785(x+1)(x-1)$
$= 0.785[\mathbf{x}(x)-\mathbf{x}(1)+\mathbf{1}(x)-\mathbf{1}(1)]$
$= 0.785[x^2+x-x-1]$
$= 0.785[x^2+(1-1)x-1]$
$= 0.785(x^2-1)$
$= 0.785x^2-0.785$
The area is $(0.785x^2-0.785)$ in^2.

117. LUGGAGE
$V = lwh$
$= (x-3)(x)(2x+2)$
$= x(x-3)(2x+2)$
$= x[\mathbf{x}(2x)+\mathbf{x}(2)-\mathbf{3}(2x)-\mathbf{3}(2)]$
$= x[2x^2+2x-6x-6]$
$= x[2x^2+(2-6)x-6]$
$= x(2x^2-4x-6)$
$= \mathbf{x}(2x^2)-\mathbf{x}(4x)-\mathbf{x}(6)$
$= 2x^{1+2}-4x^{1+1}-6x$
$= 2x^3-4x^2-6x$
The volume is $(2x^3-4x^2-6x)$ in^3.

WRITING

119-123. Answers will vary.

REVIEW

125. a. $m = \dfrac{\text{rise}}{\text{run}}$

Start at point $(-2, 0)$. Move up the axis one space (rise), then move to the right one space (run) to arrive at the blue line again.

$$m = \frac{1}{1}$$
$$= 1$$

The slope of the blue line is 1.

b. Line 2 is a vertical line. Its slope is always undefined.

c. $m = \dfrac{\text{rise}}{\text{run}}$

Start at point $(-1, 0)$. Move down the axis two spaces (rise), then move to the right three spaces (run) to arrive at the red line again.

$$m = \frac{-2}{3}$$
$$= -\frac{2}{3}$$

The slope of the red line is $-\dfrac{2}{3}$.

d. The x-axis is a horizontal line. Its slope is always zero.

CHALLENGE PROBLEMS

127. a. Find each of the following products.

i. $(x-1)(x+1) = x^2 - 1$

ii. $(x-1)(x^2+x+1)$
$$= \mathbf{x}(x^2+x+1) - \mathbf{1}(x^2+x+1)$$
$$= x^3 + x^2 + x - x^2 - x - 1$$
$$= x^3 + (1-1)x^2 + (1-1)x - 1$$
$$= x^3 - 1$$

iii. $(x-1)(x^3+x^2+x+1)$
$$= \mathbf{x}(x^3+x^2+x+1) - \mathbf{1}(x^3+x^2+x+1)$$
$$= x^4 + x^3 + x^2 + x - x^3 - x^2 - x - 1$$
$$= x^4 + (1-1)x^3 + (1-1)x^2 + (1-1)x - 1$$
$$= x^4 - 1$$

b. Write a product of two polynomials such that the result is $x^5 - 1$.

$(x-1)(x^4+x^3+x^2+x+1)$

SECTION 5.7 STUDY SET

VOCABULARY

Fill in the blanks.

1. Expressions of the form $(x+y)^2$, $(x-y)^2$, and $(x+y)(x-y)$ occur so frequently in algebra that they are called special **products**.

CONCEPTS

3. Complete each special product.

a. $(x+y)^2 = x + 2xy + y^2$

The **square** of the second term

Twice the product of the first and second terms

The square of the **first** term

b. $(x+y)(x-y) = x^2 - y^2$.

The square of the **second** term

The **square** of the first term

NOTATION

Complete each solution to find the product.

5. $(x+4)^2 = \mathbf{x^2} + 2(x)(\mathbf{4}) + \mathbf{4^2}$
$= x^2 + \mathbf{8x} + 16$

7. $(s+5)(s-5) = \mathbf{s^2} - \mathbf{5^2}$
$= s^2 - \mathbf{25}$

GUIDED PRACTICE

Find each product. See Example 1.

9. $(x+1)^2 = x^2 + 2(x)(1) + (1)^2$
$= x^2 + 2x + 1$

11. $(m-6)^2 = m^2 + 2(m)(-6) + (-6)^2$
$= m^2 - 12m + 36$

13. $(4x+5)^2 = (4x)^2 + 2(4x)(5) + (5)^2$
$= 16x^2 + 40x + 25$

15. $(7m-2)^2 = (7m)^2 + 2(7m)(-2) + (-2)^2$
$= 49m^2 - 28m + 4$

17. $(1-3y)^2 = (1)^2 + 2(1)(-3y) + (-3y)^2$
$= 1 - 6y + 9y^2$

19. $(y+0.9)^2 = y^2 + 2(y)(0.9) + (0.9)^2$
$= y^2 + 1.8y + 0.81$

21. $(a^2+b^2)^2 = (a^2)^2 + 2(a^2)(b^2) + (b^2)^2$
$= a^{2\cdot 2} + 2a^2b^2 + b^{2\cdot 2}$
$= a^4 + 2a^2b^2 + b^4$

23. $\left(s+\frac{3}{4}\right)^2 = s^2 + 2(s)\left(\frac{3}{4}\right) + \left(\frac{3}{4}\right)^2$
$= s^2 + \frac{3}{2}s + \frac{9}{16}$

Find each product. See Example 2.

25. $(x+3)(x-3) = x^2 - (3)^2$
$= x^2 - 9$

27. $(2p+7)(2p-7) = (2p)^2 - (7)^2$
$= 4p^2 - 49$

29. $(3n+1)(3n-1) = (3n)^2 - (1)^2$
$= 9n^2 - 1$

31. $\left(c+\frac{3}{4}\right)\left(c-\frac{3}{4}\right) = c^2 - \left(\frac{3}{4}\right)^2$
$= c^2 - \frac{9}{16}$

33. $(0.4-9m^2)(0.4+9m^2) = (0.4)^2 - (9m^2)^2$
$= 0.16 - 81m^{2\cdot 2}$
$= 0.16 - 81m^4$

35. $(5-6g)(5+6g) = 5^2 - (6g)^2$
$= 25 - 36g^2$

Expand each binomial. See Example 3.

37. $(x+4)^3$
$= (x+4)^2(x+4)$
$= [x^2 + 2(x)(4) + (4)^2](x+4)$
$= (x^2 + 8x + 16)(x+4)$
$= \mathbf{x^2}(x) + \mathbf{x^2}(4) + \mathbf{8x}(x) + \mathbf{8x}(4) + \mathbf{16}(x) + \mathbf{16}(4)$
$= x^3 + 4x^2 + 8x^2 + 32x + 16x + 64$
$= x^3 + (\mathbf{4} + \mathbf{8})x^2 + (\mathbf{32} + \mathbf{16})x + 64$
$= x^3 + 12x^2 + 48x + 64$

39. $(n-6)^3$

$= (n-6)^2(n-6)$

$= [n^2 + 2(n)(-6) + (-6)^2](n-6)$

$= (n^2 - 12n + 36)(n-6)$

$= \mathbf{n^2}(n) + \mathbf{n^2}(-6) - \mathbf{12n}(n) - \mathbf{12n}(-6) + \mathbf{36}(n) + \mathbf{36}(-6)$

$= n^3 - 6n^2 - 12n^2 + 72n + 36n - 216$

$= n^3 + (\mathbf{-6-12})n^2 + (\mathbf{72+36})n - 216$

$= n^3 - 18n^2 + 108n - 216$

41. $(2g-3)^3$

$= (2g-3)^2(2g-3)$

$= [(2g)^2 + 2(2g)(-3) + (-3)^2](2g-3)$

$= (4g^2 - 12g + 9)(2g-3)$

$= \mathbf{4g^2}(2g) + \mathbf{4g^2}(-3) - \mathbf{12g}(2g) - \mathbf{12g}(-3) + \mathbf{9}(2g) + \mathbf{9}(-3)$

$= 8g^3 - 12g^2 - 24g^2 + 36g + 18g - 27$

$= 8g^3 + (\mathbf{-12-24})g^2 + (\mathbf{36+18})g - 27$

$= 8g^3 - 36g^2 + 54g - 27$

43. $(a+b)^3$

$= (a+b)^2(a+b)$

$= [a^2 + 2(a)(b) + (b)^2](a+b)$

$= (a^2 + 2ab + b^2)(a+b)$

$= \mathbf{a^2}(a) + \mathbf{a^2}(b) + \mathbf{2ab}(a) + \mathbf{2ab}(b) + \mathbf{b^2}(a) + \mathbf{b^2}(b)$

$= a^3 + a^2b + 2a^2b + 2ab^2 + ab^2 + b^3$

$= a^3 + (\mathbf{1+2})a^2b + (\mathbf{2+1})ab^2 + b^3$

$= a^3 + 3a^2b + 3ab^2 + b^3$

Perform the operations.

45. $2(x^2 + 7x - 1) - 3(x^2 - 2x + 2)$

$= \mathbf{2}(x^2) + \mathbf{2}(7x) + \mathbf{2}(-1) - \mathbf{3}(x^2) - \mathbf{3}(-2x) - \mathbf{3}(2)$

$= 2x^2 + 14x - 2 - 3x^2 + 6x - 6$

$= (\mathbf{2-3})x^2 + (\mathbf{14+6})x + (\mathbf{-2-6})$

$= -x^2 + 20x - 8$

47. $(3x+4)(2x-2) - (2x+1)(x+3)$

$= \left[\mathbf{3x}(2x) - \mathbf{3x}(2) + \mathbf{4}(2x) - \mathbf{4}(2)\right] - \left[\mathbf{2x}(x) + \mathbf{2x}(3) + \mathbf{1}(x) + \mathbf{1}(3)\right]$

$= \left[6x^2 - 6x + 8x - 8\right] - \left[2x^2 + 6x + x + 3\right]$

$= \left[6x^2 + (\mathbf{-6+8})x - 8\right] - \left[2x^2 + (\mathbf{6+1})x + 3\right]$

$= \left[6x^2 + 2x - 8\right] - \left[2x^2 + 7x + 3\right]$

$= 6x^2 + 2x - 8 - 2x^2 - 7x - 3$

$= (\mathbf{6-2})x^2 + (\mathbf{2-7})x + (\mathbf{-8-3})$

$= 4x^2 - 5x - 11$

49. $-5d(4d-1)^2$

$= -5d[(4d)^2 + 2(4d)(-1) + (-1)^2$

$= -5d[16d^2 - 8d + 1]$

$= \mathbf{-5d}(16d^2) - \mathbf{5d}(-8d) - \mathbf{5d}(1)$

$= -80d^3 + 40d^2 - 5d$

51. $4d(d^2 + g^3)(d^2 - g^3) = 4d\left[(d^2)^2 - (g^3)^2\right]$

$= 4d\left[d^{2\cdot2} - g^{3\cdot2}\right]$

$= 4d\left[d^4 - g^6\right]$

$= \mathbf{4d}(d^4) + \mathbf{4d}(-g^6)$

$= 4(d^{1+4}) - 4(dg^6)$

$= 4d^5 - 4dg^6$

Find a polynomial that represents the area of the figure. Leave π in your answer.

53. $A = \frac{1}{2}bh$

$= \frac{1}{2}(2x+2)(2x-2)$

$= \frac{1}{2}[(2x)^2 - 2^2]$

$= \frac{1}{2}(4x^2 - 4)$

$= \mathbf{\frac{1}{2}}(4x^2) + \mathbf{\frac{1}{2}}(-4)$

$= 2x^2 - 2$

The area is $(2x^2 - 2)$ yd^2.

55. $A = lw$

$= (3x+1)(3x+1)$

$= (3x)^2 + 2(3x)(1) + 1^2$

$= 9x^2 + 6x + 1$

The area is $(9x^2 + 6x + 1)$ ft^2.

TRY IT YOURSELF

Perform the operations.

57. $(2v^3-8)^2=(2v^3)^2+2(2v^3)(-8)+(-8)^2$
$=4v^{3\cdot2}-32v^3+64$
$=4v^6-32v^3+64$

59. $3x(2x+3)(2x+3)$
$=3x[(2x)^2+2(2x)(3)+(3)^2]$
$=3x[4x^2+12x+9]$
$=\mathbf{3x}(4x^2)+\mathbf{3x}(12x)+\mathbf{3x}(9)$
$=12x^3+36x^2+27x$

61. $(4f+0.4)(4f-0.4)=(4f)^2-(0.4)^2$
$=16f^2-0.16$

63. $(r^2+10s)^2=(r^2)^2+2(r^2)(10s)+(10s)^2$
$=r^{2\cdot2}+20r^2s+100s^2$
$=r^4+20r^2s+100s^2$

65. $2(x+3)+4(x-2)$
$=\mathbf{2}(x)+\mathbf{2}(3)+\mathbf{4}(x)+\mathbf{4}(-2)$
$=2x+6+4x-8$
$=(\mathbf{2}+\mathbf{4})x+(\mathbf{6}-\mathbf{8})$
$=6x-2$

67. $\left(d^4+\frac{1}{4}\right)^2=(d^4)^2+2(d^4)\left(\frac{1}{4}\right)+\left(\frac{1}{4}\right)^2$
$=d^{4\cdot2}+\frac{1}{2}d^4+\frac{1}{16}$
$=d^8+\frac{1}{2}d^4+\frac{1}{16}$

69. $(d+7)(d-7)=d^2-(7)^2$
$=d^2-49$

71. $(2a-3b)^2=(2a)^2+2(2a)(-3b)+(-3b)^2$
$=4a^2-12ab+9b^2$

73. $(n+6)(n-6)=n^2-6^2$
$=n^2-36$

75. $(m+10)^2-(m-8)^2$
$=(m)^2+2(m)(10)+(10)^2$
$\quad-[(m)^2+2(m)(-8)+(-8)^2]$
$=m^2+20m+100-[m^2-16m+64]$
$=m^2+20m+100-\mathbf{1}(m^2)-\mathbf{1}(-16m)-\mathbf{1}(64)$
$=m^2+20m+100-m^2+16m-64$
$=(\mathbf{1}-\mathbf{1})m^2+(\mathbf{20}+\mathbf{16})m+(\mathbf{100}-\mathbf{64})$
$=36m+36$

77. $(2m+n)^3$
$=(2m+n)^2(2m+n)$
$=[(2m)^2+2(2m)(n)+(n)^2](2m+n)$
$=(4m^2+4mn+n^2)(2m+n)$
$=\mathbf{4m^2}(2m)+\mathbf{4m^2}(n)+\mathbf{4mn}(2m)+\mathbf{4mn}(n)$
$\quad+\mathbf{n^2}(2m)+\mathbf{n^2}(n)$
$=8m^3+4m^2n+8m^2n+4mn^2+2mn^2+n^3$
$=8m^3+(\mathbf{4}+\mathbf{8})m^2n+(\mathbf{4}+\mathbf{2})mn^2+n^3$
$=8m^3+12m^2n+6mn^2+n^3$

79. $\left(5m-\frac{6}{5}\right)^2=(5m)^2+2(5m)\left(-\frac{6}{5}\right)+\left(-\frac{6}{5}\right)^2$
$=25m^2-12m+\frac{36}{25}$

81. $(r^2-s^2)^2=(r^2)^2+2(r^2)(-s^2)+(-s^2)^2$
$=r^{2\cdot2}-2r^2s^2+s^{2\cdot2}$
$=r^4-2r^2s^2+s^4$

83. $(x-2)^2=(x)^2+2(x)(-2)+(-2)^2$
$=x^2-4x+4$

85. $(r+2)^2=r^2+2(r)(2)+(2)^2$
$=r^2+4r+4$

87. $(n-2)^4$

$= (n-2)^2(n-2)^2$

$= [(n)^2 + 2(n)(-2) + (-2)^2]$
$[(n)^2 + 2(n)(-2) + (-2)^2]$

$= (n^2 - 4n + 4)(n^2 - 4n + 4)$

$= \boldsymbol{n^2}(n^2) + \boldsymbol{n^2}(-4n) + \boldsymbol{n^2}(4)$
$- \boldsymbol{4n}(n^2) - \boldsymbol{4n}(-4n) - \boldsymbol{4n}(4)$
$+ \mathbf{4}(n^2) + \mathbf{4}(-4n) + \mathbf{4}(4)$

$= n^4 - 4n^3 + 4n^2 - 4n^3 + 16n^2 - 16n$
$+ 4n^2 - 16n + 16$

$= n^4 + (\mathbf{-4-4})n^3 + (\mathbf{4+16+4})n^2$
$+ (\mathbf{-16-16})n + 16$

$= n^4 - 8n^3 + 24n^2 - 32n + 16$

89. $5(y^2 - 2y - 6) + 6(2y^2 + 2y - 5)$

$= 5(y^2) - 5(2y) - 5(6) + 6(2y^2) + 6(2y - 6(5)$

$= 5y^2 - 10y - 30 + 12y^2 + 12y - 30$

$= (5+12)y^2 + (-10+12)y + (-30-30)$

$= 17y^2 + 2y - 60$

91. $(3x-2)^2 + (2x+1)^2$

$= (3x)^2 + 2(3x)(-2) + (-2)^2$
$+ (2x)^2 + 2(2x)(1) + (1)^2$

$= 9x^2 - 12x + 4 + [4x^2 + 4x + 1]$

$= 9x^2 - 12x + 4 + 4x^2 + 4x + 1$

$= (\mathbf{9+4})x^2 + (\mathbf{-12+4})x + (\mathbf{4+1})$

$= 13x^2 - 8x + 5$

93. $(f-8)^2 = f^2 + 2(f)(-8) + (-8)^2$
$= f^2 - 16f + 64$

95. $\left(6b + \frac{1}{2}\right)\left(6b - \frac{1}{2}\right) = (6b)^2 - \left(\frac{1}{2}\right)^2$
$= 36b^2 - \frac{1}{4}$

97. $3y(y+2) + (y+1)(y-1)$

$= [\mathbf{3y}(y+2)] + [(y+1)(y-1)]$

$= [\mathbf{3y}(y) + \mathbf{3y}(2)] + [(y^2 - 1^2)]$

$= 3y^2 + 6y + y^2 - 1$

$= (\mathbf{3+1})y^2 + 6y - 1$

$= 4y^2 + 6y - 1$

99. $(6-2d^3)^2 = 6^2 + 2(6)(-2d^3) + (-2d^3)^2$
$= 36 - 24d^3 + 4d^{3\cdot 2}$
$= 36 - 24d^3 + 4d^6$

101. $(2e+1)^3$

$= (2e+1)^2(2e+1)$

$= [(2e)^2 + 2(2e)(1) + (1)^2](2e+1)$

$= (4e^2 + 4e + 1)(2e+1)$

$= \mathbf{4e^2}(2e) + \mathbf{4e^2}(1) + \mathbf{4e}(2e) + \mathbf{4e}(1)$
$+ \mathbf{1}(2e) + \mathbf{1}(1)$

$= 8e^3 + 4e^2 + 8e^2 + 4e + 2e + 1$

$= 8e^3 + (\mathbf{4+8})e^2 + (\mathbf{4+2})e + 1$

$= 8e^3 + 12e^2 + 6e + 1$

103. $(8x+3)^2 = (8x)^2 + 2(8x)(3) + (3)^2$
$= 64x^2 + 48x + 9$

LOOK ALIKES . . .

Perform the indicated operations.

105. a. $(xy)^2 = x^{1\cdot 2}y^{1\cdot 2}$
$= x^2y^2$

b. $(x+y)^2 = x^2 + 2(x)(y) + y^2$
$= x^2 + 2xy + y^2$

107. a. $(2b^2d)^2 = 2^2b^{2\cdot 2}d^{1\cdot 2}$
$= 4b^4d^2$

b. $(2b^2 + d)^2 = (2b^2)^2 + 2(2b^2)(d) + d^2$
$= 2^2b^{2\cdot 2} + 4b^2d + d^2$
$= 4b^4 + 4bd + d^2$

APPLICATIONS

109. PLAYPENS

$$(x+6)^2 = (x)^2 + 2(x)(6) + (6)^2$$
$$= x^2 + 12x + 36$$

The area is $(x^2 + 12x + 36)$ in.^{2}

111. PAPER TOWELS

$$\pi h(R+r)(R-r) = \pi h(R^2 - r^2)$$
$$= \pi hR^2 - \pi hr^2$$

WRITING

113-115. Answers will vary.

REVIEW

117. Simplify:

$$\frac{30}{36} = \frac{5 \cdot 6}{6 \cdot 6}$$
$$= \frac{5 \cdot \overset{1}{\cancel{6}}}{6 \cdot \underset{1}{\cancel{6}}}$$
$$= \frac{5}{6}$$

119. Multiply:

$$\frac{7}{8} \cdot \frac{3}{5} = \frac{7 \cdot 3}{8 \cdot 5}$$
$$= \frac{21}{40}$$

120. Divide:

$$\frac{1}{3} \div \frac{4}{5} = \frac{1}{3} \cdot \frac{5}{4}$$
$$= \frac{1 \cdot 5}{3 \cdot 4}$$
$$= \frac{5}{12}$$

CHALLENGE PROBLEMS

121. a. Find two binomials whose product is a binomial.

$(x + 1)(x - 1) = x^2 - 1$

b. Find two binomials whose product is a trinomial.

$(x + 1)(x + 1) = x^2 + 2x + 1$

c. Find two binomials whose product is a four-term polynomial.

$(x + a)(x + b) = x^2 + bx + ax + ab$

SECTION 5.8 STUDY SET
VOCABULARY

Fill in the blanks.

1. The expression $\frac{18x^7}{9x^4}$ is a monomial divided by a **monomial**.

3. The expression $\frac{x^2-8x+12}{x-6}$ is a trinomial divided by a **binomial**.

CONCEPTS

Fill in the blanks.

5. The long division method is a series of four steps that are repeated. Put them in the correct order: subtract multiply bring down divide
   Divide, multiply, subtract, bring down

7. **Fill in the blanks:** To check an answer of a long division, we use the fact that
   divisor • **quotient** + remainder = **dividend**

NOTATION

Complete each solution.

9. $$\frac{28x^5-x^3+5x^2}{7x^2}=\frac{28x^5}{\boxed{7x^2}}-\frac{\boxed{x^3}}{7x^2}+\frac{5x^2}{\boxed{7x^2}}$$
$$=4x^{\boxed{5}-\boxed{2}}-\frac{x^{3-2}}{\boxed{7}}+5x^{\boxed{2}-\boxed{2}}$$
$$=\boxed{4x^3}-\frac{x}{7}+\frac{\boxed{5}}{\boxed{7}}$$

11. Write the polynomial $2x^2-1+5x^4$ in descending powers of x and insert placeholders for each missing term.
    $\mathbf{5x^4+0x^3+2x^2+0x-1}$

GUIDED PRACTICE

Divide the monomials. See Example 1.

13. $$\frac{x^5}{x^2}=x^{5-2}=x^3$$

15. $$\frac{12h^8}{9h^6}=\frac{4h^{8-6}}{3}=\frac{4h^2}{3}$$

17. $$\frac{-3d^4}{15d^8}=-\frac{1}{5}d^{4-8}=-\frac{1}{5}d^{-4}=-\frac{1}{5d^4}$$

19. $$\frac{10s^2}{s^3}=10s^{2-3}=10s^{-1}=\frac{10}{s}$$

21. $$\frac{8x^3y^2}{40xy^6}=\frac{1x^{3-1}y^{2-6}}{5}=\frac{1x^2y^{-4}}{5}=\frac{x^2}{5y^4}$$

23. $$\frac{-16r^3y^2}{-4r^2y^7}=4r^{3-2}y^{2-7}=4r^1y^{-5}=\frac{4r}{y^5}$$

Divide the polynomial by the monomial. See Example 2.

25. $$\frac{6x+3}{3}=\frac{6x}{3}+\frac{3}{3}=2x+1$$

27. $$\frac{a-a^3+a^4}{a^4}=\frac{a}{a^4}-\frac{a^3}{a^4}+\frac{a^4}{a^4}=a^{1-4}-a^{3-4}+a^{4-4}=a^{-3}-a^{-1}+a^0=\frac{1}{a^3}-\frac{1}{a}+1$$

29. $$\frac{6h^{12}+48h^9}{24h^{10}}=\frac{6h^{12}}{24h^{10}}+\frac{48h^9}{24h^{10}}=\frac{h^{12-10}}{4}+2h^{9-10}=\frac{h^2}{4}+2h^{-1}=\frac{h^2}{4}+\frac{2}{h}$$

31. $$\frac{9s^8-18s^5+12s^4}{3s^3}=\frac{9s^8}{3s^3}-\frac{18s^5}{3s^3}+\frac{12s^4}{3s^3}=3s^{8-3}-6s^{5-3}+4s^{4-3}=3s^5-6s^2+4s^1=3s^5-6s^2+4s$$

33. $\dfrac{7c^5 + 21c^4 - 14c^3 - 35c}{7c^2}$

$$= \frac{7c^5}{7c^2} + \frac{21c^4}{7c^2} - \frac{14c^3}{7c^2} - \frac{35c}{7c^2}$$
$$= c^{5-2} + 3c^{4-2} - 2c^{3-2} - 5c^{1-2}$$
$$= c^3 + 3c^2 - 2c^1 - 5c^{-1}$$
$$= c^3 + 3c^2 - 2c - \frac{5}{c^1}$$
$$= c^3 + 3c^2 - 2c - \frac{5}{c}$$

35. $-25x^2y^3 + 30xy^2 - 5xy$

$$= \frac{-25x^2y^3}{-5x^2y^2} + \frac{30xy^2}{-5x^2y^2} - \frac{5xy}{-5x^2y^2}$$
$$= 5x^{2-2}y^{3-2} - 6x^{1-2}y^{2-2} + x^{1-2}y^{1-2}$$
$$= 5x^0y^1 - 6x^{-1}y^0 + x^{-1}y^{-1}$$
$$= 5y - \frac{6}{x} + \frac{1}{xy}$$

Perform each division. See Examples 3 and 4.

37.
$$\begin{array}{r} x + 6 \\ x+2\overline{)x^2 + 8x + 12} \\ \underline{-(x^2 + 2x)} \\ 6x + 12 \\ \underline{-(6x + 12)} \\ 0 \end{array}$$

39.
$$\begin{array}{r} x - 2 \\ x-3\overline{)x^2 - 5x + 6} \\ \underline{-(x^2 - 3x)} \\ -2x + 6 \\ \underline{-(-2x + 6)} \\ 0 \end{array}$$

41.
$$\begin{array}{r} x + 1 + \dfrac{-1}{2x+3} \\ 2x+3\overline{)2x^2 + 5x + 2} \\ \underline{-(2x^2 + 3x)} \\ 2x + 2 \\ \underline{-(2x + 3)} \\ -1 \end{array}$$

43.
$$\begin{array}{r} 2x - 3 + \dfrac{-1}{3x-1} \\ 3x-1\overline{)6x^2 - 11x + 2} \\ \underline{-(6x^2 - 2x)} \\ -9x + 2 \\ \underline{-(-9x + 3)} \\ -1 \end{array}$$

Perform each division. See Example 5.

45.
$$\begin{array}{r} 2x - 1 \\ x+2\overline{)2x^2 + 3x - 2} \\ \underline{-(2x^2 + 4x)} \\ -x - 2 \\ \underline{-(-x - 2)} \\ 0 \end{array}$$

47.
$$\begin{array}{r} 2x + 1 \\ 5x+3\overline{)10x^2 + 11x + 3} \\ \underline{-(10x^2 + 6x)} \\ 5x + 3 \\ \underline{-(5x + 3)} \\ 0 \end{array}$$

Perform each division. See Example 6.

49.
$$\begin{array}{r} a - 5 \\ a+5\overline{)a^2 \mathbf{+ 0a} - 25} \\ \underline{-(a^2 + 5a)} \\ -5a - 25 \\ \underline{-(-5a - 25)} \\ 0 \end{array}$$

51.
$$\begin{array}{r} x + 1 \\ x-1\overline{)x^2 \mathbf{+ 0x} - 1} \\ \underline{-(x^2 - x)} \\ x - 1 \\ \underline{-(x - 1)} \\ 0 \end{array}$$

53.
$$\begin{array}{r} 2x - 3 \\ 2x+3\overline{)4x^2 \mathbf{+ 0x} - 9} \\ \underline{-(4x^2 + 6x)} \\ -6x - 9 \\ \underline{-(-6x - 9)} \\ 0 \end{array}$$

55.
$$\begin{array}{r} 9b+7 \\ 9b-7\overline{)81b^2+\mathbf{0b}-49} \\ \underline{-(81b^2-63b)} \\ 63b-49 \\ \underline{-(63b-49)} \\ 0 \end{array}$$

TRY IT YOURSELF

Perform each division.

57.
$$\begin{array}{r} y+12+\dfrac{1}{y+1} \\ y+1\overline{)y^2+13y+13} \\ \underline{-(y^2+y)} \\ 12y+13 \\ \underline{-(12y+12)} \\ 1 \end{array}$$

59.
$$\begin{aligned} \frac{15a^8b^2-10a^2b^5}{5a^3b^2} &= \frac{15a^8b^2}{5a^3b^2}-\frac{10a^2b^5}{5a^3b^2} \\ &= \frac{3a^{8-3}b^{2-2}}{1}-\frac{2a^{2-3}b^{5-2}}{1} \\ &= \frac{3a^5b^0}{1}-\frac{2a^{-1}b^3}{1} \\ &= 3a^5-\frac{2b^3}{a} \end{aligned}$$

61.
$$\begin{array}{r} 2x^2+2x+1 \\ 3x+2\overline{)6x^3+10x^2+7x+2} \\ \underline{-(6x^3+4x^2)} \\ 6x^2+7x \\ \underline{-(6x^2+4x)} \\ 3x+2 \\ \underline{-(3x+2)} \\ 0 \end{array}$$

63.
$$\begin{aligned} \frac{8x^9-32x^6}{4x^4} &= \frac{8x^9}{4x^4}-\frac{32x^6}{4x^4} \\ &= 2x^{9-4}-8x^{6-4} \\ &= 2x^5-8x^2 \end{aligned}$$

65.
$$\begin{array}{r} 3a-2 \\ 2a+3\overline{)6a^2+5a-6} \\ \underline{-(6a^2+9a)} \\ -4a-6 \\ \underline{-(-4a-6)} \\ 0 \end{array}$$

67.
$$\begin{aligned} \frac{45m^{10}}{9m^5} &= 5\cdot m^{10-5} \\ &= 5m^5 \end{aligned}$$

69.
$$\begin{array}{r} b+3 \\ 3b+2\overline{)3b^2+11b+6} \\ \underline{-(3b^2+2b)} \\ 9b+6 \\ \underline{-(9b+6)} \\ 0 \end{array}$$

71.
$$\begin{array}{r} x+3 \\ 2x-7\overline{)2x^2-x-21} \\ \underline{-(2x^2-7x)} \\ 6x-21 \\ \underline{-(6x-21)} \\ 0 \end{array}$$

73.
$$\begin{array}{r} x^2-x+1 \\ x+1\overline{)x^3+\mathbf{0x^2}+\mathbf{0x}+1} \\ \underline{-(x^3+x^2)} \\ -x^2+0x \\ \underline{-(-x^2-x)} \\ x+1 \\ \underline{-(x+1)} \\ 0 \end{array}$$

75.
$$\begin{aligned} \frac{-65rs^2}{15r^2s^5} &= \frac{-13r^{1-2}s^{2-5}}{3} \\ &= \frac{-13r^{-1}s^{-3}}{3} \\ &= -\frac{13}{3rs^3} \end{aligned}$$

77. $\dfrac{-18w^6-9}{9w^4}=\dfrac{-18w^6}{9w^4}-\dfrac{-9}{9w^4}$

$=-2w^{6-4}-\dfrac{1}{w^4}$

$=-2w^2-\dfrac{1}{w^4}$

79. $\dfrac{9m-6}{m}=\dfrac{9m}{m}-\dfrac{6}{m}$

$=9m^{1-1}-\dfrac{6}{m}$

$=9m^0-\dfrac{6}{m}$

$=9-\dfrac{6}{m}$

81.
$$
\begin{array}{r}
y^2+2y+5+\dfrac{10}{y-2} \\
y-2\overline{)y^3+\mathbf{0y^2}+y+\mathbf{0}} \\
\underline{-(y^3-2y^2)} \\
2y^2+y \\
\underline{-(2y^2-4y)} \\
5y+0 \\
\underline{-(5y-10)} \\
10
\end{array}
$$

83. $\dfrac{5x^4-10x}{25x^3}=\dfrac{5x^4}{25x^3}-\dfrac{10x}{25x^3}$

$=\dfrac{x^{4-3}}{5}-\dfrac{2x^{1-3}}{5}$

$=\dfrac{x}{5}-\dfrac{2x^{-2}}{5}$

$=\dfrac{x}{5}-\dfrac{2}{5x^2}$

85.
$$
\begin{array}{r}
x^2-2x+1 \\
4x+3\overline{)4x^3-5x^2-2x+3} \\
\underline{-(4x^3+3x^2)} \\
-8x^2-2x \\
\underline{-(-8x-6x)} \\
4x+3 \\
\underline{-(4x+3)} \\
0
\end{array}
$$

87.
$$
\begin{array}{r}
x+1+\dfrac{10}{x+5} \\
x+5\overline{)x^2+6x+15} \\
\underline{-(x^2+5x)} \\
x+15 \\
\underline{-(x+5)} \\
10
\end{array}
$$

89. $\dfrac{12x^3y^2-8x^2y-4x}{4xy}$

$=\dfrac{12x^3y^2}{4xy}-\dfrac{8x^2y}{4xy}-\dfrac{4x}{4xy}$

$=3x^{3-1}y^{2-1}-2x^{2-1}y^{1-1}-\dfrac{x^{1-1}}{y}$

$=3x^2y^1-2x^1y^0-\dfrac{x^0}{y}$

$=3x^2y-2x-\dfrac{1}{y}$

91.
$$
\begin{array}{r}
a-12+\dfrac{4}{a-5} \\
a-5\overline{)a^2-17a+64} \\
\underline{-(a^2-5a)} \\
-12a+64 \\
\underline{-(-12a+60)} \\
4
\end{array}
$$

93.
$$
\begin{array}{r}
a^2+a+1 \\
a-1\overline{)a^3+\mathbf{0a^2}+\mathbf{0a}-1} \\
\underline{-(a^3-a^2)} \\
a^2+0a \\
\underline{-(a^2-a)} \\
a-1 \\
\underline{-(a-1)} \\
0
\end{array}
$$

95.
$$
\begin{array}{r}
2x^2+x+1+\dfrac{2}{3x-1} \\
3x-1\overline{)6x^3+x^2+2x+1} \\
\underline{-(6x^3-2x^2)} \\
3x^2+2x \\
\underline{-(3x^2-x)} \\
3x+1 \\
\underline{-(3x-1)} \\
2
\end{array}
$$

97. $$\frac{8x^{17}y^{20}}{16x^{15}y^{30}} = \frac{x^{17-15}y^{20-30}}{2} = \frac{x^2y^{-10}}{2} = \frac{x^2}{2y^{10}}$$

99. $$\begin{array}{r} 3m-8 \\ 2m+5\overline{)6m^2 - \quad m - 40} \\ \underline{-(6m^2+15m)} \\ -16m-40 \\ \underline{-(-16m-40)} \\ 0 \end{array}$$

LOOK ALIKES . . .

Perform the indicated operations.

101. a. $$\frac{16x^2-16x-5}{4x} = \frac{16x^2}{4x} - \frac{16x}{4x} - \frac{5}{4x} = 4x^{2-1} - 4x^{1-1} - \frac{5}{4x} = 4x - 4 - \frac{5}{4x}$$

b. $$\begin{array}{r} 4x-5 \\ 4x+1\overline{)16x^2-16x-5} \\ \underline{-(16x^2+4x)} \\ -20x-5 \\ \underline{-(-20x-5)} \\ 0 \end{array}$$

APPLICATIONS

103. FURNACE FILTERS

$$\begin{array}{r} x-6 \\ x+4\overline{)x^2-2x-24} \\ \underline{-(x^2+4x)} \\ -6x-24 \\ \underline{-(-6x-24)} \\ 0 \end{array}$$

The length is $(x-6)$ in.

105. POOL

$$\frac{6x^2-3x+9}{3} = \frac{6x^2}{3} - \frac{3x}{3} + \frac{+9}{3} = \frac{2\cdot\cancel{3}^{1}x^2}{\cancel{3}_{1}} - \frac{\cancel{3}^{1}x}{\cancel{3}_{1}} + \frac{\cancel{3}^{1}\cdot 3}{\cancel{3}_{1}} = 2x^2 - x + 3$$

The length of one side is $(2x^2 - x + 3)$ in.

WRITING

107-109. Answers will vary.

REVIEW

111. Write an equation of the line with slope $-\frac{11}{6}$ that passes through (2, –6). Write the answer in slope–intercept form.

$$y - y_1 = m(x - x_1)$$
$$y - (-6) = -\frac{11}{6}(x-2)$$
$$y + 6 = -\frac{11}{6}(x) - \frac{11}{6}(-2)$$
$$y + 6 = -\frac{11}{6}x + \frac{11}{3}$$
$$y + 6 - \mathbf{6} = -\frac{11}{6}x + \frac{11}{3} - \mathbf{6}$$
$$y = -\frac{11}{6}x + \frac{11}{3} - \frac{\mathbf{18}}{\mathbf{3}}$$
$$y = -\frac{11}{6}x - \frac{7}{3}$$

CHALLENGE PROBLEMS

Perform each division.

113. $$\frac{6a^3-17a^2b+14ab^2-3b^3}{2a-3b}$$

$$\begin{array}{r} 3a^2 - 4ab + b^2 \\ 2a-3b\overline{)6a^3-17a^2b+14ab^2-3b^3} \\ \underline{-(6a^3-9a^2b)} \\ -8a^2b + 14ab^2 \\ \underline{-(-8a^2b + 12ab^2)} \\ 2ab^2 - 3b^3 \\ \underline{-(2ab^2-3b^3)} \\ 0 \end{array}$$

115. $(x^6+2x^4-6x^2-9)\div(x^2+3)$

$$
\begin{array}{r}
x^4 \qquad -x^2 \;-\; 3 \\
x^2+\mathbf{0x}+3\overline{\big)\, x^6+\mathbf{0x^5}+2x^4+\mathbf{0x^3}-6x^2+\mathbf{0x}-9} \\
\underline{-(x^6-0x^5+3x^4)} \qquad\qquad\qquad\qquad \\
-x^4+0x^3-6x^2 \qquad\qquad \\
\underline{-(-x^4+0x^3-3x^2)} \qquad\qquad \\
-3x^2+0x-9 \\
\underline{-(-3x^2-0x-9)} \\
0
\end{array}
$$

117. $\dfrac{a^8+a^6-4a^4+5a^2-3}{a^4+2a^2-3}$

$$
\begin{array}{r}
a^4-a^2+1 \\
a^4+2a^2-3\overline{\big)\, a^8+a^6-4a^4+5a^2-3} \\
\underline{-(a^8+2a^6-3a^4)} \qquad\qquad \\
-a^6-a^4+5a^2 \qquad \\
\underline{-(-a^6-2a^4+3a^2)} \qquad \\
a^4+2a^2-3 \\
\underline{-(a^4+2a^2-3)} \\
0
\end{array}
$$

CHAPTER 5
SECTION 5.1 REVIEW EXERCISES
Rules for Exponents

1. Identify the base and the exponent in each expression.
   a. n^{12} base n, exponent 12
   b. $(2x)^6$ base $2x$, exponent 6
   c. $3r^4$ base r, exponent 4
   d. $(y-7)^3$ base $y-7$, exponent 3

2. Write each expression in an equivalent form using an exponent.
   a. $m \cdot m \cdot m \cdot m \cdot m = \mathbf{m^5}$
   b. $-3 \cdot x \cdot x \cdot x \cdot x = \mathbf{-3x^4}$
   c. $(x+8)(x+8) = \mathbf{(x+8)^2}$
   d. $\left(\frac{1}{2}pq\right)\left(\frac{1}{2}pq\right)\left(\frac{1}{2}pq\right) = \left(\frac{1}{2}pq\right)^3$

Simplify each expression. Assume there are no divisions by 0.

3. $7^4 \cdot 7^8 = 7^{4+8}$
   $= 7^{12}$

4. $mmnn = m^2n^2$

5. $(y^7)^3 = y^{7\cdot 3}$
   $= y^{21}$

6. $(3x)^4 = 81x^4$

7. $\dfrac{b^{12}}{b^3} = b^{12-3}$
   $= b^9$

8. $-b^3b^4b^5 = -b^{12}$

9. $(-16s^3)^2s^4 = (-16)^2s^{3\cdot 2}s^4$
   $= 256s^6s^4$
   $= 256s^{6+4}$
   $= 256s^{10}$

10. $(2.1x^2y)^2 = 4.41x^4y^2$

11. $[(-9)^3]^5 = (-9)^{3\cdot 5}$
    $= (-9)^{15}$

12. $(a^5)^3(a^2)^4 = (a^{15})(a^8)$
    $= a^{23}$

13. $\left(\frac{1}{2}x^2x^3\right)^3 = \left(\frac{1}{2}\right)^3 (x^{2+3})^3$
    $= \frac{1}{8}(x^5)^3$
    $= \frac{1}{8}x^{5\cdot 3}$
    $= \frac{1}{8}x^{15}$

14. $\left(\dfrac{x^7}{3xy}\right)^2 = \left(\dfrac{x^{7-1}}{3y}\right)^2$
    $= \dfrac{x^{6\cdot 2}}{3^2y^2}$
    $= \dfrac{x^{12}}{9y^2}$

15. $\dfrac{(m-25)^{16}}{(m-25)^4} = (m-25)^{16-4}$
    $= (m-25)^{12}$

16. $\dfrac{(5y^2z^3)^3}{(yz)^5} = \dfrac{5^3y^{2\cdot 3}z^{3\cdot 3}}{y^5z^5}$
    $= \dfrac{125y^6z^9}{y^5z^5}$
    $= 125y^{6-5}z^{9-5}$
    $= 125yz^4$

17. $\dfrac{a^5a^4a^5}{a^2a} = \dfrac{a^{5+4+5}}{a^{2+1}}$
    $= \dfrac{a^{14}}{a^3}$
    $= a^{14-3}$
    $= a^{11}$

18. $\dfrac{(cd)^9}{(cd)^4} = (cd)^{9-4}$
    $= (cd)^5$
    $= c^5d^5$

Find an expression that represents the area or the volume of each figure, whichever is appropriate.

19. $V = s^3$
$= (4x^4)^3$
$= 4^3 x^{4 \cdot 3}$
$= 64x^{12}$
The volume is $64x^{12}$ in.^{3}

20. $A = s^2$
$= (y^2)^2$
$= y^{2 \cdot 2}$
$= y^4$
The area is y^4 ft^2.

SECTION 5.2 REVIEW EXERCISES
Zero and Negative Exponents

Simplify each expression. Do not use negative exponents in the answer.

21. $x^0 = 1$

22. $(3x^2y^2)^0 = 1$

23. $3x^0 = 3(1)$
$= 3$

24. $10^{-3} = \frac{1}{10^3}$
$= \frac{1}{1,000}$

25. $-5^{-2} = \frac{-1}{5^2}$
$= -\frac{1}{25}$

26. $\frac{t^4}{t^{10}} = t^{4-10}$
$= t^{-6}$
$= \frac{1}{t^6}$

27. $\frac{8}{x^{-5}} = 8x^5$

28. $-6y^4y^{-5} = -6y^{4-5}$
$= -6y^{-1}$
$= -\frac{6}{y}$

29. $\frac{7^{-2}}{2^{-3}} = \frac{2^3}{7^2}$
$= \frac{8}{49}$

30. $(x^{-3}x^{-4})^{-2} = (x^{-7})^{-2}$
$= x^{-7 \cdot -2}$
$= x^{14}$

31. $\left(\frac{-3r^4r^{-3}}{r^{-3}r^7}\right)^3 = \left(\frac{-3r^{4-3}}{r^{-3+7}}\right)^3$
$= \left(\frac{-3r^1}{r^4}\right)^3$
$= \left(\frac{-3r^{1-4}}{1}\right)^3$
$= \left(\frac{-3r^{-3}}{1}\right)^3$
$= \left(\frac{-3}{r^3}\right)^3$
$= \frac{(-3)^3}{r^{3 \cdot 3}}$
$= -\frac{27}{r^9}$

32. $\left(\frac{4z^4}{z^3}\right)^{-2} = \left(\frac{z^3}{4z^4}\right)^2$
$= \left(\frac{1}{4z^{4-3}}\right)^2$
$= \left(\frac{1}{4z}\right)^2$
$= \frac{1^2}{4^2z^2}$
$= \frac{1}{16z^2}$

33. $\dfrac{3^{-2}c^3d^3}{2^{-3}c^2d^8} = \dfrac{2^3c^{3-2}d^{3-8}}{3^2}$

$= \dfrac{8c^1d^{-5}}{9}$

$= \dfrac{8c}{9d^5}$

34. $\dfrac{t^{-30}}{t^{-60}} = t^{-30-(-60)}$

$= t^{-30+60}$

$= t^{30}$

35. $\dfrac{w(w^{-3})^{-4}}{w^{-9}} = \dfrac{w(w^{-3\cdot -4})}{w^{-9}}$

$= \dfrac{w(w^{12})}{w^{-9}}$

$= \dfrac{w^{1+12+9}}{1}$

$= w^{22}$

36. $\left(\dfrac{4}{f^4}\right)^{-10} = \left(\dfrac{f^4}{4}\right)^{10}$

$= \dfrac{f^{4\cdot 10}}{4^{10}}$

$= \dfrac{f^{40}}{4^{10}}$

SECTION 5.3 REVIEW EXERCISES

Scientific Notation

Write each number in scientific notation.

37. Move the decimal 8 places to the left.
$720{,}000{,}000 = 7.2\times 10^8$

38. Move the decimal 15 places to the left.
$9{,}370{,}000{,}000{,}000{,}000 = 9.37\times 10^{15}$

39. Move the decimal 9 places to the right.
$0.00000000942 = 9.42\times 10^{-9}$

40. Move the decimal 4 places to the right.
$0.00013 = 1.3\times 10^{-4}$

41. Move the decimal 2 places to the right.
$0.018\times 10^{-2} = 1.8\times 10^{-2+(-2)}$
$= 1.8\times 10^{-4}$

42. Move the decimal 2 places to the left.
$853\times 10^3 = 8.53\times 10^{3+2}$
$= 8.53\times 10^5$

Write each number in standard notation.

43. The exponent is 5.
Move the decimal 5 places to the right.
$1.26\times 105 = 126{,}000$

44. The exponent is -8.
Move the decimal 8 places to the left.
$3.919\times 10^{-8} = 0.00000003919$

45. The exponent is 0.
Do not move the decimal.
$2.68\times 10^0 = 2.68$

46. The exponent is 1.
Move the decimal 1 place to the right.
$5.76\times 10^1 = 57.6$

Evaluate each expression by first writing each number in scientific notation. Express the result in scientific notation and standard notation.

47. $\dfrac{(0.000012)(0.000004)}{0.00000016} = \dfrac{(1.2\times 10^{-5})(4.0\times 10^{-6})}{1.6\times 10^{-7}}$

$= \dfrac{(1.2\cdot 4.0)}{1.6}\times\dfrac{10^{-5+(-6)}}{10^{-7}}$

$= \dfrac{(1.2\cdot 4.0)}{1.6}\times\dfrac{10^{-11}}{10^{-7}}$

$= \dfrac{(1.2\cdot 4.0)}{1.6}\times 10^{-11-(-7)}$

$= 3.0\times 10^{-11+7}$

$= 3.0\times 10^{-4}$

$= 0.0003$

48. $\dfrac{(4{,}800{,}000)(20{,}000{,}000)}{600{,}000}$

$= \dfrac{(4.8\times 10^6)(2.0\times 10^7)}{6.0\times 10^5}$

$= \dfrac{(4.8\cdot 2.0)}{6}\times\dfrac{10^{13}}{10^5}$

$= \dfrac{(4.8\cdot 2.0)}{6}\times 10^{13-5}$

$= 1.6\times 10^8$

$= 160{,}000{,}000$

49. WORLD POPULATION
6.57 billion $= 6{,}570{,}000{,}000$
Move the decimal 9 places to the left.
$6{,}570{,}000{,}000 = 6.57\times 10^9$

50. ATOMS
Move the decimal 5 places to the right.
$1.0\times 10^5 = 100{,}000$

SECTION 5.4 REVIEW EXERCISES

Polynomials

51. Consider the polynomial $3x^3 - x^2 + x + 10$.

    a. How many terms does the polynomial have?
    4

    b. What is the lead term? **$3x^3$**

    c. What is the coefficient of each term?
    3, –1, 1, 10

    d. What is the constant term? **10**

52. Find the degree of each polynomial and classify it as a monomial, binomial, trinomial, or none of these.

    a. $13x^7$ **7th, monomial**
    b. $-16a^2b$ **3rd, monomial**
    c. $5^3x + x^2$ **2nd, binomial**
    d. $-3x^5 + x - 1$ **5th, trinomial**
    e. $9xy^2 + 21x^3y^3$ **6th, binomial**
    f. $4s^4 - 3s^2 + 5s + 4$ **4th, none of these**

53. Evaluate $-x^5 - 3x^4 + 3$ for $x = 0$ and $x = -2$.

$$-x^5 - 3x^4 + 3 = -(0)^5 - 3(0)^4 + 3$$
$$= 3$$

$$-x^5 - 3x^4 + 3 = -(-2)^5 - 3(-2)^4 + 3$$
$$= -(-32) - 3(16) + 3$$
$$= 32 - 48 + 3$$
$$= -16 + 3$$
$$= -13$$

54. DIVING

$$0.1875x^2 - 0.0078125x^3$$
$$= 0.1875(8)^2 - 0.0078125(8)^3$$
$$= 0.1875(64) - 0.0078125(512)$$
$$= 12 - 4$$
$$= 8$$

The amount of deflection is 8 in.

55. $y = x^2$

x	y	(x, y)
–2	$y = (-2)^2$ $= 4$	(–2, 4)
–1	$y = (-1)^2$ $= 1$	(–1, 1)
0	$y = (0)^2$ $= 0$	(0, 0)
1	$y = (1)^2$ $= 1$	(1, 1)
2	$y = (2)^2$ $= 4$	(2, 4)

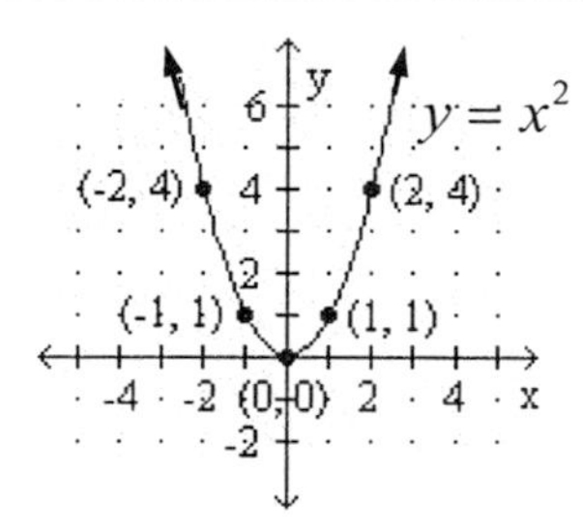

56. $y = x^3 + 1$

x	y	(x, y)
–2	$y = (-2)^3 + 1$ $= -8 + 1$ $= -7$	(–2, –7)
–1	$y = (-1)^3 + 1$ $= -1 + 1$ $= 0$	(–1, 0)
0	$y = 0^3 + 1$ $= 1$	(0, 1)
1	$y = 1^3 + 1$ $= 1 + 1$ $= 2$	(1, 2)
2	$y = 2^3 + 1$ $= 8 + 1$ $= 9$	(2, 9)

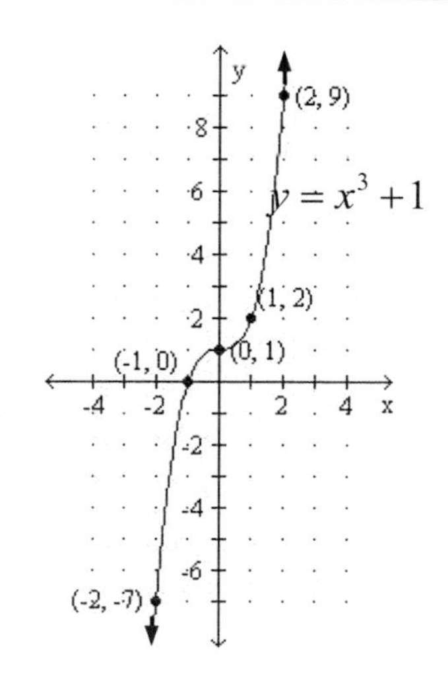

SECTION 5.5 REVIEW EXERCISES

Adding and Subtracting Polynomials

Simplify each polynomial and write the result in descending powers of one variable.

57. $6y^3+8y^4+7y^3+(-8y^4)$
$$=(8-8)y^4+(6+7)y^3$$
$$=0y^4+13y^3$$
$$=13y^3$$

58. $4a^2b+5-6a^3b-3a^2b+2a^3b+1$
$$=(-6+2)a^3b+(4-3)a^2b+(5+1)$$
$$=-4a^3b+a^2b+6$$

59. $\frac{5}{6}x^2+\frac{1}{3}y^2-\frac{1}{4}x^2-\frac{3}{4}xy+\frac{2}{3}y^2$
$$=\left(\frac{5}{6}-\frac{1}{4}\right)x^2-\frac{3}{4}xy+\left(\frac{1}{3}+\frac{2}{3}\right)y^2$$
$$=\left(\frac{5}{6}\cdot\frac{\mathbf{2}}{\mathbf{2}}-\frac{1}{4}\cdot\frac{\mathbf{3}}{\mathbf{3}}\right)x^2-\frac{3}{4}xy+\left(\frac{1}{3}+\frac{2}{3}\right)y^2$$
$$=\left(\frac{10-3}{12}\right)x^2-\frac{3}{4}xy+\left(\frac{1+2}{3}\right)y^2$$
$$=\left(\frac{7}{12}\right)x^2-\frac{3}{4}xy+\left(\frac{3}{3}\right)y^2$$
$$=\frac{7}{12}x^2-\frac{3}{4}xy+y^2$$

60. $7.6c^5-2.1c^3-0.9c^5+8.1c^4$
$$=(7.6-0.9)c^5+8.1c^4-2.1c^3$$
$$=6.7c^5+8.1c^4-2.1c^3$$

Perform the operations.

61. $(2r^6+14r^3)+(23r^6-5r^3+5r)$
$$=(2+23)r^6+(14-5)r^3+5r$$
$$=25r^6+9r^3+5r$$

62. $(7.1a^2+2.2a-5.8)-(3.4a^2-3.9a+11.8)$
$$=7.1a^2+2.2a-5.8-3.4a^2+3.9a-11.8$$
$$=(7.1-3.4)a^2+(2.2+3.9)a+(-5.8-11.8)$$
$$=3.7a^2+6.1a-17.6$$

63. $(3r^3s+r^2s^2-3rs^3-3s^4)+(r^3s-8r^2s^2-4rs^3+s^4)$
$$=(3+1)r^3s+(1-8)r^2s^2+(-3-4)rs^3+(-3+1)s^4$$
$$=4r^3s-7r^2s^2-7rs^3-2s^4$$

64. $\left(\frac{7}{8}m^4-\frac{1}{5}m^3\right)-\left(\frac{1}{4}m^4+\frac{1}{5}m^3\right)-\frac{3}{5}m^3$
$$=\frac{7}{8}m^4-\frac{1}{5}m^3-\frac{1}{4}m^4-\frac{1}{5}m^3-\frac{3}{5}m^3$$
$$=\left(\frac{7}{8}-\frac{1}{4}\right)m^4+\left(-\frac{1}{5}-\frac{1}{5}-\frac{3}{5}\right)m^3$$
$$=\left(\frac{7}{8}-\frac{1}{4}\cdot\frac{\mathbf{2}}{\mathbf{2}}\right)m^4+\left(-\frac{5}{5}\right)m^3$$
$$=\left(\frac{7}{8}-\frac{2}{8}\right)m^4+(-1)m^3$$
$$=\frac{5}{8}m^4-m^3$$

65. Step 1:
$$(2z^2+3z-7)+(-4z^3-2z-3)$$
$$=-4z^3+2z^2+(3-2)z+(-7-3)$$
$$=-4z^3+2z^2+z-10$$

Step 2:
$$(-4z^3+2z^2+z-10)-(-3z^3-4z+7)$$
$$=-4z^3+2z^2+z-10+3z^3+4z-7$$
$$=(-4+3)z^3+2z^2+(1+4)z+(-10-7)$$
$$=-z^3+2z^2+5z-17$$

66. GARDENING
$$(2x^2+x+1)-(x^2-2)=2x^2+x+1-x^2+2$$
$$=(2-1)x^2+x+(1+2)$$
$$=x^2+x+3$$
The length of the handle is (x^2+x+3) in.

67.
$$\begin{array}{r} 3x^2+5x+2 \\ +\ \underline{x^2-3x+6} \\ 4x^2+2x+8 \end{array}$$

68.
$$\begin{array}{r} 20x^3 \qquad\quad +12x \\ \underline{-(12x^3+7x^2-7x)} \end{array} \xrightarrow[\text{and add}]{\text{Change signs}} \begin{array}{r} 20x^3 \qquad\quad +12x \\ \underline{-12x^3-7x^2+\ 7x} \\ 8x^3-7x^2+19x \end{array}$$

SECTION 5.6 REVIEW EXERCISES

Multiplying Polynomials

Multiply.

69. $(2x^2)(5x) = (2 \cdot 5)(x^{2+1})$
$$= 10x^3$$

70. $(-6x^4z^3)(x^6z^2) = -6x^{4+6}z^{3+2}$
$$= -6x^{10}z^5$$

71. $5b^3 \cdot 6b^2 \cdot 4b^6 = (5 \cdot 6 \cdot 4)(b^{3+2+6})$
$$= 120b^{11}$$

72. $\frac{2}{3}h^5(3h^9 + 12h^6) = \frac{\mathbf{2}}{\mathbf{3}}\boldsymbol{h}^{\mathbf{5}}(3h^9) + \frac{\mathbf{2}}{\mathbf{3}}\boldsymbol{h}^{\mathbf{5}}(12h^6)$
$$= \left(\frac{2}{3} \cdot 3\right)h^{5+9} + \left(\frac{2}{3} \cdot 12\right)h^{5+6}$$
$$= 2h^{14} + 8h^{11}$$

73. $3n^2(3n^2 - 5n + 2)$
$$= \mathbf{3}\boldsymbol{n}^{\mathbf{2}}(3n^2) + \mathbf{3}\boldsymbol{n}^{\mathbf{2}}(-5n) + \mathbf{3}\boldsymbol{n}^{\mathbf{2}}(2)$$
$$= (3 \cdot 3)n^{2+2} + (3 \cdot -5)n^{2+1} + (3 \cdot 2)n^2$$
$$= 9n^4 - 15n^3 + 6n^2$$

74. $x^2y(y^2 - xy) = \boldsymbol{x}^{\mathbf{2}}\boldsymbol{y}(y^2) + \boldsymbol{x}^{\mathbf{2}}\boldsymbol{y}(-xy)$
$$= x^2y^{1+2} - x^{2+1}y^{1+1}$$
$$= x^2y^3 - x^3y^2$$

75. $2x(3x^4)(x+2) = (2 \cdot 3)(x^{1+4})(x+2)$
$$= 6x^5(x+2)$$
$$= \mathbf{6}\boldsymbol{x}^{\mathbf{5}}(x) + \mathbf{6}\boldsymbol{x}^{\mathbf{5}}(2)$$
$$= 6x^{5+1} + (6 \cdot 2)x^5$$
$$= 6x^6 + 12x^5$$

76. $-a^2b^2(-a^4b^2 + a^3b^3 - ab^4 + 7a)$
$$= -\boldsymbol{a}^{\mathbf{2}}\boldsymbol{b}^{\mathbf{2}}(-a^4b^2) - \boldsymbol{a}^{\mathbf{2}}\boldsymbol{b}^{\mathbf{2}}(a^3b^3)$$
$$- \boldsymbol{a}^{\mathbf{2}}\boldsymbol{b}^{\mathbf{2}}(-ab^4) - \boldsymbol{a}^{\mathbf{2}}\boldsymbol{b}^{\mathbf{2}}(7a)$$
$$= a^{2+4}b^{2+2} - a^{2+3}b^{2+3} + a^{2+1}b^{2+4} - 7a^{2+1}b^2$$
$$= a^6b^4 - a^5b^5 + a^3b^6 - 7a^3b^2$$

77. $(x+3)(x+2) = \boldsymbol{x}(x) + \boldsymbol{x}(2) + \mathbf{3}(x) + \mathbf{3}(2)$
$$= x^2 + 2x + 3x + 6$$
$$= x^2 + 5x + 6$$

78. $(2x+1)(x-1)$
$$= \mathbf{2}\boldsymbol{x}(x) + \mathbf{2}\boldsymbol{x}(-1) + \mathbf{1}(x) + \mathbf{1}(-1)$$
$$= 2x^2 - 2x + x - 1$$
$$= 2x^2 + (-2+1)x - 1$$
$$= 2x^2 - x - 1$$

79. $(3t-3)(2t+2)$
$$= \mathbf{3}\boldsymbol{t}(2t) + \mathbf{3}\boldsymbol{t}(2) - \mathbf{3}(2t) - \mathbf{3}(2)$$
$$= 6t^2 + 6t - 6t - 6$$
$$= 6t^2 - 6$$

80. $(3n^4 - 5n^2)(2n^4 - n^2)$
$$= \mathbf{3}\boldsymbol{n}^{\mathbf{4}}(2n^4) + \mathbf{3}\boldsymbol{n}^{\mathbf{4}}(-n^2) - \mathbf{5}\boldsymbol{n}^{\mathbf{2}}(2n^4) - \mathbf{5}\boldsymbol{n}^{\mathbf{2}}(-n^2)$$
$$= 6n^{4+4} - 3n^{4+2} - 10n^{2+4} + 5n^{2+2}$$
$$= 6n^8 - 3n^6 - 10n^6 + 5n^4$$
$$= 6n^8 + (-3-10)n^6 + 5n^4$$
$$= 6n^8 - 13n^6 + 5n^4$$

81. $-a^5(a^2 - b)(5a^2 + b)$
$$= -a^5\,[\boldsymbol{a}^{\mathbf{2}}(5a^2) + \boldsymbol{a}^{\mathbf{2}}(b) - \boldsymbol{b}(5a^2) - \boldsymbol{b}(b)]$$
$$= -a^5\,[5a^4 + a^2b - 5a^2b - b^2]$$
$$= -a^5\,[5a^4 - 4a^2b - b^2]$$
$$= -\boldsymbol{a}^{\mathbf{5}}(5a^4) - \boldsymbol{a}^{\mathbf{5}}(-4a^2b) - \boldsymbol{a}^{\mathbf{5}}(-b^2)$$
$$= -5a^9 + 4a^7b + a^5b^2$$

82. $6.6(a-1)(a+1) = 6.6(a^2 - 1)$
$$= 6.6a^2 - 6.6$$

83. $\left(3t - \frac{1}{3}\right)\left(6t + \frac{5}{3}\right)$
$$= \mathbf{3}\boldsymbol{t}(6t) + \mathbf{3}\boldsymbol{t}\left(\frac{5}{3}\right) + \left(-\frac{\mathbf{1}}{\mathbf{3}}\right)6t + \left(-\frac{\mathbf{1}}{\mathbf{3}}\right)\left(\frac{5}{3}\right)$$
$$= 3t(6t) + \frac{\cancel{3}^{1} \cdot 5t}{\cancel{3}_{1}} - \frac{\cancel{3}^{1} \cdot 2t}{\cancel{3}_{1}} - \frac{1 \cdot 5}{3 \cdot 3}$$
$$= 18t^2 + 5t - 2t - \frac{5}{9}$$
$$= 18t^2 + (5-2)t - \frac{5}{9}$$
$$= 18t^2 + 3t - \frac{5}{9}$$

84. $(5.5 - 6b)(2 - 4b)$
$$= \mathbf{5.5}(2) + \mathbf{5.5}(-4b) - \mathbf{6}\boldsymbol{b}(2) - \mathbf{6}\boldsymbol{b}(-4b)$$
$$= 11 - 22b - 12b + 24b^2$$
$$= 11 + (-22 - 12)b + 24b^2$$
$$= 11 - 34b + 24b^2$$

85. $(2a-3)(4a^2+6a+9)$
$= \mathbf{2a}(4a^2+6a+9) - \mathbf{3}(4a^2+6a+9)$
$= \mathbf{2a}(4a^2) + \mathbf{2a}(6a) + \mathbf{2a}(9) - \mathbf{3}(4a^2) - \mathbf{3}(6a) - \mathbf{3}(9)$
$= 8a^3 + 12a^2 + 18a - 12a^2 - 18a - 27$
$= 8a^3 + (12-12)a^2 + (18-18)a - 27$
$= 8a^3 + 0a^2 + 0a - 27$
$= 8a^3 - 27$

86.
$$\begin{array}{r} 8x^2+x-2 \\ \underline{7x^2+x-1} \\ -8x^2-x+2 \\ 8x^3+x^2-2x \\ \underline{56x^4+7x^3-14x^2} \\ 56x^4+15x^3-21x^2-3x+2 \end{array}$$

87.
$$\begin{array}{r} 4x^2-2x+1 \\ \underline{2x+1} \\ 4x^2-2x+1 \\ \underline{8x^3-4x^2+2x} \\ 8x^3 \qquad +1 \end{array}$$

88. APPLIANCES

a. Find the perimeter of the base of the dishwasher.

$P = 2w + 2l$
$= 2(x+6) + 2(2x-1)$
$= \mathbf{2}x + \mathbf{2}(6) + \mathbf{2}(2x) - \mathbf{2}(1)$
$= 2x + 12 + 4x - 2$
$= (2+4)x + (12-2)$
$= 6x + 10$

The perimeter is **(6x + 10) in.**

b. Find the area of the base of the dishwasher.

$A = wl$
$= (x+6)(2x-1)$
$= \mathbf{x}(2x) + \mathbf{x}(-1) + \mathbf{6}(2x) + \mathbf{6}(-1)$
$= 2x^2 - x + 12x - 6$
$= 2x^2 + (-1+12)x - 6$
$= 2x^2 + 11x - 6$

The area is **$(2x^2 + 11x - 6)$ in.^{2}**

c. Find the volume occupied by the dishwasher.

$V = lwh$
$= (x+6)(2x-1)(3x)$
$= [\mathbf{x}(2x) + \mathbf{x}(-1) + \mathbf{6}(2x) + \mathbf{6}(-1)](3x)$
$= [2x^2 - x + 12x - 6](3x)$
$= [2x^2 + (-1+12)x - 6](3x)$
$= [2x^2 + 11x - 6](3x)$
$= \mathbf{3x}(2x^2) + \mathbf{3x}(11x) - \mathbf{3x}(6)$
$= (3\cdot 2)x^{1+2} + (3\cdot 11)x^{1+1} - (3\cdot 6)x$
$= 6x^3 + 33x^2 - 18x$

The volume is **$(6x^3 + 33x^2 - 18x)$ in.^{3}**

SECTION 5.7 REVIEW EXERCISES

Special Products

Find each product.

89. $(a-3)^2 = a^2 + 2(a)(-3) + (-3)^2$
$= a^2 - 6a + 9$

90. $(m+2)^3 = m^3 + 6m^2 + 12m + 8$

91. $(x+7)(x-7) = (x)^2 - (7)^2$
$= x^2 - 49$

92. $(2x-0.9)(2x+0.9) = 4x^2 - 0.81$

93. $(2y+1)^2 = (2y)^2 + 2(2y)(1) + (1)^2$
$= 4y^2 + 4y + 1$

94. $(y^2+1)(y^2-1) = y^4 - 1$

95. $(6r^2+10s)^2 = (6r^2)^2 + 2(6r^2)(10s) + (10s)^2$
$= 36r^4 + 120r^2s + 100s^2$

96. $-(8a-3c)^2 = -(64a^2 - 48ac + 9c^2)$
$= -64a^2 + 48ac - 9c^2$

97. $80s(r^2+s^2)(r^2-s^2) = 80s[(r^2)^2 - (s^2)^2]$
$= 80s[r^4 - s^4]$
$= \mathbf{80s}(r^4) + \mathbf{80s}(-s^4)$
$= 80r^4s - 80s^5$

98. $4b(3b-4)^2 = 4b[(3b)^2 + 2(3b)(-4) + (-4)^2]$
$= 4b(9b^2 - 24b + 16)$
$= 36b^3 - 96b^2 + 64b$

99. $\left(t-\frac{3}{4}\right)^2 = t^2 + 2(t)\left(-\frac{3}{4}\right) + \left(-\frac{3}{4}\right)^2$
$= t^2 - \frac{3}{2}t + \frac{9}{16}$

100. $\left(x+\frac{4}{3}\right)^2 = x^2 + 2(x)\left(\frac{4}{3}\right) + \left(\frac{4}{3}\right)^2$
$= x^2 + \frac{8}{3}x + \frac{16}{9}$

Perform the operations.

101. $3(9x^2+3x+7)-2(11x^2-5x+9)$
$= \mathbf{3}(9x^2)+\mathbf{3}(3x)+\mathbf{3}(7)-\mathbf{2}(11x^2)-\mathbf{2}(-5x)-\mathbf{2}(9)$
$= 27x^2+9x+21-22x^2+10x-18$
$= (\mathbf{27}-\mathbf{22})x^2+(\mathbf{9}+\mathbf{10})x+(21-18)$
$= 5x^2+19x+3$

102. $(5c-1)^2-(c+6)(c-6)$
$= (5c)^2+2(5c)(-1)+(-1)^2-(c^2-36)$
$= 25c^2-10c+1-c^2+36$
$= (25-1)c^2-10c+(1+36)$
$= 24c^2-10c+37$

103. GRAPHIC ARTS
$[(x+3)-1][(x-1)-1]$
$= [x+(3-1)][x+(-1-1)]$
$= (x+2)(x-2)$
$= x^2-(2)^2$
$= x^2-4$
The area of the picture is (x^2-4) in.^{2}

104. Find a polynomial that represents the area of the triangle.
$A = \frac{1}{2}bh$
$= \frac{1}{2}(10x+4)(10x-4)$
$= \frac{1}{2}(100x^2-16)$
$= \frac{\mathbf{1}}{\mathbf{2}}(100x^2)+\frac{\mathbf{1}}{\mathbf{2}}(-16)$
$= 50x^2-8$
The area of the triangle is $(50x^2-8)$ in.^{2}

SECTION 5.8 REVIEW EXERCISES
Dividing Polynomials

Divide. Do not use negative exponents in the answer.

105. $\frac{16n^8}{8n^5} = 2n^{8-5}$
$= 2n^3$

106. $\frac{-14x^2y}{21xy^3} = -\frac{14x^{2-1}y^{1-3}}{21}$
$= -\frac{2\cdot \overset{1}{\cancel{7}}\, x^1y^{-2}}{3\cdot \underset{1}{\cancel{7}}}$
$= -\frac{2x}{3y^2}$

107. $\frac{a^{15}-24a^8}{6a^{12}} = \frac{a^{15}}{6a^{12}}-\frac{24a^8}{6a^{12}}$
$= \frac{a^{15-12}}{6}-4a^{8-12}$
$= \frac{a^3}{6}-4a^{-4}$
$= \frac{a^3}{6}-\frac{4}{a^4}$

108. $\frac{15a^5b+ab^2-25b}{5a^2b}$
$= \frac{15a^5b}{5a^2b}+\frac{ab^2}{5a^2b}-\frac{25b}{5a^2b}$
$= \frac{3\cdot \overset{1}{\cancel{5}}\, a^{5-2}b^{1-1}}{\underset{1}{\cancel{5}}}+\frac{a^{1-2}b^{2-1}}{5}-\frac{5\cdot \overset{1}{\cancel{5}}\, b^{1-1}}{\underset{1}{\cancel{5}}\, a^2}$
$= 3a^3b^0+\frac{a^{-1}b^1}{5}-\frac{5b^0}{a^2}$
$= 3a^3+\frac{b}{5a}-\frac{5}{a^2}$

109. $x-1\overline{)x^2-6x+5}$ with quotient $x-5$
$\underline{-(x^2-x)}$
$5x+5$
$\underline{-(-5x+5)}$
0

110. $x+3\overline{)2x^2+7x+3}$ with quotient $2x+1$
$\underline{-(2x^2+6x)}$
$x+3$
$\underline{-(x+3)}$
0

111.
$$\begin{array}{r} 5x-6+\dfrac{4}{3x+2} \\ 3x+2\overline{)15x^2-8x-8} \\ \underline{-(15x^2+10x)} \\ -18x-8 \\ \underline{-(-18x-12)} \\ 4 \end{array}$$

112.
$$\begin{array}{r} 5y-3 \\ 5y+3\overline{)25y^2+\mathbf{0y}-9} \\ \underline{-(25y^2+15y)} \\ -15y-9 \\ \underline{-(-15y-9)} \\ 0 \end{array}$$

113.
$$\begin{array}{r} 3x^2-x-4 \\ 3x+1\overline{)9x^3+\mathbf{0x^2}-13x-4} \\ \underline{-(9x^3+3x^2)} \\ -3x^2-13x \\ \underline{-(-3x^2-x)} \\ -12x-4 \\ \underline{-(-12x-4)} \\ 0 \end{array}$$

114.
$$\begin{array}{r} 3x^2+2x+1+\dfrac{2}{2x-1} \\ 2x-1\overline{)6x^3+x^2+\mathbf{0x}+1} \\ \underline{-(6x^3-3x^2)} \\ 4x^2+0x \\ \underline{-(4x^2-2x)} \\ 2x+1 \\ \underline{-(2x-1)} \\ 2 \end{array}$$

115.
$$\begin{aligned} (y+3)(3y+2) &= \mathbf{y}(3y)+\mathbf{y}(2)+\mathbf{3}(3y)+\mathbf{3}(2) \\ &= 3y^2+2y+9y+6 \\ &= 3y^2+(2+9)y+6 \\ &= 3y^2+11y+6 \end{aligned}$$

116. BEDDING
$$\begin{array}{r} 2x^2+3x-4 \\ 2x+3\overline{)4x^3+12x^2+x-12} \\ \underline{-(4x^3+6x^2)} \\ 6x^2+x \\ \underline{-(6x^2+9x)} \\ -8x-12 \\ \underline{-(-8x-12)} \\ 0 \end{array}$$

The length of the bed is $(2x^2+3x-4)$ in.

CHAPTER 5 TEST

1. **Fill in the blanks.**
   a. In the expression y^{10}, the **base** is y and the **exponent** is 10.

   b. We call a polynomial with exactly one term a **monomia**, with exactly two terms a **binomial**, and with exactly three terms a **trinomial**.

   c. The **degree** of a term of a polynomial in one variable is the value of the exponent on the variable.

   d. $(x+y)^2$, $(x-y)^2$, and $(x+y)(x-y)$ are called **special** products .

2. Use exponents to rewrite $2xxxyyyy$. $\mathbf{2x^3y^4}$

Simplify each expression. Do not use negative exponent in the answer.

3. $$\begin{aligned} y^2(yy^3) &= y^{2+1+3} \\ &= y^6 \end{aligned}$$

4. $$\begin{aligned} \left(\frac{1}{2}x^3\right)^5(x^2)^3 &= \left(\frac{1^5}{2^5}x^{3\cdot5}\right)(x^{2\cdot3}) \\ &= \left(\frac{1}{32}x^{15}\right)(x^6) \\ &= \frac{1}{32}x^{15+6} \\ &= \frac{1}{32}x^{21} \end{aligned}$$

5. $$\begin{aligned} 3.5x^0 &= 3.5(1) \\ &= 3.5 \end{aligned}$$

6. $2y^{-5}y^2 = 2y^{-5+2}$
$= 2y^{-3}$
$= \dfrac{2}{y^3}$

7. $5^{-3} = \dfrac{1}{5^3}$
$= \dfrac{1}{125}$

8. $\dfrac{(x+1)^{15}}{(x+1)^6} = (x+1)^{15-6}$
$= (x+1)^9$

9. $\dfrac{(y^{-5})^{-4}}{yy^{-2}} = \dfrac{y^{-5 \cdot -4}}{y^{1-2}}$
$= \dfrac{y^{20}}{y^{-1}}$
$= y^{20-(-1)}$
$= y^{20+1}$
$= y^{21}$

10. $\left(\dfrac{a^2b^{-1}}{4a^3b^{-2}}\right)^3 = \left(\dfrac{a^{2-3}b^{-1-(-2)}}{4}\right)^3$
$= \left(\dfrac{a^{-1}b^{-1+2}}{4}\right)^3$
$= \left(\dfrac{b}{4a}\right)^3$
$= \dfrac{b^3}{4^3a^3}$
$= \dfrac{b^3}{64a^3}$

11. $\left(\dfrac{8}{m^6}\right)^{-2} = \left(\dfrac{m^6}{8}\right)^2$
$= \dfrac{m^{6 \cdot 2}}{8^2}$
$= \dfrac{m^{12}}{64}$

12. $\dfrac{-6a}{b^{-9}} = -6ab^9$

13. $V = (10y^4)^3$
$= 10^3 y^{4 \cdot 3}$
$= 1{,}000y^{12}$
The volume of the cube is $1{,}000y^{12}$ in.^{3}

14. ELECTRICITY
Move the decimal 18 places to the left.
$6{,}250{,}000{,}000{,}000{,}000{,}000 = 6.25 \times 10^{18}$

15. The exponent is -5.
Move the decimal 5 places to the left.
$9.3 \times 10^{-5} = 0.000093$

16. $(2.3 \times 10^{18})(4.0 \times 10^{-15})$
$= (2.3 \cdot 4.0) \times (10^{18} \times 10^{-15})$
$= (2.3 \cdot 4.0) \times (10^{18+(-15)})$
$= 9.2 \times 10^3$
$= 9{,}200$

17. $x^4 + 8x^2 - 12$ **Trinomial**

Term	Coefficient	Degree
x^4	1	4
$8x^2$	8	2
-12	-12	0

Degree of the polynomial: **4**

18. **Find the degree of the polynomial**
$3x^3y + 2x^2y^3 - 5xy^2 - 6y$ **5th degree**

Complete the table of solutions and then graph the equation.

19. $y = x^2 + 2$

x	y	(x, y)
-2	$y = (-2)^2 + 2$ $= 4 + 2$ $= 6$	$(-2, 6)$
-1	$y = (-1)^2 + 2$ $= 1 + 2$ $= 3$	$(-1, 3)$
0	$y = (0)^2 + 2$ $= 0 + 2$ $= 2$	$(0, 2)$
1	$y = (1)^2 + 2$ $= 1 + 2$ $= 3$	$(1, 3)$
2	$y = (2)^2 + 2$ $= 4 + 2$ $= 6$	$(2, 6)$

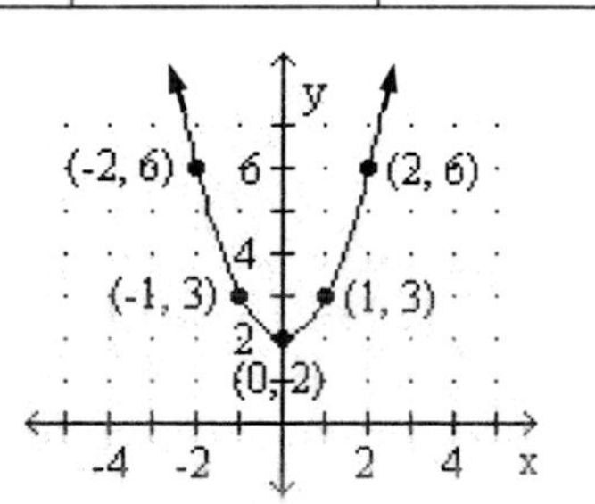

20. FREE FALL

Evaluate $-16t^2+5{,}184$ when $t=18$.

$$\begin{aligned}-16t^2+5{,}184 &= -16(\mathbf{18})^2+5{,}184\\ &= -16(324)+5{,}184\\ &= -5{,}184+5{,}184\\ &= 0\end{aligned}$$

0 ft; The rock hits the canyon floor 18 seconds after being dropped.

Simplify each polynomial.

21. $\frac{3}{5}x^2+6+\frac{1}{4}x-8-\frac{1}{2}x^2+\frac{1}{3}x$

$$\begin{aligned}&=\left(\frac{3}{5}-\frac{1}{2}\right)x^2+\left(\frac{1}{4}+\frac{1}{3}\right)x+(6-8)\\ &=\left(\frac{3}{5}\cdot\frac{\mathbf{2}}{\mathbf{2}}-\frac{1}{2}\cdot\frac{\mathbf{5}}{\mathbf{5}}\right)x^2+\left(\frac{1}{4}\cdot\frac{\mathbf{3}}{\mathbf{3}}+\frac{1}{3}\cdot\frac{\mathbf{4}}{\mathbf{4}}\right)x+(6-8)\\ &=\left(\frac{6-5}{10}\right)x^2+\left(\frac{3+4}{12}\right)x-2\\ &=\frac{1}{10}x^2+\frac{7}{12}x-2\end{aligned}$$

22. $4a^2b+5-6a^3b-3a^2b+2a^3b$

$$\begin{aligned}&=(-6+2)a^3b+(4-3)a^2b+5\\ &=-4a^3b+a^2b+5\end{aligned}$$

Perform the operations.

23. $(12.1h^3-9.9h^2+9.5)+(7.3h^3-1.2h^2-10.1)$

$$\begin{aligned}&=12.1h^3-9.9h^2+9.5+7.3h^3-1.2h^2-10.1\\ &=(12.1+7.3)h^3+(-9.9-1.2)h^2+(9.5-10.1)\\ &=19.4h^3-11.1h^2-0.6\end{aligned}$$

24. $(6b^3c-3bc)+(b^3c-2bc)-(b^3c-3bc+12)$

$$\begin{aligned}&=6b^3c-3bc+b^3c-2bc-b^3c+3bc-12\\ &=(6+1-1)b^3c+(-3-2+3)bc-12\\ &=6b^3c-2bc-12\end{aligned}$$

25.
$$\begin{array}{r}-5y^3+4y^2\qquad\quad+3\\ -(-2y^3-14y^2+17y-32)\\ \hline\end{array}$$

Change the signs of the 2nd line and add

$$\begin{array}{r}-5y^3+4y^2\qquad\quad+3\\ 2y^3+14y^2-17y+32\\ \hline -3y^3+18y^2-17y+35\end{array}$$

26. $A=2l+2w$

$$\begin{aligned}&=2(5a^2+3a-1)+2(a-9)\\ &=\mathbf{2}(5a^2)+\mathbf{2}(3a)-\mathbf{2}(1)+\mathbf{2}(a)-\mathbf{2}(9)\\ &=10a^2+6a-2+2a-18\\ &=10a^2+(6+2)a+(-2-18)\\ &=10a^2+8a-20\end{aligned}$$

The perimeter of the rectangle is $(10a^2+8a-20)$ in.

Multiply.

27. $(2x^3y^3)(5x^2y^8)=(2\cdot5)(x^{3+2})(y^{3+8})$
$$=10x^5y^{11}$$

28. $9b^3(8b^4)(-b)=(9\cdot8\cdot-1)b^{3+4+1}$
$$=-72b^8$$

29. $3y^2(y^2-2y+3)$

$$\begin{aligned}&=\mathbf{3y^2}(y^2)+\mathbf{3y^2}(-2y)+\mathbf{3y^2}(3)\\ &=3y^{2+2}-6y^{2+1}+9y^2\\ &=3y^4-6y^3+9y^2\end{aligned}$$

30. $0.6p^5(0.4p^6-0.9p^3)$

$$\begin{aligned}&=\mathbf{0.6p^5}(0.4p^6)-\mathbf{0.6p^5}(0.9p^3)\\ &=(0.6\cdot0.4)(p^{5+6})-(0.6\cdot0.9)(p^{5+3})\\ &=0.24p^{11}-0.54p^8\end{aligned}$$

31. $\frac{3}{4}s^3t^9(s^4t^8+16st)$

$$\begin{aligned}&=\mathbf{\frac{3}{4}s^3t^9}(s^4t^8)+\mathbf{\frac{3}{4}s^3t^9}(16st)\\ &=\frac{3}{4}(s^{3+4}t^{9+8})+\left(\frac{3}{4}\cdot16\right)(s^{3+1}t^{9+1})\\ &=\frac{3}{4}s^7t^{17}+12s^4t^{10}\end{aligned}$$

32. $(x-5)(3x+4)=3x^2+4x-15x-20$

$$\begin{aligned}&=3x^2+(4-15)x-20\\ &=3x^2-11x-20\end{aligned}$$

33. $\left(6t+\frac{1}{2}\right)\left(2t-\frac{3}{2}\right)$

$$= \mathbf{6t}(2t)+\mathbf{6t}\left(-\frac{3}{2}\right)+\left(\frac{\mathbf{1}}{\mathbf{2}}\right)2t+\left(\frac{\mathbf{1}}{\mathbf{2}}\right)\left(-\frac{3}{2}\right)$$

$$= 6t(2t)-\frac{\overset{1}{\cancel{2}}\cdot 3t\cdot 3}{\underset{1}{\cancel{2}}}+\frac{\overset{1}{\cancel{2}}\cdot t}{\underset{1}{\cancel{2}}}-\frac{1\cdot 3}{2\cdot 2}$$

$$= 12t^2-9t+t-\frac{3}{4}$$

$$= 12t^2-8t-\frac{3}{4}$$

34. $(3.8m-1)(2m-1)$

$$= \mathbf{3.8m}(2m)-\mathbf{3.8m}(1)-\mathbf{1}(2m)-\mathbf{1}(-1)$$
$$= 7.6m^2-3.8m-2m+1$$
$$= 7.6m^2-5.8m+1$$

35. $(a^3-6)(a^3+7)$

$$= \mathbf{a^3}(a^3)+\mathbf{a^3}(7)-\mathbf{6}(a^3)-\mathbf{6}(7)$$
$$= a^6+7a^3-6a^3-42$$
$$= a^6+(7-6)a^3-42$$
$$= a^6+a^3-42$$

36.

$$\begin{array}{r} x^2-2x+4 \\ \underline{2x-3} \\ -3x^2+6x-12 \\ \underline{2x^3-4x^2+8x} \\ 2x^3-7x^2+14x-12 \end{array}$$

37. $(1+10c)(1-10c)=(1)^2-(10c)^2$

$$=1-100c^2$$

38. $(7b^3-3t)^2=(7b^3)^2+2(7b^3)(-3t)+(-3t)^2$

$$=49b^6-42b^3t+9t^2$$

39. $(2.2a)(a+5)(a-3)$

$$= 2.2a[\mathbf{a}(a)+\mathbf{a}(-3)+\mathbf{5}(a)+\mathbf{5}(-3)]$$
$$= 2.2a[a^2-3a+5a-15]$$
$$= 2.2a[a^2+2a-15]$$
$$= \mathbf{2.2a}(a^2)+\mathbf{2.2a}(2a)+\mathbf{2.2a}(-15)$$
$$= 2.2a^3+4.4a^2-33a$$

40. $(x+y)(x-y)+(x+y)^2$

$$= x^2-y^2+x^2+2xy+y^2$$
$$= (1+1)x^2+(-1+1)y^2+2xy$$
$$= 2x^2+2xy$$

Divide.

41. $\frac{6a^2-12b^2}{24ab}=\frac{6a^2}{24ab}-\frac{12b^2}{24ab}$

$$=\frac{a^{2-1}}{4b}-\frac{b^{2-1}}{2a}$$
$$=\frac{a}{4b}-\frac{b}{2a}$$

42.

$$\begin{array}{r} x-2 \\ x+3\overline{)x^2+\ \ x-6} \\ \underline{-(x^2+3x)} \\ -2x-6 \\ \underline{-(-2x-6)} \\ 0 \end{array}$$

43.

$$\begin{array}{r} 3x^2+2x+1+\frac{2}{2x-1} \\ 2x-1\overline{)6x^3+\ x^2+\ \mathbf{0x}+1} \\ \underline{-(6x^3-3x^2)} \\ 4x^2+\ 0x \\ \underline{-(4x^2-\ 2x)} \\ 2x+1 \\ \underline{-(2x-1)} \\ 2 \end{array}$$

44.

$$\begin{array}{r} x-5 \\ x-1\overline{)x^2-6x+5} \\ \underline{-(x^2-x)} \\ -5x+5 \\ \underline{-(-5x+5)} \\ 0 \end{array}$$

The width of the rectangle is $(x-5)$ ft.

45. $(5m+1)(m-6)$

$$= \mathbf{5m}(m)+\mathbf{5m}(-6)+\mathbf{1}(m)+\mathbf{1}(-6)$$
$$= 5m^2-30m+m-6$$
$$= 5m^2+(-30+1)m-6$$
$$= 5m^2-29m-6$$

Yes, it checks.

46. Is $(a+b)^2=a^2+b^2$?

No; $(a+b)^2=a^2+2ab+b^2$

CUMULATIVE REVIEW CHAPTERS 1 - 5

1. Prime Factorization: [Section 1.2]

$$\begin{aligned}270 &= 2\cdot 135\\ &= 2\cdot 5\cdot 27\\ &= 2\cdot 5\cdot 3\cdot 9\\ &= 2\cdot 5\cdot 3\cdot 3\cdot 3\\ &= 2\cdot 3^3\cdot 5\end{aligned}$$

2. a. Commutative Property of Addition: [Section 1.4]

$a+b=\mathbf{b+a}$

b. Associative Property of Multiplication: [Section 1.6]

$(xy)z=\mathbf{x(yz)}$

Evaluate each expression. [Section 1.7]

3. $$\begin{aligned}3-4[-10-4(-5)] &= 3-4(-10+20)\\ &= 3-4(10)\\ &= 3-40\\ &= 3+(-40)\\ &= -37\end{aligned}$$

4. $$\begin{aligned}\frac{|-45|-2(-5)+1^5}{2\cdot 9-2^4} &= \frac{45-2(-5)+1}{2\cdot 9-16}\\ &= \frac{45+10+1}{18-16}\\ &= \frac{56}{2}\\ &= 28\end{aligned}$$

Simplify each expression. [Section 1.9]

5. $$\begin{aligned}27\left(\frac{2}{3}x\right) &= \frac{27\cdot 2}{3}x\\ &= \frac{3\cdot 9\cdot 2}{3}x\\ &= \frac{\overset{1}{\cancel{3}}\cdot 9\cdot 2}{\underset{1}{\cancel{3}}}x\\ &= 18x\end{aligned}$$

6. $$\begin{aligned}3x^2+2x^2-5x^2 &= (3+2-5)x^2\\ &= 0\end{aligned}$$

Solve each equation. [Section 2.2]

7. $$\begin{aligned}2-(4x+7) &= 3+2(x+2)\\ 2-4x-7 &= 3+2x+4\\ -4x-5 &= 2x+7\\ -4x-5+\mathbf{4x} &= 2x+\mathbf{4x}+7\\ -5 &= 6x+7\\ -5-\mathbf{7} &= 6x+7-\mathbf{7}\\ -12 &= 6x\\ \frac{-12}{\mathbf{6}} &= \frac{6x}{\mathbf{6}}\\ -2 &= x\end{aligned}$$

The solution is -2.

8. $$\begin{aligned}\frac{2}{5}y+3 &= 9\\ \mathbf{5}\left(\frac{2}{5}y+3\right) &= \mathbf{5}(9)\\ \mathbf{5}\left(\frac{2}{5}y\right)+\mathbf{5}(3) &= \mathbf{5}(9)\\ 2y+15 &= 45\\ 2y+15-\mathbf{15} &= 45-\mathbf{15}\\ 2y &= 30\\ \frac{2y}{\mathbf{2}} &= \frac{30}{\mathbf{2}}\\ y &= 15\end{aligned}$$

The solution is 15.

9. CANDY SALES [Section 2.3]

34% of $6.5 billions is what number?

$$\begin{aligned}0.34 \quad \cdot \quad 6.5 \quad &= x\\ 0.34(6.5) &= x\\ 2.21 &= x\end{aligned}$$

The candy sales is $2.21 billion.

10. AIR CONDITIONING [Section 2.4]

One will need to change the radius of 3 inches into part of a foot before using the data in the formula. 12 inches in a foot.

$$\begin{aligned}V &= \pi r^2 h\\ &= 3.14\left(\frac{3}{12}\right)^2(6)\\ &= 3.14(0.25)^2(6)\\ &= 3.14(0.0625)(6)\\ &= 3.14(0.375)\\ &= 1.1775\end{aligned}$$

The volume of air is about 1.2 ft^3.

11. ANGLE OF ELEVATION [Section 2.5]

Analyze

- $x°$ is the measure of one acute angle.
- $2x°$ is the measure of other acute angle.
- $90°$ is the measure of the right angle.
- What is the value of x?

Assign

Let x = the measure of one acute angle
$2x$ = the measure of other acute angle

Form

Angle 1	plus	angle 2	plus	angle 3	equals	the total of 180.
x	$+$	$2x$	$+$	90	$=$	180

Solve

$$x+2x+90=180$$
$$3x+90=180$$
$$3x+90-\mathbf{90}=180-\mathbf{90}$$
$$3x=90$$
$$\frac{3x}{\mathbf{3}}=\frac{90}{\mathbf{3}}$$
$$x=30$$

State

The value of x is 30.

Check

$$30+2(30)+90=180$$
$$30+60+90=180$$
$$180=180$$

The result checks.

12. LIVESTOCK AUCTION [Section 2.5]

Analyze

- Wants to make \$6,000.
- 4% commission.
- What is the selling price?

Assign

Let x = selling price of the hog

Form

The selling price of the hog	minus	the amount of the commission	equals	the amount made on the auction.
x	$-$	$0.04x$	$=$	$6,000$

Solve

$$x-0.04x=6,000$$
$$0.96x=6,000$$
$$\frac{0.96x}{\mathbf{0.96}}=\frac{6,000}{\mathbf{0.96}}$$
$$x=6,250$$

State

\$6,250 is the selling price.

Check

$$x-0.04x=6,000$$
$$6,250-0.04(6,250)=6,000$$
$$6,250-250=6,000$$
$$6,000=6,000$$

The result checks.

13. STOCK MARKET [Section 2.6]

Analyze

- \$45,000 is the total investment.
- Mutual fund at 12% annual interest.
- Treasury bond at 6.5% annual interest.
- \$4,300 is the first-year combined income.
- How much is invested in each account?

Assign

Let x = amount invested at 12%
$45,000-x$ = amount invested at 6.5%

Form

	Principal •	rate •	time =	interest
Mutual	x	0.12	1	$\mathbf{0.12x}$
Treasury	$45,000-x$	0.065	1	$\mathbf{0.065(45,000-x)}$
			Total:	**4,300**

The interest earned at 12%	plus	the interest earned at 6.5%	equals	the total interest.
$0.12x$	$+$	$0.065(45,000-x)$	$=$	$4,300$

Solve

$$0.12x+0.065(45,000-x)=4,300$$
$$\mathbf{1,000}[0.12x+0.065(45,000-x)]=\mathbf{1,000}(4,300)$$
$$1,000(0.12x)+1,000[0.065(45,000-x)]=1,000(4,300)$$
$$120x+65(45,000-x)=4,300,000$$
$$120x+2,925,000-65x=4,300,000$$
$$55x+2,925,000=4,300,000$$
$$55x+2,925,000-\mathbf{2,925,000}=4,300,000-\mathbf{2,925,000}$$
$$55x=1,375,000$$
$$\frac{55x}{\mathbf{55}}=\frac{1,375,000}{\mathbf{55}}$$
$$x=25,000$$

6.5% investment

$$45,000-x=45,000-25,000$$
$$=20,000$$

State

\$25,000 was invested in the mutual fund.
\$20,000 was invested in treasury bonds.

Check

The results check.

14. Solve: [Section 2.7]

$$-4x+6>17$$
$$-4x+6-\mathbf{6}>17-\mathbf{6}$$
$$-4x>11$$
$$\frac{-4x}{\mathbf{-4}}<\frac{11}{\mathbf{-4}}$$
$$x<-\frac{11}{4}$$
$$\left(-\infty,-\frac{11}{4}\right)$$

$-\frac{11}{4}$

Graph each equation.

15. $y=3x$ [Section 3.2]

x	y	(x,y)
-1	$y=3x$ $=3(-1)$ $=-3$	$(-1,-3)$
0	$y=3x$ $=3(0)$ $=0$	$(0, 0)$
1	$y=3x$ $=3(1)$ $=3$	$(1, 3)$

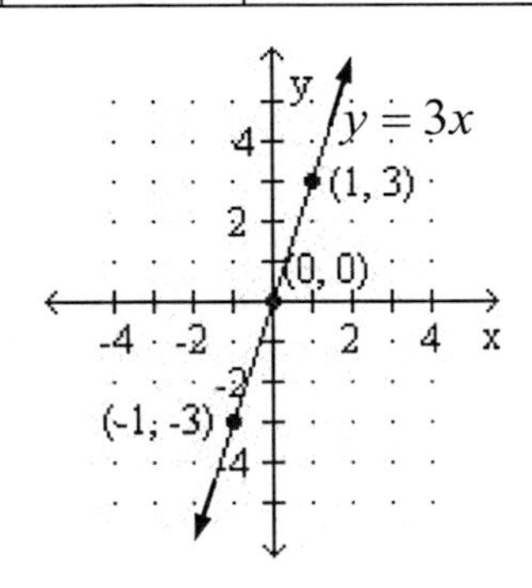

16. $x=-2$ [Section 3.3]

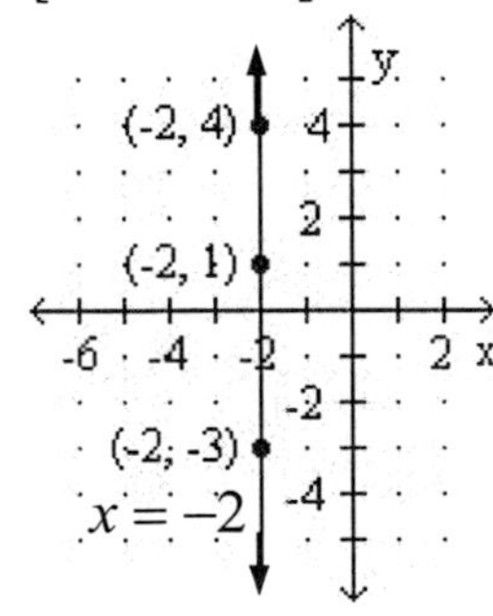

17. Find the slope of the line passing through $(6,-2)$ and $(-3,2)$. [Section 3.4]

Let $(x_1, y_1)=(6,-2)$
Let $(x_2, y_2)=(-3,2)$

$$m=\frac{y_2-y_1}{x_2-x_1}$$
$$=\frac{2-(-2)}{-3-6}$$
$$=\frac{2+2}{-3+(-6)}$$
$$=-\frac{4}{9}$$

The slope of the line is $-\frac{4}{9}$.

18. DIABETES [Section 3.4]

Select any two points that lie on the line.

Let $(x_1, y_1)=(1996,\ 8.8)$
Let $(x_2, y_2)=(2008,\ 17.8)$

$$m=\frac{y_2-y_1}{x_2-x_1}$$
$$=\frac{17.8-8.8}{2008-1996}$$
$$=\frac{9}{12}$$
$$=\frac{3}{4}$$

0.75 million people per year or

$\frac{3}{4}$ million people per year.

19. Find the slope and y-intercept of the line. Then write the equation of the line. [Section 3.5]

$$m=\frac{3}{1}$$
$$=3$$

y-intercept $=(0,-2)$

$$y=mx+b$$
$$=3x+(-2)$$
$$=3x-2$$

The equation of the line is $y=3x-2$.

20. Determine if the graphs of $y=\frac{3}{2}x-1$ and $2x+3y=10$ are parallel, perpendicular, or neither. [Section 3.5]

$y=\frac{3}{2}x-1$

$m=\frac{3}{2}$

$2x+3y=10$

$3y=-2x+10$

$y=-\frac{2}{3}x+\frac{10}{3}$

$m=-\frac{2}{3}$

The slopes are negative reciprocals. The lines are perpendicular.

Find an equation of the line with the following properties. Write the equation in slope - intercept form.

21. Passing through $(-2,10)$ with slope -4. [Section 3.6]

$$(-2,10),\quad m=-4$$
$$y-y_1=m(x-x_1)$$
$$y-10=-4[x-(-2)]$$
$$y-10=-4(x+2)$$
$$y-10=-4x-8$$
$$y-10+\mathbf{10}=-4x-8+\mathbf{10}$$
$$y=-4x+2$$

22. Is $(-2,1)$ a solution of $2x-3y\ge-6$? [Section 3.7]

$$2x-3y\ge-6$$
$$2(\mathbf{-2})-3(\mathbf{1})\overset{?}{\ge}-6$$
$$-4-3\overset{?}{\ge}-6$$
$$-7\ge-6$$

False

$(-2,1)$ is not a solution.

23. If $f(x)=2x^2+3x-9$, find $f(-5)$. [Section 3.8]

$$f(x)=2x^2+3x-9$$
$$f(\mathbf{-5})=2(\mathbf{-5})^2+3(\mathbf{-5})-9$$
$$f(-5)=2(25)+(-15)-9$$
$$f(-5)=50+(-15)+(-9)$$
$$f(-5)=35+(-9)$$
$$f(-5)=26$$

24. Is the graph of function? [Section 3.8]

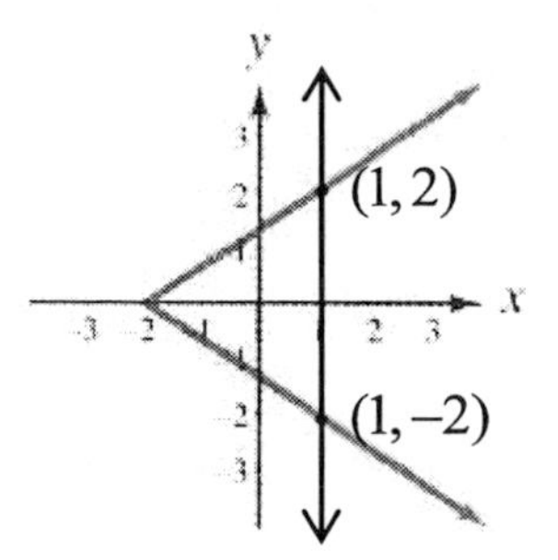

It is not a graph of a function, because the vertical line intersects the graph at points $(1,2)$ and $(1,-2)$ which have the same x-coordinate. There are many more points that demonstrate this same concept.

25. $\left(\frac{2}{3},-1\right)$, $\begin{cases} y=-3x+1 \\ 3x+3y=-2 \end{cases}$ [Section 4.1]

$y=-3x+1$

$-1\overset{?}{=}-3\left(\frac{2}{3}\right)+1$

$-1\overset{?}{=}-2+1$

$-1=-1$

True

$3x+3y=-2$

$3\left(\frac{2}{3}\right)+3(-1)\overset{?}{=}-2$

$2+(-3)\overset{?}{=}-2$

$-1=-2$

False

Since $\left(\frac{2}{3},-1\right)$ does not satisfies both equations, it is not a solution of the system.

26. Solve the system $\begin{cases} 3x+2y=14 \\ y=\frac{1}{4}x \end{cases}$ by graphing. [Section 4.1]

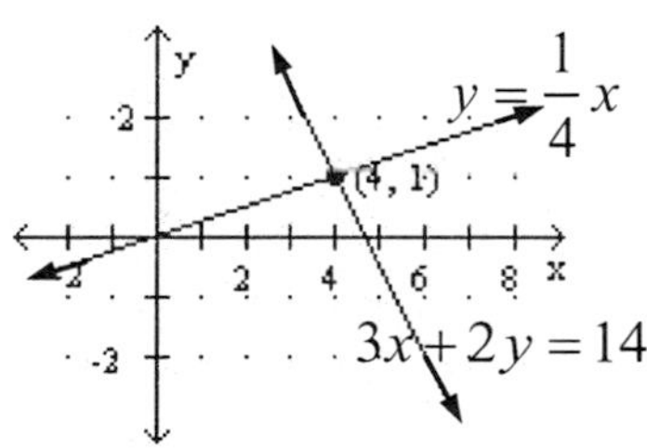

The solution is $(4,1)$.

27. Solve the system $\begin{cases} 2b-3a=18 \\ a+3b=5 \end{cases}$ by substitution. [Section 4.2]

$$\begin{cases} 2b-3a=18 \\ a+3b=5 \end{cases}$$

Solve for a in 2nd equation.

$$\begin{cases} 2b-3a=18 \\ a=5-3b \end{cases}$$

$2b-3a=18$ The 1st equation
Substitute for a.

$$2b-3(\mathbf{5-3b})=18$$
$$2b-15+9b=18$$
$$11b-15=18$$
$$11b-15+\mathbf{15}=18+\mathbf{15}$$
$$11b=33$$
$$\frac{11b}{\mathbf{11}}=\frac{33}{\mathbf{11}}$$
$$b=3$$

$a=5-3b$ The 2nd equation

$$a=5-3(3)$$
$$a=5+(-9)$$
$$a=-4$$

The solution is $(-4, 3)$.

28. Solve the system $\begin{cases} 8s+10t=24 \\ 11s-3t=-34 \end{cases}$ by elimination (addition). [Section 4.3]

$$\begin{cases} 8s+10t=24 \\ 11s-3t=-34 \end{cases}$$

Eliminate t.
Multiply both sides of the 1st equation by 3.
Multiply both sides of the 2nd equation by 10.

$$\begin{array}{r} 24s+30t=72 \\ 110s-30t=-340 \\ \hline 134s=-268 \end{array}$$

$$\frac{134s}{\mathbf{134}}=\frac{-268}{\mathbf{134}}$$
$$s=-2$$

$8s+10t=24$ The 1st equation

$$8(\mathbf{-2})+10t=24$$
$$-16+10t=24$$
$$-16+10t+\mathbf{16}=24+\mathbf{16}$$
$$10t=40$$
$$\frac{10t}{\mathbf{10}}=\frac{40}{\mathbf{10}}$$
$$t=4$$

The solution is $(-2,4)$.

29. VACATION [Section 4.4]

Analyze

- Tickets for 2 adults, 3 children cost $275.
- Tickets for 3 adults, 3 children cost $336.
- What does an adult and child ticket cost?

Assign

Let x = cost of adult ticket
y = cost of child ticket

Form

	Number •	Value =	Total Value
Adult	2	$x	**2x**
Child	3	$y	**3y**
		Total	**275**

The cost of 2 adults tickets	plus	the cost of 3 children tickets	is	275.
$2x$	$+$	$3y$	$=$	275

	Number •	Value =	Total Value
Adult	3	$x	**3x**
Child	3	$y	**3y**
		Total	**336**

The cost of 3 adults tickets	plus	the cost of 3 children tickets	is	336.
$3x$	$+$	$3y$	$=$	336

Solve

$$\begin{cases} 2x+3y=275 \\ 3x+3y=336 \end{cases}$$

Eliminate y.
Multiply both sides of the 1st equation by -1.

$$\begin{array}{r} -2x-3y=-275 \\ 3x+3y=336 \\ \hline x=61 \end{array}$$

$2x+3y=275$ The 1st equation

$$\mathbf{2(61)}+3y=275$$
$$122+3y-\mathbf{122}=275-\mathbf{122}$$
$$3y=153$$
$$\frac{3y}{\mathbf{3}}=\frac{153}{\mathbf{3}}$$
$$y=51$$

State

$61 is the cost of an adult ticket.
$51 is the cost of a child ticket.

Check

The results check.

30. Graph: $\begin{cases} y \le 2x-1 \\ x+3y>6 \end{cases}$ [Section 4.5]

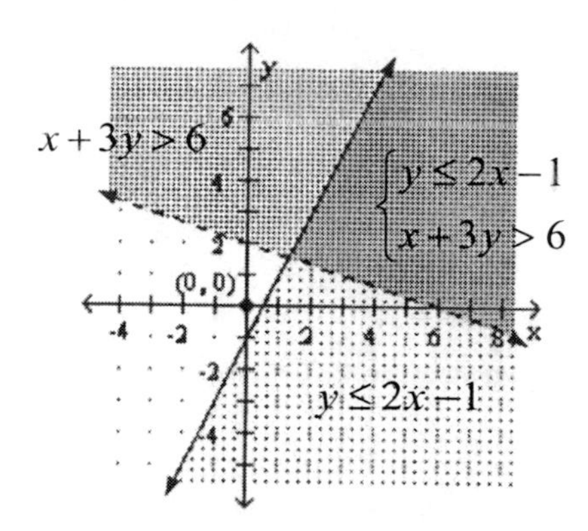

Simplify: Do not use negative exponents in the answer. [31-34 Section 5.1]

31. $(-3x^2y^4)^2 = (-3)^2(x^2)^2(y^4)^2$
$= 9x^{2\cdot 2}y^{4\cdot 2}$
$= 9x^4y^8$

32. $(v^5)^2(v^3)^4 = (v^{5\cdot 2})(v^{3\cdot 4})$
$= v^{10}v^{12}$
$= v^{10+12}$
$= v^{22}$

33. $ab^3c^4 \cdot ab^4c^2 = aa \cdot b^3b^4 \cdot c^4c^2$
$= a^{1+1} \cdot b^{3+4} \cdot c^{4+2}$
$= a^2b^7c^6$

34. $\left(\dfrac{4t^3t^4t^5}{3t^2t^6}\right)^3 = \left(\dfrac{4t^{3+4+5}}{3t^{2+6}}\right)^3$
$= \left(\dfrac{4t^{12}}{3t^8}\right)^3$
$= \left(\dfrac{4t^{12-8}}{3}\right)^3$
$= \left(\dfrac{4t^4}{3}\right)^3$
$= \dfrac{4^3t^{4\cdot 3}}{3^3}$
$= \dfrac{64t^{12}}{27}$

[35-38 Section 5.2]

35. $(2y)^{-4} = \dfrac{1}{(2y)^4}$
$= \dfrac{1}{2^4y^4}$
$= \dfrac{1}{16y^4}$

36. $\dfrac{a^4b^0}{a^{-3}} = a^{4-(-3)} \cdot 1$
$= a^{4+3}$
$= a^7$

37. $-5^{-2} = \dfrac{-1}{5^2}$
$= -\dfrac{1}{25}$

38. $\left(\dfrac{a}{x}\right)^{-10} = \left(\dfrac{x}{a}\right)^{10}$
$= \dfrac{x^{10}}{a^{10}}$

Write each number in scientific notation.

39. 615,000 [Section 5.3]
Move the decimal 5 places to the left.
$615{,}000 = 6.15 \times 10^5$

40. 0.0000013 [Section 5.3]
Move the decimal 6 places to the right.
$0.0000013 = 1.3 \times 10^{-6}$

41. Graph: $y = x^2$. [Section 5.4]

x	y	(x, y)
−3	$y = (-3)^2$ $= 9$	(−3, 9)
−2	$y = (-2)^2$ $= 4$	(−2, 4)
−1	$y = (-1)^2$ $= 1$	(−1, 1)
0	$y = (0)^2$ $= 0$	(0, 0)
1	$y = (1)^2$ $= 1$	(1, 1)
2	$y = (2)^2$ $= 4$	(2, 4)
3	$y = (3)^2$ $= 9$	(3, 9)

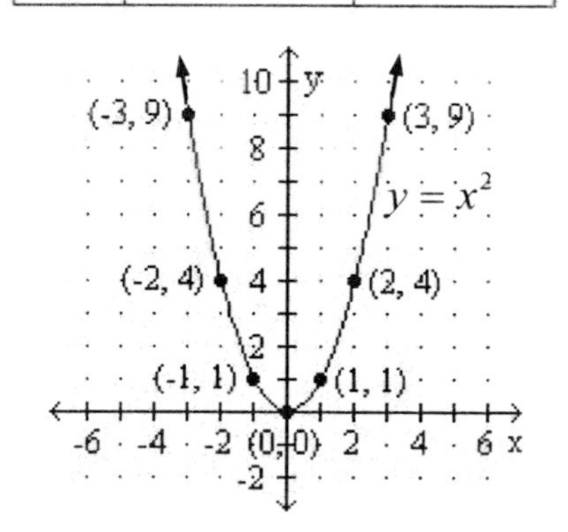

42. MUSICAL INSTRUMENT [Section 5.4]

$$0.01875x^4-0.15x^3+1.2x$$
$$=0.01875(\mathbf{2})^4-0.15(\mathbf{2})^3+1.2(\mathbf{2})$$
$$=0.01875(16)-0.15(8)+1.2(2)$$
$$=0.3-1.2+2.4$$
$$=1.5$$

The amount of deflection is 1.5 in.

Perform the operations.

43. $(4c^2+3c-2)+(3c^2+4c+2)$ [Section 5.5]

$$=(4c^2+3c^2)+(3c+4c)+(-2+2)$$
$$=(4+3)c^2+(3+4)c+(-2+2)$$
$$=7c^2+7c$$

44. Subtract: [Section 5.5]

$$\begin{array}{r} 17x^4-3x^2-65x-12 \\ -\underline{(23x^4+14x^2+3x-23)} \end{array}$$

Change the signs of the 2nd line and then add.

$$\begin{array}{r} 17x^4-3x^2-65x-12 \\ \underline{-23x^4-14x^2-3x+23} \\ -6x^4-17x^2-68x+11 \end{array}$$

45. $(2t+3s)(3t-s)$ [Section 5.6]

$$=\mathbf{2t}(3t)+\mathbf{2t}(-s)+\mathbf{3s}(3t)+\mathbf{3s}(-s)$$
$$=6t^2-2st+9st-3s^2$$
$$=6t^2+(-2+9)st-3s^2$$
$$=6t^2+7st-3s^2$$

46. $3x(2x+3)^2$ [Section 5.7]

$$=3x[(2x)^2+2(2x)(3)+(3)^2]$$
$$=3x(4x^2+12x+9)$$
$$=\mathbf{3x}(4x^2)+\mathbf{3x}(12x)+\mathbf{3x}(9)$$
$$=(3\cdot 4)x^{1+2}+(3\cdot 12)x^{1+1}+(3\cdot 9)x$$
$$=12x^3+36x^2+27x$$

47. [Section 5.8]

$$\begin{array}{r} 2x+1 \\ 5x+3\overline{)10x^2+11x+3} \\ \underline{-(10x^2+6x)} \\ 5x+3 \\ \underline{-(5x+3)} \\ 0 \end{array}$$

48. [Section 5.8]

$$\frac{2x-32}{16x}=\frac{2x}{16x}-\frac{32}{16x}$$
$$=\frac{\overset{1}{\cancel{2}}\,\overset{1}{\cancel{x}}}{\underset{1}{\cancel{2}}\cdot 8\,\underset{1}{\cancel{x}}}-\frac{2\cdot\overset{1}{\cancel{16}}}{\underset{1}{\cancel{16}}\,x}$$
$$=\frac{1}{8}-\frac{2}{x}$$

SECTION 6.1 STUDY SET

VOCABULARY

Fill in the blanks.

1. To **factor** a polynomial means to express it as a product of two (or more) polynomials.

3. To factor $m^3 + 3m^2 + 4m + 12$ by **grouping**, we begin by writing $m^2(m + 3) + 4(m + 3)$.

CONCEPTS

5. Complete each factorization.

a. $6x = 2 \cdot \mathbf{3} \cdot x$

b. $35h^2 = 5 \cdot \mathbf{7} \cdot h \cdot \mathbf{h}$

c. $18y^3z = 2 \cdot \mathbf{3} \cdot 3 \cdot \mathbf{y} \cdot y \cdot \mathbf{y} \cdot z$

7.a. Write a binomial such that the GCF of its terms is 2. **$2x + 4$ (answers may vary)**

b. Write a trinomial such that the GCF of its terms is x. **$x^3 + x^2 + x$ (answers may vary)**

Fill in the blanks to complete each factorization.

9. $2x^2 + 6x = \mathbf{2x} \cdot x + \mathbf{2x} \cdot 3$

$= \mathbf{2x}(\mathbf{x + 3})$

11. Consider the polynomial $2k - 8 + hk - 4h$.

a. How many terms does the polynomial have? **4**

b. Is there a common factor of all the terms, other than 1? **No**

c. What is the GCF of the first two terms and what is the GCF of the last two terms? **2; h**

NOTATION

Complete each factorization.

13. $8m^2 - 32m + 16 = \mathbf{8}(m^2 - 4m + 2)$

15. $b^3 - 6b^2 + 2b - 12 = \mathbf{b^2}(b - 6) + \mathbf{2}(b - 6)$

$= (\mathbf{b - 6})(b^2 + 2)$

GUIDED PRACTICE

Find the GCF of each list of numbers. See Example 1.

17. $6 = \mathbf{2} \cdot 3$
$10 = \mathbf{2} \cdot 5$
GCF is 2.

19. $18 = \mathbf{2} \cdot \mathbf{3} \cdot 3$
$24 = \mathbf{2} \cdot 2 \cdot 2 \cdot \mathbf{3}$
GCF is $2 \cdot 3 = 6$.

21. $14 = 2 \cdot \mathbf{7}$
$21 = 3 \cdot \mathbf{7}$
$42 = 2 \cdot 3 \cdot \mathbf{7}$
GCF is 7.

23. $40 = \mathbf{2} \cdot \mathbf{2} \cdot \mathbf{2} \cdot 5$
$32 = \mathbf{2} \cdot \mathbf{2} \cdot \mathbf{2} \cdot 2 \cdot 2$
$24 = \mathbf{2} \cdot \mathbf{2} \cdot \mathbf{2} \cdot 3$
GCF is $2 \cdot 2 \cdot 2 = 8$.

Find the GCF of each list of terms. See Example 2.

25. $m^4 = \mathbf{m} \cdot \mathbf{m} \cdot \mathbf{m} \cdot m$
$m^3 = \mathbf{m} \cdot \mathbf{m} \cdot \mathbf{m}$
GCF is $m \cdot m \cdot m = m^3$.

27. $15x = 3 \cdot \mathbf{5} \cdot x$
$25 = \mathbf{5} \cdot 5$
GCF is 5.

29. $20c^2 = \mathbf{2} \cdot \mathbf{2} \cdot 5 \cdot \mathbf{c} \cdot c$
$12c = \mathbf{2} \cdot \mathbf{2} \cdot 3 \cdot \mathbf{c}$
GCF is $2 \cdot 2 \cdot c = 4c$.

31. $18a^4 = 2 \cdot \mathbf{3} \cdot \mathbf{3} \cdot \mathbf{a} \cdot \mathbf{a} \cdot \mathbf{a} \cdot a$
$9a^3 = \mathbf{3} \cdot \mathbf{3} \cdot \mathbf{a} \cdot \mathbf{a} \cdot \mathbf{a}$
$27a^3 = \mathbf{3} \cdot \mathbf{3} \cdot 3 \cdot \mathbf{a} \cdot \mathbf{a} \cdot \mathbf{a}$
GCF is $3 \cdot 3 \cdot a \cdot a \cdot a = 9a^3$.

33. $24a^2 = \mathbf{2} \cdot \mathbf{2} \cdot \mathbf{2} \cdot 3 \cdot \mathbf{a} \cdot a$
$16a^3b = \mathbf{2} \cdot \mathbf{2} \cdot \mathbf{2} \cdot 2 \cdot \mathbf{a} \cdot a \cdot a \cdot b$
$40ab = \mathbf{2} \cdot \mathbf{2} \cdot \mathbf{2} \cdot 5 \cdot \mathbf{a} \cdot b$
GCF is $2 \cdot 2 \cdot 2 \cdot a = 8a$.

35. $6m^4n = 2 \cdot \mathbf{3} \cdot \mathbf{m} \cdot \mathbf{m} \cdot \mathbf{m} \cdot m \cdot \mathbf{n}$
$12m^3n^2 = 2 \cdot 2 \cdot \mathbf{3} \cdot \mathbf{m} \cdot \mathbf{m} \cdot \mathbf{m} \cdot \mathbf{n} \cdot n$
$9m^3n^3 = \mathbf{3} \cdot 3 \cdot \mathbf{m} \cdot \mathbf{m} \cdot \mathbf{m} \cdot \mathbf{n} \cdot n \cdot n$
GCF is $3 \cdot m \cdot m \cdot m \cdot n = 3m^3n$.

37. $4(x+7) = 2 \cdot 2 \cdot \mathbf{(x+7)}$
$9(x+7) = 3 \cdot 3 \cdot \mathbf{(x+7)}$
GCF is $x + 7$.

39. $4(p-t) = 2 \cdot 2 \cdot \mathbf{(p-t)}$
$p(p-t) = p \cdot \mathbf{(p-t)}$
GCF is $p - t$.

Factor out the GCF. See Example 3.

41. $3x + 6 = \mathbf{3} \cdot x + \mathbf{3} \cdot 2$
$= \mathbf{3}(x + 2)$

43. $18m - 9 = \mathbf{9} \cdot 2m - \mathbf{9} \cdot 1$
$= \mathbf{9}(2m - 1)$

45. $d^2 - 7d = \mathbf{d} \cdot d - \mathbf{d} \cdot 7$
$= \mathbf{d}(d - 7)$

47. $15c^3 + 25 = \mathbf{5} \cdot 3c^3 + \mathbf{5} \cdot 5$
$- \mathbf{5}(3c^3 + 5)$

49. $24a - 16a^2 = \mathbf{8a} \cdot 3 - \mathbf{8a} \cdot 2a$
$= \mathbf{8a}(3 - 2a)$

51. $14x^2 - 7x - 7 = \mathbf{7} \cdot 2x^2 - \mathbf{7} \cdot x - \mathbf{7} \cdot 1$
$= \mathbf{7}(2x^2 - x - 1)$

53. $t^4 + t^3 + 2t^2 = \mathbf{t^2} \cdot t^2 + \mathbf{t^2} \cdot t + \mathbf{t^2} \cdot 2$
$= \mathbf{t^2}(t^2 + t + 2)$

55. $21x^2y^3 + 3xy^2 = \mathbf{3xy^2} \cdot 7xy + \mathbf{3xy^2} \cdot 1$
$= \mathbf{3xy^2}(7xy + 1)$

Factor out –1 from each polynomial. See Example 3.

57. $-a-b=(-1)a+(-1)b$
$=(-1)(a+b)$
$=-(a+b)$

59. $-x^2-x+16=(-1)x^2+(-1)(x)+(-1)(-16)$
$=(-1)(x^2+x-16)$
$=-(x^2+x-16)$

61. $5-x=(-1)(-5)+(-1)(x)$
$=(-1)(-5+x)$
$=-(-5+x)$ or $-(x-5)$

63. $9-4a=(-1)(-9)+(-1)(4a)$
$=(-1)(-9+4a)$
$=-(-9+4a)$ or $-(4a-9)$

Factor each polynomial by factoring out the opposite of the GCF. See Example 5.

65. $-3x^2-6x=(-3x)(x)+(-3x)(2)$
$=-3x(x+2)$

67. $-4a^2b+12a^3=(-4a^2)(b)+(-4a^2)(-3a)$
$=-4a^2(b-3a)$

69. $-24x^4-48x^3+36x^2$
$=(-12x^2)2x^2+(-12x^2)(4x)+(-12x^2)(-3)$
$=-12x^2(2x^2+4x-3)$

71. $-4a^3b^2+14a^2b^2-10ab^2$
$=(-2ab^2)2a^2+(-2ab^2)(-7a)+(-2ab^2)(5)$
$=-2ab^2(2a^2-7a+5)$

Factor each expression. See Example 6.

73. $y(x+2)+3(x+2)=y(x+2)+3(x+2)$
$=(x+2)(y+3)$

75. $m(p-q)-5(p-q)=m(p-q)-5(p-q)$
$=(p-q)(m-5)$

Factor by grouping. See Example 7.

77. $2x+2y+ax+ay=(2x+2y)+(ax+ay)$
$=2(x+y)+a(x+y)$
$=(x+y)(2+a)$

79. $rs-ru+8sw-8uw=(rs-ru)+(8sw-8uw)$
$=r(s-u)+8w(s-u)$
$=(s-u)(r+8w)$

81. $7m^3-2m^2+14m-4$
$=(7m^2m-2m^2)+(2\cdot 7m-2\cdot 2)$
$=m^2(7m-2)+2(7m-2)$
$=(7m-2)(m^2+2)$

83. $5x^3-x^2+10x-2$
$=(5x^2x-x^2\cdot 1)+(2\cdot 5x-2\cdot 1)$
$=x^2(5x-1)+2(5x-1)$
$=(5x-1)(x^2+2)$

Factor by grouping. See Example 8.

85. $ab+ac+b+c=(ab+ac)+(1b+1c)$
$=a(b+c)+1(b+c)$
$=(b+c)(a+1)$

87. $rs+4s^2-r-4s=(rs+4ss)+(-r-4s)$
$=s(r+4s)-1(r+4s)$
$=(r+4s)(s-1)$

89. $2ax+2bx-3a-3b$
$=(2ax+2bx)+(-3a-3b)$
$=2x(a+b)-3(a+b)$
$=(a+b)(2x-3)$
or
$=(2ax-3a)+(2bx-3b)$
$=a(2x-3)+b(2x-3)$
$=(2x-3)(a+b)$

91. $mp-np-mq+nq$
$=(mp-np)+(-mq+nq)$
$=(mp-np)-1(mq-nq)$
$=p(m-n)-q(m-n)$
$=(m-n)(p-q)$

Factor by grouping. See Example 9.

93. $5m^3+6+5m^2+6m=5m^3+5m^2+6m+6$
$=(5m^3+5m^2)+(6m+6)$
$=(5m^2m^1+5m^2)+(6m+6)$
$=5m^2(m+1)+6(m+1)$
$=(m+1)(5m^2+6)$

95. $y^3-12+3y-4y^2=y^3-4y^2+3y-12$
$=(y^3-4y^2)+(3y-12)$
$=(y^2y^1-4y^2)+(3y-3\cdot 4)$
$=y^2(y-4)+3(y-4)$
$=(y-4)(y^2+3)$

Factor by grouping. Remember to factor out the GCF first. See Example 10.

97. $ax^3-2ax^2+5ax-10a$
$=ax^3-2ax^2+5ax-10a$
$=a(x^3-2x^2+5x-10)$
$=a[(x^2x-2x^2)+(5x-5\cdot 2)]$
$=a[x^2(x-2)+5(x-2)]$
$=a(x-2)(x^2+5)$

99. $6x^3 - 6x^2 + 12x - 12$
$$= \mathbf{6}x^3 - \mathbf{6}x^2 + \mathbf{6}\cdot 2x - \mathbf{6}\cdot 2$$
$$= \mathbf{6}(x^3 - x^2 + 2x - 2)$$
$$= \mathbf{6}[(\mathbf{x^2}x - \mathbf{x^2}\cdot 1) + (\mathbf{2}x - \mathbf{2}\cdot 1)]$$
$$= \mathbf{6}[\mathbf{x^2}(x-1) + \mathbf{2}(x-1)]$$
$$= \mathbf{6}(x-1)(\mathbf{x^2+2})$$

TRY IT YOURSELF

Factor.

101. $h^2(14+r) + 5(14+r)$
$$= h^2\mathbf{(14+r)} + 5\mathbf{(14+r)}$$
$$= \mathbf{(14+r)}(h^2+5)$$

103. $22a^3 - 33a^2 = \mathbf{11}\cdot 2\mathbf{a^2}a - \mathbf{11}\cdot 3\mathbf{a^2}$
$$= \mathbf{11a^2}(2a-3)$$

105. $ax + bx - a - b = (a\mathbf{x} + b\mathbf{x}) + (-\mathbf{1}a - \mathbf{1}b)$
$$= \mathbf{x}(a+b) - \mathbf{1}(a+b)$$
$$= (a+b)(\mathbf{x-1})$$

107. $15r^8 - 18r^6 - 30r^5$
$$= \mathbf{3}\cdot 5\mathbf{r^5}r^3 - \mathbf{3}\cdot 6\mathbf{r^5}r^1 - \mathbf{3}\cdot 10\mathbf{r^5}$$
$$= \mathbf{3r^5}(5r^3 - 6r - 10)$$

109. $27mp + 9mq - 9np - 3nq$
$$= \mathbf{3}\cdot 9mp + \mathbf{3}\cdot 3mq - \mathbf{3}\cdot 3np - \mathbf{3}\cdot nq$$
$$= \mathbf{3}(9mp + 3mq - 3np - nq)$$
$$= \mathbf{3}[(\mathbf{3}\cdot 3\mathbf{m}p + \mathbf{3}\cdot \mathbf{m}q) + (-3\mathbf{n}p - \mathbf{n}q)]$$
$$= \mathbf{3}[\mathbf{3m}(3p+q) - \mathbf{n}(3p+q)]$$
$$= \mathbf{3}(3p+q)(\mathbf{3m-n})$$

111. $-60p^2t^2 - 80pt^3 = \mathbf{-20}\cdot 3\mathbf{p}p\mathbf{t^2} - \mathbf{20}\cdot 4\mathbf{pt^2}t$
$$= \mathbf{-20pt^2}(3p+4t)$$

113. $-2x + 5 = \mathbf{(-1)}2x + \mathbf{(-1)}(-5)$
$$= \mathbf{(-1)}(2x-5)$$
$$= -(2x-5)$$

115. $6x^2 - 2xy - 15x + 5y$
$$= (\mathbf{2}\cdot 3x\mathbf{x} - \mathbf{2}\cdot \mathbf{x}y) + (\mathbf{-5}\cdot 3x + \mathbf{5}y)$$
$$= \mathbf{2x}(3x-y) - \mathbf{5}(3x-y)$$
$$= (3x-y)(\mathbf{2x-5})$$

117. $2x^3z - 4x^2z + 32xz - 64z$
$$= \mathbf{2}x^3\mathbf{z} - \mathbf{2}\cdot 2x^2\mathbf{z} + \mathbf{2}\cdot 16x\mathbf{z} - \mathbf{2}\cdot 32\mathbf{z}$$
$$= \mathbf{2z}(x^3 - 2x^2 + 16x - 32)$$
$$= \mathbf{2z}[(\mathbf{x^2}x - 2\mathbf{x^2}) + (\mathbf{16}x - \mathbf{16}\cdot 2)]$$
$$= \mathbf{2z}[\mathbf{x^2}(x-2) + \mathbf{16}(x-2)]$$
$$= \mathbf{2z}(x-2)(\mathbf{x^2+16})$$

119. $12uvw^3 - 54uv^2w^2 = \mathbf{6}\cdot 2\mathbf{uvw^2}w - \mathbf{6}\cdot 9\mathbf{uv}v\mathbf{w^2}$
$$= \mathbf{6uvw^2}(2w - 9v)$$

121. $x^3 + x^2 + x + 1 = (\mathbf{x^2}\cdot x + \mathbf{x^2}) + (x+1)$
$$= \mathbf{x^2}(x+1) + \mathbf{1}(x+1)$$
$$= (x+1)(\mathbf{x^2+1})$$

123. $-3r + 2s - 3 = \mathbf{(-1)}3r + \mathbf{(-1)}(-2s) + \mathbf{(-1)}3$
$$= \mathbf{(-1)}(3r - 2s + 3)$$
$$= -(3r - 2s + 3)$$

LOOK ALIKES . . .

125. a. $5t^3 + 6t^2 + 15t + 18$
$$= (5t^3 + 6t^2) + (15t + 18)$$
$$= (5\mathbf{t^2}t^1 + 6\mathbf{t^2}) + (\mathbf{3}\cdot 5t + \mathbf{3}\cdot 6)$$
$$= \mathbf{t^2}(5t+6) + \mathbf{3}(5t+6)$$
$$= (5t+6)(\mathbf{t^2+3})$$

b. $3t^3 + 6t^2 + 15t + 18$
$$= \mathbf{3}t^3 + (\mathbf{3}\cdot 2)t^2 + (\mathbf{3}\cdot 5)t + (\mathbf{3}\cdot 6)$$
$$= 3(t^3 + 2t^2 + 5t + 6)$$

APPLICATIONS

127. GEOMETRY

$x^3 + 4x^2 + 5x + 20$
$$= (\mathbf{x^2}\cdot x + 4\cdot \mathbf{x^2}) + (\mathbf{5}\cdot x + \mathbf{5}\cdot 4)$$
$$= \mathbf{x^2}(x+4) + \mathbf{5}(x+4)$$
$$= (x+4)(\mathbf{x^2+5})$$

The length is (x^2+5) ft.

The width is $(x+4)$ ft.

from Campus to Careers

129. **ELEMENTARY SCHOOL TEACHER**

a. Rewrite the formula in factored form.
$$V = \boldsymbol{\pi r^2}h_1 + \frac{1}{3}\boldsymbol{\pi r^2}h_2$$
$$= \boldsymbol{\pi r^2}\left(h_1 + \frac{1}{3}h_2\right)$$

The formula in factored form is
$$V = \boldsymbol{\pi r^2}\left(h_1 + \frac{1}{3}h_2\right).$$

b. Evaluate the factored formula if $h_1 = 1\frac{5}{6}$ in. $h_2 = \frac{1}{2}$ in. and $r = \frac{1}{4}$ in.

$$V = \pi r^2\left(h_1 + \frac{1}{3}h_2\right)$$
$$= 3.14\left(\frac{1}{4}\right)^2\left(1\frac{5}{6} + \frac{1}{3}\cdot\frac{1}{2}\right)$$
$$= 3.14\left(\frac{1}{16}\right)\left(\frac{11}{6} + \frac{1}{6}\right)$$
$$= 3.14\left(\frac{1}{16}\right)\left(\frac{12}{6}\right)$$
$$= 3.14\left(\frac{1}{16}\right)(2)$$
$$= 0.3925$$

20 students will each make 5 crayons which is 100 crayons. Mutiply 100 by 0.3925 and obtain the total volume of wax that will be needed. 39.25 or about 40 in.^{3} of wax.

Depending on the value of π that was used the answers will vary.

WRITING

131-133. Answers will vary.

REVIEW

135. INSURANCE COSTS

Step 1: $\$1,050 - \$925 = \$125$

This is the amount of decrease.

Step 2: 125 is what percent of 1,050?

$$125 = x \cdot 1,050$$
$$\frac{125}{\mathbf{1,050}} = \frac{1,050x}{\mathbf{1,050}}$$
$$0.119 \approx x$$
$$0.12 \approx x$$

The percent of decrease is about 12%.

CHALLENGE PROBLEMS

137. $6x^{4m}y^n + 21x^{3m}y^{2n} - 15x^{2m}y^{3n}$

$= \mathbf{3x^{2m}y^n} \cdot 2x^{2m} + \mathbf{3x^{2m}y^n} \cdot 7x^m y^n - \mathbf{3x^{2m}y^n} \cdot 5y^{2n}$

$= \mathbf{3x^{2m}y^n}(2x^{2m} + 7x^m y^n - 5y^{2n})$

SECTION 6.2 STUDY SET

VOCABULARY

Fill in the blanks.

1. The trinomial $x^2 - x - 12$ **factors** as the product of two binomials: $(x - 4)(x + 3)$.

3. The **leading** coefficient of $x^2 - 3x + 2$ is 1.

CONCEPTS

Fill in the blanks.

5. a. Before attempting to factor a trinomial, be sure that it is written in **descending** powers of a variable.
   b. Before attempting to factor a trinomial into two binomials, always factor out any **common** factors first.

7. $x^2 + 5x + 3$ cannot be factored because we cannot find two integers whose product is **3** and whose sum is **5**.

9. Check to determine whether each factorization is correct.
   a. $x^2 - x - 20 = (x + 5)(x - 4)$ **No**
   b. $4a^2 + 12a - 16 = 4(a - 1)(a + 4)$ **Yes**

11. Consider a trinomial of the form $x^2 + bx + c$.
    a. If c is positive, what can be said about the two integers that should be chosen for the factorization? **They are both positive or both negative.**
    b. If c is negative, what can be said about the two integers that should be chosen for the factorization? **One will be positive, the other negative.**

NOTATION

13. $(x + \mathbf{3})(x - \mathbf{2})$

GUIDED PRACTICE

Factor each trinomial. See Example 1.

15. The positive factors of 2 whose sum is 3 are 1 and 2.
    $x^2 + 3x + 2 = (x+2)(x+1)$

17. The positive factors of 12 whose sum is 7 are 3 and 4.
    $z^2 + 7z + 12 = (z+4)(z+3)$

Factor each trinomial. See Example 2.

19. The negative factors of 6 whose sum is -5 are -2 and -3.
    $m^2 - 5m + 6 = (m-3)(m-2)$

21. The negative factors of 28 whose sum is -11 are -4 and -7.
    $t^2 - 11t + 28 = (t-7)(t-4)$

Factor each trinomial. See Example 3.

23. Two different sign factors of -24 whose sum is $+5$ are $+8$ and -3.
    $x^2 + 5x - 24 = (x+8)(x-3)$

25. Two different sign factors of -48 whose sum is $+13$ are $+16$ and -3.
    $t^2 + 13t - 48 = (t-3)(t+16)$

Factor each trinomial. See Example 4.

27. Two different sign factors of -16 whose sum is -6 are -8 and $+2$.
    $a^2 - 6a - 16 = (a-8)(a+2)$

29. Two different sign factors of -36 whose sum is -9 are -12 and $+3$.
    $b^2 - 9b - 36 = (b-12)(b+3)$

Factor each trinomial. See Example 5.

31. Factor out a -1.
    $-x^2 - 7x - 10 = -1(x^2 + 7x + 10)$
    The positive factors of 10 whose sum is 7 are 2 and 5.
    $-x^2 - 7x - 10 = -1(x^2 + 7x + 10)$
    $= -(x+5)(x+2)$

33. Factor out a -1.
    $-t^2 - t + 30 = -1(t^2 + t - 30)$
    Two different sign factors of -30 whose sum is $+1$ are $+6$ and -5.
    $-t^2 - t + 30 = -1(t^2 + t - 30)$
    $= -(t+6)(t-5)$

35. Factor out a -1.
    $-r^2 - 3r + 54 = -1(r^2 + 3r - 54)$
    Two different sign factors of -54 whose sum is $+3$ are $+9$ and -6.
    $-r^2 - 3r + 54 = -1(r^2 + 3r - 54)$
    $= -(r+9)(r-6)$

37. Factor out a -1.
    $-m^2 + 18m - 77 = -1(m^2 - 18m + 77)$
    The negative factors of 77 whose sum is -18 are -7 and -11.
    $-m^2 + 18m - 77 = -1(m^2 - 18m + 77)$
    $= -(m-7)(m-11)$

Factor each trinomial. See Example 6.

39. The positive factors of 3 whose sum is 4 are 1 and 3.
    $a^2 + 4ab + 3b^2 = (a+3b)(a+b)$

41. Two different sign factors of -7 whose sum is -6 are -7 and $+1$.
    $x^2 - 6xy - 7y^2 = (x-7y)(x+y)$

43. Two different sign factors of -2 whose sum is $+1$ are $+2$ and -1.

$$r^2+rs-2s^2=(r+2s)(r-s)$$

45. Two negative factors of 6 whose sum is -5 are -3 and -2.

$$a^2-5ab+6b^2=(a-3b)(a-2b)$$

Factor completely. See Example 7.

47. Factor out a 2.

$$2x^2+10x+12=2(x^2+5x+6)$$

The positive factors of 6 whose sum is 5 are 2 and 3.

$$\begin{aligned}2x^2+10x+12&=2(x^2+5x+6)\\&=2(x+2)(x+3)\end{aligned}$$

49. Factor out a 6.

$$6a^2-30a+24=6(a^2-5a+4)$$

The negative factors of 4 whose sum is -5 are -1 and -4.

$$\begin{aligned}6a^2-30a+24&=6(a^2-5a+4)\\&=6(a-1)(a-4)\end{aligned}$$

51. Factor out a 5.

$$5a^2-25a+30=5(a^2-5a+6)$$

The negative factors of 6 whose sum is -5 are -2 and -3.

$$\begin{aligned}5a^2-25a+30&=5(a^2-5a+6)\\&=5(a-2)(a-3)\end{aligned}$$

53. Factor out a $-z$.

$$-z^3+29z^2-100z=-z(z^2-29z+100)$$

The negative factors of 100 whose sum is -29 are -4 and -25.

$$\begin{aligned}-z^3+29z^2-100z&=-z(z^2-29z+100)\\&=-z(z-4)(z-25)\end{aligned}$$

Write each trinomial in descending powers of one variable and factor. See Example 8.

55. $80-24x+x^2=x^2-24x+80$

The negative factors of 80 whose sum is -24 are -20 and -4.

$$\begin{aligned}80-24x+x^2&=x^2-24x+80\\&=(x-20)(x-4)\end{aligned}$$

57. $10y+9+y^2=y^2+10y+9$

The positive factors of 9 whose sum is 10 are 1 and 9.

$$\begin{aligned}10y+9+y^2&=y^2+10y+9\\&=(y+1)(y+9)\end{aligned}$$

59. $r^3-16r+6r^2=r^3+6r^2-16r$

Factor out a r.

Two different sign factors of -16 whose sum is $+6$ are $+8$ and -2.

$$\begin{aligned}r^3-16r+6r^2&=r^3+6r^2-16r\\&=r(r^2+6r-16)\\&=r(r+8)(r-2)\end{aligned}$$

61. $4r^2x+r^3+3rx^2=r^3+4r^2x+3rx^2$

Factor out a r.

The positive factors of 3 whose sum is 4 are 1 and 3.

$$\begin{aligned}4r^2x+r^3+3rx^2&=r^3+4r^2x+3rx^2\\&=r(r^2+4rx+3x^2)\\&=r(r+x)(r+3x)\end{aligned}$$

Factor each trinomial. See Example 9.

63. There are no two integers whose product is 15 and whose sum is 10, the trinomial $u^2+10u+15$ cannot be factored and is a *prime trinomial*.

65. There are no two diffferent sign integers whose product is -4 and whose sum is 2, the trinomial r^2+2r-4 cannot be factored and is a *prime trinomial*.

TRY IT YOURSELF

Choose the correct method from Section 6.1 or Section 6.2 to factor each of the following.

67. $$\begin{aligned}5x+15+xy+3y&=(\mathbf{5}x+\mathbf{5}\cdot3)+(x\mathbf{y}+3\mathbf{y})\\&=\mathbf{5}(x+3)+\mathbf{y}(x+3)\\&=(x+3)(\mathbf{5+y})\end{aligned}$$

69. $$\begin{aligned}26n^2-8n&=\mathbf{2}\cdot13\mathbf{n}n-\mathbf{2}\cdot4\mathbf{n}\\&=\mathbf{2n}(13n-4)\end{aligned}$$

71. Two different sign factors of -5 whose sum is -4 are -5 and $+1$.

$$a^2-4a-5=(a-5)(a+1)$$

73. Factor out a -1.

$$-x^2+21x+22=-1(x^2-21x-22)$$

Two different sign factors of -22 whose sum is -21 are -22 and $+1$.

$$\begin{aligned}-x^2+21x+22&=-1(x^2-21x-22)\\&=-(x-22)(x+1)\end{aligned}$$

75. $4xy-4x+28y-28$

$$\begin{aligned}&=\mathbf{4}xy-\mathbf{4}x+\mathbf{4}\cdot7y-\mathbf{4}\cdot7\\&=\mathbf{4}(xy-x+7y-7)\\&=\mathbf{4}[(x\mathbf{y}-\mathbf{x}\cdot1)+(\mathbf{7}y-\mathbf{7}\cdot1)]\\&=\mathbf{4}[\mathbf{x}(y-1)+\mathbf{7}(y-1)]\\&=\mathbf{4}(y-1)(\mathbf{x+7})\end{aligned}$$

77. Factor out a $12b^2$.

$24b^4 - 48b^3 - 36b^2$

$= \mathbf{12} \cdot 2 \cdot \mathbf{b^2} \cdot b^2 - \mathbf{12} \cdot 4 \cdot \mathbf{b^2} \cdot b - \mathbf{12} \cdot 3 \cdot \mathbf{b^2}$

$= 12b^2(2b^2 - 4b - 3)$

79. The negative factors of 18 whose sum is -9 are -3 and -6.

$r^2 - 9r + 18 = (r-3)(r-6)$

81. Factor out a $-n^2$.

$-n^4 + 28n^3 + 60n^2 = -n^2(n^2 - 28n - 60)$

Two different sign factors of -60 whose sum is -28 are -30 and $+2$.

$-n^4 + 28n^3 + 60n^2 = -n^2(n^2 - 28n - 60)$

$= -n^2(n-30)(n+2)$

83. The positive factors of 4 whose sum is 4 are 2 and 2.

$x^2 + 4xy + 4y^2 = (x+2y)(x+2y)$

$= (x+2y)^2$

85. Two different sign factors of -12 whose sum is -4 are -6 and $+2$.

$a^2 - 4ab - 12b^2 = (a-6b)(a+2b)$

87. Factor out a $4x^2$.

$4x^4 + 16x^3 + 16x^2 = 4x^2(x^2 + 4x + 4)$

The positive factors of 4 whose sum is 4 are 2 and 2.

$4x^4 + 16x^3 + 16x^2 = 4x^2(x^2 + 4x + 4)$

$= 4x^2(x+2)(x+2)$

$= 4x^2(x+2)^2$

89. The negative factors of 45 whose sum is -46 are -1 and -45.

$a^2 - 46a + 45 = (a-45)(a-1)$

91. There are no two integers whose product is 4 and whose sum is -2, the trinomial $r^2 - 2r + 4$ cannot be factored and is a *prime trinomial*.

93. $t(x+2) + 7(x+2) = (x+2)(t+7)$

95. Factor out a s^2.

$s^4 + 11s^3 - 26s^2 = s^2(s^2 + 11s - 26)$

Two different sign factors of -26 whose sum is $+11$ are $+13$ and -2.

$s^4 + 11s^3 - 26s^2 = s^2(s^2 + 11s - 26)$

$= s^2(s+13)(s-2)$

97. $15s^3 + 75 = \mathbf{15} \cdot s^3 + \mathbf{15} \cdot 5$

$= 15(s^3 + 5)$

99. $-13y + y^2 - 14 = y^2 - 13y - 14$

Two different sign factors of -14 whose sum is -13 are -14 and $+1$.

$-13y + y^2 - 14 = y^2 - 13y - 14$

$= (y-14)(y+1)$

101. Factor out a 2.

$2x^2 - 12x + 16 = 2(x^2 - 6x + 8)$

The negative factors of 8 whose sum is -6 are -4 and -2.

$2x^2 - 12x + 16 = 2(x^2 - 6x + 8)$

$= 2(x-4)(x-2)$

LOOK ALIKES...

103. The negative factors of 24 whose sum is -10 are -4 and -6.

a. $x^2 - 10x + 24 = (x-4)(x-6)$

Two different sign factors of -24 whose sum is -10 are -12 and $+2$.

b. $x^2 - 10x - 24 = (x-12)(x+2)$

APPLICATIONS

105. PETS

Factor out a x.

$x^3 + 12x^2 + 27x = x(x^2 + 12x + 27)$

The positive factors of 27 whose sum is 12 are 9 and 3.

$x^3 + 12x^2 + 27x = x(x^2 + 12x + 27)$

$= x(x+9)(x+3)$

Its length is $(x+9)$ in.

Its width is x in.

Its height is $(x+3)$ in.

WRITING

107-111. Answers will vary.

REVIEW

Simplify each expression. Write each answer without negative exponents.

113. $\dfrac{x^{12}x^{-7}}{x^3x^4} = \dfrac{x^{12-7}}{x^{3+4}}$

$= \dfrac{x^5}{x^7}$

$= \dfrac{x^{5-7}}{1}$

$= \dfrac{x^{-2}}{1}$

$= \dfrac{1}{x^2}$

115. $(x^{-3}x^{-2})^2 = (x^{-3+(-2)})^2$

$$= (x^{-5})^2$$

$$= \left(\frac{1}{x^5}\right)^2$$

$$= \frac{1}{x^{5\cdot 2}}$$

$$= \frac{1}{x^{10}}$$

CHALLENGE PROBLEMS

Factor completely.

117. The negative factors of $\frac{9}{25}$ whose sum is $-\frac{6}{5}$ are $-\frac{3}{5}$ and $-\frac{3}{5}$.

$$x^2 - \frac{6}{5}x + \frac{9}{25} = \left(x - \frac{3}{5}\right)\left(x - \frac{3}{5}\right)$$

$$= \left(x - \frac{3}{5}\right)^2$$

119. Two different sign factors of -45 whose sum is -12 are -15 and $+3$.

$$x^{2m} - 12x^m - 45 = (x^m - 15)(x^m + 3)$$

121. Find all positive integer values of c that make $n^2 + 6n + c$ factorable. **5, 8, 9**

SECTION 6.3 STUDY SET

VOCABULARY

Fill in the blanks.

1. The **leading** coefficient of $3x^2 - x - 12$ is 3.
3. The first terms of the binomial factors $(5y + 1)(y + 3)$ are **$5y$** and **y**. The second terms of the binomial factors are **1** and **3**.

CONCEPTS

5. If $10x^2 - 27x + 5$ is to be factored as the product of two binomials, what are the possible first terms of the binomial factors?

   $10x$ and x, $5x$ and $2x$

7. a. Fill in the blanks. When factoring a trinomial, we write it in **descending** powers of the variable. Then we factor out any **GCF** (including −1 if that is necessary to make the lead **coefficient** positive).

   b. What is the GCF of the terms of $6s^4 + 33s^3 + 36s^2$? **$3s^2$**

   c. Factor out −1 from $-2d^2 + 19d - 8$.

   $-(2d^2 - 19d + 8)$

A trinomial has been partially factored. Complete each statement that describes the type of integers we should consider for the blanks.

9. $5y^2 - 13y + 6 = (5y \square)(y \square)$

   Since the last term of the trinomial is positive and the middle term is negative, the integers must be **negative** factors of 6

11. $5y^2 - 7y - 6 = (5y \square)(y \square)$

   Since the last term of the trinomial is negative, the signs of the integers will be **different**.

13. Complete the key number table.

Negative factors of 12	Sum of the negative factors of 12
−1(−12)	**−13**
−2(**−6**)	**−8**
−3(−4)	**−7**

NOTATION

15. a. Suppose we wish to factor $12b^2 + 20b - 9$ by grouping. Identify a, b, and c. **12, 20, −9**

   b. What is the key number, ac? **−108**

Complete each step of the factorization of the trinomial by grouping.

17. $12t^2 + 17t + 6 = 12t^2 + 9t + 8t + 6$

$$= \mathbf{3t}(4t + 3) + \mathbf{2}(4t + 3)$$
$$= (\mathbf{4t + 3})(3t + 2)$$

GUIDED PRACTICE

Factor. See Example 1.

19. $2x^2 + 3x + 1 = (2x + 1)(x + 1)$
21. $3a^2 + 10a + 3 = (3a + 1)(a + 3)$
23. $5x^2 + 7x + 2 = (5x + 2)(x + 1)$
25. $7x^2 + 18x + 11 = (7x + 11)(x + 1)$

Factor. See Example 2.

27. $4x^2 - 8x + 3 = (2x - 3)(2x - 1)$
29. $8x^2 - 22x + 5 = (4x - 1)(2x - 5)$
31. $15t^2 - 26t + 7 = (5t - 7)(3t - 1)$
33. $6y^2 - 13y + 2 = (6y - 1)(y - 2)$

Factor. See Example 3.

35. $\underset{a}{3}x^2 - \underset{b}{2}x - \underset{c}{21}$

In $3x^2 - 2x - 21$, we have $a = 3$, $b = -2$, and $c = -21$. The key number is $ac = 3(-21) = -63$. We must find a factorization of −63 in which the sum of the factors is $b = -2$. Since the factors must have a negative product, their signs must be different. The pairs of factors are −9 and +7.

$$3x^2 - 2x - 21 = 3x^2 - 9x + 7x - 21$$
$$= (3x^2 - 9x) + (7x - 21)$$
$$= 3x(x - 3) + 7(x - 3)$$
$$= (x - 3)(3x + 7)$$

37. $\underset{a}{5}m^2 - \underset{b}{7}m - \underset{c}{6}$

In $5m^2 - 7m - 6$, we have $a = 5$, $b = -7$, and $c = -6$. The key number is $ac = 5(-6) = -30$. We must find a factorization of −30 in which the sum of the factors is $b = -7$. Since the factors must have a negative product, their signs must be different. The pairs of factors are −10 and +3.

$$5m^2 - 7m - 6 = 5m^2 - 10m + 3m - 6$$
$$= (5m^2 - 10m) + (3m - 6)$$
$$= 5m(m - 2) + 3(m - 2)$$
$$= (m - 2)(5m + 3)$$

39. $\underset{a}{7}y^2 + \underset{b}{55}y - \underset{c}{8}$

In $7y^2 + 55y - 8$, we have $a = 7$, $b = 55$ and $c = -8$. The key number is $ac = 7(-8) = -56$. We must find a factorization of –56 in which the sum of the factors is $b = 55$. Since the factors must have a negative product, their signs must be different. The pairs of factors are –1 and +56.

$$\begin{aligned} 7y^2 + 55y - 8 &= 7y^2 - y + 56y - 8 \\ &= (7y^2 - y) + (56y - 8) \\ &= y(7y - 1) + 8(7y - 1) \\ &= (7y - 1)(y + 8) \end{aligned}$$

41. $\underset{a}{11}y^2 + \underset{b}{7}y - \underset{c}{4}$

In $11y^2 + 7y - 4$, we have $a = 11$, $b = 7$ and $c = -4$. The key number is $ac = 11(-4) = -44$. We must find a factorization of –44 in which the sum of the factors is $b = 7$. Since the factors must have a negative product, their signs must be different. The pairs of factors are –4 and +11.

$$\begin{aligned} 11y^2 + 7y - 4 &= 11y^2 - 4y + 11y - 4 \\ &= (11y^2 - 4y) + (11y - 4) \\ &= y(11y - 4) + 1(11y - 4) \\ &= (11y - 4)(y + 1) \end{aligned}$$

Factor. See Example 4.

43. $\underset{a}{6}r^2 + \underset{b}{1}rs - \underset{c}{2}s^2$

In $6r^2 + 1rs - 2s^2$, we have $a = 6$, $b = 1$ and $c = -2$. The key number is $ac = 6(-2) = -12$. We must find a factorization of –12 in which the sum of the factors is $b = 1$. Since the factors must have a negative product, their signs must be different. The pairs of factors are –3 and +4.

$$\begin{aligned} 6r^2 + rs - 2s^2 &= 6r^2 - 3rs + 4rs - 2s^2 \\ &= (6r^2 - 3rs) + (4rs - 2s^2) \\ &= 3r(2r - s) + 2s(2r - s) \\ &= (2r - s)(3r + 2s) \end{aligned}$$

45. $\underset{a}{4}x^2 + \underset{b}{8}xy + \underset{c}{3}y^2$

In $4x^2 + 8xy + 3y^2$, we have $a = 4$, $b = 8$ and $c = 3$. The key number is $ac = 4(3) = 12$. We must find a factorization of 12 in which the sum of the factors is $b = 8$. Since the factors must have a positive product, their signs must be the same. The pairs of factors are +2 and +6.

$$\begin{aligned} 4x^2 + 8xy + 3y^2 &= 4x^2 + 2xy + 6xy + 3y^2 \\ &= (4x^2 + 2xy) + (6xy + 3y^2) \\ &= 2x(2x + y) + 3y(2x + y) \\ &= (2x + y)(2x + 3y) \end{aligned}$$

47. $\underset{a}{8}m^2 + \underset{b}{91}mn + \underset{c}{33}n^2$

In $8m^2 + 91mn + 33n^2$, we have $a = 8$, $b = 91$ and $c = 33$. The key number is $ac = 8(33) = 264$. We must find a factorization of 264 in which the sum of the factors is $b = 91$. Since the factors must have a positive product, their signs must be the same. The pairs of factors are +3 and +88.

$$\begin{aligned} 8m^2 + 91mn + 33n^2 &= 8m^2 + 88mn + 3mn + 33n^2 \\ &= (8m^2 + 88mn) + (3mn + 33n^2) \\ &= 8m(m + 11n) + 3n(m + 11n) \\ &= (m + 11n)(8m + 3n) \end{aligned}$$

49. $\underset{a}{15}x^2 - \underset{b}{1}xy - \underset{c}{6}y^2$

In $15x^2 - 1xy - 6y^2$, we have $a = 15$, $b = -1$ and $c = -6$. The key number is $ac = 15(-6) = -90$. We must find a factorization of –90 in which the sum of the factors is $b = -1$. Since the factors must have a negative product, their signs must be different. The pairs of factors are –10 and +9.

$$\begin{aligned} 15x^2 - xy - 6y^2 &= 15x^2 - 10xy + 9xy - 6y^2 \\ &= (15x^2 - 10xy) + (9xy - 6y^2) \\ &= 5x(3x - 2y) + 3y(3x - 2y) \\ &= (3x - 2y)(5x + 3y) \end{aligned}$$

Factor. See Example 7.

51. $$\begin{aligned} -26x + 6x^2 - 20 &= 6x^2 - 26x - 20 \\ &= 2(3x^2 - 13x - 10) \end{aligned}$$

$\underset{a}{3}x^2 - \underset{b}{13}x - \underset{c}{10}$

In $3x^2 - 13x - 10$, we have $a = 3$, $b = -13$ and $c = -10$. The key number is $ac = 3(-10) = -30$. We must find a factorization of –30 in which the sum of the factors is $b = -13$. Since the factors must have a negative product, their signs must be different. The pairs of factors are –15 and +2.

$$\begin{aligned} 2(3x^2 - 13x - 10) &= 2(3x^2 - 15x + 2x - 10) \\ &= 2[(3x^2 - 15x) + (2x - 10)] \\ &= 2[3x(x - 5) + 2(x - 5)] \\ &= 2(x - 5)(3x + 2) \end{aligned}$$

53. $15a+8a^3-26a^2=8a^3-26a^2+15a$
$$=a(8a^2-26a+15)$$

$$\underset{a}{8}a^2-\underset{b}{26}a+\underset{c}{15}$$

In $8a^2-26a+15$, we have $a=8$, $b=-26$ and $c=15$. The key number is $ac=8(15)=120$. We must find a factorization of 120 in which the sum of the factors is $b=-26$. Since the factors must have a positive product, their signs must be the same. The pairs of factors are -6 and -20.

$$\begin{aligned} a(8a^2-26a+15) &= a(8a^2-6a-20a+15) \\ &= a[(8a^2-6a)+(-20a+15)] \\ &= a[2a(4a-3)-5(4a-3)] \\ &= a(4a-3)(2a-5) \end{aligned}$$

55. $2u^2-6v^2-uv=2u^2-uv-6v^2$

$$\underset{a}{2}u^2-\underset{b}{1}uv-\underset{c}{6}v^2$$

In $2u^2-1uv-6v^2$, we have $a=2$, $b=-1$ and $c=-6$. The key number is $ac=2(-6)=-12$. We must find a factorization of -12 in which the sum of the factors is $b=-1$. Since the factors must have a negative product, their signs must be different. The pairs of factors are -4 and 3.

$$\begin{aligned} 2u^2-uv-6v^2 &= (2u^2-4uv+3uv-6v^2) \\ &= (2u^2-4uv)+(3uv-6v^2) \\ &= 2u(u-2v)+3v(u-2v) \\ &= (u-2v)(2u+3v) \end{aligned}$$

57. $36y^2-88y+32=4(9y^2-22y+8)$

$$\underset{a}{9}y^2-\underset{b}{22}y+\underset{c}{8}$$

In $9y^2-22y+8$, we have $a=9$, $b=-22$ and $c=8$. The key number is $ac=9(8)=72$. We must find a factorization of 72 in which the sum of the factors is $b=-22$. Since the factors must have a positive product, their signs must be the same. The pairs of factors are -4 and -18.

$$\begin{aligned} 4(9y^2-22y+8) &= 4(9y^2-4y-18y+8) \\ &= 4[(9y^2-4y)+(-18y+8)] \\ &= 4[y(9y-4)-2(9y-4)] \\ &= 4(9y-4)(y-2) \end{aligned}$$

TRY IT YOURSELF

Factor. If an expression is *prime*, so indicate.

59. $\underset{a}{6}t^2-\underset{b}{7}t-\underset{c}{20}$

In $6t^2-7t-20$, we have $a=6$, $b=-7$ and $c=-20$. The key number is $ac=6(-20)=-120$. We must find a factorization of -120 in which the sum of the factors is $b=-7$. Since the factors must have a negative product, their signs must be different. The pairs of factors are -15 and 8.

$$\begin{aligned} 6t^2-7t-20 &= 6t^2-15t+8t-20 \\ &= (6t^2-15t)+(8t-20) \\ &= 3t(2t-5)+4(2t-5) \\ &= (2t-5)(3t+4) \end{aligned}$$

61. $\underset{a}{15}p^2-\underset{b}{2}pq-\underset{c}{1}q^2$

In $15p^2-2pq-1q^2$, we have $a=15$, $b=-2$ and $c=-1$. The key number is $ac=15(-1)=-15$. We must find a factorization of -15 in which the sum of the factors is $b=-2$. Since the factors must have a negative product, their signs must be different. The pairs of factors are -5 and 3.

$$\begin{aligned} 15p^2-2pq-q^2 &= 15p^2-5pq+3pq-q^2 \\ &= (15p^2-5pq)+(3pq-q^2) \\ &= 5p(3p-q)+q(3p-q) \\ &= (3p-q)(5p+q) \end{aligned}$$

63. $\underset{a}{4}t^2-\underset{b}{16}t+\underset{c}{7}$

In $4t^2-16t+7$, we have $a=4$, $b=-16$ and $c=7$. The key number is $ac=4(7)=28$. We must find a factorization of 28 in which the sum of the factors is $b=-16$. Since the factors must have a positive product, their signs must be the same. The pairs of factors are -2 and -14.

$$\begin{aligned} 4t^2-16t+7 &= 4t^2-2t-14t+7 \\ &= (4t^2-2t)+(-14t+7) \\ &= 2t(2t-1)-7(2t-1) \\ &= (2t-1)(2t-7) \end{aligned}$$

65. $130r^2+20r-110=10(13r^2+2r-11)$

$$\underset{a}{13}r^2+\underset{b}{2}r-\underset{c}{11}$$

In $13r^2+2r-11$, we have $a=13$, $b=2$ and $c=-11$. The key number is $ac=13(-11)=-143$. We must find a factorization of -143 in which the sum of the factors is $b=2$. Since the factors must have a negative product, their signs must be different. The pairs of factors are -11 and 13.

$$\begin{aligned} 10(13r^2+2r-11) &= 10(13r^2-11r+13r-11) \\ &= 10[(13r^2-11r)+(13r-11)] \\ &= 10[r(13r-11)+1(13r-11)] \\ &= 10(13r-11)(r+1) \end{aligned}$$

67. $\underset{a}{8}y^2 - \underset{b}{2}y - \underset{c}{1}$

In $8y^2 - 2y - 1$, we have $a = 8$, $b = -2$ and $c = -1$. The key number is $ac = 8(-1) = -8$. We must find a factorization of -8 in which the sum of the factors is $b = -2$. Since the factors must have a negative product, their signs must be different. The pairs of factors are -4 and 2.

$$\begin{aligned}8y^2 - 2y - 1 &= 8y^2 - 4y + 2y - 1\\ &= (8y^2 - 4y) + (2y - 1)\\ &= 4y(2y - 1) + 1(2y - 1)\\ &= (2y - 1)(4y + 1)\end{aligned}$$

69. $\underset{a}{18}x^2 + \underset{b}{31}x - \underset{c}{10}$

In $18x^2 + 31x - 10$, we have $a = 18$, $b = 31$ and $c = -10$. The key number is $ac = 18(-10) = -180$. We must find a factorization of -180 in which the sum of the factors is $b = 31$. Since the factors must have a negative product, their signs must be different. The pairs of factors are -5 and 36.

$$\begin{aligned}18x^2 + 31x - 10 &= 18x^2 - 5x + 36x - 10\\ &= (18x^2 - 5x) + (36x - 10)\\ &= x(18x - 5) + 2(18x - 5)\\ &= (18x - 5)(x + 2)\end{aligned}$$

71. $$\begin{aligned}-y^3 - 13y^2 - 12y &= -y(y^2 + 13y + 12)\\ &= -y(y + 12)(y + 1)\end{aligned}$$

73. $\underset{a}{10}u^2 - \underset{b}{13}u - \underset{c}{6}$

In $10u^2 - 13u - 6$, we have $a = 10$, $b = -13$ and $c = -6$. The key number is $ac = 10(-6) = -60$. We must find a factorization of -60 in which the sum of the factors is $b = -13$. Since the factors must have a negative product, their signs must be different. The pairs of factors are none. This trinomial is prime.

$10u^2 - 13u - 6$ is prime.

75. $-6x^4 + 15x^3 + 9x^2 = -3x^2(2x^2 - 5x - 3)$

In $2x^2 - 5x - 3$, we have $a = 2$, $b = -5$ and $c = -3$. The key number is $ac = 2(-3) = -6$. We must find a factorization of -6 in which the sum of the factors is $b = -5$. Since the factors must have a negative product, their signs must be different. The pairs of factors are -6 and 1.

$$\begin{aligned}-3x^2(2x^2 - 5x - 3) &= -3x^2(2x^2 - 6x + 1x - 3)\\ &= -3x^2[(2x^2 - 6x) + (1x - 3)]\\ &= -3x^2[2x(x - 3) + 1(x - 3)]\\ &= -3x^2(x - 3)(2x + 1)\end{aligned}$$

77. $\underset{a}{6}p^2 + \underset{b}{1}pq - \underset{c}{1}q^2$

In $6p^2 + 1pq - 1q^2$, we have $a = 6$, $b = 1$ and $c = -1$. The key number is $ac = 6(-1) = -6$. We must find a factorization of -6 in which the sum of the factors is $b = 1$. Since the factors must have a negative product, their signs must be different. The pairs of factors are -2 and 3.

$$\begin{aligned}6p^2 + 1pq - 1q^2 &= 6p^2 - 2pq + 3pq - 1q^2\\ &= (6p^2 - 2pq) + (3pq - 1q^2)\\ &= 2p(3p - q) + q(3p - q)\\ &= (3p - q)(2p + q)\end{aligned}$$

79. $30r^5 + 63r^4 - 30r^3 = 3r^3(10r^2 + 21r - 10)$

In $10r^2 + 21r - 10$, we have $a = 10$, $b = 21$ and $c = -10$. The key number is $ac = 10(-10) = -100$. We must find a factorization of -100 in which the sum of the factors is $b = 21$. Since the factors must have a negative product, their signs must be different. The pairs of factors are -4 and 25.

$$\begin{aligned}&3r^3(10r^2 + 21r - 10)\\ &= 3r^3(10r^2 - 4r + 25r - 10)\\ &= 3r^3[(10r^2 - 4r) + (25r - 10)]\\ &= 3r^3[2r(5r - 2) + 5(5r - 2)]\\ &= 3r^3(5r - 2)(2r + 5)\end{aligned}$$

81. $$\begin{aligned}&16m^3n + 20m^2n^2 + 6mn^3\\ &= 2mn(8m^2 + 10mn + 3n^2)\end{aligned}$$

$\underset{a}{8}m^2 + \underset{b}{10}mn + \underset{c}{3}n^2$

In $8m^2 + 10mn + 3n^2$, we have $a = 8$, $b = 10$ and $c = 3$. The key number is $ac = 8(3) = 24$. We must find a factorization of 24 in which the sum of the factors is $b = 10$. Since the factors must have a positive product, their signs must be the same. The pairs of factors are 6 and 4.

$$\begin{aligned}&2mn(8m^2 + 10mn + 3n^2)\\ &= 2mn(8m^2 + 6mn + 4mn + 3n^2)\\ &= 2mn[(8m^2 + 6mn) + (4mn + 3n^2)]\\ &= 2mn[2m(4m + 3n) + n(4m + 3n)]\\ &= 2mn(4m + 3n)(2m + n)\end{aligned}$$

83. $\underset{a}{3}x^2 + \underset{b}{1}x + \underset{c}{6}$

In $3x^2 + 1x + 6$, we have $a = 3$, $b = 1$ and $c = 6$. The key number is $ac = 3(6) = 18$. We must find a factorization of 18 in which the sum of the factors is $b = 1$. Since the factors must have a positive product, their signs must be the same. The pairs of factors are none. This trinomial is prime.

$3x^2 + x + 6$ is prime.

85. $-12y^2-12+25y=-12y^2+25y-12$
$$=-(12y^2-25y+12)$$

$\underset{a}{12}y^2-\underset{b}{25}y+\underset{c}{12}$

In $12y^2-25y+12$, we have $a=12$, $b=-25$ and $c=12$. The key number is $ac=12(12)=144$. We must find a factorization of 144 in which the sum of the factors is $b=-25$. Since the factors must have a positive product, their signs must be the same. The pairs of factors are -9 and -16.

$$\begin{aligned}&-(12y^2-25y+12)\\&=-[12y^2-16y-9y+12]\\&=-[(12y^2-16y)+(-9y+12)]\\&=-[4y(3y-4)-3(3y-4)]\\&=-(3y-4)(4y-3)\end{aligned}$$

Choose the correct method from Sections 6.1, 6.2, or 6.3 to factor each of the following.

87. $m^2+3m-28=(m+7)(m-4)$

89. $6a^3+15a^2=3a^2(2a+5)$

91. $x^3-2x^2+5x-10=(x^3-2x^2)+(5x-10)$
$$=x^2(x-2)+5(x-2)$$
$$=(x-2)(x^2+5)$$

93. $5y^2+3-8y=5y^2-8y+3$

$\underset{a}{5}y^2-\underset{b}{8}y+\underset{c}{3}$

In $5y^2-8y+3$, we have $a=5$, $b=-8$ and $c=3$. The key number is $ac=5(3)=15$. We must find a factorization of 15 in which the sum of the factors is $b=-8$. Since the factors must have a positive product, their signs must be the same. The pairs of factors are -3 and -5.

$$\begin{aligned}5y^2-8y+3&=5y^2-5y-3y+3\\&=(5y^2-5y)+(-3y+3)\\&=5y(y-1)-3(y-1)\\&=(y-1)(5y-3)\end{aligned}$$

95. $-2x^2-10x-12=-2(x^2+5x+6)$
$$=-2(x+2)(x+3)$$

97. $12x^3y^3-18x^2y^3+15x^2y^2$
$$=3x^2y^2(4xy-6y+5)$$

99. $a^2-7ab+10b^2=(a-5b)(a-2b)$

101. $9u^6-71u^5-8u^4=u^4(9u^2-71u-8)$

$\underset{a}{9}u^2-\underset{b}{71}u-\underset{c}{8}$

In $9u^2-71u-8$, we have $a=9$, $b=-71$ and $c=-8$. The key number is $ac=9(-8)=-72$. We must find a factorization of -72 in which the sum of the factors is $b=-71$. Since the factors must have a negative product, their signs must be different. The pairs of factors are 1 and -72.

$$\begin{aligned}u^4(9u^2-71u-8)&=u^4(9u^2-72u+1u-8)\\&=u^4[(9u^2-72u)+(1u-8)]\\&=u^4[9u(u-8)+1(u-8)]\\&=u^4(u-8)(9u+1)\end{aligned}$$

APPLICATIONS from CAMPUS to CAREERS

103. ELEMENTARY SCHOOL TEACHER

$\underset{a}{4}x^2+\underset{b}{20}x-\underset{c}{11}$

In $4x^2+20x-11$, we have $a=4$, $b=20$ and $c=-11$. The key number is $ac=4(-11)=-44$. We must find a factorization of -44 in which the sum of the factors is $b=20$. Since the factors must have a negative product, their signs must be different. The pairs of factors are -2 and 22.

$$\begin{aligned}4x^2+20x-11&=4x^2+22x-2x-11\\&=(4x^2+22x)+(-2x-11)\\&=2x(2x+11)-1(2x+11)\\&=(2x+11)(2x-1)\end{aligned}$$

The length is $(2x+11)$ in.
The width is $(2x-1)$ in.

WRITING

105-107. Answers will vary.

REVIEW

Evaluate each expression.

109. $-7^2=-(7)(7)$
$$=-49$$

111. $7^0=1$

113. $\frac{1}{7^{-2}}=7^2$
$$=49$$

CHALLENGE PROBLEMS

Factor.

115. $\underset{a}{6}a^{10}+\underset{b}{5}a^{5}\underset{c}{-21}$

In $6a^{10}+5a^{5}-21$, we have $a=6$, $b=5$ and $c=-21$. The key number is $ac=6(-21)=-126$. We must find a factorization of -126 in which the sum of the factors is $b=5$. Since the factors must have a negative product, their signs must be different. The pairs of factors are 14 and -9.

$$\begin{aligned}6a^{10}+5a^{5}-21&=6a^{10}-9a^{5}+14a^{5}-21\\&=(6a^{10}-9a^{5})+(14a^{5}-21)\\&=3a^{5}(2a^{5}-3)+7(2a^{5}-3)\\&=(2a^{5}-3)(3a^{5}+7)\end{aligned}$$

117. $8x^{2}(c^{2}+c-2)-2x(c^{2}+c-2)$
$-1(c^{2}+c-2)$

$$\begin{aligned}&=(c^{2}+c-2)(8x^{2}-2x-1)\\&=(c+2)(c-1)(4x+1)(2x-1)\end{aligned}$$

SECTION 6.4 STUDY SET
VOCABULARY

Fill in the blanks.

1. $x^2 + 6x + 9$ is a **perfect** -square trinomial because it is the square of the binomial $x + 3$.

CONCEPTS

3. Consider $25x^2 + 30x + 9$.
   a. The first term is the square of **5x**.
   b. The last term is the square of **3**.
   c. The middle term is twice the product of **5x** and **3**.
5. a. $x^2 + 2xy + y^2 = (\mathbf{x} + \mathbf{y})^2$
   b. $x^2 - 2xy + y^2 = (x - y)^2$
   c. $x^2 - y^2 = (x + y)(\mathbf{x} - \mathbf{y})$
7. List the squares of the integers from 1 through 20.

   1, 4, 9, 16, 25, 36, 49, 64, 81, 100, 121, 144, 169, 196, 225, 256, 289, 324, 361, 400

NOTATION

Complete each factorization.

9. $x^2 + 10x + 25 = (x + 5)^2$
11. $x^2 - 64 = (x + 8)(x - 8)$

GUIDED PRACTICE

Determine whether each of the following is a perfect-square trinomial. See Example 1.

13. $x^2 + 18x + 81$

    $2 \cdot x \cdot 9 = 18x$ **Yes**

15. $y^2 + 2y + 4$

    $2 \cdot y \cdot 2 = 4y$ **No**

17. $9n^2 - 30n - 25$

    The last term is negative. **No**

19. $4y^2 - 12y + 9$

    $2 \cdot 2y \cdot 3 = 12y$ **Yes**

Factor. See Example 2.

21. $x^2 + 6x + 9$

    The first term x^2 is the square of $\mathbf{x}$.
    The last term 9 is the square of **3**.
    The middle term is twice the product of x and 3: $2(x)(3) = \mathbf{6x}$.

    $$x^2 + 6x + 9 = (x + 3)(x + 3) = (x + 3)^2$$

23. $b^2 + 2b + 1$

    The first term b^2 is the square of $\mathbf{b}$.
    The last term 1 is the square of **1**.
    The middle term is twice the product of b and 1: $2(b)(1) = \mathbf{2b}$.

    $$b^2 + 2b + 1 = (b + 1)(b + 1) = (b + 1)^2$$

25. $c^2 - 12c + 36$

    The first term c^2 is the square of $\mathbf{c}$.
    The last term 36 is the square of $\mathbf{-6}$.
    The middle term is twice the product of c and -6: $2(c)(-6) = \mathbf{-12c}$.

    $$c^2 - 12c + 36 = (c - 6)(c - 6) = (c - 6)^2$$

27. $9 + 4x^2 + 12x = 4x^2 + 12x + 9$

    The first term $4x^2$ is the square of $\mathbf{2x}$.
    The last term 9 is the square of **3**.
    The middle term is twice the product of $2x$ and 3: $2(2x)(3) = \mathbf{12x}$.

    $$4x^2 + 12x + 9 = (2x + 3)(2x + 3) = (2x + 3)^2$$

29. $36m^2 + 60mn + 25n^2$

    The first term $36m^2$ is the square of $\mathbf{6m}$.
    The last term $25n^2$ is the square of $\mathbf{5n}$.
    The middle term is twice the product of $6m$ and $5n$: $2(6m)(5n) = \mathbf{60mn}$.

    $$36m^2 + 60mn + 25n^2 = (6m + 5n)(6m + 5n) = (6m + 5n)^2$$

31. $81x^2 - 72xy + 16y^2$

    The first term $81x^2$ is the square of $\mathbf{9x}$.
    The last term $16y^2$ is the square of $\mathbf{-4y}$.
    The middle term is twice the product of $9x$ and $-4y$: $2(9x)(-4y) = \mathbf{-72xy}$.

    $$81x^2 - 72xy + 16y^2 = (9x - 4y)(9x - 4y) = (9x - 4y)^2$$

Factor. See Example 3.

33. $3u^2 - 18u + 27 = 3(u^2 - 6u + 9)$

The first term u^2 is the square of $\boldsymbol{u}$.
The last term 9 is the square of $-\mathbf{3}$.
The middle term is twice the product of u and -3: $2(u)(-3) = \mathbf{-6u}$.

$$3(u^2 - 6u + 9) = 3(u-3)(u-3) = 3(u-3)^2$$

35. $36x^3 + 12x^2 + x = x(36x^2 + 12x + 1)$

The first term $36x^2$ is the square of $\mathbf{6x}$.
The last term 1 is the square of $\mathbf{1}$.
The middle term is twice the product of $6x$ and 1: $2(6x)(1) = \mathbf{12x}$.

$$x(36x^2 + 12x + 1) = x(6x+1)(6x+1) = x(6x+1)^2$$

Factor. If a polynomial can't be factored, write "prime." See Example 4.

37. $x^2 - 4$

$$F^2 - L^2 = (F+L)(F-L)$$
$$\downarrow \quad \downarrow \quad \downarrow \quad \downarrow \quad \downarrow \quad \downarrow$$
$$x^2 - 2^2 = (x+2)\ (x-2)$$

39. $x^2 - 16$

$$F^2 - L^2 = (F+L)(F-L)$$
$$\downarrow \quad \downarrow \quad \downarrow \quad \downarrow \quad \downarrow \quad \downarrow$$
$$x^2 - 4^2 = (x+4)\ (x-4)$$

41. $36 - y^2$

$$F^2 - L^2 = (F+L)(F-L)$$
$$\downarrow \quad \downarrow \quad \downarrow \quad \downarrow \quad \downarrow \quad \downarrow$$
$$6^2 - y^2 = (6+y)\ (6-y)$$

43. $t^2 - 25$

$$(F^2 - L^2) = (F+L)(F-L)$$
$$\downarrow \quad \downarrow \quad \downarrow \quad \downarrow \quad \downarrow \quad \downarrow$$
$$(t^2 - 5^2) = (t+5)(t-5)$$

45. $a^2 + b^2$ is prime.

47. $y^2 - 63$ is prime.

Factor. See Example 5.

49. $25t^2 - 64$

$$F^2 - L^2 = (F+L)(F-L)$$
$$\downarrow \quad \downarrow \quad \downarrow \quad \downarrow \quad \downarrow \quad \downarrow$$
$$(5t)^2 - 8^2 = (5t+8)\ (5t-8)$$

51. $81y^2 - 1$

$$F^2 - L^2 = (F+L)\ (F-L)$$
$$\downarrow \quad \downarrow \quad \downarrow \quad \downarrow \quad \downarrow \quad \downarrow$$
$$(9y)^2 - 1^2 = (9y+1)\ (9y-1)$$

53. $9x^4 - y^2$

$$F^2 - L^2 = (F + L)\ (F - L)$$
$$\downarrow \quad \downarrow \quad \downarrow \quad \downarrow \quad \downarrow \quad \downarrow$$
$$(3x^2)^2 - y^2 = (3x^2 + y)\ (3x^2 - y)$$

55. $-49d^4 + 16c^2 = 16c^2 - 49d^4$

$$F^2 - L^2 = (F + L)\ (F - L)$$
$$\downarrow \quad \downarrow \quad \downarrow \quad \downarrow \quad \downarrow \quad \downarrow$$
$$(4c)^2 - (7d^2)^2 = (4c + 7d^2)\ (4c - 7d^2)$$

Factor. See Example 6.

57. $8x^2 - 32y^2 = 8(x^2 - 4y^2)$
$= 8(x+2y)(x-2y)$

59. $63a^2 - 7 = 7(9a^2 - 1)$
$= 7(3a+1)(3a-1)$

Factor. See Example 7.

61. $81 - s^4 = 9^2 - (s^2)^2$
$= (9+s^2)(9-s^2)$
$= (9+s^2)(3+s)(3-s)$

63. $b^4 - 256 = (b^2)^2 - (16)^2$
$= (b^2 + 16)(b^2 - 16)$
$= (b^2 + 16)(b+4)(b-4)$

TRY IT YOURSELF

Factor.

65. $a^4 - 144b^2 = (a^2)^2 - (12b)^2$
$= (a^2 + 12b)(a^2 - 12b)$

67. $9x^2y^2 + 30xy + 25$

The first term $9x^2y^2$ is the square of $\mathbf{3xy}$.
The last term 25 is the square of $\mathbf{5}$.
The middle term is twice the product of $3xy$ and 5: $2(3xy)(5) = \mathbf{30xy}$.

$$9x^2y^2 + 30xy + 25 = (3xy+5)(3xy+5) = (3xy+5)^2$$

69. $16t^4 - 16s^4 = 16(t^4 - s^4)$
$= 16[(t^2)^2 - (s^2)^2]$
$= 16(t^2 + s^2)(t^2 - s^2)$
$= 16(t^2 + s^2)(t+s)(t-s)$

71. $t^2 - 20t + 100$

The first term t^2 is the square of $\mathbf{t}$.
The last term 100 is the square of $\mathbf{-10}$.
The middle term is twice the product of t and -10: $2(t)(-10) = \mathbf{-20t}$.

$$t^2 - 20t + 100 = (t-10)(t-10) = (t-10)^2$$

73. $9y^2 - 24y + 16$

The first term $9y^2$ is the square of $\mathbf{3y}$.
The last term 16 is the square of $\mathbf{-4}$.
The middle term is twice the product of $3y$ and -4: $2(3y)(-4) = \mathbf{-24y}$.

$$9y^2 - 24y + 16 = (3y-4)(3y-4) = (3y-4)^2$$

75. $z^2 - 64 = (z+8)(z-8)$

77. $25m^4 - 25 = 25(m^4 - 1)$
$= 25[(m^2)^2 - 1]$
$= 25(m^2+1)(m^2-1)$
$= 25(m^2+1)(m+1)(m-1)$

79. $18a^5 + 84a^4b + 98a^3b^2$
$= 2a^3(9a^2 + 42ab + 49b^2)$

The first term $9a^2$ is the square of $\mathbf{3a}$.
The last term $49b^2$ is the square of $\mathbf{7b}$.
The middle term is twice the product of $3a$ and$7b$: $2(3a)(7b) = \mathbf{42ab}$.

$2a^3(9a^2 + 42ab + 49b^2)$
$= 2a^3(3a+7b)(3a+7b)$
$= 2a^3(3a+7b)^2$

81. $x^3 - 144x = x(x^2 - 144)$
$= x(x+12)(x-12)$

83. $49t^2 - 28ts + 4s^2$

The first term $49t^2$ is the square of $\mathbf{7t}$.
The last term $4s^2$ is the square of $\mathbf{-2s}$.
The middle term is twice the product of $7t$ and $-2s$: $2(7t)(-2s) = \mathbf{-28ts}$.

$$49t^2 - 28ts + 4s^2 = (7t-2s)(7t-2s) = (7t-2s)^2$$

85. $3m^4 - 3n^4 = 3(m^4 - n^4)$
$= 3[(m^2)^2 - (n^2)^2]$
$= 3(m^2+n^2)(m^2-n^2)$
$= 3(m^2+n^2)(m+n)(m-n)$

87. $25m^2 + 70m + 49$

The first term $25m^2$ is the square of $\mathbf{5m}$.
The last term 49 is the square of **7**.
The middle term is twice the product of $5m$ and 7: $2(5m)(7) = \mathbf{70m}$.

$$25m^2 + 70m + 49 = (5m+7)(5m+7) = (5m+7)^2$$

89. $-100t^2 + 20t - 1 = -(100t^2 - 20t + 1)$

The first term $100t^2$ is the square of $\mathbf{10t}$.
The last term 1 is the square of $\mathbf{-1}$.
The middle term is twice the product of $10t$ and -1: $2(10t)(-1) = \mathbf{-20t}$.

$$-(100t^2 - 20t + 1) = -(10t-1)(10t-1) = -(10t-1)^2$$

91. $6x^4 - 6x^2y^2 = 6x^2(x^2 - y^2)$
$= 6x^2(x+y)(x-y)$

93. $100a^2 + 81$ is prime.

95. $-169 + 25x^2 = -(169 - 25x^2)$
$= -(13+5x)(13-5x)$

or $-169 + 25x^2 = 25x^2 - 169$
$= (5x+13)(5x-13)$

Choose the correct method from Section 6.1, Section 6.2, Section 6.3, or Section 6.4 to factor each of the following:

97. Two different sign factors of -42 whose sum is $+1$ are $+7$ and -6.
$x^2 + x - 42 = (x+7)(x-6)$

99. $x^2 - 9 = (x+3)(x-3)$

101. $24a^3b - 16a^2b = 8a^2b(3a-2)$

103. $-2r^2 + 28r - 80 = -2(r^2 - 14r + 40)$

The negative factors of 40 whose sum is -14 are -10 and -4.
$= -2(r-10)(r-4)$

105. $x^3+3x^2+4x+12=(x^3+3x^2)+(4x+12)$

$$=(\mathbf{x^2x^1}+\mathbf{3x^2})+(\mathbf{4}x+\mathbf{4}\cdot 3)$$
$$=x^2(x+3)+4(x+3)$$
$$=(x+3)(x^2+4)$$

107. $4b^2-20b+25$

The first term $4b^2$ is the square of $2\boldsymbol{b}$.
The last term 25 is the square of $-\mathbf{5}$.
The middle term is twice the product of $2b$ and -5: $2(2b)(-5)=\mathbf{-20b}$.

$$4b^2-20b+25=(2b-5)(2b-5)$$
$$=(2b-5)^2$$

APPLICATIONS

109. GENETICS

$$p^2+2pq+q^2=(p+q)(p+q)$$
$$=(p+q)^2$$

111. PHYSICS

$$0.5gt_1^{\ 2}-0.5gt_2^{\ 2}=0.5g\left(t_1^{\ 2}-t_2^{\ 2}\right)$$
$$=0.5g\left(t_1+t_2\right)\left(t_1-t_2\right)$$

The distance is $0.5g\left(t_1+t_2\right)\left(t_1-t_2\right)$.

WRITING

113-115. Answers will vary.

REVIEW

Perform each division.

117. $\dfrac{-30c^2d^2-15c^2d-10cd^2}{-10cd}$

$$=\frac{-30c^2d^2}{-10cd}-\frac{15c^2d}{-10cd}-\frac{10cd^2}{-10cd}$$
$$=\frac{3c^{2-1}d^{2-1}}{1}+\frac{3c^{2-1}d^{1-1}}{2}+\frac{1c^{1-1}d^{2-1}}{1}$$
$$\frac{3c^1d^1}{1}+\frac{3c^1d^0}{2}+\frac{1c^0d^1}{1}$$
$$=3cd+\frac{3c}{2}+d$$

CHALLENGE PROBLEMS

119. For what value of c does $80x^2-c$ factor as $5(4x+3)(4x-3)$? **5·3·3 = 45**

Factor completely.

121. $81x^6+36x^3y^2+4y^4$

The first term $81x^6$ is the square of $\mathbf{9x^3}$.
The last term $4y^4$ is the square of $\mathbf{2y^2}$.
The middle term is twice the product of $9x^3$ and $2y^2$: $2(9x^3)(2y^2)=\mathbf{36x^3y^2}$.

$$81x^6+36x^3y^2+4y^4=(9x^3+2y^2)(9x^3+2y^2)$$
$$=(9x^3+2y^2)^2$$

123. $c^2+1.6c+0.64$

The first term c^2 is the square of $\boldsymbol{c}$.
The last term 0.64 is the square of **0.8**.
The middle term is twice the product of c and 0.8: $2(c)(0.8)=\mathbf{1.6c}$.

$$c^2+1.6c+0.64=(c+0.8)(c+0.8)$$
$$=(c+0.8)^2$$

125. $(x+5)^2-y^2=(x+5+y)(x+5-y)$

127. $c^2-\dfrac{1}{16}=c^2-\left(\dfrac{1}{4}\right)^2$

$$=\left(c+\frac{1}{4}\right)\left(c-\frac{1}{4}\right)$$

SECTION 6.5 STUDY SET

VOCABULARY

Fill in the blanks.

1. $x^3 + 27$ is the **sum** of two cubes and $a^3 - 125$ is the difference of two **cubes**.

CONCEPTS

Fill in the blanks.

3. a. $F^3 + L^3 = (\mathbf{F + L})(F^2 - FL + L^2)$

   b. $F^3 - L^3 = (F \mathbf{-} L)(\mathbf{F^2} + FL + \mathbf{L^2})$

5. $216n^3 - 125$

   ↑ This is **6n** cubed. ↑ This is **5** cubed.

7. List the first ten positive integer cubes.

   1, 8, 27, 64, 125, 216, 343, 512, 729, 1,000

9. Use multiplication to determine if the factorization is correct.

   $b^3 + 27 = (b + 3)(b^2 + 3b + 9)$

$$\begin{array}{r} b^2 + 3b + 9 \\ b + 3 \\ \hline 3b^2 + 9b + 27 \\ b^3 + 3b^2 + 9b \\ \hline b^3 + 6b^2 + 18b + 27 \end{array}$$

   No

NOTATION

Complete each factorization.

11. $a^3 + 8 = (a + 2)(a^2 - \mathbf{2a} + 4)$

13. $b^3 + 27 = (\mathbf{b + 3})(b^2 - 3b + 9)$

Give an example of each type of expression.

15. a. the sum of two cubes $\mathbf{x^3 + 8}$

    b. the cube of a sum $\mathbf{(x + 8)^3}$

GUIDED PRACTICE

Factor. See Example 1.

17. $y^3 + 125 = y^3 + 5^3$
$$= (y+5)(y^2 - y \cdot 5 + 5^2)$$
$$= (y+5)(y^2 - 5y + 25)$$

19. $a^3 + 64 = a^3 + 4^3$
$$= (a+4)(a^2 - a \cdot 4 + 4^2)$$
$$= (a+4)(a^2 - 4a + 16)$$

21. $n^3 + 512 = n^3 + 8^3$
$$= (n+8)(n^2 - n \cdot 8 + 8^2)$$
$$= (n+8)(n^2 - 8n + 64)$$

23. $8 + t^3 = 2^3 + t^3$
$$= (2+t)(2^2 - 2 \cdot t + t^2)$$
$$= (2+t)(4 - 2t + t^2)$$

25. $a^3 + 1{,}000b^3$
$$= a^3 + (10b)^3$$
$$= (a+10b)[a^2 - a \cdot 10b + (10b)^2]$$
$$= (a+10b)(a^2 - 10ab + 100b^2)$$

27. $125c^3 + 27d^3$
$$= (5c)^3 + (3d)^3$$
$$= (5c+3d)[(5c)^2 - 5c \cdot 3d + (3d)^2]$$
$$= (5c+3d)(25c^2 - 15cd + 9d^2)$$

Factor. See Example 2.

29. $a^3 - 27 = a^3 - 3^3$
$$= (a-3)(a^2 + a \cdot 3 + 3^2)$$
$$= (a-3)(a^2 + 3a + 9)$$

31. $m^3 - 343 = m^3 - 7^3$
$$= (m-7)(m^2 + m \cdot 7 + 7^2)$$
$$= (m-7)(m^2 + 7m + 49)$$

33. $216 - v^3 = 6^3 - v^3$
$$= (6-v)(6^2 + 6 \cdot v + v^2)$$
$$= (6-v)(36 + 6v + v^2)$$

35. $8s^3 - t^3 = (2s)^3 - t^3$
$$= (2s-t)[(2s)^2 + 2s \cdot t + t^2)]$$
$$= (2s-t)(4s^2 + 2st + t^2)$$

37. $1{,}000a^3 - w^3$
$$= (10a)^3 - w^3$$
$$= (10a-w)[(10a)^2 + 10a \cdot w + (w)^2]$$
$$= (10a-w)(100a^2 + 10aw + w^2)$$

39. $64x^3 - 27y^3$
$$= (4x)^3 - (3y)^3$$
$$= (4x-3y)[(4x)^2 + 4x \cdot 3y + (3y)^2]$$
$$= (4x-3y)(16x^2 + 12xy + 9y^2)$$

Factor. See Example 3.

41. $2x^3 + 2 = 2(x^3 + 1)$
$$= 2(x^3 + 1^3)$$
$$= 2(x+1)(x^2 - x \cdot 1 + 1^2)$$
$$= 2(x+1)(x^2 - x + 1)$$

43. $3d^3+81=3(d^3+27)$
$$=3(d^3+3^3)$$
$$=3(d+3)(d^2-d\cdot 3+3^2)$$
$$=3(d+3)(d^2-3d+9)$$

45. $x^4-216x=x(x^3-216)$
$$=x(x^3-6^3)$$
$$=x(x-6)(x^2+x\cdot 6+6^2)$$
$$=x(x-6)(x^2+6x+36)$$

47. $64m^3x-8n^3x$
$$=8x(8m^3-n^3)$$
$$=8x((2m)^3-n^3)$$
$$=8x(2m-n)[(2m)^2+2m\cdot n+n^2]$$
$$=8x(2m-n)(4m^2+2mn+n^2)$$

TRY IT YOURSELF

Choose the correct method from Section 6.1 through Section 6.5 and factor completely.

49. $x^2+8x+16$

The first term x^2 is the square of $\boldsymbol{x}$.
The last term 16 is the square of **4**.
The middle term is twice the product of x and 4: $2(x)(4)=\mathbf{8}\boldsymbol{x}$.

$$x^2+8x+16=(x+4)(x+4)$$
$$=(x+4)^2$$

51. $9r^2-16s^2=(3r+4s)(3r-4s)$

53. $xy-ty+sx-st=(xy-ty)+(sx-st)$
$$=y(x-t)+s(x-t)$$
$$=(x-t)(y+s)$$

55. $4p^3+32q^3=4(p^3+8q^3)$
$$=4[(p^3+(2q)^3]$$
$$=4[(p+2q)(p^2-p\cdot 2q+(2q)^2]$$
$$=4(p+2q)(p^2-2pq+4q^2)$$

57. $16c^3t^2+20c^2t^3+6ct^4$
$$=\mathbf{2}\cdot 8\mathbf{c^1}c^2\boldsymbol{t^2}+\mathbf{2}\cdot 10\mathbf{c^1}c^1t^1\boldsymbol{t^2}+\mathbf{2}\cdot 3\mathbf{c^1}t^2\boldsymbol{t^2}$$
$$=2ct^2(8c^2+10ct+3t^2)$$

$$\underset{a}{8}c^2+\underset{b}{10}ct+\underset{c}{3}t^2$$

In $8c^2+10ct+3t^2$, we have $a=8$, $b=10$ and $c=3$. The key number is $ac=8(3)=24$. We must find a factorization of 24 in which the sum of the factors is $b=10$. Since the factors must have a positive product, their signs must be the same. The pairs of factors are +4 and +6.

$$2ct^2(8c^2+10ct+3t^2)$$
$$=2ct^2[8c^2+4ct+6ct+3t^2]$$
$$=2ct^2[(8c^2+4ct)+(6ct+3t^2)]$$
$$=2ct^2[4c(2c+t)+3t(2c+t)]$$
$$=2ct^2(2c+t)(4c+3t)$$

59. $36e^4-36=36(e^4-1)$
$$=36[(e^2)^2-1)]$$
$$=36(e^2+1)(e^2-1)$$
$$=36(e^2+1)(e+1)(e-1)$$

61. $35a^3b^2-14a^2b^3+14a^3b^3$
$$=\mathbf{7}\cdot 5\boldsymbol{a^2}a^1\boldsymbol{b^2}-\mathbf{7}\cdot 2\boldsymbol{a^2}b^1\boldsymbol{b^2}+\mathbf{7}\cdot 2a^1\boldsymbol{a^2}b^1\boldsymbol{b^2}$$
$$=7a^2b^2(5a-2b+2ab)$$

63. $36r^2+60rs+25s^2$

The first term $36r^2$ is the square of $\mathbf{6}\boldsymbol{r}$.
The last term $25s^2$ is the square of $\mathbf{5}\boldsymbol{s}$.
The middle term is twice the product of $6r$ and $5s$: $2(6r)(5s)=\mathbf{60}\boldsymbol{rs}$.

$$36r^2+60rs+25s^2=(6r+5s)(6r+5s)$$
$$=(6r+5s)^2$$

LOOK ALIKES . . .

65. a. $x^2-1=(x+1)(x-1)$
b. $x^3-1=(x-1)(x^2+x+1)$

67. a. $x^2+2x=x(x+2)$
b. $x^2+2x+1=(x+1)(x+1)$
$$=(x+1)^2$$

APPLICATIONS

69. MAILING BREAKABLES

Find volume of the larger box.

$$V = lwh$$
$$= 10 \cdot 10 \cdot 10$$
$$= 1{,}000$$

The volume is 1,000 in^3.

Find volume of the smaller box.

$$V = lwh$$
$$= x \cdot x \cdot x$$
$$= x^3$$

The volume is x^3 in^3.
Subtract the samller volume from the larger volume.

$$V_L - V_S = 1{,}000 - x^3$$

The volume of packing is $(1{,}000 - x^3)$ in^3.

$$1{,}000 - x^3 = 10^3 - x^3$$
$$= (10 - x)(10^2 + 10 \cdot x + x^2)$$
$$= (10 - x)(100 + 10x + x^2)$$

WRITING

71. Answers will vary.

REVIEW

73. $\frac{7}{9} = 0.\overline{7}$, repeating decimal

75. Solve.

$$2x + 2 = \frac{2}{3}x - 2$$
$$\mathbf{3} \cdot (2x + 2) = \mathbf{3} \cdot \left(\frac{2}{3}x - 2\right)$$
$$6x + 6 = 2x - 6$$
$$6x + 6 - \mathbf{6} = 2x - 6 - \mathbf{6}$$
$$6x = 2x - 12$$
$$6x - \mathbf{2x} = 2x - 12 - \mathbf{2x}$$
$$4x = -12$$
$$\frac{4x}{\mathbf{4}} = \frac{-12}{\mathbf{4}}$$
$$x = -3$$

CHALLENGE PROBLEMS

77. a) $(x^3)^2 - 1^2$

$$= (x^3 + 1)(x^3 - 1)$$
$$= (x + 1)(x^2 - x + 1)$$
$$(x - 1)(x^2 + x + 1)$$

Factor.

79. $x^6 - y^9 = (x^2)^3 - (y^3)^3$

$$= (x^2 - y^3)[(x^2)^2 + x^2 \cdot y^3 + (y^3)^2]$$
$$= (x^2 - y^3)(x^4 + x^2y^3 + y^6)$$

81. $64x^{12} + y^{15}z^{18}$

$$= (4x^4)^3 + (y^5z^6)^3$$
$$= (4x^4 + y^5z^6)[(4x^4)^2 - 4x^4 \cdot y^5z^6 + (y^5z^6)^2]$$
$$= (4x^4 + y^5z^6)(16x^8 - 4x^4y^5z^6 + y^{10}z^{12})$$

SECTION 6.6 STUDY SET
VOCABULARY

Fill in the blanks.

1. To factor a polynomial means to express it as a **product** of two (or more) polynomials.

CONCEPTS

For each of the following polynomials, which factoring method would you use first?

3. $2x^5y - 4x^3y$

   Factor out the GCF

5. $x^2 + 18x + 81$

   Perfect square trinomial

7. $x^3 + 27$

   Sum of two cubes

9. $m^2 + 3mn + 2n^2$

   Trinomial factoring

11. What is the first question that should be asked when using the strategy of this section to factor a polynomial? **Is there a common factor?**

NOTATION

Complete each factorization.

13. $6m^3 - 28m^2 + 16m = 2m(3m^2 - \mathbf{14m} + 8)$
$= 2m(3m - 2)(\mathbf{m} - 4)$

TRY IT YOURSELF

The following is a list of random factoring problems. Factor each expression. If an expression is not factorable, write "prime." See Examples 1–5.

15. $2b^2 + 8b - 24 = 2(b^2 + 4b - 12)$

Two different sign factors of -12 whose sum is $+4$ are $+6$ and -2.

$2(b^2 + 4b - 12) = 2(b + 6)(b - 2)$

17. $8p^3q^7 + 4p^2q^3 = \mathbf{4 \cdot 2p^2pq^3q^4 + 4p^2q^3}$
$= 4p^2q^3(2pq^4 + 1)$

19. $2 + 24y + 40y^2 = 2(20y^2 + 12y + 1)$

The positive factors of 20 whose sum is 12 are 10 and 2.

$2(20y^2 + 12y + 1) = 2(2y + 1)(10y + 1)$

21. $8x^4 - 8 = 8(x^4 - 1)$
$= 8(x^2 + 1)(x^2 - 1)$
$= 8(x^2 + 1)(x + 1)(x - 1)$

23. $14c - 147 + c^2 = c^2 + 14c - 147$

Two different sign factors of -147 whose sum is $+14$ are $+21$ and -7.

$c^2 + 14c - 147 = (c + 21)(c - 7)$

25. $x^2 + 7x + 1$ is prime.

27. $-2x^5 + 128x^2 = -2x^2(x^3 - 64)$
$= -2x^2(x^3 - 4^3)$
$= -2x^2(x - 4)(x^2 + x \cdot 4 + 4^2)$
$= -2x^2(x - 4)(x^2 + 4x + 16)$

29. $a^2c + a^2d^2 + bc + bd^2$
$= (\mathbf{a^2}c + \mathbf{a^2}d^2) + (\mathbf{b}c + \mathbf{b}d^2)$
$= a^2(c + d^2) + b(c + d^2)$
$= (c + d^2)(a^2 + b)$

31. $-9x^2 + 6x - 1 = -(9x^2 - 6x + 1)$

$-(9x^2 - 6x + 1)$

The first term $9x^2$ is the square of $\mathbf{3x}$.

The last term 1 is the square of $\mathbf{-1}$.

The middle term is twice the product of $3x$ and -1: $2(3x)(-1) = \mathbf{-6x}$.

$-(9x^2 - 6x + 1) = -(3x - 1)(3x - 1)$
$= -(3x - 1)^2$

33. $-20m^3 - 100m^2 - 125m$
$= -5m(4m^2 + 20m + 25)$

$4m^2 + 20m + 25$

The first term $4m^2$ is the square of $\mathbf{2m}$.

The last term 25 is the square of **5**.

The middle term is twice the product of $2m$ and 5: $2(2m)(5) = \mathbf{20m}$.

$-5m(4m^2 + 20m + 25)$
$= -5m(2m + 5)(2m + 5)$
$= -5m(2m + 5)^2$

35. $\underset{a}{2}c^2 - \underset{b}{5}cd - \underset{c}{3}d^2$

In $2c^2 - 5cd - 3c^2$, we have $a = 2$, $b = -5$, and $c = -3$. The key number is $ac = 2(-3) = -6$. We must find a factorization of -6 in which the sum of the factors is $b = -5$. Since the factors must have a negative product, their signs must be different. The pairs of factors are -6 and 1.

$2c^2 - 5cd - 3d^2 = 2c^2 - 6cd + cd - 3d^2$
$= (2c^2 - 6cd) + (cd - 3d^2)$
$= 2c(c - 3d) + d(c - 3d)$
$= (c - 3d)(2c + d)$

37. $p^4 - 2p^3 - 8p + 16$
$$= (p^4 - 2p^3) + (-8p + 16)$$
$$= p^3(p-2) - 8(p-2)$$
$$= (p-2)(p^3 - 8)$$
$$= (p-2)(p^3 - 2^3)$$
$$= (p-2)(p-2)(p^2 + p\cdot 2 + 2^2)$$
$$= (p-2)(p-2)(p^2 + 2p + 4)$$
$$= (p-2)^2(p^2 + 2p + 4)$$

39. $a^2(x-a) - b^2(x-a) = (x-a)(a^2 - b^2)$
$$= (x-a)(a+b)(a-b)$$

41. $a^2b^2 - 144 = (ab + 12)(ab - 12)$

43. $2x^3 + 10x^2 + x + 5 = (2x^3 + 10x^2) + (x+5)$
$$= 2x^2(x+5) + 1(x+5)$$
$$= (x+5)(2x^2 + 1)$$

45. $8v^2 - 14v^3 + v^4 = v^4 - 14v^3 + 8v^2$
$$= v^2(v^2 - 14v + 8)$$

47. $18a^2 - 6ab + 42ac - 14bc$
$$= \mathbf{2}\cdot 9a^2 - \mathbf{2}\cdot 3ab + \mathbf{2}\cdot 21ac - \mathbf{2}\cdot 7bc$$
$$= 2(9a^2 - 3ab + 21ac - 7bc)$$
$$= 2\left[(9a^2 - 3ab) + (21ac - 7bc)\right]$$
$$= 2\left[(\mathbf{3}\cdot 3\mathbf{a}^1a^1 - \mathbf{3a}^1b) + (\mathbf{7}\cdot 3ac - \mathbf{7bc})\right]$$
$$= 2\left[3a(3a-b) + 7c(3a-b)\right]$$
$$= 2(3a-b)(3a+7c)$$

49. $8a^2x^3 - 2b^2x = 2x(4a^2x^2 - b^2)$
$$= 2x(2ax + b)(2ax - b)$$

51. $6x^2 - 14x + 8 = 2(3x^2 - 7x + 4)$

$\underset{a}{3x^2} \underset{b}{-7x} \underset{c}{+4}$

In $3x^2 - 7x + 4$, we have $a = 3$, $b = -7$, and $c = 4$. The key number is $ac = 3(4) = 12$. We must find a factorization of 12 in which the sum of the factors is $b = -7$. Since the factors must have a positive product, their signs must be the same. The pairs of factors are -3 and -4.

$$2(3x^2 - 7x + 4) = 2(3x^2 - 4x - 3x + 4)$$
$$= 2[(3x^2 - 4x) + (-3x + 4)]$$
$$= 2[x(3x-4) - 1(3x-4)]$$
$$= 2(3x-4)(x-1)$$

53. $4x^2y^2 + 4xy^2 + y^2 = y^2(4x^2 + 4x + 1)$

$4x^2 + 4x + 1$

The first term $4x^2$ is the square of $\mathbf{2x}$.
The last term 1 is the square of **1**.
The middle term is twice the product of $2x$ and 1: $2(2x)(1) = \mathbf{4x}$.

$$y^2(4x^2 + 4x + 1) = y^2(2x+1)(2x+1)$$
$$= y^2(2x+1)^2$$

55. $4m^5 + 500m^2 = 4m^2(m^3 + 125)$
$$= 4m^2(m^3 + 5^3)$$
$$= 4m^2(m+5)(m^2 - m\cdot 5 + 5^2)$$
$$= 4m^2(m+5)(m^2 - 5m + 25)$$

57. $a^3 - 24 - 4a + 6a^2$
$$= a^3 + 6a^2 - 4a - 24$$
$$= (\mathbf{a^2}a^1 + 6\mathbf{a^2}) + (\mathbf{-4}a - \mathbf{4}\cdot 6)$$
$$= a^2(a+6) - 4(a+6)$$
$$= (a+6)(a^2 - 4)$$
$$= (a+6)(a+2(a-2)$$

59. $4x^2 + 9y^2$ is prime.

61. $16a^5 - 54a^2 = 2a^2(8a^3 - 27)$
$$= 2a^2[(2a)^3 - 3^3]$$
$$= 2a^2[2a-3][(2a)^2 + 2a\cdot 3 + 3^2]$$
$$= 2a^2(2a-3)(4a^2 + 6a + 9)$$

63. $27x - 27y - 27z = 27(x - y - z)$

65. $xy - ty + xs - ts = (xy - ty) + (xs - ts)$
$$= y(x-t) + s(x-t)$$
$$= (x-t)(y+s)$$

67. $35x^8 - 2x^7 - x^6 = x^6(35x^2 - 2x - 1)$
$$= x^6(7x+1)(5x-1)$$

69. $5(x-2) + 10y(x-2) = (x-2)(5+10y)$
$$= (x-2)5(1+2y)$$
$$= 5(x-2)(1+2y)$$

71. $49p^2 + 28pq + 4q^2$

The first term $49p^2$ is the square of $\mathbf{7p}$.
The last term $4q^2$ is the square of $\mathbf{2q}$.
The middle term is twice the product of $7p$ and $2q$: $2(7p)(2q) = \mathbf{28pq}$.

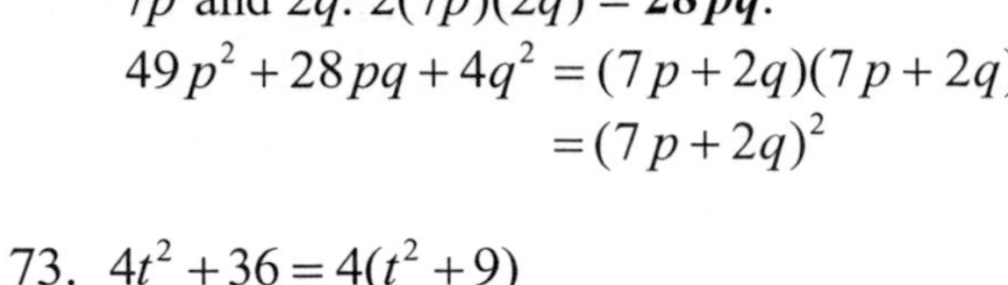

$$49p^2 + 28pq + 4q^2 = (7p+2q)(7p+2q) = (7p+2q)^2$$

73. $4t^2 + 36 = 4(t^2+9)$

75. $m^2n^2 - 9m^2 + 3n^2 - 27$

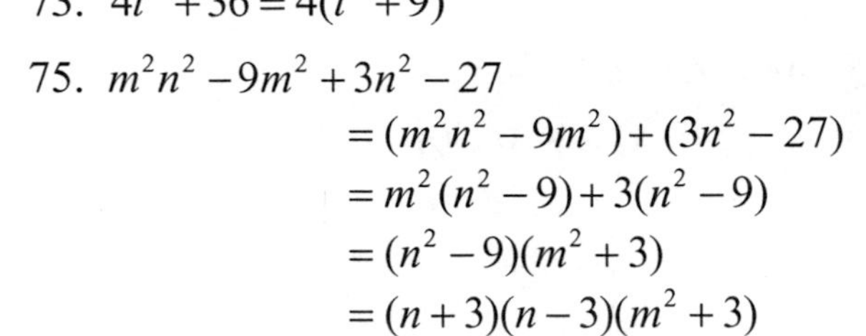

$$= (m^2n^2 - 9m^2) + (3n^2 - 27)$$
$$= m^2(n^2-9) + 3(n^2-9)$$
$$= (n^2-9)(m^2+3)$$
$$= (n+3)(n-3)(m^2+3)$$

WRITING

77-79. Answers will vary.

REVIEW

81. Graph the real numbers $-3, 0, 2,$ and $-\frac{3}{2}$ on a number line.

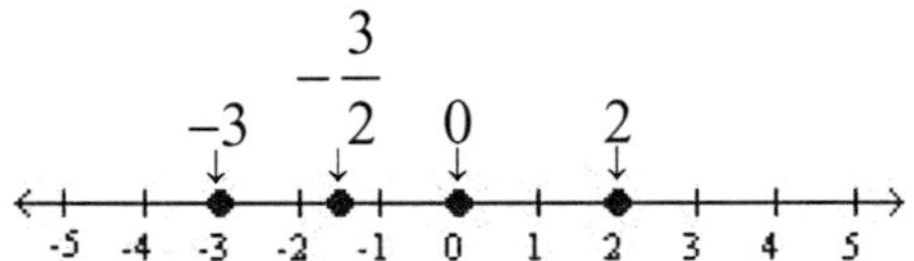

83. Graph: $y = \frac{1}{2}x + 1$.

x	y	(x, y)
-2	$y = \frac{1}{2}x+1$ $= \frac{-2}{2}+1$ $= -1+1$ $= 0$	$(-2, 0)$
0	$y = \frac{1}{2}x+1$ $= \frac{0}{2}+1$ $= 1$	$(0, 1)$
2	$y = \frac{1}{2}x+1$ $= \frac{2}{2}+1$ $= 1+1$ $= 2$	$(2, 2)$

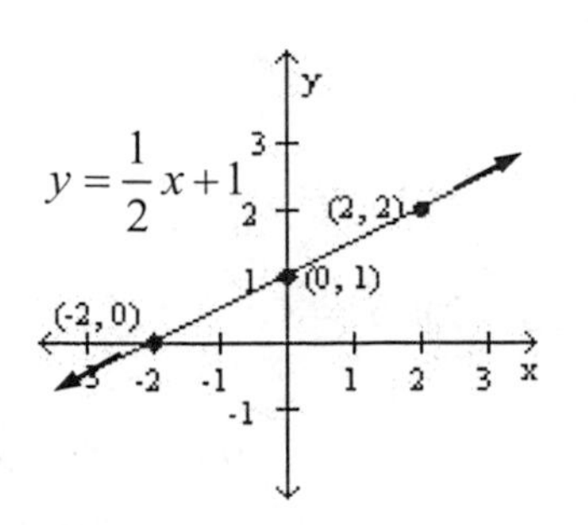

CHALLENGE PROBLEMS

Factor using rational numbers.

85. $x^6 - 4x^3 - 12 = (x^3+2)(x^3-6)$

87. $24 - x^3 + 8x^2 - 3x = 24 - 3x + 8x^2 - x^3$
$$= (24-3x) + (8x^2 - x^3)$$
$$= 3(8-x) + x^2(8-x)$$
$$= (8-x)(3+x^2)$$

89. $x^9 + y^6 = (x^3)^3 + (y^2)^3$
$$= (x^3+y^2)[(x^3)^2 - x^3 \cdot y^2 + (y^2)^2]$$
$$= (x^3+y^2)(x^6 - x^3y^2 + y^4)$$

91. $x^4 - 13x^2 + 36 = (x^2-9)(x^2-4)$
$$= (x+3)(x-3)(x+2)(x-2)$$

93. $x^2y^2 - 6xy - 16 = (xy+2)(xy-8)$

SECTION 6.7 STUDY SET
VOCABULARY

Fill in the blanks.

1. $2x^2 + 3x - 1 = 0$ and $x^2 - 36 = 0$ are examples of **quadratic** equations.

3. The **zero-factor** property states that if the product of two numbers is 0, at least one of them is 0: If $ab = 0$, then $a = \mathbf{0}$ or $b = \mathbf{0}$.

CONCEPTS

5. Which of the following are quadratic equations?
   a. $x^2 + 2x - 10 = 0$ **Yes** b. $2x - 10 = 0$ **No**
   c. $x^2 = 15x$ **Yes** d. $x^3 + x^2 + 2x = 0$ **No**

7. Set $5x + 4$ equal to 0 and solve for x. $-\frac{4}{5}$

9. What step (or steps) should be performed first before factoring is used to solve each equation?
   a. $x^2 + 7x = -6$ **Add 6 to both sides.**
   b. $x(x + 7) = 3$ **Distribute the multiplication by x and subtract 3 from both sides.**

NOTATION

Complete each solution to solve the equation.

11. $(x-1)(x+7) = 0$
    $x - 1 = \mathbf{0}$ or $\mathbf{x + 7} = 0$
    $x = 1$ | $x = \mathbf{-7}$

13. $p^2 - p - 6 = 0$
    $(\mathbf{p} - 3)(p + 2) = 0$
    $\mathbf{p - 3} = 0$ or $p + 2 = \mathbf{0}$
    $p = \mathbf{3}$ | $p = \mathbf{-2}$

GUIDED PRACTICE

Solve each equation. See Example 1.

15. $(x-3)(x-2)=0$
    $x-3=0$ or $x-2=0$
    $x=3$ | $x=2$
    The solutions are 3 and 2.

17. $(x+7)(x-7)=0$
    $x+7=0$ or $x-7=0$
    $x=-7$ | $x=7$
    The solutions are -7 and 7.

19. $6x(2x-5)=0$
    $6x=0$ or $2x-5=0$
    $\frac{6x}{\mathbf{6}}=\frac{0}{\mathbf{6}}$ | $2x=5$
    $x=0$ | $x=\frac{5}{2}$
    The solutions are 0 and $\frac{5}{2}$.

21. $-7a(3a+10)=0$
    $-7a=0$ or $3a+10=0$
    $a=0$ | $3a=-10$
    $a=-\frac{10}{3}$
    The solutions are 0 and $-\frac{10}{3}$.

23. $t(t-6)(t+8)=0$
    $t=0$ or $t-6=0$ or $t+8=0$
    $t=6$ | $t=-8$
    The solutions are 0, 6 and -8.

25. $(x-1)(x+2)(x-3)=0$
    $x-1=0$ or $x+2=0$ or $x-3=0$
    $x=1$ | $x=-2$ | $x=3$
    The solutions are 1, -2 and 3.

Solve each equation. See Example 2.

27. $x^2-13x+12=0$
    $(x-12)(x-1)=0$
    $x-12=0$ or $x-1=0$
    $x=12$ | $x=1$
    The solutions are 12 and 1.

29. $x^2-4x-21=0$
    $(x+3)(x-7)=0$
    $x+3=0$ or $x-7=0$
    $x=-3$ | $x=7$
    The solutions are -3 and 7.

31. $x^2-9x+8=0$
    $(x-8)(x-1)=0$
    $x-8=0$ or $x-1=0$
    $x=8$ | $x=1$
    The solutions are 8 and 1.

33. $a^2+8a+15=0$
    $(a+3)(a+5)=0$
    $a+3=0$ or $a+5=0$
    $a=-3$ | $a=-5$
    The solutions are -3 and -5.

Solve each equation. See Example 3.

35. $x^2-81=0$
    $(x+9)(x-9)=0$
    $x+9=0$ or $x-9=0$
    $x=-9$ | $x=9$
    The solutions are -9 and 9.

37. $t^2 - 25 = 0$

$(t+5)(t-5) = 0$

$t+5=0$ or $t-5=0$

$t=-5$ | $t=5$

The solutions are -5 and 5.

39. $4x^2 - 1 = 0$

$(2x+1)(2x-1) = 0$

$2x+1=0$ or $2x-1=0$

$2x=-1$ | $2x=1$

$x=-\frac{1}{2}$ | $x=\frac{1}{2}$

The solutions are $-\frac{1}{2}$ and $\frac{1}{2}$.

41. $9y^2 - 49 = 0$

$(3y+7)(3y-7) = 0$

$3y+7=0$ or $3y-7=0$

$3y=-7$ | $3y=7$

$y=-\frac{7}{3}$ | $y=\frac{7}{3}$

The solutions are $-\frac{7}{3}$ and $\frac{7}{3}$.

Solve each equation. See Example 4.

43. $w^2 = 7w$

$w^2 - \mathbf{7w} = 7w - \mathbf{7w}$

$w^2 - 7w = 0$

$w(w-7) = 0$

$w=0$ or $w-7=0$

| $w=7$

The solutions are 0 and 7.

45. $s^2 = 16s$

$s^2 - \mathbf{16s} = 16s - \mathbf{16s}$

$s^2 - 16s = 0$

$s(s-16) = 0$

$s=0$ or $s-16=0$

| $s=16$

The solutions are 0 and 16.

47. $4y^2 = 12y$

$4y^2 - \mathbf{12y} = 12y - \mathbf{12y}$

$4y^2 - 12y = 0$

$4y(y-3) = 0$

$4y=0$ or $y-3=0$

$y=0$ | $y=3$

The solutions are 0 and 3.

49. $3x^2 = -8x$

$3x^2 + \mathbf{8x} = -8x + \mathbf{8x}$

$3x^2 + 8x = 0$

$x(3x+8) = 0$

$x=0$ or $3x+8=0$

| $3x=-8$

| $x=-\frac{8}{3}$

The solutions are 0 and $-\frac{8}{3}$.

Solve each equation. See Example 5.

51. $3x^2 + 5x = 2$

$3x^2 + 5x - \mathbf{2} = 2 - \mathbf{2}$

$3x^2 + 5x - 2 = 0$

$(3x-1)(x+2) = 0$

$3x-1=0$ or $x+2=0$

$3x=1$ | $x=-2$

$x=\frac{1}{3}$ |

The solutions are $\frac{1}{3}$ and -2.

53. $2x^2 + x = 3$

$2x^2 + x - \mathbf{3} = 3 - \mathbf{3}$

$2x^2 + x - 3 = 0$

$(2x+3)(x-1) = 0$

$2x+3=0$ or $x-1=0$

$2x=-3$ | $x=1$

$x=-\frac{3}{2}$ |

The solutions are $-\frac{3}{2}$ and 1.

55. $5x^2 + 1 = 6x$

$5x^2 + 1 - \mathbf{6x} = 6x - \mathbf{6x}$

$5x^2 - 6x + 1 = 0$

$(5x-1)(x-1) = 0$

$5x-1=0$ or $x-1=0$

$5x=1$ | $x=1$

$x=\frac{1}{5}$ |

The solutions are $\frac{1}{5}$ and 1.

57.
$$2x^2-3x=20$$
$$2x^2-3x-\mathbf{20}=20-\mathbf{20}$$
$$2x^2-3x-20=0$$
$$(2x+5)(x-4)=0$$
$$2x+5=0 \quad \text{or} \quad x-4=0$$
$$2x=-5 \qquad x=4$$
$$x=-\frac{5}{2}$$
The solutions are $-\frac{5}{2}$ and 4.

Solve each equation. See Example 6.

59.
$$4r(r+7)=-49$$
$$4r^2+28r=-49$$
$$4r^2+28r+\mathbf{49}=-49+\mathbf{49}$$
$$4r^2+28r+49=0$$
$$(2r+7)(2r+7)=0$$
$$2r+7=0 \quad \text{or} \quad 2r+7=0$$
$$2r=-7 \qquad 2r=-7$$
$$r=-\frac{7}{2} \qquad r=-\frac{7}{2}$$
There is a repeated solution of $-\frac{7}{2}$.

61.
$$9a(a-3)=3a-25$$
$$9a^2-27a=3a-25$$
$$9a^2-27a-\mathbf{3a}=3a-25-\mathbf{3a}$$
$$9a^2-30a=-25$$
$$9a^2-30a+\mathbf{25}=-25+\mathbf{25}$$
$$9a^2-30a+25=0$$
$$(3a-5)(3a-5)=0$$
$$3a-5=0 \quad \text{or} \quad 3a-5=0$$
$$3a=5 \qquad 3a=5$$
$$a=\frac{5}{3} \qquad a=\frac{5}{3}$$
There is a repeated solution of $\frac{5}{3}$.

Solve each equation. See Example 7.

63.
$$x^3+3x^2+2x=0$$
$$x(x^2+3x+2)=0$$
$$x(x+1)(x+2)=0$$
$$x=0 \quad \text{or} \quad x+1=0 \quad \text{or} \quad x+2=0$$
$$x=-1 \qquad x=-2$$
The solutions are 0, -1 and -2.

65.
$$k^3-27k-6k^2=0$$
$$k^3-6k^2-27k=0$$
$$k(k^2-6k-27)=0$$
$$k(k-9)(k+3)=0$$
$$k=0 \quad \text{or} \quad k-9=0 \quad \text{or} \quad k+3=0$$
$$k=9 \qquad k=-3$$
The solutions are 0, 9, and -3.

TRY IT YOURSELF.

Solve each equation.

67.
$$4x^2=81$$
$$4x^2-\mathbf{81}=81-\mathbf{81}$$
$$4x^2-81=0$$
$$(2x+9)(2x-9)=0$$
$$2x+9=0 \quad \text{or} \quad 2x-9=0$$
$$2x=-9 \qquad 2x=9$$
$$x=-\frac{9}{2} \qquad x=\frac{9}{2}$$
The solutions are $-\frac{9}{2}$ and $\frac{9}{2}$.

69.
$$x^2-16x+64=0$$
$$(x-8)(x-8)=0$$
$$x-8=0 \quad \text{or} \quad x-8=0$$
$$x=8 \qquad x=8$$
There is a repeated solution of 8.

71.
$$(2s-5)(s+6)=0$$
$$2s-5=0 \quad \text{or} \quad s+6=0$$
$$2s=5 \qquad s=-6$$
$$s=\frac{5}{2}$$
The solutions are $\frac{5}{2}$ and -6.

73.
$$3b^2-30b=6b-60$$
$$3b^2-30b-\mathbf{6b}=6b-60-\mathbf{6b}$$
$$3b^2-30b-6b=-60$$
$$3b^2-30b-6b+\mathbf{60}=-60+\mathbf{60}$$
$$3b^2-30b-6b+60=0$$
$$3(b^2-10b-2b+20)=0$$
$$3[(b^2-10b)+(-2b+20)]=0$$
$$3[b(b-10)-2(b-10)]=0$$
$$3(b-10)(b-2)=0$$
$$b-10=0 \quad \text{or} \quad b-2=0$$
$$b=10 \qquad b=2$$
The solutions are 10 and 2.

75. $$\begin{aligned} k^3+k^2-20k&=0\\ k(k^2+k-20)&=0\\ k(x+5)(x-4)&=0 \end{aligned}$$

$k=0$ or $k+5=0$ or $k-4=0$

$k=-5$ | $k=4$

The solutions are 0, -5 and 4.

77. $$\begin{aligned} x^2-100&=0\\ (x+10)(x-10)&=0 \end{aligned}$$

$x+10=0$ or $x-10=0$

$x=-10$ | $x=10$

The solutions are -10 and 10.

79. $$\begin{aligned} z(z-7)&=-12\\ z^2-7z&=-12\\ z^2-7z+\mathbf{12}&=-12+\mathbf{12}\\ z^2-7z+12&=0\\ (z-3)(z-4)&=0 \end{aligned}$$

$z-3=0$ or $z-4=0$

$z=3$ | $z=4$

The solutions are 3 and 4.

81. $$\begin{aligned} 3y^2-14y-5&=0\\ (3y+1)(y-5)&=0 \end{aligned}$$

$3y+1=0$ or $y-5=0$

$3y=-1$ | $y=5$

$y=-\dfrac{1}{3}$

The solutions are $-\dfrac{1}{3}$ and 5.

83. $$\begin{aligned} (x-2)(x^2-8x+7)&=0\\ (x-2)(x-7)(x-1)&=0 \end{aligned}$$

$x-2=0$ or $x-7=0$ or $x-1=0$

$x=2$ | $x=7$ | $x=1$

The solutions are 2, 7 and 1.

85. $$\begin{aligned} (n+8)(n-3)&=-30\\ n^2+5n-24&=-30\\ n^2+5n-24+\mathbf{30}&=-30+\mathbf{30}\\ n^2+5n+6&=0\\ (n+3)(n+2)&=0 \end{aligned}$$

$n+3=0$ or $n+2=0$

$n=-3$ | $n=-2$

The solutions are -3 and -2.

87. $$\begin{aligned} x^3-6x^2&=-9x\\ x^3-6x^2+\mathbf{9x}&=-9x+\mathbf{9x}\\ x^3-6x^2+9x&=0\\ x(x^2-6x+9)&=0\\ x(x-3)(x-3)&=0 \end{aligned}$$

$x=0$ or $x-3=0$

| $x=3$

The solutions are 0 and a repeated answer of 3.

89. $$\begin{aligned} 4a^2+1&=8a+1\\ 4a^2+1-\mathbf{8a}&=8a+1-\mathbf{8a}\\ 4a^2-8a+1&=1\\ 4a^2-8a+1-\mathbf{1}&=1-\mathbf{1}\\ 4a^2-8a&=0\\ 4a(a-2)&=0 \end{aligned}$$

$4a=0$ or $a-2=0$

$a=0$ | $a=2$

The solutions are 0 and 2.

91. $$\begin{aligned} 2b(6b+13)&=-12\\ 12b^2+26b&=-12\\ 12b^2+26b+\mathbf{12}&=-12+\mathbf{12}\\ 12b^2+26b+12&=0\\ 2(6b^2+13b+6)&=0\\ 2(2b+3)(3b+2)&=0 \end{aligned}$$

$2b+3=0$ or $3b+2=0$

$2b=-3$ | $3b=-2$

$b=-\dfrac{3}{2}$ | $b=-\dfrac{2}{3}$

The solutions are $-\dfrac{3}{2}$ and $-\dfrac{2}{3}$.

93. $$\begin{aligned} 3a^3+4a^2+a&=0\\ a(3a^2+4a+1)&=0\\ a(3a+1)(a+1)&=0 \end{aligned}$$

$a=0$ or $3a+1=0$ or $a+1=0$

| $3a=-1$ | $a=-1$

$a=-\dfrac{1}{3}$

The solutions are 0, $-\dfrac{1}{3}$ and -1.

95.
$$2x^3 = 2x(x+2)$$
$$2x^3 = 2x^2 + 4x$$
$$2x^3 - \mathbf{2x^2} = 2x^2 + 4x - \mathbf{2x^2}$$
$$2x^3 - 2x^2 = 4x$$
$$2x^3 - 2x^2 - \mathbf{4x} = 4x - \mathbf{4x}$$
$$2x^3 - 2x^2 - 4x = 0$$
$$2x(x^2 - x - 2) = 0$$
$$x(x+1)(x-2) = 0$$

$x = 0$ or $x+1=0$ or $x-2=0$

$x=-1$ | $x=2$

The solutions are 0, -1 and 2.

97.
$$-15x^2 + 2 + 7x = 0$$
$$-15x^2 + 7x + 2 = 0$$
$$-(15x^2 - 7x - 2) = 0$$
$$-(5x+1)(3x-2) = 0$$

$5x+1=0$ or $3x-2=0$

$5x=-1$ | $3x=2$

$x=-\frac{1}{5}$ | $x=\frac{2}{3}$

The solutions are $-\frac{1}{5}$ and $\frac{2}{3}$.

99.
$$4p^2 - 121 = 0$$
$$(2p+11)(2p-11) = 0$$

$2p+11=0$ or $2p-11=0$

$2p=-11$ | $2p=11$

$p=-\frac{11}{2}$ | $p=\frac{11}{2}$

The solutions are $-\frac{11}{2}$ and $\frac{11}{2}$.

101.
$$d(8d-9) = -1$$
$$8d^2 - 9d = -1$$
$$8d^2 - 9d + \mathbf{1} = -1 + \mathbf{1}$$
$$8d^2 - 9d + 1 = 0$$
$$(8d-1)(d-1) = 0$$

$8d-1=0$ or $d-1=0$

$8d=1$ | $d=1$

$d=\frac{1}{8}$

The solutions are $\frac{1}{8}$ and 1.

LOOK ALIKES . . .

Factor the expression in part *a* and solve the equation in part *b*.

103. a. $x^2 + 4x - 21 = (x+7)(x-3)$

b. $x^2 + 4x - 21 = 0$

$(x+7)(x-3) = 0$

$x+7=0$ or $x-3=0$

$x=-7$ | $x=3$

The solutions are -7 and 3.

105. a. $12n^2 - 5n - 2 = (4n+1)(3n-2)$

b. $12n^2 - 5n - 2 = 0$

$(4n+1)(3n-2) = 0$

$4n+1=0$ or $3n-2=0$

$4n=-1$ | $3n=2$

$n=-\frac{1}{4}$ | $n=\frac{2}{3}$

The solutions are $-\frac{1}{4}$ and $\frac{2}{3}$.

WRITING

107-111. Answers will vary.

REVIEW

113. EXERCISE

15 min $\le t <$ 30 min

CHALLENCE PROBLEMS

Solve each equation.

115.
$$x^4 - 625 = 0$$
$$(x^2)^2 - 25^2 = 0$$
$$(x^2+25)(x^2-25) = 0$$
$$(x^2+25)(x+5)(x-5) = 0$$

$x^2+25=0$ or $x+5=0$ or $x-5=0$

$x^2=-25$ | $x=-5$ | $x=5$

not real

The solutions are -5 and 5.

117.
$$(x-3)^2 = 2x+9$$
$$x^2 - 6x + 9 = 2x + 9$$
$$x^2 - 6x + 9 - \mathbf{2x} = 2x + 9 - \mathbf{2x}$$
$$x^2 - 8x + 9 = 9$$
$$x^2 - 8x + 9 - \mathbf{9} = 9 - \mathbf{9}$$
$$x^2 - 8x = 0$$
$$x(x-8) = 0$$

$x=0$ or $x-8=0$

| $x=8$

The solutions are 0 and 8.

SECTION 6.8 STUDY SET
VOCABULARY

Fill in the blanks.

1. Integers that follow one another, such as 6 and 7, are called **consecutive** integers.
3. The longest side of a right triangle is the **hypotenuse**. The remaining two sides are the **legs** of the triangle.

CONCEPTS

5. **ii**
7. $\mathbf{20 = b(b + 5)}$
9. a. What kind of triangle is shown below?

   A right triangle

   b. What are the lengths of the legs of the triangle? **x ft; $(x + 1)$ ft**

   c. How much longer is the hypotenuse than the shorter leg? **9 ft**

NOTATION

Complete the solution to solve the equation.

11.
$$0 = -16t^2 + 32t + 48$$
$$0 = \mathbf{-16}(t^2 - 2t - 3)$$
$$0 = -16(t - 3)(t + \mathbf{1})$$
$$t - 3 = \mathbf{0} \quad \text{or} \quad t + 1 = \mathbf{0}$$
$$t = \mathbf{3} \quad \Big| \quad t = \mathbf{-1}$$

APPLICATIONS
GEOMETRY

13. FLAGS

Analyze

- Length is twice as long as it's width.
- Area is 18 sq. ft.
- Find the dimensions.

Assign

Let w = width of flag in ft

$2w$ = length of flag in ft

Form

The area of the rectangle	equals	the length	times	the width.
18	=	$2w$	$\cdot$	w

Solve

$$2w(w) = 18$$
$$2w^2 = 18$$
$$2w^2 - \mathbf{18} = 18 - \mathbf{18}$$
$$2w^2 - 18 = 0$$
$$2(w^2 - 9) = 0$$
$$2(w+3)(w-3) = 0$$

$$w + 3 = 0 \quad \text{or} \quad w - 3 = 0$$
$$\cancel{w = -3} \quad \Big| \quad w = 3$$

State

The width of the flag is 3 feet.
The length of the flag is $2(3) = 6$ feet.

Check

A rectangle with dimensions 3 feet by 6 feet has an area of 18 ft^2, and the length is twice the width. The results check.

14. BILLIARDS

Width: 5 ft; length: 10 ft

15. X-RAYS

Analyze

- Length is 2 inches more than the width.
- Area is 80 sq. in.
- Find the dimensions.

Assign

Let w = width of x-ray in inches

$w + 2$ = length of x-ray in inches

Form

The area of the rectangle	equals	the length	times	the width.
80	=	$w+2$	$\cdot$	w

Solve

$$w(w+2) = 80$$
$$w^2 + 2w = 80$$
$$w^2 + 2w - \mathbf{80} = 80 - \mathbf{80}$$
$$w^2 + 2w - 80 = 0$$
$$(w+10)(w-8) = 0$$

$$w + 10 = 0 \quad \text{or} \quad w - 8 = 0$$
$$\cancel{w = -10} \quad \Big| \quad w = 8$$

State

The width of the film is 8 inches.
The length of the film is $8 + 2 = 10$ inches.

Check

A rectangle with dimensions 8 in by 10 in has an area of 80 in^2, and the length is two inches more the width. The results check.

from CAMPUS TO CAREERS

17. ELEMENTARY TEACHERS

Analyze

- Length is 3 times the width.
- Area of wall is 90 sq. ft.
- Fire code only allows for 30% coverage.
- Find the dimensions.

Assign

Let w = width of board in feet

$3w$ = length of board in feet

First calculate 30% of the wall area.

$$90(30\%) = 90(0.30) = 27$$

Form

The area of the rectangle	equals	the length	times	the width.
27	$=$	$3w$	$\cdot$	w

Solve

$$3w(w) = 27$$
$$3w^2 = 27$$
$$3w^2 - \mathbf{27} = 27 - \mathbf{27}$$
$$3w^2 - 27 = 0$$
$$3(w^2 - 9) = 0$$
$$3(w+3)(w-3) = 0$$

$w + 3 = 0$ or $w - 3 = 0$

~~$w = -3$~~ | $w = 3$

State

The width of the board is 3 feet.

The length of the board is $3(3) = 9$ feet.

Check

A rectangle with dimensions 3 ft by 9 ft has an area of 27 ft^2, and the length is three times the width. The results check.

19. JEANS

Analyze

- Height of the triangle is 1 cm less than the base.
- Area is 15 sq. cm.
- Find the lengths of the base and height.

Assign

Let b = base of triangle in cm

$b - 1$ = height of triangle in cm

Form

The area of the triangle	equals	$\frac{1}{2}$	times	the base	times	the height.
15	$=$	$\frac{1}{2}$	$\cdot$	b	$\cdot$	$b-1$

Solve

$$\frac{1}{2}b(b-1) = 15$$
$$(\mathbf{2})\frac{1}{2}b(b-1) = (\mathbf{2})15$$
$$b(b-1) = 30$$
$$b^2 - b = 30$$
$$b^2 - b - \mathbf{30} = 30 - \mathbf{30}$$
$$b^2 - b - 30 = 0$$
$$(b+5)(b-6) = 0$$

$b + 5 = 0$ or $b - 6 = 0$

~~$b = -5$~~ | $b = 6$

State

The base is 6 cm.

The height is $6 - 1 = 5$ cm.

Check

A triangle with dimensions 5 cm by 6 cm has an area of 15 cm^2, and the height is one cm less than the base. The results check.

21. SAILBOATS

Analyze

- Length of the luff is 3 times longer than the length of the foot.
- Area is 24 sq. ft.
- Find the lengths of the luff and foot.

Assign

Let f = length of foot (base) in feet

$3f$ = length of luff (height) in feet

Form

The area of the triangle	equals one half of	the base	times	the height.
24	$=$	$\frac{1}{2} \cdot f$	$\cdot$	$3f$

Solve

$$\frac{1}{2}f(3f) = 24$$
$$\frac{3}{2}f^2 = 24$$
$$\mathbf{2} \cdot \frac{3}{2}f^2 = \mathbf{2} \cdot 24$$
$$3f^2 = 48$$
$$3f^2 - \mathbf{48} = 48 - \mathbf{48}$$
$$3f^2 - 48 = 0$$
$$3(f^2 - 16) = 0$$
$$3(f+4)(f-4) = 0$$

$f + 4 = 0$ or $f - 4 = 0$

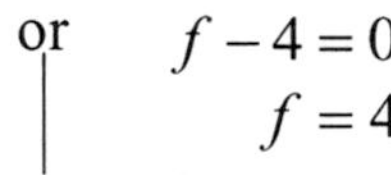
$f = -4$

$f = 4$

State

The foot is 4 ft.

The luff is $3(4) = 12$ ft.

Check

A triangle with dimensions 4 ft by 12 ft has an area of 24 ft^2, and the luff is three times the foot. The results check.

CONSECUTIVE INTEGERS

23. NASCAR

Analyze

- Product of two consecutive positive integers is 90.
- Kahne's car number is the smaller.
- Riggs' car number is the larger.
- What is the number of each car?

Assign

Let x = Kahne's car number

$x + 1$ = Riggs' car number

Form

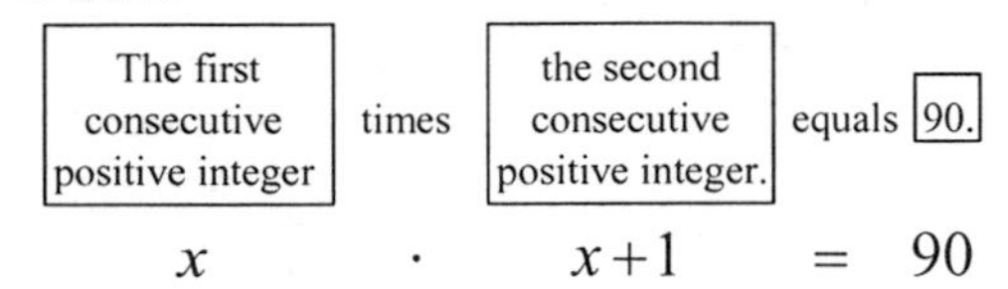

$x \cdot x+1 = 90$

Solve

$$x(x+1) = 90$$
$$x^2 + x = 90$$
$$x^2 + x - \mathbf{90} = 90 - \mathbf{90}$$
$$x^2 + x - 90 = 0$$
$$(x+10)(x-9) = 0$$

$x + 10 = 0$ or $x - 9 = 0$

$x = -10$ (crossed out)

$x = 9$

State

Kahne's car number is 9.

Riggs' car number is $9 + 1 = 10$.

Check

Since 9 and 10 are consecutive positive integers, and since $9 \cdot 10 = 90$. The results check.

25. CUSTOMER SERVICE

Analyze

- Product of two consecutive positive integers is 156.
- Ticket being served now is the smaller.
- Next ticket is the larger.
- What is the ticket number being served now?

Assign

Let $x =$ Ticket being served now
$x+1 =$ Next ticket being served

Form

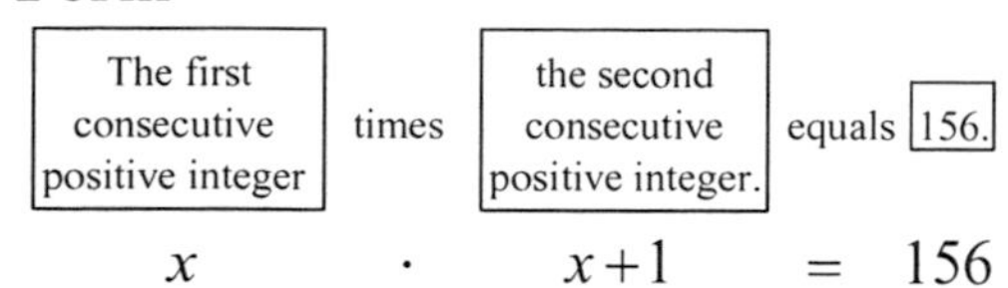

$$x \quad \cdot \quad x+1 \quad = \quad 156$$

Solve

$$x(x+1)=156$$
$$x^2+x=156$$
$$x^2+x-\mathbf{156}=156-\mathbf{156}$$
$$x^2+x-156=0$$
$$(x+13)(x-12)=0$$

$x+13=0$ or $x-12=0$

~~$x=-13$~~ | $x=12$

State

Ticket being served now is 12.

Check

Since 12 and 13 are consecutive positive integers, and since $12\cdot 13=156$. The results check.

27. PLOTTING POINTS

Analyze

- Product of two consecutive odd positive integers is 143.
- The x-coordinate is the smaller.
- The y-coordinate is the larger.
- What is the coordinate of the point?

Assign

Let $x = x$-coordinate
$x+2 = y$-coordinate

Form

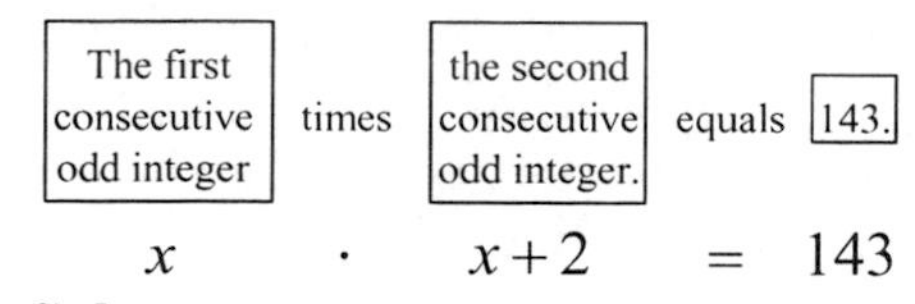

$$x \quad \cdot \quad x+2 \quad = \quad 143$$

Solve

$$x(x+2)=143$$
$$x^2+2x=143$$
$$x^2+2x-\mathbf{143}=143-\mathbf{143}$$
$$x^2+2x-143=0$$
$$(x+13)(x-11)=0$$

$x+13=0$ or $x-11=0$

~~$x=-13$~~ | $x=11$

State

The x-coordinate is 11.
The y-coordinate is 13.

Check

Since 11 and 13 are consecutive odd positive integers, and since $11\cdot 13=143$. The results check.

PYTHAGOREAN THEOREM PROBLEMS

29. HIGH-ROPES ADVENTURES COURSES

Analyze

- Anchor is 8 yds from the base of the pole. (longer leg)
- Tie off point is 6 yds up the pole. (shorter leg)
- How long is the cable? (hypotenuse)

Assign

Let $a =$ length of shorter leg in yd
$b =$ length of longer leg in yd

Form

(The length of the shorter leg)^2 plus (the length of the longer leg)^2 equals (the length of the hypotenuse.)^2

$$a^2 \quad + \quad b^2 \quad = \quad c^2$$

Solve

$$a^2+b^2=c^2$$
$$6^2+8^2=c^2$$
$$36+64=c^2$$
$$100=c^2$$
$$c^2=100$$
$$c^2-\mathbf{100}=100-\mathbf{100}$$
$$c^2-100=0$$
$$(c+10)(c-10)=0$$

$c+10=0$ or $c-10=0$

~~$c=-10$~~ | $c=10$

State

The length of cable is 10 yards long.

Check

Since $6^2+8^2=10^2$. The results check.

31. MOTO X

Analyze

- 15 ft is the longer leg.
- x ft is the shorter leg.
- 17 ft is the hypotenuse.
- What is the height of the ramp (x)?

Assign

Let x = height of the ramp in ft

Form

$$\begin{pmatrix}\text{The length of}\\ \text{the shorter leg}\end{pmatrix}^2 \text{ plus } \begin{pmatrix}\text{the length of}\\ \text{the longer leg}\end{pmatrix}^2 \text{ equals } \begin{pmatrix}\text{the length of}\\ \text{the hypotenuse.}\end{pmatrix}^2$$

$$x^2 \quad + \quad 15^2 \quad = \quad 17^2$$

Solve

$$x^2 + 15^2 = 17^2$$
$$x^2 + 225 = 289$$
$$x^2 + 225 - \mathbf{289} = 289 - \mathbf{289}$$
$$x^2 - 64 = 0$$
$$(x+8)(x-8) = 0$$

$x + 8 = 0$ or $x - 8 = 0$

~~$x = -8$~~ | $x = 8$

State

The height of the ramp is 8 ft.

Check

Since $8^2 + 15^2 = 17^2$. The results check.

33. BOATING

Analyze

- x m is the shorter leg (rise).
- Run is 7m longer than the rise.
- Ramp is 8 m longer than the rise.
- How long is each side?

Assign

Let x = length of shorter leg in meters

$x + 7$ = length of longer leg in meters

$x + 8$ = length of hypotenuse in meters

Form

$$\begin{pmatrix}\text{The length of}\\ \text{the shorter leg}\end{pmatrix}^2 \text{ plus } \begin{pmatrix}\text{the length of}\\ \text{the longer leg}\end{pmatrix}^2 \text{ equals } \begin{pmatrix}\text{the length of}\\ \text{the hypotenuse.}\end{pmatrix}^2$$

$$x^2 \quad + \quad (x+7)^2 \quad = \quad (x+8)^2$$

Solve

$$x^2 + (x+7)^2 = (x+8)^2$$
$$x^2 + x^2 + 14x + 49 = x^2 + 16x + 64$$
$$2x^2 + 14x + 49 = x^2 + 16x + 64$$
$$2x^2 + 14x + 49 - \mathbf{x^2} = x^2 + 16x + 64 - \mathbf{x^2}$$
$$x^2 + 14x + 49 = 16x + 64$$
$$x^2 + 14x + 49 - \mathbf{16x} = 16x + 64 - \mathbf{16x}$$
$$x^2 - 2x + 49 = 64$$
$$x^2 - 2x + 49 - \mathbf{64} = 64 - \mathbf{64}$$
$$x^2 - 2x - 15 = 0$$
$$(x+3)(x-5) = 0$$

$x + 3 = 0$ or $x - 5 = 0$

~~$x = -3$~~ | $x = 5$

State

The rise is 5 m.

The run is $5 + 7 = 12$ m.

The ramp is $5 + 8 = 13$ m.

Check

Since $5^2 + 12^2 = 13^2$. The results check.

QUADRATIC EQUATION MODEL PROBLEMS

35. THRILL RIDES

Analyze

- When the sunglasses hit the ground, its height will be 0 feet. To find the time that it takes for the glasses to hit the ground, we set h equal to 0, and solve the quadratic equation for t.

Assign

Let t = time in seconds

Form

$$h = -16t^2 + 64t + 80$$
$$0 = -16t^2 + 64t + 80$$

Solve

$$-16t^2 + 64t + 80 = 0$$
$$-16(t^2 - 4t - 5) = 0$$
$$-16(t+1)(t-5) = 0$$

~~$-16 = 0$~~ $t + 1 = 0$ or $t - 5 = 0$

~~$t = -1$~~ | $t = 5$

State

The equation has two solutions, -1 and 5. Since t represents time, and, in this case, time cannot be negative, we discard -1. The second solution, 5, indicates that the glasses hits the ground 5 seconds after falling off.

Check

Check this result by substituting 5 for t in $h = -16t^2 + 64t + 80$. You should get $h = 0$.

37. SOFTBALL

Analyze

- When the softball hits the ground, its height will be 0 feet. To find the time that it takes for the ball to hit the ground, we set h equal to 0, and solve the quadratic equation for t.

Assign

Let t = time in seconds

Form

$$h = -16t^2 + 63t + 4$$
$$0 = -16t^2 + 63t + 4$$

Solve

$$-16t^2 + 63t + 4 = 0$$
$$-1(16t^2 - 63t - 4) = 0$$
$$-(16t + 1)(t - 4) = 0$$

$-1 \neq 0$ (crossed out) $\quad 16t + 1 = 0 \quad$ or $\quad t - 4 = 0$

$t = -\frac{1}{16}$ (crossed out) $\quad\quad t = 4$

State

The equation has two solutions, $-\frac{1}{16}$ and 4. Since t represents time, and, in this case, time cannot be negative, we discard $-\frac{1}{16}$. The second solution, 4, indicates that the ball hits the ground 4 seconds after being thrown.

Check

Check this result by substituting 4 for t in $h = -16t^2 + 63t + 4$. You should get $h = 0$.

39. DOLPHINS

Analyze

- The height h in feet reached by a dolphin t seconds after breaking the surface of the water is given by $h = -16t^2 + 32t$.
- Height is 16 ft.

Assign

Let t = time in seconds

Form

$$h = -16t^2 + 32t$$
$$16 = -16t^2 + 32t$$

Solve

$$-16t^2 + 32t = 16$$
$$-16t^2 + 32t - \mathbf{16} = 16 - \mathbf{16}$$
$$-16t^2 + 32t - 16 = 0$$
$$-16(t^2 - 2t + 1) = 0$$
$$-16(t-1)(t-1) = 0$$

$-16 \neq 0$ (crossed out) $\quad$ or $\quad t - 1 = 0$

$t = 1$

State

The equation has two repeating solutions, 1. The solution, 1, indicates that the dolphin will touch the trainer's hand 1 second after breaking the surface of the water.

Check

Check this result by substituting 1 for t in $-16t^2 + 32t = 16$. You should get $16 = 16$.

41. CHOREOGRAPHY

Analyze

- 36 dancers are to assemble in a triangular-shaped series of rows, where each row has one more dancer than the previous row. The relationship between the number of rows r and the number of dancers d is given by $d = \frac{1}{2}r(r+1)$.
- What is the number of rows in the formation?

Assign

Let r = the number of rows

Form

$$d = \frac{1}{2}r(r+1)$$
$$36 = \frac{1}{2}r(r+1)$$

Solve

$$\frac{1}{2}r(r+1) = 36$$
$$(\mathbf{2})\frac{1}{2}r(r+1) = (\mathbf{2})(36)$$
$$r(r+1) = 72$$
$$r^2 + r = 72$$
$$r^2 + r - \mathbf{72} = 72 - \mathbf{72}$$
$$r^2 + r - 72 = 0$$
$$(r+9)(r-8) = 0$$

$r + 9 = 0 \quad$ or $\quad r - 8 = 0$

$r = -9$ (crossed out) $\quad\quad r = 8$

State

The equation has two solutions, -9 and 8. Since r represents number of rows, and, in this case, number of rows cannot be negative, we discard -9. The second solution, 8, indicates that the number of rows is 8.

Check

Check this result by substituting 8 for r in $\frac{1}{2}r(r+1) = 36$. You should get $36 = 36$.

WRITING

43-45. Answers will vary.

REVIEW

Find each special product.

47. $(5b-2)^2 = (5b)^2 + 2\cdot 5b\cdot(-2) + (-2)^2$
$= 25b^2 - 20b + 4$

49. $(s^2+4)^2 = (s^2)^2 + 2\cdot s^2\cdot(4) + 4^2$
$= s^4 + 8s^2 + 16$

51. $(9x+6)(9x-6) = (9x)^2 - (6)^2$
$= 81x^2 - 36$

CHALLENGE PROBLEMS

53. POOL BORDERS

Analyze

- $25+2w$ is the new length in meters.
- $10+2w$ is the new width in meters.
- w is the border in meters.
- Pool is 10 m by 25 m.
- Area of border pool minus area of pool is 74

Assign

Let $w =$ width of border in m

Form

The area of the outside of the pool	minus	the area of the pool	equals	the area of border.
$(2w+25)(2w+10)$	$-$	$(10)(25)$	$=$	74

Solve

$$(2w+25)(2w+10)-(10)(25)=74$$
$$4w^2+70w+250-250=74$$
$$4w^2+70w=74$$
$$4w^2+70w-\mathbf{74}=74-\mathbf{74}$$
$$4w^2+70w-74=0$$
$$2(2w^2+35w-37)=0$$
$$2(2w+37)(w-1)=0$$

$2w+37=0$ or $w-1=0$

~~$w=-\frac{37}{2}$~~ | $w=1$

State

The equation has two solutions, $-\frac{37}{2}$ and 1. Since w represents the width of the border, and, in this case, the border cannot be negative, we discard $-\frac{37}{2}$. The second solution, 1, indicates that the border is 1 m.

Check

Check this result by substituting 1 for w in $(2w+25)(2w+10)-(10)(25)=324-250=74$. You should get $74=74$.

CHAPTER 6

SECTION 6.1 REVIEW EXERCISES

The Greatest Common Factor; Factoring by Grouping

Find the prime-factorization of each number.

1. $35 = 5 \cdot 7$

2. $96 = 2 \cdot 48$
$= 2 \cdot 2 \cdot 24$
$= 2 \cdot 2 \cdot 2 \cdot 12$
$= 2 \cdot 2 \cdot 2 \cdot 2 \cdot 6$
$= 2 \cdot 2 \cdot 2 \cdot 2 \cdot 2 \cdot 3$
$= 2^5 \cdot 3$

Find the GCF of each list.

3. 28 and 35

$$\left.\begin{array}{l} 28 = 2 \cdot 2 \cdot \mathbf{7} \\ 35 = 5 \cdot \mathbf{7} \end{array}\right\} \text{GCF} = 7$$

4. $36a^4$, $54a^3$, and $126a^6$

$$\left.\begin{array}{l} 36a^4 = \mathbf{2} \cdot 2 \cdot \mathbf{3} \cdot \mathbf{3} \cdot \boldsymbol{a} \cdot \boldsymbol{a} \cdot \boldsymbol{a} \cdot a \\ 54a^3 = \mathbf{2} \cdot \mathbf{3} \cdot \mathbf{3} \cdot 3 \cdot \boldsymbol{a} \cdot \boldsymbol{a} \cdot \boldsymbol{a} \\ 126a^6 = \mathbf{2} \cdot \mathbf{3} \cdot \mathbf{3} \cdot 7 \cdot \boldsymbol{a} \cdot \boldsymbol{a} \cdot \boldsymbol{a} \cdot a \cdot a \cdot a \end{array}\right\}$$

$\text{GCF} = \mathbf{2} \cdot \mathbf{3} \cdot \mathbf{3} \cdot \boldsymbol{a} \cdot \boldsymbol{a} \cdot \boldsymbol{a} = 18a^3$

Factor.

5. $3x + 9y = \mathbf{3} \cdot x + \mathbf{3} \cdot 3y$
$= \mathbf{3}(x + 3y)$

6. $5ax^2 + 15a = \mathbf{5} \cdot \boldsymbol{a} \cdot x \cdot x + 3 \cdot \mathbf{5} \cdot \boldsymbol{a}$
$= 5a(x^2 + 3)$

7. $7s^5 + 14s^3 = \mathbf{7} \cdot s^2 \cdot \boldsymbol{s^3} + \mathbf{7} \cdot 2\boldsymbol{s^3}$
$= \mathbf{7s^3}(s^2 + 2)$

8. $\pi ab - \pi ac = \boldsymbol{\pi} \cdot \boldsymbol{a} \cdot b - \boldsymbol{\pi} \cdot \boldsymbol{a} \cdot c$
$= \pi a(b - c)$

9. $24x^3 + 60x^2 - 48x$
$= \mathbf{12} \cdot 2x^2 \cdot \boldsymbol{x} + \mathbf{12} \cdot 5x \cdot \boldsymbol{x} - \mathbf{12} \cdot 4\boldsymbol{x}$
$= \mathbf{12x}(2x^2 + 5x - 4)$

10. $x^5y^3z^2 + xy^5z^3 - xy^3z^2$
$= \boldsymbol{x} \cdot x^4 \cdot \boldsymbol{y^3} \cdot \boldsymbol{z^2} + \boldsymbol{x} \cdot \boldsymbol{y^3} \cdot y^2 \cdot \boldsymbol{z^2} \cdot z - \boldsymbol{x} \cdot \boldsymbol{y^3} \cdot \boldsymbol{z^2}$
$= xy^3z^2(x^4 + y^2z - 1)$

11. $-5ab^2 + 10a^2b - 15ab$
$= (\mathbf{-5} \cdot \boldsymbol{ab} \cdot b) + (\mathbf{-5} \cdot -2a \cdot \boldsymbol{ab}) + (\mathbf{-5} \cdot 3\boldsymbol{ab})$
$= \mathbf{-5ab}(b - 2a + 3)$

12. $4(x-2) - x(x-2) = 4(\mathbf{x-2}) - x(\mathbf{x-2})$
$= (x-2)(4-x)$

Factor out -1.

13. $-a - 7 = (\mathbf{-1})a + (\mathbf{-1})(7)$
$= (\mathbf{-1})(a + 7)$
$= -(a + 7)$

14. $-4t^2 + 3t - 1 = (\mathbf{-1})4t^2 - (\mathbf{-1})3t + (\mathbf{-1})1$
$= -(4t^2 - 3t + 1)$

Factor.

15. $2c + 2d + ac + ad = (\mathbf{2}c + \mathbf{2}d) + (\boldsymbol{a}c + \boldsymbol{a}d)$
$= \mathbf{2}(c + d) + \boldsymbol{a}(c + d)$
$= (c + d)(\mathbf{2} + \boldsymbol{a})$

16. $3xy + 18x - 5y - 30$
$= (3xy + 18x) + (-5y - 30)$
$= (\mathbf{3x}y + \mathbf{3x} \cdot 6) + [\mathbf{-5}y + (\mathbf{-5} \cdot 6)]$
$= 3x(y + 6) - 5(y + 6)$
$= (y + 6)(3x - 5)$

17. $2a^3 + 2a^2 - a - 1$
$= (\mathbf{2a^2} \cdot a + \mathbf{2a^2}) + [\mathbf{-1} \cdot a + (\mathbf{-1} \cdot 1)]$
$= 2a^2(a + 1) - 1(a + 1)$
$= (a + 1)(2a^2 - 1)$

18. $4m^2n + 12m^2 - 8mn - 24m$
$= 4m(mn + 3m - 2n - 6)$
$= 4m[(mn + 3m) + (-2n - 6)]$
$= 4m[m(n + 3) - 2(n + 3)]$
$= 4m(n + 3)(m - 2)$

SECTION 6.2 REVIEW EXERCISES

Factoring Trinomials of the Form $x^2 + bx + c$

19. What is the lead coefficient of $x^2 + 8x - 9$? **1**

20. Complete the table.

Factors of 6	Sum of the factors of 6
1(6)	**7**
2(3)	**5**
−1(−6)	**−7**
−2(−3)	**−5**

21. Two different sign factors of -24 whose sum is $+2$ are $+6$ and -4.
$$x^2+2x-24=(x+6)(x-4)$$

22. Two different sign factors of -40 whose sum is -18 are -20 and $+2$.
$$x^2-18x-40=(x-20)(x+2)$$

23. The negative factors of 45 whose sum is -14 are -5 and -9.
$$x^2-14x+45=(x-5)(x-9)$$

24. $t^2+10t+15$ is prime.

25. Factor out a -1.
$$-y^2+15y-56=-(y^2-15y+56)$$
The negative factors of 56 whose sum is -15 are -8 and -7.
$$-(y^2-15y+56)=-(y-8)(y-7)$$

26. $10y+9+y^2=y^2+10y+9$
The positive factors of 9 whose sum is $+10$ are $+9$ and $+1$.
$$y^2+10y+9=(y+9)(y+1)$$

27. Two different sign factors of -10 whose sum is $+3$ are $+5$ and -2.
$$c^2+3cd-10d^2=(c+5d)(c-2d)$$

28. $-3mn+m^2+2n^2=m^2-3mn+2n^2$
The negative factors of 2 whose sum is -3 are -2 and -1.
$$m^2-3mn+2n^2=(m-2n)(m-n)$$

29. Explain how we can check to determine whether $(x-4)(x+5)$ is the factorization of x^2+x-20.
Multiply

30. Explain why $x^2+7x+11$ is prime.
There are no two integers whose product is 11 and whose sum is 7.

Completely factor each trinomial.

31. Factor out GCF $5a^3$.
$$5a^5+45a^4-50a^3=5a^3(a^2+9a-10)$$
Two different sign factors of -10 whose sum is $+9$ are $+10$ and -1.
$$\begin{aligned}5a^5+45a^4-50a^3&=5a^3(a^2+9a-10)\\&=5a^3(a+10)(a-1)\end{aligned}$$

32. Rearrange and factor out GCF $-4x$.
$$\begin{aligned}-4x^2y-4x^3+24xy^2&=-4x^3-4x^2y+24xy^2\\&=-4x(x^2+xy-6y^2)\end{aligned}$$
Two different sign factors of -6 whose sum is $+1$ are $+3$ and -2.
$$-4x(x^2+xy-6y^2)=-4x(x+3y)(x-2y)$$

SECTION 6.3 REVIEW EXERCISES

Factoring Trinomials of the Form ax^2+bx+c

Factor each trinomial completely, if possible.

33. $\underset{a}{2}x^2\underset{b}{-5}x\underset{c}{-3}$

In $2x^2-5x-3$, we have $a=2$, $b=-5$, and $c=-3$. The key number is $ac=2(-3)=-6$. We must find a factorization of -6 in which the sum of the factors is $b=-5$. Since the factors must have a negative product, their signs must be different. The pairs of factors are -6 and $+1$.
$$\begin{aligned}2x^2-5x-3&=2x^2-6x+x-3\\&=(2x^2-6x)+(x-3)\\&=2x(x-3)+1(x-3)\\&=(x-3)(2x+1)\end{aligned}$$

34. $\underset{a}{35}y^2\underset{b}{+11}y\underset{c}{-10}$

In $35y^2+11y-10$, we have $a=35$, $b=11$, and $c=-10$. The key number is $ac=35(-10)=-350$. We must find a factorization of -350 in which the sum of the factors is $b=+11$. Since the factors must have a negative product, their signs must be different. The pairs of factors are -14 and $+25$.
$$\begin{aligned}35y^2+11y-10&=35y^2-14y+25y-10\\&=(35y^2-14y)+(25y-10)\\&=7y(5y-2)+5(5y-2)\\&=(5y-2)(7y+5)\end{aligned}$$

35. $-3x^2+13x+30=-(3x^2-13x-30)$
$$\underset{a}{3}x^2\underset{b}{-13}x\underset{c}{-30}$$
In $3x^2-13x-30$, we have $a=3$, $b=-13$ and $c=-30$. The key number is $ac=3(-30)=-90$. We must find a factorization of -90 in which the sum of the factors is $b=-13$. Since the factors must have a negative product, their signs must be different. The pairs of factors are -18 and $+5$.
$$\begin{aligned}-(3x^2-13x-30)&=-(3x^2-18x+5x-30)\\&=-[(3x^2-18x)+(5x-30)]\\&=-[3x(x-6)+5(x-6)]\\&=-(x-6)(3x+5)\end{aligned}$$

36. $-33p^2 - 6p + 18p^3 = 18p^3 - 33p^2 - 6p$
$= 3p(6p^2 - 11p - 2)$

$3p(\underset{a}{6}p^2 - \underset{b}{11}p - \underset{c}{2})$

In $6p^2 - 11p - 2$, we have $a = 6$, $b = -11$ and $c = -2$. The key number is $ac = 6(-2) = -12$. We must find a factorization of -12 in which the sum of the factors is $b = -11$. Since the factors must have a negative product, their signs must be different. The pairs of factors are -12 and $+1$.

$$\begin{aligned} 3p(6p^2 - 11p - 2) &= 3p(6p^2 - 12p + 1p - 2) \\ &= 3p[(6p^2 - 12p) + (1p - 2)] \\ &= 3p[6p(p-2) + 1(p-2)] \\ &= 3p(p-2)(6p+1) \end{aligned}$$

37. $\underset{a}{4}b^2 - \underset{b}{17}bc + \underset{c}{4}c^2$

In $4b^2 - 17bc + 4c^2$, we have $a = 4$, $b = -17$ and $c = 4$. The key number is $ac = 4(4) = 16$. We must find a factorization of 16 in which the sum of the factors is $b = -17$. Since the factors must have a positive product, their signs must be the same. The pairs of factors are -1 and -16.

$$\begin{aligned} 4b^2 - 17bc + 4c^2 &= 4b^2 - 16bc - bc + 4c^2 \\ &= (4b^2 - 16bc) + (-bc + 4c^2) \\ &= 4b(b - 4c) - c(b - 4c) \\ &= (b - 4c)(4b - c) \end{aligned}$$

38. $7y^2 + 7y - 18$ is prime.

39. ENTERTAINING

$\underset{a}{12}x^2 - \underset{b}{1}x - \underset{c}{1}$

In $12x^2 - 1x - 1$, we have $a = 12$, $b = -1$ and $c = -1$. The key number is $ac = 12(-1) = -12$. We must find a factorization of -12 in which the sum of the factors is $b = -1$. Since the factors must have a negative product, their signs must be different. The pairs of factors are -4 and $+3$.

$$\begin{aligned} 12x^2 - x - 1 &= 12x^2 - 4x + 3x - 1 \\ &= (12x^2 - 4x) + (3x - 1) \\ &= 4x(3x - 1) + 1(3x - 1) \\ &= (3x - 1)(4x + 1) \end{aligned}$$

The length is $(4x + 1)$ in.
The width is $(3x - 1)$ in.

40. In the following work, a student began to factor $5x^2 - 8x + 3$. Explain his mistake.

$(5x - \quad)(x + \quad)$

The signs of the second terms must be negative.

SECTION 6.4 REVIEW EXERCISES
Factoring Perfect - Square Trinomials and the Difference of Two Squares

Factor completely, if possible.

41. $x^2 + 10x + 25$

The first term x^2 is the square of $\boldsymbol{x}$.
The last term 25 is the square of **5**.
The middle term is twice the product of x and 5: $2(x)(5) = \mathbf{10}\boldsymbol{x}$.

$$\begin{aligned} x^2 + 10x + 25 &= (x+5)(x+5) \\ &= (x+5)^2 \end{aligned}$$

42. $9y^2 + 16 - 24y = 9y^2 - 24y + 16$

The first term $9y^2$ is the square of $\mathbf{3}\boldsymbol{y}$.
The last term 16 is the square of $\mathbf{-4}$.
The middle term is twice the product of $3y$ and -4: $2(3y)(-4) = \mathbf{-24}\boldsymbol{y}$.

$$\begin{aligned} 9y^2 - 24y + 16 &= (3y-4)(3y-4) \\ &= (3y-4)^2 \end{aligned}$$

43. $-z^2 + 2z - 1 = -(z^2 - 2z + 1)$

The first term z^2 is the square of $\boldsymbol{z}$.
The last term 1 is the square of $\mathbf{-1}$.
The middle term is twice the product of z and -1: $2(z)(-1) = \mathbf{-2}\boldsymbol{z}$.

$$\begin{aligned} -(z^2 - 2z + 1) &= -(z-1)(z-1) \\ &= -(z-1)^2 \end{aligned}$$

44. $25a^2 + 20ab + 4b^2$

The first term $25a^2$ is the square of $\mathbf{5}\boldsymbol{a}$.
The last term $4b^2$ is the square of $\mathbf{2}\boldsymbol{b}$.
The middle term is twice the product of $5a$ and $2b$: $2(5a)(2b) = \mathbf{20}\boldsymbol{ab}$.

$$\begin{aligned} 25a^2 + 20ab + 4b^2 &= (5a + 2b)(5a + 2b) \\ &= (5a + 2b)^2 \end{aligned}$$

45. $x^2 - 9$

$$\begin{array}{ccccccc} F^2 & - & L^2 & = & (F + L) & & (F - L) \\ \downarrow & & \downarrow & & \downarrow \ \downarrow & & \downarrow \ \downarrow \\ x^2 & - & 3^2 & = & (x + 3) & & (x - 3) \end{array}$$

46. $49t^2 - 121y^2$

$$\begin{array}{ccccccc} F^2 & - & L^2 & = & (F + L) & & (F - L) \\ \downarrow & & \downarrow & & \downarrow \ \downarrow & & \downarrow \ \downarrow \\ (7t)^2 & - & (11y)^2 & = & (7t + 11y) & & (7t - 11y) \end{array}$$

47. $x^2y^2 - 400$

$$\begin{array}{ccccccc} F^2 & - & L^2 & = (F & + L) & (F & - L) \\ \downarrow & & \downarrow & \downarrow & \downarrow & \downarrow & \downarrow \end{array}$$

$$(xy)^2 - 20^2 = (xy+20)(xy-20)$$

48. $8at^2 - 32a = 8a(t^2 - 4)$

$$\begin{array}{l} F^2 - L^2 = \quad (F+L)(F-L) \\ \downarrow \quad \downarrow \qquad\quad \downarrow \ \downarrow \ \downarrow \ \downarrow \end{array}$$

$$8a(t^2 - 2^2) = 8a(t+2)(t-2)$$

49. $c^4 - 256$

$$\begin{array}{l} F^2 - L^2 = (F + L)(F - L) \\ \downarrow \quad \downarrow \quad \downarrow \ \downarrow \ \downarrow \ \downarrow \end{array}$$

$$(c^2)^2 - 16^2 = (c^2+16)(c^2-16)$$

$c^2 - 16$

$$\begin{array}{l} F^2 - L^2 = (F+L)(F-L) \\ \downarrow \quad \downarrow \quad \downarrow \ \downarrow \ \downarrow \ \downarrow \end{array}$$

$$c^2 - 4^2 = (c+4)(c-4)$$

$$c^4 - 256 = (c^2+16)(c+4)(c-4)$$

50. $h^2 + 36$ is prime.

SECTION 6.5 REVIEW EXERCISES
Factoring the Sum and Difference of Two Cubes

Factor each polynomial completely.

51. $b^3 + 1 = b^3 + 1^3$

$$= (b+1)(b^2 - b\cdot 1 + 1^2)$$
$$= (b+1)(b^2 - b + 1)$$

52. $x^3 - 216 = x^3 - 6^3$

$$= (x-6)(x^2 + x\cdot 6 + 6^2)$$
$$= (x-6)(x^2 + 6x + 36)$$

53. $p^3 + 125q^3 = p^3 + (5q)^3$

$$= (p+5q)[p^2 - p\cdot 5q + (5q)^2]$$
$$= (p+5q)(p^2 - 5pq + 25q^2)$$

54. $16x^5 - 54x^2y^3$

$$= 2x^2(8x^3 - 27y^3)$$
$$= 2x^2[(2x)^3 - (3y)^3]$$
$$= 2x^2(2x-3y)[(2x)^2 + (2x)(3y) + (3y)^2]$$
$$= 2x^2(2x-3y)(4x^2 + 6xy + 9y^2)$$

SECTION 6.6 REVIEW EXERCISES
A Factoring Strategy

Factor each polynomial completely, if possible.

55. $14y^3 + 6y^4 - 40y^2 = 6y^4 + 14y^3 - 40y^2$

$$= 2y^2(3y^2 + 7y - 20)$$

$$\underset{a}{3}y^2 + \underset{b}{7}y \underset{c}{-20}$$

In $3y^2 + 7y - 20$, we have $a = 3$, $b = 7$, and $c = -20$. The key number is $ac = 3(-20) = -60$. We must find a factorization of -60 in which the sum of the factors is $b = 7$. Since the factors must have a negative product, their signs must be different. The pairs of factors are -5 and $+12$.

$$2y^2(3y^2 + 7y - 20) = 2y^2(3y^2 - 5y + 12y - 20)$$
$$= 2y^2[y(3y-5) + 4(3y-5)]$$
$$= 2y^2(3y-5)(y+4)$$

56. $5s^2t + 5s^2u^2 + 5tv + 5u^2v$

$$= 5(s^2t + s^2u^2 + tv + u^2v)$$
$$= 5[(s^2t + s^2u^2) + (tv + u^2v)]$$
$$= 5[s^2(t+u^2) + v(t+u^2)]$$
$$= 5(t+u^2)(s^2+v)$$

57. $j^4 - 16$

$$\begin{array}{l} F^2 - L^2 = (F+L)(F-L) \\ \downarrow \quad \downarrow \quad \downarrow \ \downarrow \ \downarrow \ \downarrow \end{array}$$

$$(j^2)^2 - 4^2 = (j^2+4)(j^2-4)$$

$$\begin{array}{l} F^2 - L^2 = (F+L)(F-L) \\ \downarrow \quad \downarrow \quad \downarrow \ \downarrow \ \downarrow \ \downarrow \end{array}$$

$$j^2 - 2^2 = (j+2)(j-2)$$

$$j^4 - 16 = (j^2+4)(j+2)(j-2)$$

58. $-3j^3 - 24 = -3(j^3 + 8)$

$$= -3[(j^3 + 2^3]$$
$$= -3[(j+2)(j^2 - j\cdot 2 + 2^2)]$$
$$= -3(j+2)(j^2 - 2j + 4)$$

59. $400x + 400 - m^2x - m^2$

$$= (400x + 400) + (-m^2x - m^2)$$
$$= 400(x+1) - m^2(x+1)$$
$$= (x+1)(400 - m^2)$$
$$= (x+1)(20+m)(20-m)$$

60. $12w^4 - 36w^3 + 27w^2 = 3w^2(4w^2 - 12w + 9)$

$$= 3w^2(2w-3)(2w-3)$$
$$= 3w^2(2w-3)^2$$

61. $2t^3 + 10 = \mathbf{2}t^3 + \mathbf{2}\cdot 5$

$$= 2(t^3 + 5)$$

62. $121p^2 + 36q^2$ is prime.

63. $x^2z + 64y^2z + 16xyz = x^2z + 16xyz + 64y^2z$
$= z(x^2 + 16xy + 64y^2)$

$x^2 + 16xy + 64y^2$

The first term x^2 is the square of $\boldsymbol{x}$.
The last term $64y^2$ is the square of $\mathbf{8y}$.
The middle term is twice the product of x and $8y$: $2(x)(8y) = \mathbf{16xy}$.

$z(x^2 + 16xy + 64y^2) = z(x + 8y)(x + 8y)$
$= z(x + 8y)^2$

64. $18c^3d^2 - 12c^3d - 24c^2d$
$= \mathbf{6} \cdot 3\mathbf{c^2}cd\mathbf{d} - \mathbf{6} \cdot 2\mathbf{c^2}c\mathbf{d} - \mathbf{6} \cdot 4\mathbf{c^2d}$
$= 6c^2d(3cd - 2c - 4)$

SECTION 6.7 REVIEW EXERCISES

Solving Quadratic Equations by Factoring

Solve each equation by factoring.

65. $8x(x - 6) = 0$

$8x = 0$ or $x - 6 = 0$
$x = 0$ | $x = 6$

The solutions are 0 and 6.

66. $(4x - 7)(x + 1) = 0$

$4x - 7 = 0$ or $x + 1 = 0$
$4x = 7$ | $x = -1$
$x = \frac{7}{4}$

The solutions are $\frac{7}{4}$ and -1.

67. $x^2 + 2x = 0$

$x(x + 2) = 0$
$x = 0$ or $x + 2 = 0$
$x + 2 - 2 = 0 - 2$
$x = -2$

The solutions are 0 and -2.

68. $x^2 - 9 = 0$

$(x + 3)(x - 3) = 0$
$x + 3 = 0$ or $x - 3 = 0$
$x = -3$ | $x = 3$

The solutions are -3 and 3.

69. $144x^2 - 25 = 0$

$(12x + 5)(12x - 5) = 0$

$12x + 5 = 0$ or $12x - 5 = 0$
$12x + 5 - \mathbf{5} = 0 - \mathbf{5}$ | $12x - 5 + \mathbf{5} = 0 + \mathbf{5}$
$12x = -5$ | $12x = 5$
$\frac{12x}{\mathbf{12}} = \frac{-5}{\mathbf{12}}$ | $\frac{12x}{\mathbf{12}} = \frac{5}{\mathbf{12}}$
$x = -\frac{5}{12}$ | $x = \frac{5}{12}$

The solutions are $-\frac{5}{12}$ and $\frac{5}{12}$.

70. $a^2 - 7a + 12 = 0$

$(a - 3)(a - 4) = 0$
$a - 3 = 0$ or $a - 4 = 0$
$a = 3$ | $a = 4$

The solutions are 3 and 4.

71. $2t^2 + 28t + 98 = 0$

$2(t^2 + 14t + 49) = 0$
$2(t + 7)(t + 7) = 0$

$t + 7 = 0$
$t = -7$

The solution is a repeated answer of -7.

72. $2x - x^2 = -24$

$2x - x^2 + \mathbf{x^2} = -24 + \mathbf{x^2}$
$2x = x^2 - 24$
$2x - \mathbf{2x} = x^2 - 24 - \mathbf{2x}$
$0 = x^2 - 2x - 24$
$0 = (x + 4)(x - 6)$
$x + 4 = 0$ or $x - 6 = 0$
$x = -4$ | $x = 6$

The solutions are -4 and 6.

73. $5a^2 - 6a + 1 = 0$

$(5a - 1)(a - 1) = 0$
$5a - 1 = 0$ or $a - 1 = 0$
$5a - 1 + \mathbf{1} = 0 + \mathbf{1}$ | $a - 1 + \mathbf{1} = 0 + \mathbf{1}$
$5a = 1$ | $a = 1$
$\frac{5a}{5} = \frac{1}{5}$
$a = \frac{1}{5}$

The solutions are $\frac{1}{5}$ and 1.

74.
$$2p^3 = 2p(p+2)$$
$$2p^3 = 2p^2 + 4p$$
$$2p^3 - \mathbf{2p^2} = 2p^2 + 4p - \mathbf{2p^2}$$
$$2p^3 - 2p^2 = 4p$$
$$2p^3 - 2p^2 - \mathbf{4p} = 4p - \mathbf{4p}$$
$$2p^3 - 2p^2 - 4p = 0$$
$$2p(p^2 - p - 2) = 0$$
$$2p(p+1)(p-2) = 0$$
$$p = 0 \quad \text{or} \quad p+1=0 \quad \text{or} \quad p-2=0$$
$$p = -1 \qquad p = 2$$

The solutions are 0, -1 and 2.

SECTION 6.8 REVIEW EXERCISES

Applications of Quadratic Equations

75. SANDPAPER

Analyze

- Length is 2 inches longer than it is wide.
- Area is 99 sq. in.
- Find the width.

Assign

Let w = width of sandpaper in inches
$w+2$ = length of sandpaper in inches

Form

The area of the rectangle	equals	the length	times	the width.
99	=	$w+2$	$\cdot$	w

Solve

$$w(w+2) = 99$$
$$w^2 + 2w = 99$$
$$w^2 + 2w - \mathbf{99} = 99 - \mathbf{99}$$
$$w^2 + 2w - 99 = 0$$
$$(w+11)(h-9) = 0$$
$$w+11=0 \quad \text{or} \quad w-9=0$$
$$w+11-\mathbf{11} = 0 - \mathbf{11} \qquad w = 9$$
$$\cancel{w=-11}$$

State

The width is 9 inches.
The length is $(9+2)=11$ inches.

Check

A recangle with dimensions 11 in by 9 in has an area of 99 in^2. The results check.

76. CONSTRUCTION

Analyze

- Base is 3 m longer than twice its height.
- Area is 45 sq. m.
- Find the length of the base.

Assign

Let h = height of triangle in m
$2h+3$ = base of triangle in m

Form

The area of the triangle	equals	$\frac{1}{2}$	times	the base	times	the height.
45	=	$\frac{1}{2}$	$\cdot$	$2h+3$	$\cdot$	h

Solve

$$\frac{1}{2}h(2h+3) = 45$$
$$(\mathbf{2})\frac{1}{2}h(2h+3) = (\mathbf{2})45$$
$$h(2h+3) = 90$$
$$2h^2 + 3h = 90$$
$$2h^2 + 3h - \mathbf{90} = 90 - \mathbf{90}$$
$$2h^2 + 3h - 90 = 0$$
$$(2h+15)(h-6) = 0$$
$$2h+15=0 \quad \text{or} \quad h-6=0$$
$$2h+15-\mathbf{15} = 0-\mathbf{15} \qquad h=6$$
$$2h = -15$$
$$\frac{2h}{2} = \frac{-15}{2}$$
$$\cancel{h = -\frac{15}{2}}$$

State

The height is 6 m.
The base is $2(6)+3=15$ m.

Check the Results

A triangle with dimensions 15 m by 6 m has an area of 45 cm^2, and the base is twice the height plus 3 m. The results check.

77. Fill in the blanks.

If we let $x =$ the first integer, then:

□ two consecutive integers are x and $\boxed{x+1}$

□ two consecutive even integers are x and $\boxed{x+2}$

□ two consecutive odd integers are x and $\boxed{x+2}$

78. MUSIC AWARDS

Analyze

· Product of two consecutive even integers is 120. West has been nominated less.

· The first even integer represents number of times West was nominated.

· The second even integer represents number of times Jackson was nominated.

Assign

Let $x =$ West's number of nominations

$x+2 =$ Jackson's number of nominations

Form

The first consecutive even integer	times	the second consecutive even integer.	equals	120.
x	$\cdot$	$x+2$	$=$	120

Solve

$$x(x+2)=120$$
$$x^2+2x=120$$
$$x^2+2x-\mathbf{120}=120-\mathbf{120}$$
$$x^2+2x-120=0$$
$$(x+12)(x-10)=0$$

$x+12=0$ or $x-10=0$

~~$x=-12$~~ | $x=10$

State

West has been nominated 10 times.

Jackson has been nominated $10+2=12$ times.

Check

Since 10 and 12 are consecutive even integers, and since $10\cdot 12=120$, the results check.

79. TIGHTROPE WALKERS

Analyze

· The longer leg is $(x+7)$ meters long.

· The shorter leg is x meters long.

· The hypotenuse is $(x+8)$ meters long.

· How high above the ground is the platform?

Assign

Let $x =$ length of shorter leg in m

$x+7 =$ length of longer leg in m

$x+8 =$ length of hypotenuse in m

Form

$\boxed{\text{The length of the shorter leg}}^2$	plus	$\boxed{\text{the length of the longer leg}}^2$	equals	$\boxed{\text{the length of the hypotenuse.}}^2$
x^2	$+$	$(x+7)^2$	$=$	$(x+8)^2$

Solve

$$x^2+(x+7)^2=(x+8)^2$$
$$x^2+x^2+14x+49=x^2+16x+64$$
$$2x^2+14x+49=x^2+16x+64$$
$$2x^2+14x+49-\mathbf{x^2}=x^2+16x+64-\mathbf{x^2}$$
$$x^2+14x+49=16x+64$$
$$x^2+14x+49-\mathbf{16x}=16x+64-\mathbf{16x}$$
$$x^2-2x+49=64$$
$$x^2-2x+49-\mathbf{64}=64-\mathbf{64}$$
$$x^2-2x-15=0$$
$$(x+3)(x-5)=0$$

$x+3=0$ or $x-5=0$

~~$x=-3$~~ | $x=5$

State

The platform is 5 meters above the ground.

Check

Since $5^2+12^2=13^2$, the results check.

80. BALLOONING

$h=-16t^2+1,600,$ the camera hitting the ground means h is zero.

$$h=-16t^2+1,600$$
$$0=-16t^2+1,600$$
$$0=-16(t^2-100)$$
$$0=-16(t+10)(t-10)$$

$t+10=0$ or $t-10=0$

~~$t=-10$~~ | $t=10$

The camera will hit the ground after 10 sec.

CHAPTER 6 TEST

1. Fill in the blanks.
   a. The letters GCF stand for **greatest common factor**.
   b. To factor a polynomial means to express it as a **product** of two (or more) polynomials.
   c. The **Pythagorean** theorem provides a formula relating the lengths of the three sides of a right triangle.
   d. $y^2 - 25$ is a **difference** of two squares.
   e. The trinomial $x^2 + x - 6$ factors as the product of two **binomials**: $(x + 3)(x - 2)$.
2. a. Find the prime factorizations of 45 and 30.
   $\mathbf{45 = 3^2 \cdot 5;\ 30 = 2 \cdot 3 \cdot 5}$
   b. Find the greatest common factor of $45x^4$ and $30x^3$. $\mathbf{15x^3}$

Factor. If an expression cannot be factored, write "prime."

3. $4x + 16 = 4(x + 4)$
4. $q^2 - 81 = (q + 9)(q - 9)$
5. $30a^2b^3 - 20a^3b^2 + 5ab = 5ab(6ab^2 - 4a^2b + 1)$
6. $x^2 + 9$ Prime
7. $2x(x + 1) + 3(x + 1) = (x + 1)(2x + 3)$
8. $x^2 + 4x + 3 = (x + 3)(x + 1)$
9. $-x^2 + 9x + 22 = -(x - 11)(x + 2)$
10. $60x^2 - 32x^3 + x^4 = x^2(x - 30)(x - 2)$
11. $9a - 9b + ax - bx = (a - b)(9 + x)$
12. $2a^2 + 5a - 12 = (2a - 3)(a + 4)$
13. $18x^2 + 60xy + 50y^2 = 2(3x + 5y)^2$
14. $x^3 + 8 = (x + 2)(x^2 - 2x + 4)$
15. $60m^8 - 45m^6 = 15m^6(4m^2 - 3)$
16. $3a^3 - 81 = 3(a - 3)(a^2 + 3a + 9)$
17. $16x^4 - 81 = (4x^2 + 9)(2x + 3)(2x - 3)$
18. $a^3 + 5a^2 + a + 5 = (a + 5)(a^2 + 1)$
19. $a^2 - 24 - 4a + 6a^3 = (a + 6)(a^3 - 4)$
20. $3d - 4 + 10d^2 = (5d + 4)(2d - 1)$
21. $8m^2 - 800 = 8(m + 10)(m - 10)$
22. $36n^2 - 84n + 49 = (6n - 7)^2$
23. $8r^2 - 14r + 3 = (4r - 1)(2r - 3)$
24. $t^2 - 6t + 10$ Prime
25. CHECKERS
   $(25x^2 - 40x + 16) = (5x - 4)(5x - 4)$
   The length of a side of the checkerboard is $(5x - 4)$ in.
26. Factor $x^2 - 3x - 54$. Show a check of your answer.
   $$(x - 9)(x + 6) = x^2 + 6x - 9x - 54 = x^2 - 3x - 54$$

Solve each equation.

27. $(x+3)(x-2)=0$
   $x+3=0$ or $x-2=0$
   $x=-3$ | $x=2$
   The solutions are -3 and 2.
28. $x^2 - 25 = 0$
   $(x+5)(x-5)=0$
   $x+5=0$ or $x+5=0$
   $x=-5$ | $x=5$
   The solutions are -5 and 5.
29. $36x^2 - 6x = 0$
   $6x(6x-1)=0$
   $6x=0$ or $6x-1=0$
   $x=0$ | $6x-1+1=0+1$
   $6x=1$
   $\frac{6x}{6}=\frac{1}{6}$
   $x=\frac{1}{6}$
   The solutions are 0 and $\frac{1}{6}$.
30. $x^2 + 6x = -9$
   $x^2 + 6x \mathbf{+ 9} = -9 \mathbf{+ 9}$
   $x^2 + 6x + 9 = 0$
   $(x+3)(x+3)=0$
   $x+3=0$
   $x=-3$
   The solution repeats and is -3.

31.
$$6x^2+x-1=0$$
$$(3x-1)(2x+1)=0$$

	or	
$3x-1=0$		$2x+1=0$
$3x-1+1=0+1$		$2x+1-1=0-1$
$3x=1$		$2x=-1$
$\frac{3x}{3}=\frac{1}{3}$		$\frac{2x}{2}=-\frac{1}{2}$
$x=\frac{1}{3}$		$x=-\frac{1}{2}$

The solutions are $\frac{1}{3}$ and $-\frac{1}{2}$.

32.
$$a(a-7)=18$$
$$a^2-7a=18$$
$$a^2-7a-\mathbf{18}=18-\mathbf{18}$$
$$a^2-7a-18=0$$
$$(a-9)(a+2)=0$$

	or	
$a-9=0$		$a+2=0$
$a=9$		$a=-2$

The solutions are 9, –2.

33.
$$x^3+7x^2=-6x$$
$$x^3+7x^2+6x=-6x+6x$$
$$x^3+7x^2+6x=0$$
$$x(x^2+7x+6)=0$$
$$x(x+1)(x+6)=0$$

	or		or	
$x=0$		$x+1=0$		$x+6=0$
		$x+1-1=0-1$		$x+6-6=0-6$
		$x=-1$		$x=-6$

The solutions are 0, –1, –6.

34. DRIVING SAFETY

Analyze

- Length is 3 feet longer than its width.
- Area is 54 sq. ft.
- Find the dimensions.

Assign

Let $w=$ width of blind spot in ft

$w+3=$ length of blind spot in ft

Form

The area of the rectangle	equals	the length	times	the width.
54	=	$w+3$	$\cdot$	w

Solve

$$w(w+3)=54$$
$$w^2+3w=54$$
$$w^2+3w-\mathbf{54}=54-\mathbf{54}$$
$$w^2+3w-54=0$$
$$(w+9)(w-6)=0$$

	or	
$w+9=0$		$w-6=0$
~~$w=-9$~~		$w=6$

State

The width of the blind spot is 6 feet.
The length of theblind spot is $6+3=9$ feet.

Check

A rectangle with dimensions 6 feet by 9 feet has an area of 54 ft^2, and the length is 3 feet longer than the width. The results check.

35. ROCKETRY

Analyze

- When the rocket hits the ground, its height will be 0 feet. To find the time that it takes for the ball to hit the ground, we set h equal to 0, and solve the quadratic equation for t.

Assign

Let $t=$ time in seconds

Form

$$h=-16t^2+80t$$
$$0=-16t^2+80t$$

Solve

$$-16t^2+80t=0$$
$$-16t(t-5)=0$$

	or	
$-16t=0$		$t-5=0$
$t=0$		$t=5$

State

The equation has two solutions, 0 and 5. Since t represents time, and, in this case, time cannot be zero, we discard 0. The second solution, 5, indicates that the rockets hits the ground 5 seconds after being launched.

Check

Check this result by substituting 5 for t in $h=-16t^2+80t$. You should get $h=0$.

36. ATV'S

Analyze

- Area is 33 sq. in.
- Height is 1 inch less than twice the base
- Find the lengths of the base and height.

Assign

Let x = length of base in inches

$2x-1$ = length of height in inches

Form

The area of the triangle	equals one half of	the base	times	the height.
33	$=$	$\frac{1}{2}\cdot x$	$\cdot$	$2x-1$

Solve

$$\frac{1}{2}x(2x-1)=33$$
$$\mathbf{2}\cdot\frac{1}{2}x(2x-1)=\mathbf{2}\cdot 33$$
$$x(2x-1)=66$$
$$2x^2-x=66$$
$$2x^2-x-\mathbf{66}=66-\mathbf{66}$$
$$2x^2-x-66=0$$
$$(2x+11)(x-6)=0$$

$2x+11=0$ or $x-6=0$

$2x=-11$ | $x=6$

~~$x=\frac{-11}{2}$~~

State

The base is 6 inches.

The height is $2(6)-1=11$ inches.

Check

A triangle with dimensions 6 in by 11 in has an area of 33 in^2, and the height is 1 less than twice base. The results check.

37. CONSECUTIVE NUMBERS

Analyze

- Product of two consecutive positive integers is 156.
- What are the two integers?

Assign

Let $x = 1^{st}$ consecutive positive even integer

$x+1 = 2^{nd}$ consecutive positive even integer

Form

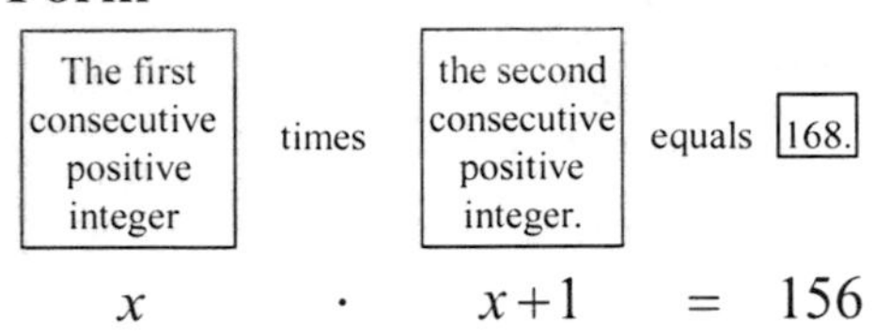

$x \quad \cdot \quad x+1 \quad = \quad 156$

Solve

$$x(x+1)=156$$
$$x^2+x=156$$
$$x^2+x-\mathbf{156}=156-\mathbf{156}$$
$$x^2+x-156=0$$
$$(x+13)(x-12)=0$$

$x+13=0$ or $x-12=0$

~~$x=-13$~~ | $x=12$

State

1st consecutive positive integer is 12.

2nd consecutive positive integer is 13.

Check

Since 12 and 13 are consecutive positive integers, and since $12\cdot 13=156$, the results check.

38. RIGHT TRIANGLE

Analyze

- The longer leg is $(x-2)$ units long.
- The shorter leg is $(x-4)$ units long.
- The hypotenuse is x units long.
- Find the length of the hypotenuse.

Assign

Let x = length of hypotenuse

$x-2$ = length of longer leg

$x-4$ = length of in shorter leg

Form

The length of the shorter leg^2	plus	the length of the longer leg^2	equals	the length of the hypotenuse.^{2}
$(x-4)^2$	$+$	$(x-2)^2$	$=$	x^2

Solve

$$(x-4)^2+(x-2)^2=x^2$$
$$x^2-8x+16+x^2-4x+4=x^2$$
$$2x^2-12x+20=x^2$$
$$2x^2-12x+20-\mathbf{x^2}=x^2-\mathbf{x^2}$$
$$x^2-12x+20=0$$
$$(x-2)(x-10)=0$$

$x-2=0$ or $x-10=0$

$x=2$ | $x=10$

State

The solution of 2 will not work for the longer leg would be 0 and the shorter leg would be -2.

The length of the hypotenuse is 10 units long.

Check

Since $4^2+8^2=10^2$, the results check.

39. What is a quadratic equation? Give an example.

A quadratic equation is an equation that can be written in the form $ax^2+bx+c=0$;
$x^2-2x+1=0$. (Answers may vary.)

40. If the product of two numbers is 0, what conclusion can be drawn about the numbers? **At least one of them is 0.**

CHAPTERS 1 - 6
CUMULATIVE REVIEW

1. HEART RATES [Section 1.1]
   185 – 150 = 35. The difference is which is 35 beats/min.

2. Find the prime factorization of 250. [Section 1.2]
$$\begin{aligned}250 &= 2\cdot 125\\ &= 2\cdot 5\cdot 25\\ &= 2\cdot 5\cdot 5\cdot 5\\ &= 2\cdot 5^3\end{aligned}$$

3. Find the quotient: [Section 1.2]
$$\begin{aligned}\frac{16}{5}\div\frac{10}{3} &= \frac{16}{5}\cdot\frac{3}{10}\\ &= \frac{16\cdot 3}{5\cdot 10}\\ &= \frac{\overset{1}{\cancel{2}}\cdot 8\cdot 3}{5\cdot \underset{1}{\cancel{2}}\cdot 5}\\ &= \frac{24}{25}\end{aligned}$$

4. Write as a decimal. [Section 1.3]
$$\frac{124}{125} = 0.992$$

5. Determine whether each statement is true or false. [Section 1.2]

   a. Every integer is a whole number. **False**
   b. Every integer is a rational number. **True**
   c. π is a real number. **True**

6. Which division is undefined, $\frac{0}{5}$ or $\frac{5}{0}$?

   [Section 1.6] $\mathbf{\frac{5}{0}}$

Evaluate each expression.

7. $3+2[-1-4(5)]$ [Section 1.7]
$$\begin{aligned}3+2[-1-4(5)] &= 3+2[-1-20]\\ &= 3+2[-1+(-20)]\\ &= 3+2(-21)\\ &= 3+(-42)\\ &= -39\end{aligned}$$

8. $\frac{|-25|-2(-5)}{9-2^4}$ [Section 1.7]
$$\begin{aligned}\frac{|-25|-2(-5)}{9-2^4} &= \frac{25+10}{9-16}\\ &= \frac{35}{9+(-16)}\\ &= \frac{35}{-7}\\ &= -5\end{aligned}$$

9. What is -3 cubed? [Section 1.7]
$$\begin{aligned}-3 \text{ cubed} &= (-3)(-3)(-3)\\ &= -27\end{aligned}$$

10. What is the value of x twenty-dollar bills? [Section 1.8] $\$20x$

11. Evaluate $\frac{-x-a}{y-b}$ for $x=-2$, $y=1$, $a=5$, and $b=2$. [Section 1.8]
$$\begin{aligned}\frac{-x-a}{y-b} &= \frac{-(-2)-5}{1-2}\\ &= \frac{2+(-5)}{1+(-2)}\\ &= \frac{-3}{-1}\\ &= 3\end{aligned}$$

12. Identify the coefficient of each term in expression $8x^2-x+9$. [Section 1.8]
    The coefficients are 8, -1, 9.

Simplify each expression. [Section 1.9]

13. $$\begin{aligned}-8y^2-5y^2+6 &= (-8-5)y^2+6\\ &= -13y^2+6\end{aligned}$$

14. $$\begin{aligned}3z+2(y-z)+y &= 3z+2y-2z+y\\ &= (2+1)y+(3-2)z\\ &= 3y+z\end{aligned}$$

Solve each equation. [15-18 Section 2.2]

15.
$$-(3a+1)+a=2$$
$$-3a-1+a=2$$
$$(-3+1)a-1=2$$
$$-2a-1=2$$
$$-2a-1+\mathbf{1}=2+\mathbf{1}$$
$$-2a=3$$
$$\frac{-2a}{\mathbf{-2}}=\frac{3}{\mathbf{-2}}$$
$$a=-\frac{3}{2}$$

16.
$$2-(4x+7)=3+2(x+2)$$
$$2-4x-7=3+2x+4$$
$$-4x+(2-7)=2x+(3+4)$$
$$-4x-5=2x+7$$
$$-4x-5-\mathbf{2x}=2x+7-\mathbf{2x}$$
$$-6x-5=7$$
$$-6x-5+\mathbf{5}=7+\mathbf{5}$$
$$-6x=12$$
$$\frac{-6x}{\mathbf{-6}}=\frac{12}{\mathbf{-6}}$$
$$x=-2$$

17.
$$\frac{3t-21}{2}=t-6$$
$$\mathbf{2}\cdot\frac{3t-21}{2}=\mathbf{2}(t-6)$$
$$3t-21=2t-12$$
$$3t-21-\mathbf{2t}=2t-12-\mathbf{2t}$$
$$t-21=-12$$
$$t-21+\mathbf{21}=-12+\mathbf{21}$$
$$t=9$$

18.
$$-\frac{1}{3}-\frac{x}{5}=\frac{3}{2}$$
$$\mathbf{30}\cdot\left(-\frac{1}{3}\right)+\mathbf{30}\cdot\left(-\frac{x}{5}\right)=\mathbf{30}\left(\frac{3}{2}\right)$$
$$-10-6x=45$$
$$-6x-10+\mathbf{10}=45+\mathbf{10}$$
$$-6x=55$$
$$\frac{-6x}{\mathbf{-6}}=\frac{55}{\mathbf{-6}}$$
$$x=-\frac{55}{6}$$

19. WATERMELON [Section 2.3]

92% of 270 is what number?

$$0.92 \cdot 270 = x$$
$$248.4=x$$
$$248\approx x$$

The approximate weight is 248 lbs.

20. Find the distance traveled by a truck traveling for 5½ hours at a rate of 60 miles per hour. [Section 2.4]

$$d=rt$$
$$=60\cdot 5\frac{1}{2}$$
$$=60\cdot 5.5$$
$$=330$$

The distance traveled is 330 miles.

21. What is the formula for simple interest? [Section 2.4] $I=Prt$

22. GEOMETRY TOOLS [Section 2.4]

$$A=\pi r^2,\quad \pi=3.141592654$$
$$\approx 3.141592654\cdot 2^2$$
$$\approx 3.141592654\cdot 4$$
$$\approx 12.56637061$$
$$\approx 12.6$$

The area is about 12.6 in^2.

23. Solve $A=P+Prt$ for t. [Section 2.4]

$$A=P+Prt$$
$$A-\boldsymbol{P}=P+Prt-\boldsymbol{P}$$
$$A-P=Prt$$
$$\frac{A-P}{\boldsymbol{Pr}}=\frac{Prt}{\boldsymbol{Pr}}$$
$$\frac{A-P}{Pr}=t$$
$$t=\frac{A-P}{Pr}$$

24. HISTORY [Section 2.5]

Analyze

- Consecutive integers represent the order of all the presidents.
- Cleveland's two terms add up to be 46.
- What are the two numbers of his 2 terms?

Assign

Let $x =$ number of his 1st term

$x+2 =$ number of his 2nd term

Form

The 1st term	plus	the 2nd term	equals	the total of 46.
x	$+$	$x+2$	$=$	46

Solve

$$x+x+2=46$$
$$2x+2=46$$
$$2x+2-\mathbf{2}=46-\mathbf{2}$$
$$2x=44$$
$$\frac{2x}{\mathbf{2}}=\frac{44}{\mathbf{2}}$$
$$x=22$$

2nd term

$$x+2=22+2$$
$$=24$$

State

He was the 22th and 24th president.

Check

$22+24=46$

The solutions check.

25. ANTIQUE SHOWS [Section 2.5]

Analyze

- 17 week tour to 3 cities.
- Will stay in LA 2 weeks longer than in LV.
- Will stay in Dallas 1 week less than twice in LV.
- How long did they spend in each city?

Assign

Let $x =$ number of weeks in Las Vegas

$x+2 =$ number of weeks in Los Angles

$2x-1 =$ number of weeks in Dallas

Form

The number of weeks in LV	plus	the number of weeks in LA	plus	the number of weeks in Dallas	equals	the total of 17.
x	$+$	$x+2$	$+$	$2x-1$	$=$	17

Solve

$$x+(x+2)+(2x-1)=17$$
$$4x+1=17$$
$$4x+1-\mathbf{1}=17-\mathbf{1}$$
$$4x=16$$
$$\frac{4x}{\mathbf{4}}=\frac{16}{\mathbf{4}}$$
$$x=4$$

LA

$$x+2=\mathbf{4}+2$$
$$=6$$

Dallas

$$2x-1=2(\mathbf{4})-1$$
$$=7$$

State

The show stays 4 weeks in Las Vegas, 6 weeks in Los Angles, and 7 weeks in Dallas.

Check

$4+6+7=17$

The solutions check.

26. PHOTOGRAPHIC CHEMICALS [Section 2.6]

Analyze

- Weak 6 liter acetic acid solution is 5%.
- Strong acetic solution is 10%.
- A mixture of the two is to be 7%.
- How many liters of 10% solution are needed?

Assign

Let x = the amount of 10% solution in liters

Form

	Amount	• Strength =	Amount of pure acid
Weak	6	0.05	**6(0.05)**
Strong	x	0.10	**0.10 x**
Mixture	$x+6$	0.07	**0.07($x+6$)**

The acetic acid in the 5% solution	plus	the acetic acid in the 10% solution	equals	the acetic acid in the 7% solution.
0.30	+	$0.10x$	=	$0.07(x+6)$

Solve

$$0.30+0.10x=0.07(x+6)$$
$$\mathbf{100}[0.30+0.10x]=\mathbf{100}[0.07(x+6)]$$
$$100(0.30)+100(0.10x)=100[0.07(x+6)]$$
$$30+10x=7(x+6)$$
$$30+10x=7x+42$$
$$30+10x-\mathbf{7x}=7x+42-\mathbf{7x}$$
$$30+3x=42$$
$$30+3x-\mathbf{30}=42-\mathbf{30}$$
$$3x=12$$
$$\frac{3x}{\mathbf{3}}=\frac{12}{\mathbf{3}}$$
$$x=4$$

State

4 liters of the 10% acetic acid solution will be needed.

Check

The acid in the 5% solution is 6(0.05), or 0.3 L.
The acid in the 10% solution is 4(0.10),or 0.4 L.
The acid in the 7% solution is 10(0.07),or 0.7 L.
Since the total was 0.3 L + 0.4 L = 0.7 L,
the solutions check.

27. DRIED FRUIT [Section 2.6]

Analyze

- Dried apple slices cost $4.60 per lb.
- Dried banana chips cost $3.40 per lb.
- 10 lb mixture sells for $4.00 per lb.
- How many pounds of each are needed?

Assign

Let x = the pounds of apples slices

$10-x$ = the pounds of banana chips

Form

Apple	x	4.60	**4.60x**
Banana	**$10-x$**	3.40	**3.40($10-x$)**
Mixture	10	4.00	4.00(10)

The value of the apple slices	plus	the value of the banana chips	equals	the value of the mixture.
$4.60x$	+	$3.40(10-x)$	=	$4.00(10)$

Solve

$$4.60x+3.40(10-x)=4.00(10)$$
$$\mathbf{100}[4.60x+3.40(10-x)]=\mathbf{100}[4.00(10)]$$
$$100(4.60x)+100[3.40(10-x)]=100[4.00(10)]$$
$$460x+340(10-x)=400(10)$$
$$460x+3{,}400-340x=4{,}000$$
$$120x+3{,}400=4{,}000$$
$$120x+3{,}400-\mathbf{3{,}400}=4{,}000-\mathbf{3{,}400}$$
$$120x=600$$
$$\frac{120x}{\mathbf{120}}=\frac{600}{\mathbf{120}}$$
$$x=5$$

State

5 lbs of apple slices will be needed.
5 lbs of banana chips will be needed.

Check

The value of the apple is 5(4.60), or $23.
The value of the banana is 5(3.40), or $17.
The value of the mixture is 10(4.00), or $40.
Since the total was $23 + $17 = $40,
the solutions check.

28. **Solve the inequality. Write the solution set in interval notation and graph it.**
[Section 2.8]

$$-\frac{x}{2}+4>5$$
$$-\frac{x}{2}+4-\mathbf{4}>5-\mathbf{4}$$
$$-\frac{x}{2}>1$$
$$\mathbf{-2}\cdot\left(-\frac{x}{2}\right)<\mathbf{-2}\cdot 1$$
$$x<-2$$
$$(-\infty,-2)$$

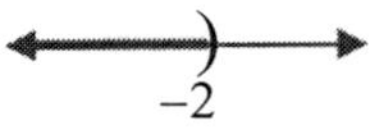

29. Is $(-2,5)$ a solution of $3x+2y=4$? [Section 3.2]

$$3x+2y=4$$
$$3(-2)+2(5)\overset{?}{=}4$$
$$-6+10\overset{?}{=}4$$
$$4=4 \text{ True}$$

$(-2, 5)$ is a solution.

30. Graph: $y=2x-3$ [Section 3.2]

x	y	(x, y)
-2	$y=2x-3$ $=2(-2)-3$ $=-4-3$ $=-7$	$(-2,-7)$
0	$y=2x-3$ $=2(0)-3$ $=0-3$ $=-3$	$(0,-3)$
2	$y=2x-3$ $=2(2)-3$ $=4-3$ $=1$	$(2, 1)$

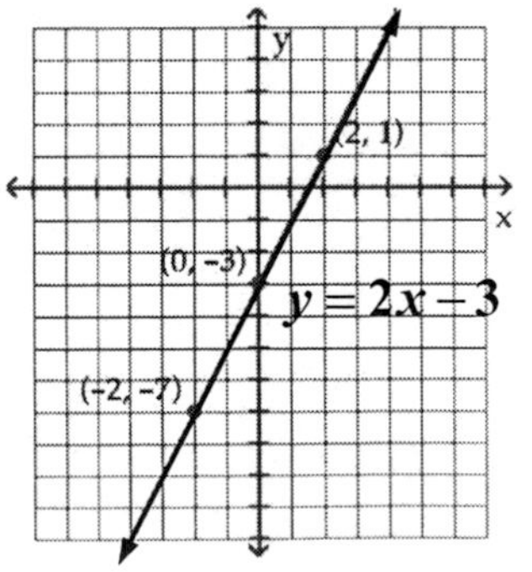

31. Is the graph of $x=3$ a vertical or horizontal line? [Section 3.3]
It is a vertical line.

32. If two lines are parallel, what can be said about their slopes? [Section 3.4]
They are the same.

33. ENCYCLOPEDIAS [Section 3.4]
Select any two points that lie on the line.
Let $(x_1, y_1)=(2005,\ 470{,}000)$
Let $(x_2, y_2)=(2010,\ 3{,}150{,}000)$

$$m=\frac{y_2-y_1}{x_2-x_1}$$
$$=\frac{3{,}150{,}000-470{,}000}{2010-2005}$$
$$=\frac{2{,}680{,}000}{5}$$
$$=536{,}000$$

The rate of change is an increase of 536,000 articles per year.

34. Find the slope and the y-intercept of the graph of $3x-3y=6$. [Section 3.5]

$$3x-3y=6$$
$$3x-3y-\mathbf{3x}=6-\mathbf{3x}$$
$$-3y=-3x+6$$
$$\frac{-3y}{\mathbf{-3}}=\frac{-3x}{\mathbf{-3}}+\frac{6}{\mathbf{-3}}$$
$$y=x-2$$

The slope is 1. The y-intercept is $(0,-2)$.

35. Find an equation of the line passing through $(-2, 5)$ and $(-3,-2)$. Write the equation in slope-intercept form. [Section 3.5]

$(-2, 5)$ and $(-3,\ -2)$
(x_1, y_1) and $(x_2,\ y_2)$

$$m=\frac{y_2-y_1}{x_2-x_1}$$
$$=\frac{-2-5}{-3-(-2)}$$
$$=\frac{-7}{-1}$$
$$=7$$

Passes through $(-2,5)$, $m=7$

$x_1=-2$ and $y_1=5$

$$y-y_1=m(x-x_1)$$
$$y-5=7[x-(-2)]$$
$$y-5=7(x+2)$$
$$y-5=7x+14$$
$$y-5+\mathbf{5}=7x+14+\mathbf{5}$$
$$y=7x+19 \quad \text{slope-intercept form}$$

36. Graph the line passing through $(-4, 1)$ that has slope -3 [Section 3.6]

$(-4,1)$, $m=-3$

Start at $(-4,1)$, go down 3, right 1

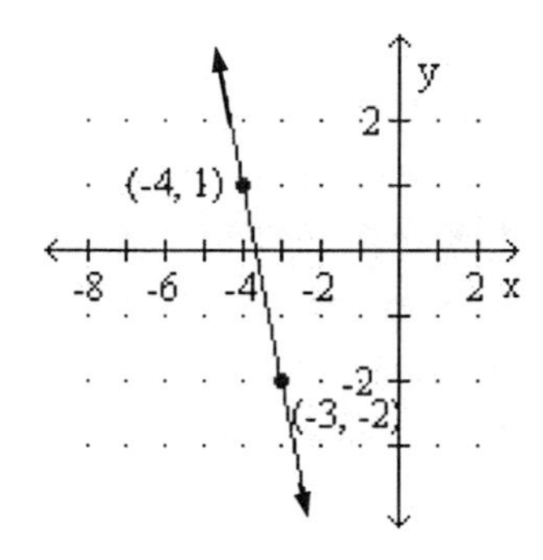

37. Graph: $8x+4y\geq -24$ [Section 3.7]

Graph as $8x+4y=-24$.

Boundary line is solid.

y-intercept:	x-intercept:
If $x=0$,	If $y=0$,
$8x+4y=-24$	$8x+4y=-24$
$\mathbf{0}+4y=-24$	$8x+2(\mathbf{0})=-24$
$4y=-24$	$8x=-24$
$\frac{4y}{\mathbf{4}}=\frac{-24}{\mathbf{4}}$	$\frac{8x}{\mathbf{8}}=\frac{-24}{\mathbf{8}}$
$y=-6$	$x=-3$

The y-int is $(0, -6)$, and the x-int is $(-3,0)$.

Select test point (0,0) and substitute into

$$8x+4y\geq -24$$
$$8(\mathbf{0})+4(\mathbf{0})\overset{?}{\geq}-24$$
$$0\geq -24$$
True

The coordinates of every point on the same side of the line as (0,0) satisfy the inequality. To indicate this, we shade the half-plane that contains the test point (0,0).

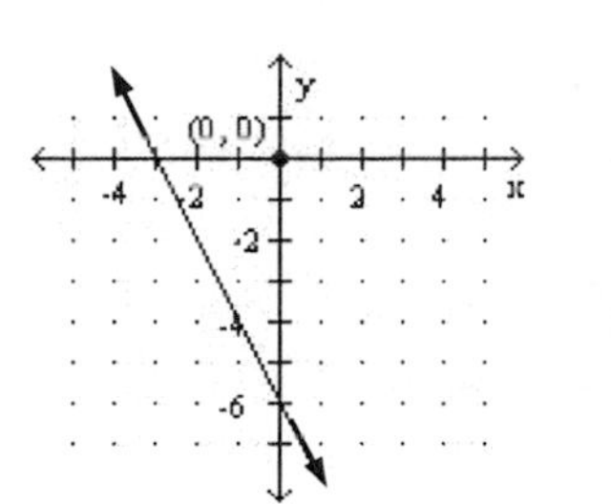

$8x+4y=-24$

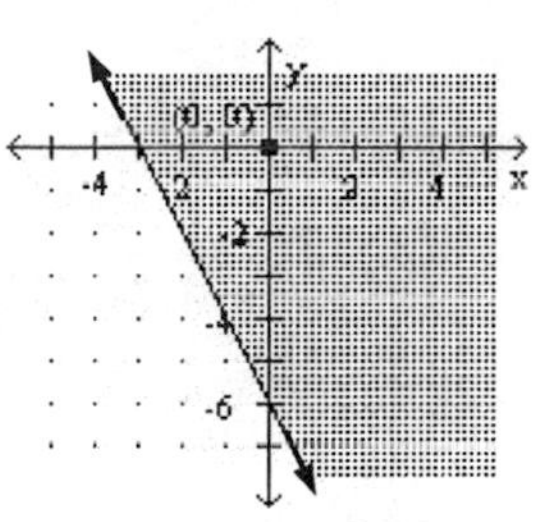

$8x+4y\geq -24$

38. If $f(x)=3x^2-2x+1$, find $f(-2)$. [Section 3.8]

$$f(x)=3x^2-2x+1$$
$$f(-\mathbf{2})=3(-\mathbf{2})^2-2(-\mathbf{2})+1$$
$$=3(4)+4+1$$
$$=12+5$$
$$=17$$

Thus, $f(-2)=17$

39. Is $\left(\frac{1}{2}, 1\right)$ a solution of the system $\begin{cases}4x-y=1\\2x+y=2\end{cases}$? [Section 4.1]

$4x-y=1$	$2x+y=2$
$4\left(\frac{1}{2}\right)-1\overset{?}{=}1$	$2\left(\frac{1}{2}\right)+1\overset{?}{=}2$
$2-1\overset{?}{=}1$	$1+1\overset{?}{=}2$
$1=1$	$2=2$
True	True

Since $(\frac{1}{2},3)$ does satisfy both equations, it is a solution of the system.

40. Solve the system $\begin{cases}3x-2y=6\\x-y=1\end{cases}$ by graphing. [Section 4.1]

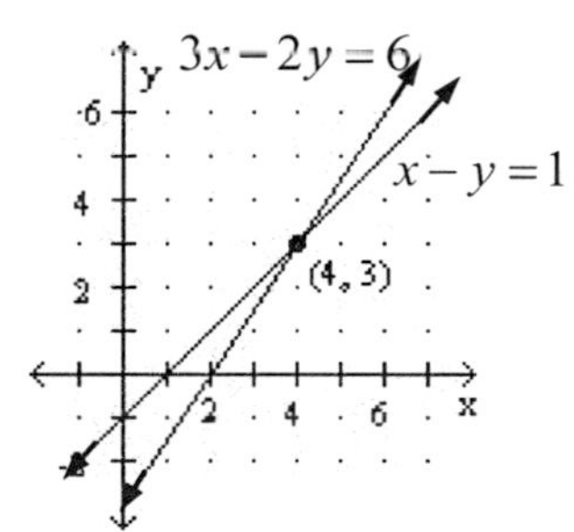

The solution is (4, 3).

41. Solve the system $\begin{cases} y = -4x + 1 \\ 4x - y = 5 \end{cases}$ by substitution. [Section 4.2]

$4x - y = 5$ The 2nd equation
Substitute for y.

$$4x - (\mathbf{-4x+1}) = 5$$
$$4x + 4x - 1 = 5$$
$$8x - 1 + \mathbf{1} = 5 + \mathbf{1}$$
$$8x = 6$$
$$\frac{8x}{\mathbf{8}} = \frac{6}{\mathbf{8}}$$
$$x = \frac{3}{4}$$

$y = -4x + 1$ The 1st equation

$$y = -4\left(\frac{\mathbf{3}}{\mathbf{4}}\right) + 1$$
$$y = -3 + 1$$
$$y = -2$$

The solution is $\left(\frac{3}{4}, -2\right)$.

Check:	Check:
$y = -4x + 1$	$4x - y = 5$
$-2 \stackrel{?}{=} -4\left(\frac{3}{4}\right) + 1$	$4\left(\frac{3}{4}\right) - (-2) \stackrel{?}{=} 5$
$-2 \stackrel{?}{=} -3 + 1$	$3 + 2 \stackrel{?}{=} 5$
$-2 = -2$	$5 = 5$
True	True

40. Solve the system $\begin{cases} 5a + 3b = -8 \\ 2a + 9b = 2 \end{cases}$ by elimination (addition) [Section 4.3]

$$\begin{cases} 5a + 3b = -8 \\ 2a + 9b = 2 \end{cases}$$

Eliminate b.

Multiply both sides of the 1st equation by -3.

$$\begin{array}{r} -15a - 9b = 24 \\ 2a + 9b = \ \ 2 \\ \hline -13a \qquad = 26 \end{array}$$
$$\frac{-13a}{\mathbf{-13}} = \frac{26}{\mathbf{-13}}$$
$$a = -2$$

$2a + 9b = 2$ The 2nd equation

$$2(\mathbf{-2}) + 9b = 2$$
$$-4 + 9b = 2$$
$$-4 + 9b + \mathbf{4} = 2 + \mathbf{4}$$
$$9b = 6$$
$$\frac{9b}{\mathbf{9}} = \frac{6}{\mathbf{9}}$$
$$b = \frac{2}{3}$$

The solution is $\left(-2, \frac{2}{3}\right)$.

43. FUNDRAISING [Section 4.4]

Analyze

- Rotary Club earned \$356 by recycling a total of 14 tons of newspaper and cardboard.
- They were paid \$31 per ton for newspaper.
- They were paid \$18 per ton for cardboard.
- How many ton of each were collected?

Assign

Let x = the tons of newspaper
y = the tons of cardboard

Form

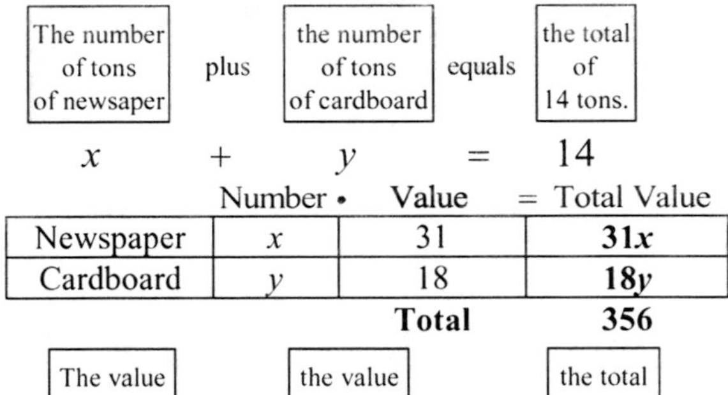

The number of tons of newsaper	plus	the number of tons of cardboard	equals	the total of 14 tons.
x	$+$	y	$=$	14

	Number •	Value	= Total Value
Newspaper	x	31	$\mathbf{31x}$
Cardboard	y	18	$\mathbf{18y}$
		Total	**356**

The value of the newspaper	plus	the value of the cardboard	equals	the total value of \$356.00.
$31x$	$+$	$18y$	$=$	356

Solve

$$\begin{cases} x + \quad y = 14 \\ 31x + 18y = 356 \end{cases}$$

Eliminate y.
Multiply both sides of the 1st equation by -18.

$$\begin{array}{r} -18x - 18y = -252 \\ 31x + 18y = 356 \\ \hline 13x \qquad = 104 \end{array}$$
$$\frac{13x}{\mathbf{13}} = \frac{104}{\mathbf{13}}$$
$$x = 8$$

$$x + y = 14 \text{ The } 1^{st} \text{ equation}$$
$$\mathbf{8} + y = 14$$
$$8 + y - \mathbf{8} = 14 - \mathbf{8}$$
$$y = 6$$

State
They collected 8 tons of newspaper.
They collected 6 tons of cardboard.

Check
The solutions check.

44. Graph $\begin{cases} 4x + 3y \geq 12 \\ y < 4 \end{cases}$ [Section 4.5]

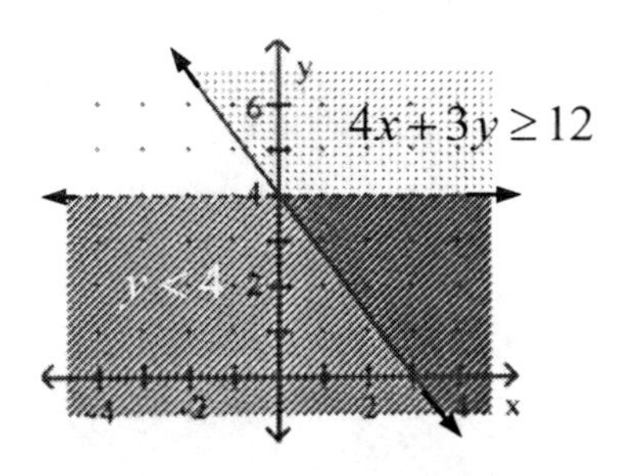

Simplify each expression. Write each answer without negative exponents.

45. [Section 5.1]

$$-y^2(4y^3) = -4y^{2+3}$$
$$= -4y^5$$

46. [Section 5.1]

$$\frac{(x^2y^5)^5}{(x^3y)^2} = \frac{x^{2\cdot5}y^{5\cdot5}}{x^{3\cdot2}y^2}$$
$$= \frac{x^{10}y^{25}}{x^6y^2}$$
$$= x^{10-6}y^{25-2}$$
$$= x^4y^{23}$$

47. [Section 5.2]

$$\frac{b^5}{b^{-2}} = b^{5-(-2)}$$
$$= b^{5+2}$$
$$= b^7$$

48. [Section 5.2]

$$2x^0 = 2 \cdot 1$$
$$= 2$$

49. Write in scientific notation. [Section 5.3]
Move the decimal 5 places to the right.
$0.00009011 = 9.011 \times 10^{-5}$

50. Write in scientific notation. [Section 5.3]
Move the decimal 6 places to the left.
$1,700,000 = 1.7 \times 10^6$

51. Find the degree of $7y^3 + 4y^2 + y + 3$. [Section 5.4]

The degree is 3.

52. Graph: $y = x^3 + 2$ [Section 5.4]

x	y	(x, y)
-3	$y = (-3)^3 + 2$ $= -27 + 2$ $= -25$	$(-3, -25)$
-2	$y = (-2)^3 + 2$ $= -8 + 2$ $= -6$	$(-2, -6)$
-1	$y = (-1)^3 + 2$ $= -1 + 2$ $= 1$	$(-1, 1)$
0	$y = (0)^3 + 2$ $= 0 + 2$ $= 2$	$(0, 2)$
1	$y = (1)^3 + 2$ $= 1 + 2$ $= 3$	$(1, 3)$
2	$y = (2)^3 + 2$ $= 8 + 2$ $= 10$	$(2, 10)$
3	$y = (3)^3 + 2$ $= 27 + 2$ $= 29$	$(3, 29)$

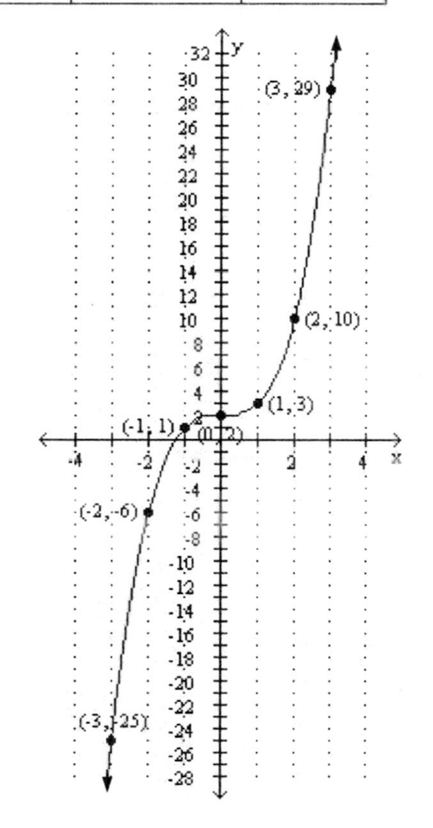

Perform the operations.

53. $(x^2-3x+8)-(3x^2+x+3)$ [Section 5.5]

$=x^2-3x+8-3x^2-x-3$

$=(1-3)x^2+(-3-1)x+(8-3)$

$=-2x^2-4x+5$

54. $4b^3(2b^2-2b)$ [Section 5.6]

$=\mathbf{4b^3}(2b^2)+\mathbf{4b^3}(-2b)$

$=4(2)b^{3+2}+4(-2)b^{3+1}$

$=8b^5-8b^4$

55. $(3x-2)(x+4)$ [Section 5.6]

$=\mathbf{3x}(x)+\mathbf{3x}(4)-\mathbf{2}(x)-\mathbf{2}(4)$

$=3x^2+12x-2x-8$

$=3x^2+10x-8$

56. $(y-6)^2$ [Section 5.7]

$=y^2+2(y)(-6)+(-6)^2$

$=y^2-12y+36$

57. $\dfrac{12a^2b^2-8a^2b-4ab}{4ab}$ [Section 5.8]

$=\dfrac{12a^2b^2}{4ab}-\dfrac{8a^2b}{4ab}-\dfrac{4ab}{4ab}$

$=3a^{2-1}b^{2-1}-2a^{2-1}b^{1-1}-a^{1-1}b^{1-1}$

$=3a^1b^1-2a^1b^0-a^0b^0$

$=3ab-2a-1$

58. [Section 5.8]

$$\begin{array}{r} 2x+1 \\ x-3\overline{)2x^2-5x-3} \\ \underline{-(2x^2-6x)} \\ x-3 \\ \underline{-(x-3)} \\ 0 \end{array}$$

59. PLAYPENS

a. perimeter of the playpen [Section 5.5]

$P=2l+2w$

$=2(x+3)+2(x+1)$

$=2x+6+2x+2$

$=(2+2)x+(6+2)$

$=4x+8$

The perimeter of the playpen is $(4x+8)$ in.

b. area of the floor of the playpen [Section 5.6]

$A=lw$

$=(x+3)(x+1)$

$=x^2+x+3x+3$

$=x^2+(1+3)x+3$

$=x^2+4x+3$

The area of the floor of the playpen is (x^2+4x+3) in^2.

c. volume of the playpen [Section 5.6]

$V=lwh$

$=(x+3)(x+1)x$

$=(x^2+x+3x+3)x$

$=(x^2+(1+3)x+3)x$

$=(x^2+4x+3)x$

$=x^3+4x^2+3x$

The volume of the playpen is (x^3+4x^2+3x) in^3.

60. Find the GCF. [Section 6.1]

$24x^5y^8=\mathbf{2}\cdot 2\cdot 2\cdot \mathbf{3}\cdot \mathbf{x\cdot x\cdot x\cdot x\cdot x}\cdot\ \mathbf{y}\cdot y^7$

$54x^6y=\mathbf{2}\cdot 3\cdot 3\cdot \mathbf{3}\cdot \mathbf{x\cdot x\cdot x\cdot x\cdot x}\cdot x\cdot \mathbf{y}$

GCF is $\mathbf{2\cdot 3\cdot x^5\cdot y}=6x^5y$.

Factor.

61. [Section 6.1]

$9b^3-27b^2=\mathbf{9\cdot b^2}b-\mathbf{9}\cdot 3\mathbf{b^2}$

$=9b^2(b-3)$

62. [Section 6.1]

$ax+bx+ay+by=(\mathbf{a}x+\mathbf{b}x)+(\mathbf{a}y+\mathbf{b}y)$

$=\mathbf{x}(a+b)+\mathbf{y}(a+b)$

$=(a+b)(x+y)$

63. u^2-3+2u [Section 6.2]

u^2+2u-3

Two different sign factors of -3 whose sum is $+2$ are $+3$ and -1.

$u^2+2u-3=(u+3)(u-1)$

64. $10x^2+x-2$ [Section 6.3]

$\underset{a}{10}x^2+\underset{b}{1}x-\underset{c}{2}$

In $10x^2+x-2$, we have $a=10$, $b=1$ and $c=-2$. The key number is $ac=10(-2)=-20$. We must find a factorization of -20 in which the sum of the factors is $b=1$. Since the factors must have a negative product, their signs must be different. The pairs of factors are

–4 and 5.

$$\begin{aligned}10x^2+x-2&=10x^2-4x+5x-2\\&=(10x^2-4x)+(5x-2)\\&=2x(5x-2)+1(5x-2)\\&=(5x-2)(2x+1)\end{aligned}$$

65. $4a^2-12a+9$ [Section 6.4]

- The first term $4a^2$ is the square of $\mathbf{2a}$.
- The last term 9 is the square of $\mathbf{-3}$.
- The middle term is twice the product of $2a$ and -3: $2(2a)(-3)=\mathbf{-12a}$.

$$\begin{aligned}4a^2-12a+9&=(2a-3)(2a-3)\\&=(2a-3)^2\end{aligned}$$

66. $9z^2-1$ [Section 6.4]

$$\begin{aligned}&\mathrm{F}^2-\mathrm{L}^2=(\mathrm{F}+\mathrm{L})(\mathrm{F}-\mathrm{L})\\&\downarrow\quad\downarrow\qquad\downarrow\ \downarrow\quad\downarrow\ \downarrow\\&(3z)^2-1^2=(3z+1)(3z-1)\end{aligned}$$

67. t^3-8 [Section 6.5]

$$\begin{aligned}&=t^3-2^3\\&=(t-2)(t^2+t\cdot2+2^2)\\&=(t-2)(t^2+2t+4)\end{aligned}$$

68. $3a^2b^2-6a^2-3b^2+6$ [Section 6.6]

$$\begin{aligned}&=3(a^2b^2-2a^2-b^2+2)\\&=3[(a^2b^2-2a^2)+(-b^2+2)]\\&=3[a^2(b^2-2)-1(b^2-2)]\\&=3(b^2-2)(a^2-1)\\&=3(b^2-2)(a+1)(a-1)\end{aligned}$$

Solve each equation.

69. $15s^2-20s=0$ [Section 6.7]

$$15s^2-20s=0$$
$$5s(3s-4)=0$$

$$5s=0 \quad\text{or}\quad 3s-4=0$$
$$s=0 \qquad\qquad 3s=4$$
$$s=\frac{4}{3}$$

The solutions are 0 and $\frac{4}{3}$.

70. $2x^2-5x=-2$ [Section 6.7]

$$2x^2-5x=-2$$
$$2x^2-5x+\mathbf{2}=-2+\mathbf{2}$$
$$2x^2-5x+2=0$$
$$(2x-1)(x-2)=0$$

$$2x-1=0 \quad\text{or}\quad x-2=0$$
$$2x=1 \qquad\qquad x=2$$
$$x=\frac{1}{2}$$

The solutions are $\frac{1}{2}$ and 2.

71. $x^3+3x^2+2x=0$ [Section 6.7]

$$x^3+3x^2+2x=0$$
$$x(x^2+3x+2)=0$$
$$x(x+1)(x+2)=0$$

$$x=0 \quad\text{or}\quad x+1=0 \quad\text{or}\quad x+2=0$$
$$x=-1 \qquad\qquad x=-2$$

The solutions are $0, -1$, and -2.

72. CAMPING

Analyze

- Length is 3 inches longer than it is wide.
- Area is 108 sq. in.
- Find the dimensions.

Assign

Let $w=$ width of surface in inches
$w+3=$ length of surface in inches

Form

The area of the rectangle	equals	the length	times	the width.
108	=	$w+3$	$\cdot$	w

Solve

$$w(w+3)=108$$
$$w^2+3w=108$$
$$w^2+3w-\mathbf{108}=108-\mathbf{108}$$
$$w^2+3w-108=0$$
$$(w+12)(h-9)=0$$

$$w+12=0 \quad\text{or}\quad w-9=0$$
$$\cancel{w=-12} \qquad\qquad w=9$$

State

The width is 9 inches.
The length is $(9+3)=12$ inches.

Check

A recangle with dimensions 9 in by 12 in has an area of 108 in^2. The solutions check.

SECTION 7.1 STUDY SET
VOCABULARY

Fill in the blanks.

1. A quotient of two polynomials, such as $\frac{x^2+x}{x^2-3x}$, is called a **rational** expression.

3. Because of the division by 0, the expression $\frac{8}{0}$ is **undefined**.

CONCEPTS

5. When we simplify $\frac{x^2+5x}{4x+20}$, the result is $\frac{x}{4}$. These equivalent expressions have the same value for all real numbers, except $x = -5$. Show that they have the same value for $x = 1$.

$$\frac{x^2+5x}{4x+20} \stackrel{?}{=} \frac{x}{4}$$

$$\frac{\mathbf{1}^2+5(\mathbf{1})}{4(\mathbf{1})+20} \stackrel{?}{=} \frac{\mathbf{1}}{4}$$

$$\frac{1+5}{4+20} \stackrel{?}{=} \frac{1}{4}$$

$$\frac{6}{24} \stackrel{?}{=} \frac{1}{4}$$

$$\frac{\overset{1}{\cancel{6}}}{\underset{1}{\cancel{6}}\cdot 4} \stackrel{?}{=} \frac{1}{4}$$

$$\frac{1}{4} = \frac{1}{4}$$

7. Simplify each expression, if possible.

a. $\frac{x-8}{x-8}$ **1** b. $\frac{x-8}{8-x}$ **−1**

c. $\frac{x+8}{8+x}$ **1** d. $\frac{x+8}{x}$ **Does not simplify**

NOTATION

Complete the solution to simplify the rational expression.

9. $$\frac{x^2+2x+1}{x^2+4x+3} = \frac{(x+1)(\mathbf{x}+1)}{(x+3)(x+\mathbf{1})}$$

$$= \frac{(x+1)\overset{1}{(\cancel{x+1})}}{(x+3)\underset{1}{(\cancel{x+1})}}$$

$$= \frac{x+1}{\mathbf{x+3}}$$

GUIDED PRACTICE

Evaluate each expression for $x = 6$. See Example 1.

11. $$\frac{x-2}{x-5} = \frac{\mathbf{6}-2}{\mathbf{6}-5} = \frac{4}{1} = 4$$

13. $$\frac{x^2-4x-12}{x^2+x-2} = \frac{\mathbf{6}^2-4(\mathbf{6})-12}{\mathbf{6}^2+\mathbf{6}-2} = \frac{36-24-12}{36+6-2} = \frac{0}{40} = 0$$

15. $$\frac{-x+1}{x^2-5x-6} = \frac{-(\mathbf{6})+1}{\mathbf{6}^2-5(\mathbf{6})-6} = \frac{-6+1}{36-30-6} = \frac{-5}{0}$$

Undefined

Evaluate each expression for $y = -3$. See Example 1.

17. $$\frac{y+5}{3y-2} = \frac{(\mathbf{-3})+5}{3(\mathbf{-3})-2} = \frac{2}{-9-2} = -\frac{2}{11}$$

19. $$\frac{-y}{y^2-y+6} = \frac{-(\mathbf{-3})}{(\mathbf{-3})^2-(\mathbf{-3})+6} = \frac{3}{9+3+6} = \frac{3}{18} = \frac{1}{6}$$

21. $\dfrac{y^2+9}{9-y^2}=\dfrac{(-\mathbf{3})^2+9}{9-(-\mathbf{3})^2}$

$=\dfrac{9+9}{9-9}$

$=\dfrac{18}{0}$

Undefined

Find all real numbers for which the rational expression is undefined. See Example 2.

23. $\dfrac{15}{x-2}$ $\quad x-2=0$; $x-2+\mathbf{2}=0+\mathbf{2}$; $x=2$

The rational expression is undefined for $x=2$.

25. $\dfrac{x+5}{8x}$ $\quad 8x=0$; $x=0$

The rational expression is undefined for $x=0$.

27. $\dfrac{15x+2}{x^2+6}$

No matter what real number is substituted for x, the denominator, x^2+6, will not be equal to 0. Thus, no real numbers make the denominator undefined.

29. $\dfrac{x+1}{2x-1}$ $\quad 2x-1=0$; $2x-1+\mathbf{1}=0+\mathbf{1}$; $2x=1$; $\dfrac{2x}{\mathbf{2}}=\dfrac{1}{\mathbf{2}}$; $x=\dfrac{1}{2}$

The rational expression is undefined for $x=\dfrac{1}{2}$.

31. $\dfrac{x^2-6x}{9}$

Since the denominator does not contain a variable. the denominator can never be equal to 0. Thus, no real numbers make the expression undefined.

33. $\dfrac{30x}{x^2-36}$ $\quad x^2-36=0$; $(x+6)(x-6)=0$

$x+6=0$; $x+6-\mathbf{6}=0-\mathbf{6}$; $x=-6$ $\quad|\quad$ $x-6=0$; $x-6+\mathbf{6}=0+\mathbf{6}$; $x=6$

The rational expression is undefined for $x=-6$ and $x=6$.

35. $\dfrac{15}{x^2+x-2}$ $\quad x^2+x-2=0$; $(x+2)(x-1)=0$

$x+2=0$; $x+2-\mathbf{2}=0-\mathbf{2}$; $x=-2$ $\quad|\quad$ $x-1=0$; $x-1+\mathbf{1}=0+\mathbf{1}$; $x=1$

The rational expression is undefined for $x=-2$ and $x=1$.

37. $\dfrac{16}{20-x}$ $\quad 20-x=0$; $20-x+\boldsymbol{x}=0+\boldsymbol{x}$; $20=x$

The rational expression is undefined for $x=20$.

Simplify each expression. See Example 3.

39. $\dfrac{45}{9a}=\dfrac{5\cdot\cancel{9}}{\cancel{9}\,a}=\dfrac{5}{a}$

41. $\dfrac{6x^4}{4x^2}=\dfrac{\cancel{2}\cdot 3x^{4-2}}{\cancel{2}\cdot 2}=\dfrac{3x^2}{2}$

Simplify. See Example 4.

43. $\dfrac{6x+3}{9}=\dfrac{3(2x+1)}{9}=\dfrac{\cancel{3}(2x+1)}{3\cdot\cancel{3}}=\dfrac{2x+1}{3}$

45. $\dfrac{x+3}{3x+9}=\dfrac{x+3}{3(x+3)}=\dfrac{\cancel{x+3}}{3\,\cancel{(x+3)}}=\dfrac{1}{3}$

47. $\dfrac{x^2-4}{x^2-6x+8}=\dfrac{(x+2)(x-2)}{(x-2)(x-4)}=\dfrac{(x+2)\,\cancel{(x-2)}}{\cancel{(x-2)}(x-4)}=\dfrac{x+2}{x-4}$

49. $\dfrac{x^2+5x+4}{x^2+4x}=\dfrac{(x+1)(x+4)}{x(x+4)}$

$=\dfrac{(x+1)\overset{1}{\cancel{(x+4)}}}{x\underset{1}{\cancel{(x+4)}}}$

$=\dfrac{x+1}{x}$

Simplify. See Example 5.

51. $\dfrac{m^2-2mn+n^2}{7m^2-7n^2}=\dfrac{(m-n)(m-n)}{7(m+n)(m-n)}$

$=\dfrac{\overset{1}{\cancel{(m-n)}}(m-n)}{7(m+n)\underset{1}{\cancel{(m-n)}}}$

$=\dfrac{m-n}{7(m+n)}$ or $\dfrac{m-n}{7m+7n}$

53. $\dfrac{4b^2+4b+1}{(2b+1)^3}=\dfrac{(2b+1)(2b+1)}{(2b+1)(2b+1)(2b+1)}$

$=\dfrac{\overset{1}{\cancel{(2b+1)}}\overset{1}{\cancel{(2b+1)}}}{\underset{1}{\cancel{(2b+1)}}\underset{1}{\cancel{(2b+1)}}(2b+1)}$

$=\dfrac{1}{2b+1}$

Simplify. See Example 6.

55. $\dfrac{10(c-3)+10}{3(c-3)+3}=\dfrac{10c-30+10}{3c-9+3}$

$=\dfrac{10c-20}{3c-6}$

$=\dfrac{10(c-2)}{3(c-2)}$

$=\dfrac{10\overset{1}{\cancel{(c-2)}}}{3\underset{1}{\cancel{(c-2)}}}$

$=\dfrac{10}{3}$

57. $\dfrac{6(x+3)-18}{3x-18}=\dfrac{6x+18-18}{3x-18}$

$=\dfrac{6x}{3(x-6)}$

$=\dfrac{2\cdot\overset{1}{\cancel{3}}\cdot x}{\underset{1}{\cancel{3}}(x-6)}$

$=\dfrac{2x}{x-6}$

Simplify. See Example 7.

59. $\dfrac{2x-7}{7-2x}=\dfrac{2x-7}{-2x+7}$

$=\dfrac{2x-7}{-1(2x-7)}$

$=\dfrac{\overset{1}{\cancel{2x-7}}}{-1\underset{1}{\cancel{(2x-7)}}}$

$=-1$

61. $\dfrac{3-4t}{8t-6}=\dfrac{-4t+3}{8t-6}$

$=\dfrac{-1(4t-3)}{2(4t-3)}$

$=\dfrac{-1\overset{1}{\cancel{(4t-3)}}}{2\underset{1}{\cancel{(4t-3)}}}$

$=-\dfrac{1}{2}$

63. $\dfrac{2-a}{a^2-a-2}=\dfrac{-a+2}{(a-2)(a+1)}$

$=\dfrac{-1(a-2)}{(a-2)(a+1)}$

$=\dfrac{-1\overset{1}{\cancel{(a-2)}}}{\underset{1}{\cancel{(a-2)}}(a+1)}$

$=-\dfrac{1}{a+1}$

65. $$\frac{25-5m}{m^2-25}=\frac{-5m+25}{(m+5)(m-5)}=\frac{-5(m-5)}{(m+5)(m-5)}=\frac{-5\,\cancel{(m-5)}}{(m+5)\,\cancel{(m-5)}}=-\frac{5}{m+5}$$

TRY IT YOURSELF

Simplify, if an expression cannot be simplified, write "Does not simplify."

67. $$\frac{a^3-a^2}{a^4-a^3}=\frac{a^2(a-1)}{a^3(a-1)}=\frac{a^{2-3}\,\cancel{(a-1)}}{\cancel{(a-1)}}=\frac{a^{-1}}{1}=\frac{1}{a}$$

69. $$\frac{4-x^2}{x^2-x-2}=\frac{-x^2+4}{x^2-x-2}=\frac{-1(x^2-4)}{x^2-x-2}=\frac{-1(x+2)(x-2)}{(x-2)(x+1)}=\frac{-1(x+2)\,\cancel{(x-2)}}{\cancel{(x-2)}\,(x+1)}=-\frac{x+2}{x+1}$$

71. $$\frac{6x-30}{5-x}=\frac{6x-30}{-x+5}=\frac{6(x-5)}{-1(x-5)}=\frac{6\,\cancel{(x-5)}}{-1\,\cancel{(x-5)}}=-6$$

73. $$\frac{x^2+3x+2}{x^2+x-2}=\frac{(x+2)(x+1)}{(x+2)(x-1)}=\frac{\cancel{(x+2)}\,(x+1)}{\cancel{(x+2)}\,(x-1)}=\frac{x+1}{x-1}$$

75. $$\frac{15x^2y}{5xy^2}=\frac{3\cdot 5\cdot x^{2-1}\cdot y^{1-2}}{5}=\frac{3\cdot\cancel{5}\cdot x^1\cdot y^{-1}}{\cancel{5}}=\frac{3x}{y}$$

77. $$\frac{x^8+9x^7}{9+x}=\frac{x^7(x+9)}{x+9}=\frac{x^7\,\cancel{(x+9)}}{\cancel{x+9}}=x^7$$

79. $$\frac{x(x-8)+16}{16-x^2}=\frac{x^2-8x+16}{-x^2+16}=\frac{(x-4)(x-4)}{-1(x^2-16)}=\frac{(x-4)(x-4)}{-1(x+4)(x-4)}=\frac{(x-4)\,\cancel{(x-4)}}{-1(x+4)\,\cancel{(x-4)}}=-\frac{x-4}{x+4}\text{ or }\frac{4-x}{4+x}$$

81. $$\frac{4c+4d}{d+c}=\frac{4(c+d)}{c+d}=\frac{4\,\cancel{(c+d)}}{\cancel{c+d}}=4$$

83. $\dfrac{3x^2-27}{2x^2-5x-3}=\dfrac{3(x^2-9)}{(2x+1)(x-3)}$

$=\dfrac{3(x+3)(x-3)}{(2x+1)(x-3)}$

$=\dfrac{3(x+3)\cancel{(x-3)}^{1}}{(2x+1)\cancel{(x-3)}_{1}}$

$=\dfrac{3(x+3)}{2x+1}$ or $\dfrac{3x+9}{2x+1}$

85. $\dfrac{-3x^2+10x+77}{x^2-4x-21}=\dfrac{-(3x^2-10x-77)}{x^2-4x-21}$

$=\dfrac{-(3x+11)(x-7)}{(x+3)(x-7)}$

$=\dfrac{-(3x+11)\cancel{(x-7)}^{1}}{(x+3)\cancel{(x-7)}_{1}}$

$=-\dfrac{3x+11}{x+3}$

87. $\dfrac{42c^3d}{18cd^3}=\dfrac{6\cdot 7\cdot c^{3-1}\cdot d^{1-3}}{3\cdot 6}$

$=\dfrac{\cancel{6}^{1}\cdot 7\cdot c^2\cdot d^{-2}}{3\cdot \cancel{6}_{1}}$

$=\dfrac{7c^2}{3d^2}$

89. $\dfrac{16a^2-1}{4a+4}=\dfrac{(4a+1)(4a-1)}{4(a+1)}$

Does not simplify.

91. $\dfrac{8u^2-2u-15}{4u^4+5u^3}=\dfrac{(2u-3)(4u+5)}{u^3(4u+5)}$

$=\dfrac{(2u-3)\cancel{(4u+5)}^{1}}{u^3\cancel{(4u+5)}}$

$=\dfrac{2u-3}{u^3}$

93. $\dfrac{(2x+3)^4}{4x^2+12x+9}$

$=\dfrac{(2x+3)^4}{(2x+3)(2x+3)}$

$=\dfrac{\cancel{(2x+3)}^{1}\cancel{(2x+3)}^{1}(2x+3)(2x+3)}{\cancel{(2x+3)}_{1}\cancel{(2x+3)}_{1}}$

$=(2x+3)^2$

95. $\dfrac{6a+3(a+2)+12}{a+2}=\dfrac{6a+3a+6+12}{a+2}$

$=\dfrac{9a+18}{a+2}$

$=\dfrac{9(a+2)}{a+2}$

$=\dfrac{9\cancel{(a+2)}^{1}}{\cancel{a+2}_{1}}$

$=9$

97. $\dfrac{15x-3x^2}{25y-5xy}=\dfrac{3x(5-x)}{5y(5-x)}$

$=\dfrac{3x\cancel{(5-x)}^{1}}{5y\cancel{(5-x)}_{1}}$

$=\dfrac{3x}{5y}$

99. $\dfrac{2x^2}{x+2}$ Does not simplify.

101. $\dfrac{18+2x}{x^2-81}=\dfrac{2(9+x)}{(x+9)(x-9)}$

$=\dfrac{2\cancel{(x+9)}^{1}}{\cancel{(x+9)}_{1}(x-9)}$

$=\dfrac{2}{x-9}$

APPLICATIONS

103. ORGAN PIPES

$$n = \frac{512}{L}$$
$$= \frac{512}{6}$$
$$= \frac{\cancel{2}^{1} \cdot 256}{\cancel{2}_{1} \cdot 3}$$
$$= \frac{256}{3}$$
$$= 85\frac{1}{3}$$

The pipe will vibrate $85\frac{1}{3}$ times per sec.

105. MEDICAL DOSAGES

$$c = \frac{4t}{t^2+1}, \text{ at } 1:00$$
$$= \frac{4(1)}{1^2+1}$$
$$= \frac{4}{2}$$
$$= 2$$

At 1:00 there is 2 mg per liter.

$$c = \frac{4t}{t^2+1}, \text{ at } 2:00$$
$$= \frac{4(2)}{2^2+1}$$
$$= \frac{8}{5}$$
$$= 1.6$$

At 2:00 there is 1.6 mg per liter.

$$c = \frac{4t}{t^2+1}, \text{ at } 3:00$$
$$= \frac{4(3)}{3^2+1}$$
$$= \frac{12}{10}$$
$$= 1.2$$

At 3:00 there is 1.2 mg per liter.

WRITING

107-111. Answers will vary.

REVIEW

State each property using the variables *a*, *b*, and when necessary, *c*.

113. a. The associative property of addition

$$(a + b) + c = a + (b + c)$$

b. The commutative property of multiplication

$$ab = ba$$

CHALLENGE PROBLEMS

Simplify each expression.

115. $$\frac{(x^2+2x+1)(x^2-2x+1)}{(x^2-1)^2}$$
$$= \frac{(x+1)(x+1)(x-1)(x-1)}{(x^2-1)(x^2-1)}$$
$$= \frac{(x+1)(x+1)(x-1)(x-1)}{(x+1)(x-1)(x+1)(x-1)}$$
$$= \frac{\cancel{(x+1)}^{1}\cancel{(x+1)}^{1}\cancel{(x-1)}^{1}\cancel{(x-1)}^{1}}{\cancel{(x+1)}_{1}\cancel{(x-1)}_{1}\cancel{(x+1)}_{1}\cancel{(x-1)}_{1}}$$
$$= 1$$

117. $$\frac{x^3-27}{x^3-9x} = \frac{(x-3)(x^2+3x+9)}{x(x^2-9)}$$
$$= \frac{(x-3)(x^2+3x+9)}{x(x+3)(x-3)}$$
$$= \frac{\cancel{(x-3)}^{1}(x^2+3x+9)}{x(x+3)\cancel{(x-3)}_{1}}$$
$$= \frac{x^2+3x+9}{x(x+3)}$$

119. $$\frac{m^3+64}{m^3+4m^2+3m+12}$$
$$= \frac{(m+4)(m^2-4m+16)}{(m^3+4m^2)+(3m+12)}$$
$$= \frac{(m+4)(m^2-4m+16)}{m^2(m+4)+3(m+4)}$$
$$= \frac{(m+4)(m^2-4m+16)}{(m+4)(m^2+3)}$$
$$= \frac{\cancel{(m+4)}^{1}(m^2-4m+16)}{\cancel{(m+4)}_{1}(m^2+3)}$$
$$= \frac{m^2-4m+16}{m^2+3}$$

SECTION 7.2 STUDY SET

VOCABULARY

Fill in the blanks.

1. The **reciprocal** of $\frac{x^2+6x+1}{10x}$ is $\frac{10x}{x^2+6x+1}$.

CONCEPTS

Fill in the blanks.

3. a. To multiply rational expressions, multiply their **numerators** and multiply their **denominators.** To divide two rational expressions, multiply the first by the **reciprocal** of the second. In symbols,

b. $\frac{A}{B}\cdot\frac{C}{D}=\frac{\mathbf{AC}}{\mathbf{BD}}$ and $\frac{A}{B}\div\frac{C}{D}=\frac{A}{B}\cdot\frac{\mathbf{D}}{\mathbf{C}}$

Simplify each expression.

5. $\frac{y\cdot y\cdot y(15-y)}{y(y-15)(y+1)}$

$$\frac{\cancel{y}\cdot y\cdot y(\cancel{15-y})}{\cancel{y}\cdot -1(\cancel{15-y})(y+1)}$$

$$-\frac{\mathbf{y^2}}{\mathbf{y+1}}$$

7. Find the product of the rational expression and its reciprocal.

$$\frac{3}{x+2}\cdot\frac{x+2}{3}=\frac{\cancel{3}}{\cancel{x+2}}\cdot\frac{\cancel{x+2}}{\cancel{3}}$$

$$=\mathbf{1}$$

NOTATION

9. What units are common to the numerator and denominator? **ft**

$$\frac{45\text{ ft}}{1}\cdot\frac{1\text{ yd}}{3\text{ ft}}$$

GUIDED PRACTICE

Multiply, and then simplify, if possible. See Example 1.

11. $\frac{3}{7}\cdot\frac{y}{2}=\frac{3y}{14}$

13. $\frac{y+2}{y}\cdot\frac{3}{y^2}=\frac{(y+2)3}{y^{1+2}}$

$$=\frac{3(y+2)}{y^3}\text{ or }\frac{3y+6}{y^3}$$

15. $\frac{35n}{12}\cdot\frac{16}{7n^2}=\frac{35n\cdot 16}{12\cdot 7n^2}$

$$=\frac{5\cdot\cancel{7}\cdot\cancel{n}\cdot\cancel{4}\cdot 4}{3\cdot\cancel{4}\cdot\cancel{7}\cdot\cancel{n}\cdot n}$$

$$=\frac{20}{3n}$$

17. $\frac{2x^2y}{3xy}\cdot\frac{3xy^2}{2}=\frac{2x^2y\cdot 3xy^2}{3xy\cdot 2}$

$$=\frac{\cancel{2}\cdot x^2\cdot\cancel{y}\cdot\cancel{3}\cdot\cancel{x}\cdot y^2}{\cancel{3}\cdot\cancel{x}\cdot\cancel{y}\cdot\cancel{2}}$$

$$=x^2y^2$$

Multiply, and then simplify, if possible. See Example 2.

19. $\frac{x+5}{5}\cdot\frac{x}{x+5}=\frac{(x+5)x}{5(x+5)}$

$$=\frac{\cancel{(x+5)}\cdot x}{5\cdot\cancel{(x+5)}}$$

$$=\frac{x}{5}$$

21. $\frac{2x+6}{x+3}\cdot\frac{3}{4x}=\frac{(2x+6)3}{(x+3)4x}$

$$=\frac{2\cdot(x+3)\cdot 3}{(x+3)\cdot 2\cdot 2\cdot x}$$

$$=\frac{\cancel{2}\cdot\cancel{(x+3)}\cdot 3}{\cancel{(x+3)}\cdot\cancel{2}\cdot 2\cdot x}$$

$$=\frac{3}{2x}$$

23. $\frac{(x+1)^2}{x+2}\cdot\frac{x+2}{x+1}=\frac{(x+1)^2(x+2)}{(x+2)(x+1)}$

$$=\frac{(x+1)\,\cancel{(x+1)}\,\cancel{(x+2)}}{\cancel{(x+2)}\,\cancel{(x+1)}}$$

$$=x+1$$

25. $$\frac{x^2-x}{x}\cdot\frac{3x-6}{3-3x}=\frac{(x^2-x)(3x-6)}{x(3-3x)}$$
$$=\frac{x(x-1)\cdot 3(x-2)}{x\cdot 3(1-x)}$$
$$=\frac{x(x-1)\cdot 3(x-2)}{x\cdot 3(-x+1)}$$
$$=\frac{\cancel{x}\cdot\cancel{(x-1)}\cdot\cancel{3}\cdot(x-2)}{\cancel{x}\cdot\cancel{3}\cdot -1\cdot\cancel{(x-1)}}$$
$$=-(x-2)\text{ or }-x+2$$

27. $$\frac{x^2+x-6}{5x}\cdot\frac{5x-10}{x+3}=\frac{(x^2+x-6)(5x-10)}{5x(x+3)}$$
$$=\frac{(x+3)(x-2)\cdot 5(x-2)}{5x(x+3)}$$
$$=\frac{\cancel{(x+3)}(x-2)\cdot\cancel{5}\cdot(x-2)}{\cancel{5}\cdot x\cdot\cancel{(x+3)}}$$
$$=\frac{(x-2)^2}{x}$$

29. $$\frac{m^2-2m-3}{2m+4}\cdot\frac{m^2-4}{m^2+3m+2}$$
$$=\frac{(m^2-2m-3)(m^2-4)}{(2m+4)(m^2+3m+2)}$$
$$=\frac{(m-3)(m+1)(m+2)(m-2)}{2(m+2)(m+2)(m+1)}$$
$$=\frac{(m-3)\cancel{(m+1)}\cancel{(m+2)}(m-2)}{2(m+2)\cancel{(m+2)}\cancel{(m+1)}}$$
$$=\frac{(m-2)(m-3)}{2(m+2)}$$

Multiply, and then simplify, if possible. See Example 3.

31. $$7m\left(\frac{5}{m}\right)=\frac{7m}{1}\left(\frac{5}{m}\right)$$
$$=\frac{7m\cdot 5}{1\cdot m}$$
$$=\frac{7\cdot\cancel{m}\cdot 5}{1\cdot\cancel{m}}$$
$$=35$$

33. $$15x\left(\frac{x+1}{5x}\right)=\frac{15x}{1}\left(\frac{x+1}{5x}\right)$$
$$=\frac{15x(x+1)}{1\cdot 5x}$$
$$=\frac{3\cdot\cancel{5}\cdot\cancel{x}\cdot(x+1)}{1\cdot\cancel{5}\cdot\cancel{x}}$$
$$=3(x+1)\text{ or }3x+3$$

35. $$12y\left(\frac{5y-8}{6y}\right)=\frac{12y}{1}\left(\frac{5y-8}{6y}\right)$$
$$=\frac{12y(5y-8)}{1\cdot 6y}$$
$$=\frac{2\cdot\cancel{6}\cdot\cancel{y}\cdot(5y-8)}{1\cdot\cancel{6}\cdot\cancel{y}}$$
$$=2(5y-8)\text{ or }10y-16$$

37. $$24\left(\frac{3a-5}{2a}\right)=\frac{24}{1}\left(\frac{3a-5}{2a}\right)$$
$$=\frac{24(3a-5)}{1\cdot 2a}$$
$$=\frac{\cancel{2}\cdot 12\cdot(3a-5)}{1\cdot\cancel{2}\cdot a}$$
$$=\frac{12(3a-5)}{a}\text{ or }\frac{36a-60}{a}$$

Divide, and then simplify, if possible. See Example 4.

39. $$\frac{2}{y}\div\frac{4}{3}=\frac{2}{y}\cdot\frac{3}{4}$$
$$=\frac{2\cdot 3}{y\cdot 2\cdot 2}$$
$$=\frac{\cancel{2}\cdot 3}{y\cdot 2\cdot\cancel{2}}$$
$$=\frac{3}{2y}$$

41. $$\frac{3a}{25}\div\frac{1}{5}=\frac{3a}{25}\cdot\frac{5}{1}$$
$$=\frac{3\cdot a\cdot 5}{5\cdot 5\cdot 1}$$
$$=\frac{3\cdot a\cdot\cancel{5}}{5\cdot\cancel{5}\cdot 1}$$
$$=\frac{3a}{5}$$

42. $$\frac{3y}{8}\div\frac{3}{2}=\frac{y}{4}$$

43. $\dfrac{x^3}{18y} \div \dfrac{x}{6y} = \dfrac{x^3}{18y} \cdot \dfrac{6y}{x}$

$$= \frac{x \cdot x^2 \cdot 6 \cdot y}{3 \cdot 6 \cdot y \cdot x}$$

$$= \frac{\cancel{x} \cdot x^2 \cdot \cancel{6} \cdot \cancel{y}}{3 \cdot \cancel{6} \cdot \cancel{y} \cdot \cancel{x}}$$

$$= \frac{x^2}{3}$$

45. $\dfrac{27p^4}{35q} \div \dfrac{9p}{21q} = \dfrac{27p^4}{35q} \cdot \dfrac{21q}{9p}$

$$= \frac{3 \cdot 9 \cdot p \cdot p^3 \cdot 3 \cdot 7 \cdot q}{5 \cdot 7 \cdot q \cdot 9 \cdot p}$$

$$= \frac{3 \cdot \cancel{9} \cdot \cancel{p} \cdot p^3 \cdot 3 \cdot \cancel{7} \cdot \cancel{q}}{5 \cdot \cancel{7} \cdot \cancel{q} \cdot \cancel{9} \cdot \cancel{p}}$$

$$= \frac{9p^3}{5}$$

Divide, and then simplify, if possible. See Example 5.

47. $\dfrac{9a-18}{28} \div \dfrac{9a^3}{35} = \dfrac{(9a-18) \cdot 35}{28 \cdot 9a^3}$

$$= \frac{9 \cdot (a-2) \cdot 5 \cdot 7}{4 \cdot 7 \cdot 9 \cdot a^3}$$

$$= \frac{\cancel{9} \cdot (a-2) \cdot 5 \cdot \cancel{7}}{4 \cdot \cancel{7} \cdot \cancel{9} \cdot a^3}$$

$$= \frac{5(a-2)}{4a^3} \text{ or } \frac{5a-10}{4a^3}$$

49. $\dfrac{x^2-4}{3x+6} \div \dfrac{2-x}{x+2} = \dfrac{x^2-4}{3x+6} \cdot \dfrac{x+2}{2-x}$

$$= \frac{(x+2)(x-2)(x+2)}{3(x+2)(-x+2)}$$

$$= \frac{\cancel{(x+2)} \cdot \cancel{(x-2)} \cdot (x+2)}{3 \cdot \cancel{(x+2)} \cdot -1 \cdot \cancel{(x-2)}}$$

$$= -\frac{x+2}{3}$$

51. $\dfrac{x^2+7x}{5x-10} \div \dfrac{(x+7)^2}{15x-30}$

$$= \frac{x(x+7)}{5(x-2)} \cdot \frac{15(x-2)}{(x+7)(x+7)}$$

$$= \frac{x\cancel{(x+7)} \cdot \cancel{5} \cdot 3\cancel{(x-2)}}{\cancel{5}\cancel{(x-2)}\cancel{(x+7)}(x+7)}$$

$$= \frac{3x}{x+7}$$

53. $\dfrac{m^2+m-20}{m} \div \dfrac{4-m}{m} = \dfrac{m^2+m-20}{m} \cdot \dfrac{m}{4-m}$

$$= \frac{(m+5) \cdot (m-4) \cdot m}{m \cdot (-m+4)}$$

$$= \frac{(m+5) \cdot \cancel{(m-4)} \cdot \cancel{m}}{\cancel{m} \cdot -1 \cdot \cancel{(m-4)}}$$

$$= -(m+5) \text{ or } -m-5$$

55. $\dfrac{t^2+5t-14}{t} \div \dfrac{t-2}{t} = \dfrac{t^2+5t-14}{t} \cdot \dfrac{t}{t-2}$

$$= \frac{(t+7)(t-2)t}{t(t-2)}$$

$$= \frac{(t+7) \cdot \cancel{(t-2)} \cdot \cancel{t}}{\cancel{t} \cdot \cancel{(t-2)}}$$

$$= t+7$$

57. $\dfrac{x^2-2x-35}{3x^2+27x} \div \dfrac{3x^2+17x+10}{18x^2+12x}$

$$= \frac{x^2-2x-35}{3x^2+27x} \cdot \frac{18x^2+12x}{3x^2+17x+10}$$

$$= \frac{(x^2-2x-35)(18x^2+12x)}{(3x^2+27x)(3x^2+17x+10)}$$

$$= \frac{(x-7)(x+5) \cdot 6x(3x+2)}{3x(x+9)(3x+2)(x+5)}$$

$$= \frac{(x-7)\cancel{(x+5)} \cdot 2 \cdot \cancel{3} \cdot \cancel{x}\cancel{(3x+2)}}{\cancel{3} \cdot \cancel{x}(x+9)\cancel{(3x+2)}\cancel{(x+5)}}$$

$$= \frac{2(x-7)}{x+9}$$

Divide, and then simplify, if possible. See Example 6.

59. $$\frac{x^2-1}{3x-3} \div (x+1) = \frac{x^2-1}{3x-3} \div \frac{(x+1)}{1}$$
$$= \frac{x^2-1}{3x-3} \cdot \frac{1}{(x+1)}$$
$$= \frac{(x+1)(x-1) \cdot 1}{3(x-1)(x+1)}$$
$$= \frac{\overset{1}{\cancel{(x+1)}}\ \overset{1}{\cancel{(x-1)}} \cdot 1}{3\ \underset{1}{\cancel{(x-1)}}\ \underset{1}{\cancel{(x+1)}}}$$
$$= \frac{1}{3}$$

61. $$\frac{n^2-10n+9}{n-9} \div (n-1) = \frac{n^2-10n+9}{n-9} \div \frac{(n-1)}{1}$$
$$= \frac{n^2-10n+9}{n-9} \cdot \frac{1}{(n-1)}$$
$$= \frac{\overset{1}{\cancel{(n-9)}} \cdot \overset{1}{\cancel{(n-1)}} \cdot 1}{\underset{1}{\cancel{(n-9)}} \cdot \underset{1}{\cancel{(n-1)}}}$$
$$= 1$$

63. $$\frac{2r-3s}{12} \div (4r^2-12rs+9s^2)$$
$$= \frac{2r-3s}{12} \div \frac{(4r^2-12rs+9s^2)}{1}$$
$$= \frac{2r-3s}{12} \cdot \frac{1}{(4r^2-12rs+9s^2)}$$
$$= \frac{(2r-3s) \cdot 1}{12(4r^2-12rs+9s^2)}$$
$$= \frac{\overset{1}{\cancel{(2r-3s)}} \cdot 1}{12\ \cancel{(2r-3s)}\ (2r-3s)}$$
$$= \frac{1}{12(2r-3s)}$$

65. $$24n^2 \div \frac{18n^3}{n-1} = \frac{24n^2}{1} \div \frac{18n^3}{n-1}$$
$$= \frac{24n^2}{1} \cdot \frac{n-1}{18n^3}$$
$$= \frac{4 \cdot \overset{1}{\cancel{6}} \cdot \overset{1}{\cancel{n^2}}\ (n-1)}{1 \cdot 3 \cdot \underset{1}{\cancel{6}} \cdot n \cdot \underset{1}{\cancel{n^2}}}$$
$$= \frac{4(n-1)}{3n}$$

Complete each unit conversion. See Examples 7 and 8.

67. $$\frac{150 \text{ yd}}{1} \cdot \frac{3 \text{ ft}}{1 \text{ yd}} = \frac{150\ \overset{1}{\cancel{\text{yd}}}}{1} \cdot \frac{3 \text{ ft}}{1\ \underset{1}{\cancel{\text{yd}}}}$$
$$= 450 \text{ ft}$$

69. $$\frac{6 \text{ pints}}{1} \cdot \frac{1 \text{ gallon}}{8 \text{ pints}} = \frac{\overset{1}{\cancel{2}} \cdot 3\ \overset{1}{\cancel{\text{pints}}}}{1} \cdot \frac{1 \text{ gallon}}{\underset{1}{\cancel{2}} \cdot 4\ \underset{1}{\cancel{\text{pints}}}}$$
$$= \frac{3}{4} \text{ gallon}$$

71. $$\frac{30 \text{ miles}}{1 \text{ hr}} \cdot \frac{1 \text{ hr}}{60 \text{ min}} = \frac{\overset{1}{\cancel{30}} \text{ miles}}{1\ \underset{1}{\cancel{\text{hr}}}} \cdot \frac{1\ \overset{1}{\cancel{\text{hr}}}}{2 \cdot \underset{1}{\cancel{30}} \text{ min}}$$
$$= \frac{1}{2} \text{ mile per min}$$

73. $$\frac{30 \text{ meters}}{1 \text{ sec}} \cdot \frac{60 \text{ sec}}{1 \text{ min}} = \frac{30 \text{ meters}}{1\ \underset{1}{\cancel{\text{sec}}}} \cdot \frac{60\ \overset{1}{\cancel{\text{sec}}}}{1 \text{ min}}$$
$$= 1{,}800 \text{ meters per min}$$

TRY IT YOURSELF

Perform the operations and simplify, if possible.

75. $$\frac{b^2-5b+6}{b^2-10b+16} \div \frac{b^2+2b}{b^2-6b-16}$$
$$= \frac{b^2-5b+6}{b^2-10b+16} \cdot \frac{b^2-6b-16}{b^2+2b}$$
$$= \frac{(b^2-5b+6)(b^2-6b-16)}{(b^2-10b+16)(b^2+2b)}$$
$$= \frac{\overset{1}{\cancel{(b-2)}}(b-3)\ \overset{1}{\cancel{(b-8)}}\ \overset{1}{\cancel{(b+2)}}}{\underset{1}{\cancel{(b-8)}}\ \underset{1}{\cancel{(b-2)}}\ b\ \underset{1}{\cancel{(b+2)}}}$$
$$= \frac{b-3}{b}$$

77. $\frac{5x-5}{25}\cdot\frac{5}{(x-1)^3}=\frac{(5x-5)5}{25(x-1)^3}$

$=\frac{\cancel{5}\cdot\cancel{5}\,\cancel{(x-1)}}{\cancel{5}\cdot\cancel{5}\,\cancel{(x-1)}(x-1)^2}$

$=\frac{1}{(x-1)^2}$

79. $\frac{6a^2}{a^2+6a+9}\cdot\frac{(a+3)^4}{4a^5}=\frac{6a^2(a+3)^4}{(a^2+6a+9)4a^5}$

$=\frac{6a^2(a+3)^4}{4a^5(a+3)^2}$

$=\frac{\cancel{2}\cdot 3\cdot a^{2-5}(a+3)^{4-2}}{\cancel{2}\cdot 2}$

$=\frac{3\cdot a^{-3}(a+3)^2}{2}$

$=\frac{3(a+3)^2}{2a^3}$

81. $\frac{36c^2-49d^2}{3d^3}\div\frac{12c+14d}{d^4}$

$=\frac{36c^2-49d^2}{3d^3}\cdot\frac{d^4}{12c+14d}$

$=\frac{(6c+7d)(6c-7d)d\cdot d^3}{3\cdot d^3\cdot 2(6c+7d)}$

$=\frac{\cancel{(6c+7d)}(6c-7d)d\cdot\cancel{d^3}}{3\cdot\cancel{d^3}\cdot 2\,\cancel{(6c+7d)}}$

$=\frac{d(6c-7d)}{6}$

83. $10h\left(\frac{5h-3}{2h}\right)=\frac{10h}{1}\left(\frac{5h-3}{2h}\right)$

$=\frac{\cancel{2}\cdot 5\,\cancel{h}(5h-3)}{1\cdot\cancel{2}\,\cancel{h}}$

$=5(5h-3)$ or $25h-15$

85. $\frac{n^2-9}{n^2-3n}\div\frac{n+3}{n^2-n}=\frac{n^2-9}{n^2-3n}\cdot\frac{n^2-n}{n+3}$

$=\frac{(n^2-9)(n^2-n)}{(n^2-3n)(n+3)}$

$=\frac{\cancel{(n+3)}\,\cancel{(n-3)}\,\cancel{n}(n-1)}{\cancel{n}\,\cancel{(n-3)}\,\cancel{(n+3)}}$

$=n-1$

87. $\frac{10r^2s}{6rs^2}\cdot\frac{3r^3}{2rs}=\frac{10r^2s\cdot 3r^3}{6rs^2\cdot 2rs}$

$=\frac{\cancel{2}\cdot 5\,\cancel{r}\cdot\cancel{r}\cdot\cancel{s}\cdot\cancel{3}\,r^3}{\cancel{2}\cdot\cancel{3}\,\cancel{r}\cdot s^2\cdot 2\,\cancel{r}\,\cancel{s}}$

$=\frac{5r^3}{2s^2}$

89. $\frac{7}{3p^3}\cdot\frac{p+2}{p}=\frac{7(p+2)}{3p^{3+1}}$

$=\frac{7(p+2)}{3p^4}$ or $\frac{7p+14}{3p^4}$

91. $\frac{5x^2+13x-6}{x+3}\div\frac{5x^2-17x+6}{x-2}$

$=\frac{5x^2+13x-6}{x+3}\cdot\frac{x-2}{5x^2-17x+6}$

$=\frac{(5x^2+13x-6)(x-2)}{(x+3)(5x^2-17x+6)}$

$=\frac{\cancel{(5x-2)}\,\cancel{(x+3)}(x-2)}{\cancel{(x+3)}\,\cancel{(5x-2)}(x-3)}$

$=\frac{x-2}{x-3}$

93. $\frac{4x^2-12xy+9y^2}{x^3y^2}\cdot\frac{x^3y}{4x^2-9y^2}$

$=\frac{(4x^2-12xy+9y^2)x^3y}{x^3y^2(4x^2-9y^2)}$

$=\frac{\cancel{x^3}\,\cancel{y}(2x-3y)\,\cancel{(2x-3y)}}{\cancel{x^3}\,\cancel{y}\,y(2x+3y)\,\cancel{(2x-3y)}}$

$=\frac{2x-3y}{y(2x+3y)}$

95. $\frac{x-2}{x}\cdot\frac{2x}{2-x}=\frac{(x-2)2x}{x(2-x)}$

$=\frac{(x-2)2x}{x(-x+2)}$

$=\frac{\overset{1}{\cancel{(x-2)}}\cdot 2\cdot\overset{1}{\cancel{x}}}{\underset{1}{\cancel{x}}\cdot -1\underset{1}{\cancel{(x-2)}}}$

$=-2$

LOOK ALIKES . . .

97. a. $\frac{3x+6}{4}\cdot\frac{4x+8}{3}=\frac{3(x+2)}{4}\cdot\frac{4(x+2)}{3}$

$=\frac{\overset{1}{\cancel{3}}\cdot\overset{1}{\cancel{4}}(x+2)(x+2)}{\underset{1}{\cancel{3}}\cdot\underset{1}{\cancel{4}}}$

$=(x+2)^2$

b. $\frac{3x+6}{4}\div\frac{4x+8}{3}=\frac{3(x+2)}{4}\div\frac{4(x+2)}{3}$

$=\frac{3(x+2)}{4}\cdot\frac{3}{4(x+2)}$

$=\frac{3\cdot 3\overset{1}{\cancel{(x+2)}}}{4\cdot 4\underset{1}{\cancel{(x+2)}}}$

$=\frac{9}{16}$

99. a. $\frac{x^2-5x+6}{2x-4}\cdot\frac{2x-6}{x-2}$

$=\frac{(x-2)(x-3)}{2(x-2)}\cdot\frac{2(x-3)}{(x-2)}$

$=\frac{\overset{1}{\cancel{2}}\overset{1}{\cancel{(x-2)}}(x-3)(x-3)}{\underset{1}{\cancel{2}}\underset{1}{\cancel{(x-2)}}(x-2)}$

$=\frac{(x-3)^2}{x-2}$

b. $\frac{x^2-5x+6}{2x-4}\div\frac{2x-6}{x-2}$

$=\frac{(x-2)(x-3)}{2(x-2)}\div\frac{2(x-3)}{(x-2)}$

$=\frac{(x-2)(x-3)}{2(x-2)}\cdot\frac{(x-2)}{2(x-3)}$

$=\frac{\overset{1}{\cancel{(x-2)}}(x-2)\overset{1}{\cancel{(x-3)}}}{2\cdot 2\underset{1}{\cancel{(x-2)}}\underset{1}{\cancel{(x-3)}}}$

$=\frac{x-2}{4}$

APPLICATIONS

101. GEOMETRY

$A=lw$

$=\left(\frac{x^2-7x}{5}\right)\left(\frac{x}{2x-14}\right)$

$=\frac{(x^2-7x)x}{5(2x-14)}$

$=\frac{x\overset{1}{\cancel{(x-7)}}x}{5\cdot 2\underset{1}{\cancel{(x-7)}}}$

$=\frac{x^2}{10}$

The area of the rectangle is $\frac{x^2}{10}$ ft^2.

103. TALKING

$\frac{12{,}000 \text{ words}}{1 \text{ day}}\cdot\frac{365 \text{ days}}{1 \text{ year}}$

$=\frac{12{,}000 \text{ words}}{1\ \underset{1}{\cancel{\text{day}}}}\cdot\frac{365\ \overset{1}{\cancel{\text{days}}}}{1 \text{ year}}$

$=4{,}380{,}000$ words per year

The average speaks 4,380,000 words per year.

105. NATURAL LIGHT

$1 \text{ yd}^2=3 \text{ ft}\cdot 3 \text{ ft}=9 \text{ ft}^2$

$\frac{72 \text{ ft}^2}{1}\cdot\frac{1 \text{ yd}^2}{9 \text{ ft}^2}=\frac{\overset{1}{\cancel{9}}\cdot 8\ \overset{1}{\cancel{\text{ft}^2}}}{1}\cdot\frac{1 \text{ yd}^2}{\underset{1}{\cancel{9}}\ \underset{1}{\cancel{\text{ft}^2}}}$

$=8 \text{ yd}^2$

The average classroom has 8 yd ^{2} of windows.

107. BEARS

$$\frac{30\text{ miles}}{1\text{ hr}}\cdot\frac{1\text{ hr}}{60\text{ min}}=\frac{\cancel{30}\text{ miles}}{1\cancel{\text{hr}}}\cdot\frac{1\cancel{\text{hr}}}{2\cdot\cancel{30}\text{ min}}$$

$$=\frac{1}{2}\text{ mile per min}$$

A bear can run $\frac{1}{2}$ mile per minute.

109. TV TRIVIA

$$\frac{160\text{ acres}}{1}\cdot\frac{1\text{ mi}^2}{640\text{ acres}}$$

$$=\frac{\cancel{160}\ \cancel{\text{acres}}}{1}\cdot\frac{1\text{ mi}^2}{4\cdot\cancel{160}\ \cancel{\text{acres}}}$$

$$=\frac{1}{4}\text{ mi}^2$$

The farm was $\frac{1}{4}$ mi^2.

WRITING

111- 113. Answers will vary.

REVIEW

115. HARDWARE

Analyze

- A leg of the triangle is 8 inches long. (longer leg)
- The width of the shelf is the other leg. (shorter leg)
- The brace is 2 inches less than twice the width of the shelf. (hypotenuse)
- Find the width of the shelf and the length of the brace.

Assign

Let x = width of the shelf in inches

$2x-2$ = length of the brace in inches

Form

$$\left(\begin{array}{c}\text{The length of}\\\text{the shorter leg}\end{array}\right)^2 \text{ plus } \left(\begin{array}{c}\text{the length of}\\\text{the longer leg}\end{array}\right)^2 \text{ equals } \left(\begin{array}{c}\text{the length of}\\\text{the hypotenuse.}\end{array}\right)^2$$

$$x^2 \quad + \quad 8^2 \quad = \quad (2x-2)^2$$

Solve

$$x^2+8^2=(2x-2)^2$$
$$x^2+64=4x^2-8x+4$$
$$x^2+64-\mathbf{x^2}=4x^2-8x+4-\mathbf{x^2}$$
$$64=3x^2-8x+4$$
$$64-\mathbf{64}=3x^2-8x+4-\mathbf{64}$$
$$0=3x^2-8x-60$$
$$0=(3x+10)(x-6)$$

$3x+10=0$ or $x-6=0$

$\cancel{x=-\frac{10}{3}}$ $\qquad x=6$

State

The width of the shelf is 6 inches.

The length of the brace is $2(6)-2=10$ inches.

Check

Since $6^2+8^2=10^2$, the answers check.

CHALLENGE PROBLEMS

Perform the operations and simplify, if possible.

117. $\frac{c^3-2c^2+5c-10}{c^2-c-2}\cdot\frac{c^3+c^2-5c-5}{c^4-25}$

$$=\frac{c^2(c-2)+5(c-2)}{(c-2)(c+1)}\cdot\frac{c^2(c+1)-5(c+1)}{(c^2-5)(c^2+5)}$$

$$=\frac{\cancel{(c-2)}\ \cancel{(c^2+5)}\ \cancel{(c+1)}\ \cancel{(c^2-5)}}{\cancel{(c-2)}\ \cancel{(c+1)}\ \cancel{(c^2-5)}\ \cancel{(c^2+5)}}$$

$$=1$$

119. $\frac{-x^3+x^2+6x}{3x^3+21x^2}\div\left(\frac{2x+4}{3x^2}\div\frac{2x+14}{x^2-3x}\right)$

$$=\frac{-x^3+x^2+6x}{3x^3+21x^2}\div\left(\frac{2x+4}{3x^2}\cdot\frac{x^2-3x}{2x+14}\right)$$

$$=\frac{-x(x^2-x-6)}{3x^2(x+7)}\div\left(\frac{2(x+2)x(x-3)}{3x^2\cdot 2(x+7)}\right)$$

$$=\frac{-x(x-3)(x+2)}{3x^2(x+7)}\cdot\left(\frac{2\cdot 3x^2(x+7)}{2x(x+2)(x-3)}\right)$$

$$=-\frac{\cancel{2}\cdot\cancel{3}\cdot\cancel{x}\cdot\cancel{x^2}\ \cancel{(x-3)}\ \cancel{(x+2)}\ \cancel{(x+7)}}{\cancel{2}\cdot\cancel{3}\ \cancel{x}\cdot\cancel{x^2}\ \cancel{(x+7)}\ \cancel{(x+2)}\ \cancel{(x-3)}}$$

$$=-1$$

SECTION 7.3 STUDY SET

VOCABULARY

Fill in the blanks.

1. The rational expressions $\frac{7}{6n}$ and $\frac{n+1}{6n}$ have the common **denominator** $6n$.

3. To **build** a rational expression, we multiply it by a form of 1. For example, $\frac{2}{n^2} \cdot \frac{8n}{8n} = \frac{16n}{8n^3}$.

CONCEPTS

Fill in the blanks.

5. To add or subtract rational expressions that have the same denominator, add or subtract the **numerators**, and write the sum or difference over the common **denominator**. In symbols,

$$\frac{A}{D} + \frac{B}{D} = \frac{A+B}{D} \quad \text{and} \quad \frac{A}{D} - \frac{B}{D} = \frac{A-B}{D}$$

7. The sum of two rational expressions is $\frac{4x+4}{5(x+1)}$. Factor the numerator and then simplify the result. $\frac{4}{5}$

9. Consider the following factorizations.
$$18x - 36 = 2 \cdot 3 \cdot 3 \cdot (x - 2)$$
$$3x - 6 = 3(x - 2)$$
   a. What is the greatest number of times the factor 3 appears in any one factorization? **Twice**
   b. What is the greatest number of times the factor $x - 2$ appears in any one factorization? **Once**

NOTATION

Complete the solution.

11. $$\frac{6a-1}{4a+1} + \frac{2a+3}{4a+1} = \frac{6a-1+(2a+\boxed{3})}{\boxed{4a+1}} = \frac{8a+\boxed{2}}{4a+1} = \frac{2\boxed{4a+1}}{4a+1} = \boxed{2}$$

GUIDED PRACTICE

Add and simplify the result, if possible. See Example 1.

13. $$\frac{9}{x} + \frac{2}{x} = \frac{9+2}{x} = \frac{11}{x}$$

15. $$\frac{x}{18} + \frac{5}{18} = \frac{x+5}{18}$$

17. $$\frac{a-5}{3a^3} + \frac{5}{3a^3} = \frac{a-5+5}{3a^3} = \frac{a}{3a^3} = \frac{\overset{1}{\cancel{a}}}{3\underset{1}{\cancel{a}} \cdot a^2} = \frac{1}{3a^2}$$

19. $$\frac{x+3}{2y} + \frac{x+5}{2y} = \frac{x+3+x+5}{2y} = \frac{2x+8}{2y} = \frac{\overset{1}{\cancel{2}}(x+4)}{\underset{1}{\cancel{2}} \cdot y} = \frac{x+4}{y}$$

Add and simplify the result, if possible. See Example 2.

21. $$\frac{2}{r^2-3r-10} + \frac{r}{r^2-3r-10} = \frac{r+2}{r^2-3r-10} = \frac{\overset{1}{\cancel{r+2}}}{\underset{1}{\cancel{(r+2)}}(r-5)} = \frac{1}{r-5}$$

23. $$\frac{3x-5}{x-2} + \frac{6x-13}{x-2} = \frac{3x-5+6x-13}{x-2} = \frac{9x-18}{x-2} = \frac{9\overset{1}{\cancel{(x-2)}}}{\underset{1}{\cancel{(x-2)}}} = 9$$

Subtract and simplify the result, if possible. See Example 3.

25. $\dfrac{2x}{25}-\dfrac{x}{25}=\dfrac{2x-x}{25}$
$=\dfrac{x}{25}$

27. $\dfrac{m-1}{6m^2}-\dfrac{5}{6m^2}=\dfrac{m-1-5}{6m^2}$
$=\dfrac{m-6}{6m^2}$

29. $\dfrac{t}{t^2+t-2}-\dfrac{1}{t^2+t-2}=\dfrac{t-1}{t^2+t-2}$
$=\dfrac{\overset{1}{\cancel{t-1}}}{\underset{1}{\cancel{(t-1)}}(t+2)}$
$=\dfrac{1}{t+2}$

31. $\dfrac{11w+6}{3w(w-9)}-\dfrac{11w}{3w(w-9)}=\dfrac{11w+6-11w}{3w(w-9)}$
$=\dfrac{2\cdot\overset{1}{\cancel{3}}}{\underset{1}{\cancel{3}}\cdot w(w-9)}$
$=\dfrac{2}{w(w-9)}$

Subtract and simplify the result, if possible. See Example 4.

33. $\dfrac{3y-2}{2y+6}-\dfrac{2y-5}{2y+6}=\dfrac{3y-2-(2y-5)}{2y+6}$
$=\dfrac{3y-2-2y+5}{2y+6}$
$=\dfrac{y+3}{2y+6}$
$=\dfrac{\overset{1}{\cancel{y+3}}}{2\underset{1}{\cancel{(y+3)}}}$
$=\dfrac{1}{2}$

35. $\dfrac{6x^2}{3x+2}-\dfrac{11x+10}{3x+2}=\dfrac{6x^2-(11x+10)}{3x+2}$
$=\dfrac{6x^2-11x-10}{3x+2}$
$=\dfrac{(2x-5)\overset{1}{\cancel{(3x+2)}}}{\underset{1}{\cancel{3x+2}}}$
$=2x-5$

37. $\dfrac{6x-5}{3xy}-\dfrac{3x-5}{3xy}=\dfrac{6x-5-(3x-5)}{3xy}$
$=\dfrac{6x-5-3x+5}{3xy}$
$=\dfrac{3x}{3xy}$
$=\dfrac{\overset{1}{\cancel{3}}\overset{1}{\cancel{x}}}{\underset{1}{\cancel{3}}\cdot\underset{1}{\cancel{x}}\cdot y}$
$=\dfrac{1}{y}$

39. $\dfrac{2-p}{p^2-p}-\dfrac{-p+2}{p^2-p}=\dfrac{2-p-(-p+2)}{p^2-p}$
$=\dfrac{2-p+p-2}{p^2-p}$
$=\dfrac{0}{p^2-p}$
$=0$

Find the LCD of each pair of rational expressions. See Example 5.

41. $\dfrac{1}{2x},\dfrac{9}{6x}$
$2x=2\cdot x$
$6x=2\cdot 3\cdot x$
$\text{LCD}=2\cdot 3\cdot x$
$=6x$

43. $\dfrac{33}{15a^3},\dfrac{9}{10a}$
$15a^3=3\cdot 5\cdot a\cdot a\cdot a$
$10a=2\cdot 5\cdot a$
$\text{LCD}=2\cdot 3\cdot 5\cdot a\cdot a\cdot a$
$=30a^3$

45. $\dfrac{35}{3a^2b},\dfrac{23}{a^2b^3}$
$3a^2b=3\cdot a\cdot a\cdot b$
$a^2b^3=a\cdot a\cdot b\cdot b\cdot b$
$\text{LCD}=3\cdot a\cdot a\cdot b\cdot b\cdot b$
$=3a^2b^3$

47. $\dfrac{8}{c}, \dfrac{8-c}{c+2}$

$c = c$

$c+2 = c+2$

$\text{LCD} = c(c+2)$

Find the LCD of each pair of rational expressions. See Example 6.

49. $\dfrac{3x+1}{3x-3}, \dfrac{3x}{4x-4}$

$3x-3 = 3(x-1)$

$4x-4 = 4(x-1)$

$\text{LCD} = 3\cdot 4\cdot(x-1)$

$= 12(x-1)$

51. $\dfrac{b-9}{4b+8}, \dfrac{b}{6}$

$4b+8 = 2\cdot 2\cdot(b+2)$

$6 = 2\cdot 3$

$\text{LCD} = 2\cdot 2\cdot 3\cdot(b+2)$

$= 12(b+2)$

53. $\dfrac{6-k}{2k+4}, \dfrac{11}{8k}$

$2k+4 = 2\cdot(k+2)$

$8k = 2\cdot 2\cdot 2\cdot k$

$\text{LCD} = 2\cdot 2\cdot 2\cdot k(k+2)$

$= 8k(k+2)$

55. $\dfrac{-2x}{x^2-1}, \dfrac{5x}{x+1}$

$x^2-1 = (x+1)(x-1)$

$x+1 = x+1$

$\text{LCD} = (x+1)(x-1)$

57. $\dfrac{4x-5}{x^2-4x-5}, \dfrac{3x+1}{x^2-25}$

$x^2-4x-5 = (x+1)(x-5)$

$x^2-25 = (x+5)(x-5)$

$\text{LCD} = (x+1)(x+5)(x-5)$

58.

59. $\dfrac{5n^2-16}{2n^2+13n+20}, \dfrac{3n^2}{n^2+8n+16}$

$2n^2+13n+20 = (2n+5)(n+4)$

$n^2+8n+16 = (n+4)(n+4)$

$\text{LCD} = (2n+5)(n+4)(n+4)$

$= (2n+5)(n+4)^2$

Build each rational expression into an equivalent expression with the given denominator. See Example 7.

61. $\dfrac{5}{r} = \dfrac{5}{r}\cdot\dfrac{\mathbf{10}}{\mathbf{10}}$

$= \dfrac{50}{10r}$

63. $\dfrac{8}{x} = \dfrac{8}{x}\cdot\dfrac{\boldsymbol{xy}}{\boldsymbol{xy}}$

$= \dfrac{8xy}{x^2y}$

65. $\dfrac{9}{4b} = \dfrac{9}{4b}\cdot\dfrac{\mathbf{3}\boldsymbol{b}}{\mathbf{3}\boldsymbol{b}}$

$= \dfrac{27b}{12b^2}$

67. $\dfrac{3x}{x+1} = \dfrac{3x}{(x+1)}\cdot\dfrac{\mathbf{(x+1)}}{\mathbf{(x+1)}}$

$= \dfrac{3x^2+3x}{(x+1)^2}$

Build each rational expression into an equivalent expression with the given denominator. See Example 8.

69. $\dfrac{x+9}{x^2+5x} = \dfrac{x+9}{x(x+5)}$

$= \dfrac{x+9}{x(x+5)}\cdot\dfrac{\boldsymbol{x}}{\boldsymbol{x}}$

$= \dfrac{x^2+9x}{x^2(x+5)}$

71. $\dfrac{t+5}{4t+8} = \dfrac{t+5}{4(t+2)}$

$= \dfrac{t+5}{4(t+2)}\cdot\dfrac{\mathbf{(t+9)}}{\mathbf{(t+9)}}$

$= \dfrac{t^2+14t+45}{4(t+2)(t+9)}$

73. $\dfrac{y+3}{y^2-5y+6} = \dfrac{y+3}{(y-2)(y-3)}$

$= \dfrac{y+3}{(y-2)(y-3)}\cdot\dfrac{\mathbf{4}\boldsymbol{y}}{\mathbf{4}\boldsymbol{y}}$

$= \dfrac{4y^2+12y}{4y(y-2)(y-3)}$

75. $\dfrac{12-h}{h^2-81}=\dfrac{12-h}{(h+9)(h-9)}$

$=\dfrac{12-h}{(h+9)(h-9)}\cdot\dfrac{\mathbf{3}}{\mathbf{3}}$

$=\dfrac{36-3h}{3(h+9)(h-9)}$

TRY IT YOURSELF

Perform the operations. Then simplify, if possible.

77. $\dfrac{3t}{t^2-8t+7}-\dfrac{3}{t^2-8t+7}=\dfrac{3t-3}{t^2-8t+7}$

$=\dfrac{3\cancel{(t-1)}}{(t-7)\cancel{(t-1)}}$

$=\dfrac{3}{t-7}$

79. $\dfrac{c}{c^2-d^2}-\dfrac{d}{c^2-d^2}=\dfrac{c-d}{c^2-d^2}$

$=\dfrac{\cancel{c-d}}{(c+d)\cancel{(c-d)}}$

$=\dfrac{1}{c+d}$

81. $\dfrac{a^2+a}{4a^2-8a}+\dfrac{2a^2-7a}{4a^2-8a}=\dfrac{a^2+a+2a^2-7a}{4a^2-8a}$

$=\dfrac{3a^2-6a}{4a^2-8a}$

$=\dfrac{3\cancel{a}\cancel{(a-2)}}{4\cancel{a}\cancel{(a-2)}}$

$=\dfrac{3}{4}$

83. $\dfrac{17a}{2a+4}-\dfrac{7a}{2a+4}=\dfrac{17a-7a}{2a+4}$

$=\dfrac{10a}{2a+4}$

$=\dfrac{\cancel{2}\cdot 5a}{\cancel{2}(a+2)}$

$=\dfrac{5a}{a+2}$

85. $\dfrac{8}{9-3x^2}-\dfrac{-6x+8}{9-3x^2}=\dfrac{8-(-6x+8)}{9-3x^2}$

$=\dfrac{8+6x-8}{9-3x^2}$

$=\dfrac{6x}{9-3x^2}$

$=\dfrac{2\cdot\cancel{3}\cdot x}{\cancel{3}(3-x^2)}$

$=\dfrac{2x}{3-x^2}$

87. $\dfrac{11n}{(n+4)(n-2)}-\dfrac{4n-1}{(n-2)(n+4)}$

$=\dfrac{11n-(4n-1)}{(n+4)(n-2)}$

$=\dfrac{11n-4n+1}{(n+4)(n-2)}$

$=\dfrac{7n+1}{(n+4)(n-2)}$

89. $\dfrac{5r-27}{3r^2-9r}+\dfrac{4r}{3r^2-9r}=\dfrac{5r-27+4r}{3r^2-9r}$

$=\dfrac{9r-27}{3r^2-9r}$

$=\dfrac{9(r-3)}{3r(r-3)}$

$=\dfrac{3\cdot\cancel{3}\cancel{(r-3)}}{\cancel{3}\cdot r\cancel{(r-3)}}$

$=\dfrac{3}{r}$

91. $\dfrac{11}{36y}+\dfrac{9}{36y}=\dfrac{11+9}{36y}$

$=\dfrac{20}{36y}$

$=\dfrac{5\cdot \cancel{4}}{9\cdot \cancel{4}y}$

$=\dfrac{5}{9y}$

93. $\dfrac{-4x}{3x^2-7x+2}-\dfrac{-3x-2}{3x^2-7x+2}=\dfrac{-4x-(-3x-2)}{3x^2-7x+2}$

$=\dfrac{-4x+3x+2}{3x^2-7x+2}$

$=\dfrac{-x+2}{3x^2-7x+2}$

$=\dfrac{-\cancel{(x-2)}}{(3x-1)\cancel{(x-2)}}$

$=-\dfrac{1}{3x-1}$

95. $\dfrac{3x^2}{x+1}-\dfrac{-x+2}{x+1}=\dfrac{3x^2-(-x+2)}{x+1}$

$=\dfrac{3x^2+x-2}{x+1}$

$=\dfrac{(3x-2)(x+1)}{x+1}$

$=\dfrac{(3x-2)\cancel{(x+1)}}{\cancel{x+1}}$

$=3x-2$

LOOK ALIKES . . .

97. a. $\dfrac{t}{12}+\dfrac{5t}{12}=\dfrac{t+5t}{12}$

$=\dfrac{\cancel{6}t}{\cancel{6}\cdot 2}$

$=\dfrac{t}{2}$

b. $\dfrac{t}{12}\cdot\dfrac{5t}{12}=\dfrac{5t^{1+1}}{12\cdot 12}$

$=\dfrac{5t^2}{144}$

c. $\dfrac{t}{12}\div\dfrac{5t}{12}=\dfrac{\cancel{t}}{\cancel{12}}\cdot\dfrac{\cancel{12}}{5\cancel{t}}$

$=\dfrac{1}{5}$

99. a. $\dfrac{m+6}{5}-\dfrac{m+2}{5}=\dfrac{m+6-(m+2)}{5}$

$=\dfrac{m+6-m-2}{5}$

$=\dfrac{4}{5}$

b. $\dfrac{m+6}{5}\cdot\dfrac{m+2}{5}=\dfrac{(m+6)(m+2)}{5\cdot 5}$

$=\dfrac{m^2+8m+12}{25}$

c. $\dfrac{m+6}{5}\div\dfrac{m+2}{5}=\dfrac{m+6}{\cancel{5}}\cdot\dfrac{\cancel{5}}{m+2}$

$=\dfrac{m+6}{m+2}$

APPLICATIONS

101. GEOMETRY

$\dfrac{5x+11}{x+2}-\dfrac{3x+5}{x+2}=\dfrac{5x+11-(3x+5)}{x+2}$

$=\dfrac{5x+11-3x-5}{x+2}$

$=\dfrac{2x+6}{x+2}$

The difference between the length and the width is $\dfrac{2x+6}{x+2}$ feet.

WRITING

103-107. Answers will vary.

REVIEW

Give the formula for . . .

109. a. simple interest

$I=Prt$

b. the area of a triangle

$A=\dfrac{1}{2}bh$

c. the perimeter of a rectangle

$P=2l+2w$

CHALLENGE PROBLEMS

Perform the operations. Simplify the results, if possible.

111. $\dfrac{3xy}{x-y}-\dfrac{x(3y-x)}{x-y}-\dfrac{x(x-y)}{x-y}$

$$=\frac{3xy-x(3y-x)-x(x-y)}{x-y}$$

$$=\frac{3xy-3xy+x^2-x^2+xy}{x-y}$$

$$=\frac{xy}{x-y}$$

113. $\dfrac{2a^2+2}{a^3+8}+\dfrac{a^3+a}{a^3+8}=\dfrac{2a^2+2+a^3+a}{a^3+8}$

$$=\frac{a^3+2a^2+a+2}{a^3+8}$$

$$=\frac{(a^3+2a^2)+(a+2)}{a^3+8}$$

$$=\frac{a^2(a+2)+1(a+2)}{(a+2)(a^2-2a+4)}$$

$$=\frac{\overset{1}{\cancel{(a+2)}}(a^2+1)}{\underset{1}{\cancel{(a+2)}}(a^2-2a+4)}$$

$$=\frac{a^2+1}{a^2-2a+4}$$

SECTION 7.4 STUDY SET
VOCABULARY

Fill in the blanks.

1. $\frac{x}{x-7}$ and $\frac{1}{x-7}$ have like denominators. $\frac{x+5}{x-7}$ and $\frac{4x}{x+7}$ have **unlike** denominators.

CONCEPTS

3. a. $\frac{x+1}{20x^2}$, $20x^2 = \mathbf{2 \cdot 2 \cdot 5 \cdot x \cdot x}$

b $\frac{3x^2-4}{x^2+4x-12}$,
$x^2+4x-12 = \mathbf{(x+6)(x-2)}$

5. $\mathbf{(x+6)(x+3)}$

Fill in the blanks.

7. To build $\frac{x}{x+2}$ so that it has a denominator of $5(x+2)$, we multiply it by 1 in the form of $\mathbf{\frac{5}{5}}$.

NOTATION

Complete the solution.

9. $$\frac{2}{5}+\frac{7}{3x} = \frac{2}{5}\cdot\frac{\mathbf{3x}}{3x}+\frac{7}{3x}\cdot\frac{\mathbf{5}}{5} = \frac{6x}{\mathbf{15x}}+\frac{35}{\mathbf{15x}} = \frac{6x+\mathbf{35}}{15x}$$

GUIDED PRACTICE

Perform the operations. Simplify, if possible. See Example 1.

11. $\left.\begin{matrix} 3=3 \\ 7=7 \end{matrix}\right\}\text{LCD} = 3\cdot 7 = 21$

$$\frac{x}{3}+\frac{2x}{7} = \frac{x}{3}\cdot\frac{\mathbf{7}}{\mathbf{7}}+\frac{2x}{7}\cdot\frac{\mathbf{3}}{\mathbf{3}} = \frac{7x}{21}+\frac{6x}{21} = \frac{7x+6x}{21} = \frac{13x}{21}$$

13. $\left.\begin{matrix} 8=2\cdot 2\cdot 2 \\ 5=5 \end{matrix}\right\}\text{LCD} = 2\cdot 2\cdot 2\cdot 5 = 40$

$$\frac{7a}{8}+\frac{4a}{5} = \frac{7a}{8}\cdot\frac{\mathbf{5}}{\mathbf{5}}+\frac{4a}{5}\cdot\frac{\mathbf{8}}{\mathbf{8}} = \frac{35a}{40}+\frac{32a}{40} = \frac{35a+32a}{40} = \frac{67a}{40}$$

Perform the operations. Simplify, if possible. See Example 2.

15. $\left.\begin{matrix} m^2=m\cdot m \\ m=m \end{matrix}\right\}\text{LCD} = m\cdot m = m^2$

$$\frac{7}{m^2}-\frac{2}{m} = \frac{7}{m^2}-\frac{2}{m}\cdot\frac{\mathbf{m}}{\mathbf{m}} = \frac{7}{m^2}-\frac{2m}{m^2} = \frac{7-2m}{m^2}$$

17. $\left.\begin{matrix} 5p^2=5\cdot p\cdot p \\ 10p=2\cdot 5\cdot p \end{matrix}\right\}\text{LCD} = 2\cdot 5\cdot p\cdot p = 10p^2$

$$\frac{3}{5p^2}-\frac{5}{10p} = \frac{3}{5p^2}\cdot\frac{\mathbf{2}}{\mathbf{2}}-\frac{5}{10p}\cdot\frac{\mathbf{p}}{\mathbf{p}} = \frac{6}{10p^2}-\frac{5p}{10p^2} = \frac{6-5p}{10p^2}$$

19. $\left.\begin{matrix} 6t=2\cdot 3\cdot t \\ 8t^3=2\cdot 2\cdot 2\cdot t\cdot t\cdot t \end{matrix}\right\}\text{LCD} = 2\cdot 2\cdot 2\cdot 3\cdot t\cdot t\cdot t = 24t^3$

$$\frac{1}{6t}-\frac{11}{8t^3} = \frac{1}{6t}\cdot\frac{\mathbf{4t^2}}{\mathbf{4t^2}}-\frac{11}{8t^3}\cdot\frac{\mathbf{3}}{\mathbf{3}} = \frac{4t^2}{24t^3}-\frac{33}{24t^3} = \frac{4t^2-33}{24t^3}$$

21. $\left.\begin{array}{l} 6c^4 = 2\cdot 3\cdot c\cdot c\cdot c\cdot c \\ 9c^2 = 3\cdot 3\cdot c\cdot c \end{array}\right\}$

$$\text{LCD} = 2\cdot 3\cdot 3\cdot c\cdot c\cdot c\cdot c = 18c^4$$

$$\frac{1}{6c^4} - \frac{8}{9c^2} = \frac{1}{6c^4}\cdot\frac{\mathbf{3}}{\mathbf{3}} - \frac{8}{9c^2}\cdot\frac{\mathbf{2c^2}}{\mathbf{2c^2}} = \frac{3}{18a^4} - \frac{16c^2}{18a^4} = \frac{3-16c^2}{18a^4}$$

Perform the operations. Simplify, if possible. See Example 3.

23. $\left.\begin{array}{l} 2a+4 = 2(a+2) \\ a^2-4 = (a+2)(a-2) \end{array}\right\}$

$$\text{LCD} = 2(a+2)(a-2)$$

$$\frac{1}{2a+4} + \frac{5}{a^2-4} = \frac{1}{2(a+2)} + \frac{5}{(a+2)(a-2)} = \frac{1}{2(a+2)}\cdot\frac{\mathbf{(a-2)}}{\mathbf{(a-2)}} + \frac{5}{(a+2)(a-2)}\cdot\frac{\mathbf{2}}{\mathbf{2}} = \frac{a-2}{2(a+2)(a-2)} + \frac{10}{2(a+2)(a-2)} = \frac{a-2+10}{2(a+2)(a-2)} = \frac{a+8}{2(a+2)(a-2)}$$

25. $\left.\begin{array}{l} 3a-2 = 3a-2 \\ 9a^2-4 = (3a+2)(3a-2) \end{array}\right\}$

$$\text{LCD} = (3a+2)(3a-2)$$

$$\frac{2}{3a-2} + \frac{5}{9a^2-4} = \frac{2}{3a-2} + \frac{5}{(3a+2)(3a-2)} = \frac{2}{(3a-2)}\cdot\frac{\mathbf{(3a+2)}}{\mathbf{(3a+2)}} + \frac{5}{(3a+2)(3a-2)} = \frac{2(3a+2)+5}{(3a+2)(3a-2)} = \frac{6a+4+5}{(3a+2)(3a-2)} = \frac{6a+9}{(3a+2)(3a-2)}$$

27. $\left.\begin{array}{l} a+2 = a+2 \\ a^2+4a+4 = (a+2)(a+2) \end{array}\right\}$

$$\text{LCD} = (a+2)(a+2)$$

$$\frac{4}{a+2} - \frac{7}{a^2+4a+4} = \frac{4}{a+2} - \frac{7}{(a+2)(a+2)} = \frac{4}{(a+2)}\cdot\frac{\mathbf{(a+2)}}{\mathbf{(a+2)}} - \frac{7}{(a+2)(a+2)} = \frac{4(a+2)-7}{(a+2)(a+2)} = \frac{4a+8-7}{(a+2)(a+2)} = \frac{4a+1}{(a+2)^2}$$

29. $\left.\begin{array}{l} 5m-5 = 5(m-1) \\ 5m^2-5m = 5m(m-1) \end{array}\right\}\text{LCD} = 5m(m-1)$

$$\frac{6}{5m^2-5m} - \frac{3}{5m-5} = \frac{6}{5m(m-1)} - \frac{3}{5(m-1)} = \frac{6}{5m(m-1)} - \frac{3}{5(m-1)}\cdot\frac{\boldsymbol{m}}{\boldsymbol{m}} = \frac{6-3m}{5m(m-1)}$$

Perform the operations. Simplify, if possible. See Example 4.

31. $\left.\begin{array}{l} t+3 = t+3 \\ t+2 = t+2 \end{array}\right\}\text{LCD} = (t+3)(t+2)$

$$\frac{9}{t+3} + \frac{8}{t+2} = \frac{9}{(t+3)}\cdot\frac{\mathbf{(t+2)}}{\mathbf{(t+2)}} + \frac{8}{(t+2)}\cdot\frac{\mathbf{(t+3)}}{\mathbf{(t+3)}} = \frac{9t+18+8t+24}{(t+3)(t+2)} = \frac{17t+42}{(t+3)(t+2)}$$

33. $\left.\begin{array}{l}2x-1=2x-1\\2x+3=2x+3\end{array}\right\}\text{LCD}=(2x-1)(2x+3)$

$$\frac{3x}{2x-1}-\frac{2x}{2x+3}$$
$$=\frac{3x}{(2x-1)}\cdot\frac{\mathbf{(2x+3)}}{\mathbf{(2x+3)}}-\frac{2x}{(2x+3)}\cdot\frac{\mathbf{(2x-1)}}{\mathbf{(2x-1)}}$$
$$=\frac{6x^2+9x-(4x^2-2x)}{(2x-1)(2x+3)}$$
$$=\frac{6x^2+9x-4x^2+2x}{(2x-1)(2x+3)}$$
$$=\frac{2x^2+11x}{(2x-1)(2x+3)}$$

35. $\left.\begin{array}{l}s+3=s+3\\s+7=s+7\end{array}\right\}\text{LCD}=(s+3)(s+7)$

$$\frac{s+7}{s+3}-\frac{s-3}{s+7}$$
$$=\frac{(s+7)}{(s+3)}\cdot\frac{\mathbf{(s+7)}}{\mathbf{(s+7)}}-\frac{(s-3)}{(s+7)}\cdot\frac{\mathbf{(s+3)}}{\mathbf{(s+3)}}$$
$$=\frac{(s+7)(s+7)-(s-3)(s+3)}{(s+3)(s+7)}$$
$$=\frac{s^2+14s+49-(s^2-9)}{(s+3)(s+7)}$$
$$=\frac{s^2+14s+49-s^2+9}{(s+3)(s+7)}$$
$$=\frac{14s+58}{(s+3)(s+7)}$$

37. $\left.\begin{array}{l}m-2=m-2\\m+5=m+5\end{array}\right\}\text{LCD}=(m-2)(m+5)$

$$\frac{3m}{m-2}-\frac{m-3}{m+5}$$
$$=\frac{3m}{(m-2)}\cdot\frac{\mathbf{(m+5)}}{\mathbf{(m+5)}}-\frac{(m-3)}{(m+5)}\cdot\frac{\mathbf{(m-2)}}{\mathbf{(m-2)}}$$
$$=\frac{3m(m+5)-(m-3)(m-2)}{(m-2)(m+5)}$$
$$=\frac{3m^2+15m-(m^2-5m+6)}{(m-2)(m+5)}$$
$$=\frac{3m^2+15m-m^2+5m-6}{(m-2)(m+5)}$$
$$=\frac{2m^2+20m-6}{(m-2)(m+5)}$$

Perform the operations. Simplify, if possible. See Example 5.

39. $\left.\begin{array}{l}s^2+5s+4=(s+1)(s+4)\\s^2+2s+1=(s+1)(s+1)\end{array}\right\}$

$\text{LCD}=(s+4)(s+1)(s+1)$

$$\frac{4}{s^2+5s+4}+\frac{s}{s^2+2s+1}$$
$$=\frac{4}{(s+1)(s+4)}\cdot\frac{\mathbf{(s+1)}}{\mathbf{(s+1)}}+\frac{s}{(s+1)(s+1)}\cdot\frac{\mathbf{(s+4)}}{\mathbf{(s+4)}}$$
$$=\frac{4s+4+s^2+4s}{(s+4)(s+1)(s+1)}$$
$$=\frac{s^2+8s+4}{(s+4)(s+1)(s+1)}$$

41. $\left.\begin{array}{l}x^2-9x+8=(x-8)(x-1)\\x^2-6x-16=(x-8)(x+2)\end{array}\right\}$

$\text{LCD}=(x-1)(x-8)(x+2)$

$$\frac{5}{x^2-9x+8}-\frac{3}{x^2-6x-16}$$
$$=\frac{5}{(x-1)(x-8)}\cdot\frac{\mathbf{(x+2)}}{\mathbf{(x+2)}}-\frac{3}{(x-8)(x+2)}\cdot\frac{\mathbf{(x-1)}}{\mathbf{(x-1)}}$$
$$=\frac{5(x+2)-3(x-1)}{(x-1)(x-8)(x+2)}$$
$$=\frac{5x+10-3x+3}{(x-1)(x-8)(x+2)}$$
$$=\frac{2x+13}{(x-1)(x-8)(x+2)}$$

43. $\left.\begin{array}{l}a^2+4a+3=(a+3)(a+1)\\a+3=a+3\end{array}\right\}$

$\text{LCD}=(a+1)(a+3)$

$$\frac{2}{a^2+4a+3}+\frac{1}{a+3}$$
$$=\frac{2}{(a+1)(a+3)}+\frac{1}{(a+3)}\cdot\frac{\mathbf{(a+1)}}{\mathbf{(a+1)}}$$
$$=\frac{2+a+1}{(a+1)(a+3)}$$
$$=\frac{\cancel{a+3}^{\,1}}{(a+1)\cancel{(a+3)}_{\,1}}$$
$$=\frac{1}{a+1}$$

45. $$\left.\begin{aligned} y^2-16&=(y+4)(y-4) \\ y^2-y-12&=(y-4)(y+3) \end{aligned}\right\}$$

$$\text{LCD}=(y+4)(y-4)(y+3)$$

$$\frac{8}{y^2-16}-\frac{7}{y^2-y-12}$$
$$=\frac{8}{(y+4)(y-4)}\cdot\frac{\mathbf{(y+3)}}{\mathbf{(y+3)}}-\frac{7}{(y-4)(y+3)}\cdot\frac{\mathbf{(y+4)}}{\mathbf{(y+4)}}$$
$$=\frac{8y+24-7y-28}{(y+4)(y-4)(y+3)}$$
$$=\frac{\overset{1}{\cancel{y-4}}}{(y+4)\underset{1}{\cancel{(y-4)}}(y+3)}$$
$$=\frac{1}{(y+4)(y+3)}$$

Perform the operations. Simplify, if possible. See Example 6.

47. $$\frac{9y}{x-4}+y=\frac{9y}{x-4}+\frac{y}{1}\cdot\frac{\mathbf{x-4}}{\mathbf{x-4}}$$
$$=\frac{9y+y(x-4)}{x-4}$$
$$=\frac{9y+xy-4y}{x-4}$$
$$=\frac{(9-4)y+xy}{x-4}$$
$$=\frac{5y+xy}{x-4}$$

49. $$\frac{8}{x}+z=\frac{8}{x}+\frac{z}{1}\cdot\frac{\mathbf{x}}{\mathbf{x}}$$
$$=\frac{8+xz}{x}$$

Perform the operations. Simplify, if possible. See Example 7.

51. $$\frac{7}{a-4}+\frac{5}{4-a}=\frac{7}{a-4}+\frac{5}{4-a}\cdot\frac{\mathbf{-1}}{\mathbf{-1}}$$
$$=\frac{7}{a-4}+\frac{-5}{-4+a}$$
$$=\frac{7}{a-4}+\frac{-5}{a-4}$$
$$=\frac{7+(-5)}{a-4}$$
$$=\frac{2}{a-4}$$

53. $$\frac{c}{7c-d}-\frac{d}{d-7c}=\frac{c}{7c-d}-\frac{d}{d-7c}\cdot\frac{\mathbf{-1}}{\mathbf{-1}}$$
$$=\frac{c}{7c-d}-\frac{-d}{-d+7c}$$
$$=\frac{c}{7c-d}-\frac{-d}{7c-d}$$
$$=\frac{c-(-d)}{7c-d}$$
$$=\frac{c+d}{7c-d}$$

TRY IT YOURSELF

Perform the operations. Then simplify, if possible.

55. $$\frac{x-7}{x^2+4x-5}-\frac{x-9}{x^2+3x-10}$$
$$=\frac{x-7}{(x+5)(x-1)}-\frac{x-9}{(x+5)(x-2)}$$
$$=\frac{(x-7)}{(x+5)(x-1)}\cdot\frac{\mathbf{(x-2)}}{\mathbf{(x-2)}}-\frac{(x-9)}{(x+5)(x-2)}\cdot\frac{\mathbf{(x-1)}}{\mathbf{(x-1)}}$$
$$=\frac{(x-7)(x-2)-(x-9)(x-1)}{(x+5)(x-1)(x-2)}$$
$$=\frac{x^2-9x+14-(x^2-10x+9)}{(x+5)(x-1)(x-2)}$$
$$=\frac{x^2-9x+14-x^2+10x-9}{(x+5)(x-1)(x-2)}$$
$$=\frac{(1-1)x^2+(-9+10)x+(14-9)}{(x+5)(x-1)(x-2)}$$
$$=\frac{x+5}{(x+5)(x-1)(x-2)}$$
$$=\frac{\overset{1}{\cancel{x+5}}}{\underset{1}{\cancel{(x+5)}}(x-1)(x-2)}$$
$$=\frac{1}{(x-1)(x-2)}$$

57. $\dfrac{3d-3}{d-9}-\dfrac{3d}{9-d}=\dfrac{3d-3}{d-9}-\dfrac{3d}{9-d}\cdot\dfrac{\mathbf{-1}}{\mathbf{-1}}$
$=\dfrac{3d-3}{d-9}-\dfrac{-3d}{-9+d}$
$=\dfrac{3d-3}{d-9}-\dfrac{-3d}{d-9}$
$=\dfrac{3d-3-(-3d)}{d-9}$
$=\dfrac{3d-3+3d}{d-9}$
$=\dfrac{6d-3}{d-9}$

59. $\dfrac{10}{x-1}+y=\dfrac{10}{x-1}+\dfrac{y}{1}\cdot\dfrac{\mathbf{(x-1)}}{\mathbf{(x-1)}}$
$=\dfrac{10+y(x-1)}{x-1}$
$=\dfrac{10+xy-y}{x-1}$
$=\dfrac{xy-y+10}{x-1}$

61. $\dfrac{b}{b+1}-\dfrac{b-1}{b+2}=\dfrac{b}{(b+1)}\cdot\dfrac{\mathbf{(b+2)}}{\mathbf{(b+2)}}-\dfrac{(b-1)}{(b+2)}\cdot\dfrac{\mathbf{(b+1)}}{\mathbf{(b+1)}}$
$=\dfrac{b(b+2)-(b-1)(b+1)}{(b+1)(b+2)}$
$=\dfrac{b^2+2b-(b^2-1)}{(b+1)(b+2)}$
$=\dfrac{b^2+2b-b^2+1}{(b+1)(b+2)}$
$=\dfrac{2b+1}{(b+1)(b+2)}$

63. $\dfrac{g}{g^2-4}+\dfrac{2}{4-g^2}=\dfrac{g}{g^2-4}+\dfrac{2}{4-g^2}\cdot\dfrac{\mathbf{-1}}{\mathbf{-1}}$
$=\dfrac{g}{g^2-4}+\dfrac{-2}{-4+g^2}$
$=\dfrac{g}{g^2-4}+\dfrac{-2}{g^2-4}$
$=\dfrac{g+(-2)}{g^2-4}$
$=\dfrac{\cancel{g-2}^{\,1}}{(g+2)\cancel{(g-2)}_{\,1}}$
$=\dfrac{1}{g+2}$

65. $\dfrac{5y}{6}+\dfrac{5y}{3}=\dfrac{5y}{6}+\dfrac{5y}{3}\cdot\dfrac{\mathbf{2}}{\mathbf{2}}$
$=\dfrac{5y}{6}+\dfrac{10y}{6}$
$=\dfrac{5y+10y}{6}$
$=\dfrac{15y}{6}$
$=\dfrac{\cancel{3}^{\,1}\cdot 5y}{\cancel{3}_{\,1}\cdot 2}$
$=\dfrac{5y}{2}$

67. $\dfrac{1}{5x}+\dfrac{7x}{x+5}=\dfrac{1}{5x}\cdot\dfrac{\mathbf{(x+5)}}{\mathbf{(x+5)}}+\dfrac{7x}{(x+5)}\cdot\dfrac{\mathbf{(5x)}}{\mathbf{(5x)}}$
$=\dfrac{x+5+35x^2}{5x(x+5)}$
$=\dfrac{35x^2+x+5}{5x(x+5)}$

69. $\dfrac{11}{5x}-\dfrac{5}{6x}=\dfrac{11}{5x}\cdot\dfrac{\mathbf{6}}{\mathbf{6}}-\dfrac{5}{6x}\cdot\dfrac{\mathbf{5}}{\mathbf{5}}$
$=\dfrac{66-25}{30x}$
$=\dfrac{41}{30x}$

71. $$\frac{x}{x+1}+\frac{x-1}{x}=\frac{x}{(x+1)}\cdot\frac{\mathbf{(x)}}{\mathbf{(x)}}+\frac{(x-1)}{x}\cdot\frac{\mathbf{(x+1)}}{\mathbf{(x+1)}}$$
$$=\frac{x^2+x^2-1}{x(x+1)}$$
$$=\frac{2x^2-1}{x(x+1)}$$

73. $$\frac{y}{y-1}-\frac{4}{1-y}=\frac{y}{y-1}-\frac{4}{1-y}\cdot\frac{\mathbf{-1}}{\mathbf{-1}}$$
$$=\frac{y}{y-1}-\frac{-4}{-1+y}$$
$$=\frac{y}{y-1}-\frac{-4}{y-1}$$
$$=\frac{y-(-4)}{y-1}$$
$$=\frac{y+4}{y-1}$$

75. $$\frac{n}{5}-\frac{n-2}{15}=\frac{n}{5}\cdot\frac{\mathbf{3}}{\mathbf{3}}-\frac{n-2}{15}$$
$$=\frac{3n}{15}-\frac{n-2}{15}$$
$$=\frac{3n-(n-2)}{15}$$
$$=\frac{3n-n+2}{15}$$
$$=\frac{2n+2}{15}$$

77. $$\frac{y+2}{5y^2}+\frac{y+4}{15y}=\frac{(y+2)}{5y^2}\cdot\frac{\mathbf{3}}{\mathbf{3}}+\frac{(y+4)}{15y}\cdot\frac{\boldsymbol{y}}{\boldsymbol{y}}$$
$$=\frac{3(y+2)+y(y+4)}{15y^2}$$
$$=\frac{3y+6+y^2+4y}{15y^2}$$
$$=\frac{y^2+7y+6}{15y^2}$$

79. $$\frac{x}{x-2}+\frac{4+2x}{x^2-4}=\frac{x}{(x-2)}\cdot\frac{\mathbf{(x+2)}}{\mathbf{(x+2)}}+\frac{4+2x}{(x+2)(x-2)}$$
$$=\frac{x(x+2)+4+2x}{(x+2)(x-2)}$$
$$=\frac{x^2+2x+4+2x}{(x+2)(x-2)}$$
$$=\frac{x^2+4x+4}{(x+2)(x-2)}$$
$$=\frac{\cancel{(x+2)}(x+2)}{\cancel{(x+2)}(x-2)}$$
$$=\frac{x+2}{x-2}$$

81. $$b-\frac{3}{a^2}=\frac{b}{1}\cdot\frac{\boldsymbol{a^2}}{\boldsymbol{a^2}}-\frac{3}{a^2}$$
$$=\frac{a^2b-3}{a^2}$$

83. $$\frac{7}{3a}+\frac{1}{a-2}=\frac{7}{3a}\cdot\frac{\mathbf{(a-2)}}{\mathbf{(a-2)}}+\frac{1}{(a-2)}\cdot\frac{\mathbf{(3a)}}{\mathbf{(3a)}}$$
$$=\frac{7(a-2)+3a}{3a(a-2)}$$
$$=\frac{7a-14+3a}{3a(a-2)}$$
$$=\frac{10a-14}{3a(a-2)}$$

85. $$\frac{3}{x^2}+\frac{17}{x}=\frac{3}{x^2}+\frac{17}{x}\cdot\frac{\boldsymbol{x}}{\boldsymbol{x}}$$
$$=\frac{3}{x^2}+\frac{17x}{x^2}$$
$$=\frac{17x+3}{x^2}$$

87. $\frac{x+2}{x+1}-5=\frac{x+2}{x+1}-\frac{5}{1}\cdot\frac{\mathbf{(x+1)}}{\mathbf{(x+1)}}$

$=\frac{x+2-5(x+1)}{x+1}$

$=\frac{x+2-5x-5}{x+1}$

$=\frac{-4x-3}{x+1}$

$=\frac{-(4x+3)}{x+1}$

$=-\frac{4x+3}{x+1}$ or $\frac{-4x-3}{x+1}$

89. $\frac{4b}{3}-\frac{5b}{12}=\frac{4b}{3}\cdot\frac{\mathbf{4}}{\mathbf{4}}-\frac{5b}{12}$

$=\frac{16b}{12}-\frac{5b}{12}$

$=\frac{16b-5b}{12}$

$=\frac{11b}{12}$

LOOK ALIKES . . .

Perform the operations and simplify, if possible.

91. a. $\frac{5}{2x}+\frac{4x}{15}=\frac{5}{2x}\cdot\frac{\mathbf{15}}{\mathbf{15}}+\frac{4x}{15}\cdot\frac{\mathbf{2x}}{\mathbf{2x}}$

$=\frac{75+8x^2}{30x}$

b. $\frac{5}{2x}\cdot\frac{4x}{15}=\frac{5\cdot 4x}{2x\cdot 15}$

$=\frac{\overset{1}{\cancel{5}}\cdot\overset{1}{\cancel{2}}\cdot 2\overset{1}{\cancel{x}}}{\underset{1}{\cancel{2}}\underset{1}{\cancel{x}}\cdot 3\cdot\underset{1}{\cancel{5}}}$

$=\frac{2}{3}$

93. a. $\frac{t}{t-5}-\frac{t}{t^2-25}=\frac{t}{(t-5)}-\frac{t}{(t+5)(t-5)}$

$=\frac{t}{(t-5)}\cdot\frac{\mathbf{(t+5)}}{\mathbf{(t+5)}}-\frac{t}{(t+5)(t-5)}$

$=\frac{t(t+5)-t}{(t-5)(t+5)}$

$=\frac{t^2+5t-t}{(t-5)(t+5)}$

$=\frac{t^2+4t}{(t-5)(t+5)}$

b. $\frac{t}{t-5}\div\frac{t}{t^2-25}=\frac{t}{(t-5)}\div\frac{t}{(t+5)(t-5)}$

$=\frac{t}{(t-5)}\cdot\frac{(t+5)(t-5)}{t}$

$=\frac{\overset{1}{\cancel{t}}}{\underset{1}{\cancel{(t-5)}}}\cdot\frac{(t+5)\overset{1}{\cancel{(t-5)}}}{\underset{1}{\cancel{t}}}$

$=t+5$

APPLICATIONS

95. $\frac{3}{2x^2}+\frac{10}{3x}=\frac{3}{2x^2}\cdot\frac{\mathbf{3}}{\mathbf{3}}+\frac{10}{3x}\cdot\frac{\mathbf{2x}}{\mathbf{2x}}$

$=\frac{9}{6x^2}+\frac{20x}{6x^2}$

$=\frac{20x+9}{6x^2}$

The total height is $\frac{20x+9}{6x^2}$ cm.

WRITING

97-99. Answers will vary.

REVIEW

101. Find the slope and y-intercept of the graph $y=8x+2$.

$y=mx+b$

$m=$ slope of line

$b=$ y-intercept, $(0, y)$

$m=8$

y-int $=(0,2)$

103. What is the slope of teh graph of $y=2$?

$y = mx + b$

$m =$ slope of line

$b =$ y-intercept, $(0, y)$

$y = 2$

This is the equation of a horizontal line.

$m = 0$

CHALLENGE PROBLEMS

Perform the operations and simplify the results, if possible.

105.
$$\left.\begin{array}{r} a-1 = a-1 \\ a+2 = a+2 \\ a^2+a-2 = (a-1)(a+2) \end{array}\right\}$$

$$\text{LCD} = (a-1)(a+2)$$

$$\frac{a}{a-1} - \frac{2}{a+2} + \frac{3(a-2)}{a^2+a-2}$$

$$= \frac{a}{(a-1)} \cdot \frac{\mathbf{(a+2)}}{\mathbf{(a+2)}} - \frac{2}{(a+2)} \cdot \frac{\mathbf{(a-1)}}{\mathbf{(a-1)}} + \frac{3(a-2)}{a^2+a-2}$$

$$= \frac{a(a+2) - 2(a-1) + 3(a-2)}{(a-1)(a+2)}$$

$$= \frac{a^2 + 2a - 2a + 2 + 3a - 6}{(a-1)(a+2)}$$

$$= \frac{a^2 + 3a - 4}{(a-1)(a+2)}$$

$$= \frac{(a-1)(a+4)}{(a-1)(a+2)}$$

$$= \frac{\overset{1}{\cancel{(a-1)}}(a+4)}{\underset{1}{\cancel{(a-1)}}(a+2)}$$

$$= \frac{a+4}{a+2}$$

107. $$\frac{1}{a+1} + \frac{a^2-7a+10}{2a^2-2a-4} \cdot \frac{2a^2-50}{a^2+10a+25}$$

$$= \frac{1}{a+1} + \frac{a^2-7a+10}{2(a^2-a-2)} \cdot \frac{2(a^2-25)}{a^2+10a+25}$$

$$= \frac{1}{a+1} + \frac{(a-2)(a-5)}{2(a-2)(a+1)} \cdot \frac{2(a+5)(a-5)}{(a+5)(a+5)}$$

$$= \frac{1}{a+1} + \frac{\overset{1}{\cancel{2}}\overset{1}{\cancel{(a-2)}}(a-5)\overset{1}{\cancel{(a+5)}}(a-5)}{\underset{1}{\cancel{2}}\underset{1}{\cancel{(a-2)}}(a+1)\underset{1}{\cancel{(a+5)}}(a+5)}$$

$$= \frac{1}{a+1} + \frac{(a-5)(a-5)}{(a+1)(a+5)}$$

$$\left.\begin{array}{r} a+1 = a+1 \\ (a+1)(a+5) = (a+1)(a+5) \end{array}\right\}$$

$$\text{LCD} = (a+1)(a+5)$$

$$= \frac{1}{a+1} + \frac{(a-5)(a-5)}{(a+1)(a+5)}$$

$$= \frac{1}{(a+1)} \cdot \frac{\mathbf{(a+5)}}{\mathbf{(a+5)}} + \frac{(a-5)(a-5)}{(a+1)(a+5)}$$

$$= \frac{1(a+5) + (a-5)(a-5)}{(a+1)(a+5)}$$

$$= \frac{a + 5 + a^2 - 10a + 25}{(a+1)(a+5)}$$

$$= \frac{a^2 - 9a + 30}{(a+1)(a+5)}$$

SECTION 7.5 STUDY SET
VOCABULARY

Fill in the blanks.

1. The expression $\frac{\frac{2}{3}-\frac{1}{x}}{\frac{x-3}{4}}$ is called a **complex** rational expression or a **complex** fraction.

CONCEPTS

Fill in the blanks.

3. Method 1: To simplify a complex fraction, write its numerator and denominator as **single** rational expression. Then perform the indicated **division** by multiplying the numerator of the complex fraction by the **reciprocal** of the denominator.

5. a. $\frac{x-3}{4}$, **Yes**

   b. $\frac{1}{12}-\frac{x}{6}$, **No**

NOTATION

Fill in the blanks to simplify each complex fraction.

7. $\frac{\frac{12}{y^2}}{\frac{4}{y^3}}=\frac{12}{y^2}\boxed{\div}\frac{4}{y^3}$

GUIDED PRACTICE

Simplify each complex fraction. See Example 1.

9. $\frac{\frac{2}{3}}{\frac{3}{4}}=\frac{2}{3}\div\frac{3}{4}$

$=\frac{2}{3}\cdot\frac{4}{3}$

$=\frac{2\cdot 4}{3\cdot 3}$

$=\frac{8}{9}$

11. $\frac{\frac{x}{2}}{\frac{6}{5}}=\frac{x}{2}\div\frac{6}{5}$

$=\frac{x}{2}\cdot\frac{5}{6}$

$=\frac{x\cdot 5}{2\cdot 6}$

$=\frac{5x}{12}$

13. $\frac{\frac{x}{y}}{\frac{1}{x}}=\frac{x}{y}\div\frac{1}{x}$

$=\frac{x}{y}\cdot\frac{x}{1}$

$=\frac{x\cdot x}{y\cdot 1}$

$=\frac{x^2}{y}$

15. $\frac{\frac{n}{8}}{\frac{1}{n^2}}=\frac{n}{8}\div\frac{1}{n^2}$

$=\frac{n}{8}\cdot\frac{n^2}{1}$

$=\frac{n\cdot n^2}{8}$

$=\frac{n^3}{8}$

17. $\frac{\frac{4a}{11}}{\frac{6a^4}{55}}=\frac{4a}{11}\div\frac{6a^4}{55}$

$=\frac{4a}{11}\cdot\frac{55}{6a^4}$

$=\frac{4a\cdot 55}{11\cdot 6a^4}$

$=\frac{\cancel{2}\cdot 2\cdot 5\cdot\cancel{11}\cdot\cancel{a}}{\cancel{11}\cdot\cancel{2}\cdot 3\cdot\cancel{a}\cdot a^3}$

$=\frac{10}{3a^3}$

19. $\frac{-\frac{x^4}{30}}{\frac{7x^2}{15}}=-\frac{x^4}{30}\div\frac{7x^2}{15}$

$=-\frac{x^4}{30}\cdot\frac{15}{7x^2}$

$=-\frac{x^4\cdot 15}{30\cdot 7x^2}$

$=-\frac{\cancel{3}\cdot\cancel{5}\cdot\cancel{x^2}\cdot x^2}{2\cdot\cancel{3}\cdot\cancel{5}\cdot 7\cdot\cancel{x^2}}$

$=-\frac{x^2}{14}$

Simplify each complex fraction. See Examples 2 or 4.

21. $\left.\begin{array}{l}2=2\\4=2\cdot 2\end{array}\right\}\text{LCD}=2\cdot 2=4$

$$\frac{\frac{1}{2}+\frac{3}{4}}{\frac{3}{2}+\frac{1}{4}}=\frac{\frac{1}{2}+\frac{3}{4}}{\frac{3}{2}+\frac{1}{4}}\cdot\mathbf{\frac{4}{4}}$$

$$=\frac{\left(\frac{1}{2}+\frac{3}{4}\right)4}{\left(\frac{3}{2}+\frac{1}{4}\right)4}$$

$$=\frac{\frac{1}{2}(4)+\frac{3}{4}(4)}{\frac{3}{2}(4)+\frac{1}{4}(4)}$$

$$=\frac{2+3}{6+1}$$

$$=\frac{5}{7}$$

23. $\left.\begin{array}{l}4=2\cdot 2\\y=y\\3=3\\2=2\end{array}\right\}\text{LCD}=2\cdot 2\cdot 3\cdot y=12y$

$$\frac{\frac{1}{4}+\frac{1}{y}}{\frac{y}{3}-\frac{1}{2}}=\frac{\frac{1}{4}+\frac{1}{y}}{\frac{y}{3}-\frac{1}{2}}\cdot\mathbf{\frac{12y}{12y}}$$

$$=\frac{\left(\frac{1}{4}+\frac{1}{y}\right)12y}{\left(\frac{y}{3}-\frac{1}{2}\right)12y}$$

$$=\frac{\frac{1}{4}(12y)+\frac{1}{y}(12y)}{\frac{y}{3}(12y)-\frac{1}{2}(12y)}$$

$$=\frac{3y+12}{4y^2-6y}\text{ or }\frac{3(y+4)}{2y(2y-3)}$$

25. $\left.\begin{array}{l}y=y\\2=2\end{array}\right\}\text{LCD}=2\cdot y=2y$

$$\frac{\frac{1}{y}-\frac{5}{2}}{\frac{3}{y}}=\frac{\frac{1}{y}-\frac{5}{2}}{\frac{3}{y}}\cdot\mathbf{\frac{2y}{2y}}$$

$$=\frac{\left(\frac{1}{y}-\frac{5}{2}\right)2y}{\left(\frac{3}{y}\right)2y}$$

$$=\frac{\frac{1}{y}(2y)-\frac{5}{2}(2y)}{\frac{3}{y}(2y)}$$

$$=\frac{2-5y}{6}$$

27. $\left.\begin{array}{l}c=c\\6=2\cdot 3\end{array}\right\}\text{LCD}=2\cdot 3\cdot c=6c$

$$\frac{\frac{4}{c}-\frac{c}{6}}{\frac{2}{c}}=\frac{\frac{4}{c}-\frac{c}{6}}{\frac{2}{c}}\cdot\mathbf{\frac{6c}{6c}}$$

$$=\frac{\left(\frac{4}{c}-\frac{c}{6}\right)6c}{\left(\frac{2}{c}\right)6c}$$

$$=\frac{\frac{4}{c}(6c)-\frac{c}{6}(6c)}{\frac{2}{c}(6c)}$$

$$=\frac{24-c^2}{12}$$

Simplify each complex fraction. See Examples 3 or 5.

29. $\text{LCD} = 3$

$$\frac{\frac{2}{3}+1}{\frac{1}{3}+1} = \frac{\frac{2}{3}+1}{\frac{1}{3}+1}\cdot\frac{\mathbf{3}}{\mathbf{3}} = \frac{\left(\frac{2}{3}+1\right)3}{\left(\frac{1}{3}+1\right)3} = \frac{\frac{2}{3}(3)+1(3)}{\frac{1}{3}(3)+1(3)} = \frac{2+3}{1+3} = \frac{5}{4}$$

31. $\text{LCD} = x$

$$\frac{\frac{1}{x}-3}{\frac{5}{x}+2} = \frac{\frac{1}{x}-3}{\frac{5}{x}+2}\cdot\frac{\boldsymbol{x}}{\boldsymbol{x}} = \frac{\left(\frac{1}{x}-3\right)x}{\left(\frac{5}{x}+2\right)x} = \frac{\frac{1}{x}(x)-3(x)}{\frac{5}{x}(x)+2(x)} = \frac{1-3x}{5+2x}$$

33. $\text{LCD} = x$

$$\frac{\frac{2}{x}+2}{\frac{4}{x}+2} = \frac{\frac{2}{x}+2}{\frac{4}{x}+2}\cdot\frac{\boldsymbol{x}}{\boldsymbol{x}} = \frac{\left(\frac{2}{x}+2\right)x}{\left(\frac{4}{x}+2\right)x} = \frac{\frac{2}{x}(x)+2(x)}{\frac{4}{x}(x)+2(x)} = \frac{2+2x}{4+2x} = \frac{\overset{1}{\cancel{2}}(1+x)}{\underset{1}{\cancel{2}}(2+x)} = \frac{x+1}{x+2}$$

35. $\text{LCD} = x$

$$\frac{\frac{3y}{x}-y}{y-\frac{y}{x}} = \frac{\frac{3y}{x}-y}{y-\frac{y}{x}}\cdot\frac{\boldsymbol{x}}{\boldsymbol{x}} = \frac{\left(\frac{3y}{x}-y\right)x}{\left(y-\frac{y}{x}\right)x} = \frac{\frac{3y}{x}(x)-y(x)}{y(x)-\frac{y}{x}(x)} = \frac{3y-xy}{xy-y} = \frac{\overset{1}{\cancel{y}}(3-x)}{\underset{1}{\cancel{y}}(x-1)} = \frac{3-x}{x-1}$$

Simplify each complex fraction. See Example 4.

37. $\left.\begin{array}{l}6 = 2\cdot 3\\ x = x\end{array}\right\}\text{LCD} = 2\cdot 3\cdot x = 6x$

$$\frac{\frac{1}{6}-\frac{2}{x}}{\frac{1}{6}+\frac{1}{x}} = \frac{\frac{1}{6}-\frac{2}{x}}{\frac{1}{6}+\frac{1}{x}}\cdot\frac{\mathbf{6x}}{\mathbf{6x}} = \frac{\left(\frac{1}{6}-\frac{2}{x}\right)6x}{\left(\frac{1}{6}+\frac{1}{x}\right)6x} = \frac{\frac{1}{6}(6x)-\frac{2}{x}(6x)}{\frac{1}{6}(6x)+\frac{1}{x}(6x)} = \frac{x-12}{x+6}$$

39. $\left.\begin{array}{l}7 = 7\\ a = a\end{array}\right\}\text{LCD} = 7\cdot a = 7a$

$$\frac{\frac{a}{7}-\frac{7}{a}}{\frac{1}{a}+\frac{1}{7}} = \frac{\frac{a}{7}-\frac{7}{a}}{\frac{1}{a}+\frac{1}{7}}\cdot\frac{\mathbf{7a}}{\mathbf{7a}} = \frac{\left(\frac{a}{7}-\frac{7}{a}\right)7a}{\left(\frac{1}{a}+\frac{1}{7}\right)7a} = \frac{\frac{a}{7}(7a)-\frac{7}{a}(7a)}{\frac{1}{a}(7a)+\frac{1}{7}(7a)} = \frac{a^2-49}{7+a} = \frac{(a-7)\overset{1}{\cancel{(a+7)}}}{\underset{1}{\cancel{7+a}}} = a-7$$

Simplify each complex fraction. See Example 5.

41. $\left.\begin{array}{l}2=2\\4=2\cdot 2\\5=5\end{array}\right\}\text{LCD}=2\cdot 2\cdot 5=20$

$$\frac{\frac{d^2}{4}+\frac{4d}{5}}{\frac{d+1}{2}}=\frac{\frac{d^2}{4}+\frac{4d}{5}}{\frac{d+1}{2}}\cdot\frac{\mathbf{20}}{\mathbf{20}}$$

$$=\frac{\left(\frac{d^2}{4}+\frac{4d}{5}\right)20}{\left(\frac{d+1}{2}\right)20}$$

$$=\frac{\frac{d^2}{4}(20)+\frac{4d}{5}(20)}{\left(\frac{d+1}{2}\right)(20)}$$

$$=\frac{5d^2+16d}{10(d+1)}$$

$$=\frac{5d^2+16d}{10d+10}$$

43. $\left.\begin{array}{l}x=x\\y=y\end{array}\right\}\text{LCD}=x\cdot y=xy$

$$\frac{\frac{2}{x}}{\frac{2}{y}-\frac{4}{x}}=\frac{\frac{2}{x}}{\frac{2}{y}-\frac{4}{x}}\cdot\frac{\boldsymbol{xy}}{\boldsymbol{xy}}$$

$$=\frac{\left(\frac{2}{x}\right)xy}{\left(\frac{2}{y}-\frac{4}{x}\right)xy}$$

$$=\frac{\left(\frac{2}{x}\right)(xy)}{\frac{2}{y}(xy)-\frac{4}{x}(xy)}$$

$$=\frac{2y}{2x-4y}$$

$$=\frac{\overset{1}{\cancel{2}}y}{\underset{1}{\cancel{2}}(x-2y)}$$

$$=\frac{y}{x-2y}$$

Simplify each complex fraction. See Example 6.

45. $\text{LCD}=x+1$

$$\frac{\frac{1}{x+1}}{1+\frac{1}{x+1}}=\frac{\frac{1}{x+1}}{1+\frac{1}{x+1}}\cdot\frac{\mathbf{(x+1)}}{\mathbf{(x+1)}}$$

$$=\frac{\left(\frac{1}{x+1}\right)(x+1)}{\left(1+\frac{1}{x+1}\right)(x+1)}$$

$$=\frac{\left(\frac{1}{x+1}\right)(x+1)}{1(x+1)+\left(\frac{1}{x+1}\right)(x+1)}$$

$$=\frac{1}{x+1+1}$$

$$=\frac{1}{x+2}$$

47. $\text{LCD}=x+2$

$$\frac{\frac{x}{x+2}}{\frac{x}{x+2}+x}=\frac{\frac{x}{x+2}}{\frac{x}{x+2}+x}\cdot\frac{\mathbf{(x+2)}}{\mathbf{(x+2)}}$$

$$=\frac{\left(\frac{x}{x+2}\right)(x+2)}{\left(\frac{x}{x+2}+x\right)(x+2)}$$

$$=\frac{\left(\frac{x}{x+2}\right)(x+2)}{\left(\frac{x}{x+2}\right)(x+2)+x(x+2)}$$

$$=\frac{x}{x+x^2+2x}$$

$$=\frac{x}{x^2+3x}$$

$$=\frac{\overset{1}{\cancel{x}}}{\underset{1}{\cancel{x}}(x+3)}$$

$$=\frac{1}{x+3}$$

TRY IT YOURSELF

Simplify each complex fraction.

49. LCD $= pq$

$$\frac{\frac{1}{p}+\frac{1}{q}}{\frac{1}{p}} = \frac{\frac{1}{p}+\frac{1}{q}}{\frac{1}{p}}\cdot\frac{\mathbf{pq}}{\mathbf{pq}}$$

$$= \frac{\left(\frac{1}{p}+\frac{1}{q}\right)pq}{\left(\frac{1}{p}\right)pq}$$

$$= \frac{\frac{1}{p}(pq)+\frac{1}{q}(pq)}{\frac{1}{p}(pq)}$$

$$= \frac{q+p}{q}$$

51. $$\frac{40x^2}{\frac{20x}{9}} = \frac{40x^2}{1}\div\frac{20x}{9}$$

$$= \frac{40x^2}{1}\cdot\frac{9}{20x}$$

$$= \frac{40\cdot 9\cdot x^2}{20x}$$

$$= \frac{2\cdot \cancel{20}\cdot 9\cdot \cancel{x}\cdot x}{\cancel{20}\,\cancel{x}}$$

$$= 18x$$

53. $$\left.\begin{aligned} 2 &= 2 \\ 4 &= 2\cdot 2 \\ c &= c \\ c^2 &= c\cdot c \end{aligned}\right\}\text{LCD} = 2\cdot 2\cdot c\cdot c = 4c^2$$

$$\frac{\frac{1}{c}+\frac{1}{2}}{\frac{1}{c^2}-\frac{1}{4}} = \frac{\frac{1}{c}+\frac{1}{2}}{\frac{1}{c^2}-\frac{1}{4}}\cdot\frac{\mathbf{4c^2}}{\mathbf{4c^2}}$$

$$= \frac{\left(\frac{1}{c}+\frac{1}{2}\right)4c^2}{\left(\frac{1}{c^2}-\frac{1}{4}\right)4c^2}$$

$$= \frac{\frac{1}{c}(4c^2)+\frac{1}{2}(4c^2)}{\frac{1}{c^2}(4c^2)-\frac{1}{4}(4c^2)}$$

$$= \frac{4c+2c^2}{4-c^2}$$

$$= \frac{2c\,\cancel{(2+c)}}{\cancel{(2+c)}(2-c)}$$

$$= \frac{2c}{2-c}$$

55. LCD $= (r+1)(r-1)$

$$\frac{\frac{1}{r+1}+1}{\frac{3}{r-1}+1} = \frac{\frac{1}{r+1}+1}{\frac{3}{r-1}+1}\cdot\frac{\mathbf{(r+1)(r-1)}}{\mathbf{(r+1)(r-1)}}$$

$$= \frac{\left(\frac{1}{r+1}+1\right)(r+1)(r-1)}{\left(\frac{3}{r-1}+1\right)(r+1)(r-1)}$$

$$= \frac{\left(\frac{1}{r+1}\right)(r+1)(r-1)+1(r+1)(r-1)}{\left(\frac{3}{r-1}\right)(r+1)(r-1)+1(r+1)(r-1)}$$

$$= \frac{(r-1)+(r+1)(r-1)}{3(r+1)+(r+1)(r-1)}$$

$$= \frac{r-1+r^2-1}{3r+3+r^2-1}$$

$$= \frac{r^2+r-2}{r^2+3r+2}$$

$$= \frac{\cancel{(r+2)}(r-1)}{\cancel{(r+2)}(r+1)}$$

$$= \frac{r-1}{r+1}$$

57. $$\frac{\frac{b^2-81}{18a^2}}{\frac{4b-36}{9a}} = \frac{b^2-81}{18a^2} \div \frac{4b-36}{9a}$$

$$= \frac{b^2-81}{18a^2} \cdot \frac{9a}{4b-36}$$

$$= \frac{(b^2-81)\cdot 9a}{18a^2(4b-36)}$$

$$= \frac{\cancel{9}\cdot \cancel{a}\,\cancel{(b-9)}(b+9)}{2\cdot \cancel{9}\,\cancel{a}\cdot a\cdot 4\,\cancel{(b-9)}}$$

$$= \frac{b+9}{8a}$$

59. $$\frac{\frac{10x}{x-3}}{\frac{6}{x-3}} = \frac{10x}{x-3} \div \frac{6}{x-3}$$

$$= \frac{10x}{x-3} \cdot \frac{x-3}{6}$$

$$= \frac{10x\cdot(x-3)}{(x-3)\cdot 6}$$

$$= \frac{\cancel{2}\cdot 5\cdot x\,\cancel{(x-3)}}{\cancel{2}\cdot 3\,\cancel{(x-3)}}$$

$$= \frac{5x}{3}$$

61. $$\left.\begin{array}{l} 8h = 2\cdot 2\cdot 2\cdot h \\ 4h = 2\cdot 2\cdot h \end{array}\right\} \text{LCD} = 2\cdot 2\cdot 2\cdot h = 8h$$

$$\frac{4-\frac{1}{8h}}{12+\frac{3}{4h}} = \frac{4-\frac{1}{8h}}{12+\frac{3}{4h}} \cdot \frac{\mathbf{8h}}{\mathbf{8h}}$$

$$= \frac{\left(4-\frac{1}{8h}\right)8h}{\left(12+\frac{3}{4h}\right)8h}$$

$$= \frac{4(8h)-\frac{1}{8h}(8h)}{12(8h)+\frac{3}{4h}(8h)}$$

$$= \frac{32h-1}{96h+6}$$

63. $$\left.\begin{array}{l} n = n \\ m = m \end{array}\right\} \text{LCD} = n\cdot m = mn$$

$$\frac{\frac{m}{n}+\frac{n}{m}}{\frac{m}{n}-\frac{n}{m}} = \frac{\frac{m}{n}+\frac{n}{m}}{\frac{m}{n}-\frac{n}{m}} \cdot \frac{\boldsymbol{mn}}{\boldsymbol{mn}}$$

$$= \frac{\left(\frac{m}{n}+\frac{n}{m}\right)mn}{\left(\frac{m}{n}-\frac{n}{m}\right)mn}$$

$$= \frac{\frac{m}{n}(mn)+\frac{n}{m}(mn)}{\frac{m}{n}(mn)-\frac{n}{m}(mn)}$$

$$= \frac{m^2+n^2}{m^2-n^2}$$

65. $$\left.\begin{array}{l} 4 = 2\cdot 2 \\ c = c \\ c^2 = c\cdot c \end{array}\right\} \text{LCD} = 2\cdot 2\cdot c\cdot c = 4c^2$$

$$\frac{\frac{2}{c^2}}{\frac{1}{c}+\frac{5}{4}} = \frac{\frac{2}{c^2}}{\frac{1}{c}+\frac{5}{4}} \cdot \frac{\mathbf{4c^2}}{\mathbf{4c^2}}$$

$$= \frac{\left(\frac{2}{c^2}\right)4c^2}{\left(\frac{1}{c}+\frac{5}{4}\right)4c^2}$$

$$= \frac{\left(\frac{2}{c^2}\right)(4c^2)}{\frac{1}{c}(4c^2)+\frac{5}{4}(4c^2)}$$

$$= \frac{8}{4c+5c^2}$$

67. $\dfrac{\frac{4t-8}{t^2}}{\frac{8t-16}{t^5}} = \dfrac{4t-8}{t^2} \div \dfrac{8t-16}{t^5}$

$= \dfrac{4t-8}{t^2} \cdot \dfrac{t^5}{8t-16}$

$= \dfrac{(4t-8)\cdot t^5}{t^2(8t-16)}$

$= \dfrac{4t^5(t-2)}{8t^2(t-2)}$

$= \dfrac{\overset{1}{\cancel{4}} \cdot \overset{1}{\cancel{t^2}} \cdot t^3 \, \overset{1}{\cancel{(t-2)}}}{2 \cdot \underset{1}{\cancel{4}} \cdot \underset{1}{\cancel{t^2}} \, \underset{1}{\cancel{(t-2)}}}$

$= \dfrac{t^3}{2}$

69. $\left.\begin{array}{l} s = s \\ s^2 = s \cdot s \\ s^3 = s \cdot s \cdot s \end{array}\right\} \text{LCD} = s \cdot s \cdot s = s^3$

$\dfrac{\frac{2}{s} - \frac{2}{s^2}}{\frac{4}{s^3} + \frac{4}{s^2}} = \dfrac{\frac{2}{s} - \frac{2}{s^2}}{\frac{4}{s^3} + \frac{4}{s^2}} \cdot \dfrac{\mathbf{s^3}}{\mathbf{s^3}}$

$= \dfrac{\left(\frac{2}{s} - \frac{2}{s^2}\right)s^3}{\left(\frac{4}{s^3} + \frac{4}{s^2}\right)s^3}$

$= \dfrac{\frac{2}{s}(s^3) - \frac{2}{s^2}(s^3)}{\frac{4}{s^3}(s^3) + \frac{4}{s^2}(s^3)}$

$= \dfrac{2s^2 - 2s}{4 + 4s}$

$= \dfrac{\overset{1}{\cancel{2}}\, s(s-1)}{2 \cdot \underset{1}{\cancel{2}}(1+s)}$

$= \dfrac{s(s-1)}{2(s+1)}$ or $\dfrac{s^2 - s}{2 + 2s}$

71. $\left.\begin{array}{l} t = t \\ t^2 = t \cdot t \end{array}\right\} \text{LCD} = t \cdot t = t^2$

$\dfrac{1 + \frac{6}{t} + \frac{8}{t^2}}{1 + \frac{1}{t} - \frac{12}{t^2}} = \dfrac{1 + \frac{6}{t} + \frac{8}{t^2}}{1 + \frac{1}{t} - \frac{12}{t^2}} \cdot \dfrac{\mathbf{t^2}}{\mathbf{t^2}}$

$= \dfrac{\left(1 + \frac{6}{t} + \frac{8}{t^2}\right)t^2}{\left(1 + \frac{1}{t} - \frac{12}{t^2}\right)t^2}$

$= \dfrac{1(t^2) + \frac{6}{t}(t^2) + \frac{8}{t^2}(t^2)}{1(t^2) + \frac{1}{t}(t^2) - \frac{12}{t^2}(t^2)}$

$= \dfrac{t^2 + 6t + 8}{t^2 + t - 12}$

$= \dfrac{(t+2)\,\overset{1}{\cancel{(t+4)}}}{(t-3)\,\underset{1}{\cancel{(t+4)}}}$

$= \dfrac{t+2}{t-3}$

73. $\text{LCD} = xy$

$\dfrac{1}{\frac{1}{x} + \frac{1}{y}} = \dfrac{1}{\frac{1}{x} + \frac{1}{y}} \cdot \dfrac{\mathbf{xy}}{\mathbf{xy}}$

$= \dfrac{(1)xy}{\left(\frac{1}{x} + \frac{1}{y}\right)xy}$

$= \dfrac{xy}{\frac{1}{x}(xy) + \frac{1}{y}(xy)}$

$= \dfrac{xy}{y + x}$

75. $\dfrac{-\dfrac{25}{16x^2}}{\dfrac{15}{32x^5}} = -\dfrac{25}{16x^2} \div \dfrac{15}{32x^5}$

$$= -\frac{25}{16x^2} \cdot \frac{32x^5}{15}$$

$$= -\frac{25 \cdot 32x^5}{16x^2 \cdot 15}$$

$$= -\frac{5 \cdot \cancel{5} \cdot 2 \cdot \cancel{16} \cdot \cancel{x^2} \cdot x^3}{3 \cdot \cancel{5} \cdot \cancel{16} \cdot \cancel{x^2}}$$

$$= -\frac{10x^3}{3}$$

77. $\text{LCD} = x-1$

$$\frac{3+\dfrac{3}{x-1}}{3-\dfrac{3}{x-1}} = \frac{3+\dfrac{3}{x-1}}{3-\dfrac{3}{x-1}} \cdot \frac{\mathbf{(x-1)}}{\mathbf{(x-1)}}$$

$$= \frac{\left(3+\dfrac{3}{x-1}\right)(x-1)}{\left(3-\dfrac{3}{x-1}\right)(x-1)}$$

$$= \frac{3(x-1)+\left(\dfrac{3}{x-1}\right)(x-1)}{3(x-1)-\left(\dfrac{3}{x-1}\right)(x-1)}$$

$$= \frac{3x-3+3}{3x-3-3}$$

$$= \frac{3x}{3x-6}$$

$$= \frac{\cancel{3}x}{\cancel{3}(x-2)}$$

$$= \frac{x}{x-2}$$

79. $\left.\begin{array}{l} d = d \\ d^2 = d \cdot d \end{array}\right\} \text{LCD} = d \cdot d = d^2$

$$\frac{1-\dfrac{9}{d^2}}{2+\dfrac{6}{d}} = \frac{1-\dfrac{9}{d^2}}{2+\dfrac{6}{d}} \cdot \frac{\mathbf{d^2}}{\mathbf{d^2}}$$

$$= \frac{\left(1-\dfrac{9}{d^2}\right)d^2}{\left(2+\dfrac{6}{d}\right)d^2}$$

$$= \frac{1(d^2)-\dfrac{9}{d^2}(d^2)}{2(d^2)+\dfrac{6}{d}(d^2)}$$

$$= \frac{d^2-9}{2d^2+6d}$$

$$= \frac{(d-3)\cancel{(d+3)}}{2d\cancel{(d+3)}}$$

$$= \frac{d-3}{2d}$$

81. $\left.\begin{array}{l} a^2b = a \cdot a \cdot b \\ ab = a \cdot b \\ ab^2 = a \cdot b \cdot b \end{array}\right\} \text{LCD} = a^2 \cdot b^2 = a^2b^2$

$$\frac{\dfrac{1}{a^2b}-\dfrac{5}{ab}}{\dfrac{3}{ab}-\dfrac{7}{ab^2}} = \frac{\dfrac{1}{a^2b}-\dfrac{5}{ab}}{\dfrac{3}{ab}-\dfrac{7}{ab^2}} \cdot \frac{\mathbf{a^2b^2}}{\mathbf{a^2b^2}}$$

$$= \frac{\left(\dfrac{1}{a^2b}-\dfrac{5}{ab}\right)a^2b^2}{\left(\dfrac{3}{ab}-\dfrac{7}{ab^2}\right)a^2b^2}$$

$$= \frac{\dfrac{1}{a^2b}(a^2b^2)-\dfrac{5}{ab}(a^2b^2)}{\dfrac{3}{ab}(a^2b^2)-\dfrac{7}{ab^2}(a^2b^2)}$$

$$= \frac{b-5ab}{3ab-7a} \text{ or } \frac{b(1-5a)}{a(3b-7)}$$

83. LCD $= 2m+1$

$$\frac{m-\frac{1}{2m+1}}{1-\frac{m}{2m+1}} = \frac{m-\frac{1}{2m+1}}{1-\frac{m}{2m+1}} \cdot \frac{\mathbf{(2m+1)}}{\mathbf{(2m+1)}}$$

$$= \frac{\left(m-\frac{1}{2m+1}\right)(2m+1)}{\left(1-\frac{m}{2m+1}\right)(2m+1)}$$

$$= \frac{m(2m+1)-\left(\frac{1}{2m+1}\right)(2m+1)}{1(2m+1)-\left(\frac{m}{2m+1}\right)(2m+1)}$$

$$= \frac{2m^2+m-1}{2m+1-m}$$

$$= \frac{2m^2+m-1}{m+1}$$

$$= \frac{\overset{1}{\cancel{(m+1)}}(2m-1)}{\underset{1}{\cancel{m+1}}}$$

$$= 2m-1$$

APPLICATIONS

85. SLOPE

$$\left.\begin{aligned} 2 &= 2 \\ 3 &= 3 \\ 4 &= 2\cdot 2 \\ 8 &= 2\cdot 2\cdot 2 \end{aligned}\right\} \text{LCD} = 2\cdot 2\cdot 2\cdot 3 = 24$$

$$m = \frac{\frac{5}{8}-\frac{1}{3}}{\frac{3}{4}-\frac{1}{2}}$$

$$= \frac{\left(\frac{5}{8}-\frac{1}{3}\right)\mathbf{24}}{\left(\frac{3}{4}-\frac{1}{2}\right)\mathbf{24}}$$

$$= \frac{\left(\frac{5}{8}\right)24-\left(\frac{1}{3}\right)24}{\left(\frac{3}{4}\right)24-\left(\frac{1}{2}\right)24}$$

$$= \frac{15-8}{18-12}$$

$$= \frac{7}{6}$$

87. ELECTRONICS

LCD $= R_1R_2$

$$\text{Total resistance} = \frac{1}{\frac{1}{R_1}+\frac{1}{R_2}}$$

$$= \frac{1(\boldsymbol{R_1R_2})}{\left(\frac{1}{R_1}+\frac{1}{R_2}\right)\boldsymbol{R_1R_2}}$$

$$= \frac{R_1R_2}{\left(\frac{1}{R_1}\right)R_1R_2+\left(\frac{1}{R_2}\right)R_1R_2}$$

$$= \frac{R_1R_2}{R_2+R_1}$$

WRITING

89-92. Answers will vary.

REVIEW

Simplify each expression. Write each answer without negative exponents.

93. $(8x)^0 = 1$

95. $$\left(\frac{4x^3}{5x^{-3}}\right)^{-2} = \left(\frac{4x^{3-(-3)}}{5}\right)^{-2}$$

$$= \left(\frac{4x^{3+3}}{5}\right)^{-2}$$

$$= \left(\frac{4x^6}{5}\right)^{-2}$$

$$= \left(\frac{5}{4x^6}\right)^{2}$$

$$= \frac{5^2}{4^2x^{6(2)}}$$

$$= \frac{25}{16x^{12}}$$

CHALLENGE PROBLEMS

Simplify.

97.
$$\left.\begin{aligned} h+1 &= h+1 \\ h+2 &= h+2 \\ h^2+3h+2 &= (h+1)(h+2) \end{aligned}\right\}$$

$$\text{LCD} = (h+1)(h+2)$$

$$\frac{\frac{h}{h^2+3h+2}}{\frac{4}{h+2}-\frac{4}{h+1}} = \frac{\left(\frac{h}{(h+1)(h+2)}\right)}{\left(\frac{4}{h+2}-\frac{4}{h+1}\right)} \cdot \frac{\mathbf{(h+1)(h+2)}}{\mathbf{(h+1)(h+2)}}$$

$$= \frac{\left(\frac{h}{(h+1)(h+2)}\right)(h+1)(h+2)}{\left(\frac{4}{h+2}\right)(h+1)(h+2)-\left(\frac{4}{h+1}\right)(h+1)(h+2)}$$

$$= \frac{h}{4(h+1)-4(h+2)}$$

$$= \frac{h}{4h+4-4h-8}$$

$$= \frac{h}{-4}$$

$$= -\frac{h}{4}$$

99. $\text{LCD} = a+1$

$$a+\frac{a}{1+\frac{a}{a+1}} = a+\frac{a}{\left(1+\frac{a}{a+1}\right)} \cdot \frac{\mathbf{(a+1)}}{\mathbf{(a+1)}}$$

$$= a+\frac{a(a+1)}{1(a+1)+\left(\frac{a}{a+1}\right)(a+1)}$$

$$= a+\frac{a(a+1)}{a+1+a}$$

$$= a+\frac{a^2+a}{2a+1}$$

$$\text{LCD} = 2a+1$$

$$= a \cdot \frac{\mathbf{(2a+1)}}{\mathbf{(2a+1)}}+\frac{a^2+a}{2a+1}$$

$$= \frac{a(2a+1)+a^2+a}{2a+1}$$

$$= \frac{2a^2+a+a^2+a}{2a+1}$$

$$= \frac{3a^2+2a}{2a+1}$$

SECTION 7.6 STUDY SET

VOCABULARY

Fill in the blanks.

1. Equations that contain one or more rational expressions, such as $\frac{x}{x+2} = 4 + \frac{10}{x+2}$, are called **rational** equations.

3. To **clear** a rational equation of fractions, multiply both sides by the LCD of all rational expressions in the equation.

CONCEPTS

5. a. $\frac{1}{x-1} = 1 - \frac{3}{x-1}$

$$\frac{1}{5-1} \stackrel{?}{=} 1 - \frac{3}{5-1}$$
$$\frac{1}{4} \stackrel{?}{=} 1 - \frac{3}{4}$$
$$\frac{1}{4} \stackrel{?}{=} \frac{4}{4} - \frac{3}{4}$$
$$\frac{1}{4} = \frac{1}{4}$$

Yes

b. $\frac{x}{x-5} = 3 + \frac{5}{x-5}$

$$\frac{5}{5-5} \stackrel{?}{=} 3 + \frac{5}{5-5}$$
$$\frac{5}{0} \stackrel{?}{=} 3 + \frac{5}{0}$$

Undefined

No

7. $\frac{x}{x-3} = \frac{1}{x} + \frac{2}{x-3}$

a. $x = \mathbf{0}$ and $x - 3 = 0$

$$x = \mathbf{3}$$

b. **0, 3**

c. **0, 3**

By what should both sides of the equation be multiplied to clear it of fractions?

9. a. $\boldsymbol{y}$

b. $\mathbf{(x+2)(x-2)}$

11. a. $4x\left(\frac{3}{4x}\right) = \frac{\overset{1}{\cancel{4}} \cdot \overset{1}{\cancel{x}} \cdot 3}{\underset{1}{\cancel{4}} \cdot \underset{1}{\cancel{x}}}$

$$= \mathbf{3}$$

b. $(x+6)(x-2)\left(\frac{3}{x-2}\right)$

$$= \frac{3(x+6)\overset{1}{\cancel{(x-2)}}}{\underset{1}{\cancel{x-2}}}$$

$$= \mathbf{3(x+6)} \text{ or } \mathbf{3x+18}$$

NOTATION

13. $\frac{2}{a} + \frac{1}{2} = \frac{7}{2a}$

$$\boxed{\mathbf{2a}}\left(\frac{2}{a} + \frac{1}{2}\right) = \boxed{\mathbf{2a}}\left(\frac{7}{2a}\right)$$
$$\boxed{\mathbf{2a}}\left(\frac{2}{a}\right) + \boxed{\mathbf{2a}}\left(\frac{1}{2}\right) = \boxed{\mathbf{2a}}\left(\frac{7}{2a}\right)$$
$$\boxed{\mathbf{4}} + a = \boxed{\mathbf{7}}$$
$$4 + a - 4 = 7 - \boxed{\mathbf{4}}$$
$$a = \boxed{\mathbf{3}}$$

GUIDED PRACTICE

Solve each equation and check the result. If an equation has no solution, so indicate.
Only a selected few of the rational equations are checked due to the complex nature of checking them. See Example 1.

15. $\left.\begin{matrix} 2 = 2 \\ 3 = 3 \\ 6 = 2 \cdot 3 \end{matrix}\right\}$ LCD $= 2 \cdot 3 = 6$

$$\frac{2}{3} = \frac{1}{2} + \frac{x}{6}$$
$$\mathbf{6}\left(\frac{2}{3}\right) = \mathbf{6}\left(\frac{1}{2} + \frac{x}{6}\right)$$
$$6\left(\frac{2}{3}\right) = 6\left(\frac{1}{2}\right) + 6\left(\frac{x}{6}\right)$$
$$2 \cdot \overset{1}{\cancel{3}}\left(\frac{2}{\underset{1}{\cancel{3}}}\right) = 3 \cdot \overset{1}{\cancel{2}}\left(\frac{1}{\underset{1}{\cancel{2}}}\right) + \overset{1}{\cancel{6}}\left(\frac{x}{\underset{1}{\cancel{6}}}\right)$$
$$4 = 3 + x$$
$$4 - \mathbf{3} = 3 + x - \mathbf{3}$$
$$1 = x$$

check: $\frac{2}{3} = \frac{1}{2} + \frac{x}{6}$

$$\frac{2}{3} \stackrel{?}{=} \frac{1}{2} + \frac{1}{6}$$
$$\frac{2}{3} \stackrel{?}{=} \frac{3}{6} + \frac{1}{6}$$
$$\frac{2}{3} \stackrel{?}{=} \frac{4}{6}$$
$$\frac{2}{3} = \frac{2}{3}$$

This checks, so the solution is 1.

17. $\left.\begin{array}{l} 2=2 \\ 4=2\cdot 2 \\ 12=2\cdot 2\cdot 3 \end{array}\right\}\text{LCD}=2\cdot 2\cdot 3=12$

$$\frac{s}{12}-\frac{s}{2}=\frac{5s}{4}$$
$$\mathbf{12}\left(\frac{s}{12}-\frac{s}{2}\right)=\mathbf{12}\left(\frac{5s}{4}\right)$$
$$12\left(\frac{s}{12}\right)-12\left(\frac{s}{2}\right)=12\left(\frac{5s}{4}\right)$$
$$\cancel{12}^{1}\left(\frac{s}{\cancel{12}_{1}}\right)-6\cdot\cancel{2}^{1}\left(\frac{s}{\cancel{2}_{1}}\right)=3\cdot\cancel{4}^{1}\left(\frac{5s}{\cancel{4}_{1}}\right)$$
$$s-6s=15s$$
$$-5s=15s$$
$$-5s+\mathbf{5s}=15s+\mathbf{5s}$$
$$0=20s$$
$$\frac{0}{\mathbf{20}}=\frac{20s}{\mathbf{20}}$$
$$0=s$$

This checks, so the solution is 0.

19. $\left.\begin{array}{l} 2=2 \\ 3=3 \\ 18=2\cdot 3\cdot 3 \end{array}\right\}\text{LCD}=2\cdot 3\cdot 3=18$

$$\frac{x}{18}=\frac{1}{3}-\frac{x}{2}$$
$$\mathbf{18}\left(\frac{x}{18}\right)=\mathbf{18}\left(\frac{1}{3}-\frac{x}{2}\right)$$
$$18\left(\frac{x}{18}\right)=18\left(\frac{1}{3}\right)-18\left(\frac{x}{2}\right)$$
$$\cancel{18}^{1}\left(\frac{x}{\cancel{18}_{1}}\right)=6\cdot\cancel{3}^{1}\left(\frac{1}{\cancel{3}_{1}}\right)-9\cdot\cancel{2}^{1}\left(\frac{x}{\cancel{2}_{1}}\right)$$
$$x=6-9x$$
$$x+\mathbf{9x}=6-9x+\mathbf{9x}$$
$$10x=6$$
$$\frac{10x}{\mathbf{10}}=\frac{6}{\mathbf{10}}$$
$$x=\frac{\cancel{2}^{1}\cdot 3}{\cancel{2}_{1}\cdot 5}$$
$$x=\frac{3}{5}$$

check: $\frac{x}{18}=\frac{1}{3}-\frac{x}{2}$

$x=\frac{3}{5}$. Convert to a decimal to make the checking easier. $\frac{3}{5}=0.6$

$$\frac{x}{18}=\frac{1}{3}-\frac{x}{2}$$
$$\frac{0.6}{18}\stackrel{?}{=}\frac{1}{3}-\frac{0.6}{2}$$
$$0.0\overline{3}3\stackrel{?}{=}0.33\overline{3}-0.3$$
$$0.03\overline{3}\approx 0.03\overline{3}$$

This checks, so the solution is $\frac{3}{5}$.

21. $\left.\begin{array}{l} 4=2\cdot 2 \\ 2=2 \\ 3=3 \end{array}\right\}\text{LCD}=2\cdot 2\cdot 3=12$

$$\frac{b}{4}+\frac{1}{2}=\frac{b}{3}-\frac{1}{4}$$
$$\mathbf{12}\left(\frac{b}{4}+\frac{1}{2}\right)=\mathbf{12}\left(\frac{b}{3}-\frac{1}{4}\right)$$
$$12\left(\frac{b}{4}\right)+12\left(\frac{1}{2}\right)=12\left(\frac{b}{3}\right)-12\left(\frac{1}{4}\right)$$
$$3\cdot\cancel{4}^{1}\left(\frac{b}{\cancel{4}_{1}}\right)+6\cdot\cancel{2}^{1}\left(\frac{1}{\cancel{2}_{1}}\right)=4\cdot\cancel{3}^{1}\left(\frac{b}{\cancel{3}_{1}}\right)-3\cdot\cancel{4}^{1}\left(\frac{1}{\cancel{4}_{1}}\right)$$
$$3b+6=4b-3$$
$$3b+6-\mathbf{3b}=4b-3-\mathbf{3b}$$
$$6=b-3$$
$$6+\mathbf{3}=b-3+\mathbf{3}$$
$$9=b$$

Check: $\frac{b}{4}+\frac{1}{2}=\frac{b}{3}-\frac{1}{4}$

$$\frac{\mathbf{9}}{4}+\frac{1}{2}\stackrel{?}{=}\frac{\mathbf{9}}{3}-\frac{1}{4}$$
$$\frac{9}{4}+\frac{2}{4}\stackrel{?}{=}\frac{36}{12}-\frac{3}{12}$$
$$\frac{11}{4}\stackrel{?}{=}\frac{33}{12}$$
$$\frac{11}{4}=\frac{11}{4}$$

This checks, so the solution is 9.

Solve each equation and check the result. If an equation has no solution, so indicate. See Example 2.

23. $\left.\begin{array}{r} k=k \\ 3k=3\cdot k \end{array}\right\}\text{LCD}=3\cdot k=3k$

$$\frac{5}{3k}+\frac{1}{k}=-2$$
$$\mathbf{3k}\left(\frac{5}{3k}+\frac{1}{k}\right)=\mathbf{3k}(-2)$$
$$3k\left(\frac{5}{3k}\right)+3k\left(\frac{1}{k}\right)=3k(-2)$$
$$\cancel{3k}\left(\frac{5}{\cancel{3k}}\right)+3\cdot\cancel{k}\left(\frac{1}{\cancel{k}}\right)=-6k$$
$$5+3=-6k$$
$$8=-6k$$
$$\frac{8}{\mathbf{-6}}=\frac{-6k}{\mathbf{-6}}$$
$$-\frac{\cancel{2}\cdot 4}{\cancel{2}\cdot 3}=k$$
$$k=-\frac{4}{3}$$

This checks, so the solution is $-\frac{4}{3}$.

25. $\left.\begin{array}{r} 4=2\cdot 2 \\ 6=2\cdot 3 \\ a=a \end{array}\right\}\text{LCD}=2\cdot 2\cdot 3\cdot a=12a$

$$\frac{1}{4}-\frac{5}{6}=\frac{1}{a}$$
$$\mathbf{12a}\left(\frac{1}{4}-\frac{5}{6}\right)=\mathbf{12a}\left(\frac{1}{a}\right)$$
$$12a\left(\frac{1}{4}\right)-12a\left(\frac{5}{6}\right)=12a\left(\frac{1}{a}\right)$$
$$3a\cdot\cancel{4}\left(\frac{1}{\cancel{4}}\right)-2a\cdot\cancel{6}\left(\frac{5}{\cancel{6}}\right)=12\cdot\cancel{a}\left(\frac{1}{\cancel{a}}\right)$$
$$3a-10a=12$$
$$-7a=12$$
$$\frac{-7a}{\mathbf{-7}}=\frac{12}{\mathbf{-7}}$$
$$a=-\frac{12}{7}$$

This checks, so the solution is $-\frac{12}{7}$.

27. $\left.\begin{array}{r} 8=2\cdot 2\cdot 2 \\ 12=2\cdot 2\cdot 3 \\ b=b \end{array}\right\}\text{LCD}=2\cdot 2\cdot 2\cdot 3\cdot b=24b$

$$\frac{1}{8}+\frac{2}{b}-\frac{1}{12}=0$$
$$\mathbf{24b}\left(\frac{1}{8}+\frac{2}{b}-\frac{1}{12}\right)=\mathbf{24b}(0)$$
$$24b\left(\frac{1}{8}\right)+24b\left(\frac{2}{b}\right)-24b\left(\frac{1}{12}\right)=24b(0)$$
$$3b\cdot\cancel{8}\left(\frac{1}{\cancel{8}}\right)+24\cancel{b}\left(\frac{2}{\cancel{b}}\right)-2b\cdot\cancel{12}\left(\frac{1}{\cancel{12}}\right)=0$$
$$3b+48-2b=0$$
$$b+48=0$$
$$b+48-\mathbf{48}=0-\mathbf{48}$$
$$b=-48$$

This checks, so the solution is -48.

29. $\left.\begin{array}{r} 5=5 \\ 10x=2\cdot 5\cdot x \\ 15=3\cdot 5 \end{array}\right\}\text{LCD}=2\cdot 3\cdot 5\cdot x=30x$

$$\frac{4}{5}-\frac{1}{10x}=\frac{7}{15}$$
$$\mathbf{30x}\left(\frac{4}{5}-\frac{1}{10x}\right)=\mathbf{30x}\left(\frac{7}{15}\right)$$
$$30x\left(\frac{4}{5}\right)-30x\left(\frac{1}{10x}\right)=30x\left(\frac{7}{15}\right)$$
$$6x\cdot\cancel{5}\left(\frac{4}{\cancel{5}}\right)-3\cdot\cancel{10x}\left(\frac{1}{\cancel{10x}}\right)=2x\cdot\cancel{15}\left(\frac{7}{\cancel{15}}\right)$$
$$24x-3=14x$$
$$24x-3-\mathbf{24x}=14x-\mathbf{24x}$$
$$-3=-10x$$
$$\frac{-3}{\mathbf{-10}}=\frac{-10x}{\mathbf{-10}}$$
$$x=\frac{3}{10}$$

This checks, so the solution is $\frac{3}{10}$.

Solve each equation and check the result. If an equation has no solution, so indicate. See Example 3.

31. LCD $= x$

$$x + \frac{8}{x} = 6$$

$$\mathbf{x}\left(x + \frac{8}{x}\right) = \mathbf{x}(6)$$

$$x(x) + x\left(\frac{8}{x}\right) = x(6)$$

$$x^2 + \overset{1}{\cancel{x}}\left(\frac{8}{\underset{1}{\cancel{x}}}\right) = 6x$$

$$x^2 + 8 = 6x$$

$$x^2 + 8 - \mathbf{6x} = 6x - \mathbf{6x}$$

$$x^2 - 6x + 8 = 0$$

$$(x-2)(x-4) = 0$$

$x - 2 = 0$ or $x - 4 = 0$

$x - 2 + 2 = 0 + 2$ | $x - 4 + 4 = 0 + 4$

$x = 2$ | $x = 4$

check: $x + \frac{8}{x} = 6$ check: $x + \frac{8}{x} = 6$

$2 + \frac{8}{2} \stackrel{?}{=} 6$ $4 + \frac{8}{4} \stackrel{?}{=} 6$

$2 + 4 \stackrel{?}{=} 6$ $4 + 2 \stackrel{?}{=} 6$

$6 = 6$ $6 = 6$

Both check, so the solutions are 2 and 4.

33. LCD $= t$

$$\frac{10}{t} - t = 3$$

$$\mathbf{t}\left(\frac{10}{t} - t\right) = \mathbf{t}(3)$$

$$t\left(\frac{10}{t}\right) - t(t) = t(3)$$

$$\overset{1}{\cancel{t}}\left(\frac{10}{\underset{1}{\cancel{t}}}\right) - t^2 = 3t$$

$$10 - t^2 = 3t$$

$$10 - t^2 - \mathbf{3t} = 3t - \mathbf{3t}$$

$$-t^2 - 3t + 10 = 0$$

$$-(t^2 + 3t - 10) = 0$$

$$-(t-2)(t+5) = 0$$

$t - 2 = 0$ or $t + 5 = 0$

$t - 2 + 2 = 0 + 2$ | $t + 5 - 5 = 0 - 5$

$t = 2$ | $t = -5$

check: $\frac{10}{t} - t = 3$ check: $\frac{10}{t} - t = 3$

$\frac{10}{2} - 2 \stackrel{?}{=} 3$ $\frac{10}{-5} - (-5) \stackrel{?}{=} 3$

$5 - 2 \stackrel{?}{=} 3$ $-2 + 5 \stackrel{?}{=} 3$

$3 = 3$ $3 = 3$

Both check, so the solutions are 2 and -5.

35. LCD $= c$

$$\frac{20}{c} + c = -9$$

$$\mathbf{c}\left(\frac{20}{c} + c\right) = \mathbf{c}(-9)$$

$$c\left(\frac{20}{c}\right) + c(c) = c(-9)$$

$$\overset{1}{\cancel{c}}\left(\frac{20}{\underset{1}{\cancel{c}}}\right) + c^2 = -9c$$

$$20 + c^2 = -9c$$

$$20 + c^2 + \mathbf{9c} = -9c + \mathbf{9c}$$

$$c^2 + 9c + 20 = 0$$

$$(c+4)(c+5) = 0$$

$c + 4 = 0$ or $c + 5 = 0$

$c + 4 - 4 = 0 - 4$ | $c + 5 - 5 = 0 - 5$

$c = -4$ | $c = -5$

Both check, so the solutions are -4 and -5.

37. LCD $= p$

$$4+\frac{15}{p}=3p$$
$$\mathbf{p}\left(4+\frac{15}{p}\right)=\mathbf{p}(3p)$$
$$p(4)+p\left(\frac{15}{p}\right)=p(3p)$$
$$4p+\overset{1}{\cancel{p}}\left(\frac{15}{\underset{1}{\cancel{p}}}\right)=3p^2$$
$$4p+15=3p^2$$
$$4p+15-\mathbf{4p}=3p^2-\mathbf{4p}$$
$$15=3p^2-4p$$
$$15-\mathbf{15}=3p^2-4p-\mathbf{15}$$
$$0=3p^2-4p-15$$
$$0=(3p+5)(p-3)$$

$$3p+5=0 \quad\text{or}\quad p-3=0$$
$$3p+5-5=0-5 \qquad p-3+3=0+3$$
$$\frac{3p}{3}=\frac{-5}{3} \qquad p=3$$
$$p=-\frac{5}{3}$$

Both check, so the solutions are $-\frac{5}{3}$ and 3

Solve each equation and check the result. If an equation has no solution, so indicate. See Example 4.

39. LCD $= x-5$

$$\frac{x}{x-5}=3+\frac{5}{x-5}$$
$$\mathbf{(x-5)}\left(\frac{x}{x-5}\right)=\mathbf{(x-5)}\left(3+\frac{5}{x-5}\right)$$
$$(x-5)\left(\frac{x}{x-5}\right)=(x-5)(3)+(x-5)\left(\frac{5}{x-5}\right)$$
$$\overset{1}{\cancel{(x-5)}}\left(\frac{x}{\underset{1}{\cancel{x-5}}}\right)=3x-15+\overset{1}{\cancel{(x-5)}}\left(\frac{5}{\underset{1}{\cancel{x-5}}}\right)$$
$$x=3x-15+5$$
$$x-\mathbf{x}=3x-10-\mathbf{x}$$
$$0=2x-10$$
$$0+\mathbf{10}=2x-10+\mathbf{10}$$
$$10=2x$$
$$\frac{10}{\mathbf{2}}=\frac{2x}{\mathbf{2}}$$
$$5=x$$

check: $$\frac{x}{x-5}=3+\frac{5}{x-5}$$
$$\frac{x}{\mathbf{5}-5}=3+\frac{5}{\mathbf{5}-5}$$
$$\frac{x}{\mathbf{0}}=3+\frac{5}{\mathbf{0}}$$

5 makes the denominators of the original equation 0. No solution. 5 is extraneous.

41. LCD $= a+2$

$$\frac{a^2}{a+2}-a=\frac{4}{a+2}$$
$$\mathbf{(a+2)}\left(\frac{a^2}{a+2}-a\right)=\mathbf{(a+2)}\left(\frac{4}{a+2}\right)$$
$$(a+2)\left(\frac{a^2}{a+2}\right)-(a+2)(a)=(a+2)\left(\frac{4}{a+2}\right)$$
$$\overset{1}{\cancel{(a+2)}}\left(\frac{a^2}{\underset{1}{\cancel{a+2}}}\right)-(a+2)(a)=\overset{1}{\cancel{(a+2)}}\left(\frac{4}{\underset{1}{\cancel{a+2}}}\right)$$
$$a^2-a(a+2)=4$$
$$a^2-a^2-2a=4$$
$$-2a=4$$
$$\frac{-2a}{\mathbf{-2}}=\frac{4}{\mathbf{-2}}$$
$$a=-2$$

check: $$\frac{a^2}{a+2}-a=\frac{4}{a+2}$$
$$\frac{a^2}{\mathbf{-2}+2}-a=\frac{4}{\mathbf{-2}+2}$$
$$\frac{a^2}{\mathbf{0}+2}-a=\frac{4}{\mathbf{0}+2}$$

-2 makes the denominators of the original equation 0. No solution. -2 is extraneous.

Solve each equation and check the result. If an equation has no solution, so indicate. See Example 5.

43. LCD $= (x+4)(x-3)$

$$\frac{x+6}{x+4}+\frac{1}{x^2+x-12}=1$$
$$\frac{x+6}{x+4}+\frac{1}{(x+4)(x-3)}=1$$
$$\mathbf{(x+4)(x-3)}\left(\frac{x+6}{x+4}+\frac{1}{(x+4)(x-3)}\right)=\mathbf{(x+4)(x-3)(1)}$$
$$(x+4)(x-3)\left(\frac{x+6}{x+4}\right)+(x+4)(x-3)\left(\frac{1}{(x+4)(x-3)}\right)=(x+4)(x-3)$$
$$(x-3)\cancel{(x+4)}\left(\frac{x+6}{\cancel{x+4}}\right)+\cancel{(x+4)}\cancel{(x-3)}\left(\frac{1}{\cancel{(x+4)}\cancel{(x-3)}}\right)=(x+4)(x-3)$$
$$(x-3)(x+6)+1=(x+4)(x-3)$$
$$x^2+3x-18+1=x^2+x-12$$
$$x^2+3x-17=x^2+x-12$$
$$x^2+3x-17-\mathbf{x^2}=x^2+x-12-\mathbf{x^2}$$
$$3x-17=x-12$$
$$3x-17-\mathbf{x}=x-12-\mathbf{x}$$
$$2x-17=-12$$
$$2x-17+\mathbf{17}=-12+\mathbf{17}$$
$$2x=5$$
$$\frac{2x}{\mathbf{2}}=\frac{5}{\mathbf{2}}$$
$$x=\frac{5}{2}$$

This checks, so the solution is $\frac{5}{2}$.

45. $\left.\begin{aligned}x+2&=x+2\\x^2+x-2&=(x+2)(x-1)\end{aligned}\right\}$ LCD $=(x+2)(x-1)$

$$\frac{2x}{x^2+x-2}+\frac{2}{x+2}=1$$
$$\mathbf{(x+2)(x-1)}\cdot\left(\frac{2x}{(x+2)(x-1)}\right)+\mathbf{(x+2)(x-1)}\cdot\left(\frac{2}{x+2}\right)=\mathbf{(x+2)(x-1)\cdot 1}$$
$$\frac{2x\cancel{(x+2)}\cancel{(x-1)}}{\cancel{(x+2)}\cancel{(x-1)}}+\frac{2\cancel{(x+2)}(x-1)}{\cancel{(x+2)}}=(x+2)(x-1)$$
$$2x+2(x-1)=(x+2)(x-1)$$
$$2x+2x-2=x^2+x-2$$
$$4x-2=x^2+x-2$$
$$4x-2-\mathbf{4x}=x^2+x-2-\mathbf{4x}$$
$$-2=x^2-3x-2$$
$$-2+\mathbf{2}=x^2-3x-2+\mathbf{2}$$
$$0=x^2-3x$$
$$0=x(x-3)$$

$x=0$ or
$$x-3=0$$
$$x-3+3=0+3$$
$$x=3$$

Both check, so the solutions are 0 and 3.

Solve each formula for the indicated variable. See Example 6.

47. LCD $=b+d$

$$h=\frac{2A}{b+d}$$
$$\mathbf{(b+d)}(h)=\mathbf{(b+d)}\left(\frac{2A}{b+d}\right)$$
$$(b+d)(h)=\cancel{(b+d)}\left(\frac{2A}{\cancel{b+d}}\right)$$
$$h(b+d)=2A$$
$$\frac{h(b+d)}{2}=\frac{\cancel{2}A}{\cancel{2}}$$
$$A=\frac{h(b+d)}{2}$$

49. LCD $=R+r$

$$I=\frac{E}{R+r}$$
$$\mathbf{(R+r)}(I)=\mathbf{(R+r)}\left(\frac{E}{R+r}\right)$$
$$(R+r)(I)=\cancel{(R+r)}\left(\frac{E}{\cancel{R+r}}\right)$$
$$I(R+r)=E$$
$$IR+Ir=E$$
$$IR+Ir-\mathbf{IR}=E-\mathbf{IR}$$
$$Ir=E-IR$$
$$\frac{\cancel{I}r}{\cancel{I}}=\frac{E-IR}{\mathbf{I}}$$
$$r=\frac{E-IR}{I}$$

51. LCD $= xyz$

$$\frac{5}{x}-\frac{4}{y}=\frac{5}{z}$$

$$(\mathbf{xyz})\left(\frac{5}{x}-\frac{4}{y}\right)=(\mathbf{xyz})\left(\frac{5}{z}\right)$$

$$\overset{1}{\cancel{x}}yz\left(\frac{5}{\underset{1}{\cancel{x}}}\right)-x\overset{1}{\cancel{y}}z\left(\frac{4}{\underset{1}{\cancel{y}}}\right)=xy\overset{1}{\cancel{z}}\left(\frac{5}{\underset{1}{\cancel{z}}}\right)$$

$$5yz-4xz=5xy$$

$$5yz-4xz+\mathbf{4xz}=5xy+\mathbf{4xz}$$

$$5yz=x(5y+4z)$$

$$\frac{5yz}{\mathbf{5y+4z}}=\frac{x\,\overset{1}{\cancel{(5y+4z)}}}{\underset{1}{\cancel{\mathbf{5y+4z}}}}$$

$$x=\frac{5yz}{5y+4z}$$

53. LCD $= rst$

$$\frac{1}{r}+\frac{1}{s}=\frac{1}{t}$$

$$(\mathbf{rst})\left(\frac{1}{r}+\frac{1}{s}\right)=(\mathbf{rst})\left(\frac{1}{t}\right)$$

$$\overset{1}{\cancel{r}}st\left(\frac{1}{\underset{1}{\cancel{r}}}\right)+r\overset{1}{\cancel{s}}t\left(\frac{1}{\underset{1}{\cancel{s}}}\right)=rs\overset{1}{\cancel{t}}\left(\frac{1}{\underset{1}{\cancel{t}}}\right)$$

$$st+rt=rs$$

$$st+rt-\mathbf{rt}=rs-\mathbf{rt}$$

$$st=r(s-t)$$

$$\frac{st}{\mathbf{s-t}}=\frac{r\,\overset{1}{\cancel{(s-t)}}}{\underset{1}{\cancel{\mathbf{s-t}}}}$$

$$r=\frac{st}{s-t}$$

Solve each formula for the indicated variable. See Example 7.

55. LCD $= n$

$$\frac{P}{n}=rt$$

$$\mathbf{n}\left(\frac{P}{n}\right)=\mathbf{n}(rt)$$

$$\overset{1}{\cancel{n}}\left(\frac{P}{\underset{1}{\cancel{n}}}\right)=nrt$$

$$P=nrt$$

57. LCD $= bd$

$$\frac{a}{b}=\frac{c}{d}$$

$$\mathbf{bd}\left(\frac{a}{b}\right)=\mathbf{bd}\left(\frac{c}{d}\right)$$

$$d\overset{1}{\cancel{b}}\left(\frac{a}{\underset{1}{\cancel{b}}}\right)=b\overset{1}{\cancel{d}}\left(\frac{c}{\underset{1}{\cancel{d}}}\right)$$

$$ad=bc$$

$$\frac{\overset{1}{\cancel{a}}d}{\underset{1}{\cancel{a}}}=\frac{bc}{a}$$

$$d=\frac{bc}{a}$$

59. LCD $= ab$

$$\frac{1}{a}+\frac{1}{b}=1$$

$$(\mathbf{ab})\left(\frac{1}{a}+\frac{1}{b}\right)=(\mathbf{ab})(1)$$

$$\overset{1}{\cancel{(a)}}b\left(\frac{1}{\underset{1}{\cancel{a}}}\right)+a\overset{1}{\cancel{(b)}}\left(\frac{1}{\underset{1}{\cancel{b}}}\right)=(ab)(1)$$

$$b+a=ab$$

$$b+a-a=ab-a$$

$$b=ab-a$$

$$b=a(b-1)$$

$$\frac{b}{b-1}=\frac{a\,\overset{1}{\cancel{(b-1)}}}{\underset{1}{\cancel{b-1}}}$$

$$a=\frac{b}{b-1}$$

61. $\left.\begin{array}{l}2=2\\6d=2\cdot 3\cdot d\end{array}\right\}\text{LCD}=2\cdot 3\cdot d=6d$

$$F=\frac{L^2}{6d}+\frac{d}{2}$$
$$(\mathbf{6d})(F)=(\mathbf{6d})\left(\frac{L^2}{6d}+\frac{d}{2}\right)$$
$$6d(F)=6d\left(\frac{L^2}{6d}\right)+3\cdot 2d\left(\frac{d}{2}\right)$$
$$6dF=L^2+3d^2$$
$$6dF-\mathbf{3d^2}=L^2+3d^2-\mathbf{3d^2}$$
$$6dF-3d^2=L^2$$
$$L^2=6dF-3d^2$$

TRY IT YOURSELF

Solve each equation and check the result. If an equation has no solution, so indicate. Only a selected few of the rational equations are checked due to the complex nature of checking them.

63. LCD $=3(x-3)$

$$\frac{1}{3}+\frac{2}{x-3}=1$$
$$\mathbf{3(x-3)}\left(\frac{1}{3}+\frac{2}{x-3}\right)=\mathbf{3(x-3)}(1)$$
$$3(x-3)\left(\frac{1}{3}\right)+3(x-3)\left(\frac{2}{x-3}\right)=3(x-3)(1)$$
$$3(x-3)\left(\frac{1}{3}\right)+3(x-3)\left(\frac{2}{x-3}\right)=3x-9$$
$$x-3+6=3x-9$$
$$x+3-x=3x-9-x$$
$$3+9=2x-9+9$$
$$\frac{2\cdot 6}{2}=\frac{2x}{2}$$
$$6=x$$

This checks, so the solution is 6.

65. $\left.\begin{array}{r}q-2=q-2\\q+1=q+1\\q^2-q-2=(q-2)(q+1)\end{array}\right\}$

LCD $=(q-2)(q+1)$

$$\frac{7}{q^2-q-2}+\frac{1}{q+1}=\frac{3}{q-2}$$
$$\frac{7}{(q-2)(q+1)}\cdot\mathbf{(q-2)(q+1)}+\frac{1}{q+1}\cdot\mathbf{(q-2)(q+1)}=\frac{3}{q-2}\cdot\mathbf{(q-2)(q+1)}$$
$$\frac{7(q-2)(q+1)}{(q-2)(q+1)}+\frac{1(q-2)(q+1)}{q+1}=\frac{3(q-2)(q+1)}{q-2}$$
$$7+q-2=3q+3$$
$$q+5-\mathbf{3}=3q+3-\mathbf{3}$$
$$q+2-\mathbf{q}=3q-\mathbf{q}$$
$$\frac{2}{2}=\frac{2q}{2}$$
$$1=q$$

This checks, so the solution is 1.

67. LCD $=(3-t)(t+3)$

$$\frac{2}{3-t}=\frac{-t}{t+3}$$
$$\mathbf{(3-t)(t+3)}\left(\frac{2}{3-t}\right)=\mathbf{(3-t)(t+3)}\left(\frac{-t}{t+3}\right)$$
$$(3-t)(t+3)\left(\frac{2}{3-t}\right)=(3-t)(t+3)\left(\frac{-t}{t+3}\right)$$
$$2(t+3)=-t(3-t)$$
$$2t+6=-3t+t^2$$
$$2t+6-\mathbf{2t}=-3t+t^2-\mathbf{2t}$$
$$6=t^2-5t$$
$$6-\mathbf{6}=t^2-5t-\mathbf{6}$$
$$0=t^2-5t-6$$
$$0=(t-6)(t+1)$$

$t-6=0$ or $t+1=0$

$t-6+6=0+6$ | $t+1-1=0-1$

$t=6$ | $t=-1$

check: $\frac{2}{3-t}=\frac{-t}{t+3}$

$$\frac{2}{3-6}\overset{?}{=}\frac{-6}{6+3}$$
$$-\frac{2}{3}\overset{?}{=}-\frac{2\cdot 3}{3\cdot 3}$$
$$-\frac{2}{3}=-\frac{2}{3}$$

check: $\frac{2}{3-t}=\frac{-t}{t+3}$

$$\frac{2}{3-(-1)}\overset{?}{=}\frac{-(-1)}{-1+3}$$
$$\frac{2}{2\cdot 2}\overset{?}{=}\frac{1}{2}$$
$$\frac{1}{2}=\frac{1}{2}$$

These check, so the solutions are -1 and 6.

69. $\left.\begin{array}{l}8=2\cdot 2\cdot 2\\ 10=2\cdot 5\\ y=y\end{array}\right\}\text{LCD}=2\cdot 2\cdot 2\cdot 5\cdot y=40y$

$$\frac{1}{8}+\frac{2}{y}=\frac{1}{y}+\frac{1}{10}$$
$$\mathbf{40y}\left(\frac{1}{8}+\frac{2}{y}\right)=\mathbf{40y}\left(\frac{1}{y}+\frac{1}{10}\right)$$
$$40y\left(\frac{1}{8}\right)+40y\left(\frac{2}{y}\right)=40y\left(\frac{1}{y}\right)+40y\left(\frac{1}{10}\right)$$
$$5\cdot\cancel{8}y\left(\frac{1}{\cancel{8}}\right)+40\cdot\cancel{y}\left(\frac{2}{\cancel{y}}\right)=40\cdot\cancel{y}\left(\frac{1}{\cancel{y}}\right)+4\cdot\cancel{10}y\left(\frac{1}{\cancel{10}}\right)$$
$$5y+80=40+4y$$
$$5y+80-4y=40+4y-4y$$
$$y+80-80=40-80$$
$$y=-40$$

This checks, so the solution is -40.

71. $\text{LCD}=x+1$

$$4-\frac{8}{x+1}=\frac{8x}{x+1}$$
$$(\mathbf{x+1})\left(4-\frac{8}{x+1}\right)=(\mathbf{x+1})\left(\frac{8x}{x+1}\right)$$
$$(x+1)(4)-(x+1)\left(\frac{8}{x+1}\right)=(x+1)\left(\frac{8x}{x+1}\right)$$
$$4x+4-\cancel{(x+1)}\left(\frac{8}{\cancel{x+1}}\right)=\cancel{(x+1)}\left(\frac{8x}{\cancel{x+1}}\right)$$
$$4x+4-8=8x$$
$$4x-4-\mathbf{4x}=8x-\mathbf{4x}$$
$$-4=4x$$
$$\frac{-\cancel{4}}{\cancel{4}}=\frac{\cancel{4}x}{\cancel{4}}$$
$$-1=x$$

check: $4-\dfrac{8}{x+1}=\dfrac{8x}{x+1}$

-1 makes the denominators of the original equation 0. No solution. -1 is extraneous.

73. $\text{LCD}=a+1$

$$\frac{5a}{a+1}-4=\frac{3}{a+1}$$
$$(\mathbf{a+1})\left(\frac{5a}{a+1}-4\right)=(\mathbf{a+1})\left(\frac{3}{a+1}\right)$$
$$(a+1)\left(\frac{5a}{a+1}\right)-(a+1)(4)=(a+1)\left(\frac{3}{a+1}\right)$$
$$\cancel{(a+1)}\left(\frac{5a}{\cancel{a+1}}\right)-(a+1)(4)=\cancel{(a+1)}\left(\frac{3}{\cancel{a+1}}\right)$$
$$5a-4(a+1)=3$$
$$5a-4a-4=3$$
$$a-4=3$$
$$a-4+\mathbf{4}=3+\mathbf{4}$$
$$a=7$$

check: $\dfrac{5a}{a+1}-4=\dfrac{3}{a+1}$

$$\frac{5(7)}{7+1}-4\stackrel{?}{=}\frac{3}{7+1}$$
$$\frac{35}{8}-4\cdot\frac{\mathbf{8}}{\mathbf{8}}\stackrel{?}{=}\frac{3}{8}$$
$$\frac{35-32}{8}\stackrel{?}{=}\frac{3}{8}$$
$$\frac{3}{8}=\frac{3}{8}$$

This checks, so the solution is 7.

75. $\text{LCD}=y+1$

$$\frac{2}{y+1}+5=\frac{12}{y+1}$$
$$(\mathbf{y+1})\left(\frac{2}{y+1}+5\right)=(\mathbf{y+1})\left(\frac{12}{y+1}\right)$$
$$(y+1)\left(\frac{2}{y+1}\right)+(y+1)(5)=(y+1)\left(\frac{12}{y+1}\right)$$
$$\cancel{(y+1)}\left(\frac{2}{\cancel{y+1}}\right)+(y+1)(5)=\cancel{(y+1)}\left(\frac{12}{\cancel{y+1}}\right)$$
$$2+5(y+1)=12$$
$$2+5y+5=12$$
$$5y+7-\mathbf{7}=12-\mathbf{7}$$
$$5y=5$$
$$\frac{5y}{\mathbf{5}}=\frac{5}{\mathbf{5}}$$
$$y=1$$

This checks, so the solution is 1.

77. LCD $= 2(x+1)$

$$\frac{3}{x+1}=\frac{x-2}{x+1}+\frac{x-2}{2}$$

$$\mathbf{2(x+1)}\left(\frac{3}{x+1}\right)=\mathbf{2(x+1)}\left(\frac{x-2}{x+1}+\frac{x-2}{2}\right)$$

$$2\,\cancel{(x+1)}\left(\frac{3}{\cancel{x+1}}\right)=2\,\cancel{(x+1)}\left(\frac{x-2}{\cancel{x+1}}\right)+\cancel{2}(x+1)\left(\frac{x-2}{\cancel{2}}\right)$$

$$6=2(x-2)+(x+1)(x-2)$$
$$6=2x-4+x^2-x-2$$
$$6-\mathbf{6}=x^2+x-6-\mathbf{6}$$
$$0=x^2+x-12$$
$$0=(x-3)(x+4)$$

$x-3=0$ or $x+4=0$

$x-3+3=0+3$ | $x+4-4=0-4$

$x=3$ | $x=-4$

This checks, so the solutions are 3 and -4.

79. LCD $=(z-3)(z+1)$

$$\frac{z-4}{z-3}=\frac{z+2}{z+1}$$

$$\mathbf{(z-3)(z+1)}\left(\frac{z-4}{z-3}\right)=\mathbf{(z-3)(z+1)}\left(\frac{z+2}{z+1}\right)$$

$$\cancel{(z-3)}(z+1)\left(\frac{z-4}{\cancel{z-3}}\right)=(z-3)\,\cancel{(z+1)}\left(\frac{z+2}{\cancel{z+1}}\right)$$

$$(z+1)(z-4)=(z-3)(z+2)$$
$$z^2-3z-4=(z-3)(z+2)$$
$$z^2-3z-4=z^2-z-6$$
$$z^2-3z-4-\mathbf{z^2}=z^2-z-6-\mathbf{z^2}$$
$$-3z-4=-z-6$$
$$-3z-4+\mathbf{3z}=-z-6+\mathbf{3z}$$
$$-4=2z-6$$
$$-4+\mathbf{6}=2z-6+\mathbf{6}$$
$$2=2z$$
$$\frac{2}{\mathbf{2}}=\frac{2z}{\mathbf{2}}$$
$$1=z$$

This checks, so the solution is 1.

81. LCD $= x$

$$\frac{3}{x}+2=3$$
$$\mathbf{x}\left(\frac{3}{x}+2\right)=\mathbf{x}(3)$$
$$x\left(\frac{3}{x}\right)+x(2)=x(3)$$
$$\cancel{x}\left(\frac{3}{\cancel{x}}\right)+2x=3x$$
$$3+2x-\mathbf{2x}=3x-\mathbf{2x}$$
$$3=x$$

check: $\frac{3}{x}+2=3$

$\frac{3}{3}+2\overset{?}{=}3$

$1+2\overset{?}{=}3$

$3=3$

This checks, so the solution is 3.

83. $\left.\begin{aligned} y+2&=y+2\\ y-2&=y-2\\ y^2-4&=(y+2)(y-2)\end{aligned}\right\}$ LCD $=(y+2)(y-2)$

$$\frac{4}{y^2-4}=\frac{1}{y-2}+\frac{1}{y+2}$$

$$\mathbf{(y+2)(y-2)}\cdot\frac{4}{(y+2)(y-2)}$$
$$=\mathbf{(y+2)(y-2)}\cdot\frac{1}{y-2}+\mathbf{(y+2)(y-2)}\cdot\frac{1}{y+2}$$

$$\frac{4\,\cancel{(y+2)}\,\cancel{(y-2)}}{\cancel{(y+2)}\,\cancel{(y-2)}}=\frac{\cancel{(y-2)}(y+2)}{\cancel{(y-2)}}+\frac{\cancel{(y+2)}(y-2)}{\cancel{(y+2)}}$$

$$4=y+2+y-2$$
$$4=2y$$
$$\frac{4}{\mathbf{2}}=\frac{2y}{\mathbf{2}}$$
$$2=y$$

check: $\frac{4}{y^2-4}=\frac{1}{y-2}+\frac{1}{y+2}$

$$\frac{4}{\mathbf{(2)}^2-4}=\frac{1}{\mathbf{2}-2}+\frac{1}{\mathbf{2}+2}$$
$$\frac{4}{\mathbf{0}}=\frac{1}{\mathbf{0}}+\frac{1}{4}$$

2 makes two denominators of the original equation 0. No solution. 2 is extraneous.

85. $\left.\begin{aligned} 3 &= 3 \\ 5d &= 5 \cdot d \\ 10d &= 2 \cdot 5 \cdot d \end{aligned}\right\} \text{LCD} = 2 \cdot 3 \cdot 5 \cdot d = 30d$

$$\frac{3}{5d} + \frac{4}{3} = \frac{9}{10d}$$

$$\mathbf{30d}\left(\frac{3}{5d} + \frac{4}{3}\right) = \mathbf{30d}\left(\frac{9}{10d}\right)$$

$$30d\left(\frac{3}{5d}\right) + 30d\left(\frac{4}{3}\right) = 30d\left(\frac{9}{10d}\right)$$

$$6 \cdot \overset{1}{\cancel{5d}}\left(\frac{3}{\underset{1}{\cancel{5d}}}\right) + 10 \cdot \overset{1}{\cancel{3}}\, d\left(\frac{4}{\underset{1}{\cancel{3}}}\right) = 3 \cdot \overset{1}{\cancel{10d}}\left(\frac{9}{\underset{1}{\cancel{10d}}}\right)$$

$$18 + 40d = 27$$

$$18 + 40d - \mathbf{18} = 27 - \mathbf{18}$$

$$\frac{\overset{1}{\cancel{40}}\, d}{\underset{1}{\cancel{\mathbf{40}}}} = \frac{9}{\mathbf{40}}$$

$$d = \frac{9}{40}$$

This checks, so the solution is $\frac{9}{40}$.

87. $\left.\begin{aligned} n+3 &= n+3 \\ n-3 &= n-3 \\ n^2-9 &= (n+3)(n-3) \end{aligned}\right\} \text{LCD} = (n+3)(n-3)$

$$\frac{n}{n^2-9} + \frac{n+8}{n+3} = \frac{n-8}{n-3}$$

$$\mathbf{(n+3)(n-3)} \cdot \frac{n}{(n+3)(n-3)} + \mathbf{(n+3)(n-3)} \cdot \frac{n+8}{n+3} = \mathbf{(n+3)(n-3)} \cdot \frac{n-8}{n-3}$$

$$\frac{n\,\overset{1}{\cancel{(n+3)}}\,\overset{1}{\cancel{(n-3)}}}{\underset{1}{\cancel{(n+3)}}\,\underset{1}{\cancel{(n-3)}}} + \frac{(n+8)\,\overset{1}{\cancel{(n+3)}}\,(n-3)}{\underset{1}{\cancel{(n+3)}}} = \frac{(n-8)(n+3)\,\overset{1}{\cancel{(n-3)}}}{\underset{1}{\cancel{(n-3)}}}$$

$$n + n^2 + 5n - 24 = n^2 - 5n - 24$$

$$n^2 + 6n - 24 - \mathbf{n^2} = n^2 - 5n - 24 - \mathbf{n^2}$$

$$6n - 24 + \mathbf{24} = -5n - 24 + \mathbf{24}$$

$$6n + \mathbf{5n} = -5n + \mathbf{5n}$$

$$11n = 0$$

$$\frac{11n}{\mathbf{11}} = \frac{0}{\mathbf{11}}$$

$$n = 0$$

check: $\frac{n}{n^2-9} + \frac{n+8}{n+3} = \frac{n-8}{n-3}$

$$\frac{0}{0^2-9} + \frac{0+8}{0+3} \overset{?}{=} \frac{0-8}{0-3}$$

$$\frac{8}{3} \overset{?}{=} \frac{-8}{-3}$$

$$\frac{8}{3} = \frac{8}{3}$$

This checks, so the solution is 0.

89. $\left.\begin{aligned} x &= x \\ x-2 &= x-2 \\ x^2-2x &= x(x-2) \end{aligned}\right\} \text{LCD} = x(x-2)$

$$\frac{3}{x-2} + \frac{1}{x} = \frac{6x+4}{x^2-2x}$$

$$\mathbf{x(x-2)} \cdot \frac{3}{x-2} + \mathbf{x(x-2)} \cdot \frac{1}{x} = \mathbf{x(x-2)} \cdot \frac{6x+4}{x(x-2)}$$

$$\frac{3x\,\overset{1}{\cancel{(x-2)}}}{\underset{1}{\cancel{(x-2)}}} + \frac{1\,\overset{1}{\cancel{x}}\,(x-2)}{\underset{1}{\cancel{x}}} = \frac{(6x+4)\,\overset{1}{\cancel{x}}\,\overset{1}{\cancel{(x-2)}}}{\underset{1}{\cancel{x}}\,\underset{1}{\cancel{(x-2)}}}$$

$$3x + x - 2 = 6x + 4$$

$$4x - 2 = 6x + 4$$

$$4x - 2 - \mathbf{4x} = 6x + 4 - \mathbf{4x}$$

$$-2 = 2x + 4$$

$$-2 - \mathbf{4} = 2x + 4 - \mathbf{4}$$

$$-6 = 2x$$

$$\frac{-6}{\mathbf{2}} = \frac{2x}{\mathbf{2}}$$

$$-3 = x$$

check:

$$\frac{3}{x-2} + \frac{1}{x} = \frac{6x+4}{x^2-2x}$$

$$\frac{3}{-3-2} + \frac{1}{-3} \overset{?}{=} \frac{6(-3)+4}{(-3)^2-2(-3)}$$

$$\frac{-3}{5} + \frac{-1}{3} \overset{?}{=} -\frac{14}{15}$$

$$\frac{-3}{5} \cdot \frac{\mathbf{3}}{\mathbf{3}} + \frac{-1}{3} \cdot \frac{\mathbf{5}}{\mathbf{5}} \overset{?}{=} -\frac{14}{15}$$

$$\frac{-9-5}{15} \overset{?}{=} -\frac{14}{15}$$

$$-\frac{14}{15} = -\frac{14}{15}$$

This checks, so the solution is -3.

91. $\left.\begin{aligned} 3 &= 3 \\ 3y-9 &= 3(y-3) \end{aligned}\right\} \text{LCD} = 3(y-3)$

$$y+\frac{2}{3}=\frac{2y-12}{3y-9}$$

$$\mathbf{3(y-3)}\cdot\left(y+\frac{2}{3}\right)=\mathbf{3(y-3)}\cdot\left(\frac{2y-12}{3y-9}\right)$$

$$y\cdot 3(y-3)+\left(\frac{2}{3}\right)\cdot 3(y-3)=\left(\frac{2y-12}{3y-9}\right)\cdot 3(y-3)$$

$$3y^2-9y+\frac{2\cdot \overset{1}{\cancel{3}}(y-3)}{\underset{1}{\cancel{3}}}=\frac{(2y-12)\,\overset{1}{\cancel{3(y-3)}}}{\underset{1}{\cancel{3(y-3)}}}$$

$$3y^2-9y+2y-6=2y-12$$

$$3y^2-7y-6-\mathbf{2y+12}=2y-12-\mathbf{2y+12}$$

$$3y^2-9y+6=0$$

$$3(y^2-3y+2)=0$$

$$3(y-2)(y-1)=0$$

$y-2=0$	or	$y-1=0$
$y-2+2=0+2$		$y-1+1=0+1$
$y=2$		$y=1$

Both check, so the solutions are 2 and 1.

93. $\left.\begin{aligned} 2 &= 2 \\ 7 &= 7 \\ 14 &= 2\cdot 7 \end{aligned}\right\} \text{LCD} = 2\cdot 7 = 14$

$$\frac{a-1}{7}-\frac{a-2}{14}=\frac{1}{2}$$

$$\mathbf{14}\left(\frac{a-1}{7}-\frac{a-2}{14}\right)=\mathbf{14}\left(\frac{1}{2}\right)$$

$$14\left(\frac{a-1}{7}\right)-14\left(\frac{a-2}{14}\right)=14\left(\frac{1}{2}\right)$$

$$2\cdot\overset{1}{\cancel{7}}\left(\frac{a-1}{\underset{1}{\cancel{7}}}\right)-\overset{1}{\cancel{14}}\left(\frac{a-2}{\underset{1}{\cancel{14}}}\right)=7\cdot\overset{1}{\cancel{2}}\left(\frac{1}{\underset{1}{\cancel{2}}}\right)$$

$$2(a-1)-(a-2)=7$$

$$2a-2-a+2=7$$

$$a=7$$

This checks, so the solution is 7.

LOOK ALIKES . . .

For each expression, perform the indicated operations and then simplify, if possible. Solve each equation and check the results.

95. a. $\frac{a}{3}+\frac{3}{5}+\frac{a}{15}=\frac{\mathbf{5}}{\mathbf{5}}\cdot\frac{a}{3}+\frac{\mathbf{3}}{\mathbf{3}}\cdot\frac{3}{5}+\frac{a}{15}$

$$=\frac{5a+9+a}{15}$$

$$=\frac{6a+9}{15}$$

$$=\frac{\overset{1}{\cancel{3}}(2a+3)}{\underset{1}{\cancel{3}}\cdot 5}$$

$$=\frac{2a+3}{5}$$

b. $\frac{a}{3}+\frac{3}{5}=\frac{a}{15}$

$$\mathbf{15}\left(\frac{a}{3}+\frac{3}{5}\right)=\mathbf{15}\cdot\frac{a}{15}$$

$$5\cdot\overset{1}{\cancel{3}}\left(\frac{a}{\underset{1}{\cancel{3}}}\right)+3\cdot\overset{1}{\cancel{5}}\left(\frac{3}{\underset{1}{\cancel{5}}}\right)=\overset{1}{\cancel{15}}\cdot\frac{a}{\underset{1}{\cancel{15}}}$$

$$5a+9=a$$

$$5a+9-\mathbf{5a}=a-\mathbf{5a}$$

$$9=-4a$$

$$\frac{9}{\mathbf{-4}}=\frac{-4a}{\mathbf{-4}}$$

$$a=-\frac{9}{4}$$

This checks, so the solution is $-\frac{9}{4}$.

97. a. $\frac{x}{x-2}-\frac{1}{x-3}$

$$=\frac{\mathbf{(x-3)}}{\mathbf{(x-3)}}\cdot\frac{x}{(x-2)}-\frac{\mathbf{(x-2)}}{\mathbf{(x-2)}}\cdot\frac{1}{(x-3)}$$

$$=\frac{x^2-3x-(x-2)}{(x-2)(x-3)}$$

$$=\frac{x^2-3x-(x-2)}{(x-2)(x-3)}$$

$$=\frac{x^2-3x-x+2}{(x-2)(x-3)}$$

$$=\frac{x^2-4x+2}{(x-2)(x-3)}$$

b. $$\frac{x}{x-2}-\frac{1}{x-3}=1$$

$$\mathbf{(x-2)(x-3)}\left(\frac{x}{x-2}-\frac{1}{x-3}\right)=\mathbf{(x-2)(x-3)}1$$

$$(x-3)(x-2)\left(\frac{x}{x-2}\right)-(x-2)(x-3)\left(\frac{1}{x-3}\right)=(x-2)(x-3)$$

$$x(x-3)-(x-2)=(x-2)(x-3)$$

$$x^2-3x-x+2=x^2-5x+6$$

$$x^2-4x+2-\mathbf{x^2}=x^2-5x+6-\mathbf{x^2}$$

$$-4x+2+\mathbf{5x}=-5x+6+\mathbf{5x}$$

$$x+2-\mathbf{2}=6-\mathbf{2}$$

$$x=4$$

This checks, so the solution is 4.

APPLICATION

99. MEDICINE

LCD $= R+B$

$$H=\frac{RB}{R+B}$$

$$\mathbf{(R+B)}\cdot H=\mathbf{(R+B)}\cdot\frac{RB}{R+B}$$

$$(R+B)\cdot H=(R+B)\cdot\frac{RB}{R+B}$$

$$H(R+B)=RB$$

$$HR+HB=RB$$

$$HR+HB-\mathbf{HR}=RB-\mathbf{HR}$$

$$HB=R(B-H)$$

$$\frac{HB}{\mathbf{B-H}}=\frac{R(B-H)}{\mathbf{(B-H)}}$$

$$\frac{HB}{B-H}=R$$

$$R=\frac{HB}{B-H}$$

101. ELECTRONICS

LCD $= rr_1r_2$

$$\frac{1}{r}=\frac{1}{r_1}+\frac{1}{r_2}$$

$$\mathbf{(rr_1r_2)}\cdot\frac{1}{r}=\mathbf{(rr_1r_2)}\cdot\left(\frac{1}{r_1}+\frac{1}{r_2}\right)$$

$$r r_1r_2\cdot\frac{1}{r}=r r_1 r_2\frac{1}{r_1}\cdot+rr_1 r_2\cdot\frac{1}{r_2}$$

$$r_1r_2=rr_2+rr_1$$

$$r_1r_2=r(r_2+r_1)$$

$$\frac{r_1r_2}{\mathbf{r_2+r_1}}=\frac{r(r_2+r_1)}{(r_2+r_1)}$$

$$\frac{r_1r_2}{r_2+r_1}=r$$

$$r=\frac{r_1r_2}{r_2+r_1}$$

WRITING

103-105. Answers will vary.

REVIEW

107. UNIFORMS

Analyze

- Cost per uniform is \$18.50
- One time set up fee is \$75.00
- Total cost is \$445.00
- Find the total number of uniforms

Assign

Let x = the total number of uniforms

Form

	Number	• Value	+ Fee =	Total value
Uniforms	x	18.50	75	445

The number of uniforms	times	the value of one unifirm	plus	the fee	equals	the total value.
x	$\cdot$	18.50	+	75	=	445

Solve

$$18.50x+75=445$$

$$18.50x+75-75=445-75$$

$$\frac{18.50x}{18.50}=\frac{370}{18.50}$$

$$x=20$$

State

The number of uniforms widgets is 20.

Check

The results check.

CHALLENGE PROBLEMS

Solve each equation and check the results. If an equation has not solution, so indicate.

109. $\text{LCD} = (x-3)$

$$\frac{x-4}{x-3} + \frac{x-2}{x-3} = x-3$$

$$(x-3)\cdot\left(\frac{x-4}{x-3}\right) + (x-3)\cdot\left(\frac{x-2}{x-3}\right) = (x-3)(x-3)$$

$$\frac{\overset{1}{\cancel{(x-3)}}(x-4)}{\underset{1}{\cancel{(x-3)}}} + \frac{\overset{1}{\cancel{(x-3)}}(x-2)}{\underset{1}{\cancel{(x-3)}}} = x^2 - 6x + 9$$

$$(x-4) + (x-2) = x^2 - 6x + 9$$

$$2x - 6 = x^2 - 6x + 9$$

$$2x - 6 - \mathbf{2x} = x^2 - 6x + 9 - \mathbf{2x}$$

$$-6 = x^2 - 8x + 9$$

$$-6 + \mathbf{6} = x^2 - 8x + 9 + \mathbf{6}$$

$$0 = x^2 - 8x + 15$$

$$0 = (x-5)(x-3)$$

	or	
$x - 5 = 0$		$x - 3 = 0$
$x - 5 + \mathbf{5} = 0 + \mathbf{5}$		$x - 3 + \mathbf{3} = 0 + \mathbf{3}$
$x = 5$		$x = 3$

check: $$\frac{x-4}{x-3} + \frac{x-2}{x-3} = x-3$$

$$\frac{\mathbf{5}-4}{\mathbf{5}-3} + \frac{\mathbf{5}-2}{\mathbf{5}-3} \stackrel{?}{=} \mathbf{5}-3$$

$$\frac{1}{2} + \frac{3}{2} \stackrel{?}{=} 2$$

$$\frac{4}{2} \stackrel{?}{=} 2$$

$$2 = 2$$

This checks, so the solution is 5.

check: $$\frac{x-4}{x-3} + \frac{x-2}{x-3} = x-3$$

$$\frac{\mathbf{3}-4}{\mathbf{3}-3} - \frac{\mathbf{3}-2}{\mathbf{3}-3} \stackrel{?}{=} \mathbf{3}-3$$

$$\frac{-1}{\mathbf{0}} - \frac{1}{\mathbf{0}} \stackrel{?}{=} 0$$

3 makes the denominators of the original equation 0. Not a solution. 3 is extraneous.

The only solution is 5.

111. $$\left.\begin{array}{l} x = x \\ x^2 = x \cdot x \end{array}\right\} \text{LCD} = x \cdot x = x^2$$

Solve

$$x^{-2} + 2x^{-1} + 1 = 0$$

$$\frac{1}{x^2} + \frac{2}{x} + 1 = 0$$

$$\left(\frac{1}{x^2} + \frac{2}{x} + 1\right)\cdot \boldsymbol{x^2} = 0 \cdot \boldsymbol{x^2}$$

$$\frac{1}{x^2}\left(x^2\right) + \frac{2}{x}\left(x^2\right) + 1\left(x^2\right) = 0$$

$$1 + 2x + x^2 = 0$$

$$x^2 + 2x + 1 = 0$$

$$(x+1)(x+1) = 0$$

$$x + 1 = 0$$

$$x + 1 - 1 = 0 - 1$$

$$x = -1$$

This checks, so the solution is -1.

SECTION 7.7 STUDY SET
VOCABULARY

Fill in the blanks.

1. In this section, problems that involve:
   - moving vehicles are called uniform **motion** problems.
   - depositing money are called **investment** problems.
   - people completing jobs are called shared-**work** problems.

CONCEPTS

3. $\frac{5+x}{8+x}=\frac{2}{3}$ **(iii)**

5. a. $\frac{\mathbf{1}}{\mathbf{45}}$ of the job per minute b. $\frac{\mathbf{x}}{\mathbf{4}}$

7. *a.*
$$d=rt$$
$$\frac{d}{\mathbf{r}}=\frac{rt}{\mathbf{r}}$$
$$\frac{d}{r}=t$$
$$\mathbf{t}=\frac{\mathbf{d}}{\mathbf{r}}$$

b.
$$I=Prt$$
$$\frac{I}{\mathbf{rt}}=\frac{Prt}{\mathbf{rt}}$$
$$\frac{I}{rt}=P$$
$$\mathbf{P}=\frac{\mathbf{I}}{\mathbf{rt}}$$

9.

	Rate	· Time	= Work completed
1st printer	$\frac{1}{15}$	x	$\frac{\mathbf{x}}{\mathbf{15}}$
2nd printer	$\frac{1}{8}$	x	$\frac{\mathbf{x}}{\mathbf{8}}$

NOTATION

11. $\frac{55}{9}=9\overline{)55}=\mathbf{6\frac{1}{9}}$ days (quotient 6; 55 − 54 = 1)

GUIDED PRACTICE

Solve each of these number problems. See Example 1.

13.

Analyze

- Begin with the fraction $\frac{2}{5}$.
- Add the same number to the numerator and denominator.
- The results is $\frac{2}{3}$.
- Find the number.

Assign

Let n = the unknown number.

Form

$$\frac{2+n}{5+n}=\frac{2}{3}$$

Solve

$$\frac{2+n}{5+n}=\frac{2}{3}$$
$$\text{LCD}=3(n+5)$$
$$\mathbf{3(n+5)}\left(\frac{2+n}{5+n}\right)=\mathbf{3(n+5)}\left(\frac{2}{3}\right)$$
$$3\cancel{(n+5)}\left(\frac{2+n}{\cancel{5+n}}\right)=\cancel{3}(n+5)\left(\frac{2}{\cancel{3}}\right)$$
$$3(n+2)=2(n+5)$$
$$3n+6=2n+10$$
$$3n+6-\mathbf{2n}=2n+10-\mathbf{2n}$$
$$n+6=10$$
$$n+6-\mathbf{6}=10-\mathbf{6}$$
$$n=4$$

State

The number is 4.

Check

$$\frac{2+n}{5+n}=\frac{2+\mathbf{4}}{5+\mathbf{4}}=\frac{6}{9}=\frac{2\cdot\cancel{3}^{1}}{3\cdot\cancel{3}_{1}}=\frac{2}{3}$$

The result checks.

15.

Analyze

- Begin with the fraction $\frac{3}{4}$.
- Double the numerator.
- Add a number to the denominator.
- The results is 1.
- Find the number.

Assign

Let n = the unknown number.

Form

$$\frac{3(2)}{4+n}=1$$

Solve

$$\frac{6}{4+n}=1$$

$$\text{LCD}=n+4$$

$$(n+4)\left(\frac{6}{n+4}\right)=(n+4)(1)$$

$$\overset{1}{\cancel{(n+4)}}\left(\frac{6}{\underset{1}{\cancel{n+4}}}\right)=n+4$$

$$6=n+4$$

$$6-\mathbf{4}=n+4-\mathbf{4}$$

$$2=n$$

State

The number is 2.

Check

$$\frac{6}{4+n}=\frac{6}{4+\mathbf{2}}=\frac{6}{6}=1$$

The result checks.

17.

Analyze

- Begin with the fraction $\frac{3}{4}$.
- Add a number to the numerator.
- Twice as much is added to the denominator.
- The results is $\frac{4}{7}$.
- Find the number.

Assign

Let n = the unknown number.

Form

$$\frac{3+n}{4+2n}=\frac{4}{7}$$

Solve

$$\frac{3+n}{4+2n}=\frac{4}{7}$$

$$\text{LCD}=7(4+2n)$$

$$\mathbf{7(4+2n)}\left(\frac{3+n}{4+2n}\right)=\mathbf{7(4+2n)}\left(\frac{4}{7}\right)$$

$$7\cancel{(4+2n)}\left(\frac{3+n}{\cancel{4+2n}}\right)=\cancel{7}(4+2n)\left(\frac{4}{\cancel{7}}\right)$$

$$7(3+n)=(4+2n)4$$

$$21+7n=16+8n$$

$$21+7n-\mathbf{7n}=16+8n-\mathbf{7n}$$

$$21=16+n$$

$$21-\mathbf{16}=16+n-\mathbf{16}$$

$$5=n$$

State

The number is 5.

Check

$$\frac{3+n}{4+2n}=\frac{3+\mathbf{5}}{4+2\cdot\mathbf{5}}=\frac{3+5}{4+10}=\frac{8}{14}=\frac{4\cdot\overset{1}{\cancel{2}}}{7\cdot\underset{1}{\cancel{2}}}=\frac{4}{7}$$

The result checks.

19.

Analyze

- The sum of a number and its reciprocal is $\frac{13}{6}$.
- Find the numbers.

Assign

Let n = the unknown number.

$\frac{1}{n}=$ the reciprocal

Form

$$n+\frac{1}{n}=\frac{13}{6}$$

Solve

$$n+\frac{1}{n}=\frac{13}{6}$$
$$\text{LCD}=6n$$
$$\mathbf{6n}\left(n+\frac{1}{n}\right)=\mathbf{6n}\left(\frac{13}{6}\right)$$
$$6n(n)+6\cancel{n}\left(\frac{1}{\cancel{n}}\right)=\cancel{6}n\left(\frac{13}{\cancel{6}}\right)$$
$$6n^2+6=13n$$
$$6n^2+6-\mathbf{13n}=13n-\mathbf{13n}$$
$$6n^2-13n+6=0$$
$$(3n-2)(2n-3)=0$$

$3n-2=0$	or	$2n-3=0$
$3n-2+2=0+2$		$2n-3+3=0+3$
$3n=2$		$2n=3$
$\frac{3n}{3}=\frac{2}{3}$		$\frac{2n}{2}=\frac{3}{2}$
$n=\frac{2}{3}$		$n=\frac{3}{2}$

State

The numbers are $\frac{2}{3}$ and $\frac{3}{2}$.

Check

$$x+\frac{1}{x}=\frac{13}{6} \qquad x+\frac{1}{x}=\frac{13}{6}$$
$$\frac{2}{3}+\frac{1}{\frac{2}{3}}\stackrel{?}{=}\frac{13}{6} \qquad \frac{3}{2}+\frac{1}{\frac{3}{2}}\stackrel{?}{=}\frac{13}{6}$$
$$\frac{2}{3}+\frac{3}{2}\stackrel{?}{=}\frac{13}{6} \qquad \frac{3}{2}+\frac{2}{3}\stackrel{?}{=}\frac{13}{6}$$
$$\frac{2}{3}\cdot\frac{\mathbf{2}}{\mathbf{2}}+\frac{3}{2}\cdot\frac{\mathbf{3}}{\mathbf{3}}\stackrel{?}{=}\frac{13}{6} \qquad \frac{3}{2}\cdot\frac{\mathbf{3}}{\mathbf{3}}+\frac{2}{3}\cdot\frac{\mathbf{2}}{\mathbf{2}}\stackrel{?}{=}\frac{13}{6}$$
$$\frac{4+9}{6}\stackrel{?}{=}\frac{13}{6} \qquad \frac{9+4}{6}\stackrel{?}{=}\frac{13}{6}$$
$$\frac{13}{6}=\frac{13}{6} \qquad \frac{13}{6}=\frac{13}{6}$$

Both results check.

APPLICATIONS

21. COOKING

Analyze

- Begin with the fraction $\frac{1}{4}$.
- Add the same number to the numerator and denominator.
- The results is $\frac{3}{4}$.
- Find the number.

Assign

Let n = the unknown number.

Form

$$\frac{1+n}{4+n}=\frac{3}{4}$$

Solve

$$\frac{1+n}{4+n}=\frac{3}{4}$$
$$\text{LCD}=4(4+n)$$
$$\mathbf{4(4+n)}\left(\frac{1+n}{4+n}\right)=\mathbf{4(4+n)}\left(\frac{3}{4}\right)$$
$$4\cancel{(4+n)}\left(\frac{1+n}{\cancel{4+n}}\right)=\cancel{4}(4+n)\left(\frac{3}{\cancel{4}}\right)$$
$$4(1+n)=3(4+n)$$
$$4+4n=12+3n$$
$$4+4n-\mathbf{3n}=12+3n-\mathbf{3n}$$
$$4+n=12$$
$$4+n-\mathbf{4}=12-\mathbf{4}$$
$$n=8$$

State

The number is 8.

Check

$$\frac{1+n}{4+n}=\frac{1+\mathbf{8}}{4+\mathbf{8}}=\frac{9}{12}=\frac{3\cdot\cancel{3}}{4\cdot\cancel{3}}=\frac{3}{4}$$

The result checks.

23. TOUR de FRANCE

Analyze

- Garin rode 80 miles.
- Armstrong rode 130 miles.
- Times are the same.
- Armstrong was 10 mph faster than Garin.
- What was each cyclist average speed?

Assign

Let r = Garin's speed in mph

$r+10$ = Armstrong's speed in mph

Form

	Rate •	Time =	Distance
Garin	r	$\frac{80}{r}$	80
Armstrong	$r+10$	$\frac{130}{r+10}$	130

Time took Garin to cycle 80 miles equals Time took Armstrong to cycle 130 miles

$$\frac{80}{r} = \frac{130}{r+10}$$

Solve

$$\frac{80}{r} = \frac{130}{r+10}$$

$$\text{LCD} = r(r+10)$$

$$\mathbf{r(r+10)}\left(\frac{80}{r}\right) = \mathbf{r(r+10)}\left(\frac{130}{r+10}\right)$$

$$\cancel{r}(r+10)\left(\frac{80}{\cancel{r}}\right) = r\,\cancel{(r+10)}\left(\frac{130}{\cancel{r+10}}\right)$$

$$80(r+10) = 130r$$

$$80r + 800 = 130r$$

$$80r + 800 - \mathbf{80r} = 130r - \mathbf{80r}$$

$$800 = 50r$$

$$\frac{800}{\mathbf{50}} = \frac{50r}{\mathbf{50}}$$

$$16 = r$$

State

Garin's speed was 16 mph.

Armstrong's speed was 26 mph.

Check

The results check.

25. PACKAGING FRUIT

Analyze

- Shorter belt is 100 feet.
- Longer belt is 300 feet.
- Times are the same.
- Longer belt 1 foot per second faster.
- What is each belt's speed?

Assign

Let r = Shorter belt's speed in fps

$r+1$ = Longer belt's speed in fps

Form

	Rate •	Time =	Distance
Shorter	r	$\frac{100}{r}$	100
Longer	$r+1$	$\frac{300}{r+1}$	300

Time for shorter to travel 100 ft equals Time for longer to travel 300 ft

$$\frac{100}{r} = \frac{300}{r+1}$$

Solve

$$\frac{100}{r} = \frac{300}{r+1}$$

$$\text{LCD} = r(r+1)$$

$$\mathbf{r(r+1)}\left(\frac{100}{r}\right) = \mathbf{r(r+1)}\left(\frac{300}{r+1}\right)$$

$$\cancel{r}(r+1)\left(\frac{100}{\cancel{r}}\right) = r \cdot \cancel{(r+1)}\left(\frac{300}{\cancel{r+1}}\right)$$

$$100(r+1) = 300r$$

$$100r + 100 = 300r$$

$$100r + 100 - \mathbf{100r} = 300r - \mathbf{100r}$$

$$100 = 200r$$

$$\frac{100}{\mathbf{200}} = \frac{200r}{\mathbf{200}}$$

$$\frac{1}{2} = r$$

State

Longer belt's speed is $1\frac{1}{2}$ feet per second.

Shorter belt's speed is $\frac{1}{2}$ foot per second.

Check

The results check.

27. BIRDS IN FLIGHT

Analyze

- Canada goose can fly 120 miles.
- Great Blue heron can fly 80 miles.
- Times are the same.
- Goose flies 10 mph faster.
- What are the speeds of both?

Assign

Let r = Heron's speed in mph

$r+10$ = Goose's speed in mph

Form

	Rate	• Time	= Distance
Heron	r	$\frac{80}{r}$	80
Goose	$r+10$	$\frac{120}{r+10}$	120

Time took heron to fly 80 miles equals Time took goose to fly 120 miles

$$\frac{80}{r} = \frac{120}{r+10}$$

Solve

$$\frac{80}{r} = \frac{120}{r+10}$$

$$\text{LCD} = r(r+10)$$

$$\mathbf{r(r+10)}\left(\frac{80}{r}\right) = \mathbf{r(r+10)}\left(\frac{120}{r+10}\right)$$

$$\cancel{r}(r+10)\left(\frac{80}{\cancel{r}}\right) = r\cdot\cancel{(r+10)}\left(\frac{120}{\cancel{r+10}}\right)$$

$$80(r+10) = 120r$$

$$80r + 800 = 120r$$

$$80r + 800 - \mathbf{80r} = 120r - \mathbf{80r}$$

$$800 = 40r$$

$$\frac{800}{\mathbf{40}} = \frac{40r}{\mathbf{40}}$$

$$20 = r$$

State

The Great Blue heron's speed is 20 mph.
The Canada goose's speed is 30 mph.

Check

The results check.

29. WIND SPEED

Analyze

- Rate of plane in still air is 255 mph.
- Flies 300 miles downwind.
- Flies 210 miles upwind.
- Times are the same for both directions.
- What is the speed of the wind?

Assign

Let r = Wind's speed in mph

Form

	Rate	• Time	= Distance
Downwind	$255+r$	$\frac{300}{255+r}$	300
Upwind	$255-r$	$\frac{210}{255-r}$	210

Time took plane to fly 300 miles downwind. equals time took plane to fly 210 miles upwind.

$$\frac{300}{255+r} = \frac{210}{255-r}$$

Solve

$$\frac{300}{255+r} = \frac{210}{255-r}$$

$$\text{LCD} = (255+r)(255-r)$$

$$\mathbf{(255+r)(255-r)}\left(\frac{300}{255+r}\right) = \mathbf{(255+r)(255-r)}\left(\frac{210}{255-r}\right)$$

$$\cancel{(255+r)}(255-r)\left(\frac{300}{\cancel{255+r}}\right) = (255+r)\cancel{(255-r)}\left(\frac{210}{\cancel{255-r}}\right)$$

$$300(255-r) = 210(255+r)$$

$$76{,}500 - 300r = 53{,}550 + 210r$$

$$76{,}500 - 300r + \mathbf{300r} = 53{,}550 + 210r + \mathbf{300r}$$

$$76{,}500 = 53{,}550 + 510r$$

$$76{,}500 - \mathbf{53{,}550} = 53{,}550 + 510r - \mathbf{53{,}550}$$

$$22{,}950 = 510r$$

$$\frac{22{,}950}{\mathbf{510}} = \frac{510r}{\mathbf{510}}$$

$$45 = r$$

State

The wind's speed is 45 mph.

Check

The result checks.

31. ROOFING HOUSES

Analyze

- Homeowner takes 7 days to roof house.
- Professional takes 4 days to roof house.
- How long will it take both of them, working together, to roof the house?

Assign

Let x = the number of days it will take both working together to roof the house.

Form

	Rate	• Time	= Work Completed
Homeowner	$\frac{1}{7}$	x	$\frac{x}{7}$
Professional	$\frac{1}{4}$	x	$\frac{x}{4}$

The part of job done by homeowner. plus The part of job done by professional. equals 1 job completed

$$\frac{x}{7} + \frac{x}{4} = 1$$

Solve

$$\frac{x}{7}+\frac{x}{4}=1$$

LCD = 28

$$\mathbf{28}\left(\frac{x}{7}+\frac{x}{4}\right)=\mathbf{28}(1)$$

$$\cancel{7}^{1}\cdot 4\left(\frac{x}{\cancel{7}_{1}}\right)+\cancel{4}^{1}\cdot 7\left(\frac{x}{\cancel{4}_{1}}\right)=28$$

$$4x+7x=28$$

$$11x=28$$

$$\frac{11x}{\mathbf{11}}=\frac{28}{\mathbf{11}}$$

$$x=\frac{28}{11}$$

$$x=2\frac{6}{11}$$

State

It takes both working together $2\frac{6}{11}$ days to roof the house.

Check

The result checks.

from CAMPUS TO CAREERS

33. RECREATION DIRECTOR

Analyze

- 1st pipe can fill pool in 12 hours.
- 2nd pipe can fill pool in 18 hours.
- Will the pool be filled for a 2:00 p.m. opening if both are turned on a 8:00 a.m. that day?

Assign

Let x = the number of hours it will take both pipes working together to fill the pool.

Form

	Rate	• Time	= Work Completed
1st pipe	$\frac{1}{12}$	x	$\frac{x}{12}$
2nd pipe	$\frac{1}{18}$	x	$\frac{x}{18}$

The part of job done by 1st pipe. plus The part of job done by 2nd pipe. equals 1 job completed

$$\frac{x}{12} + \frac{x}{18} = 1$$

Solve

$$\frac{x}{12}+\frac{x}{18}=1$$

LCD = 36

$$\mathbf{36}\left(\frac{x}{12}+\frac{x}{18}\right)=\mathbf{36}(1)$$

$$\cancel{12}^{1}\cdot 3\left(\frac{x}{\cancel{12}_{1}}\right)+\cancel{18}^{1}\cdot 2\left(\frac{x}{\cancel{18}_{1}}\right)=36$$

$$3x+2x=36$$

$$5x=36$$

$$\frac{5x}{\mathbf{5}}=\frac{36}{\mathbf{5}}$$

$$x=\frac{36}{5}$$

$$x=7\frac{1}{5}$$

State

It takes both working together $7\frac{1}{5}$ hours to fill the pool. The pool opens in 6 hours. No, the pool will not be filled in time.

Check

The result checks.

35. FILLING A POOL

Analyze

- Inlet pipe can fill pool in 4 hours.
- Outlet pipe can drain pool in 8 hours.
- How long will it take the inlet pipe to fill the pool, if the outlet pipe is left open?

Assign

Let x = the number of hours it will take the inlet pipe to fill the pool if the outlet drain is left open.

Form

	Rate	• Time	= Work Completed
Inlet pipe	$\frac{1}{4}$	x	$\frac{x}{4}$
Outlet pipe	$\frac{1}{8}$	x	$\frac{x}{8}$

The part of job done by inlet pipe.	minus	The part of job done by outlet pipe.	equals	1 job completed
$\frac{x}{4}$	$-$	$\frac{x}{8}$	$=$	1

Solve

$$\frac{x}{4}-\frac{x}{8}=1$$

$$\text{LCD}=8$$

$$8\left(\frac{x}{4}-\frac{x}{8}\right)=8(1)$$

$$2\cdot\cancel{4}\left(\frac{x}{\cancel{4}}\right)-\cancel{8}\left(\frac{x}{\cancel{8}}\right)=8$$

$$2x-x=8$$

$$x=8$$

State

It will take the inlet pipe 8 hours to fill the pool with the outlet drain left oepn.

Check

The result checks.

37. GRADING PAPERS

Analyze

- Teacher takes 30 minutes.
- Aide takes 60 minutes.
- How long will it take both of them, working together, to grade a set of papers?

Assign

Let x = the number of minutes it will take both working together to grade tests.

Form

	Rate	• Time	= Work Completed
Teacher	$\frac{1}{30}$	x	$\frac{x}{30}$
Aide	$\frac{1}{60}$	x	$\frac{x}{60}$

The part of job done by teacher.	plus	The part of job done by aide.	equals	1 job completed
$\frac{x}{30}$	$+$	$\frac{x}{60}$	$=$	1

Solve

$$\frac{x}{30}+\frac{x}{60}=1$$

$$\text{LCD}=60$$

$$60\left(\frac{x}{30}+\frac{x}{60}\right)=60(1)$$

$$\cancel{30}\cdot 2\left(\frac{x}{\cancel{30}}\right)+\cancel{60}\left(\frac{x}{\cancel{60}}\right)=60$$

$$2x+x=60$$

$$3x=60$$

$$\frac{3x}{3}=\frac{60}{3}$$

$$x=20$$

State

It takes both working together 20 minutes to grade one set of tests.

Check

The result checks.

39. PRINTERS

Analyze

- One printer takes 4 hours.
- Other printer takes 6 hours.
- How long will it take both of them, working together, to print $\frac{3}{4}$ of the schedules?

Assign

Let x = the number of hours it will take both working together to print schedules.

Form

	Rate	• Time	= Work Completed
Fast printer	$\frac{1}{4}$	x	$\frac{x}{4}$
Slow Printer	$\frac{1}{6}$	x	$\frac{x}{6}$

The part of job done by fast printer.	plus	The part of job done by slow printer.	equals	$\frac{3}{4}$ job completed
$\frac{x}{4}$	+	$\frac{x}{6}$	=	$\frac{3}{4}$

Solve

$$\frac{x}{4}+\frac{x}{6}=\frac{3}{4}$$
$$\text{LCD}=12$$
$$\mathbf{12}\left(\frac{x}{4}+\frac{x}{6}\right)=\mathbf{12}\left(\frac{3}{4}\right)$$
$$\cancel{4}\cdot 3\left(\frac{x}{\cancel{4}}\right)+\cancel{6}\cdot 2\left(\frac{x}{\cancel{6}}\right)=\cancel{4}\cdot 3\left(\frac{3}{\cancel{4}}\right)$$
$$3x+2x=9$$
$$5x=9$$
$$\frac{5x}{\mathbf{5}}=\frac{9}{\mathbf{5}}$$
$$x=1\frac{4}{5}$$

State

It takes both working together $1\frac{4}{5}$ or 1.8 hours to print $\frac{3}{4}$ of the schedules.

Check

The result checks.

41. COMPARING INVESTMENTS

Analyze

- Tax-free bonds earns \$300 interest.
- Credit union earns \$200 interest.
- Credit union rate is 2% less.
- Time for both is 1 year.
- Principals are the same amount for both.
- What are the two rates?

Assign

Let r = Bond's interest rate as a percent

$r-2$ = C Union's interest rate as a percent

Form

	Principal	• Rate	• Time	= Interest
Bonds	$\frac{300}{r}$	r	1	300
Credit	$\frac{200}{r-2}$	$r-2$	1	200

Principal invested in bonds.	equals	Principal invested with credit union.
$\frac{300}{r}$	=	$\frac{200}{r-2}$

Solve

$$\frac{300}{r}=\frac{200}{r-2}$$
$$\text{LCD}=r(r-2)$$
$$\mathbf{r(r-2)}\left(\frac{300}{r}\right)=\mathbf{r(r-2)}\left(\frac{200}{r-2}\right)$$
$$\cancel{r}\cdot(r-2)\left(\frac{300}{\cancel{r}}\right)=r\cdot\cancel{(r-2)}\left(\frac{200}{\cancel{r-2}}\right)$$
$$300(r-2)=r(200)$$
$$300r-600=200r$$
$$300r-600+\mathbf{600}=200r+\mathbf{600}$$
$$300r=200r+600$$
$$300r-\mathbf{200r}=200r+600-\mathbf{200r}$$
$$\frac{100r}{\mathbf{100}}=\frac{600}{\mathbf{100}}$$
$$r=6$$

State

Bond's rate = 6%.

Credit Union's rate $=r-2=6-2=4\%$

Check

The result checks.

43. COMPARING INVESTMENTS

Analyze

- 1st CD earns \$175 interest.
- 2nd CD earns \$200 interest.
- 2nd CD investment rate is 1% more.
- Time for both is 1 year.
- Principals are the same amount for both.
- What are the two rates?

Assign

Let $r = 1^{st}$ CD interest rate as a percent
$r+1 = 2^{nd}$ CD interest rate as a percent

Form

	Principal	• Rate	• Time	= Interest
1st CD	$\frac{\mathbf{175}}{\mathbf{r}}$	r	1	175
2nd CD	$\frac{\mathbf{200}}{\mathbf{r+1}}$	$r+1$	1	200

Principal invested in 1st CD. equals Principal invested in 2nd CD.

$$\frac{175}{r} = \frac{200}{r+1}$$

Solve

$$\frac{175}{r} = \frac{200}{r+1}$$
$$\text{LCD} = r(r+1)$$
$$\mathbf{r(r+1)}\left(\frac{175}{r}\right) = \mathbf{r(r+1)}\left(\frac{200}{r+1}\right)$$
$$\cancel{r}\cdot(r+1)\left(\frac{175}{\cancel{r}}\right) = r\cdot\cancel{(r+1)}\left(\frac{200}{\cancel{r+1}}\right)$$
$$175(r+1) = r(200)$$
$$175r + 175 = 200r$$
$$175r + 175 - \mathbf{175r} = 200r - \mathbf{175r}$$
$$175 = 25r$$
$$\frac{175}{\mathbf{25}} = \frac{25r}{\mathbf{25}}$$
$$7 = r$$

State

1st CD's rate = 7%.
2nd CD's rate $= r+1 = 7+1 = 8\%$

Check

The results check.

WRITING

45. Answers will vary.

REVIEW

47. Solve using substitution:

$$\begin{cases} x+y=4 \\ y=3x \end{cases}$$

$x+y=4$ The 1st equation
Substitute for y.

$$x+\mathbf{3x}=4$$
$$4x=4$$
$$\frac{4x}{4}=\frac{4}{4}$$
$$x=1$$

$y=3x$ The 2nd equation

$$y=3(\mathbf{1})$$
$$y=3$$

The solution is (1, 3).

49. $x+20=4x-1+2x$

$$\frac{\mathbf{21}}{\mathbf{5}}+20\overset{?}{=}4\left(\frac{\mathbf{21}}{\mathbf{5}}\right)-1+2\left(\frac{\mathbf{21}}{\mathbf{5}}\right)$$
$$\frac{21}{5}+\frac{100}{5}\overset{?}{=}\frac{84}{5}-\frac{5}{5}+\frac{42}{5}$$
$$\frac{121}{5}=\frac{121}{5}$$

True

CHALLENGE PROBLEMS

51. RIVER TOUR

Analyze

- Tour's distance is 60 miles one way.
- Rate of current is 5 mph.
- Total time is 5 hours.
- What is the boat's still water speed?

Assign

Let $r =$ boat's still water speed in mph

Form

	Rate •	Time =	Distance
Upstream	$r-5$	$\frac{60}{r-5}$	60
Downstream	$r+5$	$\frac{60}{r+5}$	60

Time took boat to go 60 miles upstream. plus Time took boat to go 60 miles downstream. equals Total time of 5 hours.

$$\frac{60}{r-5} + \frac{60}{r+5} = 5$$

Solve

$$\frac{60}{r-5}+\frac{60}{r+5}=5$$

$$\text{LCD}=(r-5)(r+5)$$

$$(r-5)(r+5)\left(\frac{60}{r-5}\right)+(r-5)(r+5)\left(\frac{60}{r+5}\right)=(r-5)(r+5)(5)$$

$$\cancel{(r-5)}(r+5)\left(\frac{60}{\cancel{r-5}}\right)+(r-5)\cancel{(r+5)}\left(\frac{60}{\cancel{r+5}}\right)=5(r^2-25)$$

$$60(r+5)+60(r-5)=5(r^2-25)$$

$$60r+300+60r-300=5r^2-125$$

$$120r=5r^2-125$$

$$5r^2-125=120r$$

$$5r^2-125-120r=120r-120r$$

$$5r^2-120r-125=0$$

$$5(r^2-24r-25)=0$$

$$5(r-25)(r+1)=0$$

$r-25=0$ or $r+1=0$

$r-25+25=0+25$ | $r+1-1=0-1$

$r=25$ | $r=-1$

State

Boat's rate should be 25 mph.

Check

The result checks.

53. SALES

Analyze

- Spent $1,200 on some radios.
- Gave away 6 radios.
- Sold each for $10 more than paid.
- How many radios did she buy?
- Broke even.

Assign

Let $x =$ number of radios bought

Form

	Quanity •	Value of 1 =	Total Value
Bought	x	$\frac{1,200}{x}$	1,200
Selling	$x-6$	$\frac{1,200}{x-6}$	1,200

Buyer's cost per radio. plus Profit per radio. equals Selling price per radio.

$$\frac{1,200}{x} + 10 = \frac{1,200}{x-6}$$

Solve

$$\frac{1,200}{x}+10=\frac{1,200}{x-6}$$

$$\text{LCD}=x(x-6)$$

$$\mathbf{x(x-6)}\left(\frac{1,200}{x}+10\right)=\mathbf{x(x-6)}\left(\frac{1,200}{x-6}\right)$$

$$\overset{1}{\cancel{x}}(x-6)\left(\frac{1,200}{\underset{1}{\cancel{x}}}\right)+x(x-6)(10)=x\,\overset{1}{\cancel{(x-6)}}\left(\frac{1,200}{\underset{1}{\cancel{x-6}}}\right)$$

$$1,200(x-6)+10x(x-6)=1,200x$$

$$1,200x-7,200+10x^2-60x=1,200x$$

$$10x^2+1,140x-7,200=1,200x$$

$$10x^2+1,140x-7,200-\mathbf{1,200x}=1,200x-\mathbf{1,200x}$$

$$10x^2-60x-7,200=0$$

$$10(x^2-6x-720)=0$$

$$10(x-30)(x+24)=0$$

$x-30=0$ or $x+24=0$

$x-30+30=0+30$ | $x+24-24=0-24$

$x=30$ | $x=-24$

State

The number of radios is 30.

Check

The result checks.

SECTION 7.8, STUDY SET

VOCABULARY

Fill in the blanks.

1. A **ratio** is the quotient of two numbers or the quotient of two quantities with the same units. A **rate** is a quotient of two quantities that have different units.

3. In $\frac{50}{3} = \frac{x}{9}$, the terms 50 and 9 are called the **extremes** and the terms 3 and x are called the **means** of the proportion.

5. Examples of **unit** prices are \$1.65 per gallon, 17¢ per day, and \$50 per foot.

CONCEPTS

7. Fill in the blanks: In a proportion, the product of the extremes is **equal** to the product of the means. In symbols, If $\frac{a}{b} = \frac{c}{d}$, then $ad = bc$.

9. SNACKS

$$\frac{\text{Number of bags} \rightarrow \mathbf{25}}{\text{Number underweight} \rightarrow \mathbf{2}} = \frac{\mathbf{1{,}000} \leftarrow \text{Number of bags}}{\boldsymbol{x} \leftarrow \text{Number underweight}}$$

11. KLEENEX

$$\frac{\text{Price}}{\text{Number of sheets}} = \frac{\text{Price}}{\text{Number of sheets}}$$

$$\frac{\mathbf{2.19}}{85} = \frac{x}{\mathbf{1}}$$

NOTATION

13. Solve for x.

$$\frac{12}{18} = \frac{x}{24}$$

$$12 \cdot 24 = 18 \cdot \boldsymbol{x}$$

$$\mathbf{288} = 18x$$

$$\frac{288}{\mathbf{18}} = \frac{18x}{\mathbf{18}}$$

$$16 = x$$

15. Fill in the blanks: The proportion $\frac{20}{1.6} = \frac{100}{8}$ can be read: 20 is to 1.6 **as** 100 is **to** 8.

GUIDED PRACTICE

Translate each ratio into a fraction in simplest form. See Example 1.

17. $\frac{4\text{ boxes}}{15\text{ boxes}} = \frac{4\cancel{\text{boxes}}}{15\cancel{\text{boxes}}} = \frac{4}{15}$

19. $\frac{18\text{ watts}}{24\text{ watts}} = \frac{3 \cdot \cancel{6}\ \cancel{\text{watts}}}{4 \cdot \cancel{6}\ \cancel{\text{watts}}} = \frac{3}{4}$

21. $\frac{30\text{ days}}{24\text{ days}} = \frac{5 \cdot \cancel{6}\ \cancel{\text{days}}}{4 \cdot \cancel{6}\ \cancel{\text{days}}} = \frac{5}{4}$

23. If 1 hour = 60 minutes, then 3 hours = 180 minutes.

$$\frac{90\text{ minutes}}{3\text{ hours}} = \frac{90\text{ minutes}}{180\text{ minutes}}$$

$$\frac{90\text{ minutes}}{180\text{ minutes}} = \frac{1 \cdot \cancel{90}\ \cancel{\text{minutes}}}{2 \cdot \cancel{90}\ \cancel{\text{minutes}}} = \frac{1}{2}$$

25. If 1 gallon = 4 quarts, then 4 gallons = 16 quarts.

$$\frac{8\text{ quarts}}{4\text{ gallons}} = \frac{8\text{ quarts}}{16\text{ quarts}}$$

$$\frac{8\text{ quarts}}{16\text{ quarts}} = \frac{1 \cdot \cancel{8}\ \cancel{\text{quarts}}}{2 \cdot \cancel{8}\ \cancel{\text{quarts}}} = \frac{1}{2}$$

27. 1 mile = 5,280 feet

$$\frac{6{,}000\text{ feet}}{1\text{ mile}} = \frac{6{,}000\text{ feet}}{5{,}280\text{ feet}}$$

$$\frac{6{,}000\text{ feet}}{5{,}280\text{ feet}} = \frac{25 \cdot \cancel{240}\ \cancel{\text{feet}}}{22 \cdot \cancel{240}\ \cancel{\text{feet}}} = \frac{25}{22}$$

Determine whether each equation is a true proportion. See Example 2.

29. $\frac{7}{3} = \frac{14}{6}$

$$7 \cdot 6 = 14 \cdot 3$$

$$42 = 42$$

Since the cross products are equal, it is a proportion.

31. $\frac{5}{8} = \frac{12}{19.4}$

$$5 \cdot 19.4 = 12 \cdot 8$$

$$97 = 96$$

Since the cross products are not equal, it is not a proportion.

Solve each proportion. See Example 3.

33. $\frac{2}{3}=\frac{x}{6}$

$2\cdot 6=3x$

$12=3x$

$\frac{12}{\mathbf{3}}=\frac{3x}{\mathbf{3}}$

$4=x$

check:

$\frac{2}{3}=\frac{x}{6}$

$\frac{2}{3}\overset{?}{=}\frac{4}{6}$

$2\cdot 6\overset{?}{=}3\cdot 4$

$12=12$

35. $\frac{63}{g}=\frac{9}{2}$

$63\cdot 2=9g$

$126=9g$

$\frac{126}{\mathbf{9}}=\frac{9g}{\mathbf{9}}$

$14=g$

check:

$\frac{63}{g}=\frac{9}{2}$

$\frac{63}{14}\overset{?}{=}\frac{9}{2}$

$63\cdot 2\overset{?}{=}9\cdot 14$

$126=126$

37. $\frac{x+1}{5}=\frac{3}{15}$

$15(x+1)=3\cdot 5$

$15x+15=15$

$15x+15-\mathbf{15}=15-\mathbf{15}$

$15x=0$

$\frac{15x}{\mathbf{15}}=\frac{0}{\mathbf{15}}$

$x=0$

check:

$\frac{x+1}{5}=\frac{3}{15}$

$\frac{0+1}{5}\overset{?}{=}\frac{3}{15}$

$\frac{1}{5}\overset{?}{=}\frac{3}{15}$

$1\cdot 15=5\cdot 3$

$15=15$

39. $\frac{5-x}{17}=\frac{13}{34}$

$34(5-x)=13\cdot 17$

$170-34x=221$

$170-34x-\mathbf{170}=221-\mathbf{170}$

$-34x=51$

$\frac{-34x}{\mathbf{-34}}=\frac{51}{\mathbf{-34}}$

$x=-\frac{3\cdot \cancel{17}^{1}}{2\cdot \cancel{17}_{1}}$

$x=-\frac{3}{2}$

check:

$\frac{5-x}{17}=\frac{13}{34}$

let $-\frac{3}{2}=-1.5$

$\frac{5-(-1.5)}{17}\overset{?}{=}\frac{13}{34}$

$\frac{6.5}{17}\overset{?}{=}\frac{13}{34}$

$6.5\cdot 34\overset{?}{=}17\cdot 13$

$221=221$

41. $\frac{15}{7b+5}=\frac{5}{2b+1}$

$15(2b+1)=5(7b+5)$

$30b+15=35b+25$

$30b+15-\mathbf{25}=35b+25-\mathbf{25}$

$30b-10=35b$

$30b-10-\mathbf{30b}=35b-\mathbf{30b}$

$-10=5b$

$\frac{-10}{\mathbf{5}}=\frac{5b}{\mathbf{5}}$

$-2=b$

check:

$\frac{15}{7b+5}=\frac{5}{2b+1}$

$\frac{15}{7(-2)+5}\overset{?}{=}\frac{5}{2(-2)+1}$

$\frac{15}{-14+5}\overset{?}{=}\frac{5}{-4+1}$

$\frac{15}{-9}\overset{?}{=}\frac{5}{-3}$

$15\cdot -3\overset{?}{=}-9\cdot 5$

$-45=-45$

43. $\frac{8x}{3}=\frac{11x+9}{4}$

$8x(4)=3(11x+9)$

$32x=33x+27$

$32x-\mathbf{33x}=33x+27-\mathbf{33x}$

$-x=27$

$(\mathbf{-1})(-x)=(\mathbf{-1})(27)$

$x=-27$

check:

$\frac{8x}{3}=\frac{11x+9}{4}$

$\frac{8(-27)}{3}\overset{?}{=}\frac{11(-27)+9}{4}$

$\frac{-216}{3}\overset{?}{=}\frac{-297+9}{4}$

$\frac{-216}{3}\overset{?}{=}\frac{-288}{4}$

$-216\cdot 4\overset{?}{=}-288\cdot 3$

$-864=-864$

Solve each proportion. See Example 4.

45. $\frac{2}{3x}=\frac{x}{6}$

$2(6)=3x(x)$

$12=3x^2$

$12-\mathbf{12}=3x^2-\mathbf{12}$

$0=3x^2-12$

$0=3(x^2-4)$

$0=3(x+2)(x-2)$

$x+2=0$ or $x-2=0$

$x+2-\mathbf{2}=0-\mathbf{2}$ | $x-2+\mathbf{2}=0+\mathbf{2}$

$x=-2$ | $x=2$

The solutions are -2 and 2.

check:

$\frac{2}{3x}=\frac{x}{6}$

$\frac{2}{3(-2)}\overset{?}{=}\frac{-2}{6}$

$-\frac{2}{6}\overset{?}{=}-\frac{2}{6}$

$-2\cdot 6\overset{?}{=}6\cdot -2$

$-12=-12$

check:

$\frac{2}{3x}=\frac{x}{6}$

$\frac{2}{3(2)}\overset{?}{=}\frac{2}{6}$

$\frac{2}{6}\overset{?}{=}\frac{2}{6}$

$2\cdot 6\overset{?}{=}2\cdot 6$

$12=12$

Both solutions check.

47. $$\frac{b-5}{3}=\frac{2}{b}$$
$$b(b-5)=2\cdot 3$$
$$b^2-5b=6$$
$$b^2-5b-\mathbf{6}=6-\mathbf{6}$$
$$b^2-5b-6=0$$
$$(b-6)(b+1)=0$$

$b-6=0$ or $b+1=0$

$b-6+\mathbf{6}=0+\mathbf{6}$ | $b+1-\mathbf{1}=0-\mathbf{1}$

$b=6$ | $b=-1$

The solutions are 6 and -1.

check:
$$\frac{b-5}{3}=\frac{2}{b}$$
$$\frac{6-5}{3}\overset{?}{=}\frac{2}{6}$$
$$\frac{1}{3}\overset{?}{=}\frac{2}{6}$$
$$1\cdot 6\overset{?}{=}3\cdot 2$$
$$6=6$$

check:
$$\frac{b-5}{3}=\frac{2}{b}$$
$$\frac{-1-5}{3}\overset{?}{=}\frac{2}{-1}$$
$$\frac{-6}{3}\overset{?}{=}\frac{2}{-1}$$
$$-6\cdot -1\overset{?}{=}3\cdot 2$$
$$6=6$$

Both solutions check.

49. $$\frac{a-4}{a}=\frac{15}{a+4}$$
$$(a-4)(a+4)=15a$$
$$a^2-16=15a$$
$$a^2-16-\mathbf{15a}=15a-\mathbf{15a}$$
$$a^2-15a-16=0$$
$$(a-16)(a+1)=0$$

$a-16=0$ or $a+1=0$

$a-16+\mathbf{16}=0+\mathbf{16}$ | $a+1-\mathbf{1}=0-\mathbf{1}$

$a=16$ | $a=-1$

The solutions are 16 and -1.

check:
$$\frac{a-4}{a}=\frac{15}{a+4}$$
$$\frac{16-4}{16}\overset{?}{=}\frac{15}{16+4}$$
$$\frac{12}{16}\overset{?}{=}\frac{15}{20}$$
$$12\cdot 20\overset{?}{=}16\cdot 15$$
$$240=240$$

check:
$$\frac{a-4}{a}=\frac{15}{a+4}$$
$$\frac{-1-4}{-1}\overset{?}{=}\frac{15}{-1+4}$$
$$\frac{-5}{-1}\overset{?}{=}\frac{15}{3}$$
$$-5\cdot 3\overset{?}{=}-1\cdot 15$$
$$-15=-15$$

Both solutions check.

51. $$\frac{t+3}{t+5}=\frac{-1}{2t}$$
$$2t(t+3)=-1(t+5)$$
$$2t^2+6t=-t-5$$
$$2t^2+6t+\boldsymbol{t}=-t-5+\boldsymbol{t}$$
$$2t^2+7t=-5$$
$$2t^2+7t+\mathbf{5}=-5+\mathbf{5}$$
$$2t^2+7t+5=0$$
$$(2t+5)(t+1)=0$$

$2t+5=0$ or $t+1=0$

$2t+5-\mathbf{5}=0-\mathbf{5}$ | $t+1-\mathbf{1}=0-\mathbf{1}$

$2t=-5$ | $t=-1$

$\frac{2t}{\mathbf{2}}=\frac{-5}{\mathbf{2}}$

$t=-\frac{5}{2}$

The solutions are $-\frac{5}{2}$ and -1.

check:
$$\frac{t+3}{t+5}=\frac{-1}{2t}$$
let $-\frac{5}{2}=-2.5$
$$\frac{-2.5+3}{-2.5+5}\overset{?}{=}\frac{-1}{2(-2.5)}$$
$$\frac{0.5}{2.5}\overset{?}{=}\frac{-1}{-5}$$
$$0.5\cdot -5\overset{?}{=}2.5\cdot -1$$
$$-2.5=-2.5$$

ch eck:
$$\frac{t+3}{t+5}=\frac{-1}{2t}$$
$$\frac{-1+3}{-1+5}\overset{?}{=}\frac{-1}{2(-1)}$$
$$\frac{2}{4}\overset{?}{=}\frac{-1}{-2}$$
$$2\cdot -2\overset{?}{=}4\cdot -1$$
$$-4=-4$$

Both solutions check.

Each pair of triangles is similar. Find the missing side length. See Example 8.

53. $\frac{4}{12}=\frac{5}{x}$

$4x=5\cdot 12$

$4x=60$

$\frac{4x}{\mathbf{4}}=\frac{60}{\mathbf{4}}$

$x=15$

The solution is 15.

check:

$\frac{4}{12}=\frac{5}{x}$

$\frac{4}{12}=\frac{5}{\mathbf{15}}$

$4\cdot 15=12\cdot 5$

$60=60$

55. $\frac{9}{12}=\frac{6}{x}$

$9x=6\cdot 12$

$9x=72$

$\frac{9x}{\mathbf{9}}=\frac{72}{\mathbf{9}}$

$x=8$

The solution is 8.

check:

$\frac{9}{12}=\frac{6}{x}$

$\frac{9}{12}\stackrel{?}{=}\frac{6}{\mathbf{8}}$

$72=72$

TRY IT YOURSELF

Solve each proportion.

57. $\frac{x-1}{x+1}=\frac{2}{3x}$

$3x(x-1)=2(x+1)$

$3x^2-3x=2x+2$

$3x^2-3x-\mathbf{2x}=2x+2-\mathbf{2x}$

$3x^2-5x=2$

$3x^2-5x-\mathbf{2}=2-\mathbf{2}$

$3x^2-5x-2=0$

$(3x+1)(x-2)=0$

$3x+1=0$ or $x-2=0$

$3x+1-\mathbf{1}=0-\mathbf{1}$ | $x-2+\mathbf{2}=0+\mathbf{2}$

$3x=-1$ | $x=2$

$\frac{3x}{\mathbf{3}}=\frac{-1}{\mathbf{3}}$

$x=-\frac{1}{3}$

The solutions are $-\frac{1}{3}$ and 2.

check:

$\frac{x-1}{x+1}=\frac{2}{3x}$

$\frac{-\frac{\mathbf{1}}{\mathbf{3}}-1}{-\frac{\mathbf{1}}{\mathbf{3}}+1}\stackrel{?}{=}\frac{2}{3\cdot -\frac{\mathbf{1}}{\mathbf{3}}}$

$\frac{-\frac{4}{3}}{\frac{2}{3}}\stackrel{?}{=}\frac{2}{-1}$

$-\frac{4}{3}\cdot -1\stackrel{?}{=}\frac{2}{3}\cdot 2$

$\frac{4}{3}=\frac{4}{3}$

check:

$\frac{x-1}{x+1}=\frac{2}{3x}$

$\frac{\mathbf{2}-1}{\mathbf{2}+1}\stackrel{?}{=}\frac{2}{3(\mathbf{2})}$

$\frac{1}{3}\stackrel{?}{=}\frac{2}{6}$

$1\cdot 6\stackrel{?}{=}3\cdot 2$

$6=6$

Both solutions check.

59. $\frac{x+1}{4}=\frac{3x}{8}$

$8(x+1)=3x(4)$

$8x+8=12x$

$8x+8-\mathbf{8x}=12x-\mathbf{8x}$

$8=4x$

$\frac{8}{\mathbf{4}}=\frac{4x}{\mathbf{4}}$

$2=x$

The solution is 2.

check:

$\frac{x+1}{4}=\frac{3x}{8}$

$\frac{\mathbf{2}+1}{4}\stackrel{?}{=}\frac{3(\mathbf{2})}{8}$

$\frac{3}{4}\stackrel{?}{=}\frac{6}{8}$

$3\cdot 8\stackrel{?}{=}4\cdot 6$

$24=24$

The solution checks.

61. $\frac{y-4}{y+1}=\frac{y+3}{y+6}$

$(y-4)(y+6)=(y+1)(y+3)$

$y^2+2y-24=y^2+4y+3$

$y^2+2y-24-\mathbf{y^2}=y^2+4y+3-\mathbf{y^2}$

$2y-24=4y+3$

$2y-24-\mathbf{2y}=4y+3-\mathbf{2y}$

$-24=2y+3$

$-24-\mathbf{3}=2y+3-\mathbf{3}$

$-27=2y$

$\frac{-27}{\mathbf{2}}=\frac{2y}{\mathbf{2}}$

$-\frac{27}{2}=y$

The solution is $-\frac{27}{2}$.

check:

$$\frac{y-4}{y+1}=\frac{y+3}{y+6}$$

let $-\frac{27}{2}=-13.5$

$$\frac{\mathbf{-13.5}-4}{\mathbf{-13.5}+1}\stackrel{?}{=}\frac{-13.5+3}{-13.5+6}$$

$$\frac{-17.5}{-12.5}\stackrel{?}{=}\frac{-10.5}{-7.5}$$

$$-17.5\cdot -7.5\stackrel{?}{=}-12.5\cdot -10.5$$

$$131.25=131.25$$

The solution checks.

63.

$$\frac{c}{10}=\frac{10}{c}$$

$$c\cdot c=10\cdot 10$$

$$c^2=100$$

$$c^2-\mathbf{100}=100-\mathbf{100}$$

$$c^2-100=0$$

$$(c+10)(c-10)=0$$

$c+10=0$ or $c-10=0$

$c+10-\mathbf{10}=0-\mathbf{10}$ | $c-10+\mathbf{10}=0+\mathbf{10}$

$c=-10$ | $c=10$

The solutions are -10 and 10.

check:

$$\frac{c}{10}=\frac{10}{c}$$

$$\frac{\mathbf{-10}}{10}\stackrel{?}{=}\frac{10}{\mathbf{-10}}$$

$$-10\cdot -10\stackrel{?}{=}10\cdot 10$$

$$100=100$$

check:

$$\frac{c}{10}=\frac{10}{c}$$

$$\frac{\mathbf{10}}{10}\stackrel{?}{=}\frac{10}{\mathbf{10}}$$

$$10\cdot 10\stackrel{?}{=}10\cdot 10$$

$$100=100$$

Both solutions check.

65.

$$\frac{m}{3}=\frac{4}{m+1}$$

$$m(m+1)=3\cdot 4$$

$$m^2+m=12$$

$$m^2+m-\mathbf{12}=12-\mathbf{12}$$

$$m^2+m-12=0$$

$$(m+4)(m-3)=0$$

$m+4=0$ or $m-3=0$

$m+4-\mathbf{4}=0-\mathbf{4}$ | $m-3+\mathbf{3}=0+\mathbf{3}$

$m=-4$ | $m=3$

The solutions are -4 and 3.

check:

$$\frac{m}{3}=\frac{4}{m+1}$$

$$\frac{\mathbf{-4}}{3}\stackrel{?}{=}\frac{4}{\mathbf{-4}+1}$$

$$\frac{-4}{3}\stackrel{?}{=}\frac{4}{-3}$$

$$-4\cdot -3\stackrel{?}{=}3\cdot 4$$

$$12=12$$

check:

$$\frac{m}{3}=\frac{4}{m+1}$$

$$\frac{\mathbf{3}}{3}\stackrel{?}{=}\frac{4}{\mathbf{3}+1}$$

$$\frac{3}{3}\stackrel{?}{=}\frac{4}{4}$$

$$3\cdot 4\stackrel{?}{=}3\cdot 4$$

$$12=12$$

Both solutions check.

67.

$$\frac{3}{3b+4}=\frac{2}{5b-6}$$

$$3(5b-6)=2(3b+4)$$

$$15b-18=6b+8$$

$$15b-18-\mathbf{6b}=6b+8-\mathbf{6b}$$

$$9b-18=8$$

$$9b-18+\mathbf{18}=8+\mathbf{18}$$

$$9b=26$$

$$\frac{9b}{\mathbf{9}}=\frac{26}{\mathbf{9}}$$

$$b=\frac{26}{9}$$

The solution is $\frac{26}{9}$.

check:

$$\frac{3}{3b+4}=\frac{2}{5b-6}$$

$$\frac{3}{3\left(\frac{\mathbf{26}}{\mathbf{9}}\right)+4}\stackrel{?}{=}\frac{2}{5\left(\frac{\mathbf{26}}{\mathbf{9}}\right)-6}$$

$$\frac{3}{\frac{26}{3}+4}\stackrel{?}{=}\frac{2}{\frac{130}{9}-6}$$

$$\frac{3}{\frac{38}{3}}\stackrel{?}{=}\frac{2}{\frac{76}{9}}$$

$$3\cdot\frac{76}{9}\stackrel{?}{=}\frac{38}{3}\cdot 2$$

$$\frac{76}{3}=\frac{76}{3}$$

The solution checks.

LOOK ALIKES . . .

Solve each equation.

69. a.

$$\frac{-2}{5}=\frac{3}{4x}$$

$$-2(4x)=3(5)$$

$$-8x=15$$

$$\frac{-8x}{\mathbf{-8}}=\frac{15}{\mathbf{-8}}$$

$$x=-\frac{15}{8}$$

b.

$$\frac{4}{x}-\frac{2}{5}=\frac{3}{4x}$$

$$\mathbf{20x}\left(\frac{4}{x}-\frac{2}{5}\right)=\mathbf{20x}\left(\frac{3}{4x}\right)$$

$$20\cancel{x}\left(\frac{4}{\cancel{x}}\right)-4x\cdot\cancel{5}\left(\frac{2}{\cancel{5}}\right)=5\cdot\cancel{4x}\left(\frac{3}{\cancel{4x}}\right)$$

$$80-8x=15$$

$$80-8x-\mathbf{80}=15-\mathbf{80}$$

$$-8x=-65$$

$$\frac{-8x}{\mathbf{-8}}=\frac{-65}{\mathbf{-8}}$$

$$x=\frac{65}{8}$$

71. a.

$$\frac{3}{a-1}=\frac{8}{a}$$

$$a(3)=8(a-1)$$

$$3a=8a-8$$

$$3a-\mathbf{8a}=8a-8-\mathbf{8a}$$

$$-5a=8$$

$$\frac{-5a}{\mathbf{-5}}=\frac{8}{\mathbf{-5}}$$

$$a=-\frac{8}{5}$$

b.

$$\frac{3}{a-1}+\frac{8}{a}=3$$

$$\boldsymbol{a(a-1)}\left(\frac{3}{a-1}+\frac{8}{a}\right)=\boldsymbol{a(a-1)}(3)$$

$$a\cdot\cancel{(a-1)}\left(\frac{3}{\cancel{a-1}}\right)+(a-1)\cdot\cancel{a}\left(\frac{8}{\cancel{a}}\right)=(a^2-a)(3)$$

$$3a+8a-8=3a^2-3a$$

$$11a-8=3a^2-3a$$

$$11a-8-\mathbf{11a}=3a^2-3a-\mathbf{11a}$$

$$-8=3a^2-14a$$

$$-8+\mathbf{8}=3a^2-14a+\mathbf{8}$$

$$0=3a^2-14a+8$$

$$0=(3a-2)(a-4)$$

$3a-2=0$ or $a-4=0$

$3a=2 \quad | \quad a-4+\mathbf{4}=0+\mathbf{4}$

$a=\frac{2}{3} \quad | \quad a=4$

The solutions are $\frac{2}{3}$ and 4.

APPLICATIONS

73. SHOPPING FOR CLOTHES

Analyze

- We know the cost of two shirts is \$25.
- How much do five shirts cost?

Assign

Let c = cost of 5 shirts

Form

2 shirts is to \$25 as 5 shirts is to \$$c$.

$$\begin{matrix} \text{2 shirts} \rightarrow \\ \text{Cost of 2 shirts} \rightarrow \end{matrix} \frac{2}{25} = \frac{5}{c} \begin{matrix} \leftarrow \text{5 shirts} \\ \leftarrow \text{Cost of 5 shirts} \end{matrix}$$

Solve

$$\frac{2}{25} = \frac{5}{c}$$
$$2c = 125$$
$$\frac{2c}{\mathbf{2}} = \frac{125}{\mathbf{2}}$$
$$c = 62.5$$

State

Five shirts cost \$62.50.

Check

$$\frac{2}{25} = \frac{5}{c}$$
$$\frac{2}{25} \stackrel{?}{=} \frac{5}{\mathbf{62.50}}$$
$$125 = 125$$

The result checks.

75. CPR

Analyze

- We know the ratio of chest compressions to breaths should be 5:2.
- How many breaths are needed for 210 compressions?

Assign

Let x = # of breaths needed

Form

5 compressions is to 2 breaths
as 210 compressions is to x breaths.

$$\begin{matrix} \text{5 compressions} \rightarrow \\ \text{2 breaths} \rightarrow \end{matrix} \frac{5}{2} = \frac{210}{x} \begin{matrix} \leftarrow \text{210 compressions} \\ \leftarrow \text{number of breaths} \end{matrix}$$

Solve

$$\frac{5}{2} = \frac{210}{x}$$
$$5x = 420$$
$$\frac{5x}{\mathbf{5}} = \frac{420}{\mathbf{5}}$$
$$x = 84$$

State

84 breaths are needed with 210 compressions.

Check

$$\frac{5}{2} = \frac{210}{x}$$
$$\frac{5}{2} \stackrel{?}{=} \frac{210}{\mathbf{84}}$$
$$420 = 420$$

The result checks.

from CAMPUS TO CAREERS

77. RECREATION DIRECTOR

a.

Analyze

- We know there is total of 966 boys and girls.
- If 504 are boys how many are girls?

Assign

Let x = number of girls

Form

The # of boys plus the # of girls is 966.

Solve

$$504 + x = 966$$
$$504 + x - \mathbf{504} = 966 - \mathbf{504}$$
$$x = 462$$

State

The are 462 girls.

Check

$$504 + x = 966$$
$$504 + 462 \stackrel{?}{=} 966$$
$$966 = 966$$

The result checks.

b.

- Find the ratio of girls to boys.

$$\frac{\text{\# of girls}}{\text{\# of boys}} = \frac{462}{504}$$
$$= \frac{11 \cdot \cancel{42}^{1}}{12 \cdot \cancel{42}_{1}}$$
$$= \frac{11}{12} \text{ or } 11:12$$

The ratio of girls to boys is 11:12.

79. COMPUTING A PAYCHECK

Analyze

- We know Billie earns \$412 for 40 hours.
- She missed 10 hours of work.
- How much did she earn?

Assign
Let x = amount earned for 30 hours

Form
40 hours is to \$412 as 30 hours is to \$$x$.

$$\begin{matrix}\text{40 hrs worked} \rightarrow \\ \text{Earned 40 hrs.} \rightarrow\end{matrix} \frac{40}{412} = \frac{30}{x} \begin{matrix}\leftarrow \text{30 hrs worked} \\ \leftarrow \text{Earned 30 hrs}\end{matrix}$$

Solve

$$\frac{40}{412} = \frac{30}{x}$$
$$40x = 12{,}360$$
$$\frac{40x}{\mathbf{40}} = \frac{12{,}360}{\mathbf{40}}$$
$$x = 309$$

State
Billie earns \$309 for 30 hours of work.

Check

$$\frac{40}{412} = \frac{30}{x}$$
$$\frac{40}{412} \stackrel{?}{=} \frac{30}{\mathbf{309}}$$
$$12{,}360 = 12{,}360$$
The result checks.

81. TWITTER

Analyze
- 7,500 tweets are sent every 10 seconds.
- How many tweets are sent in one minute?

Assign
Let x = number of tweets sent in one minute

Form
7,500 tweets is to 10 seconds as
x tweets is to 60 seconds.

$$\begin{matrix}\text{tweets} \rightarrow \\ \text{seconds} \rightarrow\end{matrix} \frac{7{,}500}{10} = \frac{x}{60} \begin{matrix}\leftarrow \text{tweets} \\ \leftarrow \text{seconds}\end{matrix}$$

Solve

$$\frac{7{,}500}{10} = \frac{x}{60}$$
$$450{,}000 = 10x$$
$$\frac{450{,}000}{\mathbf{10}} = \frac{10x}{\mathbf{10}}$$
$$x = 45{,}000$$

State
45,000 tweets are sent in one minute.

Check

$$\frac{7{,}500}{10} = \frac{x}{60}$$
$$\frac{7{,}500}{10} \stackrel{?}{=} \frac{\mathbf{45{,}000}}{60}$$
$$450{,}000 = 450{,}000$$
The result checks.

83. NUTRITION

Analyze
- We know a 10-oz milkshake contains 355 calories, 8 gm of fat and 9 gm of protein.
- What are the amounts in a 16-oz milkshake?

Assign
Let c = # of calories in a 16-oz shake

Form
335 cal is to a 10-oz shake as c cal is to a 16-oz shake.

$$\begin{matrix}\text{335 calories} \rightarrow \\ \text{10-oz shake} \rightarrow\end{matrix} \frac{335}{10} = \frac{c}{16} \begin{matrix}\leftarrow \text{number of calories} \\ \leftarrow \text{16-oz shake}\end{matrix}$$

Assign
Let f = grams of fat in a 16-oz shake

Form
8 gm is to a 10-oz shake as f gm is to a 16-oz shake.

$$\begin{matrix}\text{8 grams of fat} \rightarrow \\ \text{10-oz shake} \rightarrow\end{matrix} \frac{8}{10} = \frac{f}{16} \begin{matrix}\leftarrow \text{grams of fat} \\ \leftarrow \text{16-oz shake}\end{matrix}$$

Assign
Let p = grams of protein in a 16-oz shake

Form
9 gm is to a 10-oz shake as p gm is to a 16-oz shake.

$$\begin{matrix}\text{9 gm of protein} \rightarrow \\ \text{10-oz shake} \rightarrow\end{matrix} \frac{9}{10} = \frac{p}{16} \begin{matrix}\leftarrow \text{grams of protein} \\ \leftarrow \text{16-oz shake}\end{matrix}$$

Solve

calories

$$\frac{355}{10}=\frac{c}{16}$$
$$5680=10c$$
$$\frac{5680}{\mathbf{10}}=\frac{10c}{\mathbf{10}}$$
$$568=c$$

fat

$$\frac{8}{10}=\frac{f}{16}$$
$$128=10f$$
$$\frac{128}{\mathbf{10}}=\frac{10f}{\mathbf{10}}$$
$$12.8=f$$
$$13=f$$

protein

$$\frac{9}{10}=\frac{p}{16}$$
$$144=10p$$
$$\frac{144}{\mathbf{10}}=\frac{10p}{\mathbf{10}}$$
$$14.4=p$$
$$14=p$$

State

In a 16 - oz mikeshake there are 568 calories, 13 grams of fat and 14 grams of protein.

Check

$$\frac{355}{10}=\frac{c}{16} \qquad \frac{8}{10}=\frac{f}{16} \qquad \frac{9}{10}=\frac{p}{16}$$
$$\frac{355}{10}\stackrel{?}{=}\frac{\mathbf{568}}{16} \qquad \frac{8}{10}\stackrel{?}{=}\frac{\mathbf{12.8}}{16} \qquad \frac{9}{10}\stackrel{?}{=}\frac{\mathbf{14.4}}{16}$$
$$5680=5680 \qquad 128=128 \qquad 144=144$$

The results check.

85. MIXING FUEL

Analyze

- We know the ratio is 50 to 1.
- We know the amount of gasoline is 6 gallons and the amount of oil is 6 ounces.
- The number of ounces of gasoline is $6\cdot128=768$.
- Are the amounts correct?

Assign

Let x = the number of ounces of oil

Form

50 parts gasoline is to 1 part oil as 768 parts of gasoline is to x parts of oil.

$$\begin{matrix}\text{50 parts gas}\rightarrow \\ \text{1 part oil}\rightarrow\end{matrix}\ \frac{50}{1}=\frac{768}{x}\ \begin{matrix}\leftarrow\text{768 parts gas} \\ \leftarrow x\text{ parts oil}\end{matrix}$$

Solve

$$\frac{50}{1}=\frac{768}{x}$$
$$50x=768$$
$$\frac{50x}{\mathbf{50}}=\frac{768}{\mathbf{50}}$$
$$x=15.36$$

State

The calculated amount of 15.36 ounces is close to the stated amount of 16 ounces.

Check

$$\frac{50}{1}=\frac{768}{x}$$
$$\frac{50}{1}\stackrel{?}{=}\frac{768}{\mathbf{15.36}}$$
$$768=768$$

The result checks.

87. CAPTURE-RELEASE METHOD

Analyze

- We know 12 tagged squirrels were released.
- 35 were captured with 3 being tagged.
- How many squirrels are on the acreage?

Assign

Let x = the number of squirrels

Form

35 squirrels is to 3 tagged squirrels as x squirrels is to 12 tagged squirrels.

$$\begin{matrix}\text{35 squirrels}\rightarrow \\ \text{3 tagged}\rightarrow\end{matrix}\ \frac{35}{3}=\frac{x}{12}\ \begin{matrix}\leftarrow x\text{ squirrels} \\ \leftarrow\text{12 tagged}\end{matrix}$$

Solve

$$\frac{35}{3}=\frac{x}{12}$$
$$420=3x$$
$$\frac{420}{3}=\frac{3x}{3}$$
$$140=x$$

State

There are 140 squirrels on the acreage.

Check

$$\frac{35}{3}=\frac{x}{12}$$
$$\frac{35}{3}\stackrel{?}{=}\frac{\mathbf{140}}{12}$$
$$420=420$$

The result checks.

89. MODEL RAILROADS

Analyze

- We know the HO scale is 1:87.
- How long is a real engine in inches and feet if a model one measures 6 inches?

Assign

Let x = the length of a real engine in inches

Form

1 is to 87 as 6 inches is to x inches.

$$\begin{array}{r} \text{1 part} \rightarrow \\ \text{87 parts} \rightarrow \end{array} \frac{1}{87} = \frac{6}{x} \begin{array}{l} \leftarrow \text{6 inch model} \\ \leftarrow x \text{ inches real engine} \end{array}$$

Solve

$$\frac{1}{87} = \frac{6}{x}$$

$$x = 87(6)$$

$$x = 522$$

State

A real engine would be 522 inches long or $\frac{522 \text{ in}}{1} \cdot \frac{1 \text{ ft}}{12 \text{ in}} = 43.5$ feet.

Check

$$\frac{1}{87} = \frac{6}{x}$$

$$\frac{1}{87} \stackrel{?}{=} \frac{6}{\mathbf{522}}$$

$$522 = 522$$

The result checks.

91. BLUEPRINTS

Analyze

- We know that $\frac{1}{4}$ inch is equal to 1 foot.
- How long is the real kitchen if the length is $2\frac{1}{2}$ inches on the drawing?

Assign

Let x = the length of the real kitchen

Form

$\frac{1}{4}$ inch $= 0.25$ inches

$2\frac{1}{2}$ inches $= 2.5$ inches

0.25 inch is to 1 foot as 2.5 inches is to x feet.

$$\begin{array}{r} \text{0.25 inch} \rightarrow \\ \text{1 foot} \rightarrow \end{array} \frac{0.25}{1} = \frac{2.5}{x} \begin{array}{l} \leftarrow \text{2.5 inches} \\ \leftarrow x \text{ feet} \end{array}$$

Solve

$$\frac{0.25}{1} = \frac{2.5}{x}$$

$$0.25x = 2.5$$

$$\frac{0.25x}{0.25} = \frac{2.5}{0.25}$$

$$x = 10$$

State

The kitchen's real length is 10 feet.

Check

$$\frac{0.25}{1} = \frac{2.5}{x}$$

$$\frac{0.25}{1} \stackrel{?}{=} \frac{2.5}{\mathbf{10}}$$

$$2.5 = 2.5$$

The result checks.

For each of the following purchases, determine the better buy. See Example 7.

93. TRUMPET LESSONS
$25 for 45 minutes
45 minutes = 0.75 hour

$$\begin{array}{r} \text{cost} \rightarrow \\ 0.75 \text{ hour} \rightarrow \end{array} \frac{25}{0.75} = \frac{x}{1} \begin{array}{l} \leftarrow \text{cost} \\ \leftarrow 1 \text{ hour} \end{array}$$

$$\frac{25}{0.75} = x$$

$$33.\overline{3} = x$$

The cost per hour is $33.33.

$35 for 60 minutes,
The cost per hour is $35.00.

The lesson of $25 for 45 minutes is the better rate.

95. BUSINESS CARDS
100 cards for $9.99.

$$\begin{array}{r} \text{cost} \rightarrow \\ 100 \text{ cards} \rightarrow \end{array} \frac{9.99}{100} = \frac{x}{1} \begin{array}{l} \leftarrow \text{cost} \\ \leftarrow 1 \text{ card} \end{array}$$

$$\frac{9.99}{100} = x$$

$$0.0999 = x$$

The cost per card is $0.0999.

150 cards for $12.99.

$$\begin{array}{r} \text{cost} \rightarrow \\ 150 \text{ cards} \rightarrow \end{array} \frac{12.99}{150} = \frac{x}{1} \begin{array}{l} \leftarrow \text{cost} \\ \leftarrow 1 \text{ card} \end{array}$$

$$\frac{12.99}{150} = x$$

$$0.0866 = x$$

The cost per card is $0.0866.

The 150 cards for $12.99 is the better buy.

97. SOFT DRINKS
6 cans for $1.50.

$$\begin{array}{r} \text{cost} \rightarrow \\ 6 \text{ cans} \rightarrow \end{array} \frac{1.50}{6} = \frac{x}{1} \begin{array}{l} \leftarrow \text{cost} \\ \leftarrow 1 \text{ can} \end{array}$$

$$\frac{1.50}{6} = x$$

$$0.25 = x$$

The cost per can is $0.25.

24 cans for $6.25.

$$\begin{array}{r} \text{cost} \rightarrow \\ 24 \text{ cans} \rightarrow \end{array} \frac{6.25}{24} = \frac{x}{1} \begin{array}{l} \leftarrow \text{cost} \\ \leftarrow 1 \text{ can} \end{array}$$

$$\frac{6.25}{24} = x$$

$$0.2604 \approx x$$

The cost per can is $0.2604.

The 6-pack for $1.50 is the better buy.

99. AQUACLEAR WATER

12 8-oz bottles for $1.79.

$$\begin{array}{r} \text{cost} \rightarrow \\ 96 \text{ ounces} \rightarrow \end{array} \frac{1.79}{96} = \frac{x}{1} \begin{array}{l} \leftarrow \text{cost} \\ \leftarrow 1 \text{ ounce} \end{array}$$

$$\frac{1.79}{96} = x$$

$$0.0186 \approx x$$

The cost per ounce is about $0.0186.

24 12-oz bottles for $4.49.

$$\begin{array}{r} \text{cost} \rightarrow \\ 288 \text{ ounces} \rightarrow \end{array} \frac{4.49}{288} = \frac{x}{1} \begin{array}{l} \leftarrow \text{cost} \\ \leftarrow 1 \text{ ounce} \end{array}$$

$$\frac{4.49}{288} = x$$

$$0.0156 \approx x$$

The cost per ounce is about $0.0156.

The 24-12 oz bottles at $4.49 is the better buy.

101. HEIGHT OF A TREE

Analyze

- Similar triangles determined by the tree and its shadow and the man and his shadow.
- What is the height of the tree?

Assign

Let h = the height of the tree

Form

6 feet is to 4 feet as h feet is to 26 feet.

$$\begin{array}{r} \text{man's height} \rightarrow \\ \text{man's shadow} \rightarrow \end{array} \frac{6}{4} = \frac{h}{26} \begin{array}{l} \leftarrow \text{tree's height} \\ \leftarrow \text{tree's shadow} \end{array}$$

Solve

$$\frac{6}{4} = \frac{h}{26}$$
$$156 = 4h$$
$$\frac{156}{\mathbf{4}} = \frac{4h}{\mathbf{4}}$$
$$39 = h$$

State

A height of the tree is 39 feet.

Check

$$\frac{6}{4} = \frac{h}{26}$$
$$\frac{6}{4} \stackrel{?}{=} \frac{\mathbf{39}}{26}$$
$$156 = 156$$

The result checks.

103. SURVEYING

Analyze

- Similar triangles are determined by the layout.
- What is the width of the river?

Assign

Let w = the width of the river

Form a Proportion

20 feet is to 32 feet as w feet is to 75 feet.

$$\begin{array}{r} \text{dis to bank} \rightarrow \\ \text{dis along river} \rightarrow \end{array} \frac{20}{32} = \frac{w}{75} \begin{array}{l} \leftarrow \text{width of river} \\ \leftarrow \text{dis along river} \end{array}$$

Solve

$$\frac{20}{32} = \frac{w}{75}$$
$$1,500 = 32w$$
$$\frac{1,500}{\mathbf{32}} = \frac{32w}{\mathbf{32}}$$
$$46.875 = w$$
$$46\frac{7}{8} = x$$

State

A width of the river is $46\frac{7}{8}$ feet.

Check

$$\frac{20}{32} = \frac{x}{75}$$
$$\frac{20}{32} \stackrel{?}{=} \frac{\mathbf{46.875}}{75}$$
$$1,500 = 1,500$$

The result checks.

105. SLOPE

Analyze

- The rise to run is 14 to 21 of the given large triangle.
- What is the rise of the smaller triangle with a run of 12?

Assign

Let x = the rise of the smaller triangle

Form

14 is to 21 as x is to 12.

$$\begin{array}{r} \text{rise large} \rightarrow \\ \text{run large} \rightarrow \end{array} \frac{14}{21} = \frac{x}{12} \begin{array}{l} \leftarrow \text{rise small} \\ \leftarrow \text{run small} \end{array}$$

Solve

$$\frac{14}{21} = \frac{x}{12}$$
$$168 = 21x$$
$$\frac{168}{\mathbf{21}} = \frac{21x}{\mathbf{21}}$$
$$8 = x$$

State

The rise of the smaller triangle is 8.

Check

$$\frac{14}{21} = \frac{x}{12}$$
$$\frac{14}{21} \stackrel{?}{=} \frac{\mathbf{8}}{12}$$
$$168 = 168$$

The result checks.

WRITING

107-109. Answers will vary.

REVIEW

111. $\frac{9}{10} = 10\overline{)9.0}^{\,0.9} \Rightarrow 0.9 \cdot 100 = 90\%$

113. $30\% \text{ of } 1{,}600 = 0.30 \cdot 1{,}600$
$= 480$

CHALLENGE PROBLEMS

115. $\frac{a}{c} = \frac{b}{d}, \quad \frac{b}{a} = \frac{d}{c}, \quad \frac{c}{a} = \frac{d}{b}$

SECTION 7.9 STUDY SET

VOCABULARY

Fill in the blanks.

1. The equation $y = kx$ defines **direct** variation and the equation $y = \frac{k}{x}$ defines **inverse** variation.

Tell whether each relationship suggests direct or inverse variation.

3. Direct 5. Inverse 7. Direct

9. Inverse

CONCEPTS

Determine whether each graph represents direct variation or inverse variation.

▶ 11.

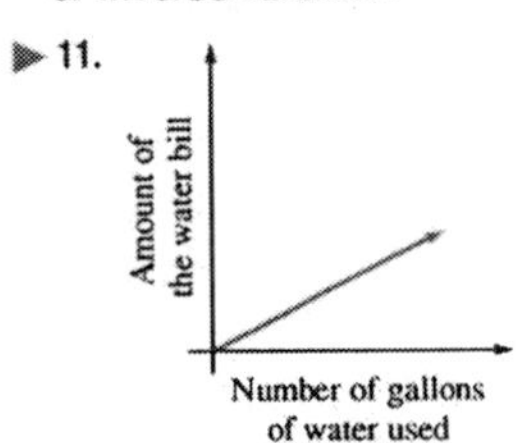

Direct

13.

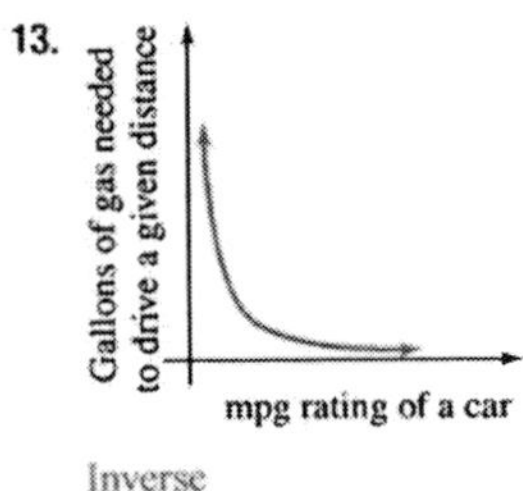

Inverse

Complete each graph by sketching either a direct variation or an inverse variation.

15.

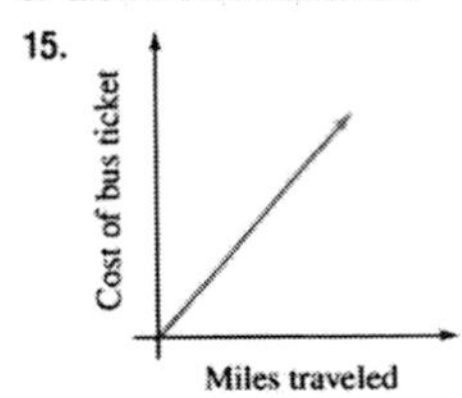

17.

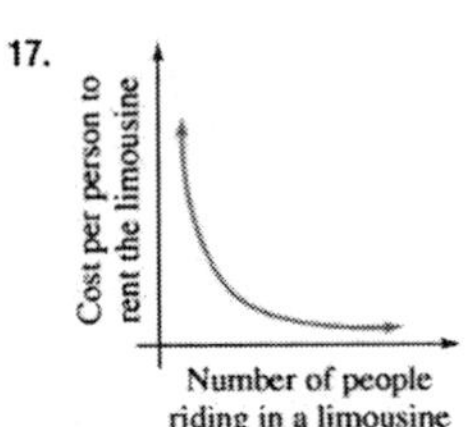

Write an equation to describe each variation. Use k for the constant of variation.

19. SWIMMING 21. GEOLOGY

$p = kd$ $d = \frac{k}{r}$

NOTATION

Determine whether the equation describes direct variation.

23. a. Yes b. No c. No d. Yes

GUIDED PRACTICE

Solve each direct variation problem. See Example 1.

25. $y = kx$ Find y when $x = 7$.

Find k:

$10 = k(2)$

$\frac{10}{2} = \frac{2k}{2}$

$5 = k$

$y = 5x$

$y = 5(7)$

$y = 35$

Thus, when $x = 7$, the value of y is 35.

27. $r = ks$ Find r when $s = 12$.

Find k:

$21 = k(7)$

$\frac{21}{7} = \frac{7k}{7}$

$3 = k$

$r = 3s$

$r = 3(12)$

$r = 36$

Thus, when $s = 12$, the value of r is 36.

29. $s = kt$ Find s when $t = 30$.

Find k:

$1.2 = k(4)$

$\frac{1.2}{4} = \frac{4k}{4}$

$0.3 = k$

$s = 0.3t$

$s = 0.3(30)$

$s = 9$

Thus, when $t = 30$, the value of s is 9.

31. $d = kt$ Find d when $t = 4$.

Find k:

$21 = k(6)$

$\frac{21}{6} = \frac{6k}{6}$

$\frac{7}{2} = k$

$d = \frac{7}{2}t$

$d = \frac{7}{2}(4)$

$d = 14$

Thus, when $t = 4$, the value of d is 14.

Solve each inverse variation problem. See Examples 3 and 4.

33. $y = \frac{k}{x}$ Find y when $x = 4$.

Find k:

$8 = \frac{k}{2}$

$8 \cdot 2 = k$

$16 = k$

$y = \frac{16}{x}$

$y = \frac{16}{4}$

$y = 4$

Thus, when $x = 4$, the value of y is 4.

35. $r = \frac{k}{t}$ Find r when $t = 200$.

Find k:

$40 = \frac{k}{10}$

$40 \cdot 10 = k$

$400 = k$

$r = \frac{400}{t}$

$r = \frac{400}{200}$

$r = 2$

Thus, when $t = 200$, the value of r is 2.

37. $p = \frac{k}{x}$

Find k:

$$6 = \frac{k}{4}$$
$$6 \cdot 4 = k$$
$$24 = k$$

Find p when $x = 1.5$.

$$p = \frac{24}{x}$$
$$p = \frac{24}{1.5}$$
$$p = 16$$

Thus, when $x = 1.5$, the value of p is 16.

39. $q = \frac{k}{s}$

Find k:

$$6 = \frac{k}{9}$$
$$6 \cdot 9 = k$$
$$54 = k$$

Find q when $s = 24$.

$$q = \frac{54}{s}$$
$$q = \frac{54}{24}$$
$$q = 2.25$$

Thus, when $s = 24$, the value of q is 2.25 or $\frac{9}{4}$.

APPLICATIONS

Solve each direct variation problem. See Example 2.

41. DRIVING

Step 1: We let d represent the distance (in miles) that a car can travel on g (gallons) of gasoline. Translating the words *distance varies directly with number of gallons,* we get the equation $d = kg$.

Step 2: To find the constant of variation, k, we substitute 360 for d and 15 for g.

$$d = kg$$
$$360 = k(15)$$
$$\frac{360}{15} = \frac{15k}{15}$$
$$24 = k$$

Step 3: We now substitute the value of k into the equation from step 1.

$$d = 24g$$

Step 4: We can find the *distance* by substituting 7 for g in the equation from step 3.

$$d = 24g$$
$$d = 24(7)$$
$$d = 168$$

The car can travel 168 miles on 7 gallons of gasoline.

43. MEDICATIONS

Step 1: We let d represent the dose (in milligrams) of medicine and the child's body weight (in p pounds). Translating the words *dose is directly proportional to the child's weight in pound,* we get the equation $d = kp$.

Step 2: To find the constant of variation, k, we substitute 124 for d and 20 for p.

$$d = kp$$
$$124 = k(20)$$
$$\frac{124}{20} = \frac{20k}{20}$$
$$6.2 = k$$

Step 3: We now substitute the value of k into the equation from step 1.

$$d = 6.2p$$

Step 4: We can find the *dose* by substituting 28 for p in the equation from step 3.

$$d = 6.2p$$
$$d = 6.2(28)$$
$$d = 173.6$$

The dose for a 28 pound child is 173.6 mg.

Solve each inverse variation problem. See Example 4.

45. TRAVELING

Step 1: We let t represent the time (in hours) that a car can travel at a certain rate of speed (r) in mph. Translating the words *distance varies indirectly as its rate of speed,*

we get the equation $t = \frac{k}{r}$.

Step 2: To find the constant of variation, k, we substitute 3 for t and 50 for r.

$$t = \frac{k}{r}$$
$$3 = \frac{k}{50}$$
$$3 \cdot 50 = k$$
$$150 = k$$

Step 3: We now substitute the value of k into the equation from step 1.

$$t = \frac{150}{r}$$

Step 4: We can find the *time* by substituting 60 for r in the equation from step 3.

$$t = \frac{150}{r}$$
$$t = \frac{150}{60}$$
$$t = 2.5$$

The trip will take 2.5 hours traveling at a rate of 60 mph.

47. ELECTRICITY

Step 1: We let a represent the current (in amps) and r the resistance (in ohms) of a circuit. Translating the words *the current of a circuit, varies inversely as the resistance,*

we get the equation $a = \frac{k}{r}$.

Step 2: To find the constant of variation, k, we substitute 30 for a and 4 for r.

$$a = \frac{k}{r}$$
$$30 = \frac{k}{4}$$
$$30 \cdot 4 = k$$
$$120 = k$$

Step 3: We now substitute the value of k into the equation from step 1.

$$a = \frac{120}{r}$$

Step 4: We can find the *amps* by substituting 15 for r in the equation from step 3.

$$a = \frac{120}{r}$$
$$a = \frac{120}{15}$$
$$a = 8$$

The current is 8 amps with a resistance of 15 ohms.

TRY IT YOURSELF

49. PULLEYS

Step 1: We let s represent the speed of the pulley (in revolutions per minute) and d the diameter of the pulley (in inches). Translating the words *the speed of a pulley is inversely proportional to its diameter,*

we get the equation $s = \frac{k}{d}$.

Step 2: To find the constant of variation, k, we substitute 120 for s and 24 for d.

$$s = \frac{k}{d}$$
$$120 = \frac{k}{24}$$
$$120 \cdot 24 = k$$
$$2{,}880 = k$$

Step 3: We now substitute the value of k into the equation from step 1.

$$s = \frac{2{,}880}{d}$$

Step 4: We can find the *speed of the smaller pulley* by substituting 16 for d in the equation from step 3.

$$s = \frac{2{,}880}{d}$$
$$s = \frac{2{,}880}{16}$$
$$s = 180$$

The speed of the smaller pulley is 180 rpm.

51. GRAVITY

Step 1: We let m represent the weight of an object on the moon (in pounds) and e the weight of the object on earth (in pounds). Translating the words *the weight of an object on the moon varies directly to its weight on earth,* we get the equation $m = ke$.

Step 2: To find the constant of variation, k, we substitute 1 for m and 6 for e.

$$m = ke$$
$$1 = k(6)$$
$$\frac{1}{\mathbf{6}} = \frac{6k}{\mathbf{6}}$$
$$\frac{1}{6} = k$$

Step 3: We now substitute the value of k into the equation from step 1.

$$m = \frac{1}{6}e$$

Step 4: We can find the *weight of the astronaut* by substituting 330 for e in the equation from step 3.

$$m = \frac{1}{6}e$$
$$m = \frac{1}{6}(330)$$
$$m = 55$$

The astronaut weights 55 pounds on the moon.

53. HOOKE'S LAW

Step 1: We let d represent the distance a spring will stretch (in centimeters) and f the force applied to the object (in kilograms). Translating the words the *distance a spring will stretch varies directly as the force applied to it,* we get the equation $d = kf$.

Step 2: To find the constant of variation, k, we substitute 24 for d and 15 for f.

$$d = kf$$
$$24 = k(15)$$
$$\frac{24}{\mathbf{15}} = \frac{15k}{\mathbf{15}}$$
$$\frac{8}{5} = k$$

Step 3: We now substitute the value of k into the equation from step 1.

$$d = \frac{8}{5}f$$

Step 4: We can find the *distance the spring will stretch* by substituting 25 for f in the equation from step 3.

$$d = \frac{8}{5}f$$
$$d = \frac{8}{5}(25)$$
$$d = 40$$

The spring will stretch 40 cm.

55. ARCHITECTURE

Step 1: We let s represent the fixed floor space, the number of sq. ft. varies inversely as f, the number of floors.
Translating the words *the number of sq. ft. varies inversely to the number of floors,* we get the equation $s = \frac{k}{f}$.

Step 2: To find the constant of variation, k, we substitute 163,000 for s and 5 for f.

$$s = \frac{k}{f}$$
$$163{,}000 = \frac{k}{5}$$
$$5 \cdot 163{,}000 = k$$
$$815{,}000 = k$$

Step 3: We now substitute the value of k into the equation from step 1.

$$s = \frac{815{,}000}{f}$$

Step 4: We can find the *sq. ft. that a 25-story building will occupy* by substituting 25 for f in the equation from step 3.

$$s = \frac{815{,}000}{f}$$
$$s = \frac{815{,}000}{25}$$
$$s = 32{,}600$$

32,600 ft^2 of floor space will occupy a 25-story building.

from CAMPUS TO CARRERS

57. RECREATION DIRECTOR

Step 1: We let v represent the value of the bike (in dollars) and a the age of the bike (in years). Translating the words *the value of a machine varies inversely as its age,* we get the equation $v=\frac{k}{a}$.

Step 2: To find the constant of variation, k, we substitute 1,600 for v and 2 for a.

$$v=\frac{k}{a}$$
$$1,600=\frac{k}{2}$$
$$2\cdot 1,600=k$$
$$3,200=k$$

Step 3: We now substitute the value of k into the equation from step 1.

$$v=\frac{3,200}{a}$$

Step 4: We can find the *value of the bike when it is 8 years old* by substituting 8 for a in the equation from step 3.

$$v=\frac{3,200}{a}$$
$$v=\frac{3,200}{8}$$
$$v=400$$

In 8 years the bike is worth \$400.

It depreciates \$1,200 over that time period. $(\$1,600-\$400=\$1,200)$

WRITING

59-61. Answers will vary.

REVIEW

Solve each equation.

63. $(a-1)(a^2+5a+6)=0$

$$(a-1)(a+2)(a+3)=0$$

$a-1=0$ or $a+2=0$ or $a+3=0$

$a-1+1=0+1$ | $a+2-2=0-2$ | $a+3-3=0-3$

$a=1$ | $a=-2$ | $a=-3$

These check, so the solutions are $1, -2$ and -3.

65. $x^3-6x^2-27x=0$

$$x(x^2-6x-27)=0$$
$$x(x+3)(x-9)=0$$

$x=0$ or $x+3=0$ or $x-9=0$

$x+3-3=0-3$ | $x-9+9=0+9$

$x=-3$ | $x=9$

These check, so the solutions are $0, -3$ and 9.

CHALLENGE PROBLEMS

67. WIND ENERGY

Step 1: We let p represent the power of the turbine (in kw) and s the speed of the wind (in mph). Translating the words the *power of the wind turbine varies directly as the cube of speed of the wind,* we get the equation $p=ks^3$.

Step 2: To find the constant of variation, k, we substitute 5.4 for p and 15 for s.

$$p=ks^3$$
$$5.4=k(15^3)$$
$$\frac{5.4}{3,375}=\frac{3,375k}{3,375}$$
$$0.0016=k$$

Step 3: We now substitute the value of k into the equation from step 1.

$$p=0.0016s^3$$

Step 4: We can find the *number of kilowatts* by substituting 30 for s in the equation from step 3.

$$p=0.0016s^3$$
$$p=(0.0016)(30^3)$$
$$p=(0.0016)(27,000)$$
$$p=43.2$$

The amount of power produced is 43.2 kw.

69. LANDING AIRCRAFT

Step 1: We let d represent the length of runway (in feet) and s the speed of the plane (in mph). Translating the words the *runway distance varies directly as the square of the touchdown speed*, we get the equation $d = ks^2$.

Step 2: To find the constant of variation, k, we substitute 1,470 for d and 70 for s.

$$d = ks^2$$
$$1{,}470 = k(70^2)$$
$$\frac{1{,}470}{4{,}900} = \frac{4{,}900k}{4{,}900}$$
$$0.3 = k$$

Step 3: We now substitute the value of k into the equation from step 1.

$$d = 0.3s^2$$

Step 4: We can find the *landing distance* by substituting 80 for s in the equation from step 3.

$$d = 0.3s^2$$
$$d = (0.3)(80^2)$$
$$d = (0.3)(6{,}400)$$
$$d = 1{,}920$$

The landing distance needed is 1,920 feet.

CHAPTER 7 REVIEW
SECTION 7.1
Simplifying Rational Expressions

1. Find undefined values.

$$\frac{x-1}{x^2-16}=\frac{x-1}{(x+4)(x-4)}$$

$$x+4=0 \quad \text{or} \quad x-4=0$$
$$x+4-\mathbf{4}=0-\mathbf{4} \quad \Big| \quad x-4+\mathbf{4}=0+\mathbf{4}$$
$$x=-4 \quad \Big| \quad x=4$$

The rational expression is undefined for $x=-4$ and $x=4$.

2. Evaluate: $\frac{x^2-1}{x-5}$ for $x=-2$.

$$\frac{x^2-1}{x-5}=\frac{(-2)^2-1}{-2-5}=\frac{4-1}{-7}=-\frac{3}{7}$$

Simplify each rational expression, if possible. Assume that no denominators are zero.

3. $$\frac{3x^2}{6x^3}=\frac{\cancel{3}\cdot \cancel{x^2}}{2\cdot\cancel{3}\cdot x\cdot\cancel{x^2}}=\frac{1}{2x}$$

4. $$\frac{5xy^2}{2x^2y^2}=\frac{5\cdot\cancel{x}\cdot\cancel{y^2}}{2x\cdot\cancel{x}\cdot\cancel{y^2}}=\frac{5}{2x}$$

5. $$\frac{x^2}{x^2+x}=\frac{x^2}{x(x+1)}=\frac{\cancel{x}\cdot x}{\cancel{x}(x+1)}=\frac{x}{x+1}$$

6. $$\frac{a^2-4}{a+2}=\frac{(a+2)(a-2)}{a+2}=\frac{\cancel{(a+2)}(a-2)}{\cancel{a+2}}=a-2$$

7. $$\frac{3p-2}{2-3p}=\frac{3p-2}{-1(-2+3p)}=\frac{\cancel{3p-2}}{-1\,\cancel{(3p-2)}}=-1$$

8. $$\frac{8-x}{x^2-5x-24}=\frac{8-x}{(x-8)(x+3)}=\frac{-1\,\cancel{(x-8)}}{\cancel{(x-8)}\,(x+3)}=-\frac{1}{x+3}$$

9. $$\frac{2x^2-16x}{2x^2-18x+16}=\frac{2x(x-8)}{2(x^2-9x+8)}=\frac{2x(x-8)}{2(x-1)(x-8)}=\frac{\cancel{2}\cdot x\,\cancel{(x-8)}}{\cancel{2}(x-1)\,\cancel{(x-8)}}=\frac{x}{x-1}$$

10. $$\frac{x^2+x-2}{x^2-x-2}=\frac{(x+2)(x-1)}{(x-2)(x+1)}$$

This will not simplify.

11. $$\frac{x^2-2xy+y^2}{(x-y)^3}=\frac{(x-y)(x-y)}{(x-y)(x-y)(x-y)}=\frac{\cancel{(x-y)}\,\cancel{(x-y)}}{(x-y)\,\cancel{(x-y)}\,\cancel{(x-y)}}=\frac{1}{x-y}$$

12. $$\frac{4(t+3)+8}{3(t+3)+6}=\frac{4t+12+8}{3t+9+6}=\frac{4t+20}{3t+15}=\frac{4(t+5)}{3(t+5)}=\frac{4\,\cancel{(t+5)}}{3\,\cancel{(t+5)}}=\frac{4}{3}$$

13. $\frac{x+1}{x} = \frac{\cancel{x}+1}{\cancel{x}} = \frac{2}{1} = 2$

x is not a common factor of the numerator and the denominator; x is a term of the numerator.

14. DOSAGES

$$C = \frac{D(A+1)}{24}$$

$$C = \frac{300(11+1)}{24} = \frac{300(12)}{24} = \frac{3,600}{24} = 150$$

The 11 year old's dosage is 150 milligrams.

SECTION 7.2

Multiplying and Dividing Rational Expressions

Multiply and simplify, if possible.

15. $\frac{3xy}{2x} \cdot \frac{4x}{2y^2} = \frac{3xy \cdot 4x}{2x \cdot 2y^2} = \frac{3 \cdot \cancel{4} \cdot x \cdot \cancel{x} \cdot \cancel{y}}{\cancel{4} \cdot \cancel{x} \cdot \cancel{y} \cdot y} = \frac{3x}{y}$

16. $56x\left(\frac{12}{7x}\right) = \frac{56x \cdot 12}{7x} = \frac{\cancel{7} \cdot 8 \cdot 12 \cdot \cancel{x}}{\cancel{7} \cdot \cancel{x}} = 96$

17. $\frac{x^2-1}{x^2+2x} \cdot \frac{x}{x+1} = \frac{(x+1)(x-1)x}{x(x+2)(x+1)} = \frac{\cancel{x}\,\cancel{(x+1)}(x-1)}{\cancel{x}(x+2)\,\cancel{(x+1)}} = \frac{x-1}{x+2}$

18. $\frac{x^2+x}{3x-15} \cdot \frac{6x-30}{x^2+2x+1} = \frac{x(x+1)6(x-5)}{3(x-5)(x+1)(x+1)} = \frac{2 \cdot \cancel{3} \cdot x\,\cancel{(x+1)}\,\cancel{(x-5)}}{\cancel{3}\,\cancel{(x-5)}\,\cancel{(x+1)}(x+1)} = \frac{2x}{x+1}$

Divide and simplify, if possible.

19. $\frac{3x^2}{5x^2y} \div \frac{6x}{15xy^2} = \frac{3x^2}{5x^2y} \cdot \frac{15xy^2}{6x} = \frac{3x^2 \cdot 15xy^2}{5x^2y \cdot 6x} = \frac{3 \cdot \cancel{3} \cdot \cancel{5}\,\cancel{x^3}\,y\,\cancel{y}}{2 \cdot \cancel{3} \cdot \cancel{5}\,\cancel{x^3}\,\cancel{y}} = \frac{3y}{2}$

20. $\frac{x^2-x-6}{1-2x} \div \frac{x^2-2x-3}{2x^2+x-1}$

$$= \frac{x^2-x-6}{1-2x} \cdot \frac{2x^2+x-1}{x^2-2x-3}$$

$$= \frac{(x-3)(x+2)(2x-1)(x+1)}{-1(-1+2x)(x-3)(x+1)}$$

$$= -\frac{\cancel{(x-3)}(x+2)\,\cancel{(2x-1)}\,\cancel{(x+1)}}{\cancel{(2x-1)}\,\cancel{(x-3)}\,\cancel{(x+1)}}$$

$$= -(x+2) \text{ or } -x-2$$

21. a. Yes b. No c. Yes d. Yes

22. TRAFFIC SIGN

$$\frac{20 \text{ miles}}{1 \text{ hour}} \cdot \frac{1 \text{ hour}}{60 \text{ minutes}} = \frac{20 \text{ miles } 1\,\cancel{\text{hour}}}{1\,\cancel{\text{hour}}\,60 \text{ minutes}} = \frac{\cancel{2} \cdot \cancel{2} \cdot \cancel{5} \text{ miles}}{3 \cdot \cancel{2} \cdot \cancel{2} \cdot \cancel{5} \text{ minutes}} = \frac{1}{3} \text{ mile per minute}$$

The conversion is $\frac{1}{3}$ mile per minute.

SECTION 7.3

Adding and Subtracting with Like Denominators; Least Common Denominators

Add or subtract and simplify, if possible.

23. $\frac{13}{15d}-\frac{8}{15d}=\frac{13-8}{15d}$

$=\frac{\overset{1}{\cancel{5}}}{3\cdot\underset{1}{\cancel{5}}d}$

$=\frac{1}{3d}$

24. $\frac{x}{x+y}+\frac{y}{x+y}=\frac{x+y}{x+y}$

$=\frac{\overset{1}{\cancel{x+y}}}{\underset{1}{\cancel{x+y}}}$

$=1$

25. $\frac{3x}{x-7}-\frac{x-2}{x-7}=\frac{3x-(x-2)}{x-7}$

$=\frac{3x-x+2}{x-7}$

$=\frac{2x+2}{x-7}$

$=\frac{2(x+1)}{x-7}$ or $\frac{2x+2}{x-7}$

26. $\frac{a}{a^2-2a-8}+\frac{2}{a^2-2a-8}=\frac{a+2}{a^2-2a-8}$

$=\frac{a+2}{(a-4)(a+2)}$

$=\frac{\overset{1}{\cancel{a+2}}}{(a-4)\underset{1}{\cancel{(a+2)}}}$

$=\frac{1}{a-4}$

Find the LCD of each pair of rational expressions.

27. $\frac{12}{x},\frac{1}{9}$

$\text{LCD}=9x$

28. $\frac{1}{2x^3},\frac{5}{8x}$

$2x^3=2\cdot x\cdot x\cdot x$

$8x=2\cdot 2\cdot 2\cdot x$

$\text{LCD}=2\cdot 2\cdot 2\cdot x\cdot x\cdot x=8x^3$

29. $\frac{7}{m},\frac{m+2}{m-8}$

$\text{LCD}=m(m-8)$

30. $\frac{x}{5x+1},\frac{5x}{5x-1}$

$\text{LCD}=(5x+1)(5x-1)$

31. $\frac{6-a}{a^2-25},\frac{a^2}{a-5}$

$a^2-25=(a+5)(a-5)$

$a-5=a-5$

$\text{LCD}=(a+5)(a-5)$

32. $\frac{4t+25}{t^2+10t+25},\frac{t^2-7}{2t^2+17t+35}$

$t^2+10t+25=(t+5)(t+5)$

$2t^2+17t+35=(2t+7)(t+5)$

$\text{LCD}=(2t+7)(t+5)^2$

Build each rational expression into an equivalent fraction having the given denominator.

33. $\frac{9}{a}\cdot\frac{\mathbf{7}}{\mathbf{7}}=\frac{63}{7a}$

34. $\frac{2y+1}{x-9}\cdot\frac{\mathbf{x}}{\mathbf{x}}=\frac{x(2y+1)}{x(x-9)}$

$=\frac{2xy+x}{x(x-9)}$

35. $\frac{b+7}{3b-15}=\frac{b+7}{3(b-5)}\cdot\frac{\mathbf{2}}{\mathbf{2}}$

$=\frac{2(b+7)}{6(b-5)}$

$=\frac{2b+14}{6(b-5)}$

36. $\dfrac{9r}{r^2+6r+5} = \dfrac{9r}{(r+1)(r+5)} \cdot \dfrac{\mathbf{r-4}}{\mathbf{r-4}}$

$= \dfrac{9r(r-4)}{(r+1)(r+5)(r-4)}$

$= \dfrac{9r^2-36r}{(r+1)(r+5)(r-4)}$

SECTION 7.4

Adding and Subtracting with Unlike Denominators

Add or subtract and simplify, if possible.

37. $\text{LCD} = 7a$

$\dfrac{1}{7} - \dfrac{1}{a} = \dfrac{1}{7} \cdot \dfrac{\mathbf{a}}{\mathbf{a}} - \dfrac{1}{a} \cdot \dfrac{\mathbf{7}}{\mathbf{7}}$

$= \dfrac{a-7}{7a}$

38. $\text{LCD} = x(x-1)$

$\dfrac{x}{x-1} + \dfrac{1}{x} = \dfrac{x}{x-1} \cdot \dfrac{\mathbf{x}}{\mathbf{x}} + \dfrac{1}{x} \cdot \dfrac{\mathbf{(x-1)}}{\mathbf{(x-1)}}$

$= \dfrac{x^2+x-1}{x(x-1)}$

39. $\left.\begin{array}{r} t^2+2t+1 = (t+1)(t+1) \\ (t+1) = t+1 \end{array}\right\}$

$\text{LCD} = (t+1)(t+1)$

$\dfrac{2t+2}{t^2+2t+1} - \dfrac{1}{t+1} = \dfrac{2t+2}{(t+1)(t+1)} - \dfrac{1}{(t+1)} \cdot \dfrac{\mathbf{(t+1)}}{\mathbf{(t+1)}}$

$= \dfrac{2t+2-(t+1)}{(t+1)(t+1)}$

$= \dfrac{2t+2-t-1}{(t+1)(t+1)}$

$= \dfrac{\overset{1}{\cancel{t+1}}}{(t+1)\underset{1}{\cancel{(t+1)}}}$

$= \dfrac{1}{t+1}$

40. $\left.\begin{array}{r} 2x = 2x \\ x^2 = x \cdot x \end{array}\right\} \text{LCD} = 2x^2$

$\dfrac{x+2}{2x} - \dfrac{2-x}{x^2} = \dfrac{x+2}{2x} \cdot \dfrac{\mathbf{x}}{\mathbf{x}} - \dfrac{2-x}{x^2} \cdot \dfrac{\mathbf{2}}{\mathbf{2}}$

$= \dfrac{x(x+2)-2(2-x)}{2x^2}$

$= \dfrac{x^2+2x-4+2x}{2x^2}$

$= \dfrac{x^2+4x-4}{2x^2}$

41. $\text{LCD} = b-1$

$\dfrac{6}{b-1} - \dfrac{b}{1-b} = \dfrac{6}{b-1} - \dfrac{b}{-1(-1+b)}$

$= \dfrac{6}{b-1} - \dfrac{-b}{b-1}$

$= \dfrac{6}{b-1} + \dfrac{b}{b-1}$

$= \dfrac{b+6}{b-1}$

42. $\text{LCD} = c$

$\dfrac{8}{c} + 6 = \dfrac{8}{c} + \dfrac{6}{1} \cdot \dfrac{\mathbf{c}}{\mathbf{c}}$

$= \dfrac{8}{c} + \dfrac{6c}{c}$

$= \dfrac{6c+8}{c}$

43. $\left.\begin{array}{r} n+3 = n+3 \\ n+7 = n+7 \end{array}\right\} \text{LCD} = (n+3)(n+7)$

$\dfrac{n+7}{n+3} - \dfrac{n-3}{n+7} = \dfrac{(n+7)}{(n+3)} \cdot \dfrac{\mathbf{(n+7)}}{\mathbf{(n+7)}} - \dfrac{(n-3)}{(n+7)} \cdot \dfrac{\mathbf{(n+3)}}{\mathbf{(n+3)}}$

$= \dfrac{(n+7)(n+7)-(n-3)(n+3)}{(n+3)(n+7)}$

$= \dfrac{n^2+14n+49-(n^2-9)}{(n+3)(n+7)}$

$= \dfrac{n^2+14n+49-n^2+9}{(n+3)(n+7)}$

$= \dfrac{14n+58}{(n+3)(n+7)}$

44. $\left.\begin{array}{r} t+2=t+2 \\ (t+2)^2=(t+2)^2 \end{array}\right\}\text{LCD}=(t+2)^2$

$$\frac{4}{t+2}-\frac{7}{(t+2)^2}=\frac{4}{(t+2)}\cdot\frac{\mathbf{(t+2)}}{\mathbf{(t+2)}}-\frac{7}{(t+2)^2}$$
$$=\frac{4(t+2)-7}{(t+2)^2}$$
$$=\frac{4t+8-7}{(t+2)^2}$$
$$=\frac{4t+1}{(t+2)^2}$$

45. $\left.\begin{array}{r} a^2-9=(a+3)(a-3) \\ a^2-a-6=(a-3)(a+2) \end{array}\right\}$

$\text{LCD}=(a+3)(a-3)(a+2)$

$$\frac{6}{a^2-9}-\frac{5}{a^2-a-6}$$
$$=\frac{6}{(a+3)(a-3)}-\frac{5}{(a-3)(a+2)}$$
$$=\frac{6}{(a+3)(a-3)}\cdot\frac{\mathbf{(a+2)}}{\mathbf{(a+2)}}-\frac{5}{(a-3)(a+2)}\cdot\frac{\mathbf{(a+3)}}{\mathbf{(a+3)}}$$
$$=\frac{6(a+2)-5(a+3)}{(a+3)(a-3)(a+2)}$$
$$=\frac{6a+12-5a-15}{(a+3)(a-3)(a+2)}$$
$$=\frac{a-3}{(a+3)(a-3)(a+2)}$$
$$=\frac{\overset{1}{\cancel{a-3}}}{(a+3)\underset{1}{\cancel{(a-3)}}(a+2)}$$
$$=\frac{1}{(a+3)(a+2)}$$

46. $\left.\begin{array}{r} 3y-6=3(y-2) \\ 4y+8=4(y+2) \end{array}\right\}\text{LCD}=12(y-2)(y+2)$

$$\frac{2}{3y-6}+\frac{3}{4y+8}$$
$$=\frac{2}{3(y-2)}+\frac{3}{4(y+2)}$$
$$=\frac{2}{3(y-2)}\cdot\frac{\mathbf{4(y+2)}}{\mathbf{4(y+2)}}+\frac{3}{4(y+2)}\cdot\frac{\mathbf{3(y-2)}}{\mathbf{3(y-2)}}$$
$$=\frac{8(y+2)+9(y-2)}{12(y-2)(y+2)}$$
$$=\frac{8y+16+9y-18}{12(y-2)(y+2)}$$
$$=\frac{17y-2}{12(y-2)(y+2)}$$

47. $\frac{-5n^3-7}{3n(n+6)}=\frac{-(5n^3+7)}{3n(n+6)}=-\frac{5n^3+7}{3n(n+6)}$

Yes, they are equivalent.

48. DIGITAL VIDEO CAMERAS

$P=2L+2W \qquad \text{LCD}=(x+6)(x-1)$

$$P=2\left(\frac{4}{x+6}\right)+2\left(\frac{3}{x-1}\right)$$
$$=\frac{8}{x+6}+\frac{6}{x-1}$$
$$=\frac{8}{(x+6)}\cdot\frac{\mathbf{(x-1)}}{\mathbf{(x-1)}}+\frac{6}{x-1}\cdot\frac{\mathbf{(x+6)}}{\mathbf{(x+6)}}$$
$$=\frac{8(x-1)+6(x+6)}{(x+6)(x-1)}$$
$$=\frac{8x-8+6x+36}{(x+6)(x-1)}$$
$$=\frac{14x+28}{(x+6)(x-1)}$$

The perimeter is $\frac{14x+28}{(x+6)(x-1)}$ units.

AREA $\quad A=LW$

$$=\left(\frac{4}{x+6}\right)\left(\frac{3}{x-1}\right)$$
$$=\frac{12}{(x+6)(x-1)}$$

The area is $\frac{12}{(x+6)(x-1)}$ square units.

SECTION 7.5

Simplifying Complex Fractions

Simplify each complex fraction.

49. $\dfrac{\frac{n^4}{30}}{\frac{7n}{15}} = \dfrac{n^4}{30} \div \dfrac{7n}{15}$

$= \dfrac{n^4}{30} \cdot \dfrac{15}{7n}$

$= \dfrac{15n^4}{30 \cdot 7n}$

$= \dfrac{\cancel{3} \cdot \cancel{5} \cdot \cancel{n} \cdot n^3}{2 \cdot \cancel{3} \cdot \cancel{5} \cdot 7 \cdot \cancel{n}}$

$= \dfrac{n^3}{14}$

50. $\dfrac{\frac{r^2-81}{18s^2}}{\frac{4r-36}{9s}} = \dfrac{r^2-81}{18s^2} \div \dfrac{4r-36}{9s}$

$= \dfrac{r^2-81}{18s^2} \cdot \dfrac{9s}{4r-36}$

$= \dfrac{9s(r+9)(r-9)}{18s^2 \cdot 4(r-9)}$

$= \dfrac{\cancel{3} \cdot \cancel{3} \cdot \cancel{s}(r+9)\cancel{(r-9)}}{2 \cdot \cancel{3} \cdot \cancel{3} \cdot 4 \cdot \cancel{s} \cdot s\cancel{(r-9)}}$

$= \dfrac{r+9}{8s}$

51. LCD $= y$

$\dfrac{\frac{1}{y}+1}{\frac{1}{y}-1} = \dfrac{\left(\frac{1}{y}+1\right)\mathbf{y}}{\left(\frac{1}{y}-1\right)\mathbf{y}}$

$= \dfrac{\frac{1}{y}(y)+1(y)}{\frac{1}{y}(y)-1(y)}$

$= \dfrac{1+y}{1-y}$

52. LCD $= 3a^2$

$\dfrac{\frac{7}{a^2}}{\frac{1}{a}+\frac{10}{3}} = \dfrac{\left(\frac{7}{a^2}\right)\mathbf{3a^2}}{\left(\frac{1}{a}+\frac{10}{3}\right)\mathbf{3a^2}}$

$= \dfrac{\frac{7}{a^2}(3a^2)}{\frac{1}{a}(3a^2)+\frac{10}{3}(3a^2)}$

$= \dfrac{21}{3a+10a^2}$

53. LCD $= (x+1)(x-1)$

$\dfrac{\frac{2}{x-1}+\frac{x-1}{x+1}}{\frac{1}{x^2-1}} = \dfrac{\frac{2}{x-1}+\frac{x-1}{x+1}}{\frac{1}{(x+1)(x-1)}} \cdot \dfrac{\mathbf{(x+1)(x-1)}}{\mathbf{(x+1)(x-1)}}$

$= \dfrac{\left(\frac{2}{x-1}\right)\cdot(x+1)(x-1)+\left(\frac{x-1}{x+1}\right)\cdot(x+1)(x-1)}{\left(\frac{1}{(x+1)(x-1)}\right)\cdot(x+1)(x-1)}$

$= \dfrac{2(x+1)+(x-1)(x-1)}{1}$

$= 2x+2+x^2-2x+1$

$= x^2+3$

54. $\left.\begin{aligned} x^2y &= x \cdot x \cdot y \\ xy &= x \cdot y \\ xy^2 &= x \cdot y \cdot y \end{aligned}\right\}$ LCD $= x^2y^2$

$\dfrac{\frac{1}{x^2y}-\frac{5}{xy}}{\frac{3}{xy}-\frac{7}{xy^2}} = \dfrac{\frac{1}{x^2y}-\frac{5}{xy}}{\frac{3}{xy}-\frac{7}{xy^2}} \cdot \dfrac{\mathbf{x^2y^2}}{\mathbf{x^2y^2}}$

$= \dfrac{\left(\frac{1}{x^2y}-\frac{5}{xy}\right)(x^2y^2)}{\left(\frac{3}{xy}-\frac{7}{xy^2}\right)(x^2y^2)}$

$= \dfrac{\left(\frac{1}{x^2y}\right)(x^2y^2)-\left(\frac{5}{xy}\right)(x^2y^2)}{\left(\frac{3}{xy}\right)(x^2y^2)-\left(\frac{7}{xy^2}\right)(x^2y^2)}$

$= \dfrac{y-5xy}{3xy-7x}$

SECTION 7.6

Solving Rational Equations

Solve each equation and check the results. If an equation has no solution, so indicate.

55. LCD $= x(x-1) \quad x \neq 0,\ 1$

$$\frac{3}{x} = \frac{2}{x-1}$$

$$\mathbf{x(x-1)}\left(\frac{3}{x}\right) = \mathbf{x(x-1)}\left(\frac{2}{x-1}\right)$$

$$\overset{1}{\cancel{x}}(x-1)\left(\frac{3}{\underset{1}{\cancel{x}}}\right) = x\,\overset{1}{\cancel{(x-1)}}\left(\frac{2}{\underset{1}{\cancel{x-1}}}\right)$$

$$3(x-1) = x(2)$$

$$3x-3 = 2x$$

$$3x-3-\mathbf{2x} = 2x-\mathbf{2x}$$

$$x-3 = 0$$

$$x-3+\mathbf{3} = 0+\mathbf{3}$$

$$x = 3$$

Check: $\frac{3}{x} = \frac{2}{x-1}$

$$\frac{3}{\mathbf{3}} \overset{?}{=} \frac{2}{\mathbf{3}-1}$$

$$\frac{3}{3} \overset{?}{=} \frac{2}{2}$$

$$1 = 1$$

The solution is 3.
The result checks.

56. LCD $= a-5$; $a \neq 5$

$$\frac{a}{a-5} = 3 + \frac{5}{a-5}$$

$$\mathbf{(a-5)}\left(\frac{a}{a-5}\right) = \mathbf{(a-5)}\left(3+\frac{5}{a-5}\right)$$

$$(a-5)\left(\frac{a}{a-5}\right) = (a-5)(3) + (a-5)\left(\frac{5}{a-5}\right)$$

$$\overset{1}{\cancel{(a-5)}}\left(\frac{a}{\underset{1}{\cancel{a-5}}}\right) = (a-5)(3) + \overset{1}{\cancel{(a-5)}}\left(\frac{5}{\underset{1}{\cancel{a-5}}}\right)$$

$$a = 3(a-5)+5$$

$$a = 3a-15+5$$

$$a = 3a-10$$

$$a-\mathbf{a} = 3a-10-\mathbf{a}$$

$$0 = 2a-10$$

$$0+\mathbf{10} = 2a-10+\mathbf{10}$$

$$10 = 2a$$

$$\frac{10}{\mathbf{2}} = \frac{2a}{\mathbf{2}}$$

$$5 = a$$

The denominator becomes zero. No solution.
5 is extraneous.

57. $\left.\begin{aligned} 3t &= 3t \\ t &= t \\ 9 &= 3\cdot 3 \end{aligned}\right\}$ LCD $= 9t$; $t \neq 0$

$$\frac{2}{3t} + \frac{1}{t} = \frac{5}{9}$$

$$\mathbf{9t}\left(\frac{2}{3t}\right) + \mathbf{9t}\left(\frac{1}{t}\right) = \mathbf{9t}\left(\frac{5}{9}\right)$$

$$3\cdot\overset{1}{\cancel{3t}}\left(\frac{2}{\underset{1}{\cancel{3t}}}\right) + 9\overset{1}{\cancel{t}}\left(\frac{1}{\underset{1}{\cancel{t}}}\right) = \overset{1}{\cancel{9}}\,t\left(\frac{5}{\underset{1}{\cancel{9}}}\right)$$

$$3(2)+9 = t(5)$$

$$6+9 = 5t$$

$$15 = 5t$$

$$\frac{15}{\mathbf{5}} = \frac{5t}{\mathbf{5}}$$

$$3 = t$$

The result checks.

58. $\left.\begin{aligned} 4a-24 &= 4(a-6) \\ 4 &= 4 \end{aligned}\right\}$

LCD $= 4(a-6)$; $a \neq 6$

$$a = \frac{3a-50}{4a-24} - \frac{3}{4}$$

$$\mathbf{4(a-6)}(a) = \mathbf{4(a-6)}\left(\frac{3a-50}{4(a-6)} - \frac{3}{4}\right)$$

$$4a(a-6) = 4(a-6)\left(\frac{3a-50}{4(a-6)}\right) - 4(a-6)\left(\frac{3}{4}\right)$$

$$4a(a-6) = \overset{1}{\cancel{4}}\,\overset{1}{\cancel{(a-6)}}\left(\frac{3a-50}{\underset{1}{\cancel{4}}\,\underset{1}{\cancel{(a-6)}}}\right) - \overset{1}{\cancel{4}}(a-6)\left(\frac{3}{\underset{1}{\cancel{4}}}\right)$$

$$4a(a-6) = 3a-50-3(a-6)$$
$$4a^2-24a = 3a-50-3a+18$$
$$4a^2-24a = -32$$
$$4a^2-24a+\mathbf{32} = -32+\mathbf{32}$$
$$4a^2-24a+32 = 0$$
$$4(a^2-6a+8) = 0$$
$$4(a-2)(a-4) = 0$$

$$\begin{array}{c|c} a-2=0 & a-4=0 \\ a-2+\mathbf{2}=0+\mathbf{2} & a-4+\mathbf{4}=0+\mathbf{4} \\ a=2 & a=4 \end{array}$$

(or)

The solutions are 2 and 4.
The results check.

59. $\left.\begin{aligned} x+2 &= x+2 \\ x+3 &= x+3 \\ (x^2+5x+6) &= (x+2)(x+3) \end{aligned}\right\}$

LCD $= (x+2)(x+3)$; $x \neq -2, -3$

$$\frac{4}{x+2} - \frac{3}{x+3} = \frac{6}{(x^2+5x+6)}$$

$$\mathbf{(x+2)(x+3)}\left(\frac{4}{x+2} - \frac{3}{x+3}\right) = \mathbf{(x+2)(x+3)}\left(\frac{6}{(x+2)(x+3)}\right)$$

$$(x+2)(x+3)\left(\frac{4}{x+2}\right) - (x+2)(x+3)\left(\frac{3}{x+3}\right) = (x+2)(x+3)\left(\frac{6}{(x+2)(x+3)}\right)$$

$$\overset{1}{\cancel{(x+2)}}(x+3)\left(\frac{4}{\underset{1}{\cancel{x+2}}}\right) - (x+2)\overset{1}{\cancel{(x+3)}}\left(\frac{3}{\underset{1}{\cancel{x+3}}}\right) = \overset{1}{\cancel{(x+2)}}\,\overset{1}{\cancel{(x+3)}}\left(\frac{6}{\underset{1}{\cancel{(x+2)}}\,\underset{1}{\cancel{(x+3)}}}\right)$$

$$4(x+3)-3(x+2) = 6$$
$$4x+12-3x-6 = 6$$
$$x+6 = 6$$
$$x+6-\mathbf{6} = 6-\mathbf{6}$$
$$x = 0$$

The solution is 0.
The result checks.

60. LCD $= 2(x+1)$; $x \neq -1$

$$\frac{3}{x+1} - \frac{x-2}{2} = \frac{x-2}{x+1}$$

$$\mathbf{2(x+1)}\left(\frac{3}{x+1} - \frac{x-2}{2}\right) = \mathbf{2(x+1)}\left(\frac{x-2}{x+1}\right)$$

$$2(x+1)\left(\frac{3}{x+1}\right) - 2(x+1)\left(\frac{x-2}{2}\right) = 2(x+1)\left(\frac{x-2}{x+1}\right)$$

$$2\overset{1}{\cancel{(x+1)}}\left(\frac{3}{\underset{1}{\cancel{x+1}}}\right) - \overset{1}{\cancel{2}}(x+1)\left(\frac{x-2}{\underset{1}{\cancel{2}}}\right) = 2\overset{1}{\cancel{(x+1)}}\left(\frac{x-2}{\underset{1}{\cancel{x+1}}}\right)$$

$$2(3)-(x+1)(x-2) = 2(x-2)$$
$$6-(x^2-x-2) = 2x-4$$
$$6-x^2+x+2 = 2x-4$$
$$-x^2+x+8 = 2x-4$$
$$-x^2+x+8-\mathbf{2x} = 2x-4-\mathbf{2x}$$
$$-x^2+x+8-2x = -4$$
$$-x^2-x+8+\mathbf{4} = -4+\mathbf{4}$$
$$-x^2-x+12 = 0$$
$$-(x^2+x-12) = 0$$
$$-(x+4)(x-3) = 0$$

$$\begin{array}{c|c} x+4=0 & x-3=0 \\ x+4-\mathbf{4}=0-\mathbf{4} & x-3+\mathbf{3}=0+\mathbf{3} \\ x=-4 & x=3 \end{array}$$

(or)

The solutions are -4 and 3.
The results check.

61. ENGINEERING

$E = 1 - \frac{T_2}{T_1}$; LCD $= T_1$

$$E = 1 - \frac{T_2}{T_1}$$

$$\boldsymbol{T_1}E = \boldsymbol{T_1}(1) - \boldsymbol{T_1}\left(\frac{T_2}{T_1}\right)$$

$$T_1E = T_1 - \overset{1}{\cancel{T_1}}\left(\frac{T_2}{\underset{1}{\cancel{T_1}}}\right)$$

$$T_1E = T_1 - T_2$$
$$T_1E - \boldsymbol{T_1} = T_1 - T_2 - \boldsymbol{T_1}$$
$$T_1E - T_1 = -T_2$$
$$T_1(E-1) = -T_2$$
$$\frac{T_1(E-1)}{\boldsymbol{E-1}} = \frac{-T_2}{\boldsymbol{E-1}}$$
$$T_1 = -\frac{T_2}{E-1} \text{ or } \frac{T_2}{1-E}$$

62. LCD $= xyz$; Solve for y.

$$\frac{1}{x}=\frac{1}{y}+\frac{1}{z}$$

$$\mathbf{xyz}\left(\frac{1}{x}\right)=\mathbf{xyz}\left(\frac{1}{y}+\frac{1}{z}\right)$$

$$xyz\left(\frac{1}{x}\right)=xyz\left(\frac{1}{y}\right)+xyz\left(\frac{1}{z}\right)$$

$$\overset{1}{\cancel{x}}yz\left(\frac{1}{\underset{1}{\cancel{x}}}\right)=x\overset{1}{\cancel{y}}z\left(\frac{1}{\underset{1}{\cancel{y}}}\right)+xy\overset{1}{\cancel{z}}\left(\frac{1}{\underset{1}{\cancel{z}}}\right)$$

$$yz=xz+xy$$

$$yz-\mathbf{xy}=xz+xy-\mathbf{xy}$$

$$yz-xy=xz$$

$$y(z-x)=xz$$

$$\frac{y(z-x)}{\mathbf{z-x}}=\frac{xz}{\mathbf{z-x}}$$

$$y=\frac{xz}{z-x}$$

SECTION 7.7

Problem Solving Using Rational Equations

63. NUMBER PROBLEM

Analyze

- Begin with the fraction $\frac{4}{5}$.
- Subtract a number from the denominator.
- Twice as much is added to the numerator.
- The results is 5.
- Find the number.

Assign

Let $n =$ the unknown number.

Form

$$\frac{4+2n}{5-n}=5$$

Solve

$$\frac{4+2n}{5-n}=5$$

$$\text{LCD}=5-n$$

$$\mathbf{(5-n)}\left(\frac{4+2n}{5-n}\right)=\mathbf{(5-n)}5$$

$$\overset{1}{\cancel{(5-n)}}\left(\frac{4+2n}{\underset{1}{\cancel{5-n}}}\right)=5(5-n)$$

$$4+2n=25-5n$$

$$4+2n+\mathbf{5n}=25-5n+\mathbf{5n}$$

$$4+7n=25$$

$$4+7n-\mathbf{4}=25-\mathbf{4}$$

$$7n=21$$

$$\frac{7n}{\mathbf{7}}=\frac{21}{\mathbf{7}}$$

$$n=3$$

State

The number is 3.

Check

$$\frac{4+2n}{5-n}=\frac{4+2(\mathbf{3})}{5-\mathbf{3}}=\frac{4+6}{2}=\frac{10}{2}=5$$

The result checks.

64. EXERCISE

Analyze

- Jogger bikes 30 miles.
- She jogs 10 miles.
- Times are the same.
- Biking was 10 mph faster than jogging.
- How fast can she jog?

Assign

Let r = jogging speed in mph

$r+10$ = biking speed in mph

Form

	Rate	• Time	= Distance
Jogging	r	$\mathbf{\frac{10}{r}}$	10
Biking	$r+10$	$\mathbf{\frac{30}{r+10}}$	30

Time it took to jog 10 miles equals time it took to bike 30 miles.

$$\frac{10}{r} = \frac{30}{r+10}$$

Solve

$$\frac{10}{r} = \frac{30}{r+10}$$

$$\text{LCD} = r(r+10)$$

$$\mathbf{r(r+10)}\left(\frac{10}{r}\right) = \mathbf{r(r+10)}\left(\frac{30}{r+10}\right)$$

$$\cancel{r}(r+10)\left(\frac{10}{\cancel{r}}\right) = r\,\cancel{(r+10)}\left(\frac{30}{\cancel{r+10}}\right)$$

$$10(r+10) = 30r$$

$$10r+100 = 30r$$

$$10r+100-\mathbf{10r} = 30r-\mathbf{10r}$$

$$100 = 20r$$

$$\frac{100}{\mathbf{20}} = \frac{20r}{\mathbf{20}}$$

$$5 = r$$

State

Her jogging speed is 5 mph.

Check

The result checks.

65. HOUSE CLEANING

The rate of work is $\frac{1}{4}$ of the job per hour.

66. HOUSE PAINTING

Analyze

- Homeowner takes 14 days.
- Professional takes 10 days.
- How long will it take both of them, working together, to paint the house?

Assign

Let x = the number of days it will take both working together to paint the house.

Form

	Rate	• Time	= Work Completed
Homeowner	$\frac{1}{14}$	x	$\mathbf{\frac{x}{14}}$
Professional	$\frac{1}{10}$	x	$\mathbf{\frac{x}{10}}$

The part of job done by homeowner plus the part of job done by professional equals 1 job completed.

$$\frac{x}{14} + \frac{x}{10} = 1$$

Solve

$$\frac{x}{14}+\frac{x}{10} = 1\text{ , LCD} = 70$$

$$\mathbf{70}\left(\frac{x}{14}+\frac{x}{10}\right) = \mathbf{70}(1)$$

$$\cancel{14}\cdot 5\left(\frac{x}{\cancel{14}}\right)+\cancel{10}\cdot 7\left(\frac{x}{\cancel{10}}\right) = 70$$

$$5x+7x = 70$$

$$12x = 70$$

$$\frac{12x}{\mathbf{12}} = \frac{70}{\mathbf{12}}$$

$$x = 5\frac{5}{6}$$

State

It takes both working together $5\frac{5}{6}$ days to paint the house.

Check

The result checks.

67. INVESTMENTS

Analyze

- S and L earns $100 interest.
- Credit Union earns $120 interest.
- Credit Union rate is 1% more.
- Time for both is 1 year.
- Principals are the same amount for both.
- What are the two rates?

Assign

Let r = S and L interest rate as a percent

$r+1$ = Credit Union interest rate as a percent

Form

	Principal	• Rate	• Time	= Interest
S and L	$\frac{100}{r}$	r	1	100
Credit Union	$\frac{120}{r+1}$	$r+1$	1	120

Principal invested in S and L equals principal invested in Credit Union.

$$\frac{100}{r} = \frac{120}{r+1}$$

Solve

$$\frac{100}{r} = \frac{120}{r+1}$$

$$\text{LCD} = r(r+1)$$

$$\mathbf{r(r+1)}\left(\frac{100}{r}\right) = \mathbf{r(r+1)}\left(\frac{120}{r+1}\right)$$

$$\overset{1}{\cancel{r}}\cdot(r+1)\left(\frac{100}{\underset{1}{\cancel{r}}}\right) = r\cdot\overset{1}{\cancel{(r+1)}}\left(\frac{120}{\underset{1}{\cancel{r+1}}}\right)$$

$$100(r+1) = r(120)$$

$$100r + 100 = 120r$$

$$100r + 100 - \mathbf{100r} = 120r - \mathbf{100r}$$

$$100 = 20r$$

$$\frac{100}{\mathbf{20}} = \frac{20r}{\mathbf{20}}$$

$$5 = r$$

State

Savings and Loan's rate is 5%.

Check

The result checks.

68. WIND SPEED

Analyze

- Rate of plane in still air is 360 mph.
- Flies 400 miles with the wind.
- Flies 320 miles against the wind.
- Times are the same for both directions.
- What is the speed of the wind?

Assign

Let r = Wind's speed in mph

Form

	Rate	• Time	= Distance
With wind	$360+r$	$\frac{400}{360+r}$	400
Against wind	$360-r$	$\frac{320}{360-r}$	320

The time it took the plane to fly 400 miles with the wind equals the time it took the plane to fly 320 miles against the wind.

$$\frac{400}{360+r} = \frac{320}{360-r}$$

Solve

$$\frac{400}{360+r} = \frac{320}{360-r}$$

$$\text{LCD} = (360+r)(360-r)$$

$$\mathbf{(360+r)(360-r)}\left(\frac{400}{360+r}\right) = \mathbf{(360+r)(360-r)}\left(\frac{320}{360-r}\right)$$

$$\overset{1}{\cancel{(360+r)}}(360-r)\left(\frac{400}{\underset{1}{\cancel{360+r}}}\right) = (360+r)\overset{1}{\cancel{(360-r)}}\left(\frac{320}{\underset{1}{\cancel{360-r}}}\right)$$

$$400(360-r) = 320(360+r)$$

$$144{,}000 - 400r = 115{,}200 + 320r$$

$$144{,}000 - 400r + \mathbf{400r} = 115{,}200 + 320r + \mathbf{400r}$$

$$144{,}000 = 115{,}200 + 720r$$

$$144{,}000 - \mathbf{115{,}200} = 115{,}200 + 720r - \mathbf{115{,}20}$$

$$28{,}800 = 720r$$

$$\frac{28{,}800}{\mathbf{720}} = \frac{720r}{\mathbf{720}}$$

$$40 = r$$

State

The wind's speed is 40 mph.

Check

The result checks.

SECTION 7.8

Proportions and Similar Triangles

Determine whether each equation is a true proportion.

69. $\frac{4}{7} = \frac{20}{34}$

$4 \cdot 34 \stackrel{?}{=} 7 \cdot 20$

$136 \neq 140$

No

70. $\frac{5}{7} = \frac{30}{42}$

$5 \cdot 42 \stackrel{?}{=} 7 \cdot 30$

$210 = 210$

Yes

Solve each proportion.

71. $\frac{3}{x} = \frac{6}{9}$

$3 \cdot 9 = 6x$

$27 = 6x$

$\frac{27}{\mathbf{6}} = \frac{6x}{\mathbf{6}}$

$\frac{9}{2} = x$

Check: $\frac{3}{x} = \frac{6}{9}$

$\frac{3}{\frac{9}{2}} \stackrel{?}{=} \frac{6}{9}$

$3 \cdot \frac{2}{9} \stackrel{?}{=} \frac{6}{9}$

$\frac{6}{9} = \frac{6}{9}$ True

$\frac{9}{2}$ is the solution.

72. $\frac{x}{3} = \frac{x}{5}$

$5x = 3x$

$5x - \mathbf{3x} = 3x - \mathbf{3x}$

$2x = 0$

$\frac{2x}{\mathbf{2}} = \frac{0}{\mathbf{2}}$

$x = 0$

Check: $\frac{x}{3} = \frac{x}{5}$

$\frac{0}{3} \stackrel{?}{=} \frac{0}{5}$

$0 = 0$ True

0 is the solution.

73. $\frac{x-2}{5} = \frac{x}{7}$

$7(x-2) = 5x$

$7x - 14 = 5x$

$7x - 14 - \mathbf{7x} = 5x - \mathbf{7x}$

$-14 = -2x$

$\frac{-14}{\mathbf{-2}} = \frac{-2x}{\mathbf{-2}}$

$7 = x$

Check: $\frac{x-2}{5} = \frac{x}{7}$

$\frac{7-2}{5} \stackrel{?}{=} \frac{7}{7}$

$\frac{5}{5} \stackrel{?}{=} \frac{7}{7}$

$1 = 1$ True

7 is the solution.

74. $\frac{2x}{x+4} = \frac{3}{x-1}$

$2x(x-1) = 3(x+4)$

$2x^2 - 2x = 3x + 12$

$2x^2 - 2x - 3x = 3x + 12 - 3x$

$2x^2 - 5x = 12$

$2x^2 - 5x - \mathbf{12} = 12 - \mathbf{12}$

$2x^2 - 5x - 12 = 0$

$(2x+3)(x-4) = 0$

$2x + 3 = 0$ or $x - 4 = 0$

$2x + 3 - \mathbf{3} = 0 - \mathbf{3}$ | $x - 4 + \mathbf{4} = 0 + \mathbf{4}$

$2x = -3$ | $x = 4$

$\frac{2x}{\mathbf{2}} = \frac{-3}{\mathbf{2}}$

$x = -\frac{3}{2}$

The solutions are $-\frac{3}{2}$ and 4.

The results check.

75. DENTISTRY

Analyze

- 3 out of 4 will develop gum disease.
- How many of his 340 adult patients will develop gum disease?

Assign

Let g = number of gum disease patients

Form

3 adults is to 4 as g adults is to 340.

3 gum disease → $\frac{3}{4} = \frac{g}{340}$ ← # of gum disease

4 patients → ← Total # of patients

Solve

$\frac{3}{4} = \frac{g}{340}$

$1{,}020 = 4g$

$\frac{1{,}020}{\mathbf{4}} = \frac{4g}{\mathbf{4}}$

$255 = g$

State

255 patients will develop gum disease.

Check

The result checks.

76. UTILITY POLES

Analyze

- Similar triangles determined by the pole and its shadow and the man and his shadow.
- What is the height of the pole?

Assign

Let h = the height of the pole

Form

6 feet is to 3.6 feet as h feet is to 12 feet.

$$\frac{6}{3.6} = \frac{h}{12} \quad \begin{array}{l}\text{man's height} \to 6,\ h \leftarrow \text{pole's height}\\ \text{man's shadow} \to 3.6,\ 12 \leftarrow \text{pole's shadow}\end{array}$$

Solve

$$\frac{6}{3.6} = \frac{h}{12}$$
$$72 = 3.6h$$
$$\frac{72}{\mathbf{3.6}} = \frac{3.6h}{\mathbf{3.6}}$$
$$20 = h$$

State

A height of the pole is 20 feet.

Check

The result checks.

77. PORCELAIN FIGURINES

Analyze the Problem

- We know the scale is 1:12.
- How tall is a real flutist in feet if a model one measures 5.5 inches?

Assign

Let x = the height of a real flutist

Form

1 is to 12 as 5.5 inches is to x inches.

$$\frac{1}{12} = \frac{5.5}{x} \quad \begin{array}{l}1 \text{ part} \to 1,\ 5.5 \leftarrow 5.5 \text{ inch model}\\ 12 \text{ parts} \to 12,\ x \leftarrow x \text{ inches real flutist}\end{array}$$

Solve

$$\frac{1}{12} = \frac{5.5}{x}$$
$$x = 66$$

State

A real flutist would be 66 inches tall or 5 feet 6 inches tall.

Check

The result checks.

78. COMPARISON SHOPPING

150 compact discs for $60.

$$\frac{60}{150} = \frac{x}{1} \quad \begin{array}{l}\text{cost} \to 60,\ x \leftarrow \text{cost}\\ 150 \text{ discs} \to 150,\ 1 \leftarrow 1 \text{ disc}\end{array}$$
$$\frac{60}{150} = x$$
$$0.4 = x$$

The cost per disc is $0.40.

250 compact discs for $98.

$$\frac{98}{250} = \frac{x}{1} \quad \begin{array}{l}\text{cost} \to 98,\ x \leftarrow \text{cost}\\ 250 \text{ discs} \to 250,\ 1 \leftarrow 1 \text{ disc}\end{array}$$
$$\frac{98}{250} = x$$
$$0.392 = x$$

The cost per disc is about $0.392.

The 250 discs for $98 is the better buy.

SECTION 7.9

Variations

Write an equation to describe each variation. Use *k* for the constant of variation.

79. FITNESS

$$c = kt$$

80. GUITARS

$$f = \frac{k}{L}$$

81. SELLING FRUIT

Step 1: We let p represent the profit (in dollars) when b (the number of baskets) are sold. Translating the words *the profit varies directly to the number of baskets sold,* we get the equation $p = kb$.

Step 2: To find the constant of variation, k, we substitute 500 for p and 300 for b.

$$p = kb$$
$$500 = k(300)$$
$$\frac{500}{300} = \frac{300k}{300}$$
$$\frac{5}{3} = k$$

Step 3: We now substitute the value of k into the equation from step 1.

$$p = \frac{5}{3}b$$

Step 4: We can find the *profit of 1,200 baskets of strawberries* by substituting 1,200 for b in the equation from step 3.

$$p = \frac{5}{3}b$$
$$p = \frac{5}{3}(1{,}200)$$
$$p = \frac{5 \cdot 400 \cdot \cancel{3}^{1}}{\cancel{3}_{1}}$$
$$p = 2{,}000$$

$2,000 is the profit on 1,200 basket of strawberries.

82. $L = \dfrac{k}{w}$

Find k:

$$30 = \frac{k}{20}$$
$$30 \cdot 20 = k$$
$$600 = k$$

The constant is 600.

83. ELECTRICITY

Step 1: We let a represent the current (in amps) and r the resistance (in ohms) of a circuit. Translating the words *the current of a circuit, varies inversely as the resistance,* we get the equation $a = \dfrac{k}{r}$.

Step 2: To find the constant of variation, k, we substitute 2.5 for a and 150 for r.

$$a = \frac{k}{r}$$
$$2.5 = \frac{k}{150}$$
$$2.5 \cdot 150 = k$$
$$375 = k$$

Step 3: We now substitute the value of k into the equation from step 1.

$$a = \frac{375}{r}$$

Step 4: We can find the *amps* by substituting 300 for r in the equation from step 3.

$$a = \frac{375}{r}$$
$$a = \frac{375}{300}$$
$$a = \frac{5}{4}$$
$$a = 1.25$$

The current is 1.25 amps with a resistance of 300 ohms.

84. The graph below shows an inverse variation.

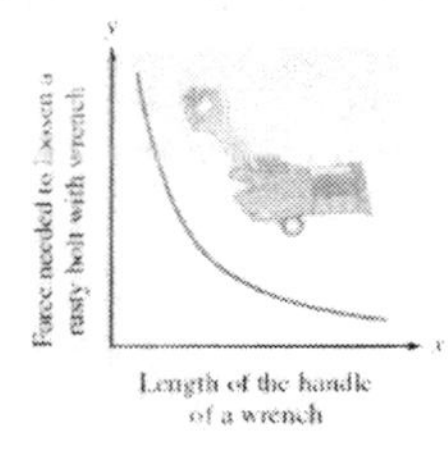

CHAPTER 7 TEST

1. **Fill in the blanks.**

a. A quotient of two polynomials, such as $\frac{x+7}{x^2+2x}$, is called a **rational** expression.

b. Two triangles with the same shape, but not necessarily the same size, are called **similar** triangles.

c. A **proportion** is a mathematical statement that two ratios or two rates are equal.

d. To **build** a rational expression, we multiply it by a form of 1. For example, $\frac{2}{5x}\cdot\frac{8}{8}=\frac{16}{40x}$.

e. To simplify $\frac{x-3}{(x+3)(x-3)}$, we remove common **factors** of the numerator and denominator.

2. MEMORY

$$n=\frac{35+5d}{d}=\frac{35+5(7)}{7}=\frac{35+35}{7}=\frac{70}{7}=10$$

After one week, 10 words will be remembered.

For what real numbers are each rational expression undefined?

3. $\frac{6x-9}{5x}$

$$5x=0$$
$$\frac{5x}{\mathbf{5}}=\frac{0}{\mathbf{5}}$$
$$x=0$$

The rational expression is undefined for $x=0$.

4. $\frac{x}{x^2+x-6}=\frac{x}{(x+3)(x-2)}$

$x+3=0$ or $x-2=0$

$x+3-\mathbf{3}=0-\mathbf{3}$ | $x-2+\mathbf{2}=0+\mathbf{2}$

$x=-3$ | $x=2$

The rational expression is undefined for $x=-3$ and $x=2$.

5. THE INTERNET

$$\frac{56\text{K bits}}{1\text{ second}}\cdot\frac{60\text{ seconds}}{1\text{ minute}}=\frac{56\text{K bits}}{1\text{ second}}\cdot\frac{60\text{ seconds}}{1\text{ minute}}=\frac{3{,}360\text{K bits}}{1\text{ minute}}$$

The modem can transmit 3,360K bits per minute or 3,360,000 bits per minute.

6. $\frac{x+5}{5}=\frac{x+5}{5}$ (crossed out)

$=x+1$

5 is not a comman factor of the numerator, and therefore can not be removed. 5 is a term of the numerator.

Simplify each rational expression.

7. $$\frac{48x^2y}{54xy^2}=\frac{8\cdot 6\cdot x\cdot x\cdot y}{9\cdot 6\cdot x\cdot y\cdot y}=\frac{8x}{9y}$$

8. $$\frac{7m-49}{7-m}=\frac{7(m-7)}{7-m}=\frac{7(m-7)}{-1(-7+m)}=-\frac{7(m-7)}{(m-7)}=-7$$

9. $$\frac{2x^2-x-3}{4x^2-9}=\frac{(2x-3)(x+1)}{(2x+3)(2x-3)}=\frac{(2x-3)(x+1)}{(2x+3)(2x-3)}=\frac{x+1}{2x+3}$$

10. $\dfrac{3(x+2)-3}{6x+5-(3x+2)}=\dfrac{3x+6-3}{6x+5-3x-2}$

$=\dfrac{3x+3}{3x+3}$

$=\dfrac{\overset{1}{\cancel{3x+3}}}{\underset{1}{\cancel{3x+3}}}$

$=1$

Find the LCD of each pair of rational expressions.

11. $\dfrac{19}{3c^2d},\dfrac{6}{c^2d^3}$; $\text{LCD}=3c^2d^3$

12. $\dfrac{4n+25}{n^2-4n-5},\dfrac{6n}{n^2-25}$

$\dfrac{4n+25}{(n-5)(n+1)},\dfrac{6n}{(n+5)(n-5)}$

$\text{LCD}=(n+5)(n-5)(n+1)$

Perform the operations. Simplify, if possible.

13. $\dfrac{12x^2y}{15xy}\cdot\dfrac{25y^2}{16x}=\dfrac{12\cdot25x^2yy^2}{15\cdot16x^2y}$

$=\dfrac{\overset{1}{\cancel{3}}\cdot\overset{1}{\cancel{4}}\cdot\overset{1}{\cancel{5}}\cdot5\,\overset{1}{\cancel{x^2}}\,\overset{1}{\cancel{y}}\,y^2}{\underset{1}{\cancel{3}}\cdot\underset{1}{\cancel{5}}\cdot\underset{1}{\cancel{4}}\cdot4\,\underset{1}{\cancel{x^2}}\,\underset{1}{\cancel{y}}}$

$=\dfrac{5y^2}{4}$

14. $\dfrac{x^2+3x+2}{3x+9}\cdot\dfrac{x+3}{x^2-4}=\dfrac{(x^2+3x+2)(x+3)}{(3x+9)(x^2-4)}$

$=\dfrac{(x+2)(x+1)(x+3)}{3(x+3)(x+2)(x-2)}$

$=\dfrac{\overset{1}{\cancel{(x+2)}}(x+1)\overset{1}{\cancel{(x+3)}}}{3\underset{1}{\cancel{(x+3)}}\underset{1}{\cancel{(x+2)}}(x-2)}$

$=\dfrac{x+1}{3(x-2)}$

15. $\dfrac{x-x^2}{3x^2+6x}\div\dfrac{3x-3}{3x^3+6x^2}=\dfrac{x-x^2}{3x^2+6x}\cdot\dfrac{3x^3+6x^2}{3x-3}$

$=\dfrac{(x-x^2)(3x^3+6x^2)}{(3x^2+6x)(3x-3)}$

$=\dfrac{x(1-x)3x^2(x+2)}{3x(x+2)3(x-1)}$

$=\dfrac{3x^2x\cdot-1(-1+x)(x+2)}{3\cdot3x(x+2)(x-1)}$

$=-\dfrac{\overset{1}{\cancel{3}}\,x^2\,\overset{1}{\cancel{x}}\,\overset{1}{\cancel{(x-1)}}\,\overset{1}{\cancel{(x+2)}}}{3\cdot\underset{1}{\cancel{3}}\,\underset{1}{\cancel{x}}\,\underset{1}{\cancel{(x+2)}}\,\underset{1}{\cancel{(x-1)}}}$

$=-\dfrac{x^2}{3}$

16. $\dfrac{a^2-16}{a-4}\div(6a+24)=\dfrac{a^2-16}{a-4}\cdot\dfrac{1}{(6a+24)}$

$=\dfrac{(a+4)(a-4)}{(a-4)6(a+4)}$

$=\dfrac{\overset{1}{\cancel{(a+4)}}\,\overset{1}{\cancel{(a-4)}}}{6\,\underset{1}{\cancel{(a-4)}}\,\underset{1}{\cancel{(a+4)}}}$

$=\dfrac{1}{6}$

17. $\dfrac{3y+7}{2y+3}-\dfrac{-3y-2}{2y+3}=\dfrac{3y+7-(-3y-2)}{2y+3}$

$=\dfrac{3y+7+3y+2}{2y+3}$

$=\dfrac{6y+9}{2y+3}$

$=\dfrac{3\,\overset{1}{\cancel{(2y+3)}}}{\underset{1}{\cancel{2y+3}}}$

$=3$

18. $\text{LCD}=10m$

$\dfrac{2n}{5m}-\dfrac{n}{2}=\dfrac{2n}{5m}\cdot\dfrac{\mathbf{2}}{\mathbf{2}}-\dfrac{n}{2}\cdot\dfrac{\mathbf{5m}}{\mathbf{5m}}$

$=\dfrac{4n-5mn}{10m}$

19. LCD $= x(x+1)$

$$\frac{x+1}{x}+\frac{x-1}{x+1}=\frac{x+1}{x}\cdot\frac{\mathbf{x+1}}{\mathbf{x+1}}+\frac{x-1}{x+1}\cdot\frac{\mathbf{x}}{\mathbf{x}}$$
$$=\frac{(x+1)(x+1)+(x-1)x}{x(x+1)}$$
$$=\frac{x^2+2x+1+x^2-x}{x(x+1)}$$
$$=\frac{2x^2+x+1}{x(x+1)}$$

20. LCD $= (a-1)$

$$\frac{a+3}{a-1}-\frac{a+4}{1-a}=\frac{a+3}{a-1}-\frac{a+4}{-1(-1+a)}$$
$$=\frac{a+3}{a-1}+\frac{a+4}{a-1}$$
$$=\frac{a+3+a+4}{a-1}$$
$$=\frac{(1+1)a+(3+4)}{a-1}$$
$$=\frac{2a+7}{a-1}$$

21. LCD $= c-4$

$$\frac{9}{c-4}+c=\frac{9}{c-4}+\frac{c}{1}$$
$$=\frac{9}{c-4}+\frac{c}{1}\cdot\frac{\mathbf{c-4}}{\mathbf{c-4}}$$
$$=\frac{9+c(c-4)}{c-4}$$
$$=\frac{9+c^2-4c}{c-4}$$
$$=\frac{c^2-4c+9}{c-4}$$

22. LCD $= (t+3)(t-3)(t+2)$

$$\frac{6}{t^2-9}-\frac{5}{t^2-t-6}$$
$$=\frac{6}{(t+3)(t-3)}-\frac{5}{(t-3)(t+2)}$$
$$=\frac{6}{(t+3)(t-3)}\cdot\frac{\mathbf{t+2}}{\mathbf{t+2}}-\frac{5}{(t-3)(t+2)}\cdot\frac{\mathbf{t+3}}{\mathbf{t+3}}$$
$$=\frac{6(t+2)-5(t+3)}{(t+3)(t-3)(t+2)}$$
$$=\frac{6t+12-5t-15}{(t+3)(t-3)(t+2)}$$
$$=\frac{t-3}{(t+3)(t-3)(t+2)}$$
$$=\frac{\overset{1}{\cancel{t-3}}}{(t+3)\underset{1}{\cancel{(t-3)}}(t+2)}$$
$$=\frac{1}{(t+3)(t+2)}$$

Simplify each complex fraction.

23. $$\frac{\frac{3m-9}{8m}}{\frac{5m-15}{32}}=\frac{3m-9}{8m}\div\frac{5m-15}{32}$$
$$=\frac{3m-9}{8m}\cdot\frac{32}{5m-15}$$
$$=\frac{3(m-3)\cdot 32}{8m\cdot 5(m-3)}$$
$$=\frac{3\cdot\overset{1}{\cancel{8}}\cdot 4\,\overset{1}{\cancel{(m-3)}}}{\underset{1}{\cancel{8}}\cdot 5m\,\underset{1}{\cancel{(m-3)}}}$$
$$=\frac{12}{5m}$$

24. LCD $= a^2s^2$

$$\frac{\frac{3}{as^2}+\frac{6}{a^2s}}{\frac{6}{a}-\frac{9}{s^2}}=\frac{\frac{3}{as^2}+\frac{6}{a^2s}}{\frac{6}{a}-\frac{9}{s^2}}\cdot\frac{\mathbf{a^2s^2}}{\mathbf{a^2s^2}}$$
$$=\frac{\left(\frac{3}{as^2}+\frac{6}{a^2s}\right)a^2s^2}{\left(\frac{6}{a}-\frac{9}{s^2}\right)a^2s^2}$$
$$=\frac{\frac{3}{as^2}(a^2s^2)+\frac{6}{a^2s}(a^2s^2)}{\frac{6}{a}(a^2s^2)-\frac{9}{s^2}(a^2s^2)}$$
$$=\frac{3a+6s}{6as^2-9a^2}$$
$$=\frac{3(a+2s)}{3a(2s^2-3a)}$$
$$=\frac{\overset{1}{\cancel{3}}(a+2s)}{\underset{1}{\cancel{3}}a(2s^2-3a)}$$
$$=\frac{a+2s}{a(2s^2-3a)}\text{ or }\frac{a+2s}{2as^2-3a^2}$$

Solve each equation. If an equation has no solution, so indicate.

25. $\text{LCD} = 3y\,;\quad y \neq 0$

$$\frac{1}{3}+\frac{4}{3y}=\frac{5}{y}$$

$$\mathbf{3y}\left(\frac{1}{3}+\frac{4}{3y}\right)=\mathbf{3y}\left(\frac{5}{y}\right)$$

$$\overset{1}{\cancel{3}}y\left(\frac{1}{\underset{1}{\cancel{3}}}\right)+\overset{1}{\cancel{3}}\overset{1}{\cancel{y}}\left(\frac{4}{\underset{1}{\cancel{3}}\underset{1}{\cancel{y}}}\right)=3\overset{1}{\cancel{y}}\left(\frac{5}{\underset{1}{\cancel{y}}}\right)$$

$$y+4=15$$

$$y+4-\mathbf{4}=15-\mathbf{4}$$

$$y=11$$

The solution is 11.

26. $\text{LCD} = n-6;\quad n \neq 6$

$$\frac{9n}{n-6}=3+\frac{54}{n-6}$$

$$\mathbf{(n-6)}\left(\frac{9n}{n-6}\right)=\mathbf{(n-6)}\left(3+\frac{54}{n-6}\right)$$

$$\overset{1}{\cancel{(n-6)}}\left(\frac{9n}{\underset{1}{\cancel{n-6}}}\right)=(n-6)(3)+\overset{1}{\cancel{(n-6)}}\left(\frac{54}{\underset{1}{\cancel{n-6}}}\right)$$

$$9n=3(n-6)+54$$

$$9n=3n-18+54$$

$$9n=3n+36$$

$$9n-\mathbf{3n}=3n+36-\mathbf{3n}$$

$$6n=36$$

$$\frac{6n}{\mathbf{6}}=\frac{36}{\mathbf{6}}$$

$$n=6$$

The denominator becomes zero.
No solution. 6 is extraneous.

27. $\text{LCD} = (q+1)(q-2);\quad q \neq -1, 2$

$$\frac{7}{q^2-q-2}+\frac{1}{q+1}=\frac{3}{q-2}$$

$$\frac{7}{(q-2)(q+1)}+\frac{1}{q+1}=\frac{3}{q-2}$$

$$\mathbf{(q-2)(q+1)}\left(\frac{7}{(q-2)(q+1)}+\frac{1}{q+1}\right)=\mathbf{(q-2)(q+1)}\left(\frac{3}{q-2}\right)$$

$$\overset{1}{\cancel{(q-2)(q+1)}}\left(\frac{7}{\underset{1}{\cancel{(q-2)(q+1)}}}\right)+(q-2)\overset{1}{\cancel{(q+1)}}\left(\frac{1}{\underset{1}{\cancel{q+1}}}\right)=\overset{1}{\cancel{(q-2)}}(q+1)\left(\frac{3}{\underset{1}{\cancel{q-2}}}\right)$$

$$7+q-2=3(q+1)$$

$$q+5=3q+3$$

$$q+5-q=3q+3-q$$

$$5=2q+3$$

$$5-\mathbf{3}=2q+3-\mathbf{3}$$

$$2=2q$$

$$\frac{2}{\mathbf{2}}=\frac{2q}{\mathbf{2}}$$

$$1=q$$

The solution is 1.

28. $\text{LCD} = 3(c-3);\quad c \neq 3$

$$\frac{2}{3}=\frac{2c-12}{3c-9}-c$$

$$\mathbf{3(c-3)}\left(\frac{2}{3}\right)=\mathbf{3(c-3)}\left(\frac{2c-12}{3(c-3)}-c\right)$$

$$\overset{1}{\cancel{3}}(c-3)\left(\frac{2}{\underset{1}{\cancel{3}}}\right)=\overset{1}{\cancel{3(c-3)}}\left(\frac{2c-12}{\underset{1}{\cancel{3(c-3)}}}\right)-3(c-3)(c)$$

$$2(c-3)=2c-12-3c(c-3)$$

$$2c-6=2c-12-3c^2+9c$$

$$2c-6=-3c^2+11c-12$$

$$2c-6-\mathbf{2c}=-3c^2+11c-12-\mathbf{2c}$$

$$-6=-3c^2+9c-12$$

$$-6+\mathbf{6}=-3c^2+9c-12+\mathbf{6}$$

$$0=-3c^2+9c-6$$

$$0=-3(c^2-3c+2)$$

$$0=-3(c-2)(c-1)$$

$c-2=0$ or $c-1=0$

$c-2+\mathbf{2}=0+\mathbf{2}$ | $c-1+\mathbf{1}=0+\mathbf{1}$

$c=2$ | $c=1$

The solutions are 1 and 2.

29. $\text{LCD} = y(y-1); \quad y \neq 0,1$

$$\frac{y}{y-1} = \frac{y-2}{y}$$

$$\mathbf{y(y-1)}\left(\frac{y}{y-1}\right) = \mathbf{y(y-1)}\left(\frac{y-2}{y}\right)$$

$$y\,\cancel{(y-1)}\left(\frac{y}{\cancel{y-1}}\right) = \cancel{y}(y-1)\left(\frac{y-2}{\cancel{y}}\right)$$

$$y^2 = (y-1)(y-2)$$

$$y^2 = y^2 - 3y + 2$$

$$y^2 - \mathbf{y^2} = y^2 - 3y + 2 - \mathbf{y^2}$$

$$0 = -3y + 2$$

$$0 + \mathbf{3y} = -3y + 2 + \mathbf{3y}$$

$$3y = 2$$

$$\frac{3y}{\mathbf{3}} = \frac{2}{\mathbf{3}}$$

$$y = \frac{2}{3}$$

The solution is $\frac{2}{3}$.

30. $\text{LCD} = (a+3)(a-3); \quad a \neq -3,3$

$$\frac{a}{a-3} + \frac{4}{a+3} = \frac{18}{a^2-9}$$

$$\mathbf{(a+3)(a-3)}\left(\frac{a}{a-3} + \frac{4}{a+3}\right) = \mathbf{(a+3)(a-3)}\left(\frac{18}{(a+3)(a-3)}\right)$$

$$(a+3)\cancel{(a-3)}\left(\frac{a}{\cancel{a-3}}\right) + \cancel{(a+3)}(a-3)\left(\frac{4}{\cancel{a+3}}\right)$$

$$= \cancel{(a+3)}\cancel{(a-3)}\left(\frac{18}{\cancel{(a+3)}\,\cancel{(a-3)}}\right)$$

$$a(a+3) + 4(a-3) = 18$$

$$a^2 + 3a + 4a - 12 = 18$$

$$a^2 + 7a - 12 = 18$$

$$a^2 + 7a - 12 - \mathbf{18} = 18 - -\mathbf{18}$$

$$a^2 + 7a - 30 = 0$$

$$(a+10)(a-3) = 0$$

$$a + 10 = 0 \quad \text{or} \quad a - 3 = 0$$

$$a = -10 \quad \Big| \quad \cancel{a = 3}$$

The solution is -10.
3 is an extraneous solution because it makes one denominator zero.

31. $H = \dfrac{RB}{R+B}$; Solve for B.

$\text{LCD} = R + B$

$$H = \frac{RB}{R+B}$$

$$\mathbf{(R+B)}H = \mathbf{(R+B)}\frac{RB}{R+B}$$

$$H(R+B) = \cancel{(R+B)}\,\frac{RB}{\cancel{R+B}}$$

$$HR + HB = RB$$

$$HR + HB - HB = RB - HB$$

$$HR = RB - HB$$

$$HR = B(R-H)$$

$$\frac{HR}{\mathbf{R-H}} = \frac{B(R-H)}{\mathbf{R-H}}$$

$$\frac{HR}{R-H} = B$$

$$B = \frac{HR}{R-H}$$

32. $\text{LCD} = rst$; Solve for s.

$$\frac{1}{r} + \frac{1}{s} = \frac{1}{t}$$

$$\mathbf{(rst)}\left(\frac{1}{r} + \frac{1}{s}\right) = \mathbf{(rst)}\left(\frac{1}{t}\right)$$

$$\cancel{r}st\left(\frac{1}{\cancel{r}}\right) + r\cancel{s}t\left(\frac{1}{\cancel{s}}\right) = rs\cancel{t}\left(\frac{1}{\cancel{t}}\right)$$

$$st + rt = rs$$

$$st + rt - \mathbf{st} = rs - \mathbf{st}$$

$$rt = s(r-t)$$

$$\frac{rt}{r-t} = \frac{s\cancel{(r-t)}}{\cancel{(r-t)}}$$

$$s = \frac{rt}{r-t}$$

33. HEALTH RISKS

$$\frac{114}{120} = \frac{\overset{1}{\cancel{6}} \cdot 19}{\underset{1}{\cancel{6}} \cdot 20} = \frac{19}{20}$$

Yes, the patient falls within the range.

34. CURRENCY EXCHANGE RATES

Let x = number of pounds

$$\frac{51}{100} = \frac{x}{3{,}500}$$
$$51 \cdot 3{,}500 = 100x$$
$$\frac{51 \cdot 3{,}500}{\mathbf{100}} = \frac{100x}{\mathbf{100}}$$
$$1{,}785 = x$$

The traveler receives 1,785 pounds.

35. TV TOWERS

Let h = the height of the tower

$$\begin{matrix}\text{man's height} \rightarrow \\ \text{man's shadow} \rightarrow\end{matrix} \frac{6}{4} = \frac{h}{114} \begin{matrix}\leftarrow \text{tower's height} \\ \leftarrow \text{tower's shadow}\end{matrix}$$
$$6 \cdot 114 = 4h$$
$$\frac{6 \cdot 114}{\mathbf{4}} = \frac{4h}{\mathbf{4}}$$
$$171 = h$$

The height of the tower is 171 feet.

36. COMPARISON SHOPPING

80 sheets for \$3.89.

$$\begin{matrix}\text{cost} \rightarrow \\ \text{80 sheets} \rightarrow\end{matrix} \frac{3.89}{80} = \frac{x}{1} \begin{matrix}\leftarrow \text{cost} \\ \leftarrow \text{1 sheet}\end{matrix}$$
$$\frac{3.89}{80} = x$$
$$0.0486 \approx x$$

The cost per sheet is about \$0.0486.

120 sheets for \$6.19.

$$\begin{matrix}\text{cost} \rightarrow \\ \text{120 sheets} \rightarrow\end{matrix} \frac{6.19}{120} = \frac{x}{1} \begin{matrix}\leftarrow \text{cost} \\ \leftarrow \text{1 card}\end{matrix}$$
$$\frac{6.19}{120} = x$$
$$0.0516 \approx x$$

The cost per sheet is about \$0.0516.

The 80 sheets for \$3.89 is the better buy.

37. CLEANING HIGHWAYS

Analyze

- Highway worker takes 7 hours.
- Helper takes 9 hours.
- How long will it take both of them, working together, to pick up trash?

Assign

Let x = the number of hours it will take both working together to pick up the trash.

Form

Rate · Time = Work Completed

	Rate	Time	Work Completed
Worker	$\frac{1}{7}$	x	$\frac{x}{7}$
Helper	$\frac{1}{9}$	x	$\frac{x}{9}$

The part of job done by worker	plus	the part of job done by helper	equals	1 job completed.
$\frac{x}{7}$	$+$	$\frac{x}{9}$	$=$	1

Solve

$$\frac{x}{7} + \frac{x}{9} = 1$$
$$\text{LCD} = 63$$
$$\mathbf{63}\left(\frac{x}{7} + \frac{x}{9}\right) = \mathbf{63}(1)$$
$$\overset{1}{\cancel{7}} \cdot 9\left(\frac{x}{\underset{1}{\cancel{7}}}\right) + \overset{1}{\cancel{9}} \cdot 7\left(\frac{x}{\underset{1}{\cancel{9}}}\right) = 63$$
$$9x + 7x = 63$$
$$16x = 63$$
$$\frac{16x}{\mathbf{16}} = \frac{63}{\mathbf{16}}$$
$$x = 3\frac{15}{16}$$

State

It takes both working together $3\frac{15}{16}$ hours to pick up the trash.

Check

The result checks.

38. PHYSICAL FITNESS

Analyze

- He roller blades 5 miles.
- He jogs 2 miles.
- Times are the same.
- Roller is 6 mph faster than jogging.
- How fast can he jog?

Assign

Let r = jogging speed in mph

$r+6$ = roller speed in mph

Form

	Rate	• Time	= Distance
Jogging	r	$\frac{2}{r}$	2
Roller Blade	$r+6$	$\frac{5}{r+6}$	5

Time took to jog 2 miles equals time took to blade 5 miles.

$$\frac{2}{r} = \frac{5}{r+6}$$

Solve

$$\frac{2}{r} = \frac{5}{r+6}$$

$$\text{LCD} = r(r+6)$$

$$r(r+6)\left(\frac{2}{r}\right) = r(r+6)\left(\frac{5}{r+6}\right)$$

$$\cancel{r}(r+6)\left(\frac{2}{\cancel{r}}\right) = r\cdot\cancel{(r+6)}\left(\frac{5}{\cancel{r+6}}\right)$$

$$2(r+6) = 5r$$

$$2r+12 = 5r$$

$$2r+12-\mathbf{2r} = 5r-\mathbf{2r}$$

$$12 = 3r$$

$$\frac{12}{\mathbf{3}} = \frac{3r}{\mathbf{3}}$$

$$4 = r$$

State

His jogging speed is 4 mph.

Check

The result checks.

39. NUMBER PROBLEM

Analyze

- Begin with the fraction $\frac{5}{8}$.
- Subtract a number from the numerator.
- Twice as much is added to the denominator.
- The results is $\frac{1}{4}$.
- Find the number.

Assign

Let n = the unknown number.

Form

$$\frac{5-n}{8+2n} = \frac{1}{4}$$

Solve

$$\frac{5-n}{8+2n} = \frac{1}{4}$$

$$\text{LCD} = 4(4+n)$$

$$\mathbf{4(4+n)}\left(\frac{5-n}{8+2n}\right) = \mathbf{4(4+n)}\frac{1}{4}$$

$$2\cdot\cancel{2}\,\cancel{(4+n)}\left(\frac{5-n}{\cancel{2}\,\cancel{(4+n)}}\right) = \cancel{4}(4+n)\frac{1}{\cancel{4}}$$

$$2(5-n) = 4+n$$

$$10-2n = 4+n$$

$$10-2n+\mathbf{2n} = 4+n+\mathbf{2n}$$

$$10 = 4+3n$$

$$10-\mathbf{4} = 4+3n-\mathbf{4}$$

$$6 = 3n$$

$$\frac{6}{\mathbf{3}} = \frac{3n}{\mathbf{3}}$$

$$2 = n$$

State

The number is 2.

Check

$$\frac{5-n}{8+2n} = \frac{5-\mathbf{2}}{8+2(\mathbf{2})} = \frac{3}{12} = \frac{1}{4}$$

The result checks.

40. **Explain what is means to clear the following equation of fractions.**

$$\frac{u}{u-1} + \frac{1}{u} = \frac{u^2+1}{u^2-u}$$

We multiply both sides of the equation by the LCD of the rational expressions appearing in the equation. The resulting equation is easier to solve.

41. **Explain the diffeence between the procedure used to simplify $\frac{1}{x}+\frac{1}{4}$ and the procedure used to solve $\frac{1}{x}+\frac{1}{4}=\frac{1}{2}$.**

To simplify $\frac{1}{x}+\frac{1}{4}$, we build each fraction to have the LCD of $4x$. To solve $\frac{1}{x}+\frac{1}{4}=\frac{1}{2}$, we multiply both sides by the LCD $4x$ to e lim inate the denominators.

42. POGO STICKS

Strategy We follow the four-step strategy for solving a direct variation problem.

Why? In the words of the problem, the phrase *varies directly* indicates that a direct variation model should be used.

Solution

Step 1 : We let f represent the force (in pounds) required to compress a spring of length l (in inches). Translating the words *force varies directly with length of spring*. We get the equation $f = kl$.

Step 2 : To find the constant of variation, k, we substitute 130 for f and 6.5 for l.

$$f = kl$$
$$130 = k(6.5)$$
$$\frac{130}{\mathbf{6.5}} = \frac{6.5k}{\mathbf{6.5}}$$
$$20 = k$$

Step 3 : We now substitute the value of k into the equation from step 1.

$$f = 20l$$

Step 4 : We can find the *force* by subst-ituting 5 for l in the equation from step 3.

$$f = kl$$
$$f = 20(5)$$
$$f = 100$$

A force of 100 pound will be needed to compress the spring 5 inches.

43.

$$r = \frac{k}{s}$$
$$40 = \frac{k}{10}$$
$$40 \cdot 10 = k$$
$$400 = k$$

$$r = \frac{k}{s}$$
$$r = \frac{400}{15}$$
$$r = \frac{80 \cdot \cancel{5}^{1}}{3 \cdot \cancel{5}_{1}}$$
$$r = \frac{80}{3} \text{ or}$$
$$r = 26\frac{2}{3}$$

Thus, when $s = 15$, the value of r is $\frac{80}{3}$ or $26\frac{2}{3}$.

44. a. The time it takes to complete a job and the number of workers on the job.
Inverse

b. The length of a beard and the time it has been growing.
Direct

CHAPTERS 1 - 7 CUMULATIVE REVIEW

1. Determine whether each statement is true or false. [Section 1.3]

   a. Every integer is a whole number. **False**

   b. 0 is not a rational number. **False**

   c. π is an irrational number. **True**

   d. The set of integers is the set of whole numbers and their opposites. **True**

2. Insert the proper symbol, < or >, in the blank to make a true statement. [Section 1.5]

$$|2-4| \boxed{?} -(-6)$$
$$|2+(-4)| \boxed{?} -(-6)$$
$$|-2| \boxed{?}\ 6$$
$$2<6$$
$$|2-4| \boxed{<} -(-6)$$

3. Evaluate: $9^2-3[45-3(6+4)]$ [Section 1.7]

$$\begin{aligned} 9^2-3[45-3(6+4)] &= 81-3[45-3(10)] \\ &= 81-3[45-30] \\ &= 81-3[15] \\ &= 81-45 \\ &= 36 \end{aligned}$$

4. Find the average (mean) test score of a student in a history class with scores of 80, 73, 61, 73, and 98. [Section 1.7]

$$\begin{aligned} \textit{Mean test score} &= \frac{80+73+61+73+98}{5} \\ &= \frac{385}{5} \\ &= 77 \end{aligned}$$

   The mean test score is 77.

5. Simplify: $8(c+7)-2(c-3)$. [Section 1.9]

$$\begin{aligned} 8(c+7)-2(c-3) &= \mathbf{8}(c)+\mathbf{8}(7)-\mathbf{2}(c)-\mathbf{2}(-3) \\ &= 8c+56-2c+6 \\ &= (8-2)c+(56+6) \\ &= 6c+62 \end{aligned}$$

6. Solve: $\frac{4}{5}d=-4$ [Section 2.1]

$$\left(\frac{\mathbf{5}}{\mathbf{4}}\right)\cdot\frac{4}{5}d=\left(\frac{\mathbf{5}}{\mathbf{4}}\right)\cdot(-4)$$
$$d=-5$$

   The solution is -5.

7. Solve: $2-3(x-5)=4(x-1)$. [Section 2.2]

$$\begin{aligned} 2-\mathbf{3}(x)-\mathbf{3}(-5) &= \mathbf{4}(x)-\mathbf{4}(1) \\ 2-3x+15 &= 4x-4 \\ -3x+17 &= 4x-4 \\ -3x+17+\mathbf{3x} &= 4x-4+\mathbf{3x} \\ 17 &= 7x-4 \\ 17+\mathbf{4} &= 7x-4+\mathbf{4} \\ 21 &= 7x \\ \frac{21}{\mathbf{7}} &= \frac{7x}{\mathbf{7}} \\ 3 &= x \end{aligned}$$

   The solution is 3.

8. GRAND KING SIZE BEDS [Section 2.3]

   1,600 is what % of 6,240?

$$\begin{aligned} 1{,}600 &= 6{,}240x \\ \frac{1{,}600}{\mathbf{6{,}240}} &= \frac{6{,}240x}{\mathbf{6{,}240}} \\ 0.256 &\approx x \\ 0.26 &\approx x \\ 26\% &\approx x \end{aligned}$$

   The increase in area is about 26%.

9. Solve: $A-c=2B+r$ for B. [Section 2.4]

$$\begin{aligned} A-c-\mathbf{r} &= 2B+r-\mathbf{r} \\ A-c-r &= 2B \\ \frac{A-c-r}{\mathbf{2}} &= \frac{2B}{\mathbf{2}} \\ \frac{A-c-r}{2} &= B \\ B &= \frac{A-c-r}{2} \end{aligned}$$

10. Change 40° C to degrees Fahrenheit. [Section 2.4]

$$F = \frac{9}{5}C + 32$$

$$F = \frac{9}{5}(\mathbf{40}) + 32$$

$$F = \frac{9}{\cancel{5}_1}(\cancel{40}^8) + 32$$

$$F = 9(8) + 32$$

$$F = 72 + 32$$

$$F = 104$$

The Fahrenheit temperature is 104°.

11. Find the volume of a pyramid that has a square base, measuring 6 feet on a side, and whose height is 20 feet. [Section 2.4]

$$V = \frac{1}{3}Bh$$

$$V = \frac{1}{3}(\mathbf{6^2})(\mathbf{20})$$

$$V = \frac{1}{3}(36)(20)$$

$$V = \frac{1}{\cancel{3}_1}(\cancel{36}^{12})(20)$$

$$V = 12(20)$$

$$V = 240$$

The volume is 240 ft^3.

12. BLENDING TEA [Section 2.6]

Analyze

- 1st tea cost \$6.40 per pound.
- 2nd tea cost \$4.00 per pound.
- 20 pound mixture worth \$5.44 per pound.
- How much of each grade of tea will be needed?

Assign

Let $x =$ the amount of 1st grade tea

$20 - x =$ the amount of 2nd grade tea

Form

	Amount	• Price	= Total Value
1st Tea	x	6.40	**6.40*x***
2nd Tea	$20 - x$	4.00	**4.00(20 – *x*)**
Mixture	20	5.44	**5.44(20)**

Value of 1st tea. plus Value of 2nd tea. equals Value of the mixture.

$$6.40x \quad + \quad 4.00(20 - x) \quad = \quad 5.44(20)$$

Solve

$$6.40x + 4.00(20 - x) = 5.44(20)$$

$$6.40x + 80 - 4.00x = 108.80$$

$$2.40x + 80 = 108.80$$

$$2.40x + 80 - \mathbf{80} = 108.80 - \mathbf{80}$$

$$2.40x = 28.80$$

$$\frac{2.40x}{\mathbf{2.40}} = \frac{28.80}{\mathbf{2.40}}$$

$$x = 12$$

State

12 lb. of the \$6.40 tea. 8 lb of the \$4 tea.

Check

$$6.40x + 4.00(20 - x) = 5.44(20)$$

$$6.40(12) + 4.00(20 - 12) \stackrel{?}{=} 5.44(20)$$

$$76.8 + 32 \stackrel{?}{=} 108.8$$

$$108.8 = 108.8 \text{ True}$$

12 pounds of \$6.40 tea and 8 pounds of \$4.00 tea are the correct solutions.

13. SPEED OF A PLANE [Section 2.6]

Analyze

- Two planes are 6,000 miles a part.
- Faster plane is 200 mph faster.
- They travel towards each other.
- They meet in 5 hours.
- What is the speed of the slower plane?

Assign

Let $x =$ the speed of slower plane in mph

$x + 200 =$ the speed of faster plane in mph

Form

	Rate	· Time	= Distance
Slower	x	5	$\mathbf{5x}$
Faster	$x+200$	5	$\mathbf{5(x+200)}$

Distance of slower.	plus	Distance of faster.	equals	Total distance.
$5x$	$+$	$5(x+200)$	$=$	6,000

Solve

$$5x + 5(x+200) = 6{,}000$$
$$5x + 5x + 1{,}000 = 6{,}000$$
$$10x + 1{,}000 = 6{,}000$$
$$10x + 1{,}000 - \mathbf{1{,}000} = 6{,}000 - \mathbf{1{,}000}$$
$$10x = 5{,}000$$
$$\frac{10x}{\mathbf{10}} = \frac{5{,}000}{\mathbf{10}}$$
$$x = 500$$

State

The speed of the slower plane is 500 mph.

Check

The result checks.

14. Solve: $7x + 2 \geq 4x - 1$ [Section 2.7]

$$7x + 2 - \mathbf{4x} \geq 4x - 1 - \mathbf{4x}$$
$$3x + 2 \geq -1$$
$$3x + 2 - \mathbf{2} \geq -1 - \mathbf{2}$$
$$3x \geq -3$$
$$\frac{3x}{\mathbf{3}} \geq \frac{-3}{\mathbf{3}}$$
$$x \geq -1$$

Interval Notation: $[-1, \infty)$

Graph: 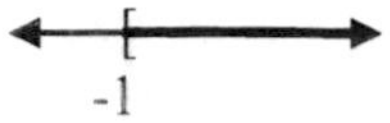

15. Graph: $y = 2x - 3$ [Section 3.2]

x	y	(x, y)
-2	$2x-3$ $=2(-2)-3$ $=-4-3$ $=-7$	$(-2,-7)$
0	$2x-3$ $=2(0)-3$ $=0-3$ $=-3$	$(0,-3)$
3	$2x-3$ $=2(3)-3$ $=6-3$ $=3$	$(3,3)$

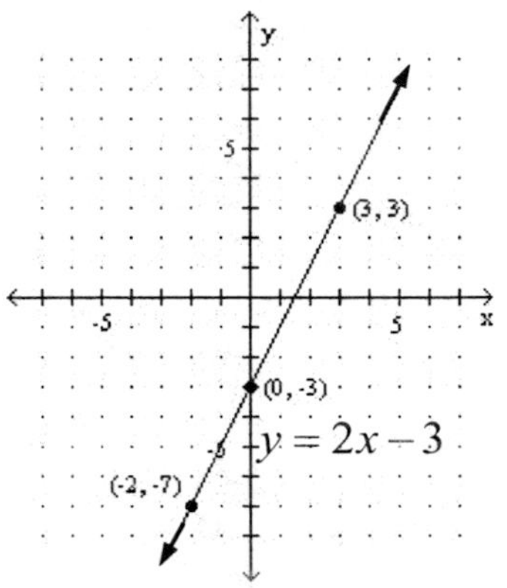

16. Find the slope of the line passing through $(-1, 3)$ and $(3, -1)$. [Section 3.4]

(x_1, y_1) (x_2, y_2)

$$m = \frac{y_2 - y_1}{x_2 - x_1}$$
$$m = \frac{-1-3}{3-(-1)}$$
$$m = \frac{-4}{4}$$
$$m = -1$$

17. CUTTING STEEL. [Section 3.4]

$(0, 0)$ and $(50, 0.4)$

(x_1, y_1) (x_2, y_2)

$$m = \frac{y_2 - y_1}{x_2 - x_1}$$
$$m = \frac{0.4 - 0}{50 - 0}$$
$$m = \frac{0.4}{50}$$
$$m = 0.008$$

The rate of change is 0.008 mm/m.

18. What is the slope of a line perpendicular to the line $y = -\frac{7}{8}x - 6$. [Section 3.5]

$y = mx + b$, where m is the slope of the line.

$-\frac{7}{8}$ is the slope of the given line.

Perpendicular slope is the negative reciprocal of the given slope.

$\frac{8}{7}$ is the slope of the perpendicular line.

$$-\frac{7}{8} \cdot \left(\frac{8}{7}\right) = -1$$

19. Write the equation of a line in slope - intercept form that has slope 3 and passes through the point (1, 5). [Section 3.6]

$$y - y_1 = m(x - x_1)$$
$$y - 5 = 3(x - 1)$$
$$y - 5 = 3x - 3$$
$$y - 5 + \mathbf{5} = 3x - 3 + \mathbf{5}$$
$$y = 3x + 2$$

20. Graph: $3x - 2y \le 6$ [Section 3.7]

- Graph the line $3x - 2y = 6$.

$$y = \frac{3}{2}x - 3$$

x	y	(x, y)
-2	$\frac{3}{2}x - 3$ $= \frac{3}{2}(-2) - 3$ $= -3 - 3$ $= -6$	$(-2, -6)$
0	$\frac{3}{2}x - 3$ $= \frac{3}{2}(0) - 3$ $= 0 - 3$ $= -3$	$(0, -3)$
2	$\frac{3}{2}x - 3$ $= \frac{3}{2}(2) - 3$ $= 3 - 3$ $= 0$	$(2, 0)$

- Test (0,0) in the inequality $3x - 2y \le 6$.

$$3(0) - 2(0) \le 6$$
$$0 \le 6$$
True

- Shade in the side that is true.

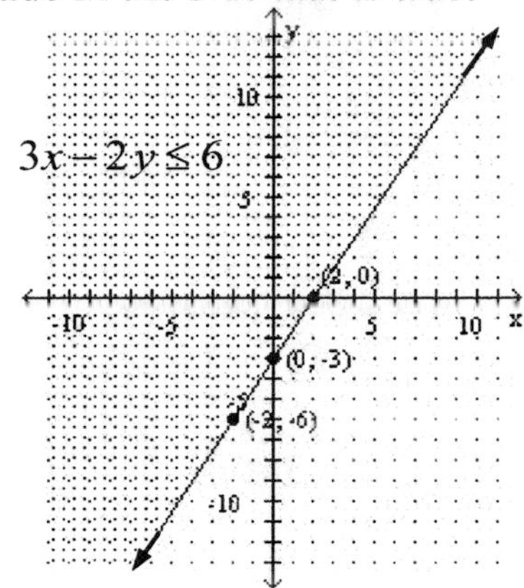

21. If $f(x) = -3x^2 - 6x$, find $f(-2)$. [Section 3.8]

$$f(x) = -3x^2 - 6x$$
$$f(\mathbf{-2}) = -3(\mathbf{-2})^2 - 6(\mathbf{-2})$$
$$f(-2) = -3(4) + 12$$
$$f(-2) = -12 + 12$$
$$f(-2) = 0$$

The output is 0.

22. Fill in the blanks. [Section 3.8]

The set of all possible input values of a function is called the **domain** and the set of all output values is called the **range**.

23. Solve the system by graphing. [Section 4.1]

$$\begin{cases} x + y = 1 \\ y = x + 5 \end{cases}$$

The method of choice to graph each line is left up to the student.

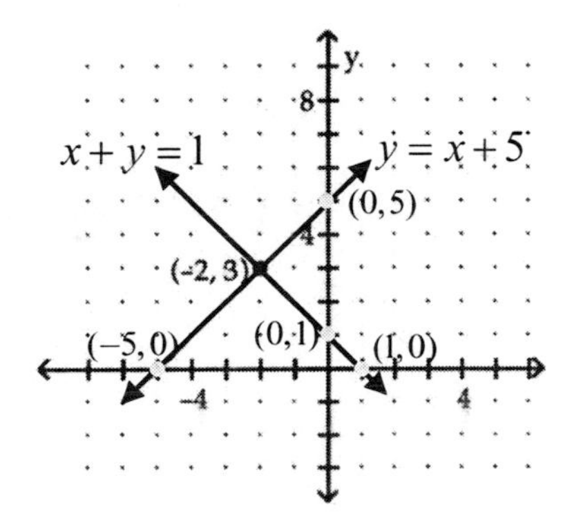

The solution is $(-2, 3)$.

24. Solve the system by substitution. [Section 4.2]

$$\begin{cases} x = 3y - 1 \\ 2x - 3y = 4 \end{cases}$$

$2x - 3y = 4$ The 2nd equation
Substitute for x.

$$2(\mathbf{3y - 1}) - 3y = 4$$
$$6y - 2 - 3y = 4$$
$$3y - 2 = 4$$
$$3y - 2 + \mathbf{2} = 4 + \mathbf{2}$$
$$3y = 6$$
$$\frac{3y}{\mathbf{3}} = \frac{6}{\mathbf{3}}$$
$$y = 2$$

$x = 3y - 1$ The 1st equation

$$x = 3(\mathbf{2}) - 1$$
$$x = 5$$

The solution is (5, 2).

Check:	Check:
$x = 3y - 1$	$2x - 3y = 4$
$5 \overset{?}{=} 3(2) - 1$	$2(5) - 3(2) \overset{?}{=} 4$
$5 = 5$	$10 - 6 \overset{?}{=} 4$
True	$4 = 4$
	True

25. Solve the system by elimination (addition). [Section 4.3]

$$\begin{cases} 2x + 3y = -1 \\ 3x + 5y = -2 \end{cases}$$

Eliminate x.
Multiply both sides of the 1st equation by 3.
Multiply both sides of the 2nd equation by -2.

$$6x + 9y = -3$$
$$\underline{-6x - 10y = 4}$$
$$-y = 1$$
$$\frac{-y}{\mathbf{-1}} = \frac{1}{\mathbf{-1}}$$
$$y = -1$$

$2x + 3y = -1$ The 1st equation

$$2x + 3(\mathbf{-1}) = -1$$
$$2x - 3 + \mathbf{3} = -1 + \mathbf{3}$$
$$2x = 2$$
$$\frac{2x}{\mathbf{2}} = \frac{2}{\mathbf{2}}$$
$$x = 1$$

The solution is $(1, -1)$

26. POKER [Section 4.4]

Analyze

- Red chips worth $5 each.
- Blue chips worth $10 each.
- Ended with $190 and 23 chips.
- How many of each chips did he have?

Assign

Let x = number of red chips
y = number of blue chips

Form

The number of red chips	plus	the number of blue chips	is	23.
x	$+$	y	$=$	23

	Number •	Value	= Total Value
Red chips	x	5	**5x**
Blue chips	y	10	**10y**
		Total	**190**

The value of red chips	plus	the value of blue chips	is	$190.
$5x$	$+$	$10y$	$=$	190

Solve

$$\begin{cases} x + y = 23 \\ 5x + 10y = 190 \end{cases}$$

Eliminate x.
Multiply both sides of the 1st equation by -5.

$$-5x - 5y = -115$$
$$\underline{5x + 10y = 190}$$
$$5y = 75$$
$$\frac{5y}{\mathbf{5}} = \frac{75}{\mathbf{5}}$$
$$y = 15$$

$x + y = 23$ The 1st equation

$$x + 15 = 23$$
$$x + 15 - \mathbf{15} = 23 - \mathbf{15}$$
$$x = 8$$

State

He had 8 red chips.
He had 15 blue chips.

Check

$x + y = 23$	$5x + 10y = 190$
$8 + 15 \overset{?}{=} 23$	$5(8) + 10(15) \overset{?}{=} 190$
$23 = 23$	$40 + 150 \overset{?}{=} 190$
True	$190 = 190$
	True

The results check.

Simplify each expression. Write each answer without using negative exponents. [Section 5.1]

27. $x^4x^3 = x^{4+3}$
$= x^7$

28. $(x^2x^3)^5 = (x^{2+3})^5$
$= (x^5)^5$
$= x^{5 \cdot 5}$
$= x^{25}$

29. $\left(\dfrac{y^3y}{2yy^2}\right)^3 = \left(\dfrac{y^{3+1}}{2y^{1+2}}\right)^3$
$= \left(\dfrac{y^4}{2y^3}\right)^3$
$= \left(\dfrac{y^{4-3}}{2}\right)^3$
$= \left(\dfrac{y^1}{2^1}\right)^3$
$= \dfrac{y^{1 \cdot 3}}{2^{1 \cdot 3}}$
$= \dfrac{y^3}{8}$

30. $\left(\dfrac{-2a}{b}\right)^5 = \dfrac{(-2)^5a^5}{b^5}$
$= -\dfrac{32a^5}{b^5}$

[Section 5.2]

31. $(a^{-2}b^3)^{-4} = a^{-2 \cdot -4}b^{3 \cdot -4}$
$= a^8b^{-12}$
$= \dfrac{a^8}{b^{12}}$

32. $\dfrac{9b^0b^3}{3b^{-3}b^4} = \dfrac{9b^{0+3}}{3b^{-3+4}}$
$= \dfrac{9b^3}{3b^1}$
$= \dfrac{\overset{1}{\cancel{3}} \cdot 3b^{3-1}}{\underset{1}{\cancel{3}}}$
$= 3b^2$

33. Write 290,000 in scientific notation. [Section 5.3]
Move the decimal 5 places to the left.
$290{,}000 = 2.9 \times 10^5$

34. What is the degree of the polynomial $5x^3 - 4x + 16$? [Section 5.4]
The degree is 3, because the degree of a polynomial is the same as the highest degree of any term of the polynomial.

35. Graph: $y = -x^3$ [Section 5.4]

x	y	(x, y)
-2	$-x^3 = -(-2)^3$ $= -(-8)$ $= 8$	$(-2, 8)$
-1	$-x^3 = -(-1)^3$ $= -(-1)$ $= 1$	$(-1, 1)$
0	$-x^3 = -(0)^3$ $= 0$	$(0, 0)$
1	$-x^3 = -(1)^3$ $= -1$	$(1, -1)$
2	$-x^3 = -(2)^3$ $= -8$	$(2, -8)$

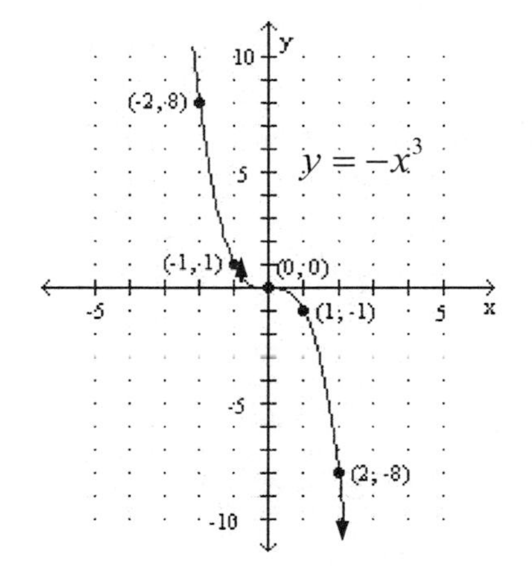

36. CONCENTRIC CIRCLES [Section 5.7] The area of the ring between the two concentric circles of radius r and R is given by the formula $A=\pi(R+r)(R-r)$. Do the multiplication on the right-hand side of the equation.

$$A=\pi(R+r)(R-r)$$
$$A=\pi(R^2-r^2)$$
$$A=\pi R^2-\pi r^2$$

Perform the operations. [Section 5.5]

37. $(3x^2-3x-2)+(3x^2+4x-3)$
$$=3x^2-3x-2+3x^2+4x-3$$
$$=(3+3)x^2+(-3+4)x+(-2-3)$$
$$=6x^2+x-5$$

38. $\left(\frac{1}{16}t^3+\frac{1}{2}t^2-\frac{1}{6}t\right)-\left(\frac{9}{16}t^3+\frac{9}{4}t^2-\frac{1}{12}t\right)$
$$=\frac{1}{16}t^3+\frac{1}{2}t^2-\frac{1}{6}t-\frac{9}{16}t^3-\frac{9}{4}t^2+\frac{1}{12}t$$
$$=\left(\frac{1}{16}-\frac{9}{16}\right)t^3+\left(\frac{1}{2}-\frac{9}{4}\right)t^2+\left(-\frac{1}{6}+\frac{1}{12}\right)t$$
$$=\left(-\frac{8}{16}\right)t^3+\left(\frac{1}{2}-\frac{9}{4}\right)t^2+\left(-\frac{1}{6}+\frac{1}{12}\right)t$$
$$=\left(-\frac{1\cdot\overset{1}{\cancel{8}}}{2\cdot\underset{1}{\cancel{8}}}\right)t^3+\left(\frac{1}{2}\cdot\frac{\mathbf{2}}{\mathbf{2}}-\frac{9}{4}\right)t^2+\left(-\frac{1}{6}\cdot\frac{\mathbf{2}}{\mathbf{2}}+\frac{1}{12}\right)t$$
$$=\left(-\frac{1}{2}\right)t^3+\left(\frac{2-9}{4}\right)t^2+\left(\frac{-2+1}{12}\right)t$$
$$=-\frac{1}{2}t^3-\frac{7}{4}t^2-\frac{1}{12}t$$

[Section 5.6]

39. $(2x^2y^3)(3x^2y^2)=(2\cdot3)(x^2x^2)(y^3y^2)$
$$=6x^{2+2}y^{3+2}$$
$$=6x^4y^5$$

40. $(2y-5)(3y+7)=6y^2+14y-15y-35$
$$=6y^2-y-35$$

41. $-4x^2z(3x^2-z)=-4x^2z(3x^2)-4x^2z(-z)$
$$=-4\cdot3x^{2+2}z-4\cdot-1x^2z^{1+1}$$
$$=-12x^4z+4x^2z^2$$

[Section 5.7]

42. $(3a-4)^2=(3a-4)(3a-4)$
$$=9a^2-12a-12a+16$$
$$=9a^2-24a+16$$

[Section 5.8]

43. $\frac{6x+9}{3}=\frac{6x}{3}+\frac{9}{3}$
$$=\frac{2\cdot\overset{1}{\cancel{3}}x}{\underset{1}{\cancel{3}}}+\frac{3\cdot\overset{1}{\cancel{3}}}{\underset{1}{\cancel{3}}}$$
$$=2x+3$$

44.
$$\begin{array}{r} x^2+2x-1 \\ 2x+3\overline{)\,2x^3+7x^2+4x-3} \\ \underline{-(2x^3+3x^2)} \\ 4x^2+4x \\ \underline{-(4x^2+6x)} \\ -2x-3 \\ \underline{-(-2x-3)} \\ 0 \end{array}$$

Factor each polynomial completely, if possible. [Section 6.1]

45. k^3t-3k^2t, GCF $=k^2t$
$$k^3t-3k^2t=\mathbf{k^2t}\cdot k-\mathbf{k^2t}\cdot3$$
$$=k^2t(k-3)$$

46. $2ab+2ac+3b+3c$
$$=(\mathbf{2}a\mathbf{b}+\mathbf{2}a\mathbf{c})+(\mathbf{3}b+\mathbf{3}c)$$
$$=\mathbf{2a}(b+c)+\mathbf{3}(b+c)$$
$$=(b+c)(\mathbf{2a}+\mathbf{3})$$

[Section 6.2]

47. $u^2-18u+81=(u-9)(u-9)$
$$=(u-9)^2$$

48. $-r^2+2+r=-r^2+r+2$
$$=-(r^2-r-2)$$
$$=-(r-2)(r+1)$$

49. $u^2+10u+15$

prime

[Section 6.3]

50. $6x^2-63-13x=6x^2-13x-63$
$$=(2x-9)(3x+7)$$

[Section 6.4]

51. $2a^2-200b^2=2(a^2-100b^2)$
$$=2(a+10b)(a-10b)$$

[Section 6.5]

52. $b^3+125=b^3+5^3$
$$=(b+5)(b^2-5b+25)$$

Solve each equation by factoring.
[Section 6.7]

53. $5x^2+x=0$

$x(5x+1)=0$

$x=0$ or $5x+1=0$

$5x+1-\mathbf{1}=0-\mathbf{1}$

$5x=-1$

$\frac{5x}{\mathbf{5}}=\frac{-1}{\mathbf{5}}$

$x=-\frac{1}{5}$

The solutions are 0 and $-\frac{1}{5}$.

54. $6x^2-5x=-1$

$6x^2-5x+\mathbf{1}=-1+\mathbf{1}$

$6x^2-5x+1=0$

$(2x-1)(3x-1)=0$

$2x-1=0$	or	$3x-1=0$
$2x-1+\mathbf{1}=0+\mathbf{1}$		$3x-1+\mathbf{1}=0+\mathbf{1}$
$2x=1$		$3x=1$
$\frac{2x}{\mathbf{2}}=\frac{1}{\mathbf{2}}$		$\frac{3x}{\mathbf{3}}=\frac{1}{\mathbf{3}}$
$x=\frac{1}{2}$		$x=\frac{1}{3}$

The solutions are $\frac{1}{2}$ and $\frac{1}{3}$.

55. COOKING [Section 6.7]

Analyze
- Cooking surface is 160 sq. in.
- Length is $w+6$.
- Width is w.
- Griddle is a rectangle.
- Find the length and width.

Assign

Let $w=$ Width of griddle in inches
$w+6=$ Length of griddle in inches

Form

Width of griddle	times	length of griddle	equals	area of griddle.
w	$\cdot$	$w+6$	$=$	160

Solve

$$w(w+6)=160$$
$$w^2+6w=160$$
$$w^2+6w-\mathbf{160}=160-\mathbf{160}$$
$$w^2+6w-160=0$$
$$(w+16)(w-10)=0$$

$w+16=0$	or	$w-10=0$
$w+16-\mathbf{16}=0-\mathbf{16}$		$w-10+\mathbf{10}=0+\mathbf{10}$
$w=-16$ (crossed out)		$w=10$

State

The width is 10 inches.
The length is 16 inches.

Check

The result checks.

56. For what values of $\frac{3x^2}{x^2-25}$ is the rational expression undefined? [Section 7.1]

$$\frac{3x^2}{x^2-25}=\frac{3x^2}{(x+5)(x-5)}$$

$$\begin{array}{c|c} x+5=0 & x-5=0 \\ x+5-\mathbf{5}=0-\mathbf{5} & x-5+\mathbf{5}=0+\mathbf{5} \\ x=-5 & x=5 \end{array}$$

(or)

The rational expression is undefined for $x=5$ and $x=-5$.

Perform the operations. Simplify, if possible. [Section 7.1]

57. $$\frac{2x^2-8x}{x^2-6x+8}=\frac{2x(x-4)}{(x-2)(x-4)}=\frac{2x\,\cancel{(x-4)}^{1}}{(x-2)\,\cancel{(x-4)}_{1}}=\frac{2x}{x-2}$$

[Section 7.2]

58. $$\frac{x^2-16}{4-x}\div\frac{3x+12}{x^3}=\frac{x^2-16}{4-x}\cdot\frac{x^3}{3x+12}$$

$$=\frac{(x^2-16)x^3}{(4-x)(3x+12)}$$

$$=\frac{x^3(x+4)(x-4)}{-1(-4+x)3(x+4)}$$

$$=-\frac{x^3\,\cancel{(x+4)}^{1}\cdot\cancel{(x-4)}^{1}}{3\,\cancel{(x-4)}_{1}\,\cancel{(x+4)}_{1}}$$

$$=-\frac{x^3}{3}$$

59. LCD $=2m+5$ [Section 7.3]

$$\frac{8m^2}{2m+5}-\frac{4m^2+25}{2m+5}=\frac{8m^2-(4m^2+25)}{2m+5}$$

$$=\frac{8m^2-4m^2-25}{2m+5}$$

$$=\frac{4m^2-25}{2m+5}$$

$$=\frac{(2m+5)(2m-5)}{2m+5}$$

$$=\frac{\cancel{(2m+5)}^{1}(2m-5)}{\cancel{2m+5}_{1}}$$

$$=2m-5$$

60. LCD $=x-3$ [Section 7.4]

$$\frac{4}{x-3}+\frac{5}{3-x}=\frac{4}{x-3}+\frac{5}{-1(-3+x)}$$

$$=\frac{4}{x-3}-\frac{5}{x-3}$$

$$=\frac{4-5}{x-3}$$

$$=\frac{-1}{x-3}$$

$$=-\frac{1}{x-3}$$

[Section 7.4]

61. $\left.\begin{array}{l} m^2+5m+6=(m+2)(m+3) \\ m^2+3m+2=(m+2)(m+1) \end{array}\right\}$

$$\text{LCD}=(m+2)(m+3)(m+1)$$

$$\frac{m}{m^2+5m+6}-\frac{2}{m^2+3m+2}$$

$$=\frac{m}{(m+2)(m+3)}-\frac{2}{(m+2)(m+1)}$$

$$=\frac{m}{(m+2)(m+3)}\cdot\frac{\mathbf{(m+1)}}{\mathbf{(m+1)}}-\frac{2}{(m+2)(m+1)}\cdot\frac{\mathbf{(m+3)}}{\mathbf{(m+3)}}$$

$$=\frac{m(m+1)-2(m+3)}{(m+2)(m+3)(m+1)}$$

$$=\frac{m^2+m-2m-6}{(m+2)(m+3)(m+1)}$$

$$=\frac{m^2-m-6}{(m+2)(m+3)(m+1)}$$

$$=\frac{(m+2)(m-3)}{(m+2)(m+3)(m+1)}$$

$$=\frac{\overset{1}{\cancel{(m+2)}}(m-3)}{\underset{1}{\cancel{(m+2)}}(m+3)(m+1)}$$

$$=\frac{(m-3)}{(m+3)(m+1)}$$

Simplify. [Section 7.5]

62. $\text{LCD}=x(x+1)$

$$\frac{2-\frac{2}{x+1}}{2+\frac{2}{x}}=\frac{2-\frac{2}{x+1}}{2+\frac{2}{x}}\cdot\frac{\mathbf{x(x+1)}}{\mathbf{x(x+1)}}$$

$$=\frac{\left(2-\frac{2}{x+1}\right)x(x+1)}{\left(2+\frac{2}{x}\right)x(x+1)}$$

$$=\frac{2x(x+1)-x(x+1)\left(\frac{2}{x+1}\right)}{2x(x+1)+x(x+1)\left(\frac{2}{x}\right)}$$

$$=\frac{2x^2+2x-x\,\overset{1}{\cancel{(x+1)}}\left(\frac{2}{\underset{1}{\cancel{x+1}}}\right)}{2x^2+2x+\overset{1}{\cancel{x}}(x+1)\left(\frac{2}{\underset{1}{\cancel{x}}}\right)}$$

$$=\frac{2x^2+2x-x(2)}{2x^2+2x+(x+1)(2)}$$

$$=\frac{2x^2+2x-2x}{2x^2+2x+2x+2}$$

$$=\frac{2x^2}{2x^2+4x+2}$$

$$=\frac{2x^2}{2(x^2+2x+1)}$$

$$=\frac{\overset{1}{\cancel{2}}x^2}{\underset{1}{\cancel{2}}(x^2+2x+1)}$$

$$=\frac{x^2}{x^2+2x+1}\text{ or }\frac{x^2}{(x+1)^2}$$

Solve each equation. [Section 7.6]

63. $\text{LCD} = 60x$; $x \neq 0$

$$\frac{7}{5x} - \frac{1}{2} = \frac{5}{6x} + \frac{1}{3}$$

$$\mathbf{60x}\left(\frac{7}{5x} - \frac{1}{2}\right) = \mathbf{60x}\left(\frac{5}{6x} + \frac{1}{3}\right)$$

$$60x\left(\frac{7}{5x}\right) - 60x\left(\frac{1}{2}\right) = 60x\left(\frac{5}{6x}\right) + 60x\left(\frac{1}{3}\right)$$

$$12 \cdot \cancel{5}\cancel{x}\left(\frac{7}{\cancel{5}\cancel{x}}\right) - 30 \cdot \cancel{2}x\left(\frac{1}{\cancel{2}}\right) = 10 \cdot \cancel{6}\cancel{x}\left(\frac{5}{\cancel{6}\cancel{x}}\right) + 20 \cdot \cancel{3}x\left(\frac{1}{\cancel{3}}\right)$$

$$12(7) - 30x = 10(5) + 20x$$

$$84 - 30x = 50 + 20x$$

$$84 - 30x + \mathbf{30x} = 50 + 20x + \mathbf{30x}$$

$$84 = 50 + 50x$$

$$84 - \mathbf{50} = 50 + 50x - \mathbf{50}$$

$$34 = 50x$$

$$\frac{34}{\mathbf{50}} = \frac{50x}{\mathbf{50}}$$

$$\frac{\cancel{2} \cdot 17}{\cancel{2} \cdot 25} = x$$

$$\frac{17}{25} = x$$

The solution is $\frac{17}{25}$.

64. $\text{LCD} = u(u-1)$; $u \neq 0, 1$

$$\frac{u}{u-1} + \frac{1}{u} = \frac{u^2+1}{u^2-u}$$

$$\mathbf{u(u-1)}\left(\frac{u}{u-1} + \frac{1}{u}\right) = \mathbf{u(u-1)}\left(\frac{u^2+1}{u^2-u}\right)$$

$$u(u-1)\left(\frac{u}{u-1}\right) + u(u-1)\left(\frac{1}{u}\right) = u(u-1)\left(\frac{u^2+1}{u(u-1)}\right)$$

$$u\cancel{(u-1)}\left(\frac{u}{\cancel{u-1}}\right) + \cancel{u}(u-1)\left(\frac{1}{\cancel{u}}\right) = \cancel{u}\cancel{(u-1)}\left(\frac{u^2+1}{\cancel{u}\cancel{(u-1)}}\right)$$

$$u^2 + u - 1 = u^2 + 1$$

$$u^2 + u - 1 - \mathbf{u^2} = u^2 + 1 - \mathbf{u^2}$$

$$u - 1 = 1$$

$$u - 1 + \mathbf{1} = 1 + \mathbf{1}$$

$$u = 2$$

The solution is 2.

65. DRAINING A TANK [Section 7.7]

Analyze

- **1st Outlet pipe can drain pool in 24 hours.**
- **2nd Outlet pipe can drain pool in 36 hours.**
- **How long will it take both pipes to drain the pool?**

Assign

Let x = the number of hours it will take both outlet pipes to drain the pool.

Form

	Rate	• Time	= Work Completed
1st Outlet pipe	$\frac{1}{24}$	x	$\frac{\mathbf{x}}{\mathbf{24}}$
2nd Outlet pipe	$\frac{1}{36}$	x	$\frac{\mathbf{x}}{\mathbf{36}}$

The part of job done by 1st outlet pipe	plus	the part of job done by 2nd outlet pipe	equals	1 job completed.
$\frac{x}{24}$	$+$	$\frac{x}{36}$	$=$	1

Solve

$$\frac{x}{24} + \frac{x}{36} = 1$$

$$\text{LCD} = 72$$

$$\mathbf{72}\left(\frac{x}{24} + \frac{x}{36}\right) = \mathbf{72}(1)$$

$$\cancel{24} \cdot 3\left(\frac{x}{\cancel{24}}\right) + \cancel{36} \cdot 2\left(\frac{x}{\cancel{36}}\right) = 72$$

$$3x + 2x = 72$$

$$5x = 72$$

$$\frac{5x}{\mathbf{5}} = \frac{72}{\mathbf{5}}$$

$$x = 14\frac{2}{5}$$

State

It will take both outlet pipes $14\frac{2}{5}$ hours to empty the pool.

Check

The result checks.

66. HEIGHT OF A TREE [Section 7.8]

Analyze

- Similar triangles determined by the tree and its shadow and the yardstick and its shadow.
- What is the height of the tree?

Assign

Let h = the height of the tree

Form

1 yard is to 2.5 feet as h feet is to 29 feet.

We will convert 1 yard to 3 feet so all units will be the same.

3 feet is to 2.5 feet as h feet is to 29 feet.

$$\begin{array}{r}\text{yardstick} \rightarrow \\ \text{yardstick shadow} \rightarrow\end{array} \frac{3}{2.5} = \frac{h}{29} \begin{array}{l}\leftarrow \text{tree's height} \\ \leftarrow \text{tree's shadow}\end{array}$$

Solve

$$\frac{3}{2.5} = \frac{h}{29}$$
$$87 = 2.5h$$
$$\frac{87}{\mathbf{2.5}} = \frac{2.5h}{\mathbf{2.5}}$$
$$34.8 = h$$

State

A height of the tree is 34.8 feet.

Check

$$\frac{3}{2.5} = \frac{h}{29}$$
$$\frac{3}{2.5} \stackrel{?}{=} \frac{\mathbf{34.8}}{29}$$
$$87 = 87$$

The result checks.

67. FORESTRY [Section 7.9]

Step 1: We let d represent the diameter (in inches) of a hardwood tree trunk and a its age in years. Translating the words *diameter varies directly with age of the tree.*
We get the equation $d = ka$.

Step 2: To find the constant of variation, k, we substitute 16 for d and 48 for a.

$$d = ka$$
$$16 = k(48)$$
$$\frac{16}{\mathbf{48}} = \frac{48k}{\mathbf{48}}$$
$$\frac{\overset{1}{\cancel{16}}}{\underset{1}{\cancel{16}} \cdot 3} = k$$
$$\frac{1}{3} = k$$

Step 3: We now substitute the value of k into the equation from step 1.

$$d = \frac{1}{3}a$$

Step 4: We can find the *diameter* by substituting 84 for a in the equation from step 3.

$$d = ka$$
$$d = \frac{1}{3}(84)$$
$$d = 28$$

The diameter of a hardwood tree that is 84 years old is 28 inches.

68. [Section 7.9]

$$y = \frac{k}{x}$$
$$8 = \frac{k}{6}$$
$$8 \cdot 6 = k$$
$$48 = k$$

$$y = \frac{k}{x}$$
$$y = \frac{48}{64}$$
$$y = \frac{\overset{1}{\cancel{2}} \cdot \overset{1}{\cancel{2}} \cdot \overset{1}{\cancel{2}} \cdot \overset{1}{\cancel{2}} \cdot 3}{\underset{1}{\cancel{2}} \cdot \underset{1}{\cancel{2}} \cdot \underset{1}{\cancel{2}} \cdot \underset{1}{\cancel{2}} \cdot 4}$$
$$y = \frac{3}{4}$$

Thus, when $x = 64$, the value of y is $\frac{3}{4}$ or 0.75.

SECTION 8.1 STUDY SET
VOCABULARY

Fill in the blanks.

1. The number b is a **square** root of the number a if $b^2 = a$.

3. The number or variable expression under a radical symbol is called the **radicand**.

5. The **Pythagorean** theorem is a formula that relates the lengths of the sides of a right triangle: $\mathbf{a^2 + b^2 = c^2}$.

CONCEPTS

Fill in the blanks.

7. Every positive number has **two** square roots, one positive and one negative. The positive square root of 25 is **5** and the negative square root of 25 is **–5**. In symbols, we write $\sqrt{25} = \mathbf{5}$ and $-\sqrt{25} = \mathbf{-5}$

9. The number 0 is the only real number with exactly one square root: $\sqrt{0} = \mathbf{0}$.

11. Identify the radicand of each radical expression.

a. $\sqrt{64}$ **64** b. $-\sqrt{36y^4}$ $\mathbf{36y^4}$

c. $2\sqrt{d}$ $\mathbf{d}$ d. $\sqrt{\frac{16}{25}}$ $\mathbf{\frac{16}{25}}$

13. Fill in the blanks: $\sqrt{29}$ must be a number between 5 and 6, because $\mathbf{\sqrt{25}} < \sqrt{29} < \mathbf{\sqrt{36}}$.

NOTATION

Complete the solution.

15. The legs of a right triangle measure 5 and 12 centimeters. Find the length of the hypotenuse.

$$a^2 + b^2 = c^2$$
$$\mathbf{5}^2 + 12^2 = c^2$$
$$25 + \mathbf{144} = c^2$$
$$\mathbf{169} = c^2$$
$$\sqrt{169} = \mathbf{c}$$
$$\mathbf{13} = c$$

The length of the hypotenuse is **13** cm.

GUIDED PRACTICE

Evaluate each square root without using a calculator.

17. $\sqrt{64} = 8$

19. $\sqrt{36} = 6$

21. $\sqrt{100} = 10$

23. $\sqrt{400} = 20$

25. $-\sqrt{81} = -9$

27. $\sqrt{1.21} = 1.1$

29. $\sqrt{169} = 13$

31. $\sqrt{\frac{9}{25}} = \frac{3}{5}$

33. $-\sqrt{\frac{1}{64}} = -\frac{1}{8}$

35. $-\sqrt{0.04} = -0.2$

37. $-\sqrt{289} = -17$

39. $\sqrt{2,500} = 50$

Use a calculator to approximate each square root to the nearest hundredths. See Example 2.

41. $\sqrt{3} \approx 1.73$

43. $\sqrt{95} \approx 9.75$

45. $\sqrt{428} \approx 20.69$

47. $2\sqrt{3} \approx 3.46$

Classify each square root as rational, irrational, or not a real number. See Example 3.

49. $\sqrt{9}$

Rational

51. $\sqrt{-21}$

Not a real number

53. $\sqrt{33}$

Irrational

55. $-\sqrt{0.25}$

Rational

Find each square root. All variables represent nonnegative real numbers. See Example 4.

57. $\sqrt{m^2} = m$

59. $\sqrt{t^4} = t^2$

61. $\sqrt{c^{10}} = c^5$

63. $\sqrt{n^{12}} = n^6$

65. $\sqrt{x^{36}} = x^{18}$

67. $\sqrt{4y^2} = 2y$

69. $\sqrt{64b^4} = 8b^2$

71. $-\sqrt{49s^{14}} = -7s^7$

Refer to the right triangle. See Example 5.

73. Find c if $a = 4$ and $b = 3$.

$$a^2 + b^2 = c^2$$
$$4^2 + 3^2 = c^2$$
$$16 + 9 = c^2$$
$$25 = c^2$$
$$\sqrt{25} = c$$
$$c = 5$$

75. Find a if $b = 15$ and $c = 17$.

$$a^2 + b^2 = c^2$$
$$a^2 + 15^2 = 17^2$$
$$a^2 + 225 = 289$$
$$a^2 + 225 - \mathbf{225} = 289 - \mathbf{225}$$
$$a^2 = 64$$
$$a = \sqrt{64}$$
$$a = 8$$

77. Find b if $a = 45$ and $c = 53$.

$$a^2 + b^2 = c^2$$
$$45^2 + b^2 = 53^2$$
$$2,025 + b^2 = 2,809$$
$$2,025 + b^2 - \mathbf{2,025} = 2,809 - \mathbf{2,025}$$
$$b^2 = 784$$
$$b = \sqrt{784}$$
$$b = 28$$

79. Find b if $c = 125$ and $a = 44$.

$$a^2 + b^2 = c^2$$
$$44^2 + b^2 = 125^2$$
$$1,936 + b^2 = 15,625$$
$$1,936 + b^2 - \mathbf{1,936} = 15,625 - \mathbf{1,936}$$
$$b^2 = 13,689$$
$$b = \sqrt{13,689}$$
$$b = 117$$

The lengths of two sides of a right triangle are given. Find the missing side length. Give an exact answer and an approximation to the nearest hundredth. See Example 6.

81. Find b if $a = 5$ cm and $c = 6$ cm

$$a^2 + b^2 = c^2$$
$$5^2 + b^2 = 6^2$$
$$25 + b^2 = 36$$
$$25 + b^2 - \mathbf{25} = 36 - \mathbf{25}$$
$$b^2 = 11$$
$$b = \sqrt{11}$$
$$b \approx 3.316 \text{ or } 3.32$$

The exact answer is $\sqrt{11}$ cm.
The approximation is about 3.32 cm.

83. Find c if $a = 12$ m and $b = 8$ m

$$a^2 + b^2 = c^2$$
$$12^2 + 8^2 = c^2$$
$$144 + 64 = c^2$$
$$208 = c^2$$
$$c = \sqrt{208}$$
$$c \approx 14.422 \text{ or } 14.42$$

The exact answer is $\sqrt{208}$ m.
The approximation is about 14.42 m.

85. Find c if $a = 9$ in. and $b = 3$ in.

$$a^2 + b^2 = c^2$$
$$9^2 + 3^2 = c^2$$
$$81 + 9 = c^2$$
$$90 = c^2$$
$$c = \sqrt{90}$$
$$c \approx 9.486 \text{ or } 9.49$$

The exact answer is $\sqrt{90}$ in.
The approximation is about 9.49 in.

87. Find a if $b = 4$ in. and $c = 6$ in.

$$a^2 + b^2 = c^2$$
$$a^2 + 4^2 = 6^2$$
$$a^2 + 16 = 36$$
$$a^2 + 16 - \mathbf{16} = 36 - \mathbf{16}$$
$$a^2 = 20$$
$$a = \sqrt{20}$$
$$a \approx 4.472 \text{ or } 4.47$$

The exact answer is $\sqrt{20}$ in.
The approximation is about 4.47 in.

For each isosceles right triangle, find the length of the legs. Give an exact answer and an approximation to the nearest hundredth. See Example 7.

89. $a^2 + b^2 = c^2$

$$x^2 + x^2 = 2^2$$
$$2x^2 = 4$$
$$\frac{2x^2}{\mathbf{2}} = \frac{4}{\mathbf{2}}$$
$$x^2 = 2$$
$$x = \sqrt{2}$$
$$x \approx 1.414 \text{ or } 1.41$$

The exact answer is $\sqrt{2}$ in.
The approximation is about 1.41 in.

Find the distance between the two points. If an answer contains a radical, give an exact answer and an approximation to the nearest hundredth. See Example 8.

91. (4, 6) and (1, 2)

$$\begin{array}{cc} \uparrow\ \uparrow & \uparrow\ \uparrow \\ (x_1, y_1) & (x_2, y_2) \end{array}$$

$$d = \sqrt{(x_2 - x_1)^2 + (y_2 - y_1)^2}$$
$$d = \sqrt{(1-4)^2 + (2-6)^2}$$
$$d = \sqrt{(-3)^2 + (-4)^2}$$
$$d = \sqrt{9+16}$$
$$d = \sqrt{25}$$
$$d = 5$$

93. (−2, −8) and (3, 4)

$$\begin{array}{cc} \uparrow\ \uparrow & \uparrow\ \uparrow \\ (x_1, y_1) & (x_2, y_2) \end{array}$$

$$d = \sqrt{(x_2 - x_1)^2 + (y_2 - y_1)^2}$$
$$d = \sqrt{[3-(-2)]^2 + [4-(-8)]^2}$$
$$d = \sqrt{(3+2)^2 + (4+8)^2}$$
$$d = \sqrt{5^2 + 12^2}$$
$$d = \sqrt{25+144}$$
$$d = \sqrt{169}$$
$$d = 13$$

95. (6, −5) and (0, −4)

$$\begin{array}{cc} \uparrow\ \uparrow & \uparrow\ \uparrow \\ (x_1, y_1) & (x_2, y_2) \end{array}$$

$$d = \sqrt{(x_2 - x_1)^2 + (y_2 - y_1)^2}$$
$$d = \sqrt{(0-6)^2 + [-4-(-5)]^2}$$
$$d = \sqrt{(-6)^2 + (-4+5)^2}$$
$$d = \sqrt{(-6)^2 + 1^2}$$
$$d = \sqrt{36+1}$$
$$d = \sqrt{37}$$
$$d \approx 6.082$$
$$d \approx 6.08$$

The exact answer is $\sqrt{37}$.
The approximation is about 6.08.

97. (9, −1) and (7, −6)

$$\begin{array}{cc} \uparrow\ \uparrow & \uparrow\ \uparrow \\ (x_1, y_1) & (x_2, y_2) \end{array}$$

$$d = \sqrt{(x_2 - x_1)^2 + (y_2 - y_1)^2}$$
$$d = \sqrt{(7-9)^2 + [-6-(-1)]^2}$$
$$d = \sqrt{(-2)^2 + (-6+1)^2}$$
$$d = \sqrt{(-2)^2 + (-5)^2}$$
$$d = \sqrt{4+25}$$
$$d = \sqrt{29}$$
$$d \approx 5.385$$
$$d \approx 5.39$$

The exact answer is $\sqrt{29}$.
The approximation is about 5.39.

APPLICATIONS

If an answer contains a radical, give an exact answer and an approximation to the nearest hundredth.

from CAMPUS TO CAREERS

99. CRIME SCENE INVESTIGATOR

a. $s = 4.5\sqrt{d}$, $d = 144$ feet

$$s = 4.5\sqrt{144}$$
$$s = 4.5(12)$$
$$s = 54$$

The speed of the car is 54 mph.

b. Yes, the driver was speeding.

$$54 - 35 = 19$$

The driver was 19 mph over.

101. BOATING

$s = 1.34\sqrt{L}$, $L = 64$ feet

$$s = 1.34\sqrt{64}$$
$$s = 1.34(8)$$
$$s = 10.72$$

The approximate maximum speed of the boat is about 10.72 knots.

103. DRAFTING

a. $h = \sqrt{2}\, l$, $l = 6$ inches

$$h = \sqrt{2}\,(6)$$
$$h = 6\sqrt{2}$$
$$h \approx 6(1.414213562)$$
$$h \approx 8.485281374$$
$$h \approx 8.49$$

The exact answer is $6\sqrt{2}$ in.
The approximation is about 8.49 in.

b. $l = \sqrt{3} \cdot \frac{h}{2}$, $h = 10$ inches

$$l = \sqrt{3} \cdot \frac{10}{2}$$
$$l = \sqrt{3} \cdot 5$$
$$l = 5\sqrt{3}$$
$$l \approx 5(1.732050808)$$
$$l \approx 8.660254038$$
$$l \approx 8.66$$

The exact answer is $5\sqrt{3}$ in.
The approximation is about 8.66 in.

105. BASEBALL

- Baseball diamond is a square.
- Both base paths (legs) are 90 feet each.
- What is the distance (hypotenuse) from home plate to 2nd base ?

$$a^2 + b^2 = c^2$$
$$90^2 + 90^2 = c^2$$
$$8{,}100 + 8{,}100 = c^2$$
$$16{,}200 = c^2$$
$$\sqrt{16{,}200} = c$$
$$c \approx \sqrt{16{,}200}$$
$$c \approx 127.279$$
$$c \approx 127.28$$

The exact answer is $\sqrt{16{,}200}$ ft.
The approximation is about 127.28 ft.

107. CARPENTRY

a. 16 in. and 30 in.

b.
$$a^2 + b^2 = c^2$$
$$16^2 + 30^2 = c^2$$
$$256 + 900 = c^2$$
$$1{,}156 = c^2$$
$$\sqrt{1{,}156} = c$$
$$34 = c$$

Its length should be 34 in.

109. FOOTBALL

- Hypotenuse (diagonal) is 6 yards.
- What is the length of one side (leg)?

a.
$$a^2 + b^2 = c^2$$
$$x^2 + x^2 = 6^2$$
$$2x^2 = 36$$
$$\frac{2x^2}{\mathbf{2}} = \frac{36}{\mathbf{2}}$$
$$x^2 = 18$$
$$x = \sqrt{18}$$
$$x \approx 4.24$$

He gains about 4.24 yd.

b. Yes; $6 \text{ yd} + 4.24 \text{ yd} > 10 \text{ yd}$.

111. DECK DESIGNS

Brace 1: $(-4, 0)$ and $(-1, -3)$

$$\begin{matrix} (-4, & 0) & & (-1, & -3) \\ \uparrow & \uparrow & & \uparrow & \uparrow \\ (x_1, & y_1) & & (x_2, & y_2) \end{matrix}$$

$$d = \sqrt{(x_2 - x_1)^2 + (y_2 - y_1)^2}$$
$$d = \sqrt{[-1 - (-4)]^2 + (-3 - 0)^2}$$
$$d = \sqrt{(-1 + 4)^2 + (-3)^2}$$
$$d = \sqrt{3^2 + (-3)^2}$$
$$d = \sqrt{9 + 9}$$
$$d = \sqrt{18}$$
$$d \approx 4.242$$

The exact length of Brace 1 is $\sqrt{18}$ m.
The approximate length is about 4.24 m.

Brace 2: $(-5, 0)$ and $(-2, -6)$

$$\begin{matrix} (-5, & 0) & & (-2, & -6) \\ \uparrow & \uparrow & & \uparrow & \uparrow \\ (x_1, & y_1) & & (x_2, & y_2) \end{matrix}$$

$$d = \sqrt{(x_2 - x_1)^2 + (y_2 - y_1)^2}$$
$$d = \sqrt{[-2 - (-5)]^2 + (-6 - 0)^2}$$
$$d = \sqrt{(-2 + 5)^2 + (-6)^2}$$
$$d = \sqrt{3^2 + (-6)^2}$$
$$d = \sqrt{9 + 36}$$
$$d = \sqrt{45}$$
$$d \approx 6.708$$

The exact length of Brace 2 is $\sqrt{45}$ m.
The approximate length is about 6.71 m.

Continued on next page

Brace 3: $(-3,-4)$ and $(-2,-2)$

$$\begin{matrix} \uparrow & \uparrow & & \uparrow & \uparrow \\ (x_1, & y_1) & & (x_2, & y_2) \end{matrix}$$

$$d=\sqrt{(x_2-x_1)^2+(y_2-y_1)^2}$$
$$d=\sqrt{[-2-(-3)]^2+[-2-(-4)]^2}$$
$$d=\sqrt{(-2+3)^2+(-2+4)^2}$$
$$d=\sqrt{1^2+2^2}$$
$$d=\sqrt{1+4}$$
$$d=\sqrt{5}$$
$$d\approx 2.236$$

The exact length of Brace 3 is $\sqrt{5}$ m.
The approximate length is about 2.24 m.

WRITING

113-115. Answers will vary.

REVIEW

117. Add:

$$(3s^2-3s-2)+(3s^2+4s-3)$$
$$=3s^2-3s-2+3s^2+4s-3$$
$$=(3+3)s^2+(-3+4)s+(-2-3)$$
$$=6s^2+s-5$$

119. Multiply:

$$(3x-2)(x+4)$$
$$=\mathbf{3x}(x)+\mathbf{3x}(4)+(\mathbf{-2})(x)+(\mathbf{-2})(4)$$
$$=3x^2+12x-2x-8$$
$$=3x^2+(12-2)x-8$$
$$=3x^2+10x-8$$

CHALLENGE PROBLEMS

121. SHORTCUTS

- Width (short leg) is 52 feet.
- Length (long leg) is 165 feet.
- How long is the diagonal (hypotenuse)?

$$a^2+b^2=c^2$$
$$52^2+165^2=c^2$$
$$2,704+27,225=c^2$$
$$29,929=c^2$$
$$\sqrt{29,929}=c$$
$$173=c$$

The diagonal is 173 feet.

$165 \text{ ft.}+52 \text{ ft.}=217 \text{ ft.}$

$217 \text{ ft.}-173 \text{ ft.}=44 \text{ ft.}$

Using the diagonal saves 44 ft.

123. Simplify:

$$\sqrt{16x^{16n}}=\sqrt{16}\cdot\sqrt{x^{16n}}$$
$$=\sqrt{4^2}\cdot\sqrt{x^{8n}\cdot x^{8n}}$$
$$=4x^{8n}$$

SECTION 8.2 STUDY SET
VOCABULARY

Fill in the blanks.

1. "To **simplify** $\sqrt{20}$" means to write it as $2\sqrt{5}$.
3. The expression $\sqrt{4 \cdot 3}$ is a square root of a **product** and $\sqrt{4}\sqrt{3}$ is a product of two **square** roots.

CONCEPTS

5. **Fill in the blanks.**
   a. The square root of the product of two positive numbers is equal to the **product** of their square roots. In symbols, $\sqrt{a \cdot b} = \sqrt{\mathbf{a}}\sqrt{\mathbf{b}}$.
   b. The square root of the quotient of two positive numbers is equal to the **quotient** of their square roots. In symbols,
   $$\sqrt{\frac{a}{b}} = \frac{\sqrt{\mathbf{a}}}{\sqrt{\mathbf{b}}}.$$
7. Which perfect square, 1, 4, 9, 16, 25, 36, 49, 64, 81, or 100, is the greatest perfect-square factor of the given number?
   a. 20 **4** b. 98 **49**
   c. 54 **9** d. 48 **16**
9. To simplify $\sqrt{40}$, which one of the following factorizations should be used? $\mathbf{\sqrt{4 \cdot 10}}$
11. Use the square root of a product property to simplify each expression.
   a. $\sqrt{81 \cdot 2} = \sqrt{\mathbf{81}}\sqrt{\mathbf{2}} = \mathbf{9\sqrt{2}}$
   b. $\sqrt{m^4 \cdot m} = \sqrt{\mathbf{m^4}}\sqrt{\mathbf{m}} = \mathbf{m^2\sqrt{m}}$

NOTATION

13. We can read $2\sqrt{6}$ as "2 **times** the square root of 6" or as "2 **radical** 6."

GUIDED PRACTICE

Simplify. See Example 1.

15. $\sqrt{20} = \sqrt{4 \cdot 5} = \sqrt{4}\sqrt{5} = 2\sqrt{5}$
17. $\sqrt{27} = \sqrt{9 \cdot 3} = \sqrt{9}\sqrt{3} = 3\sqrt{3}$
19. $\sqrt{50} = \sqrt{25 \cdot 2} = \sqrt{25}\sqrt{2} = 5\sqrt{2}$
21. $\sqrt{24} = \sqrt{4 \cdot 6} = \sqrt{4}\sqrt{6} = 2\sqrt{6}$

Simplify. See Example 2.

23. $\sqrt{500} = \sqrt{100 \cdot 5} = \sqrt{100}\sqrt{5} = 10\sqrt{5}$
25. $\sqrt{63} = \sqrt{9 \cdot 7} = \sqrt{9}\sqrt{7} = 3\sqrt{7}$

Simplify. See Example 3.

27. $\sqrt{98} = \sqrt{2 \cdot 49} = \sqrt{2 \cdot 7 \cdot 7} = \sqrt{2}\sqrt{7 \cdot 7} = \sqrt{2} \cdot 7 = 7\sqrt{2}$
29. $\sqrt{180} = \sqrt{2 \cdot 90} = \sqrt{2 \cdot 3 \cdot 30} = \sqrt{2 \cdot 3 \cdot 3 \cdot 10} = \sqrt{2 \cdot 3 \cdot 3 \cdot 2 \cdot 5} = \sqrt{2 \cdot 2 \cdot 3 \cdot 3 \cdot 5} = \sqrt{2 \cdot 2}\sqrt{3 \cdot 3}\sqrt{5} = 2 \cdot 3\sqrt{5} = 6\sqrt{5}$
31. $\sqrt{192} = \sqrt{2 \cdot 96} = \sqrt{2 \cdot 2 \cdot 48} = \sqrt{2 \cdot 2 \cdot 2 \cdot 24} = \sqrt{2 \cdot 2 \cdot 2 \cdot 2 \cdot 12} = \sqrt{2 \cdot 2 \cdot 2 \cdot 2 \cdot 2 \cdot 6} = \sqrt{2 \cdot 2 \cdot 2 \cdot 2 \cdot 2 \cdot 2 \cdot 3} = \sqrt{2 \cdot 2}\sqrt{2 \cdot 2}\sqrt{2 \cdot 2}\sqrt{3} = 2 \cdot 2 \cdot 2\sqrt{3} = 8\sqrt{3}$
33. $\sqrt{375} = \sqrt{5 \cdot 75} = \sqrt{5 \cdot 5 \cdot 15} = \sqrt{5 \cdot 5 \cdot 3 \cdot 5} = \sqrt{5 \cdot 5}\sqrt{3 \cdot 5} = 5\sqrt{15}$
35. $\sqrt{42} = \sqrt{2 \cdot 3 \cdot 7}$; Cannot be simplified
37. $\sqrt{385} = \sqrt{5 \cdot 7 \cdot 11}$; Cannot be simplified

Simplify. See Example 4.

39. $\sqrt{x^{11}} = \sqrt{x^{10} \cdot x}$
$= \sqrt{x^{10}}\sqrt{x}$
$= x^5\sqrt{x}$

41. $\sqrt{n^9} = \sqrt{n^8 \cdot n}$
$= \sqrt{n^8}\sqrt{n}$
$= n^4\sqrt{n}$

Simplify. See Example 5.

43. $4\sqrt{12x} = 4\sqrt{4 \cdot 3x}$
$= 4\sqrt{4}\sqrt{3x}$
$= 4 \cdot 2\sqrt{3x}$
$= 8\sqrt{3x}$

45. $5\sqrt{54q} = 5\sqrt{9 \cdot 6q}$
$= 5\sqrt{9}\sqrt{6q}$
$= 5 \cdot 3\sqrt{6q}$
$= 15\sqrt{6q}$

Simplify. See Example 6.

47. $\sqrt{25t^3} = \sqrt{25t^2 \cdot t}$
$= \sqrt{25t^2}\sqrt{t}$
$= 5t\sqrt{t}$

49. $\sqrt{32x^5} = \sqrt{16x^4 \cdot 2x}$
$= \sqrt{16x^4}\sqrt{2x}$
$= 4x^2\sqrt{2x}$

51. $\frac{1}{5}\sqrt{75x^7} = \frac{1}{5}\sqrt{25x^6 \cdot 3x}$
$= \frac{1}{5}\sqrt{25x^6}\sqrt{3x}$
$= \frac{1}{5} \cdot 5x^3\sqrt{3x}$
$= \frac{1}{\cancel{5}_1} \cdot \cancel{5}^1 x^3\sqrt{3x}$
$= x^3\sqrt{3x}$

53. $\frac{1}{3}\sqrt{18t^{11}} = \frac{1}{3}\sqrt{9t^{10} \cdot 2t}$
$= \frac{1}{3}\sqrt{9t^{10}}\sqrt{2t}$
$= \frac{1}{3} \cdot 3t^5\sqrt{2t}$
$= \frac{1}{\cancel{3}_1} \cdot \cancel{3}^1 t^5\sqrt{2t}$
$= t^5\sqrt{2t}$

Simplify. See Example 7.

55. $\sqrt{\frac{25}{9}} = \frac{\sqrt{25}}{\sqrt{9}}$
$= \frac{5}{3}$

57. $\sqrt{\frac{81}{64}} = \frac{\sqrt{81}}{\sqrt{64}}$
$= \frac{9}{8}$

59. $\sqrt{\frac{6}{121}} = \frac{\sqrt{6}}{\sqrt{121}}$
$= \frac{\sqrt{6}}{11}$

61. $\sqrt{\frac{75}{16}} = \frac{\sqrt{75}}{\sqrt{16}}$
$= \frac{\sqrt{25 \cdot 3}}{4}$
$= \frac{\sqrt{25}\sqrt{3}}{4}$
$= \frac{5\sqrt{3}}{4}$

Simplify. See Example 8.

63. $\sqrt{\frac{a^4}{4a}} = \sqrt{\frac{a^{4-1}}{4}}$
$= \frac{\sqrt{a^3}}{\sqrt{4}}$
$= \frac{\sqrt{a^2 \cdot a}}{2}$
$= \frac{\sqrt{a^2}\sqrt{a}}{2}$
$= \frac{a\sqrt{a}}{2}$

65. $\sqrt{\frac{r^{10}}{225r}} = \sqrt{\frac{r^{10-1}}{225}}$
$= \frac{\sqrt{r^9}}{\sqrt{225}}$
$= \frac{\sqrt{r^8 \cdot r}}{15}$
$= \frac{\sqrt{r^8}\sqrt{r}}{15}$
$= \frac{r^4\sqrt{r}}{15}$

67. $\sqrt{\frac{72x^3}{x}} = \sqrt{72x^{3-1}}$
$= \sqrt{36 \cdot 2x^2}$
$= \sqrt{36x^2}\sqrt{2}$
$= 6x\sqrt{2}$

69. $\sqrt{\frac{125n^5}{64n}} = \sqrt{\frac{125n^{5-1}}{64}}$
$= \frac{\sqrt{125n^4}}{\sqrt{64}}$
$= \frac{\sqrt{25 \cdot 5n^4}}{8}$
$= \frac{\sqrt{25n^4}\sqrt{5}}{8}$
$= \frac{5n^2\sqrt{5}}{8}$

TRY IT YOURSELF

Simplify.

71. $\sqrt{75t} = \sqrt{25 \cdot 3t}$
$= \sqrt{25}\sqrt{3t}$
$= 5\sqrt{3t}$

73. $\sqrt{\frac{48}{81}} = \frac{\sqrt{48}}{\sqrt{81}}$
$= \frac{\sqrt{16 \cdot 3}}{9}$
$= \frac{\sqrt{16}\sqrt{3}}{9}$
$= \frac{4\sqrt{3}}{9}$

75. $\sqrt{48} = \sqrt{16 \cdot 3}$
$= \sqrt{16}\sqrt{3}$
$= 4\sqrt{3}$

77. $\sqrt{4k} = \sqrt{4 \cdot k}$
$= \sqrt{4}\sqrt{k}$
$= 2\sqrt{k}$

79. $\sqrt{2d^{11}} = \sqrt{d^{10} \cdot 2d}$
$= \sqrt{d^{10}}\sqrt{2d}$
$= d^5\sqrt{2d}$

81. $\sqrt{10b} = \sqrt{2 \cdot 5b}$; Cannot be simplified.

83. $\sqrt{44} = \sqrt{4 \cdot 11}$
$= \sqrt{4}\sqrt{11}$
$= 2\sqrt{11}$

85. $\sqrt{\frac{23}{64}} = \frac{\sqrt{23}}{\sqrt{64}}$
$= \frac{\sqrt{23}}{8}$

87. $\frac{1}{6}\sqrt{72} = \frac{1}{6}\sqrt{36 \cdot 2}$
$= \frac{1}{6}\sqrt{36}\sqrt{2}$
$= \frac{1}{6} \cdot 6\sqrt{2}$
$= \frac{1}{\cancel{6}_1} \cdot \cancel{6}^1\sqrt{2}$
$= \sqrt{2}$

89. $\sqrt{\frac{75q^2}{16q^4}} = \sqrt{\frac{75}{16q^{4-2}}}$
$= \frac{\sqrt{75}}{\sqrt{16q^2}}$
$= \frac{\sqrt{25 \cdot 3}}{4q}$
$= \frac{\sqrt{25}\sqrt{3}}{4q}$
$= \frac{5\sqrt{3}}{4q}$

91. $\sqrt{\frac{20}{49}} = \frac{\sqrt{20}}{\sqrt{49}}$
$= \frac{\sqrt{4 \cdot 5}}{7}$
$= \frac{\sqrt{4}\sqrt{5}}{7}$
$= \frac{2\sqrt{5}}{7}$

93. $\sqrt{162} = \sqrt{81 \cdot 2}$
$= \sqrt{81}\sqrt{2}$
$= 9\sqrt{2}$

95. $\frac{3}{2}\sqrt{16y^5} = \frac{3}{2}\sqrt{16y^4 \cdot y}$
$= \frac{3}{2}\sqrt{16y^4}\sqrt{y}$
$= \frac{3}{2} \cdot 4y^2\sqrt{y}$
$= \frac{3}{\cancel{2}_1} \cdot 2 \cdot \cancel{2}^1 y^2\sqrt{y}$
$= 6y^2\sqrt{y}$

97. $\sqrt{50x^4} = \sqrt{25x^4 \cdot 2}$
$= \sqrt{25x^4}\sqrt{2}$
$= 5x^2\sqrt{2}$

99. $\sqrt{t^{15}} = \sqrt{t^{14} \cdot t}$
$= \sqrt{t^{14}}\sqrt{t}$
$= t^7\sqrt{t}$

LOOK ALIKES . . .

101.

a. $\sqrt{x^2} = x$

b. $\sqrt{x^3} = \sqrt{x^2 \cdot x^1}$
$= \sqrt{x^2}\sqrt{x^1}$
$= x\sqrt{x}$

c. $\sqrt{x^4} = \sqrt{x^2 \cdot x^2}$
$= \sqrt{x^2}\sqrt{x^2}$
$= x \cdot x$
$= x^2$

d. $\sqrt{x^5} = \sqrt{x^2 \cdot x^2 \cdot x^1}$
$= \sqrt{x^2}\sqrt{x^2}\sqrt{x^1}$
$= x \cdot x\sqrt{x^1}$
$= x^2\sqrt{x}$

e. $\sqrt{x^6} = \sqrt{x^2 \cdot x^2 \cdot x^2}$
$= \sqrt{x^2}\sqrt{x^2}\sqrt{x^2}$
$= x \cdot x \cdot x$
$= x^3$

f. $\sqrt{x^7} = \sqrt{x^2 \cdot x^2 \cdot x^2 \cdot x^1}$
$= \sqrt{x^2}\sqrt{x^2}\sqrt{x^2}\sqrt{x^1}$
$= x \cdot x \cdot x\sqrt{x^1}$
$= x^3\sqrt{x}$

g. $\sqrt{x^8} = \sqrt{x^2 \cdot x^2 \cdot x^2 \cdot x^2}$
$= \sqrt{x^2}\sqrt{x^2}\sqrt{x^2}\sqrt{x^2}$
$= x \cdot x \cdot x \cdot x$
$= x^4$

h. $\sqrt{x^9} = \sqrt{x^2 \cdot x^2 \cdot x^2 \cdot x^2 \cdot x^1}$
$= \sqrt{x^2}\sqrt{x^2}\sqrt{x^2}\sqrt{x^2}\sqrt{x^1}$
$= x \cdot x \cdot x \cdot x\sqrt{x^1}$
$= x^4\sqrt{x}$

103. a. $\sqrt{8}=\sqrt{4\cdot 2}$
$=\sqrt{4}\sqrt{2}$
$=2\sqrt{2}$

b. $\sqrt{18}=\sqrt{9\cdot 2}$
$=\sqrt{9}\sqrt{2}$
$=3\sqrt{2}$

c. $\sqrt{50}=\sqrt{25\cdot 2}$
$=\sqrt{25}\sqrt{2}$
$=5\sqrt{2}$

d. $\sqrt{72}=\sqrt{36\cdot 2}$
$=\sqrt{36}\sqrt{2}$
$=6\sqrt{2}$

APPLICATIONS

105. STUDYING PAST CULTURES

- Both legs are 3 feet long.
- What is the length of the diagonal?

$$\begin{aligned} a^2+b^2&=c^2\\ 3^2+3^2&=c^2\\ 9+9&=c^2\\ 18&=c^2\\ \sqrt{18}&=c\\ \sqrt{9\cdot 2}&=c\\ \sqrt{9}\sqrt{2}&=c\\ 3\sqrt{2}&=c\\ 4.24&\approx c \end{aligned}$$

The exact distance is $3\sqrt{2}$ ft.
The approximate distance is about 4.24 ft.

107. SQUARE ROOT SPIRAL

$\sqrt{4}=2,\ \sqrt{8}=2\sqrt{2},\ \sqrt{9}=3,$
$\sqrt{12}=2\sqrt{3},\ \sqrt{16}=4$

WRITING

109-111. Answers will vary.

REVIEW

Solve each system.

113. $\begin{cases} y=2x-6\\ 2x+y=6 \end{cases}$

$\begin{cases} -2x+y=-6\\ 2x+y=6 \end{cases}$

$$\begin{aligned} -2x+y&=-6\\ 2x+y&=6\\ \hline 2y&=0\\ \frac{2y}{\mathbf{2}}&=\frac{0}{\mathbf{2}}\\ y&=0 \end{aligned} \qquad \begin{aligned} 2x+y&=6\\ 2x+\mathbf{0}&=6\\ 2x&=6\\ \frac{2x}{\mathbf{2}}&=\frac{6}{\mathbf{2}}\\ x&=3 \end{aligned}$$

The solution is $(3,0)$.

115. $\begin{cases} 3x+4y=-7\\ 2x-y=-1 \end{cases}$

$\begin{cases} 3x+4y=-7\\ 8x-4y=-4 \end{cases}$

$$\begin{aligned} 3x+4y&=-7\\ 8x-4y&=-4\\ \hline 11x&=-11\\ \frac{11x}{\mathbf{11}}&=\frac{-11}{\mathbf{11}}\\ x&=-1 \end{aligned} \qquad \begin{aligned} 3x+4y&=-7\\ 3(\mathbf{-1})+4y&=-7\\ -3+4y+\mathbf{3}&=-7+\mathbf{3}\\ 4y&=-4\\ \frac{4y}{\mathbf{4}}&=\frac{-4}{\mathbf{4}}\\ y&=-1 \end{aligned}$$

The solution is $(-1,-1)$.

CHALLENGE PROBLEMS

117. **Simplify. All variables represent positive numbers.**

$$\begin{aligned} \sqrt{\frac{196a^8b^5c^9}{50abc^8}}&=\sqrt{\frac{2\cdot 98a^8b^5c^9}{2\cdot 25abc^8}}\\ &=\sqrt{\frac{\overset{1}{\cancel{2}}\cdot 2\cdot 49a^{8-1}b^{5-1}c^{9-8}}{\underset{1}{\cancel{2}}\cdot 25}}\\ &=\sqrt{\frac{2\cdot 49a^7b^4c}{25}}\\ &=\frac{\sqrt{2\cdot 49a^7b^4c}}{\sqrt{25}}\\ &=\frac{\sqrt{2\cdot 49a^6a^1b^4c}}{5}\\ &=\frac{\sqrt{49a^6b^4}\sqrt{2ac}}{5}\\ &=\frac{7a^3b^2\sqrt{2ac}}{5} \end{aligned}$$

SECTION 8.3 STUDY SET

VOCABULARY

Fill in the blanks.

1. Square roots such as $\sqrt{2}$ and $5\sqrt{2}$, that have the same radicand, are called **like** radicals.

CONCEPTS

Determine whether each pair of radicals are like radicals.

3.a. $3\sqrt{3}, 4\sqrt{3}$ **Yes** b. $9\sqrt{a}, 9\sqrt{7a}$ **No**

5. **Fill in the blanks.**

a. $5\sqrt{6}+3\sqrt{6}=(\mathbf{5}+\mathbf{3})\sqrt{6}=\mathbf{8}\sqrt{6}$

b. $2\sqrt{n}-9\sqrt{n}=(\mathbf{2}-\mathbf{9})\sqrt{n}=\mathbf{-7}\sqrt{n}$

NOTATION

Complete the solution to simplify the expression.

7.
$$\begin{aligned}9\sqrt{5}-3\sqrt{20}&=9\sqrt{5}-3\sqrt{\mathbf{4}\cdot 5}\\&=9\sqrt{5}-3\sqrt{\mathbf{4}}\sqrt{5}\\&=9\sqrt{5}-3\cdot\mathbf{2}\sqrt{5}\\&=9\sqrt{5}-\mathbf{6}\sqrt{5}\\&=\mathbf{3}\sqrt{5}\end{aligned}$$

GUIDED PRACTICE

In the following problems, all variables represent nonnegative real numbers. Perform the indicated operations, if possible, by combining like radicals. See Example 1.

9.
$$\begin{aligned}5\sqrt{7}+4\sqrt{7}&=(5+4)\sqrt{7}\\&=9\sqrt{7}\end{aligned}$$

11.
$$\begin{aligned}14\sqrt{21}-4\sqrt{21}&=(14-4)\sqrt{21}\\&=10\sqrt{21}\end{aligned}$$

13.
$$\begin{aligned}8\sqrt{n}+\sqrt{n}&=(8+1)\sqrt{n}\\&=9\sqrt{n}\end{aligned}$$

15. $\sqrt{3}+\sqrt{15}$ Does not simplify.

17.
$$\begin{aligned}\sqrt{x}-4\sqrt{x}&=(1-4)\sqrt{x}\\&=-3\sqrt{x}\end{aligned}$$

19.
$$\begin{aligned}2\sqrt{11}+3\sqrt{11}+5\sqrt{11}&=(2+3+5)\sqrt{11}\\&=10\sqrt{11}\end{aligned}$$

21.
$$\begin{aligned}4\sqrt{2}+4\sqrt{2}-4\sqrt{2}&=(4+4-4)\sqrt{2}\\&=4\sqrt{2}\end{aligned}$$

23. $7+2\sqrt{2}$; Cannot be simplified.

Perform the indicated operations, if possible, by combining like radicals. See Example 2.

25.
$$\begin{aligned}\sqrt{12}+\sqrt{27}&=\sqrt{4\cdot 3}+\sqrt{9\cdot 3}\\&=\sqrt{4}\sqrt{3}+\sqrt{9}\sqrt{3}\\&=2\sqrt{3}+3\sqrt{3}\\&=(2+3)\sqrt{3}\\&=5\sqrt{3}\end{aligned}$$

27.
$$\begin{aligned}\sqrt{18}-\sqrt{8}&=\sqrt{9\cdot 2}-\sqrt{4\cdot 2}\\&=\sqrt{9}\sqrt{2}-\sqrt{4}\sqrt{2}\\&=3\sqrt{2}-2\sqrt{2}\\&=(3-2)\sqrt{2}\\&=1\sqrt{2}\\&=\sqrt{2}\end{aligned}$$

29.
$$\begin{aligned}\sqrt{12}-\sqrt{48}&=\sqrt{4\cdot 3}-\sqrt{16\cdot 3}\\&=\sqrt{4}\sqrt{3}-\sqrt{16}\sqrt{3}\\&=2\sqrt{3}-4\sqrt{3}\\&=(2-4)\sqrt{3}\\&=-2\sqrt{3}\end{aligned}$$

31.
$$\begin{aligned}\sqrt{288}-3\sqrt{200}&=\sqrt{144\cdot 2}-3\sqrt{100\cdot 2}\\&=\sqrt{144}\sqrt{2}-3\sqrt{100}\sqrt{2}\\&=12\sqrt{2}-(3\cdot 10)\sqrt{2}\\&=12\sqrt{2}-30\sqrt{2}\\&=(12-30)\sqrt{2}\\&=-18\sqrt{2}\end{aligned}$$

33.
$$\begin{aligned}\sqrt{20}+\sqrt{45}+\sqrt{80}&=\sqrt{4\cdot 5}+\sqrt{9\cdot 5}+\sqrt{16\cdot 5}\\&=\sqrt{4}\sqrt{5}+\sqrt{9}\sqrt{5}+\sqrt{16}\sqrt{5}\\&=2\sqrt{5}+3\sqrt{5}+4\sqrt{5}\\&=(2+3+4)\sqrt{5}\\&=9\sqrt{5}\end{aligned}$$

35. $3\sqrt{200}-\sqrt{75}+\sqrt{48}$
$$\begin{aligned}&=3\sqrt{100\cdot 2}-\sqrt{25\cdot 3}+\sqrt{16\cdot 3}\\&=3\sqrt{100}\sqrt{2}-\sqrt{25}\sqrt{3}+\sqrt{16}\sqrt{3}\\&=(3\cdot 10)\sqrt{2}-5\sqrt{3}+4\sqrt{3}\\&=30\sqrt{2}+(-5+4)\sqrt{3}\\&=30\sqrt{2}-\sqrt{3}\end{aligned}$$

Perform the indicated operations, if possible, by combining like radicals. See Example 3.

37. $\sqrt{72a}-\sqrt{98a}=\sqrt{36\cdot 2a}-\sqrt{49\cdot 2a}$
$=\sqrt{36}\sqrt{2a}-\sqrt{49}\sqrt{2a}$
$=6\sqrt{2a}-7\sqrt{2a}$
$=(6-7)\sqrt{2a}$
$=-1\sqrt{2a}$
$=-\sqrt{2a}$

39. $\sqrt{18y}-\sqrt{27y}=\sqrt{9\cdot 2y}-\sqrt{9\cdot 3y}$
$=\sqrt{9}\sqrt{2y}-\sqrt{9}\sqrt{3y}$
$=3\sqrt{2y}-3\sqrt{3y}$

41. $\sqrt{2x^2}+\frac{1}{2}\sqrt{8x^2}=\sqrt{2}\sqrt{x^2}+\frac{1}{2}\sqrt{4x^2}\sqrt{2}$
$=x\sqrt{2}+\frac{1}{2}\cdot 2x\sqrt{2}$
$=x\sqrt{2}+\frac{1}{\cancel{2}}\cdot \cancel{2}x\sqrt{2}$
$=x\sqrt{2}+x\sqrt{2}$
$=(x+x)\sqrt{2}$
$=2x\sqrt{2}$

43. $\sqrt{2d^3}+\sqrt{8d^3}=\sqrt{d^2\cdot 2d}+\sqrt{4d^2\cdot 2d}$
$=\sqrt{d^2}\sqrt{2d}+\sqrt{4d^2}\sqrt{2d}$
$=d\sqrt{2d}+2d\sqrt{2d}$
$=(d+2d)\sqrt{2d}$
$=3d\sqrt{2d}$

Perform the indicated operations, if possible, by combining like radicals. See Example 4.

45. $\sqrt{4}+\sqrt{2}+3\sqrt{2}+3=2+\sqrt{2}+3\sqrt{2}+3$
$=(2+3)+(1+3)\sqrt{2}$
$=5+4\sqrt{2}$

47. $\sqrt{y^2}+3\sqrt{y}-5\sqrt{y}-15=y+3\sqrt{y}-5\sqrt{y}-15$
$=y+(3-5)\sqrt{y}-15$
$=y-2\sqrt{y}-15$

49. $2\sqrt{t^2}+5\sqrt{t}-2\sqrt{t}-5=2t+5\sqrt{t}-2\sqrt{t}-5$
$=2t+(5-2)\sqrt{t}-5$
$=2t+3\sqrt{t}-5$

51. $3\sqrt{a^2}+3\sqrt{a}-3\sqrt{a}-1=3a+3\sqrt{a}-3\sqrt{a}-1$
$=3a+(3-3)\sqrt{a}-1$
$=3a+0\sqrt{a}-1$
$=3a-1$

TRY IT YOURSELF

Perform the indicated operations, if possible.

53. $8\sqrt{6}-5\sqrt{2}-3\sqrt{6}=(8-3)\sqrt{6}-5\sqrt{2}$
$=5\sqrt{6}-5\sqrt{2}$

55. $\sqrt{24}+\sqrt{150}+\sqrt{240}=\sqrt{4\cdot 6}+\sqrt{25\cdot 6}+\sqrt{16\cdot 15}$
$=\sqrt{4}\sqrt{6}+\sqrt{25}\sqrt{6}+\sqrt{16}\sqrt{15}$
$=2\sqrt{6}+5\sqrt{6}+4\sqrt{15}$
$=(2+5)\sqrt{6}+4\sqrt{15}$
$=7\sqrt{6}+4\sqrt{15}$

57. $12+2\sqrt{5}$; Cannot be simplified.

59. $-1+2\sqrt{r}-3\sqrt{r}=-1+(2-3)\sqrt{r}$
$=-1-1\sqrt{r}$
$=-1-\sqrt{r}$

61. $-9\sqrt{21}+6\sqrt{21}=(-9+6)\sqrt{21}$
$=-3\sqrt{21}$

63. $-\sqrt{y}+\sqrt{y}=(-1+1)\sqrt{y}$
$=0\sqrt{y}$
$=0$

65. $2\sqrt{28}+\frac{1}{4}\sqrt{112}=2\sqrt{4\cdot 7}+\frac{1}{4}\sqrt{16\cdot 7}$
$=2\sqrt{4}\sqrt{7}+\frac{1}{4}\sqrt{16}\sqrt{7}$
$=4\sqrt{7}+\frac{1}{4}\cdot 4\sqrt{7}$
$=4\sqrt{7}+\frac{1}{\cancel{4}}\cdot \cancel{4}\sqrt{7}$
$=4\sqrt{7}+\sqrt{7}$
$=(4+1)\sqrt{7}$
$=5\sqrt{7}$

67. $15\sqrt{b^2}+20\sqrt{b}-3\sqrt{b}-\sqrt{16}$
$$=15b+20\sqrt{b}-3\sqrt{b}-4$$
$$=15b+(20-3)\sqrt{b}-4$$
$$=15b+17\sqrt{b}-4$$

69. $5+3\sqrt{3}+3\sqrt{3}=5+(3+3)\sqrt{3}$
$$=5+6\sqrt{3}$$

71. $\sqrt{2}+\sqrt{3}+\sqrt{5}$; Cannot be simplified.

73. $2\sqrt{45}+2\sqrt{80}=2\sqrt{9\cdot5}+2\sqrt{16\cdot5}$
$$=2\sqrt{9}\sqrt{5}+2\sqrt{16}\sqrt{5}$$
$$=(2\cdot3)\sqrt{5}+(2\cdot4)\sqrt{5}$$
$$=6\sqrt{5}+8\sqrt{5}$$
$$=(6+8)\sqrt{5}$$
$$=14\sqrt{5}$$

75. $3\sqrt{54b^2}+5\sqrt{24b^2}=3\sqrt{9b^2\cdot6}+5\sqrt{4b^2\cdot6}$
$$=3\sqrt{9b^2}\sqrt{6}+5\sqrt{4b^2}\sqrt{6}$$
$$=(3\cdot3b)\sqrt{6}+(5\cdot2b)\sqrt{6}$$
$$=9b\sqrt{6}+10b\sqrt{6}$$
$$=(9b+10b)\sqrt{6}$$
$$=19b\sqrt{6}$$

77. $\sqrt{32x^5}-\frac{2}{3}\sqrt{18x^5}=\sqrt{16x^4\cdot2x}-\frac{2}{3}\sqrt{9x^4\cdot2x}$
$$=\sqrt{16x^4}\sqrt{2x}-\frac{2}{3}\sqrt{9x^4}\sqrt{2x}$$
$$=4x^2\sqrt{2x}-\frac{2}{3}\cdot3x^2\sqrt{2x}$$
$$=4x^2\sqrt{2x}-\frac{2}{\cancel{3}_1}\cdot\cancel{3}^1x^2\sqrt{2x}$$
$$=4x^2\sqrt{2x}-2x^2\sqrt{2x}$$
$$=(4x^2-2x^2)\sqrt{2x}$$
$$=2x^2\sqrt{2x}$$

79. $2\sqrt{80}-3\sqrt{125}=2\sqrt{16\cdot5}-3\sqrt{25\cdot5}$
$$=2\sqrt{16}\sqrt{5}-3\sqrt{25}\sqrt{5}$$
$$=(2\cdot4)\sqrt{5}-(3\cdot5)\sqrt{5}$$
$$=8\sqrt{5}-15\sqrt{5}$$
$$=(8-15)\sqrt{5}$$
$$=-7\sqrt{5}$$

81. $\sqrt{48}-\sqrt{8}+\sqrt{27}-\sqrt{32}$
$$=\sqrt{16\cdot3}-\sqrt{4\cdot2}+\sqrt{9\cdot3}-\sqrt{16\cdot2}$$
$$=\sqrt{16}\sqrt{3}-\sqrt{4}\sqrt{2}+\sqrt{9}\sqrt{3}-\sqrt{16}\sqrt{2}$$
$$=4\sqrt{3}-2\sqrt{2}+3\sqrt{3}-4\sqrt{2}$$
$$=(4+3)\sqrt{3}+(-2-4)\sqrt{2}$$
$$=7\sqrt{3}-6\sqrt{2}$$

83. $6\sqrt{40y}-2\sqrt{360z}=6\sqrt{4\cdot10y}-2\sqrt{36\cdot10z}$
$$=6\sqrt{4}\sqrt{10y}-2\sqrt{36}\sqrt{10z}$$
$$=(6\cdot2)\sqrt{10y}-(2\cdot6)\sqrt{10z}$$
$$=12\sqrt{10y}-12\sqrt{10z}$$

LOOK ALIKES . . .

85. a. $18\sqrt{n}+18\sqrt{n}=(18+18)\sqrt{n}$
$$=36\sqrt{n}$$

b. $18\sqrt{n}-18\sqrt{n}=(18-18)\sqrt{n}$
$$=0\sqrt{n}$$
$$=0$$

87. a. $\sqrt{20}+\sqrt{20}=\sqrt{4\cdot5}+\sqrt{4\cdot5}$
$$=\sqrt{4}\sqrt{5}+\sqrt{4}\sqrt{5}$$
$$=2\sqrt{5}+2\sqrt{5}$$
$$=(2+2)\sqrt{5}$$
$$=4\sqrt{5}$$

b. $\sqrt{21}+\sqrt{21}=(1+1)\sqrt{21}$
$$=2\sqrt{21}$$

APPLICATIONS

89. PLAYGROUND EQUIPMENT

$$4\sqrt{180}+2(3)+10 = 4\sqrt{36\cdot 5}+6+10$$
$$= 4\sqrt{36}\sqrt{5}+16$$
$$= 4(6)\sqrt{5}+16$$
$$= 24\sqrt{5}+16$$

The exact total of pipe is $(24\sqrt{5}+16)$ ft.
The approximate total is about 70 ft.

91. CAMPING

1 pole of the parent's tent:

Step 1: $l = 0.5s\sqrt{3}$
$= (0.5)(6)\sqrt{3}$
$= 3\sqrt{3}$

1 pole of the children's tent:

Step 2: $l = 0.5s\sqrt{3}$
$= (0.5)(4)\sqrt{3}$
$= 2\sqrt{3}$

Total of 4 poles: 2 adults and 2 children

Step 3: $3\sqrt{3}+3\sqrt{3}+2\sqrt{3}+2\sqrt{3}$
$= (3+3+2+2)\sqrt{3}$
$= 10\sqrt{3}$

The exact total is $10\sqrt{3}$ ft.
The approximate total is about 17.3 ft.

WRITING

93-97. Answers will vary.

REVIEW

Simplify each expression. Write each answer without using negative exponents.

99. $3^{-2} = \frac{1}{3^2}$
$= \frac{1}{(3)(3)}$
$= \frac{1}{9}$

101. $-3^2 = -(3)(3)$
$= -9$

CHALLENGE PROBLEMS

103. Fill in the blank:

$$2y\sqrt{175y} - \sqrt{\_\_\_\_\_\_} = 0$$
$$2y\sqrt{175y} - \boxed{\sqrt{4y^2}\sqrt{175y}} = 0$$
$$2y\sqrt{175y} - \boxed{\sqrt{4y^2\cdot 175y}} = 0$$
$$2y\sqrt{175y} - \boxed{\sqrt{4\cdot 175y^2y}} = 0$$
$$2y\sqrt{175y} - \boxed{\sqrt{700y^{2+1}}} = 0$$
$$2y\sqrt{175y} - \boxed{\sqrt{700y^3}} = 0$$

The missing blank is $700y^3$.

SECTION 8.4 STUDY SET

VOCABULARY

Fill in the blanks.

1. The **denominator** of $\frac{1}{\sqrt{3}}$ is an irrational number.

3. The **conjugate** of $1+\sqrt{2}$ is $1-\sqrt{2}$.

CONCEPTS

Fill in the blanks. All variables represent nonnegative numbers.

5. a. $\sqrt{a}\cdot\sqrt{b}=\sqrt{\boldsymbol{a\cdot b}}$ b. $\frac{\sqrt{a}}{\sqrt{b}}=\sqrt{\frac{\boldsymbol{a}}{\boldsymbol{b}}}$

c. $\left(\sqrt{a}\right)^2=\boldsymbol{a}$ d. $\sqrt{b}\cdot\sqrt{b}=\boldsymbol{b}$

7. Explain why each expression is not in simplified radical form.

a. $\sqrt{\frac{3}{4}}$ **The radicand is a fraction.**

b. $\frac{1}{\sqrt{10}}$ **There is a radical in the denominator.**

NOTATION

Complete each solution.

9. $7\sqrt{2}\cdot4\sqrt{3}=7\cdot\mathbf{4}\cdot\sqrt{2}\cdot\sqrt{\mathbf{3}}$
$=28\sqrt{\mathbf{6}}$

11. $\frac{5}{\sqrt{7}}=\frac{5}{\sqrt{7}}\cdot\frac{\sqrt{7}}{\sqrt{7}}$
$=\frac{5\sqrt{7}}{\mathbf{7}}$

GUIDED PRACTICE

In the following problems, all variables represent nonnegative real numbers. Multiply and simplify, if possible. See Example 1.

13. $\sqrt{3}\cdot\sqrt{5}=\sqrt{3\cdot5}$
$=\sqrt{15}$

15. $\sqrt{5}\cdot\sqrt{10}=\sqrt{5\cdot10}$
$=\sqrt{50}$
$=\sqrt{25\cdot2}$
$=\sqrt{25}\sqrt{2}$
$=5\sqrt{2}$

17. $\sqrt{5d}\cdot\sqrt{8d}=\sqrt{5\cdot8\cdot d\cdot d}$
$=\sqrt{40d^{1+1}}$
$=\sqrt{4d^2\cdot10}$
$=\sqrt{4d^2}\sqrt{10}$
$=2d\sqrt{10}$

19. $\sqrt{5x^3}\sqrt{x^5}=\sqrt{5\cdot x^3\cdot x^5}$
$=\sqrt{5x^{3+5}}$
$=\sqrt{5x^8}$
$=\sqrt{x^8}\sqrt{5}$
$=x^4\sqrt{5}$

Multiply and simplify, if possible. See Example 2.

21. $3\sqrt{5}\cdot5=(3\cdot5)\sqrt{5}$
$=15\sqrt{5}$

23. $10\sqrt{5}\cdot\sqrt{15}=10\sqrt{5}\cdot\sqrt{15}$
$=10\sqrt{5\cdot15}$
$=10\sqrt{75}$
$=10\sqrt{25\cdot3}$
$=10\sqrt{25}\sqrt{3}$
$=(10\cdot5)\sqrt{3}$
$=50\sqrt{3}$

25. $5\sqrt{3}\cdot2\sqrt{5}=5\cdot2\cdot\sqrt{3}\cdot\sqrt{5}$
$=10\sqrt{3\cdot5}$
$=10\sqrt{15}$

27. $2\sqrt{6}\left(3\sqrt{3}\right)=2\cdot3\cdot\sqrt{6}\cdot\sqrt{3}$
$=6\sqrt{6\cdot3}$
$=6\sqrt{18}$
$=6\sqrt{9\cdot2}$
$=6\sqrt{9}\sqrt{2}$
$=6\cdot3\cdot\sqrt{2}$
$=18\sqrt{2}$

Find each power. See Example 3.

29. $\left(\sqrt{6}\right)^2=6$

31. $\left(\sqrt{y}\right)^2=y$

33. $\left(\sqrt{2b+7}\right)^2=2b+7$

35. $\left(2\sqrt{3}\right)^2 = 2^2\left(\sqrt{3}\right)^2$
$= 4 \cdot 3$
$= 12$

Multiply and simplify, if possible. See Example 4.

37. $\sqrt{2}\left(\sqrt{2}+1\right) = \sqrt{\mathbf{2}}\sqrt{2} + \sqrt{\mathbf{2}} \cdot 1$
$= 2 + \sqrt{2}$

39. $3\sqrt{3}\left(\sqrt{27}-\sqrt{2}\right) = \mathbf{3\sqrt{3}}\sqrt{27} - \mathbf{3\sqrt{3}}\sqrt{2}$
$= 3\sqrt{3 \cdot 27} - 3\sqrt{3 \cdot 2}$
$= 3\sqrt{81} - 3\sqrt{6}$
$= 3(9) - 3\sqrt{6}$
$= 27 - 3\sqrt{6}$

41. $\sqrt{x}\left(\sqrt{3x}-2\right) = \sqrt{\boldsymbol{x}}\sqrt{3x} - \sqrt{\boldsymbol{x}} \cdot 2$
$= \sqrt{x \cdot x \cdot 3} - 2\sqrt{x}$
$= \sqrt{x \cdot x}\sqrt{3} - 2\sqrt{x}$
$= x\sqrt{3} - 2\sqrt{x}$

43. $\sqrt{m}\left(\sqrt{8m}+9\right) = \sqrt{\boldsymbol{m}}\sqrt{8m} + \sqrt{\boldsymbol{m}} \cdot 9$
$= \sqrt{4 \cdot 2 \cdot m \cdot m} + 9\sqrt{m}$
$= \sqrt{4 \cdot m \cdot m}\sqrt{2} + 9\sqrt{m}$
$= 2m\sqrt{2} + 9\sqrt{m}$

Multiply and simplify, if possible. See Example 5.

45. $\left(\sqrt{2}+1\right)\left(\sqrt{2}-1\right) = \sqrt{2} \cdot \sqrt{2} - \sqrt{2} + \sqrt{2} - 1$
$= 2 + (-\sqrt{2} + \sqrt{2}) - 1$
$= 2 - 1$
$= 1$

47. $\left(\sqrt{2}-\sqrt{3}\right)\left(\sqrt{3}+\sqrt{5}\right)$
$= \sqrt{\mathbf{2}}\sqrt{3} + \sqrt{\mathbf{2}}\sqrt{5} - \sqrt{\mathbf{3}}\sqrt{3} - \sqrt{\mathbf{3}}\sqrt{5}$
$= \sqrt{2 \cdot 3} + \sqrt{2 \cdot 5} - \sqrt{3 \cdot 3} - \sqrt{3 \cdot 5}$
$= \sqrt{6} + \sqrt{10} - 3 - \sqrt{15}$

49. $\left(\sqrt{2x}+3\right)\left(\sqrt{8x}-6\right)$
$= \sqrt{\boldsymbol{2x}} \cdot \sqrt{8x} - \sqrt{\boldsymbol{2x}} \cdot 6 + \mathbf{3}\sqrt{8x} - \mathbf{3} \cdot 6$
$= \sqrt{2x \cdot 8x} - 6\sqrt{2x} + 3\sqrt{8x} - 18$
$= \sqrt{16 \cdot x \cdot x} - 6\sqrt{2x} + 3\sqrt{4 \cdot 2x} - 18$
$= \sqrt{16}\sqrt{x \cdot x} - 6\sqrt{2x} + 3\sqrt{4}\sqrt{2x} - 18$
$= 4x - 6\sqrt{2x} + 3(2)\sqrt{2x} - 18$
$= 4x - 6\sqrt{2x} + 6\sqrt{2x} - 18$
$= 4x + (-6+6)\sqrt{2x} - 18$
$= 4x + 0\sqrt{2x} - 18$
$= 4x - 18$

51. $\left(2\sqrt{7}-x\right)\left(3\sqrt{2}+x\right)$
$= \mathbf{2\sqrt{7}} \cdot 3\sqrt{2} + \mathbf{2\sqrt{7}} \cdot (x) + (\boldsymbol{-x}) \cdot 3\sqrt{2} - \boldsymbol{x} \cdot x$
$= 6\sqrt{7 \cdot 2} + 2x\sqrt{7} - 3x\sqrt{2} - x^2$
$= 6\sqrt{14} + 2x\sqrt{7} - 3x\sqrt{2} - x^2$

Find each product and simplify, if possible. See Example 6.

53. $\left(5+\sqrt{3}\right)^2 = 5^2 + 2 \cdot 5\sqrt{3} + \left(\sqrt{3}\right)^2$
$= 25 + 10\sqrt{3} + 3$
$= 28 + 10\sqrt{3}$

55. $\left(a+\sqrt{7}\right)^2 = a^2 + 2 \cdot a\sqrt{7} + \left(\sqrt{7}\right)^2$
$= a^2 + 2a\sqrt{7} + 7$

57. $\left(\sqrt{5}-\sqrt{m}\right)^2 = \left(\sqrt{5}\right)^2 + 2\left(-\sqrt{5}\right)\sqrt{m} + \left(-\sqrt{m}\right)^2$
$= 5 - 2\sqrt{5m} + m$

59. $\left(\sqrt{6}+\sqrt{3}\right)\left(\sqrt{6}-\sqrt{3}\right) = \left(\sqrt{6}\right)^2 - \left(\sqrt{3}\right)^2$
$= 6 - 3$
$= 3$

61. $\left(\sqrt{11}-y\right)\left(\sqrt{11}+y\right) = \left(\sqrt{11}\right)^2 - (y)^2$
$= 11 - y^2$

63. $\left(\sqrt{7c}-3\right)\left(\sqrt{7c}+3\right) = \left(\sqrt{7c}\right)^2 - (3)^2$
$= 7c - 9$

Simplify. See Example 7.

65. $\dfrac{\sqrt{60}}{\sqrt{6}} = \sqrt{\dfrac{60}{6}}$
$= \sqrt{10}$

67. $\frac{\sqrt{a^7}}{\sqrt{a^3}}=\sqrt{\frac{a^7}{a^3}}$
$=\sqrt{a^{7-3}}$
$=\sqrt{a^4}$
$=a^2$

69. $\frac{\sqrt{18x}}{\sqrt{25x}}=\frac{\sqrt{9\cdot 2x}}{\sqrt{25\cdot x}}$
$=\frac{\sqrt{9}\sqrt{2x}}{\sqrt{25}\sqrt{x}}$
$=\frac{3\sqrt{2x}}{5\sqrt{x}}$
$=\frac{3}{5}\sqrt{\frac{2\cancel{x}}{\cancel{x}}}$
$=\frac{3}{5}\sqrt{2}$
$=\frac{3\sqrt{2}}{5}$

71. $\frac{\sqrt{27x}}{\sqrt{75x^3}}=\frac{\sqrt{9\cdot 3x}}{\sqrt{25\cdot x^2\cdot 3x}}$
$=\frac{\sqrt{9}\sqrt{3x}}{\sqrt{25x^2}\sqrt{3x}}$
$=\frac{3\sqrt{3x}}{5x\sqrt{3x}}$
$=\frac{3}{5x}\sqrt{\frac{\cancel{3}\cancel{x}}{\cancel{3}\cancel{x}}}$
$=\frac{3}{5x}$

Rationalize each denominator and simplify. See Example 8.

73. $\frac{1}{\sqrt{3}}=\frac{1}{\sqrt{3}}\cdot\frac{\sqrt{3}}{\sqrt{3}}$
$=\frac{\sqrt{3}}{3}$

75. $\frac{4}{\sqrt{19}}=\frac{4}{\sqrt{19}}\cdot\frac{\sqrt{19}}{\sqrt{19}}$
$=\frac{4\sqrt{19}}{19}$

Rationalize each denominator and simplify. See Example 9.

77. $\sqrt{\frac{12}{5}}=\frac{\sqrt{12}}{\sqrt{5}}$
$=\frac{\sqrt{4\cdot 3}}{\sqrt{5}}$
$=\frac{2\sqrt{3}}{\sqrt{5}}\cdot\frac{\sqrt{5}}{\sqrt{5}}$
$=\frac{2\sqrt{15}}{5}$

79. $\frac{6}{\sqrt{27}}=\frac{6}{\sqrt{9\cdot 3}}$
$=\frac{6}{3\sqrt{3}}\cdot\frac{\sqrt{3}}{\sqrt{3}}$
$=\frac{6\sqrt{3}}{3\cdot 3}$
$=\frac{2\cdot\cancel{3}\sqrt{3}}{3\cdot\cancel{3}}$
$=\frac{2\sqrt{3}}{3}$

Rationalize each denominator and simplify. See Example 10.

81. $\frac{3}{\sqrt{3}-1}=\frac{3}{\sqrt{3}-1}\cdot\frac{\sqrt{3}+1}{\sqrt{3}+1}$
$=\frac{3\left(\sqrt{3}+1\right)}{\left(\sqrt{3}-1\right)\left(\sqrt{3}+1\right)}$
$=\frac{3\sqrt{3}+3}{\left(\sqrt{3}\right)^2-1^2}$
$=\frac{3\sqrt{3}+3}{3-1}$
$=\frac{3\sqrt{3}+3}{2}$

83. $\frac{5}{\sqrt{3}+\sqrt{2}}=\frac{5}{\sqrt{3}+\sqrt{2}}\cdot\frac{\sqrt{3}-\sqrt{2}}{\sqrt{3}-\sqrt{2}}$
$=\frac{5\left(\sqrt{3}-\sqrt{2}\right)}{\left(\sqrt{3}+\sqrt{2}\right)\left(\sqrt{3}-\sqrt{2}\right)}$
$=\frac{5\sqrt{3}-5\sqrt{2}}{\left(\sqrt{3}\right)^2-\left(\sqrt{2}\right)^2}$
$=\frac{5\sqrt{3}-5\sqrt{2}}{3-2}$
$=\frac{5\sqrt{3}-5\sqrt{2}}{1}$
$=5\sqrt{3}-5\sqrt{2}$

85. $\frac{\sqrt{5}+\sqrt{7}}{\sqrt{2}-\sqrt{5}}=\frac{\sqrt{5}+\sqrt{7}}{\sqrt{2}-\sqrt{5}}\cdot\frac{\sqrt{2}+\sqrt{5}}{\sqrt{2}+\sqrt{5}}$
$=\frac{\left(\sqrt{5}+\sqrt{7}\right)\left(\sqrt{2}+\sqrt{5}\right)}{\left(\sqrt{2}-\sqrt{5}\right)\left(\sqrt{2}+\sqrt{5}\right)}$
$=\frac{\sqrt{10}+\sqrt{25}+\sqrt{14}+\sqrt{35}}{\left(\sqrt{2}\right)^2-\left(\sqrt{5}\right)^2}$
$=\frac{\sqrt{10}+5+\sqrt{14}+\sqrt{35}}{2-5}$
$=\frac{\sqrt{10}+5+\sqrt{14}+\sqrt{35}}{-3}$
$=-\frac{\sqrt{10}+5+\sqrt{14}+\sqrt{35}}{3}$

87. $\frac{\sqrt{3}-\sqrt{a}}{\sqrt{5}+\sqrt{a}}=\frac{\sqrt{3}-\sqrt{a}}{\sqrt{5}+\sqrt{a}}\cdot\frac{\sqrt{5}-\sqrt{a}}{\sqrt{5}-\sqrt{a}}$
$=\frac{(\sqrt{3}-\sqrt{a})(\sqrt{5}-\sqrt{a})}{(\sqrt{5}+\sqrt{a})(\sqrt{5}-\sqrt{a})}$
$=\frac{\sqrt{15}-\sqrt{3a}-\sqrt{5a}+\sqrt{a\cdot a}}{(\sqrt{5})^2-(\sqrt{a})^2}$
$=\frac{\sqrt{15}-\sqrt{3a}-\sqrt{5a}+a}{5-a}$

TRY IT YOURSELF

Perform the operations and simplify, if possible or rationalize the denominator.

89. $\frac{3}{\sqrt{32}}=\frac{3}{\sqrt{16\cdot 2}}$
$=\frac{3}{4\sqrt{2}}$
$=\frac{3}{4\sqrt{2}}\cdot\frac{\sqrt{2}}{\sqrt{2}}$
$=\frac{3\sqrt{2}}{4\cdot 2}$
$=\frac{3\sqrt{2}}{8}$

91. $\sqrt{x}\left(\sqrt{14x}+\sqrt{2}\right)=\sqrt{x}\sqrt{14x}+\sqrt{x}\cdot\sqrt{2}$
$=\sqrt{x\cdot x\cdot 14}+\sqrt{2x}$
$=\sqrt{x\cdot x}\sqrt{14}+\sqrt{2x}$
$=x\sqrt{14}+\sqrt{2x}$

93. $\frac{\sqrt{3}}{5+\sqrt{x}}=\frac{\sqrt{3}}{5+\sqrt{x}}\cdot\frac{5-\sqrt{x}}{5-\sqrt{x}}$
$=\frac{\sqrt{3}(5-\sqrt{x})}{(5+\sqrt{x})(5-\sqrt{x})}$
$=\frac{5\sqrt{3}-\sqrt{3x}}{5^2-(\sqrt{x})^2}$
$=\frac{5\sqrt{3}-\sqrt{3x}}{25-x}$

95. $(10\sqrt{x})^2=10^2(\sqrt{x})^2$
$=100x$

97. $\sqrt{\frac{13}{7}}=\frac{\sqrt{13}}{\sqrt{7}}$
$=\frac{\sqrt{13}}{\sqrt{7}}\cdot\frac{\sqrt{7}}{\sqrt{7}}$
$=\frac{\sqrt{91}}{7}$

99. $(3p+\sqrt{5})^2=(3p)^2+2\cdot 3p\sqrt{5}+(\sqrt{5})^2$
$=9p^2+6p\sqrt{5}+5$

101. $\frac{8}{\sqrt{7}+2}=\frac{8}{\sqrt{7}+2}\cdot\frac{\sqrt{7}-2}{\sqrt{7}-2}$
$=\frac{8(\sqrt{7}-2)}{(\sqrt{7}+2)(\sqrt{7}-2)}$
$=\frac{8\sqrt{7}-16}{(\sqrt{7})^2-2^2}$
$=\frac{8\sqrt{7}-16}{7-4}$
$=\frac{8\sqrt{7}-16}{3}$

103. $\sqrt{6}\sqrt{11}=\sqrt{6\cdot 11}$
$=\sqrt{66}$

105. $\frac{\sqrt{12x^3}}{\sqrt{27x}}=\frac{\sqrt{4x^2\cdot 3x}}{\sqrt{9\cdot 3x}}$
$=\frac{\sqrt{4}\sqrt{x^2}\sqrt{3x}}{\sqrt{9}\sqrt{3x}}$
$=\frac{2x\sqrt{3x}}{3\sqrt{3x}}$
$=\frac{2x}{3}\sqrt{\frac{3x}{3x}}$
$=\frac{2x}{3}\cdot 1$
$=\frac{2x}{3}$

107. $\left(\sqrt{15}+\sqrt{2}\right)\left(\sqrt{15}-\sqrt{2}\right)=\left(\sqrt{15}\right)^2-\left(\sqrt{2}\right)^2$
$=15-2$
$=13$

109. $\dfrac{\sqrt{9}}{\sqrt{2x}}=\dfrac{3}{\sqrt{2x}}$
$=\dfrac{3}{\sqrt{2x}}\cdot\dfrac{\sqrt{\mathbf{2x}}}{\sqrt{\mathbf{2x}}}$
$=\dfrac{3\sqrt{2x}}{2x}$

111. $7\sqrt{3}\left(2\sqrt{5}\right)=7\cdot 2\sqrt{3\cdot 5}$
$=14\sqrt{15}$

113. $\dfrac{\sqrt{3}}{7+\sqrt{3}}=\dfrac{\sqrt{3}}{7+\sqrt{3}}\cdot\dfrac{\mathbf{7-\sqrt{3}}}{\mathbf{7-\sqrt{3}}}$
$=\dfrac{\sqrt{3}\left(7-\sqrt{3}\right)}{\left(7+\sqrt{3}\right)\left(7-\sqrt{3}\right)}$
$=\dfrac{7\sqrt{3}-3}{7^2-\left(\sqrt{3}\right)^2}$
$=\dfrac{7\sqrt{3}-3}{49-3}$
$=\dfrac{7\sqrt{3}-3}{46}$

115. $\sqrt{6n^4}\cdot\sqrt{10n^5}=\sqrt{6n^4\cdot 10n^5}$
$=\sqrt{60n^9}$
$=\sqrt{4\cdot n^8\cdot 15n}$
$=\sqrt{4}\sqrt{n^8}\sqrt{15n}$
$=2n^4\sqrt{15n}$

LOOK ALIKES . . .

117. a. $3\sqrt{5}\cdot 4\sqrt{5}=3\cdot 4\cdot\sqrt{5}\cdot\sqrt{5}$
$=12\cdot 5$
$=60$

b. $3\sqrt{5}+4\sqrt{5}=(3+4)\sqrt{5}$
$=7\sqrt{5}$

119. a. $\left(2\sqrt{x}+2\right)\left(\sqrt{x}-6\right)$
$=\mathbf{2\sqrt{x}}\cdot\sqrt{x}+\mathbf{2\sqrt{x}}\cdot(-6)+\mathbf{2}\sqrt{x}+\mathbf{2}\cdot(-6)$
$=2\sqrt{x\cdot x}-12\sqrt{x}+2\sqrt{x}-12$
$=2x-12\sqrt{x}+2\sqrt{x}-12$
$=2x+(-12+2)\sqrt{x}-12$
$=2x-10\sqrt{x}-12$

b. $2\sqrt{x}+2+\sqrt{x}-6=(2+1)\sqrt{x}+(2-6)$
$=3\sqrt{x}-4$

APPLICATIONS
from CAMPUS TO CAREERS

121. CRIME SCENE INVESTIGATOR

Length $=\sqrt{8^2+2^2}$
$=\sqrt{64+4}$
$=\sqrt{68}$
$=\sqrt{4\cdot 17}$
$=2\sqrt{17}$

The length is $2\sqrt{17}$ ft.

Width $=\sqrt{4^2+1^2}$
$=\sqrt{16+1}$
$=\sqrt{17}$

The width is $\sqrt{17}$ ft.

Area $=lw$
$=2\sqrt{17}\cdot\sqrt{17}$
$=2\cdot 17$
$=34$

The area is 34 ft^2.

123. TUNING GUITARS

$f=0.772\sqrt{\dfrac{T}{u}}$
$=0.772\dfrac{\sqrt{T}}{\sqrt{u}}\cdot\dfrac{\sqrt{\mathbf{u}}}{\sqrt{\mathbf{u}}}$
$=0.772\dfrac{\sqrt{Tu}}{u}$

WRITING

125-127. Answers will vary.

REVIEW

129. HOT DOG EATING CHAMPION

Analyze

- Joey can eat 54 hot dogs and buns in 10 minutes.
- How many can he eat in 30 seconds?

Assign

Let $x = \#$ of hot dogs eaten in 30 seconds

Form

54 eaten in 600 seconds as x number eaten in 30 seconds.

$$\begin{array}{r} \text{54 eaten} \rightarrow \\ \text{600 seconds} \rightarrow \end{array} \frac{54}{600} = \frac{x}{30} \begin{array}{l} \leftarrow x \text{ eaten} \\ \leftarrow \text{30 seconds} \end{array}$$

Solve

$$\frac{54}{600} = \frac{x}{30}$$
$$54 \cdot 30 = 600x$$
$$1,620 = 600x$$
$$\frac{1,620}{\mathbf{600}} = \frac{600x}{\mathbf{600}}$$
$$x = 2.7$$

State

2.7 hot dogs and buns are eaten in 30 seconds.

Check

$$\frac{54}{600} = \frac{x}{30}$$
$$\frac{54}{600} \stackrel{?}{=} \frac{\mathbf{2.7}}{30}$$
$$1,620 = 1,620$$

The result checks.

CHALLENGE PROBLEMS

131. **Rationalize the denominator of the reciprocal.**

$$\frac{\sqrt{y} - 2\sqrt{x}}{2\sqrt{x} + \sqrt{y}}$$

Reciprocal

$$\frac{2\sqrt{x} + \sqrt{y}}{\sqrt{y} - 2\sqrt{x}}$$

$$\begin{aligned}
\frac{2\sqrt{x} + \sqrt{y}}{\sqrt{y} - 2\sqrt{x}} &= \frac{2\sqrt{x} + \sqrt{y}}{\sqrt{y} - 2\sqrt{x}} \cdot \frac{\boldsymbol{\sqrt{y} + 2\sqrt{x}}}{\boldsymbol{\sqrt{y} + 2\sqrt{x}}} \\
&= \frac{\left(2\sqrt{x} + \sqrt{y}\right)\left(\sqrt{y} + 2\sqrt{x}\right)}{\left(\sqrt{y}\right)^2 - \left(2\sqrt{x}\right)^2} \\
&= \frac{2\sqrt{xy} + 4x + y + 2\sqrt{xy}}{y - 4x} \\
&= \frac{4\sqrt{xy} + 4x + y}{y - 4x}
\end{aligned}$$

SECTION 8.5 STUDY SET

VOCABULARY

Fill in the blanks.

1. $\sqrt{x+1}=3$ and $\sqrt{x}-2=10$ are examples of **radical** equations.

3. To **solve** a radical equation, we find all the values of the variable that make the equation true.

CONCEPTS

Fill in the blanks.

5. The squaring property of equality states that if two numbers are equal, their **squares** are equal. If $a = b$, then $a^2 = \boldsymbol{b^2}$.

7. Use properties of algebra to isolate each radical term on one side of the equation. **Do not solve the equation.**

a. $\sqrt{x-4}-1=2$

$\sqrt{x-4}-1+\mathbf{1}=2+\mathbf{1}$

$\sqrt{x-4}=3$

b. $8=\sqrt{x}-x$

$8+\boldsymbol{x}=\sqrt{x}-x+\boldsymbol{x}$

$\boldsymbol{x}+8=\sqrt{x}$

NOTATION

Complete each solution.

9. Solve: $\sqrt{x-3}=5$

$\left(\sqrt{x-3}\right)^{\boxed{2}}=5^{\boxed{2}}$

$\boldsymbol{x}-3=25$

$x=\mathbf{28}$

GUIDED PRACTICE

Solve each equation. See example 1.

11. $\sqrt{x}=3$

$\left(\sqrt{x}\right)^2=(3)^2$

$x=9$

Check: $\sqrt{x}=3$

$\sqrt{9}\overset{?}{=}3$

$3=3$

True

13. $\sqrt{y}=12$

$\left(\sqrt{y}\right)^2=(12)^2$

$y=144$

Check: $\sqrt{y}=12$

$\sqrt{144}\overset{?}{=}12$

$12=12$

True

15. $\sqrt{2a}=4$

$\left(\sqrt{2a}\right)^2=(4)^2$

$2a=16$

$\frac{2a}{\mathbf{2}}=\frac{16}{\mathbf{2}}$

$a=8$

Check: $\sqrt{2a}=4$

$\sqrt{2(8)}\overset{?}{=}4$

$\sqrt{16}\overset{?}{=}4$

$4=4$

True

17. $\sqrt{4n}=6$

$\left(\sqrt{4n}\right)^2=(6)^2$

$4n=36$

$\frac{4n}{\mathbf{4}}=\frac{36}{\mathbf{4}}$

$n=9$

Check: $\sqrt{4n}=6$

$\sqrt{4(9)}\overset{?}{=}6$

$\sqrt{36}\overset{?}{=}6$

$6=6$

True

Solve each equation. See Example 2.

19. $\sqrt{x+3}=2$

$\left(\sqrt{x+3}\right)^2=(2)^2$

$x+3=4$

$x+3-\mathbf{3}=4-\mathbf{3}$

$x=1$

Check: $\sqrt{x+3}=2$

$\sqrt{1+3}\overset{?}{=}2$

$\sqrt{4}\overset{?}{=}2$

$2=2$

True

21. $\sqrt{5-T}=10$

$\left(\sqrt{5-T}\right)^2=(10)^2$

$5-T=100$

$5-T-\mathbf{5}=100-\mathbf{5}$

$-T=95$

$\frac{-T}{\mathbf{-1}}=\frac{95}{\mathbf{-1}}$

$T=-95$

Check: $\sqrt{5-T}=10$

$\sqrt{5-(-95)}\overset{?}{=}10$

$\sqrt{100}\overset{?}{=}10$

$10=10$

True

23. $\sqrt{6x+19}=7$

$\left(\sqrt{6x+19}\right)^2=(7)^2$

$6x+19=49$

$6x+19-\mathbf{19}=49-\mathbf{19}$

$6x=30$

$\frac{6x}{\mathbf{6}}=\frac{30}{\mathbf{6}}$

$x=5$

Check: $\sqrt{6x+19}=7$

$\sqrt{6(5)+19}\overset{?}{=}7$

$\sqrt{30+19}\overset{?}{=}7$

$\sqrt{49}\overset{?}{=}7$

$7=7$

True

25. $\sqrt{5x-5}=5$

$\left(\sqrt{5x-5}\right)^2=(5)^2$

$5x-5=25$

$5x-5+\mathbf{5}=25+\mathbf{5}$

$5x=30$

$\frac{5x}{\mathbf{5}}=\frac{30}{\mathbf{5}}$

$x=6$

Check: $\sqrt{5x-5}=5$

$\sqrt{5(6)-5}\overset{?}{=}5$

$\sqrt{30-5}\overset{?}{=}5$

$\sqrt{35}\overset{?}{=}5$

$5=5$

True

Solve each equation. See Example 3.

27. $x=\sqrt{x^2-2x+16}$

$(x)^2=\left(\sqrt{x^2-2x+16}\right)^2$

$x^2=x^2-2x+16$

$x^2-\boldsymbol{x^2}=x^2-2x+16-\boldsymbol{x^2}$

$0=-2x+16$

$0-\mathbf{16}=-2x+16-\mathbf{16}$

$-16=-2x$

$\frac{-16}{\mathbf{-2}}=\frac{-2x}{\mathbf{-2}}$

$8=x$

Check:

$x=\sqrt{x^2-2x+16}$

$8\overset{?}{=}\sqrt{(8)^2-2(8)+16}$

$8\overset{?}{=}\sqrt{64-16+16}$

$8\overset{?}{=}\sqrt{64}$

$8=8$

True

29. $c=\sqrt{c^2-3c+39}$

$(c)^2=\left(\sqrt{c^2-3c+39}\right)^2$

$c^2=c^2-3c+39$

$c^2-\boldsymbol{c^2}=c^2-3c+39-\boldsymbol{c^2}$

$0=-3c+39$

$0+\mathbf{3c}=-3c+39+\mathbf{3c}$

$3c=39$

$\frac{3c}{\mathbf{3}}=\frac{39}{\mathbf{3}}$

$c=13$

Check:

$c=\sqrt{c^2-3c+39}$

$13\overset{?}{=}\sqrt{(13)^2-3(13)+39}$

$13\overset{?}{=}\sqrt{169-39+39}$

$13\overset{?}{=}\sqrt{169}$

$13=13$

True

31. $\sqrt{9q^2-5q+10}=3q$

$\sqrt{9q^2-5q+10}^{\,2}=(3q)^2$

$9q^2-5q+10=9q^2$

$9q^2-5q+10-\mathbf{9q^2}=9q^2-\mathbf{9q^2}$

$-5q+10=0$

$-5q+10-\mathbf{10}=0-\mathbf{10}$

$-5q=-10$

$\frac{-5q}{\mathbf{-5}}=\frac{-10}{\mathbf{-5}}$

$q=2$

Check:

$\sqrt{9q^2-5q+10}=3q$

$\sqrt{9(2)^2-5(2)+10}\overset{?}{=}3(2)$

$\sqrt{9(4)-10+10}\overset{?}{=}6$

$\sqrt{36}\overset{?}{=}6$

$6=6$

True

33. $\sqrt{4m^2+6m+6}=-2m$

$\sqrt{4m^2+6m+6}^{\,2}=(-2m)^2$

$4m^2+6m+6=4m^2$

$4m^2+6m+6-\mathbf{4m^2}=4m^2-\mathbf{4m^2}$

$6m+6=0$

$6m+6-\mathbf{6}=0-\mathbf{6}$

$6m=-6$

$\frac{6m}{\mathbf{6}}=\frac{-6}{\mathbf{6}}$

$m=-1$

Check:

$\sqrt{4m^2+6m+6}=-2m$

$\sqrt{4(-1)^2+6(-1)+6}\overset{?}{=}-2(-1)$

$\sqrt{4-6+6}\overset{?}{=}2$

$\sqrt{4}\overset{?}{=}2$

$2=2$

True

Solve each equation. See Example 4.

35. $\sqrt{x}=-6$

$\sqrt{x}^{\,2}=(-6)^2$

$x=36$

Check: $\sqrt{x}=-6$

$\sqrt{36}\overset{?}{=}-6$

$6=-6$

False

Since the result is a false statement, 36 is not a solution of the original equation and must be discarded. The original equation has no solution. The solution set is the empty set, written as $\varnothing$.

37. $\sqrt{r}+4=0$

$\sqrt{r}+4-\mathbf{4}=0-\mathbf{4}$

$\sqrt{r}=-4$

$\sqrt{r}^{\,2}=(-4)^2$

$r=16$

Check:

$\sqrt{r}+4=0$

$\sqrt{16}+4\overset{?}{=}0$

$4+4=0$

False

Since the result is a false statement, 16 is not a solution of the original equation and must be discarded. The original equation has no solution. The solution set is the empty set, written as $\varnothing$.

39. $\sqrt{2x+7}+4=1$

$\sqrt{2x+7}+4-\mathbf{4}=1-\mathbf{4}$

$\sqrt{2x+7}=-3$

$\sqrt{2x+7}^{\,2}=(-3)^2$

$2x+7=9$

$2x+7-\mathbf{7}=9-\mathbf{7}$

$2x=2$

$\frac{2x}{\mathbf{2}}=\frac{2}{\mathbf{2}}$

$x=1$

Check:

$\sqrt{2x+7}+4=1$

$\sqrt{2(1)+7}+4\overset{?}{=}1$

$\sqrt{9}+4\overset{?}{=}1$

$3+4\overset{?}{=}1$

$7=1$

False

Since the result is a false statement, 1 is not a solution of the original equation and must be discarded. The original equation has no solution. The solution set is the empty set, written as $\varnothing$.

41. $\sqrt{6-2b}-7=-9$

$\sqrt{6-2b}-7+\mathbf{7}=-9+\mathbf{7}$

$\sqrt{6-2b}=-2$

$\sqrt{6-2b}^{\,2}=(-2)^2$

$6-2b=4$

$6-2b-\mathbf{6}=4-\mathbf{6}$

$-2b=-2$

$\frac{-2b}{\mathbf{-2}}=\frac{-2}{\mathbf{-2}}$

$b=1$

Check:

$\sqrt{6-2b}-7=-9$

$\sqrt{6-2(1)}-7\overset{?}{=}-9$

$\sqrt{4}-7\overset{?}{=}-9$

$2-7\overset{?}{=}-9$

$-5=-9$

False

Since the result is a false statement, 1 is not a solution of the original equation and must be discarded. The original equation has no solution. The solution set is the empty set, written as $\varnothing$.

Solve each equation. See Example 5.

43.
$$\sqrt{3a+7}-a=3$$
$$\sqrt{3a+7}-a+\mathbf{a}=\mathbf{a}+3$$
$$\sqrt{3a+7}=a+3$$
$$\left(\sqrt{3a+7}\right)^2=(a+3)^2$$
$$3a+7=a^2+2\cdot 3a+3^2$$
$$3a+7=a^2+6a+9$$
$$a^2+6a+9=3a+7$$
$$a^2+6a+9-\mathbf{3a}=3a+7-\mathbf{3a}$$
$$a^2+3a+9=7$$
$$a^2+3a+9-\mathbf{7}=7-\mathbf{7}$$
$$a^2+3a+2=0$$
$$(a+1)(a+2)=0$$

Set each factor equal to zero and solve.

$a+1=0$	$a+2=0$
$a+1-\mathbf{1}=0-\mathbf{1}$	$a+2-\mathbf{2}=0-\mathbf{2}$
$a=-1$	$a=-2$

Check both solutions.

$\sqrt{3a+7}-a=3$	$\sqrt{3a+7}-a=3$
$\sqrt{3(-1)+7}-(-1)\stackrel{?}{=}3$	$\sqrt{3(-2)+7}-(-2)\stackrel{?}{=}3$
$\sqrt{-3+7}+1\stackrel{?}{=}3$	$\sqrt{-6+7}+2\stackrel{?}{=}3$
$\sqrt{4}+1\stackrel{?}{=}3$	$\sqrt{1}+2\stackrel{?}{=}3$
$2+1\stackrel{?}{=}3$	$1+2\stackrel{?}{=}3$
$3=3$	$3=3$
True	True

Since two true statements result when -1 and -2 are substituted for a, -1 and -2 are solutions. The solution set is $\{-1,-2\}$.

45.
$$b=\sqrt{5b+1}-1$$
$$b+\mathbf{1}=\sqrt{5b+1}-1+\mathbf{1}$$
$$b+1=\sqrt{5b+1}$$
$$(b+1)^2=\left(\sqrt{5b+1}\right)^2$$
$$b^2+2\cdot 1b+1^2=5b+1$$
$$b^2+2b+1=5b+1$$
$$b^2+2b+1-\mathbf{5b}=5b+1-\mathbf{5b}$$
$$b^2-3b+1=1$$
$$b^2-3b+1-\mathbf{1}=1-\mathbf{1}$$
$$b^2-3b=0$$
$$b(b-3)=0$$

Set each factor equal to zero and solve.

$b=0$	$b-3=0$
	$b-3+\mathbf{3}=0+\mathbf{3}$
	$b=3$

Check both solutions.

$b=\sqrt{5b+1}-1$	$b=\sqrt{5b+1}-1$
$0\stackrel{?}{=}\sqrt{5(0)+1}-1$	$3\stackrel{?}{=}\sqrt{5(3)+1}-1$
$0\stackrel{?}{=}\sqrt{0+1}-1$	$3\stackrel{?}{=}\sqrt{15+1}-1$
$0\stackrel{?}{=}\sqrt{1}-1$	$3\stackrel{?}{=}\sqrt{16}-1$
$0\stackrel{?}{=}1-1$	$3\stackrel{?}{=}4-1$
$0=0$	$3=3$
True	True

Since two true statements result when 0 and 3 are substituted for b, 0 and 3 are solutions. The solution set is $\{0,3\}$.

47.
$$y=9+\sqrt{y-3}$$
$$y-\mathbf{9}=9+\sqrt{y-3}-\mathbf{9}$$
$$y-9=\sqrt{y-3}$$
$$(y-9)^2=\left(\sqrt{y-3}\right)^2$$
$$y^2+2\cdot(-9)y+(-9)^2=y-3$$
$$y^2-18y+81=y-3$$
$$y^2-18y+81-\mathbf{y}=y-3-\mathbf{y}$$
$$y^2-19y+81=-3$$
$$y^2-19y+81+\mathbf{3}=-3+\mathbf{3}$$
$$y^2-19y+84=0$$
$$(y-7)(y-12)=0$$

Set each factor equal to zero and solve.

$y-7=0$	$y-12=0$
$y-7+\mathbf{7}=0+\mathbf{7}$	$y-12+\mathbf{12}=0+\mathbf{12}$
$y=7$	$y=12$

Check both solutions.

$y=9+\sqrt{y-3}$	$y=9+\sqrt{y-3}$
$7\stackrel{?}{=}9+\sqrt{7-3}$	$12\stackrel{?}{=}9+\sqrt{12-3}$
$7\stackrel{?}{=}9+\sqrt{4}$	$12\stackrel{?}{=}9+\sqrt{9}$
$7\stackrel{?}{=}9+2$	$12\stackrel{?}{=}9+3$
$7=11$	$12=12$
False	True

Since a true statement results when 12 is substituted for y, 12 is a solution. Since a false statement results when 7 is substituted for y, 7 is an extraneous solution. The solution set is $\{12\}$.

49. $$\sqrt{15-3t}+5=t$$
$$\sqrt{15-3t}+5-\mathbf{5}=t-\mathbf{5}$$
$$\sqrt{15-3t}=t-5$$
$$\left(\sqrt{15-3t}\right)^2=(t-5)^2$$
$$15-3t=t^2+2\cdot(-5)t+(-5)^2$$
$$15-3t=t^2-10t+25$$
$$t^2-10t+25=15-3t$$
$$t^2-10t+25+\mathbf{3t}=15-3t+\mathbf{3t}$$
$$t^2-7t+25=15$$
$$t^2-7t+25-\mathbf{15}=15-\mathbf{15}$$
$$t^2-7t+10=0$$
$$(t-2)(t-5)=0$$

Set each factor equal to zero and solve.

$t-2=0$	$t-5=0$
$t-2+\mathbf{2}=0+\mathbf{2}$	$t-5+\mathbf{5}=0+\mathbf{5}$
$t=2$	$t=5$

Check both solutions.

$\sqrt{15-3t}+5=t$	$\sqrt{15-3t}+5=t$
$\sqrt{15-3(2)}+5\stackrel{?}{=}2$	$\sqrt{15-3(5)}+5\stackrel{?}{=}5$
$\sqrt{15-6}+5\stackrel{?}{=}2$	$\sqrt{15-15}+5\stackrel{?}{=}5$
$\sqrt{9}+5\stackrel{?}{=}2$	$\sqrt{0}+5\stackrel{?}{=}5$
$3+5\stackrel{?}{=}2$	$0+5\stackrel{?}{=}5$
$8=2$	$5=5$
False	True

Since a true statement results when 5 is substituted for t, 5 is a solution. Since a false statement results when 2 is substituted for t, 2 is an extraneous solution. The solution set is $\{5\}$.

Solve each equation. See Example 6.

51. $$\sqrt{10-3x}=\sqrt{2x+20}$$
$$\left(\sqrt{10-3x}\right)^2=\left(\sqrt{2x+20}\right)^2$$
$$10-3x=2x+20$$
$$10-3x-\mathbf{2x}=2x+20-\mathbf{2x}$$
$$10-5x=20$$
$$10-5x-\mathbf{10}=20-\mathbf{10}$$
$$-5x=10$$
$$\frac{-5x}{\mathbf{-5}}=\frac{10}{\mathbf{-5}}$$
$$x=-2$$

Check the solution.
$$\sqrt{10-3x}=\sqrt{2x+20}$$
$$\sqrt{10-3(-2)}\stackrel{?}{=}\sqrt{2(-2)+20}$$
$$\sqrt{10+6}\stackrel{?}{=}\sqrt{-4+20}$$
$$\sqrt{16}\stackrel{?}{=}\sqrt{16}$$
$$4=4$$
True

The solution is -2.

The solution set is $\{-2\}$.

53. $$\sqrt{3c-8}=\sqrt{c}$$
$$\left(\sqrt{3c-8}\right)^2=\left(\sqrt{c}\right)^2$$
$$3c-8=c$$
$$3c-8-\mathbf{3c}=c-\mathbf{3c}$$
$$-8=-2c$$
$$\frac{-8}{\mathbf{-2}}=\frac{-2c}{\mathbf{-2}}$$
$$4=c$$

Check the solution.
$$\sqrt{3c-8}=\sqrt{c}$$
$$\sqrt{3(4)-8}\stackrel{?}{=}\sqrt{4}$$
$$\sqrt{12-8}\stackrel{?}{=}2$$
$$\sqrt{4}\stackrel{?}{=}2$$
$$2=2$$
True

The solution is 4.

The solution set is $\{4\}$.

55. $$5\sqrt{a}=\sqrt{10a+15}$$
$$(5)^2\left(\sqrt{a}\right)^2=\left(\sqrt{10a+15}\right)^2$$
$$25a=10a+15$$
$$25a-\mathbf{10a}=10a+15-\mathbf{10a}$$
$$15a=15$$
$$\frac{15a}{\mathbf{15}}=\frac{15}{\mathbf{15}}$$
$$a=1$$

Check the solution.
$$5\sqrt{a}=\sqrt{10a+15}$$
$$5\sqrt{1}\stackrel{?}{=}\sqrt{10(1)+15}$$
$$5(1)\stackrel{?}{=}\sqrt{10+15}$$
$$5\stackrel{?}{=}\sqrt{25}$$
$$5=5$$
True

The solution is 1.

The solution set is $\{1\}$.

57. $2\sqrt{3x+4} = \sqrt{5x+9}$

$$(2)^2\left(\sqrt{3x+4}\right)^2 = \left(\sqrt{5x+9}\right)^2$$
$$4(3x+4) = 5x+9$$
$$12x+16 = 5x+9$$
$$12x+16-\mathbf{5x} = 5x+9-\mathbf{5x}$$
$$7x+16 = 9$$
$$7x+16-\mathbf{16} = 9-\mathbf{16}$$
$$7x = -7$$
$$\frac{7x}{\mathbf{7}} = \frac{-7}{\mathbf{7}}$$
$$x = -1$$

Check the solution.

$$2\sqrt{3x+4} = \sqrt{5x+9}$$
$$2\sqrt{3(-1)+4} \stackrel{?}{=} \sqrt{5(-1)+9}$$
$$2\sqrt{-3+4} \stackrel{?}{=} \sqrt{-5+9}$$
$$2\sqrt{1} \stackrel{?}{=} \sqrt{4}$$
$$2(1) \stackrel{?}{=} 2$$
$$2 = 2$$

True

The solution is -1.

The solution set is $\{-1\}$.

TRY IT YOURSELF

Solve each equation.

59. $5\sqrt{x}-11 = 9$

$$5\sqrt{x}-11+\mathbf{11} = 9+\mathbf{11}$$
$$5\sqrt{x} = 20$$
$$\left(5\sqrt{x}\right)^2 = (20)^2$$
$$(5)^2\left(\sqrt{x}\right)^2 = 400$$
$$25x = 400$$
$$\frac{25x}{\mathbf{25}} = \frac{400}{\mathbf{25}}$$
$$x = 16$$

Check the solution.

$$5\sqrt{x}-11 = 9$$
$$5\sqrt{16}-11 \stackrel{?}{=} 9$$
$$5\cdot 4-11 \stackrel{?}{=} 9$$
$$20-11 \stackrel{?}{=} 9$$
$$9 = 9$$

True

The solution is 16.

The solution set is $\{16\}$.

61. $x = \sqrt{x^2-15}+3$

$$x-\mathbf{3} = \sqrt{x^2-15}+3-\mathbf{3}$$
$$x-3 = \sqrt{x^2-15}$$
$$(x-3)^2 = \left(\sqrt{x^2-15}\right)^2$$
$$x^2+2(-3)x+(-3)^2 = x^2-15$$
$$x^2-6x+9 = x^2-15$$
$$x^2-6x+9-\mathbf{x^2} = x^2-15-\mathbf{x^2}$$
$$-6x+9-\mathbf{9} = -15-\mathbf{9}$$
$$-6x = -24$$
$$\frac{-6x}{\mathbf{-6}} = \frac{-24}{\mathbf{-6}}$$
$$x = 4$$

Check the solution.

$$x = \sqrt{x^2-15}+3$$
$$4 \stackrel{?}{=} \sqrt{(4)^2-15}+3$$
$$4 \stackrel{?}{=} \sqrt{16-15}+3$$
$$4 \stackrel{?}{=} \sqrt{1}+3$$
$$4 \stackrel{?}{=} 1+3$$
$$4 = 4$$

True

Since a true statements result when 4 is substituted for x, 4 is a solution. The solution set is $\{4\}$.

63. $\sqrt{24+10n}-n = 4$

$$\sqrt{24+10n}-n+\mathbf{n} = \mathbf{n}+4$$
$$\sqrt{24+10n} = n+4$$
$$\left(\sqrt{24+10n}\right)^2 = (n+4)^2$$
$$24+10n = n^2+2\cdot 4n+4^2$$
$$24+10n = n^2+8n+16$$
$$24+10n-\mathbf{10n} = n^2+8n+16-\mathbf{10n}$$
$$24 = n^2-2n+16$$
$$24-\mathbf{24} = n^2-2n+16-\mathbf{24}$$
$$0 = n^2-2n-8$$
$$0 = (n+2)(n-4)$$

Set each factor equal to zero and solve.

$n+2=0$	$n-4=0$
$n+2-\mathbf{2}=0-\mathbf{2}$	$n-4+\mathbf{4}=0+\mathbf{4}$
$n=-2$	$n=4$

Check both solutions.

$\sqrt{24+10n}-n=4$	$\sqrt{24+10n}-n=4$
$\sqrt{24+10\cdot(-2)}-(-2)\stackrel{?}{=}4$	$\sqrt{24+10\cdot 4}-4\stackrel{?}{=}4$
$\sqrt{24-20}+2\stackrel{?}{=}4$	$\sqrt{24+40}-4\stackrel{?}{=}4$
$\sqrt{4}+2\stackrel{?}{=}4$	$\sqrt{64}-4\stackrel{?}{=}4$
$2+2\stackrel{?}{=}4$	$8-4\stackrel{?}{=}4$
$4=4$	$4=4$
True	True

Since two true statements result when -2 and 4 are substituted for n, -2 and 4 are solutions. The solution set is $\{-2, 4\}$.

65. $\sqrt{3t-9}=\sqrt{t+1}$

$\left(\sqrt{3t-9}\right)^2 = \left(\sqrt{t+1}\right)^2$

$3t-9=t+1$

$3t-9-\mathbf{t}=t+1-\mathbf{t}$

$2t-9=1$

$2t-9+\mathbf{9}=1+\mathbf{9}$

$2t=10$

$\frac{2t}{\mathbf{2}}=\frac{10}{\mathbf{2}}$

$t=5$

Check the solution.

$\sqrt{3t-9}=\sqrt{t+1}$

$\sqrt{3(5)-9}\overset{?}{=}\sqrt{5+1}$

$\sqrt{15-9}\overset{?}{=}\sqrt{6}$

$\sqrt{6}=\sqrt{6}$

True

The solution is 5.

The solution set is $\{5\}$.

67. $\sqrt{a}+2=9$

$\sqrt{a}+2-\mathbf{2}=9-\mathbf{2}$

$\sqrt{a}=7$

$\left(\sqrt{a}\right)^2=(7)^2$

$a=49$

Check the solution.

$\sqrt{a}+2=9$

$\sqrt{49}+2\overset{?}{=}9$

$7+2\overset{?}{=}9$

$9=9$

True

The solution is 49.

The solution set is $\{49\}$.

69. $-2=2\sqrt{x}-12$

$-2+\mathbf{12}=2\sqrt{x}-12+\mathbf{12}$

$10=2\sqrt{x}$

$(10)^2=(2\sqrt{x})^2$

$100=(2)^2\left(\sqrt{x}\right)^2$

$100=4x$

$\frac{100}{\mathbf{4}}=\frac{4x}{\mathbf{4}}$

$25=x$

Check the solution.

$-2=2\sqrt{x}-12$

$-2\overset{?}{=}2\sqrt{25}-12$

$-2\overset{?}{=}2\cdot 5-12$

$-2\overset{?}{=}10-12$

$-2=-2$

True

The solution is 25.

The solution set is $\{25\}$.

71. $1=\sqrt{x^2+x+4}-x$

$1+\mathbf{x}=\sqrt{x^2+x+4}-x+\mathbf{x}$

$x+1=\sqrt{x^2+x+4}$

$(x+1)^2=\left(\sqrt{x^2+x+4}\right)^2$

$x^2+2(1)x+(1)^2=x^2+x+4$

$x^2+2x+1=x^2+x+4$

$x^2+2x+1-\mathbf{x^2}=x^2+x+4-\mathbf{x^2}$

$2x+1=x+4$

$2x+1-\mathbf{x}=x+4-\mathbf{x}$

$x+1=4$

$x+1-\mathbf{1}=4-\mathbf{1}$

$x=3$

Check the solution.

$1=\sqrt{x^2+x+4}-x$

$1\overset{?}{=}\sqrt{(3)^2+3+4}-3$

$1\overset{?}{=}\sqrt{9+3+4}-3$

$1\overset{?}{=}\sqrt{16}-3$

$1\overset{?}{=}4-3$

$1=1$

True

Since a true statements result when 3 is substituted for x, 3 is a solution. The solution set is $\{3\}$.

73. $\sqrt{9y}=2y+1$

$\left(\sqrt{9y}\right)^2=(2y+1)^2$

$9y=(2y)^2+2\cdot 2y+1^2$

$9y=4y^2+4y+1$

$9y-\mathbf{9y}=4y^2+4y+1-\mathbf{9y}$

$0=4y^2-5y+1$

$0=(4y-1)(y-1)$

Set each factor equal to zero and solve.

$4y-1=0$

$4y-1+\mathbf{1}=0+\mathbf{1}$

$4y=1$

$\frac{4y}{\mathbf{4}}=\frac{1}{\mathbf{4}}$

$y=\frac{1}{4}$

$y-1=0$

$y-1+\mathbf{1}=0+\mathbf{1}$

$y=1$

Check both solutions.

$\sqrt{9y}=2y+1$

$\sqrt{9\cdot\frac{1}{4}}\overset{?}{=}2\cdot\frac{1}{4}+1$

$\sqrt{\frac{9}{4}}\overset{?}{=}\frac{2}{4}+1$

$\frac{3}{2}\overset{?}{=}\frac{1}{2}+\frac{2}{2}$

$\frac{3}{2}=\frac{3}{2}$

True

$\sqrt{9y}=2y+1$

$\sqrt{9\cdot 1}\overset{?}{=}2\cdot 1+1$

$\sqrt{9}\overset{?}{=}2+1$

$3=3$

True

Since two true statements result when $\frac{1}{4}$ and 1 are substituted for y, $\frac{1}{4}$ and 1 are solutions. The solution set is $\{\frac{1}{4},1\}$.

75. $\frac{1}{2}\sqrt{4t^2+2t+20}=t$

$$\mathbf{2}\cdot\frac{1}{2}\sqrt{4t^2+2t+20}=t\cdot\mathbf{2}$$

$$\sqrt{4t^2+2t+20}=2t$$

$$\left(\sqrt{4t^2+2t+20}\right)^2=(2t)^2$$

$$4t^2+2t+20=4t^2$$

$$4t^2+2t+20-\mathbf{4t^2}=4t^2-\mathbf{4t^2}$$

$$2t+20=0$$

$$2t+20-\mathbf{20}=0-\mathbf{20}$$

$$2t=-20$$

$$\frac{2t}{\mathbf{2}}=\frac{-20}{\mathbf{2}}$$

$$t=-10$$

Check the solution.

$$t=\frac{1}{2}\sqrt{4t^2+2t+20}$$

$$-10\overset{?}{=}\frac{1}{2}\sqrt{4(-10)^2+2(-10)+20}$$

$$-10\overset{?}{=}\frac{1}{2}\sqrt{4(100)-20+20}$$

$$-10\overset{?}{=}\frac{1}{2}\sqrt{400}$$

$$-10\overset{?}{=}\frac{1}{2}\cdot 20$$

$$-10=10$$

False

Since the result is a false statement, -10 is not a solution of the original equation and must be discarded. The original equation has no solution. The solution set is the empty set, written as $\varnothing$.

77. $-\sqrt{x}=-2$

$$\left(-\sqrt{x}\right)^2=(-2)^2$$

$$(-1)^2\left(\sqrt{x}\right)^2=4$$

$$x=4$$

Check the solution.

$$-\sqrt{x}=-2$$

$$-\sqrt{4}\overset{?}{=}-2$$

$$-2=-2$$

True

The solution is 4.

The solution set is $\{4\}$.

79. $10-\sqrt{s}=7$

$$10-\sqrt{s}+\sqrt{s}=7+\sqrt{s}$$

$$10=7+\sqrt{s}$$

$$10-\mathbf{7}=7+\sqrt{s}-\mathbf{7}$$

$$3=\sqrt{s}$$

$$3^2=(\sqrt{s})^2$$

$$9=s$$

Check the solution.

$$10-\sqrt{s}=7$$

$$10-\sqrt{9}\overset{?}{=}7$$

$$10-3\overset{?}{=}7$$

$$7=7$$

True

The solution is 9.

The solution set is $\{9\}$.

81. $\sqrt{4x-2}=\sqrt{3x+5}$

$$\left(\sqrt{4x-2}\right)^2=\left(\sqrt{3x+5}\right)^2$$

$$4x-2=3x+5$$

$$4x-2-\mathbf{3x}=3x+5-\mathbf{3x}$$

$$x-2=5$$

$$x-2+\mathbf{2}=5+\mathbf{2}$$

$$x=7$$

Check the solution.

$$\sqrt{4x-2}=\sqrt{3x+5}$$

$$\sqrt{4(7)-2}\overset{?}{=}\sqrt{3(7)+5}$$

$$\sqrt{28-2}\overset{?}{=}\sqrt{21+5}$$

$$\sqrt{26}=\sqrt{26}$$

True

The solution is 7.

The solution set is $\{7\}$.

83. $\sqrt{x}+1=7$

$$\sqrt{x}+1-\mathbf{1}=7-\mathbf{1}$$

$$\sqrt{x}=6$$

$$\left(\sqrt{x}\right)^2=(6)^2$$

$$x=36$$

Check the solution.

$$\sqrt{x}+1=7$$

$$\sqrt{36}+1\overset{?}{=}7$$

$$6+1\overset{?}{=}7$$

$$7=7$$

True

The solution is 36.

The solution set is $\{36\}$.

85.

$$\begin{aligned}
b &= \sqrt{2b-2}+1 \\
b-\mathbf{1} &= \sqrt{2b-2}+1-\mathbf{1} \\
b-1 &= \sqrt{2b-2} \\
(b-1)^2 &= \sqrt{2b-2}^{\,2} \\
b^2+2\cdot -1b+(-1)^2 &= 2b-2 \\
b^2-2b+1 &= 2b-2 \\
b^2-2b+1-\mathbf{2b} &= 2b-2-\mathbf{2b} \\
b^2-4b+1 &= -2 \\
b^2-4b+1+\mathbf{2} &= -2+\mathbf{2} \\
b^2-4b+3 &= 0 \\
(b-1)(b-3) &= 0
\end{aligned}$$

Set each factor equal to zero and solve.

$b-1=0$	$b-3=0$
$b-1+\mathbf{1}=0+\mathbf{1}$	$b-3+\mathbf{3}=0+\mathbf{3}$
$b=1$	$b=3$

Check both solutions.

$b=\sqrt{2b-2}+1$	$b=\sqrt{2b-2}+1$
$1\stackrel{?}{=}\sqrt{2\cdot 1-2}+1$	$3\stackrel{?}{=}\sqrt{2\cdot 3-2}+1$
$1\stackrel{?}{=}\sqrt{2-2}+1$	$3\stackrel{?}{=}\sqrt{6-2}+1$
$1\stackrel{?}{=}\sqrt{0}+1$	$3\stackrel{?}{=}\sqrt{4}+1$
$1=1$	$3\stackrel{?}{=}2+1$
True	$3=3$
	True

Since two true statements result when 1 and 3 are substituted for n, 1 and 3 are solutions. The solution set is $\{1,3\}$.

87.

$$\begin{aligned}
6+\sqrt{m}-m &= 0 \\
6+\sqrt{m}-m-\mathbf{6} &= 0-\mathbf{6} \\
\sqrt{m}-m &= -6 \\
\sqrt{m}-m+\mathbf{m} &= -6+\mathbf{m} \\
\sqrt{m} &= m-6 \\
\sqrt{m}^{\,2} &= (m-6)^2 \\
m &= m^2+2\cdot -6m+(-6)^2 \\
m &= m^2-12m+36 \\
m-\mathbf{m} &= m^2-12m+36-\mathbf{m} \\
0 &= m^2-13m+36 \\
0 &= (m-9)(m-4)
\end{aligned}$$

Set each factor equal to zero and solve.

$m-9=0$	$m-4=0$
$m-9+\mathbf{9}=0+\mathbf{9}$	$m-4+\mathbf{4}=0+\mathbf{4}$
$m=9$	$m=4$

Check both solutions.

$6+\sqrt{m}-m=0$	$6+\sqrt{m}-m=0$
$6+\sqrt{9}-9\stackrel{?}{=}0$	$6+\sqrt{4}-4\stackrel{?}{=}0$
$6+3-9\stackrel{?}{=}0$	$6+2-4\stackrel{?}{=}0$
$0=0$	$4=0$
True	False

Since a true statement results when 9 is substituted for m, 9 is a solution. Since a false statement results when 4 is substituted for m, 4 is an extraneous solution. The solution set is $\{9\}$.

89.

$$\begin{aligned}
\sqrt{2a^2-3a-4} &= a \\
\sqrt{2a^2-3a-4}^{\,2} &= (a)^2 \\
2a^2-3a-4 &= a^2 \\
2a^2-3a-4-\mathbf{a^2} &= a^2-\mathbf{a^2} \\
a^2-3a-4 &= 0 \\
(a-4)(a+1) &= 0
\end{aligned}$$

Set each factor equal to zero and solve.

$a-4=0$	$a+1=0$
$a-4+\mathbf{4}=0+\mathbf{4}$	$a+1-\mathbf{1}=0-\mathbf{1}$
$a=4$	$a=-1$

Check both solutions.

$\sqrt{2a^2-3a-4}=a$	$\sqrt{2a^2-3a-4}=a$
$\sqrt{2(\mathbf{4})^2-3(\mathbf{4})-4}\stackrel{?}{=}\mathbf{4}$	$\sqrt{2(\mathbf{-1})^2-3(\mathbf{-1})-4}\stackrel{?}{=}\mathbf{-1}$
$\sqrt{32-12-4}\stackrel{?}{=}4$	$\sqrt{2+3-4}\stackrel{?}{=}-1$
$\sqrt{16}\stackrel{?}{=}4$	$\sqrt{1}\stackrel{?}{=}-1$
$4=4$	$1=-1$
True	False

Since a true statement results when 4 is substituted for a, 4 is a solution. Since a false statement results when -1 is substituted for a, -1 is an extraneous solution. The solution set is $\{4\}$.

APPLICATIONS

91. NIAGARA FALLS

Strategy:
We will substitute 3.25 for t in the formula and solve for d.
- Time is 3.25 seconds.
- What is the height of the falls?

$$t = \frac{\sqrt{d}}{4}$$
$$\mathbf{3.25} = \frac{\sqrt{d}}{4}$$
$$\mathbf{4} \cdot 3.25 = \mathbf{4} \cdot \frac{\sqrt{d}}{4}$$
$$4(3.25) = \sqrt{d}$$
$$13 = \sqrt{d}$$
$$(13)^2 = (\sqrt{d})^2$$
$$169 = d$$

The height of the falls is 169 ft.

93. PENDULUMS

Strategy:
We will substitute 8.91 for t in the formula and solve for L.
- Time is 8.91 seconds.
- How long is the pendulum?

$$t = 1.11\sqrt{L}$$
$$\mathbf{8.91} = 1.11\sqrt{L}$$
$$\frac{8.91}{\mathbf{1.11}} = \frac{1.11\sqrt{L}}{\mathbf{1.11}}$$
$$\frac{8.91}{1.11} = \sqrt{L}$$
$$8.027 \approx \sqrt{L}$$
$$(8.027)^2 \approx (\sqrt{L})^2$$
$$64.43 \approx L$$
$$64.4 \approx L$$

The length of the pendulum is about 64.4 ft.

from CAMPUS TO CAREERS

95. CRIME SCENE INVESTIGATOR

Strategy:
We will substitute 1,100 for v in the formula and solve for h.
- Velocity is 1,100 ft/sec.
- How high is the block raised?

$$v = 891\sqrt{h}$$
$$\mathbf{1{,}100} = 891\sqrt{h}$$
$$\frac{1{,}100}{\mathbf{891}} = \frac{891\sqrt{h}}{\mathbf{891}}$$
$$\frac{1{,}100}{891} = \sqrt{h}$$
$$1.235 \approx \sqrt{h}$$
$$(1.235)^2 \approx (\sqrt{h})^2$$
$$1.525 \approx h$$
$$1.5 \approx h$$

The block will be raise about 1.5 ft.

97. HIGHWAY DESIGN

Strategy:
We will substitute 65 for s in the formula and solve for r.
- s is 65 mph.
- What radius should be applied?

$$s = 1.45\sqrt{r}$$
$$\mathbf{65} = 1.45\sqrt{r}$$
$$\frac{65}{\mathbf{1.45}} = \frac{1.45\sqrt{r}}{\mathbf{1.45}}$$
$$\frac{65}{1.45} = \sqrt{r}$$
$$44.83 \approx \sqrt{r}$$
$$(44.83)^2 \approx (\sqrt{r})^2$$
$$2{,}009.7 \approx r$$
$$2{,}010 \approx r$$

The radius is about 2,010 ft.

WRITING

99-101. Answers will vary.

REVIEW

Solve each equation.

103. $\frac{1}{2}+\frac{x}{5}=\frac{3}{4}$, LCD = 20

$$\mathbf{20}\left(\frac{1}{2}\right)+\mathbf{20}\left(\frac{x}{5}\right)=\mathbf{20}\left(\frac{3}{4}\right)$$

$$\frac{20}{2}+\frac{20x}{5}=\frac{20\cdot 3}{4}$$

$$\frac{\not{2}\cdot 10}{\not{2}}+\frac{\not{5}\cdot 4x}{\not{5}}=\frac{\not{4}\cdot 5\cdot 3}{\not{4}}$$

$$10+4x=15$$

$$10+4x-\mathbf{10}=15-\mathbf{10}$$

$$4x=5$$

$$\frac{4x}{\mathbf{4}}=\frac{5}{\mathbf{4}}$$

$$x=\frac{5}{4}$$

105. $\frac{2}{5}x+1=\frac{1}{3}+x$, LCD = 15

$$\mathbf{15}\left(\frac{2}{5}x\right)+\mathbf{15}(1)=\mathbf{15}\left(\frac{1}{3}\right)+\mathbf{15}(x)$$

$$\frac{15\cdot 2x}{5}+15=\frac{15}{3}+15x$$

$$\frac{\not{5}\cdot 3\cdot 2x}{\not{5}}+15=\frac{\not{3}\cdot 5}{\not{3}}+15x$$

$$6x+15=5+15x$$

$$6x+15-\mathbf{15}=5+15x-\mathbf{15}$$

$$6x=15x-10$$

$$6x-\mathbf{15x}=15x-10-\mathbf{15x}$$

$$-9x=-10$$

$$\frac{-9x}{\mathbf{-9}}=\frac{-10}{\mathbf{-9}}$$

$$x=\frac{10}{9}$$

CHALLENGE PROBLEMS

Solve each equation. (Hint: You have to square both sides of the equation twice.)

107. $\sqrt{n}+3=\sqrt{n+21}$

$$\left(\sqrt{n}+3\right)^2=\left(\sqrt{n+21}\right)^2$$

$$\left(\sqrt{n}\right)^2+2\cdot 3\sqrt{n}+3^2=n+21$$

$$n+6\sqrt{n}+9=n+21$$

$$n+6\sqrt{n}+9-\mathbf{9}=n+21-\mathbf{9}$$

$$n+6\sqrt{n}=n+12$$

$$n+6\sqrt{n}-\boldsymbol{n}=n-12-\boldsymbol{n}$$

$$6\sqrt{n}=-12$$

$$\left(6\sqrt{n}\right)^2=(-12)^2$$

$$6^2\left(\sqrt{n}\right)^2=144$$

$$36n=144$$

$$\frac{36n}{\mathbf{36}}=\frac{144}{\mathbf{36}}$$

$$n=4$$

Check the solution.

$\sqrt{n}+3=\sqrt{n+21}$

$\sqrt{\mathbf{4}}+3\overset{?}{=}\sqrt{\mathbf{4}+21}$

$2+3\overset{?}{=}\sqrt{25}$

$5=5$

True

The solution set is $\{4\}$.

109. $\sqrt{1-t}+\sqrt{t+9}=4$

$$\sqrt{1-t}+\sqrt{t+9}-\boldsymbol{\sqrt{t+9}}=4-\boldsymbol{\sqrt{t+9}}$$

$$\sqrt{1-t}=4-\sqrt{t+9}$$

$$\left(\sqrt{1-t}\right)^2=\left(4-\sqrt{t+9}\right)^2$$

$$1-t=4^2-2\cdot 4\sqrt{t+9}+\left(\sqrt{t+9}\right)^2$$

$$1-t=16-8\sqrt{t+9}+t+9$$

$$1-t=25-8\sqrt{t+9}+t$$

$$1-t-\boldsymbol{t}=25-8\sqrt{t+9}+t-\boldsymbol{t}$$

$$1-2t=25-8\sqrt{t+9}$$

$$1-2t-\mathbf{25}=25-8\sqrt{t+9}-\mathbf{25}$$

$$-2t-24=-8\sqrt{t+9}$$

$$\left(-2t-24\right)^2 = \left(-8\sqrt{t+9}\right)^2$$

$$(-2t)^2+2\cdot -2t(-24)+(-24)^2=(-8)^2\left(\sqrt{t+9}\right)^2$$

$$4t^2+96t+576=64(t+9)$$
$$4t^2+96t+576=64t+576$$
$$4t^2+96t+576-\mathbf{64t}=64t+576-\mathbf{64t}$$
$$4t^2+32t+576=576$$
$$4t^2+32t+576-\mathbf{576}=576-\mathbf{576}$$
$$4t^2+32t=0$$
$$4t(t+8)=0$$

Set each factor equal to zero and solve.

$4t=0$	$t+8=0$
$\frac{4t}{\mathbf{4}}=\frac{0}{\mathbf{4}}$	$t+8-\mathbf{8}=0-\mathbf{8}$
$t=0$	$t=-8$

Check both solutions.

$\sqrt{1-t}+\sqrt{t+9}=4$	$\sqrt{1-t}+\sqrt{t+9}=4$
$\sqrt{1-0}+\sqrt{0+9}\stackrel{?}{=}4$	$\sqrt{1-(-8)}+\sqrt{-8+9}\stackrel{?}{=}4$
$\sqrt{1}+\sqrt{9}\stackrel{?}{=}4$	$\sqrt{1+8}+\sqrt{1}\stackrel{?}{=}4$
$1+3\stackrel{?}{=}4$	$\sqrt{9}+\sqrt{1}\stackrel{?}{=}4$
$4=4$	$3+1\stackrel{?}{=}4$
True	$4=4$
	True

Since two true statements result when 0 and -8 are substituted for t, 0 and -8 are solutions. The solution set is $\{0,-8\}$.

111. $$\sqrt{\sqrt{x+2}}-3=0$$
$$\sqrt{\sqrt{x+2}}=3$$
$$\left(\sqrt{\sqrt{x+2}}\right)^2=(3)^2$$
$$\sqrt{x+2}=9$$
$$\left(\sqrt{x+2}\right)^2=(9)^2$$
$$x+2=81$$
$$x+2-\mathbf{2}=81-\mathbf{2}$$
$$x=79$$

SECTION 8.6 STUDY SET

VOCABULARY

Fill in the blanks.

1. We read $\sqrt[3]{8}$ as "the **cube** root of 8" and $\sqrt[4]{16}$ as "the **fourth** root of 16" and $\sqrt[5]{-1}$ as "the **fifth** root of −1."

3. "To **simplify** $\sqrt[3]{54}$" means to write it as $3\sqrt[3]{2}$.

CONCEPTS

Fill in the blanks.

5. a. $\sqrt[3]{64} = 4$ because $(\mathbf{4})^3 = 64$.

   b. $\sqrt[4]{16x^4} = 2x$ because $(\mathbf{2x})^4 = 16x^4$.

   c. $\sqrt[5]{-32} = -2$ because $(-2)^5 = \mathbf{-32}$.

7. a. $\sqrt[n]{a \cdot b} = \sqrt[n]{a}\sqrt[n]{b}$ b. $\sqrt[n]{\dfrac{a}{b}} = \dfrac{\sqrt[n]{a}}{\sqrt[n]{b}}$

9. a. To simplify $\sqrt[3]{24}$, which of the following expressions should be used? $\mathbf{\sqrt[3]{8 \cdot 3}}$

   $\sqrt[3]{12 \cdot 2}$ $\sqrt[3]{4 \cdot 6}$ $\sqrt[3]{8 \cdot 3}$ $\sqrt[3]{24 \cdot 1}$

   b. To simplify $\sqrt[4]{48}$, which of the following expressions should be used? $\mathbf{\sqrt[4]{16 \cdot 3}}$

   $\sqrt[4]{24 \cdot 2}$ $\sqrt[4]{16 \cdot 3}$ $\sqrt[4]{8 \cdot 6}$ $\sqrt[4]{48 \cdot 1}$

NOTATION

Write each exponential expression in an equivalent form involving a radical. Do not evaluate it.

11. a. $25^{1/2}$ $\mathbf{\sqrt{25}}$ b. $(-27)^{2/3}$ $\left(\sqrt[3]{-27}\right)^2$

GUIDED PRACTICE

Evaluate. See Example 1.

13. $\sqrt[3]{8} = \sqrt[3]{2^3}$
$= 2$

15. $\sqrt[3]{0} = 0$

17. $\sqrt[3]{-125} = \sqrt[3]{(-5)^3}$
$= -5$

19. $\sqrt[3]{-64} = \sqrt[3]{(-4)^3}$
$= -4$

21. $-\sqrt[3]{-1} = -\sqrt[3]{(-1)^3}$
$= -(-1)$
$= 1$

23. $-\sqrt[3]{-27} = -\sqrt[3]{(-3)^3}$
$= -(-3)$
$= 3$

25. $\sqrt[3]{729} = \sqrt[3]{9^3}$
$= 9$

27. $\sqrt[3]{\dfrac{1}{125}} = \sqrt[3]{\left(\dfrac{1}{5}\right)^3}$
$= \dfrac{1}{5}$

Evaluate. See Example 2.

29. $\sqrt[4]{16} = \sqrt[4]{2^4}$
$= 2$

31. $\sqrt[4]{256} = \sqrt[4]{4^4}$
$= 4$

33. $\sqrt[4]{-625}$
Not a real number

35. $\sqrt[5]{-243} = \sqrt[5]{(-3)^5}$
$= -3$

37. $\sqrt[5]{32} = \sqrt[5]{2^5}$
$= 2$

39. $-\sqrt[4]{\dfrac{1}{81}} = -\sqrt[4]{\left(\dfrac{1}{3}\right)^4}$
$= -\dfrac{1}{3}$

Find each root. See Example 3.

41. $\sqrt[4]{m^8} = \sqrt[4]{(m^2)^4}$
$= m^2$

43. $\sqrt[4]{x^{24}} = \sqrt[4]{(x^6)^4}$
$= x^6$

45. $\sqrt[4]{81a^4} = \sqrt[4]{3^4}\sqrt[4]{a^4}$
$= 3a$

47. $\sqrt[4]{16b^{12}} = \sqrt[4]{16}\sqrt[4]{b^{12}}$
$= \sqrt[4]{2^4}\sqrt[4]{(b^3)^4}$
$= 2b^3$

Find each root. See Example 4.

49. $\sqrt[5]{y^5} = y$

51. $\sqrt[3]{27a^6} = \sqrt[3]{(3a^2)^3}$
$= 3a^2$

53. $\sqrt[5]{x^{15}} = \sqrt[5]{(x^3)^5}$
$= x^3$

55. $\sqrt[5]{32v^{30}} = \sqrt[5]{(2v^6)^5}$
$= 2v^6$

Simplify. See Example 5.

57. $\sqrt[3]{24} = \sqrt[3]{8\cdot 3}$
$= \sqrt[3]{8}\sqrt[3]{3}$
$= 2\sqrt[3]{3}$

59. $\sqrt[3]{-128} = \sqrt[3]{-64\cdot 2}$
$= \sqrt[3]{-64}\sqrt[3]{2}$
$= -4\sqrt[3]{2}$

61. $\sqrt[4]{162} = \sqrt[4]{81\cdot 2}$
$= \sqrt[4]{81}\sqrt[4]{2}$
$= 3\sqrt[4]{2}$

63. $\sqrt[3]{72} = \sqrt[3]{8\cdot 9}$
$= \sqrt[3]{8}\sqrt[3]{9}$
$= 2\sqrt[3]{9}$

65. $\sqrt[3]{\frac{250}{27}} = \sqrt[3]{\frac{125\cdot 2}{27}}$
$= \sqrt[3]{\frac{125}{27}}\cdot\frac{\sqrt[3]{2}}{1}$
$= \frac{5}{3}\cdot\frac{\sqrt[3]{2}}{1}$
$= \frac{5\sqrt[3]{2}}{3}$

67. $\sqrt[5]{64} = \sqrt[5]{32\cdot 2}$
$= \sqrt[5]{32}\sqrt[5]{2}$
$= 2\sqrt[5]{2}$

Evaluate. See Example 6.

69. $81^{1/2} = \sqrt{81}$
$= 9$

71. $8^{1/3} = \sqrt[3]{8}$
$= 2$

73. $\left(\frac{1}{16}\right)^{1/4} = \sqrt[4]{\frac{1}{16}}$
$= \frac{1}{2}$

75. $-144^{1/2} = -1\cdot 144^{1/2}$
$= -1\cdot\sqrt{144}$
$= -1\cdot 12$
$= -12$

77. $(-125)^{1/3} = \sqrt[3]{-125}$
$= -5$

79. $\left(\frac{9}{64}\right)^{1/2} = \sqrt{\frac{9}{64}}$
$= \frac{3}{8}$

Evaluate. See Example 7.

81. $16^{3/2} = \left(\sqrt{16}\right)^3$
$= (4)^3$
$= 64$

83. $27^{2/3} = \left(\sqrt[3]{27}\right)^2$
$= (3)^2$
$= 9$

85. $(-125)^{2/3} = \left(\sqrt[3]{-125}\right)^2$
$= (-5)^2$
$= 25$

87. $4^{5/2} = \left(\sqrt{4}\right)^5$
$= (2)^5$
$= 32$

89. $\left(\frac{1}{16}\right)^{3/4} = \left(\sqrt[4]{\frac{1}{16}}\right)^3$
$= \left(\frac{1}{2}\right)^3$
$= \frac{1}{8}$

91. $-32^{3/5} = -\sqrt[5]{32}^3$
$= -(2)^3$
$= -(2)(2)(2)$
$= -8$

Evaluate. See Example 8.

93. $8^{-1/3} = \frac{1}{8^{1/3}}$
$= \frac{1}{\sqrt[3]{8}}$
$= \frac{1}{2}$

95. $25^{-3/2} = \frac{1}{25^{3/2}}$
$= \frac{1}{\left(\sqrt{25}\right)^3}$
$= \frac{1}{5^3}$
$= \frac{1}{125}$

97. $16^{-5/4} = \frac{1}{16^{5/4}}$
$= \frac{1}{\left(\sqrt[4]{16}\right)^5}$
$= \frac{1}{2^5}$
$= \frac{1}{32}$

99. $(-27)^{-2/3} = \frac{1}{(-27)^{2/3}}$

$= \frac{1}{\left(\sqrt[3]{-27}\right)^2}$

$= \frac{1}{(-3)^2}$

$= \frac{1}{9}$

101. $-81^{-3/2} = -1 \cdot 81^{-3/2}$

$= -\frac{1}{81^{3/2}}$

$= -\frac{1}{\left(\sqrt{81}\right)^3}$

$= -\frac{1}{9^3}$

$= -\frac{1}{729}$

103. $32^{-3/5} = \frac{1}{32^{3/5}}$

$= \frac{1}{\left(\sqrt[5]{32}\right)^3}$

$= \frac{1}{2^3}$

$= \frac{1}{8}$

TRY IT YOURSELF

LOOK ALIKES . . .

105. **Evaluate.**

a. $\sqrt{64} = 8$

b. $\sqrt[3]{64} = 4$

c. $\sqrt{-64}$
Not a real number.

d. $\sqrt[3]{-64} = -4$

107. **Simplify each root.**

a. $\sqrt{32} = \sqrt{16 \cdot 2}$
$= \sqrt{16}\sqrt{2}$
$= 4\sqrt{2}$

b. $\sqrt[3]{32} = \sqrt[3]{8 \cdot 4}$
$= \sqrt[3]{8}\sqrt[3]{4}$
$= 2\sqrt[3]{4}$

c. $\sqrt[4]{32} = \sqrt[4]{16 \cdot 2}$
$= \sqrt[4]{16}\sqrt[4]{2}$
$= 2\sqrt[4]{2}$

d. $\sqrt[5]{32} = 2$

APPLICATIONS

109. WINDMILLS

Strategy:
We will substitute 20 for P in the formula and solve for S.

- 20 watts of power.
- What is the speed of the wind?

$S = \sqrt[3]{\frac{P}{0.02}}$

$= \sqrt[3]{\frac{20}{0.02}}$

$= \sqrt[3]{1{,}000}$

$= 10$

The speed of the wind is 10 mph.

111. HOLIDAY DECORATIONS

Strategy:
We will substitute 24 for h and 10 for r in the formula and solve for s.

- 24 ft tall.
- 10 ft radius.
- What is the length of one string of lights?

$s = \left(r^2 + h^2\right)^{1/2}$

$= \left(10^2 + 24^2\right)^{1/2}$

$= (100 + 576)^{1/2}$

$= (676)^{1/2}$

$= \sqrt{676}$

$= 26$

The length of one string of lights is 26 ft.

113. SPEAKERS

Strategy:
We will substitute 216 for V in the formula and solve for A.

- 216 cubic inches.
- What is the area of one face the cube?

$A = V^{2/3}$

$= 216^{2/3}$

$= \left(\sqrt[3]{216}\right)^2$

$= 6^2$

$= 36$

The area of one face is 36 in^2.

WRITING

115-117. Answers will vary.

REVIEW

Graph each equation.

119. $x = 3$

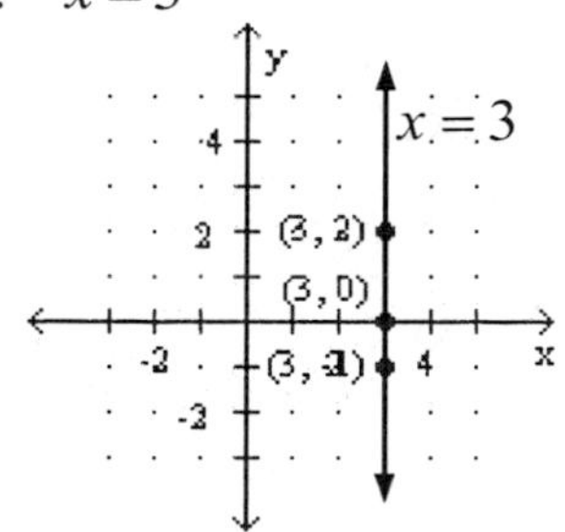

121. $2x - y = -4$

x	y	(x, y)
-2	$y = 2x + 4$ $= 2(-2) + 4$ $= -4 + 4$ $= 0$	$(-2, 0)$
0	$y = 2x + 4$ $= 2(0) + 4$ $= 0 + 4$ $= 4$	$(0, 4)$
1	$y = 2x + 4$ $= 2(1) + 4$ $= 2 + 4$ $= 6$	$(1, 6)$

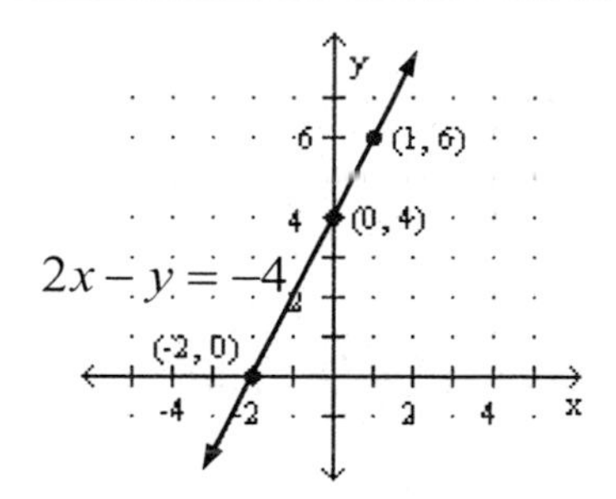

CHALLENGE PROBLEMS

123. Graph $y = \sqrt[3]{x}$

x	y	(x, y)
-27	$y = \sqrt[3]{x}$ $= \sqrt[3]{-27}$ $= -3$	$(-27, -3)$
-8	$y = \sqrt[3]{x}$ $= \sqrt[3]{-8}$ $= -2$	$(-8, -2)$
-1	$y = \sqrt[3]{x}$ $= \sqrt[3]{-1}$ $= -1$	$(-1, -1)$
0	$y = \sqrt[3]{x}$ $= \sqrt[3]{0}$ $= 0$	$(0, 0)$
1	$y = \sqrt[3]{x}$ $= \sqrt[3]{1}$ $= 1$	$(1, 1)$
8	$y = \sqrt[3]{x}$ $= \sqrt[3]{8}$ $= 2$	$(8, 2)$
27	$y = \sqrt[3]{x}$ $= \sqrt[3]{27}$ $= 3$	$(27, 3)$

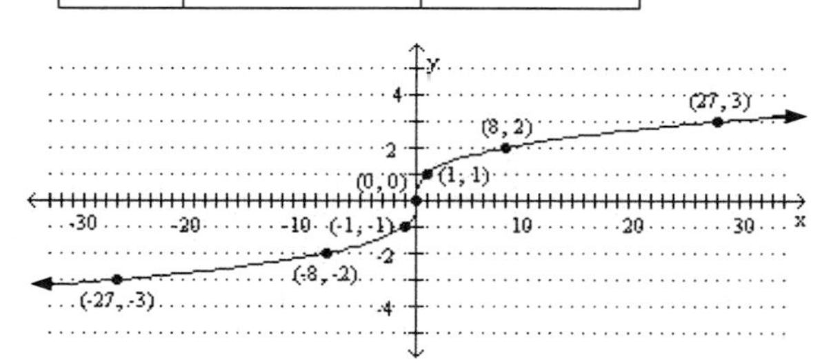

CHAPTER 8 REVIEW

SECTION 8.1

An Introduction to Square Roots

1. Fill in the blanks: $\sqrt{36} = 6$ because $(\mathbf{6})^2 = \mathbf{36}$.

2. Determine whether the statement is true or false:

$-\sqrt{4} = \sqrt{-4}$. **False**

Find each square root. ***Do not use a calculator.***

3. $\sqrt{25}$ **5** 4. $\sqrt{49}$ 7 5. $-\sqrt{144}$ **−12**

6. $\sqrt{\frac{16}{81}}$ $\mathbf{\frac{4}{9}}$ 7. $\sqrt{0.64}$ **0.8** 8. $\sqrt{1}$ **1**

Use a calculator to approximate each expression to the nearest hundredth.

9. $\sqrt{21}$ **4.58** 10. $-2\sqrt{7}$ **−5.29**

11. Determine whether each number is rational, irrational, or not a real number:

$\sqrt{-2}$ **Not a real number**

$\sqrt{68}$ **Irrational**

$\sqrt{81}$ **Rational**

$\sqrt{3}$ **Irrational**

12. ROAD SIGNS

$$v = \sqrt{2.5r}$$
$$= \sqrt{2.5(360)}$$
$$= \sqrt{900}$$
$$= 30$$

The road sign should read "30 mph".

Simplify. All variables represent positive real numbers.

13. $\sqrt{x^2}$ x 14. $\sqrt{4b^4}$ $\mathbf{2b^2}$

15. $-\sqrt{y^{12}}$ $-y^6$ 16. $\sqrt{9h^{16}}$ $\mathbf{3h^8}$

Refer to the right triangle.

17. Find c where $a = 6$ and $b = 8$.

$$c^2 = a^2 + b^2$$
$$c^2 = 6^2 + 8^2$$
$$c^2 = 36 + 64$$
$$c^2 = 100$$
$$\sqrt{c^2} = \sqrt{100}$$
$$c = 10$$

18. Find b where a = 8 and c = 17.

$$a^2 + b^2 = c^2$$
$$8^2 + b^2 = 17^2$$
$$64 + b^2 = 289$$
$$64 + b^2 - \mathbf{64} = 289 - \mathbf{64}$$
$$b^2 = 225$$
$$\sqrt{b^2} = \sqrt{225}$$
$$b = 15$$

19. THEATER SEATING

$$a^2 + b^2 = c^2$$
$$12^2 + b^2 = 13^2$$
$$144 + b^2 = 169$$
$$144 + b^2 - \mathbf{144} = 169 - \mathbf{144}$$
$$b^2 = 25$$
$$\sqrt{b^2} = \sqrt{25}$$
$$b = 5$$

The top seats are 5 ft higher.

20. TRIANGLES

$$a^2 + b^2 = c^2$$
$$x^2 + x^2 = 3^2$$
$$2x^2 = 9$$
$$\frac{2x^2}{\mathbf{2}} = \frac{9}{\mathbf{2}}$$
$$x^2 = 4.5$$
$$\sqrt{x^2} = \sqrt{4.5}$$
$$x \approx 2.121$$
$$x \approx 2.12$$

The length of one leg is about 2.12 in.

Find the distance between the points. If an answer is not exact, round to the nearest hundredth.

21. $(-7, 12)$ and $(-4, 8)$

$$(x_1, y_1) \qquad (x_2, y_2)$$

$$d = \sqrt{(x_2 - x_1)^2 + (y_2 - y_1)^2}$$
$$d = \sqrt{[-4 - (-7)]^2 + [8 - 12]^2}$$
$$d = \sqrt{(-4 + 7)^2 + (8 - 12)^2}$$
$$d = \sqrt{3^2 + (-4)^2}$$
$$d = \sqrt{9 + 16}$$
$$d = \sqrt{25}$$
$$d = 5$$

22. $(-15, -3)$ and $(-10, -16)$

$$\begin{matrix} \uparrow & \uparrow & & \uparrow & \uparrow \\ (x_1, & y_1) & & (x_2, & y_2) \end{matrix}$$

$$d = \sqrt{(x_2 - x_1)^2 + (y_2 - y_1)^2}$$
$$d = \sqrt{[-10-(-15)]^2 + [-16-(-3)]^2}$$
$$d = \sqrt{(-10+15)^2 + (-16+3)^2}$$
$$d = \sqrt{5^2 + (-13)^2}$$
$$d = \sqrt{25+169}$$
$$d = \sqrt{194}$$
$$d \approx 13.928$$
$$d \approx 13.93$$

SECTION 8.2
Simplifying Square Roots

Simplify. All variables represent positive real numbers.

23. $\sqrt{32} = \sqrt{16 \cdot 2}$
$= \sqrt{16}\sqrt{2}$
$= 4\sqrt{2}$

24. $\sqrt{x^5} = \sqrt{x^{4+1}}$
$= \sqrt{x^4}\sqrt{x^1}$
$= x^2\sqrt{x}$

25. $\sqrt{80x^2} = \sqrt{16x^2 \cdot 5}$
$= \sqrt{16x^2}\sqrt{5}$
$= 4x\sqrt{5}$

26. $-2\sqrt{63} = -2\sqrt{9}\sqrt{7}$
$= -2 \cdot 3\sqrt{7}$
$= -6\sqrt{7}$

27. $\sqrt{250t^3} = \sqrt{25t^2 \cdot 10t^1}$
$= \sqrt{25t^2}\sqrt{10t}$
$= 5t\sqrt{10t}$

28. $\sqrt{\dfrac{16}{25}} = \dfrac{\sqrt{16}}{\sqrt{25}}$
$= \dfrac{4}{5}$

29. $\sqrt{\dfrac{60}{49}} = \dfrac{\sqrt{4 \cdot 15}}{\sqrt{49}}$
$= \dfrac{\sqrt{4}\sqrt{15}}{\sqrt{49}}$
$= \dfrac{2\sqrt{15}}{7}$

30. $\sqrt{\dfrac{242x^4}{169x^2}} = \sqrt{\dfrac{242x^{4-2}}{169}}$
$= \dfrac{\sqrt{242x^2}}{\sqrt{169}}$
$= \dfrac{\sqrt{121x^2}\sqrt{2}}{13}$
$= \dfrac{11x\sqrt{2}}{13}$

31. FITNESS EQUIPMENT
- Short leg is 2 feet long.
- Long leg is 6 feet long.
- What is the length of the sit-up board?

$$a^2 + b^2 = c^2$$
$$2^2 + 6^2 = c^2$$
$$4 + 36 = c^2$$
$$40 = c^2$$
$$\sqrt{40} = \sqrt{c^2}$$
$$\sqrt{4 \cdot 10} = c$$
$$2\sqrt{10} = c$$
$$6.3 \approx c$$

The exact distance is $2\sqrt{10}$ ft.
The approximate length is 6.3 ft.

32. **Determine whether the statement is true or false:** $\sqrt{x^9} = x^3$ **False**

SECTION 8.3
Adding and Subtracting Radical Expressions

Perform the operations. All variables represent nonnegative real numbers.

33. $\sqrt{10} + \sqrt{10} = (1+1)\sqrt{10}$
$= 2\sqrt{10}$

34. $6\sqrt{x} - \sqrt{x} = (6-1)\sqrt{x}$
$= 5\sqrt{x}$

35. $\sqrt{2}+\sqrt{8}-\sqrt{18}=\sqrt{2}+\sqrt{4\cdot 2}-\sqrt{9\cdot 2}$
$=\sqrt{2}+2\sqrt{2}-3\sqrt{2}$
$=(1+2-3)\sqrt{2}$
$=0\sqrt{2}$
$=0$

36. $\sqrt{3}+4+\sqrt{27}-7=\sqrt{3}+4+\sqrt{9}\sqrt{3}-7$
$=\sqrt{3}+4+3\sqrt{3}-7$
$=(4-7)+(1+3)\sqrt{3}$
$=-3+4\sqrt{3}$

37. $5\sqrt{28}-3\sqrt{63}=5\sqrt{4\cdot 7}-3\sqrt{9\cdot 7}$
$=5\cdot 2\sqrt{7}-3\cdot 3\sqrt{7}$
$=10\sqrt{7}-9\sqrt{7}$
$=(10-9)\sqrt{7}$
$=\sqrt{7}$

38. $3\sqrt{5y^3}-5y\sqrt{20y}=3\sqrt{y^2}\sqrt{5y}-5y\sqrt{4}\sqrt{5y}$
$=3y\sqrt{5y}-5y\cdot 2\sqrt{5y}$
$=3y\sqrt{5y}-10y\sqrt{5y}$
$=(3y-10y)\sqrt{5y}$
$=-7y\sqrt{5y}$

39. Explain why we cannot add $3\sqrt{5}$ and $5\sqrt{3}$.
The radicands are different.

40. GARDENING
$7\sqrt{45}-4\sqrt{20}=7\sqrt{9}\sqrt{5}-4\sqrt{4}\sqrt{5}$
$=7\cdot 3\sqrt{5}-4\cdot 2\sqrt{5}$
$=21\sqrt{5}-8\sqrt{5}$
$=(21-8)\sqrt{5}$
$=13\sqrt{5}$
The difference in the lengths is $13\sqrt{5}$ in.

SECTION 8.4
Multiplying and Dividing Radical Expressions

Perform the operations. All variables represent nonnegative real numbers.

41. $\sqrt{2}\sqrt{3}=\sqrt{2\cdot 3}$
$=\sqrt{6}$

42. $-5\sqrt{5}^{\,2}=-5^{\,2}\ \sqrt{5}^{\,2}$
$=25\cdot 5$
$=125$

43. $3\sqrt{3x^3}\ \ 4\sqrt{6x^2}=(3\cdot 4)\ \sqrt{3x^3\cdot 6x^2}$
$=12\ \sqrt{3\cdot 6x^{3+2}}$
$=12\ \sqrt{18x^5}$
$=12\ \sqrt{9x^4\cdot 2x^1}$
$=12(3x^2)\sqrt{2x}$
$=36x^2\sqrt{2x}$

44. $\sqrt{15}+3x^{\,2}=\sqrt{15}^{\,2}+2\cdot 3x\sqrt{15}+(3x)^2$
$=15+6x\sqrt{15}+9x^2$

45. $\sqrt{2}\ \sqrt{8}-\sqrt{18}=\sqrt{2}\ \sqrt{8}-\sqrt{2}\ \sqrt{18}$
$=\sqrt{2\cdot 8}-\sqrt{2\cdot 18}$
$=\sqrt{16}-\sqrt{36}$
$=4-6$
$=-2$

46. $\sqrt{3}+\sqrt{5}\ \ \sqrt{3}-\sqrt{5}=\sqrt{3}^{\,2}-\sqrt{5}^{\,2}$
$=3-5$
$=-2$

47. $\sqrt{x}^{\,2}=x$

48. $\sqrt{t-1}^{\,2}=t-1$

49. $A=lw$
$5\sqrt{3}\ \ 2\sqrt{6}=5\cdot 2\sqrt{3\cdot 6}$
$=10\sqrt{18}$
$=10\sqrt{9\cdot 2}$
$=10(3)\sqrt{2}$
$=30\sqrt{2}$
≈ 42.42
≈ 42.4
The exact area is $30\sqrt{2}$ in^2.
The approximate area is 42.4 in^2.

50. PATIO TILES

$$\frac{1}{\sqrt{2}} = \frac{1}{\sqrt{2}} \cdot \frac{\sqrt{2}}{\sqrt{2}}$$
$$= \frac{\sqrt{2}}{2}$$

Rationalize the denominator. All variables represent positive real numbers.

51. $$\frac{9}{\sqrt{7}} = \frac{9}{\sqrt{7}} \cdot \frac{\sqrt{7}}{\sqrt{7}}$$
$$= \frac{9\sqrt{7}}{7}$$

52. $$\sqrt{\frac{3}{a}} = \frac{\sqrt{3}}{\sqrt{a}}$$
$$= \frac{\sqrt{3}}{\sqrt{a}} \cdot \frac{\sqrt{a}}{\sqrt{a}}$$
$$= \frac{\sqrt{3a}}{a}$$

53. $$\frac{16\sqrt{3}}{\sqrt{5}} = \frac{16\sqrt{3}}{\sqrt{5}} \cdot \frac{\sqrt{5}}{\sqrt{5}}$$
$$= \frac{16\sqrt{3 \cdot 5}}{\sqrt{5}\sqrt{5}}$$
$$= \frac{16\sqrt{15}}{5}$$

54. $$\frac{11}{\sqrt{75}} = \frac{11}{\sqrt{25}\sqrt{3}}$$
$$= \frac{11}{5\sqrt{3}}$$
$$= \frac{11}{5\sqrt{3}} \cdot \frac{\sqrt{3}}{\sqrt{3}}$$
$$= \frac{11\sqrt{3}}{5 \cdot 3}$$
$$= \frac{11\sqrt{3}}{15}$$

55. $$\frac{1}{\sqrt{8x}} = \frac{1}{\sqrt{4 \cdot 2x}}$$
$$= \frac{1}{2\sqrt{2x}}$$
$$= \frac{1}{2\sqrt{2x}} \cdot \frac{\sqrt{2x}}{\sqrt{2x}}$$
$$= \frac{\sqrt{2x}}{2\sqrt{2x}\sqrt{2x}}$$
$$= \frac{\sqrt{2x}}{2 \cdot 2x}$$
$$= \frac{\sqrt{2x}}{4x}$$

56. $$\frac{\sqrt{7}}{\sqrt{7}-\sqrt{2}} = \frac{\sqrt{7}}{\sqrt{7}-\sqrt{2}} \cdot \frac{\sqrt{7}+\sqrt{2}}{\sqrt{7}+\sqrt{2}}$$
$$= \frac{\sqrt{7}\ \ \sqrt{7}+\sqrt{2}}{\sqrt{7}-\sqrt{2}\ \ \sqrt{7}+\sqrt{2}}$$
$$= \frac{\sqrt{7}\sqrt{7}+\sqrt{7}\sqrt{2}}{\sqrt{7}^{\ 2} - \sqrt{2}^{\ 2}}$$
$$= \frac{7+\sqrt{14}}{7-2}$$
$$= \frac{7+\sqrt{14}}{5}$$

57. $$\frac{\sqrt{a}}{\sqrt{a}+1} = \frac{\sqrt{a}}{\sqrt{a}+1} \cdot \frac{\sqrt{a}-1}{\sqrt{a}-1}$$
$$= \frac{\sqrt{a}\ \ \sqrt{a}-1}{\sqrt{a}+1\ \ \sqrt{a}-1}$$
$$= \frac{\sqrt{a}\ \ \sqrt{a}\ \ -\sqrt{a}\ \ 1}{\sqrt{a}^{\ 2} - 1^2}$$
$$= \frac{a-\sqrt{a}}{a-1}$$

58. $\dfrac{2+\sqrt{b}}{\sqrt{b}-3}=\dfrac{2+\sqrt{b}}{\sqrt{b}-3}\cdot\dfrac{\sqrt{b}+3}{\sqrt{b}+3}$

$$=\frac{2+\sqrt{b}\quad\sqrt{b}+3}{\sqrt{b}-3\quad\sqrt{b}+3}$$

$$=\frac{2\sqrt{b}+2(3)+\sqrt{b}\sqrt{b}+\sqrt{b}(3)}{\sqrt{b}^{\;2}-(3)^2}$$

$$=\frac{2\sqrt{b}+6+b+3\sqrt{b}}{b-9}$$

$$=\frac{b+(2+3)\sqrt{b}+6}{b-9}$$

$$=\frac{b+5\sqrt{b}+6}{b-9}$$

SECTION 8.5

Solving Radical Equations

Solve each equation and check each result.

59. $\sqrt{x}=9$

$$\sqrt{x}^{\;2}=(9)^2$$

$$x=81$$

Check the solution

$$\sqrt{x}=9$$

$$\sqrt{81}\overset{?}{=}9$$

$$9=9$$

True

The solution set is 81 .

60. $\sqrt{2x+10}=10$

$$\sqrt{2x+10}^{\;2}=10^2$$

$$2x+10=100$$

$$2x+10-\mathbf{10}=100-\mathbf{10}$$

$$2x=90$$

$$\frac{2x}{\mathbf{2}}=\frac{90}{\mathbf{2}}$$

$$x=45$$

Check the solution

$$\sqrt{2x+10}=10$$

$$\sqrt{2(45)+10}\overset{?}{=}10$$

$$\sqrt{90+10}\overset{?}{=}10$$

$$\sqrt{100}\overset{?}{=}10$$

$$10=10$$

True

The solution set is 45 .

61. $\sqrt{3x+4}+5=3$

$$\sqrt{3x+4}+5-\mathbf{5}=3-\mathbf{5}$$

$$\sqrt{3x+4}=-2$$

$$\sqrt{3x+4}^{\;2}=(-2)^2$$

$$3x+4=4$$

$$3x+4-\mathbf{4}=4-\mathbf{4}$$

$$3x=0$$

$$\frac{3x}{\mathbf{3}}=\frac{0}{\mathbf{3}}$$

$$x=0$$

Check the solution

$$\sqrt{3x+4}+5=3$$

$$\sqrt{3(0)+4}+5\overset{?}{=}3$$

$$\sqrt{4}+5\overset{?}{=}3$$

$$2+5\overset{?}{=}3$$

$$7=3$$

False

No Solution.

62.

$$\sqrt{b+12}=3\sqrt{b+4}$$
$$\sqrt{b+12}^{\;2}=3^2\;\sqrt{b+4}^{\;2}$$
$$b+12=9(b+4)$$
$$b+12=9b+36$$
$$b+12-\mathbf{b}=9b+36-\mathbf{b}$$
$$12=8b+36$$
$$12-\mathbf{36}=8b+36-\mathbf{36}$$
$$-24=8b$$
$$\frac{-24}{\mathbf{8}}=\frac{8b}{\mathbf{8}}$$
$$-3=b$$

Check the solution

$$\sqrt{b+12}=3\sqrt{b+4}$$
$$\sqrt{-3+12}\overset{?}{=}3\sqrt{-3+4}$$
$$\sqrt{9}\overset{?}{=}3\sqrt{1}$$
$$3=3$$
True

The solution set is -3 .

63.

$$\sqrt{p^2+8p+13}=p+3$$
$$\sqrt{p^2+8p+13}^{\;2}=(p+3)^2$$
$$p^2+8p+13=p^2+2\cdot 3p+3^2$$
$$p^2+8p+13=p^2+6p+9$$
$$p^2+8p+13-\mathbf{p^2}=p^2+6p+9-\mathbf{p^2}$$
$$8p+13=6p+9$$
$$8p+13-\mathbf{6p}=6p+9-\mathbf{6p}$$
$$2p+13=9$$
$$2p+13-\mathbf{13}=9-\mathbf{13}$$
$$2p=-4$$
$$\frac{2p}{2}=\frac{-4}{\mathbf{2}}$$
$$p=-2$$

Check the solution

$$\sqrt{p^2+8p+13}=p+3$$
$$\sqrt{(-2)^2+8(-2)+13}\overset{?}{=}(-2)+3$$
$$\sqrt{4-16+13}\overset{?}{=}1$$
$$\sqrt{1}\overset{?}{=}1$$
$$1=1$$
True

The solution set is -2 .

64.

$$\sqrt{24+10y}-y=4$$
$$\sqrt{24+10y}-y+\mathbf{y}=4+\mathbf{y}$$
$$\sqrt{24+10y}=y+4$$
$$\sqrt{24+10y}^{\;2}=(y+4)^2$$
$$24+10y=y^2+8y+16$$
$$y^2+8y+16=10y+24$$
$$y^2+8y+16-\mathbf{10y}=10y+24-\mathbf{10y}$$
$$y^2-2y+16=24$$
$$y^2-2y+16-\mathbf{24}=24-\mathbf{24}$$
$$y^2-2y-8=0$$
$$(y-4)(y+2)=0$$

Set each factor equal to zero and solve.

$y-4=0$	$y+2=0$
$y-4+\mathbf{4}=0+\mathbf{4}$	$y+2-\mathbf{2}=0-\mathbf{2}$
$y=4$	$y=-2$

Check both solutions.

$\sqrt{24+10y}-y=4$	$\sqrt{24+10y}-y=4$
$\sqrt{24+10(4)}-4\overset{?}{=}4$	$\sqrt{24+10(-2)}-(-2)\overset{?}{=}4$
$\sqrt{24+40}-4\overset{?}{=}4$	$\sqrt{24-20}+2\overset{?}{=}4$
$\sqrt{64}-4\overset{?}{=}4$	$\sqrt{4}+2\overset{?}{=}4$
$8-4\overset{?}{=}4$	$2+2\overset{?}{=}4$
$4=4$	$4=4$
True	True

The solution set is $-2, 4$.

65.

$$5+\sqrt{3n+3}=n$$
$$5+\sqrt{3n+3}-\mathbf{5}=n-\mathbf{5}$$
$$\sqrt{3n+3}=n-5$$
$$\sqrt{3n+3}^{\;2}=(n-5)^2$$
$$3n+3=n^2+2\cdot -5n+(-5)^2$$
$$3n+3=n^2-10n+25$$
$$n^2-10n+25=3n+3$$
$$n^2-10n+25-\mathbf{3n}=3n+3-\mathbf{3n}$$
$$n^2-13n+25=3$$
$$n^2-13n+25-\mathbf{3}=3-\mathbf{3}$$
$$n^2-13n+22=0$$
$$(n-2)(n-11)=0$$

Set each factor equal to zero and solve.

$n-2=0$	$n-11=0$
$n-2+\mathbf{2}=0+\mathbf{2}$	$n-11+\mathbf{11}=0+\mathbf{11}$
$n=2$	$n=11$

Check both solutions.

$n = 5+\sqrt{3n+3}$	$n = 5+\sqrt{3n+3}$
$2 \overset{?}{=} 5+\sqrt{3\cdot 2+3}$	$11 \overset{?}{=} 5+\sqrt{3\cdot 11+3}$
$2 \overset{?}{=} 5+\sqrt{6+3}$	$11 \overset{?}{=} 5+\sqrt{33+3}$
$2 \overset{?}{=} 5+\sqrt{9}$	$11 \overset{?}{=} 5+\sqrt{36}$
$2 \overset{?}{=} 5+3$	$11 \overset{?}{=} 5+6$
$2 = 8$	$11 = 11$
False	True

Since a true statement results when 11 is substituted for n, 11 is a solution. Since a false statement results when 2 is substituted for n, 2 is an extraneous solution. The solution set is $\{11\}$.

66.
$$\sqrt{a}-2 = a-8$$
$$\sqrt{a}-2+\mathbf{2} = a-8+\mathbf{2}$$
$$\sqrt{a} = a-6$$
$$\sqrt{a}^{\,2} = (a-6)^2$$
$$a = a^2-12a+36$$
$$a^2-12a+36 = a$$
$$a^2-12a+36-\boldsymbol{a} = a-\boldsymbol{a}$$
$$a^2-13a+36 = 0$$
$$(a-9)(a-4) = 0$$

Set each factor equal to zero and solve.

$a-9=0$	$a-4=0$
$a-9+\mathbf{9}=0+\mathbf{9}$	$a-4+\mathbf{4}=0+\mathbf{4}$
$a=9$	$a=4$

Check both solutions.

$\sqrt{a}-2 \overset{?}{=} a-8$	$\sqrt{a}-2 \overset{?}{=} a-8$
$\sqrt{9}-2 \overset{?}{=} 9-8$	$\sqrt{4}-2 \overset{?}{=} 4-8$
$3-2 \overset{?}{=} 1$	$2-2 \overset{?}{=} -4$
$1 = 1$	$0 = -4$
True	False

Since a true statement results when 9 is substituted for a, 9 is a solution. Since a false statement results when 4 is substituted for a, 4 is an extraneous solution. The solution set is $\{9\}$.

67. FERRIS WHEELS

Strategy:

We will substitute 2 for t in the formula and solve for d.

- Time is 2 seconds.
- What is the height of the Ferris wheel?

$$t = \sqrt{\frac{d}{16}}$$
$$\mathbf{2} = \sqrt{\frac{d}{16}}$$
$$2^2 = \left(\sqrt{\frac{d}{16}}\right)^2$$
$$4 = \frac{d}{16}$$
$$\mathbf{16}\cdot 4 = \mathbf{16}\cdot\frac{d}{16}$$
$$64 = \frac{\overset{1}{\cancel{16}}\,d}{\underset{1}{\cancel{16}}}$$
$$64 = d$$

The height of the Ferris wheel is 64 ft.

68. THE HORIZON
$$d = \sqrt{1.5h}$$
$$12 = \sqrt{1.5h}$$
$$(12)^2 = \sqrt{1.5h}$$
$$144 = 1.5h$$
$$\frac{144}{\mathbf{1.5}} = \frac{1.5h}{\mathbf{1.5}}$$
$$96 = h$$

The estimated height of the tower is 96 ft.

SECTION 8.6

Higher-Order Roots and Rational Exponents

Evaluate each root.

69. $\sqrt[3]{-27} = \sqrt[3]{(-3)^3} = -3$

70. $-\sqrt[3]{125} = -5$

71. $\sqrt[4]{81} = \sqrt[4]{(3)^4} = 3$

72. $\sqrt[5]{32} = 2$

73. $\sqrt[3]{0} = 0$

74. $\sqrt[5]{-1} = -1$

75. $\sqrt[3]{\frac{1}{64}} = \sqrt[3]{\left(\frac{1}{4}\right)^3} = \frac{1}{4}$

76. $\sqrt[4]{256} = 4$

Simplify. All variables represent nonnegative real numbers.

77. $\sqrt[3]{x^3} = x$

78. $\sqrt[3]{27y^6} = \sqrt[3]{\left(3y^2\right)^3} = 3y^2$

79. $\sqrt[4]{16a^{12}} = \sqrt[4]{(2a^3)^4}$
$= 2a^3$

80. $\sqrt[5]{b^{20}} = \sqrt[5]{(b^4)^5}$
$= b^4$

81. $\sqrt[3]{54} = \sqrt[3]{27 \cdot 2}$
$= \sqrt[3]{27} \cdot \sqrt[3]{2}$
$= 3\sqrt[3]{2}$

82. $\sqrt[4]{80} = \sqrt[4]{16}\sqrt[4]{5}$
$= 2\sqrt[4]{5}$

83. $\sqrt[3]{-\frac{16}{343}} = \frac{\sqrt[3]{-16}}{\sqrt[3]{343}}$
$= \frac{\sqrt[3]{-8 \cdot 2}}{\sqrt[3]{7^3}}$
$= -\frac{2\sqrt[3]{2}}{7}$

84. $\sqrt[3]{\frac{56}{125}} = \frac{\sqrt[3]{56}}{\sqrt[3]{125}}$
$= \frac{\sqrt[3]{8}\sqrt[3]{7}}{\sqrt[3]{125}}$
$= \frac{2\sqrt[3]{7}}{5}$

Evaluate.

85. $49^{1/2} = \sqrt{49}$
$= 7$

86. $(-1{,}000)^{1/3} = \sqrt[3]{-1{,}000}$
$= -10$

87. $36^{3/2} = \sqrt{36}^{\,3}$
$= (6)^3$
$= 216$

88. $\left(\frac{8}{27}\right)^{2/3} = \frac{8^{2/3}}{27^{2/3}}$
$= \frac{\sqrt[3]{8}^{\,2}}{\sqrt[3]{27}^{\,2}}$
$= \frac{2^2}{3^2}$
$= \frac{4}{9}$

89. $4^{-3/2} = \frac{1}{4^{3/2}}$
$= \frac{1}{\sqrt{4}^{\,3}}$
$= \frac{1}{2^3}$
$= \frac{1}{8}$

90. $-81^{5/4} = -\sqrt[4]{81}^{\,5}$
$= -(3^5)$
-243

91. DENTISTRY
$A = h^{-3/2}$
$= 16^{-3/2}$
$= \frac{1}{16^{3/2}}$
$= \frac{1}{\sqrt{16}^{\,3}}$
$= \frac{1}{4^3}$
$= \frac{1}{64}$

$\frac{1}{64}$ of the original dose is still in the system.

92. Explain why $(-64)^{1/2}$ is not a real number.
No real number squared is – 64.

CHAPTER 8 TEST

1. **Fill in the blanks.**
   a. The symbol $\sqrt{\ }$ is called a **radical** symbol.
   b. The **radicand** of $\sqrt{25a^2}$ is $25a^2$.
   c. The **Pythagorean** theorem relates the lengths of the sides of a right triangle.
   d. We read $\sqrt[3]{8}$ as "the **cube** **root** of 8."
   e. For the expression $\sqrt[5]{32}$, the **index** is 5.

2. Fill in the blanks: $\sqrt{16} = 4$ because $(4)^2 =$ **16**.

Find each square root.

3. $\sqrt{100} = 10$

4. $-\sqrt{\frac{64}{9}} = -\frac{\sqrt{64}}{\sqrt{9}}$
$= -\frac{8}{3}$

5. $\sqrt{0.25} = 0.5$

6. $\sqrt{1} = 1$

7. ELECTRONICS
Strategy:
We will substitute 980 for P and 20 for R in the formula and solve for I.
- Watts is 980.
- Ohms is 20.
- What is the current in amps?

$$I = \sqrt{\frac{P}{R}}$$
$$= \sqrt{\frac{980}{20}}$$
$$= \sqrt{49}$$
$$= 7$$

The current is 7 amps.

8. LADDERS
- Ladder (hypotenuse) is 26 feet long.
- Long leg is 24 feet long.
- How far is the ladder from the wall?

$$a^2 + b^2 = c^2$$
$$a^2 + 24^2 = 26^2$$
$$a^2 + 576 = 676$$
$$a^2 + 576 - \mathbf{576} = 676 - \mathbf{576}$$
$$a^2 = 100$$
$$\sqrt{a^2} = \sqrt{100}$$
$$a = 10$$

The ladder is 10 ft from the wall.

9. Determine whether each number is rational, irrational, or not a real number:

$\sqrt{19}$ **Irrational**

$\sqrt{-16}$ **Not a real number**

$\sqrt{144}$ **Rational**

10. Find the distance between $(-4,5)$ and $(2,11)$. Express the result in simplified radical form.

$$(-4, 5) \text{ and } (2, 11)$$
$$\uparrow\ \uparrow \qquad \uparrow\ \uparrow$$
$$(x_1, y_1) \quad (x_2, y_2)$$

$$d = \sqrt{(x_2 - x_1)^2 + (y_2 - y_1)^2}$$
$$d = \sqrt{[2-(-4)]^2 + (11-5)^2}$$
$$d = \sqrt{(2+4)^2 + (6)^2}$$
$$d = \sqrt{6^2 + 36}$$
$$d = \sqrt{36+36}$$
$$d = \sqrt{72}$$
$$d = \sqrt{36 \cdot 2}$$
$$d = 6\sqrt{2}$$
$$d \approx 8.49$$

Simplify each expression. All variables represent positive real numbers.

11. $\sqrt{4x^2} = 2x$

12. $\sqrt{54x^3} = \sqrt{9}\sqrt{6}\sqrt{x^2}\sqrt{x^1}$
$$= \sqrt{9x^2}\sqrt{6x}$$
$$= 3x\sqrt{6x}$$

13. $\sqrt{\frac{50}{49}} = \frac{\sqrt{50}}{\sqrt{49}}$
$$= \frac{\sqrt{25}\sqrt{2}}{\sqrt{49}}$$
$$= \frac{5\sqrt{2}}{7}$$

14. $\sqrt{\frac{18a^6}{2a}} = \sqrt{\frac{\overset{1}{\cancel{2}} \cdot 9a^{6-1}}{\underset{1}{\cancel{2}}}}$
$$= \sqrt{9a^5}$$
$$= \sqrt{9a^4}\sqrt{a}$$
$$= 3a^2\sqrt{a}$$

Perform each operation and simplify. All variables represent nonnegative real numbers.

15. $8\sqrt{5} + \sqrt{5} = (8+1)\sqrt{5}$
$$= 9\sqrt{5}$$

16. $16\sqrt{11} - 4\sqrt{11} - 9\sqrt{11} = (16-4-9)\sqrt{11}$
$$= 3\sqrt{11}$$

17. $\sqrt{12b^4} + \sqrt{27b^4} = \sqrt{4b^4 \cdot 3} + \sqrt{9b^4 \cdot 3}$
$$= 2b^2\sqrt{3} + 3b^2\sqrt{3}$$
$$= (2b^2 + 3b^2)\sqrt{3}$$
$$= 5b^2\sqrt{3}$$

18. $10\sqrt{3} \cdot \sqrt{2} = 10\sqrt{3 \cdot 2}$
$$= 10\sqrt{6}$$

19. $\sqrt{3}\ \ \sqrt{8}+\sqrt{6}\ =\sqrt{3}\ \sqrt{8}\ +\sqrt{3}\ \sqrt{6}$
$=\sqrt{3\cdot 8}+\sqrt{3\cdot 6}$
$=\sqrt{24}+\sqrt{18}$
$=\sqrt{4}\sqrt{6}+\sqrt{9}\sqrt{2}$
$=2\sqrt{6}+3\sqrt{2}$

20. $x\sqrt{50x}-\sqrt{200x^3}$
$=\ x\sqrt{25}\sqrt{2x}-\sqrt{100x^2}\sqrt{2x}$
$=5x\sqrt{2x}-10x\sqrt{2x}$
$=(5x-10x)\sqrt{2x}$
$=-5x\sqrt{2x}$

21. $\sqrt{2}+\sqrt{3}\ \ \sqrt{2}-\sqrt{3}\ =\ \sqrt{2}\ ^2-\ \sqrt{3}\ ^2$
$=2-3$
$=-1$

22. $-2\sqrt{8x}\ \ 3\sqrt{12x^4}\ =\ -2\cdot 3\ \ \sqrt{8x}\sqrt{12x^4}$
$=-6\sqrt{8x\cdot 12x^4}$
$=-6\sqrt{96x^5}$
$=-6\sqrt{16x^4}\sqrt{6x}$
$=-6(4x^2)\sqrt{6x}$
$=-24x^2\sqrt{6x}$

23. $5\sqrt{x}-1\ \ \sqrt{x}-4$
$=5\sqrt{x}\cdot\sqrt{x}+5\sqrt{x}\cdot(-4)-1\sqrt{x}-1\cdot(-4)$
$=5\sqrt{x\cdot x}-20\sqrt{x}-\sqrt{x}+4$
$=5x+(-20-1)\sqrt{x}+4$
$=5x-21\sqrt{x}+4$

24. $2\sqrt{3t}\ ^2=(2)^2\ \sqrt{3t}\ ^2$
$=4(3t)$
$=12t$

25. $\sqrt{x+1}\ ^2=x+1$

26. $\sqrt{x}+1\ ^2=\ \sqrt{x}\ ^2+2\cdot 1\sqrt{x}+1^2$
$=x+2\sqrt{x}+1$

27. SEWING
Perimeter $=a+b+c$
$=\sqrt{8}+\sqrt{32}+\sqrt{40}$
$=\sqrt{4\cdot 2}+\sqrt{16\cdot 2}+\sqrt{4\cdot 10}$
$=\sqrt{4}\sqrt{2}+\sqrt{16}\sqrt{2}+\sqrt{4}\sqrt{10}$
$=2\sqrt{2}+4\sqrt{2}+2\sqrt{10}$
$=(2+4)\sqrt{2}+2\sqrt{10}$
$=6\sqrt{2}+2\sqrt{10}$
The total inches is $(6\sqrt{2}+2\sqrt{10})$ in.

28. $a^2+b^2=c^2$
$x^2+x^2=16^2$
$2x^2=256$
$\frac{2x^2}{2}=\frac{256}{2}$
$x^2=128$

$\sqrt{x^2}=\sqrt{128}$
$x=\sqrt{64\cdot 2}$
$x=8\sqrt{2}$
$x\approx 11.313$
$x\approx 11.31$

The length of a leg is $8\sqrt{2}$ ft ≈ 11.31 ft.

Rationalize each denominator. All variables represent positive real numbers.

29. $\frac{2}{\sqrt{7}}=\frac{2}{\sqrt{7}}\cdot\frac{\sqrt{7}}{\sqrt{7}}$
$=\frac{2\sqrt{7}}{\sqrt{7}\sqrt{7}}$
$=\frac{2\sqrt{7}}{7}$

30. $\frac{\sqrt{3}}{\sqrt{x}-8}=\frac{\sqrt{3}}{\sqrt{x}-8}\cdot\frac{\sqrt{x}+8}{\sqrt{x}+8}$
$=\frac{\sqrt{3}\ \ \sqrt{x}+8}{\sqrt{x}-8\ \ \sqrt{x}+8}$
$=\frac{\sqrt{3x}+8\sqrt{3}}{\sqrt{x}\ ^2-8^2}$
$=\frac{\sqrt{3x}+8\sqrt{3}}{x-64}$

Solve each equation and check the result.

31. $\sqrt{x}=15$
$\sqrt{x}\ ^2=(15)^2$
$x=225$

Check:
$\sqrt{x}=15$
$\sqrt{225}\stackrel{?}{=}15$
$15=15$
True

The solution set is 225 .

32.
$$\sqrt{2-x}-2=6$$
$$\sqrt{2-x}-2+\mathbf{2}=6+\mathbf{2}$$
$$\sqrt{2-x}=8$$
$$\sqrt{2-x}^{\,2}=(8)^2$$
$$2-x=64$$
$$2-x+\boldsymbol{x}=64+\boldsymbol{x}$$
$$2=64+x$$
$$2-\mathbf{64}=64+x-\mathbf{64}$$
$$-62=x$$

Check the solution
$$\sqrt{2-x}-2=6$$
$$\sqrt{2-(-62)}-2\overset{?}{=}6$$
$$\sqrt{2+62}-2\overset{?}{=}6$$
$$\sqrt{64}-2\overset{?}{=}6$$
$$8-2\overset{?}{=}6$$
$$6=6$$
True

The solution set is -62 .

33.
$$\sqrt{3x+9}=2\sqrt{x+1}$$
$$\sqrt{3x+9}^{\,2}=2\sqrt{x+1}^{\,2}$$
$$3x+9=2^2\ \sqrt{x+1}^{\,2}$$
$$3x+9=4(x+1)$$
$$3x+9=4x+4$$
$$3x+9-\mathbf{3}\boldsymbol{x}=4x+4-\mathbf{3}\boldsymbol{x}$$
$$9=x+4$$
$$9-\mathbf{4}=x+4-\mathbf{4}$$
$$5=x$$

Check the solution
$$\sqrt{3x+9}=2\sqrt{x+1}$$
$$\sqrt{3(5)+9}\overset{?}{=}2\sqrt{5+1}$$
$$\sqrt{15+9}\overset{?}{=}2\sqrt{6}$$
$$\sqrt{24}\overset{?}{=}2\sqrt{6}$$
$$\sqrt{4}\sqrt{6}\overset{?}{=}2\sqrt{6}$$
$$2\sqrt{6}=2\sqrt{6}$$
True

The solution set is 5 .

34.
$$x=\sqrt{x-1}+1$$
$$x-\mathbf{1}=\sqrt{x-1}+1-\mathbf{1}$$
$$x-1=\sqrt{x-1}$$
$$(x-1)^2=\sqrt{x-1}^{\,2}$$
$$x^2+2\cdot(-1)x+(-1)^2=x-1$$
$$x^2-2x+1=x-1$$
$$x^2-2x+1-\boldsymbol{x}=x-1-\boldsymbol{x}$$
$$x^2-3x+1=-1$$
$$x^2-3x+1+\mathbf{1}=-1+\mathbf{1}$$
$$x^2-3x+2=0$$
$$(x-2)(x-1)=0$$

Set each factor equal to zero and solve.

$x-2=0$	$x-1=0$
$x=2$	$x=1$

Check both solutions.

$x=\sqrt{x-1}+1$	$x=\sqrt{x-1}+1$
$2\overset{?}{=}\sqrt{2-1}+1$	$1\overset{?}{=}\sqrt{1-1}+1$
$2\overset{?}{=}\sqrt{1}+1$	$1\overset{?}{=}\sqrt{0}+1$
$2\overset{?}{=}1+1$	$1\overset{?}{=}0+1$
$2=2$	$1=1$
True	True

Since two true statements result when 2 and 1 are substituted for x, 2 and 1 are solutions. The solution set is $\{2,1\}$.

35.
$$\sqrt{7t+4}+1=0$$
$$\sqrt{7t+4}+1-\mathbf{1}=0-\mathbf{1}$$
$$\sqrt{7t+4}=-1$$
$$\sqrt{7t+4}^{\,2}=(-1)^2$$
$$7t+4=1$$
$$7t+4-4=1-4$$
$$7t=-3$$
$$\frac{7t}{\mathbf{7}}=\frac{-3}{\mathbf{7}}$$
$$t=-\frac{3}{7}$$

Check:
$$\sqrt{7t+4}+1=0$$
$$\sqrt{7\cdot-\frac{\mathbf{3}}{\mathbf{7}}+4}+1\overset{?}{=}0$$
$$\sqrt{-3+4}+1\overset{?}{=}0$$
$$\sqrt{1}+1\overset{?}{=}0$$
$$2=0$$
False

Since the result is a false statement, $-\frac{3}{7}$ is not a solution of the original equation and must be discarded. The original equation has no solution. The solution set is the empty set, written as $\varnothing$.

36.
$$\sqrt{m}-5=m-11$$
$$\sqrt{m}-5+\mathbf{5}=m-11+\mathbf{5}$$
$$\sqrt{m}=m-6$$
$$\left(\sqrt{m}\right)^2=(m-6)^2$$
$$m=m^2+2\cdot(-6)m+(-6)^2$$
$$m=m^2-12m+36$$
$$m^2-12m+36-\mathbf{m}=m-\mathbf{m}$$
$$m^2-13m+36=0$$
$$(m-9)(m-4)=0$$

Set each factor equal to zero and solve.

$m-9=0$	$m-4=0$
$m=9$	$m=4$

Check both solutions.

$\sqrt{m}-5=m-11$	$\sqrt{m}-5=m-11$
$\sqrt{9}-5\overset{?}{=}9-11$	$\sqrt{4}-5\overset{?}{=}4-11$
$3-5\overset{?}{=}-2$	$2-5\overset{?}{=}-7$
$-2=-2$	$-3=-7$
True	False

Since a true statement results when 9 is substituted for m, 9 is a solution. Since a false statement results when 4 is substituted for m, 4 is an extraneous solution. The solution set is $\{9\}$.

37. CARPENTRY
- Height up the wall (leg) is 3feet.
- Length along the floor (leg) is 4 feet.
- What is the reading of the tape measure (hypotenuse)?

$$a^2+b^2=c^2$$
$$3^2+4^2=c^2$$
$$9+16=c^2$$
$$25=c^2$$
$$25-\mathbf{25}=c^2-\mathbf{25}$$
$$0=c^2-25$$
$$0=(c+5)(c-5)$$

~~$c+5=0$~~
~~$c=-5$~~

$$c-5=0$$
$$c=5$$

The measurement should be 5 ft.
He is applying the Pythagorean theorem.

38. CURVING TEST SCORES
- Original score is s.
- New score is N.
- Find the original test score if the "curved" grade is 80.

$$N=10\sqrt{s}$$
$$80=10\sqrt{s}$$
$$\frac{80}{\mathbf{10}}=\frac{10\sqrt{s}}{\mathbf{10}}$$
$$8=\sqrt{s}$$
$$8^2=\left(\sqrt{s}\right)^2$$
$$64=s$$

The original test score was 64.

39. Explain why we cannot add $3\sqrt{2}$ and $2\sqrt{3}$.
They are not like radicals, the radicands are different.

40. Explain why $\sqrt{-49}$ is not a real number.
No real number squared is equal to -49.

Find each root, if possible. All variables represent nonnegative real numbers.

41. $\sqrt[3]{-125}=\sqrt[3]{(-5)^3}$
$=-5$

42. $\sqrt[4]{-256}$
Not a real number.

43. $\sqrt[3]{\frac{1}{64}}=\sqrt[3]{\left(\frac{1}{4}\right)^3}$
$=\frac{1}{4}$

44. $\sqrt[5]{-32x^{10}}=\sqrt[5]{(-2)^5(x^2)^5}$
$=-2x^2$

Simplify. All variables represent nonnegative real numbers.

45. $\sqrt[4]{81x^4}=\sqrt[4]{3^4x^4}$
$=3x$

46. $\sqrt[3]{88}=\sqrt[3]{8\cdot 11}$
$=\sqrt[3]{8}\sqrt[3]{11}$
$=2\sqrt[3]{11}$

Evaluate.

47. $144^{\frac{1}{2}}=\sqrt{144}$
$=12$

48. $8^{\frac{2}{3}}=\left(\sqrt[3]{8}\right)^2$
$=2^2$
$=4$

49. $(-27)^{-4/3}=\frac{1}{(-27)^{4/3}}$
$=\frac{1}{\left(\sqrt[3]{-27}\right)^4}$
$=\frac{1}{(-3)^4}$
$=\frac{1}{81}$

50. $-9^{-1/2}=-\frac{1}{9^{1/2}}$
$=-\frac{1}{\sqrt{9}}$
$=-\frac{1}{3}$

CHAPTERS 1 - 8 CUMULATIVE REVIEW

1. Determine whether each statement is true or false. [Section 1.3]

   a. All whole numbers are integers. **True**

   b. π is a rational number. **False**

   c. A real number is neither rational or irrational. **True**

2. Evaluate: [Section 1.7]

$$\frac{-3(3+2)^2-(-5)}{17-|-22|}=\frac{-3(5)^2+5}{17-22}$$
$$=\frac{-3(25)+5}{17+(-22)}$$
$$=\frac{-75+5}{-5}$$
$$=\frac{-70}{-5}$$
$$=14$$

3. Simplify: [Section 1.9]

$$3p-6(p+z)+p=3p-6p-6z+p$$
$$=(3-6+1)p-6z$$
$$=-2p-6z$$

4. Solve: [Section 2.2]

$$2-(4x+7)=3+2(x+2)$$
$$2-4x-7=3+2x+4$$
$$-4x-5=2x+7$$
$$-4x-5-\mathbf{2x}=2x+7-\mathbf{2x}$$
$$-6x-5=7$$
$$-6x-5+\mathbf{5}=7+\mathbf{5}$$
$$-6x=12$$
$$\frac{-6x}{\mathbf{-6}}=\frac{12}{\mathbf{-6}}$$
$$x=-2$$

5. BACKPACKS: [Section 2.3]

   - r is the percent (20%).
   - b is the weight (85 lb) of girl.
   - What is the amount (A) to be carried?

   $A=rb$

   $A=(0.20)(85)$

   $A=17$

   Total weight should be no more than 17 lb.

6. SURFACE AREA: [Section 2.4]

   - 202 in^2 is the total surface area.
   - 9 inches is the length.
   - 5 inches is the width.
   - What is the height?

$$A=2lw+2wh+2lh$$
$$202=2(9)(5)+2(5)h+2(9)h$$
$$202=90+10h+18h$$
$$202=90+28h$$
$$202-\mathbf{90}=90+28h-\mathbf{90}$$
$$112=28h$$
$$\frac{112}{\mathbf{28}}=\frac{28h}{\mathbf{28}}$$
$$4=h$$

   The height is 4 in.

7. SEARCH AND RESCUE: [Section 2.6]

Analyze

- 1st team travels north at 2 mph.
- 2nd team travels south at 4 mph.
- Both teams start from same place at the same time.
- How long will it take them to search a distance of 21 miles between them?

Assign

Let t = time in hours they are 21 miles apart

Form

	rate	• time	= distance
North	2	t	$\mathbf{2t}$
South	4	t	$\mathbf{4t}$
		Total:	**21**

The distance traveled north	plus	the distance traveled south	equals	the total distance.
$2t$	$+$	$4t$	$=$	21

Solve

$$2t+4t=21$$
$$6t=21$$
$$\frac{6t}{\mathbf{6}}=\frac{21}{\mathbf{6}}$$
$$t=3.5$$

State

After 3.5 hours they will be 21 miles apart.

Check

The distance traveled north is 2(3.5), or 7 mi.
The distance traveled south is 4(3.5), or 14 mi.
Since the total was 7 mi + 14 mi = 21 mi, the result checks.

8. BLENDING COFFEE [Section 2.6]

Analyze

- Regular coffee sells for \$8 per lb.
- Gourmet coffee sells for \$14 per lb.
- A blend is to sell for \$10 per lb.
- Start with 40 lb of gourmet coffee.
- How many pounds of regular coffee are needed?

Assign

Let x = the lb of regular coffee needed

Form

	Number •	Value =	Total value
Gourmet	40	14	**560**
Regular	x	8	$\mathbf{8x}$
Blend	$\mathbf{x+40}$	10	$\mathbf{10(x+40)}$

The value of the gourmet	plus	the value of the regular	equals	the total value of the blend.
560	+	$8x$	=	$10(x+40)$

Solve

$$560+8x=10(x+40)$$
$$560+8x=10x+400$$
$$560+8x-\mathbf{8x}=10x+400-\mathbf{8x}$$
$$560-2x+400$$
$$560-\mathbf{400}=2x+400-\mathbf{400}$$
$$160=2x$$
$$\frac{160}{\mathbf{2}}=\frac{2x}{\mathbf{2}}$$
$$80=x$$

State

80 lb of the regular coffee will be needed.

Check

The value of the gourmet is 40(\$14), or \$560.
The value of the regular is 80(\$8), or \$640.
The value of the blend is 120(\$10), or \$1,200.
Since the total was $\$560+\$640=\$1,200$, the result checks.

9. Solve: $3-3x\geq 6+x$ [Section 2.7]

$$3-3x\geq 6+x$$
$$3-3x-\mathbf{x}\geq 6+x-\mathbf{x}$$
$$3-4x\geq 6$$
$$3-4x-\mathbf{3}\geq 6-\mathbf{3}$$
$$-4x\geq 3$$
$$\frac{-4x}{\mathbf{-4}}\leq\frac{3}{\mathbf{-4}}$$
$$x\leq -\frac{3}{4}$$
$$\left(-\infty,-\frac{3}{4}\right]$$

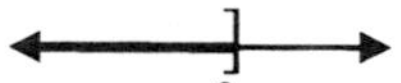

10. Is $(-6,-7)$ a solution of $4x-3y=-4$? [Section 3.1]

$$4x-3y=-4$$
$$4(\mathbf{-6})-3(\mathbf{-7})\overset{?}{=}-4$$
$$-24+21\overset{?}{=}-4$$
$$-3=-4$$
False

Graph each equation.

11. $y=\frac{1}{2}x$ [Section 3.2]

x	y	(x, y)
−2	$y=\frac{1}{2}x$ $=\frac{1}{2}(-2)$ $=-1$	(−2, −1)
0	$y=\frac{1}{2}x$ $=\frac{1}{2}(0)$ $=0$	(0, 0)
2	$y=\frac{1}{2}x$ $=\frac{1}{2}(2)$ $=1$	(2, 1)

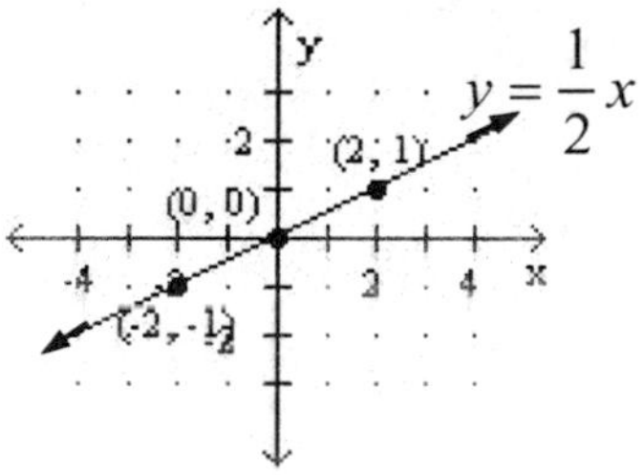

12. $3x - 4y = 12$ [Section 3.3]

y-intercept:	x-intercept:
If $x = 0$,	If $y = 0$
$3(\mathbf{0}) - 4y = 12$	$3x - 4(\mathbf{0}) = 12$
$-4y = 12$	$3x = 12$
$y = -3$	$x = 4$

The y-intercept is (0, –3), and the x-intercept is (4, 0).

Check Point

$$3x - 4y = 12$$
$$3(1) - 4y = 12$$
$$-4y = 9$$
$$y = -\frac{9}{4}$$
$$\left(1, -\frac{9}{4}\right)$$

13. $x = 5$ [Section 3.3]

x	y	(x, y)
5	–3	(5,–3)
5	0	(5,0)
5	2	(5,2)

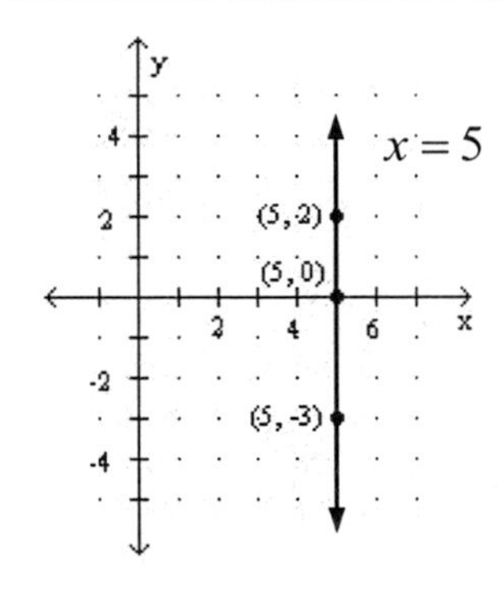

14. $y = 2x^2 - 3$ [Section 5.4]

x	y	(x, y)
–3	$y = 2(-3)^2 - 3$ $= 2(9) - 3$ $= 18 + (-3)$ $= 15$	(–3, 15)
–2	$y = 2(-2)^2 - 3$ $= 2(4) - 3$ $= 8 + (-3)$ $= 5$	(–2, 5)
–1	$y = 2(-1)^2 - 3$ $= 2(1) - 3$ $= 2 + (-3)$ $= -1$	(–1, –1)
0	$y = 2(0)^2 - 3$ $= 2(0) - 3$ $= 0 + (-3)$ $= -3$	(0, –3)
1	$y = 2(1)^2 - 3$ $= 2(1) - 3$ $= 2 + (-3)$ $= -1$	(1, –1)
2	$y = 2(2)^2 - 3$ $= 2(4) - 3$ $= 8 + (-3)$ $= 5$	(2, 5)
3	$y = 2(3)^2 - 3$ $= 2(9) - 3$ $= 18 + (-3)$ $= 15$	(3, 15)

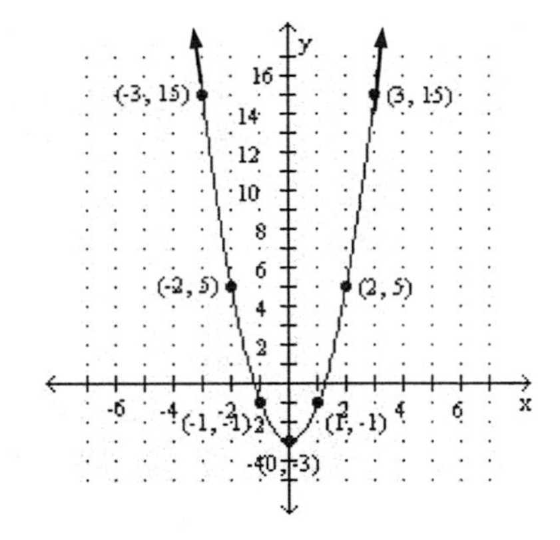

15. SKYPE [Section 3.4]

Find the slope of the line passing through (2007, 217) and (2010, 559).

$$(x_1, y_1) \qquad (x_2, y_2)$$
$$m = \frac{y_2 - y_1}{x_2 - x_1}$$
$$m = \frac{559 - 217}{2010 - 2007}$$
$$m = \frac{342}{3}$$
$$m = 114$$

An increase of 114 million subscribers/yr.

16. What is the slope of the line defined by each equation? [Section 3.5]

$y = mx + b$, where m is the slope of the line.

a. $y = 3x - 7$

$m = 3$

b.
$$2x + 3y = -10$$
$$2x + 3y - \mathbf{2x} = \mathbf{-2x} - 10$$
$$3y = -2x - 10$$
$$\frac{3y}{\mathbf{3}} = \frac{-2x}{\mathbf{3}} - \frac{10}{\mathbf{3}}$$
$$y = -\frac{2}{3}x - \frac{10}{3}$$
$$m = -\frac{2}{3}$$

17. Write the equation of the line passing through (– 2, 5) and (4, 8) in slope-intercept form. [Section 3.6]

$$(-2,5) \text{ and } (4,\ 8)$$
$$(x_1, y_1) \text{ and } (x_2, y_2)$$
$$m = \frac{y_2 - y_1}{x_2 - x_1}$$
$$= \frac{8-5}{4-(-2)}$$
$$= \frac{3}{4+2}$$
$$= \frac{3}{6}$$
$$= \frac{1}{2}$$

Passes through (4,8) , $m = \dfrac{1}{2}$

$x_1 = 4$ and $y_1 = 8$

$$y - y_1 = m(x - x_1)$$
$$y - 8 = \frac{1}{2}(x-4)$$
$$y - 8 = \frac{\mathbf{1}}{\mathbf{2}}x - \frac{\mathbf{1}}{\mathbf{2}}(4)$$
$$y - 8 = \frac{1}{2}x - 2$$
$$y - 8 + \mathbf{8} = \frac{1}{2}x - 2 + \mathbf{8}$$
$$y = \frac{1}{2}x + 6$$

slope-intercept form

18. If $f(x) = 2x^2 - 3x + 1$, find $f(-3)$. [Section 3.8]

$$f(x) = 2x^2 - 3x + 1$$
$$f(-3) = 2(\mathbf{-3})^2 - 3(\mathbf{-3}) + 1$$
$$= 2(9) + 9 + 1$$
$$= 18 + 9 + 1$$
$$= 28$$
$$f(-3) = 28$$

19. BOATING. [Section 3.8]
Yes, the graph represents a function.

20. Solve the system $\begin{cases} x + y = 4 \\ y = x + 6 \end{cases}$ by graphing.
[Section 4.1] The solution is $(-1, 5)$.

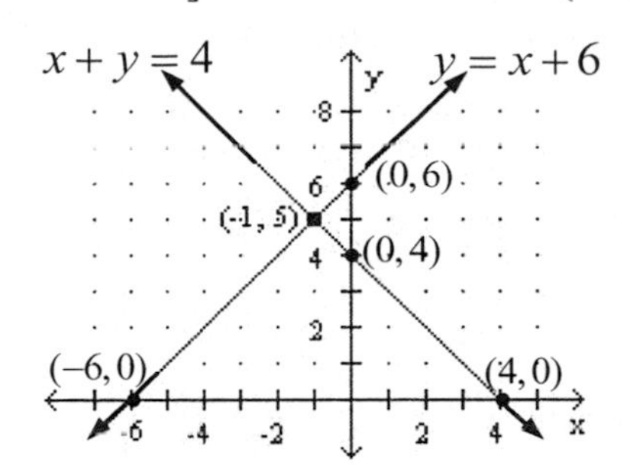

Solve each system of equations.

21. $\begin{cases} x = y + 4 \\ 2x + y = 5 \end{cases}$ [Section 4.2]

$2x + y = 5$ The 2nd equation
Substitute for x.

$$2(\mathbf{y+4}) + y = 5$$
$$2y + 8 + y = 5$$
$$3y + 8 = 5$$
$$3y + 8 - \mathbf{8} = 5 - \mathbf{8}$$
$$3y = -3$$
$$\frac{3y}{\mathbf{3}} = \frac{-3}{\mathbf{3}}$$
$$y = -1$$

$x = y + 4$ The 1st equation
$x - (\mathbf{-1}) + 4$
$x = 3$
The solution is (3, 1).

22. $\begin{cases} 3s + 4t = 5 \\ 2s - 3t = -8 \end{cases}$ [Section 4.3]

Eliminate t.
Multiply both sides of the 1st equation by 3.
Multiply both sides of the 2nd equation by 4.

$$9s + 12t = 15$$
$$8s - 12t = -32$$
$$17s = -17$$
$$\frac{17s}{\mathbf{17}} = \frac{-17}{\mathbf{17}}$$
$$s = -1$$

$3s + 4t = 5$ The 1st equation
$$3(\mathbf{-1}) + 4t = 5$$
$$-3 + 4t = 5$$
$$-3 + 4t + \mathbf{3} = 5 + \mathbf{3}$$
$$4t = 8$$
$$\frac{4t}{\mathbf{4}} = \frac{8}{\mathbf{4}}$$
$$t = 2$$

The solution is $(-1, 2)$.

23. FINANCIAL PLANNING [Section 4.4]

Analyze

- Investing $6,000.
- Savings account paying 6% annual interest.
- Development plan paying 12% annually.
- Combined interest for first year was $540.
- How much was invested at each rate?

Assign

Let x = amount invested in savings account

y = amount invested in mini-mall

Form

The amount invested in savings	plus	the amount invested in the mini-mall	is	$6,000.
x	$+$	y	$=$	6,000

	Principal ·	Rate ·	Time =	Interest
Savings	x	6%	1 yr	**$0.06x$**
Mini-mall	y	12%	1 yr	**$0.12y$**

Total Interest = $540

The interest earned by savings	plus	the interest earned by mini-mall	is	$540.
$0.06x$	$+$	$0.12y$	$=$	540

Solve

$$\begin{cases} x+y=6,000 \\ 0.06x+0.12y=540 \end{cases}$$

Eliminate x.

Multiply both sides of the 1st equation by -6.

Multiply both sides of the 2nd equation by 100.

$$\begin{aligned} -6x-6y &= -36,000 \\ 6x+12y &= 54,000 \\ \hline 6y &= 18,000 \\ \frac{6y}{\mathbf{6}} &= \frac{18,000}{\mathbf{6}} \\ y &= 3,000 \end{aligned}$$

$$\begin{aligned} x+y &= 6,000 \quad \text{The 1st equation} \\ x+\mathbf{3,000} &= 6,000 \\ x+3,000-3,000 &= 6,000-3,000 \\ x &= 3,000 \end{aligned}$$

State

$3,000 was invested at 6%.

$3,000 was invested at 12%.

Check

The results check.

24. Graph: $\begin{cases} 3x+2y\ge 6 \\ x+3y\le 6 \end{cases}$. [Section 4.5]

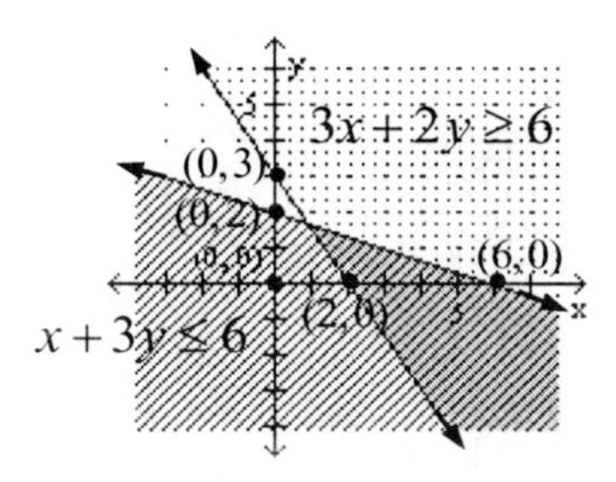

Simplify. Use only positive exponents in your answers.

25. [Section 5.1]

$$\begin{aligned} \left(x^5\right)^2\left(x^7\right)^3 &= \left(x^{5\cdot 2}\right)\left(x^{7\cdot 3}\right) \\ &= \left(x^{10}\right)\left(x^{21}\right) \\ &= x^{10+21} \\ &= x^{31} \end{aligned}$$

26. [Section 5.1]

$$\begin{aligned} \left(\frac{a^3b}{c^4}\right)^5 &= \frac{a^{3\cdot 5}b^{1\cdot 5}}{c^{4\cdot 5}} \\ &= \frac{a^{15}b^5}{c^{20}} \end{aligned}$$

27. [Section 5.2]

$$\begin{aligned} 4^{-3}\cdot 4^{-2}\cdot 4^5 &= 4^{-3+(-2)+5} \\ &= 4^0 \\ &= 1 \end{aligned}$$

28. [Section 5.2]

$$\begin{aligned} \left(2a^{-2}b^3\right)^{-4} &= 2^{1\cdot -4}a^{-2\cdot -4}b^{3\cdot -4} \\ &= 2^{-4}a^8b^{-12} \\ &= \frac{a^8}{2^4b^{12}} \\ &= \frac{a^8}{16b^{12}} \end{aligned}$$

29. ASTRONOMY [Section 5.3]

$$\begin{aligned} &(3.0\times 10^{16})(1.6\times 10^2) \\ &= (3.0\times 1.6)(10^{16}\times 10^2) \\ &= 4.8\times 10^{16+2} \\ &= 4.8\times 10^{18} \end{aligned}$$

The distance to Betelgeuse is 4.8×10^{18} m.

30. Write 0.0000000043 in scientific notation. [Section 5.3]

Move the decimal 9 places to the right.

$0.0000000043 = 4.3\times10^{-9}$

Perform the operations.

31. $(3a^2-2a+4)-(a^2-3a+7)$ [Section 5.5]

$$\begin{aligned}&= 3a^2-2a+4-a^2+3a-7\\&=(3a^2-a^2)+(-2a+3a)+(4-7)\\&=(3-1)a^2+(-2+3)a+(4-7)\\&=2a^2+a-3\end{aligned}$$

32. $0.3p^5(0.4p^4-6p^2)$ [Section 5.6]

$$\begin{aligned}&=\mathbf{0.3p^5}(0.4p^4)-\mathbf{0.3p^5}(6p^2)\\&=0.3(0.4)p^{5+4}-0.3(6)p^{5+2}\\&=0.12p^9-1.8p^7\end{aligned}$$

33. $(-3t+2s)(2t-3s)$ [Section 5.6]

$$\begin{aligned}&=-\mathbf{3t}(2t)-\mathbf{3t}(-3s)+\mathbf{2s}(2t)+\mathbf{2s}(-3s)\\&=-6t^2+9st+4st-6s^2\\&=-6t^2+(9+4)st-6s^2\\&=-6t^2+13st-6s^2\end{aligned}$$

34. $(4b-8)^2$ [Section 5.7]

$$\begin{aligned}&=(4b)^2+2(4b)(-8)+(-8)^2\\&=16b^2-64b+64\end{aligned}$$

35. $\left(6b+\frac{1}{2}\right)\left(6b-\frac{1}{2}\right)$ [Section 5.7]

$$\begin{aligned}&=(6b)^2-\left(\frac{1}{2}\right)^2\\&=36b^2-\frac{1}{4}\end{aligned}$$

36. $(2x^2+3x-2)\div(x+2)$ [Section 5.8]

$$\begin{array}{r}2x-1\\x+2\overline{)\,2x^2+3x-2}\\\underline{-(2x^2+4x)}\\-x-2\\\underline{-(-x-2)}\\0\end{array}$$

Factor completely, if possible.

37. $12x^2y-6xy^2+9xy^3$ [Section 6.1]

$$\begin{aligned}&=(\mathbf{3}\cdot4\mathbf{xxy}-\mathbf{3}\cdot2\mathbf{xy}y+\mathbf{3}\cdot3\mathbf{xy}y^2)\\&=3xy(4x-2y+3y^2)\end{aligned}$$

38. $2x^2+2xy-3x-3y$ [Section 6.1]

$$\begin{aligned}&=(2x^2+2xy)+(-3x-3y)\\&=(\mathbf{2x}\cdot x+\mathbf{2x}\cdot y)+(\mathbf{(-3)}\cdot x+\mathbf{(-3)}\cdot y)\\&=\mathbf{2x}(x+y)+\mathbf{(-3)}(x+y)\\&=(x+y)(2x-3)\end{aligned}$$

39. $x^2+7x+10$ [Section 6.2]

The positive factors of 10 whose sum is 7 are 2 and 5.

$x^2+7x+10=(x+2)(x+5)$

40. $\underset{a}{6}a^2-\underset{b}{7}a-\underset{c}{20}$ [Section 6.3]

In $6a^2-7a-20$, we have $a=6$, $b=-7$, and $c=-20$. The key number is $ac=6(-20)=-120$. We must find a factorization of -120 in which the sum of the factors is $b=-7$. Since the factors must have a negative product, their signs must be different. The pairs of factors are -15 and 8.

$$\begin{aligned}6a^2-7a-20&=6a^2-15a+8a-20\\&=(6a^2-15a)+(8a-20)\\&=3a(2a-5)+4(2a-5)\\&=(2a-5)(3a+4)\end{aligned}$$

41. $6+3x^2+x=\underset{a}{3}x^2+\underset{b}{1}x+\underset{c}{6}$ [Section 6.3]

In $3x^2+1x+6$, we have $a=3$, $b=1$ and $c=6$. The key number is $ac=3(6)=18$. We must find a factorization of 18 in which the sum of the factors is $b=1$. Since the factors must have a positive product, their signs must be the same. The pairs of factors are none. This trinomial is prime.

$3x^2+x+6$ is prime.

42. $25a^2-70ab+49b^2$ [Section 6.4]

- The first term $25a^2$ is the square of $\mathbf{5a}$.
- The last term $49b^2$ is the square of $-\mathbf{7b}$.
- The middle term is twice the product of $5a$ and $-7b$: $2(5a)(-7b)=\mathbf{-70ab}$.

$$\begin{aligned}25a^2-70ab+49b^2&=(5a-7b)(5a-7b)\\&=(5a-7b)^2\end{aligned}$$

43. [Section 6.5]

$$\begin{aligned}a^3+8b^3&=(a)^3+(2b)^3\\&=(a+2b)[a^2-a\cdot2b+(2b)^2]\\&=(a+2b)(a^2-2ab+4b^2)\end{aligned}$$

44. [Section 6.6]

$$\begin{aligned}2x^5 - 32x &= 2x(x^4 - 16)\\ &= 2x(x^2+4)(x^2-4)\\ &= 2x(x^2+4)(x+2)(x-2)\end{aligned}$$

Solve each equation.

45. $x^2 + 3x + 2 = 0$ [Section 6.7]

$(x+1)(x+2) = 0$

$x+1=0$ or $x+2=0$

$x=-1$ | $x=-2$

The solutions are -1 and -2.

46. $5x^2 = 10x$ [Section 6.7]

$5x^2 - \mathbf{10x} = 10x - \mathbf{10x}$

$5x^2 - 10x = 0$

$5x(x-2) = 0$

$5x = 0$ or $x-2=0$

$x=0$ | $x=2$

The solutions are 0 and 2.

47. $6x^2 - x = 2$ [Section 6.7]

$6x^2 - x - \mathbf{2} = 2 - \mathbf{2}$

$6x^2 - x - 2 = 0$

$(3x-2)(2x+1) = 0$

$3x-2=0$ or $2x+1=0$

$3x = 2$ | $2x = -1$

$x = \frac{2}{3}$ | $x = -\frac{1}{2}$

The solutions are $\frac{2}{3}$ and $-\frac{1}{2}$.

48. $a^2 - 25 = 0$ [Section 6.7]

$(a+5)(a-5) = 0$

$a+5=0$ or $a-5=0$

$a=-5$ | $a=5$

The solutions are -5 and 5.

49. CHILDREN'S STICKERS [Section 6.8]

Analyze

- Width is 1cm less than the length.
- Rectangular sticker has area of is 20 sq. cm.
- Find the length.

Assign

Let l = length in cm

$l - 1$ = width in cm

Form

The length	times	the width	equals	the area of the sticker.
l	$\cdot$	$l-1$	$=$	20

Solve

$$l(l-1) = 20$$
$$l^2 - l = 20$$
$$l^2 - l - \mathbf{20} = 20 - \mathbf{20}$$
$$l^2 - l - 20 = 0$$
$$(l+4)(l-5) = 0$$

$l+4=0$ or $l-5=0$

~~$l = -4$~~ | $l = 5$

State

The length of the sticker is 5 cm.

The width of the sticker is $5-1=4$ cm.

Check

A rectangle with dimensions 4 cm by 5 cm has an area of 20 cm^2, and the width is one cm less than the length. The result checks.

50. Simplify: [Section 7.1]

$$\frac{x^2+2x+1}{x^2-1} = \frac{(x+1)(x+1)}{(x+1)(x-1)} = \frac{\overset{1}{\cancel{(x+1)}}(x+1)}{\underset{1}{\cancel{(x+1)}}(x-1)} = \frac{x+1}{x-1}$$

Perform the operations. Simplify, if possible.

51. $\frac{p^2-p-6}{3p-9} \div \frac{p^2+6p+9}{p^2-9}$ [Section 7.2]

$$= \frac{p^2-p-6}{3p-9} \cdot \frac{p^2-9}{p^2+6p+9}$$
$$= \frac{(p-3)(p+2)(p+3)(p-3)}{3(p-3)(p+3)(p+3)}$$
$$= \frac{\cancel{(p-3)}^1(p+2)\cancel{(p+3)}^1(p-3)}{3\cancel{(p-3)}_1\cancel{(p+3)}_1(p+3)}$$
$$= \frac{(p+2)(p-3)}{3(p+3)}$$

52. [Section 7.2]

$$\frac{12x^2}{7-x} \cdot \frac{x-7}{20x^3} = \frac{12x^2}{-1(-7+x)} \cdot \frac{x-7}{20x^3}$$
$$= -\frac{12x^2(x-7)}{20x^3(x-7)}$$
$$= -\frac{\cancel{4}^1 \cdot 3\cancel{x^2}^1\cancel{(x-7)}^1}{\cancel{4}_1 \cdot 5\cancel{x^2}_1 \cdot x\cancel{(x-7)}_1}$$
$$= -\frac{3}{5x}$$

53. [Section 7.3]

$$\frac{13}{15a} - \frac{8}{15a} = \frac{13-8}{15a}$$
$$= \frac{5}{15a}$$
$$= \frac{\cancel{5}^1}{3 \cdot \cancel{5}_1 a}$$
$$= \frac{1}{3a}$$

54. $\frac{x+2}{x+5} - \frac{x-3}{x+7}$ [Section 7.4]

$$= \frac{(x+2)}{(x+5)} \cdot \frac{\mathbf{(x+7)}}{\mathbf{(x+7)}} - \frac{(x-3)}{(x+7)} \cdot \frac{\mathbf{(x+5)}}{\mathbf{(x+5)}}$$
$$= \frac{(x+2)(x+7)-(x-3)(x+5)}{(x+5)(x+7)}$$
$$= \frac{(x^2+9x+14)-(x^2+2x-15)}{(x+5)(x+7)}$$
$$= \frac{x^2+9x+14-x^2-2x+15}{(x+5)(x+7)}$$
$$= \frac{(1-1)x^2+(9-2)x+(14+15)}{(x+5)(x+7)}$$
$$= \frac{7x+29}{(x+5)(x+7)}$$

55. [Section 7.4]

$$\frac{1}{6b^4} - \frac{8}{9b^2} = \frac{1}{6b^4} \cdot \frac{\mathbf{3}}{\mathbf{3}} - \frac{8}{9b^2} \cdot \frac{\mathbf{2b^2}}{\mathbf{2b^2}}$$
$$= \frac{3-16b^2}{18b^4}$$

56. $\dfrac{\frac{1}{x}+\frac{1}{y}}{\frac{1}{x}-\frac{1}{y}}$, LCD $= xy$ [Section 7.5]

$$\frac{\frac{1}{x}+\frac{1}{y}}{\frac{1}{x}-\frac{1}{y}} = \frac{\frac{1}{x}+\frac{1}{y}}{\frac{1}{x}-\frac{1}{y}} \cdot \frac{\mathbf{xy}}{\mathbf{xy}}$$
$$= \frac{\left(\frac{1}{x}+\frac{1}{y}\right)xy}{\left(\frac{1}{x}-\frac{1}{y}\right)xy}$$
$$= \frac{\frac{1}{x}(xy)+\frac{1}{y}(xy)}{\frac{1}{x}(xy)-\frac{1}{y}(xy)}$$
$$= \frac{\frac{\cancel{x}^1 y}{\cancel{x}_1} + \frac{x\cancel{y}^1}{\cancel{y}_1}}{\frac{\cancel{x}^1 y}{\cancel{x}_1} - \frac{x\cancel{y}^1}{\cancel{y}_1}}$$
$$= \frac{y+x}{y-x}$$

57. Solve: $\frac{7}{a^2-a-2}+\frac{1}{a+1}=\frac{3}{a-2}$

[Section 7.6]

$$\left.\begin{aligned} a+1&=a+1\\ a-2&=a-2\\ a^2-a-2&=(a+1)(a-2)\end{aligned}\right\}\text{LCD}=(a+1)(a-2)$$

$$a\neq -1,2$$

$$\frac{7}{a^2-a-2}+\frac{1}{a+1}=\frac{3}{a-2}$$

$$\frac{7}{a^2-a-2}\cdot(a+1)(a-2)+\frac{1}{a+1}\cdot(a+1)(a-2)=\frac{3}{a-2}\cdot(a+1)(a-2)$$

$$\frac{7\,\cancel{(a+1)}\,\cancel{(a-2)}}{\cancel{(a+1)}\,\cancel{(a-2)}}+\frac{\cancel{(a+1)}(a-2)}{\cancel{(a+1)}}=\frac{3(a+1)\,\cancel{(a-2)}}{\cancel{(a-2)}}$$

$$7+(a-2)=3(a+1)$$
$$7+a-2=3a+3$$
$$a+5=3a+3$$
$$a+5-\mathbf{a}=3a+3-\mathbf{a}$$
$$5=2a+3$$
$$5-\mathbf{3}=2a+3-\mathbf{3}$$
$$2=2a$$
$$\frac{2}{\mathbf{2}}=\frac{2a}{\mathbf{2}}$$
$$1=a$$

Checking:

$$\frac{7}{a^2-a-2}+\frac{1}{a+1}=\frac{3}{a-2}$$
$$\frac{7}{(1)^2-(1)-2}+\frac{1}{1+1}\stackrel{?}{=}\frac{3}{1-2}$$
$$\frac{7}{1-1-2}+\frac{1}{1+1}\stackrel{?}{=}\frac{3}{1-2}$$
$$\frac{7}{-2}+\frac{1}{2}\stackrel{?}{=}\frac{3}{-1}$$
$$\frac{-7}{2}+\frac{1}{2}\stackrel{?}{=}-3$$
$$\frac{-6}{2}\stackrel{?}{=}-3$$
$$-3=-3$$

True

58. FILLING A POOL [Section 7.7]

Analyze

- 1st pipe can fill pool in 5 hours.
- 2nd pipe can fill pool in 4 hours.
- How long will it take the two pipes, working together, to fill the pool?

Assign

Let x = the number of hours it will take both pipes working together to fill the pool.

Form

	Rate	• Time	= Work Completed
1st pipe	$\frac{1}{5}$	x	$\frac{\boldsymbol{x}}{\mathbf{5}}$
2nd pipe	$\frac{1}{4}$	x	$\frac{\boldsymbol{x}}{\mathbf{4}}$

The part of job done by 1st pipe. plus The part of job done by 2nd pipe. equals 1 job completed

$$\frac{x}{5}\quad+\quad\frac{x}{4}\quad=\quad 1$$

Solve

$$\frac{x}{5}+\frac{x}{4}=1$$
$$\text{LCD}=20$$
$$\mathbf{20}\left(\frac{x}{5}+\frac{x}{4}\right)=\mathbf{20}(1)$$
$$\cancel{5}\cdot4\left(\frac{x}{\cancel{5}}\right)+\cancel{4}\cdot5\left(\frac{x}{\cancel{4}}\right)=20$$
$$4x+5x=20$$
$$9x=20$$
$$\frac{9x}{\mathbf{9}}=\frac{20}{\mathbf{9}}$$
$$x=\frac{20}{9}$$
$$x=2\frac{2}{9}$$

State

It takes both working together $2\frac{2}{9}$ hours to fill the pool.

Check

The result checks.

59. ONLINE SALES [Section 7.8]

Analyze

- We know 9 sales for every 500 hits.
- How many out of 360,000 hits were sales?

Assign

Let x = number of total sales

9 sales is to 500 hits as x sales is to 360,000 hits.

Form

$$\begin{array}{l} \text{9 sales} \rightarrow \\ \text{500 hits} \rightarrow \end{array} \frac{9}{500} = \frac{x}{360{,}000} \begin{array}{l} \leftarrow \text{total number of sales} \\ \leftarrow \text{360,000 total hits} \end{array}$$

Solve

$$\frac{9}{500} = \frac{x}{360{,}000}$$

$$3{,}240{,}000 = 500x$$

$$\frac{3{,}240{,}000}{\mathbf{500}} = \frac{500x}{\mathbf{500}}$$

$$6{,}480 = x$$

State

6,480 sales were made out of 360,000 hits.

Check

$$\frac{9}{500} = \frac{x}{360{,}000}$$

$$\frac{9}{500} \stackrel{?}{=} \frac{\mathbf{6{,}480}}{360{,}000}$$

$$3{,}240{,}000 = 3{,}240{,}000$$

60. CAFETERIAS [Section 7.9]

Step 1: We let D represent the time (in days) that it takes to consume the punch. Translating the words *dispenser of punch varies inversely as the number of children.* We get the equation $D = \frac{k}{c}$.

Step 2: To find the constant of variation, k, we substitute 3 for D and 45 for c.

$$D = \frac{k}{c}$$

$$3 = \frac{k}{45}$$

$$3 \cdot 45 = k$$

$$135 = k$$

Step 3: We now substitute the value of k into the equation from step 1.

$$D = \frac{135}{c}$$

Step 4: We can find the *number of days* by substituting 30 for c in the equation from step 3.

$$D = \frac{135}{c}$$

$$D = \frac{135}{30}$$

$$D = \frac{9 \cdot \cancel{15}^{1}}{2 \cdot \cancel{15}_{1}}$$

$$D = 4\frac{1}{2}$$

The punch will last $4\frac{1}{2}$ days for 30 children.

61. CARGO SPACE [Section 8.1]

- Height (shorter leg) of opening is 48 in.
- Width (longer leg) of opening is 55 in.
- How wide (hypotenuse) of piece of ply wood will fit?

$$c^2 = a^2 + b^2$$

$$c^2 = 48^2 + 55^2$$

$$c^2 = 2{,}304 + 3{,}025$$

$$c^2 = 5{,}329$$

$$c = \sqrt{5{,}329}$$

$$c = 73$$

Its width would be 73 in.

62. Explain why $\sqrt{-4}$ is not a real number. [Section8.1]

No real number squared is -4.

Simplify. All variables represent nonnegative real numbers.

63. $\sqrt{64} = 8$ [Section 8.1]

64. [Section 8.2]

$$\sqrt{32x^3} = \sqrt{16x^2 \cdot 2x}$$

$$= \sqrt{16x^2}\sqrt{2x}$$

$$= 4x\sqrt{2x}$$

65. [Section 8.6]

$$-\sqrt[3]{-27m^6} = -\sqrt[3]{(-3)^3(m^2)^3}$$
$$= -(-3m^2)$$
$$= 3m^2$$

66. [Section 8.6]

$$\sqrt[4]{81t^4} = \sqrt[4]{(3)^4(t)^4}$$
$$= 3t$$

Perform each operation and simplify. All variables represent nonnegative real numbers.

67. $\sqrt{48} - \sqrt{8} + \sqrt{27} - \sqrt{32}$ [Section 8.3]

$$= \sqrt{16 \cdot 3} - \sqrt{4 \cdot 2} + \sqrt{9 \cdot 3} - \sqrt{16 \cdot 2}$$
$$= \sqrt{16}\sqrt{3} - \sqrt{4}\sqrt{2} + \sqrt{9}\sqrt{3} - \sqrt{16}\sqrt{2}$$
$$= 4\sqrt{3} - 2\sqrt{2} + 3\sqrt{3} - 4\sqrt{2}$$
$$= (4\sqrt{3} + 3\sqrt{3}) + (-2\sqrt{2} - 4\sqrt{2})$$
$$= (4+3)\sqrt{3} + (-2-4)\sqrt{2}$$
$$= 7\sqrt{3} - 6\sqrt{2}$$

68. $\left(\sqrt{y} - 4\right)\left(\sqrt{y} - 5\right)$ [Section 8.4]

$$= \sqrt{\mathbf{y}} \cdot \sqrt{y} + \sqrt{\mathbf{y}} \cdot (-5) - \mathbf{4}\sqrt{y} - \mathbf{4} \cdot (-5)$$
$$= \sqrt{y \cdot y} - 5\sqrt{y} - 4\sqrt{y} + 20$$
$$= y + (-5-4)\sqrt{y} + 20$$
$$= y - 9\sqrt{y} + 20$$

Rationalize the denominator. All variables represent positive real numbers. [Section 8.4]

69. $$\frac{4}{\sqrt{5}} = \frac{4}{\sqrt{5}} \cdot \frac{\sqrt{\mathbf{5}}}{\sqrt{\mathbf{5}}}$$
$$= \frac{4\sqrt{5}}{5}$$

70. $$\sqrt{\frac{17}{2x}} = \frac{\sqrt{17}}{\sqrt{2x}}$$
$$= \frac{\sqrt{17}}{\sqrt{2x}} \cdot \frac{\sqrt{\mathbf{2x}}}{\sqrt{\mathbf{2x}}}$$
$$= \frac{\sqrt{17 \cdot 2x}}{2x}$$
$$= \frac{\sqrt{34x}}{2x}$$

71. Solve: [Section 8.5]

$$\sqrt{6x+19} - 5 = 2$$
$$\sqrt{6x+19} - 5 + \mathbf{5} = 2 + \mathbf{5}$$
$$\sqrt{6x+19} = 7$$
$$\left(\sqrt{6x+19}\right)^2 = (7)^2$$
$$6x + 19 = 49$$
$$6x + 19 - \mathbf{19} = 49 - \mathbf{19}$$
$$6x = 30$$
$$\frac{6x}{\mathbf{6}} = \frac{30}{\mathbf{6}}$$
$$x = 5$$

Check:

$$\sqrt{6x+19} - 5 = 2$$
$$\sqrt{6 \cdot \mathbf{5} + 19} - 5 \stackrel{?}{=} 2$$
$$\sqrt{30+19} - 5 \stackrel{?}{=} 2$$
$$\sqrt{49} - 5 \stackrel{?}{=} 2$$
$$7 - 5 \stackrel{?}{=} 2$$
$$2 = 2$$

True

The solution is 5.

72. Evaluate: $16^{3/2}$. [Section 8.6]

$$16^{3/2} = \left(\sqrt{16}\right)^3$$
$$= (4)^3$$
$$= 64$$

SECTION 9.1 STUDY SET

VOCABULARY

Fill in the blank.

1. $x^2 - 15 = 0$ is an example of a **quadratic** equation.

CONCEPTS

Fill in the blanks.

3. The square root property of equations: If $x^2 = c$, then $x = \sqrt{c}$ or $x = -\sqrt{c}$.

5. Use a property of equality to isolate the variable on the left side of the equation.

   a. $x+9=\pm\sqrt{2}$ $\quad$ $\boldsymbol{x = -9 \pm \sqrt{2}}$

   b. $6x = 3\pm\sqrt{2}$ $\quad$ $\boldsymbol{x = \dfrac{3\pm\sqrt{2}}{6}}$

7. Is $2\sqrt{5}$ a solution of $x^2 = 20$?

$$x^2 = 20$$
$$(2\sqrt{5})^2 \stackrel{?}{=} 20$$
$$2^2(\sqrt{5})^2 \stackrel{?}{=} 20$$
$$4\cdot 5 \stackrel{?}{=} 20$$
$$20 = 20$$

Yes

NOTATION

9. Write the statement $x=\sqrt{6}$ or $x=-\sqrt{6}$ using a $\pm$ symbol (double-sign notation). $\boldsymbol{x = \pm\sqrt{6}}$

GUIDED PRACTICE

Use the square root property to solve each equation, if possible. See Example 1.

11. $$x^2 - 36 = 0$$
$$x^2 - 36 + \mathbf{36} = 0 + \mathbf{36}$$
$$x^2 = 36$$
$$x = \sqrt{36} \quad \text{or} \quad x = -\sqrt{36}$$
$$x = 6 \quad\quad x = -6$$
$$x = \pm 6$$

The solutions are ± 6.

13. $$x^2 = \frac{49}{16}$$
$$x = \sqrt{\frac{49}{16}} \quad \text{or} \quad x = -\sqrt{\frac{49}{16}}$$
$$x = \frac{7}{4} \quad\quad x = -\frac{7}{4}$$
$$x = \pm\frac{7}{4}$$

The solutions are $\pm\frac{7}{4}$.

15. $$5x^2 = 125$$
$$\frac{5x^2}{\mathbf{5}} = \frac{125}{\mathbf{5}}$$
$$x^2 = 25$$
$$x = \sqrt{25} \quad \text{or} \quad x = -\sqrt{25}$$
$$x = 5 \quad\quad x = -5$$
$$x = \pm 5$$

The solutions are ± 5.

17. $$x^2 - 6 = 0$$
$$x^2 - 6 + \mathbf{6} = 0 + \mathbf{6}$$
$$x^2 = 6$$
$$x = \sqrt{6} \quad \text{or} \quad x = -\sqrt{6}$$
$$x = \pm\sqrt{6}$$

The solutions are $\pm\sqrt{6}$.

19. $$m^2 = 20$$
$$m = \sqrt{20} \quad \text{or} \quad m = -\sqrt{20}$$
$$m = \pm\sqrt{20}$$
$$m = \pm\sqrt{4\cdot 5}$$
$$m = \pm\sqrt{4}\sqrt{5}$$
$$m = \pm 2\sqrt{5}$$

The solutions are $\pm 2\sqrt{5}$.

21. $$t^2 = 72$$
$$t = \pm\sqrt{72}$$
$$t = \pm\sqrt{36\cdot 2}$$
$$t = \pm 6\sqrt{2}$$

The solutions are $\pm 6\sqrt{2}$.

23. $$2x^2 + 8 = 23$$
$$2x^2 + 8 - \mathbf{8} = 23 - \mathbf{8}$$
$$2x^2 = 15$$
$$\frac{2x^2}{\mathbf{2}} = \frac{15}{\mathbf{2}}$$
$$x^2 = \frac{15}{2}$$
$$x = \pm\sqrt{\frac{15}{2}}$$
$$x = \pm\frac{\sqrt{15}}{\sqrt{2}} \cdot \frac{\sqrt{2}}{\sqrt{2}}$$
$$x = \pm\frac{\sqrt{30}}{2}$$
The solutions are $\pm\frac{\sqrt{30}}{2}$.

25. $x^2 = -81$

$x^2 = \pm\sqrt{-81}$

No real-number solutions.

Use the square root property to solve each equation. See Example 2.

27. $$(x+1)^2 = 25$$
$$x+1 = \sqrt{25} \text{ or } x+1 = -\sqrt{25}$$
$$x+1 = \pm\sqrt{25}$$
$$x+1 = \pm 5$$
$$x+1-\mathbf{1} = -\mathbf{1} \pm 5$$
$$x = -1 \pm 5$$

$x = -1+5$ or	$x = -1-5$
$x = 4$	$x = -6$
Check:	Check:
$(x+1)^2 = 25$	$(x+1)^2 = 25$
$(\mathbf{4}+1)^2 \overset{?}{=} 25$	$(\mathbf{-6}+1)^2 \overset{?}{=} 25$
$5^2 \overset{?}{=} 25$	$(-5)^2 \overset{?}{=} 25$
$25 = 25$	$25 = 25$
True	True

The solutions are 4 and -6.

29. $$(x-2)^2 = 8$$
$$x-2 = \pm\sqrt{8}$$
$$x-2 = \pm\sqrt{4 \cdot 2}$$
$$x-2 = \pm 2\sqrt{2}$$
$$x-2+\mathbf{2} = \mathbf{2} \pm 2\sqrt{2}$$
$$x = 2 \pm 2\sqrt{2}$$
The solutions are $2 \pm 2\sqrt{2}$.

31. $$(s+9)^2 = 63$$
$$s+9 = \pm\sqrt{63}$$
$$s+9 = \pm\sqrt{9 \cdot 7}$$
$$s+9 = \pm 3\sqrt{7}$$
$$s+9-\mathbf{9} = -\mathbf{9} \pm 3\sqrt{7}$$
$$s = -9 \pm 3\sqrt{7}$$
The solutions are $-9 \pm 3\sqrt{7}$.

33. $$(5c-10)^2 - 6 = 0.$$
$$(5c-10)^2 - 6 + \mathbf{6} = 0 + \mathbf{6}$$
$$(5c-10)^2 = 6$$
$$5c - 10 = \pm\sqrt{6}$$
$$5c - 10 + \mathbf{10} = \mathbf{10} \pm \sqrt{6}$$
$$5c = 10 \pm \sqrt{6}$$
$$\frac{5c}{\mathbf{5}} = \frac{10 \pm \sqrt{6}}{\mathbf{5}}$$
$$c = \frac{10 \pm \sqrt{6}}{5}$$
The solutions are $\frac{10 \pm \sqrt{6}}{5}$.

Use the square root property to solve each equation. Approximate each solution to the nearest tenth. See Example 3.

35. $x^2 + 2x + 1 = 10$
$$(x+1)^2 = 10$$
$$x+1 = \pm\sqrt{10}$$
$$x+1-\mathbf{1} = -\mathbf{1} \pm \sqrt{10}$$
$$x = -1 \pm \sqrt{10}$$
The solutions are $-1 \pm \sqrt{10}$ and 2.2, –4.2

37. $x^2 - 18x + 81 = 7$
$$(x-9)^2 = 7$$
$$x - 9 = \pm\sqrt{7}$$
$$x - 9 + \mathbf{9} = \mathbf{9} \pm \sqrt{7}$$
$$x = 9 \pm \sqrt{7}$$
The solutions are $9 \pm \sqrt{7}$ and 11.6, 6.4

39. $a^2 - 6a + 9 = 40$
$$(a-3)^2 = 40$$
$$a - 3 = \pm\sqrt{40}$$
$$a - 3 = \pm\sqrt{4 \cdot 10}$$
$$a - 3 = \pm 2\sqrt{10}$$
$$a - 3 + \mathbf{3} = \mathbf{3} \pm 2\sqrt{10}$$
$$a = 3 \pm 2\sqrt{10}$$
The solutions are $3 \pm 2\sqrt{10}$ and 9.3, -3.3

41. $m^2+4m+4=75$
$$(m+2)^2=75$$
$$m+2=\pm\sqrt{75}$$
$$m+2=\pm\sqrt{25\cdot 3}$$
$$m+2=\pm 5\sqrt{3}$$
$$m+2-\mathbf{2}=\mathbf{-2}\pm 5\sqrt{3}$$
$$m=-2\pm 5\sqrt{3}$$
The solutions are $-2\pm 5\sqrt{3}$ and 6.7, −10.7

TRY IT YOURSELF

Use the square root property to solve each equation, if possible.

43. $(x+12)^2=27$
$$x+12=\pm\sqrt{27}$$
$$x+12=\pm\sqrt{9\cdot 3}$$
$$x+12=\pm 3\sqrt{3}$$
$$x+12-\mathbf{12}=\mathbf{-12}\pm 3\sqrt{3}$$
$$x=-12\pm 3\sqrt{3}$$
The solutions are $-12\pm 3\sqrt{3}$.

45. $m^2=98$
$$m=\pm\sqrt{98}$$
$$m=\pm\sqrt{49\cdot 2}$$
$$m=\pm 7\sqrt{2}$$
The solutions are $\pm 7\sqrt{2}$.

47. $4x^2=400$
$$\frac{4x^2}{\mathbf{4}}=\frac{400}{\mathbf{4}}$$
$$x^2=100$$
$x=\sqrt{100}$ or $x=-\sqrt{100}$
$x=10$ | $x=-10$
$$x=\pm 10$$
The solutions are ± 10.

49. $(3x+1)^2-18=0$
$$(3x+1)^2-18+\mathbf{18}=0+\mathbf{18}$$
$$(3x+1)^2=18$$
$$3x+1=\pm\sqrt{9\cdot 2}$$
$$3x+1=\pm 3\sqrt{2}$$
$$3x+1-\mathbf{1}=\mathbf{-1}\pm 3\sqrt{2}$$
$$3x=-1\pm 3\sqrt{2}$$
$$\frac{3x}{\mathbf{3}}=\frac{-1\pm 3\sqrt{2}}{\mathbf{3}}$$
$$x=\frac{-1\pm 3\sqrt{2}}{3}$$
The solutions are $\frac{-1\pm 3\sqrt{2}}{3}$.

51. $b^2-12b+36=2$
$$(b-6)^2=2$$
$$b-6=\pm\sqrt{2}$$
$$b-6+\mathbf{6}=\mathbf{6}\pm\sqrt{2}$$
$$b=6\pm\sqrt{2}$$
The solutions are $6\pm\sqrt{2}$.

53. $b^2-17=0$
$$b^2-17+\mathbf{17}=0+\mathbf{17}$$
$$b^2=17$$
$b=\sqrt{17}$ or $b=-\sqrt{17}$
$$b=\pm\sqrt{17}$$
The solutions are $\pm\sqrt{17}$.

55. $6r^2-3=4$
$$6r^2-3+\mathbf{3}=4+\mathbf{3}$$
$$6r^2=7$$
$$\frac{6r^2}{\mathbf{6}}=\frac{7}{\mathbf{6}}$$
$$r^2=\frac{7}{6}$$
$$r=\pm\sqrt{\frac{7}{6}}$$
$$r=\pm\frac{\sqrt{7}}{\sqrt{6}}\cdot\frac{\mathbf{\sqrt{6}}}{\mathbf{\sqrt{6}}}$$
$$r=\pm\frac{\sqrt{42}}{6}$$
The solutions are $\pm\frac{\sqrt{42}}{6}$.

57. $(y-15)^2-8=0.$
$$(y-15)^2-8+\mathbf{8}=0+\mathbf{8}$$
$$(y-15)^2=8$$
$$y-15=\pm\sqrt{8}$$
$$y-15=\pm\sqrt{4\cdot 2}$$
$$y-15=\pm 2\sqrt{2}$$
$$y-15+\mathbf{15}=\mathbf{15}\pm 2\sqrt{2}$$
$$y=15\pm 2\sqrt{2}$$
The solutions are $15\pm 2\sqrt{2}$.

59. $t^2=\frac{1}{144}$
$$t=\pm\sqrt{\frac{1}{144}}$$
$$t=\pm\frac{\sqrt{1}}{\sqrt{144}}$$
$$t=\pm\frac{1}{12}$$
The solutions are $\pm\frac{1}{12}$.

61. $4(t-7)^2-12=0$
$$4(t-7)^2-12+\mathbf{12}=0+\mathbf{12}$$
$$4(t-7)^2=12$$
$$\frac{4(t-7)^2}{\mathbf{4}}=\frac{12}{\mathbf{4}}$$
$$(t-7)^2=3$$
$$t-7=\pm\sqrt{3}$$
$$t-7+\mathbf{7}=\mathbf{7}\pm\sqrt{3}$$
$$t=7\pm\sqrt{3}$$
The solutions are $7\pm\sqrt{3}$.

63. $h^2+25=0$
$$h^2+25-\mathbf{25}=0-\mathbf{25}$$
$$h^2=-25$$
$$h=\sqrt{-25}$$
No real-number solutions.

65. $5x^2+1=18$
$$5x^2+1-\mathbf{1}=18-\mathbf{1}$$
$$5x^2=17$$
$$\frac{5x^2}{\mathbf{5}}=\frac{17}{\mathbf{5}}$$
$$x^2=\frac{17}{5}$$
$$x=\pm\sqrt{\frac{17}{5}}$$
$$x=\pm\frac{\sqrt{17}}{\sqrt{5}}\cdot\frac{\sqrt{\mathbf{5}}}{\sqrt{\mathbf{5}}}$$
$$x=\pm\frac{\sqrt{85}}{5}$$
The solutions are $\pm\frac{\sqrt{85}}{5}$.

67. $(x+2)^2=81$
$$x+2=\pm\sqrt{81}$$
$$x+2=\pm 9$$
$$x+2-\mathbf{2}=\mathbf{-2}\pm 9$$
$$x=-2\pm 9$$

$x=-2+9$	or	$x=-2-9$
$x=7$		$x=-11$

Check:	Check:
$(x+2)^2=81$	$(x+2)^2=81$
$(7+2)^2\overset{?}{=}81$	$(-11+2)^2\overset{?}{=}81$
$9^2\overset{?}{=}81$	$(-9)^2\overset{?}{=}81$
$81=81$	$81=81$
True	True

The solutions are 7 and -11.

69. $x^2-14=0$
$$x^2-14+\mathbf{14}=0+\mathbf{14}$$
$$x^2=14$$
$$x=\pm\sqrt{14}$$
The solutions are $\pm\sqrt{14}$.

71. $(8y+9)^2=44$

$$8y+9=\pm\sqrt{44}$$
$$8y+9=\pm\sqrt{4\cdot 11}$$
$$8y+9=\pm 2\sqrt{11}$$
$$8y+9-\mathbf{9}=-\mathbf{9}\pm 2\sqrt{11}$$
$$8y=-9\pm 2\sqrt{11}$$
$$\frac{8y}{\mathbf{8}}=\frac{-9\pm 2\sqrt{11}}{\mathbf{8}}$$
$$y=\frac{-9\pm 2\sqrt{11}}{8}$$

The solutions are $\frac{-9\pm 2\sqrt{11}}{8}$.

73. $\frac{1}{2}a^2+6=4$

$$\frac{1}{2}a^2+6-\mathbf{6}=4-\mathbf{6}$$
$$\frac{1}{2}a^2=-2$$
$$\mathbf{2}\cdot\frac{1}{2}a^2=-2\cdot\mathbf{2}$$
$$a^2=-4$$
$$a=\sqrt{-4}$$

No real-number solutions.

LOOK ALIKES . . .

Solve the equation in part *a* using factoring. Solve the equation in part *b* using the square root property.

75. a. $b^2-9=0$

$$(b+3)(b-3)=0$$

$b+3=0$ or $b-3=0$

$b+3-\mathbf{3}=0-\mathbf{3}$ | $b-3+\mathbf{3}=0+\mathbf{3}$

$b=-3$ | $b=3$

$$b=\pm 3$$

The solutions are ± 3.

b. $b^2-8=0$

$$b^2-8+\mathbf{8}=0+\mathbf{8}$$
$$b^2=8$$

$b=\sqrt{8}$ or $b=-\sqrt{8}$

$=\sqrt{4\cdot 2}$ | $=-\sqrt{4\cdot 2}$

$=2\sqrt{2}$ | $=-2\sqrt{2}$

$$b=\pm 2\sqrt{2}$$

The solutions are $\pm 2\sqrt{2}$.

77. a. $x^2-14x-49=0$

$$(x-7)(x-7)=0$$
$$x-7=0$$
$$x=7$$

7 is a repeated solution.

b. $x^2-14x-49=6$

$$(x-7)^2=6$$
$$\sqrt{(x-7)^2}=\pm\sqrt{6}$$
$$x-7=\pm\sqrt{6}$$
$$x-7+\mathbf{7}=\mathbf{7}\pm\sqrt{6}$$
$$x=7\pm\sqrt{6}$$

The solutions are $7\pm\sqrt{6}$.

APPLICATIONS

For Exercises 79 - 84, use the formula $d=16t^2$, where d is the distance an object falls (in feet) and t is the time (in seconds) that is has been falling. See Example 4.

79. LIGHTHOUSES

$$d=16t^2$$
$$144=16t^2$$
$$\frac{144}{\mathbf{16}}=\frac{16t^2}{\mathbf{16}}$$
$$\frac{144}{16}=t^2$$
$$t^2=9$$
$$t=\pm\sqrt{9}$$
$$t=\pm 3$$
$$t=3$$

It will take the object 3 seconds to hit the ground.

81. SCIENCE HISTORY

$$d=16t^2$$
$$183=16t^2$$
$$\frac{183}{\mathbf{16}}=\frac{16t^2}{\mathbf{16}}$$
$$\frac{183}{16}=t^2$$
$$t^2=11.4375$$
$$t=\pm\sqrt{11.4375}$$
$$t\approx\pm 3.38$$
$$t\approx 3.4$$

It will take the object about 3.4 seconds to hit the ground.

83. DAREDEVILS

$$d = 16t^2$$
$$200 = 16t^2$$
$$\frac{200}{\mathbf{16}} = \frac{16t^2}{\mathbf{16}}$$
$$\frac{200}{16} = t^2$$
$$t^2 = 12.5$$
$$t = \pm\sqrt{12.5}$$
$$t \approx \pm 3.53$$
$$t \approx 3.5$$

It will take him about 3.5 seconds to hit the water.

85. PRO WRESTLING

$$A = s^2$$
$$400 = s^2$$
$$\pm\sqrt{400} = s$$
$$20 = s$$

The length of one side is 20 ft.

87. ESCAPE VELOCITY

- $g = 78{,}545$
- $R = 3{,}960$
- What is the escape velocity?

$$\frac{v^2}{2g} = R$$
$$\frac{v^2}{2 \cdot 78{,}545} = 3{,}960$$
$$\frac{v^2}{157{,}090} = 3{,}960$$
$$\frac{v^2}{157{,}090} \cdot \frac{\mathbf{157{,}090}}{1} = 3{,}960 \cdot \mathbf{157{,}090}$$
$$v^2 = 622{,}076{,}400$$
$$v = \pm\sqrt{622{,}076{,}400}$$
$$v \approx \pm 24{,}941.4$$
$$v \approx 24{,}941$$

The escape velocity is about 24,941 mi/hr.

from CAMPUS TO CAREERS

89. GRAPHIC DESIGNER

- x = width in inches
- $1.618x$ = length in inches
- Area is 275 in^2.
- Find the length and width.

$$A = lw$$
$$275 = 1.618x \cdot x$$
$$1.618x^2 = 275$$
$$\frac{1.618x^2}{\mathbf{1.618}} = \frac{275}{\mathbf{1.618}}$$
$$x^2 \approx 169.9629$$
$$x \approx \pm\sqrt{169.9629}$$
$$x \approx \pm 13.03$$
$$x = 13$$

$$\text{length} = 1.618x$$
$$= 1.618(13)$$
$$\approx 21.03$$
$$\approx 21$$

The width is 13 inches and the length is 21 inches.

WRITING

91-93. Answers will vary.

REVIEW

Solve each equation.

95.
$$\sqrt{5x-6} = 2$$
$$\left(\sqrt{5x-6}\right)^2 = 2^2$$
$$5x - 6 = 4$$
$$5x - 6 + \mathbf{6} = 4 + \mathbf{6}$$
$$5x = 10$$
$$\frac{5x}{\mathbf{5}} = \frac{10}{\mathbf{5}}$$
$$x = 2$$

Check:
$$\sqrt{5x-6} = 2$$
$$\sqrt{5(2)-6} \stackrel{?}{=} 2$$
$$\sqrt{10-6} \stackrel{?}{=} 2$$
$$\sqrt{4} \stackrel{?}{=} 2$$
$$2 = 2$$
True

The solution is 2.

97.

$$2\sqrt{x} = \sqrt{5x-16}$$
$$\left(2\sqrt{x}\right)^2 = \left(\sqrt{5x-16}\right)^2$$
$$2^2\left(\sqrt{x}\right)^2 = 5x-16$$
$$4x = 5x-16$$
$$4x+\mathbf{16} = 5x-16+\mathbf{16}$$
$$4x+16 = 5x$$
$$4x+16-\mathbf{4x} = 5x-\mathbf{4x}$$
$$16 = x$$

Check:

$$2\sqrt{x} = \sqrt{5x-16}$$
$$2\sqrt{16} \stackrel{?}{=} \sqrt{5(16)-16}$$
$$2(4) \stackrel{?}{=} \sqrt{80-16}$$
$$8 \stackrel{?}{=} \sqrt{64}$$
$$8 = 8$$

True

The solution is 16.

CHALLENGE PROBLEMS

Solve each equation.

99.

$$25\left(x+\frac{1}{3}\right)^2 = 144$$
$$\frac{25\left(x+\frac{1}{3}\right)^2}{\mathbf{25}} = \frac{144}{\mathbf{25}}$$
$$\left(x+\frac{1}{3}\right)^2 = \frac{144}{25}$$
$$x+\frac{1}{3} = \pm\frac{\sqrt{144}}{\sqrt{25}}$$
$$x+\frac{1}{3} = \pm\frac{12}{5}$$
$$x+\frac{1}{3}-\frac{\mathbf{1}}{\mathbf{3}} = -\frac{\mathbf{1}}{\mathbf{3}}\pm\frac{12}{5}$$
$$x = -\frac{1}{3}\pm\frac{12}{5}$$
$$x = -\frac{1}{3}\cdot\frac{\mathbf{5}}{\mathbf{5}}\pm\frac{12}{5}\cdot\frac{\mathbf{3}}{\mathbf{3}}$$
$$x = \frac{-5\pm 36}{15}$$

$$x = \frac{-5+36}{15} \quad \text{or} \quad x = \frac{-5-36}{15}$$
$$x = \frac{31}{15} \quad\quad x = -\frac{41}{15}$$

The solutions are $\frac{31}{15}$ and $-\frac{41}{15}$.

SECTION 9.2 STUDY SET

VOCABULARY

Fill in the blanks.

1. When we add 9 to $x^2 + 6x$, we say that we have completed the **square** on $x^2 + 6x$.

CONCEPTS

3. Find one-half of the given number and square the result.

   a. 6 **9** b. –5 $\mathbf{\frac{25}{4}}$

5. Fill in the blank: To complete the square on $x^2 + 8x$, add the square of **one-half** of the coefficient of x.

7. What is the first step to solve the equation by completing the square? **Do not solve.**

   a. $x^2 + 9x + 7 = 0$ **Subtract 7 from both sides**

   b. $4x^2 + 5x - 16 = 0$ **Divide both sides by 4**

9. Determine whether each statement is true or false.

   a. Any quadratic equation can be solved by the factoring method. **False**

   b. Any quadratic equation can be solved by the completing the square. **True**

NOTATION

11. Translate to mathematical symbols: *the square of one-half of nine*. Then evaluate the expression.

$$\left(\frac{1}{2}\cdot 9\right)^2 = \left(\frac{9}{2}\right)^2$$

$$= \frac{81}{4}$$

GUIDED PRACTICE

Complete the square and factor the resulting perfect square trinomial. See Example 1.

13. $x^2 + 2x$

To complete the square:

$\frac{1}{2}(2) = 1$

and $(1)^2 = 1$.

Add 1 to the binomial.

$$x^2 + 2x + \mathbf{1} = (x+1)^2$$

15. $x^2 - 4x$

To complete the square:

$\frac{1}{2}(-4) = -2$

and $(-2)^2 = 4$.

Add 4 to the binomial.

$$x^2 - 4x + 4 = (x-2)^2$$

17. $a^2 - 7a$

To complete the square:

$\frac{1}{2}(-7) = -\frac{7}{2}$

and $\left(-\frac{7}{2}\right)^2 = \frac{49}{4}$.

Add $\frac{49}{4}$ to the binomial.

$$a^2 - 7a + \frac{49}{4} = \left(a - \frac{7}{2}\right)^2$$

19. $x^2 + x$

To complete the square:

$\frac{1}{2}(1) = \frac{1}{2}$

and $\left(\frac{1}{2}\right)^2 = \frac{1}{4}$.

Add $\frac{1}{4}$ to the binomial.

$$x^2 + x + \frac{1}{4} = \left(x + \frac{1}{2}\right)^2$$

21. $b^2 - \frac{2}{3}b$

To complete the square:

$\frac{1}{2}\left(-\frac{2}{3}\right) = -\frac{1}{3}$

and $\left(-\frac{1}{3}\right)^2 = \frac{1}{9}$.

add $\frac{1}{9}$ to the binomial.

$$b^2 - \frac{2}{3}b + \frac{1}{9} = \left(b - \frac{1}{3}\right)^2$$

23. $x^2 - \frac{5}{2}x$

To complete the square:

$\frac{1}{2}\left(-\frac{5}{2}\right) = -\frac{5}{4}$

and $\left(-\frac{5}{4}\right)^2 = \frac{25}{16}$.

Add $\frac{25}{16}$ to the binomial.

$$x^2 - \frac{5}{2}x + \frac{25}{16} = \left(x - \frac{5}{4}\right)^2$$

Solve each equation by completing the square. See Example 2.

25. $x^2+4x=5$

To complete the square:

$\frac{1}{2}(4)=2$

and $(2)^2=4$.

Add 4 to both sides of the equation.

$$x^2+4x+\mathbf{4}=5+\mathbf{4}$$
$$(x+2)^2=9$$
$$x+2=\pm\sqrt{9}$$
$$x+2=\pm 3$$
$$x+2-\mathbf{2}=-\mathbf{2}\pm 3$$
$$x=-2\pm 3$$

$x=-2+3$
$x=1$

or

$x=-2-3$
$x=-5$

The solutions are -5 and 1.

27. $g^2-2g=15$

To complete the square:

$\frac{1}{2}(-2)=-1$

and $(-1)^2=1$.

Add 1 to both sides of the equation.

$$g^2-2g+\mathbf{1}=15+\mathbf{1}$$
$$(g-1)^2=16$$
$$g-1=\pm\sqrt{16}$$
$$g-1=\pm 4$$
$$g-1+\mathbf{1}=\mathbf{1}\pm 4$$
$$g=1\pm 4$$

$g=1+4$
$g=5$

or

$g=1-4$
$g=-3$

The solutions are -3 and 5.

29. $x^2+6x=-8$

To complete the square:

$\frac{1}{2}(6)=3$

and $(3)^2=9$.

Add 9 to both sides of the equation.

$$x^2+6x+\mathbf{9}=-8+\mathbf{9}$$
$$(x+3)^2=1$$
$$x+3=\pm\sqrt{1}$$
$$x+3=\pm 1$$
$$x+3-\mathbf{3}=-\mathbf{3}\pm 1$$
$$x=-3\pm 1$$

$x=-3+1$
$x=-2$

or

$x=-3-1$
$x=-4$

The solutions are -2 and -4.

31. $k^2-8k=-12$

To complete the square:

$\frac{1}{2}(-8)=-4$

and $(-4)^2=16$.

Add 16 to both sides of the equation.

$$k^2-8k+\mathbf{16}=-12+\mathbf{16}$$
$$(k-4)^2=4$$
$$k-4=\pm\sqrt{4}$$
$$k-4=\pm 2$$
$$k-4+\mathbf{4}=\mathbf{4}\pm 2$$
$$k=4\pm 2$$

$k=4+2$
$k=6$

or

$k=4-2$
$k=2$

The solutions are 2 and 6.

Solve each equation by completing the square. Approximate the solutions to the nearest hundredth. See Example 3.

33. $s^2-4s-3=0$

$$s^2-4s-3+\mathbf{3}=0+\mathbf{3}$$
$$s^2-4s=3$$

To complete the square:

$\frac{1}{2}(-4)=-2$

and $(-2)^2=4$.

Add 4 to both sides of the equation.

$$s^2-4s+\mathbf{4}=3+\mathbf{4}$$
$$(s-2)^2=7$$
$$s-2=\pm\sqrt{7}$$
$$s-2+\mathbf{2}=\mathbf{2}\pm\sqrt{7}$$
$$s=2\pm\sqrt{7}$$

$s=2+\sqrt{7}$
$s\approx 2+2.645$
$s\approx 4.65$

or

$s=2-\sqrt{7}$
$s\approx 2-2.645$
$s\approx -0.65$

The exact solutions are: $2\pm\sqrt{7}$.
The approximate solutions are: $4.65, -0.65$.

35. $$x^2-2x-17=0$$
$$x^2-2x-17+\mathbf{17}=0+\mathbf{17}$$
$$x^2-2x=17$$

To complete the square:

$\frac{1}{2}(-2)=-1$

and $(-1)^2=1$.

Add 1 to both sides of the equation.

$$x^2-2x+\mathbf{1}=17+\mathbf{1}$$
$$(x-1)^2=18$$
$$x-1=\pm\sqrt{18}$$
$$x-1+\mathbf{1}=\mathbf{1}\pm\sqrt{18}$$
$$x=1\pm\sqrt{18}$$
$$x=1\pm\sqrt{9}\sqrt{2}$$
$$x=1\pm3\sqrt{2}$$

$x=1+3\sqrt{2}$	or	$x=1-3\sqrt{2}$
$x\approx1+4.242$	|	$x\approx1-4.242$
$x\approx5.24$		$x\approx-3.24$

The exact solutions are: $1\pm3\sqrt{2}$.
The approximate solutions are: $5.24,-3.24$.

37. $$x^2+8x-6=0$$
$$x^2+8x-6+\mathbf{6}=0+\mathbf{6}$$
$$x^2+8x=6$$

To complete the square:

$\frac{1}{2}(8)=4$

and $(4)^2=16$.

Add 16 to both sides of the equation.

$$x^2+8x+\mathbf{16}=6+\mathbf{16}$$
$$(x+4)^2=22$$
$$x+4=\pm\sqrt{22}$$
$$x+4-\mathbf{4}=\mathbf{-4}\pm\sqrt{22}$$
$$x=-4\pm\sqrt{22}$$

$x=-4+\sqrt{22}$	or	$x=-4-\sqrt{22}$
$x\approx-4+4.690$	|	$x\approx-4-4.690$
$x\approx0.69$		$x\approx-8.69$

The exact solutions are: $-4\pm\sqrt{22}$.
The approximate solutions are: $0.69,-8.69$.

39. $$x^2+6x+4=0$$
$$x^2+6x+4-\mathbf{4}=0-\mathbf{4}$$
$$x^2+6x=-4$$

To complete the square:

$\frac{1}{2}(6)=3$

and $(3)^2=9$.

Add 9 to both sides of the equation.

$$x^2+6x+\mathbf{9}=-4+\mathbf{9}$$
$$(x+3)^2=5$$
$$x+3=\pm\sqrt{5}$$
$$x+3-\mathbf{3}=\mathbf{-3}\pm\sqrt{5}$$
$$x=-3\pm\sqrt{5}$$

$x=-3+\sqrt{5}$	or	$x=-3-\sqrt{5}$
$x\approx-3+2.236$	|	$x\approx-3-2.236$
$x\approx-0.76$		$x\approx-5.24$

The exact solutions are: $-3\pm\sqrt{5}$.
The approximate solutions are: $-0.76,-5.24$.

Solve each equation by completing the square. Approximate the solutions to the nearest hundredth. See Example 4.

41. $$x^2-7x=5$$

To complete the square:

$\frac{1}{2}(-7)=\frac{-7}{2}$

and $\left(\frac{-7}{2}\right)^2=\frac{49}{4}$.

Add $\frac{49}{4}$ to both sides of the equation.

$$x^2-7x+\mathbf{\frac{49}{4}}=5+\mathbf{\frac{49}{4}}$$
$$\left(x-\frac{7}{2}\right)^2=\frac{5}{1}\cdot\mathbf{\frac{4}{4}}+\frac{49}{4}$$
$$\left(x-\frac{7}{2}\right)^2=\frac{20}{4}+\frac{49}{4}$$
$$\left(x-\frac{7}{2}\right)^2=\frac{69}{4}$$
$$x-\frac{7}{2}=\pm\frac{\sqrt{69}}{\sqrt{4}}$$
$$x-\frac{7}{2}=\pm\frac{\sqrt{69}}{2}$$
$$x-\frac{7}{2}+\mathbf{\frac{7}{2}}=\mathbf{\frac{7}{2}}\pm\frac{\sqrt{69}}{2}$$
$$x=\frac{7\pm\sqrt{69}}{2}$$

$$x=\frac{7+\sqrt{69}}{2} \quad \text{or} \quad x=\frac{7-\sqrt{69}}{2}$$
$$x\approx\frac{7+8.306}{2} \quad\quad x\approx\frac{7-8.306}{2}$$
$$x\approx\frac{15.306}{2} \quad\quad x\approx\frac{-1.306}{2}$$
$$x\approx 7.653 \quad\quad x\approx -0.653$$
$$x\approx 7.65 \quad\quad x\approx -0.65$$

The exact solutions are: $\frac{7\pm\sqrt{69}}{2}$.

The approximate solutions are: 7.65, –0.65.

43. $b^2-5b=10$

To complete the square:

$\frac{1}{2}(-5)=\frac{-5}{2}$

and $\left(\frac{-5}{2}\right)^2=\frac{25}{4}$.

Add $\frac{25}{4}$ to both sides of the equation.

$$b^2-5b+\mathbf{\frac{25}{4}}=10+\mathbf{\frac{25}{4}}$$
$$\left(b-\frac{5}{2}\right)^2=\frac{10}{1}\cdot\mathbf{\frac{4}{4}}+\frac{25}{4}$$
$$\left(b-\frac{5}{2}\right)^2=\frac{40}{4}+\frac{25}{4}$$
$$\left(b-\frac{5}{2}\right)^2=\frac{65}{4}$$
$$b-\frac{5}{2}=\pm\frac{\sqrt{65}}{\sqrt{4}}$$
$$b-\frac{5}{2}=\pm\frac{\sqrt{65}}{2}$$
$$b-\frac{5}{2}+\mathbf{\frac{5}{2}}=\mathbf{\frac{5}{2}}\pm\frac{\sqrt{65}}{2}$$
$$b=\frac{5\pm\sqrt{65}}{2}$$

$$b=\frac{5+\sqrt{65}}{2} \quad \text{or} \quad b=\frac{5-\sqrt{65}}{2}$$
$$b\approx\frac{5+8.062}{2} \quad\quad b\approx\frac{5-8.062}{2}$$
$$x\approx\frac{13.062}{2} \quad\quad x\approx\frac{-3.062}{2}$$
$$x\approx 6.531 \quad\quad x\approx -1.531$$
$$x\approx 6.53 \quad\quad x\approx -1.53$$

The exact solutions are: $\frac{5\pm\sqrt{65}}{2}$.

The approximate solutions are: 6.53, –1.53.

45. $t^2+3t=20$

To complete the square:

$\frac{1}{2}(3)=\frac{3}{2}$

and $\left(\frac{3}{2}\right)^2=\frac{9}{4}$.

Add $\frac{9}{4}$ to both sides of the equation.

$$t^2+3t+\mathbf{\frac{9}{4}}=20+\mathbf{\frac{9}{4}}$$
$$\left(t+\frac{3}{2}\right)^2=\frac{20}{1}\cdot\mathbf{\frac{4}{4}}+\frac{9}{4}$$
$$\left(t+\frac{3}{2}\right)^2=\frac{80}{4}+\frac{9}{4}$$
$$\left(t+\frac{3}{2}\right)^2=\frac{89}{4}$$
$$t+\frac{3}{2}=\pm\sqrt{\frac{89}{4}}$$
$$t+\frac{3}{2}=\pm\frac{\sqrt{89}}{2}$$
$$t+\frac{3}{2}-\mathbf{\frac{3}{2}}=-\mathbf{\frac{3}{2}}\pm\frac{\sqrt{89}}{2}$$
$$t=\frac{-3\pm\sqrt{89}}{2}$$

$$t=\frac{-3+\sqrt{89}}{2} \quad \text{or} \quad t=\frac{-3-\sqrt{89}}{2}$$
$$t\approx\frac{-3+9.433}{2} \quad\quad t\approx\frac{-3-9.433}{2}$$
$$t\approx\frac{6.433}{2} \quad\quad t\approx\frac{-12.433}{2}$$
$$t\approx 3.216 \quad\quad t\approx -6.216$$
$$t\approx 3.22 \quad\quad t\approx -6.22$$

The exact solutions are: $\frac{-3\pm\sqrt{89}}{2}$.

The approximate solutions are: 3.22, –6.22.

47. $x^2+x=9$

To complete the square:

$\frac{1}{2}(1)=\frac{1}{2}$

and $\left(\frac{1}{2}\right)^2=\frac{1}{4}$.

Add $\frac{1}{4}$ to both sides of the equation.

$$x^2+x+\mathbf{\frac{1}{4}}=9+\mathbf{\frac{1}{4}}$$
$$\left(x+\frac{1}{2}\right)^2=\frac{9}{1}\cdot\mathbf{\frac{4}{4}}+\frac{1}{4}$$
$$\left(x+\frac{1}{2}\right)^2=\frac{36}{4}+\frac{1}{4}$$
$$\left(x+\frac{1}{2}\right)^2=\frac{37}{4}$$
$$x+\frac{1}{2}=\pm\sqrt{\frac{37}{4}}$$
$$x+\frac{1}{2}=\pm\frac{\sqrt{37}}{2}$$
$$x+\frac{1}{2}-\mathbf{\frac{1}{2}}=-\mathbf{\frac{1}{2}}\pm\frac{\sqrt{37}}{2}$$
$$x=\frac{-1\pm\sqrt{37}}{2}$$

	or	
$x=\frac{-1+\sqrt{37}}{2}$		$x=\frac{-1-\sqrt{37}}{2}$
$x\approx\frac{-1+6.082}{2}$		$x\approx\frac{-1-6.082}{2}$
$x\approx\frac{5.082}{2}$		$x\approx\frac{-7.082}{2}$
$x\approx 2.541$		$x\approx -3.541$
$x\approx 2.54$		$x\approx -3.54$

The exact solutions are: $\frac{-1\pm\sqrt{37}}{2}$.

The approximate solutions are: 2.54, −3.54.

Solve each equation by completing the square. See Example 5.

49. $2x^2-7x-3=0$

$$\frac{2x^2}{\mathbf{2}}-\frac{7x}{\mathbf{2}}-\frac{3}{\mathbf{2}}=\frac{0}{\mathbf{2}}$$
$$x^2-\frac{7}{2}x-\frac{3}{2}=0$$
$$x^2-\frac{7}{2}x-\frac{3}{2}+\mathbf{\frac{3}{2}}=0+\mathbf{\frac{3}{2}}$$
$$x^2-\frac{7}{2}x=\frac{3}{2}$$

To complete the square:

$\frac{1}{2}\left(-\frac{7}{2}\right)=-\frac{7}{4}$

and $\left(-\frac{7}{4}\right)^2=\frac{49}{16}$.

Add $\frac{49}{16}$ to both sides of the equation.

$$x^2-\frac{7}{2}x+\mathbf{\frac{49}{16}}=\frac{3}{2}+\mathbf{\frac{49}{16}}$$
$$\left(x-\frac{7}{4}\right)^2=\frac{3}{2}\cdot\mathbf{\frac{8}{8}}+\frac{49}{16}$$
$$\left(x-\frac{7}{4}\right)^2=\frac{24}{16}+\frac{49}{16}$$
$$\left(x-\frac{7}{4}\right)^2=\frac{73}{16}$$
$$x-\frac{7}{4}=\pm\sqrt{\frac{73}{16}}$$
$$x-\frac{7}{4}=\pm\frac{\sqrt{73}}{4}$$
$$x-\frac{7}{4}+\mathbf{\frac{7}{4}}=\mathbf{\frac{7}{4}}\pm\frac{\sqrt{73}}{4}$$
$$x=\frac{7\pm\sqrt{73}}{4}$$

The solutions are $\frac{7\pm\sqrt{73}}{4}$.

51. $4a^2 - 9a + 1 = 0$

$$\frac{4a^2}{\mathbf{4}} - \frac{9a}{\mathbf{4}} + \frac{1}{\mathbf{4}} = \frac{0}{\mathbf{4}}$$

$$a^2 - \frac{9}{4}a + \frac{1}{4} = 0$$

$$a^2 - \frac{9}{4}a + \frac{1}{4} - \frac{\mathbf{1}}{\mathbf{4}} = 0 - \frac{\mathbf{1}}{\mathbf{4}}$$

$$a^2 - \frac{9}{4}a = -\frac{1}{4}$$

To complete the square:

$$\frac{1}{2}\left(-\frac{9}{4}\right) = -\frac{9}{8}$$

and $\left(-\frac{9}{8}\right)^2 = \frac{81}{64}$.

Add $\frac{81}{64}$ to both sides of the equation.

$$a^2 - \frac{9}{4}a + \frac{\mathbf{81}}{\mathbf{64}} = -\frac{1}{4} + \frac{\mathbf{81}}{\mathbf{64}}$$

$$\left(a - \frac{9}{8}\right)^2 = -\frac{1}{4}\cdot\frac{\mathbf{16}}{\mathbf{16}} + \frac{81}{64}$$

$$\left(a - \frac{9}{8}\right)^2 = -\frac{16}{64} + \frac{81}{64}$$

$$\left(a - \frac{9}{8}\right)^2 = \frac{65}{64}$$

$$a - \frac{9}{8} = \pm\sqrt{\frac{65}{64}}$$

$$a - \frac{9}{8} = \pm\frac{\sqrt{65}}{8}$$

$$a - \frac{9}{8} + \frac{\mathbf{9}}{\mathbf{8}} = \frac{\mathbf{9}}{\mathbf{8}} \pm \frac{\sqrt{65}}{8}$$

$$a = \frac{9 \pm \sqrt{65}}{8}$$

The solutions are $\frac{9 \pm \sqrt{65}}{8}$.

53. $5b^2 + 3b - 4 = 0$

$$\frac{5b^2}{\mathbf{5}} + \frac{3b}{\mathbf{5}} - \frac{4}{\mathbf{5}} = \frac{0}{\mathbf{5}}$$

$$b^2 + \frac{3}{5}b - \frac{4}{5} = 0$$

$$b^2 + \frac{3}{5}b - \frac{4}{5} + \frac{\mathbf{4}}{\mathbf{5}} = 0 + \frac{\mathbf{4}}{\mathbf{5}}$$

$$b^2 + \frac{3}{5}b = \frac{4}{5}$$

To complete the square:

$$\frac{1}{2}\left(\frac{3}{5}\right) = \frac{3}{10}$$

and $\left(\frac{3}{10}\right)^2 = \frac{9}{100}$.

Add $\frac{9}{100}$ to both sides of the equation.

$$b^2 + \frac{3}{5}b + \frac{\mathbf{9}}{\mathbf{100}} = \frac{4}{5} + \frac{\mathbf{9}}{\mathbf{100}}$$

$$\left(b + \frac{3}{10}\right)^2 = \frac{4}{5}\cdot\frac{\mathbf{20}}{\mathbf{20}} + \frac{9}{100}$$

$$\left(b + \frac{3}{10}\right)^2 = \frac{80}{100} + \frac{9}{100}$$

$$\left(b + \frac{3}{10}\right)^2 = \frac{89}{100}$$

$$b + \frac{3}{10} = \pm\sqrt{\frac{89}{100}}$$

$$b + \frac{3}{10} = \pm\frac{\sqrt{89}}{10}$$

$$b + \frac{3}{10} - \frac{\mathbf{3}}{\mathbf{10}} = -\frac{\mathbf{3}}{\mathbf{10}} \pm \frac{\sqrt{89}}{10}$$

$$b = \frac{-3 \pm \sqrt{89}}{10}$$

The solutions are $\frac{-3 \pm \sqrt{89}}{10}$.

55.

$$3x^2+5x-5=0$$
$$\frac{3x^2}{\mathbf{3}}+\frac{5x}{\mathbf{3}}-\frac{5}{\mathbf{3}}=\frac{0}{\mathbf{3}}$$
$$x^2+\frac{5}{3}x-\frac{5}{3}=0$$
$$x^2+\frac{5}{3}x-\frac{5}{3}+\frac{\mathbf{5}}{\mathbf{3}}=0+\frac{\mathbf{5}}{\mathbf{3}}$$
$$x^2+\frac{5}{3}x=\frac{5}{3}$$

To complete the square:

$$\frac{1}{2}\left(\frac{5}{3}\right)=\frac{5}{6}$$

and $\left(\frac{5}{6}\right)^2=\frac{25}{36}$.

Add $\frac{25}{36}$ to both sides of the equation.

$$x^2+\frac{5}{3}x+\frac{\mathbf{25}}{\mathbf{36}}=\frac{5}{3}+\frac{\mathbf{25}}{\mathbf{36}}$$
$$\left(x+\frac{5}{6}\right)^2=\frac{5}{3}\cdot\frac{\mathbf{12}}{\mathbf{12}}+\frac{25}{36}$$
$$\left(x+\frac{5}{6}\right)^2=\frac{60}{36}+\frac{25}{36}$$
$$\left(x+\frac{5}{6}\right)^2=\frac{85}{36}$$
$$x+\frac{5}{6}=\pm\sqrt{\frac{85}{36}}$$
$$x+\frac{5}{6}=\pm\frac{\sqrt{85}}{6}$$
$$x+\frac{5}{6}-\frac{\mathbf{5}}{\mathbf{6}}=-\frac{\mathbf{5}}{\mathbf{6}}\pm\frac{\sqrt{85}}{6}$$
$$x=\frac{-5\pm\sqrt{85}}{6}$$

The solutions are $\frac{-5\pm\sqrt{85}}{6}$.

Solve each equation by completing the square. See Example 6.

57.

$$4x^2-13=24x$$
$$4x^2-13-\mathbf{24x}=24x-\mathbf{24x}$$
$$4x^2-24x-13=0$$
$$4x^2-24x-13+\mathbf{13}=0+\mathbf{13}$$
$$4x^2-24x=13$$
$$\frac{4x^2}{\mathbf{4}}-\frac{24x}{\mathbf{4}}=\frac{13}{\mathbf{4}}$$
$$x^2-6x=\frac{13}{4}$$

To complete the square:

$$\frac{1}{2}(-6)=-3$$

and $(-3)^2=9$.

Add 9 to both sides of the equation.

$$x^2-6x+\mathbf{9}=\frac{13}{4}+\mathbf{9}$$
$$(x-3)^2=\frac{13}{4}+\frac{9}{1}\cdot\frac{\mathbf{4}}{\mathbf{4}}$$
$$(x-3)^2=\frac{13}{4}+\frac{36}{4}$$
$$(x-3)^2=\frac{49}{4}$$
$$x-3=\pm\sqrt{\frac{49}{4}}$$
$$x-3=\pm\frac{7}{2}$$
$$x-3+\mathbf{3}=\mathbf{3}\pm\frac{7}{2}$$
$$x=\frac{3}{1}\cdot\frac{\mathbf{2}}{\mathbf{2}}\pm\frac{7}{2}$$
$$x=\frac{6\pm7}{2}$$

$x=\frac{6+7}{2}$ or $x=\frac{6-7}{2}$

$x=\frac{13}{2}$ | $x=-\frac{1}{2}$

The solutions are $\frac{13}{2}$ and $-\frac{1}{2}$.

59.

$$9a^2-5=-18a$$
$$9a^2-5+\mathbf{18a}=-18a+\mathbf{18a}$$
$$9a^2+18a-5=0$$
$$9a^2+18a-5+\mathbf{5}=0+\mathbf{5}$$
$$9a^2+18a=5$$
$$\frac{9a^2}{9}+\frac{18a}{9}=\frac{5}{9}$$
$$a^2+2a=\frac{5}{9}$$

To complete the square:

$\frac{1}{2}(2)=1$

and $(1)^2=1$.

Add 1 to both sides of the equation.

$$a^2+2a+\mathbf{1}=\frac{5}{9}+\mathbf{1}$$
$$(a+1)^2=\frac{5}{9}+\frac{1}{1}\cdot\frac{\mathbf{9}}{\mathbf{9}}$$
$$(a+1)^2=\frac{5}{9}+\frac{9}{9}$$
$$(a+1)^2=\frac{14}{9}$$
$$a+1=\pm\sqrt{\frac{14}{9}}$$
$$a+1=\pm\frac{\sqrt{14}}{3}$$
$$a+1-\mathbf{1}=-\mathbf{1}\pm\frac{\sqrt{14}}{3}$$
$$a=\frac{-1}{1}\cdot\frac{\mathbf{3}}{\mathbf{3}}\pm\frac{\sqrt{14}}{3}$$
$$a=\frac{-3\pm\sqrt{14}}{3}$$

The solutions are $\frac{-3\pm\sqrt{14}}{3}$.

61.

$$4t^2+11=48t$$
$$4t^2+11-\mathbf{48t}=48t-\mathbf{48t}$$
$$4t^2-48t+11=0$$
$$4t^2-48t+11-\mathbf{11}=0-\mathbf{11}$$
$$4t^2-48t=-11$$
$$\frac{4t^2}{\mathbf{4}}-\frac{48t}{\mathbf{4}}=\frac{-11}{\mathbf{4}}$$
$$t^2-12t=-\frac{11}{4}$$

To complete the square:

$\frac{1}{2}(-12)=-6$

and $(-6)^2=36$.

Add 36 to both sides of the equation.

$$t^2-12t+\mathbf{36}=-\frac{11}{4}+\mathbf{36}$$
$$(t-6)^2=\frac{-11}{4}+\frac{36}{1}\cdot\frac{\mathbf{4}}{\mathbf{4}}$$
$$(t-6)^2=\frac{-11}{4}+\frac{144}{4}$$
$$(t-6)^2=\frac{133}{4}$$
$$t-6=\pm\sqrt{\frac{133}{4}}$$
$$t-6=\pm\frac{\sqrt{133}}{2}$$
$$t-6+\mathbf{6}=\mathbf{6}\pm\frac{\sqrt{133}}{2}$$
$$t=\frac{6}{1}\cdot\frac{\mathbf{2}}{\mathbf{2}}\pm\frac{\sqrt{133}}{2}$$
$$t=\frac{12\pm\sqrt{133}}{2}$$

The solutions are $\frac{12\pm\sqrt{133}}{2}$.

63.

$$16x^2+3=-64x$$
$$16x^2+3+\mathbf{64x}=-64x+\mathbf{64x}$$
$$16x^2+64x+3=0$$
$$16x^2+64x+3-\mathbf{3}=0-\mathbf{3}$$
$$16x^2+64x=-3$$
$$\frac{16x^2}{\mathbf{16}}+\frac{64x}{\mathbf{16}}=\frac{-3}{\mathbf{16}}$$
$$x^2+4x=-\frac{3}{16}$$

To complete the square:

$$\frac{1}{2}(4)=2$$

and $(2)^2=4$.

Add 4 to both sides of the equation.

$$x^2+4x+\mathbf{4}=\frac{-3}{16}+\mathbf{4}$$
$$(x+2)^2=\frac{-3}{16}+\frac{4}{1}\cdot\frac{\mathbf{16}}{\mathbf{16}}$$
$$(x+2)^2=\frac{-3}{16}+\frac{64}{16}$$
$$(x+2)^2=\frac{61}{16}$$
$$x+2=\pm\sqrt{\frac{61}{16}}$$
$$x+2=\pm\frac{\sqrt{61}}{4}$$
$$x+2-\mathbf{2}=-\mathbf{2}\pm\frac{\sqrt{61}}{4}$$
$$x=\frac{-2}{1}\cdot\frac{\mathbf{4}}{\mathbf{4}}\pm\frac{\sqrt{61}}{4}$$
$$x=\frac{-8\pm\sqrt{61}}{4}$$

The solutions are $\frac{-8\pm\sqrt{61}}{4}$.

Solve each equation by completing the square. Approximate the solutions to the nearest hundredth. See Example 6.

65.

$$2x^2=-6x-1$$
$$2x^2+\mathbf{6x}=-6x-1+\mathbf{6x}$$
$$2x^2+6x=-1$$
$$\frac{2x^2}{\mathbf{2}}+\frac{6x}{\mathbf{2}}=\frac{-1}{\mathbf{2}}$$
$$x^2+3x=-\frac{1}{2}$$

To complete the square:

$$\frac{1}{2}(3)=\frac{3}{2}$$

and $\left(\frac{3}{2}\right)^2=\frac{9}{4}$.

Add $\frac{9}{4}$ to both sides of the equation.

$$x^2+3x+\frac{\mathbf{9}}{\mathbf{4}}=-\frac{1}{2}+\frac{\mathbf{9}}{\mathbf{4}}$$
$$\left(x+\frac{3}{2}\right)^2=\frac{-1}{2}\cdot\frac{\mathbf{2}}{\mathbf{2}}+\frac{9}{4}$$
$$\left(x+\frac{3}{2}\right)^2=\frac{-2}{4}+\frac{9}{4}$$
$$\left(x+\frac{3}{2}\right)^2=\frac{7}{4}$$
$$x+\frac{3}{2}=\pm\sqrt{\frac{7}{4}}$$
$$x+\frac{3}{2}=\pm\frac{\sqrt{7}}{2}$$
$$x+\frac{3}{2}-\frac{\mathbf{3}}{\mathbf{2}}=-\frac{\mathbf{3}}{\mathbf{2}}\pm\frac{\sqrt{7}}{2}$$
$$x=\frac{-3\pm\sqrt{7}}{2}$$

$x=\frac{-3+\sqrt{7}}{2}$	or	$x=\frac{-3-\sqrt{7}}{2}$
$x\approx\frac{-3+2.645}{2}$		$x\approx\frac{-3-2.645}{2}$
$x\approx\frac{-0.355}{2}$		$x\approx\frac{-5.645}{2}$
$x\approx-0.177$		$x\approx-2.822$
$x\approx-0.18$		$x\approx-2.82$

The solutions are approximately -0.18 and -2.82.

67. $$3x^2 - 4 = -2x$$
$$3x^2 - 4 + \mathbf{2x} = -2x + \mathbf{2x}$$
$$3x^2 + 2x - 4 = 0$$
$$3x^2 + 2x - 4 + \mathbf{4} = 0 + \mathbf{4}$$
$$3x^2 + 2x = 4$$
$$\frac{3x^2}{\mathbf{3}} + \frac{2x}{\mathbf{3}} = \frac{4}{\mathbf{3}}$$
$$x^2 + \frac{2}{3}x = \frac{4}{3}$$

To complete the square:

$$\frac{1}{2}\left(\frac{2}{3}\right) = \frac{1}{3}$$

and $\left(\frac{1}{3}\right)^2 = \frac{1}{9}$.

Add $\frac{1}{9}$ to both sides of the equation.

$$x^2 + \frac{2}{3}x + \frac{\mathbf{1}}{\mathbf{9}} = \frac{4}{3} + \frac{\mathbf{1}}{\mathbf{9}}$$
$$\left(x + \frac{1}{3}\right)^2 = \frac{4}{3} \cdot \frac{\mathbf{3}}{\mathbf{3}} + \frac{1}{9}$$
$$\left(x + \frac{1}{3}\right)^2 = \frac{12}{9} + \frac{1}{9}$$
$$\left(x + \frac{1}{3}\right)^2 = \frac{13}{9}$$
$$x + \frac{1}{3} = \pm\sqrt{\frac{13}{9}}$$
$$x + \frac{1}{3} = \pm\frac{\sqrt{13}}{3}$$
$$x + \frac{1}{3} - \frac{\mathbf{1}}{\mathbf{3}} = -\frac{\mathbf{1}}{\mathbf{3}} \pm \frac{\sqrt{13}}{3}$$
$$x = \frac{-1 \pm \sqrt{13}}{3}$$

	or	
$x = \frac{-1+\sqrt{13}}{3}$		$x = \frac{-1-\sqrt{13}}{3}$
$x \approx \frac{-1+3.605}{3}$		$x \approx \frac{-1-3.605}{3}$
$x \approx \frac{2.605}{3}$		$x \approx \frac{-4.605}{3}$
$x \approx 0.868$		$x \approx -1.535$
$x \approx 0.87$		$x \approx -1.54$

The solutions are approximately 0.87 and -1.54.

69. $$4x^2 + 12x = 6$$
$$\frac{4x^2}{\mathbf{4}} + \frac{12x}{\mathbf{4}} = \frac{6}{\mathbf{4}}$$
$$x^2 + 3x = \frac{3}{2}$$

To complete the square:

$$\frac{1}{2}(3) = \frac{3}{2}$$

and $\left(\frac{3}{2}\right)^2 = \frac{9}{4}$.

Add $\frac{9}{4}$ to both sides of the equation.

$$x^2 + 3x + \frac{\mathbf{9}}{\mathbf{4}} = \frac{3}{2} + \frac{\mathbf{9}}{\mathbf{4}}$$
$$\left(x + \frac{3}{2}\right)^2 = \frac{3}{2} \cdot \frac{\mathbf{2}}{\mathbf{2}} + \frac{9}{4}$$
$$\left(x + \frac{3}{2}\right)^2 = \frac{6}{4} + \frac{9}{4}$$
$$\left(x + \frac{3}{2}\right)^2 = \frac{15}{4}$$
$$x + \frac{3}{2} = \pm\sqrt{\frac{15}{4}}$$
$$x + \frac{3}{2} = \pm\frac{\sqrt{15}}{2}$$
$$x + \frac{3}{2} - \frac{\mathbf{3}}{\mathbf{2}} = -\frac{\mathbf{3}}{\mathbf{2}} \pm \frac{\sqrt{15}}{2}$$
$$x = \frac{-3 \pm \sqrt{15}}{2}$$

	or	
$x = \frac{-3+\sqrt{15}}{2}$		$x = \frac{-3-\sqrt{15}}{2}$
$x \approx \frac{-3+3.872}{2}$		$x \approx \frac{-3-3.872}{2}$
$x \approx \frac{0.872}{2}$		$x \approx \frac{-6.872}{2}$
$x \approx 0.436$		$x \approx -3.436$
$x \approx 0.44$		$x \approx -3.44$

The solutions are approximately 0.44 and -3.44.

71.
$$6m^2 - 8m - 3 = 0$$
$$\frac{6m^2}{\mathbf{6}} - \frac{8m}{\mathbf{6}} - \frac{3}{\mathbf{6}} = \frac{0}{\mathbf{6}}$$
$$m^2 - \frac{4}{3}m - \frac{1}{2} = 0$$
$$m^2 - \frac{4}{3}m - \frac{1}{2} + \frac{\mathbf{1}}{\mathbf{2}} = 0 + \frac{\mathbf{1}}{\mathbf{2}}$$
$$m^2 - \frac{4}{3}m = \frac{1}{2}$$

To complete the square:

$$\frac{1}{2}\left(-\frac{4}{3}\right) = -\frac{2}{3}$$

and $\left(-\frac{2}{3}\right)^2 = \frac{4}{9}$.

Add $\frac{4}{9}$ to both sides of the equation.

$$m^2 - \frac{4}{3}m + \frac{\mathbf{4}}{\mathbf{9}} = \frac{1}{2} + \frac{\mathbf{4}}{\mathbf{9}}$$
$$\left(m - \frac{2}{3}\right)^2 = \frac{1}{2} \cdot \frac{\mathbf{9}}{\mathbf{9}} + \frac{4}{9} \cdot \frac{\mathbf{2}}{\mathbf{2}}$$
$$\left(m - \frac{2}{3}\right)^2 = \frac{9}{18} + \frac{8}{18}$$
$$\left(m - \frac{2}{3}\right)^2 = \frac{17}{18}$$
$$m - \frac{2}{3} = \pm\sqrt{\frac{17}{18}}$$
$$m - \frac{2}{3} = \pm\frac{\sqrt{17}}{3\sqrt{2}}$$
$$m - \frac{2}{3} = \pm\frac{\sqrt{17}}{3\sqrt{2}} \cdot \frac{\sqrt{\mathbf{2}}}{\sqrt{\mathbf{2}}}$$
$$m - \frac{2}{3} = \pm\frac{\sqrt{34}}{6}$$
$$m - \frac{2}{3} + \frac{\mathbf{2}}{\mathbf{3}} = \frac{\mathbf{2}}{\mathbf{3}} \pm \frac{\sqrt{34}}{6}$$
$$m = \frac{2}{3} \cdot \frac{\mathbf{2}}{\mathbf{2}} \pm \frac{\sqrt{34}}{6}$$
$$m = \frac{4 \pm \sqrt{34}}{6}$$

	or	
$m = \frac{4+\sqrt{34}}{6}$		$m = \frac{4-\sqrt{34}}{6}$
$m \approx \frac{4+5.830}{6}$		$m \approx \frac{4-5.830}{6}$
$m \approx \frac{9.830}{6}$		$m \approx \frac{-1.831}{6}$
$m \approx 1.638$		$m \approx -0.305$
$m \approx 1.64$		$m \approx -0.31$

The solutions are approximately 1.64 and -0.31.

TRY IT YOURSELF

Solve each equation by completing the square.

73.
$$x^2 = -7x - 2$$
$$x^2 + \mathbf{7x} = -7x - 2 + \mathbf{7x}$$
$$x^2 + 7x = -2$$

To complete the square:

$$\frac{1}{2}(7) = \frac{7}{2}$$

and $\left(\frac{7}{2}\right)^2 = \frac{49}{4}$.

Add $\frac{49}{4}$ to both sides of the equation.

$$x^2 + 7x + \frac{\mathbf{49}}{\mathbf{4}} = -2 + \frac{\mathbf{49}}{\mathbf{4}}$$
$$\left(x + \frac{7}{2}\right)^2 = \frac{-2}{1} \cdot \frac{\mathbf{4}}{\mathbf{4}} + \frac{49}{4}$$
$$\left(x + \frac{7}{2}\right)^2 = \frac{-8}{4} + \frac{49}{4}$$
$$\left(x + \frac{7}{2}\right)^2 = \frac{41}{4}$$
$$x + \frac{7}{2} = \pm\sqrt{\frac{41}{4}}$$
$$x + \frac{7}{2} = \pm\frac{\sqrt{41}}{2}$$
$$x + \frac{7}{2} - \frac{\mathbf{7}}{\mathbf{2}} = -\frac{\mathbf{7}}{\mathbf{2}} \pm \frac{\sqrt{41}}{2}$$
$$x = \frac{-7 \pm \sqrt{41}}{2}$$

The solutions are $\frac{-7 \pm \sqrt{41}}{2}$.

75.
$$x^2 - 2x - 4 = 0$$
$$x^2 - 2x - 4 + \mathbf{4} = 0 + \mathbf{4}$$
$$x^2 - 2x = 4$$

To complete the square:
$\frac{1}{2}(-2) = -1$
and $(-1)^2 = 1$.
Add 1 to both sides of the equation.

$$x^2 - 2x + \mathbf{1} = 4 + \mathbf{1}$$
$$(x-1)^2 = 5$$
$$x - 1 = \pm\sqrt{5}$$
$$x - 1 = \pm\ \sqrt{5}$$
$$x - 1 + \mathbf{1} = \mathbf{1} \pm \sqrt{5}$$
$$x = 1 \pm \sqrt{5}$$

The solutions are $1 \pm \sqrt{5}$.

77.
$$9x^2 - 36x - 1 = 0$$
$$9x^2 - 36x - 1 + \mathbf{1} = 0 + \mathbf{1}$$
$$9x^2 - 36x = 1$$
$$\frac{9x^2}{\mathbf{9}} - \frac{36x}{\mathbf{9}} = \frac{1}{\mathbf{9}}$$
$$x^2 - 4x = \frac{1}{9}$$

To complete the square:
$\frac{1}{2}(-4) = -2$
and $(-2)^2 = 4$.
Add 4 to both sides of the equation.

$$x^2 - 4x + \mathbf{4} = \frac{1}{9} + \mathbf{4}$$
$$(x-2)^2 = \frac{1}{9} + \frac{4}{1} \cdot \frac{\mathbf{9}}{\mathbf{9}}$$
$$(x-2)^2 = \frac{1}{9} + \frac{36}{9}$$
$$(x-2)^2 = \frac{37}{9}$$
$$x - 2 = \pm\sqrt{\frac{37}{9}}$$
$$x - 2 = \pm\frac{\sqrt{37}}{3}$$
$$x - 2 + \mathbf{2} = \mathbf{2} \pm \frac{\sqrt{37}}{3}$$
$$x = \frac{2}{1} \cdot \frac{\mathbf{3}}{\mathbf{3}} \pm \frac{\sqrt{37}}{3}$$
$$x = \frac{6 \pm \sqrt{37}}{3}$$

The solutions are $\frac{6 \pm \sqrt{37}}{3}$.

79.
$$x^2 - 12x = -35$$

To complete the square:
$\frac{1}{2}(-12) = -6$
and $(-6)^2 = 36$.
Add 36 to both sides of the equation.

$$x^2 - 12x + \mathbf{36} = -35 + \mathbf{36}$$
$$(x-6)^2 = 1$$
$$x - 6 = \pm\sqrt{1}$$
$$x - 1 + \mathbf{6} = \mathbf{6} \pm 1$$
$$x = 6 \pm 1$$

$x = 6 + 1$ or $x = 6 - 1$

$x = 7$ | $x = 5$

The solutions are 7 and 5.

81. $$n^2 - 9n = 5$$

To complete the square:

$$\frac{1}{2}(-9) = -\frac{9}{2}$$

and $\left(-\frac{9}{2}\right)^2 = \frac{81}{4}$.

Add $\frac{81}{4}$ to both sides of the equation.

$$n^2 - 9n + \mathbf{\frac{81}{4}} = 5 + \mathbf{\frac{81}{4}}$$

$$\left(n - \frac{9}{2}\right)^2 = \frac{5}{1} \cdot \mathbf{\frac{4}{4}} + \frac{81}{4}$$

$$\left(n - \frac{9}{2}\right)^2 = \frac{20}{4} + \frac{81}{4}$$

$$\left(n - \frac{9}{2}\right)^2 = \frac{101}{4}$$

$$n - \frac{9}{2} = \pm\sqrt{\frac{101}{4}}$$

$$n - \frac{9}{2} = \pm\frac{\sqrt{101}}{2}$$

$$n - \frac{9}{2} + \mathbf{\frac{9}{2}} = \mathbf{\frac{9}{2}} \pm \frac{\sqrt{101}}{2}$$

$$n = \frac{9 \pm \sqrt{101}}{2}$$

The solutions are $\frac{9 \pm \sqrt{101}}{2}$.

83. $$3n^2 - 8n = -4$$

$$\frac{3n^2}{3} - \frac{8n}{3} = \frac{-4}{3}$$

$$n^2 - \frac{8}{3}n = \frac{-4}{3}$$

To complete the square:

$$\frac{1}{2}\left(-\frac{8}{3}\right) = -\frac{4}{3}$$

and $\left(-\frac{4}{3}\right)^2 = \frac{16}{9}$.

Add $\frac{16}{9}$ to both sides of the equation.

$$n^2 - \frac{8}{3}n + \mathbf{\frac{16}{9}} = \frac{-4}{3} + \mathbf{\frac{16}{9}}$$

$$\left(n - \frac{4}{3}\right)^2 = \frac{-4}{3} \cdot \mathbf{\frac{3}{3}} + \frac{16}{9}$$

$$\left(n - \frac{4}{3}\right)^2 = \frac{-12}{9} + \frac{16}{9}$$

$$\left(n - \frac{4}{3}\right)^2 = \frac{4}{9}$$

$$n - \frac{4}{3} = \pm\sqrt{\frac{4}{9}}$$

$$n - \frac{4}{3} = \pm\frac{2}{3}$$

$$n - \frac{4}{3} + \mathbf{\frac{4}{3}} = \mathbf{\frac{4}{3}} \pm \frac{2}{3}$$

$$n = \frac{4 \pm 2}{3}$$

$n = \frac{4+2}{3}$ or $n = \frac{4-2}{3}$

$n = \frac{6}{3}$ | $n = \frac{2}{3}$

$n = 2$

The solutions are 2 and $\frac{2}{3}$.

85. $$x^2 - 2x - 5 = 0$$

$$x^2 - 2x - 5 + \mathbf{5} = 0 + \mathbf{5}$$

$$x^2 - 2x = 5$$

To complete the square:

$$\frac{1}{2}(-2) = -1$$

and $(-1)^2 = 1$.

Add 1 to both sides of the equation.

$$x^2 - 2x + \mathbf{1} = 5 + \mathbf{1}$$

$$(x-1)^2 = 6$$

$$x - 1 = \pm\sqrt{6}$$

$$x - 1 + \mathbf{1} = \mathbf{1} \pm \sqrt{6}$$

$$x = 1 \pm \sqrt{6}$$

The solutions are $1 \pm \sqrt{6}$.

87. $$a^2 - 4a + 7 = 0$$

$$a^2 - 4a + 7 - \mathbf{7} = 0 - \mathbf{7}$$

$$a^2 - 4a = -7$$

To complete the square:

$\frac{1}{2}(-4)=-2$

and $(-2)^2=4$.

Add 4 to both sides of the equation.

$$a^2-4a+\mathbf{4}=-7+\mathbf{4}$$
$$(a-2)^2=-3$$
$$a-2=\pm\sqrt{-3}$$

No real-number solutions.

89.
$$\frac{1}{2}t^2-1=-\frac{5}{4}t$$
$$\mathbf{2}\cdot\frac{1}{2}t^2-\mathbf{2}\cdot 1=\mathbf{2}\cdot-\frac{5}{4}t$$
$$t^2-2=-\frac{5}{2}t$$
$$t^2-2+\mathbf{2}=-\frac{5}{2}t+\mathbf{2}$$
$$t^2+\frac{\mathbf{5}}{\mathbf{2}}\mathbf{t}=-\frac{5}{2}t+2+\frac{\mathbf{5}}{\mathbf{2}}\mathbf{t}$$
$$t^2+\frac{5}{2}t=2$$

To complete the square:

$\frac{1}{2}\left(\frac{5}{2}\right)=\frac{5}{4}$

and $\left(\frac{5}{4}\right)^2=\frac{25}{16}$.

Add $\frac{25}{16}$ to both sides of the equation.

$$t^2+\frac{5}{2}t+\frac{\mathbf{25}}{\mathbf{16}}=2+\frac{\mathbf{25}}{\mathbf{16}}$$
$$\left(t+\frac{5}{4}\right)^2=2\cdot\frac{\mathbf{16}}{\mathbf{16}}+\frac{25}{16}$$
$$\left(t+\frac{5}{4}\right)^2=\frac{32+25}{16}$$
$$\left(t+\frac{5}{4}\right)^2=\frac{57}{16}$$
$$t+\frac{5}{4}=\pm\sqrt{\frac{57}{16}}$$
$$t+\frac{5}{4}=\pm\frac{\sqrt{57}}{4}$$
$$t+\frac{5}{4}-\frac{\mathbf{5}}{\mathbf{4}}=-\frac{\mathbf{5}}{\mathbf{4}}\pm\frac{\sqrt{57}}{4}$$
$$t=\frac{-5\pm\sqrt{57}}{4}$$

The solutions are $\frac{-5\pm\sqrt{57}}{4}$.

91.
$$3x^2-18x-21=0$$
$$3x^2-18x-21+\mathbf{21}=0+\mathbf{21}$$
$$3x^2-18x=21$$
$$\frac{3x^2}{\mathbf{3}}-\frac{18x}{\mathbf{3}}=\frac{21}{\mathbf{3}}$$
$$x^2-6x=7$$

To complete the square:

$\frac{1}{2}(-6)=-3$

and $(-3)^2=9$.

Add 9 to both sides of the equation.

$$x^2-6x+\mathbf{9}=7+\mathbf{9}$$
$$(x-3)^2=16$$
$$x-3=\pm\sqrt{16}$$
$$x-3=\pm 4$$
$$x-3+\mathbf{3}=\mathbf{3}\pm 4$$
$$x=3\pm 4$$
$$x=7 \text{ and } -1$$

The solutions are 7, -1.

93.
$$2x^2+24x+44=0$$
$$2x^2+24x+44-\mathbf{44}=0-\mathbf{44}$$
$$2x^2+24x=-44$$
$$\frac{2x^2}{\mathbf{2}}+\frac{24x}{\mathbf{2}}=\frac{-44}{\mathbf{2}}$$
$$x^2+12x=-22$$

To complete the square:

$\frac{1}{2}(12)=6$

and $(6)^2=36$.

Add 36 to both sides of the equation.

$$x^2+12x+\mathbf{36}=-22+\mathbf{36}$$
$$(x+6)^2=14$$
$$x+6=\pm\sqrt{14}$$
$$x+6-\mathbf{6}=\mathbf{-6}\pm\sqrt{14}$$
$$x=-6\pm\sqrt{14}$$

The solutions are $-6+\sqrt{14}$, $-6-\sqrt{14}$.

LOOK ALIKES...

Solve each equation in part *a* using factoring. Solve the equation in part *b* by completing the square.

95. a. $y^2+5y+4=0$

$(y+1)(y+4)=0$

$y+1=0$ or $y+4=0$

$y=-1$ | $y=-4$

The solutions are $-1,-4$.

b. $y^2+5y+3=0$

$y^2+5y+3-\mathbf{3}=0-\mathbf{3}$

$y^2+5y=-3$

To complete the square:

$\frac{1}{2}(5)=\frac{5}{2}$

and $\left(\frac{5}{2}\right)^2=\frac{25}{4}$.

Add $\frac{25}{4}$ to both sides of the equation.

$$y^2+5y+\frac{\mathbf{25}}{\mathbf{4}}=-3+\frac{\mathbf{25}}{\mathbf{4}}$$

$$\left(y+\frac{5}{2}\right)^2=-3\cdot\frac{\mathbf{4}}{\mathbf{4}}+\frac{25}{4}$$

$$\left(y+\frac{5}{2}\right)^2=\frac{-12+25}{4}$$

$$\left(y+\frac{5}{2}\right)^2=\frac{13}{4}$$

$$y+\frac{5}{2}=\pm\sqrt{\frac{13}{4}}$$

$$y+\frac{5}{2}=\pm\frac{\sqrt{13}}{2}$$

$$y+\frac{5}{2}-\frac{\mathbf{5}}{\mathbf{2}}=-\frac{\mathbf{5}}{\mathbf{2}}\pm\frac{\sqrt{13}}{2}$$

$$y=\frac{-5\pm\sqrt{13}}{2}$$

The solutions are $\frac{-5\pm\sqrt{13}}{2}$.

97. a. $2a^2-8a=0$

$2a(a-4)=0$

$2a=0$ or $a-4=0$

$a=0$ | $a=4$

The solutions are $0,4$.

b. $2a^2-8a=1$

$$\frac{2a^2}{\mathbf{2}}-\frac{8a}{\mathbf{2}}=\frac{1}{\mathbf{2}}$$

$$a^2-4a=\frac{1}{2}$$

To complete the square:

$\frac{1}{2}(-4)=-2$

and $(-2)^2=4$.

Add 4 to both sides of the equation.

$$a^2-4a+\mathbf{4}=\frac{1}{2}+\mathbf{4}$$

$$(a-2)^2=\frac{1}{2}+4\cdot\frac{\mathbf{2}}{\mathbf{2}}$$

$$(a-2)^2=\frac{1+8}{2}$$

$$(a-2)^2=\frac{9}{2}$$

$$a-2=\pm\sqrt{\frac{9}{2}}$$

$$a-2=\pm\frac{3}{\sqrt{2}}$$

$$a-2=\pm\frac{3}{\sqrt{2}}\cdot\frac{\sqrt{\mathbf{2}}}{\sqrt{\mathbf{2}}}$$

$$a-2=\pm\frac{3\sqrt{2}}{2}$$

$$a-2+\mathbf{2}=\mathbf{2}\pm\frac{3\sqrt{2}}{2}$$

$$a=2\cdot\frac{\mathbf{2}}{\mathbf{2}}\pm\frac{3\sqrt{2}}{2}$$

$$a=\frac{4\pm3\sqrt{2}}{2}$$

The solutions are $\frac{4\pm3\sqrt{2}}{2}$.

APPLICATIONS

99. GEOMETRY

The number of small squares is **4**.

Area: $x^2 + 4x + \mathbf{4} = (x + \mathbf{2})^2$

WRITING

101-103. Answers will vary.

REVIEW

LOOK ALIKES . . .

Perform the indicated operation and simplify when possible.

105. $\frac{x+3}{x-3} \cdot \frac{x+1}{x^2-9} = \frac{(x+3)(x+1)}{(x-3)(x+3)(x-3)}$

$$= \frac{\overset{1}{\cancel{(x+3)}}(x+1)}{(x-3)\underset{1}{\cancel{(x+3)}}(x-3)}$$

$$= \frac{x+1}{(x-3)^2}$$

107. $\frac{x+3}{x-3} + \frac{x+1}{x^2-9}$

$$= \frac{x+3}{x-3} + \frac{x+1}{(x+3)(x-3)}$$

$$= \frac{(x+3)}{(x-3)} \cdot \frac{\mathbf{(x+3)}}{\mathbf{(x+3)}} + \frac{x+1}{(x+3)(x-3)}$$

$$= \frac{(x+3)(x+3) + (x+1)}{(x+3)(x-3)}$$

$$= \frac{x^2 + 2 \cdot 3x + 9 + x + 1}{(x+3)(x-3)}$$

$$= \frac{x^2 + 6x + 9 + x + 1}{(x+3)(x-3)}$$

$$= \frac{x^2 + 7x + 10}{(x+3)(x-3)}$$

$$= \frac{x^2 + 7x + 10}{x^2 - 9}$$

CHALLENGE

Solve each equation by completing the square.

109. $0.2x^2 + 0.4x + 0.1 = 0$

$$\frac{0.2x^2}{\mathbf{0.2}} + \frac{0.4x}{\mathbf{0.2}} + \frac{0.1}{\mathbf{0.2}} = \frac{0}{\mathbf{0.2}}$$

$$x^2 + 2x + 0.5 = 0$$

$$x^2 + 2x + 0.5 - \mathbf{0.5} = 0 - \mathbf{0.5}$$

$$x^2 + 2x = -0.5$$

$$x^2 + 2x = -\frac{1}{2}$$

To complete the square:
$\frac{1}{2}(2) = 1$
and $(1)^2 = 1$.
Add 1 to both sides of the equation.

$$x^2 + 2x + \mathbf{1} = -\frac{1}{2} + \mathbf{1}$$

$$(x+1)^2 = -\frac{1}{2} + 1$$

$$(x+1)^2 = \frac{-1}{2} + \frac{1}{1} \cdot \frac{\mathbf{2}}{\mathbf{2}}$$

$$(x+1)^2 = \frac{-1+2}{2}$$

$$(x+1)^2 = \frac{1}{2}$$

$$x + 1 = \pm\frac{\sqrt{1}}{\sqrt{2}}$$

$$x + 1 = \pm\frac{1}{\sqrt{2}} \cdot \frac{\sqrt{\mathbf{2}}}{\sqrt{\mathbf{2}}}$$

$$x + 1 = \pm\frac{\sqrt{2}}{2}$$

$$x + 1 - \mathbf{1} = \mathbf{-1} \pm \frac{\sqrt{2}}{2}$$

$$x = \frac{-1}{1} \cdot \frac{\mathbf{2}}{\mathbf{2}} \pm \frac{\sqrt{2}}{2}$$

$$x = \frac{-2 \pm \sqrt{2}}{2}$$

The solutions are $\frac{-2 \pm \sqrt{2}}{2}$.

111.

$$x(x+3)-\frac{1}{2}=-2$$
$$x^2+3x-\frac{1}{2}=-2$$
$$x^2+3x-\frac{1}{2}+\mathbf{\frac{1}{2}}=-2+\mathbf{\frac{1}{2}}$$
$$x^2+3x=-2+\frac{1}{2}$$
$$x^2+3x=\frac{-2}{1}\cdot\mathbf{\frac{2}{2}}+\frac{1}{2}$$
$$x^2+3x=\frac{-4+1}{2}$$
$$x^2+3x=\frac{-3}{2}$$

To complete the square:

$\frac{1}{2}(3)=\frac{3}{2}$

and $\left(\frac{3}{2}\right)^2=\frac{9}{4}$.

Add $\frac{9}{4}$ to both sides of the equation.

$$x^2+3x+\mathbf{\frac{9}{4}}=-\frac{3}{2}+\mathbf{\frac{9}{4}}$$
$$\left(x+\frac{3}{2}\right)^2=-\frac{3}{2}+\mathbf{\frac{9}{4}}$$
$$\left(x+\frac{3}{2}\right)^2=\frac{-3}{2}\cdot\mathbf{\frac{2}{2}}+\frac{9}{4}$$
$$\left(x+\frac{3}{2}\right)^2=\frac{-6+9}{4}$$
$$\left(x+\frac{3}{2}\right)^2=\frac{3}{4}$$
$$x+\frac{3}{2}=\pm\frac{\sqrt{3}}{\sqrt{4}}$$
$$x+\frac{3}{2}=\pm\frac{\sqrt{3}}{2}$$
$$x+\frac{3}{2}-\mathbf{\frac{3}{2}}=-\mathbf{\frac{3}{2}}\pm\frac{\sqrt{3}}{2}$$
$$x=\frac{-3\pm\sqrt{3}}{2}$$

The solutions are $\frac{-3\pm\sqrt{3}}{2}$.

SECTION 9.3 STUDY SET

VOCABULARY

Fill in the blank.

1. The general **quadratic** equation is $ax^2 + bx + c = 0$, where $a \neq 0$.

CONCEPTS

3. Write each equation in $ax^2 + bx + c = 0$ form.
   a. $x^2 + 2x = -5$ b. $3x^2 = -2x + 1$
   $\mathbf{x^2 + 2x + 5 = 0}$ $\mathbf{3x^2 + 2x - 1 = 0}$

5. Divide both sides of $2x^2 - 4x + 8 = 0$ by 2, and then find a, b, and c.

$$2x^2 - 4x + 8 = 0$$
$$\frac{2x^2}{\mathbf{2}} - \frac{4x}{\mathbf{2}} + \frac{8}{\mathbf{2}} = \frac{0}{\mathbf{2}}$$
$$x^2 - 2x + 4 = 0$$
$$\mathbf{a = 1,\ b = -2,\ c = 4}$$

7. a. How many terms does the expression $10 \pm 15\sqrt{2}$ have? **2**
   b. What common factor do the terms have? **5**

9. A student used the quadratic formula to solve an equation and obtained

$$x = \frac{-3 \pm \sqrt{15}}{2}$$

   a. How many solutions does the equation have? **2**
   b. What are they? $\mathbf{\frac{-3+\sqrt{15}}{2}, \frac{-3-\sqrt{15}}{2}}$
   c. Approximate each to the nearest hundredth. **0.44, –3.44**

NOTATION

Complete the solution.

11. Solve: $x^2 - 5x - 6 = 0$.

$$x = \frac{-b \pm \sqrt{b^2 - 4ac}}{2a}$$
$$x = \frac{-(\mathbf{-5}) \pm \sqrt{(-5)^2 - 4(1)(\mathbf{-6})}}{2(\mathbf{1})}$$
$$x = \frac{5 \pm \sqrt{25 + \mathbf{24}}}{2}$$
$$x = \frac{5 \pm \sqrt{\mathbf{49}}}{2}$$
$$x = \frac{\mathbf{5 \pm 7}}{2}$$
$$x = \frac{5+7}{2} = \mathbf{6} \text{ or } x = \frac{5+7}{2} = \mathbf{-1}$$

GUIDED PRACTICE

Use the quadratic formula to solve each equation. See Example 1.

13. $x^2 + 5x + 6 = 0$

$$1x^2 + 5x + 6 = 0$$
$$a = 1,\ b = 5,\ c = 6$$
$$x = \frac{-b \pm \sqrt{b^2 - 4ac}}{2a}$$
$$x = \frac{-5 \pm \sqrt{5^2 - 4(1)(6)}}{2(1)}$$
$$x = \frac{-5 \pm \sqrt{25 - 24}}{2}$$
$$x = \frac{-5 \pm \sqrt{1}}{2}$$
$$x = \frac{-5 \pm 1}{2}$$
$$x = \frac{-5+1}{2} \quad \text{or} \quad x = \frac{-5-1}{2}$$
$$x = \frac{-4}{2} \quad\quad x = \frac{-6}{2}$$
$$x = -2 \quad\quad x = -3$$

The solutions are -2 and -3.

15. $x^2 + 7x + 12 = 0$

$$1x^2 + 7x + 12 = 0$$
$$a = 1,\ b = 7,\ c = 12$$
$$x = \frac{-b \pm \sqrt{b^2 - 4ac}}{2a}$$
$$x = \frac{-7 \pm \sqrt{7^2 - 4(1)(12)}}{2(1)}$$
$$x = \frac{-7 \pm \sqrt{49 - 48}}{2}$$
$$x = \frac{-7 \pm \sqrt{1}}{2}$$
$$x = \frac{-7 \pm 1}{2}$$
$$x = \frac{-7+1}{2} \quad \text{or} \quad x = \frac{-7-1}{2}$$
$$x = \frac{-6}{2} \quad\quad x = \frac{-8}{2}$$
$$x = -3 \quad\quad x = -4$$

The solutions are -3 and -4.

17. $4x^2+3x-1=0$

$a=4,\ b=3,\ c=-1$

$$x=\frac{-b\pm\sqrt{b^2-4ac}}{2a}$$
$$x=\frac{-3\pm\sqrt{3^2-4(4)(-1)}}{2(4)}$$
$$x=\frac{-3\pm\sqrt{9+16}}{8}$$
$$x=\frac{-3\pm\sqrt{25}}{8}$$
$$x=\frac{-3\pm5}{8}$$

$x=\frac{-3+5}{8}$ or $x=\frac{-3-5}{8}$

$x=\frac{2}{8}$ | $x=\frac{-8}{8}$

$x=\frac{1}{4}$ | $x=-1$

The solutions are $\frac{1}{4}$ and -1.

19. $6x^2+5x-6=0$

$a=6,\ b=5,\ c=-6$

$$x=\frac{-b\pm\sqrt{b^2-4ac}}{2a}$$
$$x=\frac{-5\pm\sqrt{5^2-4(6)(-6)}}{2(6)}$$
$$x=\frac{-5\pm\sqrt{25+144}}{12}$$
$$x=\frac{-5\pm\sqrt{169}}{12}$$
$$x=\frac{-5\pm13}{12}$$

$x=\frac{-5+13}{12}$ or $x=\frac{-5-13}{12}$

$x=\frac{8}{12}$ | $x=\frac{-18}{12}$

$x=\frac{2}{3}$ | $x=-\frac{3}{2}$

The solutions are $\frac{2}{3}$ and $-\frac{3}{2}$.

21. $3x^2-5x-2=0$

$a=3,\ b=-5,\ c=-2$

$$x=\frac{-b\pm\sqrt{b^2-4ac}}{2a}$$
$$x=\frac{-(-5)\pm\sqrt{(-5)^2-4(3)(-2)}}{2(3)}$$
$$x=\frac{5\pm\sqrt{25+24}}{6}$$
$$x=\frac{5\pm\sqrt{49}}{6}$$
$$x=\frac{5\pm7}{6}$$

$x=\frac{5+7}{6}$ or $x=\frac{5-7}{6}$

$x=\frac{12}{6}$ | $x=\frac{-2}{6}$

$x=2$ | $x=-\frac{1}{3}$

The solutions are 2 and $-\frac{1}{3}$.

23. $5x^2-13x-6=0$

$a=5,\ b=-13,\ c=-6$

$$x=\frac{-b\pm\sqrt{b^2-4ac}}{2a}$$
$$x=\frac{-(-13)\pm\sqrt{(-13)^2-4(5)(-6)}}{2(5)}$$
$$x=\frac{13\pm\sqrt{169+120}}{10}$$
$$x=\frac{13\pm\sqrt{289}}{10}$$
$$x=\frac{13\pm17}{10}$$

$x=\frac{13+17}{10}$ or $x=\frac{13-17}{10}$

$x=\frac{30}{10}$ | $x=\frac{-4}{10}$

$x=3$ | $x=-\frac{2}{5}$

The solutions are 3 and $-\frac{2}{5}$.

Use the quadratic formula to solve each equation. Approximate the solutions to nearest hundredth. See Example 2.

25.
$$x^2+1=-3x$$
$$x^2+1+\mathbf{3x}=-3x+\mathbf{3x}$$
$$1x^2+3x+1=0$$
$$a=1,\ b=3,\ c=1$$
$$x=\frac{-b\pm\sqrt{b^2-4ac}}{2a}$$
$$x=\frac{-3\pm\sqrt{3^2-4(1)(1)}}{2(1)}$$
$$x=\frac{-3\pm\sqrt{9-4}}{2}$$
$$x=\frac{-3\pm\sqrt{5}}{2}$$

	or	
$x=\frac{-3+\sqrt{5}}{2}$		$x=\frac{-3-\sqrt{5}}{2}$
$x\approx\frac{-3+2.236}{2}$		$x\approx\frac{-3-2.236}{2}$
$x\approx\frac{-0.763}{2}$		$x\approx\frac{-5.236}{2}$
$x\approx-0.381$		$x\approx-2.618$
$x\approx-0.38$		$x\approx-2.62$

The exact solutions are: $\frac{-3\pm\sqrt{5}}{2}$.

The approximate solutions are: $-0.38, -2.62$.

27.
$$x^2-4=-7x$$
$$x^2-4+\mathbf{7x}=-7x+\mathbf{7x}$$
$$1x^2+7x-4=0$$
$$a=1,\ b=7,\ c=-4$$
$$x=\frac{-b\pm\sqrt{b^2-4ac}}{2a}$$
$$x=\frac{-7\pm\sqrt{7^2-4(1)(-4)}}{2(1)}$$
$$x=\frac{-7\pm\sqrt{49+16}}{2}$$
$$x=\frac{-7\pm\sqrt{65}}{2}$$

	or	
$x=\frac{-7+\sqrt{65}}{2}$		$x=\frac{-7-\sqrt{65}}{2}$
$x\approx\frac{-7+8.062}{2}$		$x\approx\frac{-7-8.062}{2}$
$x\approx\frac{1.062}{2}$		$x\approx\frac{-15.062}{2}$
$x\approx0.531$		$x\approx-7.531$
$x\approx0.53$		$x\approx-7.53$

The exact solutions are: $\frac{-7\pm\sqrt{65}}{2}$.

The approximate solutions are: $0.53, -7.53$.

29.
$$3x^2-x=3$$
$$3x^2-x-\mathbf{3}=3-\mathbf{3}$$
$$3x^2-x-3=0$$
$$a=3,\ b=-1,\ c=-3$$
$$x=\frac{-b\pm\sqrt{b^2-4ac}}{2a}$$
$$x=\frac{-(-1)\pm\sqrt{(-1)^2-4(3)(-3)}}{2(3)}$$
$$x=\frac{1\pm\sqrt{1+36}}{6}$$
$$x=\frac{1\pm\sqrt{37}}{6}$$

	or	
$x=\frac{1+\sqrt{37}}{6}$		$x=\frac{1-\sqrt{37}}{6}$
$x\approx\frac{1+6.082}{6}$		$x\approx\frac{1-6.082}{6}$
$x\approx\frac{7.082}{6}$		$x\approx\frac{-5.082}{6}$
$x\approx1.180$		$x\approx-0.847$
$x\approx1.18$		$x\approx-0.85$

The exact solutions are: $\frac{1\pm\sqrt{37}}{6}$.

The approximate solutions are: $1.18, -0.85$.

31. $$7x^2 - 3x = 1$$
$$7x^2 - 3x - \mathbf{1} = 1 - \mathbf{1}$$
$$7x^2 - 3x - 1 = 0$$
$$a = 7,\ b = -3,\ c = -1$$
$$x = \frac{-b \pm \sqrt{b^2 - 4ac}}{2a}$$
$$x = \frac{-(-3) \pm \sqrt{(-3)^2 - 4(7)(-1)}}{2(7)}$$
$$x = \frac{3 \pm \sqrt{9 + 28}}{14}$$
$$x = \frac{3 \pm \sqrt{37}}{14}$$

	or	
$x = \frac{3 + \sqrt{37}}{14}$		$x = \frac{3 - \sqrt{37}}{14}$
$x \approx \frac{3 + 6.082}{14}$		$x \approx \frac{3 - 6.082}{14}$
$x \approx \frac{9.082}{14}$		$x \approx \frac{-3.082}{14}$
$x \approx 0.648$		$x \approx -0.220$
$x \approx 0.65$		$x \approx -0.22$

The exact solutions are: $\frac{3 \pm \sqrt{37}}{14}$.
The approximate solutions are: $0.65, -0.22$.

33. $$4x^2 = 7x - 2$$
$$4x^2 - \mathbf{7x} + \mathbf{2} = 7x - 2 - \mathbf{7x} + \mathbf{2}$$
$$4x^2 - 7x + 2 = 0$$
$$a = 4,\ b = -7,\ c = 2$$
$$x = \frac{-b \pm \sqrt{b^2 - 4ac}}{2a}$$
$$x = \frac{-(-7) \pm \sqrt{(-7)^2 - 4(4)(2)}}{2(4)}$$
$$x = \frac{7 \pm \sqrt{49 - 32}}{8}$$
$$x = \frac{7 \pm \sqrt{17}}{8}$$

	or	
$x = \frac{7 + \sqrt{17}}{8}$		$x = \frac{7 - \sqrt{17}}{8}$
$x \approx \frac{7 + 4.123}{8}$		$x \approx \frac{7 - 4.123}{8}$
$x \approx \frac{11.123}{8}$		$x \approx \frac{2.876}{8}$
$x \approx 1.390$		$x \approx 0.359$
$x \approx 1.39$		$x \approx 0.36$

The exact solutions are: $\frac{7 \pm \sqrt{17}}{8}$.
The approximate solutions are: $1.39, 0.36$.

35. $$2x^2 = 7 - 9x$$
$$2x^2 + \mathbf{9x} - \mathbf{7} = 7 - 9x + \mathbf{9x} - \mathbf{7}$$
$$2x^2 + 9x - 7 = 0$$
$$a = 2,\ b = 9,\ c = -7$$
$$x = \frac{-b \pm \sqrt{b^2 - 4ac}}{2a}$$
$$x = \frac{-9 \pm \sqrt{9^2 - 4(2)(-7)}}{2(2)}$$
$$x = \frac{-9 \pm \sqrt{81 + 56}}{4}$$
$$x = \frac{-9 \pm \sqrt{137}}{4}$$

	or	
$x = \frac{-9 + \sqrt{137}}{4}$		$x = \frac{-9 - \sqrt{137}}{4}$
$x \approx \frac{-9 + 11.704}{4}$		$x \approx \frac{-9 - 11.704}{4}$
$x \approx \frac{2.704}{4}$		$x \approx \frac{-20.704}{4}$
$x \approx 0.676$		$x \approx -5.176$
$x \approx 0.68$		$x \approx -5.18$

The exact solutions are: $\frac{-9 \pm \sqrt{137}}{4}$.
The approximate solutions are: $0.68, -5.18$.

Use the quadratic formula to solve each equation, if possible. See Example 3.

37. $2n^2+10n+11=0$

$a=2,\ b=10,\ c=11$

$$n=\frac{-b\pm\sqrt{b^2-4ac}}{2a}$$

$$n=\frac{-10\pm\sqrt{10^2-4(2)(11)}}{2(2)}$$

$$n=\frac{-10\pm\sqrt{100-88}}{4}$$

$$n=\frac{-10\pm\sqrt{12}}{4}$$

$$n=\frac{-10\pm\sqrt{4}\sqrt{3}}{4}$$

$$n=\frac{-10\pm2\sqrt{3}}{4}$$

$$n=\frac{2\left(-5\pm\sqrt{3}\right)}{4}$$

$$n=\frac{\overset{1}{\cancel{2}}\left(-5\pm\sqrt{3}\right)}{\underset{1}{\cancel{2}}\cdot2}$$

$$n=\frac{-5\pm\sqrt{3}}{2}$$

The exact solutions are: $\frac{-5\pm\sqrt{3}}{2}$.

39. $2d^2-6d+1=0$

$a=2,\ b=-6,\ c=1$

$$d=\frac{-b\pm\sqrt{b^2-4ac}}{2a}$$

$$d=\frac{-(-6)\pm\sqrt{(-6)^2-4(2)(1)}}{2(2)}$$

$$d=\frac{6\pm\sqrt{36-8}}{4}$$

$$d=\frac{6\pm\sqrt{28}}{4}$$

$$d=\frac{6\pm\sqrt{4}\sqrt{7}}{4}$$

$$d=\frac{6\pm2\sqrt{7}}{4}$$

$$d=\frac{2\left(3\pm\sqrt{7}\right)}{4}$$

$$d=\frac{\overset{1}{\cancel{2}}\left(3\pm\sqrt{7}\right)}{\underset{1}{\cancel{2}}\cdot2}$$

$$d=\frac{3\pm\sqrt{7}}{2}$$

The exact solutions are: $\frac{3\pm\sqrt{7}}{2}$.

41. $3x^2-8x+2=0$

$a=3,\ b=-8,\ c=2$

$$x=\frac{-b\pm\sqrt{b^2-4ac}}{2a}$$

$$x=\frac{-(-8)\pm\sqrt{(-8)^2-4(3)(2)}}{2(3)}$$

$$x=\frac{8\pm\sqrt{64-24}}{6}$$

$$x=\frac{8\pm\sqrt{40}}{6}$$

$$x=\frac{8\pm\sqrt{4}\sqrt{10}}{6}$$

$$x=\frac{8\pm2\sqrt{10}}{6}$$

$$x=\frac{2\left(4\pm\sqrt{10}\right)}{6}$$

$$x=\frac{\overset{1}{\cancel{2}}\left(4\pm\sqrt{10}\right)}{\underset{1}{\cancel{2}}\cdot3}$$

$$x=\frac{4\pm\sqrt{10}}{3}$$

The exact solutions are: $\frac{4\pm\sqrt{10}}{3}$.

43.
$$4x^2 - 12x = -1$$
$$4x^2 - 12x + \mathbf{1} = -1 + \mathbf{1}$$
$$4x^2 - 12x + 1 = 0$$
$$a = 4,\ b = -12,\ c = 1$$
$$x = \frac{-b \pm \sqrt{b^2 - 4ac}}{2a}$$
$$x = \frac{-(-12) \pm \sqrt{(-12)^2 - 4(4)(1)}}{2(4)}$$
$$x = \frac{12 \pm \sqrt{144 - 16}}{8}$$
$$x = \frac{12 \pm \sqrt{128}}{8}$$
$$x = \frac{12 \pm \sqrt{64}\sqrt{2}}{8}$$
$$x = \frac{12 \pm 8\sqrt{2}}{8}$$
$$x = \frac{4\left(3 \pm 2\sqrt{2}\right)}{8}$$
$$x = \frac{\overset{1}{\cancel{4}}\left(3 \pm 2\sqrt{2}\right)}{2 \cdot \underset{1}{\cancel{4}}}$$
$$x = \frac{3 \pm 2\sqrt{2}}{2}$$
The exact solutions are: $\frac{3 \pm 2\sqrt{2}}{2}$.

45.
$$x^2 = 1 - 2x$$
$$x^2 + \mathbf{2x} - \mathbf{1} = 1 - 2x + \mathbf{2x} - \mathbf{1}$$
$$1x^2 + 2x - 1 = 0$$
$$a = 1,\ b = 2,\ c = -1$$
$$x = \frac{-b \pm \sqrt{b^2 - 4ac}}{2a}$$
$$x = \frac{-2 \pm \sqrt{2^2 - 4(1)(-1)}}{2(1)}$$
$$x = \frac{-2 \pm \sqrt{4 + 4}}{2}$$
$$x = \frac{-2 \pm \sqrt{8}}{2}$$
$$x = \frac{-2 \pm \sqrt{4}\sqrt{2}}{2}$$
$$x = \frac{-2 \pm 2\sqrt{2}}{2}$$
$$x = \frac{2\left(-1 \pm \sqrt{2}\right)}{2}$$
$$x = \frac{\overset{1}{\cancel{2}}\left(-1 \pm \sqrt{2}\right)}{\underset{1}{\cancel{2}}}$$
$$x = -1 \pm \sqrt{2}$$
The exact solutions are: $-1 \pm \sqrt{2}$.

47.
$$a^2 + 4a = 3$$
$$a^2 + 4a - \mathbf{3} = 3 - \mathbf{3}$$
$$1a^2 + 4a - 3 = 0$$
$$a = 1,\ b = 4,\ c = -3$$
$$a = \frac{-b \pm \sqrt{b^2 - 4ac}}{2a}$$
$$a = \frac{-4 \pm \sqrt{4^2 - 4(1)(-3)}}{2(1)}$$
$$a = \frac{-4 \pm \sqrt{16 + 12}}{2}$$
$$a = \frac{-4 \pm \sqrt{28}}{2}$$
$$a = \frac{-4 \pm \sqrt{4}\sqrt{7}}{2}$$
$$a = \frac{-4 \pm 2\sqrt{7}}{2}$$
$$a = \frac{2\left(-2 \pm \sqrt{7}\right)}{2}$$
$$a = \frac{\overset{1}{\cancel{2}}\left(-2 \pm \sqrt{7}\right)}{\underset{1}{\cancel{2}}}$$
$$a = -2 \pm \sqrt{7}$$
The exact solutions are: $-2 \pm \sqrt{7}$.

Use the quadratic formula to solve each equation, if possible. See Example 4.

49.
$$3m^2+5=2m$$
$$3m^2+5-\mathbf{2m}=2m-\mathbf{2m}$$
$$3m^2-2m+5=0$$
$$a=3,\ b=-2,\ c=5$$
$$m=\frac{-b\pm\sqrt{b^2-4ac}}{2a}$$
$$m=\frac{-(-2)\pm\sqrt{(-2)^2-4(3)(5)}}{2(3)}$$
$$m=\frac{2\pm\sqrt{4-60}}{6}$$
$$m=\frac{2\pm\sqrt{-56}}{6}$$
Since $\sqrt{-56}$ is not a real number, $3m^2-2m+5=0$ has no real-number solutions.

51.
$$7x^2-x=-8$$
$$7x^2-x+\mathbf{8}=-8+\mathbf{8}$$
$$7x^2-x+8=0$$
$$a=7,\ b=-1,\ c=8$$
$$x=\frac{-b\pm\sqrt{b^2-4ac}}{2a}$$
$$x=\frac{-(-1)\pm\sqrt{(-1)^2-4(7)(8)}}{2(7)}$$
$$x=\frac{1\pm\sqrt{1-224}}{14}$$
$$x=\frac{1\pm\sqrt{-223}}{14}$$
Since $\sqrt{-223}$ is not a real number, $7x^2-x+8=0$ has no real-number solutions.

TRY IT YOURSELF

Use the most efficient method to solve each equation. Give the exact solutions, and approximations to the nearest hundredth when appropriate.

53. $(2y-1)^2=25$ Use square root method.
$$2y-1=\pm\sqrt{25}$$
$$2y-1=\pm 5$$
$$2y-1+\mathbf{1}=\mathbf{1}\pm5$$
$$2y=1\pm5$$
$$\frac{2y}{\mathbf{2}}=\frac{1\pm5}{\mathbf{2}}$$
$$y=\frac{1\pm5}{2}$$

$y=\frac{1+5}{2}$	or	$y=\frac{1-5}{2}$
$y=\frac{6}{2}$		$y=\frac{-4}{2}$
$y=3$		$y=-2$

The solutions are 3 and -2.

55. $2x^2+x=5$ Use Quadratic Formula.
$$2x^2+x-\mathbf{5}=5-\mathbf{5}$$
$$2x^2+x-5=0$$
$$a=2,\ b=1,\ c=-5$$
$$x=\frac{-b\pm\sqrt{b^2-4ac}}{2a}$$
$$x=\frac{-1\pm\sqrt{1^2-4(2)(-5)}}{2(2)}$$
$$x=\frac{-1\pm\sqrt{1+40}}{4}$$
$$x=\frac{-1\pm\sqrt{41}}{4}$$

$x=\frac{-1+\sqrt{41}}{4}$	or	$x=\frac{-1-\sqrt{41}}{4}$
$x\approx\frac{-1+6.403}{4}$		$x\approx\frac{-1-6.403}{4}$
$x\approx\frac{5.403}{4}$		$x\approx\frac{-7.403}{4}$
$x\approx1.350$		$x\approx-1.850$
$x\approx1.35$		$x\approx-1.85$

The exact solutions are: $\frac{-1\pm\sqrt{41}}{4}$.

The approximate solutions are: 1.35, -1.85.

57. $x^2-12x+35=0$ Use Factoring.

$$(x-5)(x-7)=0$$

$$x-5=0 \quad \text{or} \quad x-7=0$$

$$x=5 \qquad x=7$$

The solutions are 5 and 7.

59. $r^2+9r+1=0$ Use Quadratic Formula.

$$1r^2+9r+1=0$$

$$a=1,\ b=9,\ c=1$$

$$r=\frac{-b\pm\sqrt{b^2-4ac}}{2a}$$

$$r=\frac{-9\pm\sqrt{9^2-4(1)(1)}}{2(1)}$$

$$r=\frac{-9\pm\sqrt{81-4}}{2}$$

$$r=\frac{-9\pm\sqrt{77}}{2}$$

$$r=\frac{-9+\sqrt{77}}{2} \quad \text{or} \quad r=\frac{-9-\sqrt{77}}{2}$$

$$r\approx\frac{-9+8.774}{2} \qquad r\approx\frac{-9-8.774}{2}$$

$$r\approx\frac{-0.225}{2} \qquad r\approx\frac{-17.774}{2}$$

$$r\approx-0.112 \qquad r\approx-8.887$$

$$r\approx-0.11 \qquad r\approx-8.89$$

The exact solutions are: $\frac{-9\pm\sqrt{77}}{2}$.

The approximate solutions are: $-0.11, -8.89$.

61. $5x^2-2x=8$ Use Quadratic Formula.

$$5x^2-2x-\mathbf{8}=8-\mathbf{8}$$

$$5x^2-2x-8=0$$

$$a=5,\ b=-2,\ c=-8$$

$$x=\frac{-b\pm\sqrt{b^2-4ac}}{2a}$$

$$x=\frac{-(-2)\pm\sqrt{(-2)^2-4(5)(-8)}}{2(5)}$$

$$x=\frac{2\pm\sqrt{4+160}}{10}$$

$$x=\frac{2\pm\sqrt{164}}{10}$$

$$x=\frac{2\pm\sqrt{4\cdot41}}{10}$$

$$x=\frac{2\pm2\sqrt{41}}{10}$$

$$x=\frac{\cancel{2}^{1}\left(1\pm\sqrt{41}\right)}{\cancel{2}_{1}\cdot5}$$

$$x=\frac{1\pm\sqrt{41}}{5}$$

$$x=\frac{1+\sqrt{41}}{5} \quad \text{or} \quad x=\frac{1-\sqrt{41}}{5}$$

$$x\approx\frac{1+6.403}{5} \qquad x\approx\frac{1-6.403}{5}$$

$$x\approx\frac{7.403}{5} \qquad x\approx\frac{-5.403}{5}$$

$$x\approx1.480 \qquad x\approx-1.080$$

$$x\approx1.48 \qquad x\approx-1.08$$

The exact solutions are: $\frac{1\pm\sqrt{41}}{5}$.

The approximate solutions are: $1.48, -1.08$.

63. $x^2+3x=9$ Use Quadratic Formula.

$$x^2+3x-\mathbf{9}=9-\mathbf{9}$$

$$1x^2+3x-9=0$$

$$a=1,\ b=3,\ c=-9$$

$$x=\frac{-b\pm\sqrt{b^2-4ac}}{2a}$$

$$x=\frac{-3\pm\sqrt{3^2-4(1)(-9)}}{2(1)}$$

$$x=\frac{-3\pm\sqrt{9+36}}{2}$$

$$x=\frac{-3\pm\sqrt{45}}{2}$$

$$x=\frac{-3\pm\sqrt{9\cdot5}}{2}$$

$$x=\frac{-3\pm3\sqrt{5}}{2}$$

$$x=\frac{-3+3\sqrt{5}}{2} \quad \text{or} \quad x=\frac{-3-3\sqrt{5}}{2}$$

$$x\approx\frac{-3+6.708}{2} \qquad x\approx\frac{-3-6.708}{2}$$

$$x\approx\frac{3.708}{2} \qquad x\approx\frac{-9.708}{2}$$

$$x\approx1.854 \qquad x\approx-4.854$$

$$x\approx1.85 \qquad x\approx-4.85$$

The exact solutions are: $\frac{-3\pm3\sqrt{5}}{2}$.

The approximate solutions are: $1.85, -4.85$.

65. $7x^2+6x+4=0$ Use Quadratic Formula.

$a=7,\ b=6,\ c=4$

$$x=\frac{-b\pm\sqrt{b^2-4ac}}{2a}$$

$$x=\frac{-6\pm\sqrt{6^2-4(7)(4)}}{2(7)}$$

$$x=\frac{-6\pm\sqrt{36-112}}{14}$$

$$x=\frac{-6\pm\sqrt{-76}}{14}$$

Since $\sqrt{-76}$ is not a real number, $7x^2+6x+4=0$ has no real-number solutions.

67. $4c^2+16c=0$ Use Factoring.

$4c(c+4)=0$

$4c=0$ or $c+4=0$

$c=0$ | $c=-4$

The solutions are 0 and -4.

69. $3y^2=18$ Use square root method.

$$\frac{3y^2}{\mathbf{3}}=\frac{18}{\mathbf{3}}$$

$y^2=6$

$y=\pm\sqrt{6}$

The exact solutions are: $\pm\sqrt{6}$.
The approximate solutions are: ±2.45.

71. $x^2-4x=29$ Use Quadratic Formula.

$x^2-4x-\mathbf{29}=29-\mathbf{29}$

$1x^2-4x-29=0$

$a=1,\ b=-4,\ c=-29$

$$x=\frac{-b\pm\sqrt{b^2-4ac}}{2a}$$

$$x=\frac{-(-4)\pm\sqrt{(-4)^2-4(1)(-29)}}{2(1)}$$

$$x=\frac{4\pm\sqrt{16+116}}{2}$$

$$x=\frac{4\pm\sqrt{132}}{2}$$

$$x=\frac{4\pm\sqrt{4\cdot33}}{2}$$

$$x=\frac{4\pm2\sqrt{33}}{2}$$

$$x=\frac{\overset{1}{\cancel{2}}\left(2\pm1\sqrt{33}\right)}{\underset{1}{\cancel{2}}}$$

$x=2\pm\sqrt{33}$

$x=2+\sqrt{33}$ or $x=2-\sqrt{33}$

$x\approx2+5.744$ | $x\approx2-5.744$

$x\approx7.744$ | $x\approx-3.744$

$x\approx7.74$ | $x\approx-3.74$

The exact solutions are: $2\pm\sqrt{33}$.
The approximate solutions are: 7.74, –3.74.

73. $t^2-24t+144=0$ Use Factoring.

$(t-12)(t-12)=0$

$t-12=0$

$t=12$

The solution is 12, a repeated solution.

75. $x^2-63=0$ Use square root method.

$x^2-63+\mathbf{63}=0+\mathbf{63}$

$x^2=63$

$x=\pm\sqrt{63}$

$x=\pm\sqrt{9\cdot7}$

$x=\pm3\sqrt{7}$

$x=3\sqrt{7}$ or $x=-3\sqrt{7}$

$x\approx7.937$ | $x\approx-7.937$

$x\approx7.94$ | $x\approx-7.94$

The exact solutions are: $\pm3\sqrt{7}$.
The approximate solutions are: ±7.94.

APPLICATIONS
from CAMPUS TO CAREERS

77. GRAPHIC DESIGNER

Analyze

- Area of the triangle is 80 in^2.
- Base is 6 in longer than the height.
- What are the lengths of base and height?
- Area formula of a triangle is: $A = \frac{1}{2}bh$.

Assign

Let $h =$ height of the triangle in inches.
$h + 6 =$ base of triangle in inches.

Form

$\frac{1}{2}$	times	the base	times	the height	equals	the area of the triangle.
$\frac{1}{2}$	$\cdot$	h	$\cdot$	$h+6$	$=$	80

Solve

$$\frac{1}{2}x(x+6) = 80$$
$$(\mathbf{2})\frac{1}{2}x(x+6) = (\mathbf{2})80$$
$$x(x+6) = 160$$
$$x^2 + 6x = 160$$
$$x^2 + 6x - \mathbf{160} = 160 - \mathbf{160}$$
$$x^2 + 6x - 160 = 0$$
$$(x+16)(x-10) = 0$$

$x + 16 = 0$ or $x - 10 = 0$

~~$x = -16$~~ $\quad x = 10$

State

The height is 10 inches.
The base is $x + 6 = 10 + 6 = 16$ inches.

Check

The result checks.

79. EARTHQUAKES

Analyze

- Hypotenuse is 52 inches (one board).
- Length is l inches.
- Height is 10 inches less than the length.
- Find the length and height to nearest tenth.
- Pythagorean theroem: $a^2 + b^2 = c^2$.

Assign

Let $l =$ length of window (longer leg) in inches.
$l - 10 =$ height of window (shorter leg) in inches.

Form

length^2	plus	height^2	equals	hypotenuse.^2
l^2	$+$	$(l-10)^2$	$=$	52^2

Solve

$$l^2 + (l-10)^2 = 52^2$$
$$l^2 + l^2 - 20l + 100 = 2,704$$
$$2l^2 - 20l + 100 = 2,704$$
$$2l^2 - 20l + 100 - \mathbf{2,704} = 2,704 - \mathbf{2,704}$$
$$2l^2 - 20l - 2,604 = 0$$
$$2(l^2 - 10l - 1,302) = 0$$
$$a = 1,\ b = -10,\ c = -1,302$$
$$l = \frac{-b \pm \sqrt{b^2 - 4ac}}{2a}$$
$$l = \frac{-(-10) \pm \sqrt{(-10)^2 - 4(1)(-1,302)}}{2(1)}$$
$$l = \frac{10 \pm \sqrt{100 + 5,208}}{2}$$
$$l = \frac{10 \pm \sqrt{5,308}}{2}$$

$l = \frac{10 + \sqrt{5,308}}{2}$ or $l = \frac{10 - \sqrt{5,308}}{2}$

$l \approx \frac{10 + 72.856}{2}$ $\quad l \approx \frac{10 - 72.856}{2}$

$l \approx \frac{82.856}{2}$ $\quad$ ~~$l \approx \frac{-62.856}{2}$~~

$l \approx 41.42$

$l \approx 41.4$

State

The length is 41.4 inches.
The height is $41.4 - 10 = 31.4$ inches.

Check

The results check.

81. COMICS

Analyze

- Area of the strip is 100 cm^2.
- Length is 4 cm more than twice the width.
- What are the dimensions to the nearest tenth?
- Formula: $A = lw$.

Assign

Let w = width of strip in cm.

$2w+4$ = length of strip in cm.

Form

Length	times	width	equals	the area of the rectangle.
$2w+4$	$\cdot$	w	$=$	100

Solve

$$w(2w+4)=100$$
$$2w^2+4w=100$$
$$2w^2+4w-\mathbf{100}=100-\mathbf{100}$$
$$2w^2+4w-100=0$$
$$2(w^2+2w-50)=0$$
$$a=1,\ b=2,\ c=-50$$
$$w=\frac{-b\pm\sqrt{b^2-4ac}}{2a}$$
$$w=\frac{-(2)\pm\sqrt{(2)^2-4(1)(-50)}}{2(1)}$$
$$w=\frac{-2\pm\sqrt{4+200}}{2}$$
$$w=\frac{-2\pm\sqrt{204}}{2}$$

$w=\frac{-2\pm\sqrt{204}}{2}$ or $w=\frac{-2-\sqrt{204}}{2}$

$w\approx\frac{-2+14.28}{2}$ | $w\approx\frac{-2-14.28}{2}$

$w\approx\frac{12.28}{2}$ | ~~$w\approx\frac{-16.28}{2}$~~

$w\approx 6.14$

$w\approx 6.1$

State

The width is 6.1 cm.
The length is $2(6.1)+4=12.2+4=16.2$ cm.

Check

The result checks.

83. INVESTING

Analyze

- $A=\$5,724.50$
- $P=\$5,000$
- What is the rate?

Assign

Let r = the rate as a percent

Form

$$A=P(1+r)^2$$

Solve

$$A=P(1+r)^2$$
$$5,724.50=5,000(1+r)^2$$
$$\frac{5,724.50}{\mathbf{5,000}}=\frac{\cancel{5,000}(1+r)^2}{\cancel{\mathbf{5,000}}}$$
$$1.1449=(1+r)^2$$
$$(1+r)^2=1.1449$$
$$1+r=\pm\sqrt{1.1449}$$
$$1+r-\mathbf{1}=-\mathbf{1}\pm\sqrt{1.1449}$$
$$r=-1\pm\sqrt{1.1449}$$

$r=-1+\sqrt{1.1449}$ or $r=-1-\sqrt{1.1449}$

$r=-1+1.07$ | $r=-1-1.07$

$r=0.07$ | ~~$r=-2.07$~~

State

The rate is $0.07=7\%$.

Check

The result checks.

WRITING

85-89. Answers will vary.

REVIEW

Solve each equation for the specified variable.

91. $A=p+prt$; for r

$$A=p+prt$$
$$A-\mathbf{p}=p+prt-\mathbf{p}$$
$$A-p=prt$$
$$\frac{A-p}{\mathbf{pt}}=\frac{\cancel{p}r\cancel{t}}{\cancel{\mathbf{p}}\cancel{\mathbf{t}}}$$
$$r=\frac{A-p}{pt}$$

93. $\frac{1}{r}=\frac{1}{r_1}+\frac{1}{r_2}$; for r

$\text{LCD}=rr_1r_2$

$$\frac{1}{r}=\frac{1}{r_1}+\frac{1}{r_2}$$

$$\frac{1}{r}\cdot(rr_1r_2)=\left(\frac{1}{r_1}+\frac{1}{r_2}\right)\cdot(rr_1r_2)$$

$$\frac{1}{\not{r}}\cdot\frac{\not{r}r_1r_2}{1}=\frac{1}{\not{r_1}}\cdot\frac{r\not{r_1}r_2}{1}+\frac{1}{\not{r_2}}\cdot\frac{rr_1\not{r_2}}{1}$$

$$r_1r_2=rr_2+rr_1$$

$$r_1r_2=r(r_2+r_1)$$

$$\frac{r_1r_2}{r_2+r_1}=\frac{r\,\cancel{(r_2+r_1)}}{\cancel{(r_2+r_1)}}$$

$$\frac{r_1r_2}{r_2+r_1}=r$$

$$r=\frac{r_1r_2}{r_2+r_1}$$

97. $\frac{(n+5)(2n+1)}{2}=3.5$

$$\mathbf{2}\cdot\frac{(n+5)(2n+1)}{2}=\mathbf{2}\cdot 3.5$$

$$\frac{\cancel{2}}{1}\cdot\frac{(n+5)(2n+1)}{\cancel{2}}=2\cdot 3.5$$

$$(n+5)(2n+1)=7$$

$$2n^2+10n+n+5=7$$

$$2n^2+11n+5=7$$

$$2n^2+11n+5-\mathbf{7}=7-\mathbf{7}$$

$$2n^2+11n-2=0$$

$$a=2,\ b=11,\ c=-2$$

$$n=\frac{-b\pm\sqrt{b^2-4ac}}{2a}$$

$$n=\frac{-11\pm\sqrt{11^2-4(2)(-2)}}{2(2)}$$

$$n=\frac{-11\pm\sqrt{121+16}}{4}$$

$$n=\frac{-11\pm\sqrt{137}}{4}$$

CHALLENGE PROBLEMS

Use the quadratic formula to solve each equation.

95. $\frac{2}{3}x^2-\frac{1}{3}=\frac{5}{9}x$

$$\frac{\mathbf{9}}{\mathbf{1}}\cdot\frac{2}{3}x^2-\frac{\mathbf{9}}{\mathbf{1}}\cdot\frac{1}{3}=\frac{\mathbf{9}}{\mathbf{1}}\cdot\frac{5}{9}x$$

$$\frac{\cancel{9}}{1}\cdot\frac{2}{\cancel{3}}x^2-\frac{\cancel{9}}{1}\cdot\frac{1}{\cancel{3}}=\frac{\cancel{9}}{1}\cdot\frac{5}{\cancel{9}}x$$

$$6x^2-3=5x$$

$$6x^2-3-\mathbf{5x}=5x-\mathbf{5x}$$

$$6x^2-5x-3=0$$

$$a=6,\ b=-5,\ c=-3$$

$$x=\frac{-b\pm\sqrt{b^2-4ac}}{2a}$$

$$x=\frac{-(-5)\pm\sqrt{(-5)^2-4(6)(-3)}}{2(6)}$$

$$x=\frac{5\pm\sqrt{25+72}}{12}$$

$$x=\frac{5\pm\sqrt{97}}{12}$$

The exact solutions are $\frac{5\pm\sqrt{97}}{12}$.

Solve each equation and approximate each solution to the nearest tenth.

99. $2.4x^2-9.5x+6.2=0$ Use Quadratic Formula.

$$a=2.4,\ b=-9.5,\ c=6.2$$

$$x=\frac{-b\pm\sqrt{b^2-4ac}}{2a}$$

$$x=\frac{-(-9.5)\pm\sqrt{(-9.5)^2-4(2.4)(6.2)}}{2(2.4)}$$

$$x=\frac{9.5\pm\sqrt{90.25-59.52}}{4.8}$$

$$x=\frac{9.5\pm\sqrt{30.73}}{4.8}=\frac{\mathbf{10}(9.5)\pm\sqrt{\mathbf{100}(30.73)}}{\mathbf{10}(4.8)}$$

$$x=\frac{95\pm\sqrt{3073}}{48}\quad\text{or}\quad\frac{9.5\pm 5.54}{4.8}$$

$x=\frac{9.5+5.5}{4.8}$	or $x=\frac{9.5-5.5}{4.8}$
$x\approx\frac{15}{4.8}$	$x\approx\frac{4}{4.8}$
$x\approx 3.12$	$x\approx 0.83$
$x\approx 3.1$	$x\approx 0.8$

The approximate solutions are: 3.1, 0.8.

101. DECKING

Analyze

- Dimensions of the pool 24 ft by 14 ft.
- Area of the decking is 368 ft^2.
- What is width of the decking?
- Length of pool and decking is $2x+24$.
- Width of pool and decking is $2x+14$.

Assign

Let $x =$ width of decking in ft.

Form

The area (lw) of the pool and decking	minus	the area (lw) of the pool	equals	the area of the decking.
$(2x+24)(2x+14)$	$-$	$(24)(14)$	$=$	368

Solve

$$(2x+24)(2x+14)-(24)(14)=368$$
$$4x^2+28x+48x+336-336=368$$
$$4x^2+76x+336-336=368$$
$$4x^2+76x=368$$
$$4x^2+76x-\mathbf{368}=368-\mathbf{368}$$
$$4x^2+76x-368=0$$
$$4(x^2+19x-92)=0$$
$$4(x+23)(x-4)=0$$

$x+23=0$ or $x-4=0$

~~$x=-23$~~ | $x=4$

State

The width of decking is 4 ft.

Check

The result checks.

SECTION 9.4 STUDY SET

VOCABULARY

Fill in the blanks.

1. $9 + 2i$ is an example of a **complex** number. The **real** part is 9 and the **imaginary** part is 2.

CONCEPTS

Fill in the blanks.

3. a. $i = \sqrt{-1}$ b. $i \cdot i = i^2 = -1$

5. a. $5i + 3i = 8i$ b. $5i - 3i = 2i$

7. To write the quotient $\frac{2-3i}{6-i}$ as a complex number in standard form, we multiply it by $\frac{6+i}{6+i}$.

9. Determine whether each statement is true or false.
   a. Every real number is a complex number. **True**
   b. $2 + 7i$ is an imaginary number. **True**
   c. $\sqrt{-16}$ is a real number. **False**
   d. In the complex number system, all quadratic equations have solutions. **True**

NOTATION

11. Write each expression so it is clear that i is not within the radical symbol.
   a. $\sqrt{7}i$; $i\sqrt{7}$ b. $2\sqrt{3}i$; $2i\sqrt{3}$

GUIDED PRACTICE

Write each expression in terms of *i*. See Example 1.

13. $\sqrt{-9} = \sqrt{-1 \cdot 9}$
$= \sqrt{-1}\sqrt{9}$
$= i \cdot 3$
$= 3i$

15. $\sqrt{-7} = \sqrt{-1 \cdot 7}$
$= \sqrt{-1}\sqrt{7}$
$= i\sqrt{7}$
or $\sqrt{7}i$

17. $\sqrt{-24} = \sqrt{-1 \cdot 24}$
$= \sqrt{-1}\sqrt{24}$
$= i\sqrt{4 \cdot 6}$
$= i\sqrt{4}\sqrt{6}$
$= i \cdot 2\sqrt{6}$
$= 2i\sqrt{6}$
or $2\sqrt{6}i$

19. $-\sqrt{-32} = -\sqrt{-1 \cdot 32}$
$= -\sqrt{-1}\sqrt{32}$
$= -i\sqrt{16 \cdot 2}$
$= -i\sqrt{16}\sqrt{2}$
$= -i \cdot 4\sqrt{2}$
$= -4i\sqrt{2}$
or $-4\sqrt{2}i$

21. $5\sqrt{-81} = 5\sqrt{-1 \cdot 81}$
$= 5\sqrt{-1}\sqrt{81}$
$= 5i \cdot 9$
$= 45i$

23. $\sqrt{-\frac{25}{9}} = \sqrt{-1 \cdot \frac{25}{9}}$
$= \frac{\sqrt{-1 \cdot 25}}{\sqrt{9}}$
$= \frac{\sqrt{-1}\sqrt{25}}{\sqrt{9}}$
$= \frac{5i}{3}$
$= \frac{5}{3}i$

Write each number in the form *a* + *bi*. See Example 2.

25. $12 = 12 + 0i$

27. $\sqrt{-100} = \sqrt{-1 \cdot 100}$
$= \sqrt{-1}\sqrt{100}$
$= 10i$
$= 0 + 10i$

29. $6 + \sqrt{-16} = 6 + \sqrt{-1 \cdot 16}$
$= 6 + \sqrt{-1}\sqrt{16}$
$= 6 + 4i$

31. $-9 - \sqrt{-49} = -9 - \sqrt{-1 \cdot 49}$
$= -9 - \sqrt{-1}\sqrt{49}$
$= -9 - 7i$

Perform the operations. Write all answers in the form *a* + *bi*. See Example 3.

33. $(3 + 4i) + (5 - 6i) = (3 + 5) + (4i - 6i)$
$= 8 - 2i$

35. $(7-3i)-(4+2i)=(7-3i)+(-4-2i)$
$=(7-4)+(-3i-2i)$
$=3-5i$

37. $(14-4i)-9i=14+(-4i-9i)$
$=14-13i$

39. $15+(-3-9i)=(15-3)-9i$
$=12-9i$

Perform the operations. Write all answers in the form *a* + *bi*. See Example 4.

41. $3(2-i)=\mathbf{3}\cdot 2-\mathbf{3}\cdot i$
$=6-3i$

43. $-5i(5-5i)=\mathbf{-5i}\cdot 5-\mathbf{5i}\cdot -5i$
$=-25i+25i^2$
$=-25i+25(\mathbf{-1})$
$=-25i-25$
$=-25-25i$

45. $(3-2i)(2+3i)=\mathbf{3}\cdot 2+\mathbf{3}\cdot 3i-\mathbf{2i}\cdot 2-\mathbf{2i}\cdot 3i$
$=6+9i-4i-6i^2$
$=6+5i-6(\mathbf{-1})$
$=6+5i+6$
$=(6+6)+5i$
$=12+5i$

47. $(4+i)(3-i)=\mathbf{4}\cdot 3+\mathbf{4}\cdot -i+\mathbf{i}\cdot 3-\mathbf{i}\cdot i$
$=12-4i+3i-i^2$
$=12-i-1(\mathbf{-1})$
$=12-i+1$
$=(12+1)-i$
$=13-i$

Write each quotient in the form *a* + *bi*. See Example 5.

49. $\dfrac{5}{2-i}=\dfrac{5}{2-i}\cdot\dfrac{\mathbf{2+i}}{\mathbf{2+i}}$
$=\dfrac{10+5i}{4-i^2}$
$=\dfrac{10+5i}{4-(\mathbf{-1})}$
$=\dfrac{10+5i}{4+1}$
$=\dfrac{10+5i}{5}$
$=\dfrac{10}{5}+\dfrac{5}{5}i$
$=2+i$

51. $\dfrac{-4i}{7-2i}=\dfrac{-4i}{7-2i}\cdot\dfrac{\mathbf{7+2i}}{\mathbf{7+2i}}$
$=\dfrac{-28i-8i^2}{49-4i^2}$
$=\dfrac{-28i-8(\mathbf{-1})}{49-4(\mathbf{-1})}$
$=\dfrac{-28i+8}{49+4}$
$=\dfrac{8-28i}{53}$
$=\dfrac{8}{53}-\dfrac{28}{53}i$

53. $\dfrac{2+3i}{2-3i}=\dfrac{2+3i}{2-3i}\cdot\dfrac{\mathbf{2+3i}}{\mathbf{2+3i}}$
$=\dfrac{4+2\cdot 6i+9i^2}{4-9i^2}$
$=\dfrac{4+12i+9(\mathbf{-1})}{4-9(\mathbf{-1})}$
$=\dfrac{4+12i-9}{4+9}$
$=\dfrac{-5+12i}{13}$
$=-\dfrac{5}{13}+\dfrac{12}{13}i$

55. $\dfrac{4-3i}{7-i}=\dfrac{4-3i}{7-i}\cdot\dfrac{\mathbf{7+i}}{\mathbf{7+i}}$
$=\dfrac{28+4i-21i-3i^2}{49-i^2}$
$=\dfrac{28-17i-3(\mathbf{-1})}{49-(\mathbf{-1})}$
$=\dfrac{28-17i+3}{49+1}$
$=\dfrac{31-17i}{50}$
$=\dfrac{31}{50}-\dfrac{17}{50}i$

Solve each equation. Write all solutions in the form *a* + *bi*. See Example 6.

57.
$$x^2+9=0$$
$$x^2+9-\mathbf{9}=0-\mathbf{9}$$
$$x^2=-9$$
$$x=\pm\sqrt{-9}$$
$$x=\pm\sqrt{-1\cdot 9}$$
$$x=\pm\sqrt{-1}\sqrt{9}$$
$$x=\pm 3i$$
The solutions are $0\pm 3i$.

59.
$$d^2+8=0$$
$$d^2+8-\mathbf{8}=0-\mathbf{8}$$
$$d^2=-8$$
$$d=\pm\sqrt{-8}$$
$$d=\pm\sqrt{-1\cdot 8}$$
$$d=\pm\sqrt{-1}\sqrt{4\cdot 2}$$
$$d=\pm 2i\sqrt{2}$$
The solutions are $0\pm 2i\sqrt{2}$.

61.
$$(x+3)^2=-1$$
$$x+3=\pm\sqrt{-1}$$
$$x+3=\pm\sqrt{-1}$$
$$x+3=\pm i$$
$$x+3-\mathbf{3}=\mathbf{-3}\pm i$$
$$x=-3\pm i$$
The solutions are $-3\pm i$.

63.
$$(x-11)^2=-75$$
$$x-11=\pm\sqrt{-75}$$
$$x-11=\pm\sqrt{-1}\sqrt{75}$$
$$x-11=\pm\sqrt{-1}\sqrt{25\cdot 3}$$
$$x-11=\pm 5i\sqrt{3}$$
$$x-11+\mathbf{11}=\mathbf{11}\pm 5i\sqrt{3}$$
$$x=11\pm 5i\sqrt{3}$$
The solutions are $11\pm 5i\sqrt{3}$.

Solve each equation. Write all solutions in the form *a* + *bi*. See Example 7.

65.
$$1x^2-3x+4=0$$
$$a=1,\ b=-3,\ c=4$$
$$x=\frac{-b\pm\sqrt{b^2-4ac}}{2a}$$
$$x=\frac{-(-3)\pm\sqrt{(-3)^2-4(1)(4)}}{2(1)}$$
$$x=\frac{3\pm\sqrt{9-16}}{2}$$
$$x=\frac{3\pm\sqrt{-7}}{2}$$
$$x=\frac{3\pm\sqrt{-1\cdot 7}}{2}$$
$$x=\frac{3\pm\sqrt{-1}\sqrt{7}}{2}$$
$$x=\frac{3\pm i\sqrt{7}}{2}$$
$$x=\frac{3}{2}\pm\frac{\sqrt{7}}{2}i$$
The solutions are $\frac{3}{2}\pm\frac{\sqrt{7}}{2}i$.

67.
$$2x^2+x+1=0$$
$$a=2,\ b=1,\ c=1$$
$$x=\frac{-b\pm\sqrt{b^2-4ac}}{2a}$$
$$x=\frac{-1\pm\sqrt{1^2-4(2)(1)}}{2(2)}$$
$$x=\frac{-1\pm\sqrt{1-8}}{4}$$
$$x=\frac{-1\pm\sqrt{-7}}{4}$$
$$x=\frac{-1\pm\sqrt{-1\cdot 7}}{4}$$
$$x=\frac{-1\pm\sqrt{-1}\sqrt{7}}{4}$$
$$x=\frac{-1\pm i\sqrt{7}}{4}$$
$$x=-\frac{1}{4}\pm\frac{\sqrt{7}}{4}i$$
The solutions are $-\frac{1}{4}\pm\frac{\sqrt{7}}{4}i$.

TRY IT YOURSELF

Perform the operations. Write all answers in the form *a* + *bi*.

69. $\dfrac{3}{5+i}=\dfrac{3}{5+i}\cdot\dfrac{\mathbf{5-i}}{\mathbf{5-i}}$

$$=\frac{15-3i}{25-i^2}$$
$$=\frac{15-3i}{25-(\mathbf{-1})}$$
$$=\frac{15-3i}{25+1}$$
$$=\frac{15-3i}{26}$$
$$=\frac{15}{26}-\frac{3}{26}i$$

71. $(6-i)+(9+3i)=(6+9)+(-i+3i)$

$$=15+2i$$

73. $(-3-8i)-(-3-9i)=(-3-8i)+(3+9i)$

$$=(-3+3)+(-8i+9i)$$
$$=0+i$$

75. $(2+i)(2+3i)=\mathbf{2}\cdot 2+\mathbf{2}\cdot 3i+\boldsymbol{i}\cdot 2+\boldsymbol{i}\cdot 3i$

$$=4+6i+2i+3i^2$$
$$=4+(6+2)i+3(\mathbf{-1})$$
$$=4+8i-3$$
$$=(4-3)+8i$$
$$=1+8i$$

77. $\dfrac{3-2i}{3+2i}=\dfrac{3-2i}{3+2i}\cdot\dfrac{\mathbf{3-2i}}{\mathbf{3-2i}}$

$$=\frac{9-2\cdot 6i+4i^2}{9-4i^2}$$
$$=\frac{9-12i+4(\mathbf{-1})}{9-4(\mathbf{-1})}$$
$$=\frac{9-12i-4}{9+4}$$
$$=\frac{5-12i}{13}$$
$$=\frac{5}{13}-\frac{12}{13}i$$

79. $(10-9i)+(-1+i)=(10-1)+(-9i+i)$

$$=9-8i$$

81. $-4(3+4i)=\mathbf{-4}\cdot 3-\mathbf{4}\cdot 4i$

$$=-12-16i$$

83. $(2+i)^2=2^2+\mathbf{2}\cdot 2i+i^2$

$$=4+4i+i^2$$
$$=4+4i+1(\mathbf{-1})$$
$$=4+4i-1$$
$$=(4-1)+4i$$
$$=3+4i$$

Solve each equation. Write all solutions in the form *a* + *bi*.

85. $2x^2+x=-5$

$$2x^2+x+\mathbf{5}=-5+\mathbf{5}$$
$$2x^2+x+5=0$$
$$a=2,\ b=1,\ c=5$$
$$x=\frac{-b\pm\sqrt{b^2-4ac}}{2a}$$
$$x=\frac{-1\pm\sqrt{1^2-4(2)(5)}}{2(2)}$$
$$x=\frac{-1\pm\sqrt{1-40}}{4}$$
$$x=\frac{-1\pm\sqrt{-39}}{4}$$
$$x=\frac{-1\pm\sqrt{-1\cdot 39}}{4}$$
$$x=\frac{-1\pm\sqrt{-1}\sqrt{39}}{4}$$
$$x=\frac{-1\pm i\sqrt{39}}{4}$$
$$x=-\frac{1}{4}\pm\frac{\sqrt{39}}{4}i$$

The solutions are $-\dfrac{1}{4}\pm\dfrac{\sqrt{39}}{4}i$.

87. $(x-4)^2 = -45$

$$x-4 = \pm\sqrt{-45}$$
$$x-4 = \pm\sqrt{-1}\sqrt{45}$$
$$x-4 = \pm\sqrt{-1}\sqrt{9\cdot 5}$$
$$x-4 = \pm 3i\sqrt{5}$$
$$x-4+\mathbf{4} = \mathbf{4} \pm 3i\sqrt{5}$$
$$x = 4 \pm 3i\sqrt{5}$$

The solutions are $4 \pm 3i\sqrt{5}$.

89. $b^2 + 2b + 2 = 0$

$$1b^2 + 2b + 2 = 0$$
$$a = 1,\ b = 2,\ c = 2$$
$$b = \frac{-b \pm \sqrt{b^2 - 4ac}}{2a}$$
$$b = \frac{-2 \pm \sqrt{2^2 - 4(1)(2)}}{2(1)}$$
$$b = \frac{-2 \pm \sqrt{4-8}}{2}$$
$$b = \frac{-2 \pm \sqrt{-4}}{2}$$
$$b = \frac{-2 \pm \sqrt{-1\cdot 4}}{2}$$
$$b = \frac{-2 \pm \sqrt{-1}\sqrt{4}}{2}$$
$$b = \frac{-2 \pm i\sqrt{4}}{2}$$
$$b = \frac{-2 \pm 2i}{2}$$
$$b = -\frac{2}{2} \pm \frac{2}{2}i$$
$$b = -1 \pm i$$

The solutions are $-1 \pm i$.

91. $x^2 = -36$

$$x = \pm\sqrt{-36}$$
$$x = \pm\sqrt{-1\cdot 36}$$
$$x = \pm\sqrt{-1}\sqrt{36}$$
$$x = \pm 6i$$

The solutions are $0 \pm 6i$.

93. $x^2 = -\frac{16}{9}$

$$x = \pm\sqrt{-\frac{16}{9}}$$
$$x = \pm\frac{\sqrt{-16}}{\sqrt{9}}$$
$$x = \pm\frac{\sqrt{-1}\sqrt{16}}{\sqrt{9}}$$
$$x = \pm\frac{4i}{3}$$
$$x = \pm\frac{4}{3}i$$

The solutions are $0 \pm \frac{4}{3}i$.

95. $3x^2 + 2x + 1 = 0$

$$a = 3,\ b = 2,\ c = 1$$
$$x = \frac{-b \pm \sqrt{b^2 - 4ac}}{2a}$$
$$x = \frac{-2 \pm \sqrt{2^2 - 4(3)(1)}}{2(3)}$$
$$x = \frac{-2 \pm \sqrt{4-12}}{6}$$
$$x = \frac{-2 \pm \sqrt{-8}}{6}$$
$$x = \frac{-2 \pm \sqrt{-1\cdot 8}}{6}$$
$$x = \frac{-2 \pm \sqrt{-1}\sqrt{4\cdot 2}}{6}$$
$$x = \frac{-2 \pm 2i\sqrt{2}}{6}$$
$$x = -\frac{\overset{1}{\cancel{2}}}{\underset{1}{\cancel{2}}\cdot 3} \pm \frac{\overset{1}{\cancel{2}}\sqrt{2}}{\underset{1}{\cancel{2}}\cdot 3}i$$
$$x = -\frac{1}{3} \pm \frac{\sqrt{2}}{3}i$$

The solutions are $-\frac{1}{3} \pm \frac{\sqrt{2}}{3}i$.

APPLICATIONS

97. ELECTRICITY

$V = V_1 + V_2$
$V = (10.31 - 5.97i) + (8.14 + 3.79i)$
$V = (10.31 + 8.14) + (-5.97i + 3.79i)$
$V = 18.45 - 2.18i$
The total voltage is $18.45 - 2.18i$.

WRITING

99-101. Answers will vary.

REVIEW

Rationalize the denominator.

103. $$\frac{1}{\sqrt{7}} = \frac{1}{\sqrt{7}} \cdot \frac{\sqrt{7}}{\sqrt{7}} = \frac{1\sqrt{7}}{\sqrt{7}\sqrt{7}} = \frac{\sqrt{7}}{7}$$

105. $$\frac{8}{\sqrt{x}-2} = \frac{8}{\sqrt{x}-2} \cdot \frac{\sqrt{x}+2}{\sqrt{x}+2} = \frac{8\sqrt{x}+16}{x-4}$$

CHALLENGE PROBLEMS

Perform the operations.

107. $$\left(2\sqrt{2}+i\sqrt{2}\right)\left(3\sqrt{2}-i\sqrt{2}\right)$$
$$= \mathbf{2\sqrt{2}} \cdot 3\sqrt{2} + \mathbf{2\sqrt{2}} \cdot -i\sqrt{2} + \boldsymbol{i\sqrt{2}} \cdot 3\sqrt{2} + \boldsymbol{i\sqrt{2}} \cdot -i\sqrt{2}$$
$$= 6\sqrt{4} - 2i\sqrt{4} + 3i\sqrt{4} - i^2\sqrt{4}$$
$$= 6 \cdot 2 - 2i \cdot 2 + 3i \cdot 2 - (-1) \cdot 2$$
$$= 12 - 4i + 6i + 2$$
$$= (12+2) + (-4+6)i$$
$$= 14 + 2i$$

SECTION 9.5 STUDY SET

VOCABULARY

Fill in the blank.

1. $y = 3x^2 + 5x - 1$ is a **quadratic** equation in two variables. Its graph is a cup-shaped figure called a **parabola**.

3. Points where a parabola intersects the x-axis are called the x-**intercepts** of the graph and the point where a parabola intersects the y-axis is called the y-**intercept** of the graph.

CONCEPTS

Fill in the blanks.

5. The graph of $y = ax^2 + bx + c$ opens downward when a **<** 0 and upward when a > **0**.

7. a. To find the y-intercepts of a graph, substitute **0** for x in the given equation and solve for y.
   b. To find the x-intercepts of a graph, substitute 0 for y in the given equation and solve for x.

9. a. What do we call the curve shown in the graph? **Parabola**
   b. What are the x-intercepts of the graph? **(1, 0), (3, 0)**
   c. What is the y-intercept of the graph? **(0, –3)**
   d. What is the vertex? **(2, 1)**
   e. Draw the axis of symmetry on the graph. **It is a vertical line through (2, 1).**

11. The vertex of a parabola is (1, –3), its y-intercept is (0, –2), and it passes through the point (3, 1). Draw the axis of symmetry and use it to help determine two other points on the parabola. **(2, –2), (–1, 1)**

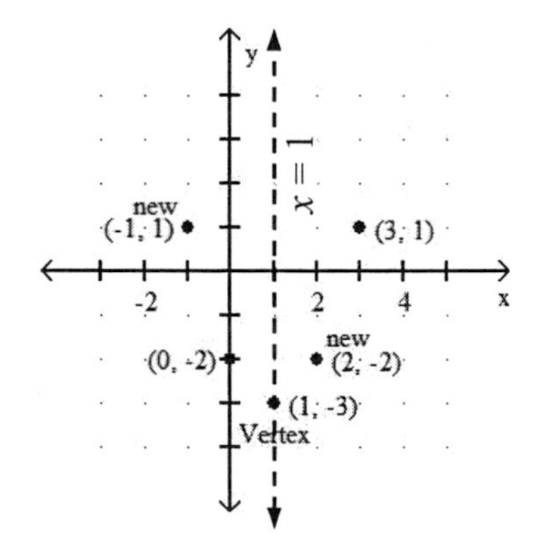

13. Examine the graph of $y = x^2 + 2x - 3$. How many real-number solutions does the equation $x^2 + 2x - 3 = 0$ have? Find them. **Two; –3, 1**

NOTATION

15. Consider the equation $y = 2x^2 + 4x - 8$.
   a. What are a, b, and c? **2, 4, –8**
   b. Find $\dfrac{-b}{2a} = \dfrac{-4}{2(2)} = \mathbf{-1}$

GUIDED PRACTICE

Find the x- and y-intercepts of the graph of the quadratic equation. See Example 1.

17. $y = x^2 - 6x + 8$

To find the x-intercepts, we let $y = 0$ and solve the resulting quadratic equation.

$$y = x^2 - 6x + 8$$
$$0 = x^2 - 6x + 8$$
$$0 = (x-2)(x-4)$$
$$x - 2 = 0 \quad \text{or} \quad x - 4 = 0$$
$$x = 2 \qquad\qquad x = 4$$

The x-intercepts are $(2,0)$ and $(4,0)$.

To find the y-intercept, we let $x = 0$ and find y.

$$y = x^2 - 6x + 8$$
$$y = 0^2 - 6(0) + 8$$
$$y = 8$$

The y-intercept is $(0,8)$.

Note that the y-coordinate of the y-intercept is the same as the value of the constant term c on the right side of the equation.

19. $y = -x^2 - 10x - 21$

To find the x-intercepts, we let $y = 0$ and solve the resulting quadratic equation.

$$y = -x^2 - 10x - 21$$
$$\mathbf{-1}(0) = \mathbf{-1}(-x^2 - 10x - 21)$$
$$0 = x^2 + 10x + 21$$
$$0 = (x+3)(x+7)$$
$$x + 3 = 0 \quad \text{or} \quad x + 7 = 0$$
$$x = -3 \qquad\qquad x = -7$$

The x-intercepts are $(-3,0)$ and $(-7,0)$.

To find the y-intercept, we let $x = 0$ and find y.

$$y = -x^2 - 10x - 21$$
$$y = -0^2 - 10(0) - 21$$
$$y = -21$$

The y-intercept is $(0,-21)$.

Find the vertex of the graph of each quadratic equation. See Example 2.

21. $y = 2x^2 - 4x + 1$

$a = 2,\ b = -4$

To find the x-coordinate of the vertex, we substitute the values for a and b into the formula $x = \frac{-b}{2a}$.

$$x = \frac{-b}{2a} = \frac{-(-4)}{2(2)} = \frac{4}{4} = 1$$

The x-coordinate of the vertex is 1.

To find the y-coordinate, we substitute 1 for x in the original equation.

$$y = 2x^2 - 4x + 1$$
$$y = 2(\mathbf{1})^2 - 4(\mathbf{1}) + 1$$
$$y = 2 - 4 + 1$$
$$y = -1$$

The y-coordinate of the vertex is -1.

The vertex of the parabola is $(1, -1)$.

23. $y = -1x^2 + 6x - 8$

$a = -1,\ b = 6$

To find the x-coordinate of the vertex, we substitute the values for a and b into the formula $x = \frac{-b}{2a}$.

$$x = \frac{-b}{2a} = \frac{-(6)}{2(-1)} = \frac{-6}{-2} = 3$$

The x-coordinate of the vertex is 3.
To find the y-coordinate, we substitute 3 for x in the original equation.

$$y = -x^2 + 6x - 8$$
$$y = -(\mathbf{3})^2 + 6(\mathbf{3}) - 8$$
$$y = -9 + 18 - 8$$
$$y = 1$$

The y-coordinate of the vertex is 1.
The vertex of the parabola is $(3,1)$.

Graph each quadratic equation by finding the vertex, the x- and y-intercepts, and the axis of symmetry of its graph. See Examples 3 and 4.

25. $y = 1x^2 + 2x - 3$

$a = 1,\ b = 2$

Find the vertex.

$$x = \frac{-b}{2a} = \frac{-(2)}{2(1)} = \frac{-2}{2} = -1$$

$x = -1$

$$y = x^2 + 2x - 3$$
$$= (-\mathbf{1})^2 + 2(-\mathbf{1}) - 3$$
$$= 1 - 2 - 3$$
$$= -4$$

The vertex of the parabola is $(-1, -4)$.

Find the x-intercepts. Let $y = 0$. Solve the quadratic.

$$y = x^2 + 2x - 3$$
$$0 = x^2 + 2x - 3$$
$$0 = (x+3)(x-1)$$

$x + 3 = 0$ or $x - 1 = 0$

$x = -3$ | $x = 1$

The x-intercepts are $(-3, 0)$ and $(1, 0)$.

Find the y-intercept. It is the constant.

$y = x^2 + 2x - 3$

The y-intercept is $(0, -3)$.

The axis of symmetry is the vertical line passing through the vertex: $x = -1$.

Plot two additional points, one by substitution and the second one by the property of symmetry.

x	y	(x, y)
2	$y = x^2 + 2x - 3$ $= 2^2 + 2(2) - 3$ $= 5$	$(2,5)$

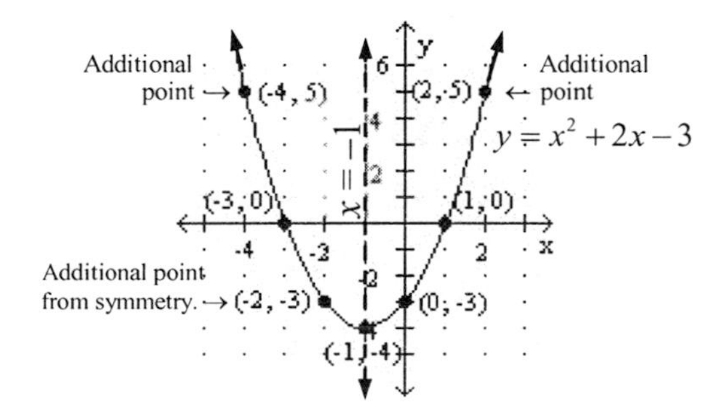

27. $y = 2x^2 + 8x + 6$

$a = 2,\ b = 8$

Find the vertex.

$$x = \frac{-b}{2a} = \frac{-(8)}{2(2)} = \frac{-8}{4} = -2$$

$x = -2$

$$y = 2x^2 + 8x + 6 = 2(-2)^2 + 8(-2) + 6 = 8 - 16 + 6 = -2$$

The vertex of the parabola is $(-2, -2)$.

Find the x-intercepts. Let $y = 0$. Solve the quadratic.

$$y = 2x^2 + 8x + 6$$
$$0 = 2(x^2 + 4x + 3)$$
$$0 = 2(x+3)(x+1)$$

$x + 3 = 0$ or $x + 1 = 0$

$x = -3$ | $x = -1$

The x-intercepts are $(-3, 0)$ and $(-1, 0)$.

Find the y-intercept. It is the constant.

$y = 2x^2 + 8x + 6$

The y-intercept is $(0, 6)$.

The axis of symmetry is the vertical line passing through the vertex. $x = -2$.

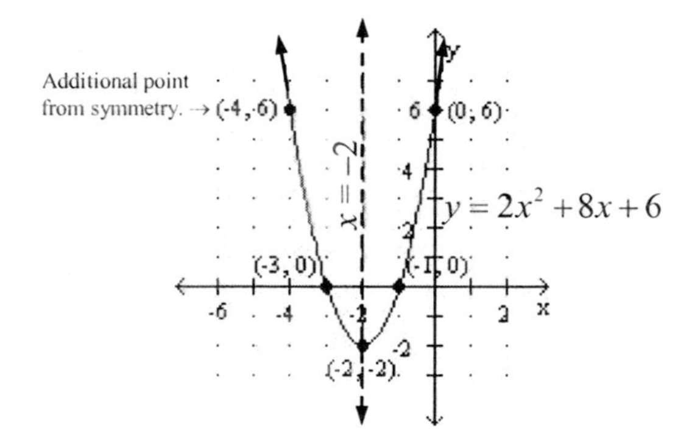

29. $y = -1x^2 + 2x + 3$

$a = -1,\ b = 2$

Find the vertex.

$$x = \frac{-b}{2a} = \frac{-(2)}{2(-1)} = \frac{-2}{-2} = 1$$

$x = 1$

$$y = -x^2 + 2x + 3 = -(1)^2 + 2(1) + 3 = -1 + 2 + 3 = 4$$

The vertex of the parabola is $(1, 4)$.

Find the x-intercepts. Let $y = 0$. Solve the quadratic.

$$y = -x^2 + 2x + 3$$
$$0 = -(x^2 - 2x - 3)$$
$$0 = -(x-3)(x+1)$$

$x - 3 = 0$ or $x + 1 = 0$

$x = 3$ | $x = -1$

The x-intercepts are $(3, 0)$ and $(-1, 0)$.

Find the y-intercept. It is the constant.

$y = -x^2 + 2x + 3$

The y-intercept is $(0, 3)$.

The axis of symmetry is the vertical line passing through the vertex. $x = 1$.

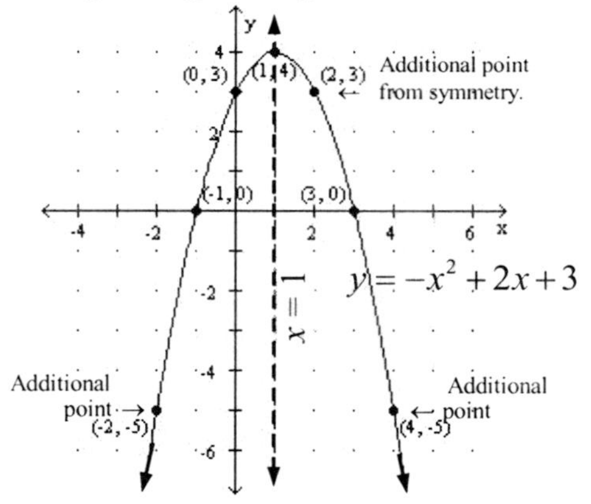

31. $y = -1x^2 + 5x - 4$

$a = -1,\ b = 5$

Find the vertex.

$$x = \frac{-b}{2a} = \frac{-(5)}{2(-1)} = \frac{-5}{-2} = \frac{\mathbf{5}}{2}$$

$$x = \frac{5}{2}$$

$$y = -x^2 + 5x - 4$$
$$= -\left(\frac{\mathbf{5}}{\mathbf{2}}\right)^2 + 5\left(\frac{\mathbf{5}}{\mathbf{2}}\right) - 4$$
$$= -\frac{25}{4} + \frac{25}{2} - \frac{4}{1}$$
$$= -\frac{25}{4} + \frac{25}{2}\cdot\frac{\mathbf{2}}{\mathbf{2}} - \frac{4}{1}\cdot\frac{\mathbf{4}}{\mathbf{4}}$$
$$= \frac{-25 + 50 - 16}{4}$$
$$= \frac{9}{4}$$

The vertex of the parabola is $\left(\frac{5}{2}, \frac{9}{4}\right)$.

Find the x-intercepts. Let $y = 0$. Solve the quadratic.

$y = -x^2 + 5x - 4$

$0 = -(x^2 - 5x + 4)$

$0 = -(x-1)(x-4)$

$x - 1 = 0$ or $x - 4 = 0$

$x = 1$ | $x = 4$

The x-intercepts are $(1, 0)$ and $(4, 0)$.

Find the y-intercept. It is the constant.

$y = -x^2 + 5x - 4$

The y-intercept is $(0, -4)$.

The axis of symmetry is the vertical line passing through the vertex. $x = \frac{5}{2}$.

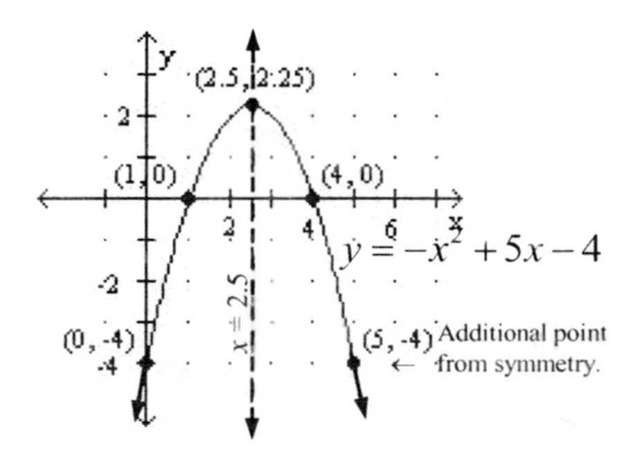

33. $y = 1x^2 - 2x + 0$

$a = 1,\ b = -2$

Find the vertex.

$$x = \frac{-b}{2a} = \frac{-(-2)}{2(1)} = \frac{2}{2} = 1$$

$x = 1$

$$y = x^2 - 2x$$
$$= (\mathbf{1})^2 - 2(\mathbf{1})$$
$$= 1 - 2$$
$$= -1$$

The vertex of the parabola is $(1, -1)$.

Find the x-intercepts. Let $y = 0$. Solve the quadratic.

$y = x^2 - 2x$

$0 = x(x - 2)$

$x = 0$ or $x - 2 = 0$

$x = 2$

The x-intercepts are $(0, 0)$ and $(2, 0)$.

Find the y-intercept. It is the constant.

$y = x^2 - 2x + 0$

The y-intercept is $(0, 0)$.

The axis of symmetry is the vertical line passing through the vertex. $x = 1$.

Plot two additional points, one by substitution and the second one by the property of symmetry.

x	y	(x, y)
-1	$y = x^2 - 2x$ $= (-1)^2 - 2(-1)$ $= 3$	$(-1, 3)$

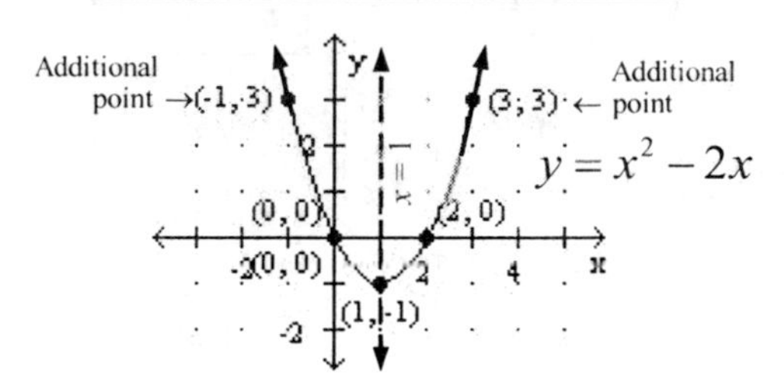

35. $y = 1x^2 + 4x + 4$

$a = 1,\ b = 4$

Find the vertex.

$$x = \frac{-b}{2a} = \frac{-(4)}{2(1)} = -2$$

$$\begin{aligned} x &= -2 \\ y &= x^2 + 4x + 4 \\ &= (-2)^2 + 4(-2) + 4 \\ &= 4 - 8 + 4 \\ &= 0 \end{aligned}$$

The vertex of the parabola is $(-2, 0)$.

Find the x-intercepts. Let $y = 0$. Solve the quadratic.

$$\begin{aligned} y &= x^2 + 4x + 4 \\ 0 &= (x+2)(x+2) \\ x + 2 &= 0 \\ x &= -2 \end{aligned}$$

The x-intercept is $(-2, 0)$.

Find the y-intercept. It is the constant.

$y = x^2 + 4x + 4$

The y-intercept is $(0, 4)$.

The axis of symmetry is the vertical line passing through the vertex. $x = -2$.

Plot two additional points, one by substitution and the second one by the property of symmetry.

x	y	(x, y)
-3	$y = x^2 + 4x + 4$ $= (-3)^2 + 4(-3) + 4$ $= 1$	$(-3, 1)$

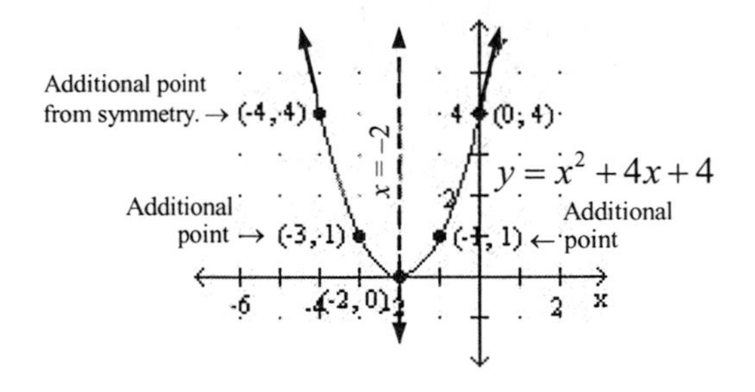

37. $y = -1x^2 - 4x + 0$

$a = -1,\ b = -4$

Find the vertex.

$$x = \frac{-b}{2a} = \frac{-(-4)}{2(-1)} = \frac{4}{-2} = -2$$

$$\begin{aligned} x &= -2 \\ y &= -x^2 - 4x \\ &= -(-2)^2 - 4(-2) \\ &= -4 + 8 \\ &= 4 \end{aligned}$$

The vertex of the parabola is $(-2, 4)$.

Find the x-intercepts. Let $y = 0$. Solve the quadratic.

$$\begin{aligned} y &= -x^2 - 4x \\ 0 &= -x(x+4) \end{aligned}$$

$x = 0$ or $x + 4 = 0$

$x = -4$

The x-intercepts are $(0, 0)$ and $(-4, 0)$.

The axis of symmetry is the vertical line passing through the vertex. $x = -2$.

Find the y-intercept. It is the constant.

$y = -x^2 - 4x + 0$

The y-intercept is $(0, 0)$.

Plot two additional points, one by substitution and the second one by the property of symmetry.

x	y	(x, y)
-1	$y = -x^2 - 4x$ $= (-1)^2 - 4(-1)$ $= 1 + 4$ $= 5$	$(-1, 5)$

(-2, 4)
(-4, 0)
(0, 0)
$x = -2$
$y = -x^2 - 4x$
(-5, -5) ← Additional point
(1, -5) ← Additional point
y
x

39. $y = 2x^2 + 3x - 2$

$a = 2,\ b = 3$

Find the vertex.

$$x = \frac{-b}{2a} = \frac{-(3)}{2(2)} = -\frac{3}{4}$$

$$x = -\frac{3}{4}$$
$$y = 2x^2 + 3x - 2$$
$$= 2\left(-\frac{3}{4}\right)^2 + 3\left(-\frac{3}{4}\right) - 2$$
$$= \frac{9}{8} - \frac{9}{4}\cdot\frac{\mathbf{2}}{\mathbf{2}} - \frac{2}{1}\cdot\frac{\mathbf{8}}{\mathbf{8}}$$
$$= \frac{9 - 18 - 16}{8}$$
$$= -\frac{25}{8}$$

The vertex of the parabola is $\left(-\frac{3}{4}, -\frac{25}{8}\right)$.

Find the x-intercepts. Let $y = 0$. Solve the quadratic.

$$y = 2x^2 + 3x - 2$$
$$0 = 2x^2 + 3x - 2$$
$$0 = (2x - 1)(x + 2)$$

$2x - 1 = 0$ or $x + 2 = 0$

$2x = 1$ | $x = -2$

$x = \frac{1}{2}$

The x-intercepts are $\left(\frac{1}{2}, 0\right)$ and $(-2, 0)$.

Find the y-intercept. It is the constant.

$y = 2x^2 + 3x - 2$

The y-intercept is $(0, -2)$.

The axis of symmetry is the vertical line passing through the vertex. $x = -\frac{3}{4}$.

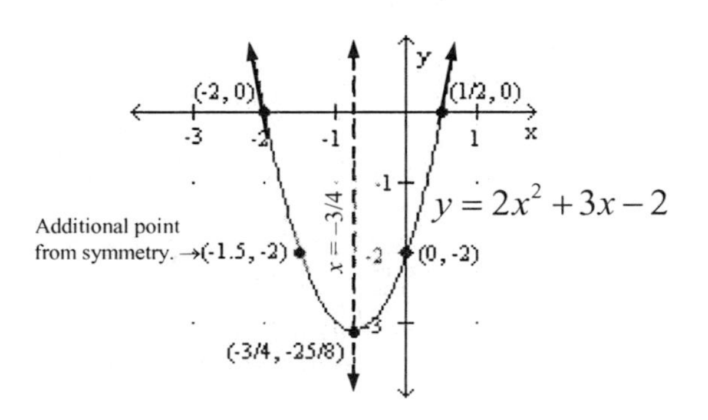

41. $y = 4x^2 - 12x + 9$

$a = 4,\ b = -12$

Find the vertex.

$$x = \frac{-b}{2a} = \frac{-(-12)}{2(4)} = \frac{3}{2}$$

$$x = \frac{3}{2}$$
$$y = 4x^2 - 12x + 9$$
$$= 4\left(\frac{3}{2}\right)^2 - 12\left(\frac{3}{2}\right) + 9$$
$$= 9 - 18 + 9$$
$$= 0$$

The vertex of the parabola is $\left(\frac{3}{2}, 0\right)$.

Find the x-intercepts. Let $y = 0$. Solve the quadratic.

$$y = 4x^2 - 12x + 9$$
$$0 = 4x^2 - 12x + 9$$
$$0 = (2x - 3)(2x - 3)$$
$$2x - 3 = 0$$
$$2x = 3$$
$$x = \frac{3}{2}$$

The x-intercept is $\left(\frac{3}{2}, 0\right)$.

Find the y-intercept. It is the constant.

$y = 4x^2 - 12x + 9$

The y-intercept is $(0, 9)$.

The axis of symmetry is the vertical line passing through the vertex. $x = \frac{3}{2}$.

Plot two additional points, one by substitution and the second one by the property of symmetry.

x	y	(x, y)
1	$y = 4x^2 - 12x + 9$ $= 4(1)^2 - 12(1) + 9$ $= 4 - 12 + 9$ $= 1$	$(1, 1)$

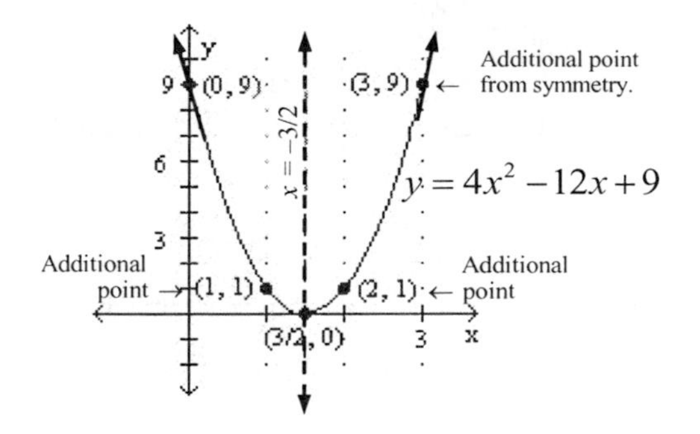

Graph each quadratic equation by finding the vertex, the x- and y-intercepts, and the axis of symmetry of its graph. See Example 5.

43. $y = 1x^2 - 4x - 1$

$a = 1,\ b = -4$

Find the vertex.

$$x = \frac{-b}{2a} = \frac{-(-4)}{2(1)} = \frac{4}{2} = 2$$

$x = 2$

$$y = x^2 - 4x - 1 = (2)^2 - 4(2) - 1 = 4 - 8 - 1 = -5$$

The vertex of the parabola is $(2, -5)$.

Find the x-intercepts. Let $y = 0$. Solve the quadratic.

$y = x^2 - 4x - 1$

$0 = 1x^2 - 4x - 1$

$a = 1,\ b = -4,\ c = -1$

$$x = \frac{-b \pm \sqrt{b^2 - 4ac}}{2a}$$

$$x = \frac{-(-4) \pm \sqrt{(-4)^2 - 4(1)(-1)}}{2(1)}$$

$$x = \frac{4 \pm \sqrt{16 + 4}}{2}$$

$$x = \frac{4 \pm \sqrt{20}}{2}$$

$$x = \frac{4 \pm \sqrt{4}\sqrt{5}}{2}$$

$$x = \frac{4 \pm 2\sqrt{5}}{2}$$

$$x = \frac{\overset{1}{\cancel{2}}\left(2 \pm \sqrt{5}\right)}{\underset{1}{\cancel{2}}}$$

$$x = 2 \pm \sqrt{5}$$

The x-intercepts are $(2 + \sqrt{5}, 0)$ and $(2 - \sqrt{5}, 0)$.

Find the y-intercept. It is the constant.

$y = x^2 - 4x - 1$

The y-intercept is $(0, -1)$.

The axis of symmetry is the vertical line passing through the vertex. $x = 2$.

Plot two additional points, one by substitution and the second point by the property of symmetry.

x	y	(x, y)
-1	$y = x^2 - 4x - 1$ $= (-1)^2 - 4(-1) - 1$ $= 1 + 4 - 1$ $= 4$	$(-1, 4)$

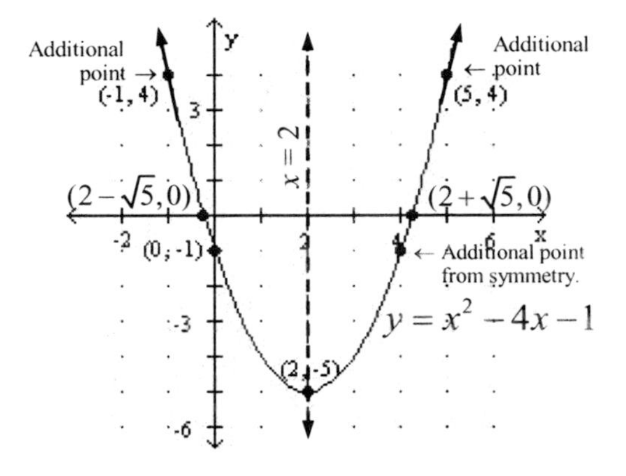

45. $y = -1x^2 - 2x + 2$

$a = -1,\ b = -2$

Find the vertex.

$$x = \frac{-b}{2a} = \frac{-(-2)}{2(-1)} = \frac{2}{-2} = -1$$

$x = -1$

$$y = -x^2 - 2x + 2 = -(-1)^2 - 2(-1) + 2 = -1 + 2 + 2 = 3$$

The vertex of the parabola is $(-1, 3)$.

Find the x-intercepts. Let $y = 0$. Solve the quadratic.

$y = -x^2 - 2x + 2$

$0 = -1x^2 - 2x + 2$

$a = -1,\ b = -2,\ c = 2$

$$x = \frac{-b \pm \sqrt{b^2 - 4ac}}{2a}$$

$$x = \frac{-(-2) \pm \sqrt{(-2)^2 - 4(-1)(2)}}{2(-1)}$$

$$x = \frac{2 \pm \sqrt{4 + 8}}{-2}$$

$$x = \frac{2 \pm \sqrt{12}}{-2}$$

$$x = \frac{2 \pm \sqrt{4}\sqrt{3}}{-2}$$

$$x = \frac{2 \perp 2\sqrt{3}}{-2}$$

$$x = \frac{\overset{1}{\cancel{2}}\left(1 \pm \sqrt{3}\right)}{-1 \cdot \underset{1}{\cancel{2}}}$$

$$x = -1 \pm \sqrt{3}$$

The x-intercepts are $(-1+\sqrt{3}, 0)$ and $(-1-\sqrt{3}, 0)$.

Find the y-intercept. It is the constant.

$y = -x^2 - 2x + 2$

The y-intercept is $(0, 2)$.

The axis of symmetry is the vertical line passing through the vertex. $x = -1$.

Plot two additional points, one by substitution and the second point by the property of symmetry.

x	y	(x, y)
2	$y = -x^2 - 2x + 2$ $= -(2)^2 - 2(2) + 2$ $= -4 - 4 + 2$ $= -6$	$(2, -6)$

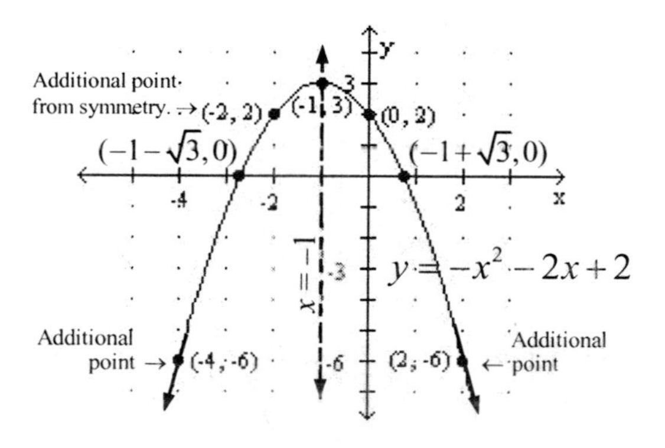

47. $y = -1x^2 - 6x - 4$

$a = -1,\ b = -6$

Find the vertex.

$$x = \frac{-b}{2a} = \frac{-(-6)}{2(-1)} = \frac{6}{-2} = -3$$

$x = -3$

$$y = -x^2 - 6x - 4 = -(-3)^2 - 6(-3) - 4 = -9 + 18 - 4 = 5$$

The vertex of the parabola is $(-3, 5)$.

Find the x-intercepts. Let $y = 0$. Solve the quadratic.

$y = -x^2 - 6x - 4$

$0 = -1x^2 - 6x - 4$

$a = -1,\ b = -6,\ c = -4$

$$x = \frac{-b \pm \sqrt{b^2 - 4ac}}{2a}$$

$$x = \frac{-(-6) \pm \sqrt{(-6)^2 - 4(-1)(-4)}}{2(-1)}$$

$$x = \frac{6 \pm \sqrt{36 - 16}}{-2}$$

$$x = \frac{6 \pm \sqrt{20}}{-2}$$

$$x = \frac{6 \pm \sqrt{4}\sqrt{5}}{-2}$$

$$x = \frac{6 \pm 2\sqrt{5}}{-2}$$

$$x = \frac{\overset{1}{\cancel{2}}\left(3 \pm \sqrt{5}\right)}{-1 \cdot \underset{1}{\cancel{2}}}$$

$$x = -3 \pm \sqrt{5}$$

The x-intercepts are $(-3+\sqrt{5}, 0)$ and $(-3-\sqrt{5}, 0)$.

Find the y-intercept. It is the constant.

$y = -x^2 - 6x - 4$

The y-intercept is $(0, -4)$.

The axis of symmetry is the vertical line passing through the vertex. $x = -3$.

Plot two additional points, one by substitution and the second point by the property of symmetry.

x	y	(x, y)
-2	$y=-x^2-6x-4$ $=-(-2)^2-6(-2)-4$ $=-4+12-4$ $=4$	$(-6,-4)$

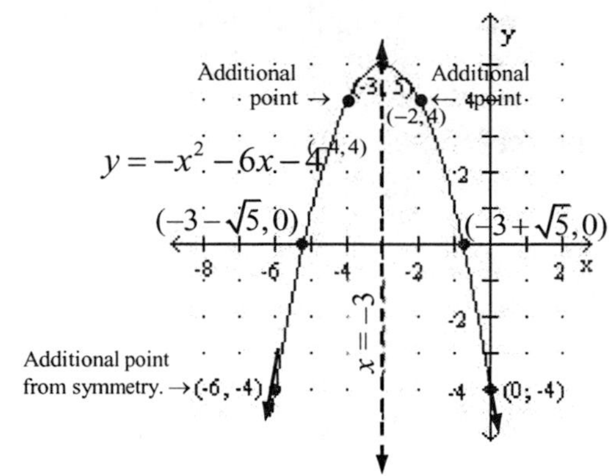

49. $y=x^2-6x+10$
    $a=1,\ b=-6$

Find the vertex.

$$x=\frac{-b}{2a}=\frac{-(-6)}{2(1)}=\frac{6}{2}=3$$

$x=3$

$$y=x^2-6x+10=(3)^2-6(3)+10=9-18+10=1$$

The vertex of the parabola is $(3,1)$.

Find the x-intercepts. Let $y=0$. Solve the quadratic.

$y=x^2-6x+10$
$0=1x^2-6x+10$
$a=1,\ b=-6,\ c=10$

$$x=\frac{-b\pm\sqrt{b^2-4ac}}{2a}$$

$$x=\frac{-(-6)\pm\sqrt{(-6)^2-4(1)(10)}}{2(1)}$$

$$x=\frac{6\pm\sqrt{36-40}}{2}$$

$$x=\frac{6\pm\sqrt{-4}}{-2}$$

$$x=\frac{6\pm\sqrt{-1\cdot 4}}{-2}$$

$$x=\frac{6\pm 2i}{-2}$$

There are no x-intercepts because the equation does not have any real number solutions.

Find the y-intercept. It is the constant.

$y=x^2-6x+10$

The y-intercept is $(0,10)$.

The axis of symmetry is the vertical line passing through the vertex. $x=3$.

Plot two additional points, one by substitution and the second point by the property of symmetry.

x	y	(x, y)
1	$y=x^2-6x+10$ $=(1)^2-6(1)+10$ $=1-6+10$ $=5$	$(1,5)$

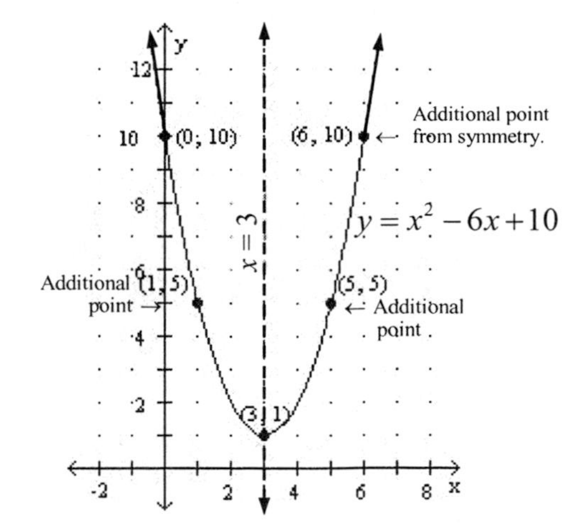

APPLICATION

51. BIOLOGY

It is a vertical line through the body of the butterfly.

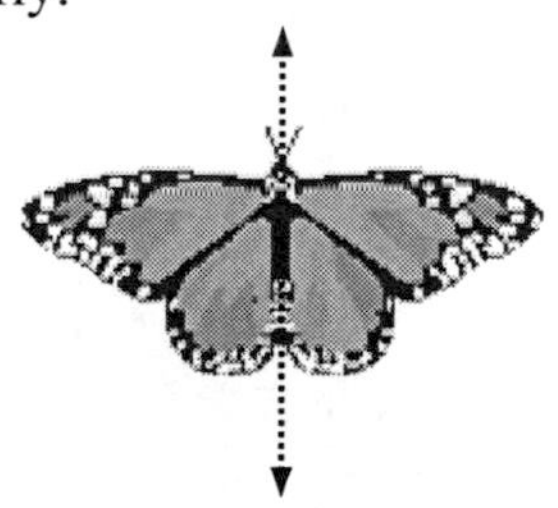

53. COST ANALYSIS

The ordered pair (30, 100) indicates that the cost to manufacture a carburetor is lowest ($100) for a production run of 30 units.

55. TRAMPOLINE

a) The ordered pair (0.5, 14) indicates that she is **14 feet** above the ground $\frac{1}{2}$ second after bounding upward.

b) The ordered pairs (0.25, 9) and (1.75, 9) indicate that she is 9 feet above the ground at **0.25 seconds** and **1.75 seconds**.

c) The vertex (1, 18) indicates the maximum number of feet above the ground. She gets **18 feet** above the ground after **1 second**.

WRITING

57-61. Answers will vary.

REVIEW

Simplify each expression.

63.
$$\begin{aligned}\sqrt{8}-\sqrt{50}+\sqrt{72} &= \sqrt{4\cdot 2}-\sqrt{25\cdot 2}+\sqrt{36\cdot 2}\\ &= \sqrt{4}\sqrt{2}-\sqrt{25}\sqrt{2}+\sqrt{36}\sqrt{2}\\ &= 2\sqrt{2}-5\sqrt{2}+6\sqrt{2}\\ &= (2-5+6)\sqrt{2}\\ &= 3\sqrt{2}\end{aligned}$$

65.
$$\begin{aligned}3\sqrt{z}\left(\sqrt{4z}-\sqrt{z}\right) &= 3\sqrt{z}\left(\sqrt{4}\sqrt{z}-\sqrt{z}\right)\\ &= 3\sqrt{z}\left(2\sqrt{z}-\sqrt{z}\right)\\ &= 3\sqrt{z}\left(\sqrt{z}\right)\\ &= 3\sqrt{z^2}\\ &= 3z\end{aligned}$$

CHALLENGE PROBLEMS

67. Graph $y=x^2-x-2$ and use the graph to solve the quadratic equation $-2=x^2-x-2$.

The graph of $y=x^2-x-2$ passes through $(0,\ -2)$ and $(1,\ -2)$.
This means x^2-x-2 has the value -2 when $x=0$ and for $x=1$.

Proof by solving:

$$\begin{aligned}x^2-x-2&=-2\\ x^2-x-2+\mathbf{2}&=-2+\mathbf{2}\\ x^2-x&=0\\ x(x-1)&=0\end{aligned}$$

Set $x=0$ and $x-1=0$

$$x=1$$

Thus, the solutions of $-2=x^2-x-2$ are 0 and 1.

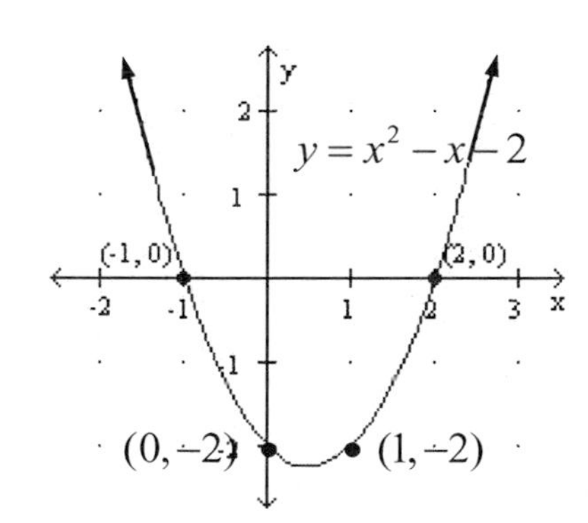

CHAPTER 9 REVIEW

SECTION 9.1

Solving Quadratic Equations: The Square Root Property

Use the square root property to solve each equation.

1. $x^2 = 64$

$$x = \pm\sqrt{64}$$

$$x = \sqrt{64} \quad \text{or} \quad x = -\sqrt{64}$$

$$x = 8 \quad\quad\quad x = -8$$

$$x = \pm 8$$

The solutions are ± 8.

2. $t^2 - 8 = 0$

$$t^2 - 8 + \mathbf{8} = 0 + \mathbf{8}$$

$$t^2 = 8$$

$$t = \pm\sqrt{8}$$

$$t = \pm\sqrt{4 \cdot 2}$$

$$t = \pm 2\sqrt{2}$$

The solutions are $\pm 2\sqrt{2}$.

3. $2x^2 - 1 = 149$

$$2x^2 - 1 + \mathbf{1} = 149 + \mathbf{1}$$

$$2x^2 = 150$$

$$\frac{2x^2}{\mathbf{2}} = \frac{150}{\mathbf{2}}$$

$$x^2 = 75$$

$$x = \pm\sqrt{75}$$

$$x = \pm\sqrt{25 \cdot 3}$$

$$x = \pm 5\sqrt{3}$$

The solutions are $\pm 5\sqrt{3}$.

4. $(x-1)^2 = 25$

$$x - 1 = \pm\sqrt{25}$$

$$x - 1 + \mathbf{1} = \mathbf{1} \pm 5$$

$$x = 1 \pm 5$$

$$x = 1 + 5 \quad \text{or} \quad x = 1 - 5$$

$$x = 6 \quad\quad\quad x = -4$$

The solutions are 6 and -4.

5. $(9x-8)^2 = 40$

$$9x - 8 = \pm\sqrt{40}$$

$$9x - 8 = \pm\sqrt{4 \cdot 10}$$

$$9x - 8 = \pm 2\sqrt{10}$$

$$9x - 8 + \mathbf{8} = +\mathbf{8} \pm 2\sqrt{10}$$

$$9x = 8 \pm 2\sqrt{10}$$

$$\frac{9x}{\mathbf{9}} = \frac{8 \pm 2\sqrt{10}}{\mathbf{9}}$$

$$x = \frac{8 \pm 2\sqrt{10}}{9}$$

The solutions are $\frac{8 \pm 2\sqrt{10}}{9}$.

6. $4(x-2)^2 - 9 = 0$

$$4(x-2)^2 - 9 + \mathbf{9} = 0 + \mathbf{9}$$

$$4(x-2)^2 = 9$$

$$\frac{4(x-2)^2}{\mathbf{4}} = \frac{9}{\mathbf{4}}$$

$$(x-2)^2 = \frac{9}{4}$$

$$x - 2 = \pm\sqrt{\frac{9}{4}}$$

$$x - 2 = \pm\frac{3}{2}$$

$$x - 2 + \mathbf{2} = \mathbf{2} \pm \frac{3}{2}$$

$$x = 2 \pm \frac{3}{2}$$

$$x = \frac{2}{1} \cdot \frac{\mathbf{2}}{\mathbf{2}} \pm \frac{3}{2}$$

$$x = \frac{4 \pm 3}{2}$$

$$x = \frac{4+3}{2} \quad \text{or} \quad x = \frac{4-3}{2}$$

$$x = \frac{7}{2} \quad\quad\quad x = \frac{1}{2}$$

The solutions are $\frac{7}{2}$ and $\frac{1}{2}$.

7. $p^2 - 20p + 100 = 9$

$$(p-10)^2 = 9$$
$$p - 10 = \pm\sqrt{9}$$
$$p - 10 = \pm 3$$
$$p - 10 + \mathbf{10} = \mathbf{10} \pm 3$$
$$p = 10 \pm 3$$

$p = 10 + 3$ or $p = 10 - 3$

$p = 13$ | $p = 7$

The solutions are 13 and 7.

8. $9m^2 + 6m + 1 = 6$

$$(3m+1)^2 = 6$$
$$3m + 1 = \pm\sqrt{6}$$
$$3m + 1 - \mathbf{1} = -\mathbf{1} \pm \sqrt{6}$$
$$3m = -1 \pm \sqrt{6}$$
$$\frac{3m}{\mathbf{3}} = \frac{-1 \pm \sqrt{6}}{\mathbf{3}}$$
$$m = \frac{-1 \pm \sqrt{6}}{3}$$

The solutions are $\frac{-1 \pm \sqrt{6}}{3}$.

Use the square root property to find all real-number solutions of each equation. Round each solution to the nearest hundredth.

9. $x^2 = 12$

$$x = \pm\sqrt{12}$$
$$x = \pm\sqrt{4 \cdot 3}$$
$$x = \pm 2\sqrt{3}$$

$x = \sqrt{12}$ or $x = -\sqrt{12}$

$x \approx 3.464$ | $x \approx -3.464$

$x \approx 3.46$ | $x \approx -3.46$

$$x \approx \pm 3.46$$

To the nearest hundredths, the approximate solutions are ± 3.46.

10. $(x-1)^2 = 55$

$$x - 1 = \pm\sqrt{55}$$
$$x - 1 + \mathbf{1} = \mathbf{1} \pm \sqrt{55}$$
$$x = 1 \pm \sqrt{55}$$

$x = 1 + \sqrt{55}$ or $x = 1 - \sqrt{55}$

$x \approx 1 + 7.416$ | $x \approx 1 - 7.416$

$x \approx 1 + 7.42$ | $x \approx 1 - 7.42$

$x \approx 8.42$ | $x \approx -6.42$

To the nearest hundredths, the approximate solutions are 8.42 and -6.42.

11. $m^2 + 36 = 0$

$$m^2 + 36 - \mathbf{36} = 0 - \mathbf{36}$$
$$m^2 = -36$$
$$m = \pm\sqrt{-36}$$

There are no real-number solutions.

12. $(2x-3)^2 = -8$

$$2x - 3 = \pm\sqrt{-8}$$

There are no real-number solutions.

13. CLIFF DIVERS

$$d = 16t^2$$
$$148 = 16t^2$$
$$16t^2 = 148$$
$$\frac{16t^2}{\mathbf{16}} = \frac{148}{\mathbf{16}}$$
$$t^2 = \frac{148}{16}$$
$$t = \pm\sqrt{\frac{148}{16}}$$
$$t = \pm\sqrt{9.25}$$
$$t \approx \pm 3.04$$
$$t \approx 3.0$$

It will take the diver approximately 3.0 second to enter the water.

14. RUBIK'S CUBE

$$A = s^2$$
$$s^2 = A$$
$$s^2 = \frac{81}{16}$$
$$s = \pm\sqrt{\frac{81}{16}}$$
$$s = \pm\frac{\sqrt{81}}{\sqrt{16}}$$
$$s = \pm\frac{9}{4}$$

The length of one side of a Rubik's cube is $2\frac{1}{4}$ inches.

SECTION 9.2

Solving Quadratic Equations: Completing the Square

Complete the square to make each expression a perfect square trinomial. Then factor.

15. x^2+4x

To complete the square:

$\frac{1}{2}(4)=2$

and $(2)^2=4$.

Add 4 to the binomial.

$x^2+4x+\mathbf{4}$

$(x+2)^2$

16. t^2-5t

To complete the square:

$\frac{1}{2}(-5)=-\frac{5}{2}$

and $\left(-\frac{5}{2}\right)^2=\frac{25}{4}$.

Add $\frac{25}{4}$ to the binomial.

$t^2-5t+\frac{25}{4}$

$\left(t-\frac{5}{2}\right)^2$

Solve each quadratic equation by completing the square.

17. $x^2-8x+15=0$

$x^2-8x+15-\mathbf{15}=0-\mathbf{15}$

$x^2-8x=-15$

To complete the square:

$\frac{1}{2}(-8)=-4$

and $(-4)^2=16$.

Add 16 to both sides of the equation.

$x^2-8x+\mathbf{16}=-15+\mathbf{16}$

$(x-4)^2=1$

$x-4=\pm\sqrt{1}$

$x-4+\mathbf{4}=\mathbf{4}\pm 1$

$x=4\pm 1$

$x=4+1$

$x=5$

or

$x=4-1$

$x=3$

The solutions are 5 and 3.

18. $x^2=-5x+14$

$x^2+\mathbf{5x}=-5x+14+\mathbf{5x}$

$x^2+5x=14$

To complete the square:

$\frac{1}{2}(5)=\frac{5}{2}$

and $\left(\frac{5}{2}\right)^2=\frac{25}{4}$.

Add $\frac{25}{4}$ to both sides of the equation.

$x^2+5x+\mathbf{\frac{25}{4}}=14+\mathbf{\frac{25}{4}}$

$\left(x+\frac{5}{2}\right)^2=14+\frac{25}{4}$

$\left(x+\frac{5}{2}\right)^2=\frac{14}{1}\cdot\mathbf{\frac{4}{4}}+\frac{25}{4}$

$\left(x+\frac{5}{2}\right)^2=\frac{56+25}{4}$

$\left(x+\frac{5}{2}\right)^2=\frac{81}{4}$

$x+\frac{5}{2}=\pm\sqrt{\frac{81}{4}}$

$x+\frac{5}{2}=\pm\frac{\sqrt{81}}{\sqrt{4}}$

$x+\frac{5}{2}=\pm\frac{9}{2}$

$x+\frac{5}{2}-\mathbf{\frac{5}{2}}=-\mathbf{\frac{5}{2}}\pm\frac{9}{2}$

$x=\frac{-5\pm 9}{2}$

$x=\frac{-5+9}{2}$ or $x=\frac{-5-9}{2}$

$x=\frac{4}{2}$ | $x=\frac{-14}{2}$

$x=2$ | $x=-7$

The solutions are 2 and -7.

19. $x^2+2x=5$

To complete the square:

$\frac{1}{2}(2)=1$

and $(1)^2=1$.

Add 1 to both sides of the equation.

$x^2+2x+\mathbf{1}=5+\mathbf{1}$

$(x+1)^2=6$

$x+1=\pm\sqrt{6}$

$x+1-\mathbf{1}=-\mathbf{1}\pm\sqrt{6}$

$x=-1\pm\sqrt{6}$

The solutions are $-1\pm\sqrt{6}$.

20. $4x^2 - 16x = 7$

$$\frac{4x^2}{\mathbf{4}} - \frac{16x}{\mathbf{4}} = \frac{7}{\mathbf{4}}$$

$$x^2 - 4x = \frac{7}{4}$$

To complete the square:

$\frac{1}{2}(-4) = -2$

and $(-2)^2 = 4$.

Add 4 to both sides of the equation.

$$x^2 - 4x + \mathbf{4} = \frac{7}{4} + \mathbf{4}$$

$$(x-2)^2 = \frac{7}{4} + \frac{4}{1} \cdot \frac{\mathbf{4}}{\mathbf{4}}$$

$$(x-2)^2 = \frac{7+16}{4}$$

$$(x-2)^2 = \frac{23}{4}$$

$$x - 2 = \pm\sqrt{\frac{23}{4}}$$

$$x - 2 = \pm\frac{\sqrt{23}}{\sqrt{4}}$$

$$x - 2 + \mathbf{2} = \mathbf{2} \pm \frac{\sqrt{23}}{2}$$

$$x = 2 \pm \frac{\sqrt{23}}{2}$$

$$x = \frac{2}{1} \cdot \frac{\mathbf{2}}{\mathbf{2}} \pm \frac{\sqrt{23}}{2}$$

$$x = \frac{4 \pm \sqrt{23}}{2}$$

The solutions are $\frac{4 \pm \sqrt{23}}{2}$.

21. $2x^2 - 2x - 1 = 0$

$$2x^2 - 2x - 1 + \mathbf{1} = 0 + \mathbf{1}$$

$$2x^2 - 2x = 1$$

$$\frac{2x^2}{\mathbf{2}} - \frac{2x}{\mathbf{2}} = \frac{1}{\mathbf{2}}$$

$$x^2 - x = \frac{1}{2}$$

To complete the square:

$\frac{1}{2}(-1) = -\frac{1}{2}$

and $\left(-\frac{1}{2}\right)^2 = \frac{1}{4}$.

Add $\frac{1}{4}$ to both sides of the equation.

$$x^2 - x + \frac{\mathbf{1}}{\mathbf{4}} = \frac{1}{2} + \frac{\mathbf{1}}{\mathbf{4}}$$

$$\left(x - \frac{1}{2}\right)^2 = \frac{1}{2} \cdot \frac{\mathbf{2}}{\mathbf{2}} + \frac{1}{4}$$

$$\left(x - \frac{1}{2}\right)^2 = \frac{2+1}{4}$$

$$x - \frac{1}{2} = \pm\sqrt{\frac{3}{4}}$$

$$x - \frac{1}{2} = \pm\frac{\sqrt{3}}{2}$$

$$x - \frac{1}{2} + \frac{\mathbf{1}}{\mathbf{2}} = \frac{\mathbf{1}}{\mathbf{2}} + \frac{\sqrt{3}}{2}$$

$$x = \frac{1 \pm \sqrt{3}}{2}$$

The solutions are $\frac{1 \pm \sqrt{3}}{2}$.

22.

$$3x^2 + 5x + 2 = 0$$
$$3x^2 + 5x + 2 - \mathbf{2} = 0 - \mathbf{2}$$
$$3x^2 + 5x = -2$$
$$\frac{3x^2}{\mathbf{3}} + \frac{5x}{\mathbf{3}} = \frac{-2}{\mathbf{3}}$$
$$x^2 + \frac{5}{3}x = -\frac{2}{3}$$

To complete the square:

$$\frac{1}{2}\left(\frac{5}{3}\right) = \frac{5}{6}$$

and $\left(\frac{5}{6}\right)^2 = \frac{25}{36}$.

Add $\frac{25}{36}$ to both sides of the equation.

$$x^2 + \frac{5}{3}x + \frac{\mathbf{25}}{\mathbf{36}} = \frac{-2}{3} + \frac{\mathbf{25}}{\mathbf{36}}$$
$$\left(x + \frac{5}{6}\right)^2 = \frac{-2}{3} \cdot \frac{\mathbf{12}}{\mathbf{12}} + \frac{25}{36}$$
$$\left(x + \frac{5}{6}\right)^2 = \frac{-24 + 25}{36}$$
$$\left(x + \frac{5}{6}\right)^2 = \frac{1}{36}$$
$$x + \frac{5}{6} = \pm\sqrt{\frac{1}{36}}$$
$$x + \frac{5}{6} = \pm\frac{\sqrt{1}}{\sqrt{36}}$$
$$x + \frac{5}{6} = \pm\frac{1}{6}$$
$$x + \frac{5}{6} - \frac{\mathbf{5}}{\mathbf{6}} = -\frac{\mathbf{5}}{\mathbf{6}} \pm \frac{1}{6}$$
$$x = \frac{-5 \pm 1}{6}$$

	or	
$x = \frac{-5+1}{6}$		$x = \frac{-5-1}{6}$
$x = \frac{-4}{6}$		$x = \frac{-6}{6}$
$x = -\frac{2}{3}$		$x = -1$

The solutions are $-\frac{2}{3}$ and -1.

Solve each quadratic equation by completing the square. Approximate the solutions to the nearest hundredth.

23.

$$x^2 + 4x + 1 = 0$$
$$x^2 + 4x + 1 - \mathbf{1} = 0 - \mathbf{1}$$
$$x^2 + 4x = -1$$

To complete the square:

$$\frac{1}{2}(4) = 2$$

and $(2)^2 = 4$.

Add 4 to both sides of the equation.

$$x^2 + 4x + \mathbf{4} = -1 + \mathbf{4}$$
$$(x + 2)^2 = 3$$
$$x + 2 = \pm\sqrt{3}$$
$$x + 2 - \mathbf{2} = -\mathbf{2} \pm \sqrt{3}$$
$$x = -2 \pm \sqrt{3}$$

The exact solutions are $-2 \pm \sqrt{3}$.

	or	
$x = -2 + \sqrt{3}$		$x = -2 - \sqrt{3}$
$x \approx -2 + 1.732$		$x \approx -2 - 1.732$
$x \approx -2 + 1.73$		$x \approx -2 - 1.73$
$x \approx -0.27$		$x \approx -3.73$

The approximate solutions are -0.27 and -3.73.

24. $x^2 - 7x = 5$

To complete the square:

$\frac{1}{2}(-7) = \frac{-7}{2}$

and $\left(\frac{-7}{2}\right)^2 = \frac{49}{4}$.

Add $\frac{49}{4}$ to both sides of the equation.

$$x^2 - 7x + \mathbf{\frac{49}{4}} = 5 + \mathbf{\frac{49}{4}}$$
$$\left(x - \frac{7}{2}\right)^2 = 5 + \frac{49}{4}$$
$$\left(x - \frac{7}{2}\right)^2 = \frac{5}{1} \cdot \mathbf{\frac{4}{4}} + \frac{49}{4}$$
$$\left(x - \frac{7}{2}\right)^2 = \frac{20 + 49}{4}$$
$$\left(x - \frac{7}{2}\right)^2 = \frac{69}{4}$$
$$x - \frac{7}{2} = \pm\sqrt{\frac{69}{4}}$$
$$x - \frac{7}{2} = \pm\frac{\sqrt{69}}{\sqrt{4}}$$
$$x - \frac{7}{2} + \mathbf{\frac{7}{2}} = \mathbf{\frac{7}{2}} \pm \frac{\sqrt{69}}{2}$$
$$x = \frac{7 \pm \sqrt{69}}{2}$$

The exact solutions are $\frac{7 \pm \sqrt{69}}{2}$.

	or	
$x = \frac{7 + \sqrt{69}}{2}$		$x = \frac{7 - \sqrt{69}}{2}$
$x \approx \frac{7 + 8.306}{2}$		$x \approx \frac{7 - 8.306}{2}$
$x \approx \frac{15.306}{2}$		$x \approx \frac{-1.306}{2}$
$x \approx 7.653$		$x \approx -0.653$
$x \approx 7.65$		$x \approx -0.65$

The approximate solutions are 7.65 and -0.65.

SECTION 9.3

Solving Quadratic Equations: The Quadratic Formula

Write each equation in $ax^2 + bx + c = 0$ form and find *a*, *b*, and *c*.

25. $x^2 + 2x = -5$

$$x^2 + 2x + \mathbf{5} = -5 + \mathbf{5}$$
$$x^2 + 2x + 5 = 0$$
$$1x^2 + 2x + 5 = 0$$
$$a = 1,\ b = 2,\ c = 5$$

26. $6x^2 = 2x + 1$

$$6x^2 - \mathbf{2x} = 2x + 1 - \mathbf{2x}$$
$$6x^2 - 2x = 1$$
$$6x^2 - 2x - \mathbf{1} = 1 - \mathbf{1}$$
$$6x^2 - 2x - 1 = 0$$
$$a = 6,\ b = -2,\ c = -1$$

Use the quadratic formula to find all real-number solutions of each equation.

27. $x^2 - 2x - 15 = 0$

$$1x^2 - 2x - 15 = 0$$
$$a = 1,\ b = -2,\ c = -15$$
$$x = \frac{-b \pm \sqrt{b^2 - 4ac}}{2a}$$
$$x = \frac{-(-2) \pm \sqrt{(-2)^2 - 4(1)(-15)}}{2(1)}$$
$$x = \frac{2 \pm \sqrt{4 + 60}}{2}$$
$$x = \frac{2 \pm \sqrt{64}}{2}$$
$$x = \frac{2 \pm 8}{2}$$

	or	
$x = \frac{2 + 8}{2}$		$x = \frac{2 - 8}{2}$
$x = \frac{10}{2}$		$x = \frac{-6}{2}$
$x = 5$		$x = -3$

The solutions are 5 and -3.

28.

$$6x^2 = 7x + 3$$
$$6x^2 - \mathbf{7x} = 7x + 3 - \mathbf{7x}$$
$$6x^2 - 7x = 3$$
$$6x^2 - 7x - \mathbf{3} = 3 - \mathbf{3}$$
$$6x^2 - 7x - 3 = 0$$
$$a = 6,\ b = -7,\ c = -3$$
$$x = \frac{-b \pm \sqrt{b^2 - 4ac}}{2a}$$
$$x = \frac{-(-7) \pm \sqrt{(-7)^2 - 4(6)(-3)}}{2(6)}$$
$$x = \frac{7 \pm \sqrt{49 + 72}}{12}$$
$$x = \frac{7 \pm \sqrt{121}}{12}$$
$$x = \frac{7 \pm 11}{12}$$

$$x = \frac{7+11}{12} \quad \text{or} \quad x = \frac{7-11}{12}$$
$$x = \frac{18}{12} \quad\quad x = \frac{-4}{12}$$
$$x = \frac{3}{2} \quad\quad x = -\frac{1}{3}$$

The solutions are $\frac{3}{2}$ and $-\frac{1}{3}$.

29.

$$p^2 - 4 = 2p$$
$$p^2 - 4 - \mathbf{2p} = 2p - \mathbf{2p}$$
$$1p^2 - 2p - 4 = 0$$
$$a = 1,\ b = -2,\ c = -4$$
$$p = \frac{-b \pm \sqrt{b^2 - 4ac}}{2a}$$
$$p = \frac{-(-2) \pm \sqrt{(-2)^2 - 4(1)(-4)}}{2(1)}$$
$$p = \frac{2 \pm \sqrt{4 + 16}}{2}$$
$$p = \frac{2 \pm \sqrt{20}}{2}$$
$$p = \frac{2 \pm \sqrt{4 \cdot 5}}{2}$$
$$p = \frac{2 \pm 2\sqrt{5}}{2}$$
$$p = \frac{\overset{1}{\cancel{2}}(1 \pm \sqrt{5})}{\underset{1}{\cancel{2}}}$$
$$p = 1 \pm \sqrt{5}$$

The solutions are $1 \pm \sqrt{5}$.

30.

$$x^2 + 7 = 6x$$
$$x^2 + 7 - \mathbf{6x} = 6x - \mathbf{6x}$$
$$1x^2 - 6x + 7 = 0$$
$$a = 1,\ b = -6,\ c = 7$$
$$x = \frac{-b \pm \sqrt{b^2 - 4ac}}{2a}$$
$$x = \frac{-(-6) \pm \sqrt{(-6)^2 - 4(1)(7)}}{2(1)}$$
$$x = \frac{6 \pm \sqrt{36 - 28}}{2}$$
$$x = \frac{6 \pm \sqrt{8}}{2}$$
$$x = \frac{6 \pm \sqrt{4 \cdot 2}}{2}$$
$$x = \frac{6 \pm 2\sqrt{2}}{2}$$
$$x = \frac{\overset{1}{\cancel{2}}(3 \pm \sqrt{2})}{\underset{1}{\cancel{2}}}$$
$$x = 3 \pm \sqrt{2}$$

The solutions are $3 \pm \sqrt{2}$.

31.
$$3x^2+3x=1$$
$$3x^2+3x-\mathbf{1}=1-\mathbf{1}$$
$$3x^2+3x-1=0$$
$$a=3,\ b=3,\ c=-1$$
$$x=\frac{-b\pm\sqrt{b^2-4ac}}{2a}$$
$$x=\frac{-3\pm\sqrt{3^2-4(3)(-1)}}{2(3)}$$
$$x=\frac{-3\pm\sqrt{9+12}}{6}$$
$$x=\frac{-3\pm\sqrt{21}}{6}$$
The solutions are $\frac{-3\pm\sqrt{21}}{6}$.

32.
$$5x^2+x=1$$
$$5x^2+x-\mathbf{1}=1-\mathbf{1}$$
$$5x^2+x-1=0$$
$$a=5,\ b=1,\ c=-1$$
$$x=\frac{-b\pm\sqrt{b^2-4ac}}{2a}$$
$$x=\frac{-1\pm\sqrt{1^2-4(5)(-1)}}{2(5)}$$
$$x=\frac{-1\pm\sqrt{1+20}}{10}$$
$$x=\frac{-1\pm\sqrt{21}}{10}$$
$$x=\frac{-1\pm\sqrt{21}}{10}$$
The solutions are $\frac{-1\pm\sqrt{21}}{10}$.

33. $7x^2-x+2=0$
$$a=7,\ b=-1,\ c=2$$
$$x=\frac{-b\pm\sqrt{b^2-4ac}}{2a}$$
$$x=\frac{-(-1)\pm\sqrt{(-1)^2-4(7)(2)}}{2(7)}$$
$$x=\frac{1\pm\sqrt{1-56}}{14}$$
$$x=\frac{1\pm\sqrt{-55}}{14}$$
There are no real-number solutions.

34.
$$2x^2+6x=5$$
$$2x^2+6x-\mathbf{5}=5-\mathbf{5}$$
$$2x^2+6x-5=0$$
$$a=2,\ b=6,\ c=-5$$
$$x=\frac{-b\pm\sqrt{b^2-4ac}}{2a}$$
$$x=\frac{-6\pm\sqrt{6^2-4(2)(-5)}}{2(2)}$$
$$x=\frac{-6\pm\sqrt{36+40}}{4}$$
$$x=\frac{-6\pm\sqrt{76}}{4}$$
$$x=\frac{-6\pm\sqrt{4\cdot 19}}{4}$$
$$x=\frac{-6\pm 2\sqrt{19}}{4}$$
$$x=\frac{\overset{1}{\cancel{2}}(-3\pm\sqrt{19})}{\underset{1}{\cancel{2}}\cdot 2}$$
$$x=\frac{-3\pm\sqrt{19}}{2}$$
The solutions are $\frac{-3\pm\sqrt{19}}{2}$.

Use the most efficient method to find all real-number solutions of each equation.

35. $4x^2 + 16x = 0$

$$4x(x+4) = 0$$

$$x = 0 \quad \text{or} \quad x + 4 = 0$$

$$x = -4$$

The solutions are 0 and -4.

36. $(y+3)^2 = 16$

$$y + 3 = \pm\sqrt{16}$$

$$y + 3 = \pm 4$$

$$y + 3 - \mathbf{3} = \mathbf{-3} \pm 4$$

$$y = -3 \pm 4$$

$$y = -3 + 4 \quad \text{or} \quad y = -3 - 4$$

$$y = 1 \quad | \quad y = -7$$

The solutions are 1 and -7.

37. $3g^2 - 81 = 0$

$$3g^2 - 81 + \mathbf{81} = 0 + \mathbf{81}$$

$$3g^2 = 81$$

$$\frac{3g^2}{\mathbf{3}} = \frac{81}{\mathbf{3}}$$

$$g^2 = 27$$

$$g = \pm\sqrt{27}$$

$$g = \pm\sqrt{9 \cdot 3}$$

$$g = \pm 3\sqrt{3}$$

The solutions are $\pm 3\sqrt{3}$.

38. $3x^2 - 6x = -1$

$$\frac{3x^2}{\mathbf{3}} - \frac{6x}{\mathbf{3}} = \frac{-1}{\mathbf{3}}$$

$$x^2 - 2x = -\frac{1}{3}$$

To complete the square:

$\frac{1}{2}(-2) = -1$

and $(-1)^2 = 1$.

Add 1 to both sides of the equation.

$$x^2 - 2x + \mathbf{1} = \frac{-1}{3} + \mathbf{1}$$

$$(x-1)^2 = \frac{-1}{3} + 1 \cdot \frac{\mathbf{3}}{\mathbf{3}}$$

$$(x-1)^2 = \frac{-1+3}{3}$$

$$(x-1)^2 = \frac{2}{3}$$

$$x - 1 = \pm\sqrt{\frac{2}{3}}$$

$$x - 1 = \pm\frac{\sqrt{2}}{\sqrt{3}} \cdot \frac{\sqrt{\mathbf{3}}}{\sqrt{\mathbf{3}}}$$

$$x - 1 = \pm\frac{\sqrt{6}}{3}$$

$$x - 1 + \mathbf{1} = \mathbf{+1} \pm \frac{\sqrt{6}}{3}$$

$$x = 1 \cdot \frac{\mathbf{3}}{\mathbf{3}} \pm \frac{\sqrt{6}}{3}$$

$$x = \frac{3 \pm \sqrt{6}}{3}$$

The solutions are $\frac{3 \pm \sqrt{6}}{3}$.

39. $2x^2 + 2x - 5 = 0$

$a = 2,\ b = 2,\ c = -5$

$$x = \frac{-b \pm \sqrt{b^2 - 4ac}}{2a}$$

$$x = \frac{-2 \pm \sqrt{2^2 - 4(2)(-5)}}{2(2)}$$

$$x = \frac{-2 \pm \sqrt{4 + 40}}{4}$$

$$x = \frac{-2 \pm \sqrt{44}}{4}$$

$$x = \frac{-2 \pm \sqrt{4 \cdot 11}}{4}$$

$$x = \frac{-2 \pm 2\sqrt{11}}{4}$$

$$x = \frac{\overset{1}{\cancel{2}}(-1 \pm \sqrt{11})}{\underset{1}{\cancel{2}} \cdot 2}$$

$$x = \frac{-1 \pm \sqrt{11}}{2}$$

The solutions are $\frac{-1 \pm \sqrt{11}}{2}$.

40.
$$a^2 = 4a - 4$$
$$a^2 - \mathbf{4a} = 4a - 4 - \mathbf{4a}$$
$$a^2 - 4a = -4$$
$$a^2 - 4a + \mathbf{4} = -4 + \mathbf{4}$$
$$a^2 - 4a + 4 = 0$$
$$(a-2)(a-2) = 0$$
$$a - 2 = 0$$
$$a = 2$$
There is a repeated solution of 2.

41.
$$(2x-5)^2 = 64$$
$$2x - 5 = \pm\sqrt{64}$$
$$2x - 5 = \pm 8$$
$$2x - 5 + \mathbf{5} = \mathbf{5} \pm 8$$
$$2x = 5 \pm 8$$
$$\frac{2x}{\mathbf{2}} = \frac{5 \pm 8}{\mathbf{2}}$$
$$x = \frac{5 \pm 8}{2}$$

$$x = \frac{5+8}{2} \quad \text{or} \quad x = \frac{5-8}{2}$$
$$x = \frac{13}{2} \qquad\qquad x = -\frac{3}{2}$$

The solutions are $\frac{13}{2}$ and $-\frac{3}{2}$.

42.
$$a^2 - 2a + 5 = 0$$
$$1a^2 - 2a + 5 = 0$$
$$a = 1,\ b = -2,\ c = 5$$
$$x = \frac{-b \pm \sqrt{b^2 - 4ac}}{2a}$$
$$x = \frac{-(-2) \pm \sqrt{(-2)^2 - 4(1)(5)}}{2(1)}$$
$$x = \frac{2 \pm \sqrt{4 - 20}}{2}$$
$$x = \frac{2 \pm \sqrt{-16}}{2}$$
There are no real-number solutions.

43.
$$3x^2 + 2x - 2 = 0$$
$$a = 3,\ b = 2,\ c = -2$$
$$x = \frac{-b \pm \sqrt{b^2 - 4ac}}{2a}$$
$$x = \frac{-2 \pm \sqrt{2^2 - 4(3)(-2)}}{2(3)}$$
$$x = \frac{-2 \pm \sqrt{4 + 24}}{6}$$
$$x = \frac{-2 \pm \sqrt{28}}{6}$$
$$x = \frac{-2 \pm \sqrt{4 \cdot 7}}{6}$$
$$x = \frac{-2 \pm 2\sqrt{7}}{6}$$
$$x = \frac{\cancel{2}^{1}(-1 \pm \sqrt{7})}{\cancel{2}_{1} \cdot 3}$$
$$x = \frac{-1 \pm \sqrt{7}}{3}$$

The exact solutions are $\frac{-1 \pm \sqrt{7}}{3}$.

$$x = \frac{-1+\sqrt{7}}{3} \quad \text{or} \quad x = \frac{-1+\sqrt{7}}{3}$$
$$x \approx \frac{-1+2.645}{3} \qquad x \approx \frac{-1-2.645}{3}$$
$$x \approx \frac{1.645}{3} \qquad x \approx \frac{-3.645}{3}$$
$$x \approx 0.548 \qquad x \approx -1.215$$
$$x \approx 0.55 \qquad x \approx -1.22$$

The approximate solutions are 0.55 and -1.22.

44. SECURITY GATES

Analyze

- The length (longer leg) is 14 ft longer than the width (shorter leg).
- The diagonal (hypotenuse) is 26 ft.
- Find the width and length.

Assign

Let $w =$ width in ft

$w+14 =$ length in ft

Form

$$\left(\text{The length of the shorter leg}\right)^2 \text{ plus } \left(\text{the length of the longer leg}\right)^2 \text{ equals } \left(\text{the length of the hypotenuse}\right)^2$$

$$w^2 \quad + \quad (w+14)^2 \quad = \quad 26^2$$

Solve

$$w^2+(w+14)^2=26^2$$
$$w^2+w^2+28w+196=676$$
$$2w^2+28w+196=676$$
$$2w^2+28w+196-\mathbf{676}=676-\mathbf{676}$$
$$2w^2+28w-480=0$$
$$2(w^2+14w-240)=0$$
$$2(w+24)(w-10)=0$$

$w+24=0$ or $w-10=0$

~~$w=-24$~~ | $w=10$

State

The width of the gate is 10 ft and the length is $w+14=10+14=24$ ft.

Check

The results check.

45. GRAND CANYON

Analyze

- Let t equal the time it takes for the rock to hit the bottom of the canyon.

Assign

Let $t =$ the time in seconds

Form

$$-16t^2+8t+5{,}040=0$$

Solve

$$-16t^2+8t+5{,}040=0$$
$$-8(2t^2-t-630)=0$$
$$-8(2t+35)(t-18)=0$$

$2t+35=0$ or $t-18=0$

$2t=-35$ | $t=18$

~~$t=-\frac{35}{2}$~~

State

It will take the rock 18 seconds to hit the bottom of the canyon.

Check

The result checks.

46. GEOMETRY

Analyze

- Area of the triangle is 30 in^2.
- The base is $x+2$ inches long.
- The height is x inches tall.
- What is the height?

Assign

Let $x =$ height of triangle in inches.

$x+2 =$ base of triangle in inches.

Form

$$\boxed{\tfrac{1}{2}} \text{ times } \boxed{\text{the base}} \text{ times } \boxed{\text{the height}} \text{ equals } \boxed{\text{the area of the triangle.}}$$

$$\frac{1}{2} \quad \cdot \quad x+2 \quad \cdot \quad x \quad = \quad 30$$

Solve

$$\frac{1}{2}x(x+2)=30$$
$$(\mathbf{2})\frac{1}{2}x(x+2)=(\mathbf{2})30$$
$$x(x+2)=60$$
$$x^2+2x=60$$
$$x^2+2x-\mathbf{60}=60-\mathbf{60}$$
$$x^2+2x-60=0$$

$$x=\frac{-b+\sqrt{b^2-4ac}}{2a} \quad \text{or} \quad x=\frac{-b-\sqrt{b^2-4ac}}{2a}$$
$$x=\frac{-(2)+\sqrt{(2)^2-4(1)(-60)}}{2(1)} \quad \Big| \quad x=\frac{-(2)-\sqrt{(2)^2-4(1)(-60)}}{2(1)}$$
$$x=\frac{-2+\sqrt{244}}{2} \quad \Big| \quad \cancel{x=\frac{-2-\sqrt{244}}{2}}$$
$$x\approx 6.81$$
$$x\approx 6.8$$

State

The height of the triangle is about 6.8 in.

Check

The result checks.

SECTION 9.4
Complex Numbers

Write each expression in terms of *i*.

47. $\sqrt{-25} = \sqrt{-1 \cdot 25}$
$= \sqrt{-1}\sqrt{25}$
$= i \cdot 5$
$= 5i$

48. $\sqrt{-18} = \sqrt{-1 \cdot 18}$
$= \sqrt{-1}\sqrt{18}$
$= i\sqrt{9 \cdot 2}$
$= i \cdot 3\sqrt{2}$
$= 3i\sqrt{2}$

49. $-\sqrt{-49} = -\sqrt{-1 \cdot 49}$
$= -\sqrt{-1}\sqrt{49}$
$= -i \cdot 7$
$= -7i$

50. $\sqrt{-\frac{9}{64}} = \frac{\sqrt{-9}}{\sqrt{64}}$
$= \frac{i\sqrt{9}}{\sqrt{64}}$
$= \frac{3i}{8}$
$= \frac{3}{8}i$

51. Complete the diagram.

Complex numbers	
Real numbers	**Imaginary numbers**

52. Determine whether each statement is true or false.
   a. Every real number is a complex number. **True**
   b. $3 - 4i$ is an imaginary number. **True**
   c. $\sqrt{-4}$ is a real number. **False**
   d. i is a real number. **False**

Give the complex conjugate of each number.

53. $3 + 6i$ **$3 - 6i$**　54. $-1 - 7i$ **$-1 + 7i$**
55. 19i **0 – 19i**　56. –i **0 + i**

Perform the operations. Write all answers in the form *a + bi*.

57. $(3+4i)+(5-6i) = (3+5)+(4i-6i)$
$= 8-2i$

58. $(7-3i)-(4+2i) = (7-3i)+(-4-2i)$
$= (7-4)+(-3i-2i)$
$= 3-5i$

59. $3i(2-i) = \mathbf{3i}(2) - \mathbf{3i}(i)$
$= 6i - 3i^2$
$= 6i - 3(\mathbf{-1})$
$= 3 + 6i$

60. $(2+3i)(3-i) = \mathbf{2}\cdot 3 - \mathbf{2}\cdot i + \mathbf{3i}\cdot 3 - \mathbf{3i}\cdot i$
$= 6 - 2i + 9i - 3i^2$
$= 6 + 7i - 3(\mathbf{-1})$
$= 6 + 7i + 3$
$= (6+3) + 7i$
$= 9 + 7i$

61. $\frac{2+3i}{2-3i} = \frac{2+3i}{2-3i} \cdot \frac{\mathbf{2+3i}}{\mathbf{2+3i}}$
$= \frac{(2+3i)(2+3i)}{(2-3i)(2+3i)}$
$= \frac{4+2\cdot 6i+9i^2}{4-9i^2}$
$= \frac{4+12i+9(\mathbf{-1})}{4-9(\mathbf{-1})}$
$= \frac{4+12i-9}{4+9}$
$= \frac{-5+12i}{13}$
$= -\frac{5}{13} + \frac{12}{13}i$

62. $\frac{3}{5+i} = \frac{3}{5+i} \cdot \frac{\mathbf{5-i}}{\mathbf{5-i}}$
$= \frac{3(5-i)}{(5+i)(5-i)}$
$= \frac{15-3i}{25-i^2}$
$= \frac{15-3i}{25-(\mathbf{-1})}$
$= \frac{15-3i}{25+1}$
$= \frac{15-3i}{26}$
$= \frac{15}{26} - \frac{3}{26}i$

Solve each equation. Write all solutions in the form a + b*i*.

63. $$x^2+9=0$$
$$x^2+9-\mathbf{9}=0-\mathbf{9}$$
$$x^2=-9$$
$$x=\pm\sqrt{-9}$$
$$x=\pm\sqrt{-1}\sqrt{9}$$
$$x=\pm 3i$$
$$x=0\pm 3i$$
The solutions are $0\pm 3i$.

64. $$3x^2=-16$$
$$\frac{3x^2}{\mathbf{3}}=\frac{-16}{\mathbf{3}}$$
$$x^2=-\frac{16}{3}$$
$$x=\pm\sqrt{-\frac{16}{3}}$$
$$x=\pm\frac{\sqrt{-16}}{\sqrt{3}}$$
$$x=\pm\frac{\sqrt{-1}\sqrt{16}}{\sqrt{3}}$$
$$x=\pm\frac{4i}{\sqrt{3}}$$
$$x=\pm\frac{4i}{\sqrt{3}}\cdot\frac{\sqrt{\mathbf{3}}}{\sqrt{\mathbf{3}}}$$
$$x=\pm\frac{4i\sqrt{3}}{3}$$
$$x=0\pm\frac{4\sqrt{3}}{3}i$$
The solutions are $0\pm\frac{4\sqrt{3}}{3}i$.

65. $$(p-2)^2=-24$$
$$p-2=\pm\sqrt{-24}$$
$$p-2=\pm\sqrt{-1}\sqrt{24}$$
$$p-2=\pm\sqrt{-1}\sqrt{4\cdot 6}$$
$$p-2=\pm 2i\sqrt{6}$$
$$p-2+\mathbf{2}=\mathbf{2}\pm 2i\sqrt{6}$$
$$p=2\pm 2i\sqrt{6}$$
The solutions are $2\pm 2i\sqrt{6}$.

66. $$(q+3)^2=-54$$
$$q+3=\pm\sqrt{-54}$$
$$q+3=\pm\sqrt{-1\cdot 54}$$
$$q+3=\pm\, i\sqrt{9\cdot 6}$$
$$q+3-\mathbf{3}=-\mathbf{3}\pm 3i\sqrt{6}$$
$$q=-3\pm 3i\sqrt{6}$$
The solutions are $-3\pm 3i\sqrt{6}$.

67. $$x^2+2x=-2$$
To complete the square:
$$\frac{1}{2}(2)=1$$
and $1^2=1$.
Add 1 to both sides of the equation.
$$x^2+2x+\mathbf{1}=+\mathbf{1}-2$$
$$(x+1)^2=-1$$
$$x+1=\pm\sqrt{-1}$$
$$x+1=\pm\, i\sqrt{1}$$
$$x+1=\pm\, i$$
$$x+1-\mathbf{1}=-\mathbf{1}\pm i$$
$$x=-1\pm i$$
The solutions are $-1\pm i$.

68. $$2x^2-3x+2=0$$
$$a=2,\ b=-3,\ c=2$$
$$x=\frac{-b\pm\sqrt{b^2-4ac}}{2a}$$
$$x=\frac{-(-3)\pm\sqrt{(-3)^2-4(2)(2)}}{2(2)}$$
$$x=\frac{3\pm\sqrt{9-16}}{4}$$
$$x=\frac{3\pm\sqrt{-7}}{4}$$
$$x=\frac{3\pm\sqrt{-1}\sqrt{7}}{4}$$
$$x=\frac{3\pm i\sqrt{7}}{4}$$
$$x=\frac{3}{4}\pm\frac{\sqrt{7}}{4}i$$
The solutions are $\frac{3}{4}\pm\frac{\sqrt{7}}{4}i$.

SECTION 9.5

Graphing Quadratic Equations

69. Refer to the figure.
    a. What are the x-intercepts of the parabola?
       (–3, 0), (1, 0)
    b. What is the y-intercept of the parabola?
       (0, –3)
    c. What is the vertex of the parabola?
       (–1, –4)
    d. Draw the axis of symmetry of the parabola on the graph.

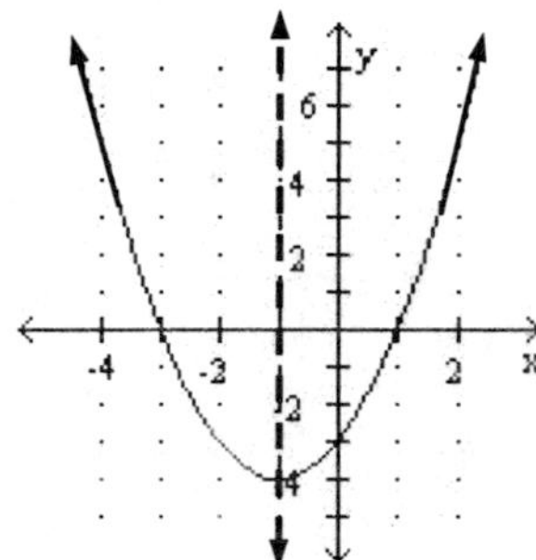

It is a vertical line through (−1, −4).

70. The point (0, –3) lies on the parabola graphed above. Use symmetry to determine the coordinates of another point that lies on the parabola? **(–2, –3)**

Find the vertex of the graph of each quadratic equation and tell which direction the parabola opens. *Do not draw the graph.*

71. $y = 2x^2 - 4x + 7$
    $a = 2,\ b = -4$

    Find the vertex.

    $x = \frac{-b}{2a} = \frac{-(-4)}{2(2)} = \frac{4}{4} = 1$

    $x = 1$
    $y = 2x^2 - 4x + 7$
    $= 2(1)^2 - 4(1) + 7$
    $= 2 - 4 + 7$
    $= 5$

    The vertex of the parabola is $(1, 5)$.

    It opens upward because of the positive coefficient of the x^2 term.

72. $y = -3x^2 + 18x - 11$
    $a = -3,\ b = 18$

    Find the vertex.

    $x = \frac{-b}{2a} = \frac{-(18)}{2(-3)} = \frac{-18}{-6} = 3$

    $x = 3$
    $y = -3x^2 + 18x - 11$
    $= -3(3)^2 + 18(3) - 11$
    $= -27 + 54 - 11$
    $= 16$

    The vertex of the parabola is (3,16).

    It opens downward because of the negative coefficient of the x^2 term.

Find the x- and y-intercepts of the graph of each quadratic equation.

73. $y = x^2 + 6x + 5$

    Find the x-intercepts. Let $y = 0$.
    Solve the quadratic.

    $y = x^2 + 6x + 5$
    $0 = (x + 5)(x + 1)$
    $x + 5 = 0$ or $x + 1 = 0$
    $x = -5$ | $x = -1$

    The x-intercepts are $(-5, 0)$ and $(-1, 0)$.

    Find the y-intercept. It is the constant.

    $y = x^2 + 6x + 5$

    The y-intercept is $(0, 5)$.

74. $y = x^2 + 2x + 3$

Find the x-intercepts. Let $y = 0$.
Solve the quadratic.

$$y = x^2 + 2x + 3$$
$$0 = x^2 + 2x + 3$$
$$a = 1,\ b = 2,\ c = 3$$
$$x = \frac{-b \pm \sqrt{b^2 - 4ac}}{2a}$$
$$x = \frac{-2 \pm \sqrt{2^2 - 4(1)(3)}}{2(1)}$$
$$x = \frac{-2 \pm \sqrt{4 - 12}}{2}$$
$$x = \frac{-2 \pm \sqrt{-8}}{-2}$$

There are no x-intercepts because the equation does not have any real number solutions.

Find the y-intercept. It is the constant.

$$y = x^2 + 2x + 3$$

The y-intercept is $(0, 3)$.

Graph each quadratic equation by finding the vertex, the x- and y-intercepts, and the axis of symmetry of its graph.

75. $y = x^2 + 2x - 3$

$$y = 1x^2 + 2x - 3$$
$$a = 1,\ b = 2$$

Find the vertex.

$$x = \frac{-b}{2a} = \frac{-(2)}{2(1)} = -1$$

$x = -1$

$$y = x^2 + 2x - 3 = (-1)^2 + 2(-1) - 3 = 1 - 2 - 3 = -4$$

The vertex of the parabola is $(-1, -4)$.

Find the x-intercepts. Let $y = 0$.
Solve the quadratic.

$$y = x^2 + 2x - 3$$
$$0 = (x + 3)(x - 1)$$

$x + 3 = 0$ or $x - 1 = 0$

$x = -3$ | $x = 1$

The x-intercepts are $(-3, 0)$ and $(1, 0)$.

Find the y-intercept. It is the constant.

$$y = x^2 + 2x - 3$$

The y-intercept is $(0, -3)$.

The axis of symmetry is the vertical line passing through the vertex: $x = -1$.

Plot two additional points, one by substitution and the second one by the property of symmetry.

x	y	(x, y)
2	$y = x^2 + 2x - 3$ $= (2)^2 + 2(2) - 3$ $= 4 + 4 - 3$ $= 5$	$(2, 5)$

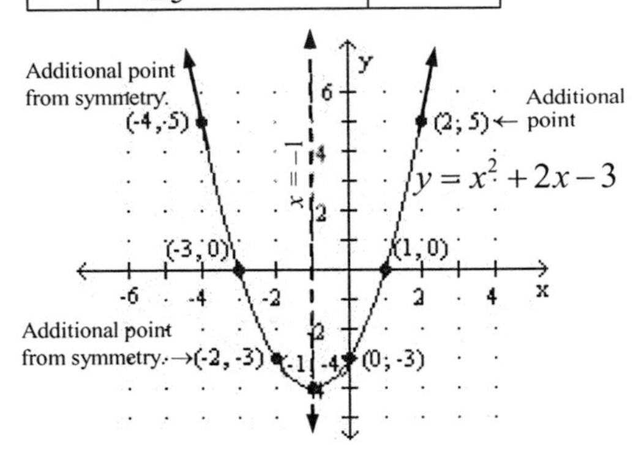

76. $y = -2x^2 + 4x - 2$

$$a = -2,\ b = 4$$

Find the vertex.

$$x = \frac{-b}{2a} = \frac{-(4)}{2(-2)} = 1$$

$x = 1$

$$y = -2x^2 + 4x - 2 = -2(1)^2 + 4(1) - 2 = -2 + 4 - 2 = 0$$

The vertex of the parabola is $(1, 0)$.

Find the x-intercepts. Let $y = 0$.
Solve the quadratic.

$$y = -2x^2 + 4x - 2$$
$$0 = -2(x^2 - 2x + 1)$$
$$0 = -2(x - 1)(x - 1)$$
$$x - 1 = 0$$
$$x = 1$$

The x-intercept is $(1, 0)$.

Find the y-intercept. It is the constant.

$y = -2x^2 + 4x - 2$

The y-intercept is $(0, -2)$.

The axis of symmetry is the vertical line passing through the vertex: $x = 1$.

Plot two additional points, one by substitution and the second one by the property of symmetry.

x	y	(x, y)
3	$y = -2x^2 + 4x - 2$ $= -2(3)^2 + 4(3) - 2$ $= -18 + 12 - 2$ $= -8$	$(3, -8)$

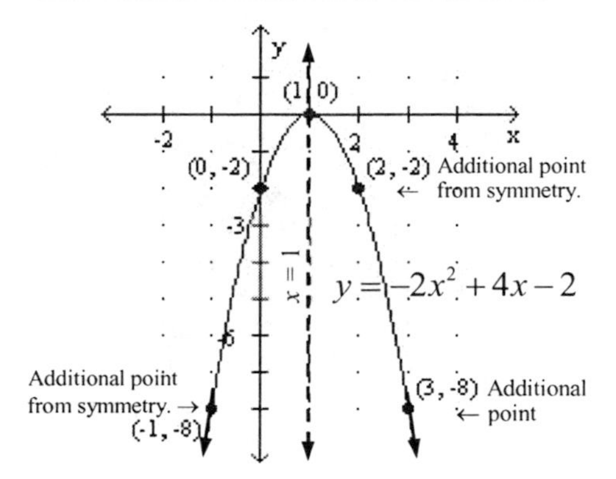

77. $y = -x^2 - 2x + 5$

$y = -1x^2 - 2x + 5$

$a = -1,\ b = -2$

Find the vertex.

$$x = \frac{-b}{2a} = \frac{-(-2)}{2(-1)} = \frac{2}{-2} = -1$$

$x = -1$

$$\begin{aligned} y &= -x^2 - 2x + 5 \\ &= -(-1)^2 - 2(-1) + 5 \\ &= -1 + 2 + 5 \\ &= 6 \end{aligned}$$

The vertex of the parabola is $(-1, 6)$.

Find the x-intercepts. Let $y = 0$. Solve the quadratic.

$y = -x^2 - 2x + 5$

$0 = -1x^2 - 2x + 5$

$a = -1,\ b = -2,\ c = 5$

$$x = \frac{-b \pm \sqrt{b^2 - 4ac}}{2a}$$

$$x = \frac{-(-2) \pm \sqrt{(-2)^2 - 4(-1)(5)}}{2(-1)}$$

$$x = \frac{2 \pm \sqrt{4 + 20}}{-2}$$

$$x = \frac{2 \pm \sqrt{24}}{-2}$$

$$x = \frac{2 \pm \sqrt{4}\sqrt{6}}{-2}$$

$$x = \frac{2 \pm 2\sqrt{6}}{-2}$$

$$x = \frac{\overset{1}{\cancel{2}}\left(1 \pm \sqrt{6}\right)}{-1 \cdot \underset{1}{\cancel{2}}}$$

$$x = -1 \pm \sqrt{6}$$

The x-intercepts are $(-1 + \sqrt{6}, 0)$ and $(-1 - \sqrt{6}, 0)$.

Find the y-intercept. It is the constant.

$y = -x^2 - 2x + 5$

The y-intercept is $(0, 5)$.

The axis of symmetry is the vertical line passing through the vertex. $x = -1$.

Plot two additional points, one by substitution and the second point by the property of symmetry.

x	y	(x, y)
2	$y = -x^2 - 2x + 5$ $= -(2)^2 - 2(2) + 5$ $= -4 - 4 + 5$ $= -3$	$(2, -3)$

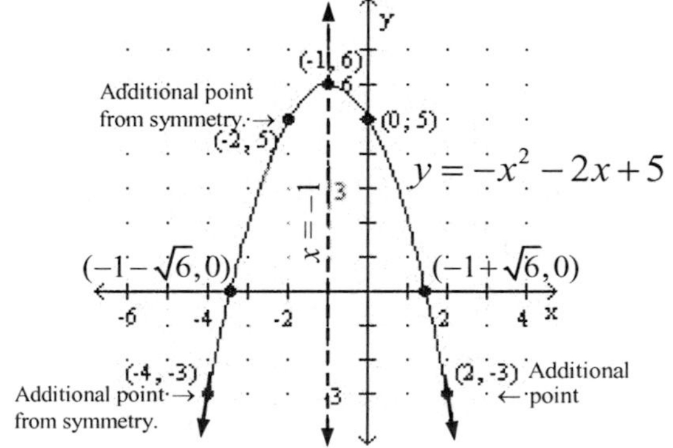

78. $y = x^2 + 4x - 1$

$y = 1x^2 + 4x - 1$

$a = 1,\ b = 4$

Find the vertex.

$$x = \frac{-b}{2a} = \frac{-(4)}{2(1)} = -2$$

$x = -2$

$y = x^2 + 4x - 1$

$= (-2)^2 + 4(-2) - 1$

$= 4 - 8 - 1$

$= -5$

The vertex of the parabola is $(-2, -5)$.

Find the x-intercepts. Let $y = 0$. Solve the quadratic.

$y = x^2 + 4x - 1$

$0 = 1x^2 + 4x - 1$

$a = 1,\ b = 4,\ c = -1$

$$x = \frac{-b \pm \sqrt{b^2 - 4ac}}{2a}$$

$$x = \frac{-4 \pm \sqrt{4^2 - 4(1)(-1)}}{2(1)}$$

$$x = \frac{-4 \pm \sqrt{16 + 4}}{2}$$

$$x = \frac{-4 \pm \sqrt{20}}{2}$$

$$x = \frac{-4 \pm \sqrt{4}\sqrt{5}}{2}$$

$$x = \frac{-4 \pm 2\sqrt{5}}{2}$$

$$x = \frac{\overset{1}{\cancel{2}}\left(-2 \pm \sqrt{5}\right)}{\underset{1}{\cancel{2}}}$$

$$x = -2 \pm \sqrt{5}$$

The x-intercepts are $(-2 + \sqrt{5}, 0)$ and $(-2 - \sqrt{5}, 0)$.

Find the y-intercept. It is the constant.

$y = x^2 + 4x - 1$

The y-intercept is $(0, -1)$.

The axis of symmetry is the vertical line passing through the vertex. $x = -2$.

Plot two additional points, one by substitution and the second point by the property of symmetry.

x	y	(x, y)
1	$y = x^2 + 4x - 1$ $= (1)^2 + 4(1) - 1$ $= 1 + 4 - 1$ $= 4$	$(1, 4)$

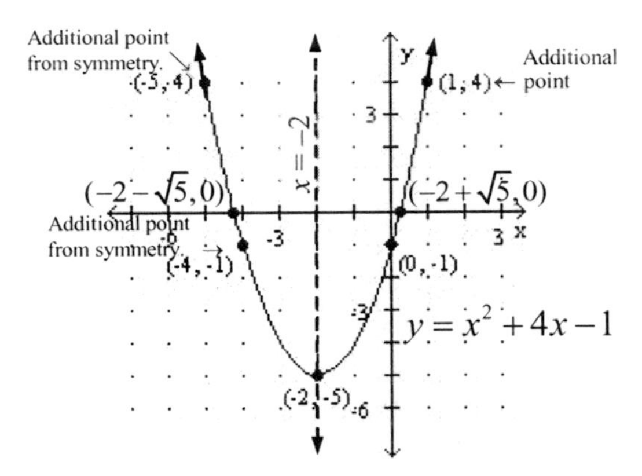

79. The graphs of three quadratic equations in two variables are shown. Fill in the blanks.

$x^2 + x - 2 = 0$ has **2** real-number solution(s).
Give the solution(s): $\mathbf{-2, 1}$

$2x^2 + 12x + 18 = 0$ has **1** repeated real-number solution(s).
Give the solution(s): $\mathbf{-3}$

$-x^2 + 4x - 5 = 0$ has **no** real-number solution(s).

80. What important information can be obtained from the vertex of the parabola in the graph below? **The maximum profit of $16,000 is obtained from the sale of 400 units.**

CHAPTER 9 TEST

1. Fill in the blanks.

a. A **quadratic** equation can be written in the form $ax^2 + bx + c = 0$, where a, b, and c represent real numbers and $a \neq 0$.

b. $x^2 + 8x + 16$ is a perfect- **square** trinomial because $x^2 + 8x + 16 = (x + 4)^2$

c. When we add 25 to $x^2 + 10x$, we say we have **completed** the square on $x^2 + 10x$.

d. We read $3 \pm \sqrt{2}$ as "three **plus** or **minus** the square root of two."

e. The **leading** coefficient of $3x^2 + 8x - 9$ is 3 and the **constant** term is -9 .

2. Write the statement $x = \sqrt{5}$ or $x = -\sqrt{5}$ using double-sign notation. $\boldsymbol{x = \pm\sqrt{5}}$.

Solve each equation by the square root method.

3. $x^2 = 17$

$$x = \pm\sqrt{17}$$

The solutions are $\pm\sqrt{17}$.

4. $r^2 - 48 = 0$

$$r^2 - 48 + \mathbf{48} = 0 + \mathbf{48}$$
$$r^2 = 48$$
$$r = \pm\sqrt{48}$$
$$r = \pm\sqrt{16 \cdot 3}$$
$$r = \pm 4\sqrt{3}$$

The solutions are $\pm 4\sqrt{3}$.

5. $(x-2)^2 = 3$

$$x - 2 = \pm\sqrt{3}$$
$$x - 2 + \mathbf{2} = \mathbf{2} \pm \sqrt{3}$$
$$x = 2 \pm \sqrt{3}$$

The solutions are $2 \pm \sqrt{3}$.

6. $4y^2 - 20 = 5$

$$4y^2 - 20 + \mathbf{20} = 5 + \mathbf{20}$$
$$4y^2 = 25$$
$$\frac{4y^2}{\mathbf{4}} = \frac{25}{\mathbf{4}}$$
$$y^2 = \frac{25}{4}$$
$$y = \pm\frac{\sqrt{25}}{\sqrt{4}}$$
$$y = \pm\frac{5}{2}$$

The solutions are $\pm\frac{5}{2}$.

7. $t^2 = \frac{1}{49}$

$$t = \pm\sqrt{\frac{1}{49}}$$
$$t = \pm\frac{1}{7}$$

The solutions are $\pm\frac{1}{7}$.

8. $x^2 + 16x + 64 = 24$

$$(x+8)^2 = 24$$
$$x + 8 = \pm\sqrt{24}$$
$$x + 8 = \pm\sqrt{4 \cdot 6}$$
$$x + 8 = \pm 2\sqrt{6}$$
$$x + 8 - \mathbf{8} = \mathbf{-8} \pm 2\sqrt{6}$$
$$x = -8 \pm 2\sqrt{6}$$

The solutions are $-8 \pm 2\sqrt{6}$.

9. Explain why the equation $m^2 + 49 = 0$ has no real-number solutions.

$m = \pm\sqrt{-49}$ and $\sqrt{-49}$ is not a real number.

10. $n^2 - 32 = 0$

$$\left(\mathbf{4\sqrt{2}}\right)^2 - 32 \stackrel{?}{=} 0$$
$$16(2) - 32 \stackrel{?}{=} 0$$
$$32 - 32 \stackrel{?}{=} 0$$
$$0 = 0$$

True

$4\sqrt{2}$ is a solution.

Complete the square and factor the resulting perfect-square trinomial.

11. $x^2 - 14x$

To complete the square:

$$\frac{1}{2}(-14) = -7$$

and $(-7)^2 = 49$.

Add 49 to the binomial.

$$x^2 - 14x + \mathbf{49}$$
$$(x-7)^2$$

12. $c^2 - 7c$

To complete the square:

$\frac{1}{2}(-7) = -\frac{7}{2}$

and $\left(-\frac{7}{2}\right)^2 = \frac{49}{4}$.

Add $\frac{49}{4}$ to the binomial.

$$c^2 - 7c + \mathbf{\frac{49}{4}}$$

$$(c - \frac{7}{2})^2$$

13. $x^2 + x$

To complete the square:

$\frac{1}{2}(1) = \frac{1}{2}$

and $\left(\frac{1}{2}\right)^2 = \frac{1}{4}$.

Add $\frac{1}{4}$ to the binomial.

$$x^2 + x + \mathbf{\frac{1}{4}}$$

$$(x + \frac{1}{2})^2$$

14. $a^2 - \frac{5}{3}a$

To complete the square:

$\frac{1}{2}\left(-\frac{5}{3}\right) = -\frac{5}{6}$

and $\left(-\frac{5}{6}\right)^2 = \frac{25}{36}$.

Add $\frac{25}{36}$ to the binomial.

$$a^2 - \frac{5}{3}a + \mathbf{\frac{25}{36}}$$

$$\left(a - \frac{5}{6}\right)^2$$

15. $a^2 + 2a - 4 = 0$

$$a^2 + 2a - 4 + \mathbf{4} = 0 + \mathbf{4}$$
$$a^2 + 2a = 4$$

To complete the square:

$\frac{1}{2}(2) = 1$

and $(1)^2 = 1$.

Add 1 to both sides of the equation.

$$a^2 + 2a + \mathbf{1} = 4 + \mathbf{1}$$
$$(a+1)^2 = 5$$
$$a + 1 = \pm\sqrt{5}$$
$$a + 1 - \mathbf{1} = -\mathbf{1} \pm \sqrt{5}$$
$$a = -1 \pm \sqrt{5}$$

The exact solutions are $-1 \pm \sqrt{5}$.

	or	
$a = -1 + \sqrt{5}$		$a = -1 - \sqrt{5}$
$a \approx -1 + 2.236$		$a \approx -1 - 2.236$
$a \approx -1 + 2.24$		$a \approx -1 - 2.24$
$a \approx 1.24$		$a \approx -3.24$

The approximate solutions are 1.24 and -3.24.

16. $a^2 + a = 3$

To complete the square:

$\frac{1}{2}(1) = \frac{1}{2}$

and $\left(\frac{1}{2}\right)^2 = \frac{1}{4}$.

Add $\frac{1}{4}$ to both sides of the equation.

$$a^2 + a + \mathbf{\frac{1}{4}} = 3 + \mathbf{\frac{1}{4}}$$
$$\left(a + \frac{1}{2}\right)^2 = 3 \cdot \mathbf{\frac{4}{4}} + \frac{1}{4}$$
$$\left(a + \frac{1}{2}\right)^2 = \frac{12 + 1}{4}$$
$$a + \frac{1}{2} = \pm\sqrt{\frac{13}{4}}$$
$$a + \frac{1}{2} = \pm\frac{\sqrt{13}}{2}$$
$$a + \frac{1}{2} - \mathbf{\frac{1}{2}} = -\mathbf{\frac{1}{2}} \pm \frac{\sqrt{13}}{2}$$
$$a = \frac{-1 \pm \sqrt{13}}{2}$$

The solutions are $\frac{-1 \pm \sqrt{13}}{2}$.

17. $m^2 - 4m + 10 = 0$

$$m^2 - 4m + 10 - \mathbf{10} = 0 - \mathbf{10}$$

$$m^2 - 4m = -10$$

To complete the square:

$\frac{1}{2}(-4) = -2$

and $(-2)^2 = 4$.

Add 4 to both sides of the equation.

$$m^2 - 4m + \mathbf{4} = -10 + \mathbf{4}$$

$$(m-2)^2 = -6$$

$$m - 2 = \pm\sqrt{-6}$$

There are no real-number solutions.

18. $2x^2 = 3x + 2$

$$2x^2 - \mathbf{3x} = 3x + 2 - \mathbf{3x}$$

$$2x^2 - 3x = 2$$

$$\frac{2x^2}{\mathbf{2}} - \frac{3x}{\mathbf{2}} = \frac{2}{\mathbf{2}}$$

$$x^2 - \frac{3}{2}x = 1$$

To complete the square:

$\frac{1}{2}\left(-\frac{3}{2}\right) = -\frac{3}{4}$

and $\left(-\frac{3}{4}\right)^2 = \frac{9}{16}$.

Add $\frac{9}{16}$ to both sides of the equation.

$$x^2 - \frac{3}{2}x + \frac{\mathbf{9}}{\mathbf{16}} = 1 + \frac{\mathbf{9}}{\mathbf{16}}$$

$$\left(x - \frac{3}{4}\right)^2 = 1 \cdot \frac{\mathbf{16}}{\mathbf{16}} + \frac{9}{16}$$

$$\left(x - \frac{3}{4}\right)^2 = \frac{16+9}{16}$$

$$x - \frac{3}{4} = \pm\sqrt{\frac{25}{16}}$$

$$x - \frac{3}{4} = \pm\frac{\sqrt{25}}{\sqrt{16}}$$

$$x - \frac{3}{4} = \pm\frac{5}{4}$$

$$x - \frac{3}{4} + \frac{\mathbf{3}}{\mathbf{4}} = \frac{\mathbf{3}}{\mathbf{4}} \pm \frac{5}{4}$$

$$x = \frac{3 \pm 5}{4}$$

$x = \frac{3+5}{4}$ or $x = \frac{3-5}{4}$

$x = \frac{8}{4}$ | $x = \frac{-2}{4}$

$x = 2$ | $x = -\frac{1}{2}$

The solutions are 2 and $-\frac{1}{2}$.

Use the quadratic formula to solve each equation.

19. $2x^2 - 5x - 12 = 0$

$a = 2,\ b = -5,\ c = -12$

$$x = \frac{-b \pm \sqrt{b^2 - 4ac}}{2a}$$

$$x = \frac{-(-5) \pm \sqrt{(-5)^2 - 4(2)(-12)}}{2(2)}$$

$$x = \frac{5 \pm \sqrt{25+96}}{4}$$

$$x = \frac{5 \pm \sqrt{121}}{4}$$

$$x = \frac{5 \pm 11}{4}$$

$x = \frac{5+11}{4}$ or $x = \frac{5-11}{4}$

$x = \frac{16}{4}$ | $x = \frac{-6}{4}$

$x = 4$ | $x = -\frac{3}{2}$

The solutions are 4 and $-\frac{3}{2}$.

20. $5x^2 + 11x = -3$

$5x^2 + 11x + \mathbf{3} = -3 + \mathbf{3}$

$5x^2 + 11x + 3 = 0$

$a = 5\ b = 11,\ c = 3$

$$x = \frac{-b \pm \sqrt{b^2 - 4ac}}{2a}$$

$$x = \frac{-11 \pm \sqrt{11^2 - 4(5)(3)}}{2(5)}$$

$$x = \frac{-11 \pm \sqrt{121 - 60}}{10}$$

$$x = \frac{-11 \pm \sqrt{61}}{10}$$

The solutions are $\frac{-11 \pm \sqrt{61}}{10}$.

21. $4n^2 - 12n + 1 = 0$

$a = 4,\ b = -12,\ c = 1$

$$n = \frac{-b \pm \sqrt{b^2 - 4ac}}{2a}$$

$$n = \frac{-(-12) \pm \sqrt{(-12)^2 - 4(4)(1)}}{2(4)}$$

$$n = \frac{12 \pm \sqrt{144 - 16}}{8}$$

$$n = \frac{12 \pm \sqrt{128}}{8}$$

$$n = \frac{12 \pm \sqrt{64 \cdot 2}}{8}$$

$$n = \frac{12 \pm 8\sqrt{2}}{8}$$

$$n = \frac{\overset{1}{\cancel{4}}(3 \pm 2\sqrt{2})}{\underset{1}{\cancel{4}} \cdot 2}$$

$$n = \frac{3 \pm 2\sqrt{2}}{2}$$

The solutions are $\frac{3 \pm 2\sqrt{2}}{2}$.

22. $7t^2 = -6t - 4$

$7t^2 + \mathbf{6t} = -6t - 4 + \mathbf{6t}$

$7t^2 + 6t = -4$

$7t^2 + 6t + \mathbf{4} = -4 + \mathbf{4}$

$7t^2 + 6t + 4 = 0$

$a = 7,\ b = 6,\ c = 4$

$$t = \frac{-b \pm \sqrt{b^2 - 4ac}}{2a}$$

$$t = \frac{-6 \pm \sqrt{(6)^2 - 4(7)(4)}}{2(7)}$$

$$t = \frac{-6 \pm \sqrt{36 - 112}}{14}$$

$$t = \cancel{\frac{-6 \pm \sqrt{-76}}{14}}$$

There are no real-number solutions.

23. $3x^2 - 2x - 2 = 0$

$a = 3,\ b = -2,\ c = -2$

$$x = \frac{-b \pm \sqrt{b^2 - 4ac}}{2a}$$

$$x = \frac{-(-2) \pm \sqrt{(-2)^2 - 4(3)(-2)}}{2(3)}$$

$$x = \frac{2 \pm \sqrt{4 + 24}}{6}$$

$$x = \frac{2 \pm \sqrt{28}}{6}$$

$$x = \frac{2 \pm \sqrt{4 \cdot 7}}{6}$$

$$x = \frac{2 \pm 2\sqrt{7}}{6}$$

$$x = \frac{\overset{1}{\cancel{2}}(1 \pm \sqrt{7})}{\underset{1}{\cancel{2}} \cdot 3}$$

$$x = \frac{1 \pm \sqrt{7}}{3}$$

The exact solutions are $\frac{1 \pm \sqrt{7}}{3}$.

	or	
$x = \frac{1 + \sqrt{7}}{3}$		$x = \frac{1 - \sqrt{7}}{3}$
$x \approx \frac{1 + 2.646}{3}$		$x \approx \frac{1 - 2.646}{3}$
$x \approx \frac{3.646}{3}$		$x \approx \frac{-1.646}{3}$
$x \approx 1.215$		$x \approx -0.548$
$x \approx 1.22$		$x \approx -0.55$

The approximate solutions are 1.22 and -0.55.

24.

$$x^2 - 2x - 4 = 0$$
$$\left(\mathbf{1}+\sqrt{\mathbf{5}}\right)^2 - 2\left(\mathbf{1}+\sqrt{\mathbf{5}}\right) - 4 \stackrel{?}{=} 0$$
$$(\mathbf{1})^2 + 2\left(\sqrt{\mathbf{5}}\right) + \left(\sqrt{\mathbf{5}}\right)^2 - \mathbf{2}(1) - \mathbf{2}\left(\sqrt{5}\right) - 4 \stackrel{?}{=} 0$$
$$1 + 2\sqrt{5} + 5 - 2 - 2\sqrt{5} - 4 \stackrel{?}{=} 0$$
$$(1+5-2-4) + (2\sqrt{5} - 2\sqrt{5}) \stackrel{?}{=} 0$$
$$0 = 0$$

True

$1+\sqrt{5}$ is a solution.

25. ARCHERY

$$A = \pi r^2$$
$$5{,}026 = \pi r^2$$
$$\pi r^2 = 5{,}026$$
$$\frac{\pi r^2}{\boldsymbol{\pi}} = \frac{5{,}026}{\boldsymbol{\pi}}$$
$$r^2 = \frac{5{,}026}{\pi}$$
$$r = \pm\sqrt{\frac{5{,}026}{\pi}}$$
$$r \approx \pm\sqrt{1{,}599.8}$$
$$r \approx \pm 39.9$$
$$r \approx 40$$

The radius of the target rounded to the nearest cm is 40 cm.

26. ST. LOUIS

The distance d that an object falls (in feet) is related to the time t (in seconds) that is has been falling by the formula:

$$d = 16t^2$$
$$630 = 16t^2$$
$$\frac{630}{\mathbf{16}} = \frac{16t^2}{\mathbf{16}}$$
$$\frac{630}{16} = t^2$$
$$t^2 = 39.375$$
$$t = \pm\sqrt{39.375}$$
$$t \approx 6.27$$
$$t \approx 6.3$$

It will take the tool about 6.3 seconds to hit the ground.

27. NEW YORK CITY

Analyze

- Area of the rectangular screen is 2,665 ft^2.
- The length is 17 ft less than twice its width.
- Find the width and length of the sign.

Assign

Let x = width in ft.

$2x - 17$ = height in ft.

Form

The width of the screen	times	the height of the screen	equals	the area of the screen.
x	$\cdot$	$2x-17$	$=$	$2{,}665$

Solve

$$x(2x-17) = 2{,}665$$
$$2x^2 - 17x = 2{,}665$$
$$2x^2 - 17x - \mathbf{2{,}665} = 2{,}665 - \mathbf{2{,}665}$$
$$2x^2 - 17x - 2{,}665 = 0$$
$$a = 2,\ b = -17,\ c = -2{,}665$$
$$x = \frac{-b \pm \sqrt{b^2 - 4ac}}{2a}$$
$$x = \frac{-(-17) \pm \sqrt{(-17)^2 - 4(2)(-2{,}665)}}{2(2)}$$
$$x = \frac{17 \pm \sqrt{289 + 21{,}320}}{4}$$
$$x = \frac{17 \pm \sqrt{21{,}609}}{4}$$
$$x = \frac{17 \pm 147}{4}$$

$$x = \frac{17+147}{4} \quad \text{or} \quad x = \frac{17-147}{4}$$
$$x = \frac{164}{4} \qquad\qquad x = \frac{-130}{4}$$
$$x = 41 \qquad\qquad \cancel{x = -32.5}$$

State

The width of the srceen is 41 ft.

The height is $2x - 17 = 2(\mathbf{41}) - 17 = 65$ ft.

Check

The results check.

28. GEOMETRY

Analyze

- Hypotenuse is 8 ft.
- Long leg is 4 ft longer than short leg.
- Find the length of the two legs to nearest tenth.
- Pythagorean theroem: $a^2 + b^2 = c^2$.

Assign

Let l = length of short leg in ft.
$l + 4$ = length of long leg in ft.

Form

$\boxed{\text{short leg}}^2$ plus $\boxed{\text{long leg}}^2$ equals $\boxed{\text{hypotenuse.}}^2$

$l^2 \quad + \quad (l+4)^2 \quad = \quad 8^2$

Solve

$$l^2 + (l+4)^2 = 8^2$$
$$l^2 + l^2 + 8l + 16 = 64$$
$$2l^2 + 8l + 16 = 64$$
$$2l^2 + 8l + 16 - \mathbf{64} = 64 - \mathbf{64}$$
$$2l^2 + 8l - 48 = 0$$
$$2(l^2 + 4l - 24) = 0$$
$$a = 1,\ b = 4,\ c = -24$$
$$l = \frac{-b \pm \sqrt{b^2 - 4ac}}{2a}$$
$$l = \frac{-4 \pm \sqrt{4^2 - 4(1)(-24)}}{2(1)}$$
$$l = \frac{-4 \pm \sqrt{16 + 96}}{2}$$
$$l = \frac{-4 \pm \sqrt{112}}{2}$$

$l = \frac{-4 + \sqrt{112}}{2}$ or $l = \frac{-4 - \sqrt{112}}{2}$

$l \approx \frac{-4 + 10.58}{2}$ | $l \approx \frac{-4 - 10.58}{2}$

$l \approx \frac{6.58}{2}$ | $l \approx \frac{-16.58}{2}$ (crossed out)

$l \approx 3.29$

$l \approx 3.3$

State

The width is 3.3 feet.
The length is $3.3 + 4 = 7.3$ feet.

Check

The results check.

Use the most efficient method to solve each equation.

29. $x^2 - 4x = -2$

To complete the square:

$\frac{1}{2}(-4) = -2$

and $(-2)^2 = 4$.

Add 4 to both sides of the equation.

$$x^2 - 4x + \mathbf{4} = -2 + \mathbf{4}$$
$$(x-2)^2 = 2$$
$$x - 2 = \pm\sqrt{2}$$
$$x - 2 + \mathbf{2} = \mathbf{2} \pm \sqrt{2}$$
$$x = 2 \pm \sqrt{2}$$

The solutions are $2 \pm \sqrt{2}$.

30. $(3b+1)^2 = 16$

$$3b + 1 = \pm\sqrt{16}$$
$$3b + 1 = \pm 4$$
$$3b + 1 - \mathbf{1} = -\mathbf{1} \pm 4$$
$$3b = -1 \pm 4$$
$$\frac{3b}{\mathbf{3}} = \frac{-1 \pm 4}{\mathbf{3}}$$
$$b = \frac{-1 \pm 4}{3}$$

$b = \frac{-1+4}{3}$ or $b = \frac{-1-4}{3}$

$b = \frac{3}{3}$ | $b = -\frac{5}{3}$

$b = 1$

The solutions are 1 and $-\frac{5}{3}$.

31. $u^2 - 24 = 0$

$$u^2 - 24 + \mathbf{24} = 0 + \mathbf{24}$$
$$u^2 = 24$$
$$u = \pm\sqrt{24}$$
$$u = \pm\sqrt{4 \cdot 6}$$
$$u = \pm 2\sqrt{6}$$

The solutions are $\pm 2\sqrt{6}$.

32. $$6n^2 - 36n = 0$$
$$6n(n-6) = 0$$
$$6n = 0 \quad \text{or} \quad n - 6 = 0$$
$$n = 0 \quad \Big| \quad n = 6$$
The solutions are 0 and 6.

Write each expression in terms of *i*.

33. $$\sqrt{-100} = \sqrt{-1 \cdot 100} = i\sqrt{100} = i10 \text{ or } 10i$$

34. $$-\sqrt{-18} = -\sqrt{-1 \cdot 18} = -i\sqrt{9 \cdot 2} = -3i\sqrt{2} \text{ or } -3\sqrt{2}i$$

Perform the operations. Write all answers in the form *a + bi*.

35. $$(8+3i)+(-7-2i) = (8-7)+(3i-2i) = 1+i$$

36. $$(5+3i)-(6-9i) = (5+3i)+(-6+9i) = (5-6)+(3i+9i) = -1+12i$$

37. $$(2-4i)(3+2i) = \mathbf{2} \cdot 3 + \mathbf{2} \cdot 2i - \mathbf{4i} \cdot 3 - \mathbf{4i} \cdot 2i$$
$$= 6+4i-12i-8i^2$$
$$= 6-8i-8(\mathbf{-1})$$
$$= 6-8i+8$$
$$= (6+8)-8i$$
$$= 14-8i$$

38. $$\frac{3-2i}{3+2i} = \frac{3-2i}{3+2i} \cdot \frac{\mathbf{3-2i}}{\mathbf{3-2i}}$$
$$= \frac{9+2 \cdot -6i + 4i^2}{9-4i^2}$$
$$= \frac{9-12i+4(\mathbf{-1})}{9-4(\mathbf{-1})}$$
$$= \frac{9-12i-4}{9+4}$$
$$= \frac{5-12i}{13}$$
$$= \frac{5}{13} - \frac{12}{13}i$$

Solve each equation. Write all solutions in the form *a + bi*.

39. $$x^2 + 100 = 0$$
$$x^2 + 100 - \mathbf{100} = 0 - \mathbf{100}$$
$$x^2 = -100$$
$$x = \pm\sqrt{-100}$$
$$x = \pm\sqrt{-1 \cdot 100}$$
$$x = \pm\sqrt{-1}\sqrt{100}$$
$$x = \pm 10i$$
$$x = 0 \pm 10i$$
The solutions are $0 \pm 10i$.

40. $$(a+3)^2 = -1$$
$$a+3 = \pm\sqrt{-1}$$
$$a+3 = \pm i$$
$$a+3-\mathbf{3} = \mathbf{-3} \pm i$$
$$a = -3 \pm i$$
The solutions are $-3 \pm i$.

41. $$n^2 = -\frac{25}{16}$$
$$n = \pm\sqrt{-\frac{25}{16}}$$
$$n = \pm\sqrt{-1 \cdot \frac{25}{16}}$$
$$n = \pm\sqrt{-1}\sqrt{\frac{25}{16}}$$
$$n = \pm\frac{5}{4}i$$
$$n = 0 \pm \frac{5}{4}i$$
The solutions are $0 \pm \frac{5}{4}i$.

42. $3x^2+2x+1=0$

$a=3,\ b=2,\ c=1$

$$x=\frac{-b\pm\sqrt{b^2-4ac}}{2a}$$

$$x=\frac{-2\pm\sqrt{2^2-4(3)(1)}}{2(3)}$$

$$x=\frac{-2\pm\sqrt{4-12}}{6}$$

$$x=\frac{-2\pm\sqrt{-8}}{6}$$

$$x=\frac{-2\pm\sqrt{-1\cdot 8}}{6}$$

$$x=\frac{-2\pm i\sqrt{8}}{6}$$

$$x=\frac{-2\pm i\sqrt{4\cdot 2}}{6}$$

$$x=\frac{-2\pm 2i\sqrt{2}}{6}$$

$$x=\frac{\overset{1}{\cancel{2}}(-1\pm i\sqrt{2})}{\underset{1}{\cancel{2}}\cdot 3}$$

$$x=\frac{-1\pm i\sqrt{2}}{3}$$

The solutions are $\frac{-1\pm i\sqrt{2}}{3}$ or $-\frac{1}{3}\pm\frac{\sqrt{2}}{3}i$.

43. ADVERTISING

The most air conditioners sold in a week (18) occurred when 3 ads were run.

44. **Fill in the blanks.**

The graph of $y=ax^2+bx+c$ *opens downward when* $a<0$ *and upward when* $a>0$.

Graph each quadratic equation by finding the vertex, the x- and y-intercepts, and the axis of symmetry of its graph.

45. $y=x^2+6x+5$

$y=1x^2+6x+5$

$a=1,\ b=6$

Find the vertex.

$$x=\frac{-b}{2a}=\frac{-(6)}{2(1)}=-3$$

$x=-3$

$y=x^2+6x+5$

$=(-3)^2+6(-3)+5$

$=9-18+5$

$=-4$

The vertex of the parabola is $(-3,-4)$.

Find the x-intercepts. Let $y=0$. Solve the quadratic.

$y=x^2+6x+5$

$0=(x+5)(x+1)$

$x+5=0$ or $x+1=0$

$x=-5$ $\quad$ $x=-1$

The x-intercepts are $(-5,0)$ and $(-1,0)$.

Find the y-intercept. It is the constant.

$y=x^2+6x+5$

The y-intercept is $(0,5)$.

The axis of symmetry is the vertical line passing through the vertex: $x=-3$.

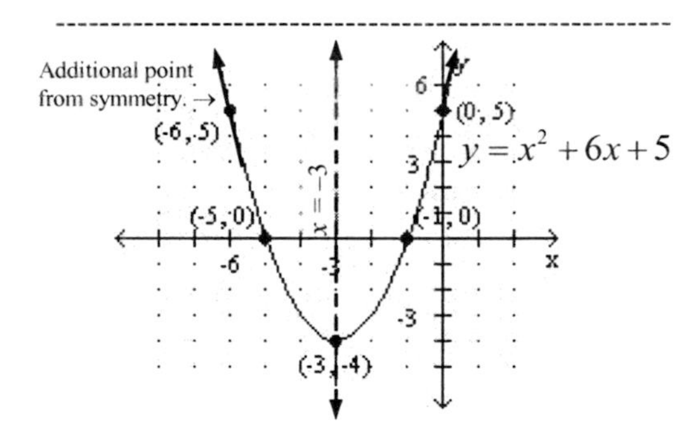

46. $y=-x^2+6x-7$

$y=-1x^2+6x-7$

$a=-1,\ b=6$

Find the vertex.

$$x=\frac{-b}{2a}=\frac{-(6)}{2(-1)}=3$$

$x=3$

$y=-x^2+6x-7$

$=-(3)^2+6(3)-7$

$=-9+18-7$

$=2$

The vertex of the parabola is $(3,2)$.

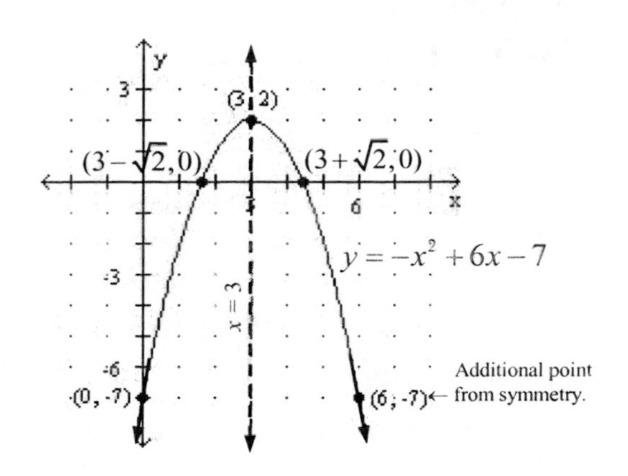

Find the x-intercepts. Let $y=0$. Solve the quadratic.

$y=-x^2+6x-7$

$0=-1x^2+6x-7$

$a=-1,\ b=6,\ c=-7$

$$x=\frac{-b\pm\sqrt{b^2-4ac}}{2a}$$

$$x=\frac{-6\pm\sqrt{6^2-4(-1)(-7)}}{2(-1)}$$

$$x=\frac{-6\pm\sqrt{36-28}}{-2}$$

$$x=\frac{-6\pm\sqrt{8}}{-2}$$

$$x=\frac{-6\pm\sqrt{4}\sqrt{2}}{-2}$$

$$x=\frac{-6\pm2\sqrt{2}}{-2}$$

$$x=\frac{\overset{1}{\cancel{2}}\left(-3\pm\sqrt{2}\right)}{-1\cdot\underset{1}{\cancel{2}}}$$

$x=3\pm\sqrt{2}$

The x-intercepts are $(3+\sqrt{2},0)$ and $(3-\sqrt{2},0)$.

Find the y-intercept. It is the constant.

$y=-x^2+6x-7$

The y-intercept is $(0,-7)$.

The axis of symmetry is the vertical line passing through the vertex: $x=3$.

CHAPTERS 1-9 CUMULATIVE REVIEW

1. Determine whether each statement is true or false. [Section 1.3]
   a. Every rational number can be written as a ratio of two integers. **True**
   b. The set of real numbers corresponds to all points on the number line. **True**
   c. The whole numbers and their opposites form the set of integers. **True**

2. DRIVING SAFETY In cold-weather climates, salt is spread on roads to keep snow and ice from bonding to the pavement. This allows snowplows to remove built-up snow quickly. According to the graph, when is the accident rate the highest? [Section 1.4]

   2 hours before salt is spread

3. Evaluate: [Section 1.7]

$$\begin{aligned} -4+2[-7-3(-9)] &= -4+2[-7+27] \\ &= -4+2[20] \\ &= -4+40 \\ &= 36 \end{aligned}$$

4. Evaluate: [Section 1.7]

$$\begin{aligned} \left|\frac{4}{5}\cdot 10-12\right| &= \left|\frac{4}{\cancel{5}_1}\cdot\frac{2\cdot\cancel{5}^1}{1}-12\right| \\ &= |8-12| \\ &= |-4| \\ &= 4 \end{aligned}$$

5. Evaluate: $(x-a)^2+(y-b)^2$ for $x=-2$, $y=1$, $a=5$, and $b=-3$. [Section 1.8]

$$\begin{aligned} (x-a)^2+(y-b)^2 &= (-2-5)^2+[1-(-3)]^2 \\ &= [-2+(-5)]^2+(1+3)^2 \\ &= (-7)^2+(4)^2 \\ &= 49+16 \\ &= 65 \end{aligned}$$

6. Simplify: [Section 1.9]

$$\begin{aligned} 3p-6(p-9)+p &= 3p-6p+54+p \\ &= (3-6+1)p+54 \\ &= -2p+54 \end{aligned}$$

7. Solve: $\frac{5}{6}k=10$ and check the results. [Section 2.2]

$$\begin{aligned} \frac{5}{6}k &= 10 \\ \mathbf{\frac{6}{5}}\cdot\frac{5}{6}k &= \mathbf{\frac{6}{5}}\cdot\frac{10}{1} \\ k &= \frac{6\cdot 10}{5} \\ k &= \frac{6\cdot 2\cdot\cancel{5}^1}{\cancel{5}_1} \\ k &= 12 \end{aligned}$$

Check:

$$\begin{aligned} \frac{5}{6}k &= 10 \\ \frac{5}{6}\cdot\mathbf{\frac{12}{1}} &\stackrel{?}{=} 10 \\ \frac{5\cdot 2\cdot\cancel{6}^1}{\cancel{6}_1} &\stackrel{?}{=} 10 \\ 10 &= 10 \end{aligned}$$

True

The solution is 12.

8. Solve: $-(3a+1)+a=2$ and check the results. [Section 2.2]

$$\begin{aligned} -(3a+1)+a &= 2 \\ -3a-1+a &= 2 \\ (-3+1)a-1 &= 2 \\ -2a-1+\mathbf{1} &= 2+\mathbf{1} \\ -2a &= 3 \\ \frac{-2a}{\mathbf{-2}} &= \frac{3}{\mathbf{-2}} \\ a &= -\frac{3}{2} \end{aligned}$$

Check:

$$\begin{aligned} -(3a+1)+a &= 2 \\ -\left(3\cdot\mathbf{\frac{-3}{2}}+1\right)+\left(\mathbf{-\frac{3}{2}}\right) &\stackrel{?}{=} 2 \\ -\left(\frac{-9}{2}+1\cdot\mathbf{\frac{2}{2}}\right)-\frac{3}{2} &\stackrel{?}{=} 2 \\ -\left(\frac{-9+2}{2}\right)-\frac{3}{2} &\stackrel{?}{=} 2 \\ -\left(\frac{-7}{2}\right)-\frac{3}{2} &\stackrel{?}{=} 2 \\ \frac{7}{2}-\frac{3}{2} &\stackrel{?}{=} 2 \\ \frac{4}{2} &\stackrel{?}{=} 2 \\ 2 &= 2 \end{aligned}$$

True

The solution is $-\frac{3}{2}$.

9. LOOSE CHANGE [Section 2.3]

- $60 in coins
- Voucher is worth $54.12.
- What percent was the pocessing fee?

$$\begin{aligned} 60-60r &= 54.12 \\ 60-60r-\mathbf{60} &= 54.12-\mathbf{60} \\ -60r &= -5.88 \\ \frac{-60r}{\mathbf{-60}} &= \frac{-5.88}{\mathbf{-60}} \\ r &= 0.098 \end{aligned}$$

The processing fee was 9.8%.

10. Solve for r. [Section 2.4]

$$T = 2r + 2t$$
$$T - \mathbf{2t} = 2r + 2t - \mathbf{2t}$$
$$T - 2t = 2r$$
$$\frac{T - 2t}{\mathbf{2}} = \frac{2r}{\mathbf{2}}$$
$$\frac{T - 2t}{2} = r$$
$$r = \frac{T - 2t}{2}$$

11. SELLING A HOME [Section 2.5]

Analyze

- Wants to make \$330,000 off the sale.
- Charged 4% commision of the sale.
- What is the selling price?

Assign

Let x = the selling price of the house

Form

The selling price of the house	minus	the amount of the commission	equals	the asking amount.
x	$-$	$0.04x$	$=$	330,000

Solve

$$x - 0.04x = 330,000$$
$$0.96x = 330,000$$
$$\frac{0.96x}{\mathbf{0.96}} = \frac{330,000}{\mathbf{0.96}}$$
$$x = 343,750$$

State

The selling price was \$343,750.

Check

$$x - 0.04x = 330,000$$
$$343,750 - 0.04(343,750) \stackrel{?}{=} 330,000$$
$$343,750 - 13,750 \stackrel{?}{=} 330,000$$
$$330,000 = 330,000$$

The result checks.

12. BUSINESS LOANS [Section 2.6]

Analyze

- \$28,000 is the total amount of the 2 loans.
- 1st business loan at 7% annual interest.
- 2nd business loan at 10% annual interest.
- \$2,560 is the first-year combined income.
- How much was each loan?

Assign

Let x = amount invested at 7%

$28,000 - x$ = amount invested at 10%

Form

	Principal	• rate	• time =	interest
1st loan	x	0.07	1	**$0.07x$**
2nd loan	$28,000 - x$	0.10	1	**$0.10(28,000 - x)$**
			Total:	**2,560**

The interest earned at 7%	plus	the interest earned at 10%	equals	the total interest.
$0.07x$	$+$	$0.10(28,000 - x)$	$=$	2,560

Solve

$$0.07x + 0.10(28,000 - x) = 2,560$$
$$\mathbf{100}[0.07x + 0.10(28,000 - x)] = \mathbf{100}(2,560)$$
$$100(0.07x) + 100[0.10(28,000 - x)] = 100(2,560)$$
$$7x + 10(28,000 - x) = 256,000$$
$$7x + 280,000 - 10x = 256,000$$
$$-3x + 280,000 = 256,000$$
$$-3x + 280,000 - \mathbf{280,000} = 256,000 - \mathbf{280,000}$$
$$-3x = -24,000$$
$$\frac{-3x}{\mathbf{-3}} = \frac{-24,000}{\mathbf{-3}}$$
$$x = 8,000$$

Amount of loan at 10%

$$28,000 - x = 28,000 - 8,000$$
$$= 20,000$$

State

\$8,000 was the amount of first loan.

\$20,000 was the amount of second loan.

Check

$$\left.\begin{array}{l} 0.07(\$8,000) = \$560 \\ 0.10(20,000) = \$2,000 \end{array}\right\} \$2,560$$

The results check.

13. Solve: [Section 2.7]

$$5x+7<2x+1$$
$$5x+7-\mathbf{2x}<2x+1-\mathbf{2x}$$
$$3x+7<1$$
$$3x+7-\mathbf{7}<\mathbf{1}-\mathbf{7}$$
$$3x<-6$$
$$\frac{3x}{\mathbf{3}}<\frac{-6}{\mathbf{3}}$$
$$x<-2$$
$$(-\infty,-2)$$

−2

14. Check to determine whether $(-5,-3)$ a solution of $2x-3y=-1$. [Section 3.1]

$$2x-3y=-1$$
$$2(-\mathbf{5})-3(-\mathbf{3})\stackrel{?}{=}-1$$
$$-10+9\stackrel{?}{=}-1$$
$$-1=-1$$

True

$(-5,-3)$ is a solution of $2x-3y=-1$.

Graph each equation or inequality.

15. $y=-x+2$ [Section 3.2]

x	y	(x, y)
−2	$y=-x+2$ $=-(-2)+2$ $=2+2$ $=4$	$(-2, 4)$
0	$y=-x+2$ $=-0+2$ $=2$	$(0, 2)$
3	$y=-x+2$ $=-(3)+2$ $=-3+2$ $=-1$	$(3, -1)$

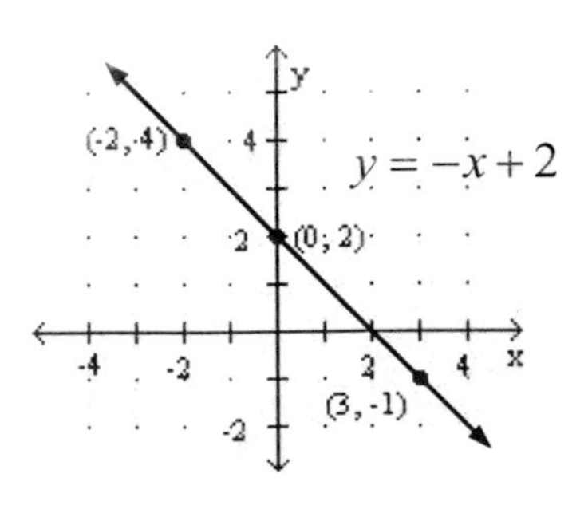

16. $2y-2x=6$ [Section 3.3]

y-intercept:
If $x=0$,
$$2y+2(\mathbf{0})=6$$
$$2y=6$$
$$y=3$$

x-intercept:
If $y=0$
$$2(\mathbf{0})-2x=6$$
$$-2x=6$$
$$x=-3$$

The y-intercept is $(0,3)$, and the x-intercept is $(-3, 0)$.

Check Point
$$2y-2x=6$$
$$2y-2(-\mathbf{1})=6$$
$$2y+2-\mathbf{2}=6-\mathbf{2}$$
$$2y=4$$
$$y=2$$
$$(-1, 2)$$

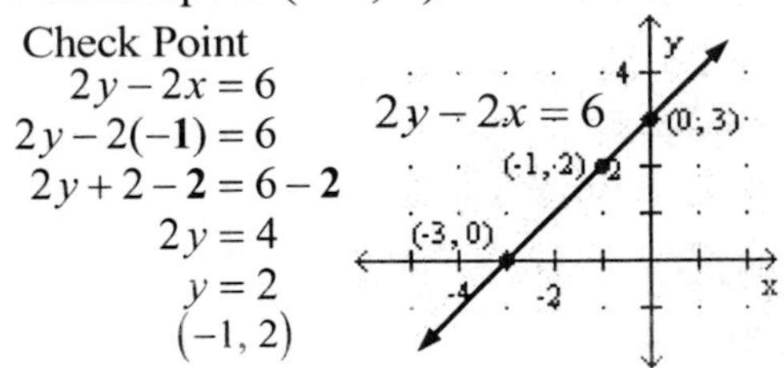

17. $y=-3$ [Section 3.3]

x	y	(x, y)
−3	−3	$(-3,-3)$
0	−3	$(0,-3)$
2	−3	$(2,-3)$

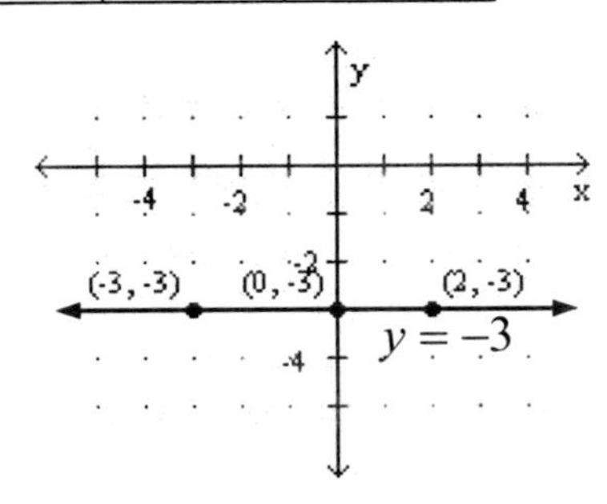

18. $y<3x$, Boundary line is dashed. [Section 3.7]

The line passes through the origin so select test point $(1,1)$ and substitute into

$$y<3x$$
$$\mathbf{1}\stackrel{?}{<}3(\mathbf{1})$$
$$1<3$$

True

The coordinates of every point on the same side of the line as $(1,1)$ satisfy the inequality. To indicate this, we shade the half-plane that contains the test point $(1,1)$.

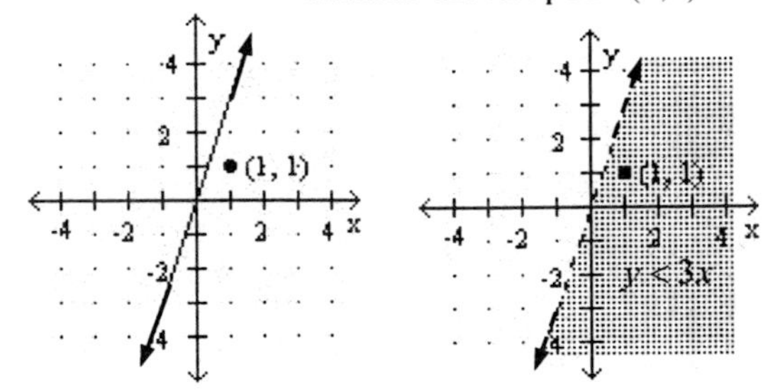

19. Find the slope of the line passing through $(-2,-2)$ and $(-12,-8)$. [Section 3.4]
(x_1, y_1) (x_2, y_2)

$$m = \frac{y_2 - y_1}{x_2 - x_1}$$
$$m = \frac{-8-(-2)}{-12-(-2)}$$
$$m = \frac{-8+2}{-12+2}$$
$$m = \frac{-6}{-10}$$
$$m = \frac{3}{5}$$

20. TV NEWS [Section 3.4]
Find the slope of the line passing through (1995, 35,500,000) and (2009, 22,900,000).

$$m = \frac{y_2 - y_1}{x_2 - x_1}$$
$$m = \frac{22,900,000 - 35,500,000}{2009 - 1995}$$
$$m = \frac{-12,600,000}{14}$$
$$m = -900,000$$

A decrease of 900,000 viewers per year.

21. What is the slope of the line defined by $4x + 5y = 6$? [Section 3.5]
$y = mx + b$, where m is the slope of the line.

$$4x + 5y = 6$$
$$4x + 5y - \mathbf{4x} = -\mathbf{4x} + 6$$
$$5y = -4x + 6$$
$$\frac{5y}{\mathbf{5}} = \frac{-4x}{\mathbf{5}} + \frac{6}{\mathbf{5}}$$
$$y = -\frac{4}{5}x + \frac{6}{5}$$
$$m = -\frac{4}{5}$$

22. Write the equation of the line whose graph has slope $m = -2$ and y-intercept (0, 1). [Section 3.5]

$$y = mx + b$$
$$y = -2x + 1$$

23. Are the graphs of $y = 4x + 9$ and $x + 4y = -10$ parallel, perpendicular, or neither? [Section 3.5]

$y = 4x + 9$	$x + 4y = -10$
$m = 4$	$x + 4y - \mathbf{x} = -10 - \mathbf{x}$
	$4y = -x - 10$
	$\frac{4y}{\mathbf{4}} = \frac{-x}{\mathbf{4}} - \frac{10}{\mathbf{4}}$
	$y = -\frac{1}{4}x - \frac{10}{4}$
	$y = -\frac{1}{4}x - \frac{5}{2}$
	$m = -\frac{1}{4}$

The two slopes are negative reciprocal of each other. The graphs are perpendicular.

24. Write the equation of the line whose graph has slope $m = \frac{1}{4}$ and passes through the point (8, 1). Answer in intercept form. [Section 3.6]
Passes through (8, 1)

$$m = \frac{1}{4},\ x_1 = 8 \text{ and } y_1 = 1$$
$$y - y_1 = m(x - x_1)$$
$$y - 1 = \frac{1}{4}(x - 8)$$
$$y - 1 = \frac{\mathbf{1}}{\mathbf{4}}(x) - \frac{\mathbf{1}}{\mathbf{4}}(8)$$
$$y - 1 = \frac{1}{4}x - 2$$
$$y - 1 + \mathbf{1} = \frac{1}{4}x - 2 + \mathbf{1}$$
$$y = \frac{1}{4}x - 1 \text{ slope-intercept form}$$

25. Graph the line passing through (–2, –1) and having slope $\frac{4}{3}$. [Section 3.6]

Plot $(-2,-1)$ first.

Step 1: Use $\frac{\text{rise}}{\text{run}} = \frac{4}{3}$ to locate (1,3).

Start at $(-2,-1)$ and then count up the y-axis 4 spaces, then count 3 spaces to the right on the x-axis and draw a dot.

Step 2: Use $\frac{-\text{rise}}{-\text{run}} = \frac{-4}{-3}$ to locate $(-5,-5)$.

Start at $(-2,-1)$ and then count down the y-axis 4 spaces, then count 3 spaces to the left on the x-axis and draw a dot.

Step 3: Connect the dots for the graph.

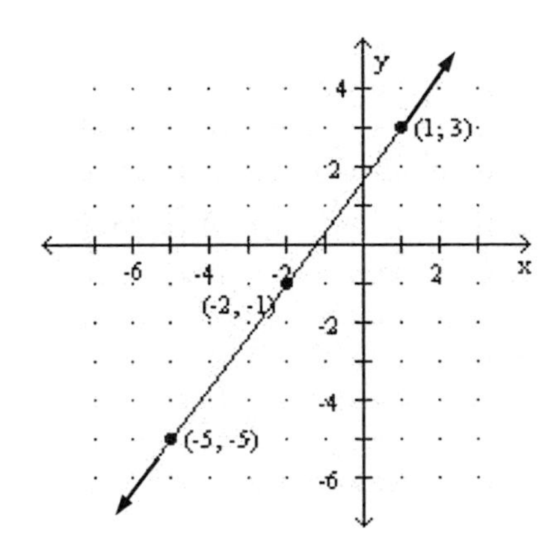

26. If $f(x) = 3x^2 + 3x - 8$, find $f(-1)$. [Section 3.8]

$$f(x) = 3x^2 + 3x - 8$$
$$f(\mathbf{-1}) = 3(\mathbf{-1})^2 + 3(\mathbf{-1}) - 8$$
$$= 3(1) - 3 - 8$$
$$= -8$$
$$f(-1) = -8$$

27. Find the domain and range of the relation {(1, 8), (4, –3), (–4, 2), (5, 8)}. [Section 3.8]
domain: {–4, 1, 4, 5}; range: {–3, 2, 8}

28. Is this the graph of a function? [Section 3.8]
Yes, because a vertical line will cut the graph at only one point.

29. Solve using the graphing method. [Section 4.1]

$$\begin{cases} x + y = 1 \\ y = x + 5 \end{cases}$$

$x + y = 1$	$x + y = 1$
y – intercept: let $x = 0$	x – intercept: let $y = 0$
$x + y = 1$	$x + y = 1$
$0 + y = 1$	$x + 0 = 1$
$y = 1$	$x = 1$
y-intercept $= (0,1)$	x-intercept $= (1,0)$

$y = x + 5$	$y = x + 5$
y – intercept: let $x = 0$	x – intercept: let $y = 0$
$y = x + 5$	$y = x + 5$
$y = 0 + 5$	$0 = x + 5$
$y = 5$	$-5 = x$
y-intercept $= (0,5)$	x-intercept $= (-5,0)$

The solution is $(-2, 3)$.

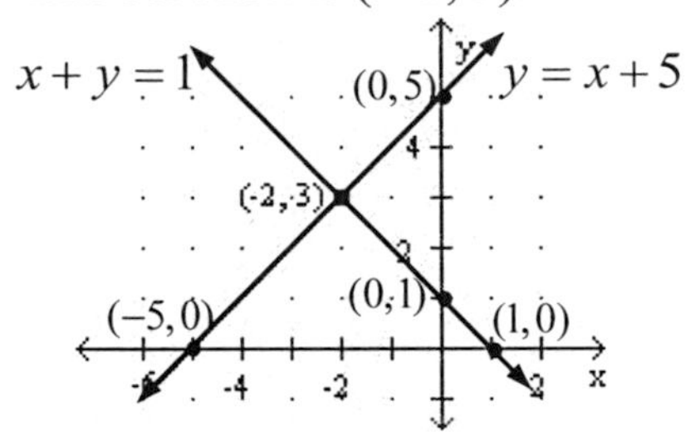

30. Solve using the substitution method. [Section 4.2]

$$\begin{cases} y = 2x + 5 \\ x + 2y = -5 \end{cases}$$

$x + 2y = -5$ The 2nd equation Substitute for y.

$$x + 2(\mathbf{2x + 5}) = -5$$
$$x + 4x + 10 = -5$$
$$5x + 10 = -5$$
$$5x + 10 - \mathbf{10} = -5 - \mathbf{10}$$
$$5x = -15$$
$$\frac{5x}{\mathbf{5}} = \frac{-15}{\mathbf{5}}$$
$$x = -3$$

$y = 2x + 5$ The 1st equation

$$y = 2(\mathbf{-3}) + 5$$
$$y = -6 + 5$$
$$y = -1$$

The solution is $(-3, -1)$.

31. Solve using the elimination (addition) method. [Section 4.3]

$$\begin{cases} \frac{3}{5}s + \frac{4}{5}t = 1 \\ -\frac{1}{4}s + \frac{3}{8}t = 1 \end{cases}$$

Clear both equations of fractions.

$$\begin{cases} \mathbf{5}\left(\frac{3}{5}s\right) + \mathbf{5}\left(\frac{4}{5}t\right) = \mathbf{5}(1) \\ \mathbf{8}\left(-\frac{1}{4}s\right) + \mathbf{8}\left(\frac{3}{8}t\right) = \mathbf{8}(1) \end{cases}$$

$$\begin{cases} 3s + 4t = 5 \\ -2s + 3t = 8 \end{cases}$$

Eliminate s.

Multiply both sides of the 1st equation by 2.

Multiply both sides of the 2nd equation by 3.

$$6s + 8t = 10$$
$$-6s + 9t = 24$$
$$17t = 34$$
$$\frac{17t}{\mathbf{17}} = \frac{34}{\mathbf{17}}$$
$$t = 2$$

$3s + 4t = 5$ The 1st equation

$$3s + 4(\mathbf{2}) = 5$$
$$3s + 8 = 5$$
$$3s + 8 - \mathbf{8} = 5 - \mathbf{8}$$
$$3s = -3$$
$$\frac{3s}{\mathbf{3}} = \frac{-3}{\mathbf{3}}$$
$$s = -1$$

The solution is $(-1, 2)$.

32. AVIATION [Section 4.4]

Analyze

- Flying with the wind plane flew 3,000 mi in 5 hr.
- Against same wind, the trip took 6 hrs.
- Find the speed of the plane in still air.

Assign

Let $x =$ speed of plane in still air in mph

$y =$ speed of air current in mph

Form

	Rate •	Time =	Distance
With wind	$x + y$	5	$\mathbf{5(x + y)}$

Total Distance = 3,000

The rate of the plane with the wind	times	5 hours traveled	is	3,000 miles.
$(x + y)$	$\cdot$	5	=	3,000

	Rate •	Time =	Distance
Against wind	$x - y$	6	$\mathbf{6(x - y)}$

Total Distance = 3,000

The rate of the plane against the wind	times	6 hours traveled	is	3,000 miles.
$(x - y)$	$\cdot$	6	=	3,000

Solve

$$\begin{cases} 5(x + y) = 3,000 \\ 6(x - y) = 3,000 \end{cases}$$

Distribute in both equations.

$$\begin{cases} 5x + 5y = 3,000 \\ 6x - 6y = 3,000 \end{cases}$$

Eliminate y.

Multiply both sides of the 1st equation by 6.

Multiply both sides of the 2nd equation by 5.

$$30x + 30y = 18,000$$
$$30x - 30y = 15,000$$
$$60x = 33,000$$
$$\frac{60x}{\mathbf{60}} = \frac{33,000}{\mathbf{60}}$$
$$x = 550$$

State

The speed of the plane in still air is 550 mph.

Check

The result checks.

33. MIXING CANDY [Section 4.4]

Analyze

- Merchant mixes hard candy with soft candy to obtain a 48 pound mixture.
- Mixture sells at $4.50 per pound.
- Hard candy sells for $4.00 a pound.
- Soft candy sells for $6.00 a pound.
- How many lbs of each should be used?

Assign

Let x = amount of $4.00/lb hard candy in lbs
y = amount of $6.00/lb soft candy in lbs

Form

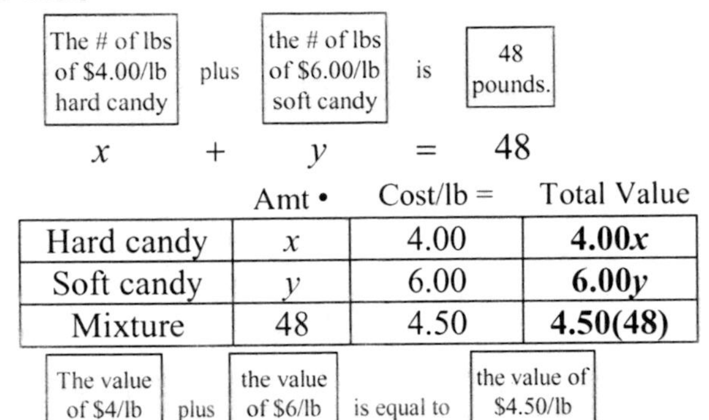

The # of lbs of $4.00/lb hard candy	plus	the # of lbs of $6.00/lb soft candy	is	48 pounds.
x	$+$	y	$=$	48

	Amt •	Cost/lb =	Total Value
Hard candy	x	4.00	**4.00*x***
Soft candy	y	6.00	**6.00*y***
Mixture	48	4.50	**4.50(48)**

The value of $4/lb hard candy	plus	the value of $6/lb soft candy	is equal to	the value of $4.50/lb mixture.
$4x$	$+$	$6y$	$=$	4.50(48)

Solve

$$\begin{cases} x+y=48 \\ 4x+6y=216 \end{cases}$$

Eliminate y.
Multiply both sides of the 1st equation by -6.

$$\begin{aligned} -6x-6y &= -288 \\ 4x+6y &= \ \ 216 \\ \hline -2x \qquad &= -72 \end{aligned}$$

$$\frac{-2x}{\mathbf{-2}}=\frac{-72}{\mathbf{-2}}$$

$$x=36$$

The amount of soft candy.

$$x+y=48 \quad \text{The 1st equation}$$

$$\mathbf{36}+y=48$$

$$36+y-\mathbf{36}=48-\mathbf{36}$$

$$y=12$$

State

36 lb of $4.00/lb hard candy will be needed.
12 lb of $6.00/lb soft candy will be needed.

Check

The results check.

34. Solve the system of linear inequalities. [Section 4.5]

$$\begin{cases} 3x+4y>-7 \\ 2x-3y\ge 1 \end{cases}$$

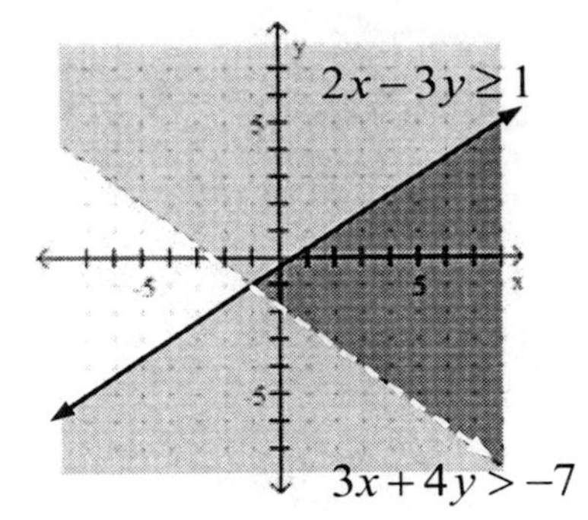

Simplify each expression. Write each answer without using parentheses or negative exponents.

35. [Section 5.1]

$$\begin{aligned} y^3(y^2y^4) &= y^3(y^{2+4}) \\ &= y^3(y^6) \\ &= y^{3+6} \\ &= y^9 \end{aligned}$$

36. [Section 5.1]

$$\begin{aligned} \left(\frac{b^2}{3a}\right)^3 &= \frac{b^{2\cdot 3}}{3^3a^3} \\ &= \frac{b^6}{27a^3} \end{aligned}$$

37. [Section 5.2]

$$\begin{aligned} \frac{10a^4a^{-2}}{5a^2a^0} &= \frac{2\cdot \overset{1}{\cancel{5}}\, a^{4+(-2)}}{\underset{1}{\cancel{5}}\, a^{2+0}} \\ &= \frac{2a^2}{a^2} \\ &= \frac{2a^{2-2}}{1} \\ &= 2a^0 \\ &= 2\cdot 1 \\ &= 2 \end{aligned}$$

38. [Section 5.2]

$$\left(\frac{21x^{-2}y^2z^{-2}}{7x^3y^{-1}}\right)^{-2} = \left(\frac{3\cdot \overset{1}{\cancel{7}}\, x^{-2-3}y^{2-(-1)}z^{-2}}{\underset{1}{\cancel{7}}}\right)^{-2}$$

$$= \left(\frac{3x^{-5}y^{2+1}z^{-2}}{1}\right)^{-2}$$

$$= (3x^{-5}y^3z^{-2})^{-2}$$

$$= (3^{-2}x^{-5\cdot -2}y^{3\cdot -2}z^{-2\cdot -2})$$

$$= 3^{-2}x^{10}y^{-6}z^4$$

$$= \frac{x^{10}z^4}{3^2y^6}$$

$$= \frac{x^{10}z^4}{9y^6}$$

39. FIVE-CARD POKER [Section 5.3]

Express 2.6×10^6 using standard notation.

The exponent is 6.

Move the decimal 6 places to the right.

$2.6 \times 10^6 = 2{,}600{,}000$

40. Write 0.00073 in scientific notation. [Section 5.3]

Move the decimal 4 places to the right.

$0.00073 = 7.3 \times 10^{-4}$

41. Graph: $y = x^3 - 2$ [Section 5.4]

x	y	(x, y)
-3	$y=(-3)^3-2$ $=-27+(-2)$ $=-29$	$(-3, -29)$
-2	$y=(-2)^3-2$ $=-8+(-2)$ $=-10$	$(-2, -10)$
-1	$y=(-1)^3-2$ $=-1+(-2)$ $=-3$	$(-1, -3)$
0	$y=(0)^3-2$ $=0+(-2)$ $=-2$	$(0, -2)$
1	$y=(1)^3-2$ $=1+(-2)$ $=-1$	$(1, -1)$
2	$y=(2)^3-2$ $=8+(-2)$ $=6$	$(2, 6)$
3	$y=(3)^3-2$ $=27+(-2)$ $=25$	$(3, 25)$

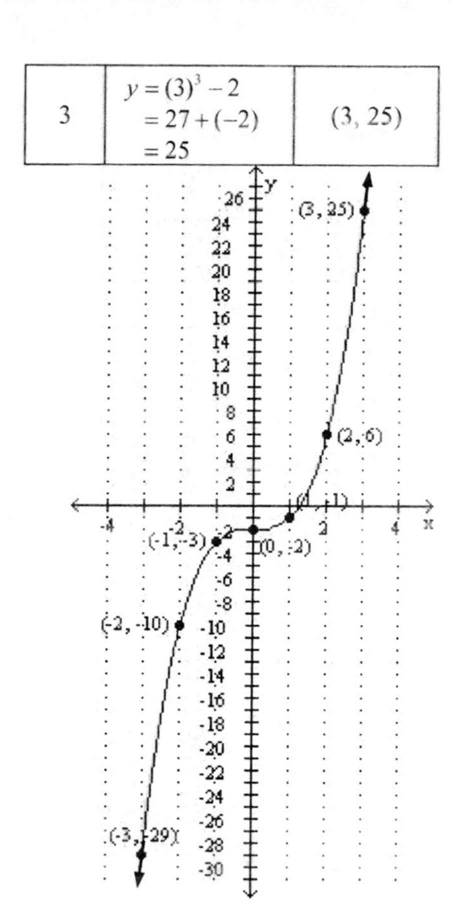

42. Write a polynomial that represents the perimeter of the rectangle. [Section 5.5]

$$P = 2l + 2w$$
$$= 2(x^3 + 3x) + 2(2x^3 - x)$$
$$= \mathbf{2}(x^3) + \mathbf{2}(3x) + \mathbf{2}(2x^3) - \mathbf{2}(x)$$
$$= 2x^3 + 6x + 4x^3 - 2x$$
$$= (2+4)x^3 + (6-2)x$$
$$= 6x^3 + 4x$$

The perimeter is $(6x^3 + 4x)$ units.

Perform the operations.

43. [Section 5.5]

$$4(4x^3 + 2x^2 - 3x - 8) - 5(2x^3 - 3x + 8)$$

$$= \mathbf{4}(4x^3) + \mathbf{4}(2x^2) - \mathbf{4}(3x) - \mathbf{4}(8) - \mathbf{5}(2x^3) - \mathbf{5}(-3x) - \mathbf{5}(8)$$

$$= 16x^3 + 8x^2 - 12x - 32 - 10x^3 + 15x - 40$$

$$= (16-10)x^3 + 8x^2 + (-12+15)x + (-32-40)$$

$$= 6x^3 + 8x^2 + 3x - 72$$

44. [Section 5.6]

$$(-2a^3)(3a^2) = (-2\cdot 3)(a^{3+2})$$
$$= -6a^5$$

45. [Section 5.6]

$(2b-1)(3b+4)$

$$= \mathbf{2b}(3b) + \mathbf{2b}(4) - \mathbf{1}(3b) - \mathbf{1}(4)$$
$$= 6b^2 + 8b - 3b - 4$$
$$= 6b^2 + (8-3)b - 4$$
$$= 6b^2 + 5b - 4$$

46. [Section 5.6]

$(3x+y)(2x^2 - 3xy + y^2)$

$$= \mathbf{3x}(2x^2) + \mathbf{3x}(-3xy) + \mathbf{3x}(y^2) + \mathbf{y}(2x^2) + \mathbf{y}(-3xy) + \mathbf{y}(y^2)$$
$$= 6x^3 - 9x^2y + 3xy^2 + 2x^2y + -3xy^2 + y^3$$
$$= 6x^3 + (-9+2)x^2y + (3-3)xy^2 + y^3$$
$$= 6x^3 - 7x^2y + 0xy^2 + y^3$$
$$= 6x^3 - 7x^2y + y^3$$

47. [Section 5.7]

$$(2x+5y)^2 = (2x)^2 + 2(2x \cdot 5y) + (5y)^2$$
$$= 4x^2 + 20xy + 25y^2$$

48. [Section 5.7]

$$(9m^2 - 1)(9m^2 + 1) = (9m^2)^2 - (1)^2$$
$$= 81m^4 - 1$$

49. [Section 5.8]

$$\frac{12a^3b - 9a^2b^2 + 3ab}{6a^2b}$$
$$= \frac{2 \cdot \overset{1}{\cancel{6}}\, a^3b}{\underset{1}{\cancel{6}}\, a^2b} - \frac{3 \cdot \overset{1}{\cancel{3}}\, a^2b^2}{2 \cdot \underset{1}{\cancel{3}}\, a^2b} + \frac{\overset{1}{\cancel{3}}\, ab}{2 \cdot \underset{1}{\cancel{3}}\, a^2b}$$
$$= \frac{2a^3b}{a^2b} - \frac{3a^2b^2}{2a^2b} + \frac{ab}{2a^2b}$$
$$= 2a^{3-2}b^{1-1} - \frac{3}{2}a^{2-2}b^{2-1} + \frac{1}{2}a^{1-2}b^{1-1}$$
$$= 2a^1b^0 - \frac{3}{2}a^0b^1 + \frac{1}{2}a^{-1}b^0$$
$$= 2a - \frac{3}{2}b + \frac{1}{2a}$$

50. $x-3\overline{)2x^2 - 3 - 5x}$ [Section 5.8]

$$\begin{array}{r} 2x+1 \\ x-3\overline{)2x^2 - 5x - 3} \\ \underline{-(2x^2 - 6x)} \\ x - 3 \\ \underline{-(x-3)} \\ 0 \end{array}$$

Factor each expression completely.

51. [Section 6.1]

$6a^2 - 12a^3b + 36ab$

$$= \mathbf{6a} \cdot a - \mathbf{6a} \cdot 2a^2b + \mathbf{6a} \cdot 6b$$
$$= 6a(a - 2a^2b + 6b)$$

52. [Section 6.1]

$2x + 2y + ax + ay$

$$= (2x + 2y) + (ax + ay)$$
$$= (\mathbf{2} \cdot x + \mathbf{2} \cdot y) + (\mathbf{a} \cdot x + \mathbf{a} \cdot y)$$
$$= \mathbf{2}(x+y) + \mathbf{a}(x+y)$$
$$= (x+y)(2+a)$$

53. [Section 6.2]

$x^2 - 6x - 16$

Two different sign factors of -16 whose sum is -6 are -8 and $+2$.

$x^2 - 6x - 16 = (x-8)(x+2)$

54. [Section 6.3]

$30y^5 + 63y^4 - 30y^3$

$3y^3(\underset{a}{10}y^2 + \underset{b}{21}y - \underset{c}{10})$

In $10y^2 + 21y - 10$, we have $a = 10$, $b = 21$ and $c = -10$. The key number is $ac = 10(-10) = -100$. We must find a factorization of -100 in which the sum of the factors is $b = 21$. Since the factors must have a negative product, their signs must be different. The pairs of factors are -4 and $+25$.

$3y^3(10y^2 + 21y - 10)$

$$= 3y^3[10y^2 - 4y + 25y - 10]$$
$$= 3y^3[(10y^2 - 4y) + (25y - 10)]$$
$$= 3y^3[2y(5y-2) + 5(5y-2)]$$
$$= 3y^3(5y-2)(2y+5)$$

55. [Section 6.4]

$$t^4 - 16 = (t^2+4)(t^2-4)$$
$$= (t^2+4)(t+2)(t-2)$$

56. [Section 6.5]

$$b^3 + 125 = (b)^3 + (5)^3$$
$$= (b+5)[b^2 - b \cdot 5 + (5)^2]$$
$$= (b+5)(b^2 - 5b + 25)$$

Solve each equation by factoring.

57. [Section 6.7]

$$3x^2 + 8x = 0$$
$$x(3x+8) = 0$$

$x = 0$ or $3x + 8 = 0$
$3x = -8$
$x = -\frac{8}{3}$

The solutions are 0 and $-\frac{8}{3}$.

58. [Section 6.7]

$$15x^2 - 2 = 7x$$
$$15x^2 - 2 - \mathbf{7x} = 7x - \mathbf{7x}$$
$$15x^2 - 7x - 2 = 0$$
$$(3x-2)(5x+1) = 0$$

$3x - 2 = 0$ or $5x + 1 = 0$
$3x = 2$ | $5x = -1$
$x = \frac{2}{3}$ | $x = -\frac{1}{5}$

The solutions are $\frac{2}{3}$ and $-\frac{1}{5}$.

59. GEOMETRY [Section 6.8]

Analyze

- Height of the triangle is x in.
- Base of the triangle is $x + 4$ in.
- Area is 22.5 sq. in.
- Find the height.

Assign

Let x = height of triangle in inches
$x + 4$ = base of triangle in inches

Form

The area of the triangle	equals	$\frac{1}{2}$	times	the base	times	the height.
22.5	=	$\frac{1}{2}$	$\cdot$	$x+4$	$\cdot$	x

Solve

$$\frac{1}{2}x(x+4) = 22.5$$
$$(\mathbf{2})\frac{1}{2}x(x+4) = (\mathbf{2})22.5$$
$$x(x+4) = 45$$
$$x^2 + 4x = 45$$
$$x^2 + 4x - \mathbf{45} = 45 - \mathbf{45}$$
$$x^2 + 4x - 45 = 0$$
$$(x+9)(x-5) = 0$$

$x + 9 = 0$ or $x - 5 = 0$
~~$x = -9$~~ | $x = 5$

State

The height is 5 inches.
The base is $5 + 4 = 9$ inches.

Check

The results check.

60. For what value is $\frac{x}{x+8}$ undefined? [Section 7.1]

$$x + 8 = 0$$
$$x = -8$$

It is undefined for $x = -8$.

Simplify each expression.

61. [Section 7.1]

$$\frac{3x^2 - 27}{x^2 + 3x - 18} = \frac{3(x^2 - 9)}{(x+6)(x-3)}$$
$$= \frac{3(x+3)\overset{1}{\cancel{(x-3)}}}{(x+6)\underset{1}{\cancel{(x-3)}}}$$
$$= \frac{3(x+3)}{x+6}$$

62. [Section 7.1]

$$\frac{a-15}{15-a} = \frac{a-15}{-1(-15+a)}$$
$$= \frac{\overset{1}{\cancel{a-15}}}{-1\underset{1}{\cancel{(a-15)}}}$$
$$= -1$$

Perform the operations and simplify when possible.

63. [Section 7.2]

$$\frac{x^2-x-6}{2x^2+9x+10} \div \frac{x^2-25}{2x^2+15x+25}$$

$$= \frac{x^2-x-6}{2x^2+9x+10} \cdot \frac{2x^2+15x+25}{x^2-25}$$

$$= \frac{(x^2-x-6)(2x^2+15x+25)}{(2x^2+9x+10)(x^2-25)}$$

$$= \frac{(x-3)(x+2)(2x+5)(x+5)}{(2x+5)(x+2)(x+5)(x-5)}$$

$$= \frac{(x-3)\,\overset{1}{\cancel{(x+2)}}\,\overset{1}{\cancel{(2x+5)}}\,\overset{1}{\cancel{(x+5)}}}{\underset{1}{\cancel{(2x+5)}}\,\underset{1}{\cancel{(x+2)}}\,\underset{1}{\cancel{(x+5)}}(x-5)}$$

$$= \frac{x-3}{x-5}$$

64. [Section 7.3]

$$\frac{1}{s^2-4s-5} + \frac{s}{s^2-4s-5} = \frac{s+1}{s^2-4s-5}$$

$$= \frac{s+1}{(s+1)(s-5)}$$

$$= \frac{\overset{1}{\cancel{s+1}}}{\underset{1}{\cancel{(s+1)}}(s-5)}$$

$$= \frac{1}{s-5}$$

65. [Section 7.4]

$$\frac{x+5}{xy} - \frac{x-1}{x^2y}$$

$$\left.\begin{aligned} xy &= x \cdot y \\ x^2y &= x \cdot x \cdot y \end{aligned}\right\} \text{LCD} = x \cdot x \cdot y = x^2y$$

$$\frac{x+5}{xy} - \frac{x-1}{x^2y} = \frac{(x+5)}{xy} \cdot \frac{\mathbf{x}}{\mathbf{x}} - \frac{x-1}{x^2y}$$

$$= \frac{x(x+5)}{x^2y} - \frac{x-1}{x^2y}$$

$$= \frac{x(x+5)-(x-1)}{x^2y}$$

$$= \frac{x^2+5x-x+1}{x^2y}$$

$$= \frac{x^2+4x+1}{x^2y}$$

66. [Section 7.4]

$$\frac{x}{x-2} + \frac{3x}{x^2-4}$$

$$\left.\begin{aligned} x-2 &= x-2 \\ x^2-4 &= (x+2)(x-2) \end{aligned}\right\} \text{LCD} = (x+2)(x-2)$$

$$\frac{x}{x-2} + \frac{3x}{x^2-4}$$

$$= \frac{x}{(x-2)} \cdot \frac{\mathbf{(x+2)}}{\mathbf{(x+2)}} + \frac{3x}{(x+2)(x-2)}$$

$$= \frac{x(x+2)+3x}{(x+2)(x-2)}$$

$$= \frac{x^2+2x+3x}{(x+2)(x-2)}$$

$$= \frac{x^2+5x}{(x+2)(x-2)} \text{ or } \frac{x^2+5x}{x^2-4}$$

Simplify each complex fraction.

67. [Section 7.5]

$$\frac{\frac{9m-27}{m^6}}{\frac{2m-6}{m^8}} = \frac{9m-27}{m^6} \div \frac{2m-6}{m^8}$$

$$= \frac{9m-27}{m^6} \cdot \frac{m^8}{2m-6}$$

$$= \frac{m^8(9m-27)}{m^6(2m-6)}$$

$$= \frac{m^8 \cdot 9(m-3)}{m^6 \cdot 2(m-3)}$$

$$= \frac{9m^2 \cdot \overset{1}{\cancel{m^6}}\,\overset{1}{\cancel{(m-3)}}}{2\,\underset{1}{\cancel{m^6}}\,\underset{1}{\cancel{(m-3)}}}$$

$$= \frac{9m^2}{2}$$

68. [Section 7.5]

$$\frac{\frac{5}{y}+\frac{4}{y+1}}{\frac{4}{y}-\frac{5}{y+1}}$$

$\text{LCD} = y(y+1)$

$$\frac{\frac{5}{y}+\frac{4}{y+1}}{\frac{4}{y}-\frac{5}{y+1}} = \frac{\frac{5}{y}+\frac{4}{y+1}}{\frac{4}{y}-\frac{5}{y+1}} \cdot \frac{\mathbf{y(y+1)}}{\mathbf{y(y+1)}}$$

$$= \frac{\left(\frac{5}{y}+\frac{4}{y+1}\right)y(y+1)}{\left(\frac{4}{y}-\frac{5}{y+1}\right)y(y+1)}$$

$$= \frac{\frac{5\cancel{y}(y+1)}{\cancel{y}}+\frac{4y\cancel{(y+1)}}{\cancel{y+1}}}{\frac{4\cancel{y}(y+1)}{\cancel{y}}-\frac{5y\cancel{(y+1)}}{\cancel{y+1}}}$$

$$= \frac{5(y+1)+4y}{4(y+1)-5y}$$

$$= \frac{5y+5+4y}{4y+4-5y}$$

$$= \frac{9y+5}{4-y}$$

Solve each equation.

69. [Section 7.6]

$$\frac{2p}{3}-\frac{1}{p}=\frac{2p-1}{3}$$

$\text{LCD} = 3p \;, \quad p \neq 0$

$$\frac{2p}{3}-\frac{1}{p}=\frac{2p-1}{3}$$

$$\mathbf{3p}\left(\frac{2p}{3}-\frac{1}{p}\right)=\mathbf{3p}\left(\frac{2p-1}{3}\right)$$

$$3p\left(\frac{2p}{3}\right)-3p\left(\frac{1}{p}\right)=3p\left(\frac{2p-1}{3}\right)$$

$$\cancel{3}p\left(\frac{2p}{\cancel{3}}\right)-3\cancel{p}\left(\frac{1}{\cancel{p}}\right)=\cancel{3}p\left(\frac{2p-1}{\cancel{3}}\right)$$

$$2p^2-3=2p^2-p$$

$$2p^2-3-\mathbf{2p^2}=2p^2-p-\mathbf{2p^2}$$

$$-3=-p$$

$$\frac{-3}{\mathbf{-1}}=\frac{-p}{\mathbf{-1}}$$

$$3=p$$

The solution is 3.

70. [Section 7.6]

$$\frac{7}{q^2-q-2}+\frac{1}{q+1}=\frac{3}{q-2}$$

$$\left.\begin{array}{c} q-2=q-2 \\ q+1=q+1 \\ q^2-q-2=(q-2)(q+1) \end{array}\right\}$$

$\text{LCD} = (q-2)(q+1) \quad , \; q \neq -1, 2$

$$\frac{7}{q^2-q-2}+\frac{1}{q+1}=\frac{3}{q-2}$$

$$\frac{7}{(q-2)(q+1)}\cdot\mathbf{(q-2)(q+1)}+\frac{1}{q+1}\cdot\mathbf{(q-2)(q+1)}=\frac{3}{q-2}\cdot\mathbf{(q-2)(q+1)}$$

$$\frac{7\cancel{(q-2)}\cancel{(q+1)}}{\cancel{(q-2)}\cancel{(q+1)}}+\frac{1(q-2)\cancel{(q+1)}}{\cancel{q+1}}=\frac{3\cancel{(q-2)}(q+1)}{\cancel{q-2}}$$

$$7+q-2=3q+3$$

$$q+5-\mathbf{3}=3q+3-\mathbf{3}$$

$$q+2-\mathbf{q}=3q-\mathbf{q}$$

$$2=2q$$

$$\frac{2}{\mathbf{2}}=\frac{2q}{\mathbf{2}}$$

$$1=q$$

The solution is 1.

71. **[Section 7.6]**

Solve the formula $\frac{1}{a}+\frac{1}{b}=1$ for a.

$\frac{1}{a}+\frac{1}{b}=1$; LCD $= ab$

$$\frac{1}{a}+\frac{1}{b}=1$$

$$(ab)\left(\frac{1}{a}+\frac{1}{b}\right)=(ab)(1)$$

$$\cancel{(a)}b\left(\frac{1}{\cancel{a}}\right)+a\cancel{(b)}\left(\frac{1}{\cancel{b}}\right)=(ab)(1)$$

$$b+a=ab$$

$$b+a-\mathbf{a}=ab-\mathbf{a}$$

$$b=ab-a$$

$$b=a(b-1)$$

$$\frac{b}{\mathbf{b-1}}=\frac{a\cancel{(b-1)}}{\cancel{b-1}}$$

$$\frac{b}{b-1}=a$$

$$a=\frac{b}{b-1}$$

72. ROOFING [Section 7.7]

Analyze

- Homeowner takes 7 days to roof house.
- Professional takes 4 days to roof house.
- How long will it take both of them, working together, to roof the house?

Assign

Let x = the number of days it will take both working together to roof the house.

Form

	Rate	• Time	= Work Completed
Homeowner	$\frac{1}{7}$	x	$\frac{x}{7}$
Professional	$\frac{1}{4}$	x	$\frac{x}{4}$

The part of job done by homeowner.	plus	The part of job done by professional.	equals	1 job completed
$\frac{x}{7}$	$+$	$\frac{x}{4}$	$=$	1

Solve

$$\frac{x}{7}+\frac{x}{4}=1$$

$$\text{LCD}=28$$

$$28\left(\frac{x}{7}+\frac{x}{4}\right)=28(1)$$

$$\cancel{7}\cdot 4\left(\frac{x}{\cancel{7}}\right)+\cancel{4}\cdot 7\left(\frac{x}{\cancel{4}}\right)=28$$

$$4x+7x=28$$

$$11x=28$$

$$\frac{11x}{\mathbf{11}}=\frac{28}{\mathbf{11}}$$

$$x=\frac{28}{11}$$

$$x=2\frac{6}{11}$$

State

It takes both working together $2\frac{6}{11}$ days to roof the house.

Check

The result checks.

73. LOSING WEIGHT [Section 7.8]

Analyze

· We know it takes 350 days to lose 10 pounds.

· How long will it take to lose 25 pounds?

Assign

Let d = # of days to lose 25 pounds

10 lbs is to 350 days as 25 lbs is to d # of days.

Form

$$\begin{matrix}\text{10 lbs}\rightarrow \\ \text{350 days}\rightarrow\end{matrix}\frac{10}{350}=\frac{25}{d}\begin{matrix}\leftarrow \text{25 lbs} \\ \leftarrow d \text{ days}\end{matrix}$$

Solve

$$\frac{10}{350}=\frac{25}{d}$$
$$10d=8,750$$
$$\frac{10d}{\mathbf{10}}=\frac{8,750}{\mathbf{10}}$$
$$d=875$$

State

It will take 875 days to lose 25 pounds.

Check

The result checks.

74. SIMILAR TRIANGLES [Section 7.8]

Analyze

Similar triangles are given with the smaller one with sides of 6 and 4. The larger one has sides of x and 26 respectively.

What is the length of the x side?

Assign

Let x = the length of the missing side

6 is to 4 as x is to 26.

Form

$$\begin{matrix}\overline{AB}\text{ of small triangle}\rightarrow \\ \overline{BC}\text{ of small triangle}\rightarrow\end{matrix}\frac{6}{4}=\frac{x}{26}\begin{matrix}\leftarrow \overline{DE}\text{ of large triangle} \\ \leftarrow \overline{EF}\text{ of large triangle}\end{matrix}$$

Solve

$$\frac{6}{4}=\frac{x}{26}$$
$$156=4x$$
$$\frac{156}{\mathbf{4}}=\frac{4x}{\mathbf{4}}$$
$$39=x$$

State

A length of the x side is 39 units.

Check

The result checks.

75. Suppose w varies directly as x. If $w = 1.2$ when $x = 4$, find w when $x = 30$. [Section 7.9]

$w = kx$	Find w when $x = 30$.
Find k:	$w = 0.3x$
$1.2 = k(4)$	$w = 0.3(30)$
$\frac{1.2}{\mathbf{4}}=\frac{4k}{\mathbf{4}}$	$w = 9$
$0.3 = k$	

Thus, when $x = 30$, the value of w is 9.

76. GEARS [Section 7.9]

The speed of a gear varies inversely with the number of teeth. If a gear with 10 teeth makes 3 revolutions per second, how many revolutions per second will a gear with 25 teeth make?

$r=\frac{k}{t}$	Find r when $t = 25$.
Find k:	$r=\frac{k}{t}$
$3=\frac{k}{10}$	$r=\frac{30}{25}$
$10\cdot 3 = k$	$r = 1.2$
$30 = k$	

The number of revolutions is 1.2 rpm.

Simplify each radical expression. All variables represent positive numbers.

77. [Section 8.1]

$$\sqrt{100x^2}=\sqrt{100}\sqrt{x^2}$$
$$=10x$$

78. [Section 8.2]

$$-\sqrt{18b^3}=-\sqrt{9}\sqrt{2}\sqrt{b^2}\sqrt{b}$$
$$=-3b\sqrt{2b}$$

Perform the indicated operation.

79. [Section 8.3]

$$3\sqrt{24}+\sqrt{54}=3\sqrt{4\cdot 6}+\sqrt{9\cdot 6}$$
$$=3\cdot 2\sqrt{6}+3\sqrt{6}$$
$$=6\sqrt{6}+3\sqrt{6}$$
$$=(6+3)\sqrt{6}$$
$$=9\sqrt{6}$$

80. [Section 8.4]

$$\left(\sqrt{2}+1\right)\left(\sqrt{2}-3\right)=\mathbf{\sqrt{2}}\cdot\sqrt{2}-\mathbf{\sqrt{2}}\cdot 3+\mathbf{1}\sqrt{2}-\mathbf{1}\cdot 3$$
$$=2-3\sqrt{2}+\sqrt{2}-3$$
$$=(2-3)+(-3+1)\sqrt{2}$$
$$=-1-2\sqrt{2}$$

Rationalize the denominator. [81 & 82 Section 8.4]

81. $$\frac{8}{\sqrt{10}}=\frac{8}{\sqrt{10}}\cdot\frac{\mathbf{\sqrt{10}}}{\mathbf{\sqrt{10}}}$$
$$=\frac{8\sqrt{10}}{10}$$
$$=\frac{\overset{1}{\not{2}}\cdot 4\sqrt{10}}{\underset{1}{\not{2}}\cdot 5}$$
$$=\frac{4\sqrt{10}}{5}$$

82. $$\frac{\sqrt{2}}{3-\sqrt{a}}=\frac{\sqrt{2}}{(3-\sqrt{a})}\cdot\frac{\mathbf{(3+\sqrt{a})}}{\mathbf{(3+\sqrt{a})}}$$
$$=\frac{\sqrt{2}\left(3+\sqrt{a}\right)}{\left(3-\sqrt{a}\right)\left(3+\sqrt{a}\right)}$$
$$=\frac{3\sqrt{2}+\sqrt{2a}}{3^2-\left(\sqrt{a}\right)^2}$$
$$=\frac{3\sqrt{2}+\sqrt{2a}}{9-a}$$

Solve each equation. [83 & 84 Section 8.5]

83. $$\sqrt{6x+1}+2=7$$
$$\sqrt{6x+1}+2-\mathbf{2}=7-\mathbf{2}$$
$$\sqrt{6x+1}=5$$
$$\left(\sqrt{6x+1}\right)^2=(5)^2$$
$$6x+1=25$$
$$6x+1-\mathbf{1}=25-\mathbf{1}$$
$$6x=24$$
$$\frac{6x}{\mathbf{6}}=\frac{24}{\mathbf{6}}$$
$$x=4$$

The solution is 4.

Check:
$$\sqrt{6x+1}+2=7$$
$$\sqrt{6(4)+1}+2\stackrel{?}{=}7$$
$$\sqrt{24+1}+2\stackrel{?}{=}7$$
$$\sqrt{25}+2\stackrel{?}{=}7$$
$$5+2\stackrel{?}{=}7$$
$$7=7$$
True

84. $$\sqrt{3t+7}=t+3$$
$$\left(\sqrt{3t+7}\right)^2=(t+3)^2$$
$$3t+7=t^2+2\cdot 3t+(3)^2$$
$$3t+7=t^2+6t+9$$
$$t^2+6t+9=3t+7$$
$$t^2+6t+9-\mathbf{3t}=3t+7-\mathbf{3t}$$
$$t^2+3t+9=7$$
$$t^2+3t+9-\mathbf{7}=7-\mathbf{7}$$
$$t^2+3t+2=0$$
$$(t+2)(t+1)=0$$

Set each factor equal to zero and solve.

$t+2=0$	$t+1=0$
$t=-2$	$t=-1$

Check both solutions.

$\sqrt{3t+7}=t+3$	$\sqrt{3t+7}=t+3$
$\sqrt{3(\mathbf{-2})+7}\stackrel{?}{=}\mathbf{-2}+3$	$\sqrt{3(\mathbf{-1})+7}\stackrel{?}{=}\mathbf{-1}+3$
$\sqrt{-6+7}\stackrel{?}{=}1$	$\sqrt{-3+7}\stackrel{?}{=}2$
$\sqrt{1}\stackrel{?}{=}1$	$\sqrt{4}\stackrel{?}{=}2$
$1=1$	$2=2$
True	True

The solutions are -2 and -1.

Simplify each radical expression. All variables represent positive numbers. [Section 8.6]

85. $$\sqrt[3]{\frac{27m^3}{8n^6}}=\frac{\sqrt[3]{27m^3}}{\sqrt[3]{8n^6}}$$
$$=\frac{\sqrt[3]{27}\sqrt[3]{m^3}}{\sqrt[3]{8}\sqrt[3]{n^6}}$$
$$=\frac{3m}{2n^2}$$

86. $\sqrt[4]{16}=2$

Evaluate each expression. [Section 8.6]

87. $$25^{3/2}=\left(\sqrt{25}\right)^3$$
$$=(5)^3$$
$$=125$$

88. $(-8)^{-4/3} = \dfrac{1}{(-8)^{4/3}}$

$$= \frac{1}{\left(\sqrt[3]{-8}\right)^4}$$

$$= \frac{1}{(-2)^4}$$

$$= \frac{1}{16}$$

Solve each equation. [Section 9.1]

89. $t^2 = 75$

$$t = \pm\sqrt{75}$$
$$t = \pm\sqrt{25 \cdot 3}$$
$$t = \pm 5\sqrt{3}$$

The solutions are $\pm 5\sqrt{3}$.

90.
$$(6y+5)^2 - 72 = 0$$
$$(6y+5)^2 - 72 + \mathbf{72} = 0 + \mathbf{72}$$
$$(6y+5)^2 = 72$$
$$6y+5 = \pm\sqrt{72}$$
$$6y+5 = \pm\sqrt{36 \cdot 2}$$
$$6y+5 = \pm\sqrt{36} \cdot \sqrt{2}$$
$$6y+5 = \pm 6\sqrt{2}$$
$$6y+5-\mathbf{5} = \mathbf{-5} \pm 6\sqrt{2}$$
$$6y = -5 \pm 6\sqrt{2}$$
$$\frac{6y}{\mathbf{6}} = \frac{-5 \pm 6\sqrt{2}}{\mathbf{6}}$$
$$y = \frac{-5 \pm 6\sqrt{2}}{6}$$

The solutions are $\dfrac{-5 \pm 6\sqrt{2}}{6}$.

91. STORAGE CUBES [Section 9.1]

- Each side (leg) is x inches.
- The diagonal (hypotenuse) is 15 inches.
- What is the height of the entire storage unit?
- Use $a^2 + b^2 = c^2$ because it is a right triangle.

$$a^2 + b^2 = c^2$$
$$x^2 + x^2 = 15^2$$
$$2x^2 = 225$$
$$\frac{2x^2}{\mathbf{2}} = \frac{225}{\mathbf{2}}$$
$$x^2 = \frac{225}{2}$$
$$x^2 = 112.5$$
$$x = \pm\sqrt{112.5}$$
$$x = \pm 10.60$$
$$x = 10.6$$

The length of one side of the cube is 10.6 in. Now multiply 10.6 by 2 to obtain the height of the entire storage unit.

$10.6(2) = 21.2$

The total height of the storage unit is 21.2 in.

92. Solve $x^2 + 8x + 12 = 0$ by completing the square. [Section 9.2]

$$x^2 + 8x + 12 = 0$$
$$x^2 + 8x + 12 - \mathbf{12} = 0 - \mathbf{12}$$
$$x^2 + 8x = -12$$

To complete the square:

$\frac{1}{2}(8) = 4$

and $(4)^2 = 16$.

Add 16 to both sides of the equation.

$$x^2 + 8x + \mathbf{16} = -12 + \mathbf{16}$$
$$(x+4)^2 = 4$$
$$x + 4 = \pm\sqrt{4}$$
$$x + 4 = \pm 2$$
$$x + 4 - \mathbf{4} = \mathbf{-4} \pm 2$$
$$x = -4 \pm 2$$

$$x = -4 + 2$$
$$x = -2$$

or

$$x = -4 - 2$$
$$x = -6$$

The solutions are -2 and -6.

93. Solve $4x^2 - x - 2 = 0$ using the quadratic formula. Give the exact solutions, and then approximate each to the nearest hundredth. [Section 9.3]

$$4x^2 - x - 2 = 0$$
$$4x^2 - 1x - 2 = 0$$
$$a = 4,\ b = -1,\ c = -2$$
$$x = \frac{-b \pm \sqrt{b^2 - 4ac}}{2a}$$
$$x = \frac{-(-1) \pm \sqrt{(-1)^2 - 4(4)(-2)}}{2(4)}$$
$$x = \frac{1 \pm \sqrt{1+32}}{8}$$
$$x = \frac{1 \pm \sqrt{33}}{8}$$

The exact solutions are $\frac{1 \pm \sqrt{33}}{8}$.

	or	
$x = \frac{1+\sqrt{33}}{8}$		$x = \frac{1-\sqrt{33}}{8}$
$x \approx \frac{1+5.744}{8}$		$x \approx \frac{1-5.744}{8}$
$x \approx \frac{6.744}{8}$		$x \approx \frac{-4.744}{8}$
$x \approx 0.843$		$x \approx -0.593$
$x \approx 0.84$		$x \approx -0.59$

The approximate solutions are $0.84, -0.59$.

94. QUILTS [Section 9.3]

Analyze

- Length of the quilt is 11 ft less than twice its width.
- Area of the quilt is 12,865 ft^2.
- Find its width and length.

Assign

Let x = width of the quilt in ft.
$2x - 11$ = length of the quilt in ft.

Form

The length of the quilt	times	the width of the quilt	equals	the area of the quilt.
$(2x-11)$	$\cdot$	x	$=$	12,865

Solve

$$x(2x-11) = 12{,}865$$
$$2x^2 - 11x = 12{,}865$$
$$2x^2 - 11x - \mathbf{12{,}865} = 12{,}865 - \mathbf{12{,}865}$$
$$2x^2 - 11x - 12{,}865 = 0$$
$$a = 2,\ b = -11,\ c = -12{,}865$$
$$x = \frac{-b \pm \sqrt{b^2 - 4ac}}{2a}$$
$$x = \frac{-(-11) \pm \sqrt{(-11)^2 - 4(2)(-12{,}865)}}{2(2)}$$
$$x = \frac{11 \pm \sqrt{121 + 102{,}920}}{4}$$
$$x = \frac{11 \pm \sqrt{103{,}041}}{4}$$
$$x = \frac{11 \pm 321}{4}$$

	or	
$x = \frac{11+321}{4}$		$x = \frac{11-321}{4}$
$x = \frac{332}{4}$		~~$x = \frac{-310}{4}$~~
$x = 83$		

State

The width of the quilt is 83 ft.
The length of the quilt is $2(83) - 11 = 155$ ft.

Check

$83 \text{ ft} \cdot 155 \text{ ft} = 12{,}865 \text{ ft}^2$
The result checks.

Write each expression in terms of *i*. [Section 9.4]

95. $\sqrt{-49} = \sqrt{-1 \cdot 49}$
$= \sqrt{-1}\sqrt{49}$
$= i \cdot 7$
$= 7i$

96. $\sqrt{-54} = \sqrt{-1 \cdot 54}$
$= i\sqrt{54}$
$= i\sqrt{9 \cdot 6}$
$= i\sqrt{9}\sqrt{6}$
$= 3i\sqrt{6}$

Perform the operations. Express each answer in the form *a + bi*. [97 – 100, Section 9.4]

97. $(2+3i)-(1-2i) = (2+3i)+(-1+2i)$
$= (2-1)+(3+2)i$
$= 1+5i$

98. $(7-4i)+(9+2i) = (7+9)+(-4+2)i$
$= 16-2i$

99. $(3-2i)(4-3i)$
$= \mathbf{3} \cdot 4 + \mathbf{3} \cdot -3i - \mathbf{2i} \cdot 4 - \mathbf{2i} \cdot -3i$
$= 12 - 9i - 8i + 6i^2$
$= 12 + (-9-8)i + 6(-\mathbf{1})$
$= 12 - 17i - 6$
$= (12-6) - 17i$
$= 6 - 17i$

100. $\frac{3-i}{2+i} = \frac{3-i}{2+i} \cdot \frac{\mathbf{2-i}}{\mathbf{2-i}}$
$= \frac{6-3i-2i+i^2}{4-i^2}$
$= \frac{6-5i+(-\mathbf{1})}{4-(-\mathbf{1})}$
$= \frac{5-5i}{4+1}$
$= \frac{5-5i}{5}$
$= \frac{\overset{1}{\cancel{5}}(1-i)}{\underset{1}{\cancel{5}}}$
$= 1-i$

Solve each equation. Express the solutions in the form *a + bi*. [101 & 102, Section 9.4]

101. $x^2+16=0$
$x^2+16-\mathbf{16} = 0-\mathbf{16}$
$x^2 = -16$
$x = \pm\sqrt{-16}$
$x = \pm\sqrt{-1 \cdot 16}$
$x = \pm\sqrt{-1}\sqrt{16}$
$x = \pm 4i$
The solutions are $0 \pm 4i$.

102. $x^2-4x=-5$
$x^2-4x+\mathbf{5} = -5+\mathbf{5}$
$1x^2-4x+5=0$
$a=1,\ b=-4,\ c=5$
$x = \frac{-b \pm \sqrt{b^2-4ac}}{2a}$
$x = \frac{-(-4) \pm \sqrt{(-4)^2-4(1)(5)}}{2(1)}$
$x = \frac{4 \pm \sqrt{16-20}}{2}$
$x = \frac{4 \pm \sqrt{-4}}{2}$
$x = \frac{4 \pm \sqrt{-1 \cdot 4}}{2}$
$x = \frac{4 \pm 2i}{2}$
$x = \frac{\overset{1}{\cancel{2}}(2 \pm i)}{\underset{1}{\cancel{2}}}$
$x = 2 \pm i$
The solutions are $2 \pm i$.

103. Graph the quadratic equation $y = 2x^2 + 8x + 6$. Find the vertex, the x- and y-intercepts, and the axis of symmetry of the graph. [Section 9.5]

$$y = 2x^2 + 8x + 6$$
$$a = 2,\ b = 8$$

Find the vertex.

$$x = \frac{-b}{2a} = \frac{-(8)}{2(2)} = \frac{-8}{4} = -2$$

$x = -2$

$$\begin{aligned} y &= 2x^2 + 8x + 6 \\ &= 2(\mathbf{-2})^2 + 8(\mathbf{-2}) + 6 \\ &= 8 - 16 + 6 \\ &= -2 \end{aligned}$$

The vertex of the parabola is $(-2, -2)$.

Find the x-intercepts. Let $y = 0$. Solve the quadratic.

$$y = 2x^2 + 8x + 6$$
$$0 = 2(x^2 + 4x + 3)$$
$$0 = 2(x + 3)(x + 1)$$

$x + 3 = 0$ or $x + 1 = 0$

$x = -3$ | $x = -1$

The x-intercepts are $(-3, 0)$ and $(-1, 0)$.

Find the y-intercept. It is the constant.

$$y = 2x^2 + 8x + 6$$

The y-intercept is $(0, 6)$.

The axis of symmetry is the vertical line passing through the vertex. $x = -2$.

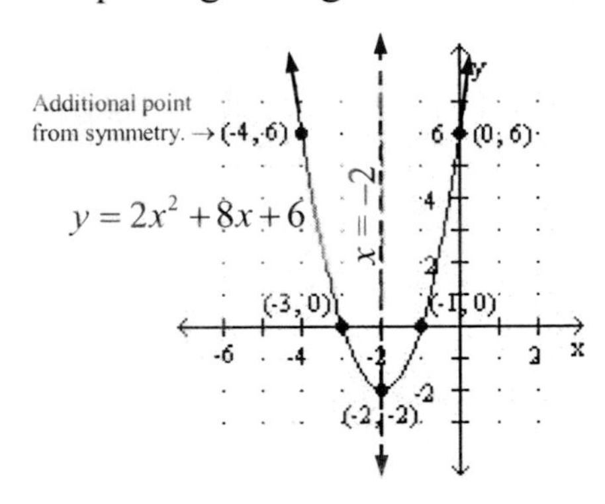

104. POWER OUTPUT [Section 9.5]

At 4,000 rpm the engine reaches its maximum power output.

CPSIA information can be obtained
at www.ICGtesting.com
Printed in the USA
FFOW01n0001231216
30676FF

9 781111 989026